OXYGEN RADICALS IN BIOLOGY AND MEDICINE

BASIC LIFE SCIENCES

Alexander Hollaender, Founding Editor

Recent volumes in the series:

Volume 36 ANEUPLOIDY: Etiology and Mechanisms
Edited by Vicki L. Dellarco, Peter E. Voytek, and Alexander Hollaender

Volume 37 GENETIC ENGINEERING OF ANIMALS: An Agricultural Perspective
Edited by J. Warren Evans and Alexander Hollaender

Volume 38 MECHANISMS OF DNA DAMAGE AND REPAIR: Implications for Carcinogenesis and Risk Assessment
Edited by Michael G. Simic, Lawrence Grossman, and Arthur C. Upton

Volume 39 ANTIMUTAGENESIS AND ANTICARCINOGENESIS MECHANISMS
Edited by Delbert M. Shankel, Philip E. Hartman, Tsuneo Kada, and Alexander Hollaender

Volume 40 EXTRACHROMOSOMAL ELEMENTS IN LOWER EUKARYOTES
Edited by Reed B. Wickner, Alan Hinnebusch, Alan M. Lambowitz, I. C. Gunsalus, and Alexander Hollaender

Volume 41 TAILORING GENES FOR CROP IMPROVEMENT: An Agricultural Perspective
Edited by George Bruening, John Harada, Tsune Kosuge, and Alexander Hollaender

Volume 42 EVOLUTION OF LONGEVITY IN ANIMALS: A Comparative Approach
Edited by Avril D. Woodhead and Keith H. Thompson

Volume 43 PHENOTYPIC VARIATION IN POPULATIONS: Relevance to Risk Assessment
Edited by Avril D. Woodhead, Michael A Bender, and Robin C. Leonard

Volume 44 GENETIC MANIPULATION OF WOODY PLANTS
Edited by James W. Hanover and Daniel E. Keathley

Volume 45 ENVIRONMENTAL BIOTECHNOLOGY: Reducing Risks from Environmental Chemicals through Biotechnology
Edited by Gilbert S. Omenn

Volume 46 BIOTECHNOLOGY AND THE HUMAN GENOME: Innovations and Impact
Edited by Avril D. Woodhead and Benjamin J. Barnhart

Volume 47 PLANT TRANSPOSABLE ELEMENTS
Edited by Oliver Nelson

Volume 48 HUMAN ACHONDROPLASIA: A Multidisciplinary Approach
Edited by Benedetto Nicoletti, Steven E. Kopits, Elio Ascani, and Victor A. McKusick

Volume 49 OXYGEN RADICALS IN BIOLOGY AND MEDICINE
Edited by Michael G. Simic, Karen A. Taylor, John F. Ward, and Clemens von Sonntag

---

OXYGEN RADICALS IN BIOLOGY AND MEDICINE

Edited by

Michael G. Simic

National Bureau of Standards
Gaithersburg, Maryland

Karen A. Taylor

International Life Sciences Institute
Washington, D.C.

John F. Ward

University of California, San Diego
La Jolla, California

and

Clemens von Sonntag

Max-Planck-Institut für Strahlenchemie
Mülheim a.d. Ruhr, Federal Republic of Germany

Assistant Editor

Janet N. Hauber

PLENUM PRESS • NEW YORK AND LONDON

Library of Congress Cataloging in Publication Data

International Congress on Oxygen Radicals (4th: 1987: University of California at San Diego)
Oxygen radicals in biology and medicine.
(Basic life sciences; v. 49)
"Proceedings of the Fourth International Congress on Oxygen Radicals, held June 27-July 3, 1987, at the University of California at San Diego, La Jolla, California"—T.p. verso
Includes bibliographies and index.
1. Active oxygen—Physiological effect—Congresses. 2. Pathology, Molecular—Congresses. 3. Free radical reactions—Congresses. 4. Oxidation, Physiological—Congresses. I. Simic, Michael G. II. Title. III. Series. [DNLM: 1. Free Radicals—congresses. 2. Oxygen—congresses. W3 BA255 v. 49 / QV 312 I6105 1987]
QP535.01I473 1987 574.19'214 88-25494

Proceedings of the Fourth International Congress on Oxygen Radicals, held June 27–July 3, 1987, at the University of California at San Diego, La Jolla, California

ISBN-13: 978-1-4684-5570-0 e-ISBN-13: 978-1-4684-5568-7
DOI: 10.1007/978-1-4684-5568-7

Softcover reprint of the hardcover 1st edition 1988

A Division of Plenum Publishing Corporation
233 Spring Street, New York, N.Y. 10013

OXYGEN RADICALS IN BIOLOGY AND MEDICINE

is dedicated to the memory of
Dr. Alexander Hollaender
(1898–1986)

Honorary Chairman of the
Fourth International Congress on Oxygen Radicals

PREFACE

This book is based on the papers presented at the "Fourth International Congress on Oxygen Radicals (4-ICOR)," held June 27 - July 3, 1987, at the University of California, La Jolla. The chapters deal with the phenomena associated with highly reactive oxygen species (hydroxy, peroxy, alkoxy, aroxy, and superoxide radicals, as well as singlet oxygen) and their peroxidation products (hydrogen peroxide, hydroperoxides, peroxides, and epoxides) as they relate to the fields of chemistry, food technology, nutrition, biology, pharmacology, and medicine. The kinetics, energetics, and mechanistic aspects of the reactions of these species and the interrelationship of oxygen radicals (or any other free radicals) and peroxidized products have been emphasized. Special attention is focused on the mechanisms of the generation of free radicals and peroxy products in biosystems and on the adverse effects of these radicals and products in humans. The topics span the continuum from the simple chemistry of model systems to the complex considerations of clinical medicine.

The book also explores the mechanisms of agents that protect against free radicals and peroxy products _in vitro_ and _in vivo_. These agents include antioxidants used in materials, food antioxidants, physiological antioxidants, and antioxienzymes (SOD, glutathione peroxidase, and catalases). The use of these inhibitors to prevent damage to organs being prepared for transplantation, thereby maintaining the quality of transplanted organs and/or extending their "shelf-life," also is examined.

The recognition of free radical processes and a better understanding of their mechanisms forecasts a revolution in the life sciences and in medicine. It is hoped that this book not only will inform readers but also will inspire them to continue expanding our ability to deal with these elusive transient species _in vivo_ by providing the intellectual stimuli for imaginative research and original ideas for the development of preventive medicine.

Michael G. Simic
4-ICOR Chairman

SPONSORS

The sponsorship, financial assistance, and encouragement of the following organizations is gratefully acknowledged:

Sponsors

National Bureau of Standards
National Heart, Lung, and Blood Institute
International Life Sciences Institute
National Cancer Institute
American Industrial Health Council
Institute for Research on Aging
Oxygen Society

Benefactors

Grunenthal GMBH

Major Contributors

Burroughs Wellcome Company
Henkel Corporation
The Upjohn Company
Shell Oil Company

Contributors

Johnson & Johnson
The Council for Tobacco Research
Pharmacia-Chiron Partnership
Glaxo, Inc.
Bristol-Meyers Company
Alcon Laboratories
Sandoz Pharmaceutical Company
Nabisco Brands, Inc.
E.I. duPont de Nemours & Company
Pfizer, Inc.
Hoffmann-La Roche, Inc.
Calreco, Inc.
Eastman Kodak Co.

ACKNOWLEDGMENTS

The editors wish to thank the many individuals whose contributions and assistance made possible the 4th International Congress on Oxygen Radicals (4-ICOR) and publication of the proceedings in this book.

For their help in planning and implementing the 4-ICOR program, we acknowledge the organizing committee (Steven D. Aust, Benon H.J. Bielski, Christopher S. Foote, Norman I. Krinsky, Chris E. Kuyatt, William E.M. Lands, Etsuo Niki, and Helmut Sies), chairmen of the previous ICOR meetings (Ajit Singh, E. Lawrence Powers, Michael A.J. Rodgers, and Wolf Bors), and chairmen of scientific sessions (Klaus-Dieter Asmus, William A. Pryor, Lester Packer, Britton Chance, Helmut Sies, Irwin Fridovich, Irving H. Goldberg, Gregory B. Bulkley, Peter A. Ward, E. Lawrence Powers, John W. Finley, Clemens von Sonntag, Robin L. Willson, Etsuo Niki, Edward A. Dratz, Erik van Kuijk, John F. Ward, Daniel M. Ziegler, Michael A. Trush, J. Edwin Seegmiller, Denham Harman, Hermann Esterbauer, Sampath Parthasarathy, Bernard M. Babior, Joe M. McCord, Earl R. Stadtman, Kelvin J.A. Davies, Ambler Thompson, Stefan L. Marklund, John E. Biaglow, Michael A.J. Rodgers, Alex Sevanian, Steven Aust, Thomas L. Robertson, and William B. Weglicki).

We are exceptionally grateful to Ernest Ambler, Director of NBS, and Raymond G. Kammer, Deputy Director of NBS, for their farsighted support of interdisciplinary research and the 4-ICOR meeting.

The support and encouragement of Claude Lenfant, Director of the National Heart, Lung, and Blood Institute, Leopold Flohé, Director of Research and Development, Grünethal GMBH, and Donald E. Stevenson, Director of Toxicology, Health, Safety and Environment, Shell Oil Company, are deeply appreciated.

Special thanks are extended to Professors Etsuo Niki and Toshikazu Yoshikawa for their crucial support in the organization of the program and the preparation of this book.

The technical and management assistance of Janet N. Hauber, Kathy C. Stang, Sara R. Torrence, Lene Hartman, Cassandra Beck, Terry W. Shuggars, Gloria G. Wiersma, Eva M. Wood, and Christopher Soares is greatly appreciated.

The press coordination of Michael Baum was invaluable in publicizing oxygen radicals and their impact on biology and medicine.

M.G.S.
K.A.T.
J.F.W.
C.v.S.

CONTENTS

INTRODUCTION

GENERATION OF OXY RADICALS

DETECTION AND MEASUREMENTS

CHEMILUMINESCENCE

FREE RADICALS AND OXYGEN SPECIES IN LIPIDS

FREE RADICALS AND OXYGEN SPECIES IN PROTEINS

FREE RADICALS AND OXYGEN SPECIES IN LIPOPROTEINS

FREE RADICALS AND OXYGEN SPECIES IN DNA

P-450

OXIDATIVE STRESS

ANTIOXIDANT DEFENSE

ENZYMATIC DEFENSE

PHARMACOLOGY AND TOXICOLOGY

ISCHEMIA - REPERFUSION

CARDIOVASCULAR SYSTEM

GENERAL MEDICINE

INTRODUCTION TO PEROXIDATION AND ANTIOXIDATION MECHANISMS

Michael G. Simic and Karen A. Taylor*

National Bureau of Standards
Gaithersburg, Maryland 20899, U.S.A.

*International Life Sciences Institute
Washington, D.C. 20036, U.S.A.

INTRODUCTION

The propagation and maintenance of life requires substantial energy. The evolution of life on Earth led to utilization of oxygen as the major mediator of energy release from organic molecules, which are the source of energy for most organisms. The participation of oxygen in energy release is not essential, as demonstrated in anaerobic and sulfur bacteria that function without it. Why is oxygen-mediated energy release favored by most organisms? The major reasons for the evolutionary selection of oxygen are (1) its ready availability, (2) the high energy yield from oxidation, (3) the reversibility of the process (oxygen is consumed by energy-releasing processes such as cellular respiration and evolved in energy-trapping processes such as photosynthesis), and (4) the easy distribution of oxygen due to its gaseous state and its solubility in biocomponents under normal conditions.

The outstanding benefits from oxygen-mediated "oxidative energetics," as everything else in life, have a darker side.[1] The dark side of oxygen is manifested either directly or in less obvious ways. It has long been known that breathing pure oxygen (100% O_2, 1 atm.) instead of air (20% O_2, 1 atm.) is detrimental to living organisms because uncontrolled and undesirable oxidative processes are enhanced by higher oxygen concentrations in tissues. A less obvious, and until recently unknown, detrimental facet of oxygen is the formation of oxy radicals *in vivo*.[2] The reasons for the substantial lag between the discovery of free radicals in model systems and of free radical processes *in vivo* were the generally short lifetimes of radicals and the difficulties associated with direct detection and monitoring of free radicals *in vivo*. It is ironic that the recognition of free radicals *in vivo* often resulted from the discovery of biological defense systems and agents capable of neutralizing free radicals and their products in cells and intercellular regions. A good example is the discovery of superoxide dismutase,[3] which catalyzes rapid recombination of superoxide radicals,

$$2\ \cdot O_2^- + 2H^+ \xrightarrow{SOD} H_2O_2 \qquad (1)$$

Reaction (1) illustrates the close relationship between oxy radicals and their peroxidized products; this relationship is critical in the overall assessment of the toxicity of oxygen.

An historical overview reveals that the modern concept of free radicals is a product of this century, although the rudimentary idea was first considered by French and German organic chemists in the 1800s. Haber and Weiss dealt with the inter-relationship between H_2O_2, $\cdot OH$ radicals, and $\cdot O_2^-$ in the early 1930s. Michaelis also discussed free radicals in biochemistry during the 1930s. The generation of $\cdot O_2^-$ by xanthine-xanthine oxidase was demonstrated by Fridovich and McCord much later, in the 1960s, and the participation of free radicals in the arachidonic acid cascade was recognized by Samuelson in the 1970s. Most major advances in this area have been in the last two decades due to the development of pulse radiolysis,[4] laser photolysis,[5] and highly sensitive analytical instrumentation. The implications of free radical processes are beginning to touch every biologically and medically oriented discipline, such as replicative inactivation of DNA,[6] mutation,[7] carcinogenesis,[8] atherosclerosis,[9] arthritis,[10] and aging.[11] What began as an obscure idea outside the mainstream of chemistry has now become essential for understanding the basic processes of life, conception, growth, aging, disease, and death.[12]

Basic reactions leading to the formation of oxy radicals and their peroxy products are presented, and approaches[13] to detection, monitoring, and measurement of free radical parameters (reaction rate constants, k; pK_a values; and redox potentials, E_7 at pH = 7) that define their reactivities and properties are briefly described. The detailed accounts and specific effects of oxy species in systems ranging from the simplest model systems to the complex human organism are described in this book.

FREE RADICALS

At the turn of the century, radicals (i.e., groups of atoms such as $-CH_3$) in free form were recognized by Gomberg,[14] who discovered the relatively stable triphenylmethyl radical, $\cdot C\phi_3$. The current, broader definition of a free radical is a molecular or atomic specie with an unpaired electron, e.g., $\cdot CH_3$ (methyl), $\cdot H$ (hydrogen atom), $\cdot OH$ (hydroxyl), and $\cdot CH_2CH_3$ (ethyl). The unpaired electron of a free radical is represented by a dot on the atom or a group in which it predominantly resides.

In aromatic, unsaturated, and more complex free radicals, the unpaired electron partially resides at different atoms or groups due to resonance. The most familiar resonant radical is the bisallylic radical in polyunsaturated fatty acids,

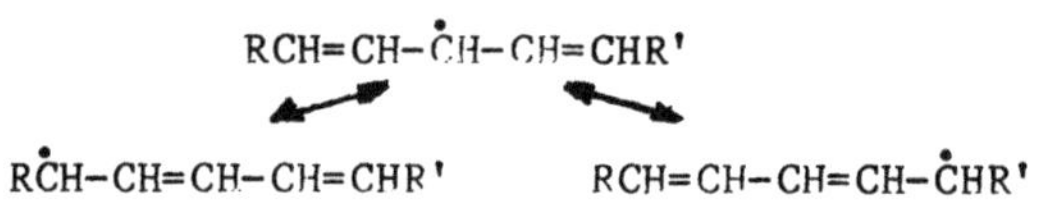

Most free radicals are very reactive with oxygen, with rate constants of the order of $k \sim 10^9\ M^{-1}\ s^{-1}$ or less.[15] There are radicals, however, that do not react with oxygen, e.g., oxy radicals.

OXY RADICALS

Oxy radicals are a class of free radicals with the unpaired electron residing predominantly on an oxygen atom. The best known examples are superoxide radical, $\cdot OO^-$ (or $\cdot O_2^-$), hydroxy radical, $\cdot OH$, and various peroxy

Table 1. Classes of Oxy Radicals and Related Oxygen Species

Species	Name	Origin
$^3O_2$	Triplet oxygen	Stable atmospheric form
$^1O_2$	Singlet oxygen	$^3O_2$, Peroxidation
RH	Parent molecule	
R•	Free radical	RH
ROO•	Peroxy radical	R• + O_2, ROOH
$\bullet O_2H$	Hydroperoxy radical	$\bullet O_2^- + H^+$
$\bullet O_2^-$	Superoxide radical	$O_2 + e^-$
•OH	Hydroxy radical	H_2O_2, H_2O
H_2O_2	Hydrogen peroxide	$\bullet O_2^-$, Biogeneration
RO•	Alkoxy radical	ROOH, ROOR
ROOH	Hydroperoxide	ROO•, $^1O_2$
ROOR	Peroxide	Peroxidation
$\overset{O}{>C{-}C<}$	Epoxide	Unsaturated ROO• or ROO•
ArO•	Phenoxy & aroxy radicals	Phenolic antioxidants

radicals, ROO•. Distinctive classes of oxy radicals are shown in Table 1, as well as other nonradical species participating in peroxidation (autoxidation) processes. There are two systems of nomenclature for oxy radicals. The one in this chapter uses the suffix "y" (e.g., oxy, hydroxy, and peroxy), whereas the alternative nomenclature uses the ending "yl" (e.g., oxyl, hydroxyl, and peroxyl).

GENERATION OF FREE RADICALS

Free radicals are generated via numerous processes in the atmosphere, environment, foods, chemical systems, and living organisms.

One of the most convenient modes of free radical generation, especially for studying their properties, is ionizing radiations such as X- and γ-rays and high-energy electrons. In aqueous systems radiation splits water,

$$H_2O \rightsquigarrow \bullet OH + e^-_{aq} + H^+ \qquad (2)$$

whereas in organic media it breaks chemical bonds, e.g.,

$$RH \rightsquigarrow \bullet R + \bullet H \qquad (3)$$

In 1914 Debierne[16] proposed that •OH radicals were formed by radiation, not long after the discovery of ionizing radiation by Roentgen in 1895 and of radioactivity by Becquerel in 1896.

A few years earlier, in 1894, Fenton,[17] who did not recognize that the process involved an •OH radical generation, initiated oxidation of organic molecules by ferrous-hydrogen peroxide mixtures. The explanation of the process was provided by Haber and Weiss in 1932,[18]

$$H_2O_2\cdot + Fe^{2+} \rightarrow \cdot OH + OH^- + Fe^{3+} \qquad (4)$$

Hence, reaction (4) is known as the Haber-Weiss reaction and the mixture of Fe(II) and H_2O_2 as Fenton reagent; the entire process is known as the Fenton-Haber-Weiss process, which gives each of them credit for their contributions to understanding this •OH-generating system.

The Haber-Weiss reaction has been a focus of interest, especially in the last two decades. Since only very low levels of free ferrous ions are found in biosystems, complexes of iron that undergo reaction (4) have been investigated. In addition, the recycling of ferrous ion, e.g., with ascorbate, has been of interest because even minute levels of ferrous ion may perpetuate the process of free radical formation. In general, ferric ion or its complexes are rapidly reduced by ascorbate,

$$Fe(III) + AH^- \rightarrow Fe(II) + \cdot A^- + H^+ \qquad (5)$$

The ascorbate radical, $\cdot A^-$, has a low redox potential[19] and can reduce a variety of oxidants, e.g.,

$$Fe(III) + \cdot A^- \rightarrow Fe(II) + A \qquad (6)$$

The generation of alkoxy radicals from hydroperoxides is similar to the Haber-Weiss reaction,

$$ROOH + Fe(II) \rightarrow RO\cdot + OH^- + Fe(III) \qquad (7)$$

Reactions (4) and (5) are of particular interest because metabolic and free radical generation of H_2O_2 and ROOH, as well as singlet oxygen in certain instances, occurs in biosystems,

$$^1O_2 + RH \rightarrow ROOH \qquad (8)$$

Additional modes of generating free radicals in biosystems[15] involve ultrasound,

$$H_2O \xrightarrow{1\ MHz} \cdot OH + \cdot H \qquad (9)$$

photochemical generation of triplet states,

$$S + h\nu \rightarrow {}^3S \qquad (10)$$

$$^3S + RH \rightarrow SH + \cdot R \qquad (11)$$

bioreduction of halogenated compounds in the liver, e.g.,[20]

$$CCl_4 + e^- \rightarrow \cdot CCl_3 + Cl^- \qquad (12)$$

and biochemical and enzymatic processes, e.g.,

$$O_2 + \text{xanthine} \xrightarrow{\text{xanthine oxidase}} \cdot O_2^- + \text{urate} \qquad (13)$$

DETECTION AND MONITORING OF FREE RADICALS

There are two basic approaches in detecting and monitoring free radicals: direct and indirect.

Direct Detection and Monitoring

Because of their short lifetimes, direct detection of free radicals requires, in most cases, time-resolved physicochemical techniques.[4,5] ESR (or EPR) is a direct technique based on the absorption of microwaves by the unpaired electron in a magnetic field. Absorption and emission of light, conductivity, raman scattering, and other physico-chemical techniques, although good monitoring techniques, require additional proof that the signal they measure originates from a free radical, and not from a transient molecular specie.

In order to study free radicals in model systems, two techniques are prominent, pulse-radiolysis[4] and laser photolysis,[5] both of them capable of time resolution on the order of picoseconds.

In pulse radiolysis,[4] free radicals are generated by a short pulse (nanoseconds to microseconds) of high-energy electrons (2-10 MeV) in water or organic media. Hence, the initial radicals, e.g., reactions (2) and (3), are generated within the time frame of the electron pulse in a suitable reaction cell at nanomolar to micromolar concentrations, as required. At micromolar concentrations most radicals disappear in a microsecond or longer. If a solute is added to a medium, the initial radical (e.g., •OH) reacts with it (if the solute is reactive) to give a second generation of free radicals. Under suitable conditions in the presence of selected solutes, a cascade of free radical reactions can be created,

$$\bullet OH \xrightarrow{RH} \bullet R \xrightarrow{O_2} ROO\bullet \xrightarrow{ArOH} ArO\bullet \xrightarrow{AH^-} \bullet A^-$$

Scheme 1. A cascade of free radicals generated by pulse radiolysis of aqueous solutions containing organic substrates, RH; oxygen; an antioxidant, ArOH; and ascorbate, AH^-.

The cascade is monitored via changes in the absorption spectra (λ_{max} for •OH, ~ 200 nm; •R, ~ 230 nm; ROO•, 245 nm; ArO•, 400 nm; and $\bullet A^-$, 360 nm).

The principles of laser photolysis[5] are similar to pulse radiolysis, with one major difference: whereas ionizing radiations ionize the medium (solutes at $< 10^{-1}$ M are ionized only to a negligible extent), photons interact with the highly absorbing solutes. In monophotonic processes, free radicals are generated via reactions (10) and (11), for example, whereas in biphotonic processes solutes are ionized in the same manner as by ionizing radiations, i.e., reaction (3).

Consequently, pulse radiolysis and laser photolysis are extremely convenient for the study of free radical properties such as acid-base (pK values of free radicals),[21] redox (redox potential at pH = 7, E_7, or any other pH), and reactivities (rate constants, k, of various reactions of free radicals with like radicals, different radicals, or solutes).[22-24] It is important to recognize that major advances in free radical biology and medicine will come only through comprehensive understanding of free radical properties.

Indirect Detection and Monitoring

It is unfortunate that both pulse radiolysis and laser photolysis are less adaptable to *in vivo* studies. Hence, *in vivo*, specific reaction products of free radicals must be studied. These biomarkers must be relatively stable and should not be generated by any other processes. Unfortunately, very few highly specific biomarkers have been identified at present.

Another approach uses compounds that are reactive with free radicals, resulting in the formation of a reaction product that is a stable free radical, which can be detected easily by ESR. This method, spin trapping, has been quite successful for detection of free radicals in the heart.[25,26] Since the concentration of spin traps cannot be sufficiently high *in vivo* without perturbing the living system, the absence of an ESR signal does not necessarily mean the absence of free radicals since other competing reactions may occur.

PROTECTION FROM FREE RADICALS

Free radicals are extremely reactive in general and can inflict considerable damage to biomolecules, such as DNA, enzymes, membranes, and proteins, which may lead to serious biological consequences, such as cancer, damage to organs, and numerous other pathologic disorders.

Protection from the damaging effect of free radicals[27] is accomplished by low-molecular-weight physiological constituents, such as glutathione (GSH), vitamin E (E-OH), and β-carotene (Car), and by enzymes, such as catalases (Cat), superoxide dismutases (SOD), and glutathione peroxidases (GP), to name just a few of the best known examples.

Protection may occur by intervention at different stages of free radical processes. One way is to intervene at the pre-radical stage by eliminating H_2O_2 and ROOH with Cat or GP or by inactivating metal ions with complexing agents. Once free radicals are generated, they can be scavenged by different agents (e.g., DMSO, mannitol, SOD, E-OH, and GSH).

Rather than describing them by the stage in the free radical process in which they act, subclasses of protection agents are traditionally subdivided according to a more traditional scheme: scavengers, repair agents, antioxidants, and antioxienzymes (antioxidant enzymes).

Free Radical Scavengers

The most popular free radical scavenger in biosystems has been DMSO due to its very high solubility in biomaterials. The reaction of DMSO with free radicals is a substitution reaction,[28]

$$(CH_3)_2SO + \cdot R \rightarrow R(CH_3)SO + \cdot CH_3 \qquad (14)$$

DMSO does not eliminate free radicals; it just replaces a radical with a methyl radical, provided that the first radical is reasonably reactive with DMSO. DMSO, for example, is an extremely efficient •OH radical scavenger,

$$(CH_3)_2SO + \cdot OH \rightarrow (OH)(CH_3)SO + \cdot CH_3 \qquad (15)$$

Based on these properties deuterated DMSO might be proposed as a marker for free radical generation. In the first step a deuterated methyl radical would be generated,

$$(CD_3)_2SO + \cdot R \rightarrow R(CD_3)SO + \cdot CD_3 \qquad (16)$$

which, on abstraction from neighboring C-H or S-H bonds, yields a specific, easily recognizable product — deuterated methane,

$$\cdot CD_3 + RH \rightarrow CHD_3 + \cdot R \qquad (17)$$

Mannitol (Man) has been used to scavenge •OH radicals in numerous biosystems since it reacts rapidly with •OH,[22]

$$Man + \cdot OH \rightarrow \cdot Man(-H) + H_2O \qquad (18)$$

$$k = 2 \times 10^9 \ M^{-1} \ s^{-1}$$

In this reaction the highly reactive •OH radical is replaced by the less reactive C-centered mannitol radical, •Man(-H). A shortcoming of mannitol is its limited solubility in biosystems.

In general, scavenging of a free radical may be effected at different stages of a free radical cascade,

$$\cdot R_{in} \rightarrow \cdot R'' \rightarrow \rightarrow \rightarrow \cdot R_{fin} \rightarrow P \qquad (19)$$

where $\cdot R_{in}$ is the free radical initially generated and $\cdot R_{fin}$ is the final radical in the cascade before the product, P, is formed. In competition with reaction (19) is the damage to a biomolecule caused by one of the radicals in the cascade, e.g.,

$$\cdot R_{in} + Bio \rightarrow \cdot Bio\text{-}R_{in} \qquad (20)$$

It is difficult, however, to single out the particular radical in the cascade that participates in reaction (20). Because free radical cascades occur normally in biosystems and since it is not known which radical in the cascade is capable of damaging a biomolecule, preventing the damage by scavenging or inactivating a particular radical in the cascade does not imply that the particular radical is the damaging one. If the damage to a biomolecule is effected by a daughter radical in the cascade, scavenging or inactivation of a precursor radical also would effectively prevent the damage.

Repair Agents

Sulfhydryl compounds have a weak S-H bond strengths (~ 85 kcal/mol) and are capable of repairing C-centered radicals, which usually have bond strengths > 90 kcal/mol, e.g.,

$$GSH + \cdot CH_3 \rightarrow GS\cdot + CH_4 \qquad (21)$$

The rate constant for abstraction of H by a radical could be as high as $k = 10^9 \ M^{-1} \ s^{-1}$ for certain aromatic sulfhydryls.[29] Most RSH compounds, however, do not react very rapidly with C-centered or peroxy radicals. It is important to remember that the reaction of repair agents with peroxy radicals leads to the formation of hydroperoxides and not to the restitution of the original undamaged molecule.

Antioxidants

Historically, there have been two classes of antioxidants: (a) inhibitors of initiation of free radical processes (e.g., metal-complexing agents that prevent the Haber-Weiss reaction) and (b) chain-breaking antioxidants (e.g., BHA, BHT, and vitamin E).[9]

In general, chain-breaking antioxidants are good electron donors[29] capable of inactivating peroxy radicals. One class of common antioxidants is the phenolic antioxidants, Ar-OH. The following reactions of phenolic antioxidants with peroxy radicals can be discerned,

$$ROO\cdot + Ar\text{-}OH \rightarrow ROO^- + Ar\text{-}OH^+ \qquad (22)$$

followed by,

$$ROO^- + H^+ \rightleftharpoons ROOH \qquad (23)$$

$$pK_a \sim 12.8$$

and

$$Ar\text{-}OH^+ \rightarrow Ar\text{-}O\cdot + H^+ \qquad (24)$$

Usually, only the overall process is indicated,

$$ROO\cdot + Ar\text{-}OH \rightarrow ROOH + Ar\text{-}O\cdot \qquad (25)$$

which may give the false impression that antioxidants are H-atom donors.

Antioxienzymes

These enzymes may eliminate the precursors of free radicals. The best known examples are catalases and glutathione peroxidases,[2]

$$2H_2O_2 \xrightarrow{Cat} O_2 + 2H_2O \qquad (26)$$

$$ROOH + 2GSH \xrightarrow{GP} ROH + H_2O + GSSG \qquad (27)$$

In contrast, SOD eliminates superoxide radicals as shown in reaction (1).

FUTURE TRENDS

Free radical biology and medicine are expanding at an accelerated rate, as indicated by the increasing number of conferences, specialized journals, and books dealing with the subject. Future trends in the development of the field are of particular interest to all of us because the field, once fully developed, will profoundly affect the quality of our lives.

There is still a great chasm between mechanistic understanding of free radical processes and applications to biosystems. What has delayed the transfer of these basic concepts to fields such as biology and medicine? Two major constraining factors, which have been pointed out in this chapter, are the elusive nature of short-lived free radicals, especially in biosystems, and the lack of suitable noninvasive detection and monitoring techniques. A more comprehensive and mechanistic understanding of free radical processes and oxy products under complex physiological conditions in vitro and in vivo is critical but has not yet been achieved. The mechanisms of free radical processes, as well as methods for detection, monitoring, and measurement, are still challenging areas that require further development. It is through resolution of these mechanistic aspects of free radical processes and through the investigation of interrelationships between oxy products and free radicals that we will gain the ability to manipulate these processes for the benefit of all.

With the current increased interest in oxy radicals, however, it is crucial that the assignments of biomedical phenomena to free radical processes be thoroughly scrutinized, systematically cross-checked, and, above all, based on highly specific biomarkers and mechanistic understanding to ensure rapid progress in the field. It is hoped that the papers included in Oxygen Radicals in Biology and Medicine will advance the field toward these goals.

ACKNOWLEDGMENTS

The authors thank David Bergtold, John Biaglow, and Lisa Karam for their helpful comments and suggestions and Gloria Wiersma for technical execution of the paper.

REFERENCES

1. J. D. Balantine, "Pathology of Oxygen Toxicity," Academic Press, New York (1982).
2. C. Rotilio, ed., "Superoxide and SOD in Biology and Medicine," Elsevier, New York (1986).
3. J. M. McCord and I. Fridovich, Superoxide dismutase, J. Biol. Chem. 244:6049 (1969).
4. C. von Sonntag, "The Chemical Basis of Radiation Biology," Taylor & Francis, New York (1987).
5. R. V. Bensasson, E. J. Land, and T. G. Truscott, "Flash Photolysis and Pulse Radiolysis," Pergamon Press, New York (1983).
6. E. J. Hall, "Radiobiology for the Radiobiologist," Harper and Row, Hagerstown, 1978.
7. F. Hutchinson, Chemical changes induced in DNA by ionizing radiation, Prog. Nucleic Acid Res. Mol. Biol. 32:115 (1985).
8. P. Cerutti, O. F. Nygaard, and M. G. Simic, eds., "Anticarcinogenesis and Radiation Protection," Plenum Press, New York (1987).
9. M. G. Simic and M. Karel, eds., "Autoxidation in Food and Biological Systems," Plenum Press, New York (1980).
10. B. Halliwell and J. M. C. Gutteridge, "Free Radicals in Biology and Medicine," Clarendon Press, Oxford (1985).
11. D. Armstrong, R. S. Sohal, R. G. Cutler, and T. F. Slater, eds., "Free Radicals in Molecular Biology, Aging, and Disease," Raven Press, New York (1985).
12. W. A. Pryor, ed., "Free Radicals in Biology," vol. 1-4, Academic Press, New York (1976-1984).
13. C. Rice-Evans and B. Halliwell, eds., "Free Radicals: A Search for New Methodology," Richelieu Press, Paris (1988).
14. M. Gomberg, An instance of trivalent carbon: Triphenylmethyl, J. Am. Chem. Soc. 22:757 (1900).
15. M. G. Simic and K. A. Taylor, Free radical mechanisms of oxidation reactions, in: "Warmed-Over Flavor of Meat," A. J. St. Angelo and M. E. Bailey, eds., Academic Press, Orlando (1987).

16. I. G. Draganic and Z. D. Draganic, "The Radiation Chemistry of Water," Academic Press, New York (1971).
17. H. J. H. Fenton, Oxidation of tartaric acid in the presence of iron, J. Chem. Soc. Trans. 65:899 (1894).
18. F. Haber and J. Weiss, Uber die Katalyse des Hydroperoxydes, Naturewiss. 20:948 (1932).
19. S. Steenken and P. Neta, One-electron redox potentials of phenols. Hydroxy- and aminophenols and related compounds of biological interest, J. Phys. Chem. 86:3661 (1982).
20. J. M. McCord and I. Fridovich, The reduction of cytochrome-c by milk xanthine oxidase, J. Biol. Chem. 243:5753 (1968).
21. E. Hayon and M. G. Simic, Acid-base properties of free radicals in solutions, Acc. Chem. Res. 7:114 (1974).
22. Selected Specific Rates of Reactions of Transients from Water in Aqueous Solutions. U.S. Government Printing Office, Washington, D.C.
    I. Hydrated Electron. M. Anbar, M. Bambenek, and A. B. Ross, eds. (1973). NSRDS-NBS 43, Hydrated Electron, Supplemental Data (1975). NSRDS-NBS 43.
    II. Hydrogen Atom. M. Anbar, Farhataziz, and A. B. Ross, eds. (1975). NSRDS-NBS 51.
    III. Hydroxyl and Perhydroxyl Radicals. Farhatiziz and A. B. Ross, eds. (1977). NSRDS-NBS 59.
23. A. B. Ross and P. Neta, eds., "Rate Constants for Reactions of Aliphatic Carbon-Centered Radicals in Aqueous Solutions," NSRDS-NBS 70 (1982).
24. Landolt-Bernstein, New Series, Group II, vol. 13, "Radical Reaction Rates in Liquids," H. Fischer, ed., Springer-Verlag, New York.
25. J. L. Zweier, J. T. Flaherty, and M. L. Weisfeldt, Direct measurement of free radical generation following reperfusion of ischemic myocardium, Proc. Natl. Acad. Sci. U.S.A. 84:1404 (1987).
26. C. M. Arroyo, J. H. Kramer, R. H. Leiboff, G. W. Mergner, B. F. Dickens, and W. B. Weglicki, Spin trapping of oxygen and carbon-centered free radicals in ischemic canine myocardium. Free Radical Biol. Med. 3:313 (1987).
27. B. N. Ames, Dietary carcinogens and anticarcinogens, Science 221:1256 (1983).
28. M. C. R. Symons, Unstable intermediates. Part CLXIII. Methyl radical anion adducts: Radiation effects on dimethyl sulphoxide, dimethyl sulphone, and their solutions in water and methanol, J. Chem. Soc., Perkin Trans. 2:908 (1976).
29. M. G. Simic, E. P. L. Hunter, and S. V. Jovanovic, Electron vs. H-atom transfer in chemical repair, in: "Anticarcinogenesis and Radiation Protection," P. Cerutti, O. F. Nygaard, and M. G. Simic, eds., Plenum Press, New York (1987).

THE REDOX THERMODYNAMICS FOR DIOXYGEN SPECIES (O_2, $O_2^{-}\cdot$, $HOO\cdot$, HOOH, AND HOO^-) AND MONOOXYGEN SPECIES (O, $O^{-}\cdot$, $\cdot OH$, AND ^{-}OH) IN WATER AND APROTIC SOLVENTS

Donald T. Sawyer

Department of Chemistry
Texas A&M University
College Station, Texas 77843, U.S.A.

Biological systems activate dioxygen (O_2) by reduction via electron transfer and H-atom transfer to give intermediates that may be further activated by transition metals in proteins and metabolic cofactors. The reaction thermodynamics for these processes are influenced by the solution matrix and its acidity. Thus, the redox thermodynamics of O_2 are directly dependent upon proton activity,

$$O_2 + 4H^+ + 4e^- \rightarrow 2\ H_2O \qquad E^{\circ\prime} \tag{1}$$

which in turn depends upon the reaction matrix. Table 1 summarizes the pK_a' values for a series of Brønsted acids in several aprotic solvents and water.[1] In acetonitrile the activity values for pK_a' range from -8.8 for $(H_3O)ClO_4$ to 30.4 for H_2O. This means that the formal potential ($E^{\circ\prime}$) for reaction 1 in acetonitrile (MeCN) is +1.75 V vs. NHE in the presence of $1M(H_3O)ClO_4$ and -0.56 V in the presence of 1M $(Bu_4N)OH$.

Another limiting factor with respect to chemical-energy flux for oxidative metabolism and respiration is the solubility of O_2. Because of its non-polar character dioxygen is much more soluble in organic solvents than in H_2O (Table 2).[2]

The reduction potentials for O_2 and various intermediate species in H_2O at pH 0, 7, and 14 are summarized in Figure 1;[3-5] similar data for O_2 in MeCN at pH -8.8, 10.0, and 30.4 are presented in Figure 2.[2,6,7]

Table 1. Effective pK_a' Values for Brønsted Acids in Aprotic Solvents and Water[1]

	Solvent [0.5M $(Et_4N)ClO_4$]				
Brønsted Acid	MeCN	DMF	Me_2SO	pyr	H_2O
$(H_3O)ClO_4$	-8.8	0.7	2.6	4.6	0.0
$MePhSO_3H$	-3.8	-	-	-	-
$(PyrH)ClO_4$	1.8	3.1	4.4	5.7	4.9
2,4-$(NO_2)_2$PhOH	4.3	4.5	4.9	5.5	-
$(NH_4)ClO4$	5.9	9.6	11.7	7.3	8.7
$(Et_3NH)Cl$	10.0	9.9	12.7	7.6	10.1
PhC(O)OH	7.9	11.5	13.6	11.6	3.2
2-Pyr-C(O)OH	8.6	-	-	-	-
PhOH	16.0	19.4	20.8	20.1	9.2
p-EtOPhOH	19.3	21.5	23.8	21.8	9.6
H_2O	30.4	34.7	36.7	30.5	14.8

MeCN, acetonitrile; DMF, dimethylformamide; Pyr, pyridine

Table 2. Solubilities of O_2 (1 atm) in Various Solvents[2]

Solvent	$[O_2]_{1\ atm}$, mM
H_2O	1.0
Me_2SO	2.1
DMF	4.8
Pyr	4.9
MeCN	8.1
Hydrocarbons	~10
Fluorocarbons	~25

The reduction manifolds for O_2 (Figures 1 and 2) indicate that the limiting step (in terms of reduction potential) is the first electron transfer to O_2, and that an electron source adequate for the reduction of O_2 will produce all of the other reduced forms of dioxygen ($O_2^-\cdot$, $HO_2\cdot$, H_2O_2, HO_2^-, $\cdot OH$) via reduction, hydrolysis, and disproportionation steps (Scheme 1).[8,9] Thus, the most direct means to activate O_2 is the addition of an electron (or hydrogen atom), which results in significant fluxes of several reactive oxygen species.

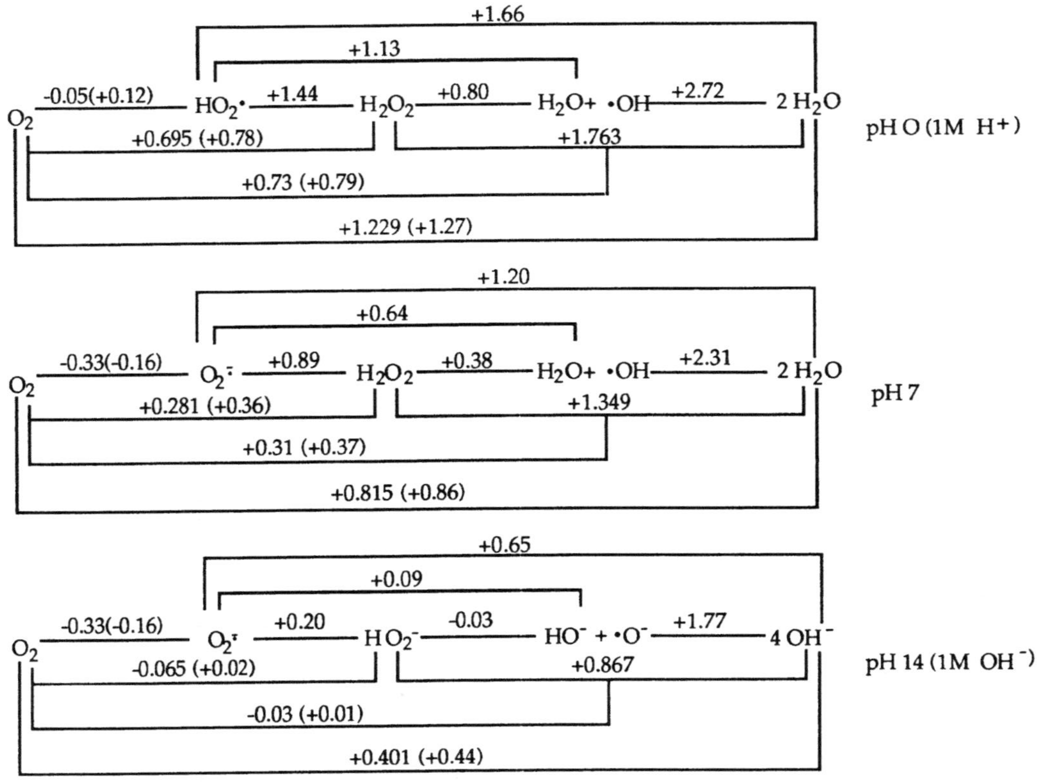

Figure 1. Standard reduction potentials for dioxygen species in water [O_2 at 1 atm] (formal potentials for O_2 at unit activity).

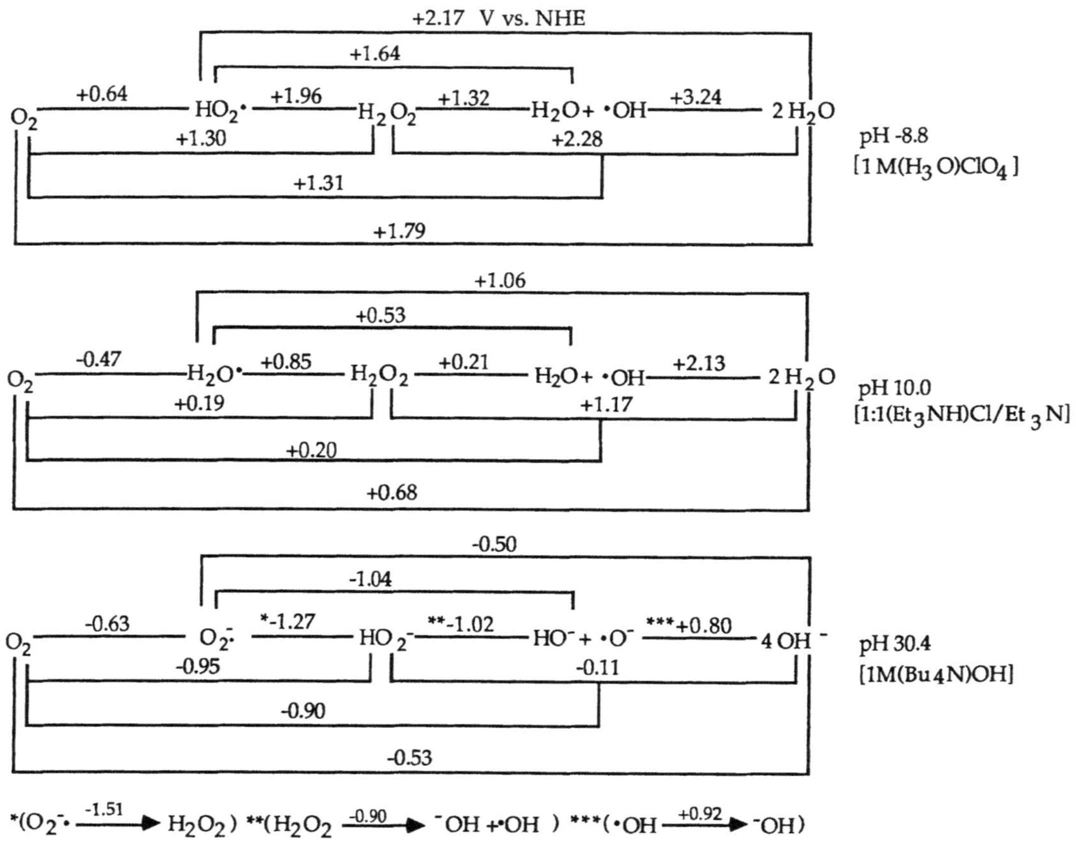

Figure 2. Formal reduction potentials for dioxygen species in acetonitrile (O_2 at unit activity).

REACTIVITY OF $O_2^{-}\cdot$ AND $HO_2\cdot$

The dominant characteristic of $O_2^{-}\cdot$ in any medium is its ability to act as a strong Brønsted base via formation of $HO_2\cdot$,[10,11] which reacts with allylic hydrogens, itself, or a second $O_2^{-}\cdot$ (Scheme 1).

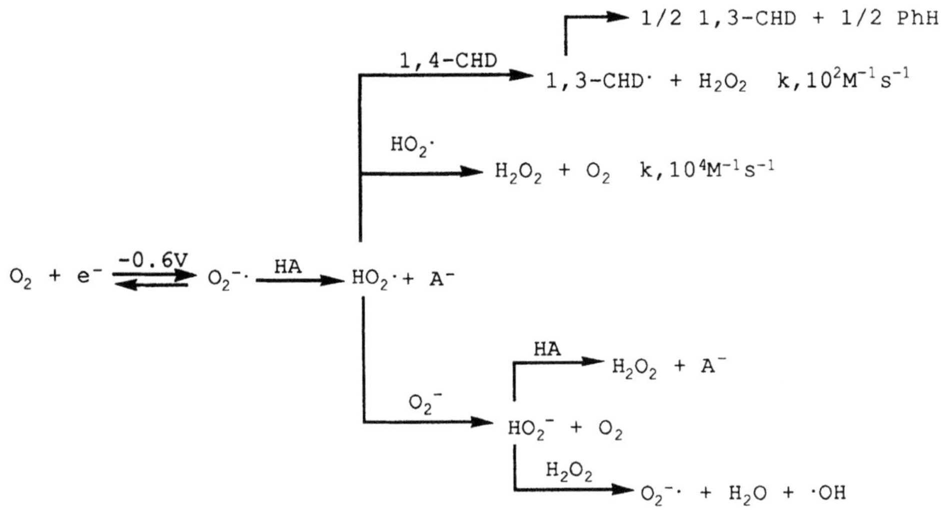

Scheme 1 (Me_2SO)

Within water, superoxide ion is rapidly converted to dioxygen and peroxide

$$2\ O_2^{-}\cdot + H_2O \rightarrow O_2 + HO_2^{-} + HO^{-} \qquad K,\ 2.5\times10^{8}\ M \qquad (2)$$

Such a proton-driven disproportionation process means that $O_2^{-}\cdot$ can deprotonate acids much weaker than water (up to $pK_a \approx 23$).[12]

The data of Figures 1 and 2 indicate that $O_2^{-}\cdot$ is a moderate one-electron reducing agent [cytochrome c(Fe^{III}) is reduced in H_2O[13,14] and iron(III) porphyrins in dimethylformamide].

$$Fe^{III}TPP^{+} + O_2^{-}\cdot \rightarrow Fe^{II}TPP + O_2 \qquad \Delta E,\ +0.7\ V \qquad (3)$$

Superoxide ion reacts with proton sources to form $HO_2\cdot$, which disproportionates via a second $O_2^{-}\cdot$ or itself (Scheme 1). However, with limiting fluxes of protons to control the rate of $HO_2\cdot$ formation from $O_2^{-}\cdot$, the rate of decay of $HO_2\cdot$ is enhanced by reaction with the allylic hydrogens of excess 1,4-cyclohexadiene (1,4-CHD).[15] Because $HO_2\cdot$ disproportionation is a second-order process, low concentrations favor hydrogen-atom abstraction from 1,4-CHD. This is especially so for Me_2SO, in which the rate of disproportionation for $HO_2\cdot$ is the slowest (PhCl>MeCN>H_2O>DMF>Me_2SO).

REDOX THERMODYNAMICS FOR ATOMIC OXYGEN

The electron-transfer potentials for the reduction of ground-state atomic oxygen in water and in acetonitrile are summarized in Tables 3 and 4.[3,4,15] In addition, selected values for the reduction of the zero-valent oxygen in O_3, $HOIO_3$, and HOCl are included.[3] Because the valence (charge number) of oxygen in water is zero, these reduction processes represent the facilitated reduction of protons via stabilization of the product H-atoms through strong covalent-bond formation with atomic oxygen. Thus, the standard-state reduction of protons,

$$2H^+ + 2e^- \rightarrow H_2 \qquad E°,\ 0.00V \text{ vs. NHE} \qquad (4)$$

is shifted to a much more favored process in the presence of atomic oxygen

$$O_{(g)} + 2H^+ + 2e^- \rightarrow H_2O \qquad E°,\ +2.42V \qquad (5)$$

Likewise, the electron-transfer reduction of H_2O [with zero-valent (neutral charge) hydrogen and oxygen], which must overcome the stabilization of the strong O-H bonds, results in a -1 charge for oxygen rather than proton reduction

$$H_2O + e^- \rightarrow 1/2\ H_2 + {}^-OH \qquad E°,\ -0.82V \text{ vs. NHE} \qquad (6)$$

In contrast, an equivalent charge-density change for the neutral oxygen in ·OH is strongly favored[7]

$$\cdot OH + e^- \rightarrow {}^-OH \qquad E°,\ +1.89V \qquad (7)$$

and reduction of the combination of the zero-charge oxygen of water with -1-charge $O^-\cdot$ yields two -1-charge hydroxide ions[4]

$$O^-\cdot + H_2O + e^- \rightarrow 2\ {}^-OH \qquad E°,\ +1.77V \qquad (8)$$

Although the chlorine atom is among the strongest one-electron oxidants known[4]

$$Cl\cdot + e^- \rightarrow Cl^- \qquad E°,\ +2.41V \qquad (9)$$

when it is stabilized by the covalent bond of the Cl_2 molecule and of the ClO^- ion its propensity to change valence is reduced[3,4]

$$Cl_2 + e^- \rightarrow Cl_2^-\cdot \qquad E°,\ +0.63V \qquad (10)$$

$$Cl_2 + 2e^- \rightarrow 2Cl^- \qquad E°,\ +1.36V \qquad (11)$$

$ClO^- + e^- \rightarrow Cl^- + O^{-}\cdot$ $E°$, +0.02V   (12)

$ClO^- + H_2O + 2e^- \rightarrow Cl^- + 2\ {}^-OH$ $E°$, +0.89V   (13)

Thus, Cl_2 can transfer an electron from an ^-OH because of the stabilization of $\cdot OH$ via bond formation with an $H\cdot$ from another ^-OH and of the resulting $O^-\cdot$ via bond formation with a $Cl\cdot$ [$\Delta E°$, +0.47V (Eq. (11) - Eq. (13))]

$Cl_2 + 2OH^- \rightarrow ClO^- + Cl^- + H_2O$ (14)

NUCLEOPHILIC ATTACK VIA SINGLE-ELECTRON TRANSFER

The facile nucleophilic displacement of chloride ion from alkyl chlorine compounds (all 0-charge atoms) by ^-OH (-1-charge O) to give an alcohol (with all neutral-charge atoms) is a charge-transfer process via electron transfer from the ^-OH to the chlorine atom of RCl[16,17]

$n\text{-}BuCl + {}^-OH \rightarrow n\text{-}BuOH + Cl^-$ (15)

Likewise the nucleophilic addition of ^-OH (-1-charge O) to CO_2 gives bicarbonate ion with the charge of the oxygen of the C-OH group changed to zero via electron transfer (and charge transfer) to a neutral carbonyl oxygen

$O{=}C{=}O + {}^-OH \rightarrow HO\text{-}C({=}O)\text{-}O^-$ (16)

The dissolution of $Na_2O(s)$ (-2-charge O) in water occurs via a charge transfer (and electron transfer) from O^{2-} to the neutral oxygen of H_2O to give ^-OH (-1-charge O)

$Na_2O(s) + H_2O \rightarrow 2Na^+ + 2\ {}^-OH$ (17)

The data of Figures 1 and 2 and of Tables 3 and 4 provide persuasive evidence that the electron-transfer and atom-transfer chemistry for dioxygen species and monoxygen species is strongly dependent upon proton activity, anion solvation, and relative X-OO and Y-O covalent bond energies. Hence, the solvation energy of ^-OH is 22.5 kcal less in acetonitrile than in water, which causes the $\cdot OH/{}^-OH$ redox potential to be +0.92V vs NHE in MeCN instead of +1.89V vs NHE in H_2O.

Table 3. Redox Potentials for Atomic Oxygen Species in Aqueous Media

	E°, V vs. NHE
pH 0:	
$O_{(g)} + 2H^+ + 2e^- \rightarrow H_2O$	+2.42
$O_{(g)} + H^+ + e^- \rightarrow \cdot OH$	+2.12
$\cdot OH + H^+ + e^- \rightarrow H_2O$	+2.72
$O_{3(g)} + 2H^+ + 2e^- \rightarrow O_{2(g)} + H_2O$	+2.07
$O_3 + e^- \rightarrow O_3^{-}\cdot$	+0.95
$HOOH + 2H^+ + 2e^- \rightarrow 2H_2O$	+1.76
$HOIO_3 + H^+ + 2e^- \rightarrow IO_3^- + H_2O$	+1.6
$HOCl + 2H^+ + 2e^- \rightarrow HCl + H_2O$	+1.49
pH 7:	
$O_{(g)} + 2H^+ + 2e^- \rightarrow H_2O$	+2.01
$O_{(g)} + H^+ + e^- \rightarrow \cdot OH$	+1.71
$\cdot OH + H^+ + e^- \rightarrow H_2O$	+2.31
$O_{3(g)} + 2H^+ + 2e^- \rightarrow O_{2(g)} + H_2O$	+1.66
$O_{3(g)} + e^- \rightarrow O_3^{-}\cdot$	+0.95
$HOOH + 2H^+ + 2e^- \rightarrow 2H_2O$	+1.35
$IO_4^- + 2H^+ + 2e^- \rightarrow IO_3^- + H_2O$	+1.2
pH 14:	
$O_{(g)} + H_2O + 2e^- \rightarrow 2\ ^-OH$	+1.60
$O_{(g)} + H_2O + e^- \rightarrow \cdot OH + ^-OH$	+1.31
$O_{(g)} + e^- \rightarrow O^{-}\cdot$	+1.43
$\cdot OH + e^- \rightarrow ^-OH$	+1.89
$HOH + e^- \rightarrow H\cdot + ^-OH$	-2.93
$O^{-}\cdot + H_2O + e^- \rightarrow 2\ ^-OH$	+1.77
$O_{3(g)} + H_2O + 2e^- \rightarrow O_{2(g)} + 2\ ^-OH$	+1.25
$O_{3(g)} + e^- \rightarrow O_3^{-}\cdot$	+0.95
$O_3^{-}\cdot + H_2O + e^- \rightarrow O_{2(g)} + 2\ ^-OH$	+1.55
$HOO^- + H_2O + 2e^- \rightarrow 3\ ^-OH$	+0.87
$IO_4^- + H_2O + 2e^- \rightarrow IO_3^- + 2\ ^-OH$	+0.8
$ClO^- + H_2O + 2e^- \rightarrow Cl^- + 2\ ^-OH$	+0.89
$ClO^- + e^- \rightarrow Cl^- + O^{-}\cdot$	+0.02

Table 4. Redox Potentials for Atomic Oxygen Species in MeCN [H+ ≡ 1M $(H_3O)ClO_4$; ^-OH ≡ 1M $(Bu_4N)OH(MeOH)$]

	E°, V vs. NHE
Acid, pH(-8.8):	
$O_{(g)} + H^+ + e^- \rightarrow \cdot OH$	+2.64
$\cdot OH + H^+ + e^- \rightarrow H_2O$	+3.24
$O_{(g)} + 2H^+ + 2e^- \rightarrow H_2O$	+2.94
$O_{3(g)} + 2H^+ + 2e^- \rightarrow O_{2(g)} + H_2O$	+2.59
$HOOH + 2H^+ + 2e^- \rightarrow 2H_2O$	+2.28
$HOCl + 2H^+ + 2e^- \rightarrow HCl + H_2O$	+2.0
Base, pH 30.4:	
$O_{(g)} + H_2O + e^- \rightarrow \cdot OH + {}^-OH \rightarrow O^{-}\cdot\ (H_2O)$	+0.34
$O^{-}\cdot + H_2O + e^- \rightarrow 2\ {}^-OH$	+0.80
$\cdot OH + e^- \rightarrow {}^-OH$	+0.92
$HOH + e^- \rightarrow H\cdot + {}^-OH$	-3.90
$O_{(g)} + H_2O + 2e^- \rightarrow 2\ {}^-OH$	+0.63
$O_3 + H_2O + 2e^- \rightarrow O_2 + 2\ {}^-OH$	+0.28
$ClO^- + H_2O + 2e^- \rightarrow Cl^- + 2\ {}^-OH$	-0.08
$HOO^- + H_2O + 2e^- \rightarrow 3\ {}^-OH$	-0.10

ACKNOWLEDGMENTS

This work was supported by the National Science Foundation under grant No. CHE-8516247.

REFERENCES

1. W.C. Barrette, Jr., H.W. Johnson, Jr., and D.T. Sawyer, Voltammetric evaluation of the effective acidities (pK_a') for Brønsted acids in aprotic solvents, Anal. Chem. 56:1890 (1984).
2. D.T. Sawyer, G. Chiericato, Jr., C.T. Angelis, E.J. Nanni, Jr., and T. Tsuchiya, Effects of media and electrode materials on the electrochemical reduction of dioxygen, Anal. Chem. 54:1720(1982).
3. R. Parsons, "Handbook of Electrochemical Constants," Butterworths Scientific Publications, London (1959).
4. H.A. Schwarz and R.W. Dodson, Equilibrium between hydroxyl radicals and thallium(II) and the oxidation potential of $OH^-(aq)$, J. Phys. Chem. 88:3643 (1984).
5. J. Wilshire and D.T. Sawyer, Redox chemistry of dioxygen species, Acc. Chem. Res. 12:105 (1979).
6. P. Cofré and D.T. Sawyer, Electrochemical reduction of dioxygen to perhydroxyl ($HO_2\cdot$) in aprotic solvents that contain Brønsted acids, Anal. Chem. 58:1057 (1986).

7. P. Cofré and D.T. Sawyer, Redox chemistry of hydrogen peroxide in anhydrous acetonitrile, Inorg. Chem. 25:2089 (1986).
8. D.T. Sawyer, J.L. Roberts, Jr., T. Tsuchiya, and G.S. Srivatsa, Generation of activated oxygen species by electron-transfer reduction of dioxygen in the presence of protons, chlorinated hydrocarbons, methyl viologen, and transition metal ions, in: "Oxygen Radicals in Chemistry and Biology," W. Bors, M. Saran and D. Tait, eds., Walter de Gruyter and Co., Berlin, (1984).
9. J.L. Roberts, Jr., M.M. Morrison, and D.T. Sawyer, Base-induced generation of superoxide ion and hydroxyl radical from hydrogen peroxide, J. Am. Chem. Soc. 100:329 (1978).
10. J.L. Roberts, Jr. and D.T. Sawyer, Activation of superoxide ion by reactions with protons. Electrophiles, secondary amines, radicals, and reduced metal ions, Israel J. Chem. 23:430 (1983).
11. D.-H. Chin, G. Chiericato, Jr., E.J. Nanni, Jr., and D.T. Sawyer, Proton-induced disproportionation of superoxide ion in aprotic media, J. Am. Chem. Soc. 104:1296 (1982).
12. M.J. Gibian, D.T. Sawyer, T. Ungerman, R.Tangpoonpholvivat, and M.M. Morrison, Reactivity of superoxide ion with carbonyl compounds in aprotic solvents," J. Am. Chem. Soc. 101:640 (1979).
13. I. Fridovich, "The Biology of Superoxide and of Superoxide Dismutases--In Brief," in: "Oxygen and Oxy Radicals," M.A.J. Rodgers and E.L. Powers, eds., Academic Press, New York, (1981).
14. J.A. Fee, Is Superoxide Toxic and are Superoxide Dismutases Essential for Aerobic Life?," in: "Oxygen and Oxy Radicals," M.A.J. Rodgers and E.L. Powers, eds., Academic Press, New York, (1981).
15. P.K.S. Tsang, P. Cofré, and D.T. Sawyer, Electrochemical oxidation of hydroxide ion in acetonitrile and its facilitation by transition-metal complexes, Inorg. Chem. 26:3604 (1987).
16. J.L. Roberts, Jr. and D.T. Sawyer, Facile degradation by superoxide ion of carbon tetrachloride, chloroform, methylene chloride, and p,p'-DDT in aprotic media, J. Am. Chem. Soc. 103:712 (1981).
17. J.L. Roberts, Jr., T.S. Calderwood, and D.T. Sawyer, Oxygenation by superoxide ion of CCl_4, $FCCl_3$, $HCCl_3$, p,p'-DDT, and related trichloromethyl substrates ($RCCl_3$) in aprotic solvents, J. Am. Chem. Soc. 105:7691 (1983).

ELECTRON SPIN RESONANCE INVESTIGATIONS OF OXYGEN-CENTERED FREE RADICALS IN BIOLOGICAL SYSTEMS

Ronald P. Mason and Kim M. Morehouse

Laboratory of Molecular Biophysics
National Institute of Environmental Health Sciences
PO Box 12233
Research Triangle Park, North Carolina 27709, U.S.A.

INTRODUCTION

Oxygen-centered free radicals have been detected directly with ESR in a variety of biological processes such as lipid autoxidation,[1] the enzymatic formation of superoxide by xanthine oxidase,[2] and hydroxyl radical formation by the γ-irradiation of ice.[3] Since the common isotope of oxygen is spinless, no nuclear hyperfine interaction is possible, and the g-value and any hydrogen hyperfine coupling provide the only criteria for distinguishing an oxygen-centered free radical from other free radicals. The greatest limitation on the direct detection of oxygen-centered free radicals, aside from their high reactivity, is their paramagnetic properties.

Although the hydroxyl radical has been observed with ESR in ice, its spectrum has never been detected in the liquid phase. Acording to Symons,[4] this is largely due to the fact that the unpaired electron is in an orbitally degenerate state, which causes the hydroxyl radical g-tensor to vary from 0 to 4. In the absence of an asymmetric hydrogen bonding (such as in ice) to lift this degeneracy, the ESR spectrum is so broad that the hydroxyl radical cannot be detected in a fluid state. Similar reasoning has been advanced to explain the failure to detect alkoxyl radicals in solution.[4] The g-anisotropy of the peroxyl radical is much less, and tert-peroxyl radicals have been detected in hematin/tert-hydroperoxide solutions.[5] The reactivity of secondary peroxyl free radicals, such as lipid peroxyl free radicals, is much greater due their decay by the Russell mechanism,[6]

$$2\ LOO^{\cdot} \rightarrow LOH + L{=}O + {}^{1}O_2 \qquad (1)$$

and internal cyclization.[7] As a result, the direct ESR investigation of the autoxidation of unsaturated lipids is limited to low temperatures.

The limited use of direct ESR in the investigation of oxygen-centered free radicals can be traced to the general necessity for low temperatures, which may affect the results. In addition, for enzymatic systems the technically complex freeze-quench technique must be used.

These limitations of direct ESR approaches have led to the increasing use of the spin trapping technique. In this approach, a reactive free radical reacts with a double bond of a diamagnetic compound, the radical trap, to form a less reactive free radical, the radical adduct. The most useful radical trap for oxygen-centered free radicals is 5,5-dimethyl-1-pyrroline N-oxide (DMPO), which has been used extensively to study superoxide[8] and hydroxyl radical[9] formation in biological systems.

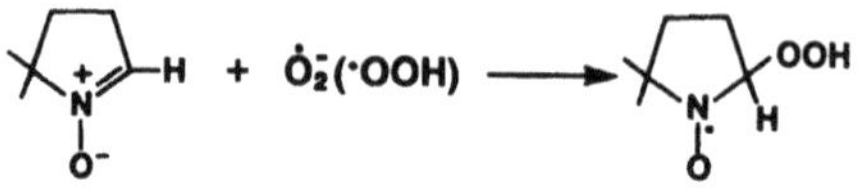

(2)

$$DMPO + \dot{O}_2^-(\cdot OOH) \longrightarrow DMPO/\dot{O}_2^-(\cdot OOH) \quad (3)$$

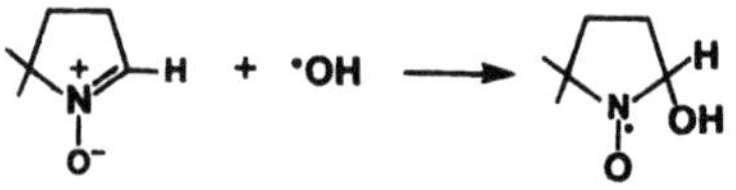

(4)

$$DMPO + \cdot OH \longrightarrow DMPO/\cdot OH \quad (5)$$

The DMPO/superoxide and DMPO/hydroxyl radical adducts were first assigned in 1974 using chemical evidence for their formation by Harbour et al.[10] We have reviewed the chemical evidence for the structure of these radical adducts and shown them to be oxygen-centered by using [$^{17}O_2$]oxygen.[11] The first report of spin trapping an oxygen-centered radical in a biological system was the trapping of the light-induced superoxide radical anion in chloroplasts by Harbour et al.[12] Since that first report, the use of spin trapping to detect oxygen-centered radicals in biological systems has multipled. Although peroxyl[5] and even alkoxyl[13] radical adducts of DMPO have been reported in biological systems, most investigations have focused on the formation of radical adducts of superoxide and hydroxyl radicals. The rest of this review will center on the problems involved in proving that the adducts formed are not artifactual, rather than on the vast literature.

If the superoxide radical is trapped, then the formation of the radical adduct must be inhibited by superoxide dismutase.[14]

$$DMPO + \dot{O}_2^-(\cdot OOH) \xrightarrow[]{SOD \;\; X} DMPO/\dot{O}_2^-(\cdot OOH) \quad (6)$$

Often the formation of the DMPO/·OH radical adduct is also inhibited by superoxide dismutase, which implies that (1) the hydroxyl radical formation is dependent on superoxide (perhaps by the reduction of ferric iron for the Fenton reaction) or (2) superoxide is trapped first and the resulting superoxide radical adduct is reduced to form the hydroxyl radical adduct.[14]

$$DMPO/\cdot OOH \xrightarrow{reduce} DMPO/\cdot OH \quad (7)$$

This is the reduction of a hydroperoxide to an alcohol. Glutathione peroxidase catalyzes this reaction.[15] A third possibility is that the DMPO/superoxide radical adduct decomposes to the hydroxyl radical, which may then be trapped.[16] Despite these reports, the ESR spectrum of the DMPO/superoxide radical adduct has been obtained from incubations of NADPH oxidase without obvious DMPO/·OH radical adduct contamination,[17] implying that the formation of DMPO/·OH from DMPO/superoxide radical adduct is not spontaneous, but requires some additional reactant.[18] In any case, a good indication that something artifactual is occurring is the inability of catalase to inhibit hydroxyl radical adduct formation, a most unlikely event. In short, DMPO is not a useful trap for the hydroxyl radical if superoxide is directly involved in the hydroxyl radical formation.

In addition, due to air oxidation of the hydroxylamine hydrolysis product of DMPO, the DMPO/hydroxyl radical adduct is a detectable impurity when DMPO is added to buffer,[19] so weak signals are suspect.

$$\text{DMPO} + {}^{-}\text{OH} \xrightarrow{\text{slow}} \text{DMPO}/{}^{-}\text{OH} \xrightarrow{[\text{O}]} \text{DMPO}/{}^{\cdot}\text{OH} \qquad (8)$$

This type of nucleophilic reaction can give background signals in the trapping of many free radicals.

The best way to prove the existence of the hydroxyl radical in a system is to perform a competition experiment with a hydroxyl radical scavenger. Dimethylsulfoxide reacts with the hydroxyl radical to form the readily trapped methyl radical.[16]

$$^{\cdot}\text{OH} + \text{DMSO} \longrightarrow {}^{\cdot}\text{CH}_3 \qquad (9)$$

$$\text{DMPO} + {}^{\cdot}\text{CH}_3 \longrightarrow \text{DMPO}/{}^{\cdot}\text{CH}_3 \qquad (10)$$

Ethanol[16] and formate, which form the carbon-centered free radicals upon hydroxyl radical attack, can also be used in these competition experiments.[20]

$$^{\cdot}\text{OH} + \text{DMPO} \xrightarrow{k_1} \text{DMPO}/{}^{\cdot}\text{OH} \qquad (11)$$

$$^{\cdot}\text{OH} + \text{HCO}_2^{-} \xrightarrow{k_2} {}^{\cdot}\text{CO}_2^{-} + \text{H}_2\text{O} \qquad (12)$$

$$^{\cdot}\text{CO}_2^{-} + \text{DMPO} \xrightarrow{k_3} \text{DMPO}/{}^{\cdot}\text{CO}_2^{-} \qquad (13)$$

$$^{\cdot}\text{OH} + \text{CH}_3\text{CH}_2\text{OH} \xrightarrow{k_4} \text{CH}_3\dot{\text{C}}\text{HOH} + \text{H}_2\text{O} \qquad (14)$$

$$\text{CH}_3\dot{\text{C}}\text{HOH} + \text{DMPO} \xrightarrow{k_5} \text{DMPO}/\text{CH}_3\dot{\text{C}}\text{HOH} \qquad (15)$$

Most artifacts leading to DMPO/˙OH radical adduct formation will be excluded by the use of hydroxyl radical scavengers if both the scavenger-derived radical adduct is detected and a corresponding decrease in the DMPO/˙OH radical adduct concentration is found.

This approach has been extended with a quantitative kinetic criterion to exclude species which are sufficiently reactive to oxidize the "specific" hydroxyl radical scavengers.[21]

Measurement of the initial rates of formation of the DMPO/˙OH and DMPO/scavenger radical adducts removes the effects of the differential radical adduct decay rates.[22] Using this approach, the relative efficiency of two hydroxyl radical scavengers is quantitatively predictable from the known rate constants for the reactions of the hydroxyl radical with these scavengers, that is k_2 and k_4. From the ratio of the rates of formation of these two radical adducts we can calculate k_2/k_4.

$$\frac{k_2}{k_4} = \frac{d(DMPO/{}^{\bullet}CO_2^-)/dt}{d(DMPO/CH_3\dot{C}HOH)/dt} \times \frac{[CH_3CH_2OH]}{[HCO_2^-]} \tag{16}$$

In this expression, the ratio of k_2/k_4 from spin trapping should agree with the rate constants for the reaction of the hydroxyl radical with these scavengers as determined from pulse radiolysis.

We have applied this methodology to a system consisting of NADPH, ferredoxin reductase and ferredoxin.[23] Ferredoxin is a very low potential electron donor, and several groups have proposed that it can reduce hydrogen peroxide to the hydroxyl radical. The well-known ESR spectrum of the DMPO/hydroxyl radical adduct is obtained from an incubation containing ferredoxin reductase, ferredoxin, and NADPH in the presence of 0.1 mM diethylenetriaminepentaacetic acid (DTPA) (Fig. 1A). Omission of any of these components results in greatly decreased radical adduct formation (Fig. 1B-D). Oxygen is required, and superoxide dismutase has no effect. The DMPO/hydroxyl radical adduct formation was inhibited by the addition of catalase, which implies that the hydroxyl radical arises from the reduction of hydrogen peroxide. In fact, only in biochemical systems where the DMPO/$^{\bullet}$OH radical adduct is superoxide dismutase-insensitive and inhibited by catalase can a proof of hydroxyl radical formation be made. Unfortunately these criteria exclude from investigation xanthine oxidase, neutrophils, and most other systems of interest.

We used the kinetic criterion by measuring the relative rate of formation of the DMPO/$^{\bullet}$OH, DMPO/$^{\bullet}CO_2^-$, and DMPO/$CH_3\dot{C}HOH$ radical adducts. After obtaining initial rates of the three species, we compared our results for the ratio of rate constants with the pulse radiolysis literature and obtained agreement within a factor of 1.5 to 2.5, which we consider to be probably as good as can be expected.[23]

At this point you may think we have proven that ferrodoxin reduces hydrogen peroxide to the hydroxyl radical, but this is not true. The effect of various metal chelators was studied to investigate the dependence of the production of the hydroxyl radical by this enzymatic system on adventitious transition metals. When no chelator was present, or when Desferal was used in place of DTPA (Fig. 1E), a very large decrease of the DMPO/hydroxyl radical adduct was detected. When EDTA was added in place of DTPA, DMPO/$^{\bullet}$OH radical adduct formation was as rapid as with DTPA. Even when the buffer was pretreated with Chelex 100, the addition of EDTA caused rapid DMPO/hydroxyl radical adduct formation.

These results suggest that transition metals chelated with DTPA or EDTA are responsible for the formation of hydroxyl radicals. Furthermore, they indicate that very low concentrations of transition metals are required to saturate this reaction. From our data, we conclude that reduced ferredoxin is responsible for reducing the chelated transition metals, which then reduce hydrogen peroxide to form the hydroxyl radical. Apparently phosphate buffer contains high enough concentrations of contaminating transition metals to facilitate this reaction. Presumably if buffer, NADPH, ferredoxin reductase, ferredoxin, and transition metal chelators could be obtained free of contaminating transition metals, no DMPO/hydroxyl radical adduct would be detected, even in the presence of EDTA or DTPA. In conclusion, no matter how low the potential of the biochemical electron donor, Fenton chemistry prevails, and even the very low potential ferredoxin does not form the hydroxyl radical directly. Using other approaches and other systems, Borg

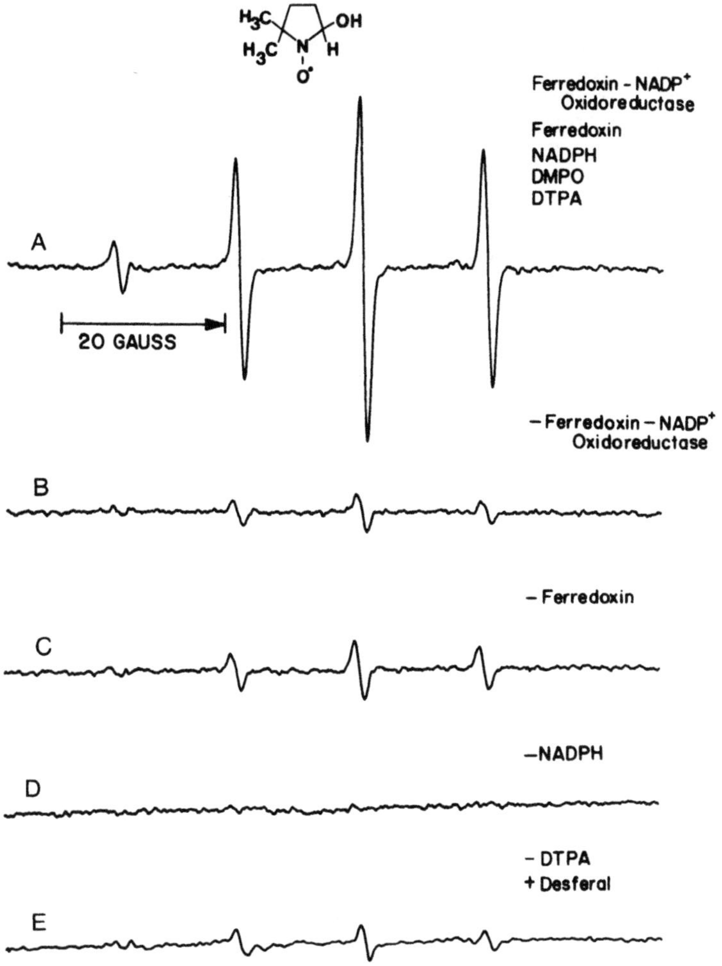

Figure 1: ESR spectra of the DMPO/hydroxyl radical adduct formed during the enzymatic reduction of molecular oxygen. A) An aerobic incubation containing 0.02 U/ml ferredoxin:NADP$^+$ oxidoreductase, 50 μg/ml ferredoxin, 0.5 mM NADPH and 100 mM DMPO in 100 mM potassium phosphate buffer (pH 7.4) containing 0.1 mM DTPA. B) Same as A, but in the absence of oxidoreductase. C) Same as A, but in the absence of ferredoxin. D) Same as A, but in the absence of NADPH. E) Same as A, with 0.1 mM Desferal replacing DTPA.

et al.[24] and Winterbourn and Sutton[25] have come to similar conclusions concerning the inevitability of the chemical Fenton reaction in biochemical systems.

REFERENCES

1. J. Yanex, C. L. Sevilla, D. Becker, and M. D. Sevilla, Low-temperature autoxidation in unsaturated lipids: An electron spin resonance study, J. Phys. Chem. 91:487 (1987).

2. P. F. Knowles, J. F. Gibson, F. M. Pick, and R. C. Bray, Electron-spin-resonance evidence for enzymic reduction of oxygen to a free radical, the superoxide ion, Biochem. J. 111:53 (1969).
3. J. K. Kochi, Oxygen radicals, in: "Free Radicals", Volume II, J. K. Kochi, ed., John Wiley & Sons, New York (1973).
4. M. C. R. Symons, Electron spin resonance spectra of organic oxy radicals. The radical $(CH_3)_3CO_3$, J. Am. Chem. Soc. 91:5924 (1969).
5. B. Kalyanaraman, C. Mottley, and R. P. Mason, A direct electron spin resonance and spin-trapping investigation of peroxyl free radical formation by hematin/hydroperoxide systems, J. Biol. Chem. 258:3855 (1983).
6. K. U. Ingold, Peroxy radicals, Acct. Chem. Res. 2:1 (1969).
7. N. A. Porter, L. S. Lehman, B. A. Weber, and K. J. Smith, Unified mechanism for polyunsaturated fatty acid autoxidation. Competition of peroxy radical hydrogen atom abstraction, β-scission, and cyclization, J. Am. Chem. Soc. 103:6447 (1981).
8. P. J. Thornalley and J. V. Bannister, The spin trapping of superoxide radicals, in: "CRC Handbook of Methods for Oxygen Radical Research," R. A. Greenwald, ed., CRC Press, Boca Raton (1985).
9. G. R. Buettner, Spin trapping of hyroxyl radical, in: "CRC Handbook of Methods for Oxygen Radical Research," R. A. Greenwald, ed., CRC Press, Boca Raton (1985).
10. J. R. Harbour, V. Chow, and J. R. Bolton, An electron spin resonance study of the spin adducts of OH and HO_2 radicals with nitrones in the ultraviolet photolysis of aqueous hydrogen peroxide solutions, Can. J. Chem. 52:3549 (1974).
11. C. Mottley, H. D. Connor, and R. P. Mason, [^{17}O]oxygen hyperfine structure for the hydroxyl and superoxide radical adducts of the spin traps, DMPO, PBN and 4-POBN, Biochem. Biophys. Res. Commun. 141:622 (1986).
12. J. R. Harbour and J. R. Bolton, Superoxide formation in spinach chloroplasts: Electron spin resonance detection by spin trapping, Biochem. Biophys. Res. Commun. 64:803 (1975).
13. B. G. Taffe, N. Takahashi, T. W. Kensler, and R. P. Mason, Generation of free radicals from organic hydroperoxide tumor promoters in isolated mouse keratinocytes, J. Biol. Chem. 262:12143 (1987).
14. E. Finkelstein, G. M. Rosen, E. J. Rauckman, and J. Paxton, Spin trapping of superoxide, Mol. Pharmacol. 16:676 (1979).
15. G. M. Rosen and B. A. Freeman, Detection of superoxide generated by endothelial cells, Proc. Natl. Acad. Sci. USA 81:7269 (1984).
16. E. Finkelstein, G. M. Rosen, and E. J. Rauckman, Production of hydroxyl radical by decomposition of superoxide spin-trapped adducts, Mol. Pharmacol. 21:262 (1982).
17. J. V. Bannister, P. Bellavite, M. C. Serra, P. J. Thornalley, and F. Rossi, An EPR study of the production of superoxide radicals by neutrophil NADPH oxidase, FEBS Lett. 145:323 (1982).
18. P. J. Thornalley, Theory and biological applications of the electron spin resonance technique of spin trapping, Life Chem. Reports 4:57 (1986).
19. R. A. Floyd and B. B. Wiseman, Spin-trapping free radicals in the autooxidation of 6-hydroxydopamine, Biochim. Biophys. Acta 586:196 (1979).
20. J. R. Harbour and J. R. Bolton, The involvement of the hydroxyl radical in the destructive photooxidation of chlorophylls in vivo and in vitro, Photochem. Photobiol. 28:231 (1978).
21. J. W. Lown and H-H. Chen, Evidence for the generation of free hydroxyl radicals from certain quinone antitumor antibiotics upon reductive activation in solution, Can. J. Chem. 59:390 (1981).

22. A. L. Castelhano, M. J. Perkins, and D. Griller, Spin trapping of hydroxyl in water: Decay kinetics for the ·OH and $CO_2^{-\cdot}$ adducts to 5,5-dimethyl-1-pyrroline-N-oxide, Can. J. Chem. 61:298 (1983).
23. K. M. Morehouse and R. P. Mason, The transition metal-mediated formation of the hydroxyl free radical during the reduction of molecular oxygen by ferredoxin-ferredoxin:$NADP^+$ oxidoreductase, J. Biol. Chem. 263:1204 (1988).
24. D. C. Borg, K. M. Schaich, and A. Forman, Autoxidative cytotoxicity: Is there metal-independent formation of hydroxyl radicals? Are there "crypto-hydroxyl" radicals?, in: "Oxygen Radicals in Chemistry and Biology," W. Bors, M. Saran, and D. Tait, eds., Walter de Gruyter, Berlin (1984).
25. C. C. Winterbourn and H. C. Sutton, Hydroxyl radical production from hydrogen peroxide and enzymatically generated paraquat radicals: Catalytic requirements and oxygen dependence, Arch. Biochem. Biophys. 235:116 (1984).

SUPEROXIDE CHEMISTRY IN NON-AQUEOUS MEDIA

Aryeh A. Frimer

Department of Chemistry
Bar-Ilan University
Ramat-Gan 52100, Israel

INTRODUCTION

The focal role played by molecular oxygen and oxygen species in living systems seems to have been realized almost from the very discovery of oxygen more than two centuries ago.[1] The continuing interest in oxygen stems from the complicated role it plays not only in the "breath of life" but also more interestingly in oxygen toxicity - what might poetically be called the "breath of death." One of the active oxygen species implicated in this detrimental process is the ubiquitous one-electron reduction product of molecular oxygen, superoxide anion radical[2-4], $O_2^{\cdot-}$

Although most superoxide activity is expected in the hydrophilic areas of the cell, a study of the chemistry of $O_2^{\cdot-}$ in aprotic media was deemed of great importance for the following reasons. First, as Irwin Fridovich[5] has so poignantly noted, "If we are going to try to understand how $O_2^{\cdot-}$ does its "dirty work", we must first learn what it is capable of doing." Second, studies carried out in aqueous media involve not only $O_2^{\cdot-}$, but also its conjugate acid HOO· , its acid catalyzed disproportionation products HOO^- and H_2O_2, and the Haber-Weiss reaction product HO· [see equations 1-3 below].[6]

$$O_2^{\cdot-} + H^+ \rightleftharpoons HOO\cdot \quad pK_a\,(HO_2\cdot) = 4.69 \qquad (1)$$

$$HOO\cdot + O_2^{\cdot-} \longrightarrow HOO^- + O_2 \quad k_2 = 10^8\ M^{-1}s^{-1} \qquad (2)$$

$$HOOH + O_2^{\cdot-} \xrightarrow{Fe^{+3}} HO\cdot + HO^- + O_2 \qquad (3)$$

$$2O_2^{\cdot-} \longrightarrow {}^3O_2 + O_2^{-2} \quad \Delta G > 28\ Kcal/mol$$

In aprotic media, on the other hand, solvent-induced acid-catalyzed disproportionation (equations 1 and 2) is essentially absent, while highly unfavorable energetics rule out spontaneous disproportionation.[7] Third, aprotic organic solvents are more likely to resemble the hydrophobic phospholipid bilayer of membranes or the active sites in many enzymes. Finally, organic chemists were curious what "magic" could be wrought with this relatively new and unexplored reagent.

GENERATION OF SUPEROXIDE IN APROTIC MEDIA

Two basic approaches are presently employed for preparing superoxide solutions in aprotic media. The first involves in situ electrolytic reduction of molecular oxygen, which permits the controlled generation of low concentrations ($< 10^{-2}$ M) of pure $O_2^{\overline{\cdot}}$ and is well suited for mechanistic studies.[8,9] Alternatively, superoxide salts can serve as well-defined sources of $O_2^{\overline{\cdot}}$. Thus, tetramethylammonium superoxide has been synthesized and is quite soluble in a number of organic solvents.[10,11] More convenient, however, is the commercially available potassium superoxide (KO_2) which can be solubilized in aprotic organic solvents (e.g., benzene, toluene, pyridine, acetonitrile, DMSO, DMF, THF, and triethylamine) through the agency of crown ethers.[12] This latter method generates high enough levels of $O_2^{\overline{\cdot}}$ ($\sim$ 0.1 M) to permit product studies and synthesis.

A priori, the crown ether-solubilized $O_2^{\overline{\cdot}}$ might either be paired (equation 4, path a) with the potassium cation-crown ether complex [$C(K^+)$] or exist as a free ion (path b).

$$C(K^+) + O_2^{\overline{\cdot}} \xleftarrow[\text{b}]{\text{path}} C + KO_2 \xrightarrow[\text{a}]{\text{path}} C(K^+)O_2^{\overline{\cdot}} \qquad (4)$$

Ionic pairing (path a) is the expected state of aggregation in a non-polar aprotic medium such as benzene, and this is indeed what is found.[13]

A caveat regarding choice of solvent is in order at this juncture. Dimethylsulfoxide has been widely used for superoxide studies, originally because KO_2 is somewhat soluble in this highly polar solvent, even in the absence of crown-ether. However, as we will discuss shortly, $O_2^{\overline{\cdot}}$ is an extremely efficient base, with the ability to deprotonate even weak oxygen and carbon acids with pK_a's above 30. DMSO with a pK_a of 23 [14] is probably not inert to $O_2^{\overline{\cdot}}$. Indeed, many of the purported superoxide reactions carried out in this solvent may in fact result from the dimsyl anion or the hydroperoxy radical formed concomitantly[13,15-17] (equation 5).

$$CH_3SOCH_3 + O_2^{\overline{\cdot}} \longrightarrow CH_3SOCH_2^{\ominus} + HOO\cdot \qquad (5)$$

Similar warnings are appropriate for the slightly acidic (pK_a = 25)[14] solvent acetonitrile.[18,21] Indeed, the lifetime of electrogenerated $O_2^{\overline{\cdot}}$ in CH_3CN is reportedly ten times shorter than in pyridine.[19] Anomolous results have been reported for tetrahydrofuran. For example, $O_2^{\overline{\cdot}}$ mediated oxidation of hydroxyflavones and indoles[22-24] observed in THF simply does not occur in the truly inert solvents benzene or toluene. The problematic nature of THF was in fact noted two decades ago by two pioneers in the superoxide field, Le Berre and Berguer,[25] who discussed the instability of THF superoxide solutions. Finally, halogenated solvents are often unsuitable since they, too, interact with superoxides.[26-30]

MODES OF REACTION

The data obtained over the past decade has led scientists to suggest four basic modes of action for $O_2^{\overline{\cdot}}$ in aprotic media, including electron transfer, nucleophilic substitution, deprotonation, and perhaps hydrogen abstraction.[31,32] It is important to remember, however, that subsequent to each of these primary modes, secondary oxidative processes can, and generally do, take over. Let us now examine each of these modes in detail.

Electron transfer. Electron transfer is one of the most common modes of $O_2^{\cdot-}$ in biological systems and is the essence of the dismutation process (equation 2). This is, of course, not surprising considering that the redox potential of the $O_2/O_2^{\cdot-}$ couple (vs. NHE) is -0.33V in water and -0.6V in organic solvents.[2] This quarter volt gap between aqueous and aprotic media cannot be attributed to differences in the dielectric constants of the media since the electrochemical potential of oxygen is relatively insensitive to the differing dielectric constants of a variety of aprotic solvents.[33-36] To understand this and many other phenomena related to $O_2^{\cdot-}$ activity (e.g., nucleophilicity and basicity [vide infra]), we must recall that $O_2^{\cdot-}$ is a small, hard, nonpolarizable anion. In aqueous/protic media it will be highly and tightly solvated and hence thermodynamically more stable than in aprotic media in which such solvation mechanisms are generally absent.

Superoxide plays the role of reductant with metals.[37] For example, in the extremely important metal catalyzed Haber-Weiss reaction[38] (equation 6; R = H).

$$O_2^{\cdot-} + ROOH \xrightarrow{Fe^{+3}} O_2 + RO\cdot + HO^- \quad (6)$$

$$O_2^{\cdot-} + Fe^{+3} \longrightarrow Fe^{+2} + O_2 \quad (7)$$

$$Fe^{+2} + ROOH \longrightarrow Fe^{+3} + RO^- + HO \quad (8)$$

Superoxide serves to reduce the metal cation (Fe^{+3}) to a lower oxidation state (Fe^{+2}; equation 7), which in turn oxidizes the peroxide linkage in a Fenton reaction (equation 8). Hence, the metal-catalyzed Haber-Weiss reaction is, in fact, a superoxide-driven Fenton process.

Superoxide will not reduce peroxides directly,[31,32,39] nor does it interact with simple olefins (including arachadonic acid and other PUFA) or aromatic compounds. It will, however, transfer electrons to good electron acceptors[31,32,40] such as quinones,[31,32] diketones,[31,31,41,42] chalcones,[40,43] nitrobenzenes,[40,44] as well as electron-poor cyano- and nitroolefins[40] (equation 9-13). Recently Gibian and Russo[45] have demonstrated that the rapid superoxide-mediated cis-trans isomerization of 1,3-di-t-butylpropenone occurs via electron transfer as well (equation 14).

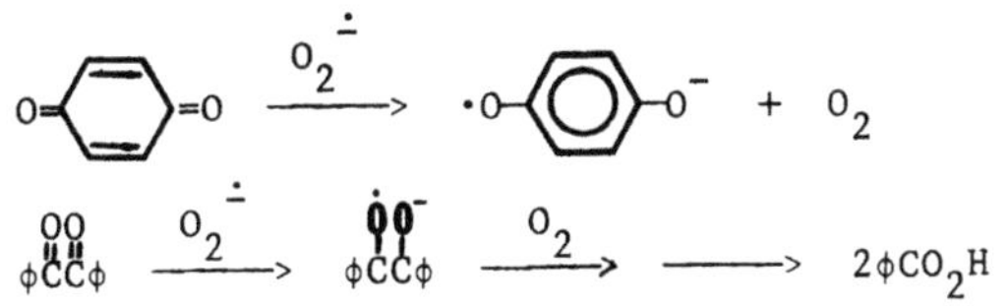

(9)
(10)

Figure 1

$$Ar^{1}\overset{O}{\overset{\|}{C}}CH{=}CHAr^{2} \xrightarrow{O_2^{\overset{\cdot}{-}}} [Ar^{1}\overset{O}{\overset{\|}{C}}CH{=}CHAr^{2}]^{\overset{\cdot}{-}} \xrightarrow{O_2} \longrightarrow \quad (11)$$

$$Ar^{1}CO_2H, Ar^{2}CO_2H, Ar^{2}CH_2CO_2H$$

$$XC_6H_4NO_2 \xrightarrow{O_2^{\overset{\cdot}{-}}} [XC_6H_4NO_2]^{\overset{\cdot}{-}} \xrightarrow{O_2} \longrightarrow HOC_6H_4NO_2 \quad (12)$$

X = F, Cl, Br, I, NO_2

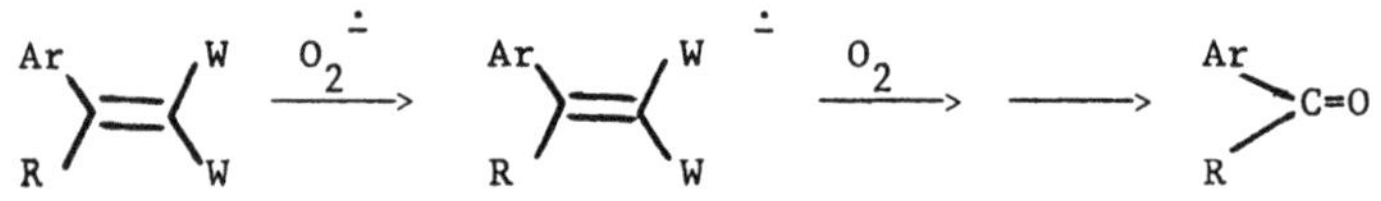

(13)

R = H, alkyl, aryl W = NO_2, CN

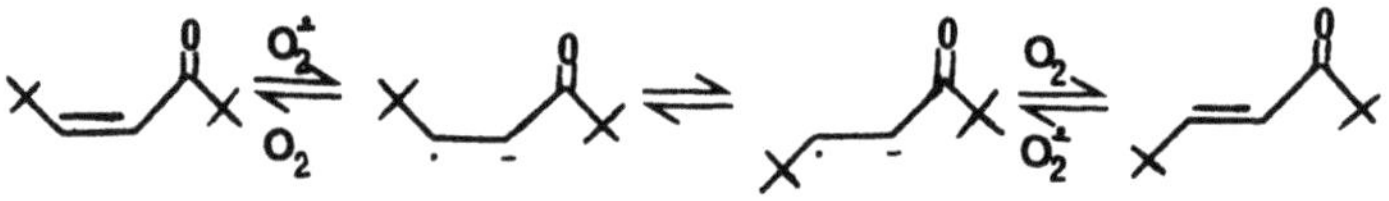

(14)

Figure 2

Of some interest is the question of the $O_2/O_2^{\overset{\cdot}{-}}$ self-exchange rate[37] (equation 15) for which widely variant values ranging from 10^{-8} - 10^{7} $M^{-1}S^{-1}$ have been obtained in aqueous media.

$$O_2^{\overset{\cdot}{-}} + O_2 \longrightarrow O_2 + O_2^{\overset{\cdot}{-}} \quad (15)$$

One relevant observation[40] for aprotic media is that KO_2^{36} (O^{18}-labelled potassium superoxide) solubilized in benzene with 18-crown-6 did not undergo observable exchange with atmospheric oxygen, O_2^{32}. The reason for the absence of this thermoneutral process may be attributed to the very different solvation and bond lengths of neutral dioxygen and the superoxide anion. These factors result in a high energy of activation for self-exchange.

In all the above cases in which $O_2^{\overset{\cdot}{-}}$ transfers an electron to a reducible substrate, including the dismutation process (equation 2), the question arises whether the molecular oxygen generated is of singlet ($^1O_2$) or triplet ($^3O_2$) multiplicity[31,32] (see scheme below).

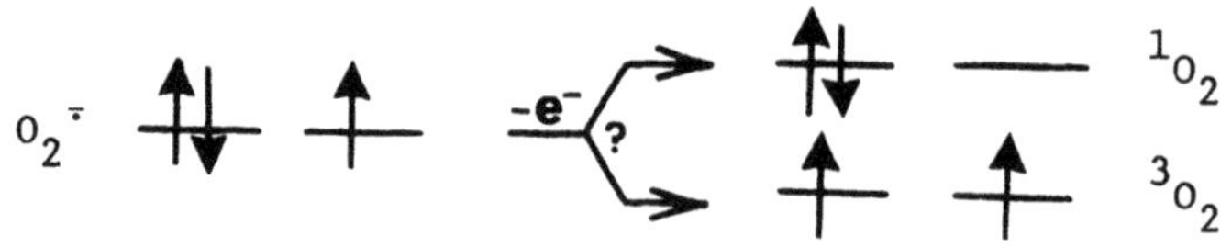

The biological importance of this question is that $O_2^{\overset{\cdot}{-}}$ could possibly serve as a "dark" pathway precursor to the biologically toxic $^1O_2$. Nevertheless, despite extensive research on all aspects of this question, there has been essentially no experimental substantiation for the formation of $^1O_2$ in these processes. The sole exceptions are the reactions of superoxide with selected radical cations[31,32] and oxidants.[46] The generation of $^1O_2$ has also been observed in the reaction of $O_2^{\cdot-}$ with diaryl peroxides[47] and carbon tetrachloride[48]; however, this is probably the result of the decomposition of unstable peroxy intermediates[28] formed by nucleophilic attack of $O_2^{\cdot-}$ on these substrates (vide infra).

Nucleophilic attack. In aprotic media, $O_2^{\cdot-}$ is an extremely vigorous nucleophile.[31,32,49] This "supernucleophilicity" has been rationalized in terms of the α-effect and is related to the destabilizing effect of the vicinal pairs of non-bonding electrons.[31,32] Superoxide reacts with halides and sulfonates by an S_N2 process to produce hydroperoxides, peroxides, alcohols, or carbonyl compounds, depending on the substrate, reaction conditions, and work-up procedures, as outlined below.

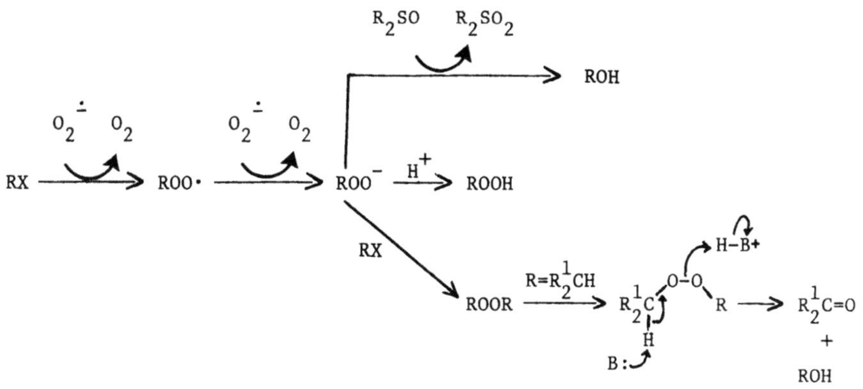

Figure 3

Similarly, esters including linoleates and acyl halides undergo nucleophilic attack, yielding the corresponding diacyl peroxides[39] or carboxylic acids.

In aqueous media, $O_2^{\cdot-}$ reacts by this mode sluggishly, if at all, as a result of the inhibitory effect of the tight hydration sphere surrounding this small, charged anion. It is plausible, however, that in the hydrophobic environment of the phospholipid bilayer, superoxide might effect membrane cleavage via hydrolysis of the ester linkages. Indeed, Goldstein and Weissman[50] have shown that liposomes exposed to fluxes of enzymatically generated superoxide gradually become leaky; however, both catalase and superoxide dismutase inhibit this process, suggesting a role for hydroxyl radical.

Basicity. The pK_a of HOO· , determined by aqueous radiolytic studies, is 4.69 (equation 1).[6] Its conjugate base, $O_2^{\cdot-}$, should be as weakly basic as acetate. Nevertheless, in aprotic media superoxide has proven to be a powerful mediator of base-catalyzed processes. This greater basicity has been rationalized from three perspectives. First, $O_2^{\cdot-}$ in aprotic media lacks the "stifling" solvation sphere that is commonplace for small hard anions in aqueous and protic media. Furthermore, in the poorly solvating aprotic media the equilibrium described by equation 1 should be shifted away from the charged species $O_2^{\cdot-}$ towards its electroneutral counterpart HOO· . Indeed, Sawyer[51] has reported a higher pK_a value for HOO· - around 12 in DMF, and this value should be even higher in apolar solvents. More importantly, however, unlike the low fluxes of superoxide obtained when it is generated enzymatically or radiolytically,

$O_2^{\cdot-}$ is longer lived and generally present in much higher concentration in aprotic media. In the presence of excess $O_2^{\cdot-}$, the facile disproportionation reaction (equation 2) between $O_2^{\cdot-}$ and its conjugate acid (HOO•) drives the unfavorably deprotonation (equation 1) to the right.

Qualitatively, we have consistently found over the past 13 years of research that the rate of a base-catalyzed autoxidative process depends on the exact nature of the base utilized and decreases in the order t-butoxide > superoxide > hydroxide. This is consistent with the work of Le Berre and Berguer[25] and Stanley[52] who found that the relative rate of oxygen evolution from the reaction of $O_2^{\cdot-}$ with methanol, propanol, heptanol, and isopropanol was 20:7:5:1. Substantial reaction with t-butanol could only be obtained when the alcohol was in large excess. It should be noted that the pK_a of n-alkyl alcohols is approximately 33 in DMF;[34] yet, these weakly acidic oxygen acids are efficiently deprotonated by $O_2^{\cdot-}$.

Similarly, weak carbon acids are also effectively deprotonated. Recently, Frimer et al.[13] carried out an in-depth kinetic study of the superoxide-mediated oxidation of substituted diphenylmethanes (pK_a values of up to ∿ 35; equation 16). The rate of reaction was found to be first order in superoxide and substrate. Furthermore, the excellent correlation of the rate constant with σ^- and the magnitudes of ρ (3.96) and the primary isotope effect ($k_H/k_D \approx 2.25$) all require a proton (not a hydrogen atom) transfer in the rate-determining step. The resulting benzylic anion undergoes subsequent oxygenation to the corresponding ketone (equation 16).

$$Ar_2CH_2 \xrightarrow{O_2^{\cdot-}} Ar_2CH^{\ominus} \xrightarrow{O_2} Ar_2CHOO^{\ominus} \longrightarrow Ar_2C{=}O \qquad (16)$$

In light of superoxide's basicity, it is not surprising that it induces elimination reactions as well as base-catalyzed autoxidative processes. In the latter category fall a variety of substrates[31,32] including ketones enones, enols[53], steroids[54], phenols (including α-tocopherol[55]) and arylmethyl) amines[56] to mention a few. Recently, Sawyer[57,58] has argued that certain substrates containing two heteroatom-hydrogen bonds in proximity such as 3,5-di-tert-butyl catechol, ascorbic acid, dihydrophenazine and dihydrolumiflavin undergo reaction with $O_2^{\cdot-}$ via a sequential transfer of a proton and a hydrogen atom. The latter step is mediated by the incipient hydroperoxy radical formed in the former step (equation 17). A dissenting view is presented by Afanas'ev.[59]

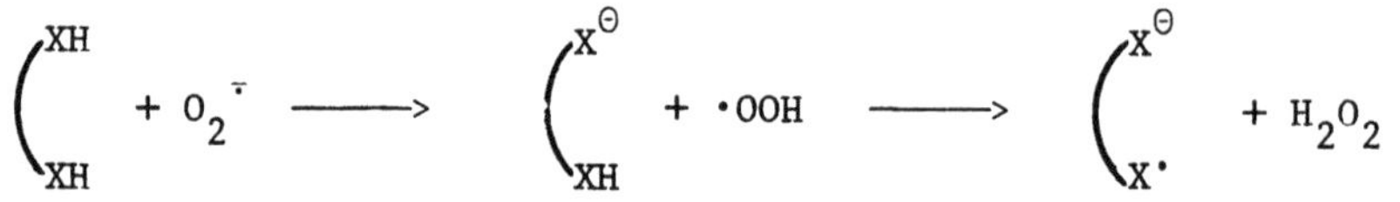

(17)

Radical reactions. Radical reactions are the final modes of action anticipated (though rarely, if ever, observed) for superoxide anion radical including hydrogen atom abstraction; addition to olefins, often initiating polymerization; and radical-radical coupling. Surprisingly, $O_2^{\cdot-}$ turns out to be a terrible radical. Regarding superoxide's ability to abstract atoms, thermodynamic calculations[60] suggest that $O_2^{\cdot-}$ is only likely to do so from those very rare substrates bearing R-H bonds approaching 66 Kcal. Indeed, the reinvestigation of systems originally assumed to be initiated by hydrogen atom abstraction mediated by $O_2^{\cdot-}$ (equation 18, path a) have nearly all turned out to be proton abstraction followed by oxidation of the resulting anion either by the hydroperoxy radical or molecular oxygen (equation 18, path b).[13,31,32] [In aqueous media, superoxide can serve as an oxidant and thereby initiate radical processes.[2-4]]

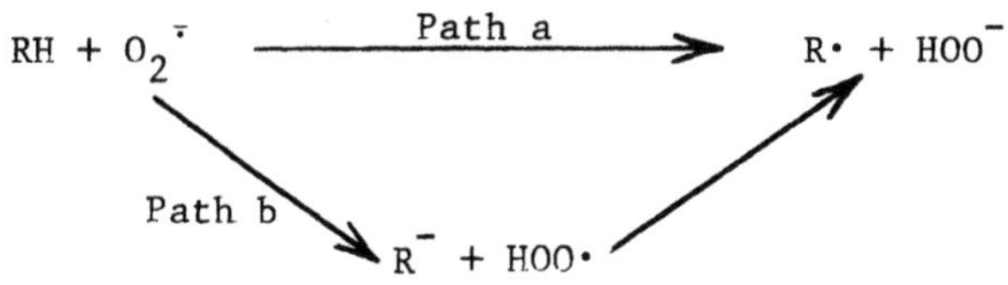

(18)

We have noted above that superoxide is generally unreactive with olefins, and here again thermodynamic calculations confirm that the radical addition of superoxide to olefinic double bonds is more than 18 Kcal endothermic.[32] $O_2^{\overline{\cdot}}$ does not initiate free radical chain polymerization and actually inhibits styrene polymerization.[52,61] Finally, radical-radical coupling, so common and rapid with normal radicals, is quite rare for $O_2^{\overline{\cdot}}$, with the noted exception of superoxide anion-radical cation couplings.[31,32,42]

RELEVANCE TO BIOLOGICAL SYSTEMS

We close by noting that the relevance of this organic chemistry to biological systems is presently being explored by Frimer and coworkers. As noted above, in aprotic media superoxide mediates the conversion of 1-halo-2,4-dinitrobenzenes to the corresponding phenols (equation 12).[40] In preliminary studies, Frimer, Hameiri-Buch, and Aljadeff[62] have incorporated these substrates (X=F,Cl,Br, or I) into the bilayer of dimyristoylphosphatidylcholine bath-sonicated liposomes. Fluxes of superoxide were generated using the xanthine oxidase-acetaldehyde system and the slow formation of phenol was followed by U.V. at ∿ 360. Research is underway to determine the mechanism and kinetics of this process.

REFERENCES

1. A. A. Frimer and I. Rosenthal in Foreword to "Active Oxygen - Part A", Israel J. Chem., 23:398 (1983).
2. See the articles of I. Fridovich and J. Fee and the subsequent discussion in "Oxygen and Oxy-Radicals in Chemistry and Biology", M. A. J. Rodgers and E. L. Powers, Eds., Academic Press, New York (1981), pp. 197-239.
3. See the exchange of correspondence between I. Fridovich, and D. T. Sawyer and J. S. Valentine, Accounts Chem. Res., 15:200 (1982).
4. R. M. Baum, Superoxide theory of oxygen toxicity is center of heated debate, Chem. Eng. News April 9, 1984 p.20-26 and subsequent Letters to the Editor on June 4, 1984, p.50 and July 2, 1984, p.47.
5. Comments of I. Fridovich in J. D. Spikes and H. M. Swartz, International conference on singlet oxygen and related species in chemistry and biology - review and general discussion, Photochem. Photobiol., 28:921 (1978).
6. B. H. J. Bielsky, Reevaluation of the spectral and kinetic properties of HO_2 and $O_2^{\overline{\cdot}}$ free radicals, Photochem. Photobiol., 28:645 (1978).
7. A. D. Goolsby and D. T. Sawyer, The electrochemical reduction of superoxide ion and the oxidation of hydroxide ion in dimethylsulfoxide, Anal. Chem., 40:83 (1968).
8. A. D. T. Sawyer and M. J. Gibian, The chemistry of superoxide ion, Tetrahedron, 35:1471 (1979); b. J. Wilshire and D. T. Sawyer, Redox chemistry of dioxygen species, Accounts Chem. Res., 12:105 (1979).
9. S. Torii, The new role of electroreductive mediators in electroorganic synthesis, Synthesis, 873 (1986); see especially 883 ff.
10. A. D. McElroy and J. S. Hasman, Synthesis of tetramethylammonium

superoxide, Inorg. Chem., 40:1798 (1964).
11. J. W. Peters and C. S. Foote, Chemistry of superoxide ion. II. Reaction of hydroperoxides, J. Amer. Chem. Soc., 98:873 (1976).
12. J. S. Valentine and A. B. Curtis, A convenient preparation of solutions of superoxide anion and the reaction of superoxide anion with a copper (II) complex, J. Amer. Chem. Soc., 97:224 (1975).
13. A. A. Frimer, T. Farkash-Solomon and G. Aljadeff, The Mechanism of the superoxide anion radical [$O_2^{\cdot-}$] mediated oxidation of diarylmethanes, J. Org. Chem., 51:2093 (1986).
14. A. Streitweiser and C. H. Heathcock,"Introduction to Organic Chemistry", Collier Macmillan, N.Y. (1976), appendix IV, p.1190.
15. I. Rosenthal and A. A. Frimer, The reaction between tetracyclone and superoxide anion radical, Tetrahedron Lett., 3731 (1975).
16. R. L. Arudi, A. D. Allen and B. H. J. Bielski, Some observations on the chemistry of KO_2 - DMSO solutions, FEBS Lett., 35:265 (1981).
17. N. Kornblum and S. Singaram, Conversion of nitriles to amides and esters to acids by the action of sodium superoxide, J. Org. Chem. 44:4727 (1979).
18. K. Shibata, Y. Saito, K. Urano and M. Matsui, Reactions of Schiff bases with superoxide ion in acetonitrile, Bull. Chem. Soc., Japan, 59:3323 (1986).
19. M. Tezuka, H. Hamada, Y. Ohkatsu and T. Osa, Reactivity of electrogenerated superoxide ion. II. Correlation between stability and reactivity of superoxide ion, Denki Kagaku, 44:17 (1976).
20. Y. Sawaki and Y. Ogata, Mechanism of the reaction of nitriles with alkaline hydrogen peroxide. Reactivity of peroxycarboximidic acid and application to superoxide ion reaction, Bull. Chem. Soc. Japan 54:793 (1981).
21. S. Oae, T. Takata and Y. H. Kim, Reaction of organic sulfur compounds with hyperoxide anion ($O_2^{\cdot-}$). IV. Evidence for formation of peroxy sulfur intermediates: oxidation of sulfoxides, phosphines and olefins with intermediary peroxysulfur species, Bull. Chem. Soc. Japan, 54:2712 (1981).
22. E. Balogh-Hergovich and G. Speier, Ring cleavage and ring expansion of indoles by superoxide ion, Tetrahedron Lett. 23:4473 (1982).
23. E. Balogh-Hergovich and G. Speier, The reactions of indoles with superoxide ion, Oxidat. Commun., 4:337 (1983);
24. M. M. A. El-Sukkary and G. Speier, Oxygenation of 3-hydroxyflavones by superoxide anion, J. Chem. Soc. Chem. Commun., 745 (1981) and G. Speier, personal communication (1984).
25. A. Le Berre and Y. Berguer, Reactions des superoxydes alcalins avec des composes organiqueś. I. Reactions radicalaires, Bull. Soc. Chim. Fr., 2363 (1966); see especially the second footnote to Table II.
26. J. L. Roberts, Jr., T. C. Calderwood, and D. T. Sawyer, Oxygenation by superoxide ion of CCl_4, $FCCl_3$, $HCCl_3$ P,P'-DDT, and related trichloromethyl substrates ($RCCl_3$) in aprotic solvents, J. Amer. Chem. Soc., 105:7691 (1983).
27. S. T. Parrington and G. B. Kenion, Superoxide: base, nucleophile, radical or electron transfer reagent, J. Chem. Soc., Chem. Commun., 731 (1982).
28. J. R. Kanofsky, Singlet oxygen production in superoxide ion - halocarbon systems, J. Amer. Chem. Soc., 108:2977 (1986).
29. C. A. Long and B. H. J. Bielsky, Rate of reaction of superoxide radical with chloride-containing species, J. Phys. Chem., 84:555 (1980).
30. D. K. Akutagawa, N. Furukawa and S. Oae, Oxidation of S,S-Diaryl-N-(p-tolylsulfonyl)sulfilimines and N-unsubstituted S,S-diaryl-sulfilimines with potassium hyperoxide anion radical ($O_2^{\cdot-}$) in the presence of 1-bromopropane, benzoyl chloride, carbon tetrachloride, chloroform or dichloromethane in aprotic media, Bull. Chem. Soc. Japan 57:1104 (1984).

31. A. A. Frimer, The organic chemistry of superoxide anion radical, in "The Chemistry of Functional Groups: Peroxides", S. Patai, ed., Wiley, Chichester (1983) pp 429-461.
32. A. A. Frimer, Organic reactions involving the superoxide anion, in "Superoxide Dismutase", L. W. Oberley ed., Chemical Rubber Co., Boca Raton, FL. (1982), Vol. 2, pp 83-125.
33. J. A. Fee and J.S. Valentine. Chemical and physical properties of superoxide, in "Superoxide and Superoxide Dismutase", A. M. Michelson, J. M. McCord and I. Fridovich eds., Academic Press, N.Y. (1977) p.19.
34. D. T. Sawyer and E. J. Nanni, Jr., Redox chemistry of O_2 and peroxides, in "Oxygen and oxy-radicals in chemistry and biology", M. A. J. Rodgers and E. L. Powers, eds., Academic Press, N.Y. (1981) p.15-44.
35. J. Chevalet, F. Rouelle, L. Gierst and J. P. Lambert, Electrogeneration and some properties of superoxide ion in aqueous solutions, J. Electroanal. Chem., 39:201 (1972).
36. R. Dietz, M. E. Peover and P. Rothbaum, Electrode reactions of organic compounds with electrochemically generated $O_2^{\cdot-}$ ion, Chem. Ing. Tech. 42:185 (1970).
37. M. S. McDavell, J. H. Espenson and A. Bakac, Kinetics of aqueous outer-sphere electron-transfer reactions of superoxide ion. Implications concerning $O_2/O_2^{\cdot-}$ self exchange rate constant, Inorg. Chem., 23:2232 (1984).
38. W. H. Koppenol, Thermodynamics of the Fenton-driven Haber-Weiss and related reactions in "Oxy-Radicals and their Scavenger Systems", G. Cohen and R. A. Greenwald, eds., Elsevier, New York (1983) Vol. I, pp 84-88.
39. See however, A.R. Forrester and V. Purushotham, Mechanism of reactions of carboxylic acid derivatives with superoxide, hydrolysis of esters by superoxide, J. Chem. Soc. Chem. Commun. 1505 and J. Chem. Soc. Perkin I, 945 (1987) who present evidence for an electron transfer mechanism in the superoxide mediated "saponification" of o-phenylbenzoyl peroxide.
40. A. A. Frimer and I. Rosenthal, Chemical reactions of superoxide anion radical in aprotic solvents, Photochem. Photobiol., 28:711 (1978).
41. See however, D. T. Sawyer, J. J. Stamp and K. A. Menton, Reactivity of superoxide ion with ethyl pyruvate, α-diketones and benzil in DMF, J. Org. Chem., 48:3733 (1983) who suggest a nucleophilic addition by $O_2^{\cdot-}$.
42. See discussion in J. L. Roberts and D. T. Sawyer, Activation of Superoxide ion by reactions with protons, electrophiles, secondary amines, radicals, and reduced metal ions, Israel J. Chem. 23:430 (1983).
43. D. C. Neckers and G. Hauck, Reactions of diphenylcyclopropenone and tetracyclones with potassium superoxide, J. Org. Chem., 48:4691 (1983).
44. See however, M. Gareil, J. Pinson and J. M. Saveant. Nucleophilic substitution of aromatic halides by electrogenerated superoxide ions, Nouveau J. Chim., 5:311 (1981) who suggest a nucleophilic addition by $O_2^{\cdot-}$.
45. M. J. Gibian and S. Russo, Electron transfer between superoxide ion and an α,β-unsaturated ketone, J. Org. Chem. 49:4304 (1984).
46. E. J. Corey, M. M. Mehrotra and A. U. Khan, Antiarthritic gold compounds effectively quench electronically excited singlet oxygen, Science, 236:68 (1987).
47. W. C. Danen and R. L. Arudi, Generation of singlet oxygen in the reaction of superoxide anion radical with diacyl peroxides, J. Amer. Chem. Soc., 100:3944 (1978).
48. A. U. Khan, Direct spectral evidence of the generation of singlet molecular oxygen ($^1\Delta_g$) in the reaction of potassium superoxide with water, J. Amer. Chem. Soc., 103:6516 (1981).
49. The original kinetic studies were carried out in DMSO, whose problematic nature we have discussed. Nevertheless, two runs were repeated in pyridine with similar results; See reference 6a Table 1.

50. I. M. Goldstein and G. Weissmann, Effects of the generation of superoxide anion on permeability of liposomes, Biochem. Biophys. Res. Commun., 75:604 (1977).
51. D. -H. Chin, G. Chiercato, Jr., E. J. Nanni, Jr., and D. T. Sawyer Proton-induced disproportionation of superoxide ion in aprotic media, J. Am. Chem. Soc., 104:1296 (1982).
52. J. P. Stanley, Reactions of superoxides and peroxides, J. Org. Chem., 45:1413 (1980).
53. A. A. Frimer, G. Aljadeff and P. Gilinsky-Sharon, Reaction of coumarins with superoxide anion radical: facile entry to o-coumarinic acid systems, Israel J. Chem., 27:39 (1986).
54. A. A. Frimer, P. Gilinsky-Sharon, J. Hameiri and G. Aljadeff, Superoxide-, tert-butoxide-, and hydroxide-mediated autoxidation of 3-oxo-Δ^4-steroids in aprotic media, J. Org. Chem., 47:2818 (1982).
55. S. Matsumoto, M. Matsuo and Y. Iitaka, Oxygenations of Vitamin E (α-tocopherol) and its model compound in the presence of potassium superoxide suspended in THF and unusual acyloin rearrangements, J. Org. Chem. 51:1435 (1986).
56. A. A. Frimer, G. Aljadeff and J. Ziv, Reaction of (arylmethyl) amines with superoxide anion radical in aprotic media. Insights into cytokynin senescence inhibition, J. Org. Chem. 48:1700 (1983).
57. D. T. Sawyer, T. S. Calderwood, C. L. Johlman and C. I. Wilkins, Oxidation by superoxide ion of catechols, ascorbic acid, dihydrophenazine and reduced flavins to their respective anion radicals. A common mechanism via a sequential proton-hydrogen atom transfer, J. Org. Chem., 50:1409 (1985) and references cited therein.
58. Cf. J. M. Fukuto, E. W. DiStefano, J. N. Burstyn, J. S. Valentine and A. K. Cho, Mechanism of oxidation of N-hydroxyphentermine by superoxide, Biochemistry 24:4161 (1985).
59. I. B. Afanas'ev, V. V. Grabovetskii and N. S. Kuprianova, Kinetics and mechanism of the reactions of superoxide in solution. Part 5. Kinetics and mechanism of the interaction of superoxide with Vitamin E and ascorbic acid, J. Chem. Soc. Perkin II, 281 (1987).
60. J. F. Liebman and J. S. Valentine, Concerning the radical character of superoxide. The H-O bond energy of HO_2. Israel J. Chem., 23:439 (1983).
61. G. Feroci and S. Roffia, Study of the reactions between styrene and electrogenerated superoxide ion in dimethylformamide, J. Electroanal. Chem. 81:387 (1977).
62. A. A. Frimer, J. Hameiri-Buch and G. Aljadeff, unpublished results (1987).

ACKNOWLEDGEMENTS

We would like to thank the U.S.-Israel Binational Science Foundation (Grant 85-33) for its kind support of much of the research described herein.

PEROXY RADICALS AND SINGLET OXYGEN ($^1O_2$) FROM THE ADDITION OF SUPEROXIDE ION ($O_2^{-}\cdot$) TO CCl_4, CF_3CCl_3, $PhCCl_3$, N-BuBr AND N-BuCl IN ACETONITRILE

Shigenobu Matsumoto, Hiroshi Sugimoto, and
Donald T. Sawyer*

Department of Chemistry
Texas A&M University
College Station, Texas 77843, U.S.A.

The initial product from the reaction of RX(R=Cl_3C, F_3CCl_2, $PhCCl_2$, PhC(O), nBu, X=Cl; R=nBu, X=Br) with $O_2^{-}\cdot$ is ROO·, which can be reduced by a second $O_2^{-}\cdot$ to form ROO^- (a reactive nucleophile) or can dimerize to form ROOOOR. The latter dissociates homolytically to ROOR and O_2. The longer-lived forms of tetraoxide react with (1) 1,4-cyclohexadiene-(1,4-CHD) via dehydrogenation to give PhH, ROOR, and HOOH, (2) diphenylisobenzofuran via dioxygenation to give dibenzoylbenzene and ROOR, and (3) rubrene via dioxygenation to give its endoperoxide and ROOR. Thus the reactivity of ROOOOR parallels that of singlet oxygen ($^1O_2$). Because of its diffusion-controlled dimerization, the primary product from the RX/$O_2^{-}\cdot$ reaction ($CCl_3COO\cdot$ for $CCl_4/O_2^{-}\cdot$) does not exhibit any reactivity with the allyl hydrogens of 1,4-CHD. Hence, at millimolar concentrations this peroxy radical (the suspected cytotoxin from the aerobic activation of CCl_4) does not exist long enough to react with allylic hydrogens. The biological hazard of CCl_4 may be due to the transient formation of $Cl_3COOOOCCl_3$.

The combination of $O_2^{-}\cdot$[1] with excess RX results in a rapid stoichiometric (1:1) reaction to give ROOR, ROO^-, O_2, and Cl^- as the major products. Table 1 summarizes (1) the reactions of $O_2^{-}\cdot$ with RX, (2) the conversion efficiencies of 1,4-CHD by the intermediate products of the primary reactions of RX with $O_2^{-}\cdot$, and (3) the overall products from the combination of $O_2^{-}\cdot$ with

*To whom correspondence should be addressed

Table 1. Reactivity and Products from the Addition of $O_2^{-}\cdot$ (10 mM) to Excess RX and 1,4-Cyclohexadiene (1,4-CHD) in Acetonitrile

RX	Initial mol-ratios $O_2^{-}\cdot$/RX/1,4-CHD			1,4-CHD/PhH[a] Conversion Efficiency %	Products[b]
CCl_4	1	10	20	0	O_2, Cl^-, ($Cl_3COOCCl_3$)[c]
F_3CCCl_3	1	50	20	8	O_2, Cl^-, PhH, ($F_3CCCl_2OOCCl_2CF_3$)
$PhCCl_3$	1	50	20	42	PhH, O_2, Cl^-, ($PhCCl_2OOCCl_2Ph$)
n-BuBr	1	50	20	18	PhH, BuOOBu, O_2, Br^-
n-BuCl	1	50	20	<3	PhH, BuOOBu, O_2, Cl^-

[a]100% represents one PhH per two $O_2^{-}\cdot$ ions ($2\ O_2^{-}\cdot + 2\ RX \rightarrow [ROOOOR] + 2X^-$).
[b]Products assayed via GC, NMR, HPLC, and GC-MS.
[c]Inferred from other products [$Cl_2C{=}O$, CO_2, $HOC(O)O^-$, Cl^-].

Table 2. Reactivity and Products from the Addition of $O_2^{-}\cdot$ (10 mM) to Excess RX and $^1O_2$-trapping Agents in Acetonitrile

RX/trap	$O_2^{-}\cdot$/RX/trap			conversion efficiency, %	
CCl_4/DPIBF	5	6000	1	11	DBB
PhC(O)Cl/DPIBF	5	10	1	15	DBB
CCl_4/rubrene	5	6000	1	1	rubrene-endoperoxide(tr)
$PhCCl_3$/cis-PhCH=CHPh	1	50	20	0	$PhCCl_2OOCCl_2Ph$

large excesses of RX and 1,4-CHD. For $O_2^{-}\cdot$ concentrations from 20 mM to 50 mM there are not any combinations of $O_2^{-}\cdot$/RX/1,4-CHD in MeCN that yield detectable amounts of 1,3-CHD. For each primary reactant, the conversion efficiency for 1,4-CHD to PhH is less than unity and independent of O_2 concentration. In contrast the reaction of HOO· with 1,4-CHD in MeCN[2] and Me_2SO[3] has second-order rate constants of $3.5 \times 10^2\ M^{-1}s^{-1}$ and $1.6 \times 10^2\ M^{-1}s^{-1}$, respectively; yields approximately equimolar amounts of 1,3-CHD and PhH; and occurs via the abstraction by HOO· of an allylic hydrogen atom from 1,4-CHD to give (1,3-CHD·) and HOOH. The absence of significant amounts of 1,3-CHD, ROOH, and ROH in the products from the reactions in Table's 1 and 2 indicates that ROO· plus 1,4-CHD, under the experimental conditions, does not react at a competitive rate ($k_{ox} < 10^2\ M^{-1}s^{-1}$).[2-4]

When $O_2^{-}\cdot$ is combined with excess CCl_4 [or an equimolar amount of PhC(O)Cl] in the presence of diphenylisobenzofuran (DPIBF) or rubrene the substrates are dioxygenated by the reactive intermediate (Table 2). The time domain for the reaction with DPIBF appears to be consistent with the lifetime (< 0.1 s) for the transient intermediate of the $O_2^{-}\cdot/CCl_4$ reaction, which implicates it as the source of the dioxygen for formation of dibenzoylbenzene(DBB) from DPIBF and of rubrene-endoperoxide from rubrene. The short lifetime of this intermediate precludes a net reaction with 1,4-CHD (Table 1), which appears to have a much slower rate. These results and the absence of significant reactivity by the intermediate with cis-stilbene parallel the chemistry of $^1O_2$.[5,6] There is copious emission of white light from the apparent concerted homolytic dissociation of the ROOOOR intermediate to ROOR and $^1O_2$ (the specific weak emission due to $(^1O_2)_2$ cannot be discerned).

The present results demonstrate that the lifetime of the ROO· formed from the reaction of $O_2^{-}\cdot$ with excess RX is too short in MeCN (at millimolar concentrations) for it to react with the allylic hydrogens of 1,4-CHD. Because the latter is believed to be the mode of toxicity for CCl_4 in mammalian liver,[6,7] other reaction paths need to be considered. The rapid formation of ROOOOR species from $O_2^{-}\cdot$ in the presence of excess

RX and their ability to dehydrogenate 1,4-CHD and to dioxygenate highly conjugated aromatic systems with reactivities analogous to $^1O_2$ represent a serious biological hazard. Because HOOH and ROOR are co-products, their toxicity needs to be considered as part of the hazard from ingestion of halogenated hydrocarbons.

ACKNOWLEDGMENT

This work was supported by the Welch Foundation under Grant No. A-1042.

REFERENCES

1. K. Yamaguchi, T.S. Calderwood, and D.T. Sawyer, Corrections and additional insights to the synthesis and characterization of tetramethylammonium superoxide [$(Me_4N)O_2$], Inorg. Chem. 25:1289 (1986).
2. J.A. Howard and K.U. Ingold, Absolute rate constants for hydrocarbon autoxidation. V. The hydroperoxy radical in chain propagation and termination, Can. J. Chem. 45:785 (1967).
3. D.T. Sawyer, M.S. McDowell, and K.S. Yamaguchi, Reactivity of Perhydroxyl(HOO·) in Aprotic Media, Chem. Res. Toxicology 1:000 (1988).
4. G.W. Burton, D.O. Foster, B. Perly, T.F. Slater, I.C.P. Smith, and K.U. Ingold, Biological antioxidants, Phil. Trans. R. Soc. Lond. B311:565 (1985).
5. C.S. Foote, Light, oxygen, and toxicity, in: Pathology of Oxygen, Academic Press, New York (1982).
6. T.F. Slater, Activation of carbon tetrachloride: Chemical principles and biological significance, in: "Free Radicals Lipid Peroxidation, and Cancer," D.C.H. McBrien and T.F. Slater, eds., Academic Press, New York (1982).
7. T.F. Slater, K.H. Cheeseman, and K.O. Ingold, Carbon tetrachloride toxicity as a model for studying free-radical mediated liver injury," Phil. Trans. R. Soc. Lond. B311:633 (1985).

THE UNIQUENESS BOTH OF SUPEROXIDE TOXICITY AND OF THE PROTECTIVE ROLE OF SUPEROXIDE DISMUTASE

Gidon Czapski and Sara Goldstein

Department of Physical Chemistry
The Hebrew University of Jerusalem
Jerusalem 91904, Israel

INTRODUCTION

The role of superoxide radical ($\cdot O_2^-$) in biological processes was established after Fridovich and McCord discovered the enzyme copper-zinc superoxide dismutase (SOD),[1] which catalyzes superoxide dismutation very efficiently. Later, enzymes with similar catalytic activity containing iron and manganese at the active site were discovered.

It has been demonstrated that many copper compounds, from bare copper ions to simple and more complicated low molecular weight copper complexes, catalyze superoxide dismutation with similar efficiency to that of the native enzyme.[2] This feature raised a big debate between Fridovich and Fee a decade ago about the real biolgical role of SOD.[3] The latter claimed that because many copper compounds can substitute SOD, the real function of the enzyme had not yet been discovered. This dispute is long over as sufficient supporting evidence left no doubt about the function of SOD as catalyst for superoxide dismutation. However, in this presentation we will show what is unique *in vivo* about SOD as compared to the other copper compounds which have SOD activity only *in vitro*.

The deleterious role of superoxide in biological systems has been well established. It is also clear that in many but not in all cases, the toxicity is expressed or at least enhanced in the presence of metal compounds. The role of the metal compounds is usually attributed to their ability to catalyze the Haber-Weiss reaction producing the deleterious hydroxyl radical which can cause damage[2,4,5]:

$$M^{n+} + \cdot O_2^- \longrightarrow M^{(n-1)+} + O_2 \qquad (1)$$

$$M^{(n-1)+} + H_2O_2 \longrightarrow M^{n+} + OH^- + \cdot OH \qquad (2)$$

$$\text{net: } \cdot O_2^- + H_2O_2 \longrightarrow O_2 + OH^- + \cdot OH \qquad (3)$$

According to this mechanism, the role of superoxide is to reduce the metal compound; it is surprising that superoxide is more toxic than many other reducing agents present in the cell at concentrations far exceeding that of superoxide, such as glutathione, vitamin C, NAD(P)H, etc. One may also wonder how we survive if the above mechanism is valid in living cells. In the second part of this presentation we will propose a working hypothesis which may explain the unique toxicity of superoxide as compared to the other reductants present in vivo.

DISCUSSION

What Is Unique about SOD as Compared to Other Copper Compounds?

There are two main reasons why copper compounds which have SOD activity in vitro may lose it in vivo: 1) The cell components may sequester the metal out of its complex resulting in a compound with no SOD activity. Similarly, the copper compound may form ternary complexes with some of the cell components which also do not have SOD activity. We have shown that Cu(II)-phenanthroline, which catalyzes superoxide dismutation very efficiently, loses this activity in the presence of DNA due to the formation of ternary complexes which react very slowly with superoxide as compared to the free complex.[6] 2) The superoxide dismutation catalyzed by SOD and by many other copper compounds is believed to involve alternate reduction and oxidation of the copper in so-called the ping-pong mechanism[7-10]:

$$Cu(II) + \cdot O_2^- \rightleftharpoons Cu(I) + O_2$$

$$Cu(I) + \cdot O_2^- + 2H^+ \longrightarrow Cu(II) + H_2O_2 \qquad (4)$$

According to this mechanism the dismutation rate is given by eq. 5:

$$-d[\cdot O_2^-]/dt = k[Cu(II)]_o[\cdot O_2^-] \qquad (5)$$

If $k_{-1}[O_2] < k_4[\cdot O_2^-]$, reaction (-1) can be neglected and the catalytic rate constant is given by eq. 6:

$$k = k_{cat} = 2k_1k_4/(k_1 + k_4) \qquad (6)$$

However, if $k_{-1}[O_2] > k_4[\cdot O_2^-]$ then we have to take into account reaction (-1) and we obtain eq. 7:

$$k = k_r = 2k_1k_4/(k_1 + k_4 + k_{-1}[O_2]/[\cdot O_2^-]) \qquad (7)$$

It turns out that the second and most important reason for the difference between SOD and the other copper compounds arises from the fact that the reoxidation of the reduced enzyme by molecular oxygen is slow (k_{-1} = 0.44 $M^{-1}sec^{-1}$)[7] as compared to most copper compounds where the reoxidation is much faster ($k_{-1} = 10^3$-10^5 $M^{-1}sec^{-1}$).[8,9,11] The concentration of superoxide in vivo is orders of magnitude lower than that generated under pulse radiolysis experiments (0.01-0.1 nM as compared to 1-10 μM); therefore in vivo k would depend on k_{-1}, while under pulse radiolysis experiments it would not (Table 1).

Therefore, in living cells only in the case of SOD reoxidation by oxygen can be neglected, while for the rest of the compounds this does not apply and eq. 7 describes their behaviour. As a result, the SOD activity of some of the copper compounds may be less than 1% of that determined by pulse radiolysis. It turns out that if one wishes to synthesize an effective SOD substitute to operate in vivo, one has to search for a copper compound for which the reoxidation by oxygen and other oxidants present in the cell is slow, has a high stability constant and does not form ternary complexes with the cell components.

Table 1: Relative SOD Activity (k_r/k_{cat}) as a Function of $[\cdot O_2^-]$ and k_{-1} in aerated solutions

$[\cdot O_2^-]$, M	k_{-1}, $M^{-1}sec^{-1}$ 1	10	10^2	10^3	10^4	10^5
10^{-6}-10^{-5}	1	1	1	1	1	1
10^{-10}	1	1	0.9	0.5	0.1	0.013
10^{-11}	1	0.9	0.5	0.1	0.013	0.0013

Why Is Superoxide More Toxic Than Any Other Reductant Present in Vivo?

The toxicity of superoxide as described by reactions 1 and 2 implies that the only role of superoxide is to reduce the metal compound. This mechanism has been modified to the site-specific mechanism, where it is assumed that these same reactions take place with the metal bound to the biological target.[2,12] This modification explains only some of the experimental results, but not the unique toxicity of superoxide because the cell contains many reducing agents that can substitute for superoxide in reaction 1. It is known that these reductants may be toxic in the presence of hydrogen peroxide and copper compounds,[12] but in normal conditions the cell survives, while the presence of superoxide is toxic.

We propose that the uniqueness of superoxide may arise from its ability to oxidize metal compounds rather than reduce them:

$$M^{n+} + \cdot O_2^- + 2H^+ \longrightarrow M^{(n+1)+} + H_2O_2 \qquad (8)$$

The oxidized metal may be Cu(III), Fe(IV) or Mn(III), which are all known to be highly oxidizing species. These species may attack the biological target directly or if they are bound to the target, an intramolecular electron-transfer reaction may cause degradation of the binding site. These entities may also be in equilibrium with OH radical and the damage may occur through this radical. This mechanism may account for the uniqueness of superoxide as vitamin C, glutathione or NADH cannot substitute superoxide in this proposed mechanism.

Some support for the role of Cu(III), Fe(IV) and Mn(III) in biological damage was given before. It has been shown that Cu(III) complexes of proteins, peptides and bleomycin[13-15] and complexes of Fe(IV) amino-carboxylate[16] undergo intramolecular electron-transfer reaction causing degradation of the ligand, while in some systems Mn(III) oxidizes other targets.[17] Our working hypothesis does not eliminate the formation of $\cdot$OH radical or other oxidizing species through the Haber-Weiss mechanism. It is possible that in some systems our proposed mechanism operates, while in others the Haber-Weiss mechanism or both mechanisms operate.

CONCLUSIONS

We have shown that many copper compounds which have SOD activity *in vitro* would not substitute SOD *in vivo* mainly because their reduced form reacts rapidly with molecular oxygen as compared to the native enzyme. The unique toxicity of superoxide as compared to other biological reductants may arise from its ability to oxidize metal compounds, forming highly oxidizing species that may cause biological damage.

REFERENCES

1. J. M. McCord and I. Fridovich, Superoxide dismutase: An enzyme function for erythocuprein, J. Biol. Chem. 244:6049 (1969).
2. S. Goldstein and G. Czapski, The role of metal ions and their complxes in enhancing damage in biological systems or in protecting these systems from the toxicity of O_2^-, J. Free Rad. Biol. Med. 2:3 (1986).
3. I. Fridovich and J. A. Fee in: "Oxygen and Oxy-Radicals in Chemistry and Biology," M. A. Rodgers and E. L. Powers, eds, Academic Press, (1981).
4. R. C. Lynch and I. Fridovich, Autoinactivation of xanthine oxidase. The role of superoxide radical and hydrogen peroxide, Biochim. Biophys. Acta 571:195 (1979).
5. G. Czapski and Y. Ilan, On the generation of of the hydroxyl agent from superoxide radical: can the Haber-Weiss reaction be the source of OH radicals? Photochem. Photobiol. 28:651 (1978).
6. S. Goldstein and G. Czapski, Mechanism of the reactions of some copper complexes in the presence of DNA with O_2^-, H_2O_2 and molecular oxygen, J. Am. Chem. Soc. 108:2244 (1986).
7. D. Klug-Roth, I. Fridovich and J. Rabani, Pulse radiolytic investigation of superoxde catalyzed disproporнation. Mechanism for bovine superoxide dismutase, J. Am. Chem. Soc. 95:2786 (1973).
8. J. Weinstein and B. H. J. Bielski, Reaction of superoxide radical with copper(II)-histidine complexes, J. Am. Chem. Soc. 102:4916 (1980).
9. S. Goldstein and G. Czapski, Kinetic of oxidation of cuprous complexes of substituted phenanthroline and 2,2'-bipyridyl by molecular oxygen and by hydrogen peroxide in aqueous solution, Inorg. Chem. 24:1087 (1985).
10. R. Brigelius, R. Spottl, W. Bors, E. Lengfelder, M. Saran and U. Weser, SOD actvity of low molecular weight Cu^{2+}-chelates studied by pulse radiolysis, FEBS Lett. 47:72 (1974).
11. A. Zuberbuhler, Helv. Chim. Acta 53:473,669 (1970).
12. A. Samuni, J. Aronovich, D. Godinger, M. Chevion and G. Czapski, On the cytotoxicity of vitamin C and metal ions. A site specific mechanism, Eur. J. Biochem. 137:119 (1983).
13. M. Anbar and A. Levitzki, Copper-induced radiolytic deactivation of α-amylase and catalas, Radiat. Res. 27:32 (1966).
14. J. S. Rybka, J. L. Kartz, T. A. Neubecker and D. W. Margerum, Reactions of copper(III) tetraglycine, Inorg. Chem. 19:2791 (1980).
15. S. Goldstein and G. Czapski, Mechanism of reduction of bleomycin-Cu(II) by CO_2^- and oxidation of bleomycin-Cu(I) by H_2O_2 in the absence and presence of DNA, Int. J. Rad. Biol. 51:259 (1987).
16. J. D. Rush and W. H. Koppenol, The reaction between ferrous polyamino-carboxylate complexes and hydrogen peroxide: an investigation of the reaction intermediates by stopped flow spectrophotometry, J. Inorg. Biochem. 29:199 (1987).
17. B. Halliwell, Manganese ions, oxidation reactions and the superoxide radical, NeuroToxicology 5:113 (1984).

PEROXYL RADICALS IN AQUEOUS MEDIA

C. von Sonntag

Max-Planck-Institut für Strahlenchemie
D-4330 Mülheim/Ruhr, Stiftstr. 34-36, W. Germany

INTRODUCTION

Oxygen is almost always present in our environment, and radicals that are formed by biochemical processes or by UV or ionizing radiation are likely to be converted into the corresponding peroxyl radicals. The living cell contains about 70% water, hence, the aqueous medium predominates. Many biologically important molecules such as DNA have an aqueous surrounding.

Radiation techniques provide the most powerful tools for the study of the fate of peroxyl radicals in aqueous solution.[1] In the radiolysis of N_2O-saturated aqueous solutions, hydroxyl radicals are formed, as shown in reactions 1 and 2. In the presence of oxygen (N_2O/O_2 [4:1 v/v]-saturated) $^\bullet OH$ radicals react with an added substrate either by H-abstraction (e.g., reaction 3) or addition to a C=C double bond, while the H-atoms are scavenged by the oxygen (reaction 4). Oxygen converts the substrate radicals into the corresponding peroxyl radicals (reaction 5).

$$H_2O \xrightarrow[\text{radiation}]{\text{ionizing}} {}^\bullet OH,\ e^-_{aq},\ H^\bullet,\ H_2O_2,\ H_2,\ H^+ \quad (1)$$

$$e^-_{aq} + N_2O \longrightarrow {}^\bullet OH + N_2 + OH^- \quad (2)$$

$$^\bullet OH + RH \longrightarrow H_2O + R^\bullet \quad (3)$$

$$H^\bullet + O_2 \longrightarrow HO_2^\bullet \quad (4)$$

$$R^\bullet + O_2 \longrightarrow RO_2^\bullet \quad (5)$$

The yields of $^\bullet OH$ and $HO_2^\bullet$ radicals under these conditions are well known ($G(^\bullet OH) = 0.6\ \mu mol\ J^{-1}$, $G(HO_2^\bullet) = 0.06\ \mu mol\ J^{-1}$), and product studies can be based upon these values. In addition kinetic studies can be done by applying the ionizing radiation as a pulse for a fraction of a microsecond and monitoring the formation and subsequent decay of the peroxyl radicals either by photometry or conductometry. The latter technique is especially useful whenever $HO_2^\bullet$ radicals ($pK(HO_2^\bullet) = 4.7$) are released.

Often, the product of the bimolecular decay of uncharged peroxyl radicals is an organic acid, and its formation can also be followed by this technique. The combination of product studies and pulse radiolysis has yielded results that enable us to give a detailed picture of the different pathways by which organic peroxyl radicals can decay in an aqueous environment.

FORMATION OF PEROXYL RADICALS

Often the radical $R^\bullet$ formed in reaction 3 has a considerable UV absorption while the peroxyl radical $RO_2^\bullet$ has not. Thus, reaction 5 can be followed spectroscopically after a short (≤ 1 μs) pulse of ionizing radiation. A case in point is the radical derived from glycine anhydride (Fig. 1). In the inset of Fig. 1. a plot of the observed speudo-first-order rate constant of the decay of the absorption of radical $R^\bullet$ is plotted as a function of the oxygen concentration. From the slope the bimolecular rate constant k_5 can be calculated. This rate constant is usually close to diffusion-controlled, in the given example the value is 2×10^9 $dm^3 mol^{-1} s^{-1}$. There are, however, examples where this rate constant is below 10^9 $dm^3 mol^{-1} s^{-1}$, and some radicals appear not to react with oxygen at all, although they have a considerable spin density at carbon.[1]

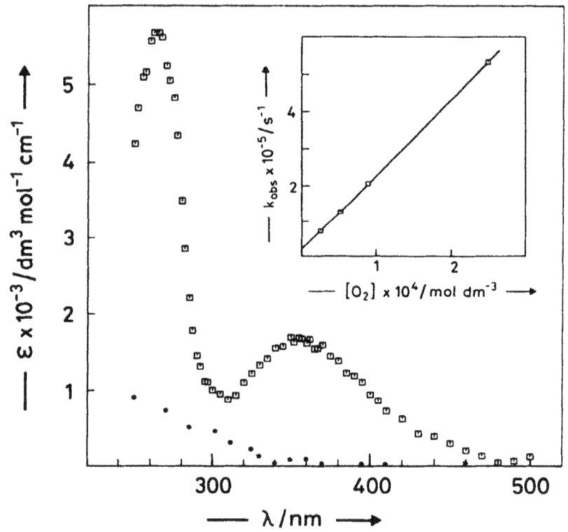

Fig. 1. UV spectra of the radical derived from glycine anhydride (squares) and its corresponding peroxyl radical (small asterisks), pH 6.5. Inset: Observed first-order decay rate constant of the glycine anhydride radical as a function of the oxygen concentration.

FORMATION OF $HO_2^\bullet/O_2^{\bullet-}$

In aqueous solutions one of the commonest reactions is the elimination of $HO_2^\bullet/O_2^{\bullet-}$. We may distinguish four cases:

1. The reaction of oxygen with a given radical yields $O_2^{\bullet-}$

immediately, i.e., an intermediate peroxyl radical, if formed at all, must have a lifetime of less than 1 μs. If the experiment were performed at elevated oxygen pressures to make oxygen addition faster, short-lived peroxyl radicals as intermediates might still be detected, but as things stand these reactions are indistinguishable from electron-transfer processes. Cases reported to date are the reactions of oxygen with $CO_2^{\cdot-}$, $(CH_3)_2NCH_2^{\cdot}$,[2] $^{\cdot}CH(OH)_2$,[3] and the C(1) radical of glucose[4] (e.g., reaction 6).

$$(CH_3)_2NCH_2^{\cdot} + O_2 \longrightarrow (CH_3)_2\overset{+}{N}{=}CH_2 + O_2^{\cdot-} \quad (6)$$

2. Oxygen addition is much faster than the subsequent elimination of $O_2^{\cdot-}$. A case in point is the tertiary radical derived from acetaldehyde dimethyl acetal[5] (reaction 7, $k_7 = 7 \cdot 10^4\ s^{-1}$).

$$CH_3C(OCH_3)_2O_2^{\cdot} \longrightarrow CH_3\overset{+}{C}(OCH_3)_2 + O_2^{\cdot-} \quad (7)$$

3. A spontaneous $HO_2^{\cdot}$ elimination has been observed with many peroxyl radicals derived from alcohols, carbohydrates and amino acids (e.g., reaction 8).[6,7]

$$^{\cdot}O_2C(CH_3)_2OH \longrightarrow (CH_3)_2C{=}O + HO_2^{\cdot} \quad (8)$$

A cyclic five-membered transition state has been envisaged[8] for this process. A peroxyl radical that is also prone to spontaneous $HO_2^{\cdot}$ elimination is the cyclohexadienyl peroxyl radical.[5]

4. Whenever such a spontaneous $HO_2^{\cdot}$ elimination is slow, the reaction may be speeded up by the deprotonation of the OH- or NH-group in α-position to the peroxyl function (e.g., reactions 9 and 10). In fact there are cases in which virtually only

$$^{\cdot}O_2CH_2OH + OH^- \rightleftharpoons {}^{\cdot}O_2CH_2O^- + H_2O \longrightarrow CH_2O + O_2^{\cdot-} \quad (9/10)$$

the OH^--induced reaction occurs while under most experimental conditions the spontaneous elimination can be neglected (in the case of the peroxyl radical derived from glycine anhydride, $k \leq 1\ s^{-1}$).[5]

Other variations of $HO_2^{\cdot}/O_2^{\cdot-}$ elimination reactions are likely to be found on more detailed study of polymers. Their peroxyl radicals have intrinsically long lifetimes, hence, they stand a good chance to undergo such reactions, in competition with a bimolecular decay, even if the $HO_2^{\cdot}/O_2^{\cdot-}$ elimination is very slow.[9]

To illustrate the measurement of the kinetics of $HO_2^{\cdot}/O_2^{\cdot-}$ formation the $O_2^{\cdot-}$ buildup from the OH^--induced decay of the peroxyl radical derived from glycine anhydride is shown in Fig. 2. In Fig. 2 (top) the $O_2^{\cdot-}$ buildup (reflected in the formation of nitroform anion upon reaction of tetranitromethane with $O_2^{\cdot-}$)[1,5] was followed by measuring the changes of absorbance at 350 nm, while in Fig. 2 (bottom) use has been made of the change in conductivity when in basic solution $O_2^{\cdot-}$ ($\lambda^\circ = 60\ \Omega^{-1}mol^{-1}cm^2$)

replaces OH^- ($\lambda° = 170\ \Omega^{-1} mol^{-1} cm^2$). Direct observation of the evolution of $O_2^{\bar{\cdot}}$ from the peroxyl radical optical spectra do not furnish very precise information because these spectra are relatively weak and uncharacteristic (cf. Fig. 3).

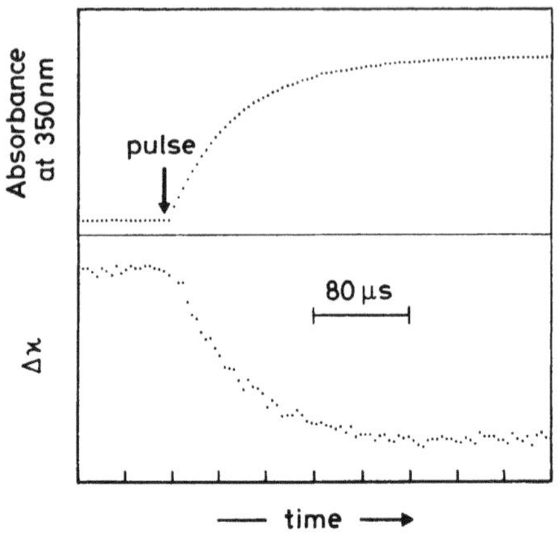

Fig. 2. Hydroxide ion-induced $O_2^{\bar{\cdot}}$ elimination of the peroxyl radical derived from glycine anhydride at pH 9.8. Top: Measured by following the change in absorbance at 350 nm (nitroform anion formation in the presence of 6×10^{-5} mol dm^{-3} tetranitrometane).[5] Bottom: Measured by following the change in conductivity.

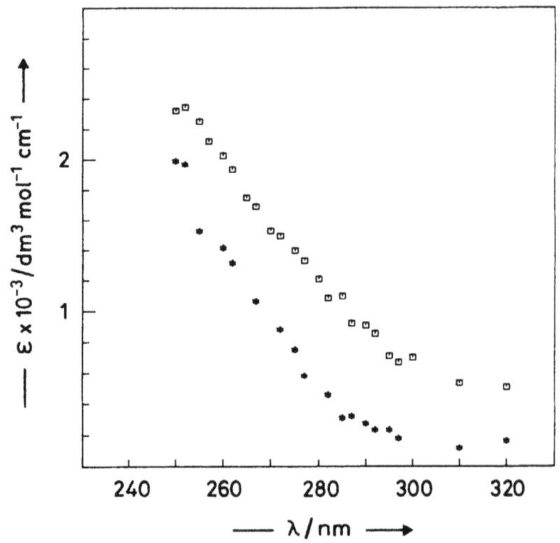

Fig. 3. Absorption spectra of the peroxyl radical derived from glycine anhydride (bottom) and that of the product(s) of the OH^--induced $O_2^{\bar{\cdot}}$ elimination, pH 11.4. The latter spectrum is mainly due to $O_2^{\bar{\cdot}}$.

BIMOLECULAR DECAY OF PEROXYL RADICALS

Many peroxyl radicals are incapable of eliminating $HO_2^{\bullet}/O_2^{\bullet-}$, or do so only slowly. These then decay bimolecularly. From studies in organic solvents at very low temperatures, we know that a tetroxide is formed as a reversible intermediate (cf. reference[10]). Such studies cannot be done in aqueous solutions because of the low temperatures required for equilibrium to be established. Hence, the tetroxide intermediate shown in reaction 11 is adduced by analogy. It must be short-lived, because in all pulse radiolysis studies peroxyl radicals which are incapable of fast $HO_2^{\bullet}/O_2^{\bullet-}$ elimination decay strictly by second order. These studies place an upper limit of about 40 μs on the lifetime of the postulated tetroxide intermediate at room temperature, although its lifetime is likely to be much shorter than this. With many peroxyl radicals that have been studied, the bimolecular decay is close to diffusion-controlled ($2k \approx 10^9$ $dm^3 mol^{-1} s^{-1}$).[1] Hence, the decay of the postulated tetroxide intermediate into two peroxyl radicals (reaction 12) is of minor, or at most, comparable, importance relative to its decay into non-radical products. With primary and secondary peroxyl radicals, there are a number of different routes. Tertiary peroxyl radicals, however, cannot undergo all these reactions. The various decay routes known so far are shown together in Scheme 1.

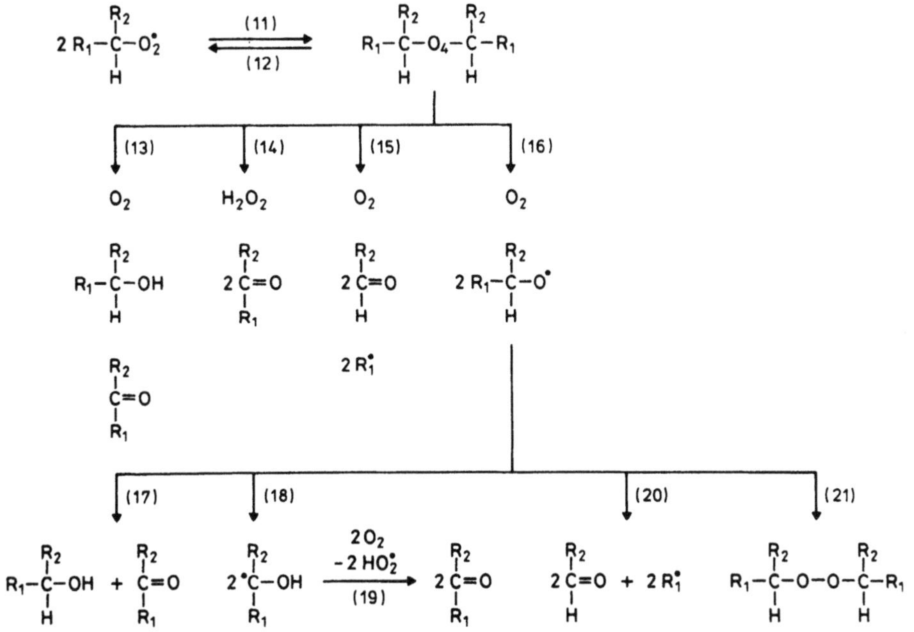

Scheme 1

Reaction 13 depicts the well-known Russell mechanism,[11] which leads to one molecule each of oxygen, a carbonyl compound and an alcohol. It is written as a concerted reaction with a six-membered transition state. It is the most commonly suggested decay route, but there are many examples which clearly show that it often plays only a minor role.

In fact, hydroxymethylperoxyl radicals decay bimolecularly, but give rise to two molecules of formic acid and one molecule of hydrogen peroxide[12]. Such a process is shown in reaction 14. A transition state involving two five-membered rings may be written, but accepting the possibility of water molecules participating other transition states can be suggested as well. This process is a concerted reaction; its free-radical analogue is discussed below.

The third concerted reaction is fragmentation reaction 15. This reaction is usually written with free oxyl radicals as intermediates. In the case of cyclopentylperoxyl radicals, where this process predominates, it has been clearly shown, however, that no free oxyl radicals are involved in this fragmention.[13] In other cases such an "unzipping reaction" is also quite likely, but it is impossible to distinguish it from the free-radical analogue.

Apart from these three concerted reactions there is often a fragmentation into oxygen and two oxyl radicals (reaction 16). This route opens up further free-radical pathways.

The two oxyl radicals that are formed may disproportionate in the solvent cage (reaction 17). This reaction is the free-radical analogue of the Russell mechanism, and it is not always possible to separate the two. In the Russell mechanism a considerable primary kinetic isotope effect ensues when the H-atom involved in the transfer is substituted by a D-atom, but not in oxyl radical formation. Such a case has indeed been observed.[14] To what extent a kinetic isotope effect has to be taken into account in the disproportionation reaction 17 is an open question. This effect might be of some importance for the interpretation of such studies, because there are some competing reactions of the oxyl radical.

In aqueous solution, oxyl radicals with an H-atom in α-position undergo a very rapid 1,3-H shift, forming α-hydroxyalkyl radicals (reaction 18) (reference 15 and references therein). Their subsequent reactions will lead to $HO_2^{\cdot}$ formation (reaction 19, see also reactions 8-10) as discussed above. In many cases the $HO_2^{\cdot}$ elimination is much faster than the bimolecular decay of the peroxyl radicals forming the tetroxide intermediate. For this reason the buildup of $HO_2^{\cdot}/O_2^{\cdot-}$ is kinetically of second order and hence can be distinguished easily from the spontaneous and base-induced $HO_2^{\cdot}/O_2^{\cdot-}$ elimination, which is kinetically of first or pseudo-first order. Since $HO_2^{\cdot}/O_2^{\cdot-}$ can disproportionate yielding hydrogen peroxide and oxygen, this decay sequence is the free-radical analogue of the concerted pathway 14.

In competition with reaction 14, and certainly of major importance in the case of tertiary peroxyl radicals, there is fragmentation of a neighbouring C-C bond (reaction 20). It appears that the aqueous environment speeds up the rate of fragmentation by as much as three orders of magnitude compared with the same reaction in the gas phase or in non-polar solution.[1] As mentioned above, it is not always possible to distinguish this route from the concerted process (reaction 15).

The two oxyl radicals formed in reaction 16 may combine to yield an organic peroxide. This reaction take place essentially only in the solvent cage because reactions 18 and 20 are very fast ($t_{1/2} \leq 1\ \mu s$), i.e., any oxyl radicals that become separated will be very likely to follow these routes rather than terminate (reactions 17 and 21). This is probably the main reason why in aqueous solutions organic peroxides are usually very minor products.[16]

As indicated above there are many reactions that lead to $HO_2^{\bullet}/O_2^{\bullet-}$ formation. These radicals can interact with the organic peroxyl radicals; the major product is a hydroperoxide (reaction 22).[9]

$$RO_2^{\bullet} + HO_2^{\bullet}\ (O_2^{\bullet-} + H^+) \longrightarrow RO_2H + O_2 \qquad (22)$$

There is another source of organic hydroperoxides. Substrates with loosely-bound H-atoms can donate an H-atom to peroxyl radicals, and the well-known autoxidation chain reaction sets in. This reaction usually requires rather high substrate concentrations because the rate constant of H-abstraction (reaction 23) is generally low (k_{23} ranging from 0.1 to 10^3 $dm^3 mol^{-1} s^{-1}$).

$$RO_2^{\bullet} + HS \longrightarrow RO_2H + S^{\bullet} \qquad (23)$$

At low substrate concentrations and at low radical initiation rates (i.e., low steady-state concentrations of peroxyl radicals) an intramolecular H-abstraction reaction can occur. Such a reaction is most pronounced with polymer-derived peroxyl radicals. In poly(U) this process is one important step in the sequence of reactions leading to strand breakage.[17]

REFERENCES

1. C. von Sonntag, "The Chemical Basis of Radiation Biology" London: Taylor and Francis (1987).
2. S. Das, M.N. Schuchmann, H.-P. Schuchmann, and C. von Sonntag, The production of the superoxide radical anion by the OH radical-induced oxidation of trimethylamine in oxygenated aqueous solution. The kinetics of the hydrolysis of (hydroxymethyl)dimethylamine. Chem. Ber. 120:319 (1987).
3. E. Bothe and D. Schulte-Frohlinde, Reaction of dihydroxymethylradical with molecular oxygen in aqueous solutions. Z. Naturforsch. 35b:1035 (1980).
4. E. Bothe, D. Schulte-Frohlinde, and C. von Sonntag, Kinetics

of $HO_2^{\cdot}$ elimination from peroxyl radicals derived from glucose and polyhydric alcohols. J. Chem. Soc. Perkin Trans. II :416 (1978).
5. S. Das, O. J. Mieden, X.-M. Pan, M. Repas, M. N. Schuchmann, H.-P. Schuchmann, C. von Sonntag, and H. Zegota, following paper.
6. E. Bothe, M. N. Schuchmann, D. Schulte-Frohlinde, and C. von Sonntag, $HO_2^{\cdot}$ elimination from α-hydroxyalkylperoxyl radicals in aqueous solution. Photochem. Photobiol. 28:639 (1978).
7. S. Abramovitch and J. Rabani, Pulse radiolytic investigations of peroxy radicals in aqueous solutions of acetate and glycine. J. Phys. Chem. 80:1562 (1976).
8. E. Bothe, G. Behrens, and D. Schulte-Frohlinde, Mechanism of the first order decay of 2-hydroxypropyl-2-peroxyl radicals and of $O_2^{\cdot-}$ formation in aqueous solution. Z. Naturforsch. 32b:886 (1977).
9. M. N. Schuchmann and C. von Sonntag, Hydroxyl radical-induced oxidation of diisopropyl ether in oxygenated aqueous solution. A product and pulse radiolysis study. Z. Naturforsch. 42b:495 (1987).
10. J. E. Bennett, D.M. Brown, and B. Mile, Studies by electron spin resonance of the reactions of alkylperoxy radicals. Part 2. Equilibrium between alkylperoxy radicals and tetroxide molecules. Trans. Faraday Soc. 66:397 (1970).
11. G. A. Russell, Deuterium-isotope effects in the autoxidation of aralkyl hydrocarbons. Mechanism of the interaction of peroxy radicals. J. Am. Chem. Soc. 79:3871 (1957).
12. E. Bothe and D. Schulte-Frohlinde, The bimolecular decay of the α-hydroxymethylperoxyl radicals in aqueous solution. Z. Naturforsch. 33b:786 (1978).
13. H. Zegota, M.N. Schuchmann, and C. von Sonntag, Cyclopentylperoxyl and cyclohexylperoxyl radicals in aqueous solution. A study by product analysis and pulse radiolysis. J. Phys. Chem. 88:5589 (1984).
14. M. N. Schuchmann and C. von Sonntag, Hydroxyl radical-induced oxidation of 2-methyl-2-propanol in oxygenated aqueous solution. A product and pulse radiolysis study. J. Phys. Chem. 83:780 (1979).
15. H.-P.Schuchmann and C. von Sonntag, Photolysis at 185 nm of dimethyl ether in aqueous solution: Involvement of the hydroxymethyl radical. J. Photochem. 16:289 (1981).
16. H.-P.Schuchmann and C. von Sonntag, Methylperoxyl radicals: A study of the γ-radiolysis of methane in oxygenated aqueous solutions. Z. Naturforsch. 39b:217 (1984).
17. E. Bothe, G. Behrens, E. Böhm, B. Sethuram, and D. Schulte-Frohlinde, Hydroxyl radical-induced strand break formation of poly(U) in the presence of oxygen. Comparison of the rates as determined by conductivity, ESR and rapid-mix experiments with a thiol. Int. J. Radiat. Biol. 49:57 (1986).

ASPECTS OF THE $HO_2^{\bullet}$ ELIMINATION REACTION FROM ORGANIC PEROXYL RADICALS: SOME RECENT EXAMPLES

S. Das, O.J. Mieden, X.-M. Pan, M. Repas,
M.N. Schuchmann, H.-P. Schuchmann, C. von Sonntag
and H. Zegota

Max-Planck-Institut für Strahlenchemie
D-4330 Mülheim/Ruhr
Stiftstr. 34-36
W. Germany

INTRODUCTION

It is known that in aqueous solution some organic radicals upon reaction with oxygen form $HO_2^{\bullet}/{\cdot}O_2^-$ without a detectable peroxyl radical intermediate. Cases in point are the radicals derived from formate,[1] formaldehyde hydrate[2] and glucose (at C(1)).[3] Many other radicals carrying OH or NH functions in α-position to the radical site form peroxyl radicals whose rate of spontaneous $HO_2^{\bullet}$ elimination can be measured. Among these are the peroxyl radicals derived from alcohols, carbohydrates (for a review see reference 4) and amino acids.[5] However, in some cases this elimination reaction is very slow and $HO_2^{\bullet}$ is practically only eliminated in a base-catalysed reaction. In this case a peroxyl radical anion is a likely intermediate which subsequently loses ${\cdot}O_2^-$. Examples are the peroxyl radicals derived from pyrimidines.[6] In continuation of these studies we have been looking at the OH radical-induced oxidation of trimethylamine, acetaldehyde dimethylacetal, glycine anhydride and benzene. Peroxyl radicals derived from these substrates were generated radiolytically as described in the preceding paper by C. von Sonntag. The elimination of $HO_2^{\bullet}/{\cdot}O_2^-$ from these peroxyl radicals has been determined pulse-radiolytically due to the rapid reaction of ${\cdot}O_2^-$ with tetranitromethane,[7] or on the basis of its conductivity ($pK_a(HO_2^{\bullet})$ = 4.7).[8] The latter two reactions are not given by organic peroxyl radicals.

RESULTS AND DISCUSSION

Trimethylamine

The (dimethylamino)methyl radical formed by H-abstraction from trimethylamine at pH > 9 (reaction 1, see reference 9) reacts with oxygen with a rate constant of $3.5 \times 10^9\ dm^3 mol^{-1} s^{-1}$, yielding $\cdot O_2^-$ (reaction 2) without a peroxyl radical stage being detectable ($t_{1/2} < 1\ \mu s$). The dimethylimonium ion adds OH^- and forms (hydroxymethyl)dimethylamine (reaction 3, $k_3 = 2.8 \times 10^8\ dm^3 mol^{-1}\ s^{-1}$). This is hydrolysed to dimethylamine and formaldehyde hydrate (reaction 4, $k_{obs} = 4\ s^{-1}$). Details have been published.[10]

$$N(CH_3)_3 + {}^{\cdot}OH \longrightarrow N(CH_3)_2CH_2^{\cdot} + H_2O \quad (1)$$

$$N(CH_3)_2CH_2^{\cdot} + O_2 \longrightarrow {}^{+}N(CH_3)_2{=}CH_2 + \cdot O_2^- \quad (2)$$

$$^{+}N(CH_3)_2{=}CH_2 + OH^- \longrightarrow N(CH_3)_2CH_2OH \quad (3)$$

$$N(CH_3)_2CH_2OH + H_2O \longrightarrow NH(CH_3)_2 + H_2C(OH)_2 \quad (4)$$

Acetaldehyde Dimethylacetal

Hydroxyl radicals react with acetaldehyde dimethylacetal by abstracting mainly the acetalic and O-methyl H atoms (reactions 5 and 6). The 1,1-dimethoxyethyl radical can be distinguished from the methoxyethoxymethyl radical by the fact that the former reacts with tetranitromethane (TNM) via electron transfer to give the nitroform anion (NF^-) directly, whilst the latter reacts with TNM to form an adduct with $\lambda_{max} \leq 300$ nm, which then decays ($k = 4.4 \times 10^4\ s^{-1}$) to give NF^-. Similar results have been reported in the case of hydroxyl radicals reacting with dimethoxymethane.[11] Both radicals also react rapidly with oxygen. The methoxyethoxymethyl peroxyl radical has the properties of an ether-derived peroxyl radical. If such radicals eliminate $HO_2^{\cdot}/\cdot O_2^-$ at all they do so extremely slowly ($k \leq 1\ s^{-1}$).[12,13] The very rapid $HO_2^{\cdot}/\cdot O_2^-$ elimination observed in the present system therefore pertains to reaction 7 ($k_7 = 7 \times 10^4\ s^{-1}$). The resulting carbocation reacts rapidly with water and eventually yields methyl acetate (reactions 8 and 9). The yield of $\cdot O_2^-$ formation from reaction 7 was determined pulse-radiolytically by conductivity detection of $H^+/\cdot O_2^-$ ($G = 0.18\ \mu mol\ J^{-1}$) and by optical detection of NF^- ($G = 0.19\ \mu mol\ J^{-1}$) derived from the fast reaction of $\cdot O_2^-$ with TNM. These two values correspond to the G value of the 1,1-dimethoxyethyl radical as determined in the absence of oxygen (see above) and the G value of methyl acetate determined by gas chromatography upon γ-radiolysis.

$$CH_3CH(OCH_3)_2 + {}^{\cdot}OH \longrightarrow CH_3C^{\cdot}(OCH_3)_2 + H_2O \quad (5)$$

$$CH_3CH(OCH_3)_2 + {}^{\cdot}OH \longrightarrow CH_3CH(OCH_3)OCH_2^{\cdot} + H_2O \quad (6)$$

$$CH_3C(OCH_3)_2O_2^{\cdot} \longrightarrow CH_3C(OCH_3)_2^{+} + \cdot O_2^- \quad (7)$$

$$CH_3C(OCH_3)_2^{+} + H_2O \longrightarrow CH_3C(OCH_3)_2OH + H^+ \quad (8)$$

$$CH_3C(OCH_3)_2OH \longrightarrow CH_3COOCH_3 + CH_3OH \quad (9)$$

Glycine Anhydride

Glycine anhydride is the smallest peptide. Owing to its structural symmetry, H-abstraction by radicals such as OH produces only one kind of carbon-centered radical which facilitates the study of this system. In a previous study the rate constant of oxygen addition to this radical was determined to be $k = 1.2 \times 10^9$ $dm^3mol^{-1}s^{-1}$.[14] In neutral and acid solution, the termination of the resulting peroxyl radicals has a rate constant of $2k = 7 \times 10^8$ $dm^3mol^{-1}s^{-1}$; under these conditions spontaneous $HO_2^{\bullet}$ elimination is not observed ($k \leq 1$ s^{-1}). However, upon its deprotonation (reaction 10) $\cdot O_2^-$ is eliminated (reaction 12, $k_{12} = 1.6 \times 10^5$ s^{-1}). From the OH^- concentration dependence of the observed $\cdot O_2^-$ elimination rate constant and assuming $k_{10} = 10^{10}$ dm^3 $mol^{-1}s^{-1}$, the pK_a value of the peroxyl radical can be estimated at 10.6. In the light of our results, the earlier value[14] of 7.5 appears unlikely. The value 10.6 is close to that of the C(6)-peroxyl radical derived from 5,6-dihydrouracil[6] which is calculated at 10.0 with the same assumption.

$$-COCH(O_2^{\bullet})NH- + OH^- \rightleftarrows -COCH(O_2^{\bullet})N^- + H_2O \qquad (10/11)$$

$$-COCH(O_2^{\bullet})N^- \longrightarrow -COCH{=}N- + \cdot O_2^- \qquad (12)$$

Benzene

Hydroxyl radicals react with benzene by addition, forming the hydroxycyclohexadienyl radical which reacts with oxygen ($k = 5 \times 10^8$ $dm^3mol^{-1}s^{-1}$).[15] One of the many products (65 % based on the OH radical yield) is phenol. Our pulse radiolysis experiments indicate that it is mainly formed by unimolecular $HO_2^{\bullet}$ elimination ($k \approx 900$ s^{-1}) from the hydroxycyclohexadienylperoxyl radicals, of which four different forms are conceivable. It is not yet clear which one is the precursor. It has further been established that muconaldehyde is not formed ($G \leq 0.001$ μmol J^{-1}).

CONCLUSIONS

Our examples indicate that $HO_2^{\bullet}$-elimination from organic peroxyl radicals in aqueous solutions is a widespread phenomenon. It must play an important part in biological systems. Especially with macromolecular peroxyl radicals such as those derived from nucleic acids[13,16] which have a long lifetime with respect to bimolecular termination, such processes are favoured even though the unimolecular decay rate constants might be low.

REFERENCES

1. E. J. Hart, Mechanism of the γ-ray-induced chain oxidation of aqueous ferrous sulfate-formic acid-oxygen solutions, J. Am. Chem. Soc. 74:4174 (1952).
2. E. Bothe and D. Schulte-Frohlinde, Reaction of dihydroxymethylradical with molecular oxygen in aqueous solutions, Z. Naturforsch. 35b:1035 (1980).
3. E. Bothe, D. Schulte-Frohlinde, and C. von Sonntag, Kinetics of $HO_2^{\bullet}$ elimination from peroxyl radicals derived from glucose and polyhydric alcohols, J. Chem. Soc. Perkin Trans. II 416 (1978).
4. E. Bothe, M. N. Schuchmann, D. Schulte-Frohlinde, and C. von Sonntag, $HO_2^{\bullet}$ elimination from α-hydroxyalkylperoxyl radicals in

aqueous solution, Photochem. Photobiol. 28:639 (1978).
5. S. Abramovitch and J. Rabani, Pulse radiolytic investigations of peroxy radicals in aqueous solutions of acetate and glycine, J. Phys. Chem. 80:1562 (1976).
6. M. I. Al-Sheikhly, A. Hissung, H.-P. Schuchmann, M. N. Schuchmann, C. von Sonntag, A. Garner, and G. Scholes, Radiolysis of dihydrouracil and dihydrothymine in aqueous solutions containing oxygen; first- and second-order reactions of organic peroxyl radicals; the role of isopyrimidines as intermediates, J. Chem. Soc. Perkin Trans. II 601 (1984).
7. J. Rabani, W. A. Mulac, and M. S. Matheson, The pulse radiolysis of aqueous tetranitromethane. I. Rate constants and the extinction coefficient of e^-_{aq}. II. Oxygenated solutions, J. Phys. Chem. 69:53 (1965).
8. B. H. J. Bielski, D. E. Cabelli, R. L. Arudi, and A. B. Ross, Reactivity of HO_2/O_2^- radicals in aqueous solution, J. Phys. Chem. Ref. Data 14:1041 (1985).
9. S. Das and C. von Sonntag, Oxidation of trimethylamine by OH radicals in aqueous solution, as studied by pulse radiolysis, ESR and product analysis. The reactions of the alkylamine radical cation, the aminoalkyl radical and the protonated aminoalkyl radical, Z. Naturforsch. 41b:505 (1986).
10. S. Das, M. N. Schuchmann, H.-P. Schuchmann, and C. von Sonntag, The production of the superoxide radical anion by the OH radical-induced oxidation of trimethylamine in oxygenated aqueous solution. The kinetics of the hydrolysis of (hydroxymethyl)dimethylamine, Chem. Ber. 120:319 (1987).
11. J. Eibenberger, D. Schulte-Frohlinde, and S. Steenken, One-electron oxidation of α-monoalkoxyalkyl radicals by tetranitromethane via an intermediate adduct. Influence of radical structure on rate of decomposition of the adduct, J. Phys. Chem. 84:704 (1980).
12. M. N. Schuchmann and C. von Sonntag, Hydroxyl radical induced oxidation of diethyl ether in oxygenated aqueous solution. A product and pulse radiolysis study, J. Phys. Chem. 86:1995 (1982).
13. M. N. Schuchmann and C. von Sonntag, Hydroxyl radical-induced oxidation of diisopropyl ether in oxygenated aqueous solution. A product and pulse radiolysis study, Z. Naturforsch. 42b:495 (1987).
14. E. Hayon and M. Simic, Acid-base properties of organic peroxy radicals, $^{\bullet}OORH$, in aqueous solution, J. Am. Chem. Soc. 95:6681 (1973).
15. L. M. Dorfman, I. A. Taub, and R. E. Bühler, Pulse radiolysis studies. I. Transient spectra and reaction-rate constants in irradiated aqueous solutions of benzene, J. Chem. Phys. 36:3051 (1962).
16. C. von Sonntag, "The Chemical Basis of Radiation Biology", Taylor and Francis, London (1987).

KINETIC STUDIES OF ALKOXYL RADICALS*

J.C. Scaiano

Division of Chemistry
National Research Council
Ottawa, Canada K1A 0R6

INTRODUCTION

A large number of relative reactivities for alkoxyl radicals was reported in the sixties and early seventies.[1-3] However, absolute rates proved very difficult to obtain. This resulted largely from the fact that alkoxyl radicals are invisible to ESR techniques, which at the time were one of the main sources of kinetic data for free radicals. Further, at ambient temperatures alkoxyl radicals are normally too short-lived for conventional ESR methods.

In the late seventies we developed experimental methods for the determination of absolute rate constants for alkoxyl radical reactions using laser photolysis techniques.[4-5] Application of these methods led to a large volume of kinetic data that allows a basic understanding of the reactivity of alkoxyl radicals, while providing sufficient information to transform earlier competitive data into absolute reaction rates.[6]

METHODOLOGY

Most of our studies on alkoxyl radicals have concentrated on the *tert*-butoxyl radical, which can be readily generated by photolysis of di-*tert*-butyl peroxide. This peroxide is commercially available and sufficiently unreactive towards hydrogen abstraction that it can be used as a convenient co-solvent for these studies. Pulsed laser irradiation was normally carried out with a nitrogen laser (337 nm) and occasionally with an excimer (308 nm) or frequency-doubled ruby laser (347 nm).

---

* Issued as NRCC 28009

The radicals generated by a laser pulse can decay by a variety of reactions, including β-cleavage, reaction with the solvent (usually hydrogen abstraction), or reaction with the substrate of interest. Reactions 1-4 illustrate these processes.

$$(t\text{-BuO-})_2 \xrightarrow{h\nu} 2\ t\text{-BuO}\cdot \quad (1)$$

$$t\text{-BuO}\cdot \xrightarrow{k_\beta} CH_3COCH_3 + \dot{C}H_3 \quad (2)$$

$$t\text{-BuO}\cdot + S(\text{solvent}) \xrightarrow{k_s} t\text{-BuOH} + S\cdot \quad (3)$$

$$t\text{-BuO}\cdot + XH \xrightarrow{k_x} t\text{-BuOH} + X\cdot \quad (4)$$

If X• has a convenient absorption in the visible or near UV regions, the growth of the signal due to X• will follow pseudo-first order kinetics and will be determined by eq. 5,

$$k_{exptl} = k_o + k_x[XH] \quad (5)$$

where $k_o = k_\beta + k_s[S]$ and k_{exptl} is the pseudo-first order rate constant for the formation of X• at a particular concentration of XH. For example, if the substrate is benzhydrol, hydrogen abstraction leads to the diphenyl ketyl radical, $Ph_2\dot{C}OH$ (λ_{max} ~540 nm);[5] Figure 1 illustrates the absorption spectrum of $Ph_2\dot{C}OH$ and a representative transient absorption trace monitored at 540 nm.

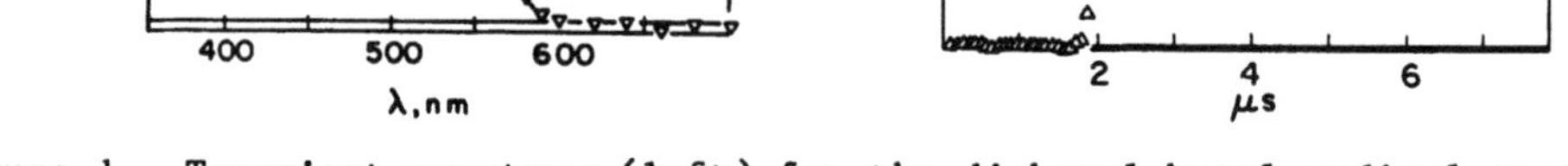

Figure 1. Transient spectrum (left) for the diphenyl ketyl radical produced by reaction of *tert*-butoxyl radicals with benzhydrol, and build-up of the diphenyl ketyl radical (right) formed in reaction 4 and monitored at 540 nm.

The rate constant of interest, k_x, is then obtained from the slope of a plot of k_{exptl} vs. [XH]. For example, in the case of Ph_2CHOH one obtains a rate constant of $6.9 \times 10^6\ M^{-1}s^{-1}$ at room temperature in a 1:2 mixture of benzene:di-*tert*-butyl peroxide (v/v).[4,5] With this technique it is relatively straightforward to determine $\underline{k}_x$ for any substrate that yields a radical with convenient absorption properties. For example, α-tocopherol yields a phenoxyl radical with λ_{max} = 425 nm; in this system we determined $\underline{k}_x = 3.8 \times 10^9\ M^{-1}s^{-1}$.[7]

The method, as described above, may not be as general as the previous paragraph may suggest. Numerous substrates of interest do not lead to a radical X• that can be readily monitored in the visible or near UV regions. For example, hydrocarbons such as cyclohexane will not lead to any detectable absorption signal at $\lambda > 300$ nm. We also note that the tert-butoxyl radical is essentially transparent at wavelengths longer than those corresponding to the excitation lasers employed. Thus, many systems of interest involve a combination of reactants and products, all of which are in practice "invisible" to the technique of laser flash photolysis. We have developed a technique that allows the measurement of absolute rate constants despite of the limitation mentioned above.

In a reaction system containing two substrates, XH and RH, such that X• can be observed directly, but R• is invisible to the technique, the build-up of the signal due to X• will follow pseudo-first order kinetics, such that the experimental rate constant, $\underline{k}_{exptl}$ will be given by

$$\underline{k}_{exptl} = \underline{k}_o + \underline{k}_x[XH] + \underline{k}_R[RH] \qquad (6)$$

where $\underline{k}_R$ is the absolute rate constant for reaction 7, and all other parameters have the same meaning as eq. 5.

$$t\text{-BuO}\cdot + RH \xrightarrow{\underline{k}_R} \text{products} \qquad (7)$$

Thus, a plot of $\underline{k}_{exptl}$ vs. [RH], keeping the concentration of XH constant, yields $\underline{k}_R$ from the slope. The idea of a method that can be used to measure absolute rate constants for a reaction that is invisible to the technique employed is generally received with some degree of skepticism. Indeed, there is a hidden limitation in the application of this method: The rate constant obtained, $\underline{k}_R$, is an overall, or molecular rate constant, which includes all possible sites and modes of attack. In this sense, $\underline{k}_R$ carries no mechanistic information, but rather, it is simply the rate constant with which RH scavenges the alkoxyl radicals. The mode and site of attack can usually be established by other methods, such as product studies or ESR spectroscopy.

REACTIVITY OF *Tert*-BUTOXYL RADICALS

The reactivity of oxygen-centered radicals generally follows the order HO• > alkoxyl > peroxyl > phenoxyl.[6] The technique described in the previous section has been used to measure the rate constants for a wide range of reactions; representative values have been summarized in Table 1.

It should be noted that when abstraction/addition competitions are possible *tert*-butoxyl radicals show a clear preference for hydrogen abstraction. Toluene, cumene[5,9,10] and olefins bearing allylic hydrogens exemplify this behaviour.

Substrates having labile O-H bonds are highly reactive in nonpolar solvents, but their reactivity is moderated in hydrogen bonding media. *Tert*-butyl hydroperoxide and the phenols included in Table 1 illustrate the reactivities in non-polar media.[5,13] In the case of phenol, its rate constants are 3.3×10^8, 2.2×10^7, and 4.7×10^6 $M^{-1}s^{-1}$ in benzene, methanol, and pyridine, respectively.[13] Abstraction from O-H bonds is more rapid than

Table 1. Rate Constants for Reactions of *Tert*-butoxyl Radicals at or Near Room Temperature in Nonpolar Media.[4,5,7-15]

Substrate	10^{-6} k_R[a]	Substrate	10^{-6} k_R[a]
Cyclopentane	0.88	Methanol	0.29
Cyclohexane	1.6	Ethanol	1.1
Toluene	0.23	2-Propanol	1.8
Ethylbenzene	1.05	1-Phenylethanol	1.8
Mesitylene	0.83	Benzhydrol	6.9
Diphenylmethane	1.3	Trimethylamine	110
Norbornane	0.58	Triethylamine	180
Cyclopentene	6.9	Quinuclidine	6.0
Cyclohexene	5.8	DABCO	28
Cyclooctene	2.6	Diethylaniline	130
Norbornene	1.09	Tetrahydropyrrol	95
Cyclopentadiene	7.2	Piperidine	79
1,4-Cyclohexadiene	53	n-Propylamine	17
1,3-Cyclohexadiene	42	*Tert*-butylamine	3.3
1,5-Cyclooctadiene	8.9	Phenol	330
1,3-Cyclooctadiene	3.6	4-Methoxyphenol	1600
Norbornadiene	3.2	4-Methylphenol	550
1-Octene	1.5	4-Chlorophenol	290
2-Methyl-1,2-pentene	14.5	4-Cyanophenol	97
2,3-Dimethyl-1,2-butene	15	α-Tocopherol	3800
Diethyl ether	3.9	*Tert*-butyl hydroperoxide	250
Diisopropyl ether	1.2	Propionaldehyde	89
Ethylene oxide	1.1	Benzaldehyde	68
Oxacyclobutane	4.0	Ethylformate	0.51
Tetrahydrofuran	8.3	Tri-*n*-butylgermane	92
Tetrahydropyran	2.7	Tri-*n*-butylstannane	190
1,4-Dioxane	1.5	Triethylsilane	5.7

[a] In units of $M^{-1}s^{-1}$

abstraction from a C-H bond of comparable energy; this is believed to result from the small magnitude of the non-bonded repulsive term between two oxygen atoms (e.g., in RO-H-OOR).

Table 1 also includes a series of substrates bearing allylic C-H bonds. Although fatty acids have not been included in the table, we have carried out a detailed study of several representative substrates.[16] It was observed that their reactivity is simply a function of the number of secondary, allylic, and doubly allylic hydrogens available. Equation 8 can be used to estimate the reactivity of polyunsaturated fatty acids.

$$k_R = [0.072\ (\#H_{sec}) + 0.518\ (\#H_{allylic}) + 2.716\ (\#H_{doubly\ allylic})] \times 10^6\ M^{-1}s^{-1} \qquad (8)$$

Stereoelectronic effects play an important role in the reactions of alkoxyl radicals. In particular, this has been demonstrated for numerous examples involving ethers and amines (see Table I).[11,12] An interesting example is provided by the triethylamine-quinuclidine pair.[12] Both have three CH_2 groups bonded to the nitrogen atom. While in the case of triethylamine it is possible to delocalize the unpaired electron following H-abstraction, the same stabilization is severely restricted in quinuclidine since the lone pair on nitrogen is unfavorably located with respect to the C-H bonds in the rigid molecular frame.

Isotope effects have been examined in several reactions of tert-butoxyl radicals. Representative values of k_H/k_D are 3.3 for 2-propanol-d_8, 2.8 for tetrahydrofuran-d_8, 1.8 for benzaldehyde-d_6, 2.9 for cyclohexene-d_{10}, and 1.23 for tri-n-butylstannane-d_1.[5,10,11,15]

Most of the examples mentioned above involve the cleavage of O-H or C-H bonds. Table 1 also includes a few examples (Sn-H, Si-H, N-H, etc) that involve other bonds.

OTHER ALKOXYL RADICALS: GENERATION AND REACTIVITY

In principle, all alkoxyl radicals can be readily generated photochemically by irradiation of the appropriate peroxide; however, in many cases these substrates are not available in adequate quantities or cannot be safely handled at the concentrations required for these experiments. The sensitized decomposition of the peroxides,[17] or of the corresponding hyponitrites (RON_2OR)[18] offer a convenient photochemical source of alkoxyl radicals.

Table 2. Kinetic Parameters for the β-cleavage of Alkoxyl Radicals in Nonpolar Media.[6,18,19]

Radical	k_β/s^{-1} [a]	E_a(kcal/mol)	$\log(A/s^{-1})$
t-BuO•	2.9×10^4	-	-
$Ph(CH_3)_2CO\bullet$	9.2×10^5	8.60	12.36
$PhCH_2CH_2O\bullet$	2.3×10^7	4.83	10.89

[a] At 20-27°C

The intermolecular reactivity of alkoxyl radicals is not very dependent upon substitution, although minor steric effects are detectable. For example, the reaction of $PhCH_2CH_2O\bullet$ with benzhydrol is 3.6 times faster than that of t-BuO• with benzhydrol.[18]

Major substituent effects are frequently observed in the β-cleavage of alkoxyl radicals. Thus, $PhCH_2CH_2O\bullet$ cleaves more readily than t-BuO• (see Table 2) reflecting the stability of the benzyl radicals produced as a product.[18]

The rate constants for the β-cleavage of alkoxyl radicals are very sensitive to the polarity of the solvent, in particular, ability to stabilize the carbonyl compounds formed as products. Thus, the β-cleavage of tert-butoxyl is greatly accelerated by $SnCl_4$,[20] whereas, in the case of $PhCH_2CH_2O\bullet$ in chlorobenzene, addition of 0.43 M ethanol changes the rate constant from 1.87×10^6 to 1.07×10^7 s^{-1} at -35°C.[18] Product studies have shown a similar solvent dependence in the competition between β-cleavage (e.g., reaction 2) and hydrogen abstraction (e.g., reaction 4).[21]

The knowledge of rate constants for β-cleavage (Table 2), is extremely useful since competitive studies (e.g., determination of k_β/k_x for reactions 2 and 4) are straightforward and only require chromatographic analysis for the alcohol and ketone. These data can be readily converted to absolute rate constants using the data in Table 2.

ALKOXYL RADICALS AS PRECURSORS FOR OTHER RADICALS

The high reactivity of alkoxyl radicals makes them convenient precursors for other radicals, an approach that has been widely used in laser photolysis studies. In particular, in the case of silyl radicals we have carried out extensive studies using reaction 1 followed by 10 as the radical source.[22-24]

$$\text{t-BuO}\bullet + R_3SiH \longrightarrow \text{t-BuOH} + R_3Si\bullet \qquad (10)$$

The reactivity of $R_3Si\cdot$ radicals can then be examined using the same techniques described herein for alkoxyl radicals.

REFERENCES

1. C. Walling, Some aspects of the chemistry of alkoxyl radicals, Pure Appl. Chem. 15:69 (1967).
2. K.U. Ingold, Rate constants for some reactions of oxy radicals, Pure Appl. Chem. 5:49 (1967).
3. J.A. Howard, Absolute rate constants for reaction of oxyl radicals, Adv. Free-Radical Chem. 4:49 (1972).
4. R.D. Small, Jr., and J.C. Scaiano, Absolute rates of hydrogen abstraction by tert-butoxy radicals. J. Am. Chem. Soc. 100:296 (1978).
5. H. Paul, R.D. Small, Jr., and J.C. Scaiano, Hydrogen abstraction by tert-butoxy radicals. A laser photolysis and electron spin resonance study, J. Am. Chem. Soc. 100:4520 (1978).
6. J.A. Howard and J.C. Scaiano, "Radical Reaction Rates in Liquids; Oxyl, Peroxyl and Related Radicals", Landolt-Börnstein, New Series Volume 13, Subvolume d, Springer-Verlag, Berlin, 1985.
7. C. Evans, unpublished results.
8. C. Chatgilialoglu, K.U. Ingold, J. Lusztyk, A.S. Nazran, and J.C. Scaiano, Formation, decay, and spectral characterization of some alkyl- and aryl-substituted carbon-, silicon-, germanium- and tin-centered radicals, Organometallics 2:1332 (1983).
9. P.C. Wong, D. Griller, and J.C. Scaiano, A kinetic study of the reactions of tert-butoxyl with alkenes: hydrogen abstration vs. addition, J. Am. Chem. Soc. 104:5106 (1982).
10. M.V. Encinas and J.C. Scaiano, Reaction of benzophenone triplets with allylic hydrogens. A laser flash photolysis study, J. Am. Chem. Soc. 103:6393 (1981).
11. V. Malatesta and J.C. Scaiano, Absolute rate constants for the reactions of tert-butoxyl with ethers: Importance of the stereoelectronic effect, J. Org. Chem. 47:1455 (1982).
12. D. Griller, J.A. Howard, P.R. Marriott, and J.C. Scaiano, Absolute rate constants for the reactions of tert-butoxyl, tert-butoxylperoxyl, and benzophenone triplet with amines: The importance of a stereoelectronic effect, J. Am. Chem. Soc. 103:619 (1981).
13. P.K. Das, M.V. Encinas, S. Steenken, and J.C. Scaiano, Reaction of tert-butoxy radicals with phenols. Comparison with the reactions of carbonyl triplets, J. Am. Chem. Soc. 103:4162 (1981).
14. C. Chatgilialoglu, L. Lunazzi, D. Macciantelli, and G. Placucci, Absolute rate constants for hydrogen abstration from aldehydes and conformational studies of the corresponding aromatic acyl radicals, J. Am. Chem. Soc. 106:5252 (1984).
15. J.C. Scaiano, Detection of trialkylstannyl radicals using laser flash photolysis, J. Am. Chem. Soc. 102:5399 (1980).
16. R.D. Small, Jr., J.C. Scaiano, and L.K. Patterson, Radical processes in lipids. A laser photolysis study of t-butoxy radical reactivity toward fatty acids. Photochem. Photobiol. 29:49 (1979).
17. J.C. Scaiano and G.G. Wubbels, Photosensitized dissociation of di-tert-butyl peroxide. Energy transfer to a repulsive excited state, J. Am. Chem. Soc. 103:640 (1981).
18. G.D. Mendenhall, L.C. Stewart, and J.C. Scaiano, Laser photolysis study of the reactions of alkoxyl radicals generated in the photosensitized decomposition of organic hyponitrites, J. Am. Chem. Soc. 104:5109 (1982).

19. A. Baignée, J.A. Howard, J.C. Scaiano, and L.C. Stewart, Absolute rate constants for reactions of cumyloxy in solution, J. Am. Chem. Soc. 105:6120 (1983).
20. A.G. Davies, B.P. Roberts, and J.C. Scaiano, The reaction of tert-butoxyl radicals with stannic chloride, J. Organometal. Chem. 42:C27 (1972).
21. C. Walling and P.J. Wagner, Positive halogen compounds. X. solvent effects in the reactions of t-butoxy radicals, J. Am. Chem. Soc. 86:3368 (1964).
22. C. Chatgilialoglu, K.U. Ingold, and J.C. Scaiano, Absolute rate constants for the addition of triethylsilyl radicals to the carbonyl group, J. Am. Chem. Soc. 104:5119 (1982).
23. C. Chatgilialoglu, K.U. Ingold, and J.C. Scaiano, Absolute rate constants for the reaction of triethylsilyl radicals with organic halides, J. Am. Chem. Soc. 104:5123 (1982).
24. C. Chatgilialoglu, K.U. Ingold, and J.C. Scaiano, Absolute rate constants for the addition of triethylsilyl radicals to various unsaturated compounds, J. Am. Chem. Soc. 105:3292 (1983).

RADICAL-INDUCED DEGRADATION OF ORGANIC HALOGEN AND SULFUR COMPOUNDS IN OXYGENATED AQUEOUS SOLUTIONS

Klaus-Dieter Asmus, Manohar Lal,* Jörg Mönig,** and Christian Schöneich

Hahn-Meitner-Institut Berlin, Bereich Strahlenchemie
Glienicker Str. 100, 1000 Berlin 39, Federal Republic of Germany

*Bhabha Atomic Research Centre, Trombay. Bombay-400085, India

**GSF Institut für Tieflagerung, Theodor-Heuss-Allee 4
3300 Braunschweig, Federal Republic of Germany

INTRODUCTION

Halogenated and sulfur-containing organic compounds are two groups of chemicals that have long attracted the attention of scientists interested in free radicals. Numerous studies on aliphatic halogen substrates, for example, CCl_4 or halothane ($CF_3CHClBr$), have revealed a striking similarity between the product patterns obtained from metabolism and from free radical-induced degradation in vitro.[1] This finding renders free radical studies potentially useful tools for understanding the chemistry of metabolic pathways.

The significance of sulfur-containing compounds in biologic processes does not need an elaborate introduction. Although the roles of sulfhydryl and disulfide compounds, to name just the most important substrates, are well established, there are still many important questions and controversial results. One seemingly simple problem has recently become the focus of renewed investigations, namely, the determination of absolute rate constants for the addition of molecular oxygen to thiyl radicals.[2] This question may be related to the formation and properties of other peroxyl radicals; therefore, joint discussion of halogenated and sulfur-containing radicals seems reasonable.

The experimental data presented in this paper were obtained via radiation chemical methods, including time-resolved pulse radiolysis studies and product analysis from ^{60}Co-γ-radiolysis. For details the reader is referred to the original literature.[3,4]

ADDITION OF O_2 TO ALIPHATIC HALOGENATED $-\dot{C}\cdot$ RADICALS

Carbon-centered aliphatic radicals are generally known to undergo addition of molecular oxygen in a fast process to yield the corresponding peroxyl radicals

$$-\overset{|}{\underset{|}{C}}\cdot \; + \; O_2 \; \rightarrow \; -\overset{|}{\underset{|}{C}}-OO\cdot \qquad (1)$$

Second-order rate constants for these reactions in the liquid phase are typically on the order of 10^9 M^{-1} s^{-1}, i.e., they indicate essentially diffusion-controlled processes. Values up to several orders of magnitude lower are obtained in the presence of strong electron-withdrawing substituents.[5] The electrophilic character of the molecular oxygen provides an explanation of this finding.

On the basis of this information it may be expected that halogen substitution would also exert enough electron-withdrawing power to reduce the electron density at the radical site and, thus, to reduce its reactivity towards O_2. It is therefore surprising that halogen substitution does not significantly influence the rate of oxygen addition, as shown by the rate constants in Table 1.

OXIDIZING PROPERTIES OF HALOGENATED PEROXYL RADICALS

Although halogen substitution has no effect on the rate constants of O_2 addition to C-centered radicals, it has a dramatic effect on the redox properties of the respective peroxyl radicals. Whereas $CH_3OO\cdot$ exhibits virtually no oxidizing properties, $CCl_3OO\cdot$ is a very good oxidant.[8] A short list of selected rate constants for the general reaction of halogenated peroxyl radicals with an electron donor, D,

$$RXOO\cdot \; + \; D \; \rightarrow \; RXOO^- \; + \; D^{+}\cdot \qquad (2)$$

where X is a halogen, reveals values on the order of up to 10^8 $M^{-1}s^{-1}$ (Table 2). This indicates that these electron transfer reactions, although efficient and fast, are not entirely diffusion controlled.

At present, the exact oxidation potentials of halogenated peroxyl radicals are still unknown. It is possible, however, to estimate the one-electron oxidation potential of the $CCl_3OO\cdot/CCl_3OO^-$ couple as $\geq +1.4$ V (vs. NHE)[9,10] by comparison with other radical oxidants.

Table 1. Rate Constants for the Addition of Molecular Oxygen to Halogenated Carbon-Centered Radicals

Radical R•	k (R• + O_2), M^{-1} s^{-1}	Ref.
$\dot{C}Cl_3$	3.3×10^9	6
$CF_3\dot{C}HCl$	1.3×10^9	7
$CH_3\dot{C}Cl_2$	$1\text{-}5 \times 10^9$	*
$CH_3\dot{C}HCl$	0.9×10^9	*

*M. Lal, C. Schöneich, C. Hahn, J. Mönig, and K.-D. Asmus, unpublished results.

Table 2. Rate Constants for the Reaction of Halogenated Peroxyl Radicals with Electron Donors

RXOO•	D	k_2, M^{-1} s^{-1}	Ref.
$CCl_3OO\cdot$	metiazinic acid	1.3×10^9	6
	methionine	2.9×10^7	9
	Me_2S	1.3×10^8	9
$CF_3CHClOO\cdot$	methionine	1.4×10^6	9
	ascorbate	6.1×10^8	11
	α-tocopherol	9.4×10^7	11
$CH_3CCl_2OO\cdot$	ascorbate	4.6×10^8	*
	propyl gallate	8.0×10^7	*
	promethazine	3.7×10^7	*
	chloropromazine	7.4×10^7	*

*M. Lal, C. Schöneich, C. Hahn, J. Mönig, and K.-D. Asmus, unpublished results

REACTION PRODUCTS FROM HALOGENATED PEROXYL RADICALS

The main molecular reaction products that arise from nonhalogenated peroxyl radicals are alcohols and aldehydes or ketones. These products are formed via a variety of molecular and radical pathways.[12-14] An example of a molecular pathway is the so-called "Russell" mechanism, which is based on a cyclic rearrangement within an intermediate tetroxide. The radical pathways, on the other hand, involve an alkoxy radical as the most important transient.

The stable reaction products that result from the degradation of halogenated peroxyl radicals are basically formed via these same two pathways. Another important consideration is the possible simultaneous presence of halogen and oxygen functions at the same carbon atom. This substitution pattern is known to generally undergo fast hydrolysis and/or hydrogen halide elimination namely,

$$-\overset{X}{\underset{|}{\overset{|}{C}}}-OH \xrightarrow{-HX} -\overset{|}{C}=O \qquad (3)$$

"alcohol" aldehyde/ketone

and

$$-\overset{X}{\overset{|}{C}}=O \xrightarrow{\text{hydrolysis}} -\overset{OH}{\overset{|}{C}}=O \qquad (4)$$

"aldehyde" acid

Thus, the stable products from the degradation of α-halogen peroxyl radicals are at the next higher oxidation state than their nonhalogenated analogues. To illustrate this pattern, Table 3 lists aldehydes and acids that have been identified in the reductive (e.g., by e^-_{aq}) or •OH radical-induced degradation of a number of selected halogenated compounds.[1,4,6,11] The biological implications of the aldehyde and acid generation in these radical-mediated processes should be stressed.

Table 3. Peroxyl Radicals and Products from Various Halogenated Compounds

Compound (plus e^-_{aq} or •OH)	α-Halogen peroxyl radical	Aldehyde	Acid	
CCl_4	$CCl_3OO\cdot$	---	CO_2	/ HCl
CH_3CCl_3	$CH_3CCl_2OO\cdot$	---	CH_3COOH	/ HCl
CH_3CHCl_2	$CH_3CHClOO\cdot$	CH_3CHO	CH_3COOH	/ HCl
CH_2ClCH_2Cl	$CH_2ClCHClOO\cdot$	CH_2ClCHO	$CH_2ClCOOH$	/ HCl
CH_2BrCH_2Br	$CH_2BrCHBrOO\cdot$	CH_2BrCHO	$CH_2BrCOOH$	/ HBr
CF_3CH_2I	$CF_3CHIOO\cdot$	CF_3CHO	CF_3COOH	/ HI
$CF_3CHClBr$	$CF_3CHClOO\cdot$	CF_3CHO	CF_3COOH	/ HCl,HBr
CHF_2OCF_2CHFCl	$CHF_2OCF_2CHFOO\cdot$	CHF_2OCF_2CHO	CHF_2OCF_2COOH	/ HCl,HF

CARBON-CARBON BOND CLEAVAGE

Another significant finding is possible carbon-carbon bond cleavage in peroxidized radicals. Except for 1,1,1-trifluoroethane derivatives, relatively high yields of HCHO, HCOOH, and/or CO_2 (up to 30 of the initial peroxyl radical yield) are found (M. Lal, C. Schöneich, C. Hahn, J. Mönig, and K.-D. Asmus, unpublished results). Quantitative analysis and complete material balance studies are still in progress. It appears, however, that these fragments, together with the products listed in the preceding section, fully account for a complete material balance.

The mechanism of fragmentation is still an unsolved question. Two plausible routes are briefly outlined. The first one is essentially based on a radical cleavage from an alkoxy radical.

$$-\overset{|}{\underset{|}{C}}-\overset{|}{\underset{|}{C}}-O\cdot \quad \rightarrow \quad -\overset{|}{\underset{|}{C}}\cdot \; + \; -\overset{|}{C}=O \qquad (5)$$

The subsequent addition of O_2 and hydrolysis, depending on the degree of halogen substitution at the carbonyl site, then leads to stable fragment products. A second proposed pathway involves formation of intramolecular dioxetane from the initial peroxyl radical or from an intermediate hydroperoxide

$$R-\overset{|}{\underset{|}{C}}-\overset{|}{\underset{|}{C}}-OO\cdot \quad \rightarrow \quad R\cdot \qquad (6a)$$

$$+ \quad \begin{array}{c} -\overset{|}{\underset{|}{C}}-\overset{|}{\underset{|}{C}}- \\ O-O \end{array}$$

$$R-\overset{|}{\underset{|}{C}}-\overset{|}{\underset{|}{C}}-OOH \quad \rightarrow \quad H^+/R^- \qquad (6b)$$

followed by homolytic scission of the dioxetane

$$\begin{array}{c} -\overset{|}{\underset{|}{C}}-\overset{|}{\underset{|}{C}}- \\ O-O \end{array} \quad \rightarrow \quad 2 \;\; -C=O \qquad (7)$$

The nature of the stable products will again depend on the degree of halogen substitution in the "carbonyl" fragment It should be noted that reaction (6b) has been demonstrated to occur in the photo-oxidation of ethylene via a CH_2BrCH_2OOH intermediate.[15]

O_2 ADDITION TO THIYL RADICALS

Although carbon-centered radicals apparently react very efficiently with molecular oxygen the situation is quite different with sulfur-centered radicals.[2,16,17] Among the five major types, namely,

$$RS\cdot,\quad (RSSR)^{-}\cdot,\quad (RSSR)^{+}\cdot,\quad R_2S^{+}\cdot,\quad (R_2S)_2{}^{+}\cdot$$

the cationic species do not react with oxygen which is considered to be due to the pronounced electrophilicity of O_2. The disulfide radical anion readily transfers an electron to O_2 but does not add oxygen. The latter process seems only to occur with thiyl radicals

$$RS\cdot + O_2 \rightarrow RSOO\cdot \qquad (8)$$

to yield a thioperoxyl radical.

Although a number of measurements of the rate constants for his reaction have been made, there is still serious concern about how fast this addition actually is. To illustrate this problem, two instructive rate constants are now examined, namely, $k_8 = 8 \times 10^9\ M^{-1}s^{-1}$ and $4 \times 10^7\ M^{-1}s^{-1}$, published for the oxygen addition to the thiyl radicals from cysteine (CysS•)[18] and penicillamine (PenS•),[2] respectively. First, the significant difference between these two rate constants is remarkable because they refer to structurally very similar species

CysS•:

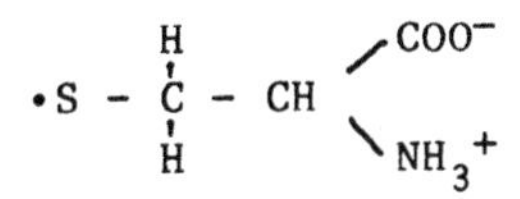

PenS•:

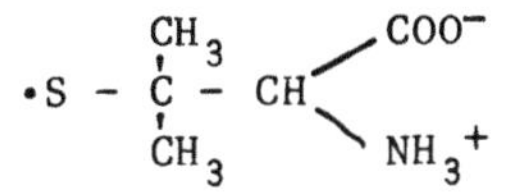

Second, it is surprising that the addition of oxygen to CysS• occurs faster than its addition to PenS• Electron-density considerations would suggest opposite results because the two electron-releasing methyl groups should facilitate electrophilic addition to PenS• Finally, it should be noted that the rate constant for the CysS• reaction is faster than that for oxygen addition to a carbon-centered radical. This does not correspond to the much lower nucleophilicity of thiyl radicals as compared to carbon-centered radicals. In fact, RS• radicals, like RO• homologues, are electrophiles and act as reasonably good oxidants towards a large number of electron donors via the general reaction

$$RS\cdot + D \rightarrow RS^{-} + D^{+}\cdot \qquad (9)$$

Although the published data on rate constants are controversial, nevertheless, they seem to be based on thoroughly conducted experiments and considerations. However, most rate constant determined so far have been based on quite complex and possibly not fully understood competitive reaction schemes.

A completely new approach based on RS• generation via reductive dissociation of the corresponding disulfides, RSSR, and a competition between reactions 8 and 9 has recently allowed the determination and re-evaluation of rate constants for oxygen addition to thiyl radicals.[2] Using the electron donor ascorbate and measuring the thiolate yields formed via reaction 9, rate constants have been obtained that are considerably below the diffusion limits. Some of our data are listed in Table 4.[2] These findings are further substantiated by a most recent study by Akhlaq and von Sonntag (private communication). Using nitroform as electron acceptor, they found $8.6 \times 10^7\ M^{-1}s^{-1}$ and $2.2 \times 10^8\ M^{-1}s^{-1}$ for the reaction of GS• and CysS•, respectively. All these results corroborate the electrophilicity of both RS• and O_2.

Table 4. Rate Constants for Oxygen Addition to Thiyl Radicals[2]

Thiyl radical	$k(RS\cdot + O_2)$, $M^{-1}s^{-1}$
CysS•	6.1×10^7
PenS•	2.5×10^7
homo-CysS•	8.0×10^7
CysS•[a]	2.8×10^7
$CH_2(COO^-)CH_2CH_2S\cdot$	2.2×10^8
GS•	3.0×10^7

[a]CysS• and GS• denote thiyl radicals from cysteamine and glutathione, respectively.

This new competitive scheme offers an advantage over the earlier studies because it is possible to vary the identity of the electron donor, D, to allow independent cross-checks. The present data may, however, need adjustment to higher values if the oxygen addition is not irreversible but rather is a reversible process, i.e.,

$$RS\cdot + O_2 \rightleftarrows RSOO\cdot \tag{10}$$

Such a suggestion has, in fact, been made in case of the thiyl radical from glutathione.[17]

In principle, a reversibility is not unrealistic considering the redox similarities between the RS• and RO• radicals. Generally, an O_2 addition to the oxy radical is extremely inefficient, but an equilibrium

$$RO\cdot + O_2 \rightleftarrows RO\text{-}OO\cdot \tag{11}$$

is indicated at low temperatures. Any discussion of reversible oxygen addition should include the chemical properties of the peroxyl radicals. Thus, it is noted that -C-OO• radicals, which are formed via an irreversible process, can undergo a large variety of consecutive reactions.[12-14] For -O-OO• radicals, however, no reaction is known except for the back reaction of equilibrium 11. The -S-OO• radicals seem to assume an intermediate position, perhaps somewhat closer to the carbo peroxyl radicals. An irreversible reaction pathway out of an equilibrium is provided, for example, by the slightly oxidizing properties of RSOO• radicals.[2,17]

At present, it is not possible to accommodate all available experimental data on $RS\cdot/O_2$ systems into an unambiguous quantitative framework. We seem to be left with two possibilities. Either, the oxygen addition is irreversible; then the corresponding rate constants are considerably lower than for a diffusion-controlled reaction. Higher rate constants may only apply if, alternatively, oxygen addition is a reversible process. In conclusion, interaction of thiyl radicals with molecular oxygen is considerably less efficient (at least from the kinetic point of view) than previously believed, irrespective of which of the two possibilities applies.

Our discussion demonstrates the need for further investigations on the kinetics of the formation and decay of thio peroxyl radicals including studies on the chemical properties and possibly the physical identity of these species. In summary, it is clear that oxygen addition to thiyl radicals is not an isolated problem but rather should be viewed in connection with peroxyl radicals in general.

REFERENCES

1. K.-D. Asmus, D. Bahnenann, K. Krischer, M. Lal, and J. Mönig, One-electron induced degradation of halogenated methanes and ethanes in oxygenated and anoxic aqueous solutions, Life Chemistry Reports, 3:1 (1985).
2. J. Mönig, K.-D. Asmus, L. G. Forni, and R. L. Willson, On the reaction of molecular oxygen with thiyl radicals. A reexamination, Int. J. Radiat. Biol. 52:589 (1987).
3. K.-D. Asmus, Pulse radiolysis methodology, in: "Methods in Enzymology," L. E. Packer, ed., vol 105, Academic Press, New York (1984).
4. M. Lal, J. Mönig, and K.-D. Asmus, Acid formation in the radical mediated degradation of chlorinated ethanes in aqueous environment. A radiation chemical study, J. Chem. Soc., Perkin Trans. 2:1639 (1987).
5. Landolt-Bornstein, "Numerical Data and Functional Relationships in Science and Technology. Radical Reaction Rates. New Series," H. Fischer, ed., vol 13 Springer-Verlag, Berlin (1984).
6. J. Mönig, D. Bahnemann, and K.-D. Asmus, One-electron reduction of CCl_4 in oxygenated aqueous solutions: A CCl_3O_2 free radical mediated formation of Cl^- and CO_2, Chem. Biol. Interact. 47:15 (1983).
7. J. Mönig, K.-D. Asmus, M. Schaeffer, T. F. Slater, and R. L. Willson, Electron transfer reactions of halothane-derived peroxyl free radicals, $CF_3CHClO_2\cdot$: Measurement of absolute rate constants, J. Chem. Soc., Perkin Trans. 2:1144 (1983)
8. J. E. Packer, T. F. Slater, and R. L. Willson, Reactions of the carbon tetrachloride-related peroxy free radical ($CCl_3O_2\cdot$) with amino acids: Pulse radiolysis evidence, Life Sci. 2:2617 (1978).
9. J. Mönig, H. Gobl, and K.-D. Asmus, Free radical one-electron versus hydroxyl radical induced oxidation. Reaction of trichloromethyl-peroxyl radicals with simple and substituted aliphatic sulphides in aqueous solutions, J. Chem. Soc., Perkin Trans. 2:647 (1985).
10. M. Bonifacic and K.-D. Asmus, One-electron redox potentials of $RSSR^{+}\cdot$ -RSSR couples from dimethyl disulfide and lipoic acid, J. Chem. Soc., Perkin Trans, 2:1805 (1986)
11. J. Mönig and K.-D. Asmus, Formation and reactions of halothane peroxy free radicals, in: "Oxygen Radicals in Chemistry and Biology," W. Bors, H. Saran, and D. Tait eds., W. de Gruyter & Co, Berlin (1984).
12. H. Zegota, H. N. Schuchmann, and C. von Sonntag, Elucidation of the mechanisms of peroxyl radical reactions in aqueous solutions using the pulse radiolysis technique, J. Radioanal. Nucl. Chem., Articles 101/102:199 (1986).
13. K.-D. Asmus, Formation and properties of oxygen radicals as studied by radiation chemical methods, in: "Reactive Oxygen Species in Chemistry, Biology and Medicine" NATO Advanced Study Institute Series, Plenum Press, New York, in press.
14. J. A. Howard, Peroxides, in: "The Chemistry of Functional Groups," S. Patai, ed., John Wiley & Sons, New York (1983).
15. W. Adam and W. J. Baader, Effects of methylation on the thermal stability and chemiluminescence properties of 1,2-dioxetanes, J. Am. Chem. Soc. 107:410 (1985).
16. K.-D. Asmus, Sulfur-centered free radicals. in: "Radioprotectors and Anticarcinogens," O. F. Nygaard and M. G. Simic, eds., Academic Press, New York (1983).
17. H. Tamba, G. Simone and H. Quintiliani, Interactions of thiyl free radicals with oxygen, a pulse radiolysis study, Int. J. Radiat. Biol. 50:595 (1986).
18. J. P. Barton amd J. E. Packer, The radiolysis of oxygenated cysteine solutions at neutral pH. The role of $RSSR^-$ and O_2^-, Int. J. Radiat. Phys. Chem. 2:159 (1970).

THIYL FREE RADICAL METABOLITES OF THIOL DRUGS AND GLUTATHIONE

Kathryn T. Knecht, Carolyn Mottley*, and Ronald P. Mason*

Curriculum in Toxicology
University of North Carolina
Chapel Hill, North Carolina 27514, U.S.A.

*Laboratory of Molecular Biophysics
National Institute of Environmental Health Sciences
PO Box 12233
Research Triangle Park, North Carolina 27709, U.S.A.

Lactoperoxidase, a prototypical mammalian peroxidase, is part of an antimicrobial system found in secreted fluids.[1] It is believed to act by oxidation of thiocyanate.[2] Not surprisingly, it will also metabolize thiol moieties, oxidizing them to free radical intermediates. Thiol compounds are important in pharmacology and toxicology, and we have examined the production of highly reactive thiyl free radicals by this mechanism.

A range of thiol compounds has been studied. Cysteine, a radioprotector,[3] shows mutagenicity in the Ames test[4] and toxicity in cell culture.[5] D-Penicillamine, a methylated analog of cysteine, is used to treat Wilson's disease and rheumatoid arthritis.[6] It produces a number of serious side effects, predominantly immune related.[7] The cysteine-containing tripeptide glutathione is an endogenous cellular protectant which shows mutagenicity in the Ames assay.[4] N-acetylcysteine is a mucolytic agent and acetaminophen antidote[8] with minimal toxicity.

Cysteine, incubated as described in the legend to Figure 1, forms a thiyl radical which can be trapped by the spin trap DMPO. The ESR signal is shown in Figure 1A. Adding H_2O_2 increases the signal slightly (Figure 1B). A requirement for H_2O_2 is shown by the greatly diminished signal intensity in the presence of catalase (Figure 1C). The reaction requires native enzyme (Figure 1D).

In addition, the signal is nearly abolished by incubation in nitrogen-bubbled (oxygen-depleted) buffer (Figure 2A), unless H_2O_2 is added (Figure 2B). Catalase in this system eliminates the residual signal (Figure 2C), suggesting the radical formation in Figure 2A could have been due to incomplete oxygen depletion. The requirement for native enzyme again demonstrates that the signal is enzyme mediated (Figure 2D).

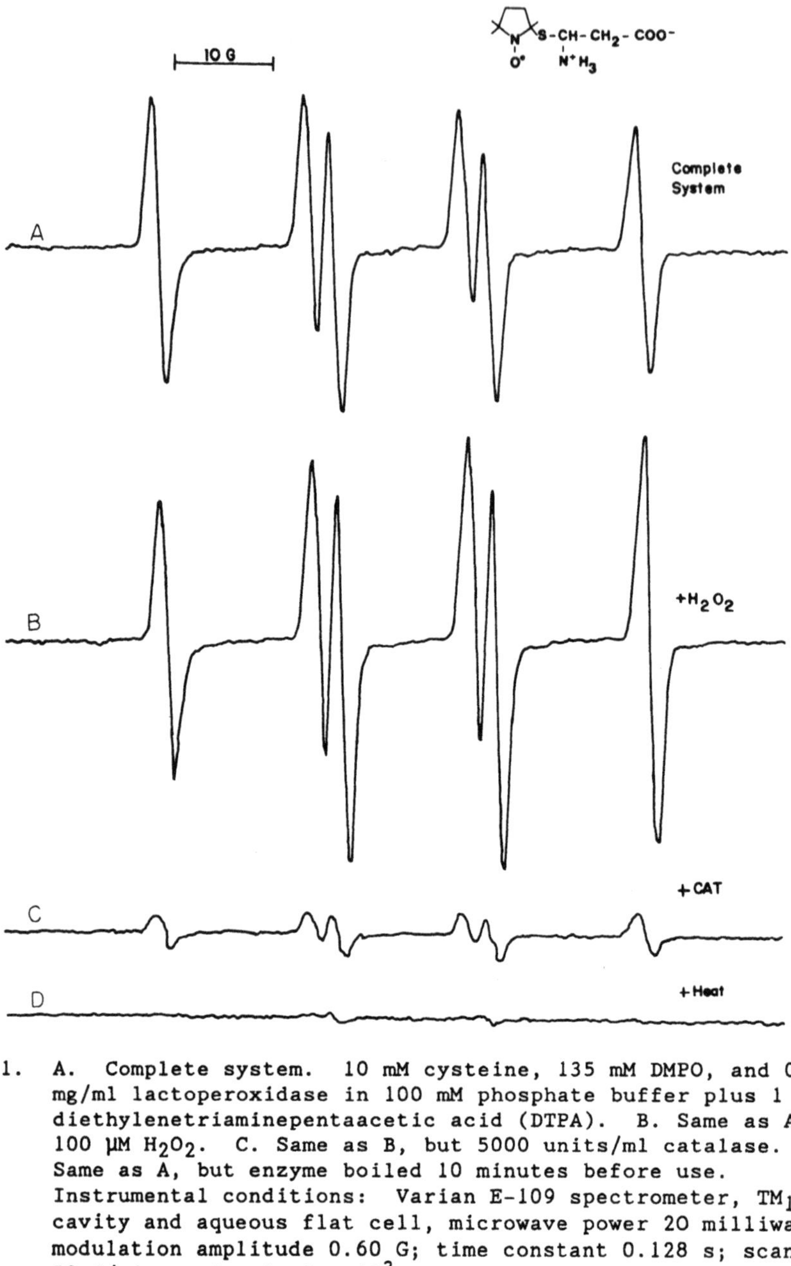

Figure 1. A. Complete system. 10 mM cysteine, 135 mM DMPO, and 0.1 mg/ml lactoperoxidase in 100 mM phosphate buffer plus 1 mM diethylenetriaminepentaacetic acid (DTPA). B. Same as A, but 100 μM H_2O_2. C. Same as B, but 5000 units/ml catalase. D. Same as A, but enzyme boiled 10 minutes before use. Instrumental conditions: Varian E-109 spectrometer, TM_{110} cavity and aqueous flat cell, microwave power 20 milliwatts; modulation amplitude 0.60 G; time constant 0.128 s; scan rate 50 G/min; and gain 8 x 10^3.

D-penicillamine behaves in a similar fashion to cysteine for radical formation, enzyme and H_2O_2 requirements, both in aerobic and in anaerobic systems (data not shown).

Production of the glutathione thiyl radical is H_2O_2 dependent even in the presence of oxygen. The reaction is still enzyme mediated (Figure 3). N-acetyl cysteine data is similar to those seen with glutathione (not shown). N-acetylation of cysteine apparently impairs its oxidation.

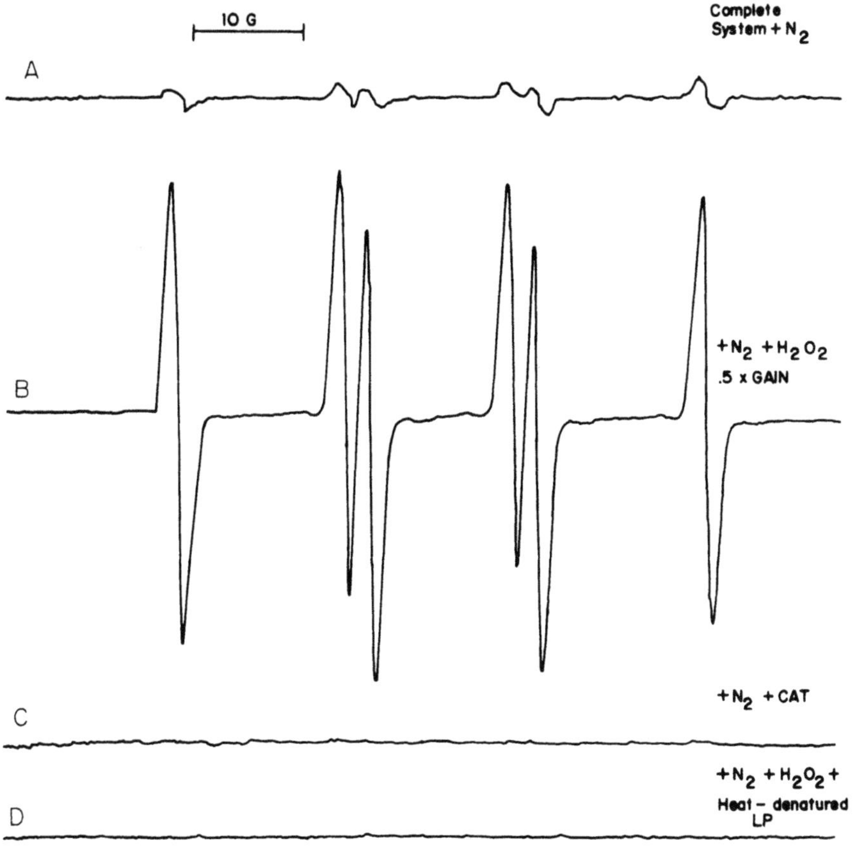

Figure 2. A. Same as 1A, but bubbled 10 min with N_2 before addition of enzyme. B. Same as A, but 100 μM H_2O_2 added last minute of N_2 bubbling. C. Same as A, but 5000 units/ml catalase added. D. Same as A, but enzyme boiled 10 minutes before use. Instrumental conditions are the same as for Figure 1. Gain 8 x 10^3 except B, gain 4 x 10^3.

$$RS^{\cdot} + RSH\ (RS^-) \rightarrow [RSSR]^{\dot{-}} + H^+ \quad (1)$$

$$[RSSR]^{\dot{-}} + O_2 \rightarrow RSSR + \dot{O}_2^- \quad (2)$$

$$2\dot{O}_2^- + 2H^+ \rightarrow H_2O_2 + O_2 \quad (3)$$

The requirement for exogenous H_2O_2 in glutathione or N-acetyl cysteine radical formation contrasts with results seen with cysteine and penicillamine. The latter thiols do require H_2O_2 for radical production, as shown by catalase inhibition. However, they seem to generate sufficient peroxide during the course of aerobic incubation to sustain radical formation without exogenous H_2O_2.[9,10]

This capability would allow cysteine and penicillamine but not glutathione or N-acetyl-cysteine to maintain radical production in low-H_2O_2 environments *in vivo*, possibly contributing to the greater relative toxicities seen with the former two compounds.

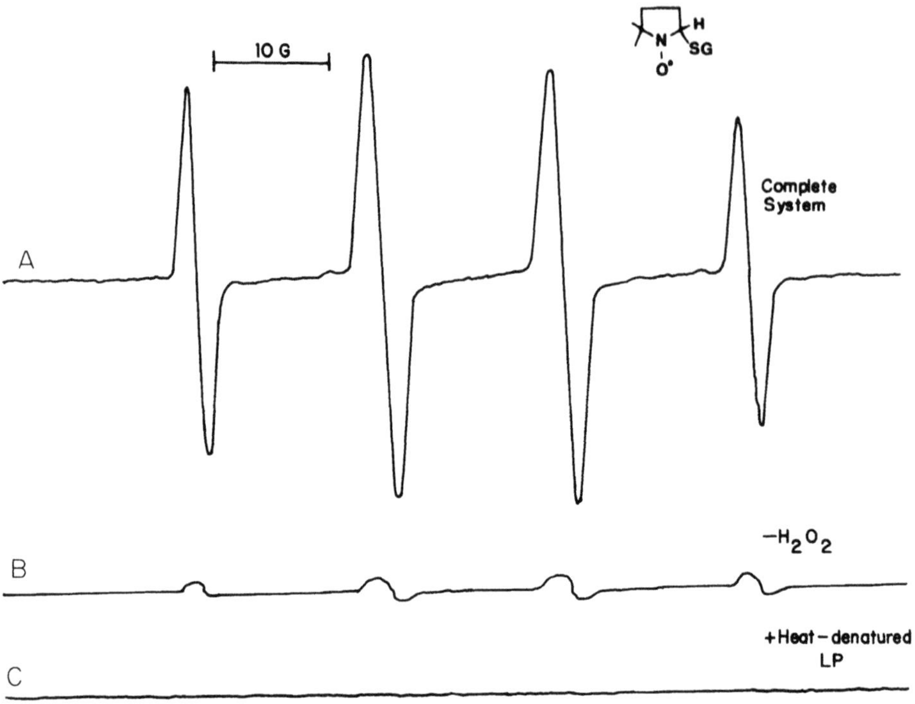

Figure 3. A. Complete system. 10 mM reduced glutathione, 135 mM DMPO, 100 μM H_2O_2, and 0.1 mg/ml lactoperoxidase in 100 mM phosphate buffer plus 1 mM DTPA. B. Same as A, - H_2O_2. C. Same as A, but lactoperoxidase boiled 10 minutes before use. Instrumental conditions are the same as for Figure 1, except gain 4 x 10^3.

The addition of 100 μM H_2O_2 to cysteine incubated aerobically produces only a slight increase in signal intensity (Figure 1A,B). However, addition of this same quantity of H_2O_2 to the anaerobic cysteine incubation produces a signal twice the height of either of those seen in air. A similar enhancement of H_2O_2 effect by oxygen depletion was seen with penicillamine. This may be due to the combination of thiyl radicals with oxygen instead of the spin trap, either directly to form the sulfur peroxyl radical $RSOO^{\cdot}$[11] or indirectly via the disulfide anion radical as described above.[9,10]

ACKNOWLEDGMENTS

KTK was supported by NIH Grant T32 ES 07126.

REFERENCES

1. E.L. Thomas, K.P. Bates, and M.M. Jefferson, Peroxidase antimicrobial system of human saliva: Requirements for accumulation of hypothiocyanite, J. Dent. Res. 60:785 (1981).
2. T.M. Aune and E.L. Thomas, Accumulation of hypothiocyanite ion during peroxidase-catalyzed oxidation of thiocyanate ion, Eur. J. Biochem. 80:209 (1977).

3. P.C. Jocelyn, in: "Biochemistry of the SH Group," Academic Press, New York (1972).
4. H. Glatt, M. Protić-Sabljić, and F. Oesch, Mutagenicity of glutathione and cysteine in the Ames test, Science 220:961 (1983).
5. G. Saez, P.J. Thornalley, H.A.O. Hill, R. Hems, and J.V. Bannister, The production of free radicals during the autooxidation of cysteine and their effect on isolated rat hepatocytes, Biochim. Biophys. Acta 719:24 (1982).
6. J. Peisach and W.E. Blumberg, A mechanism for the action of penicillamine in the treatment of Wilson's disease, Mol. Pharm. 5: 200 (1969).
7. H.B. Stein, A.C. Patterson, R.C. Offer, C.J. Atkins, A. Teufel, and H.S. Robinson, Adverse effects of D-penicillamine in rheumatoid arthritis, Ann. Intern. Med. 92:24 (1980).
8. J.B. Gerber, J.S. Macdonald, R.D. Harbison, J.-P.Villeneuve, A.J.J. Wood, and A.S. Nies, Effect of N-acetylcysteine on hepatic covalent binding of paracetamol, Lancet i:657 (1977).
9. L.S. Harman, C. Mottley, and R.P. Mason, One- and two-electron oxidation of reduced glutathione by peroxidases, J. Biol. Chem. 259:5606 (1984).
10. C. Mottley, K. Toy, and R.P. Mason, Oxidation of thiol drugs and biochemicals by the lactoperoxidase/hydrogen peroxide system, Mol. Pharm. 31:417 (1987).
11. M. Tamba, G. Simone, and M. Quintiliani, Interactions of thiyl free radicals with oxygen: a pulse radiolysis study, Int. J. Radiat. Biol. 50:595 (1986).

OXIDATION OF QUINONES BY H_2O_2: FORMATION OF EPOXY- AND HYDROXYQUINONE ADDUCTS AND ELECTRONICALLY EXCITED STATES

Anders Brunmark and Enrique Cadenas

Department of Pathology II, University of Linköping
S-581 85 Linköping, Sweden

INTRODUCTION

The cellular activation of quinonoid compounds (diketones with the structure O=C-(C=C-)$_n$C=O requires their reduction by either a one- or two-electron transfer process which can be accomplished by several NADPH or NADH reductases.[1,2] The redox features of quinone chemistry have been well characterized[3] and their extrapolation to biological processes has furnished much useful information on the toxicological aspects of quinonoid compounds.[4,5]

The addition-substitution chemistry of quinones, on the other hand, has been described extensively.[6] This study is concerned mainly with the O_2-addition and sulphur-addition reactions to quinones. The former is contemplated here as the O_2 addition to and H_2O_2 oxidation of the quinone, similar to the reaction of sodium hydroperoxide with naphthoquinones[6] and yielding the corresponding quinone epoxide product. The latter, the sulphur addition, is focused on the 1,4-reductive addition by GSH to quinones or quinone epoxides. The product of this reaction is a hydroquinone-glutathionyl conjugate which can also undergo autoxidation reactions with production of $O_2^{\cdot -}$ and eventually H_2O_2, as expected from the formation of a semiquinone-glutathionyl adduct intermediate identified by EPR studies.[7] Therefore, in addition to the enzymic reductive activation of quinone compounds, the conjugation of the latter with GSH could render them equally active in autoxidation processes and thus contribute to cellular oxidative injury.

OXIDATION OF QUINONES BY H_2O_2: QUINONE EPOXIDE FORMATION

The reaction of the simplest unsubstituted quinone, *p*-benzoquinone with equimolar amounts of H_2O_2 in slightly-alcaline aqueous buffer (pH 7.6) is accompanied by a decrease in the absorption peak of *p*-benzoquinone in the ultraviolet region (244 nm) and a parallel increase at 212 and 484 nm.[8] The former absorption peak is attributed to the formation of 2,3-epoxy-*p*-benzoquinone, whereas the latter absorption peak is attributed to 2-OH-*p*-benzoquinone, its accumulation initially sustained by the H_2O-addition reaction to *p*-benzoquinone.[9] The NMR and UV absorption spectra of synthesized 2,3-epoxy-*p*-benzoquinone (upon reaction of sodium perborate with *p*-benzoquinone at pH 8.5 for 15 s) agreed with those obtained from the reaction with H_2O_2 with *p*-benzoquinone.[8]

The epoxide retained the redox properties of the parent quinone compound, thus making its identification possible with HPLC with electrochemical detection.[8] Figure 1A shows a chromatogram of the products formed during 20-min incubation of *p*-benzoquinone with H_2O_2. Peaks b and c co-chromatographed with the peaks observed with the pure and synthesized 2,3-epoxy-*p*-benzoquinone and *p*-benzoquinone, respectively. Peak a was eluted almost with the void volume and its identification was not possible, although it likely might correspond to 2-OH-*p*-benzoquinone. The half-wave potentials (calculated from the corresponding hydrodynamic chromatograms) as well as the retention times of peaks a-c are indicated in Figure 1A. Figure 1B illustrates the time course of substrate (*p*-benzoquinone and H_2O_2) consumption and product (2,3-epoxy- and 2-OH-*p*-benzoquinone) formation; the generation of the former product proceeded at a rate of 315 μM x min^{-1}, whereas that of the latter at a rate of 8.9 μM x min^{-1}.

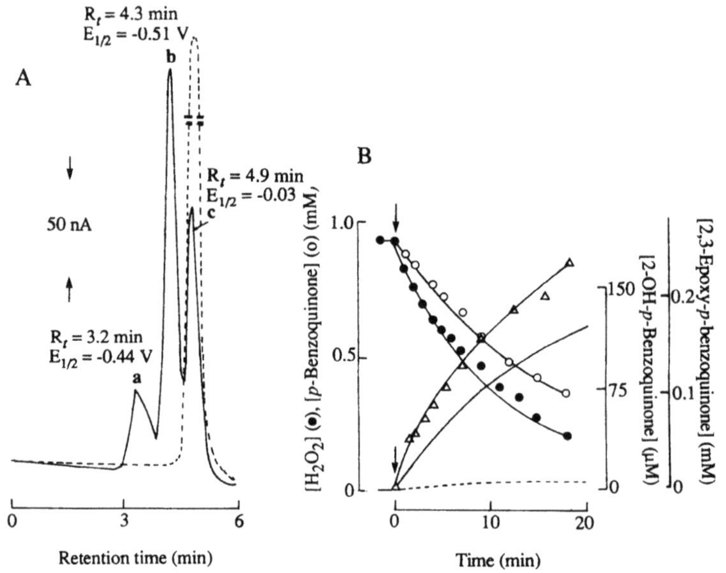

Figure 1. Formation of 2,3-epoxy-*p*-benzoquinone and 2-OH-*p*-benzoquinone during the reaction of *p*-benzoquinone with H_2O_2. (A) HPLC with electrochemical detection chromatogram of *p*-benzoquinone (---) and the products (—) formed during 20-min incubation of equimolar amounts of *p*-benzoquinone and H_2O_2. Peaks a, b, and c and their respective half-wave potentials ($E_{1/2}$) and retention times (R_t) are indicated in the figure. (B) Time course of reactants [H_2O_2 and *p*-benzoquinone] consumption and products [2,3-epoxy-*p*-benzoquinone (Δ) and 2-OH-*p*-benzoquinone (—)] formation. The production of 2-OH-*p*-benzoquinone in the absence of H_2O_2 is indicated by the dotted line (---). Assay conditions as in reference 8.

The H_2O_2-mediated oxidation of *p*-benzoquinone described here was extended to other *p*-benzo- and 1,4-naphthoquinone compounds with different degrees of substituion.[10] Reactions 1 and 2 illustrate the formation of 2,3-epoxy-*p*-benzoquinone and 2,3-epoxy-1,4-naphthoquinone upon oxidation of the unsubstituted parent quinone by H_2O_2.

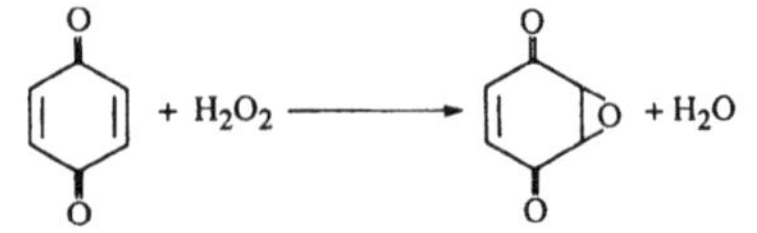

[1]

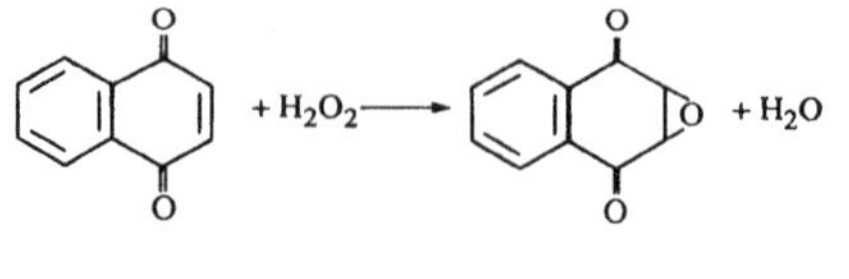

[2]

The presence of methyl substituents close to the double bond diminishes the rate of H_2O_2-mediated formation of quinone epoxides; accordingly, the yield of quinone epoxide formation decreases with increasing substitution of the quinone compound. The rates of H_2O_2-mediated epoxidation of the *p*-benzoquinone series decreases in the following order: *p*-benzoquinone > 2-methyl-*p*-benzoquinone > 2,5-dimethyl-*p*-benzoquinone (with second order rate constants for the corresponding quinone epoxide formation of 0.28, 0.13, and 0.09 $M^{-1}s^{-1}$, respectively). The rate of epoxidation for the 1,4-naphthoquinone series decreases in the following order: 1,4-naphthoquinone > 2-methyl-1,4-naphthoquinone (menadione) > 2,3-dimethyl-1,4-naphthoquinone (with second order rate constants of 0.26-, 0.07-, and 0.002 $M^{-1}s^{-1}$, respectively).[10]

The restraining role of methyl substitution against epoxidation may be explained in terms of either steric hindrance by the bulky methyl group or changes in the electron density in the quinone. The H_2O_2-mediated oxidation of these *p*-benzo- and 1,4-napthoquinone series (as illustrated in reactions 1 and 2) seems not to be dependent on the redox potential of the reacting quinone.[10]

FORMATION OF ELECTRONICALLY EXCITED STATES DURING THE OXIDATION OF *p*-BENZOQUINONE BY H_2O_2

In addition to the two molecular products (2,3-epoxy- and 2-OH-*p*-benzoquinone), electronically excited states are also generated during the oxidation of *p*-benzoquinone by H_2O_2 in slightly alkaline aqueous buffer. The formation of excited states is presumably brought about in a pathway involving the free radical intermediates implied in the H_2O-addition reaction to *p*-benzoquinone, i.e. in the formation of 2-OH-*p*-benzoquinone.

The features of excited state generation are partly summarized in Figure 2:

[a] The relaxation of the excited state to the ground state is accompanied by photoemission with a maximum at about 485-520 nm, corresponding to a free energy of about 53 kcal/mole (Figure 2B).

[b] The emission is not dependent on the presence of O_2 in the reaction mixture; it proceeds and, moreover, it is slightly (15-20%) enhanced under anaerobiosis.[8]

[c] Emission is efficiently inhibited by the free radical scavenger (+)-cyanidanol-3 (Figure 2A) which is known to react very rapidly with HO• and somewhat less rapidly with other carbon- or O_2-centered radicals. The inhibition of photoemission by (+)-cyanidanol-3 cannot be accounted for by a reaction with either *p*-benzosemiquinone or 2-OH-*p*-benzosemiquinone. These reactions would lead to the formation of *p*-benzohydroquinone and 2-OH-*p*-benzohydroquinone, respectively, products which were not detected during the course of the reaction. [d] The time courses and pH dependencies of 2-OH-*p*-benzoquinone formation and photoemission are similar, thus suggesting the participation of the hydroxyquinone compound in the photoemissive process.[8]

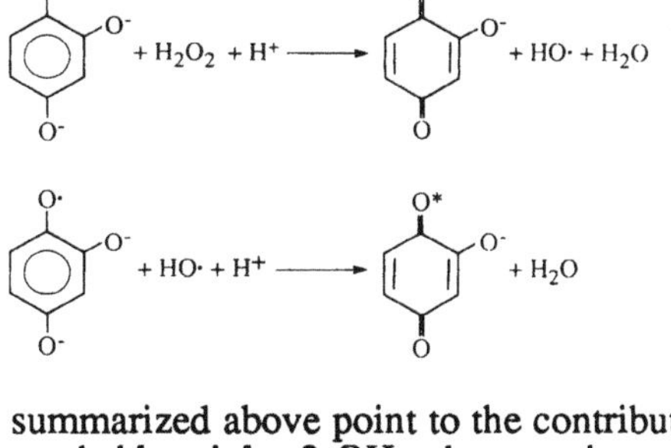

[3]

[4]

The characteristics summarized above point to the contribution of an excited triplet ketone of n,π* character, probably triplet 2-OH-*p*-benzoquinone, to the photoemission observed. The mechanism for the formation of a triplet 2-OH-*p*-benzoquinone could be formally accounted for within a two-step process. First, the reduction of H_2O_2 to HO• and H_2O by 2-OH-*p*-benzosemiquinone (reaction 3). It was suggested that any semiquinone/quinone couple with a reduction potential between -0.33 volts and 0.46 volts could theoretically bring about a non-metal, or organic Fenton reaction.[11] Second, the oxidation of 2-OH-*p*-benzosemiquinone by HO• to yield excited triplet 2-OH-*p*-benzoquinone and H_2O

(reaction 4). The calculated photoemission yield of this reaction is 7.1×10^{-8} photons/2-OH-*p*-benzoquinone molecule formed. A similar mechanism for the formation of excited states by electron transfer was proposed for the bacterial diaphorase-catalyzed redox transitions involving *p*-benzoquinone,[12] the non-enzymic oxidation of hydroquinone derivatives,[13] and the enzymic oxidation of cathecol.[14]

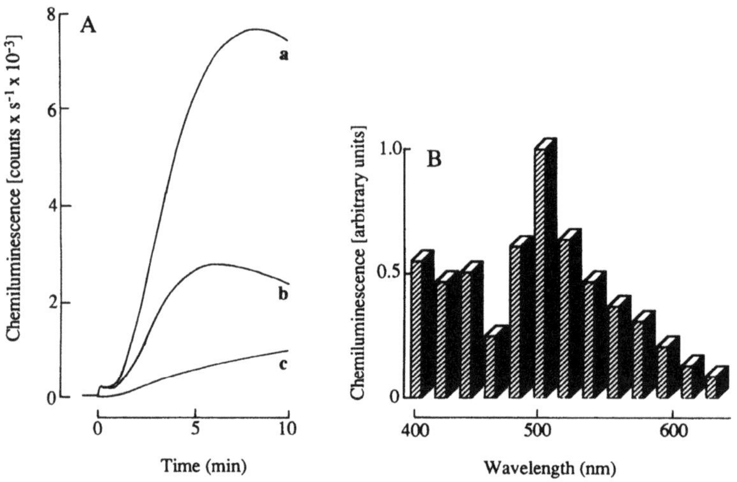

Figure 2. (A) Time course of photoemission during the reaction of *p*-benzoquinone with H_2O_2. Assay conditions: 1 mM *p*-benzoquinone in 0.1 M potassium phosphate buffer, pH 7.65, was supplemented with 1 mM H_2O_2 to initiate the reaction. Trace a: no further additions; trace b, plus 200 μM GSH; trace c, plus 5 μM (+)-cyanidanol-3. (B) Spectral distribution of photoemission under anaerobiosis and as in trace a.

EFFECT OF GSH ON 2,3-EPOXY-*p*-BENZOQUINONE AND EXCITED STATE FORMATION DURING THE OXIDATION OF *p*-BENZOQUINONE BY H_2O_2

Quinones are known to undergo nucleophilic attack by SH-containing molecules to yield a substituted hydroquinone.[6] The reaction between GSH and *p*-benzoquinone results in a rapid loss of total glutathione (GSH + 2GSSG)[8] accompanied by a parallel decrease in the HPLC/electrochemical detection signal corresponding to *p*-benzoquinone (analyzed with reductive-applied potential; Figure 3A; traces a and b) and simultaneous appearance of a signal detected with oxidative-applied potential (Figure 3A; trace c). The latter peak has a shorter retention time (R_t = 3.2 min) than the unsubstituted *p*-benzohydroquinone (R_t = 4.2 min) and a $E_{1/2}$ = +0.2 volts, and it corresponds to the glutathionyl-hydroquinone adduct. GSH and *p*-benzoquinone are totally consumed when present in equimolar amounts, suggesting a ratio of the unit for the glutathionyl-substituted hydroquinone formation, thus withstanding the formation of a sole product of conjugation.[8]

2,3-Epoxy-*p*-benzoquinone also reacts with GSH, although at slower rates than *p*-benzoquinone. The 1,4-reductive addition involving the quinone epoxide leads to epoxide ring rearrangement and reduction of the quinone. The product formed is expected to be a 2-OH-3-glutathionyl-*p*-benzohydroquinone, rapidly oxidizing to 2-OH-3-glutathionyl-*p*-benzoquinone (Figure 3B). The former product is more prone to autoxidation than the parent compound lacking the -OH substitutent, as was recently reported for various hydroxy derivatives of *p*-benzoquinones and 1,4-naphthoquinones.[10]

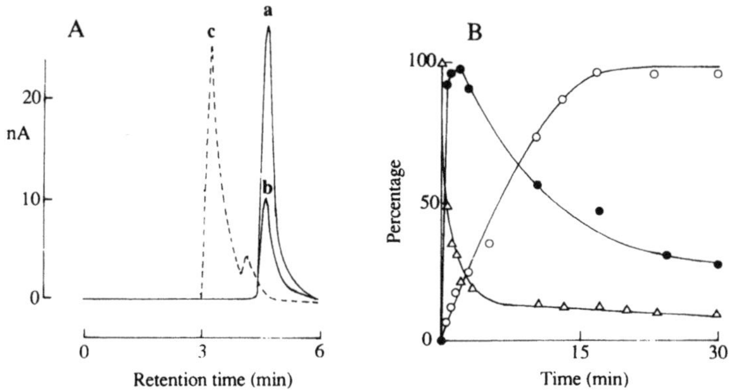

Figure 3. 1,4-Reductive addition of GSH to *p*-benzoquinone and 2,3-epoxy-*p*-benzoquinone.(A) HPLC/electrochemical detection chromatogram indicating *p*-benzoquinone (trace a) and its decrease upon reaction with GSH (trace b) and formation of a product detected with oxidative-applied potential (trace c) (see in the text) corresponding to the glutathionyl-hydroquinone conjugate. (B) Time course of 2,3-epoxy-*p*-benzoquinone consumption (Δ) and formation of products detected with reductive- and oxidative-applied potential, probably corresponding to 2-OH-*p*-benzoquinone (o) and 2-OH-*p*-benzohydroquinone (•), respectively.

Reactions 5 and 6 account for these observations in terms of the formation of a product resulting from the 1,4-reductive addition of GSH to *p*-benzoquinone and 2,3-epoxy-*p*-benzoquinone, respectively.

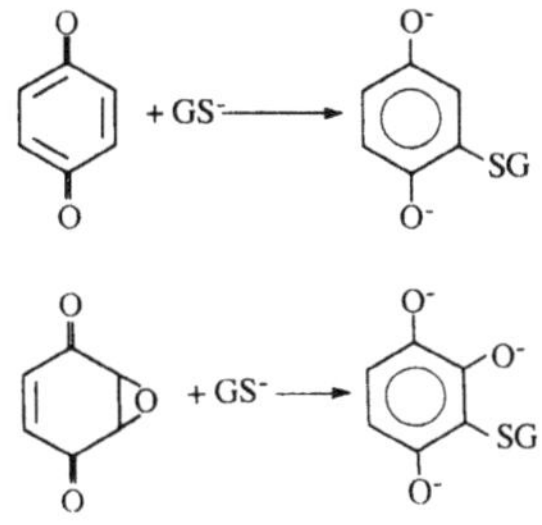

[5]

[6]

CONCLUSIONS

The oxidation of *p*-benzoquinone by H_2O_2 in slightly alkaline aqueous buffer yields as main molecular products quinone epoxide and 2-OH-quinone.[8] The accumulation of the latter, though significantly enhanced in the presence of H_2O_2 (Figure 1B), is sustained by the H_2O-addition reactions to quinones.[9] The *p*-benzoquinone epoxide formed does not contribute to emission in this experimental model, despite the report that nucleophilic attack to epoxides yields excited carbonyls.[15] The quenching of photoemission by GSH (Figure 2A) cannot be attributed to the withdrawal, upon a 1,4-reductive addition mechanism, of quinone available to react with H_2O_2. Amounts of GSH smaller than quinone present are required to inhibit emission,[8] thus suggesting a quenching of the excited state by the conjugate, involving either the S-C bond or the hydroquinone moiety.[16]

Triplet quinones, as investigated by laser flash absorption spectroscopy, are reactive species capable of oxidation of hydroquinones with formation of semiquinones [$^3Q^* + QH_2 \longrightarrow 2\ QH^\cdot$][17] or of undergoing reduction with different amines by either H-atom or electron transfer.[18] These reactive forms of quinones might be of toxicological interest and contribute to the tissue oxidative injury caused by quinone intracellular metabolism.[4]

ACKNOWLEDGMENTS

Supported by grants 7679 and 4481 from the Swedish Medical Research Council.

REFERENCES

1. I. Yamazaki, Free radicals and enzyme-substrate reactions, in: "Free Radicals in Biology," W.A. Pryor, ed., vol. III, p. 193, Academic Press, New York (1977).
2. H. Nohl, W. Jordan, and R.J. Youngman, Quinones in biology: Functions in electron transfer and oxygen activation, Adv. Free Radical Biol. Med. 2: 211 (1986).
3. J.Q. Chambers, Electrochemistry of quinones, in: "The Chemistry of Quinonoid Compounds," S. Patai, ed., p. 737, John Wiley, New York (1974).
4. M.T. Smith, C.G. Evans, H. Thor, and S. Orrenius, Quinone-induced oxidative injury to cells and tissues, in: "Oxidative Stress", H. Sies, ed., p.73, Academic Press, London (1985).
5. L. Rossi, G.A. Moore, S. Orrenius, and P.J. O'Brien, Quinone toxicity in hepatocytes without oxidative stress, Arch. Biochem. Biophys. 251: 25 (1987).
6. K.T. Finley, The addition and substitution chemistry of quinones, in: "The Chemistry of Quinonoid Compounds," S. Patai, ed., p. 877, John Wiley, New York (1974).
7. N. Takahashi, J. Schreiber, V. Fischer, and R.P. Mason, Formation of glutathione-conjugated semiquinones by the reaction of quinones with glutathione: an ESR study, Arch. Biochem. Biophys. 252: 41 (1987).
8. A. Brunmark and E. Cadenas. Electronically-excited state generation during the reaction of *p*-benzoquinone with H_2O_2. Relation to product formation: 2-OH- and 2,3-epoxy-*p*-benzoquinone. The effect of GSH. Free Radical Biol. Med. 3: 169 (1987).
9. W. Flaig, and J.C. Salfeld, Nachweiss der Bildung von Hydroxy-p-Benzochinon als Zwischenprodukt bei der Autoxydation von Hydrochinon. Naturwissenschaften 47: 516 (1960).
10. A. Brunmark, E. Cadenas, J. Segura-Aguilar, C. Lind, and L. Ernster. DT-Diaphorase-catalyzed two-electron reduction of various *p*-benzoquinone and 1,4-naphthoquinone epoxides, Free Radical Biol. Med. 4: 000 (1988).
11. W.H. Koppenol and J. Butler, Energetics of interconversion reaction of oxyradicals, Adv. Free Radical Biol. Med. 1: 91 (1985).
12. M. Ginsberg and E. Cadenas, Electronically-excited state generation during diaphorase-catalyzed benzoquinone reduction, Photobiochem. Photobiophys. 9: 223 (1985).
13. J. Stauff and P. Bartolmes, Chemilumineszenz bei der oxidativen Bildung von Triplett Zuständen des Antrasemichinon- und Anthrachinon-2-sulfonats, Angew. Chem. 82: 321 (1970).
14. M. Villablanca and G. Cilento, Enzymatic generation of electronically-excited states by electron transfer, Photochem. Photobiol. 42: 591 (1985).
15. J. Stauff, I. Simo, and W.L. Tsai, Chemilumineszenz von Epoxides, Naturforschung 33c: 769 (1978).
16. M.V. Encinas, E.A. Lissi, and A.F. Olea, Quenching of triplet benzophenone by vitamins E and C and by sulfur-containing aminoacids and peptides, Photochem. Photobiol. 42: 347 (1985).
17. E. Amouyal and R. Bensasson. Duroquinone triplet reduction in cyclohexane, ethanol and H_2O, and by durohydroquinone, J. Chem. Soc.,Faraday Trans. I, 72: 1274 (1976).
18. E. Amouyal and R. Bensasson, Interaction of duroquinone lowest triplet with amines, J. Chem. Soc., Faraday Trans. I, 73: 1561 (1977).

FROM NITRIC OXIDE TO DESFERAL:

NITROGEN FREE RADICALS AND IRON IN OXIDATIVE INJURY

Robin L. Willson

Department of Biochemistry
Brunel University of West London
Uxbridge, Middlesex, UB8 3PH U.K.

INTRODUCTION

During the last decade the possible role of free radicals in the development of disease, particularly cancer and inflammation, and in tissue injury following heart attack, transplant surgery, or certain types of chemical poisoning has attracted considerable interest.

Since much of the research in this area was undoubtedly stimulated by the identification of the enzyme superoxide dismutase in 1969, it is not surprising that much of the interest has focussed on the role of the superoxide radical ($O_2^{\bullet -}$) itself or other oxygen-centred radicals such as the hydroxyl radical ($OH^\bullet$), the hydroperoxy radical ($HO_2^\bullet$) or various organic peroxy radicals ($RO_2^\bullet$).

Yet a large variety of nitrogen free radicals also exist whose chemistry and biology is equally fascinating; indeed a significant number are thought to be important toxic intermediates. Many nitrogen radicals are very long lived, and it is therefore perhaps not surprising that several, including nitric oxide, $NO^\bullet$, nitrogen dioxide, $NO_2^\bullet$, Fremy's salt and porphyrexide, were amongst the first free radicals to be characterised.[1,2]

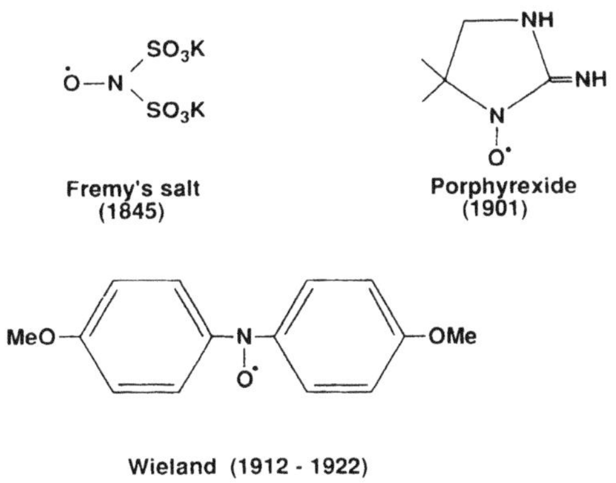

Figure 1

In this paper I will briefly illustrate just a few of the various types of nitrogen free radicals that have been detected in biological systems and then focus on some aspects of the free radical biochemistry of the stable organic nitroxide free radicals which are proving so useful in the laboratory as indicators of free radical involvement (spin-trapping experiments). Such nitroxides have been shown to be effective antioxidants and radiosensitisers and may also be important intermediates in the metabolism of a number of carcinogens, cocaine and possibly the iron binding drug, desferrioxamine (Desferal).

NITROGEN FREE RADICALS

In the present context the term nitrogen free radical refers to those free radicals in which the unpaired electron is centered on the nitrogen atom or, alternatively, is situated in a delocalised electronic orbital covering several atoms one of which is a nitrogen. Some, such as the azyl radical ($N_3^\bullet$) and the hydrazyl radical ($NH_2NH^\bullet$) can be considered as inorganic. Examples of some organic nitrogen radicals of biological relevance are shown below.

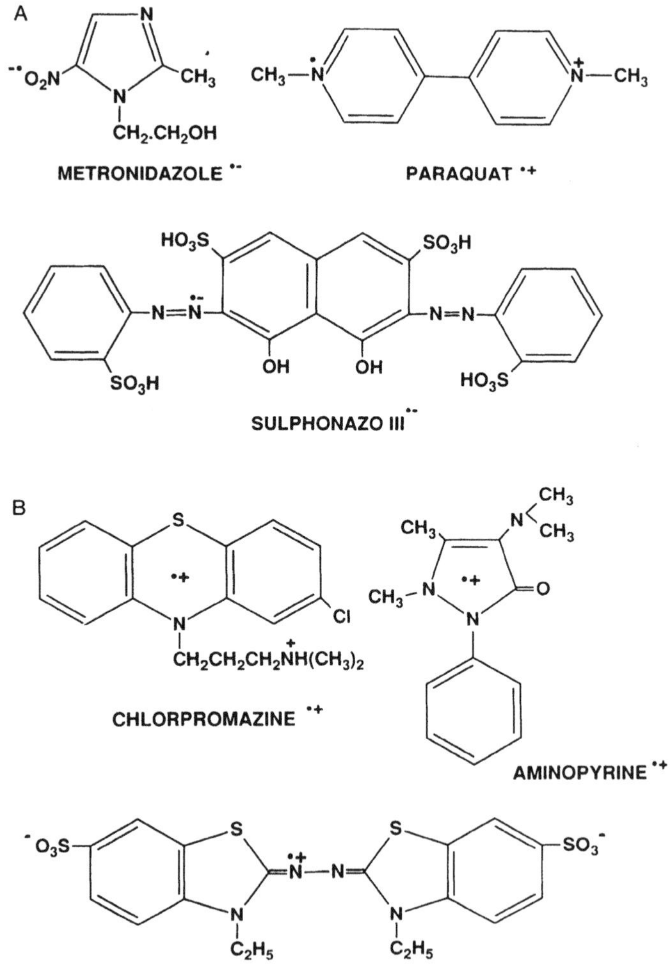

Figure 2

The electron spin resonance spectra of the radical cations derived from the herbicide methyl viologen (paraquat),[3] the radical anion of the diazo compound sulphazo III[4] and the radical anion of the drug metronidazole,[5] widely used in the treatment of anaerobic infections, have been measured during the incubation of the parent compounds with liver microsomes. The electron spin resonance spectra of the metronidazole and paraquat radicals have also been observed in incubates of the drugs with Trichomonas foetus[6] and lung epithelial cells,[7] respectively. Their optical absorption spectra have been determined by pulse radiolysis.[8,9]

The optical absorption spectrum of the radical cation of the salivary peroxidase substrate ABTS (2,2-azino bis-[3]-ethylbenzthiazoline-6-sulfonate) has also been measured by pulse radiolysis.[10] The optical absorption spectra of the corresponding radicals from the drugs aminopyrine[11,12] and chlorpromazine[13] have been determined by pulse radiolysis and their electron spin resonance spectra measured during incubation of the drugs with hydrogen peroxide and horseradish peroxidase.[14-16] In the case of chlorpromazine the optical absorption spectrum has also been obtained on oxidation of the drug with ferric or persulphate ions.[17]

NITROXIDE FREE RADICALS

Nitroxide free radicals can be considered as derivatives of nitric oxide, whose free radical stability has been attributed to its hybrid structure. Unlike many other nitrogen-centred radicals, organic nitroxides do not normally dimerize at room temperature due, it is

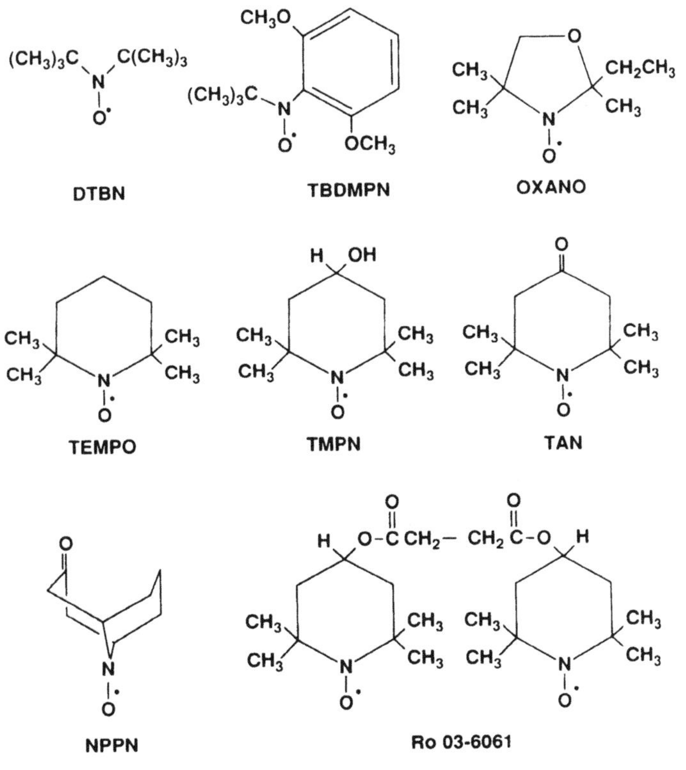

Figure 3

believed, to an inherently stable electronic arrangement around the nitrogen and oxygen atoms. Steric effects arising from the groups attached to the nitrogen atom apparently play little or no part in preventing dimerization although they are thought to be involved in preventing disproportionation or fragmentation reactions occurring.[2]

The above nitroxides can be obtained either by

a) oxidation of hydroxylamines

$$\text{e.g.,}\quad C_6H_5NHOH - e \longrightarrow C_6H_5\underset{O^{\bullet}}{N}H \qquad (1)$$

b) reduction of nitroso compounds

$$\text{e.g.,}\quad C_6H_5NO^{\bullet} + e \longrightarrow C_6H_5\underset{O^{\bullet}}{N}H \qquad (2)$$

c) radical addition to nitroso compounds

$$\text{e.g.,}\quad (CH_3)_3CNO + CH_2OH \longrightarrow (CH_3)_3C\underset{O^{\bullet}}{N}\text{-}CH_2OH \qquad (3)$$

d) radical addition to nitrones

$$\text{e.g.,}\quad C_6H_5CH{=}\underset{O^{-}}{N^{+}}\text{-}C(CH_3)_3 + CCl_3^{\bullet} \longrightarrow C_6H_5CH(CCl_3)\underset{O^{\bullet}}{N}\text{-}C(CH_3)_3 \qquad (4)$$

The above radical addition reactions with nitroso compounds such as nitrosobenzene and 2-methyl-2-nitrosopropane (TBN, t-nitroso butane) and nitrones such as 5,5, dimethyl-l-pyrroline-l-oxide (DMPO) or phenyl-t-butyl nitrone (PBN) are the basis underlying the use of these compounds as spin traps. They convert highly reactive short-lived radicals into less reactive longer-lived radicals identifiable by electron spin resonance.[18-20]

Nitroxides can be chemically reduced to the corresponding hydroxylamine or amine, the products and rates of reaction depending on the particular reactants. Of those biologically relevant reducing agents, ascorbate, cysteine, glutathione, NADPH and associated enzyme systems have been variously reported to afford reduction.

$$R_2NO^{\bullet} + e \xrightarrow{H^{+}} R_2NOH \qquad (5)$$

Nitroxides also react rapidly with a wide variety of organic free radicals. Covalent linking, oxidation or reduction can occur, depending on the particular radical. Triacetoneamine N-oxyl has been shown to be bound to macromolecules in irradiated cells presumably by a radical-radical reaction.[21,22] The absolute rate of reaction of this and similar reactions of nitroxides with organic radicals have been determined by pulse radiolysis using either direct spectroscopic, competition, or conductivity methods.[23-29]

$$\text{thymine (OH)}^{\bullet} + TAN^{\bullet} \longrightarrow \text{thymine (OH)-TAN} \qquad (6)$$

$$CO_2^{\bullet -} + TAN^{\bullet} \longrightarrow CO_2 + TAN^{-} \qquad (7)$$

$$Br_2^{\bullet -} + TAN \longrightarrow 2Br^{-} + TAN^{+} \qquad (8)$$

The free radical reactivity of nitroxides is also reflected in the various observations of their being able to be effective inhibitors of the thermal oxidative destruction of polyformaldehyde, polypropylene, polyamides and rubber, as well as of various polymerization reactions. They are thought to combine with propagating alkyl radicals, R•, and on occasions with peroxy radicals, RO_2•.[30-33]

NITROXIDES AND RADIOSENSITIZATION

It is still widely thought that the radiation resistance of hypoxic cells may be partly responsible for the failure to control some cancers. Although several theories have been put forward to explain why cells are much more sensitive to radiation in the presence of oxygen as compared to its absence, the original fixation-repair hypothesis developed in the early 1960s still serves as a useful model. A radical R• formed on a vital molecule through the loss of a hydrogen atom can either react with a thiol such as glutathione and be repaired or react with oxygen forming a peroxy radical whose subsequent reactions do not lead to the reformation of the original molecule.

$$R^\bullet + R'SH \longrightarrow RH + R'S^\bullet \quad \text{repair} \tag{9}$$

$$R^\bullet + O_2 \longrightarrow RO_2^\bullet \quad \text{damage} \tag{10}$$

In the light of the above, it is not suprising, therefore, that when the search for other compounds which could sensitise like oxygen began in the late 1950s, attention soon turned to the stable free radical nitric oxide and later the more manageable, and probably less toxic, organic nitroxides. Like oxygen, they might compete for the vital free radical and so prevent its repair.[18,34-36]

$$R^\bullet + R'SH \longrightarrow RH + R'S^\bullet \tag{11}$$

$$R^\bullet + \text{nitroxide}^\bullet \longrightarrow \text{R-nitroxide} \tag{12}$$

Organic nitroxides subsequently shown to sensitise hypoxic cells in vitro include di-t-butyl nitroxide (DTBN), t-butyl-2,6-dimethoxy-phenyl nitroxide (TBDMPN), tetramethyl pentamethylene nitroxide (TEMPO), triacetoneamine-N-oxyl, (TAN) norpseudopelletierine-N-oxyl (NPPN), and the bi-radical bis(2,2,6,6-tetramethyl-1-oxyl-4-piperidinyl) succinate (Ro-03-6061).[21,22,26,37-40]

At about the same time that these investigations began, however, it was found that N-ethylmaleimide could also sensitize bacteria.[41] This was attributed to its ability to react with thiols, so lowering their concentration and thus enabling traces of oxygen to compete more favourably in the above reactions.

$$R'SH + NEM \longrightarrow \text{products} \tag{13}$$

The possibility that the nitroxides might also be acting in this way was strengthened by the finding that exposure of cells to NO or various nitroxides led to significant removal of SH groups, the rate of depletion depending on the particular nitroxide. Although associated rapid-mixing experiments still showed that fast processes, probably free radical, were involved in radiosensitisation, the possibility that biochemical reactions were at least partly responsible could not be ruled out.

Perhaps not surprisingly, in the light of the above, the nitroxides have so far proved ineffective in vivo as radiosensitisers.[42,43] On the positive side, however, their reactions with organic free radicals and with cellular reducing compounds have provided clues as to how nitroxides generated intracellularly from foreign compounds might behave.

As long ago as 1948 the nitroxide formed either by the reduction of nitrobenzene or the oxidation of phenyl hydroxylamine was proposed as the toxic intermediate responsible for methaemaglobinaemia caused by exposure to these and related chemicals.[44]

$$ArNO_2 \dashrightarrow ArNO_2^- \dashrightarrow ArNO \dashrightarrow ArNHO^\bullet \qquad (14)$$

$$\leftarrow\!\!-\!- \; ArNHOH \leftarrow\!\!-\!- ArNH^\bullet \leftarrow\!\!-\!- ArNH_2 \qquad (15)$$

$$ArNHO^\bullet + Fe\ II\ Hb \xrightarrow{H^+} ArNHOH + Fe\ III\ Hb \qquad (16)$$

More recently a number of nitroxide free radicals have been detected in microsomal suspensions of various carcinogenic amines.[45,46] Initially it was thought that the spectra observed were due to the directly related nitroxide free radicals formed by oxidation in the P450 system. Recent results, however, suggest (a) that such oxidations by microsomes may be mediated by the superoxide radical[47,48] or a peroxy radical[49,50] and (b) that the nitroxide formed is not that produced directly from the corresponding hydroxylamine but one derived from a larger substituted hydroxylamine formed from the reaction of the corresponding nitroso compound with an unsaturated membrane fatty acid.[51] The formation of such larger compound hydroxylamines was originally proposed to explain the ESR signals observed on incubation of nitrosobenzene with 2,3-dimethyl-2-butene.[52] The hydroxylamine produced subsquently reacted with oxygen or nitrosobenzene to form the nitroxide.

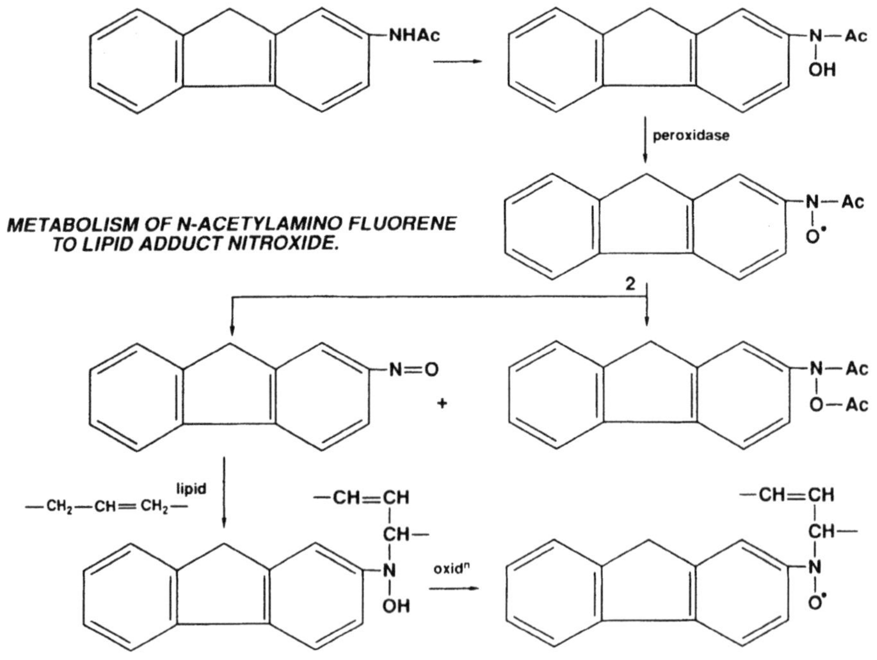

Figure 4

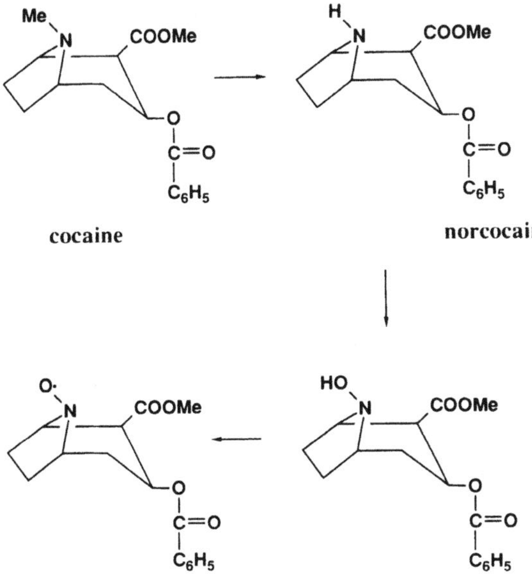

cocaine norcocaine

norcocaine nitroxide N-hydroxynorcocaine

Figure 5

In the case of 2-acetyl aminofluorene the nitroso compound itself is thought to be formed from the carcinogen by N-hydroxylation[53] followed by oxidation to the nitroxide which then reacts biomolecularly to form nitroso fluorene and N-acetyl O-acetyl aminofluorene.[54]

A similar mechanism involving an N-hydroxy compound has been proposed to explain the hepatoxicity of cocaine. It is now thought that the drug is first demethylated by microsomal enzymes to norcocaine and then oxidized by mixed-function oxidases first to N-hydroxy norcocaine and then to the nitroxide. The nitroxide is then thought to undergo a covalent binding by some unexplained mechanism to microsomal protein, and it is this covalent binding that is thought to be responsible for the acute hepatic necrosis observed.[55-58]

The formation of the nitroxide has been shown to be species-dependent and to have a half-life of less than 4 minutes in microsomal suspensions. Glutathione protects the liver against cocaine-induced toxicity. In vitro it has been shown to react with the nitroxide, probably by an iron-catalysed reaction, since no reaction is observed if the iron-chelating agent DETAPAC is present in the phosphate buffered solution.

IRON AND THE DESFERAL NITROXIDE

During the last decade, interest in the role of iron in free radical-mediated tissue injury has grown considerably, and many investigations continue to take place into the free radical biochemistry of iron-chelating agents. In this respect the drug desferrioxamine, widely used since 1962 in the treatment of hemochromatosis and iron poisoning, has attracted much attention both as a laboratory tool and as a possible ameliorator of rheumatoid disease and reperfusion injury.

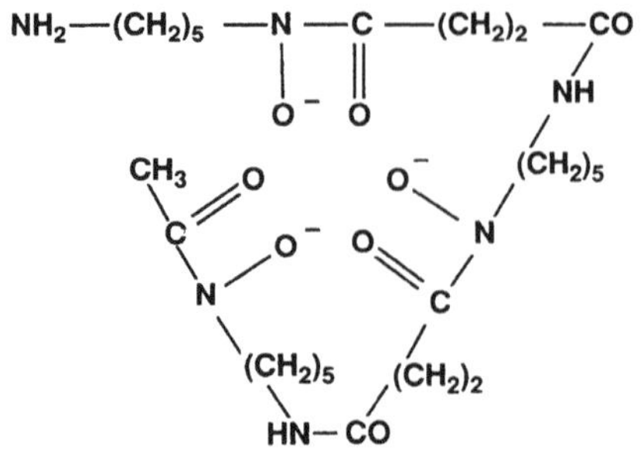

Figure 6

Such applications of metal-chelating drugs, however, are fraught with difficulties. Visual and auditory neurotoxicity have been observed in patients receiving desferrioxamine. In the treatment of inflammatory joint disease, continued use of such a drug may cause unacceptable levels of side effects in what generally would be non-life-threatening disorders. Indeed, in a recent pilot study, coma and serious retinal pigmentation occurred when desferrioxamine was administered to a rheumatoid patient being simultaneously treated with prochlorperazine.[59,60]

Similarly, as a tool in mechanistic free radical biochemistry, the drug may not just act as an iron chelator but may also scavenge free radicals directly.[61,62] As might be expected, the drug does react rapidly with the hydroxyl radical and recent studies on the inhibition of cytochrome c reduction by the xanthine-xanthine oxidase system have indicated that it also can react, albeit more slowly, with superoxide free radical, $O_2^{\cdot-}$.[63-65] Experiments in our own laboratory in which superoxide has been generated in irradiated formate solutions have confirmed that this reaction can occur significantly. Furthermore, in doing so a relatively stable but nevertheless reactive nitroxide free radical is formed in the process and is sufficiently long-lived to be observed directly by electron spin resonance.[66]

The nitroxide signal observed was similar to that detected in a solution containing t-butyl hydroperoxide, and under the radiation conditions had a half-life of approximately 10 minutes. The signal, which is attributed to the reaction of $O_2^{\cdot-}$ with one of the three ionised hydroxamic acid groups present in the molecule, was not observed if the drug was replaced by DPTA or EDTA or if superoxide dismutase, Trolox C, ascorbate, GSH, cysteine, or methionine were included in the irradiated system, in agreement with these antioxidants being able to compete effectively for $O_2^{\cdot-}$.

$$O_2^{\cdot-} + DesNO^- \longrightarrow O_2^{2-} + DesNO^\cdot \qquad (17)$$

$$O_2^{\cdot-} + Antioxidant \longrightarrow products \qquad (18)$$

The free radical signal was also absent if glutathione or the water-soluble derivative of vitamin E, Trolox C, was added to the Desferal solutions post-irradiation, in agreement with these compounds reacting readily with the nitroxide.

$$DesNO^\cdot + GSH \longrightarrow DesNOH + GS^\cdot \qquad (19)$$

$$DesNO^\cdot + TxOH \longrightarrow DesNOH + TxO^\cdot \qquad (20)$$

These ESR studies had, in fact, been prompted by related enzyme studies which had shown that a product of $O_2^{\cdot -}$ with Desferal, unlike $O_2^{\cdot -}$ itself, could inactivate alcohol dehydrogenase. Indeed, on a concentration basis, the toxin, now considered to be the nitroxide, was more effective at inactivating the enzyme than the hydroxyl free radical. Again, the above antioxidants could protect the enzyme if included at the time of irradiation, and in the case of GSH, cysteine, and methionine, even if present with the enzyme when it was mixed with the irradiated Desferal solution. Furthermore, recent studies have shown that YADH can also be inactivated by the xanthine-xanthine oxidase-formate system in the presence of Desferal but not EDTA or DETPA. SOD again protects.[67]

$$O_2^{\cdot -} + \text{YADH} \longrightarrow \text{no reaction} \qquad (21)$$

$$\text{DesNO}^{\cdot} + \text{YADH} \longrightarrow \text{inactivation} \qquad (22)$$

This reaction of $O_2^{\cdot -}$ with Desferal is supported by the earlier observation that the nitroxide OXANO can be produced from the reaction of the radical with the related hydroxamic acid.[68]

$$O_2^{\cdot -} + \text{OXANOH} \longrightarrow O_2^{2-} + \text{OXANO}^{\cdot} + H^+ \qquad (23)$$

It does seem that, contrary to previous suggestions,[64] the $O_2^{\cdot -}$ scavenging ability of Desferal or related hydroxylamine derivatives could have a profound effect in biological systems, even in the presence of high concentrations of hydroxyl radical scavengers. Although formate itself may not be present in a biological cell in sufficiently high concentration, the possibility that other compounds are present which react with OH to form a radical which then reacts with oxygen forming $O_2^{\cdot -}$ or an organic peroxy radical, cannot be ruled out.[69,70]

Clearly there is still much to be learned about the free radical biochemistry of nitrogen radicals in general and organic nitroxides in particular.

ACKNOWLEDGEMENTS

Financial support from the Cancer Research Campaign. the Medical Research Council and the National Foundation for Cancer Research is gratefully acknowledged. This presentation on nitrogen free radicals would have been impossible without the collaborations of Professor K-D Asmus, Dr D Bahnemann, Dr J Monig and Dr S Nigam, Hahn-Meitner Institut Berlin; Professor J E Packer, University of Auckland, New Zealand; Dr V O Mora-Arellano, Universidad de Oriente, Venezuela and Professor T F Slater, Dr M Davies, Ms R Donkor, Mrs C Dunster, Dr L G Forni, Dr C A Gee, Mrs S Hunter, Ms S Jonas, Dr J Mahood, Dr K Mondon, Mr K Pisharath, Dr A J F Searle, Mr C Selby and Dr B S Wolfenden of Brunel University.

REFERENCES

1. W. A. Waters, "Chemistry of free radicals", Oxford University Press (1948).
2. A. R. Forrester, J. M. Hay, and R. H. Thomson, "Organic chemistry of stable free radicals", Academic Press, London & New York (1968).
3. R. P. Mason and J. L. Holtzman, The role of catalytic superoxide formation in the O_2 inhibition of nitroreductase, Biochem. Biophys.Res.Comm. 67:1267-1274 (1975).

4. R. P. Mason, Inhibition of azo reductase by oxygen, Mol.Pharmacol. 14:665-671 (1978).
5. E. Perez-Reyes, B. Kalyanaraman, and R. P. Mason, The reductive metabolism of metronidazole and ronidazole by aerobic liver microsomes, Mol.Pharmacol. 17:239-244 (1980).
6. S. N. J. Moreno, R. P. Mason, R. P. A. Muniz, F. S. Cruz, and R. Docampo, Generation of free radicals from metronidazole and other nitroimidazoles by Tritrichomonas foetus, J.Biol.Chem. 258:4051-4054 (1983).
7. J. K. Horton, R. Brigelius, R. P. Mason, and J. R. Bend, Paraquat uptake into freshly isolated rabbit lung epithelial cells and its reduction to the paraquat radical under anaerobic conditions, Mol. Pharmacol. 29:484-488 (1986).
8. R. L. Willson, B. C. Gilbert, P. D. R. Marshall, and R. O. C. Norman, Metronidazole ('Flagyl'): A pulse radiolysis and e.s.r. study, Int.J.Radiat.Biol. 26:427-434 (1974)
9. J. A. Farrington, M. Ebert, E. J. Land, and K. Fletcher, Bipyridylium quaternary salts and related compounds, Biochim.Biophys.Acta 314: 372-381 (1973).
10. B. S. Wolfenden and R. L. Willson, Radical-cations as reference chromogens in kinetic studies of one-electron transfer reactions: pulse radiolysis studies of 2,2'-Azinobis-(3-ethylbenzthiazoline-6-sulphonate), J.Chem.Soc.Perkin.Trans.II 805-812 (1982).
11. I. Wilson, P. Wardman, G. M. Cohen, and M. D'Arcy Docherty, Reductive role of glutathione in the redox cycling of oxidizable drugs, Biochem.Pharmacol. 35:21-22 (1986).
12. L. G. Forni, V. O. Mora-Arellano, J. E. Packer, and R. L. Willson, Aminopyrine and antipyrine free radical cations pulse radiolysis studies of one-electron transfer reactions, J.Chem.Soc.Perkin Trans.II In Press.
13. D. Bahnemann and K-D Asmus, Free radical induced one-electron oxidation of the phenothiazines chlorpromazine and promethazine, J.Chem.Soc.Perkin Trans.II 1661-1668 (1983).
14. B. W. Griffin, Free radical intermediate in the N-demethylation of aminopyrine by horseradish peroxidase-hydrogen peroxide, FEBS Letts. 74:139-143 (1977).
15. B. W. Griffin and P. L. Ting, Mechanism of N-demethylation of aminopyrine by hydrogen peroxide catalyzed by horseradish peroxidase, metmyoglobin and protohemin, Biochemistry 17:2206-2211 (1978).
16. L. H. Piette, G. Bulow, and I. Yamazaki, Electron-paramagnetic-resonance studies of the chlorpromazine free radical formed during enzymic oxidation by peroxidase-hydrogen peroxide, Biochim. Biophys.Acta 88:120-129 (1964).
17. D. C. Borg and G. C. Cotzias, Interaction of trace metals with phenothiazine drug derivatives, II. Formation of free radicals, Proc.N.A.S. 48:623-642 (1962).
18. S. Forshult, C. Lagercrontz, and K. Torsell, Use of nitroso compounds as scavengers for the study of short-lived free radicals in organic reactions, Acta Chem.Scand. 23:522-530 (1969).
19. M. J. Perkins, Essay in free radical chemistry, spec. pub. No. 27, 97 Chem. Soc. London (1970).
20. E. G. Janzen. Spin trapping, Acc.Chem.Res. 4:31 (1971).
21. T. Brustad, Radiosensitization of anoxic cells of E. coli K-12 by organic nitroxyl free radicals, Int.J.Radiat.Biol. 22:443-456 (1972).
22. E. Wold and T. Brustad, Binding of nitroxyls to radiation-induced DNA-transients in E. coli K-12 under in vivo conditions and its relevance for radiosensitization, Int.J.Radiat.Biol. 25:225-233 (1974).
23. P. T. Emmerson and R. L. Willson, Pulse radiolysis of aqueous

solutions thymine and triacetoneamine N-oxyl, J.Phys.Chem. 72: 3669-3671 (1968).
24. R. L. Willson and P.T. Emmerson, Reaction of triacetoneamine-N-oxyl with radiation-induced radicals from DNA and from deoxyribonucleotides in aqueous solution, in "Radiation protection and sensitisation", H.L. Moroson and M. Quintiliani, ed., Taylor and Francis, London (1970) .
25. R. L. Willson, Pulse radiolysis studies on reaction of triacetoneamine-N-oxyl with radiation-induced free radicals, Trans.Farad.Soc. 67:3008-3019 (1971).
26. P. T. Emmerson, E. M. Fielden, and I. Johansen, A possible steric factor in the sensitization of anoxic bacteria to X-rays by N-oxyl radicals, Int.J.Radiat.Biol. 19:229-236 (1971).
27. K-D. Asmus, S. Nigam, and R. L. Willson, Kinetics of nitroxyl radical reactions: A pulse-radiolysis conductivity study, Int.J.Radiat. Biol. 29:211-219 (1976).
28. S. Nigam, K-D. Asmus, and R. L. Willson, Electron transfer and addition reactions of free nitroxyl radicals with radiation induced radicals, Trans.Farad.Soc. 2324 (1976)
29. P. O'Neill, T. C. Jenkins, and E. M. Fielden, Interaction of oxygen and nitroxyls with radiation-induced radicals of DNA and related bases in aqueous solution, Rad.Res. 82:55-64 (1980).
30. O. L. Harle and J. R. Thomas, Detection of free radical intermediates in the action of oxidation inhibitors, J.Amer.Chem.Soc. 79:2973-2974 (1957).
31. I. T. Brownlie and K. U. Ingold, The inhibited autoxidation of styrene. Part VII. Inhibition by nitroxides and hydroxylamines, Can.J.Chem. 45:2427-2432 (1967).
32. J. T. Weil, J. Van der Veen, H. S. Olcott, Stable nitroxides as lipid antioxidants, Nature 219:168-169 (1968).
33. J. Van der Veen, J. T. Weil, T. E. Kennedy, H. S. Olcott, Aliphatic hydroxylamines as lipid antioxidants, Lipid 5:509-512 (1970).
34. P. Howard-Flanders, Effect of nitric oxide on the radiosensitivity of bacteria, Nature 180:1191-1192 (1957).
35. J. P. Lynch and P. Howard-Flanders, Effect of pretreatment with nitric oxide and N-ethylmaleimide on the level of sulphydryl compounds in bacteria and on their sensitivity to x-irradiation under anoxia, Nature 194:1247-1249 (1964).
36. P. T. Emmerson and P Howard-Flanders, Preferential sensitization of anoxic bacteria to x-rays by organic nitroxide-free radicals, Radiation Research 26:54-62 (1965).
37. P. T. Emmerson, Sensitization of anoxic recombination-deficient mutants of escherichia coli K12 to X-rays by triacetoneamine N-oxyl, Radiation Research 36:410-417 (1968).
38. D. A. Agnew and L. D. Skarsgard, Sensitization of anoxic mammalian cells to radiation by triacetoneamine-N-oxyl. Effect of pre- and post-irradiation treatment, Radiation Research 51:97-109 (1972).
39. N. M. Blackett, W. E. Wooliscroft, E. M. Fielden, and S. C. Lillicrap, Radiation modifying effect of the free radical norpseudopelletierene-N-oxyl on normal bone marrow stem cells in vitro and in vivo, Radiation Research 58:361-372 (1974).
40. B. C. Cooke, E. M. Fielden, M. Johnson, and C. E. Smithen, Polyfunctional radiosensitizers, Radiation Research 65:152-162 (1976).
41. B. A. Bridges, Sensitization of eschericha coli to gama radiation by N-ethyl maleimide, Nature 188:415 (1960).
42. H. B. Hewitt and E R Blake, Studies of the toxicity and radiosensitising activity of triacetoneamine-N-oxyl in mice, Br.J.Radiol. 43:91-96 (1970).
43. S. Hornsey, Sensitization of hypoxic cells in vivo with triacetoneamine N-oxyl (TAN), Int.J.Radiat.Biol. 22:91-94 (1972).

44. W. Heubner, Giftung aromatischer Nitroverbingdungen, Arch.exp.Path. Pharmak. 205:310 (1948).
45. A. Stier and I. Reitz, Radical production of amine oxidation by liver microsomes, Xenobiotica 1:499-500 (1971).
46. A. Stier, I. Reitz, and E. Sackmann, Radical accumulation in liver microsomal membranes during biotransformation of aromatic amines and nitro compounds, Arch. Pharmacol. 274:189-191 (1972).
47. E. J. Rauckman, G. M. Rosen, and B. B. Kitchell, Superoxide Radical as an intermediate in the oxidation of hydroxylamines by mixed function amine oxidase, Mol. Pharmacol. 15:131-137 (1979).
48. G. M. Rosen and E. J. Rauckman, Formation and reduction of a nitroxide radical by liver microsomes, Biochem.Pharmacol. 26:675-678 (1977).
49. R. A. Floyd, L. M. Soong, R. N. Walker, and M. Stuart, Lipid hydroperoxide activation of N-hydroxy-N-acetylaminofluorene via a free radical route, Cancer Res. 36:2761-2767 (1976).
50. R. A. Floyd, L. M. Soong, and P. L. Culver, Horseradish peroxidase/hydrogen peroxide-catalyzed oxidation of the carcinogen N-hydroxy-N-acetyl-2-aminofluorene as effected by cyanide and ascorbate, Cancer Res. 36:1510-1519 (1976).
51. R. A. Floyd, L. M. Soong, M. A. Stuart, and D. L. Reigh, Free radicals and carcinogenesis, Arch.Biochem.Biophys. 185:450-457 (1978).
52. A. B. Sullivan, Electron spin resonance studies of a stable arylnitroso-olefin adduct free radical, J.Org.Chem. 31:2811-2817 (1966)
53. J. W. Cramer, J. A. Miller, and E. C. Miller, N-hydroxylation: a new metabolic reaction observed in the rat with the carcinogen 2-acetylaminofluorene, J.Biol.Chem. 235:885-888 (1960).
54. H. Bartsch and E. Hecker, On the metabolic activation of the carcinogen N-hydroxy-N-2-acetylaminofluorene, Biochim.Biophys.Acta 567-578 (1971)
55. M. L. Thompson, L. Shuster, and K. Shaw, Cocaine-induced hepatic necrosis in mice, Biochem.Pharmacol. 28:2389-2395 (1979).
56. M. A. Evans, Microsomal activation of N-hydroxy norcocaine to a reactive nitroxide, Toxicologist 1:1 (1981).
57. M. A. Evans and M. E. Johnson, The role of a reactive nitroxide radical in cocaine-induced hepatic necrosis, Fed.Proc. 40:638 (1981).
58. E. J. Rauckman, G. M. Rosen and J. Cavagnaro, Norcocaine nitroxide a potential hepatotoxic metabolite of cocaine, Molec.Pharmacol. 21:458-463 (1982)
59. D. R. Blake, P. Winyard, J. Lunec, A. Williams, P. A. Good, S. J. Crewes. J. M. C. Gutteridge, D. Rowley, B. Halliwell, A. Cornish, and R. C. Hider, Cerebral and ocular toxicity induced by desferrioxamine, Quart.J.Med. 56:345-355 (1985).
60. N. F. Olivieri, J. R. Bungic, E. Chew, T. Gallant, R. Harrison, N. Keenan, W. Logan, D. Mitchell, G. Ricci, B. Skarf, M. Taylor, and M. Freedman, Visual and auditory neurotoxicity patients receiving subcutaneous desferrioxamine infusions, New England J. Med. 314:869-873 (1986).
61. R. L. Willson, Iron and hydroxyl free radicals in enzyme inactivation and cancer, in: NFCR Cancer Symposia "Free radicals, lipid peroxidation and cancer", D. C. H. McBrien and T. F. Slater, ed., Academic Press, London (1982).
62. J. Butler and B. Halliwell, Reaction of iron-EDTA chelates with a superoxide radical, Arch.Biochem.Biophys. 218:174-178 (1982).
63. J. Sinaceur, C. Ribiere, J. Nordmann, and R. Nordmann, Desferrioxamine, a scavenger of superoxide radicals? Biochem. Pharmacol. 33:1693-1694 (1984).

64. B. Halliwell, Use of desferrioxamine as a "probe" for iron-dependent formation of hydroxyl radicals, Biochem.Pharmacol. 34:229-233 (1985).
65. C. Ribiere, D. Sabourault, J. Sinaceur, R. Nordmann, C. Houce-Levin, and C. Ferradini, Radiolysis study of the reaction of desferrioxamine with $O_2^{\bullet -}$ free radicals, in "Superoxide and superoxide dismutase in chemistry, biology and medicine, G. Rotilio, ed., Elsevier Science, pp.47-49 (1986).
66. M. J. Davies, R. Donkor, C. A. Dunster, C. A. Gee, S. Jonas, and R. L. Willson, Desferrioxamine (Desferal) and superoxide free radicals: formation of an enzyme-damaging nitroxide, Biochem.J. 246:725-729 (1987).
67. R. Donkor and R. L. Willson, (1987) To be published.
68. G. M. Rosen, E. Finkelstein, and E. J. Rauckman, A method for the detection of superoxide in biological systems, Arch.Biochem. Biophys. 215:367-378 (1982).
69. C. A. Gee, K. J. Kittridge, and R. L. Willson, Peroxy free radicals, enzymes and radiation damage: sensitisation of oxygen and protection by superoxide dismutase and antioxidants, Brit.J. Radiol. 58:251-256 (1985).
70. R. L. Willson, Organic peroxy free radicals as ultimate agents in oxygen toxicity, in: "Oxidative stress", Academic Press Inc. London (1985) pp. 41-72.

SINGLET OXYGEN: CURRENT MECHANISTIC ASPECTS OF PHOTO-OXIDATION

A.A. Gorman

Chemistry Department
University of Manchester
Manchester M13 9PL, U.K.

INTRODUCTION

Two principal biological photo-oxidation processes are well-recognised, termed "Type I" and "Type II." The former is characterised by primary interaction of substrate with the sensitizer excited state to give radicals whose subsequent reaction with oxygen leads to products. The "Type II" classification is restricted to oxidation processes involving the intermediacy of the lowest electronically excited state of molecular oxygen, $O_2(^1\Delta_g)$. The principal mode of formation of the latter, both in vivo and in vitro, is via electronic energy transfer from a sensitizer triplet state to ground state molecular oxygen. In this paper it is shown that the oxygen quenching of triplet states also can lead to radicals which are produced via a process which competes with singlet oxygen formation. In addition, current results concerning the mechanism of reaction of the latter species with organic/biological substrates will be discussed.

PRODUCTION AND CHARACTERISATION OF THE ANTHRALIN RADICAL

Anthralin (AnH; *1*), 1,8-dihydroxyanthrone, is of considerable importance in the topical treatment of psoriasis and other skin diseases typified by enhanced cell proliferation.[1,2] Discussion concerning its mechanism of action[3] has centred around the radical *2* (An$^\bullet$) and the corresponding peroxy radical *3*. The former has been characterised as a result of a comprehensive investigation of the photochemistry of AnH.

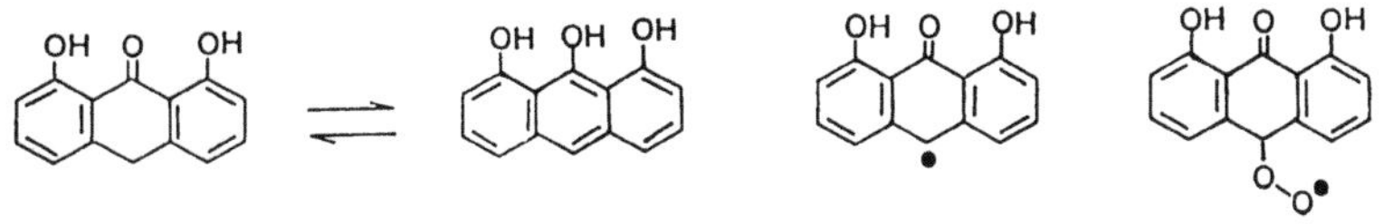

1a *1b* *2* *3*

Nanosecond and picosecond laser experiments have shown that in benzene $^1AnH^*$ is very short-lived, < 30 ps, and that deactivation involves neither triplet formation nor intramolecular H-transfer. Pulse radiolysis of deaerated benzene solutions of AnH results in 'immediate' formation of $^3AnH^*$ which has been fully characterised. Although the latter's quenching by oxygen proceeds with a typical rate constant, $2.2 \times 10^{9}\ l\ mol^{-1}s^{-1}$, it yields not only singlet oxygen (64%) but also a species absorbing in the red at 720 nm (15%). This has been identified as An˙, and control experiments have shown that it is produced via a secondary reaction of a non-ionic intermediate other than singlet oxygen, which is formed by reaction of oxygen with AnH . The most likely route is summarised in equation 1 (cf reference 4). Conclusive identification of An˙ came from the photochemistry

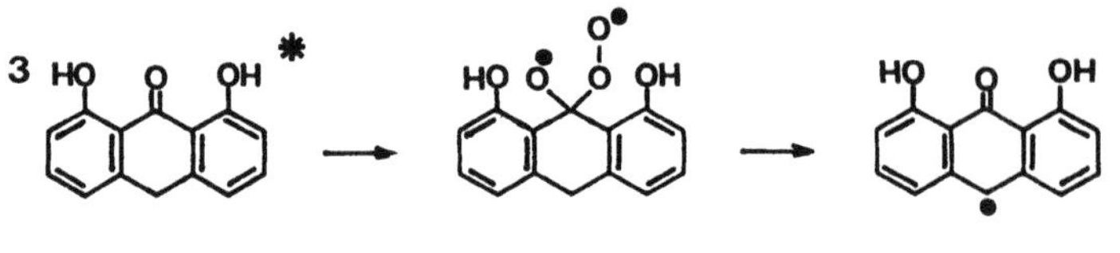

(1)

of the dehydro-dimer of AnH (*4*; An–An) and reaction of AnH with the azide radical as described below:

(a) Both ^{1}An–An* and ^{3}An–An*, formed by laser excitation and pulse radiolysis respectively, fragment spontaneously to give a radical pair (Figure 1). In the case of the triplet, cage escape occurs to give free An˙, identical in all respects with the 720 nm species formed from the reaction of AnH with oxygen. Recombination (microseconds/milliseconds) gives an isomer of *4*, probably *5* (Figure 1). For the singlet radical pair 'immediate' cage recombination gives *5*. The latter slowly (in seconds) isomerises to *4*.

(b) Pulse radiolytic production of $N_3^{\bullet}$ [5] in aqueous acetonitrile containing AnH yields An˙ via electron transfer from the enol *1b* followed by deprotonation. The high stability of An˙ is demonstrated by the fact that vitamin E model radicals, produced in the same manner, are 'repaired' by AnH which is converted to An˙. The latter is insensitive to oxygen.

MECHANISM OF REACTION OF SINGLET OXYGEN WITH ORGANIC/BIOLOGICAL SUBSTRATES

The quenching of singlet oxygen may be physical or chemical in nature, corresponding in living systems to biological protection and biological damage, respectively. It has recently been proposed[6] that such processes proceed via the intermediacy of a reversibly formed exciplex (Figure 2) and that the balance between physical and chemical quenching is determined solely by the relative values of k_{isc} and k_p. The limiting values for the experimental rate constant for quenching, k_q^p, are given by equation 2,

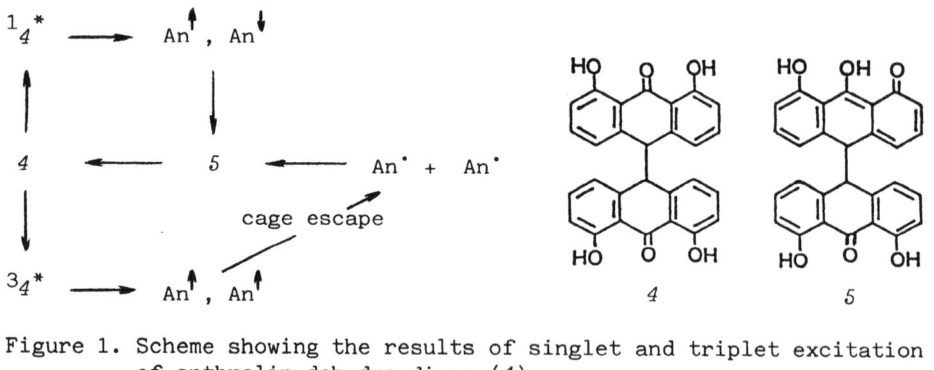

Figure 1. Scheme showing the results of singlet and triplet excitation of anthralin dehydro-dimer (*4*).

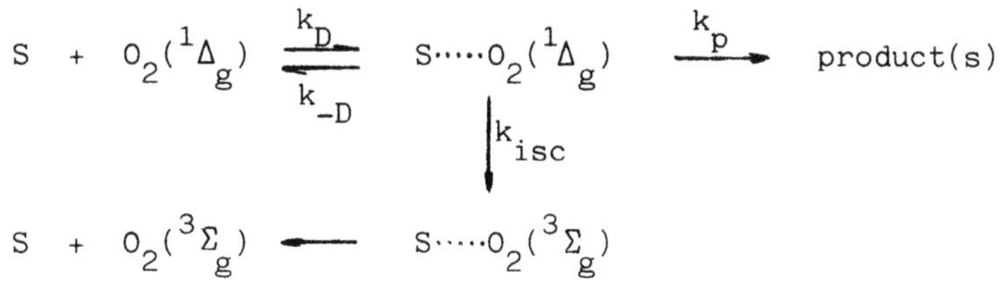

Figure 2. Scheme for chemical and/or physical deactivation of $O_2(^1\Delta_g)$.

$k_{-D} << (k_{isc} + k_p)$ and equation 3, $k_{-D} >> k_{isc}$ (100% physical quenching) or $k_{-D} >> k_p$ (100% chemical quenching), where K is the equilibrium constant for exciplex formation. In the latter situation the experimental enthalpy of activation (equation 4) can be negative if $\Delta H°$, the enthalpy of formation of the exciplex is more negative than the positive activation enthalpy of the second step.

$$k_q = k_D \text{ (diffusion limit)} \tag{2}$$

$$k_q = Kk_{isc} \text{ or } Kk_p \text{ (pre-equilibrium limit)} \tag{3}$$

$$\Delta H^{\ddagger}_{exp} = (\Delta H° + \Delta H^{\ddagger}_{isc}) \text{ or } (\Delta H° + \Delta H^{\ddagger}_{p}) \tag{4}$$

In order to confirm the mechanism outlined in Figure 2 we have performed experiments aimed at identifying both the pre-equilibrium and diffusion limits for both physical and chemical quenchers of singlet oxygen by means of temperature variation in toluene. Conversion from one limit to

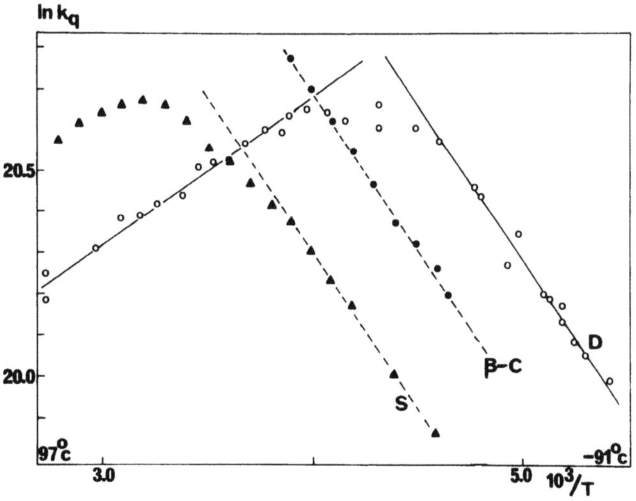

Figure. 3. Arrhenius plots for the quenching in toluene of singlet oxygen luminescence by β-carotene (β-C), strychnine (S) and 1,3-diphenylisobenzofuran (D). The ln k_q values for β-carotene have been lowered by 2.0 units to facilitate comparison.

the other over an accessible temperature range required 'fast' quenchers. Those chosen were strychnine (physical; 9 x 10^8 l $mol^{-1}s^{-1}$ at RT)[7] and 1,3-diphenylisobenzofuran (chemical; 8 x 10^8 l $mol^{-1}s^{-1}$ at RT).[8] These were compared with β-carotene which quenches (1.1 x 10^{10} l $mol^{-1}s^{-1}$ at RT) by triplet energy transfer at a close to diffusion-controlled rate.[9] Rate constants for quenching of singlet oxygen were determined by laser excitation of tetraphenylporphyrin at 532 nm in aerated toluene, both with and without quencher, over a 180°C temperature range. Singlet oxygen rate parameters were determined by time-resolved measurement of its infrared luminescence at 1270 nm. Arrhenius plots are shown in Figure 3. For both the physical and chemical quenchers both the pre-equilibrium and diffusion limits have been defined. The slopes at low temperature are clearly identical to that for β-carotene in the diffusion-controlled situation. These results provide compelling evidence that the mechanism outlined in Figure 2 operates generally for singlet oxygen quenchers. In a biological situation the balance between protection and damage is therfore determined by the fate of the exciplex.

ACKNOWLEDGMENTS

Experiments were performed at the Christie Hospital and Holt Radium Institute, Manchester. This presentation was made possible as a result of collaborations involving J. M. Bruce, I. Hamblett, C. W. Kerr, C. Lambert, S. P. McNeeney and M. C. Standen. Partial support for this work came from SERC grants GR/C 25623 and GR/D 52169.

REFERENCES

1. H. P. Baden, ed., "The Chemotherapy of Psoriasis," Pergamon Press, Oxford (1984).
2. H. H. Roenigk and H. I. Marbach, eds., "Psoriasis," Marcel Dekker, New York (1983).
3. K. K. Mustakallio, J. Martinmaa, R. Vilvara, and J. Halmekoski, Free radicals and the treatment of psoriasis with special reference to dithranol, Med. Biol. 62:155 (1984).
4. A. A. Gorman and M. A. J. Rodgers, The quenching of aromatic ketone triplets by oxygen: Competing singlet oxygen and biradical formation, J. Amer. Chem. Soc. 108:5074 (1986).
5. J. Butler, E. J. Land, A. J. Swallow, and W. Prutz, The azide radical and its reactions with tryptophan and tyrosine, Rad. Phys. Chem. 23:265 (1984).
6. A. A. Gorman, I. R. Gould, I. Hamblett, and M. C. Standen, Reversible exciplex formation between singlet oxygen, $^1\Delta_g$, and vitamin E. Solvent and temperature effects, J. Amer. Chem. Soc. 106:6956 (1984).
7. A. A. Gorman, I. Hamblett, K. Smith, and M. C. Standen, Strychnine: a fast physical quencher of singlet oxygen ($^1\Delta_g$), Tetrahedron Lett. 25:581 (1984).
8. A. A. Gorman, G. Lovering, and M. A. J. Rodgers, The entropy-controlled reactivity of singlet oxygen ($^1\Delta_g$) towards furans and indoles in toluene. A variable temperature study by pulse radiolysis, J. Amer. Chem. Soc. 101:3050 (1979).
9. A Farmilo and F. Wilkinson, On the mechanism of quenching of singlet oxygen in solution, Photochem. Photobiol. 18:447 (1973).

HOMOLYSIS OF SUBSTITUTED ANISOLES:

SUBSTITUENT EFFECTS ON PHENOXYL RADICAL STABILITIES

Stephen E. Stein and Mahendra M. Suryan

Chemical Kinetics Division
National Bureau of Standards
Gaithersburg, MD 20899, U.S.A.

INTRODUCTION

The effect of a substituent on reaction thermodynamics (energetics) is perhaps the most fundamental of the "substituent effects." In free radical reactions, other factors, most notably "polar effects," are typically deduced from deviations in thermodynamic/rate correlations. A detailed understanding of radical substituent effects therefore requires the availability of thermodynamic information.

These thermodynamic effects can be expressed as differences in bond dissociation energies (relative bond strengths), ΔBDE in substituted and unsubstituted molecules, as illustrated below for phenoxyl radicals.

$$\Delta BDE = BDE(XC_6H_4O\text{-}CH_3) - BDE(C_6H_5O\text{-}CH_3) \qquad (1)$$

Much of the recent work on substituent effects in free radicals has focused on benzyl radicals. However, the small effects on benzyl radicals are hard to measure precisely. On the other hand, thermolysis rates of reactions yielding phenoxyl radicals have been reported to be extremely sensitive to hydroxy and methoxy substituents.[1-3] Phenoxyl radicals may therefore be attractive candidates for this study. Moreover, an extensive literature exists on the properties of these species.

In the present work we report substituent effects on the phenoxy-methyl bond strength in anisole. A total of 14 substituents, a number of them in each of the available ring positions, were investigated. Substituent effects on $O\text{-}CH_3$ bond strengths in anisoles are similar to effects on O-H bond strengths on phenols. Present results can therefore be used to interpret to O-H bond strengths in phenols. Related substituent effects influence the effectiveness of phenolic antioxidants such as vitamin E[4] and BHT.[5] We also examine the correlations of the relative O-methyl bond strengths with the Hammett substituent constants and the available hyperfine coupling data on phenoxyl radicals.

METHODOLOGY

Phenoxy-methyl bond strengths were calculated by determining the dissociation rates of substituted anisoles in the gas phase by the method of very low-pressure pyrolysis (VLPP). This is a well-suited method for these measurements because the method provides rather precise (±1%) relative rate constants and, at the low pressures of these experiments, complications from bimolecular reactions are avoided.[6] Anisoles are especially suited for study by this method since the low O-methyl bond strengths (ca. 64 kcal mol^{-1}) in these compounds cause them to homolyze under relatively mild (VLPP) conditions (900-1200K).

The thermolysis products of substituted anisoles were analyzed using a quadrupole mass spectrometer. The results indicated that all substituted anisoles we studied thermally decomposed by an initial O-methyl bond scission forming methyl and substituted phenoxyl radicals. The latter subsequently decomposed to CO and C_5H_5 by ring rupture.

Unimolecular reaction rate constants, k_{uni}, were calculated by the equation $(k_{uni}/k_e) = (I_o-I)/I$, where I_o and I are the mass spectral intensities of the parent species before and after the reaction, respectively, and k_e is the escape rate constant, dependent on the reactor geometry, temperature and molecular weight of the species. The value of k_e for our reactor is $3.96\sqrt{(T/M)}$.

The rate constants so determined were translated to the high pressure values by means of RRKM theory,[8] and subsequently the Arrhenius parameters were extracted. The differences in the energies of activation were essentially the differences in O-methyl bond strengths.

RESULTS AND DISCUSSION

The results consolidated in Table 1 indicate that the O-methyl bond strengths vary over a wide range depending on both the nature and the position of the substituent. In the ortho position, where all substituents weakened the bonds, effects were the largest (-0.2 to -7.4 kcal). Para effects fell over a range (+1.1 to -4.0 kcal), while the meta effects were evenly distributed over a quite narrow range, +1.1 to -1.0 kcal.

In the following discussion we presume that substituent effects on the $O-CH_3$ bond are the same as for the phenolic O-H bond. Any differences can be expressed entirely in terms of stabilities of the parent molecules. Specifically, this difference is equal to the enthalpy ΔH, for the transfer of the substituent (X) from anisole to phenol: $PhOH + XPhOCH_3 \rightarrow XPhOH + PhOCH_3$. The enthalpy of this reaction is indeed small.[9,10]

In the discussion to follow, effects of specific substituents are considered first, in the order presented in Table 1. We then examine the results as a whole, including comparisons with substituent constants, ESR data and phenolic H-abstraction rates.

Hydroxy and Amino Anisoles

The effects of an o-hydroxy and an o-amino substituent is exceptional, weakening the $O-CH_3$ bond strength by nearly 7 kcal mol^{-1}. In earlier studies also, Vuori,[1,2] Virk[3] and coworkers found o-hydroxyanisole

Table 1. Differences in the $O-CH_3$ Bond Dissociation Energies of Substituted Anisoles Relative to that of Anisole (63.5 kcal mol^{-1}), as measured by VLPP experiments

Substituent	Substituent Location			
X	Ortho	Meta	Para	Other
OH	-7.1	**0.3**	-2.5	
		2.6[a]	-3.7[a]	
NH_2	**-7.4**	**-0.4**	**-2.9**	
OCH_3	**-4.2**	**-1.0**	**-3.9**	
			-4.3[b]	
CH_3	**-2.6**	**-0.5**	**-1.9**	
			-1.8[b]	
CH_2OH	-2.4			
F	-1.9	**0.9**	**-1.1**	
Penta-F[c]				-2.7
Cl	-2.2			
Br	-1.7			
$COCH_3$	-1.4	**0.3**	**0.6**	
$CHCH_2$	-2.4			
CN	-0.2	**1.1**	**0.3**	
NO_2	-1.3	**-0.5**	**1.2**	
2,3Benzo				-5.9
				-3.8[a]
3,4Benzo				-2.5
				-1.8[b]

Values in boldface are from present study.
[a,b] From ref. 7.
[a] Via Bronsted relationship.
[b] from equilibrium constant.
[c] Refers to 1,2,3,4,5-pentafluoroanisole.

(guaiacol) to be exceptionally labile. Based on this observation Virk postulated a concerted reaction pathway for the decomposition of guaiacol, forming o-quinone and methane directly, without the intermediacy of free radicals. On the other hand, by measuring the total spin concentrations in the pyrolysis products of isomeric hydroxyanisoles, Vuori[1] concluded that homolysis--not a concerted reaction--was the primary reaction pathway. Therefore, to resolve these mechanistic disparities, we determined the amount of methane and ethane in the pyrolysis products of hydroxyanisoles and hydroxyphenetole. The results presented showed that the concerted path made little contribution.

Methoxyanisoles

In contrast to a hydroxy and an amino, a methoxy substituent weakened the bond strength by 4 kcal in either the ortho or the para positions. This may imply that little steric interactions may be present in both o-hydroxyanisole and guaiacol. In the meta position, methoxy lowers the bond strength by 1 kcal vs. slight increase by hydroxy.

Methylanisoles

A methyl substituent weakens the O-methyl bond in anisole from all positions, and 0.7 kcal mol^{-1} more in the ortho than in the para position. This may be attributed to relaxation of steric interaction in the molecule. Qualitatively similar effects were found for C-methyl bonds in methyl-substituted ethylbenzenes[11] although the magnitudes of the effects in o- and p-methyl anisoles are about 1.5 kcal mol^{-1} larger.

Hydroxymethylanisoles

Since the most obvious explanation for the bond weakening caused by o-hydroxy and amino substituents is a strong hydrogen bond in the radical, we determined the influence of another potential hydrogen bonding substituent, the o-hydroxymethyl group. The H-atom in the hydroxyl group in this substituent may form a hydrogen bond with the oxygen via a six-membered ring. However, the observed bond weakening was only 2.4 kcal mol^{-1}, nearly the same as found for an ortho methyl group (2.6 kcal mol^{-1}).

Haloanisoles

Para substitution by either a fluorine or chlorine atom causes a 1 kcal mol^{-1} weakening of the O-methyl bond while ortho substitution by these halogens, and bromine as well, weaken the bond by about 2 kcal mol^{-1}. This modest o-/p- difference is presumably due to relaxation of repulsion in the o-substituted molecules. Meta substitution by fluorine strengthens the bond by nearly 1 kcal mol^{-1} while chlorine at this position has virtually no effect. Substitution of five fluorine atoms weakens the O-methyl bond by 2.7 kcal mol^{-1}, nearly the same as would be predicted by additivity (3.1 kcal mol^{-1}).

Cyano, Acetyl, Nitro Substituents

These electron-withdrawing groups had relatively small effects on bond strengths. In contrast to the other substituents, in the para position the O-methyl bonds were strengthened, although the magnitude of this effect was greater than 1 kcal only for nitro. The largest meta effect of any substituent studied was observed for the cyano group, which strengthened the O-methyl bond by 1.1 kcal. Ortho substitution by these substituents caused modest bond weakening, this presumably due to a relaxation of repulsion in the molecule.

Vinyl Anisole and the Methoxynaphthalenes

In these molecules, the O-methyl bond is expected to be weakened by additional delocalization of the odd electron onto the "substituents." The methoxynaphthalenes are, in effect, benzo-substituted anisoles.
The decline in bond strength in the series anisole, 2-methoxynaphthalene, 1-methoxynaphthalene reflects the increase in relative resonance stabilization energies in the radicals.[12] As for the methyl substituent, the benzo-group stabilized the phenoxy radical substantially more than the benzyl radical[13] (an extra stabilization of 5.9 kcal mol^{-1} for 1-methoxynaphthalene versus 2.9 kcal mol^{-1} for 1-ethylnaphthalene). Bond weakening caused by the o-vinyl group presumably results from a combination of steric relief and additional conjugation in the radical.

Comparison with Previous Results

Mahoney and DaRooge have determined a series of substituent effects on O-H bond strengths in phenols.[7] Their published values for the relative O-H bond strengths of substituted phenols are included in Table 1. Each of their values was obtained by one of two methods: 1) from an equilibrium constant derived from forward and reverse rate constants for H-transfer reactions or 2) from a H-abstraction rate constant and the Bronsted relationship (log k $\propto$ ΔBDE). As Table 1 indicates, the values for p-methyl, 3,4-benzo, and p-methoxy phenols obtained from equilibrium measurements agree well with our values for the anisoles (average deviation = 0.4 kcal). However, ΔBDEs for 2,3-benzo, m-hydroxy, and p-hydroxy substituted phenols derived form the Bronsted relationship are substantially different (average deviation = 1.8 kcal). We feel that this difference is likely due to the failure of the Bronsted relation in H-abstraction reactions.

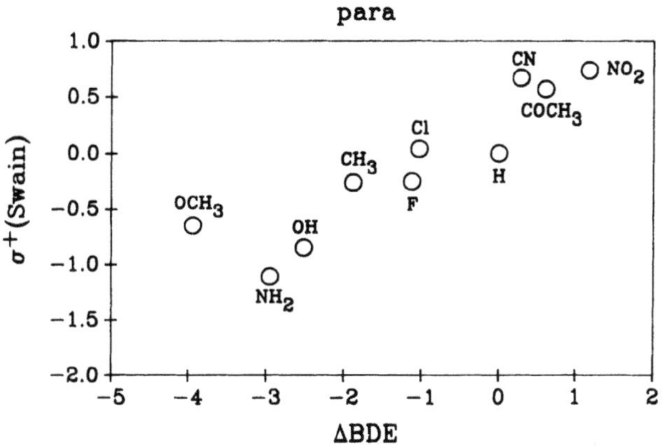

Figure 1. Plot of electrophilic substituent constant σ^+ vs. ΔBDE.

Substituent Constants and ΔBDE

In this section we examine the relation between ΔBDE and two substituent constants, σ^+ (from solvolysis of cumyl chlorides in 90% acetone[14] and σ_R (Taft's resonance term[15]).

ΔBDE is plotted against σ^+ in Figure 1. While the two values are clearly correlated (corr. constant = 0.91), there are sizable deviations. ΔBDE is a major but not the sole contributor to σ^+. Separate contributions to σ^+ can arise from interactions of net charge with the substituent. In fact, differences in trends of ΔBDE and σ^+ can be rationalized if this extra stabilization of charge is due to a charge-dipole interaction. Specifically, the larger deviations in ΔBDE vs. σ^+ trends seem to correlate with trends in the effective dipole moment of the substituent.

Consider, for instance, methoxy and hydroxy substituents. While their σ^+ values are nearly the same, the p-methoxy group has a significantly greater bond-weakening effect than does the p-hydroxy group. It seems plausible that the higher bond dipole of OH increases the ability of this substituent to stabilize positive charge, compensating for its inherently weaker resonance contribution to σ^+.

The relatively small magnitude of the bond strengthening effects of the electron withdrawing groups in comparison with differences in σ^+ values can also be rationalized in this manner (the positive charge being destabilized by these substituents).

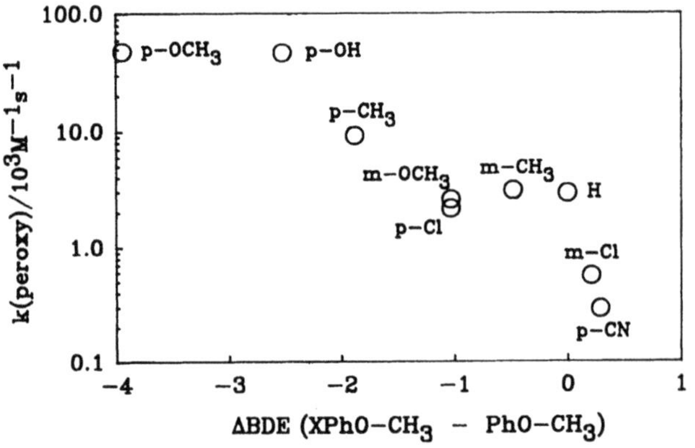

Figure 2. Plot of the rate constant k_{abs}, for the phenolic H-abstraction reaction by peroxy radical of a number of phenols[16] vs. ΔBDE of the corresponding anisoles.

ΔBDE also correlates well with σ_R; the correlation constant of 0.92 is virtually the same as for the σ^+/ΔBDE comparison. This is not surprising since σ+ and σ_R are themselves strongly correlated (correlation constant = 0.96 for the substituents examined in this work). However, if the methyl and methoxy substituents are ignored, we obtain a remarkable correlation of ΔBDE and σ_R of 0.99. Unfortunately, no other single term in Taft's scheme can account for the apparently exceptional behavior of both methyl and methoxy. It is curious that from a practical point of view (antioxidants), the ill-behaved substituents are also the most important.

Phenolic H-Abstraction Rates in Relation to ΔBDE

Extensive kinetic studies of the effect of substituents on rates of H-transfer to oxygen centered radicals have been reported by Ingold,[16] Simonyi,[17] and their coworkers. Excellent linear correlations between log

k_{abs} and σ^+ for para substituents have been reported by both groups. In Figure 2 Ingold's rates are plotted versus ΔBDE. Reaction rates correlate distinctly better with σ^+ (corr. const. = 0.99) than with ΔBDE (corr. const. = 0.92), immediately suggesting that polar effects are operative in these reactions.

For ortho substituents, there seems to be little correlation between H-abstraction rates and ΔBDE. Simonyi and coworkers[17] have determined the H-transfer rates by polyvinyl acetate radicals from o- and p-hydroxyphenols to be approximately equal while for H-transfer to the 2-allyl benzyl radical; Franz and coworkers have found p-hydroxyphenol to be slightly more reactive than o-hydroxyphenol.[18] Our results indicate that the O-H bond in the ortho isomer is 5 kcal weaker than the para isomer.

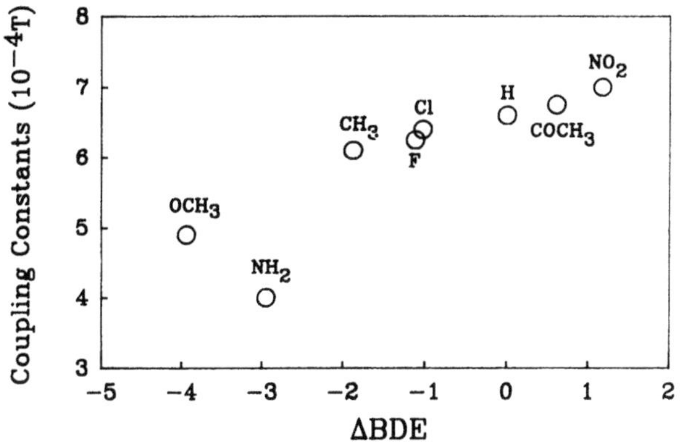

Figure 3. Plot of the ortho H-hfcc of para substituted phenoxy radicals[19] vs. ΔBDE for the corresponding substituted anisoles.

ESR Spin Density and ΔBDE

Since the above calculations suggest that ΔBDE values for para substituents are due primarily to differences in the stability of the radicals, one might expect a correlation between ESR spectral parameters of p-substituted phenoxyl radicals and ΔBDE. Plots of ΔBDE versus the proton hyperfine coupling constants (hfcc) for phenoxy radicals measured by Dixon and co-workers are given for p-substituents in Figure 3. For o-substituents, hfcc values at each of the two different conjugated positions are used. For p-substituents, values at only one was measurable. Perhaps surprisingly, the correlation was at least as good for o-substituents as for p-substituents. Both ΔBDE and hfcc effects for meta substituents were too small to make a meaningful comparison of these properties.

Dixon and coworkers found that substituent-induced changes in spin density at positions conjugated to the substituent were almost exactly compensated by opposite changes in the neighboring non-conjugated positions. Hence, the substituent affected the degree of **alternation** of hfcc values rather than their absolute values, thereby obscuring the physical connection between hfcc and substituent stabilization energies for phenoxyl radicals.

On the Origin of the Large Ortho-Amino and Ortho-Hydroxyl Effects

The most obvious explanation for the large ortho effect of both hydroxyl and amino is the presence of a strong bond between the H-atom on the substituent and the phenoxyl oxygen in the radical. We find such a strong bond surprising since, given the high degree of ring delocalization in the phenoxyl radical, the oxygen atom might be expected to more closely related to a carbonyl oxygen than a true oxygen-centered radical like methoxy. We would not have expected a carbonyl group to form particularly strong H-bonds. Nevertheless, we can find no better explanation.

The present experiments measure, in effect, the change in Gibbs energy caused by the substituent. Since any H-bonding in the radical would reduce the reaction entropy (and therefore the A-factor for dissociation), the actual bonding enthalpy should be even larger than the reported ΔBDE. Assuming that this bonding increases the barrier of rotation from 0 to ca. 8 kcal mol^{-1}, and then using Pitzer's tables,[20] we estimate that an additional 3 kcal mol^{-1} should be added to the reaction enthalpy from this effect. Thus, the H-bonding energy in the o-hydroxyphenoxy and o-aminophenoxy radicals is estimated to be ca. 10 kcal mol^{-1} greater than in the parent molecules. These are very strong hydrogen bonds.

CONCLUSIONS

We have determined the effects of a number of substituents on the O-CH_3 bond in anisoles. The range of ΔBDE (kcal mol^{-1}) for ortho substituents was -0.2 to -7.4, for para 1.2 to -3.9 and for meta 1.2 to -1.1. In the ortho and para positions, electron-donating substituents ($\sigma^+ > 0$) significantly weakened bonds, electron-accepting groups ($\sigma^+ < 0$) had significantly less effect but strengthened bonds somewhat when in the para position. Meta substituents showed no clear systematic trends.

An exceptionally large bond-weakening was observed for the o-NH_2 and the o-OH groups. It is tentatively proposed that this effect is due to an unexpectedly strong H-bond in the substituted phenoxyl radical. We calculate that such a bond is ca. 10 kcal mol^{-1} stronger than in the molecule.

Phenolic H-abstraction rates for p-substituents show some correlation with ΔBDE, but are correlated much better by σ^+ substituent constants. This is interpreted as consequence of significant polar effects in the H-abstraction reactions.

ΔBDE values for both ortho and para substituents are well correlated by ESR hyperfine coupling constants of substituted phenoxy radicals. However, unlike seemingly related effects observed for substituted benzyl radicals, stabilizing substituents decrease the apparent alternation of hfcc's (spin density) in the radical rather than simply withdrawing density from the other sites.

ΔBDE correlates poorly with existing measures of substituent effects for carbon-centered radicals. This is somewhat surprising since in terms of odd-electron density, phenoxyl radicals are essentially carbon centered radicals.

ACKNOWLEDGMENTS

The authors thank the Gas Research Institute for the financial support.

REFERENCES

1. A. Vuori, Pyrolysis studies of some simple coal related aromatic methyl ethers, Fuel 65:1575 (1986).
2. A. Vuori and J. B. Bredenberg, Thermolysis of substituted anisoles, Prepr. Pap. ACS Div. Fuel Chem. 30:366 (1981).
3. M. T. Klein and P.S. Virk, Model pathways for gas release from lignites, Prepr. Pap. ACS Div. Fuel Chem. 25:80 (1980).
4. G. W. Burton and K. U. Ingold, Vitamin E: Application of the principles of physical organic chemistry to the exploration of its structure and function, Acc. Chem. Res. 19:194 (1986).
5. M. W. Ranney, "Antioxidants, Recent Developments," Chemical Technology Review No. 127, Noyes Data Corporation, New Jersey (1979).
6. D. M. Golden, G. N. Spokes and S. W. Benson, Very low-pressure pyrolysis (VLPP); A versatile kinetic tool, Angew. Chem. Int. Ed. 12:534 (1973).
7. L. R. Mahoney and M. A. DaRooge, The kinetic behavior and thermochemical properties of phenoxy radicals, J. Am. Chem. Soc. 97:4722 (1975).
8. P. J. Robinson and K. A. Holbrook, "Unimolecular Reactions," Wiley Interscience (1971).
9. A. Pross and L. Random, A theoretical approach to substituent interactions in substituted benzenes, Progress in Physical Organic Chemistry, 13:1 (1981).
10. A. Finch, P. J. Gardner, and D. Wu, Studies on nitrophenols. Part IV. The standard enthalpies of combustion and formation of 1,2-dihydroxybenzene, 1,2- and 1,4-nitrophenol, Thermochem. Acta. 66:333 (1983).
11. B. D. Barton and S. E. Stein, Pyrolysis of alkyl benzenes. Relative stabilities of methyl-substituted benzyl radicals, J. Phys. Chem. 84:2141 (1980).
12. W. C. Herndon, Bond orders and ionization potentials from resonance theory, J. Am. Chem. Soc. 98:887 (1976), and the references cited therein.
13. D. F. McMillen and D. M. Golden, Hydrocarbon bond dissociation energies, Ann. Rev. Phys. Chem. 33:493 (1982).
14. Y. Okamoto and H. C. Brown, Electrophilic substituent constants, J. Org. Chem. 22:485 (1957).
15. R. W. Taft and I. C. Lewis, Theoretical approach to substituent effects, J. Am. Chem. Soc. 80:485 (1958).
16. H. A. Howard and K. U. Ingold, The inhibited autoxidation of styrene, Can. J. Chem. 41:1744 (1963).
17. M. Simonyi, I. Fitos, J. Kardos, I. Kovacs, I. Lukovitz, and J. Pospisil, Isotope effects in hydrogen atom abstraction from substituted phenols by polyvinyl acetate radicals in vinyl acetate: Evidence for tunneling, J. Chem. Soc. Far. Trans. 73:1286 (1977).
18. J. A. Franz, M. S. Alnajjar, R. D. Barrows, D. L. Kaisaki, D. M. Camaioni, and N. K. Suleiman, Reactions of the 2-allylbenzyl radical: relative and absolute rate constants for abstraction of hydrogen atom from thiophenol, dicyclohexylphosphine, phenols and arylalkyl donors, J. Org. Chem. 51:1446 (1986).

19. M. T. Dixon, M. Moghimi, and D. J. Murphy, Substituent effects in the ESR spectra of phenoxyl radicals, J. Chem. Soc. 1713 (1974).
20. G. N. Lewis, M. Randall, K. S. Pitzer, and L. Brewer, "Thermodynamics," p446, 2nd Ed., McGraw-Hill Book Co., New York (1961).

REDOX PROPERTIES OF OXY AND ANTIOXIDANT RADICALS

Slobodan V. Jovanovic and Michael G. Simic*

Boris Kidric Institute of Nuclear Sciences
P.O. Box 522
11001 Beograd, Yugoslavia

*Center for Radiation Research
National Bureau Standards
Gaithersburg, Maryland 20899, U.S.A.

INTRODUCTION

Defense mechanisms against damaging effects of oxy radicals are elaborate and numerous,[1-3] which suggests that these mechanisms may have evolved over a long period of time with the development of biological and environmental oxidations which are inadvertently associated with generation of oxy radicals. What are the criteria that determined the choice of defense mechanisms? On the molecular level the efficiency of a particular defense mechanism depends on the reactivities of its subcomponents with oxy radicals, and its ability to reverse or prevent the damaging consequences of oxy radical reactions. Antioxidants are an important part of the defense system. Mechanistic aspects of some well-known and some potential physiological antioxidants are discussed.

GENERATION OF FREE RADICALS

Free radicals can be conveniently generated and studied by radiation chemical techniques. In dilute aqueous solutions the energy of ionizing radiation is deposited in water and the following reactive species are formed,[4]

$$H_2O \rightsquigarrow \cdot OH\ (2.8),\ e_{aq}^-\ (2.7),\ \text{and}\ H\ (0.55) \qquad (1)$$

Numbers in parentheses indicate radiation chemical yields in G-units (G-unit = number of molecules formed per 100 eV of radiation energy absorbed). These primary radicals may be converted into desired secondary radicals in the reactions with appropriate solutes. Numerous such reactions are known.[5-11]

REDOX POTENTIALS

The one-electron redox potential of a redox couple consisting of a molecule and corresponding free radical may be determined by the technique of pulse radiolysis. If free radicals can be selectively generated from a molecule with an unknown redox potential and a molecule which is a primary or a secondary redox standard and if the two redox couples are in equilibrium, the unknown redox potential may be obtained from the parameters of the redox equilibrium. The difference in redox potentials of the redox couples is calculated from the equilibrium constant, using the Nernst equation,

$$\Delta E = 0.059 \log K \tag{2}$$

where K is the equilibrium constant. The unknown redox potential is then derived from the redox potential difference and the known redox potential of the standard. For example, the following redox equilibria with the secondary redox standards p-methoxyphenol, p-MeOϕOH, with E_7 = 0.6 V,[12] and ascorbate, AH^-, with E_7 = 0.28 V,[12] were used to determine the redox potentials of the two dimethoxyphenols,[13]

$$3,5\text{-}(MeO)_2\phi O\cdot + p\text{-}MeO\phi OH \rightleftharpoons p\text{-}MeO\phi O\cdot + 3,5\text{-}(MeO)_2\phi OH \tag{3}$$

$$k_f = 6.6\times10^6\ M^{-1}s^{-1},\ k_r = 6.5\times10^5\ M^{-1}s^{-1},$$

$$K_{kin} = 10,\ K_{abs} = 5,\ \Delta E = 0.05\ V$$

where $K_{kin} = k_f/k_r$, $K_{abs} = A(p\text{-}MeO\phi O\cdot)/A(3.5\text{-}(MeO)_2\phi O\cdot)$, and A = absorbance of a particular radical at the equilibrium (provided absorption spectra of radicals are not overlapping). Similarly,

$$3,5\text{-}(MeO)_2\phi O\cdot + 3,4\text{-}(MeO)_2\phi OH \rightleftharpoons 3,4\text{-}(MeO)_2\phi O\cdot + 3,5\text{-}(MeO)_2\phi OH \tag{4}$$

$$k_f = 2.85\times10^7\ M^{-1}s^{-1},\ k_r = 1.2\times10^5\ M^{-1}s^{-1},$$

$$K_{kin} = 233,\ K_{abs} = 200,\ \Delta E = 0.14\ V$$

$$3,4\text{-}(MeO)_2\phi O\cdot + AH^- \rightleftharpoons 3,4\text{-}(MeO)_2\phi OH + \cdot A^- + H^+ \tag{5}$$

$$k_f = 3.3\times10^7\ M^{-1}s^{-1},\ k_r = 1.1\times10^4\ M^{-1}s^{-1},$$

$$K_{kin} = 3000,\ K_{abs} = 2500,\ \Delta E = 0.21\ V$$

From equilibrium (3), the one-electron redox potential of the 3,5-dimethoxyphenol is determined to be 0.65 V at pH 7. Redox equilibria (4) and (5) give E_7 = 0.51 V for the one-electron redox potential of the 3,4-dimethoxyphenol and also confirm the redox potential of 3,5-dimethoxyphenol. Such cross-checks of redox potentials are a valuable confirmation of the accuracy of absolute redox potential values.

MECHANISM OF OXY RADICAL REACTIONS

As indicated in Table 1, there are several broad classes of oxy radicals, which differ in structure, reactivities, and one-electron redox potentials. Hydroxy, •OH, alkoxy, •OR, and arylalkoxy, •ORAr, radicals predominantly add to double bonds and aromatic nuclei, and at somewhat lower rates abstract H-atoms from C-H bonds despite their high oxidation potential. Aryloxy and vinyloxy radicals, ArO•, of which phenoxy-type radicals, ϕO•, are the best known, have the unpaired electron highly delocalized. This delocalization reflects strongly the mechanisms of their reactions. The possibility of forming a new bond is

Table 1. Redox Properties of Oxy Radicals

Oxy Radical	Redox Properties	E_7, V	Ref.
Hydroxy, •OH	Strong oxidant Addition to double bonds and aromatic rings suggest they are electrophiles	2.6	14
Alkyloxy, •OR	Oxidant Addition to double bonds and aromatic rings suggest they are electrophiles		
Arylalkoxy, •ORAr	Oxidant Addition to double bonds and aromatic rings suggest they are electrophiles		
Aryloxy, •OAr, and vinyloxy	Oxidant	< 0.9[a]	15
Peroxy, ZROO•	Oxidant	< 1.1[b]	

[a] The upper limit of 0.9 V is the redox potential of phenol.
[b] This value is obtained for trichloromethylperoxy radical, which is the strongest oxidant among peroxy radicals (see next section).

lower in resonance-stabilized aryloxy and vinyloxy radicals. Hence, aryloxy radicals react predominantly by one-electron transfer, while addition to aromatic nuclei and abstraction of H atoms is unfavorable.

Peroxy radicals, ZROO•, are moderately strong oxidants. The oxidation potential of peroxy radicals depends on how electrophilic Z is. Stronger electron-withdrawing substituent(s) in Z result in a higher oxidation potential of corresponding peroxy radical, ZROO•. For example, if hydrogens are substituted by chlorines, the redox potential of the peroxy radical would rise from ~0.6 V [15] for methylperoxy, $H_3COO\cdot$, to > 1.1 V for trichloromethylperoxy radical, $Cl_3COO\cdot$. This indicates that the unpaired electron is delocalized among halogens, generating slightly positive oxy structure ($^-Cl_3COO^+$).[16]

REDOX PROPERTIES OF TRICHLOROMETHYLPEROXY RADICAL

Trichloromethylperoxy radical, $Cl_3COO\cdot$, may easily be generated by pulse radiolysis of an aqueous solution of 2-propanol, acetone and carbon tetrachloride.[21] This radical is a strong oxidant. It rapidly reacts with various organic substrates (see Table 2), $k = 10^6 - 10^9\ M^{-1}s^{-1}$. The reactions of $Cl_3COO\cdot$ radical with an organic substrate, AH, may be viewed as an outer-sphere electron transfer processes, as depicted below in reactions (6)-(9),

Table 2. Reactivities of Trichloromethylperoxy Radical with Organic Solutes[a]

Solute, AH	$k(Cl_3COO^{\bullet} + AH)$ $M^{-1}s^{-1}$	Ref.	$E_7(\bullet A/AH)$ V	Ref.
1-Methyluracil	$< 10^5$	13	1.6	13
dAMP	$< 10^5$	13	1.3	13
dGMP	$\sim 6 \times 10^5$	13	1.05	18
Histidine	2.5×10^6	17	0.93	13
Tryptophan	8.7×10^7	20	0.64	21
Vitamin E	5.0×10^8	19	0.48	13
Isobarbituric acid	6.5×10^7	13	0.46	13
DOPA	$\sim 5 \times 10^3$	13	0.46	13
Uric acid	3.0×10^8	13	0.59	13
5-Trp-OH	1.5×10^8	13	0.57	13
Ascorbate	2.0×10^8	19	0.28	12

[a]Determined by pulse radiolysis in aqueous solutions at pH 7, 20°C. Redox potentials are accurate to ± 0.05 V.

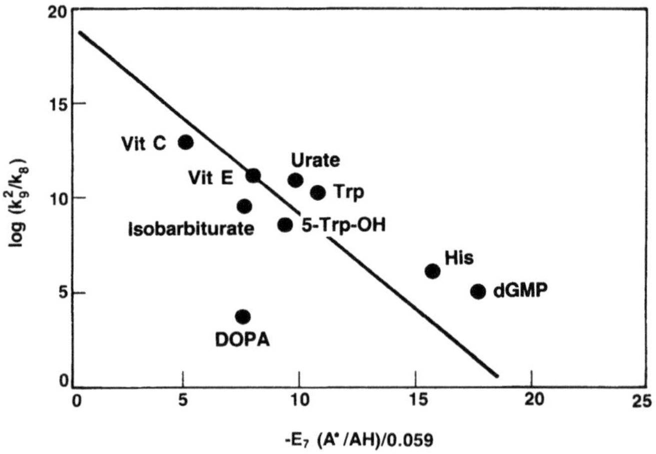

Figure 1. Evaluation of redox potential of trichloromethylperoxy radical. The points are calculated by using the modified equation (9) (see text for details). A full line is drawn through experimental points, assuming that the mechanism may be described by the Marcus cross relation.

$$Cl_3COO\cdot + AH = Cl_3COO^- + \cdot AH^+ \quad (6)$$

$$\cdot AH^+ = \cdot A + H^+ \quad (7)$$

$$\cdot AH^+ + AH = AH + \cdot AH^+ \quad (8)$$

$$Cl_3COO\cdot + Cl_3COO^- = Cl_3COO\cdot + Cl_3COO^- \quad (9)$$

The Marcus cross relation[22] relates the rate of electron transfer, reaction (6), with the difference in the redox potentials of the redox couples and their self-exchange rates,

$$k_6 \sim (k_8 k_9 K_6)^{1/2} \quad (10)$$

Figure 1 depicts the use of modified equation (10) for rate constants of reactions of $Cl_3COO\cdot$ with various substrates,[13,17-21] and the data on redox potentials of substrates and the self-exchange rates.[12,13,17,18,21]

The redox potential of the $Cl_3COO\cdot$ radical, $E_7 \geq 1.1$ V, is calculated from the intercept on the abscissa in Figure 1. This is in a good agreement with the value $E_7 < 1.3$ V, estimated from the lack of reactivity of $Cl_3COO\cdot$ with deoxyadenosine-5'-monophosphate ($E_7 = 1.3$ V), and pyrimidine DNA bases ($E_7 > 1.3$ V).

Most reactions of $Cl_3COO\cdot$ radical may be rationalized by one-electron transfer mechanism described by the Marcus relation, as shown in Figure 1. There are, however, a few reactions whose parameters did not obey the formalism of the Marcus cross relation. In the case of

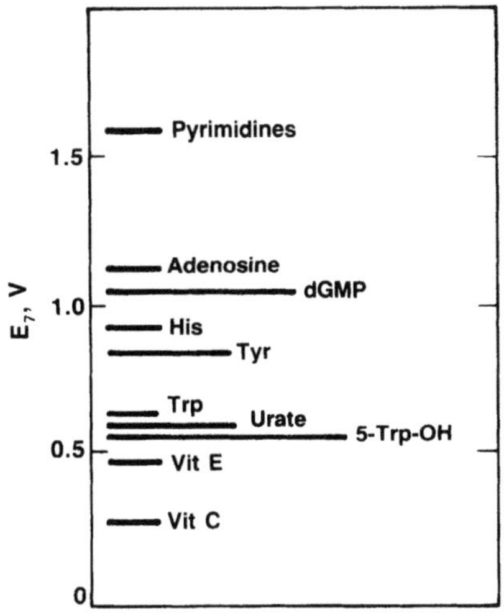

Figure 2. One-electron redox potentials of most vulnerable components of DNA and proteins and some natural antioxidants. The values are taken from references 11,12,15,16,19 and pertain to neutral aqueous solutions at 20°C.

dGMP and histidine, observed deviations might be due to the experimental difficulties in obtaining accurate rates of electron transfer, whereas the large deviation of the oxidation of 3,4-dihydroxyphenylalanine, DOPA, might be attributed to the involvement of some additional reactions.

REDOX PROPERTIES OF ANTIOXIDANT RADICALS

An efficient inactivation of oxy radical would depend on the ability of an antioxidant to intercept the radical before it damages any other component of a biosystem. In addition, the redox properties of antioxidant radicals should be low enough not to oxidize biomolecules. For example, histidine is an efficient scavenger of $Cl_3COO\cdot$ radicals,

$$Cl_3COO\cdot + His \rightarrow \cdot His(-H) + Cl_3COOH \qquad (11)$$

$$k_{11} = 2.5 \times 10^6 \ M^{-1}s^{-1} \ [ref.17]$$

On the other hand, because of its relatively high redox potential, E_7 = 0.93 V,[17] the histidinyl radical, •His(-H), may oxidize tyrosine (E_7 = 0.85 V [21]). Figure 2 summarizes the redox potentials of biological molecules which are likely to be damaged by one-electron oxidation, together with the redox potentials of natural antioxidants which act as electron donors, i.e., repairing agents of certain classes of oxidative damage to DNA bases and amino acids.

Low redox potentials of naturally occurring antioxidants, such as vitamin E, vitamin C, uric acid, and 5-hydroxyindole derivatives, enable them to efficiently scavenge oxy radicals (see Table 2 and previous section) and also to repair many biomolecules damaged by various oxy radicals. The redox potential of vitamin E, E_7(vitamin E) = 0.48 V,[13] is lower than the redox potential of the weakest peroxy radical, $E_7(H_3COO\cdot) \sim 0.6$ V.[22] Hence, vitamin E is an excellent inhibitor of chain propagation of lipid peroxidation by peroxy radicals, and a potential repair agent of certain amino acid radicals, E_7(tryptophan) = 0.64 V,[21] and E_7(tyrosine) = 0.85 V.[21]

5-Hydroxyindole derivatives, such as 5-hydroxyindole-3-acetic acid ($HO\text{-}In\text{-}CH_2CO_2^-$), 5-hydroxytryptophan (Trp-OH) and 5-hydroxytryptamine ($HO\text{-}In\text{-}CH_2CH_2\text{-}NH_3^+$) are particularly interesting as a new class of endogenous antioxidants, i.e., electron donors, since their redox (E_7 = 0.57 V)[25] and electronic properties (HIA has negatively charged side chain, Trp-OH is zwitterionic, and serotonin is positive at pH 7) make them versatile as antioxidants in diverse cellular microenvironments. For example, double-negatively charged DNA-guanyl radical, DNA-$\dot{G}$(-H), which is believed to be formed upon direct ionization of DNA,[26,27] is efficiently repaired by 5-hydroxyindole derivatives,[18]

$$DNA\text{-}\dot{G}(-H) + HO\text{-}In\text{-}CH_2CO_2^- \rightarrow \cdot O\text{-}In\text{-}CH_2CO_2^- + DNA\text{-}G \qquad (12)$$

$$k_{12} = 1.0 \times 10^8 \ M^{-1}s^{-1}$$

$$DNA\text{-}\dot{G}(-H) + Trp\text{-}OH \rightarrow Trp\text{-}O\cdot + DNA\text{-}G \qquad (13)$$

$$k_{13} = 5.4 \times 10^8 \ M^{-1}s^{-1}$$

$$DNA\text{-}\dot{G}(\text{-}H) + HO\text{-}In\text{-}CH_2CH_2\text{-}NH_3^+ \rightarrow \cdot O\text{-}In\text{-}CH_2CH_2NH_3^+ + DNA\text{-}G \quad (14)$$

$$k_{14} = 5.2\times10^9\ M^{-1}s^{-1}$$

Although the difference in redox potentials between these three derivatives is negligible, because of the electrostatic effect the rate of repair by serotonin is almost two orders of magnitude higher than the rate for 5-hydroxyindole-3-acetic acid.

CONCLUSIONS

The last three reactions (12-14) demonstrate unambiguously that in addition to favorable redox properties of the repair agents, factors reflecting the microenvironment of the bio-radical play an important role in the efficacy of repair. Understanding the kinetics, energetics, and mechanisms also allows synthesis of antioxidants with tailor-made properties for efficient repair of oxidative damage in biosystems.

ACKNOWLEDGMENTS

Thanks are due to Karen A. Taylor for editorial assistance and Terry Shuggars and Eva M. Wood for technical execution of the paper.

REFERENCES

1. W.A. Pryor, "Free Radicals in Biology," ed., vol. 1-7, Academic Press, New York (1976-1984).
2. O.F. Nygaard and M.G. Simic, eds., "Radioprotectors and Anticarcinogens," Academic Press, New York (1983).
3. M.G. Simic and M. Karel, eds., "Autoxidation in Food and Biological Systems," Plenum Press, New York (1980).
4. A.J. Swallow, "Radiation Chemistry - An Introduction," Longmans, London (1973).
5. Farhataziz and A.B. Ross, "Selected Specific Rates of Reactions of Transients from Water in Aqueous Solutions. III. Hydroxyl Radical and Perhydroxyl Radical and Their Radical Ions," NBS-NSRDS 59, Government Printing Office, Washington (1977).
6. M. Anbar, M. Bambenek and A. Ross, "Selected Specific Rates of Reactions of Transients from Water in Aqueous Solutions. I. Hydrated Electron," NBS-NSRDS 43, Government Printing Office, Washington (1977).
7. M. Anbar, Farhataziz, and A. Ross, "Selected Specific Rates of Reactions of Transients from Water in Aqueous Solutions. II. Hydrogen Atom," NBS-NSRDS 51, Government Printing Office, Washington (1975).
8. A.B. Ross and P. Neta, "Rate Constants for Reactions of Aliphatic Carbon-Centered Radicals in Aqueous Solution," NBS-NSRDS 70, Government Printing Office, Washington (1982).
9. A.B. Ross, and P. Neta, "Rate Constants for Reactions of Inorganic Radicals in Aqueous Solution," NBS-NSRDS 65, Government Printing Office, Washington (1979).
10. G.V. Buxton and R.M. Sellers, "Compilation of Rate Constants for the Reactions of Metal Ions in Unusual Valency States," NBS-NSRDS 62, Government Printing Office, Washington (1978).

11. C. von Sonntag, "The Chemical Basis of Radiation Biology," Taylor & Francis, New York (1987).
12. S. Steenken and P. Neta, "One-electron redox potentials of phenols," J. Phys. Chem. 86:3661-3667 (1982).
13. S.V. Jovanovic and M.G. Simic, unpublished.
14. H.A. Schwarz and R.W. Dodson, "Equilibrium between hydroxyl radicals and thallium(II) and the oxidation potential of OH(aq)," J. Phys. Chem. 88:3643-3647 (1984).
15. R.E. Huie and P. Neta, "Kinetics of one-electron transfer reactions involving ClO_2 and NO_2," J. Phys. Chem. 90:1193-1198 (1986).
16. E.P.L. Hunter and M.G. Simic, Kinetics of peroxy radical reactions with antioxidants, "Oxy Radicals and Their Scavenger Systems," vol. 1, G. Cohen and R.A. Greenwald, eds., Elsevier, New York, (1983).
17. I. Cudina and S.V. Jovanovic, "Free radical inactivation of enzymes," Radiat. Phys. Chem., in press.
18. S.V. Jovanovic and M.G. Simic, "The DNA-guanyl radical," submitted to Biophys. Biochim. Acta.
19. J.E. Packer, T.F. Slater, and R.L. Willson, "Direct observation of a free radical interaction between vitamin E and vitamin C," Nature (London) 278:737-738 (1979).
20. J.E. Packer, J.S. Mahood, R.L. Willson, and B. Wolfenden, "Reactions of the trichloromethylperoxy free radical ($Cl_3COO^{\cdot}$)," Int. J. Radiat. Biol. 39:135-141 (1981).
21. S.V. Jovanovic, A. Harriman, and M.G. Simic, "Electron-transfer reactions of tryptophan and tyrosine derivatives," J. Phys. Chem. 90:1935-1939 (1986).
22. R.E. Huie and P. Neta, "Rate constants for one-electron oxidation by methylperoxyl radicals in aqueous solutions," Int. J. Chem. Kinet. 18:1185-1191 (1986).
23. J.E. Packer, R.L. Willson, D. Bahneman, and K.-D. Asmus, "Electron transfer reactions of halogenated aliphatic-peroxyl radicals," J. Chem. Soc., Perkin Trans. II:296-299 (1980).
24. R.A. Marcus, "Chemical and electrochemical electron-transfer theory," Annu. Rev. Phys. Chem. 15:155-196 (1964).
25. M.G. Simic, S. Steenken, and S.V. Jovanovic, unpublished.
26. J. Huttermann, K. Voit, H. Oloff, W. Kohnlein, A. Graslund, and A. Rupprecht, "Specific formation of electron gain and loss centres in X-irradiated oriented fibres of DNA at low temperatures," Faraday Discuss. Chem. Soc. 78:135-149 (1984).
27. P.J. Boon, P.M. Cullis, M.C.R. Symmons, and B.W. Wren, "Effects of ionizing radiation on deoxyribonucleic acid and related systems. Part 1. The role of oxygen," J. Chem. Soc., Perkin Trans. II:1393-1399 (1984).

THE ROLE OF METALS IN AQUEOUS SUPEROXIDE CHEMISTRY

Benon H. J. Bielski

Chemistry Department
Brookhaven National Laboratory
Upton, New York 11973 USA

A focal point in the research of metal catalyzed reactions involving oxygen derived species/radicals ($\cdot OH$, $HO_2\cdot/O_2^{\overline{\cdot}}$, H_2O_2, etc.) is the search for the true nature and mode of formation of the oxidizing agent(s) responsible for the biological damage observed in systems containing both hydrogen peroxide and superoxide radicals. There appears to be a general agreement that one of the most deleterious species is the OH radical, generated either in the "Metal-Catalyzed Haber-Weiss Reaction" (reactions 1-3, Mechanism I)[1]

$$O_2^{\overline{\cdot}} + Fe^{3+} \longrightarrow O_2 + Fe^{2+} \qquad (1)$$

$$H_2O_2 + Fe^{2+} \longrightarrow \cdot OH + OH^- + Fe^{3+} \qquad (2)$$

$$O_2^{\overline{\cdot}} + H_2O_2 \longrightarrow \cdot OH + OH^- + O_2 \qquad (3)$$

or a "Reductant-Driven Fenton-Type Reaction" (reactions 4 and 2, Mechanism II; L_m = ligand; reductants can be AH_2, QH_2, $O_2^{\overline{\cdot}}$, etc.).

$$2L_mM^{n+} + QH_2 \longrightarrow 2L_mM^{(n-1)+} + 2H^+ + Q \qquad (4)$$

$$L_mM^{(n-1)+} + H_2O_2 \longrightarrow L_mM^{n+} + \cdot OH + OH^- \qquad (2)$$

The subtle difference between these two mechanisms is that Mechanism I is merely a specific case of a system (Mechanism II) that generates $\cdot OH$ radicals using a variety of metals (Fe, Cu, Mn, Ti, etc.) and reductants.[2-10]

While there is ample evidence that the Fenton reaction, which is carried out in the very acidic pH range, generates free OH radicals (reaction 2), this is not believed to be the case for Fenton-type reactions carried out near neutrality or in the basic pH range.[4-10] The latter studies have been carried out by necessity with metal complexes because the corresponding parent metal ions form insoluble hydroxides at pH > 2-3. The most frequently used metal complex in industrial oxidation/hydroxylation processes is FeEDTA.[11] When the iron ion is chelated, the oxidation-reduction potential is changed,[4] for example

$Fe^{2+} \longrightarrow Fe^{3+} + e^-$ $\qquad E_o = -0.771$ V

$FeEDTA^{2-} \longrightarrow FeEDTA^- + e^-$ $\qquad E_o = +0.177$ V

Because of the change in potential, the iron complex of EDTA reacts with hydrogen peroxide approximately 200 times more rapidly than with the Fe^{2+}-aquo complex;[12] a definite advantage from the catalytic point of view.

A negative attribute of this class of chelates is that they are organic in nature and, hence, scavenge OH radicals very efficiently (k's ≈ 10^9 M^{-1}).[13] In our laboratory preliminary pulse radiolysis studies of the interaction of free OH radicals with Fe(III)DTPA under anaerobic conditions show that the overall mechanism is complex. It involves inter- as well as intra-electron/H-atom transfer reactions, the final major product of which is a reduced Fe(II)DTPA complex. These results are in agreement with reports that metals chelated with organic ligands can undergo reduction when exposed to OH radicals.[14-18] Our present work supports, in particular, an earlier hypothesis that the OH radical generated in the reaction between Fe(II)DTPA and H_2O_2 reacts with DTPA and leads, in presence of molecular oxygen, to the formation of $DTPAO_2\bullet$, which in turn could react with another Fe(II)DTPA/DTPA molecule or any other substrate in the medium.[19] Hence, if the Fenton-type reaction generates an OH radical, many reaction pathways are a priori possible. Which pathway(s) will dominate will depend entirely upon the concentrations and rates of reaction of the respective reactants in the system. The Fenton-type reaction may involve two concurrent reactions (5 and 6) in which

$$L_mFe^{2+} + H_2O_2 \longrightarrow L_mFe^{3+} + \bullet OH + OH^- \qquad (5)$$

$$L_mFe^{2+} + H_2O_2 \longrightarrow [L_mFe{=}O]^{2+} + H_2O \qquad (6)$$

The mechanism by which $[Fe{=}O]^{2+}$ is generated (reaction 6) most likely involves some form of transient Fe(II) or Fe(III) peroxy complex.[20]

The acceptance of two concurrent reactions (5 and 6) for the Fenton-type reaction is obviously not a solution, but rather a temporary compromise waiting to be resolved. This mechanism is the result of research efforts that span many decades. A comparison of product distribution for the Fenton/Fenton-type reactions, carried out with a number of substrates in acidic and neutral/basic solutions showed that although in the acidic region the oxidizing agent was indeed the OH radical (the product distribution resembled closely results obtained by free OH radicals generated by ionizing radiation), in the neutral and basic region the products showed that a more stereospecific or selective mechanism is operational in addition to the more random attack by the OH radical. It was concluded that the oxidizing agent responsible for this enhanced specificity may be the ferryl iron.[5-10,21]

That Fe(IV) is formed during catalysis of many iron-containing oxygenases or in hemoproteins is well established. There are also a number of successful studies of Fe(IV) complexes in aprotic solvents[22,23] and numerous patents[11] that describe industrial processes in which Fe-EDTA complexes serve as catalysts in the oxidation/hydroxylation of organic compounds by either molecular oxygen (reactions 7-9) or hydrogen peroxide (reactions 5 and 6).[11]

$$L_mFe^{2+} + O_2 \longrightarrow L_mFe^{2+}O_2 \quad (7)$$

$$L_mFe^{2+}O_2 + L_mFe^{2+} \longrightarrow L_mFe^{2+}O_2Fe^{2+}L_m \quad (8)$$

$$L_mFe^{2+}O_2Fe^{2+}L_m \longrightarrow 2[L_mFe{=}O]^{2+} \quad (9)$$

Although there is little direct evidence that iron complexes undergo reactions 7-9 in aqueous solutions, such chemistry has been well established for some cobalt complexes.[24]

A direct comparison of results from different laboratories is difficult at present, because experiments were not carried out under identical conditions. The charting of overall mechanistic trends however (as illustrated on the above FeDTPA systems), is not only possible, but very useful in planning experiments and research programs. A strategy we have been following in our laboratory, because of the availability of the pulse radiolysis and various stopped-flow techniques, is the generation of the Fe(IV) oxidation state by oxidation of Fe(III) salts/complexes with the free OH radical.[25] A different pathway to hypervalent iron states that we are also pursuing is the reduction of the relatively stable ferrate ion (FeO_4^{2-}) by the hydrated electron or the CO_2^- radical, leading to formation of the perferryl ion (FeO_4^{3-}).[25,26] Our results indicate that although ferryl and perferryl iron have different spectral characteristics, their chemical behaviour is very similar. Therefore, perferryl is one of our models in the study of the chemistry ferryl carries out with organic substrates.

Thus far the Fe(IV) state has been observed in our studies only in presence of inorganic ligands (OH^-, $P_2O_7^{4-}$, and HCO_3^-/CO_3^{2-}):

$$\bullet OH + Fe(OH)_4^- \xrightarrow{10^8\ M^{-1}\ s^{-1}} Fe(IV) + H_2O \quad (10)$$

$$\bullet OH + HCO_3^-/CO_3^{2-} \xrightarrow{4.0 \times 10^8\ M^{-1}\ s^{-1}} H_2O/OH^- + CO_3^{\bullet-} \quad (11)$$

$$CO_3^{\bullet-} + L_mFe^{3+} \xrightarrow{2.4 \times 10^8\ M^{-1}\ s^{-1}} L_mFe(IV) + CO_3^{2-} \quad (12)$$

L_m can be OH^-, HCO_3^-/CO_3^{2-} or a combination of all three; the true nature of the Fe(IV) complex in carbonate solutions has not yet been determined.

$$\bullet OH + [FeOH(P_2O_7)_2]^{6-} \xrightarrow{5 \times 10^7\ M^{-1}\ s^{-1}} [Fe(IV)(P_2O_7)_2]^{5-} \quad (13)$$

While the spectra of the various Fe(IV) complexes are very similar (λ_{max} = 430 nm, ε = 900 ± 100 M^{-1} cm^{-1}), their decay kinetics vary with pH and the nature of the ligand ($t_{1/2} \approx$ 0.1 to 1.0 sec).[25-27]

The Iron(V) species is relatively stable in alkaline solutions, where it has a $t_{1/2} \approx$ 170 ms. Its UV spectrum shows a shoulder between 260 and 300 nm (ε = 4-5.0 x 10^3 M^{-1} cm^{-1}), while in the visible range it has a maximum near 490 nm (ε = 630 M^{-1} cm^{-1}).[26]

$$FeO_4^{2-} + e_{aq}^- \xrightarrow{2 \times 10^{10}\ M^{-1}\ s^{-1}} FeO_4^{3-} \quad (14)$$

$$FeO_4^{2-} + CO_2^{\bullet-} \xrightarrow{3.5 \times 10^8\ M^{-1}\ s^{-1}} FeO_4^{3-} + CO_2 \quad (15)$$

Preliminary experiments with formate show that $k(Fe(V) + HCOO^-)/k(FeVI) + HCOO^-) = 10^5$. This high increase in reactivity suggests that industrial oxidation processes utilizing Fe(VI) may be more efficient if carried out in presence of one-electron reducing agents. This, of course, applies only to oxidation reactions since in hydroxylation reactions the different hypervalent oxidation states may lead to different products. For example, the hydroxylation of phenol with Fe(VI) yields almost exclusively parasubstitution (p-hydroquinone and p-quinone, 98%). When the same reaction is carried out in presence of ascorbic acid, which reduces Fe(VI) to Fe(V)/Fe(IV) at a rate of $k \approx 10^6\ M^{-1}\ s^{-1}$, the product distribution is: (65% o-QH_2 and o-Q), (27% p-QH_2 and p-Q) and 8% resorcinol. This distribution varies with the ratio of [phenol]/[ascorbate].[28]

These preliminary results suggest that if hypervalent iron states are formed in Fenton-type reactions, they may be, from a biological point of view, just as deleterious as OH radicals because of their relatively longer lifetimes, higher specificity in oxidation and hydroxylation reactions, and the capacity to abstract H atoms. That Fe(VI) oxidizes various organic compounds stereospecifically has been known for some time.[26,29] While Fe(VI) abstracts H atoms at a relatively slow rate, the results described above show that this rate increases by orders of magnitude as one descends from Fe(VI) to Fe(V). Preliminary competition studies suggest that in reactions with organic compounds, Fe(IV) reacts as fast or faster than does Fe(V).

Another subject of current interest is the catalytic dismutation of $HO_2\bullet/O_2^{\bar{\bullet}}$ in the presence of certain metal ions/complexes and superoxide dismutases (SOD). As is well known, $HO_2\bullet/O_2^{\bar{\bullet}}$ can oxidize or reduce a metal ion/complex to a different oxidation state or can undergo a metal catalyzed disproportionation that usually involves a cyclic pathway between two oxidation states of the metal.[2] Which pathway dominates in a given system will depend upon such factors as the redox potential of the metal complex, nature of the ligand, pH, ionic strength and temperature.

Since metal complexes mimic superoxide dismutases <u>in vitro</u>, it is not surprising that they have become a favored research tool with which researchers probe the catalytic mechanisms of these enzymes. Currently available evidence suggests that while $O_2^{\bar{\bullet}}$ forms only mononuclear complexes,

$$L_mM^{n+} + O_2^{\bar{\bullet}} \longrightarrow L_mMOO^{(n-1)+} \qquad (16)$$

the $HO_2\bullet$ radical forms, in addition, dinuclear complexes.[30]

$$HO_2\bullet + L_mM^{n+} \longrightarrow L_mMOOH^{n+} \qquad (17)$$

$$L_mMOOH^{n+} + L_mM^{n+} \longrightarrow [L_mMOOHML_m]^{2n+} \qquad (18)$$

Formation of dinuclear complexes is apparently possible because of an electron transfer from the metal ion to the $HO_2\bullet$ radical with subsequent addition of another metal ion:[31]

$$HO_2\bullet + L_mM^{n+} \longrightarrow [L_m^{(n+1)+}HO_2^-] \qquad (19)$$

$$[L_mM^{(n+1)+}HO_2^-] + L_mM^{n+} \longrightarrow [L_mM^{(n+1)}HO_2^-M^{n+}L_m] \qquad (20)$$

In aqueous solutions the mononuclear complexes of $HO_2\cdot$ and $O_2^{\bar{\cdot}}$ are in equilibrium with each other (21,-21). Their pKs may depend upon the nature of the ligand (L_m) as for example in the $L_mMn(II)-HO_2\cdot/O_2^{\bar{\cdot}}$ complexes, for which the values range from 2.5 to 3.5 ± 0.5:[32]

$$L_mMOOH^{n+} \rightleftharpoons L_mMOO^{(n-1)+} + H^+ \qquad (21,-21)$$

A subject of speculation is whether superoxide radicals form transient adducts with superoxide dismutases, e.g., $SOD-Cu(II)-O_2^-$, and if they do, what effect they have on the structural integrity of the active site, and thus on the efficiency of the catalytic cycle. The catalytic dismutation of superoxide radicals by simple Cu(II) complexes can be described by the following mechanism:[33,34]

$$L_mCu^{2+} + O_2^{\bar{\cdot}} \longleftrightarrow L_mCuO_2^+ \longleftrightarrow L_mCu^+ + O_2 \qquad (22)$$

$$L_mCu^+ + O_2^{\bar{\cdot}} + 2H^+ \longrightarrow L_mCu^{2+} + H_2O_2 \qquad (23)$$

$$L_mCuO_2^+ + O_2^{\bar{\cdot}} + 2H^+ \longrightarrow L_mCu^{2+} + H_2O_2 \qquad (24)$$

A survey of the reports in which $Cu^{2+}-HO_2\cdot/O_2^{\bar{\cdot}}$ complexes have been postulated suggests that their formation and stability depends, similarly to the Mn(II) complexes, upon the nature of the ligands and pH of the medium.[32-34] For example, an early study of the reaction of superoxide radicals with Cu(II)-histidine complexes showed that only one complex, $(CuHist_2H)^{3+}$, of six that are known to exist in the pH range between 1 and 10 dismutates $HO_2\cdot/O_2^{\bar{\cdot}}$ with an optimum rate at pH 4.7.[35] Despite great efforts, no evidence could be established for the formation of a transient $Cu(II)-Hist-HO_2\cdot/O_2^{\bar{\cdot}}$ complex. A reinvestigation of this system in presence of high oxygen concentrations (up to 0.16 M) confirmed the earlier findings.[36]

A suggestion in a recent theoretical treatise[37] that arginine-141 may promote formation of a $SOD-Cu(II)-O_2^{\bar{\cdot}}$ through its stabilizing effect upon the $Cu^{2+}-O_2^{\bar{\cdot}}$ moiety prompted us to study the reaction of superoxide radical with Cu(II)-arginine complexes. Arginine functions as a bidentate ligand, chelating Cu(II) with the carboxylic and α-amino group. Assuming that it maintains its square planar configuration in aqueous solutions, there is a chance that the terminal amino group acts in a fashion postulated for the $SOD-Cu^{2+}$ and stabilizes the $Cu(II)-Arg-O_2^{\bar{\cdot}}$ complex. To enhance the chances of observing such a complex, the studies were carried out in the presence of high oxygen concentrations (up to 0.16 M), affecting the equilibrium conditions illustrated in reaction[22]. Measured dismutation rates of O_2^- under similar experimental conditions showed a decrease in rate with increasing oxygen concentration. Also, a detailed analysis of the kinetic runs demonstrates that the dismutation of O_2^- proceeds by two processes, a pseudo-first order and a first order reaction. Although we were not able to separate reaction 23 from 24, the overall behavior of this system can be described by reactions 22-24. These results demonstrate formation of transient superoxide adducts with Cu(II)-arginine, but they should not be used as a model for the interpretation of the SOD-Cu catalyzed dismutation mechanism of $O_2^{\bar{\cdot}}$. Similar studies at high oxygen concentrations with the $SOD-Cu^{2+}$ enzyme itself failed to reveal formation of an O_2^- adduct.[38] Theoretical approximations of the dissociation constants for the $SOD-Cu^{2+}-O_2^{\bar{\cdot}}$ and $Arg-Cu^{2+}-O_2^{\bar{\cdot}}$, which take into consideration that the two systems have different redox potentials, suggest that oxygen concentrations higher than those attainable under present experimental conditions would be required to reveal the presence of an $O_2^{\bar{\cdot}}$ adduct.[38] That SOD-Cu^{2+} is a much more sophisticated catalyst is apparent from the fact that whereas its k_{cat} changes little with pH in the range between ~ 5.0-10,[39] the copper complexes are active only in relatively narrow pH ranges with maximal activities near pH 5. Although the SOD-Cu(II)-$O_2^{\bar{\cdot}}$ complex remains, for the time being, a subject of speculation,[36,37,40] positive evidence for the formation of such a copper protein adduct has been accumulating in studies with ceruloplasmin.[41]

ACKNOWLEDGEMENTS

This work was supported by NIH Grant R01GM23658-10 and carried out at Brookhaven National Laboratory which is operated under contract DE-AC02-76CH00016 with the U.S. Department of Energy.

REFERENCES

1. F. Haber and J. Weiss, Proc. R. Soc. (London), A147:332 (1934).
2. W. Bors, M. Saran, and G. Czapski, The relevance of the superoxide anion radical in biological systems, in: "Proceedings of the International Symposium on Superoxide and Superoxide Dismutases," W. H. Bannister and J. V. Bannister, eds., Elsevier/North-Holland, New York, pp. 1-31 and references therein (1979).
3. S. D. Aust, L. A. Morehouse, and C. E. Thomas, Role of metals in oxygen radical reactions", J. Free. Rad. Biol. Med. 1:3 and references therein (1985).
4. W. H. Koppenol and J. Butler, Energetics of interconversion reactions of oxyradicals, Adv. Free Rad. Biol. Med. 1:91 and references therein (1985).
5. R. R. Grinstead, Oxidation of salicylate by model peroxidase catalyst iron-ethylenediaminetetraacetato-iron(III) acid, J. Am. Chem. Soc. 82:3472 (1960).
6. C. Walling, Fenton's Reagent Revisited, Acc. Chem. Res. 8:125 (1975).
7. G. A. Hamilton, J. W. Hanifin, Jr., and J. P. Friedman, The hydroxylation of aromatic compounds by hydrogen peroxide in the presence of catalytic amounts of ferric ion and catechol. Product studies, mechanism, and relation to some enzymic reaction, J. Am. Chem. Soc. 88:5269 (1966).
8. M. Grootveld and B. Halliwell, An aromatic hydroxylation assay for hydroxyl radicals utilizing high-performance HPLC. Use to investigate the effect of EDTA on the fenton reaction, Free. Rad. Res. Comms. 1:243 (1986).
9. J. T. Groves and M. Van Der Puy, Stereospecific aliphatic hydroxylation by iron-hydrogen peroxide. Evidence for a stepwise process, J. Am. Chem. Soc. 98:5290 (1976).
10. A. E. Cahil and H. Taube, The use of heavy oxygen in the study of reactions of hydrogen peroxide, J. Am. Chem. Soc. 74:2312 (1952).
11. Chelating Agents in Oxidation-Reduction Reactions, Technical Information-Literature Code G-1, Organic Chemical Division, W. R. Grace & Co. (1985).
12. O. K. Borggaard, O. Faver, and V. S. Andersen, Polarographic study of the rate of oxidation of iron(II) chelates by hydrogen peroxide, Acta Chem. Scand. 25:3541 (1971).
13. Farhataziz and A. B. Ross, "Selected specific rates of reactions of transients from water in aqueous solution. III. Hydroxyl radical and perhydroxyl radical and their radical ions." NSRDS-NBS 59 (1977).

14. S. N. Bhattacharyya and K. P. Kundu, The radiation chemistry of aqueous solutions of ferric ethylenediaminetetraacetate, Int. J. Radiat. Phys. Chem. 3:1 (1971).
15. B. K. Sharma and K. Sahul, ^{60}Co γ-Radiolysis of iron(III) nitrilotriacetate in aqueous solutions, Radiat. Phys. Chem. 20:341 (1982).
16. K. P. Kundu and N. Matsuura, Gamma-radiolysis of ferric ethylene diamine tetra-acetate in neutral aqueous solution, Int. J. Radiat. Phys. Chem. 7:565 (1975).
17. M. Masarwa, H. Cohen, and D. Meyerstein, The effect of nitrilotriacetate on the mechanism of reduction of copper(II) ions by α-hydroxyalkyl free radicals via complexes with copper-carbon bonds as intermediates. A pulse-radiolytic study. Inorg. Chem. 25:4897 (1986).
18. G. R. A. Johnson and N. B. Nazhat, Kinetics and mechanism of the reaction of the bis(1,10-phenanthroline) copper(I) ion with hydrogen peroxide in aqueous solution, J. Am. Chem. Soc. 109:1990 (1987).
19. G. Cohen, D. Lewis, and P. M. Sinet, Oxygen consumption during the Fenton-type reaction between hydrogen peroxide and a ferrous-chelate (Fe^{2+}-DTPA), J. Inorg. Biochem. 15:143 (1981).
20. J. D. Rush and W. H. Koppenol, The Reaction of ferrous EDTA with hydrogen peroxide: Evidence against hydroxyl radical formation, J. Free Rad. Biol. Med. 1:281 (1985).
21. G. W. Klein, K. Bhatia, V. Madhavan, and R. H. Schuler, Reaction of OH with benzoic acid. Isomer distribution in the radical intermediates, J. Phys. Chem. 79:1767 (1975).
22. H. Sugimoto and D. T. Sawyer, Iron(II)-induced activation of hydrogen peroxide to ferryl ion (FeO^{2+}) and singlet oxygen ($^1O_2$) in acetonitrile: Monoxygenations, dehydrogenations of organic substrates, J. Am. Chem. Soc. 106:4283 (1984).
23. P. N. Balasubramanian and T. C. Bruice, Oxygen transfer involving non-heme iron. The reaction of (EDTA)Fe^{III} with m-chloroperbenzoic acid, J. Am. Chem. Soc. 108:5495 (1986).
24. R. D. Jones, D. A. Summerville, and F. Basolo, Synthetic oxygen carriers related to biological systems, Chem. Rev. 79:139 (1979).
25. J. D. Rush and B. H. J. Bielski, Pulse radiolysis studies of alkaline Fe(III) and Fe(VI) solutions. Observation of transient iron complexes with intermediate oxidation states, J. Am. Chem. Soc. 108:523 (1986).
26. B. H. J. Bielski and M. J. Thomas, Studies of hypervalent iron in aqueous solutions. Part I. Radiation induced reduction of Fe(VI) to Fe(V) by CO_2^-, J. Am. Chem. Soc. 109:7761 (1987).
27. J. Melton and B. H. J. Bielski (unpublished results).
28. J. Cyr and B. H. J. Bielski (unpublished results).
29. J. D. Carr, P. B. Kelter, A. Tabatabai, D. Spichal, J. Erickson, and C. W. McLaughlin, Properties of ferrate(VI) in aqueous solution: An alternate oxidant in wastewater treatment, Proc. Conf. Water Chlorination Chem. Environ. Impact & Health Effect, R. L. Jolley ed.; Lewis Chelsea, pub.; pp 1285-1298 and references therein (1985).
30. B. H. J. Bielski, Fast kinetic studies of dioxygen-derived species and their metal complexes, Phil. Trans. R. Soc. Lond. B311:473 and references therein (1985).
31. G. G. Jayson, B. J. Parson, and A. J. Swallow, Oxidation of ferrous ions by perhydroxyl radicals, J. Chem. Soc., Faraday Trans. 1 236 (1973).

32. D. E. Cabelli and B. H. J. Bielski, Pulse radiolysis study of the kinetics and mechanisms of the reactions between manganese(II) complexes and HO_2/O_2^- radicals. 2. The phosphate complex and an overview, J. Phys. Chem. 88:6291 (1984).
33. D. Klug-Roth and J. Rabani, Pulse radiolytic studies on reactions of aqueous superoxide radicals with copper(II) complexes, J. Phys. Chem. 80:588 (1976).
34. G. V. Buxton, J. C. Green, and R. M. Sellers, Oxidation of copper(I)-olefin complexes in aqueous solution by oxygen and hydrogen peroxide, J. Chem. Soc., Dalton Trans. 2160 (1976).
35. J. Weinstein and B. H. J. Bielski, Reaction of superoxide radicals with copper(II)-histidine complexes, J. Am. Chem. Soc. 102:4916 (1980).
36. D. E. Cabelli, B. H. J. Bielski, and J. Holcman, The interaction between copper(II)-arginine complexes and HO_2/O_2^- radicals, a pulse radiolysis study, J. Am. Chem. Soc. 109:3665 (1987).
37. R. Osman and H. Basch, On the mechanism of action of superoxide dismutase: A theoretical study, J. Am. Chem Soc. 106:5710 (1984).
38. J. Holcman, private communication.
39. L. W. Oberly, ed., "Superoxide Dismutase", vol. 1 and 2, CRC Press, Boca Raton, FL (1982).
40. M. C. R. Symons and J. M. Stephenson, Mechanism of the reaction of bovine copper-zinc superoxide dismutase. J. Chem. Soc., Faraday Trans. 1 79:2983 (1983).
41. A. Plonka, J. Mayer, D. Metodiewa, J. L. Gebicki, A. Zgirski, and M. Grabska, Superoxide radical dismutation by copper proteins, J. Rad. Nuc. Chem. 101:221 and references therein (1986).

FENTON CHEMISTRY REVISITED: AMINO ACID OXIDATION

E.R. Stadtman and B.S. Berlett

Laboratory of Biochemistry
National Heart, Lung, and Blood Institute
National Institutes of Health
Bethesda, Maryland 20892, U.S.A.

INTRODUCTION

Many mixed-function oxidation (MFO) systems catalyze the O_2/Fe(II)-dependent oxidative inactivation of enzymes.[1-5] Such oxidation is likely implicated in the regulation of enzyme degradation[2,5-7] in the accumulation of altered forms of enzymes during aging[8,9] in neutrophil function.[10] Results of mechanistic studies indicate that MFO systems generate H_2O_2 and Fe(II) which interact at metal-binding sites on the protein to generate active oxygen species (•OH, perferryl ion, singlet oxygen); the activated oxygen reacts in situ with the side chains of proximal amino acid residues (especially with histidine, arginine, proline, and lysine residues) and converts them to carbonyl derivatives.[1,3,11] Protein oxidation by MFO systems is thus attributable to Fenton chemistry that occurs at Fe(II) binding sites on the protein.

Zs-Nagy and Floyd[12] demonstrated that •OH generation by a Fenton reagent comprised of H_2O_2/Fe(III)/ADP reacts with free amino acids. The present investigation was undertaken to examine the mechanism of amino acid oxidation by the Fenton reagent with the hope that it would contribute to an understanding of the mechanism of protein oxidation by MFO systems.

EXPERIMENTAL RESULTS

Leucine Oxidation

The oxidation of leucine (and other amino acids) is almost completely described by a combination of reactions 1-5. The amounts of carbonyl compounds formed in reactions 1 and 2 were measured as the 2,4-dinitrophenylhydrazone derivatives before ($RCHO + RCOCOO^-$) and after ($RCOCOO^-$ only) extraction with diethylether. Isovaleraldehyde and α-ketoisocaproate were identified as the major carbonyl compounds formed from leucine, and their concentrations were verified by HPLC chromatography (Stadtman, Berlett, Tsai, and Levine, unpublished procedures). The production of isovaleric acid in the oxidation of leucine (reactions 3 and 5) was monitored continuously by manometric measurement of the acid-mediated evolution of CO_2 which occurs when the reaction is carried out in a bicarbonate-CO_2 buffer (pH 7.6) in a Warburg apparatus, fitted with double side-arm vessels. The small amount of isovaleraldehydeoxime formed during the oxidation of leucine (reaction 4) was identified by HPL chromatography

$$RCH\overset{+}{N}H_3CO\overset{-}{O} + H_2O_2 \longrightarrow RCHO + \overset{+}{N}H_4 + HCO_3^- \tag{1}$$

$$RCH\overset{+}{N}H_3CO\overset{-}{O} + H_2O_2 \longrightarrow \overset{+}{N}H_4 + RCOCO\overset{-}{O} + H_2O \tag{2}$$

$$RCH\overset{+}{N}H_3CO\overset{-}{O} + 2H_2O_2 \longrightarrow \overset{+}{N}H_4 + RCO\overset{-}{O} + CO_2 + 2H_2O \tag{3}$$

$$RCH\overset{+}{N}H_3CO\overset{-}{O} + 2H_2O_2 \longrightarrow RCH{=}NOH + CO_2 + 3H_2O \tag{4}$$

$$RCHO + H_2O_2 \longrightarrow RCO\overset{-}{O} + H^+ + H_2O \tag{5}$$

$$2H_2O_2 \longrightarrow 2H_2O + O_2 \tag{6}$$

and by its ability to inhibit the oxidation of ethanol by liver alcohol dehydrogenase (Stadtman, Berlett, Tsai, and Levine, unpublished data). As shown in Figure 1A, the production of oxygen (reaction 6) is stimulated by both ADP and Fe(II) and is relatively independent of the presence leucine. However, the evolution of CO_2, which is a measure of the amount of isovaleric acid formed by reactions 3 and 5, is completely dependent on the presence of leucine (Figure 1B); it is stimulated by Fe(II) and much more so by the presence of both Fe(II) and ADP. That CO_2 evolution is an expression of acid formation was confirmed by showing that the amount of ether-extractable fatty acid produced is equivalent to the amount of CO_2 evolved. The ether-extractable acid was identified as isovaleric acid by HPLC chromatography.

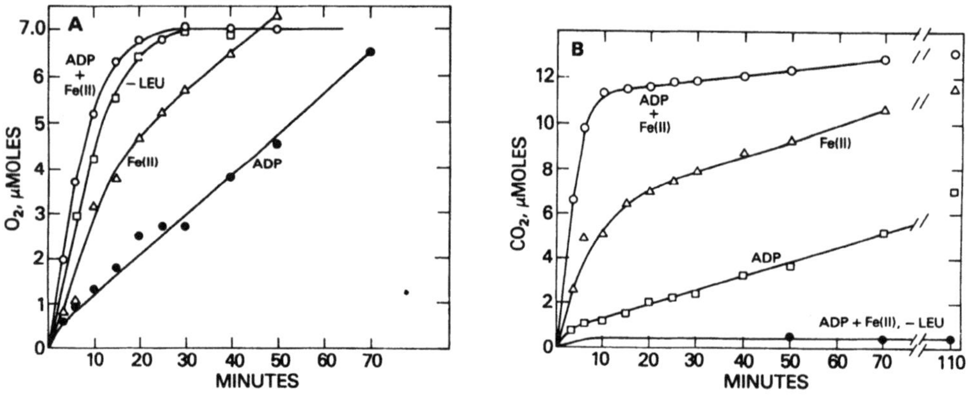

Figure 1. Manometric measurement of leucine oxidation and oxygen generation by Fenton reagents. The reaction mixture in each of two double side-arm Warburg flasks contained 23.5 mM $NaHCO_3$, and as indicated, 30 mM leucine, 100 μM $FeSO_4$, 250 μM ADP, and 30 mM H_2O_2; gas phase, 5% CO_2/95% N_2. Initially, the H_2O_2 was in one side arm of each vessel. After 15 min shaking, 37 °C, oxsorbant was introduced into the vented side arm of just one of the two vessels; after 5 min, the H_2O_2 was mixed with the other reactants. The change in pressure in the vessel without oxsorbant is due to both the evolution of CO_2 and the production of O_2; in the presence of oxsorbant the change in pressure is attributable to CO_2 only. The difference in pressure changes between the two flasks is a measure of O_2 production.

The Bicarbonate Effect

The requirement for HCO_3^- in the oxidation of leucine to carbonyl derivative (reactions 1 and 2) is illustrated by the data in Figure 2. Bicarbonate may exert its effect by facilitating the oxidation-reduction of Fe(II)/Fe(III) since bicarbonate stimulates the auto-oxidation of Fe(II) (Figure 3A) and the reduction of Fe(III) by hydroxylamine (Figure 3B). It is unlikely that the bicarbonate effect is due to the formation of peroxycarbonate or peroxybicarbonate derivatives since acetate which can also form a peroxy derivative has no effect on amino acid oxidation. The attractive possibility that bicarbonate exerts its effect via the formation of bicarbonate or CO_2 radicals remains to be investigated.

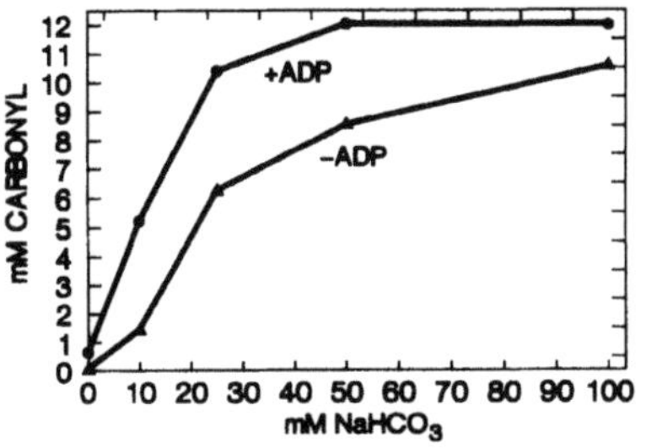

Figure 2. Effect of HCO_3^- and ADP on leucine oxidation. Reaction mixtures contained 100 mM Hepes buffer (pH 7.55), 50 mM leucine, 20 mM H_2O_2, 100 µM $FeSO_4$, and the indicated amounts of $NaHCO_3$, in the presence, (-●-●-) or absence (-▲-▲-) of 250 µM ADP. After 5 min at 25°C, the concentration of carbonyl compounds was measured.

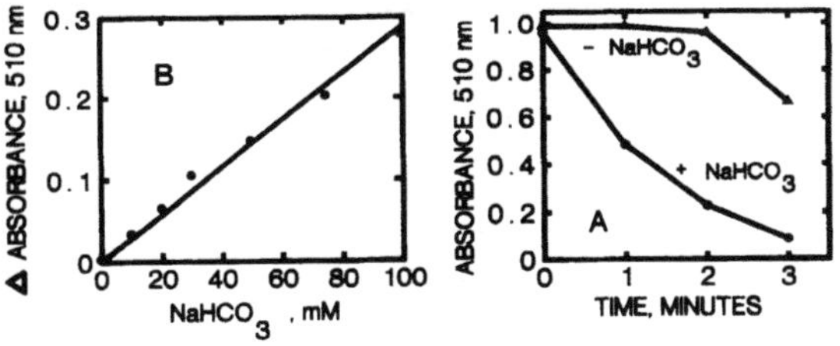

Figure 3. Effect of HCO_3^- on the auto-oxidation of Fe(II) (A) and reduction of Fe(III) by NH_2OH (B). A. Mixtures (0.5 ml, 37°C) contained 50 mM Hepes (pH 7.4), 32 mM $NaHCO_3$ (where indicated), and 50 µM $FeSO_4$. At times indicated, the concentration of Fe(II) was measured by reaction with 1,10-orthophenanthroline (absorbance at 510 nm). B. Mixtures contained 100 mM Hepes (7.4), 100 µM $FeCl_3$, 100 µM o-phenanthroline, and $NaHCO_3$ as indicated. Reaction started by addition of 100 µM NH_2OH. The ordinate indicates the change in absorbance at 510 mM after 1 min.

Amino acids react with HCO_3^- to form carbamino derivatives, which form stable salts with divalent cations. Accordingly, the iron salts of carbamino derivatives could be implicated in the bicarbonate effect.

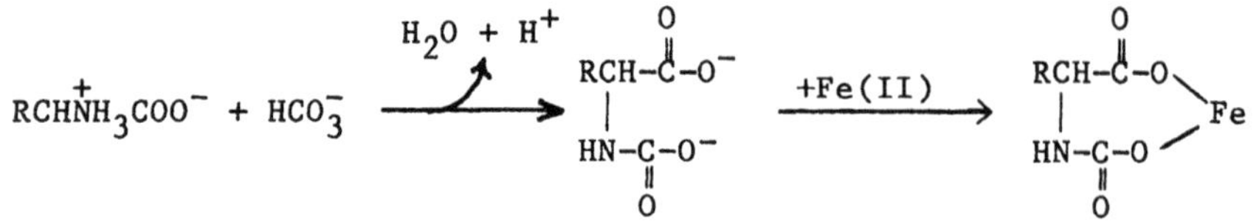

(7)

Results of preliminary spectrophotometric studies indicate that a ternary complex is formed between HCO_3^-, Fe(III) and leucine (unpublished data). The further interaction of such a complex with another chelator might alter the redox potential of iron in the quaternary complex and thereby facilitate site-specific peroxidation of the amino acid moiety of the complex; this would account for the failure of radical scavengers to inhibit the oxidation of amino acids (see below).

Effect of Iron Chelators

It is evident from the data in Figure 4 that chelators are able to stimulate the oxidation of leucine to carbonyl compounds when they are present at substoichiometric concentrations, but inhibit completely the oxidation when they are present at concentrations in excess of that required to chelate all of the iron present. A similar relationship between chelator concentration and other iron-dependent reactions has been reported previously.[13] Our results suggest that at least two forms of iron are required for maximal activity, i.e., the iron-chelate complex and in addition, either free iron or a second kind of iron complex (perhaps an iron amino acid bicarbonate complex).

When a chelator is in excess of the concentration needed to chelate all of the iron, amino acid oxidation is delayed. After a lag during which the excess amount of chelator is destroyed (unpublished data), amino acid oxidation occurs at a rate equal to that obtained in the absence of excess chelator (Figure 5). Curiously, in the absence of unchelated iron, peroxidation of excess chelator occurs to the complete exclusion of amino acid oxidation.

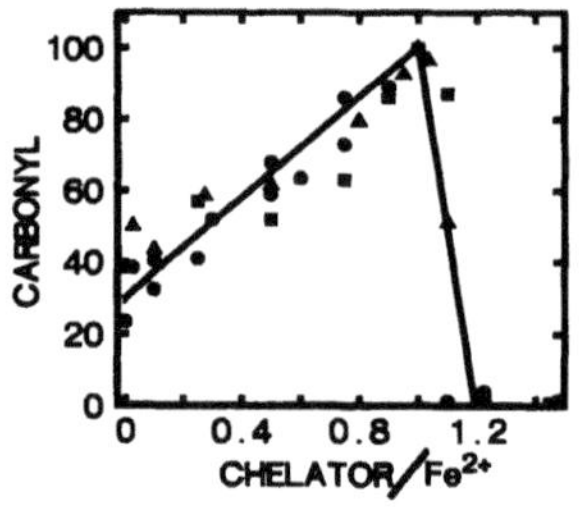

Figure 4. Effect of chelator/Fe(II) ratio on oxidation of leucine. Mixtures contained, initially, 23.5 mM $NaHCO_3$, 50 μM leucine, and chelator as indicated. After gassing with 95% N_2/5% CO_2, 100 μM $FeSO_4$ and then 30 mM H_2O_2 were added and the samples incubated 30 min at 37°C, then assayed for carbonyl compounds. To facilitate comparisons, data were normalized to peak activity = 100. Chelators were: Desferal, ▲, DETAPAC, ■ , EDTA, ●.

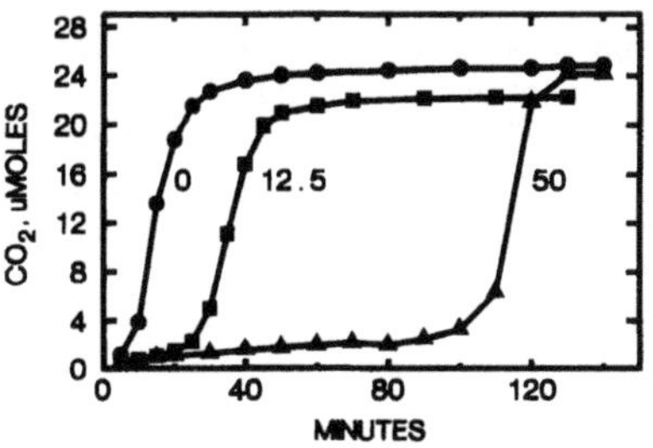

Figure 5. Effect of excess EDTA on oxidation of leucine. Conditions were as in Figure 1 except ADP was replaced with EDTA. The numbers on curves refer to the amount of EDTA that is in excess of the 100 μM Fe(II) added. Thus 0, 12.5, and 50 refer to EDTA concentrations of 100, 112, and 150 μM, respectively.

Effect of Radical Scavengers

Of several radical scavengers tested, mannitol, urea, formate, ethanol, α-tocopheral, and dimethylsulfoxide do not inhibit the oxidation of amino acids by the Fenton system. The insensitivity of oxidation to radical scavengers indicates either that free radicals are not involved or that the oxidation involves peroxidation of an amino acid-Fe(II)-chelate complex to yield active oxygen which reacts <u>in situ</u> with the amino acid moiety of the complex. The latter interpretation is supported by spectrophotometric studies showing that a highly oxidizable ternary complex between Fe(II), amino acid (AA), and ferrozine (FZ) is formed.

When Fe(II) is added to ferrozine (FZ), a colored complex, $(FZ)_3 \cdot Fe(II)$, is formed exhibiting an absortpion maximum at 560 nm (ε = 27.9 $mM^{-1}cm^{-1}$); this complex is resistant to oxidation by H_2O_2 (data not shown). When Fe(II) is added to a mixture of FZ and amino acid (AA), a ternary complex, $(FZ)_n \cdot Fe(II) \cdot AA$, ($\varepsilon_{630nm}$ = 15.17 $mM^{-1}cm^{-1}$) is formed. This complex is rapidly oxidized by H_2O_2 to colorless products. In the absence of H_2O_2, the peak absorbance at 630 nm decreases and the absorbance at 560 nm increases due to conversion of the ternary complex to the $(FZ)_3 \cdot Fe(II)$ complex in a pseudo first-order process which is nearly proportional to the amino acid concentration (unpublished data). These results are consistent with the following mechanism.

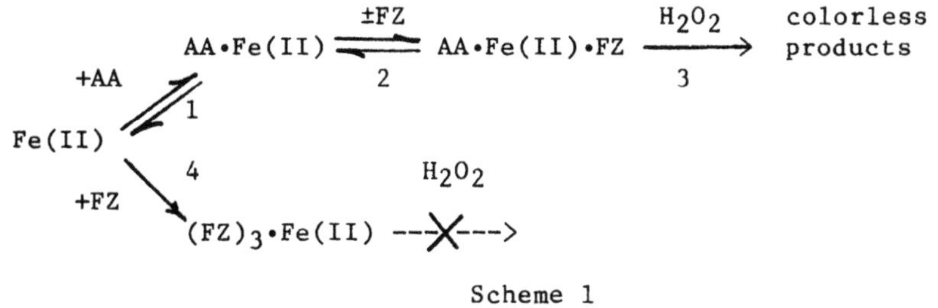

Scheme 1

At high concentrations of AA, reactions 1 and 2 occur much more rapidly than reaction 4 (Scheme 1); however, with time, the Fe(II) produced through reversibility of reactions 1 and 2 will react slowly with FZ (reaction 4, Scheme 1) to generate the thermodynamically more stable $(FZ)_3 \cdot Fe(II)$ complex. Qualitatively identical spectral changes are exhibited by all of the common amino acids. They differ only in the stability of the ternary complex. For example, the apparent first-order rate constants for the conversion of $AA \cdot F_Z$ to $(FZ)_2Fe(II)$ + AA are 0.05, 0.69, and 0.94 when AA = histidine, leucine, and glycine, respectively.

REFERENCES

1. L. Fucci, C. N. Oliver, M. J. Coon, and E. R. Stadtman, Inactivation of key metabolic enzymes by mixed-function oxidation reactions: Possible implication in protein turnover and ageing, Proc. Natl. Acad. Sci. U.S.A. 80:1521 (1983).
2. C. N. Oliver, R. L. Levine, and E. R. Stadtman, Regulation of glutamine synthetase degradation, in: "Experiences in Biochemical Perception," L. N. Ornston and S. G. Sligar, eds., Academic Press, New York (1982).
3. R. L. Levine, Oxidative modification of glutamine synthetase. II. Characterization of the ascorbate model system, J. Biol. Chem. 258:11828 (1983).
4. K. Kim, S. G. Rhee, and E. R. Stadtman, Nonenzymatic cleavage of proteins by reactive oxygen species generated by dithiothreitol and iron, J. Biol. Chem. 260:15394 (1985).
5. R. L. Levine, C. N. Oliver, R. M. Fulks, and E. R. Stadtman, Turnover of bacterial glutamine synthetase: Oxidative inactivation precedes proteolysis, Proc. Natl. Acad. Sci. U.S.A. 78:2120 (1981).
6. A. J. Rivett, Preferential degradation of the oxidatively modified form of glutamine synthetase by intracellular mammalian proteases, J. Biol. Chem. 260:300 (1985).
7. A. J. Rivett, J. E. Roseman, C. N. Oliver, R. L. Levine, and E. R. Stadtman, Covalent modification of proteins by mixed-function oxidation: Recognition by intracellular proteases, in: "Intracellular Protein Catabolism," E. A. Khairallah, J. S. Bond, and J. W. C. Bird, eds., Alan R. Liss, New York (1985).
8. C. N. Oliver, B. Ahn, M. E. Wittenberger, R. L. Levine, and E. R. Stadtman, Age-related alterations of enzymes may involve mixed-function oxidation reactions, in: "Modification of Proteins during Aging," R. C. Adelman and E. R. Dekker, eds., Alan R. Liss, New York (1985).
9. C. N. Oliver, B.-W. Ahn, E. J. Moerman, S. Goldstein, and E. R. Stadtman, Age-related changes in oxidized proteins, J. Biol. Chem. 262:5488 (1987).
10. C. N. Oliver, Inactivation of enzymes and oxidative modification of proteins by stimulated neutrophils, Arch. Biochem. Biophys. 253: 62 (1987).
11. C. N. Oliver, B. Ahn, M. E. Wittenberger, and E. R. Stadtman, Oxidative inactivation of enzymes: Implication in protein turnover and aging, in: "Cellular Regulation and Malignant Growth," S. Ebashi, ed., Japan Scientific Societies Press, Tokyo/Springer-Verlag, Berlin (1985).
12. I. Zs-Nagy and R. A. Floyd, Hydroxyl free radical reaction with amino acids and proteins studied by electron spin resonance spectroscopy and spin-trapping, Biochim. Biophys. Acta 790:238 (1984).
13. J. M. C. Gutteridge, R. Richmond, and B. Halliwell, Inhibition of the iron-catalyzed formation of hydroxyl radicals from superoxide and of lipid peroxidation by desferrioxamine, Biochem. J. 184:469 (1979).

IRON REDOX REACTIONS AND LIPID PEROXIDATION

Steven D. Aust

Biotechnology Center
Utah State University
Logan, Utah 84322-4430, U.S.A.

INTRODUCTION

The peroxidation of membrane polyunsaturated fatty acids is usually proposed to be initiated by the hydroxyl radical generated by the Haber-Weiss reaction, or a superoxide-driven Fenton reaction. However, it has been impossible to show that either liver microsomal NADPH-dependent lipid peroxidation or xanthine oxidase-dependent lipid peroxidation involves the hydroxyl radical, despite the fact that the hydroxyl radical is quite capable of promoting lipid peroxidation[1]. Catalase has no effect on microsomal lipid peroxidation[2] and xanthine oxidase-dependent lipid peroxidation is often stimulated by catalase[1]. Lipid peroxidation can be initiated simply by the addition of Fe(II). If the iron does not oxidize rapidly, lipid peroxidation only occurs after a lag period. The lag is not affected by either SOD or catalase but can be abolished by the addition of Fe(III) or oxidants that catalyze oxidation of a portion of the Fe(II). These and other data led us to propose that lipid peroxidation is catalyzed by some type of Fe(II):Fe(III) complex[3,4]. The purpose of this communication is to discuss the proposal that lipid peroxidation is indeed catalyzed by some type of Fe(II):Fe(III) complex in light of the redox chemistry or iron. Scheme 1 can be used to lead this discussion.

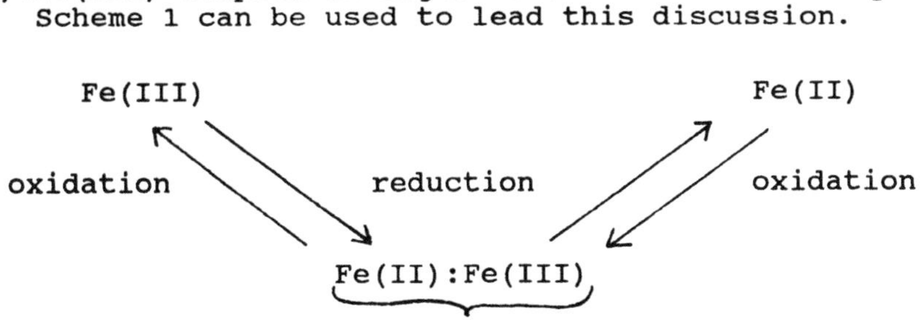

Scheme 1

The scheme shows that lipid peroxidation using Fe(III) requires reduction of the iron, whereas lipid peroxidation using Fe(II) requires its oxidation. The redox activity of iron is influenced by both pH and chelation. These may be important both thermodynamically and kinetically when considering rates of iron reduction or oxidation, especially by enzymes. Since both reduction and oxidation, including autoxidation, may occur simultaneously it will be important to consider not only the concentration of each species of iron but also the concentration of both reductants (plural) and oxidants (plural).

Fe(II)-DEPENDENT LIPID PEROXIDATION

According to the proposed mechanism, Fe(II)-dependent lipid peroxidation should eventually occur upon the autoxidation of some of the iron. Such is the case as demonstrated by the fact that the addition of ADP-chelated Fe(II) to liposomes results in lipid peroxidation after a lag period (Figure 1). The lag period is essentially unaffected by superoxide dismutase (SOD) or catalase, suggesting that oxidation by molecular oxygen (and lipid peroxides as they are formed) was responsible for the formation of the necessary Fe(III).

Oxidation of Fe(II) to provide the necessary Fe(III) can also be accomplished by H_2O_2 (Figure 2). Interestingly, the reaction of Fe(II) and H_2O_2 (Fenton

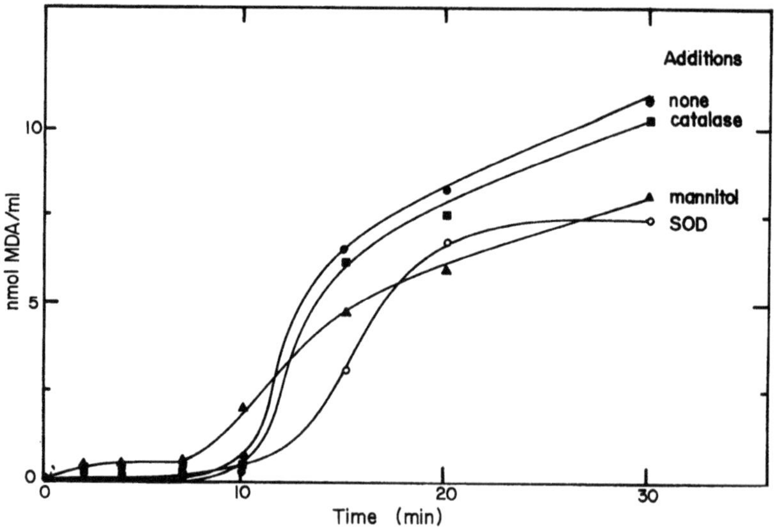

Figure 1. Fe(II) dependent peroxidation of phospholipid liposomes. Reaction mixtures (2.5 ml final volume) contained phospholipid liposomes (1 μmol lipid phosphate/ml), 1.7 mM ADP and 0.1 mM $FeCl_2$ in 30 mM NaCl, pH 7.0. Where indicated 500 mM mannitol, 100 u/ml SOD or 100 u/ml catalase were added. Aliquots were taken periodically and assayed for malondialdehyde (MDA) by the thiobarbituric acid assay.

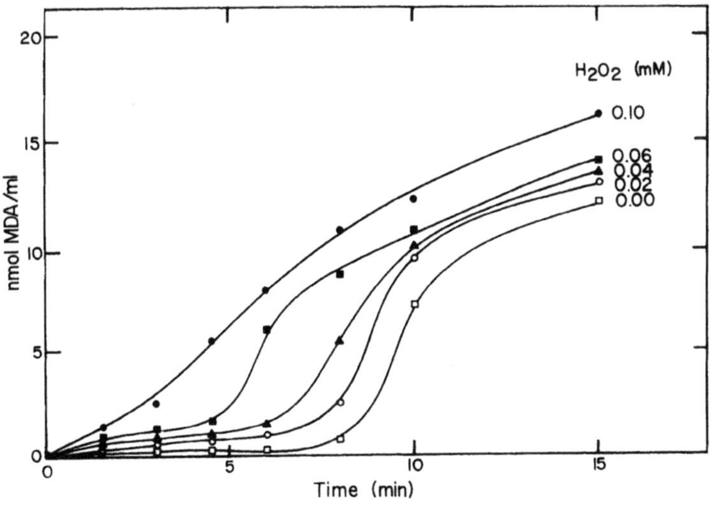

Figure 2. Effect of H_2O_2 on Fe(II) dependent peroxidation of phospholipid liposomes. The reactions were as described in the legend for Figure 1 except that H_2O_2 was added to the final concentration indicated.

reaction) also results in the production of the hydroxyl radical but we have never observed any correlation between rates of lipid peroxidation and hydroxyl radical production.

In general, we have found that the sensitivity of lipid peroxidation to excess oxidation of Fe(II) usually increases with decreasing concentration of iron. For example, when ferritin is the source of iron for lipid peroxidation, in which iron is slowly released by reduction by superoxide, for example, and never reaches more than micromolar concentrations, activity is dependent upon the addition of catalase. This is true for both xanthine oxidase[5] and paraquat-dependent[6] lipid peroxidation. The rate of iron oxidation approaches the rate of release of Fe(II) from ferritin such that the addition of catalase limits the rate of Fe(II) oxidation to give the appropriate Fe(II):Fe(III) ratio required for lipid peroxidation. Thus, with xanthine oxidase and ferritin it is necessary to add catalase in order to observe lipid peroxidation.

Fe(III)-DEPENDENT LIPID PEROXIDATION

Most model lipid peroxidation systems start with Fe(III) plus some type of reducing system, the most common being xanthine oxidase or liver microsomes. The former is a superoxide-dependent system, whereas the latter is a direct iron-reduction system. The former is therefore sensitive to SOD and is inhibited by H_2O_2. The latter involves NADPH-cytochrome P450 reductase and cytochrome P450. Since iron reduction is essentially direct, liver microsomal lipid peroxidation is insensitive to SOD and catalase, which is fortunate because microsomes are usually contaminated with these enzymes. Microsomes are also contaminated with ferritin, and in the absence of contaminating or added iron, or redox cycling chemicals such as paraquat, iron can be released by the small amounts of superoxide generated by microsomes if contamination by SOD has been eliminated[7]. The direct enzymatic reduction of iron is further complicated by the iron chelator because the enzymes may reduce some but not all chelates. Also the rate of iron reduction by the enzymes must be balanced against the rate of iron autoxidation, also influenced by the chelator. Thus ADP-chelated iron supports microsomal lipid peroxidation even though it is very slowly reduced by microsomes because it is also very slowly oxidized. EDTA, on the other hand, will inhibit microsomal lipid peroxidation when it chelates all the redox active iron because its high rate of autoxidation limits the accumulation of Fe(II). However, EDTA will stimulate microsomal lipid peroxidation if iron is present in excess of EDTA because EDTA-Fe(III) is an excellent electron acceptor for NADPH-cytochrome P450 reductase and EDTA-Fe(II) will reduce other chelates of iron, such as ADP-Fe(III), which readily promote lipid peroxidation.

The reactions of iron in xanthine oxidase-dependent lipid peroxidation are shown in scheme 2. The rate constants for each reaction are for unchelated iron, a condition that probably does not exist, but are included for the purpose of this discussion. In general superoxide is less discriminating with respect to the iron chelates it will reduce than are the mixed-function oxidases of microsomes and possibly other iron- reducing enzymes. One would expect that only electrostatic factors would be involved in the reduction of iron chelates by superoxide (given the appropriate redox potentials). However, it must be recognized that certain ferrous chelates are also oxidized by superoxide. This appears to be the case for citrate-Fe(II). Thus citrate-chelated iron has relatively good SOD-like activity[8] and it is almost impossible to promote lipid peroxidation with citrate-Fe(III)[9] making citrate a very good antioxidant with respect to lipid peroxidation.

Note that it is important when studying superoxide-dependent lipid peroxidation to also consider the oxidation

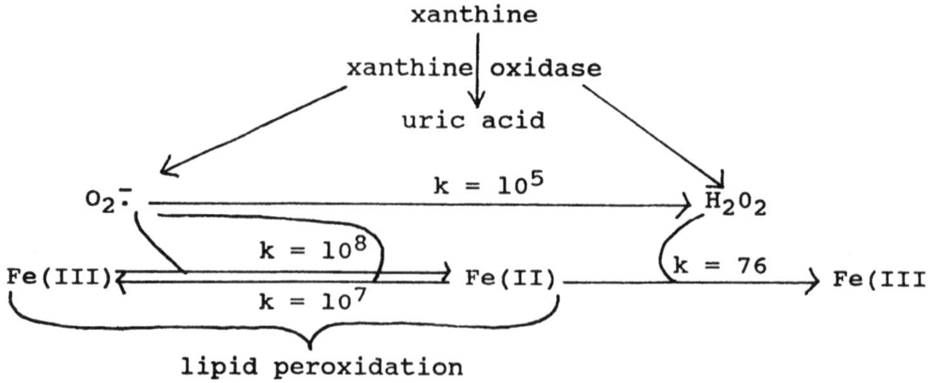

Scheme 2

of Fe(II) by H_2O_2. H_2O_2 is produced both by superoxide dismutation and directly by enzymes like xanthine oxidase. Although slow, the rate of Fe(II) oxidation by H_2O_2 can compete with the net rate of Fe(III) reduction by $O_2^{\cdot-}$ because the ratio of the rates of iron reduction to oxidation by $O_2^{\cdot-}$ is only 10. But, since these rates are dependent upon the iron chelator, it is necessary to determine the actual rate of iron reduction in the system being used. Since it is difficult to establish the individual rates of Fe(III) reduction versus Fe(II) oxidation by superoxide, we generally determine the "steady state" concentration of reduced iron. This is exemplified by the data in Table 1 which compares the rate of iron reduction and lipid peroxidation by citrate- versus ADP-chelated iron. This discovery led us to show that citrate-Fe(III) has SOD-like activity and that citrate-Fe(III) will rarely promote lipid peroxidation[8].

LIPID PEROXIDATION BY OTHER Fe(III) REDUCTANTS

It is also important to recognize that Fe(III) can be reduced by a variety of chemicals, including cellular reductants, and thereby promote lipid peroxidation. Unfortunately the reduction of iron by reductants and subsequent iron autoxidation is often mistakenly referred to as autoxidation, and frequently investigators attempt to show that the partially reduced species of oxygen, produced by this "autoxidation," are responsible for lipid peroxidation. Such is not the case. The direct reaction of molecular oxygen with unactivated organic compounds is spin forbidden. Instead chemicals such as ascorbic acid, glutathione, cysteine, dithiothreitol, etc. reduce iron, which subsequently autoxidizes. The reduced iron, in combination with the remaining Fe(III), catalyzes lipid peroxidation. A classic example of this is ascorbic acid, which is an excellent iron reductant. Ascorbic acid is known to be a prooxidant at low concentration and an antioxidant at higher concentrations. This concept can be rationalized if it is recognized that lipid peroxidation requires both Fe(II) and Fe(III) and at high concentrations of reductant, where the rate of Fe(III) reduction far exceeds the rate of Fe(II) oxidation, no lipid peroxidation will occur. Maximum rates of lipid peroxidation will occur when the rates of reduction and oxidation are balanced to provide the optimum ratio of Fe(II) to Fe(III)[10]. Thus it is important to consider all possible iron reductants and

Table 1. Iron Reduction and Lipid Peroxidation by Xanthine Oxidase and Citrate Versus ADP Chelated Iron

	Iron Reduction	Malondialdehyde
	nmoles/min/ml	
Citrate-Fe(III)	0.5	0.1
ADP-Fe(III)	6.1	1.2

oxidants when considering the final rate of lipid peroxidation.

EFFECT OF pH ON Fe(II) AUTOXIDATION AND LIPID PEROXIDATION

The final factor that needs to be discussed is the effects of pH because the rate of Fe(II) autoxidation increases with increasing pH. Therefore, the effects of pH on lipid peroxidation are complex. At low pH lipid peroxidation is frequently slow because most of the iron will be as Fe(II) and oxidants are required to provide some of the necessary Fe(III). At high pH lipid peroxidation can also be minimal because Fe(II) autoxidation converts all of the iron to the inactive Fe(III). This phenomenon is best exemplifed by starting with a Fe(II) chelate and studying its rate of autoxidation versus its ability to catalyze lipid peroxidation as a function of pH (Figure 3). At low pH Fe(II) autoxidation is extremely slow and no lipid peroxidation occurs. At high pH autoxidation is very rapid and again very little lipid peroxidation occurs. But at moderate rates of Fe(II) autoxidation, ie pH 7-8, lipid peroxidation reaches a maximum rate when just sufficient Fe(III) is produced to combine with Fe(II) to catalyze lipid peroxidation. These systems can be manipulated using various iron chelates and various reductants or oxidants to obtain almost any lipid peroxidation system so desired.

BIOLOGICAL SIGNIFICANCE

In biology, iron metabolism is tightly controlled, presumably because it is so redox active if "improperly" chelated. Obviously the iron in transferrin, lactoferrin,

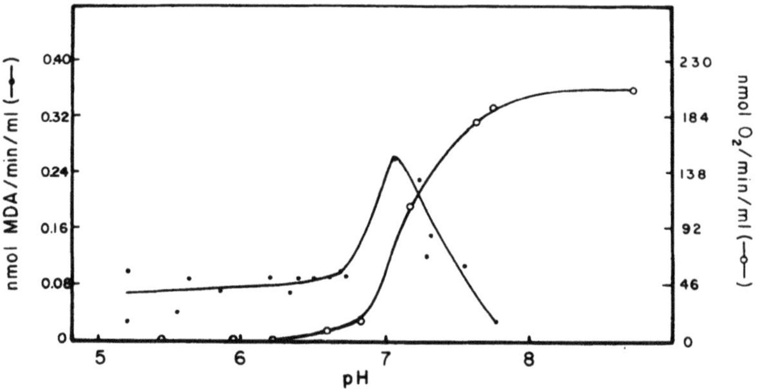

Figure 3. Effect of pH on AMP-Fe(II) autoxidation and peroxidation of phospholipid liposomes. Oxygen consumption, measured for AMP-Fe(II) (1.7:0.1 mM) at 37°C in 30 mM NaCl, was used to assess Fe(II) autoxidation at the indicated pH. Lipid peroxidation was assessed under similar conditions with 1 μmol phospholipid phosphate/ml by assaying for malondialdehyde (MDA) by the thiobarbituric acid assay.

or ferritin is inactive, although it can be released by superoxide or organic radicals. Thus perhaps in various toxicities or pathologies, iron is released from these proteins. Iron also is released from certain hemoproteins by a reaction with hydrogen peroxide or organic peroxides[11]. Obviously, chelation of this iron upon its release is of utmost interest because its could prevent tissue damage caused by lipid peroxidation.

The low molecular weight chelate or chelates of iron are also of interest. Although none has been identified, they have at least been observed[12-14]. They must represent a small but very important pool of iron. They presumably serve to transfer iron from ferritin to iron- requiring enzymes, such as ferrochelatase, aconitase, ribonucleotide reductase, prolyl hydroxylase, etc. Interestingly, although perhaps artifactual because the studies were obviously not conducted with the natural chelator, these systems are all saturated with very low concentrations of iron, and the iron must always be supplied (but not necessary active) in the ferrous form. It is interesting to note that a low molecular weight chelate of Fe(II), perhaps reduced by cellular reductants, should not catalyze lipid peroxidation if our hypothesis is correct, unless the reducing system was compromised (by loss of cellular glutathione, NADPH or ascorbate, for examples) or by excess oxidation (by H_2O_2, for example), situations that are frequently referred to as "oxidative stress." Thus, superoxide may be more deleterious because it is a potential source of an oxidant (H_2O_2), and we should be more concerned with H_2O_2 and other oxidants, or enzymes involved in the metabolism of these oxidants.

REFERENCES

1. M. Tien, B.A. Svingen, and S.D. Aust, An investigation into the role of hydroxyl radical in xanthine oxidase-dependent lipid peroxidation, Arch. Biochem. Biophys 216:142 (1982).
2. L.A. Morehouse, M. Tien, J.R. Bucher, and S.D. Aust, The effect of hydrogen peroxide on the initiation of microsomal lipid peroxidation, Biochem. Pharmacol. 32:123 (1983).
3. J.R. Bucher, M. Tien, and S.D. Aust, The requirement for ferric in the initiation of lipid peroxidation by chelated ferrous iron, Biochem. Biophys. Res. Commun. 111:777 (1983).
4. G. Minotti and S.D. Aust, The requirement for iron (III) in the initiation of lipid peroxidation by iron (II) and hydrogen peroxide, J. Biol. Chem. 262:1098 (1986).
5. C.E. Thomas, L.A. Morehouse, and S.D. Aust, Ferritin and superoxide-dependent lipid peroxidation, J. Biol. Chem. 260:3275 (1985).
6. M. Saito, C.E. Thomas, and S.D. Aust, Paraquat and ferritin-dependent lipid peroxidation, J. Free Rad. Biol. Med. 1:179 (1985).
7. C.E. Thomas and S.D. Aust, Rat liver microsomal NADPH-dependent release of iron from ferritin and lipid peroxidation, J. Free Rad. Biol. Med. 1:293 (1985).

8. G. Minotti and S.D. Aust, Superoxide-dependent redox cycling of citrate-Fe^{3+}: Evidence for a superoxide dismutase-like activity, Arch. Biochem. Biophys. 253:257 (1987).
9. T.C. Pederson, J.A. Buege, and S.D. Aust, Microsomal electron transport: The role of NADPH-cytochrome c reductase in liver microsomal lipid peroxidation, J. Biol. Chem. 248:7134 (1973).
10. M. Tien, J.R. Bucher, and S.D. Aust, Thiol-dependent lipid peroxidation, Biochem. Biophys. Res. Commun. 107:279 (1982).
11. J.M.C. Gutteridge, Iron promoters of the Fenton reaction and lipid peroxidation can be released from haemoglobin by peroxides, FEBS Letters 201:291 (1986).
12. R.L. Jones, C.M. Peterson, R.W. Grady, and A. Cerami, Low molecular weight iron-binding factor from mammalian tissue that potentiates bacterial growth, J. Exp. Med. 151:418 (1980).
13. R.L. Jones, R.W. Grady, M.P. Sorette, and A. Cerami, Host-associated iron transfer factor in normal humans and patients with transfusion siderosis, J. Lab. Clin. Med. 107:431 (1986).
14. A. van der Ende, A. du Maine, C.F. Simmons, A.L. Schwartz, and G.J. Straus, Iron metabolism in BeWo chorion carcinoma cells. Transferrin-mediated uptake and release of iron, J. Biol. Chem. 262:8910 (1987).

THE GENERATION OF FERRYL OR HYDROXYL RADICALS DURING INTERACTION OF HEMEPROTEINS WITH HYDROGEN PEROXIDE

Joseph Kanner, Stella Harel, and Avior Menahem Salan

Department of Science, Volcani Center
Bet Dagan, Israel

INTRODUCTION

Previous evidence for hydroxyl radical generation in ferric ion-derived Fenton reactions was presented by Walling et al.[1] and Gutteridge[2]. Methemoglobin (MetHb) and metmyoglobin (MetMb), the ferric states of these proteins, are activated by H_2O_2 producing a short-lived intermediate composed of one oxidizing equivalent on the heme, forming a ferryl, and one on the globin, i.e. giving an oxo-ferryl hemeglobin radical[3]. Studies on metmyoglobin-peroxide show that although it is not identical with compounds I or II or horseradish peroxidase (HRP), it has some structural features in common with both[4].

Recently, Sadrzadeh et al.[5] postulated that hemoglobin in the presence of superoxide anione or H_2O_2 promotes hydroxyl radicals formation and works as a biological Fenton reagent. This work was most recently criticized by Gutteridge[6] who suggested that iron released from hemoglobin during the interaction with H_2O_2 is the true generator of ·OH in model systems containing these reagents.

The present study was conducted to evaluate the generation of ferryl or ·OH during interaction of ferric hemeproteins with H_2O_2, using as a probe membranal lipid peroxidation and oxidation of 2-keto-4-thiomethyl butyric acid (KTBA) to ethylene and its inhibition by hydroxyl radical scavengers, antioxienzymes and chelators.

MATERIALS AND METHODS

All the reagents were of the highest purity commercially available. The hemeproteins and ascorbic acid were treated with Chelex-100. Isolation of the microsomal fraction from turkey dark muscle tissue was done by a procedure described by Apgar and Hultin[7]. Microsomal lipid peroxidation was done by a system containing glucose oxidase, 0.04 units, and glucose, 400 μM. The glucose oxidase system was adjusted for H_2O_2 generation of 1 nmol/ml/min at pH 7.0 and 25°C. Thiobarbituric acid-reactive compounds (TBARS) and proteins were determined as previously described[8]. Degradation of KTBA 1 mM (in 50 mM acetate buffer, pH 7.0) to ethylene was conducted in a glass tube closed with a serum cap and incubated in a shaking bath at 25°C. The ethylene production in the reaction was identified with the use of a 6-ft Poropak Q column and a flame ionization detector in a Packard gas chromatograph.

The results are means of triplicates of two separate experiments; in the figures each error bar I denotes the standard deviation.

RESULTS AND DISCUSSION

The results demonstrated that free ferric ions catalyzed KTBA oxidation most probably by hydroxyl radicals. Superoxide dismutase and ceruloplasmin, which totally failed to inhibit H_2O_2- activated hemeproteins catalysis, inhibit very efficiently the free ferric ion-derived Fenton reaction (Table 1). These results could be explained only if ferric ions during interaction with H_2O_2 produce superoxide and ferrous ions. Our data supported those accepted by Gutteridge[9] concerning oxidation of deoxyribose caused by H_2O_2 and ferric EDTA. The partial inhibition of KTBA oxidation by H_2O_2-activated hemeproteins with hydroxyl radical scavengers could be explained by the interaction of the hydroxyl radical scavengers with free radical intermediates of degradated KTBA[10]. MetMb degraded by a high concentration of H_2O_2 releases iron (results not shown) and it oxidized KTBA by a mechanism resembling that of the free iron.

Table 1. KTBA Oxidation by Several Catalysts Activated by H_2O_2 Generated Enzymatically[a]

Treatments[b]		$FeCl_3$		Ethylene μmol/30 min							
				MetMb[c] destroyed		MetMb		MetHb		HRP	
Control, glucose/glucose oxidase		0.99	(-)[d]	1.11	(-)	2.76	(-)	3.12	(-)	4.11	(-)
+ Mannitol	(100 mM)	0.22	(78)	0.40	(64)	1.72	(48)	2.38	(24)	2.92	(65)
+ Formate	(50 mM)	0.04	(96)	0.30	(73)	1.41	(49)	0.91	(71)	2.50	(40)
+ DMSO	(200 mM)	0.35	(65)	-	-	2.60	(6)	2.37	(24)	3.71	(10)
+ SOD	(200 U)	0.47	(57)	0.46	(58)	2.82	(-)	3.09	(-)	4.32	(-)
+ Ceruloplasmin	(10 U)	0.02	(98)	0.80	(28)	2.68	(3)	3.23	(-)	4.21	(-)
+ Catalase	(200 U)	0.25	(75)	0.20	(82)	0.34	(88)	0.54	(83)	1.60	(61)
+ EDTA	(33 μM)	0.40	(60)	0.62	(45)	2.70	(-)	3.20p	(-)	4.09	(-)
+ DETA	(33 μM)	0.14	(85)	0.78	(30)	2.65	(4)	2.65	(-)	4.10	(-)

[a]H_2O_2 was generated by glucose oxidase system at a rate of 1 nmol/ml/min at pH 7.0 and 25°C.
[b]The concentration of all catalyzers was 30 μM.
[c]MetMb (30 μM) was destroyed by 600 μM of H_2O_2 and incubated for 18 h, reaction was done with more 200 μM H_2O_2 without glucose oxidase.
[d]Percent inhibition.

EDTA is a known enhancer of the superoxide iron-derived Fenton reaction[9]. However, in several systems where the target molecule for oxidation was chelate iron (KTBA chelate iron[11]), EDTA was found to reduce the oxidation,[9,12], probably removing iron from site-specific areas to free solution (Figure 1). Membranal lipid peroxidation by ferric ions in the presence of ascorbic acid was inhibited by a low concentration of EDTA. As EDTA could not remove iron from hemeproteins, it is not surprising that it did not affect KTBA oxidation and membranal lipid peroxidation by H_2O_2-activated metmyoglobin and methemeglobin (Figure 2).

The iron in cytochrome C is covalently bound to the polypeptide chain by all six coordinations. Iron cytochrome C could interact with H_2O_2[13] and with hydroperoxides[14] without forming a complex as other hemeproteins do. We assume that KTBA oxidation by cytochrome C is generated by ·OH partially produced by iron bonded to the heme. Hydroxyl radicals generated in such a system will most probably interact in part with the heme pocket leading to its oxidative degradation and release of iron, as was actually found in our study (results not shown). Free iron was released from the molecule affected KTBA oxidation and this, in turn, was inhibited by EDTA (Figure 1).

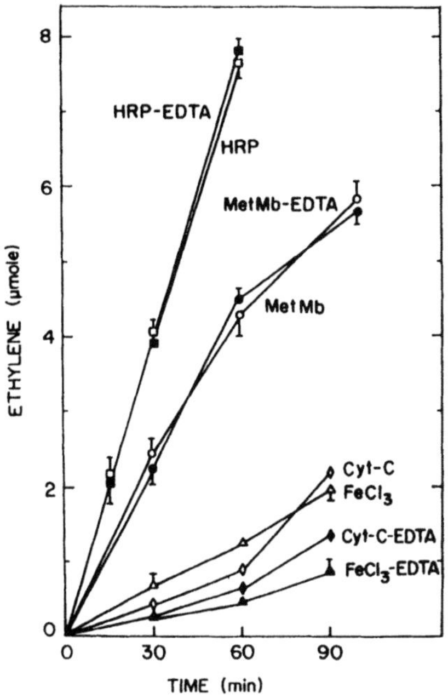

Figure 1. The rate of ethylene generation during KTBA oxidation by several hemeproteins activated by H_2O_2 produced enzymatically at a rate of 1 nmol/ml/min at pH 7.0, 25°C. MetMb or other reagents, 30 μM; EDTA, 33 μM.

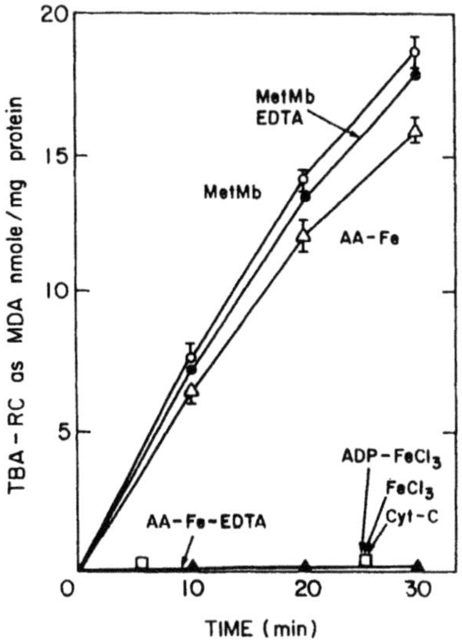

Figure 2. The rate of membranal lipid peroxidation by several catalyzers activated by H_2O_2 generated enzymatically as in Figure 1. MetMb, 30 μM; AA-Fe, ascorbic acid 200 μM and $FeCl_3$, 10 μM; ADP-$FeCl_3$, 33-30 μM; Cyc C, 30 μM; $FeCl_3$, 30 μM; EDTA, 33 μM.

CONCLUSIONS

Our results demonstrated that systems generated H_2O_2 at low concentrations of activated hemeproteins, such as myoglobin or hemoglobin, and this catalysed oxidation of target molecules mostly by ferryl species. At high H_2O_2 concentration, hemeproteins release iron and the oxidation of a target molecule is also affected by ·OH.

REFERENCES

1. C. Walling, Fenton's reaction revisited, Acc. Chem. Res. 8:125 (1975).
2. J. M. C. Gutteridge, Superoxide dismutase inhibits the superoxide-driven Fenton reaction at two different levels. Implications for a wider protective role, FEBS Lett. 185:
3. J. B. Fox, R. A. Nicholas, S. A. Ackerman, and C. E. Swift, A multiple wavelength analysis of the reaction between hydrogen peroxide and metmyoglobin. Biochemistry 13:5178 (1974).
4. M. Chance, L. Powers, C. Kumar, and B. Chance, X-ray absorption studies of myoglobin peroxide reveal functional differences between globins and heme enzymes, Biochemistry 12:1259 (1986).
5. S. M. H. Sadrzadeh, E. Graf, S. Scott Panter, P. E. Hollaway, and J. W. Eaton, Hemoglobin, biologic Fenton reagent. J. Biol. Chem. 259:14354 (1984).
6. J. M. C. Gutteridge, Iron promoters of the Fenton reaction and lipid peroxidation can be released from hemoglobin by peroxides, FEBS Lett. 201:291 (1986).
7. M. E. Apgar and H. O. Hultin, Lipid peroxidation in fish muscle microsomes in the frozen state, Cryobiology 19:154 (1982)
8. J. Kanner and S. Harel, Initiation of membranal lipid peroxidation by activated metmyoglobin and methemoglobin, Arch. Biochem. Biophys. 237:314 (1985).
9. J. M. C. Gutteridge, Reactivity of hydroxyl and hydroxyl-like radicals discriminated by release of thiobarbituric acid-reactive material from deoxysugars nucleosides and benzoate, Biochem. J. 224:761 (1984).
10. C. C. Winterbourn, The ability of scavengers to distinguish HO· production in the iron-catalyzed Haber-Weiss reaction: Composition of four assays for HO·, Free Rad. Biol. Med. 3:33 (1987).
11. G. D. Lawrence, Ethylene formation from methionine and its analogs. in: "Handbook of Methods for Oxygen Radical Research," R.A. Greenwald, ed, CRC Press Inc., Boca Raton, Florida (1985).
12. E. Graf, J. R. Mahoney, R. G. Bryant, and J. W. Eaton, Iron catalyzed hydroxyl radical formation. Stringent requirements for free iron coordination site, J. Biol. Chem. 259:3620 (1984).
13. P. L. Vandewalle and N. O. Petersen, Oxidation of reduced cytochrome C by hydrogen peroxide. FEBS Lett. 210:195 (1987).
14. A. L. Tappel, Oxidative fat rancidity in food products. I. Linoleate oxidation catalyzed by hemin, hemoglobin and cytochrome C, Food Res. 18:560 (1953).

THE PRODUCTION OF HYDROXYL AND SEMIQUINONE FREE RADICALS DURING THE AUTOXIDATION OF REDOX ACTIVE FLAVONOIDS

William F. Hodnick, B. Kalyanaraman,* Chris A. Pritsos and Ronald S. Pardini

Allie M. Lee Lab for Cancer Research and the Natural Products Lab
Department of Biochemistry
University of Nevada-Reno, Reno, Nevada 89557, U.S.A.

*National Biomedical Electron Spin Resonance Center
Department of Radiology
Medical College of Wisconsin
8701 Watertown Plank Road
Milwaukee, Wisconsin 53226, U.S.A.

INTRODUCTION

Catechols and other polyphenolic compounds have been shown to autoxidize and generate reactive oxygen species. A few examples of such compounds include epinephrine,[1] 6-hydroxydopamine,[2] and pyrogallol.[3] We have previously reported that members of a class of polyphenolics, the flavonoids (Figure 1), behave in a similar manner. In particular four flavonoids, myricetin (3,5,7,3',4',5'-hexahydroxyflavone), quercetagetin (3,5,6,7,3',4'-hexahydroxyflavone), delphinidin chloride (3,5,7,3',4',5'-hexahydroxyflavylium chloride) and quercetin 3,5,7,3'4'-pentahydroxyflavone) induce cyanide-insensitive respiration in isolated beef heart mitochondrial preparations. Additionally, these compounds autoxidize in aqueous solutions at pH 7.5, as determined by measuring oxygen consumption and oxygen-dependent spectral changes. It was demonstrated that both the respiratory bursts and the autoxidation result in the production of superoxide ($\bullet O_2^-$) and hydrogen peroxide (H_2O_2).[4]

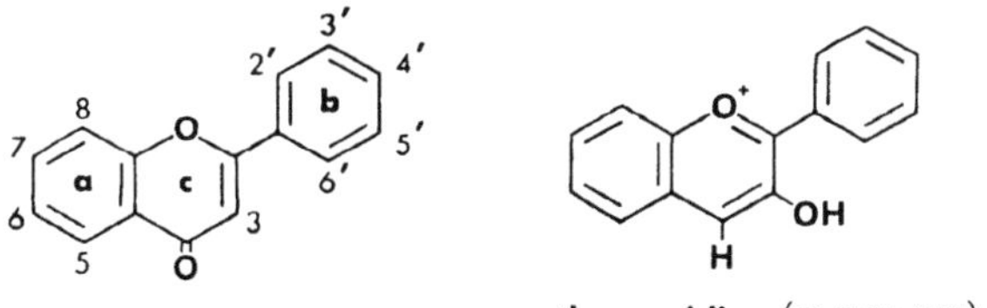

Figure 1. Structure and numbering system of flavonoids.

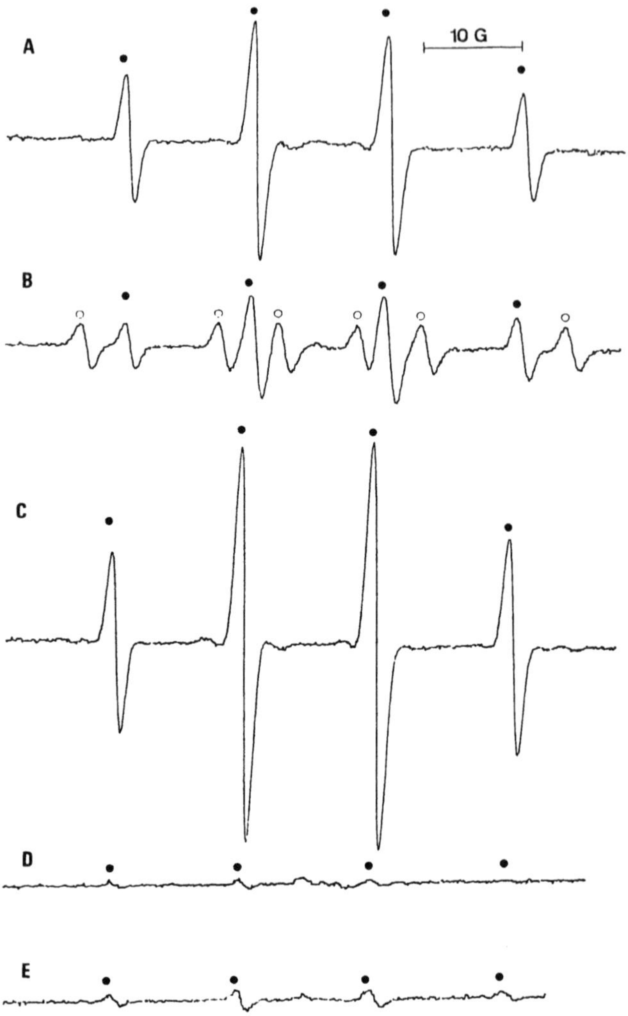

Figure 2. Production of hydroxyl radical during the autoxidation of quercetin in aqueous buffer (pH 7.5).

A) Quercetin + DMPO + Fe(III)-EDTA in air
B) A + EtOH
C) A + SOD or $FeCl_3$ (xs)
D) A + catalase
E) Quercetin + DMPO + Fe(III)-desferroximine

● - DMPO-OH
○ - DMPO-EtOH

RESULTS AND DISCUSSION

The presence of both $\cdot O_2^-$ and H_2O_2 suggests that hydroxyl radical ($\cdot$OH) could be formed via the metal-ion-catalyzed Haber-Weiss reaction. The production of $\cdot$OH was monitored using electron spin resonance (ESR) and spin trapping techniques. As exemplified by quercetin (Figure 2), in the presence of ethylenediaminetetraacetic acid (EDTA) and 5,5,-dimethyl-1-pyrroline-1-oxide (DMPO), all four of the above-mentioned flavonoids gave the ESR signal for the DMPO-OH spin adduct and the DMPO-EtOH adduct in the presence of excess ethanol, indicating the production of •OH. The addition of $FeCl_3$ or superoxide dismutase (SOD) to the reaction mixture resulted in a dramatic increase in the DMPO-OH signal, while catalase completely abrogated it. Without chelator (EDTA) there was no signal, and the presence of diethylenetriaminepentaacetic acid (DETAPAC) or desferroximine greatly diminished the signal demonstrating the need for a redox-active chelate. The addition of Fe(III)-EDTA to flavonoid solutions under anaereobic conditions produced time-dependent auxochromic shifts in their absorption spectra and resulted in the reduction of the Fe(III) to Fe(II)-EDTA. These data suggest that the flavonoids autoxidize to produce $\cdot O_2^-$, which dismutates to produce H_2O_2. In the presence of Fe(III)-EDTA the flavonoid can directly reduce the Fe(III) to Fe(II), resulting in the production of •OH through Fenton chemistry.

o-Semiquinone radicals are intermediates of the one-electron oxidation of catechols and other related compounds. Therefore, it seemed likely that such species would be produced by the autoxidation of flavonoids. However, at acidic and neutral pH, o-semiquinones and related species formed from catechols are fairly unstable. The steady-state radical concentration can be considerably enhanced by the addition of di- or trivalent metal ions which complex with the free radicals, a procedure termed spin stabilization.[5] Therefore, using Mg^{2+} as the stabilizing metal ion, semiquinone species have been detected using ESR-spin stabilization techniques (Figure 3).

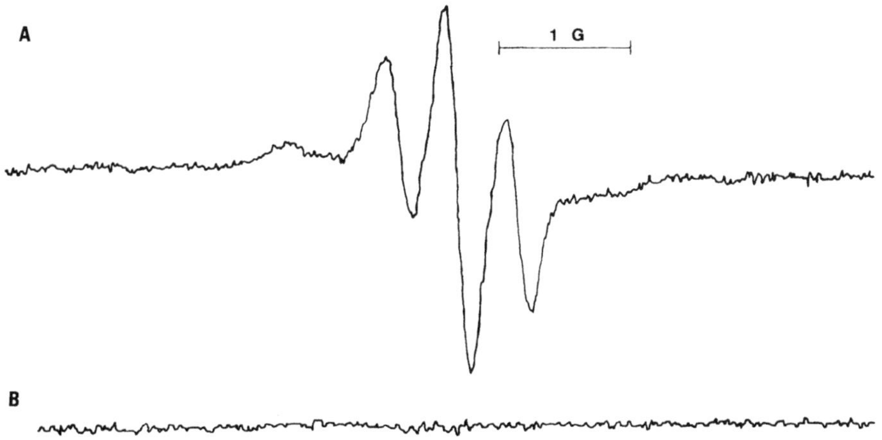

Figure 3. ESR spectra of spin-stablilized o-semiquinone species produced during the autoxidation of myricetin in aqueous buffer (pH 7.8).

A) Myricetin + Mg^{2+}
B) without Magnesium

The autoxidation of flavonoids has previously been reported but, in general, has been shown to require the presence of several hydroxyl groups and to proceed only in alkaline solution producing both quinoid species and trace amounts of H_2O_2.[6] An exception is that reported by Hathway and Seakins of the autoxidation of flavans related to catechin at pH 8.[7] The four flavonoids discussed herein underwent autoxidation and interacted with isolated mitochondria to produce radicals at a biologically significant pH, which suggests the potential to induce oxidative stress in biological systems. Flavonoids have been reported to exhibit cytotoxic and antineoplastic activity.[8] More recently a synthetic flavonoid derivative, flavone acetic acid (FAA), has been reported to possess antitumor activity against several solid tumors in mice.[9] Flavonoids have also been shown to be toxic to insects and as secondary plant metabolites to serve as defensive chemicals (allelochemicals) against herbivorous insects.[10,11] The production of these oxy- and o-semiquinone radicals may contribute to the cytotoxic and allelopathic actions as well as several other reported biological activities of the flavonoids. Consistent with this hypothesis, flavonoids were found to be cytotoxic in a clonogenic assay of cultured EMT6 mouse mammary carcinoma cells under aerobic conditions but essentially were nontoxic to hypoxic cells, demonstrating an oxygen requirement for their cytotoxic action.

REFERENCES

1. H.P. Misra and I. Fridovich, The role of superoxide anion in the autoxidation of epinephrine and a simple assay for superoxide dismutase, J. Biol. Chem. 247:3170 (1972).
2. R.E. Heikkila and G. Cohen, 6-Hydroxydopamine: Evidence for superoxide radical as an oxidative intermediate, Science 181:456 (1973).
3. S. Marklund and G. Marklund, Involvement of superoxide anion radical in the autoxidation of pyrogallol and a convenient assay for superoxide dismutase, Eur. J. Biochem. 47:469 (1974).
4. W.F. Hodnick, F.S. Kung, W.J. Roettger, C.W. Bohmont and R.S. Pardini, Inhibition of mitochondrial respiration and production of toxic oxygen radicals by flavonoids: A structure-activity study, Biochem. Pharmacol. 35:2345 (1986).
5. B. Kalyanaraman, C.C. Felix, and R.C. Sealy, Electron spin resonance-spin stabilization of semiquinones produced during oxidation of epinephrine and its analogues, J. Biol. Chem. 259:54 (1984).
6. O.R. Gottlieb, Flavonols, in: "The Flavonoids" Part 1, J.B. Harborne, T.J. Mabry, and H. Mabry, eds., Academic Press, New York (1975).
7. D.E. Hathway and J.W.T. Seakins, Autoxidation of polyphenols. Part III. Autoxidation in neutral aqueous solution of flavans related to catechin, J. Chem. Soc. p. 1562 (1957).
8. J.P. Brown, A review of the genetic effects of naturally occurring flavonoids, anthraquinones and related compounds, Mutation Res. 75:243 (1980).
9. T. Corbett, M. Bissery, A. Wozniak, J. Plowman, L. Polin, E. Tapazoglou, J. Dieckman, and F. Valeriote, Solid tumor activity of flavone acetic acid (FAA), Proc. Am. Assoc. Cancer Res. 27:281 (Abstract 1116) (1986).
10. P.A. Hedin and S.K. Waage, Roles of flavonoids in plant resistance to insects, in: "Plant Flavonoids in Biology and Medicine: Biochemical, Pharmacological and Structure-Activity Relationships," V. Cody, E. Middleton, Jr., and J.B. Harborne, eds., Prog. Clin. Biol. Res. 213, Alan R. Liss, Inc., New York, p. 87 (1986).
11. S. Ahmad, C.A. Pritsos, S.M. Bowen, K.E. Kirkland, G.T. Blomquist, and R.S. Pardini, Activities of enzymes that detoxify superoxide anion and related toxic oxy radicals in Trichoplusia ni, Arch. Insect Biochem. Physiol. 6:85 (1987).

FENTON REACTIONS IN LIPID PHASES:

IRON SOLUBILITY, ·OH PRODUCTION, AND INITIATION OF LIPID OXIDATION

K.M. Schaich and D.C. Borg

Medical Department
Brookhaven National Laboratory
Upton, New York 11973, U.S.A.

INTRODUCTION

There has been much discussion in this meeting regarding the involvement of hydroxyl radicals (·OH), oxidizing lipids, and metal ions in oxidative damage in living tissue and in food systems. However, the causal relationships linking these three species to each other and to specific damage in cells have not yet been clarified. Indeed, the literature is replete with contradictory claims regarding their activity and effects.

We feel that most studies seeking to determine the mechanistic interrelationships of these agents have missed a critical site of action, namely, the membranes. Because ·OH are extraordinarily reactive, they do not survive for more than a few collisions following their formation, and thus their reactions are intrinsically site specific, i.e., characterized by the potential substrates for ·OH oxidation that are within but ten or a few tens of angstroms of where the ·OH are produced.[1,2] This implies that Fenton-generated ·OH must react at the sites where iron is bound, and for effective initiation of lipid oxidation this means on the membrane surface or within the lipid bilayer. Fenton reactions forming ·OH from iron and H_2O_2 in the suspending medium are ineffective or may even be protective if the concentrations of H_2O_2 or of reactive reducing equivalents are rate limiting.

Although there is substantial inferential evidence to support the lipid phases of membranes as the site for both Fenton-type ·OH formation in the primordial initiation of lipid oxidation and for subsequent reactions of lipid oxyl radicals with proteins and DNA, there is little direct experimental evidence for this notion. Hence, we reasoned that it was important to determine experimentally whether Fenton-like formation of ·OH can occur in a lipid phase.

Simple model systems consisting of bulk lipids with Fenton reactants (H_2O_2, Fe^{2+} or Fe^{3+}, and a reducing agent) partitioned into them have been used to study various aspects of Fenton reactions in lipid phases. Although bulk lipids are weak surrogates for structurally organized membranes, they do provide experimental access to environments with hydrophobicity comparable to membrane interiors, and we expect that comparisons of solubilities and binding to different classes may provide some insight regarding metal localization in membranes.

RESULTS AND DISCUSSION

Our present research using this technique to measure solubilities of Fenton reactants shows that H_2O_2 partitioning into bulk lipid depends on H_2O_2 concentrations in the aqueous phase and that greater than micromolar levels are achievable in oleic acid. Partitioning of millimolar solutions of Fe^{2+} and Fe^{3+} complexes and chelates into oleic acid and methyl linoleate, quantified by atomic absorption spectroscopy (graphite furnace with deuterium arc background correction), indicated solubilities of about micromolar (ppm) in oleic acid and substantially lower (ppb) in methyl linoleate (Table 1).[3] Although systematic studies of iron distribution as a function of iron concentration have yet to be done, it is likely the data represent saturation values rather than partition coefficients. Optical absorption spectra indicate that iron ADP and iron desferrioxamine (i.e., ferrioxamine) transfer as chelates to oleic acid. EDTA and DTPA chelates have very low solubility in these lipids.

Table 1. Partitioning of Iron into Oleic Acid and Methyl Linoleate

	Concentrations (μM) partitioned into			
	Oleic Acid		Methyl Linoleate	
Iron complex	Ferric	Ferrous	Ferric	Ferrous
Ammonium Sulfate	5.90[a]	6.00	0.28	0.30
Sulfate	3.50	3.50	-[b]	-
Chloride	3.50	3.50	-	-
Citrate	0.09	3.40	-	-
ADP[c]	>18	6.30	0.64	0.19
EDTA[c]	0.40	0.54	0.02	~0[d]
DTPA[c]	0.11	~0	~0	~0
DFO[c]	>18	-	0.56	-

[a] Averages of four or more determinations

[b] Not determined

[c] Prepared with iron ammonium sulfate salts

[d] Too low for accurate determination

To determine whether Fenton reactions actually occur in lipid phases, we separately partitioned H_2O_2, iron compounds, and hydroxylamine (as a lipid-soluble reducing agent when necessary) into neat fatty acids or esters. All solutions and lipids were sparged and then maintained under argon. To start the reactions, aliquots from each of the lipid phases were mixed, the spin-trap DMPO (5,5-dimethyl-1-pyrroline-N-oxide) was added, and radical production was monitored by electron paramagnetic resonance (EPR). When Fenton reactions, either with Fe^{2+} or with Fe^{3+} and a reducing agent, were run in protic oleic acid, the well-known four-line signal of the ·OH adduct was produced with DMPO, although the nitrogen and hydrogen hyperfine splitting constants were not equal as they are in water (Figure 1).[1,3] If oleic acid is incubated with iron before DMPO is added, the peroxyl radical is trapped (Figure 1), and if the Fenton reaction or iron incubation is allowed to proceed for some time, complex spectra are obtained with alkyl, alkoxyl, and peroxyl radical adducts. Thus the reaction system behaves as would be expected from classical chemistry of lipid oxidation.[1,4]

Monitoring the changes in ·OH adduct concentrations with time by EPR showed that the form of iron complexation affected the efficiency of Fenton

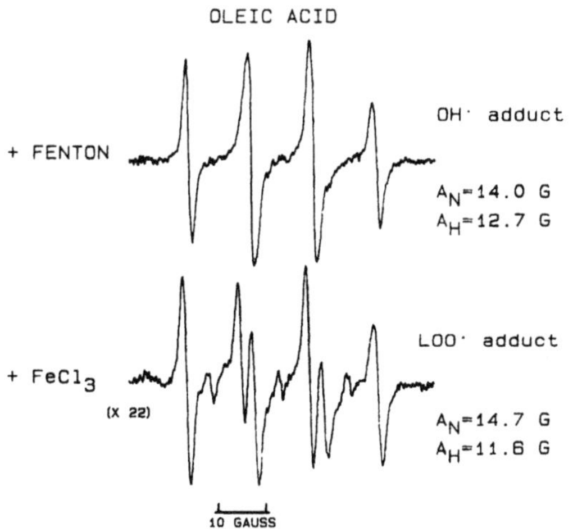

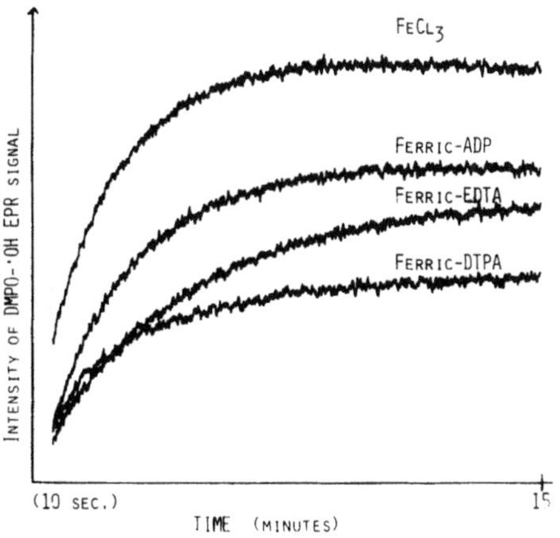

Figure 1 (left). EPR spectra of DMPO adducts in oleic acid. Top, ·OH adduct formed in Fenton reaction; bottom, lipid peroxyl (LOO·) adduct formed in iron-initiated autoxidation of oleic acid.

Figure 2 (right). ·OH adduct formation in lipid-phase Fenton reactions driven by various iron chelates. Reactants: ferric compound, H_2O_2, hydroxylamine, and DMPO in oleic acid.

reactions in lipid phases (Figure 2). ADP chelates supported Fenton kinetics in oleic acid at rates approaching that of $FeCl_3$, while EDTA and DTPA chelates reacted much more slowly and provided lower total adduct concentrations. There are at least three factors affecting these observed kinetics: 1) the absolute concentrations of chelates partitioned into the oleic acid are not the same, 2) EDTA and DTPA are known to react rapidly with ·OH and may have competed with DMPO for the ·OH as they were formed close to the chelates, or 3) the redox characteristics of the three chelates may be different in oleic acid than in water.

When the reactions were run in aprotic lauric, oleic, or linoleic methyl esters instead of the more protic acids, DMPO gave signals consisting almost entirely of lipid peroxyl radical adducts, with weak, if any, contributions from hydroxyl radicals.[1,3] Especially with linoleate, the signals changed with time,[3] indicating changes of radical species and varying proportions of different radicals (cf. Figure 3). These results suggest either that the ·OH adduct is not stable in aprotic media or that ·OH reactions with aprotic lipids are much faster than those with protic lipids, and they also raise a more general question: What are the preferred targets for reaction with ·OH in membranes, where most fatty acids are present as esters?

Have these results answered the question of whether Fenton reactions occur within lipid phases and initiate lipid oxidation? Clearly the chemistry occurs: all the makings for compartmentalized Fenton reactions are present in the model systems, and ·OH is produced. Oxidized species of lipid also can be spin trapped, indicating that lipid oxidation occurs, and changing envelopes of spin-trapped lipid species over time strongly suggest that the characteristic chain reactions of lipid peroxidation have been established with the traces of oxygen that remain. Lipid diene conjugation and peroxide values confirm initiation of autoxidation (unpublished data).

Hence intramembranous formation of ·OH also appears feasible in vivo provided that appropriately complexed nonheme iron is available and endogenous reducing sources can replace the hydroxylamine used in the models.

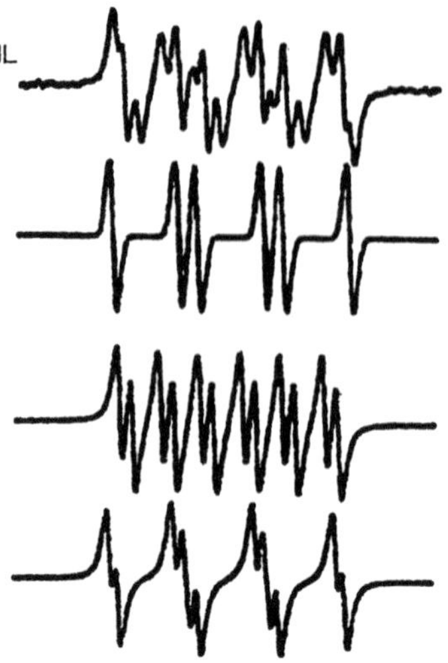

Figure 3. Typical complex EPR spectrum of free radical adducts formed during Fenton reactions in methyl linoleate (ML). A, B, and C are components of this spectrum, determined by simulations and other computer analyses. A: A_N=13.4, A_H=10.4 G (species not identified); B: A_N=12.82, A_H=6.65, A_H=2.2 G (LO· adduct); and C: A_N=12.62, A_H=10.27, A_H=1.47 G (LOO· adduct).

However, whether chain propagation through lipid bilayers or free radical damage to embedded macromolecules will predominate in biological membranes depends upon the character of the nearby lipid and upon the microenvironment of the redox-active metal's binding site.

REFERENCES

1. D. C. Borg and K. M. Schaich, Cytotoxicity from coupled redox cycling of autoxidizing xenobiotics and metals, Isr. J. Chem. 24:38 (1984).

2. S. Goldstein and G. Czapski, The role and mechanism of metal ions and their complexes in enhancing damage in biological systems or in protecting these systems from the toxicity of $O_2^{\cdot-}$, J. Free Rad. Biol. Med. 2:3 (1986).

3. K. M. Schaich and D. C. Borg, Fenton reactions in lipid phases, Lipids 23:in press (1988).

4. D. C. Borg, K. M. Schaich, and A. Forman, Autoxidative cytotoxicity: Is there metal-independent formation of hydroxyl radicals? Are there "crypto-hydroxyl" radicals?, in: "Oxygen Radicals in Chemistry and Biology," W. Bors, M. Saran, and D. Tait, eds., Walter de Gruyter, Berlin (1984).

THE ROLE OF HEMOGLOBIN IN GENERATING OXYRADICALS

Joseph M. Rifkind, Lu Zhang, Jane M. Heim, and Abraham Levy

Laboratory of Cellular and Molecular Biology
Gerontology Research Center
National Institute on Aging
National Institutes of Health
Baltimore, Maryland 21224, U.S.A.

INTRODUCTION

The largest fraction of dioxygen in mammalian systems is associated with hemoglobin. The primary function of hemoglobin is the transport of oxygen from the lungs to the tissues, which requires reversible oxygen binding. Nevertheless, a slow rate of autoxidation does take place. It has furthermore been demonstrated that autoxidation coincides with the formation of superoxide.[1,2]

$$HbO_2 \rightarrow Hb^+ + \dot{O}_2^- \qquad (1)$$

Lynch and Fridovich[3] have found that superoxide is readily transported through the erythrocyte anion channel.[4] In these experiments[4] superoxide dismutase (SOD), either inside or outside of resealed ghosts, readily reacted with superoxide formed within the sealed ghosts. It was therefore assumed that no oxyradicals would escape out of intact erythrocytes which contain high levels of superoxide dismutase, catalase and glutathione peroxidase. Recent studies have, however, shown that superoxide does leak out of sickle erythrocytes[5] where there is evidence for enhanced oxidative instability. In these experiments the level of superoxide which was found to leak out of normal erythrocytes was not significant. However, these results suggest that under conditions of enhanced oxidative stress, superoxide may be generated even by normal erythrocytes (Figure 1).

Although low oxygen pressure is expected to lower the concentration of oxyradicals generated by neutrophils[6] or mitochondra,[7] there is evidence that reduced oxygen pressure enhances autoxidation of hemoglobin[8,9] and may therefore produce enhanced levels of superoxide within erythrocytes. In this paper we have explored the possibility that hypoxia acts as a source of oxidative stress in the erythrocyte. It is shown that this oxidative stress results in the leakage of superoxide out of the erythrocyte. The significance of this finding in explaining tissue damage under hypoxic conditions is discussed.

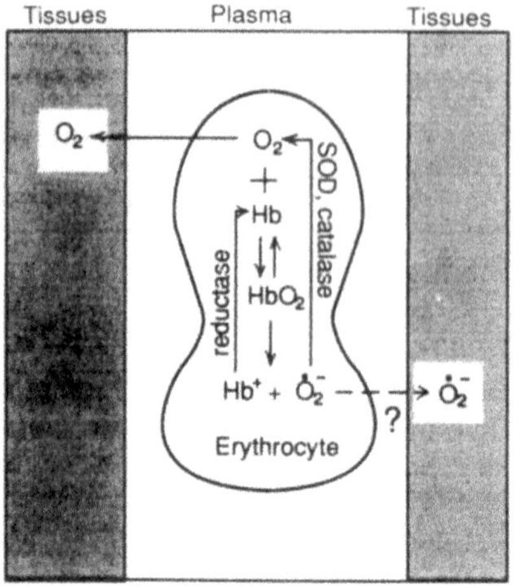

Figure 1. A schematic presentation of the erythrocyte showing how superoxide may leak out of the cell under oxidative stress initiating a process which results in oxyradical tissue damage.

METHODS

Erythrocytes were obtained from fresh human blood washed in isotonic media to remove the plasma, buffy coat and any traces of hemoglobin. In order to measure the leakage of superoxide out of the cells, the washed erythrocytes were suspended in PBS pH 7.4, which contained 1000 U/ml of catalase and 5×10^{-4} M cytochrome C. The catalase was added to eliminate any effect associated with the possible formation of H_2O_2. The suspended erythrocytes were incubated at 37°C in a Tonometer (Model 237) from Instrumentation Laboratory while being equilibrated with gas at specified partial pressures of oxygen. At various time intervals an aliquot was removed and the cells spun down. The reduction[10] of cytochrome C in the suspending medium was monitored by using HPLC to separate and quantitate the various components present in the supernatant. The separation was performed on a cation exchange column (poly Cat A obtained from poly LC, Columbia, MD) using a 30 minute (20-90%) slightly concave gradient (#7 on the Waters Model 660 Solvent programmer) with Buffer A:0.04 M Bis Tris pH 6.5 and Buffer B:0.04 M Bis Tris pH 6.8, 0.6 M NaCl. Under these conditions it was possible to separate reduced cytochrome C and oxidized cytochrome C as well as hemoglobin. Since oxyhemoglobin slowly reduces[11] cytochrome C, samples which indicated more than a negligible level of hemoglobin were discarded. No reduction was found when cytochrome C was incubated in the same medium without erythrocytes. For oxidation studies samples of hemolyzed cells were incubated in the same way as the cells. The cells were hemolyzed with 0.1% Triton X and stored at 4°C. The hemolyzed cells were discarded after 2 days or earlier if the level of oxidation in the stored cells increased by 1%. The oxidation was monitored by determining the ratio of the absorbance at 500 nm and 576 nm, where methemoglobin and oxyhemoglobin, respectively, have absorbtion maxima in the visible. By using computer-stored spectra with absorbance readings of $\pm$ 0.001 it was possible to determine the oxidation with a precision of $\pm$ 0.2%.

RESULTS

The leakage of superoxide from intact erythrocytes was measured by following the reduction of cytochrome C incubated at 37°C (Figure 2). In these incubations the hemoglobin concentration is 50-fold higher than that of the cytochrome C. The 5-10% reduction of cytochrome C detected per hour therefore corresponds approximately to one molecule of cytochrome C reduced per 1000 hemoglobin molecules. The reduction of cytochrome C in this system can be attributed to leakage of either superoxide or other reducing agents which can leak out of the erythrocyte without lysis. In Figure 2 the same erythrocyte suspension was used for the 150 mm Hg and the 3.5 mm Hg experiments. Therefore, the increased rate of reduction at low oxygen pressure in Figure 2 demonstrates that the rate of reduction is influenced by an oxygen dependent process. Figure 3 with 1000 μ/ml SOD on the outside of the cell shows that this oxygen pressure dependence is eliminated in the presence of SOD. The enhanced rate of cytochrome C reduction at 3.5 mm Hg pO_2 is thus attributed to enhanced leakage of superoxide out of the cell. From a comparison of the slopes at a value of 3.5 mm Hg and 150 mm Hg in Figure 2, an enhanced leakage of 13 μM of superoxide/per hour can be determined.

The primary source for superoxide radicals within the erythrocyte is the autoxidation of hemoglobin. We, therefore, studied the rate of oxidation of hemoglobin in hemolyzed cells as a function of oxygen pressure (Figure 4). These results indicate low rates of oxidation when samples are equilibrated with pure nitrogen (0 mm Hg) or 150 mm Hg and higher value at intermediate oxygen pressures. The maximum rate of oxidation is found in the region of 2 mm Hg. The range for the rates of

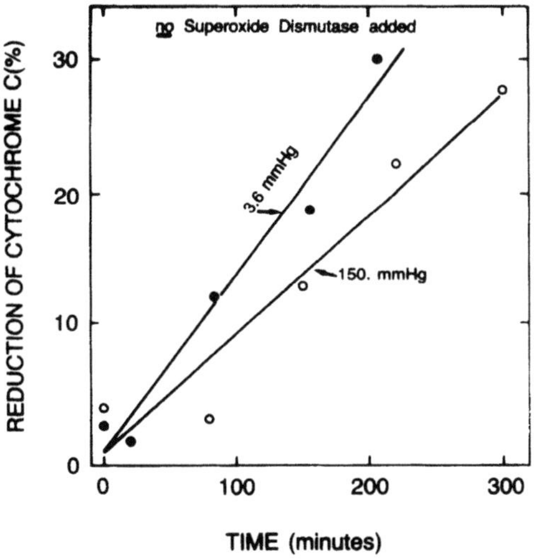

Figure 2. The reduction of cytochrome C during a 37°C incubation with intact erythrocytes; 1000 μ/ml catalase added to the medium. Oxygen pressure during incubation: (0), 150 mm Hg; (•), 3.6 mm Hg.

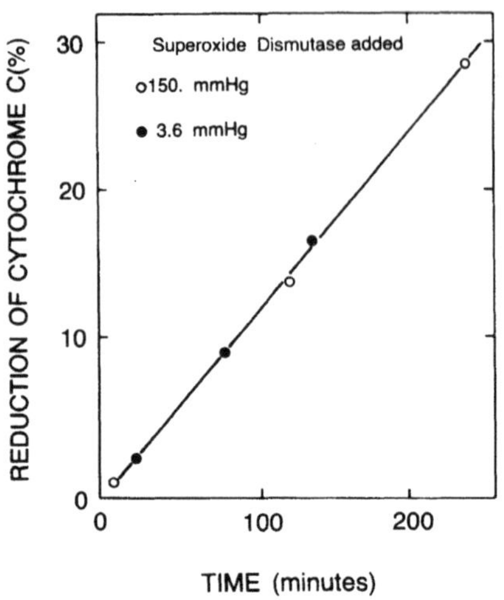

Figure 3. The reduction of cytochrome C during a 37°C incubation with intact erythrocytes. All conditions the same as for Figure 2 except 1000 μ/ml superoxide dismutase added to the medium.

oxidation in these experiments corresponds to 1.4% per hour at 2 mm Hg and ~0.3% per hour at atmospheric pressure. These levels are appreciably lower than those found with hemoglobin stored at -60°C where the methemoglobin reductase system is no longer functional. An enhanced rate of oxidation at low oxygen pressures is however found irrespective of the hemoglobin preparation indicative of an oxygen dependence of the autoxidation process and not the enzyme reduction process.

DISCUSSION

The finding that low oxygen pressure results in both enhanced leakage of superoxide and enhanced rates of autoxidation implies that the enhanced leakage is due to higher levels of superoxide formation which originates from a greater susceptibility for hemoglobin autoxidation. The results of Figure 2 and 4 imply that both intracellular methemoglobin reductase and SOD are not able to fully compensate for the oxidative stress experienced by the cell at low oxygen pressure. The additional cytochrome C reduced at low oxygen pressure is, however, equivalent to only 7% of the extra hemoglobin oxidized at low oxygen pressure. These results imply that the cellular mechanisms for protecting against superoxide are at least an order of magnitude more effective than those responsible for maintaining reduced hemoglobin (Figure 1). The superoxide which leaks out of the cell is nevertheless a potential source for cellular and tissue damage (Figure 1). When it is considered that blood is in intimate contact with all body tissues, the contribution of erythrocytes to the oxidative damage associated with hypoxic or ischemic stress must be recognized.

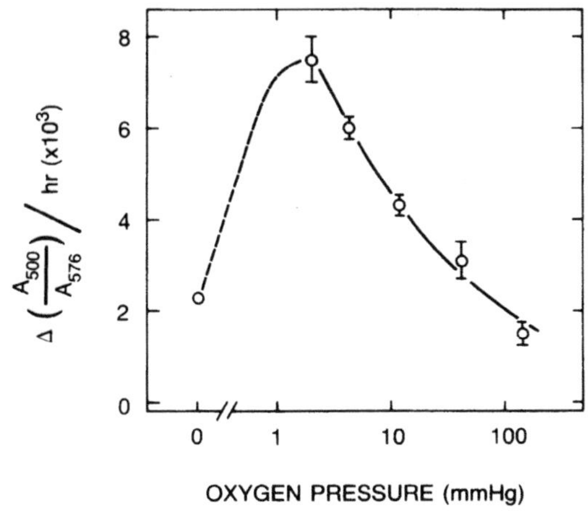

Figure 4. Oxidation of hemoglobin in hemolyzed cells at 37°C as a function of the oxygen pressure. The extent of oxidation is determined from the ratio of the absorbance at 500 nm (peak in methemoglobin) to that at 576 nm (the peak in oxyhemoglobin). Changes are determined by a least square analysis of the data. The errors indicated in the figure are the standard deviation of the slope from the least squares analysis.

REFERENCES

1. H. P. Misra and I. Fridovich, The generation of superoxide radical during the autoxidation of hemoglobin, J. Biol. Chem. 247:6960 (1972).
2. R. Wever, B. Oudega, and B. F. van Gelder, Generation of superoxide radicals during the autoxidation of mammalian oxyhemoglobin, Biochim. Biophys. Acta 302:475 (1973).
3. R. E. Lynch and I. Fridovich, Effects of superoxide on the erythrocyte membrane, J. Biol. Chem. 253:1838 (1978).
4. R. E. Lynch and I. Fridovich, Permeation of the erythrocyte stroma by superoxide radical, J. Biol. Chem. 253:4697 (1978).
5. R. P. Herbel, J. W. Eaton, M. Balasingam, and M. H. Steinberg, Spontaneous oxygen radical generation by sickle erythrocytes, J. Clin. Invest. 70:1253 (1982).
6. S. J. Weiss, The role of superoxide in the destruction of erythrocyte targets by human neutrophils, J. Biol. Chem. 255:9912 (1980).
7. H. Nohl, D. Hegner, and K.-H. Summer, The mechanism of toxic action of hyperbaric oxygenation on the mitochondria of rat-heart cells, Biochem. Pharmacol. 30:1753 (1981).
8. A. Mansouri and K. H. Winterhalter, Nonequivalence of chains in hemoglobin oxidation, Biochemistry 12:946 (1973).
9. W. J. Wallace, R. A. Houtchens, J. C. Maxwell, and W. S. Caughey, Mechanism of autooxidation for hemoglobins and myoglobins, J. Biol. Chem. 257:4966 (1982).

10. J. M. McCord and I. Fridovich, Superoxide dismutase: An enzymic function for erythrocuprein (hemocuprein), J. Biol. Chem. 244:6049 (1969).
11. A. Tomoda, A. Tsuji, and Y. Yoneyama, Mechanism of hemoglobin oxidation by ferricytochrome C under aerobic and anaerobic conditions, J. Biol. Chem. 255:7978 (1980).

SPECIFIC DETECTION OF LIPID HYDROPEROXIDES USING HPLC-EC METHOD

S. Matsushita, J. Terao, K. Yamada, and
S.S. Shibata

Research Institute for Food Science
Kyoto University
Uji, Kyoto 611, Japan

INTRODUCTION

Recently, the assay of lipid hydroperoxides has gained increasing attention since lipid peroxidation products are believed to be involved in the process of aging, mutagenicity and many kinds of disease. For this reason, several methods have been developed for the detection of lipid hydroperoxides in biological materials. These include the thiobarbituric acid method, UV measurement, fluorometric assay, chemiluminescense measurement, and enzymatic assays. However, these methods still have some limitations in their application.

HPLC has been used to separate hydroperoxide isomers from secondary products and also to analyze hydroperoxides in oxidized oils. In those cases, hydroperoxides were detected by the UV absorption of the conjugated dienes which occur in their structures as a result of the formation of hydroperoxides. This technique does not detect the hydroperoxy group directly and may be non-specific when used for oxidized oils. Moreover, hydroperoxides which do not contain conjugated dienes cannot be detected with this method. Therefore, an alternative method is necessary to detect hydroperoxy groups directly after HPLC separation.

Previously we attempted a polarographic study of lipid hydroperoxides [1,2] and found that polarography with a dropping mercury electrode could be used to determine the extent of lipid peroxidation (Figure 1). Recently, an electro-chemical (EC) detector[3] for HPLC equipped with a glassy carbon electrode has been developed which we have examined for its ability to detect lipid peroxides (HPLC-EC).

Lipid hydroperoxides are supposed can be reduced or oxidized:

Reduction $ROOH + 2e^- + 2H^+ \dashrightarrow ROH + H_2O$

Oxidation $ROOH \dashrightarrow R^+ + O_2 + 2e^- + H^+$

Reduction forms hydroxy compounds, and oxidation the release of oxygen.

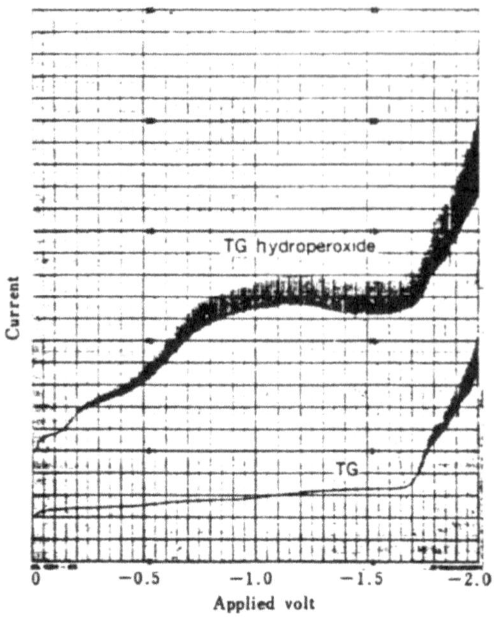

Figure 1. Polarograms of triacylglycerol and triacylglycerol hydroperoxides.

DETECTION OF PHOSPHOLIPID HYDROPEROXIDES

The main phospholipids in biological materials, phosphatidylcholine (PC) and phosphatidylethanolamine (PE), have been separated by HPLC[4]. When both phospholipids (PC and PE), their hydroperoxides (PC-OOH and PE-OOH), and their hydroxy derivatives (PC-OH and PE-OH) were analyzed by reverse-phase HPLC with an ODS column and with methanol elution, unoxidized PC and PE were separated from their hydroperoxides and hydroxy derivatives, but their hydroperoxides and hydroxy derivatives could not be separated from each other. With normal-phase HPLC using a Zorbax SIL column and linear gradient elution of water in acetonitrile, the PC and PE derivatives could be separated, although within each series unoxidized, -OOH, and -OH could not be separated[5]. Hydroperoxides and hydroxy derivatives monitored by their absorbance at 234 nm were attributed to conjugated dienes specific to the peroxidation of polyunsaturated fatty acids. Unoxidized lipids were monitored with absorption at 203 nm (nonspecific end-absorption).

In an experiment with rats which had been treated with CCl_4, reverse-phase HPLC showed large peaks attributed to peroxidation of phospholipids in the liver. Normal-phase HPLC analysis confirmed that both PC and PE in the liver phospholipids were peroxidized after CCl_4 treatment. Therefore reverse-phase and normal-phase HPLC can complement each other serving as direct and sensitive methods for the determination of lipid peroxide levels in biological sources. However, UV detection cannot distinguish phospholipid-OOH from their -OH derivatives and the phospholipid-OOH and -OH derivatives are scarcely separated from each other. Therefore, the phospholipid-OOH and -OH derivatives are difficult to distinguish from each other using the HPLC-UV system.

HPLC-EC detection was examined using reverse-phase separation[6]. This technique is suitable for separating small amounts of oxidized products from an abundance of unoxidized phospholipids in biological materials, though PC and PE oxidation products cannot be separated.

A Shimadzu liquid chromatograph LC-4A with a YMC A-212-C-8 column (6 x 150 mm, 5 μm particles size) was used. A Shimadzu SPD-2AS variable wavelength UV detector and an IRICA E-502 amperometric detector containing a glassy carbon electrode were used as the UV and EC detectors, respectively. UV and EC detectors were connected in series. The chromatograms were recorded by Shimadzu Chromatopack C-R1A (for the UV detector) and C-R3A (for the EC detector) processors. Prior to analysis it was necessary to stabilize the electrode during a 2 hr run.

The sample was dissolved in a mixture of methanol and chloroform (97:3, v/v), and 150 μl of the solution was injected onto the HPLC column. The column was eluted isocratically with methanol/water/ acetonitrile (90.5:7:2.5, v/v/v) containing 0.25 mM sodium perchlorate at a flow rate of 1.8 ml/min. The eluent was monitored at 235 nm by the UV detector and -300 mV vs Ag/AgCl on the EC detector. The mobile phase was kept free from oxygen by continuous bubbling of helium. No attempt was made to remove oxygen from the sample solution.

Under these analytical conditions, PC-OOH of rat liver gave prominent peaks on the chromatogram monitored by the EC detector and similarly on the UV detector (Figure 2). However, no peaks for PC-OH were detected by EC. Unoxidized eluted later than PC-OOH and PC-OH on the chromatogram (detected by nonspecific end absorption at 203 nM). Unoxidized PC was not detected by the EC detector.

Similar results were obtained for rat liver PE, PE-OOH and PE-OH (Figure 3). Only PE-OOH gave prominent peaks on the chromatogram monitored by the EC detector.

For both PC and PE, a linear relationship was observed between the amounts of their hydroperoxides and the peak areas in the chromatograms obtained by UV and EC detections in the range from 1 to 25 nmol (correlation coefficient: 0.990 for PC-OOH by EC detection, 0.999 for PC-OOH by UV detection, 0.823 for PE-OOH by EC detection, 0.980 for PE-OOH by UV detection) (Figure 4). The slope of the regression lines indicates that the sensitivity of EC detection is more than 10 times higher than that of UV detection for both PC-OOH and PE-OOH. The lowest detectable level using the EC detector was determined to be 200 pmol for both PC-OOH and PE-OOH.

From the above results, it is concluded that the combination of reverse-phase HPLC and EC detector can serve as a sensitive and specific method for the quantitative analysis of phospholipid hydroperoxides.

The application of this method to lipid peroxidation products in vivo was examined. Phospholipids of rat liver after CCl_4 administration were carefully prepared as described in Figure 5[4] and analyzed using this technique. Phospholipids (0.1 mg) were dissolved in a mixture of methanol/chloroform (97:3, v/v, 150 μl) and injected onto the HPLC column. The column was eluted with methanol/water/acetonitrile (87.6:/10:/2.4, v/v/v). The other conditions were the same as those described above.

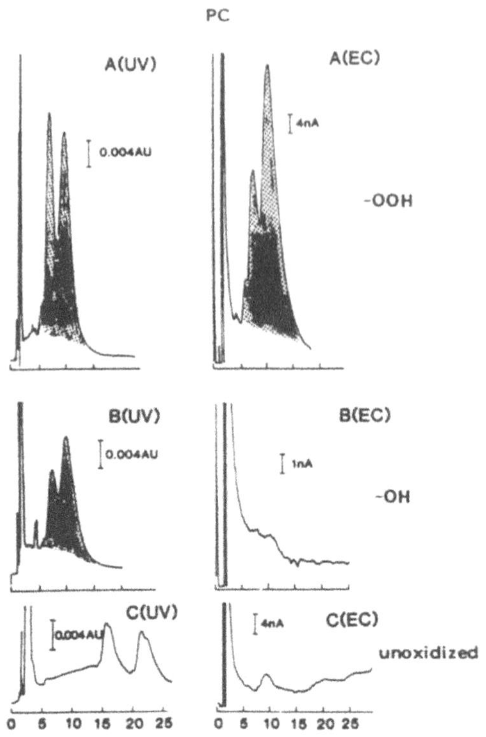

Figure 2. Chromatograms of hydroperoxy and hydroxy derivatives of rat liver PC and unoxidized PC. (A) hydroperoxy derivatives (40 nmol.) (B) hydroxy derivatives (20 nmol), and (C) unoxidized PC (20 nmol) were injected onto the column.

Phospholipids obtained from the control rat liver gave no major peak on the chromatogram of the UV detector. A sample containing standard phospholipid hydroperoxides showed two prominent peaks at 9-14 mins (Figure 6). In the case of EC detector, there was only one prominent peak at around 8 min from the control sample. Two additional peaks were observed at 9-14 min when standard phospholipid hydroperoxides were added. These results show that the fractions at 9-14 min contained phospholipid hydroperoxides.

The samples from CCl_4-injected rats (after 1 and 8 hr) gave two prominent peaks around 9-16 min by the UV detector (Figure 7). The retention times of the peaks corresponded to those of phospholipid hydroperoxides. However, there were no peaks on the chromatogram of the EC detector. These results suggest that phospholipid hydroperoxides formed in the liver are reduced to their hydroxy derivatives [7] without any release of hydroperoxy fatty acids from the phospholipid molecules.

DETECTION OF FATTY ACID HYDROPEROXIDES

Linoleic acid hydroperoxides showed very similar chromatograms as monitored by UV and by EC detections after separation by reverse-phase HPLC with a YMC packed ODS column (6 x 150 nm, 5 μm particle size) and eluting with methanol/water/acetic acid (76:24:0.1-v/v/v) containing 30

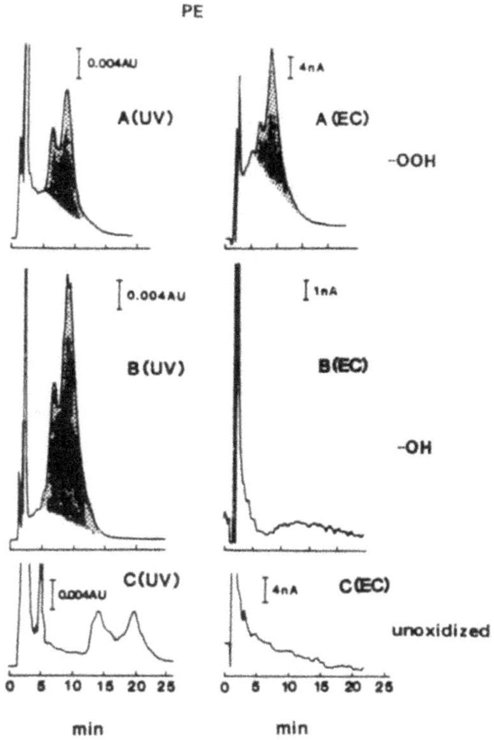

Figure 3. Chromatograms of hydroperoxy and hydroxy derivatives of rat liver PE and unoxidized PE. (A) hydroperoxy derivatives (20 nmol), (B) hydroxy derivatives (60 nmol), and (C) unoxidized PE were injected onto the column.

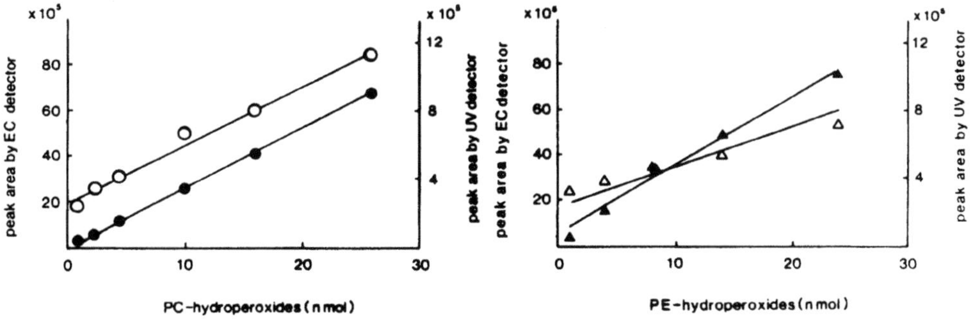

Figure 4. Calibration curves of PC and PE hydroperoxides. •,▲ UV detection and o,Δ EC detection.

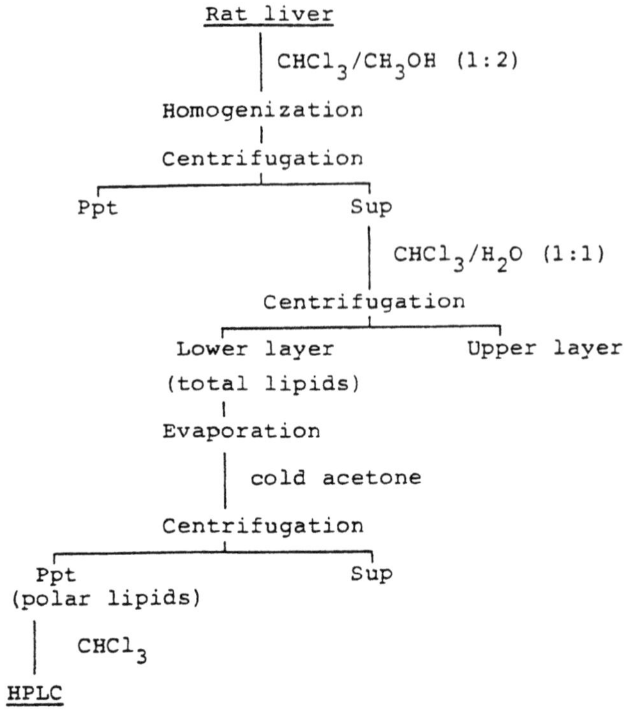

Figure 5. Phospholipid preparation from rat liver.

mM lithium acetate. The flow rate was 1.6 ml/min and the effluent was monitored at 235 nm by the UV detector and -300 mV vs Ag/AgCl on the EC detector. The mobile phase was kept free of oxygen by continuous helium bubbling and no attempt was made to remove oxygen from the sample solution.

When linoleic acid hydroperoxides were applied to the reverse-phase HPLC overlapping peaks that occurred at around 18-24 min are hypothesized to be the 13-*cis*, *trans*, 13-*trans*, *trans*, 9-*cis*, *trans* and 9-*trans*, *trans* isomers (Figure 8). On the other hand, linoleic acid hydroxy derivatives eluting at the same times as the linoleic hydroperoxides again were detected only by UV absorbance.

Calibration curves for the determination of lipid hydroperoxides by UV and EC detections showed good correlations between the peak areas and amounts of hydroperoxides in the range of 1 to 10 nmol (Figure 9). The correlation coefficient was 0.98. From the regression lines obtained by UV and EC detections, the sensitivity of the EC detector was around 50% of that of the UV detector.

In the case of arachidonic acid hydroperoxides, similar results were obtained.

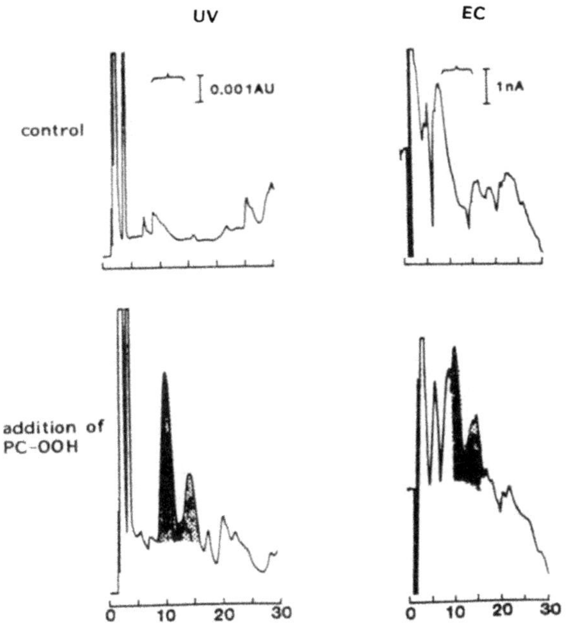

Figure 6. Chromatograms of unoxidized rat liver phospholipids and the addition of standard hydroperoxides.

DETECTION OF TRIACYLGLYCEROL HYDROPEROXIDES

The experimental conditions used above, that is, the reducing side of the EC detector, were not suitable for the analysis of triacylglycerol hydroperoxides; the solvent is unable to provide protons necessary for the completion of the reactions. Therefore, the oxidative side of the electrode potential was used.

The HPLC-EC system was essentially the same as that described above, with slight modification. After the UV detector, a side branch with a constant-flow pump was connected to supply the solvent containing the supporting electrolyte. The HPLC analysis was performed on a YMC packed ODS column (6 x 150 mm, 5 μm particle size) for reverse-phase HPLC. A solution of 2-propanol/acetonitrile (1:3, v/v) was used as the eluting solvent. The flow rate was maintained at 1.8 ml/min. The eluent was monitored at 235 and 203 nm. From the side branch, the same elution solvent containing 100 mM sodium perchlorate was added to the eluent from the UV detector at a flow rate of 0.45 ml/min. The final concentration of the supporting electrolyte became 20 mM. The electric potential of the EC cell was fixed at +900 mV vs Ag/AgCl. The mobile phase was kept free of oxygen by continuous helium bubbling and no attempt was made to remove oxygen from the sample solution.

Trilinoleoylglycerol hydroperoxides appeared as three overlapping peaks (assigned to 13-_trans_, _trans_, a mixture of 9-_cis_, _trans_ and 13-_cis_, _trans_, and 9-_trans_, _trans_ isomers) at around 13 min elution time on both chromatograms (UV and EC detectors) (Figure 10). The two calibration curves of trilinoleoylglycerol hydroperoxides were linear in

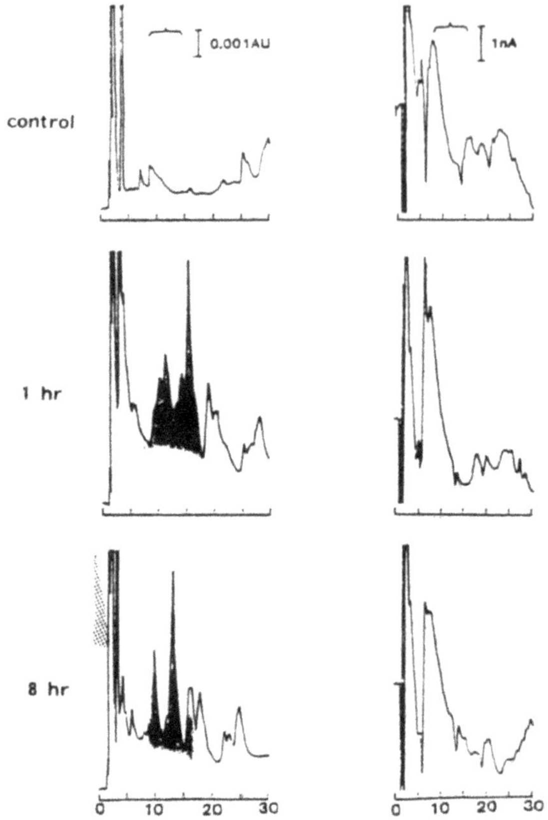

Figure 7. Chromatograms of the phospholipid fraction of CCl_4 injected rat liver.

the range 10 to 25 nmol (correlation coefficient 0.999 for UV and 0.993 for EC) (Figure 11). However, the sensitivity of EC detection was only one-tenth of that of UV detection.

Trilinoleoylglycerol hydroperoxides appeared as three overlapping peaks (assigned to 13-trans, trans, a mixture of 9-cis, trans and 13-cis, trans, and 9-trans, trans isomers) at around 13 min elution time on both chromatograms (UV and EC detectors) (Figure 10). The two calibration curves of trilinoleoylglycerol hydroperoxides were linear in the range 10 to 25 nmol (correlation coefficient 0.999 for UV and 0.993 for EC) (Figure 11) However, the sensitivity of EC detection was only one-tenth of that of UV detection.

The usefulness of this method was verified by analyzing oxidized soybean oil (Figure 12). Hydroperoxide peaks were clearly observed in the chromatograms obtained by the two detection methods.

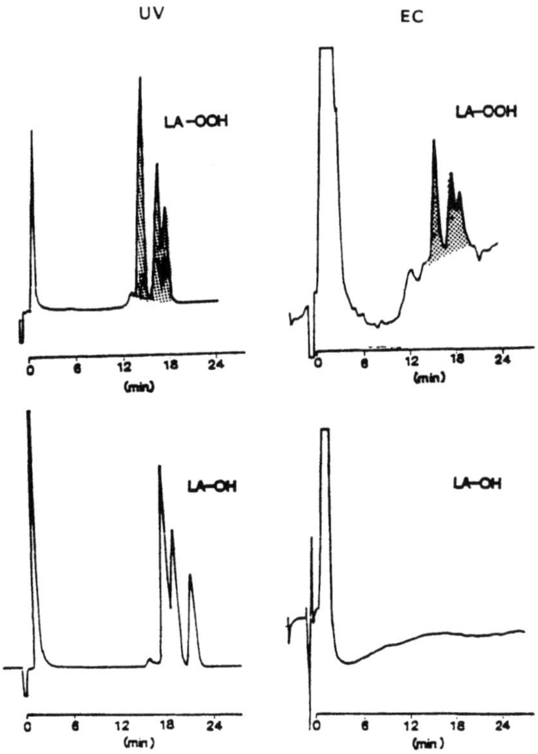

Figure 8. Chromatogram of linoleic acid hydroperoxides obtained by UV and EC detections.

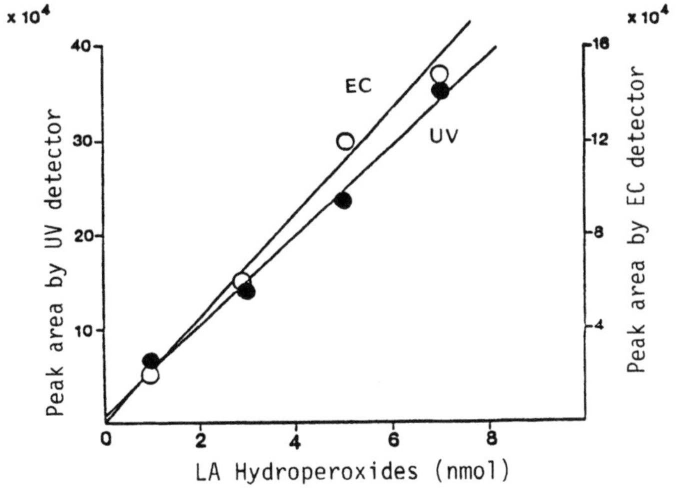

Figure 9. Calibration curves of linoleic acid hydroperoxides by UV and EC detections. • UV detection, o EC detection.

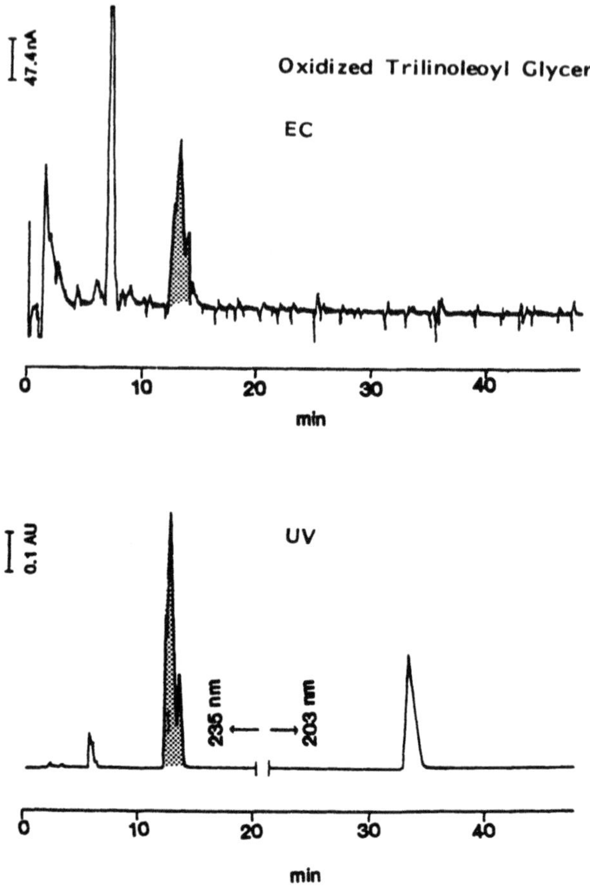

Figure 10. Chromatogram of oxidized and unoxidized trilinoleoylglycerol (TL). TL hydroperoxides (250 nmol) and non-oxidized TL (100 nmol) were applied to the column.

CONCLUSIONS

From the results obtained above, we conclude that the EC method may be useful for analyzing lipid hydroperoxides specifically in biological materials as well as foods. However, there are still some difficulties in the operation: (i) Two hours preoperation before starting the analysis is necessary to stabilize the electrode. (ii) Base line-drift or base line-noise occurs often. Further studies will be necessary to improve the EC detection of lipid hydroperoxides.

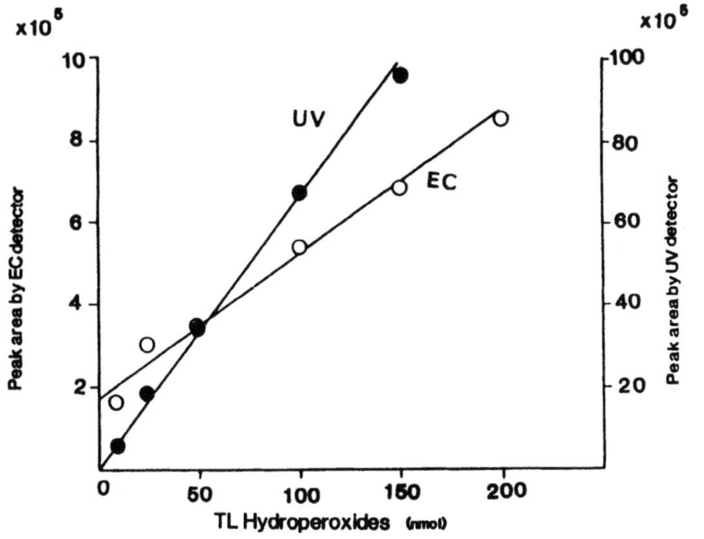

Figure 11. Calibration curves of trilinoleoylglycerol hydroperoxides obtained by UV and EC detections. • UV detection. o EC detection.

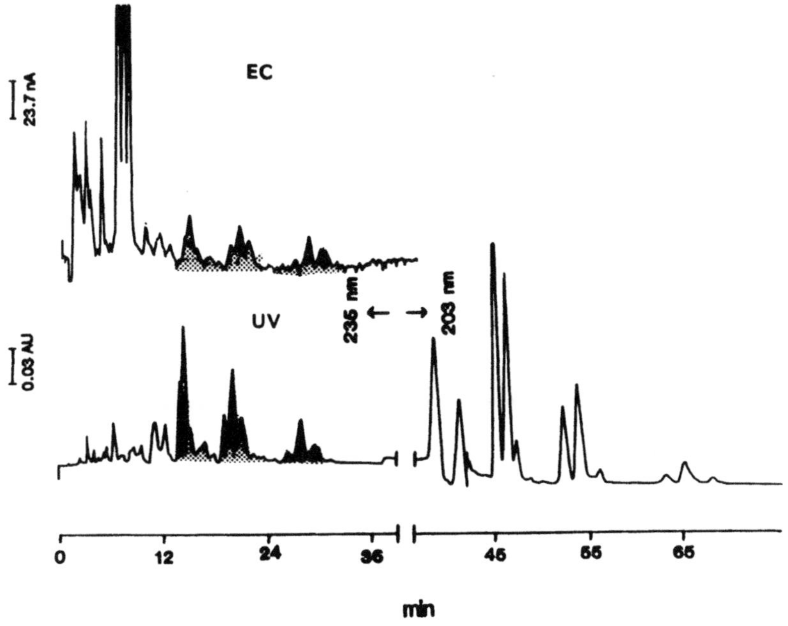

Figure 12. Chromatogram of oxidized soybean oil.

REFERENCES

1. P.T. Gamage, T.Mori, and S. Matsushita, Effects of linoleic acid hydroperoxides and their secondary products on the growth of Escherichia coli, Agric. Biol. Chem. 35:33 (1971).
2. K. Sohde, S. Izutani, and S. Matsushita, Preparation of triglyceride hydroperoxides, Agric. Biol. Chem. 37:1979 (1973).
3. P.T. Kissinger, C. Refshauge, R. Dreiling, and R.N. Adams, An electrochemical detector for liquid chromatography with picogram sensitivity, Anal. Lett., 6:465 (1973).
4. J. Terao, I. Asano, and S. Matsushita, High performance liquid chromatographic determination of phospholipid peroxidation products of rat liver after carbon tetrachloride administration. Arch. Biochem. Biophys. 235-326-333 (1984).
5. J. Terao, I. Asano, and S. Matsushita, Preparation of hydroperoxy and hydroxy derivatives of rat liver phosphatidylcholine and phosphatidylethanolamine, Lipids 20:312 (1985).
6. K. Yamada, J. Terao, and S. Matsushita, Electrochemical detection of phospholipid hydroperoxides in reverse-phase high performance liquid chromatography. Lipids 22:125 (1987).
7. H. Huges, C.U. Smith, E.C. Horning, and J.R. Mitchell, High-performance liquid chromatography and gas chromatography-mass spectrometry determination of specific lipid peroxidation products in vivo, Anal. Biochem. 130:431 (1983).

QUANTITATIVE DETERMINATION OF FATTY ACID HYDROPEROXIDES IN VIVO

Hartmut Frank, Manfred Wiegand, and Dietmar Thiel

Institut für Toxikologie
Universität Tübingen
Wilhelmstrasse 56
D-7400 Tübingen, FRG

INTRODUCTION

The toxicity of tetrachloromethane, acetaminophen, adriamycin and alcohol is believed to result from oxidative degradation of polyunsaturated fatty acids of lipid membranes, usually referred to as "lipid peroxidation"[1]; its participation in atherosclerosis and carcinogenicity has also been suggested[2].

The hypothetical mechanism of lipid peroxidation has been deduced from microsomal experiments and is based upon determination of (a) thiobarbituric acid-reactive substances (TBAR)[3], (b) the conjugation of double bonds resulting in an increase of absorbance at 235 nm[4] and (c) in vivo exhalation of ethane and pentane[5]. Evidently, all these phenomena are associated with the final stages of a complex sequence of radical reactions; it is conceivable that (b) and (c) may also arise from radical reactions without participation of oxygen. The primary products involved in "lipid peroxidation", the lipid hydroperoxides, should be more reliable as semiquantitative indicators for the involvement of oxygen.

METHODS

The main problem in quantitative determination of lipid hydroperoxides is their chemical instability. They can be analyzed by liquid chromatography[6,7], but for determination of trace amounts in complex biological matrices, the detection modes of LC are not sufficiently sensitive and/or selective.

When fatty acid hydroperoxides are reduced to the corresponding hydroxides, they can be analyzed by gas chromatography/mass spectrometry[8]. This approach enables the sensitive qualitative analysis of fatty acid hydroperoxides and hydroxides in lipid samples from various tissues[9,10], but hydroperoxides and hydroxides cannot be differentiated and cannot be determined quantitatively and independently. However, this is feasible with hydroperoxide standards labeled in the hydroperoxide moiety[8]. [$^{18}O_2$]Hydroperoxides may be synthesized via lipoxygenase-catalyzed addition of $^{18}O_2$[11].

The hydroperoxides derived from linoleic acid are the most abundant after microsomal lipid peroxidation[8]. With a higher number of double bonds

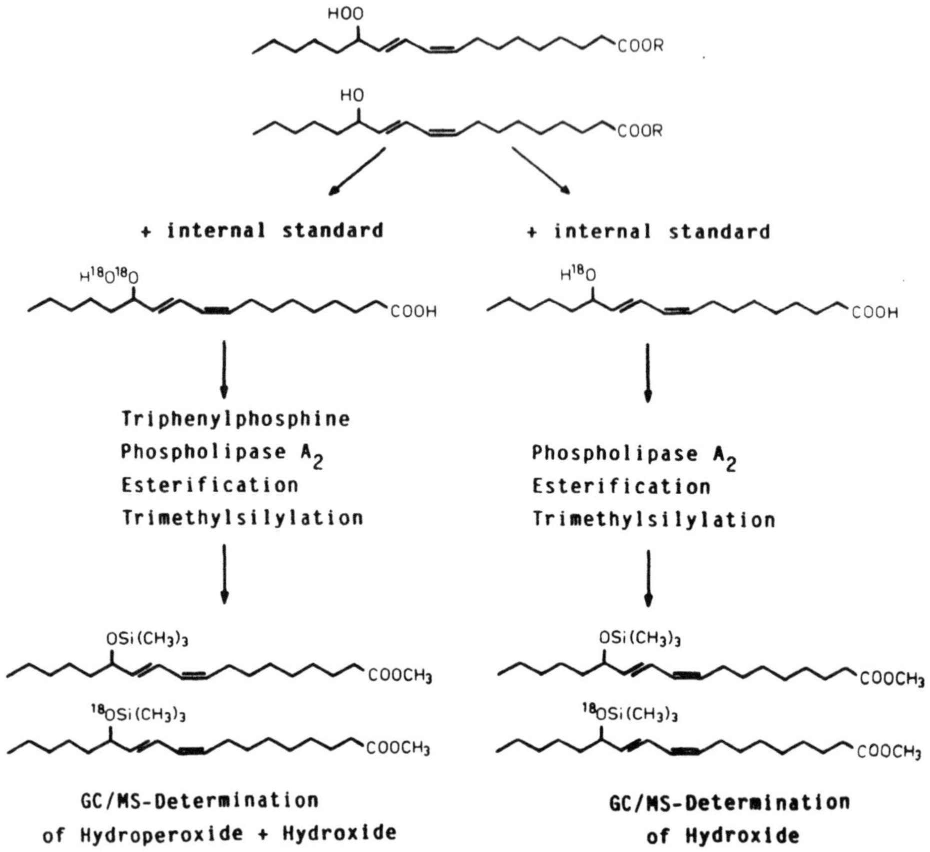

Figure 1. Schematic presentation of the preparation of a lipid sample for quantitative determination of linoleic acid hydroperoxides and hydroxides in membrane lipids by isotope-dilution and gas chromatography/mass spectrometry.

the respective hydroperoxides are less stable; only the 5- and 15-isomer of arachidonic acid hydroperoxide could be found, none of docosahexaenoic acid.

The sample preparation is shown in Figure 1. Although the recoveries may be incomplete on all stages, the internal standards compensate for degradation and losses during work-up. Multiple ion detection enables an accurate determination of hydroperoxides and hydroxides. It is important to verify the respective retention times, as identical mass spectrometric fragments may arise from different oxidized fatty acids. The sum of the hydroperoxide and hydroxides and, in a second aliquot, the hydroxides alone are determined. The relative amounts of all isomers are determined with hydroxy fatty acid [$^{2}H_3$]methylesters as internal standards.

RESULTS

The amounts of hydroperoxides formed upon incubation of microsomes with Fe^{2+}/ADP under an oxygen partial pressure of 20 kPa are compared to the amounts in microsomes from male Sprague-Dawley rats exposed to 500

Table 1. 13-Hydroperoxy and 13-Hydroxy Octadecadienoic Acid in Rat Liver Microsomes, nmol/mg Microsomal Protein ± s.d. (n = 5)

	13-Hydroperoxide + 13-Hydroxide	13-Hydroxide	Difference
Control	0.09 ± 0.03	0.10 ± 0.03	-
In vivo CCl_4	0.23 ± 0.04	0.12 ± 0.02	0.11
In vitro ADP/Fe^{2+}	6.9 ± 1.1	1.0 ± 0.3	5.9

µmol/kg bodyweight tetrachloromethane, half of which is metabolized[11] (Table 1).

13-Hydroxy octadecadienoic acid is present in microsomes from untreated animals, but the amounts in both aliquots, reduced and non-reduced, are the same; obviously, 13-hydroperoxide is not present. After exposure of the animals to tetrachloromethane, the sum of hydroperoxide and hydroxide in the reduced sample is about 0.14 nmol/mg greater than in the controls, but not the amount of hydroxide in the non-reduced sample. Aerobic incubation of microsomes from control animals with Fe^{2+}/ADP for 15 minutes entails formation of 6.9 nmol hydroxide plus hydroperoxide/mg protein, 1.0 nmol of which is hydroxide. Thus, the formation of fatty acid hydroperoxides induced by tetrachloromethane in vivo is marginal compared to the amount formed upon aerobic incubation of microsomes with ferrous salt.

The hydroperoxides have been correlated quantitatively to other parameters usually monitored in this context (Table 2). Metabolism of 250 µmol CCl_4/kg body weight entails an increase in linoleic acid[13]; nevertheless, trichloromethyl octadecadienoic and octadecenoic acid[14] and the corresponding hydroperoxide are found. Arachidonic acid and docosahexaenoic acid are lowered; only 0.5% of the lost amount of the former is recovered as trichloromethyl fatty acid, about 0.1% as hydroperoxide and 0.025% as pentane. For docosahexaenoic acid, ethane could be found to about 0.1% of the lost amount.

Table 2. Effects of CCl_4-Metabolism (Relative to Control Animals) on Polyunsaturated Fatty Acids of the Hepatic Endoplasmic Reticulum, Hydroperoxy, Hydroxy, and Trichloromethyl Fatty Acids, Alkane Exhalation and Thiobarbituric Acid-Reactive Material (TBAR) in Male Sprague-Dawley Rats (nmol/kg Body Weight ± s.d., n = 3)

CCl_4-Dose	500 000
CCl_4 Metabolized	250 000 ± 20 000
Δ Linoleic acid	+ 50 000 ± 23 000
Trichloromethylated (all isomers)	1 300 ± 750
Hydroperoxides (all isomers)	435 ± 90
Hydroxides (all isomers)	60 ± 70
Δ Arachidonic acid	- 120 000 ± 41 000
Trichloromethylated (all isomers)	700 ± 550
Hydroperoxides (5-, 15-isomers)	135 ± 30
Hydroxides (5-, 15-isomers)	5 ± 20
Pentane	30 ± 14
Δ Docosahexaenoic acid	- 30 000 ± 14 000
Trichloromethylated	traces
Hydroperoxides	not detected
Hydroxides	not detected
Ethane	28 ± 15
TBAR	not detected

CONCLUSIONS

Quantitative determination of fatty acid hydroperoxides in microsomal lipids shows that incubation of microsomes with Fe^{2+}/ADP and exposure of male Sprague-Dawley rats to CCl_4 are not comparable. 99% of the lost polyunsaturated fatty acids are not accounted for. It is likely they are buried in an analytically intractable fraction of crosslinked material.

REFERENCES

1. R.O. Recknagel, E.A. Glende, and A.M. Hruszkewycz, Chemical mechanisms in carbon tetrachloride toxicity, in: "Free Radicals", W.A. Pryor, ed., Vol III, pp 97-132 Academic Press, New York (1977).
2. K. Yagi, ed., "Lipid peroxides in Biology and Medicine", Academic Press, New York (1982).
3. K.M. Wilbur, F. Bernheim, and O.W. Shapiro, The thiobarbituric acid reagent as a test for the oxidation of unsaturated fatty acids by various agents, Arch. Biochem. Biophys. 24:305 (1949).
4. C.A. Riley, G. Cohen, and M. Lieberman, Ethane evolution: A new index of lipid peroxidation, Science 183:208 (1974).
5. R.O. Recknagel and A.K. Ghoshal, Quantitative estimation of peroxidative degeneration of rat liver microsomal and mitochondrial lipids after carbon tetrachloride poisoning, Exp. Mol. Pathol. 5:413 (1966).
6. H.W. Chan and G. Levett, Autoxidation of methyl linoleate. Separation of isomeric mixtures of methyl linoleate hydroperoxides and methyl hydroxy linoleates, Lipids 12:99 (1977).
7. N.A. Porter, J.Logan, and V. Kontoyannidou, Preparation and purification of arachidonic acid hydroperoxides of biological importance, J. Org. Chem. 44:3177 (1979).
8. H. Frank, M. Wiegand, D. Thiel, and H. Remmer, Isomers of linoleic acid hydroperoxides from peroxidation of microsomal lipids, in: "Oxygen Radicals in Chemistry and Biology", N. Bors, M. Saran, and D. Tait, eds., pp 267-272, de Gruyter, Berlin (1984).
9. F.J.G.M. van Kuijk, D.W. Thomas, R.J. Stephens, and E.A. Dratz, Gas chromatography-mass spectrometry method for determination of phospholipid peroxides, J. Free Rad. Biol. Med. 1:215 (1985).
10. H. Frank, M. Wiegand, M. Strecker, and D. Thiel, Monohydroperoxides of linoleic acid in endoplasmic lipids of rats exposed to tetrachloromethane, Lipids 22:689 (1987).
11. C.P.A. van Os, M. Vente, and J.F.G. Vliegenthart, A NMR shift method for determination of the enantiomeric composition of hydroperoxides formed by lipoxygenase, Biochim. Biophys. Acta 574:103 (1979).
12. H. Dürk, C. Klessen, and H. Frank, Tetrachloromethane metabolism in vivo under normoxia and hypoxia; biochemical and histopathological effects relative to alkane exhalation, Arch. Toxicol. 60:115 (1987).
13. H. Frank, H. Dürk, and D. Thiel, Fatty acids in liver microsomal lipids exposed to hypoxia, tetrachloromethane, or both, Biochim. Biophys. Acta 922:54 (1987).
14. B. Link, H. Dürk, D. Thiel, H. Frank, Binding of trichloromethyl radicals to lipids of the hepatic endoplasmic reticulum during tetrachloromethane metabolism, Biochem. J. 223:577 (1984).

DETECTION OF LIPID PEROXIDATION PRODUCTS IN LIPIDS AND TISSUES BY GAS CHROMATOGRAPHY-MASS SPECTROMETRY

F.J.G.M. van Kuijk, D.W. Thomas*, R.J. Stephens*, and E.A. Dratz**

Department of Biochemistry MF University of Nijmegen
P.O. Box 9101 6500 HB Nijmegen, The Netherlands

*Life Sciences Division SRI International Menlo Park
California 94025, U.S.A.

**Chemistry Department Montana State University Bozeman
Montana 59717, U.S.A.

INTRODUCTION

Formation of oxygen radicals and lipid peroxidation have been suggested to play a key role in various types of tissue degeneration and pathology such as heart disease, aging, cancer, and retinal degeneration.[1,2] Membrane phospholipids and triglycerides contain the principal deposits of unsaturated fatty acids and are thought to be the primary sites of the lipid peroxidation process. The oxidation of membrane phospholipids has been hypothesized to cause an increase in the permeability of cell membranes and/or to inhibit membrane ion pumps and to lead to cell death. This loss of membrane barrier function is thought to lead to edema, disturbances in electrolyte balance, and elevation of intracellular calcium which contributes to the malfunctioning of cells.

Detection of phospholipid peroxidation products has been limited due to a lack of sensitive, direct methods. Hydroperoxides or secondary oxidation (decomposition) products of hydroperoxides have rarely been chemically identified, and most assays that have been used to date are rather nonspecific and insensitive.[3-10] Very few methods for measurement of lipid peroxidation have been based on direct identification of oxidized lipids.[11-16] In the course of this study we have developed three new, high sensitivity techniques for determination of phospholipid peroxides and aldehydic decomposition products of lipid peroxidation (Table 1).[17-19] These new methods use gas chromatographic separation and usually use mass spectrometric detection of lipid peroxidation products.

METHODS AND DISCUSSION

Reference phospholipid hydroperoxides were synthesized by photooxidation of commercially available and natural lipids. For analysis by gas chromatography-mass spectrometry (GC-MS), the phospholipid or triglyceride hydroperoxides were usually first reduced with sodium

Table 1. Summary of Techniques Developed in the Course of This Study

	Methyl Ester	PFB Ester	PFB Oxime
Substrate specificity	Fatty acid hydroperoxides Hydroxy fatty acids 9-Oxononanoate		4-Hydroxyalkenals Alkenals Alkanals
Detection limit	10 ng	1-10 pg	10-100 pg
MS Detection[a]	EI	NICI	NICI
Main Ion	Fragments	M-PFB	M-PFB-HOTMS
Application	Adipose	All tissues	All tissues
GC Detection[b]	FID	ECD	ECD
Reference	17	18	19

[a]EI = Electron ionization, NICI = Negative Ion Chemical Ionization.
[b]FID = Flame Ionization Detection, ECD = Electron Capture Dectection.

borohydride to form more stable hydroxy derivatives. The second step of derivatization for detection of phospholipid or triglyceride peroxides was based on transesterification at room temperature into either fatty acid methyl esters or fatty acid pentafluorobenzyl esters.[17,18] Further derivatization involved formation of trimethylsilyl (TMS) ethers of the hydroxy groups.

Methyl ester derivatives and their detection with electron ionization (EI) proved to be useful for obtaining structural information on the lipid peroxides, with a sensitivity limited to 10 ng of oxidation product on the GC column.[17] This allowed detection of the different positional isomers from photooxidized unsaturated fatty acid chains.[12,17] It was found that the nonconjugated isomers, which are specific for singlet oxygen mediated photooxidation reactions, could be easily separated and distinguished from the conjugated isomers. The conjugated isomers are formed upon autooxidation, as well as photooxidation.[17]

This new analytical method was applied to adipose from vitamin E- and selenium-deficient rats. Large amounts of oxidized linoleate were detected, whereas essentially no oxidized oleate was found, even though both fatty acids were present in comparable amounts.[20] Only conjugated derivatives were found in the antioxidant-deficient adipose, which was confirmed by identification of the EI fragmentation patterns by mass spectrometry.[20] On the other hand, only trace amounts of oxidation products were detected in vitamin E- and selenium-supplemented adipose with a more sensitive GC-MS method discussed below. These findings indicate that antioxidant deficiency has greatly enhanced the occurrence of autooxidative processes in these animals.

Detection of oxidized phospholipids in small samples of other biological tissues required a more sensitive derivatization approach. This was accomplished by a novel transesterification procedure for conversion of fatty acid chains in lipids to fatty acid pentafluorobenzyl (PFB) esters.[18]

The pentafluorobenzyl (PFB) esters yield only one ion [M minus PFB (M-PFB)] in the mass spectrometer if detected by negative ion chemical ionization (NICI).[18] This lack of fragmentation and the higher thermal stability of the PFB esters (leading to much better recoveries of oxidation products of polyunsaturated fatty acids in the gas chromatograph, as compared to methyl esters), allowed detection of 1-10 picogram of the oxidation product. In standards of photooxidized rat retinal lipids, all of the mono-oxidized products of the major unsaturated fatty acid esters were detected, as well as some double oxidation products.

Oxidation products of oleate (18:1), linoleate (18:2), arachidonate (20:4) and docosahexaenoate (22:6) were detected in reference photooxidized rat retinal lipids with both GC-MS methods for detection of fatty acid esters.[17,18]

The PFB ester method allowed us to demonstrate for the first time the presence of in vivo oxidized linoleate and docosahexaenoate in the retinal phospholipids of both vitamin E-deficient and supplemented animals.[21] In addition, evidence was obtained that some of the oxidized fatty acids in the retina were derived from photooxidative processes. This supports the hypothesis that a lipid peroxidation mechanism is involved in light damage to the retina. Differences in lipid peroxide levels for both of these dietary groups could not be established at this point due to the lack of suitable stable isotope internal standards for this assay.[21]

High detection sensitivity could also be obtained with PFB oxime derivatives of 4-hydroxyalkenals.[19] The 4-hydroxyalkenals are thought to be some of the most important breakdown products of lipid peroxides because of their strong biological activity.[19] These PFB oxime derivatives yield a mixture of specific and nonspecific fragments under NICI conditions. The most abundant specific fragment obtained was M minus $C_7F_5H_2$ minus HOTMS (M - $C_7F_5H_2$ - HOTMS).[19] The fact that more fragments are formed compared to the PFB esters explains the lower detection sensitivity for PFB oximes (Table 1).

A deuterated internal standard of 4-hydroxynonenal was synthesized from a precursor provided by Dr. Hermann Esterbauer, which allowed quantitation by employing specific ion monitoring and ratios of peak areas of the protonated and deuterated fragments. A fivefold increase in 4-hydroxynonenal content was found in retinas from vitamin E-deficient as compared to vitamin E-supplemented rats or dogs. It is speculated that these relatively small differences are caused by the high reactivity of the 4-hydroxyalkenal compounds towards sulfhydryl groups.[22,23] The oxime method employed for analysis liberates the aldehydes from Schiff base linkages, but the 4-hydroxyalkenals bound to protein sulfhydryls are not recovered with the current method. Solution of this problem requires further development of the method.

Since 4-hydroxynonenal is not completely recovered, it may be preferable to assay less reactive aldehydes, such as 9-oxononanoate[24] to obtain evidence of toxic aldehyde production by lipid peroxidation. This is also a major decomposition product of linoleate hydroperoxides but is not released from the parent lipids like 4-hydroxyalkenals since it remains esterified in the phospholipids. Upon transesterification, a methyl ester derivative or a PFB ester derivative will be formed with the methods we employ. Usually after protection of the aldehyde as the methyloxime, the latter derivative has been used to estimate the compound in rat tissues. In preliminary studies it was found that the 9-oxononanoate content was about ten times higher than the 4-hydroxynonenal content. This suggests that most 4-hydroxynonenal is irreversibly bound to sulfhydryl groups.[22,23] Since 9-oxononanoate is much less reactive than 4-hydroxyalkenals or lipid hydroperoxides, it may well serve as a highly useful quantitative indicator of the occurrence of lipid peroxidation in small biological samples.

The methods developed can also be used in a GC system without a mass spectrometer. In the case of PFB derivatives, very sensitive detection (1 pg) can be obtained with an electron capture detector (ECD). A major difference with the GC assay is that it can be carried out with one internal standard, which must be separated chromatographically from other commonly occurring oxidized fatty acids and need not be isotopically labeled. A commercially available synthetic (16:0)(16:1) phospholipid that is photooxidized can serve as an internal standard since the 16:1 content of most biological tissues is very low (Handelman, van Kuijk, Adler, and Dratz, unpublished results).

REFERENCES

1. J. L. Marx, Oxygen free radicals linked to many diseases, Science 235:529 (1987).
2. G. B. Bulkley, The role of oxygen free radicals in human disease processes, Surgery 94:407 (1983).
3. T. F. Slater, Overview of methods used for detecting lipid peroxidation, in: "Methods in Enzymology," volume 105, L. A. Packer, ed., Academic Press, Orlando (1984).
4. R. P. Bird and H. H. Draper, Comparative studies on different methods of malonaldehyde determination, in: "Methods in Enzymology," volume 105, L. A. Packer, ed., Academic Press, Orlando (1984).
5. H.-Y. Lee and A. S. Csallany, Measurement of free and bound malondialdehyde in vitamin E deficient and supplemented rat liver tissues, Lipids 22:104 (1987).
6. R. O. Recknagel and J. R. Glende, Spectrophotometric detection of lipid conjugated dienes, in: "Methods in Enzymology," volume 105, L. A. Packer, ed., Academic Press, Orlando (1984).
7. T. Asakawa and S. Matsushita, A colorimetric microdetermination of peroxide values utilizing aluminum chloride as the catalyst, Lipids 15:965 (1980).
8. R. L. Heath and A. L. Tappel, A new sensitive assay for the measurement of hydroperoxides, Anal. Biochem. 76:184 (1976).
9. F. J. G. M. van Kuijk, G. J. Handelman, and E. A. Dratz, Consecutive action of phospholipase A_2 and glutathione peroxidase is required for reduction of phospholipid hydroperoxides and provides a convenient method to determine peroxide values in membranes, J. Free Rad. Biol. Med. 1:421 (1985).
10. J. Terao, I. Asano, and S. Matsushita, High-performance liquid chromatographic determination of phospholipid peroxidation products of rat liver after carbon tetrachloride administration, Arch. Biochem. Biophys. 235:326 (1984).
11. M. Tsuchida, T. Miura, K. Mizutani, and K. Aibara, Fluorescent substances in mouse and human sera as a parameter of in vivo lipid peroxidation, Biochim. Biophys. Acta 834:196 (1985).
12. E. N. Frankel, W. E. Neff, and T. R. Bessler, Analysis of autooxidized fats by gas chromatography-mass spectrometry: V. Photosensitized oxidation, Lipids 14:961 (1979).
13. H. Hughes, C. V. Smith, E. C. Horning, and J. R. Mitchel, High performance liquid chromatography and gas chromatography-mass spectrometry determination of specific lipid peroxidation products in vivo, Anal. Biochem. 130:431 (1983).
14. H. Hughes, C. V. Smith, J. O. Tsokos-Kuhn, and J. R. Mitchel, Quantitation of lipid peroxidation products by gas chromatography-mass spectrometry, Anal. Biochem. 152:107 (1986).

15. J. Lang, C. Celotto, and H. Esterbauer, Quantitative determination of the lipid peroxidation product 4-hydroxynonenal by high performance liquid chromatography, Anal. Biochem. 150:369 (1985).
16. H. Frank, M. Wiegand, M. Strecker, and D. Thiel, Monohydroperoxides of linoleic acid in endoplasmic lipids of rats exposed to tetrachloromethane, Lipids 22:689 (1987).
17. F. J. G. M. van Kuijk, D. W. Thomas, R. J. Stephens, and E. A. Dratz, Gas chromatography-mass spectrometry method for determination of phospholipid peroxides: I. Transesterification to form methyl esters, J. Free Rad. Biol. Med. 1:215 (1985).
18. F. J. G. M. van Kuijk, D. W. Thomas, R. J. Stephens, and E. A. Dratz, Gas chromatography-mass spectrometry method for determination of phospholipid peroxides: II. Transesterification to form pentafluorobenzyl esters and detection with picogram sensitivity, J. Free Rad. Biol. Med. 1:387 (1985).
19. F. J. G. M. van Kuijk, D. W. Thomas, R. J. Stephens, and E. A. Dratz, Occurrence of 4-hydroxyalkenals in rat tissues determined as pentafluorobenzyloxime derivatives by gas chromatography-mass spectrometry, Biochem. Biophys. Res. Comm. 139:144 (1986).
20. F. J. G. M. van Kuijk, D. W. Thomas, R. J. Stephens, and E. A. Dratz, Lipid peroxidation products in rat adipose associated with vitamin E deficiency measured by gas chromatography-mass spectrometry, in: "Biological Effects of Lipid Peroxidation," A. Sevanian, ed., in press.
21. R. J. Stephens, D. S. Negi, S. M. Short, F. J. G. M. van Kuijk, E. A. Dratz, and D. W. Thomas, Lipid peroxidation and phototoxic degeneration, in: "Oxygen Radicals in Biology and Medicine," M. G. Simic, K. A. Taylor, J. A. Ward, C. von Sonntag, eds., Plenum Publishing Corp., New York (1988).
22. M. Poot, A. Verkerk, J. F. Koster, H. Esterbauer, and J. F. Jongkind, Influence of cumene hydroperoxide and 4-hydroxynonenal on the glutathione metabolism during in vitro aging of human skin fibroblasts, Eur. J. Biochem. 162:287 (1987).
23. A. Benedetti, H. Esterbauer, M. Ferrali, R. Fulceri, and M. Comporti, Evidence for aldehydes bound to liver microsomal protein following CCl_4 or $BrCCl_3$ poisoning, Biochim. Biophys. Acta 711:345 (1982).
24. E. N. Frankel, Lipid oxidation: Mechanisms, products and biological significance, J. Am. Oil Chem. Soc. 61:1908 (1984).

DIRECT CHEMICAL IONIZATION MASS SPECTROMETRY OF UNDERIVATIZED ARACHIDONIC ACID METABOLITES: APPLICATION TO LEUKOTRIENE BIOSYNTHESIS

Josef Gut, Gene C. Jamieson,* and James R. Trudell

Department of Anesthesia
Stanford University School of Medicine
Stanford, California 94305-5117, U.S.A.

*Salutar
428 Oakmead Parkway
Sunnyvale, California 94086, U.S.A.

INTRODUCTION

Leukotrienes are a group of endogenous compounds active in hypersensitivity and inflammations. Few reports describe the generation and metabolism of leukotrienes (in particular LTB_4) by other than specialized cells (i.e., macrophages, neutrophils, basophils). However, hepatocytes, too, seem capable of biosynthesis and inactivation of leukotrienes. Our work on the possible involvement of HPETEs or leukotrienes in hepatotoxic processes required a mass spectral analysis method (MS) that allowed the structural elucidation of minimal amounts of these compounds after separation on HPLC.

MS plays an important role in the elucidation of the structure of leukotrienes and their precursors. Low volatility and thermal instability of the compounds previously has required lengthy derivatizations to the silyl ether and methyl ester forms before gas chromatography/mass spectrometry (GS/MS).[1] Recently, GC/MS techniques were successfully applied in the analysis of lipid peroxidation products, including precursors of leukotrienes, after derivatization to fatty acid pentafluorobenzyl esters.[2]

Newer on- or off-line techniques employed in leukotriene analysis include thermospray liquid chromatography/mass spectrometry (LC/MS) and fast atom bombardment (FAB). Thermospray LC/MS was successfully used in the structural analysis of 5-HETE, LTB_4, and other hydroxy-metabolites of docosahexaenoic acid, and the formation of 11-, 9- and 5-HETEs by human inflammatory cells was recently confirmed using the same technique. FAB also has allowed successful analyses of underivatized LTC_4, LTD_4 and LTE_4. However, these techniques suffer from limitations such as restriction to MS-compatible HPLC-eluent systems, high background, and the requirement of considerable amounts of material to be analyzed.[1]

Direct chemical ionization (DCI) mass spectral analysis has been widely applied to analysis of medium molecular weight polar compounds without prior derivatizations. Here, we describe a novel implementation of this technique to obtain information on both the molecular ion and diagnostic fragment ions of underivatized 5-HPETE and LTA_4-free acid. Also, the conversion of LTA_4 to LTB_4 during incubation with human liver microsomes is confirmed.

EXPERIMENTAL

Instrument and Technique

Mass spectra were obtained by direct chemical ionization as described by Jamieson et al.[3] The method involves rapid sample heating on a polyimide-coated, fused silica fiber with direct exposure to the reagent gas chemical ionization plasma. One microliter of the compound, dissolved in an appropriate solvent (i.e., methanol), was applied to the probe tip and the solvent was evaporated. The probe, which also introduced the reagent gas, was inserted into the mass spectrometer chemical ionization source. The sample was ballistically heated by radiant heating from the ionization chamber and vaporized. The molecules were ionized as they vaporized into the plume of the chemical ionization reagent gas plasma. Experiments were performed using a Hewlett Packard 5985A mass spectrometer scanned from 200 to 360 amu every 50 msec. Selected ion monitoring (SIM) increases the sensitivity of this technique and makes possible quantification in the range of 1 µg to 10 pg.

Biological generation of LTB_4 from LTA_4

Human liver microsomes were kept under argon in 10 mM sodium borate, pH 8.5, containing 20% glycerol (v/v) and 5 mg BSA/ml at room temperature. LTA_4, previously prepared in acetonitrile/10 mM sodium borate, pH 10.0 (40:60, v/v) containing 5 mg BSA/ml was added to 20 µM final concentration and the incubation continued for 2 min. Reactions were stopped by adding 10 µl of acetic acid; products were extracted into 3 ml diethyl ether, taken to dryness under argon, re-dissolved in 100 µl eluent, and separated on Ultrasphere ODS in methanol/water/acetic acid 75:25:0.02, v/v). Material eluting with the same retention time as synthetic LTB_4 (i.e., 13.5 min) was collected and a UV spectrum recorded; the solution was taken to dryness under argon, re-suspended in methanol, and subjected to DCIMS analysis.

RESULTS AND DISCUSSION

Direct chemical ionization (DCI) mass spectra of underivatized 5-HPETE, LTA_4 and LTB_4 were obtained using a direct chemical ionization probe and ammonia as reagent gas. We recorded spectra of 100 ng to 1 µg amounts of 5-HPETE and LTA_4, as well as of 10 ng of biologically produced LTB_4, obtained after incubation of human liver microsomes with LTA_4. Using the selected ion monitoring mode resulted in the detection of as little as 10 pg of synthetic LTB_4 and 100 pg of 5-HPETE.[4,5]

Diagnostically different fragmentations of compounds of identical molecular weight (such as 5- and 15-HPETE and LTB_4) demonstrated the capacity of DCIMS to provide information that is essential in the mass spectral analysis of these compounds. An ammonia molecular adduct $(M + NH_4)^+$ was obtained with 5-HPETE at m/z = 354 (Figure 1). Loss of water from the ammonium adduct yielded a weak signal at m/z 336 $(M + NH_4 - H_2O)^+$. Other prominent fragment ions include m/z = 319, $(M + H - H_2O)^+$; m/z = 303, $(M + H - H_2O_2)^+$. We suggest that the base peak, m/z = 233, arises by loss of C_5H_{10}, the terminal 5 carbon atoms on the fatty acid chain, from the m/z 303 fragment ion. M/z = 279 is due to a trace of dibutylphthalate in the sample.

Leukotriene A_4 (LTA_4) is the highly unstable allylic epoxide intermediate in the biosynthetic pathway of leukotrienes. In order to obtain pure

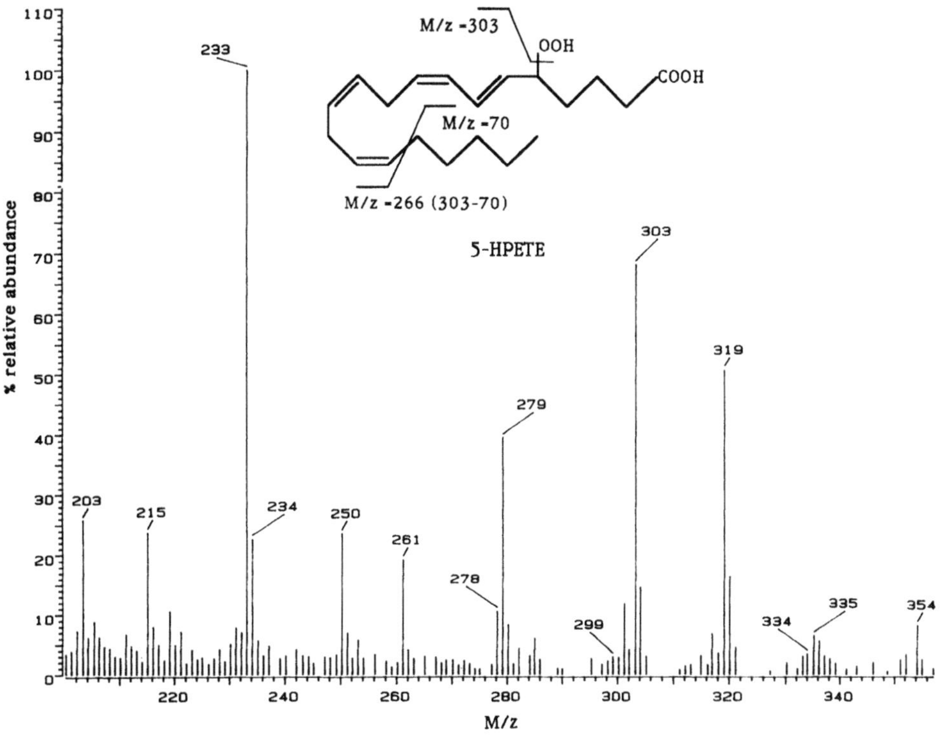

Figure 1. Ammonia DCI spectrum of 100 ng 5-HPETE.

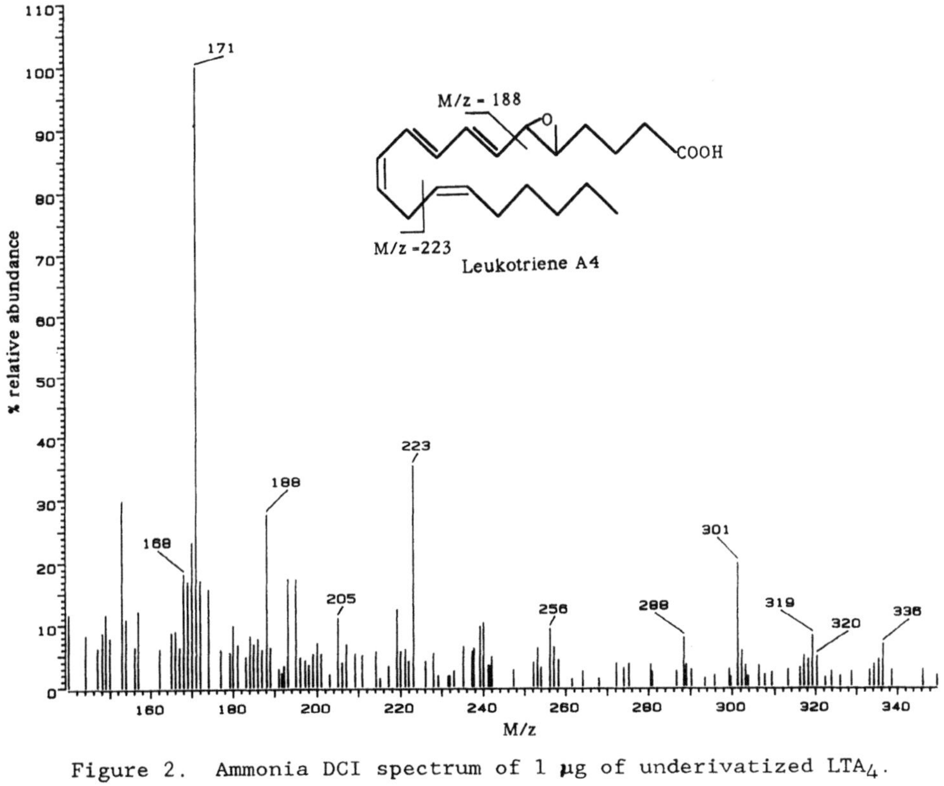

Figure 2. Ammonia DCI spectrum of 1 µg of underivatized LTA_4.

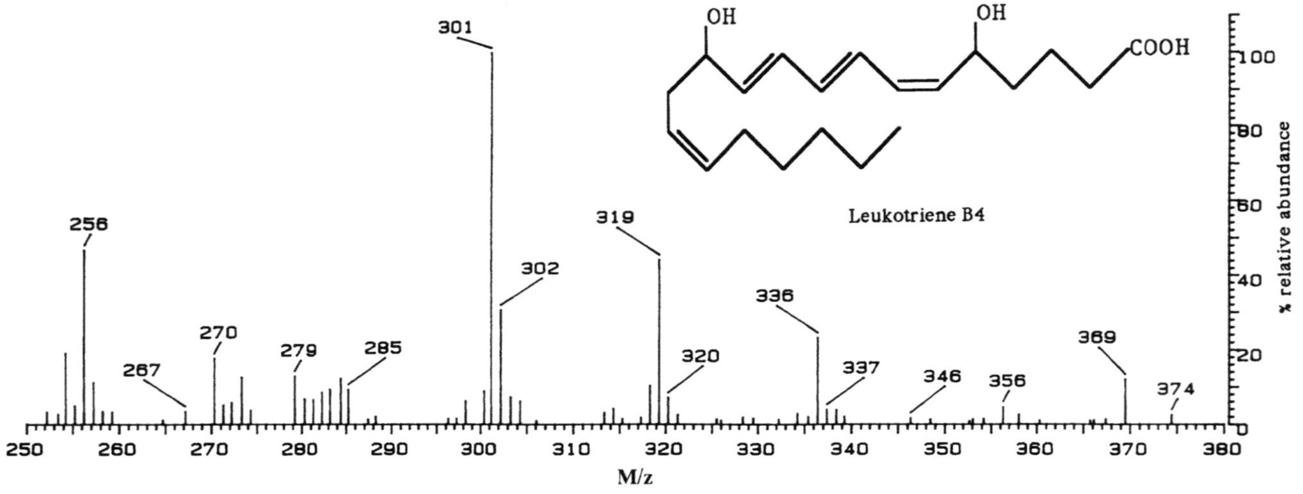

Figure 3. Ammonia DCI spectrum of 10 ng biologically generated LTB_4.

LTA_4, we performed chromatography of the LTA_4 free acid on RP-HPLC under alkaline conditions using a completely volatile solvent system. An ammonia molecular adduct $(M + NH_4)^+$, m/z = 336, was detected in the DCI mass spectrum of this preparatively chromatographed, non-derivatized LTA_4 (Figure 2). A protonated molecule ion at $(M + H)^+$, m/z = 319, was also seen. Loss of water from the latter ion was attributed to $(M + H - H_2O)^+$, m/z = 301. A McLafferty rearrangement of the epoxide is suggested to result in opening of the epoxide ring and migration of the C_7 - C_8 and C_9 - C_{10} double bonds followed by cleavage at C_9 - C_{10}, providing the base peak m/z = 171, $(M + H - C_{11}H_{18})^+$. Cleavage of the allylic C_{13} - C_{14} bond gives rise to m/z = 223, $(M + H - C_7H_{12})^+$. The fragment ion at m/z = 188 can be attributed to the end of the hydrocarbon chain C_7 through C_{20}, $C_{13}H_{20}$, arising from the break of the allylic C_6 - C_7 bond.

During our work on the involvement of leukotriene B_4 or precursors in hepatotoxic processes, material with retention times on HPLC identical to synthetic LTB_4 that was extracted from human liver microsomes after incubation with LTA_4 was analyzed by DCI-MS. The major peaks in the ammonia DCI mass spectrum of 10 ng underivatized LTB_4 (Figure 3) were m/z = 354, $(M + NH_4)^+$; m/z = 336, $(M + NH_4 - H_2O)^+$; m/z = 319, $(M + H - H_2O)^+$; and a base peak of m/z = 301, $(M + H - 2H_2O)^+$.

Mass spectrometry is not a technique to use as a single means to identify leukotrienes. MS allows the structural identification of fatty acid derivatives and leukotrienes, although it is not usually useful in determining stereochemistry.[1,5] In separate experiments, it has been shown that the biologically generated LTB_4 has a binding affinity to receptors on human neutrophils similar to authentic LTB_4 in a highly stereoselective displacement assay.[4] Only the combination of the mass spectral findings together with the high activity of this material in this receptor displacement assay for LTB_4 allows the assignment of the structure of LTB_4 to the biological material.

REFERENCES

1. R. C. Murphy, Eicosanoid mass spectrometry, Spectra 10:11 (1985).

2. F. J. G. M. van Kuijk, E. W. Thomas, R. J. Stevens, and E. A. Dratz, Gas chromatography-mass spectrometry method for determination of phospholipid peroxides, J. Free Rad. Biol. Med. 1:387 (1986).

3. G. C. Jamieson, C. C. Reuter, and W. L. Fitch, Polyimide-coated fused silica as an extended probe for in-beam chemical ionization mass spectrometry, Anal. Chem. 57:121 (1985).

4. J. Gut, D. W. Goldman, G. C. Jamieson, and J. R. Trudell, Conversion of leukotriene A_4 to leukotriene B_4: Catalysis by human liver microsomes under anaerobic conditions, Arch. Biochem. Biophys. 259:497 (1987).

5. J. Gut, G. C. Jamieson, and J. R. Trudell, Quantification of 5- and 15-HPETEs by direct chemical ionization mass spectrometry, Free Rad. Biol. Med. 3:323 (1987).

CHEMILUMINESCENCE - HIGH PERFORMANCE LIQUID CHROMATOGRAPHY OF PHOSPHATIDYLCHOLINE HYDROPEROXIDE IN HUMAN BLOOD PLASMA

Teruo Miyazawa, Keiichi Yasuda, Kenshiro Fujimoto, and Takashi Kaneda*

Department of Food Chemistry, Faculty of Agriculture
Tohoku University, Sendai 980, Japan

*Kohriyama Women's College, Kohriyama 963, Japan

INTRODUCTION

Normal phase HPLC (high performance liquid chromatography) with a UV detection has been used frequently for analyzing phospholipid classes. (1, 2) However, hydroperoxide and its hydroxy derivative usually give almost the same retention time as that for corresponding unoxidized species. (3) We have shown previously that chemiluminescence (CL) detection is very advantageous in its specificity and sensitivity for hydroperoxides. (4, 5) To analyze the phosphatidylcholine hydroperoxide (PCOOH), we newly devised the chemiluminescence - HPLC (CL-HPLC) system, in which luminol-cytochrome c solution is used as a hydroperoxide-specific CL reagent. We applied this new method to confirm the presence of PCOOH in blood plasma of healthy donors.

MATERIALS AND METHODS

Scheme 1 shows CL reaction which incorporated into the present CL-HPLC system. Lipid hydroperoxides react with heme compounds and produce oxidants, and then luminol is oxidized under alkaline conditions and emits CL. Cytochrome c-luminol mixture is most favorable for producing strong CL on reacting with lipid hydroperoxides. (4, 5)

Figure 1 shows a shematic diagram of CL-HPLC for analyzing PCOOH. Normal phase silica gel column was equipped in HPLC and the eluate was mixed with the CL reagent. The mixture was further lead to quartz spiral coils (200 µl) to measure CL with a photomultiplier which functioned as a single photon counting methods. (6, 7) Each 5 ml of blood collected from 14 healthy laboratory volunteers was centrifuged, and 2 ml of heparinized plasma was obtained. The plasma total lipid was extracted quantitatively with a mixture of chloroform and methanol (2:1) containing butyl hydroxytoluene (BHT, 0.002%) as antioxidant. The extraction was repeated three times. After evaporation, the plasma total lipid was diluted with 40 µl of chloroform-methanol (2:1), and

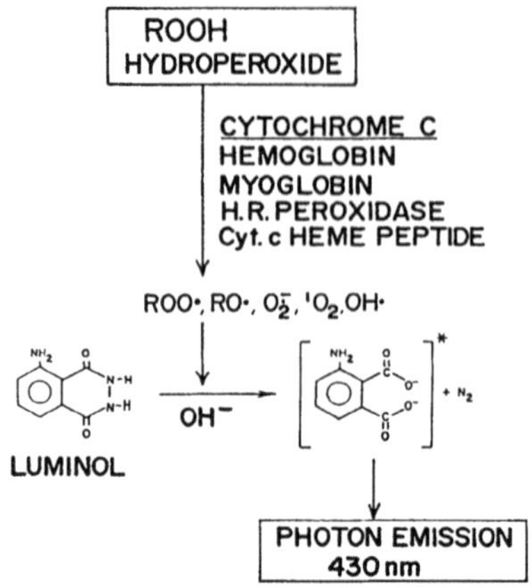

Scheme 1. Luminol-dependent CL on reacting lipid hydroperoxides with cytochrome c.

20 µl was assayed in CL-HPLC. A calibration curve was made on PCOOH prepared by photooxidation of egg yolk PC.

RESULTS AND DISCUSSION

Figure 2 shows a typical chromatogram of authentic PCOOH in CL-HPLC. By the UV detection, the PCOOH peak was not detected at the picomole levels. Hydroxy PC prepared by $NaBH_4$ reduction of PCOOH showed a peak line with a conjugated diene detection but not with CL detection. (4) Phospholipid hydroperoxides such as PCOOH and PEOOH (phosphatidylethanolamine hydroperoxide) were separatively and specifically detectable and differentiated from the unoxidized lipids and also from their hydroxy derivatives on normal phase HPLC by the CL detection. (4) The retention times of other phospholipids were clearly different from PC under these conditions. For the calibration curve,

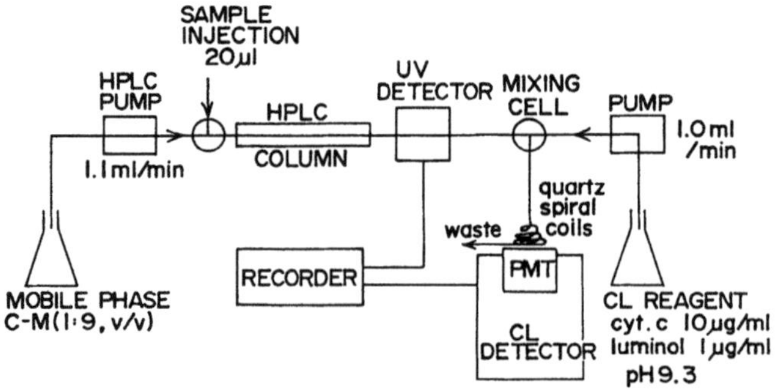

Figure 1. Schematic diagram of CL-HPLC. (PMT, photomultiplier.)

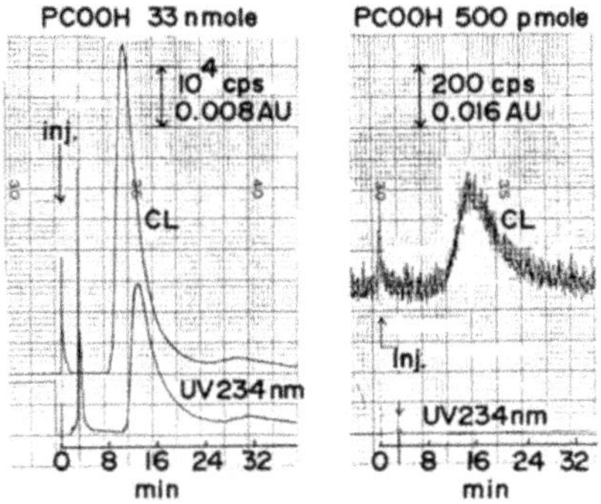

Figure 2. CL-HPLC of authentic PCOOH. (CL, chromatogram obtained with CL detection. UV, chromatogram obtained with the conjugated diene absorption.)

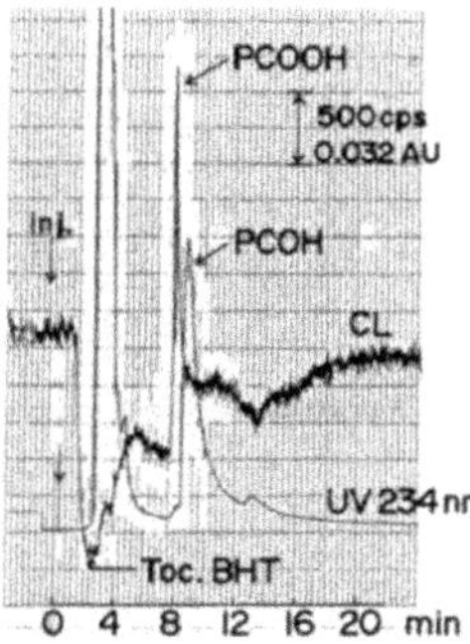

Figure 3. CL-HPLC of PCOOH in plasma total lipid obtained from a healthy female donor.

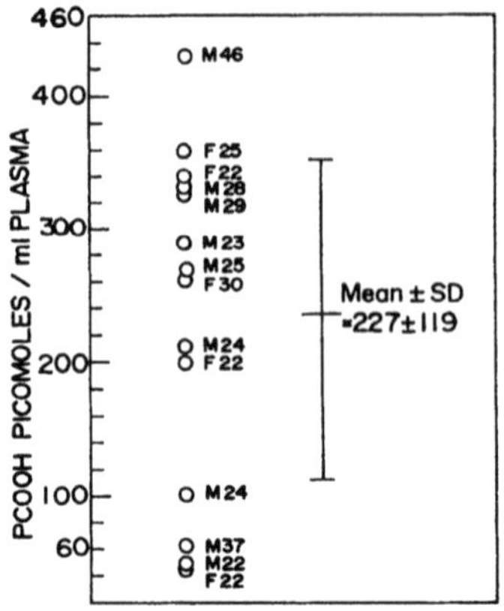

Figure 4. PCOOH levels of blood plasma of 14 healthy donors as estimated by CL-HPLC. (M, male; F, female; figure, age.)

a simple proportional relationship was obtained between the integrated CL counts of PCOOH peak and the concentrations, in the wide range from 50 picomoles to 2000 picomoles. Figure 3 is a typical chromatogram of PCOOH for plasma total lipid of a healthy female donor. As shown, a single PCOOH peak was successfully found with the CL detection.

Figure 4 shows PCOOH levels in blood plasma of 14 healthy donors estimated by CL-HPLC. The PCOOH is in the range from 50 to 430 picomoles per ml of plasma, and was detected for all the blood samples examined. The mean value is 227 picomoles per ml of plasma.

REFERENCES

1. W. S. M. Geurts Van Kessel, W. M. A. Hax, R. A. Damel and J. De Gier, High performance liquid chromatographic separation and direct ultraviolet detection of phospholipids, Biochim. Biophys. Acta 486 :524 (1977).
2. W. M. A. Hax and W. S. M. Geurts Van Kessel, High-performance liquid chromatographic separation and photometric detection of phospholipids, J. Chromatogr. 142 :735 (1977).
3. J. Terao, I. Asano and S. Matsushita, Preparation of hydroperoxy and hydroxy derivatives of rat liver phosphatidylcholine and phosphatidylethanolamine, Lipids 20 :312 (1985).
4. T. Miyazawa, K. Yasuda and K. Fujimoto, Chemiluminescence - high performance liquid chromatography of phosphatidylcholine hydroperoxide, Anal. Lett. 20 :915 (1987).
5. T. Miyazawa, K. Fujimoto and T. Kaneda, Detection of picomole levels in lipid hydroperoxides by a chemiluminescence assay, Agric. Biol. Chem. 51 :2569 (1987).
6. H. Inaba, A. Yamagishi, C. Takyu, B. Yoda, Y. Goto, T. Miyazawa, T. Kaneda and A. Saeki, Development of an ultra-high sensitive photon counting system and its application to biomedical measurements, Optics and Lasers in Engineering 3 :125 (1982).
7. T. Kaneda and T. Miyazawa, Lipid peroxides and nutrition, Wld. Rev. Nutr. Diet. 50 :186 (1987).

METHOD FOR TRACING OXYGEN-18 IN VIVO:

APPLICATION TO OZONE DOSIMETRY IN ANIMALS

Mourad Aissa and Gary E. Hatch*

Northrop Services, Inc. – Environmental Sciences
Research Triangle Park, NC 27709, U.S.A.
*Inhalation Toxicology Division of U.S. Environmental Protection Agency
Research Triangle Park, NC 27711, U.S.A.

INTRODUCTION

The lack of a long-lived radioactive isotope of oxygen has impeded tracing studies involving oxygen labels. The Environmental Protection Agency (EPA) has investigated new methods for tracing oxygen-18 in vivo [1,2] to detect reaction products in inhaled ozone (O_3) and other pollutants in tissues and for investigating mechanisms of toxicity. This report describes a technique for exposing experimental animals to oxygen-18 labeled ozone ($^{18}O_3$), then converting oxygen present in respiratory tissues to CO_2 prior to isotope ratio mass spectrometry to determine the enrichment of oxygen-18 (^{18}O) in the tissue. Since O_3 toxicity is believed to arise from either direct reaction of O_3 with tissue constituents or from initiation of free radical-induced autoxidations, the ^{18}O-tracing techniques should be applicable to detecting these as well as other types of oxygen reactions. It was hoped that this approach would be more generally applicable than approaches in which a specific reaction product was extracted then analyzed for the presence of the label.

MATERIALS AND METHODS

In the methods used, animals were exposed for 1 h to 1.0 ppm $^{18}O_3$. The $^{18}O_3$ was produced using a mixture of nitrogen and 5% $^{18}O_2$, which was introduced into a silent arc O_3 generator. The formed $^{18}O_3$ was then diluted with purified air before passing into the chamber. Immediately following the exposures of the animals, tissues were removed, lyophilized, and 2- to 3-mg samples weighed in silver cups and placed into a CNOS analyzer (Carlo Erba Inc., Peabody, MA). Figure 1 shows a schematic of the method for determining ^{18}O content of the tissues. The elemental oxygen content was determined by the elemental analyzer from the yield of CO resulting from tissue pyrolysis.[3] The effluent CO was oxidized to CO_2 by passing it through a column of iodine pentoxide at 120°C. The CO_2 was condensed into a glass tube using a series of cryogenic traps, then its isotopic composition was determined by isotope ratio mass spectrometry.[4] Benzoic acid standards containing different amounts of ^{18}O were prepared by isotopic exchange with ^{18}O-labeled water[5] and were directly decarboxylated (500°C, 10 min) to CO_2 to determine their isotopic composition.[6] These standards were used to determine parameters of a mathematical model for correcting isotopic values obtained from the CO_2. Corrections were necessary because of memory effects (i.e., oxygen atoms from previous samples) and blank effects (oxygen atoms from quartz tubes or possibly from impurities in the helium stream used by the elemental analyzer). Mathematical models were based on nonlinear regression of isotopic data obtained from samples of different mass and isotopic composition.

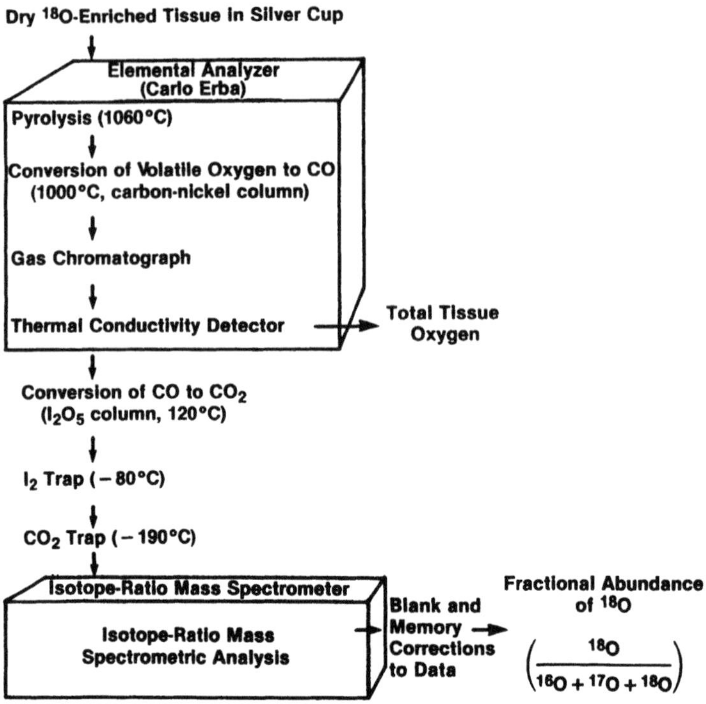

Figure 1. Schematic diagram of the procedure for sample preparation and analysis used to trace ^{18}O in tissue sample.

RESULTS

The mathematical correction model is an isotopic and mass balance equation that relates the ^{18}O fractional abundance of the ith sample, $F_{s,i}$, to the fractional abundances of the formed CO, the I_2O_5 i reaction column, and the blank ($F_{CO,\,i}$, F_I, F_B). $N_{s,i}$ and N_B are, respectively, the number of oxygen atoms present in the ith sample and the number of oxygen atoms that originate from the blank. Parameters a and b represent the exponential nature of the memory effect.

$$F_{s,i} = F_{CO,i} + \frac{N_B(F_{CO,i} - F_B) + \sum_{j=1}^{i-1} N_{s,i-j} a e^{-jb} (F_{CO,i} - F_{s,i-j})}{N_{s,i}\,(1 - \sum_{j=1}^{j-1} a e^{-jb})}$$

As seen in Table 1, preliminary data from animals exposed to $^{18}O_3$ indicated measurable enrichments in ^{18}O in respiratory tissue including lung, trachea, and nasal epithelium. In contrast, blood of the same animals showed no measurable ^{18}O enrichment. The oxygen percentages and the ^{18}O enrichments shown are equivalent to 0.1 to 1.0 nM of ^{18}O per milligram of dry tissue. No enrichment in tissue ^{18}O was observed after a similar exposure of animals to air containing a 60-ppm excess concentration of $^{18}O_2$. Therefore, the high reactivity of the O_3 molecule appeared to be responsible for the uptake of the ^{18}O rather than metabolic incorporation of the label from oxygen.

TABLE 1. Excess Oxygen-18 in Tissues of Animals Exposed to 1.0 ppm $^{18}O_3$

Experiment	Animal	Tissue	n	Tissue Oxygen, Percent of Dry Weight	^{18}O Enrichment µmoles ^{18}O/mole Total Oxygen[b]
1	Mouse	Lung	5	19.3 ± 0.5	14.8 ± 7.8
2	Rat	Lung	5	18.6 ± 1.8	6.6 ± 6.4
		Trachea	4	21.3 ± 1.2	12.4 ± 8.9
		Nasopharynx	4	19.5 ± 2.3	7.3 ± 1.5
		Blood	4	22.3 ± 0.3	−1.1 ± 0.0
3	Rabbit	Lung	2	19.9 ± 0.1	15.6 ± 3.5
		Alveolar Macrophage	2	14.6 ± 0.4	45.2 ± 3.0
		Lung Surfactant Pellet	2	10.1 ± 0.8	55.0 ± 6.5

[a] Means ± S.D. are shown for n animals. Exposure duration was 1 h.
[b] Relative to the same tissue from animals exposed to air.

CONCLUSIONS

In summary, techniques involving pyrolysis of tissues to CO_2 followed by isotope ratio mass spectrometry appeared to be effective in tracing reaction products of inhaled $^{18}O_3$ in vivo. Such techniques appear to have the advantage of tracing the oxygen molecule in any tissue sample or fraction without prior purification to individual components. Determination of the relative concentration of O_3 products in DNA, lipid, and cytosolic protein components should be possible using these techniques. It should also be possible to determine the extent of autoxidation reactions occurring during O_3 and other chemical exposures by simultaneously exposing animals to both $^{18}O_2$ and to the free-radical generating species.

REFERENCES

1. J.M. Hayes and J. Santrock, Isotopic study of the inhalation toxicology of oxidants. National Technical Information Service No. PB86-109485. EPA #600/1-85/013 (1985).
2. J. Santrock and J.M. Hayes, Adaptation of the Unterzaucher procedure for determination comof oxygen-18 in organic substances. Anal. Chem. 59:119 (1987).
3. E. Pella and B. Colmbo, Improved instrumental determination of oxygen in organic pounds. Anal. Chem. 44:1563 (1972).
4. J. Santrock, S.A. Studley, and J.M. Hayes, Isotopic analyses based on the mass spectrum of carbon dioxide. Anal. Chem. 57:1444 (1985).
5. I. Roberts and H.C. Urey, Kinetics of the exchange between benzoic acid and water. J. Am. Chem. Soc. 61:2580 (1939).
6. J. Santrock and J.M. Hayes, Pyrolytic decarboxylation of aromatic acids as a means of isotopic analysis. Anal. Chem. 57:1441 (1985).

Disclaimer: The research described in this article has been reviewed by the Health Effects Research Laboratory, U.S. Environmental Protection Agency, and approved for publication. Approval does not signify that the contents necessarily reflect the views and policies of the Agency nor does mention of trade names or commercial products constitute endorsement or recommendation for use.

MALONDIALDEHYDE DERIVATIVES IN URINE

H.H. Draper and M. Hadley

Department of Nutritional Sciences
College of Biological Science
University of Guelph, Guelph, Ontario, Canada N1G 2W1

INTRODUCTION

Malondialdehyde (MDA) excretion in rat urine is responsive to peroxidation of dietary lipids and to factors which increase lipid peroxidation *in vivo*: vitamin E deficiency and administration of iron, CCl_4 or adriamycin. Urinary MDA is also increased by administration of the lipolytic hormones ACTH and epinephrine and by fasting, particularly when the tissues have been enriched by feeding highly unsaturated fish oil fatty acids.[1,2]

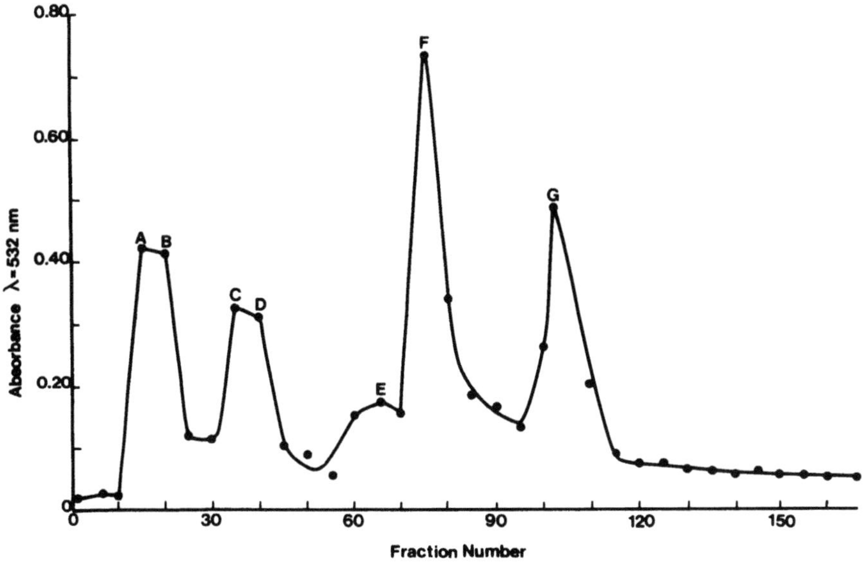

Fig. 1. Elution profile of MDA derivatives obtained by anion exchange chromatography of urine from rats fed a vitamin E-deficient diet containing cod liver oil. MDA was determined spectrophotometrically as the thiobarbituric acid derivative.[3]

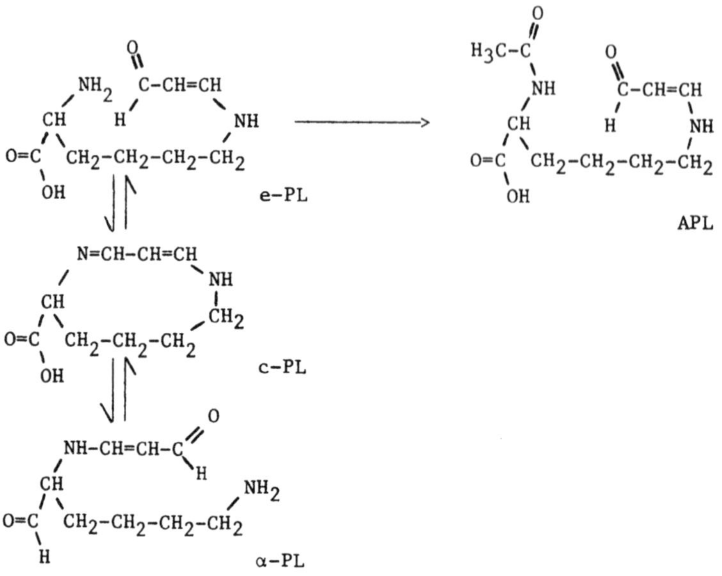

Fig. 2. Structures of lysine-MDA adducts and of the acetylated lysine adduct excreted in urine.

Anion exchange chromatography of rat urine revealed the presence of several acid-hydrolyzable MDA derivatives (Fig. 1). One of two major metabolites (1F) was purified by Sephadex chromatography and reverse phase HPLC, and shown to be the acetylated lysine-MDA adduct N-α-acetyl-ε-(2-propenal)lysine (APL).[4] The other (1A) was identified as N-ε-(2-propenal)-lysine (ε-PL) in a cyclized form (c-PL) (Fig. 2).

The reaction of MDA with lysine yields enaminals at the α- and ε-position, each of which, when isolated and stored in aqueous solution, formed a mixture of all three compounds. Cyclization of ε-PL was found to occur during anion exchange chromatography of urine. Neither α-PL nor c-PL appears to be a excretory form of MDA.

The possible occurrence in urine of MDA adducts of the phospholipid amino bases serine and ethanolamine (Fig. 3) was investigated. Synthetic 1:1 serine-MDA and ethanolamine-MDA enaminals were prepared.[4] These adducts were found to co-chromatograph with two urinary MDA metabolites (Fig. 1E and 1B, respectively) throughout the isolation procedure. Acid hydrolysis of

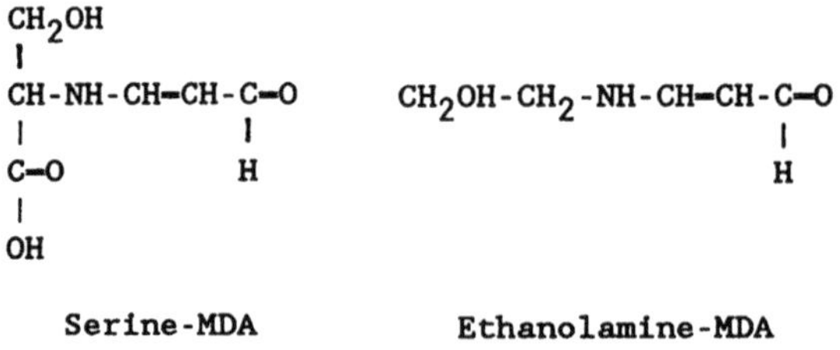

Fig. 3. Structures of the enaminals formed by reactions of MDA with serine and ethanolamine.

purified metabolite 1E yielded serine and MDA in a 1:1 molar ratio.[5] Metabolite 1B was similarly identified as a 1:1 adduct of ethanolamine and MDA (unpublished results).

Free MDA (Fig. 1G) is evident in the urine of rats fed diets containing highly peroxidized oils, but is present in negligible amounts in the urine of rats fed a natural diet. The identity of metabolites 1C and 1D is being investigated.

Excretion of the two lysine adducts of MDA by rats is decreased by feeding a peroxide-free diet or by short-term fasting; however, small amounts of these metabolites are still found in the urine, indicating that they are formed by lipid peroxidation *in vivo* as well as in the diet. ε-PL has been identified as the main form of MDA absorbed from the intestine, where it is apparently formed by a reaction between free MDA derived from the oxidative decomposition of dietary polyunsaturated fatty acids and the ε-amino groups of lysine residues of food proteins, followed by its release during proteolysis.[6] It is partially acetylated before excretion, probably by an enzyme in the liver, where N-acetylation is a common detoxification reaction. The serine and ethanolamine adducts appear to be entirely of endogenous origin.

Fractionation of human urine revealed a preponderance of the two lysine derivatives, plus small amounts of the adducts of serine and ethanolamine and of compounds corresponding to C and D in rat urine (Fig. 1) (unpublished results). Work on the identity of these compounds is in progress. Free MDA was undetectable.

The excretion of total MDA in urine, determined using a specific HPLC procedure,[1] is a sensitive index of MDA formation in the diet and tissues of humans as well as rats. Increases in MDA excretion have been observed in humans following the ingestion of cod liver oil unsupplemented with antioxidants.[7] The excretion of serine-MDA and ethanolamine-MDA adducts may be a useful index of the oxidative decomposition of polyunsaturated fatty acids *in vivo*. Identification of the remaining MDA metabolites detected in urine may provide clues to other biochemical events involved in lipid peroxidation *in vivo*. The form(s) in which MDA is generated in response to specific catalysts of lipid peroxidation is of interest from the standpoint of their possible toxicological significance.

ACKNOWLEDGMENTS

This research was supported by the Natural Sciences and Engineering Research Council of Canada and the National Cancer Institute of Canada.

REFERENCES

1. H.H. Draper, L. Polensek, M. Hadley and L.G. McGirr, Urinary malondialdehyde as an indicator of lipid peroxidation in the diet and in the tissues, *Lipids* 19:836 (1984).
2. S.N. Dhanakoti and H.H. Draper, Response of urinary malondialdehyde to factors which stimulate lipid peroxidation *in vivo*, *Lipids* 22:643 (1987).
3. R.O. Sinnhuber and T.C. Yu, 2-Thiobarbituric acid method for the measurement of rancidity in fishery products. II. The quantitative determination of malonaldehyde, *Food Technol.* 12:9 (1958).
4. L.G. McGirr, M. Hadley and H.H. Draper, The identification of N-α-acetyl-ε-(2-propenal)lysine as a urinary metabolite of malondialdehyde, *J. Biol. Chem.* 260:15427 (1985).

5. M. Hadley and H.H. Draper, Identification of N-2-propenal serine as a urinary metabolite of malondialdehyde, Federation J. (in press).
6. L.A. Piché, P.D. Cole, M. Hadley, R. van den Bergh and H.H. Draper, Identification of N-ε-(2-propenal)lysine as the main form of malondialdehyde in food digesta, Carcinogenesis (in press).
7. L.A. Piché, H.H. Draper and P.D. Cole, Malondialdehyde excretion by subjects consuming cod liver oil versus a concentrate of n-3 fatty acids, Lipids (in press).

ELECTRON SPIN RESONANCE OF THE SEMIQUINONE OF THE ANTITUMOR ANTIBIOTIC STREPTONIGRIN AND ITS METAL COMPLEXES

Helena S. Soedjak, Barney L. Bales, and Joseph Hajdu

Departments of Chemistry and Physics and Astronomy
California State University at Northridge
Northridge, California 91330, U.S.A.

INTRODUCTION

Mechanistic elucidation of the redox chemistry of aminoquinone antibiotics exhibiting potent antitumor activity has become one of the most timely problems of cancer chemotherapy today[1,2]. Specifically, streptronigrin (Figure 1a), is currently regarded as one of the most effective agents for the treatment of a broad spectrum of human cancers[3], yet its clinical application is greatly limited due to severe cytotoxicity of the drug. It has been reported that streptonigrin is activated *in vivo* by one-electron reduction forming

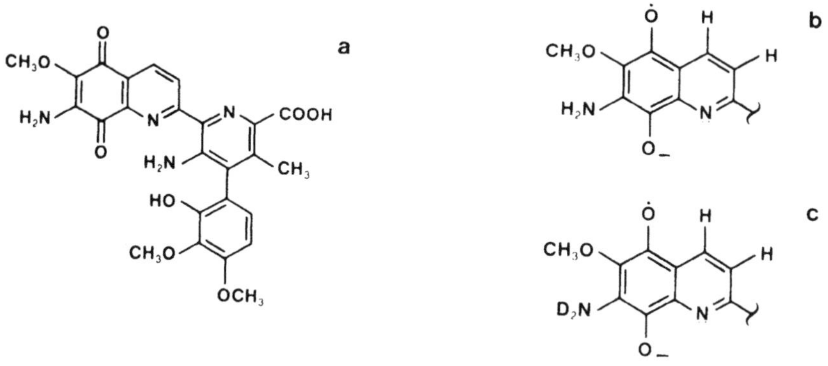

Figure 1. Structure of streptonigrin (a), streptonigrin semiquinone (b) and deuterated semiquinone (c).

the semiquinone[4-6] and that the mechanism of antitumor action of streptonigrin depends on participation of metal ions[7].

Although evidence has been presented for the formation of radicals in bacterial systems[4,5], microsomal preparations[6], as well as enzymological studies[8], identification and characterization of the free radical intermediate remain to be accomplished[9]. Metal ions have been shown to enhance the bactericidal activity of streptonigrin through complexation[10], yet the structure and the chemistry of the complexes remain to be elucidated. In the present work we have used ESR spectroscopy to prove that the free radical intermediate is the semiquinone and to study its structure. Also we provide ESR evidence of the formation of the metal complex formed with Zn^{+2} and Cd^{+2}.

Reduction of 2.0 mM streptonigrin solution in 0.1 M LiOH/MeOH with excess solid $NaBH_4$ (<0.2 mg) followed by air oxidation produces a deep-green solution (eq. 1) whose ESR spectrum is shown in Figure 2a. When samples prepared under these reaction conditions are kept in sealed capillary tubes, the signal persists for about 1 day; however, exposure to oxygen results in disappearance of the ESR spectrum within 5 min.

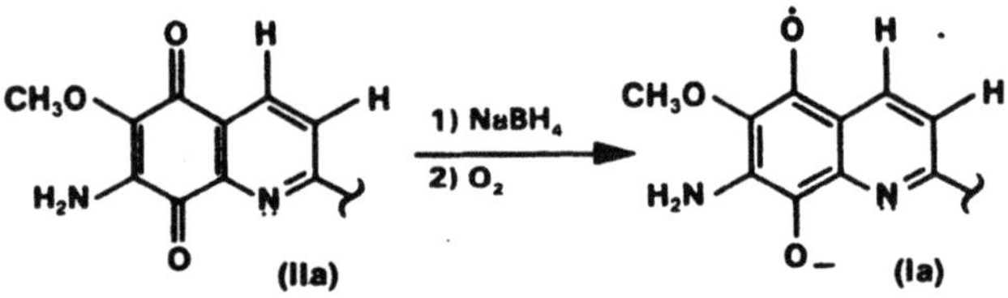

The 18-line well-resolved spectrum (Figure 2a) appears intriguingly simple due to accidentally commensurate coupling constants. This same simplicity, however, leads to ambiguities in making the assignments, since we wish to make no *a priori* assumptions about the number of interacting nuclei. With no restrictions on the number of nitrogens and protons interacting with the unpaired electron, 11 satisfactory combinations are generated by computer simulation. In order to determine the correct assignments, the exchangeable protons were deuterated in streptonigrin and the free radical generated as described above. Experimentally, Streptonigrin was either predeuterated in dioxane-deuterium oxide followed by freeze drying or the fully protonated

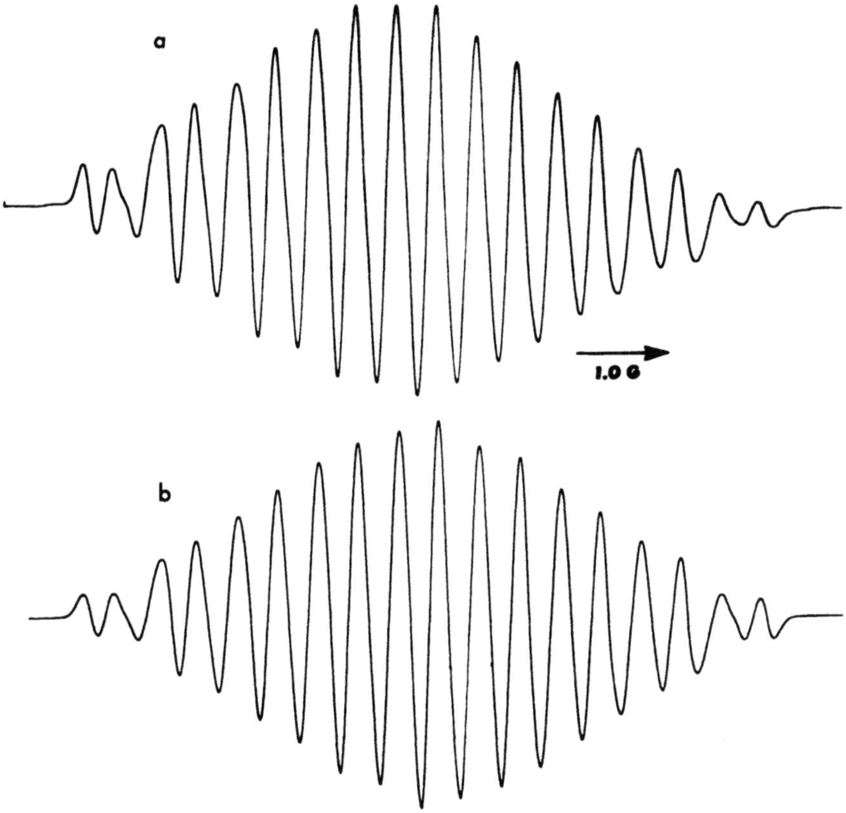

Figure 2. (a) ESR spectrum of streptonigrin semiquinone in MeOH/LiOH. Microwave power, 1.0 mW; modulation amplitude, 0.25 G; 12 G sweep in 500 s with a 320 ms time constant. Reduction of the modulation amplitude resulted in no further resolution. (b) Computer simulation using the parameters in Table 1 and a line width of 0.18 G.

streptonigrin was simply dissolved in deuteromethanol. In these reaction mixtures, 0.1 LiOD and $NaBD_4$ were used. The ESR spectrum, shown in Figure 3, turned out to be the the same irrespective of the method of preparation. This first-derivative spectrum, quite unusual to say the least, is reminiscent of a mixture of absorption and emission spectra with broad lines. The simulation turned out to be a very sensitive function of all of the parameters. Reducing two of the proton couplings by a factor of 6.5[13] and leaving all others constant unequivocally ruled out all of the 11 parameter sets consistent with the spectrum in Figure 1a except one. The computer-simulated spectra of the protonated and the deuterated radical are presented in Figures 2b and 3b respectively and the corresponding coupling constants are listed in Table 1. The simulations use a line shape that is a linear combination of Lorentzian (30% doubly-integrated intensity) and Gaussian(70%).

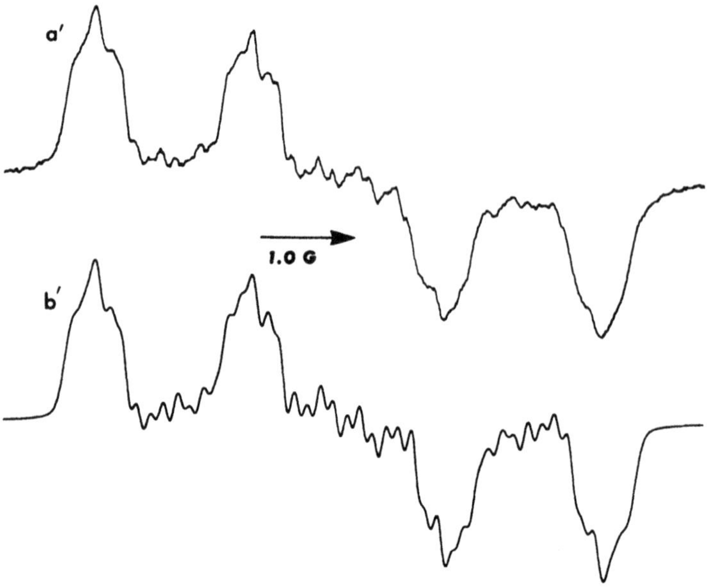

Figure 3. (a) ESR spectrum of deuterated streptonigrin semiquinone in MeOD/LiOD. Microwave power, 2.0 mW; modulation amplitude, o.08 G; 12 G sweep in 500 s with a 320 ms time constant. (b) computer simulation using the parameters in Table 1 and a line width of o.18 G.

Table 1. ESR Parameters[a] for the Streptonigrin Semiquinone in MeOH/LiOH and Deuterated Streptonigrin Semiquinone in MeOD/LiOD, T = 25 °C.

Position		MeOH	MeOD	MeOH/Cd^{2+}	MeOH/Zn^{2+}
	a^{N}_{NH}	= 1.34	1.34	1.19	1.08
	a^{H}_{NH}	= 0.89	0.137[b]	1.00	1.09
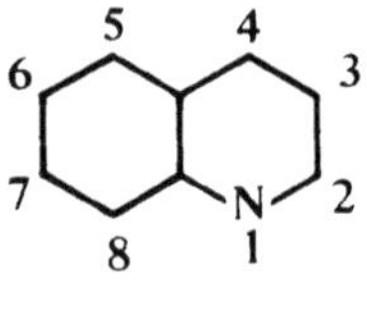	$a^{H}_{OCH_3}$	< 0.05	< 0.05	< 0.05	< 0.05
	a^{H}_{4}	= 1.76	1.76	1.96	2.00
	a^{H}_{3}	= 0.31	0.31	0.43	0.46
	a^{N}_{1}	= 0.46	0.46	0.61	0.67
				g = 2.0043 ± 0.0001	

[a]Values in Gauss. Estimated error ±0.02 G, a^{H} . [b]a^{D}_{ND} , ± 0.005 G.

The assignments in Table 1 clearly show for the first time that the free radical is the semiquinone (eq. 1) and in agreement with those reported for synthetic aminosemiquinones[12] and related systems[14] indicate that the unpaired electron is delocalized over the 5,8-quinoline semiquinone system. The assignment is consistent with a pattern established earlier[15] that in heterocyclic semiquinones, the ring containing the nitrogen atoms has very little effect on the spin density.

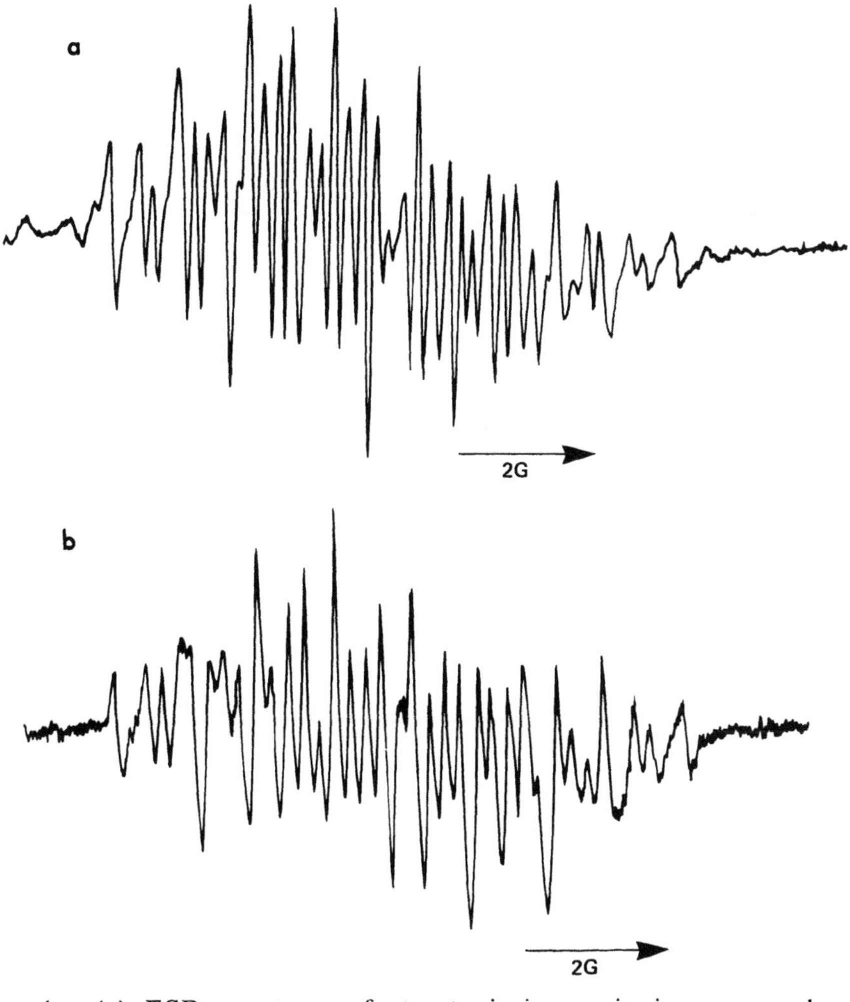

Figure 4. (a) ESR spectrum of streptonigrin semiquinone complexed with Cd^{+2} in MeOH/LiOH. Microwave power, 2.0 mW; modulation amplitude, 0.32 G; 12 G sweep in 500 s with a 320 ms time constant. (b) ESR spectrum of streptonigrin semiquinone complexed with Zn^{+2} in MeOH/LiOH.

Figure 4 shows preliminary results of experiments designed to demonstrate the complexation of metals with the semiquinone. Equimolar solutions of 2.0 mM streptonigrin and Zn^{+2} (from anhydrous zinc chloride) or Cd^{+2} (from cadmium acetate dihydrate) were reduced by $NaBH_4$ and the ESR spectra recorded as described above. The spectra in Figure 4a show lines in the wings due to hyperfine coupling to $^{111}Cd^{+2}$ and $^{113}Cd^{+2}$ in natural abundance (12.75 and 12.26% respectively). The mixture of the two magnetic nuclear isotopes together with nonmagnetic $^{112}Cd^{+2}$ gives a superposition that is complicated but a $^{113}Cd^{+2}$ coupling constant near 4 G may be inferred. Figure 4b when compared with Figure 2a shows that the spin density in the semiquinone is altered upon complexation since the nonmagnetic Zn^{+2} does not lead to further hyperfine structure. Detailed structural determination of the metal complexes is currently underway with the use of isotropically substituted $^{113}Cd^{+2}$.

Preparation and characterization of the semiquinone and its metal complexes should provide an exciting new direction of research toward elucidation of the structural, chemical, and biological properties of this important antitumor antibiotic.

ACKNOWLEDGMENTS

We are grateful to Dr. John D. Douros of Drug Research and Development, Chemotherapy, NCI for a generous gift of streptonigrin. This work was supported by grants from the N. I. H.,the California State University Foundation Northridge (through the Research and Grants Committee) to BLB and JH, and the Research Corporation to JH.

REFERENCES

1. P.A. Andrews, S. S. Pan, N. R. Bachur, Electrochemical Reductive Activation of Mitomycin C.,.J. Am. Chem. Soc. 108:4158(1986).

2. S. J. Danishefsky, M. Egbertson,On the Characterization of Intermediates in the Mitomycin Activation Cascade: A Practical Synthesis of an Aziridinomitosene, J. Am Chem. Soc. 108:4648(1986).

3. H. L. Davis, D. D. Von Hoff, J. T. Henny, M. Rózencweig,The Role of Antitumor Antibiotics in Current Oncologic Practice,Cancer Chemother. Pharmacol. 1:83(1978).

4. J. R. White, H. H. Dearman,Generation of Free Radicals From Phenazine Methosulfate, Streptonigrin, and Riboflavin in Bacterial Suspensions, Proc. Natl. Acad. Sci. (USA) 59:887(1965).

5. K. Ishizu, H. H. Dearman, M. T. Huang, J. R. White, Electron Paramagnetic Resonance Observations in Biogenic Semiquinone and 5-methyl Phenazinium Radicals, Biochim. Biophys. Acta 165:283(1968).

6. N. R. Bachur, S. L. Gordon, H. Kon,NADH Cytochrome P-450 Reductase Activation of quinone Anticancer Agents to Free Radicals, Proc. Natl. Acad. Sci. (USA) 76:954(1979).

7. B. K. Bhuyan,Phleomycin, Xanthomycin, Streptonigrin, Nogalomycin, and Aurantin, in: "Antibiotics", D. Gottlieb, P. D. Shaw eds. Springer Verlag, New York(1967).

8. H. L. White, J. R. White,Lethal Action and Metabolic Effects of Streptonigrin in Escherichia coli, Mol. Pharmacol. 4:549(1968).

9. It should be pointed out that these reports merely presented the recorded ESR spectra as experimental evidence for the formation of radicals, providing no assignments or correlation between the obtained spectra and the putative semiquinone structure (cf. ref. 4-6).

10. J. R. White, Streptonigrin-Transition Metal Complexes: Binding to DNA and Biological Activity, Biochem. Biophys. Res. Commun. 77:387(1977).

11. J. R. Bolton, A. Carrington, J. Dos Santos-Veiga,Electron Spin Resonance Studies of Semiquinones and related N-Heterocyclic Molecules in Acid Solution, J. Mol. Phys. 5:465(1962).

12. J. D. R. Clay, D. J. Murphy, Electron Spin Resonance Spectra of Aminosemiquinones and Diaminosemiquinones, J. Chem. Soc. [Faraday 2] 77:1589(1981).

13. A. Carrington, A. D. McLachlan, "Introduction to Magnetic Resonance", Harper and Row, New York(1967).

14. J. D. R. Clay, D. J. Murphy, The Electron Spin Resonance Spectra of Semiquinones From Some Heterocyclic Quinones, J. Chem. Soc. [Perkin 1] 1781(1984).

15. W. T. Dixon, Electron Spin Resonance Spectroscopy and Free Radical Reactions, Annu. Rep. Chem. Soc. Sect. B. 151(1973).

16. J. Hajdu, Interactions of Metal Ions with Streptonigrin and Biological Properties of the Complexes, in:"Metal Ions in Biological Systems", H. Siegel ed., M. Dekker, Inc., New york(1985).

THE DETECTION OF SINGLET OXYGEN IN BIOCHEMICAL SYSTEMS USING 1268 nm CHEMILUMINESCENCE

Jeffrey R. Kanofsky

Medical Service
Edward Hines, Jr., Veterans Administration Hospital
Hines, Illinois 60141, U.S.A.

Department of Medicine
Loyola University Stritch School of Medicine
Maywood, Illinois 60153, U.S.A.

INTRODUCTION

The production of singlet oxygen has been proposed to contribute to the deleterious effects of a number of biochemical processes including lipid peroxidation, the generation of superoxide anion and the neutrophil respiratory burst.[1-3] While a variety of methods have been developed to detect singlet oxygen in biochemical systems, the measurement of singlet oxygen emission at 1268 nm has proved to be particularly useful.[4,5] This paper will discuss theoretical and practical factors of relevance to the technique and then review data obtained using this method from a number of biochemical systems.

EXPERIMENTAL METHODS

The $^1\Delta_g$ state of oxygen is the only electronically excited oxygen species to have a significant half-life in the biochemical environment (of the order of 1 μs). Transition from the $^1\Delta_g$ state to the ground state ($^3\Sigma_g$) results in 1268 nm emission.[6] The detection of singlet oxygen in biochemical systems by searching for 1268 nm emission was first suggested by Khan and Kasha.[7] Advances in infrared detector technology have now made possible the measurement of the weak 1268 radiation produced by the low concentrations of singlet oxygen present in biochemical systems.

Several investigators have looked for dimol singlet oxygen emission at 634 and 703 nm in biochemical systems because photomultiplier tubes, highly sensitive to light at these wavelengths, are readily available.[8-10] The dimol emission bands result from the interaction of two molecules of singlet oxygen and consequently vary with the square of the singlet oxygen oxygen concentration.[11] When compared to the use of 1268 nm emission, however, this approach has two disadvantages. First, the quadratic dependence of the dimol emission intensity upon the singlet oxygen concentration makes quantitative measurements of singlet oxygen

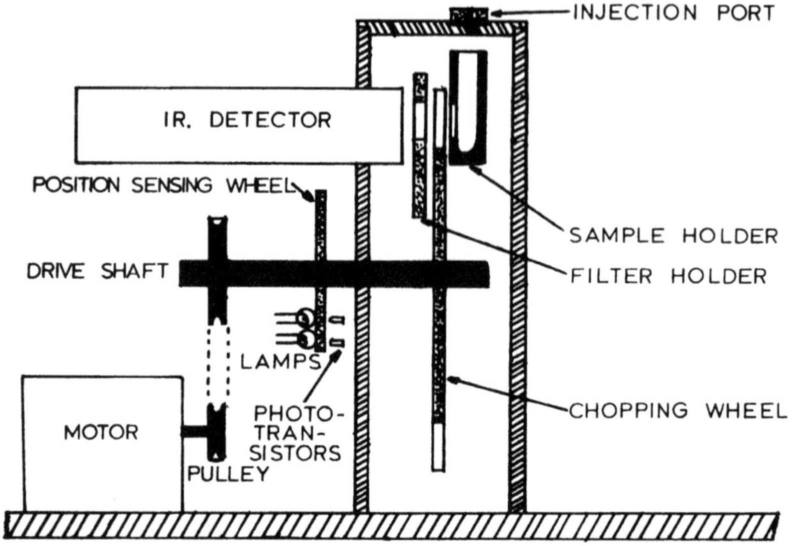

Figure 1. Chemiluminescence spectrometer for measuring near infrared emission from biochemical systems. An injection port permits measurement of light emission immediately after a reactant is added. The lamps, position sensing wheel and phototransistors provide a reference signal for the lock-in amplifier. For other details see text.

yields more difficult. Second, interference from other chemiluminescence processes is more often a problem with dimol emission than monomol emission because the absolute intensity of the dimol bands in biochemical systems is several orders of magnitude lower than the monomol band.[12]

Figure 1 shows a chemiluminescence spectrometer designed to measure near-infrared emission from biochemical systems.[4] A liquid-nitrogen-cooled germanium diode (E0817L, North Coast Optical Systems and Sensors, Santa Rosa, California) provides high sensitivity to near infrared radiation from 1000 to 1700 nm. A set of 7 interference filters with 100 nm bandwidths and center wavelengths from 1070 to 1680 nm selects the wavelength of interest. A chopping wheel modulates the light output of the sample at 37 Hz before reaching the germanium diode. Synchronous amplification of the E0817L output with an Evans Associates model 4110 lock-in amplifier (Berkeley, California) substantially improves the signal to noise ratio.

Quantitative measurements of singlet oxygen production in biochemical systems require a singlet oxygen calibration standard. The hydrogen peroxide plus hypochlorous acid reaction, which stoichiometrically produces singlet oxygen, is generally used for this purpose.[13-15] Calibration of the E0817L detection system using this standard demonstrates good linearity over five to six orders of magnitude.[15]

Since there is no theoretical reason that a species other than singlet oxygen cannot emit at 1268 nm, it is desirable to obtain additional evidence to support the assignment of 1268 nm emission to singlet oxygen. Three types of measurements are useful: (1) spectral analysis of the light emission showing a single band near 1268 nm; (2) a large increase in light emission when light water solvent is replaced by deuterium oxide (the half-life of singlet oxygen is 15 times longer in deuterium oxide than in light water); and (3) a decrease in the light intensity caused by known singlet oxygen quenchers or traps.[15]

BIOCHEMICAL SYSTEMS

Peroxidases

Initial interest in possible singlet oxygen production from peroxidases came from studies of human polymorphonuclear leukocytes.[1] The polymorphonuclear leukocyte is a phagocytic cell responsible for the destruction of bacteria. The production of potent oxidants by a metabolic process called the respiratory burst kills microorganisms and contributes to the tissue damage associated with inflammation. Allen et al. were the first to demonstrate that the leukocyte respiratory burst was accompanied by the production of light at visible wavelengths.[1] These authors proposed that singlet oxygen was produced by the leukocyte and could explain both the neutrophil chemiluminescence and the bacterial killing.[1] A plausible reaction sequence for singlet oxygen production by leukocytes is as follows[1,16-18]:

$$O_2 + NADPH \xrightarrow{\text{NADPH oxidase}} O_2^{\cdot -} + NADP + H^+ \quad (1)$$

$$2\,O_2^{\cdot -} + 2\,H^+ \xrightarrow{\text{Superoxide dismutase}} H_2O_2 + O_2 \quad (2)$$

$$H^+ + H_2O_2 + Cl^- \xrightarrow{\text{Myeloperoxidase}} HOCl + H_2O \quad (3)$$

$$H_2O_2 + HOCl \longrightarrow H_2O + H^+ + Cl^- + O_2(^1\Delta_g) \quad (4)$$

There is good biochemical evidence for reactions 1 through 3.[16-18] In addition, Reaction 4 is known to stoichiometrically produce singlet oxygen.[13]

Several peroxidase-hydrogen-peroxide-halide systems have now been shown to produce 1268 nm chemiluminescence.[4,5,14,15,19-21] Spectral analysis of the emission, enhancement of the light production in deuterium oxide and inhibition of light production by the singlet oxygen quenchers azide anion and histidine confirm the assignment of this emission to singlet oxygen.[4,5,14,15,19-21] Lactoperoxidase, chloroperoxidase and myeloperoxidase have all been shown to produce singlet oxygen.[4,5,14,15,19-21] As shown in Figure 2, horseradish peroxidase, which slowly oxidizes bromide ion,[22] also produces singlet oxygen.

Reactions 3 and 4 predict that one molecule of singlet oxygen will be produced for every two molecules of hydrogen peroxide consumed. Under optimal conditions, the oxidation of bromide ion by lactoperoxidase, chloroperoxidase or myeloperoxidase produces near stoichiometric amounts of singlet oxygen.[14,15,20-21] The chloroperoxidase-hydrogen-peroxide-chloride system is equally efficient.[14] In contrast, the myeloperoxidase-hydrogen-peroxide-chloride system (the one present in the human polymorphonuclear leukocyte) is always inefficient.[20] In acid pH, the yield is only 15 to 20 percent, while at neutral pH the yield is extremely small due to rapid enzyme inactivation.[20]

Singlet oxygen production from intact polymorphonuclear leukocytes has not been detected using either 1268 nm emission or cholesterol trap-

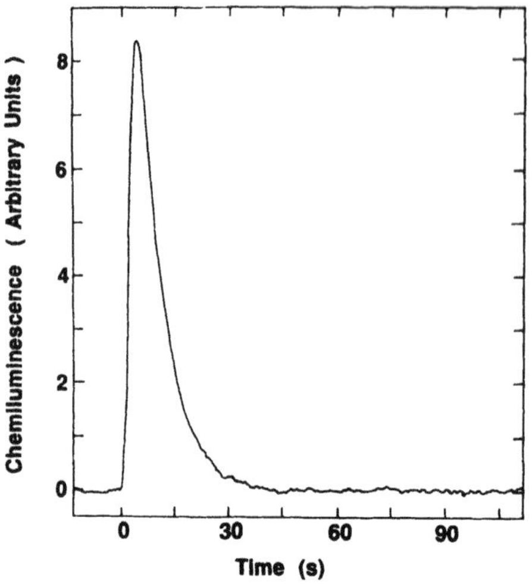

Figure 2. Time course of 1268 nm chemiluminescence for the horseradish-peroxidase-hydrogen-peroxide-bromide system. Conditions: 10 μM horseradish peroxidase, 2 mM hydrogen peroxide, 200 mM sodium bromide, pH 4.0, 100 mM sodium acetate buffer, light water solvent.

ping.[23,24] The hydrogen peroxide plus hypochlorous acid reaction is apparently too slow to compete with the consumption of hypochlorous acid by a variety of biomolecules. The production of chlorinated oxidants, such as chloramines, greatly exceeds any singlet oxygen production by the intact leukocyte.[25,26]

Soybean Lipoxygenase Isozymes

Lipoxygenases are another enzymatic source of singlet oxygen. Under appropriate conditions, the oxidation of linoleic acid by either soybean lipoxygenase-1 or soybean lipoxygenase-3 produces modest quantities of singlet oxygen.[27] These enzymatic oxidations are believed to produce carbon-centered radicals and peroxyl radicals as intermediates.[28,29] The reaction of two peroxyl radicals via a Russell mechanism constitutes a plausible mechanism for the observed singlet oxygen production.[30,31]

$$2RR'CHOO^{\cdot} \longrightarrow RR'CHOH + RR'CO + O_2(^1\Delta g) \qquad (5)$$

The biochemical requirements for singlet oxygen production by soybean lipoxygenase isozymes correlate with those for the co-oxidation of dyes such as β-carotene.[27,32,33] Lipoxygenase-3 requires high concentrations of hydroperoxide for either singlet oxygen production or for co-oxidation.[33] Lipoxygenase-1 has moderate co-oxidizing activity only under anaerobic conditions.[32] Significant quantities of singlet oxygen are produced by lipoxygenase-1 only at low oxygen concentrations.[27] At high oxygen concentrations, radical production by lipoxygenase-1 is suppressed. Under anaerobic conditions, carbon-centered radicals have no oxygen to form peroxyl radicals. The addition of organic solvent to the lipoxygenase-1 system greatly increases the yield of singlet oxygen. This may result from a decrease in the stability of the enzyme-substrate complex causing increased radical production. Lipoxygenase-2 has low cooxidation activity and does not produce detectable quantities of singlet oxygen.[27,33]

Bleomycin

Bleomycin is an antineoplastic antibiotic used to treat several types of human cancers. Its clinical use is limited by a number of toxic effects including a syndrome of pulmonary fibrosis and inflammation. Both the cytotoxic action of bleomycin against cancer cells and its toxicity against patients are believed related to the production of potent oxidants by a bleomycin-transition metal complex. Oxidative DNA cleavage is easily demonstrated in model systems.[34]

The Fe(III)-bleomycin complex catalyzes the decomposition of hydroperoxides of unsaturated fatty acids to produce modest quantities of singlet oxygen.[35] When hydrogen peroxide or a number of saturated hydroperoxides are used as substrates, no singlet oxygen is produced.[35] The hydroperoxide requirements for singlet oxygen production by Fe(III)-bleomycin complex are the same as those for heme-containing enzymes.[35]

The reaction of peroxyl radicals via a Russell mechanism is a plausible explanation for the singlet oxygen produced by Fe(III)-bleomycin.[30,31,35] The work of Dix et al. provides an explanation for the fact that only unsaturated hydroperoxides produce singlet oxygen in the Fe(III)-bleomycin system.[36] These authors have studied the peroxyl radical production from the hematin-catalyzed decomposition of hydroperoxides and found peroxyl radicals are generated only from unsaturated hydroperoxides.[36] These authors postulate the initial formation of an alkoxyl radical which rearranges to a carbon-centered radical. This rearrangement requires the presence of a carbon-carbon double band. The carbon-centered radical then reacts with oxygen to form a peroxyl radical.

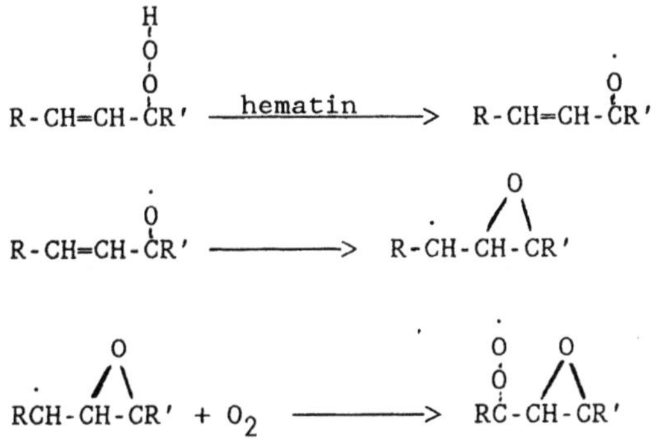

(6)

(7)

(8)

The relationship between singlet oxygen production by Fe(III)-bleomycin complex and the therapeutic or toxic effects of bleomycin will require further study.

CONCLUSION

At the current time, measurements of 1268 nm emission have been reported in only a few biochemical systems. Many other potential biochemical pathways to singlet oxygen have yet to be explored. Particularly interesting is the recent suggestion of Weiss et al. that human eosinophils may produce singlet oxygen by the selective oxidation of bromide with eosinophil peroxidase.[37]

ACKNOWLEDGEMENTS

This work is supported by Grant GM32974 from the National Institutes of Health and by the Veterans Administration Research Service. I wish to thank William Wardman for technical assistance in performing experiments and Anita Osis for assistance in preparation of the manuscript.

REFERENCES

1. R. C. Allen, R. L. Stjernholm, and R.H. Steele, Evidence for the generation of electronic excitation state(s) in human polymorphonuclear leukocytes and its participation in bacterial activity, Biochem. Biophys. Res. Commun. 47:679 (1972).

2. A. U. Khan, Singlet molecular oxygen from superoxide anion and sensitized flourescence of organic molecules, Science 168:476 (1970).

3. R. M. Howes and R. H. Steele, Microsomal (μS) chemiluminescence (CL) induced by NADPH and its relation to lipid peroxidation. Res. Commun. Chem. Pathol. Pharmacol. 2:619 (1971).

4. J. R. Kanofsky, Singlet oxygen production by lactoperoxidase, evidence from 1270 nm chemiluminescence, J. Biol. Chem. 258:5991 (1983).

5. A. U. Khan, P. Gebauer, and L.P. Hager, Chloroperoxidase generation of singlet Δ molecular oxygen observed directly by spectroscopy in the 1- to 1.6 μm region, Proc. Natl. Acad. Sci. USA 80:5195 (1983).

6. R. J. Browne and E. A. Ogryzlo, Chemiluminescence from the reaction of chlorine with hydrogen peroxide, Proc. Chem. Soc. 117(1964).

7. A. U. Khan and M. Kasha, Direct spectroscopic observation of singlet oxygen emission at 1268 nm excited by sensitizing dyes of biological interest in liquid solution. Proc. Natl. Acad. Sci. USA 76:6047 (1979).

8. E. Cadenas and H. Sies, Singlet oxygen formation detected by low-level chemiluminescence during enzymatic reduction of prostaglandin G2 to H2, Hoppe-Seyler's Z. Physiol. Chem. 364:519 (1983).

9. E. Cadenas, R. P. Daniele, and B. Chance, Low level chemiluminescence of alveolar macrophages, Spectral evidence for singlet oxygen generation, FEBS Letters 123:225 (1981).

10. B. D. Cheson, R. S. Christensen, R. Sperling, B. E. Koehler, and B. M. Babior, The origin of the chemiluminescence of phagocytosing granulocytes, J. Clin. Invest. 58:789 (1976).

11. R. J. Browne and E. A. Ogryzlo, The yield of singlet oxygen in the reaction of chlorine with hydrogen peroxide, Can. J. Chem. 43:2915 (1965).

12. C. S. Foote, Detection of singlet oxygen in complex systems: A critique, in: "Biochemical and Clinical Aspects of Oxygen," W. S. Caughey, ed., Academic Press, New York (1979).

13. A. M. Held, D. J. Halko, and J. K. Hurst, Mechanism of chlorine oxidation of hydrogen peroxide, J. Am. Chem. Soc. 100:5732 (1978).

14. J. R. Kanofsky, Singlet oxygen production by chloroperoxidase-hydrogen peroxide-halide systems, J. Biol. Chem. 259:5596 (1983).

15. J. R. Kanofsky, Singlet oxygen production by lactoperoxidase: Halide dependence and quantitation of yield, J. Photochem. 25:105 (1984).

16. B. M. Babior, R. S. Kipnes, and J. T. Curnutte, The production by leukocytes of superoxide, a potential bactericidal agent, J. Clin. Invest. 52:741 (1973).

17. J. E. Harrison and J. Schultz, Studies on the chlorinating activity of myeloperoxidase, J. Biol. Chem. 251:1371 (1976).

18. C. C. Winterbourn, Comparative reactivities of various biological compounds with myeloperoxidase-hydrogen peroxide-chloride, and similarity of the oxidant to hypochlorite, Biochim, Biophys. Acta. 840:204 (1985).

19. A. U. Khan, Enzyme system generation of singlet ($^1\Delta g$) molecular oxygen observed directly by 1.0-1.8 μm luminescence spectroscopy, J. Am. Chem. Soc. 105:7195 (1983).

20. J. R. Kanofsky, J. Wright, G. E. Miles-Richardson, and A. I. Tauber, Biochemical requirements for singlet oxygen production by purified human myeloperoxidase, J. Clin. Invest. 74:1489 (1984).

21. A. U. Khan, Myeloperoxidase singlet molecular oxygen generation detected by direct infrared electronic emission, Biochem. Biophys. Res. Commun. 122:668 (1984).

22. P. L. Ashley and B. W. Griffin, Chloroperoxidase-catalyzed halogenation of antipyrine, a drug substrate of liver microsomal cytochrome P-450, Arch. Biochem. Biophys. 210:167 (1981).

23. J. R. Kanofsky and A. I. Tauber, Non-physiologic production of singlet oxygen by human neutrophils and by the myeloperoxidase-H_2O_2-halide system, Blood 62:82a (1983).

24. C. S. Foote, R. B. Abakerli, R. L. Clough, and R. I. Lehrer, On the question of singlet oxygen production in polymorphonuclear leukocytes, in: "Bioluminescence and Chemiluminescence," M. A. Deluca and W. D. McElroy, eds., Academic Press, Inc., New York (1981).

25. C. S. Foote, T. E. Goyne, and R. I. Lehrer, Assessment of chlorination by human neutrophils, Nature 301:715 (1983).

26. S. T. Test, M. B. Lampert, P. J. Ossanna, J. G. Thoene, and S. J. Weiss, Generation of nitrogen-chlorine oxidants by human phagocytes, J. Clin. Invest. 74:1341 (1984).

27. J. R. Kanofsky and B. Axelrod, Singlet oxygen production by soybean lipoxygenase isozymes, J. Biol. Chem. 261:1099 (1986).

28. J. J. M. C. de Groot, G. A. Veldink, J. F. G. Vliegenthart, J. Boldingh, R. Wever, and B. F. van Gelder, Demonstration by EPR spectroscopy of the functional role of iron in soybean lipoxygenase-1, Biochem. Biophys. Acta 377:71 (1975).

29. P. Schieberle, W. Grosch, H. Kexel, and H.-L. Schmidt, A study of oxygen scrambling in the enzymic and non-enzymic oxidation linoleic acid, Biochim. Biophys. Acta 666:322 (1981).

30. G. A. Russell, Deuterium-isotope effects in the autooxidation of aralkyl hydrocarbons. Mechanism of interaction of peroxy radicals, J. Am. Chem. Soc. 79:3871 (1957).

31. J. A. Howard and K. U. Ingold, The self-reaction of sec-butylperoxy radicals. Confirmation of the Russell mechanism, J. Am. Chem. Soc. 90:1056 (1968).

32. B. Axelrod, Lipoxygenase, in: "Food Related Enzymes," J. R. Whitaker, ed., Adv. in Chem. Series 136:324 (1974).

33. C. S. Ramadoss, E. K. Pistorius, and B. Axelrod, Coupled oxidation of carotene by lipoxygenase requires two isoenzymes, Arch. Biochem. Biophys. 190:549 (1978).

34. E. A. Sausville, J. Peisach, and S. B. Horwitz, A role for ferrous iron and oxygen in the degradation of DNA by bleomycin, Biochem. Biophys. Res. Commun. 73:814 (1976).

35. J. R. Kanofsky, Singlet oxygen production by bleomycin, J. Biol. Chem. 261:13546 (1986).

36. T. A. Dix, R. Fontana, A. Panthani, and L. J. Marnett, Hematin-catalyzed epoxidation of 7,8-dihydroxy-7,8-dihydrobenzo[a]pyrene by polyunsaturated fatty acid hydroperoxides, J. Biol. Chem. 260:5358 (1985).

37. S. J. Weiss, S. T. Test, C. M. Eckmann, D. Roos, and S. Ragiani, Brominating oxidants generated by human eosinophils, Science 234:200 (1986).

PHAGOCYTE DIOXYGENATION REACTIONS YIELDING CHEMILUMINESCENCE: THE MAXIMUM MULTIPLICITY AND SPIN CONSERVATION RULES RELATIVE TO OXYGEN REACTIVITY

Robert C. Allen

U.S. Army Institute of Surgical Research
Fort Sam Houston, Texas 78234-6200, U.S.A.

INTRODUCTION

Considering the diversity of microbes and microbial defenses, the microbicidal activities of phagocytes are astonishingly broad spectrum, lethal, and controlled.[1] Microbicidal action is a consequence of the metabolic mechanisms whereby phagocytes alter the quantum state of O_2, and in so doing, realize the thermodynamic potential for direct O_2 reactivity. Chemiluminescence is an energy product of the resulting dioxygenation reactions and, under proper test conditions, can be employed for the detection and quantification of such reactions.[2-4]

The present confusion regarding the role of oxygen "radicals" in biologic defense and injury results from an incomplete consideration of oxygen chemistry. Use of the term "radical" as synonymous with "highly reactive" is common and apparently growing. Are radicals by nature highly reactive? Imagine living in an environment that is ~21% Cl_2. Chlorine gas is a nonradical diatomic molecule. Its reaction with biological molecules is expectedly exothermic and, in fact, it is an excellent sterilizing agent at far lower concentrations. Now consider that we live in an atmosphere that is ~21% O_2. Molecular oxygen is a diradical diatomic molecule. Furthermore, the reaction of O_2 with biological molecules is of relatively greater exothermicity, yet O_2 is well tolerated. In fact, the great electronegativity of oxygen insures efficient energy liberation from metabolically derived reducing equivalents. Can O_2, a diradical, really be less reactive than nonradical Cl_2? Yes! The reactivity of a radical is dependent on its coreactant. Radicals tend to react with radicals.

QUANTUM CHEMISTRY OF OXYGEN

Hund's maximum multiplicity rule states that for a set of degenerate orbitals, maximum stability (lowest energy) is achieved when one electron occupies each orbital before two electrons occupy any given orbital of the set, and that the electrons of singly occupied orbitals will have parallel spins. In its ground electronic state, O_2 has one electron occupying each of its two π^* (* = antibonding) orbitals, and each of these electrons has the same quantum spin number. Spin multiplicity is defined as the absolute value of the product of the sum of the electron spins, S, multiplied times two, plus one; $|2S|+1$. The electron spin quantum number, s, is described as either +1/2 or -1/2. For ground state O_2, S is equal to 1/2 + 1/2 or 1, and the spin multiplicity is $|2(1)|+1 = 3$ or triplet. Triplet multiplicity defines the diradical, paramagnetic properties of O_2.[5-8]

With regard to radical properties, how does triplet O_2 compare with superoxide? Perhydroxylic acid (hydrodioxylic acid, $\cdot O_2H$) and its conjugate base the superoxide anion ($\cdot O_2^-$) are products of one equivalent reduction of O_2. In keeping with the Aufbau principle, this electron is added to the π^* orbital, and consequently one of the two π^* orbitals of superoxide is fully filled and the other half filled. The value of S is therefore 1/2, and the spin multiplicity is doublet; $|2(1/2)|+1 = 2$. Consequently, the superoxide anion has less radical and paramagnetic character than ground state O_2. The measured magnetic susceptibility in Bohr magnetons is approximately 2.8 for O_2 and 2.0 for $\cdot O_2^-$.[9] With the exception of molecules such as flavoproteins and metalloproteins that are involved in one equivalent redox reactions, biological molecules are typically nonradical and of singlet multiplicity.

SPIN CONSERVATION AND ABSOLUTE REACTION RATE THEORY

Reaction exothermicity provides information as to the direction and magnitude of chemical change. However, the factors that govern rate are best considered in the context of absolute reaction rate equation:

$$\text{Reaction Rate} = \kappa(kT/h)\, K^*\, e^{-Ea/RT}$$

where κ is the transmission coefficient, (kT/h) is a frequency constant (i.e., k is the Boltzman entropy constant, T is the absolute temperature and h is Planck's constant), and K^* is the equilibrium constant between the normal and activated states of the reactants. Ea is the activation energy per mole at absolute zero (approximately equal to the experimental energy of activation obtained by the classical Arrhenius relationship) and R is the gas constant. Thus, reaction rate is influenced by the relative magnitudes of the partition functions, the energy of activation (Ea), and the transmission coefficient κ. Not all species which have acquired the necessary free energy to cross the potential energy barrier will do so. The transmission coefficient κ defines the ratio of the number of complexes which yield product to the number having reached the activated state at the top of the barrier.

The Wigner spin conservation rule states that systems resist any change in spin angular momentum, and thus in multiplicity. By extrapolation, κ should approximate unity when the reactants, the reaction complex and products have the same symmetry properties and the same quantum numbers for S and L (the total orbital angular momentum). On the other hand, reactions involving a change in multiplicity probably always have a κ of less than 10^{-4}.[7,10] Based on reactant multiplicity, the Wigner-Witmer correlation rules also allow determination of the reaction complex multiplicity as well as the multiplicities of the resulting products.

BIOCHEMICAL CONSEQUENCES OF SPIN CONSERVATION

Direct oxygenation of singlet multiplicity biological molecules by either triplet O_2 or doublet $\cdot O_2^-$ violates the Wigner spin conservation rule. As such, the value of κ will be many orders of magnitude lower than unity. The diradical character of O_2 actually protects biological systems from combustive oxygenation.

Phagocyte microbicidal action is accomplished through metabolically driven mechanisms that overcome the multiplicity barrier to direct O_2 reactivity.[4] The utilization of metabolically derived reducing potential for univalent reduction of O_2 to $\cdot O_2^-$ decreases multiplicity from triplet to doublet. Based on this reasoning, superoxide was proposed as an intermediate in granulocyte metabolism, and NADPH oxidase was proposed as the generator enzyme.[2,11] This mechanism was postulated at least one year prior to the experimental report of Babior et al.[12] In keeping with the quantum principles here described, no direct microbicidal function was attributed to this $\cdot O_2^-$ radical. In addition, phagocyte generation of $\cdot O_2H$ might also be required for acidification of the phagolysosome. Reduction of O_2 by one reducing equivalent yields $\cdot O_2H$ with a pKa of 4.8. As such, generation of this acid will drive the pH toward 4.8, the pH of the phagolysosome.[3]

The rate of nonenzymatic dismutation of superoxide at pH 4.8 is ~10^8 M^{-1} s^{-1}.[13] In the absence of anionic repulsion, radicals react with radicals. In accord with the Wigner-Witmer rules, this doublet-doublet annihilation should proceed via a singlet reaction complex, i.e., via overlap of the radical orbitals, and as such, the products of reaction are singlet multiplicity H_2O_2 and singlet multiplicity O_2.[14]

Is superoxide dismutase (SOD) important, and if so, why? Consistent with the Wigner-Witmer rules, the SOD-catalyzed disproportionation of superoxide proceeds stepwise yielding mixed multiplicity products, i.e., singlet H_2O_2 and triplet ground state O_2.[15] The cyclic redox activity of such metalloenzymes provides an effective means of catalysis. The same rule applies to the "Haber-Weiss reaction". The direct reaction of doublet $\cdot O_2^-$ with singlet H_2O_2 has an unfavorable κ. A more likely mechanism would require stepwise reduction involving a metal intermediate. Univalent reduction of the π^* orbital of H_2O_2 results in the excess antibonding character responsible for cleavage of the oxygen-oxygen bond, yielding singlet multiplicity H_2O and doublet multiplicity $\cdot OH$.

The same octet rule that establishes $\cdot O_2^-$ as a relatively good univalent reductant also establishes $\cdot OH$ as a potent univalent dehydrogenating agent; i.e., the ionization potential of $\cdot O_2^-$ will be low and that of $\cdot OH$ will be high. Driven by its very high electron affinity, $\cdot OH$ will univalently dehydrogenate virtually any organic substrate.[16] In keeping with the correlation rules, the reaction complex will be doublet, and the products are singlet multiplicty H_2O and doublet multiplicity organic radical.

Radicals react with radicals, and doublet organic radicals will readily react with triplet O_2 via a doublet complex to yield a doublet multiplicity radical product. As such, doublet radical reactions can <u>chain propagate</u> by reaction with triplet O_2. Superoxide can serve an important function as <u>terminator</u> of radical chain reactions. The reaction of doublet $\cdot OH$ with doublet $\cdot O_2^-$ has a rate constant of 10^{10} M^{-1} s^{-1}.[17] Based on the Wigner-Witmer rules, this radical-radical annihilation will have a singlet complex and yield singlet O_2 and a singlet H_2O. Likewise, $\cdot O_2^-$ can serve as a true oxygenation agent when reacting with doublet multiplicity organic substrates. This radical-radical annihilation will also proceed via the favored singlet reaction complex, yielding a singlet multiplicity dioxygenated product. The disintegration of such products, e.g., dioxetanes and endoperoxides, yields n,π^* excited carbonyls that can relax by photon emission, i.e., chemiluminescence.

The reactive potential of singlet multiplicity H_2O_2 is realized via myeloperoxidase.[1,18] This acid optimum enzyme is a $H_2O_2{:}Cl^-$ oxido-reductase yielding singlet HOCl and singlet H_2O as products. Furthermore, HOCl can react with H_2O_2 via a singlet reaction complex to yield singlet O_2 and singlet H_2O.[19] Depending on the reaction conditions, H_2O_2, HOCl or singlet O_2 can react with singlet multiplicity organic substrates to yield singlet multiplicity products. In the case of dioxetanes and endoperoxides, disintegration yields n,π* carbonyls which relax via photon emission.

REFERENCES

1. S. J. Klebanoff and R.C. Clark, "The Neutrophil: Function and Clinical Disorders," North-Holland, Amsterdam (1978).
2. R. C. Allen, R. L. Stjernholm, and R. H. Steele, Evidence of the generation of (an) electronic excitation state(s) in human polymorphonuclear leukocytes and its participation in bactericidal activity, Biochem. Biophys. Res. Commun. 47:679 (1972).
3. R. C. Allen, Reduced, radical, and excited state oxygen in leukocyte microbicidal activity, Front. Biol. 48:197 (1979).
4. R. C. Allen, Phagocytic leukocyte oxygenation activity and chemiluminescence: A kinetic approach to analysis, Meth. Enzymol. 133:449 (1986).
5. H. Eyring, The activated complex in chemical reactions, J. Chem. Phys. 3:107 (1935).
6. G. Herzberg, "Molecular Spectra and Molecular Structure I. Spectra of Diatomic Molecules," Van Nostrand Reinhold, New York (1950).
7. K. L. Laidler, "The Chemical Kinetics of Excited States," University Press, Oxford (1955).
8. L. Salem, "Electrons in Chemical Reactions. First Principles," Wiley-Interscience, New York (1982).
9. E. W. Neuman, Potassium superoxide and the three electron bond, J. Chem. Phys. 2:31 (1934).
10. A. E. Stearn and H. Eyring, Nonadiabatic reactions. The decomposition of N_2O, J. Chem. Phys. 3:778 (1935).
11. R. C. Allen, R. L. Stjernholm, R. R. Benerito, and R. H. Steele, Functionality of electronic excitation states in human microbicidal activity, in: "Chemiluminescence and Bioluminescence," M.J. Cormier, D. M. Hercules, and J. Lee eds., Plenum, New York, p. 498 (1973).
12. B. M. Babior, R. S. Kipnes, and J. T. Curnutte, Biological defense mechanisms: The production by leukocytes of superoxide, a potential bactericidal agent, J. Clin. Invest. 52:741 (1973).
13. D. Behar, G. Czapski, J. Rabani, L. M. Dorfman, and H. A. Schwarz, The acid dissociation constant and decay kinetics of the perhydroxyl radical, J. Phys. Chem. 74:3209 (1970).
14. A. U. Khan, Singlet molecular oxygen from superoxide and sensitized fluorescence of organic molecules, Science 168:476 (1970).
15. J. M. McCord and I. Fridovich, Superoxide dismutase. An enzymatic function for erythrocuprein, J. Biol. Chem. 244:6049 (1969).
16. L. M. Dorfman and G. L. Adams, Reactivity of the hydroxyl radical in aqueous solution, Natl. Bureau Stds., NSRDS-46 (1973).
17. K. Sehested, O. L. Rasmussen, and H. Fricke, Rate constants of OH with HO_2, O_2^-, $H_2O_2^+$ from hydrogen peroxide formation in pulse-irradiated oxygenated water, J. Phys. Chem. 72:626 (1968).
18. R. C. Allen, Halide dependence of the myeloperoxidase-mediated antimicrobial system of the polymorphonuclear leukocyte in the phenomenon of electronic excitation, Biochem. Biophys. Res. Commun. 63:675 (1975).
19. A. U. Khan and M. Kasha, Red chemiluminescence of molecular oxygen in aqueous solution, J. Chem. Phys. 39:2105 (1963).

BRYOSTATIN STIMULATION OF HUMAN NEUTROPHIL

LUMINOL AND DBA (LUCIGENIN) CHEMILUMINESCENCE

J.T. Warren, A.H. Esa, G.R. Pettit, and W.S. May

The Johns Hopkins University School of Hygiene and Public Health, Baltimore, Maryland 21205, USA, University of Arizona, Phoenix, AZ

INTRODUCTION

Bryostatins are macrocyclic lactones derived from the marine bryozoan *Bugula neritina* and are currently being investigated for their apparent anti-neoplastic activities.[1,2] In addition, certain bryostatins have been shown to share certain properties with the tumor promoting compound 12-0-tetradecanoylphorbol 13-acetate (TPA). These properties include the activation of human neutrophils to generate superoxide anion and the release of specific granules[3]. *In vitro*, certain bryostatins also appear to bind and activate the TPA receptor protein kinase-C (PK-C), a process which results in the phosphorylation of nearly identical protein substrates phosphorylated by TPA[3].

In these studies, we asked whether bryostatin 1 stimulated human neutrophil chemiluminescence (CL), and if so, what the nature of this response was. In addition, as an example of using neutrophil CL in structure-activity studies, we tested each of eight bryostatin congeners for their relative CL-generating potential.

METHODS

The chemiluminigenic probes luminol (10μM) or DBA (lucigenin, 100 μM) were incubated at 37°C with 5 million isolated human neutrophils in Dulbecco's PBS buffer (DPBS, pH 7.4) containing 1 g/l glucose and 5 mM $CaCl_2$ and $MgCl_2$ in 1 ml CL cocktails for a minimum of 10 min in an LKB luminometer. Relative CL intensity (light intensity in mV x sec) was monitored at preset intervals (i.e., 1 or 2 min) for 10 sec each. Once stable, baseline CL values were obtained, stimulant was added, and CL was monitored for up to 15 min or until near-baseline CL was reattained. Control cocktails contained 0.01% DMSO, that concentration of DMSO in matched stimulant-added cocktails. TPA, superoxide dismutase (SOD), H-7 (1-[5-isoquinolinesulfonyl]-2-methylpiperazine dihydrochloride), indomethacin, and NDGA (nordihydroguaiaretic acid) were obtained commercially. Serum-treated zymosan (STZ) was prepared according to Roos et al.[4]

RESULTS AND DISCUSSION

Bryostatin-stimulated CL

Bryostatin 1 (Bryo) stimulated human neutrophils produced both DBA- (Figure 1A) and luminol- (Figure 1B) enhanced CL at levels comparable to those produced by 162 nM (100 ng/ml) TPA. As is commonly found for the TPA response, the DBA response elicited by Bryo was inhibitable (though not completely) by SOD (100 U/ml), but not by azide (10 μM). The kinetics of the Bryo response was nearly identical to that of the TPA response, the DBA CL peaking at about 10 min then quickly decreasing to baseline by 20-25 min. In addition, the corresponding luminol response to Bryo was azide sensitive (Figure 1B), indicating that the Bryo-stimulated oxidative burst results in the production of both superoxide anion and myeloperoxidase (MPO)-catalyzed formation of hypochlorous acid, the predominant luminol-sensitive oxidant produced by human neutrophils.[5] In similar studies (data not shown), the PK-C inhibitor, H-7 (100 μM), also inhibited the Bryo DBA CL response by >50% . Lower H-7 concentrations (10 and 1.0μM) failed to inhibit this activity. H-7 also inhibited (>50%) both the PMA-(162nM) and STZ-(0.1 mg/ml) stimulated DBA CL activity. Interestingly, the lipoxygenase inhibitor, NDGA (10 μM), inhibited the Bryo DBA CL response as well (>70%), whereas the cyclooxygenase inhibitor, indomethacin (10μM), had no effect.

Sequential Stimulation of CL

The signal transduction pathway of bryostatin activation of human neutrophil CL was further investigated in sequential stimulation studies with the two well characterized neutrophil secretagogues TPA and STZ and bryostatin 1. Here, neutrophils were first stimulated with either bryostatin or a secretagogue, and DBA CL was monitored until the response peaked and then reached near baseline levels. Then, these cells were restimulated with either the same or a different stimulant. In previous experiments, we found that if the same stimulant was used in the restimlation phase, then the CL response was refractory (e.g., would not generate CL). However, restimulation with an agent that appeared to use a different signal transduction pathway of CL generation (than that activated during the first phase) resulted in a second CL response. Figure 2 represents data of this sort in experiments with bryostatin 1 (Bryo). Note that while Bryo is refractory to both itself (closed triangle) and to STZ (closed diamond), and TPA is refractory to Bryo (open triangle), STZ was not refractory to Bryo (open diamond), nor was TPA to STZ (closed square). Therefore, Bryo appears to share, in part, the activation pathway of STZ in addition to the pathway activated by TPA.

CL in Structure-activity Analyses of Bryostatins

Finally, a series of bryostatin congeners (Bryo 1, 3, 4, 8, 9, 10, 11 and 13) were analyzed for their relative activation of human neutrophil DBA CL (Figure 3). The structures of these Bryo are known, and their relative effects as activators of PK-C are under investigation. Based on the CL kinetics we obtained, three groups of Bryos emerged: (i) 1, 3, and 4, (ii) 8, 9, 10, and 11 and (iii) 13. Group (i) Bryos clearly peaked sooner and, in general, at higher CL levels, though Bryo 4 peaked at a level consistent with the intermediate Group (ii) Bryos. Only Bryo 11 reached the peak CL levels reached by the Group (i) Bryos, though at a later time point consistent with the other Group (ii) Bryos. Bryo 13 was the only inactive congener. It will be interesting to see if the degree and kinetics of the DBA CL responses generated by these three groups of Bryos correlates with their respective potentials to activate PK-C.

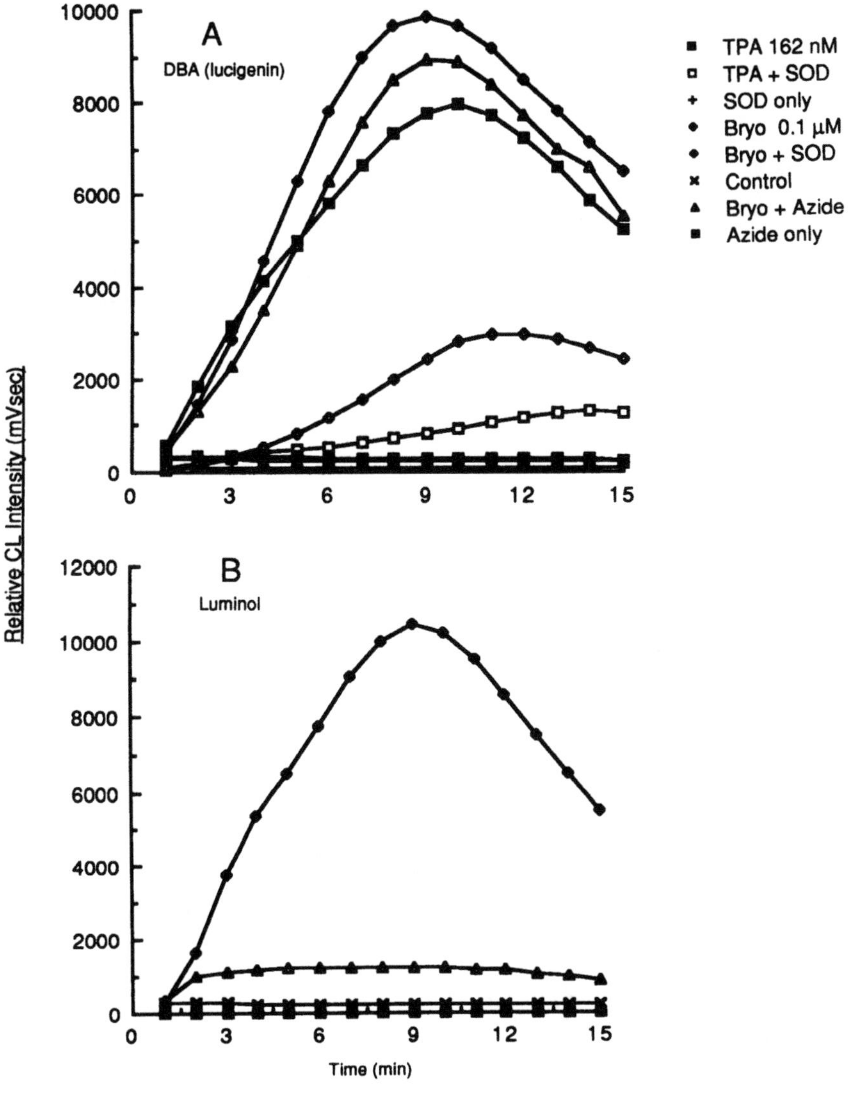

Figure 1. Bryostatin-stimulated human neutrophil CL. (A) DBA (lucigenin) neutrophil CL; (b) luminol neutrophil CL; TPA, 12-0-tetradecan-oylphorbol 13-acetate; SOD, superoxide dismutase; Bryo, Bryostatin 1.

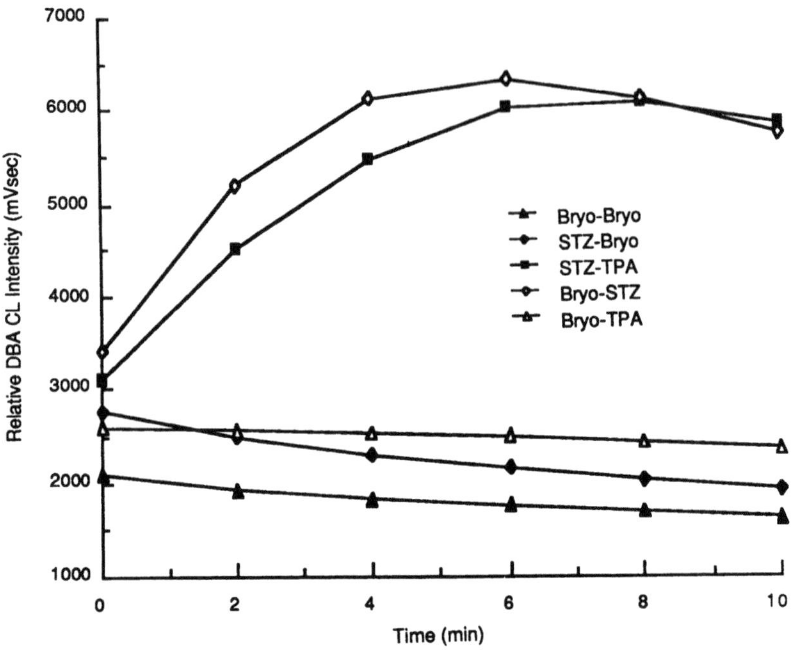

Figure 2. Sequential stimulation of neutrophil DBA CL, restimulation phase. Bryo, Bryostatin 1 (0.1 μM); STZ, serum-treated zymosan (0.1 mg/ml); TPA, 12-O-tetradecanoylphorbol 13-acetate (162 nM).

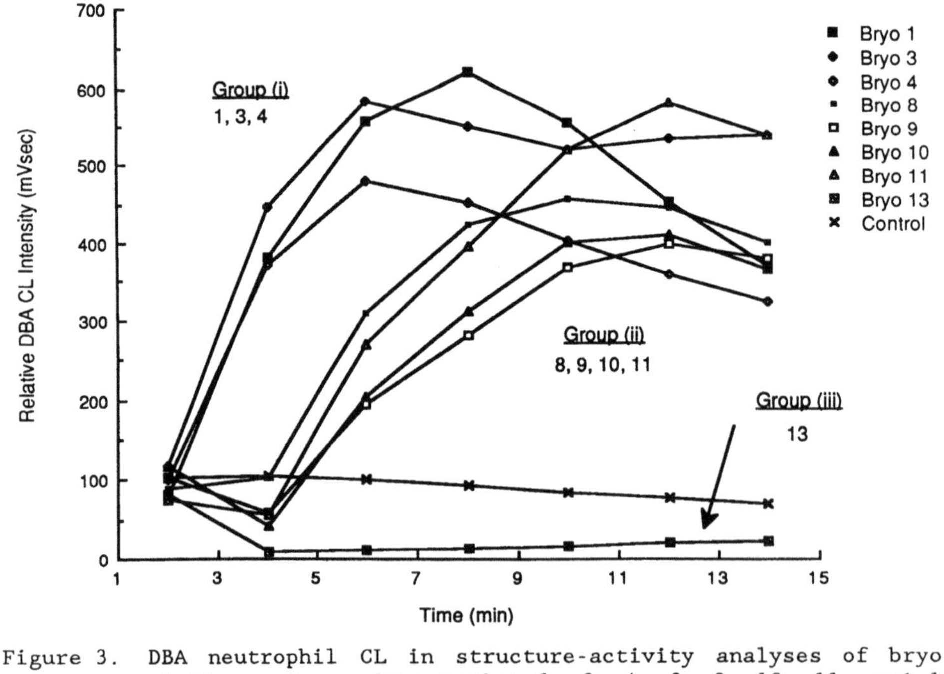

Figure 3. DBA neutrophil CL in structure-activity analyses of bryostatins. Bryo, Bryostatin; 1, 3, 4, 8, 9, 10, 11, and 13 each refer to structurally different Bryo congeners; control, 0.01% DMSO.

In summary, bryostatins can stimulate human neutrophil CL with either of two commonly used chemiluminigenic probes, DBA (lucigenin) and luminol, and these responses are inhibitable by SOD and azide, respectively. Our results using inhibitors of arachidonate metabolism, in CL studies with bryostatin 1, indicate that these metabolites may play a role in the generation of bryostatin CL and that the lipoxygenase pathway may be more active in this process. Sequential stimulation studies indicate that bryostatin shares pathways of CL activation with both TPA and STZ. Our CL system appears to be a useful method for examining potential structure-activity relationships among the different bryostatins.

REFERENCES

1. G.R. Pettit, C.L. Hearld, Y. Kamano, D. Gust and R. Aoyagi, The structure of Bryostatin 2 from the marine bryoszoan Bugula neritina, J. Nat. Prod. 46:528 (1983.
2. G.R. Pettit, C.L. Herald and Y. Kamano, The structure of the Bugula neritina (marine, bryozoa) antineoplastic component Bryostatin 3, J. Org. Chem. 48:5354 (1983).
3. R.L. Berkow and A.S. Kraft, Bryostatin, a non-phorbol macrocyclic lactone, activates intact human polymorphonuclear leukocytes and binds to the phorbol ester receptor, Biochem. Biophys. Res. Comm. 131:1109 (1985).
4. D. Roos, A.A.M. Bot, M.L.J. van schaik, M. de Boer and M.R. Daha, Interaction between human neutrophils and zymosan particles: the role of opsonins and divalent cations, J. Immunol. 126:433 (1981).
5. K. Cheung, A.C. Archibald and M.F. Robinson, The origin of chemilumminescence produced by neutrophils stimulated by opsonized zymosan, J. Immunol. 130:2324 (1983).

LUCIFERASE REVERSE MICELLAR BIOLUMINESCENCE: KINETICS AND EMISSION SPECTRA

Ambler Thompson, W. H. Biggley, and Howard H. Seliger

Department of Biology and McCollum-Pratt Institute
The Johns Hopkins University Baltimore, Maryland 21218
U.S.A.

INTRODUCTION

Firefly luciferase is a hydrophobic enzyme that which catalyses the oxidation of luciferin by molecular oxygen and requires the enzymatic cofactors ATP and magnesium ions. This enzymatic reaction is accompanied by intense luminescence. Luciferase is generally considered a membrane-associated protein. Most of the current extensive knowledge of luciferase enzymology has been been carried out in aqueous systems in which the lifetime of bioluminescence following the injection of ATP is on the order of seconds. Therefore, the kinetics of its reaction in membrane-mimetic systems such as the colloidal reverse micelle system would be expected to be quite different. Following the initial report of Belyaeva et. al.[1], we decided to investigate the effect of pH on the spectral and kinetic properties of the luciferase bioluminescence reaction in colloidal reverse micelles[2] formed from the interaction of the nonionic surfactant Brij 96, water and cyclohexane. Previous aqueous luciferase investigations have shown that as the pH is decreased from 8.0, the bioluminescence emission maxima shifts from 562 nm, the same as observed in the *in vivo* bioluminescence, to > 600 nm, with a corresponding decrease in the bioluminescence quantum efficiency.

EXPERIMENTAL

Purified firefly (*Photinus pyralis*) luciferase, luciferin, ATP and Brij 96 (poly-10-hydroxyethylene oleate) where obtained from Sigma Chemical Company. Typical reverse micelle bioluminescence reactions contained 20 μg luciferase (0.4 nmole assuming a molecular weight of 5×10^4), 520 nmole ATP, 55 nmole luciferin, and 2 μmole $MgCl_2$ in 0.24 ml of 50 mM sodium phosphate buffer at varying pH. Measurement of the final pH was carried out on the aqueous phase. The true pH of the aqueous phase inside the reverse micelles could not be determined experimentally, but might be inferred from a comparison with the aqueous bioluminescence emission spectra previously reported.

The Procedure for initiation of the bioluminescence reaction and formation of the reverse micellar gel was as follows: all enzymatic reactants except ATP were added to 1.0 ml of 0.25 M Brij 96 in cyclohexane, then a small (10 μl) aliquot containing ATP was added and the tube was mixed by vigorous shaking by hand and placed in the photometer or the spectrometer.

Kinetics of the luciferase reverse micelle bioluminescence were measured by an EMI sidewindow phototube operated in the DC mode. Intensity data was acquired at a 450 Hz sampling rate and analysed with a Hewlett Packard 216 computer. The bioluminesence emission spectra were measured with a 1-m f/3 grating spectrometer, designed by W. Fastie, calibrated for spectral sensitivity with an NBS standard lamp.

RESULTS and DISCUSSION

In aqueous solution, luciferase bioluminescence reaches a maxima in approximately 0.3 s after an injection of ATP. This luminescent flash typically decays to a baseline emission in less than 5-10 s depending on the purity of the luciferase preparation. In contrast, the luciferase reverse micelle reaction reaches a maxima within 5-8 minutes after addition of ATP, approximately 10^3 times slower than the aqueous reaction. The reverse micelle bioluminescence reaction subsequently

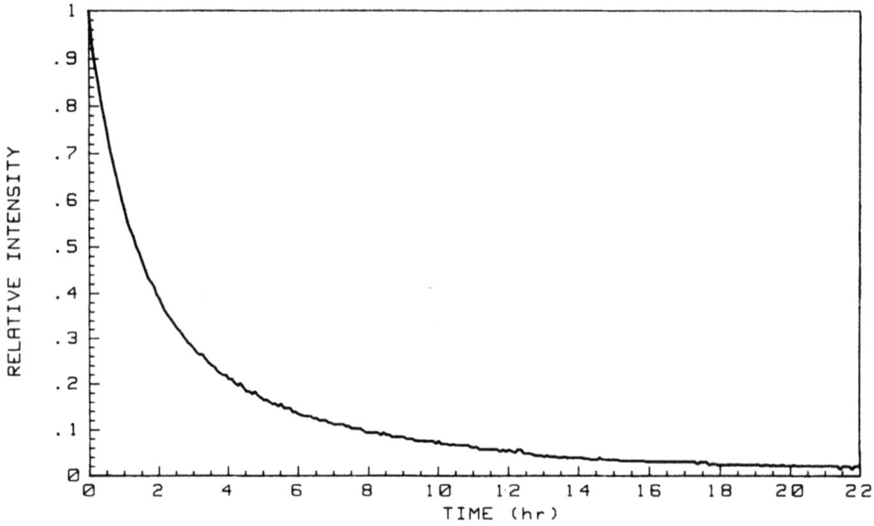

Figure 1. Kinetics of the reverse micelle bioluminescense reaction.

decays (Figure 1) over many hours with kinetics that appear to fit the sum of two exponential decays. These decay rates were relatively independent of buffer pH with the faster decaying component (k_1 = 0.0134 min^{-1}) comprising 67% of the initial emission and the slower decaying component (k_2 = 0.002 min^{-1}) the remaining 33% of the initial emission.

Marked shifts in the pH profiles of enzymatic reactions entrained in reverse micellar environments have previously been observed. This has been proposed to be due to several factors: the surfactant may cause a local shift in the pH; the acid-base properties of the groups on the solubilized enzyme may alter as a result of the change in the microenvironment; and lastly, solubilization may cause conformational changes in the enzyme that effect a change in the pK of ionizable groups in the enzyme active site. The large effect of pH on the color of the bioluminescence emission of luciferase should be an effective indicator of whether there has been a shift in pH due to microenvironmental influences. The effect of reverse micelle buffer pH on the bioluminescence emission spectra (Figure 2) paralleled the effect noted in earlier aqueous luciferase reactions, where, as the pH was decreased from 8.1 to 5.54, the emission maxima shifted from 562 nm to 616 nm with an approximately 70% decrease in the overall quantum efficiency of the bioluminescence reaction. Therefore, if there is a microenvironment effect on the pH of the water and luciferase entrained in the reverse micelles, it is small, on the order of 0.2-0.3 pH unit. The explanation for the pronounced red shift in the emission spectra with acidic pH is the protonation of the 6'-hydroxy constituent at

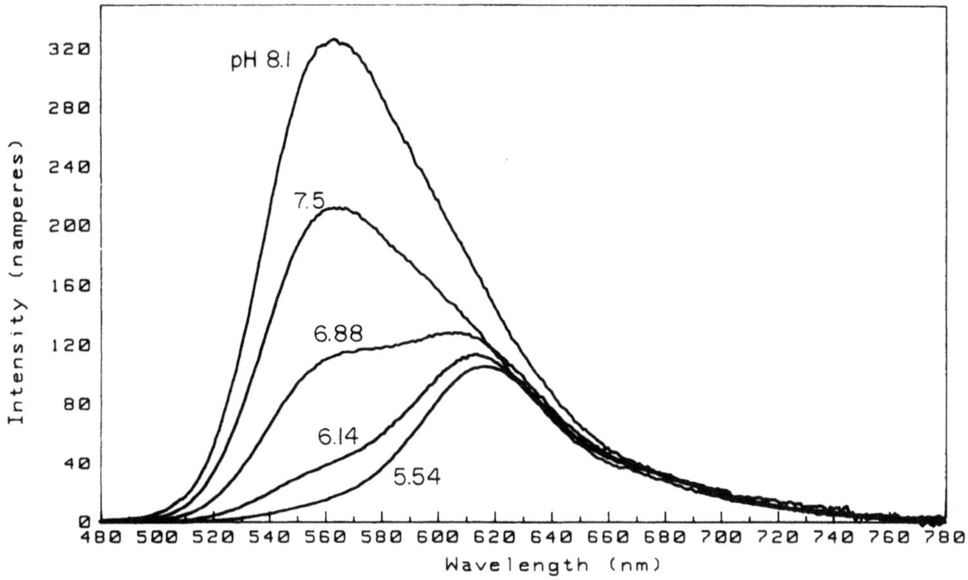

Figure 2. Effect of pH on reverse micelle bioluminescence emission spectra.

decreased pH. It is thought that the formation of a dianion is responsible for the natural yellow-green bioluminescence.

By taking the rate constants of the mixed exponential decay and the integral spectral emission in absolute units of photons/sec, the total light yield from the luciferase reverse micelle reaction was calculated to be on the order of 3 x 10^{15} photons at pH 8.1. Since the number of luciferase molecules present in the reaction was 2.4 x10^{12}, the minimum turnover number or number of photons per luciferase molecule was approximately 12. Previous aqueous luciferase measurements have indicated that oxyluciferin, the product of luciferase oxidation remains tightly bound to luciferase[3], thus inhibiting enzymatic turnover. This is a conservative estimate since it assumes that all of the luciferase molecules in the reverse micellar gel remained active and were not denatured by this treatment.

The wide spectral range of this long-lived reverse micelle bioluminescence makes it particularly useful as a single photon source for calibration of low-level spectrometers. The long lifetime of this luciferase emission makes correction for decay of the luminescence minimal, unlike the luminol reaction, and when used in conjunction with the blue luminol chemiluminescence reaction, calibrations of almost the entire visible spectrum can be accomplished. A large number of the enzymes that emit chemiluminescence when oxidizing substrate molecules are hydrophobic, membrane-associated enzymes. Since reverse micelles are membrane-mimetic model systems, the chemiluminescent reactions of these enzymes should be investigated in reverse micelles as a model closer to cellular environment than aqueous _in vitro_ reactions.

REFERENCES

1. E. I. Belyaeva, L. Y. Brouko, N. N. Ugarouc, N. L. Klyachko, A. V. Levashov, K. Martinek, and I. V. Berez. _Doklady Academii Nauk SSR_ _273_:494 (1983).
2. K. Martinek, A. V. Levashov, N. Klyachko, Y. L. Khmelnitski and I.V. Berezin, _Eur J. Biochem_. _155_: 453 (1986).
3. H. H. Seliger and R. A. Morton, "Photophysiology," A. C. Geise, ed., Vol. 4, pp.315-353. Academic Press, New York, (1968).

PLATELET CHEMILUMINISCENCE INDUCED BY LINOLEIC AND ARACHIDONIC ACIDS

T. Metsä-Ketelä, A. Montfoort,* and T. Kunnas

Department of Biomedical Sciences
University of Tampere
Tampere, Finland
*Department of Pathology I
Erasmus University
Rotterdam, The Netherlands

INTRODUCTION

Mill et al.[1] demonstrated that linoleic acid (18:2N6) and other long-chain fatty acids can induce platelet low-level chemiluminescence. It turned out that only those fatty acids which are susceptible either to enzymatic or autoxidation, e.g., polyunsaturated fatty acids, emit light.

Wörner[2] demonstrated that luminol enhanced chemiluminescence of arachidonic acid consisted of two peaks. By using more or less spesific cyclooxygenase (CO) and lipoxygenase (LO) inhibitors he concluded that the two peaks were generated by different enzymes.

In the present study we have further studied this chemiluminescence.

MATERIALS AND METHODS

Platelets

Blood was obtained from The Finnish Red Cross Blood Bank and from Dutch normal donors who had not taken aspirin for 14 days. The blood was mixed immediately with dextrose-citrate buffer, pH 6.5, in a ratio of nine parts of blood to one part of anticoagulant. Platelets were isolated essentially as described by Mills et al.[1] The platelet-rich plasma was centrifuged for 20 min at 135 x g at room temperature. Washed platelets were prepared by adding 1.5 ml of 10 % EDTA, pH 7.4, to 9 ml of platelet-rich plasma and subsequently centrifuging for 15 min at 1085 x g at 4°C. The platelets were resuspended in Hanks balanced salt solution (HBBS), pH 7.4 (10^8 platelets/ml). The washed platelets were examined for contaminating neutrophils (normally less than 0.1 %).

Fatty Acids

Arachidonic acid and linoleic acid were converted into their Na salts by addition of very small amounts of 8 N NaOH to a few microliters of an alcoholic stock solution. After evaporation of the alcohol 0.9 % NaCl was added to obtain the correct concentration. The purity and the concentrations of the fatty acids were determined by capillary GC as described before.[3] The hydroperoxy derivatives of arachidonic acid (AA-OOH) and linoleic acid (LA-OOH) were prepared according to Verhagen et al.[4] using a pure preparation of soybean lipoxygenase I which was a generous gift from prof. J.F.G. Vliegenthart (University of Utrecht, The Netherlands). The hydroperoxy fatty acid concentration was determined spectrophotometrically at 234 nm using a molar extinction coefficient of 25.000/mole x cm.

Chemiluminescence

The chemiluminescence experiments were carried out in a LKB-Wallac Luminometer 1251 equiped with a mixing device and temperature controlled. The experiments were carried out at 37°C; 550 µl of washed platelets (55 x 10^6 cells) were incubated with 50 µl of fatty acid salts solution and 20 µl of luminol (3-aminophtalhydrazide) in a stock solution of 10 mM in 0.1 M borate buffer (pH 9.0).

Thromboxane Measurements

For thromboxane B-2 measurements the platelet suspensions were acidified with HCl to pH 3 and heated at 56°C for 10 min. The thromboxane B-2 was measured by radioimmunoassay.[5]

RESULTS

Incubation of washed human platelets with arachidonic acid and linoleic acid produced signals which reached a maximum in about 30 s (Figure 1). Sometimes we observed two peaks, the first peak having its maximum in about 6 s, but in most of the samples only one peak could be observed.

The light emitted upon incubation of platelets with the same concentrations of the hydroperoxy fatty acids was only 10-20 % of the chemiluminescence with intact arachidonic and linoleic acid (Figure 2).

The arachidonic and linoleic acid -induced chemiluminiscence response depended on the concentration of the fatty acids (Figure 3). The response produced by linoleic acid was greater than that obtained by arachidonic acid. Interesting is the fact that the chemiluminescence response in case of arachidonic acid is supressed by aspirin while in the case of linoleic acid the signal is enhanced (Figure 3).

Arachidonic acid was the only fatty acid which could stimulate the formation of thromboxane B-2 (Figure 4). It also produced platelet aggregation in final concentrations up to 0.06 mM. At higher concentrations both fatty acids and especially their hydroperoxides gage rise to lyses of cells as indicated by the release of LDH.

DISCUSSION

The results indicate that intact arachidonic and linoleic acids are metabolized by human platelets. The main metabolic pathway to arachidonic is the CO pathway.[6] The high Km (450 µM) of linoleic acid for CO and the elevation of the chemiluminescence response by aspirin could possibly indicate that linoleic acid is metabolized by lipoxygenase. However, linoleic acid is a poor substrate for the main LO in platelets, e.g. 12-lipoxygenase.[7] In the study by Nugteren,[7] arachidonic acid was reported to be a 20-fold better substrate for 12-LO than linoleic acid. However in our chemiluminescence experiments linoleic acid produced a higher signal than arachidonic acid when platelets were incubated with the same fatty acid concentration. Recently, Wong et al.[8] reported that platelets also contain a 15-LO which could be responsible for the chemiluminescence response. A higher chemiluminescence signel for linoleic acid compared to arachidonic acid was also reported by Mills et al.[1]

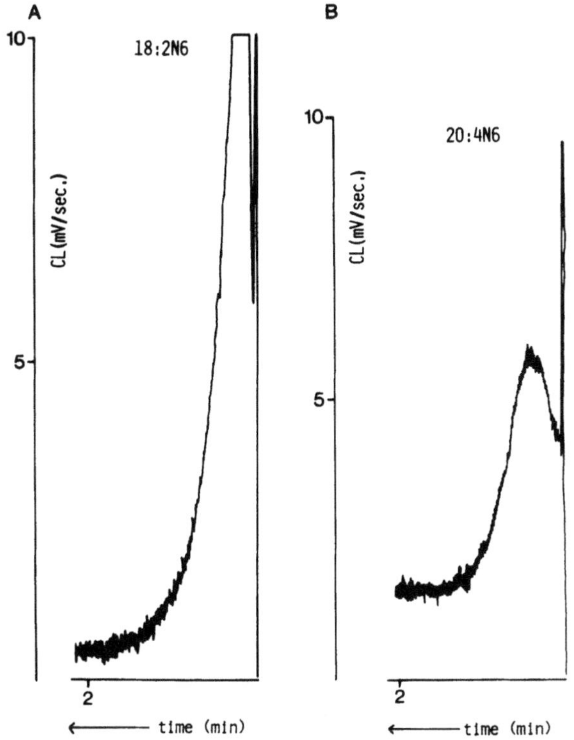

Figure 1. The chemiluminescence response of human washed platelets to linoleic acid (A) and arachidonic acid (B). 55 x 10^6 platelets suspended in HBBS were preincubated with 300 μM luminol for 10 min at 37°C before addition of the fatty acid Na salt. The total volume of the mixture was 620 μl.

Platelets preincubated with aspirin and platelets isolated from blood of individuals who had taken aspirin showed reduced chemiluminescence upon incubation with arachidonic acid. These results agree with the data described by Mills et al.[1] and Wörner.[3] However the chemiluminescence signal produced by linoleic acid in our study was very much enhanced by aspirin.

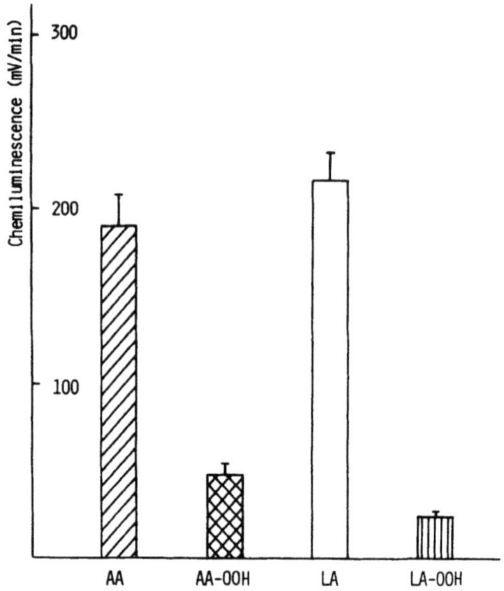

Figure 2. The chemiluminescence responses of washed human platelets to arachidonic acid, linoleic acid, arachidonic acid hydroperoxide and linoleic acid hydroperoxide (0.1 mM final concentrations). The chemiluminescence was integrated during the first minute of the response. Means and standard errors of six experiments are given.

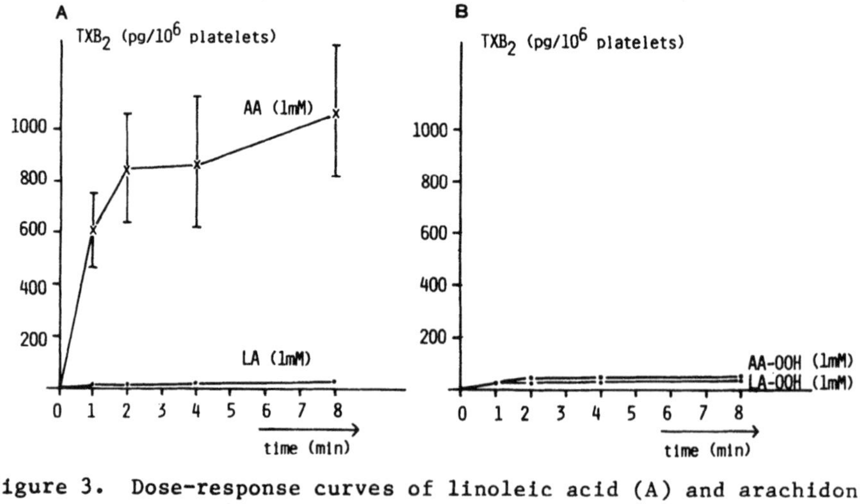

Figure 3. Dose-response curves of linoleic acid (A) and arachidonic acid (B) induced chemiluminescence responses of human platelets and the effect of 0.1 mM (final concentration) aspirin (ASA).

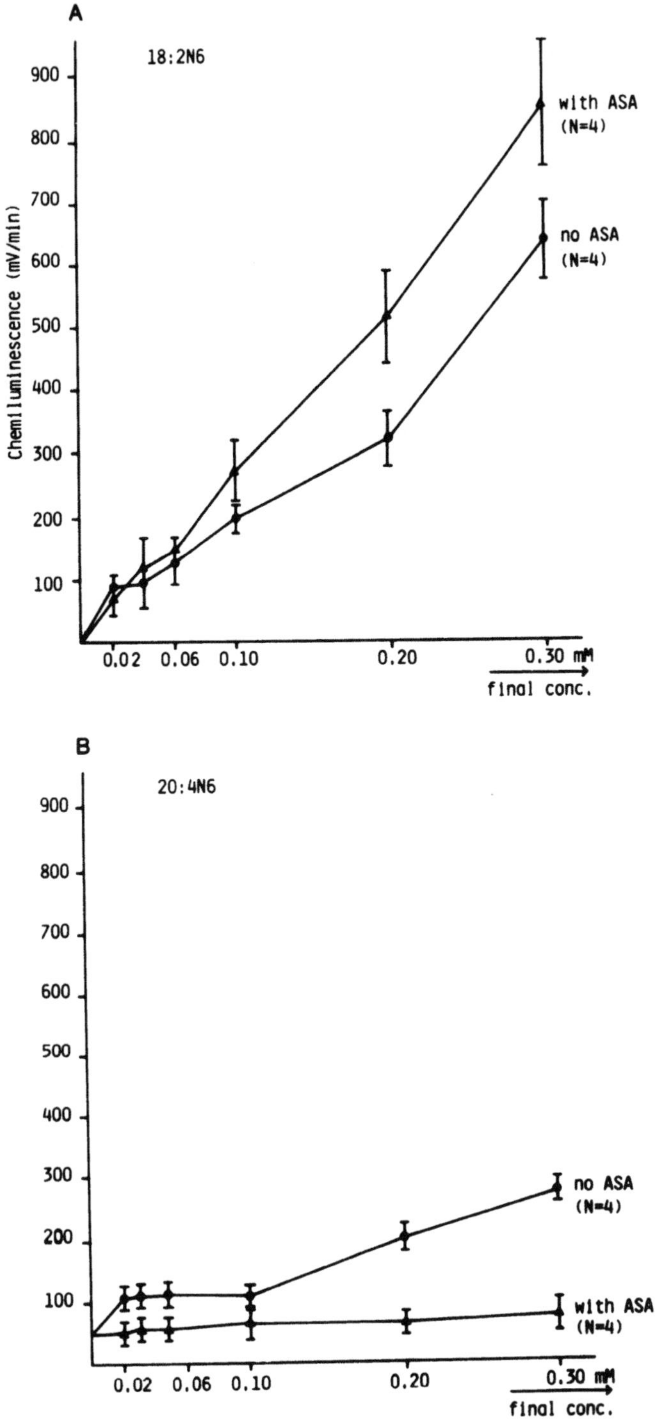

Figure 4. Effect of arachidonic and linoleic acids (A) and their hydroperoxides (B) on thromboxane B-2 formation by human platelets.

REFERENCES

1. E. L. Mills, J. M. Gerrard, D. Filipovich, J. D. White, and P. G. Qui, The chemiluminescence response of human platelets, J. Clin. Invest., 61:807 (1978).
2. P. Wörner, Arachidonic acid-induced chemiluminescence of human platelets: contribution of the prostaglandin and lipoxygenase pathways, Thromb. Haemost., 46:584 (1981).
3. A. Montfoort, C. C. M. Rutten-van Beysterveld, and M. R. Wortelboer, Molecular species of diacylphosphatidylethanolamine in rat and mouse given the same diet, Biochem. Int., 6:569 (1983).
4. J. Verhagen, G. A. Veldink, M. R. Egmond, J. F. G. Vliegenthart, J. Bolding, and J. van der Star, Stedy-state kinetics of the anaerobic reaction of soybean lipoxygenase-1 with linoleic acid and 13-L-hydroperoxylinoleic acid, Biochim. Biophys. Acta, 529:369 (1978).
5. E. Seppälä, O. Pora, and T. Metsä-Ketelä, A modified method for extraction and purification of prostaglandin with resin XAD-2, Prostaglandins Leukotrienes Med., 14:235 (1984).
6. J. B. Smith, and A. L. Willis, Aspirin selectively inhibits prostaglandin production in human platelets, Nature New Biol., 231:235 (1971).
7. D. H. Nugteren, Arachidonate lipoxygenase in blood platelets, Biochim. Biophys. Acta, 380:299 (1975).
8. P. Y. K. Wong, P. Westerlund, M. Hamberg, E. Granström, P. H. W. Chao, and Samuelsson, 15-lipoxygenase in human platelets, J. Biol. Chem., 260:9162 (1985).

LOW LEVEL CHEMILUMINESCENCE FROM ISOLATED RAT HEPATOCYTES, INTACT LUNG AND INTESTINE *in situ*

Julio F. Turrens, Cecilia Giulivi,* Claudia Pinus,* Emilio Roldan,* Carla Lavagno,* and Alberto Boveris*

Department of Biochemistry
College of Medicine
University of South Alabama
Mobile, AL 36688, U.S.A.
* Catedra de Fisicoquimica
Facultad de Farmacia y Bioquimica
University of Buenos Aires
Junin 956, Buenos Aires, Argentina

INTRODUCTION

The spontaneous emission of chemiluminescence (CL) appears to reflect changes in the intracellular steady-state concentration of oxygen radicals.[1,2] This report presents some applications of the CL technique for determining changes in the intracellular metabolism of free radicals in isolated cells and intact organs *in situ*.

MATERIALS AND METHODS

The changes in CL were determined using an EMI 9658 photomultiplier, responsive in the 300 to 900 nm region, as previously described.[1] Myeloperoxidase,[3] SOD[4] and catalase[5] activities were determined as indicated elsewhere. Glutathione concentration was determined fluorometrically following the method of Hissin and Hilf.[6]

Rat hepatocytes were isolated as described elsewhere.[7] The determinations of changes in CL from intact organs were carried out using anesthetized rats (pentobarbital 60 mg/kg, i.p.). For studies involving lung CL, the rats were connected to an animal respirator (80 cycles/min, 4 ml/cycle), prior to opening the chest. The intestine (jejune, approximately 12 cm^2) was exposed through a small abdominal opening and subjected to ischemia for 2 to 5 min. In every case the rest of the animal was covered with aluminum foil to avoid detection of light coming from other organs (particularly skin).

RESULTS AND DISCUSSION

Three different scenarios were designed for studying the relationship between changes in CL and oxidative stress in systems of increasing complexity: a) inhibition of antioxidant enzymes in liver homogenates or isolated liver cells; b) paraquat-mediated oxygen toxicity to the lung and c) reperfusion of ischemic rat intestine. In every case, the oxidative injury correlated with an increased CL, confirming that this is a suitable technique for determining changes in the steady-state concentration of oxygen radicals.

Spontaneous CL from isolated hepatocytes increased in different proportions upon addition of antioxidant enzymes inhibitors. Inhibition of catalase activity by aminotriazole increased spontaneous CL (8 cps/mg protein) twofold at concentrations very similar to those required for inhibiting catalase activity (70 mM, data not shown). Figure 1 shows that cyanide, an inhibitor of both catalase and CuZn SOD, increased CL eightfold when both enzymes were inhibited (0.5 mM KCN). Diethyldithiocarbamate, a quite specific inhibitor of CuZn SOD also stimulated fivefold the emission of CL (Figure 1), indicating that an increased steady state concentration of superoxide anion may be responsible for most of the increased CL. The inhibition of SOD increased the emission of CL 2.5-fold more than the inhibition of catalase.

Sulphydryl reagents such as phorone, iodoacetamide and diethylmaleate, stimulated the emission of CL by liver homogenates in different proportions. Although there was a correlation between concentration of reduced glutathione (GSH) and CL (Figure 2), there was no linear relationship between the two variables. Iodoacetamide reduced the GSH concentration in liver homogenates to a similar extent than diethylmaleate, stimulating the emission of CL 2.5-fold more than the second inhibitor. This result may reflect the differences in sensitivity of different metabolic pathways towards each sulphydryl reagent rather than just simple variations in the intracellular concentration of reduced glutathione.

The CL technique was also used to study two different forms of oxidative stress to intact organs in situ. The first system was the paraquat-dependent lung injury. Rats were pre-injected with either 30 or 60 mg/kg paraquat (i.v.), and the changes in CL were monitored at different times. Spontaneous CL from rat lung increased 2.5-fold (control 100 $\pm$ 25 cps/pair lungs) 6 h or 30 h after a single i.v. injection of 30 or 60 mg/kg paraquat, respectively (Figure 3). The role of polymorphonuclear cells migration to the lungs was estimated by determining myeloperoxidase in perfused lungs after paraquat treatment. There was a significant

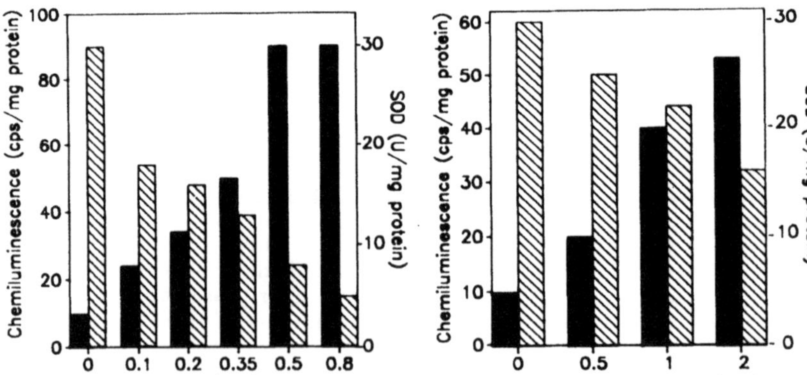

Figure 1: Changes in spontaneous CL (filled bars) from isolated hepatocytes upon inhibition of SOD activity (hatched bars). KCN also inhibits catalase at very low concentrations (10 μM) but its effect on CL seems to reflect the inhibitory action on SOD or the combined inhibition of both enzymes.

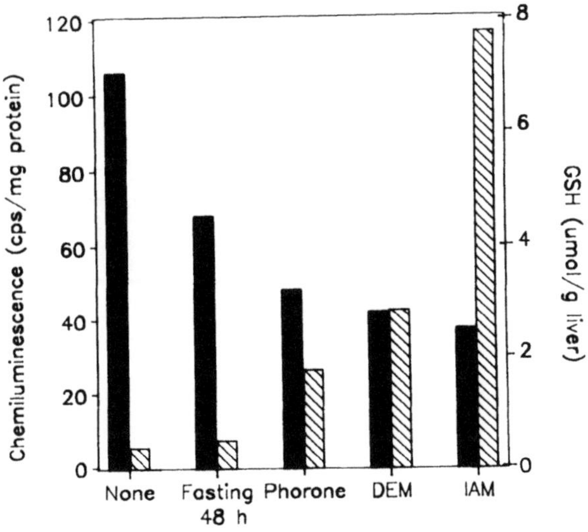

Figure 2. Changes in CL (filled bars) upon depletion of reduced glutathione (hatched bars) using different sulphydryl reagents. Every set of bars represents CL and reduced glutathione concentration in fresh homogenates from liver. The results correspond to the mean from five different samples.

increase (fivefold) in lung myeloperoxidase activity 6 h and 30 h after injection of 60 or 30 mg paraquat/kg, respectively. These results suggest that part of the augmented CL may have come from activated phagocytes migrating to the injured areas.

Another oxidative process of biological importance is that occurring during reperfusion of ischemic tissues. In this case we studied the changes in CL of rat intestine during the early stages of reperfusion. The results indicate that during reperfusion, there is a 40% increase in the emission of CL, which can be completely inhibited by pretreatment of rats with the xanthine oxidase inhibitor allopurinol (Table 1). The results are in accordance with the proposed model involving a conversion of the enzyme xanthine dehydrogenase into the oxidase form during ischemia, which in turn generates oxygen radicals during reperfusion.[8] The role of polymorphonuclear cells activation during reperfusion was not studied in this model.

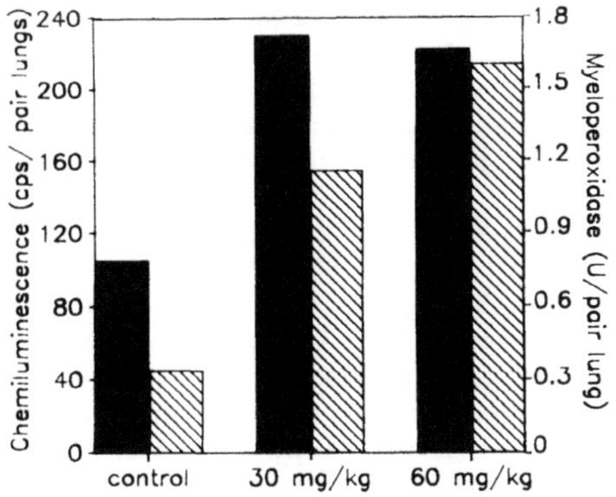

Figure 3. Changes in rat lung CL (filled bars) and myeloperoxidase activity (hatched bars) 30 or 6 h after pretreatment with a single i.v. injection of either 30 or 60 mg paraquat/kg, respectively.

Table 1. Changes in Spontaneous Intestine CL upon Reperfusion after 3 and 5 min of Ischemia

Treatment	n	Time during Ischemia (min)	% Change in CL[a] (treated/control) (X ± S.E.M.)
Ischemia	3	5	48 ± 8
Ischemia	5	5	26 ± 6
Reperfusion	3	8	140 ± 10
Reperfusion	5	8	125 ± 7
Reperfusion + allopurinol[b]	3	5	95 ± 1
Reperfusion + allopurinol	5	5	91 ± 4

[a] The emission of CL prior to ischemia was utilized as the control value for every animal, in order to avoid variability among different animals. Previous studies determined that the spontaneous CL from control rat intestine (10 ± 1 cps/cm^2) remained constant during more than 20 minutes, indicating that any change in CL during ischemia and reperfusion was a consequence of the treatment.

[b] Allopurinol was injected i.p. in two doses of 40 and 100 mg/kg body weight 24 h and 3 h before the experiment.

The results presented in this paper indicate that CL is a suitable method for monitoring changes in the steady-state concentration of free radicals in intact tissues. More research needs to be carried out in order to identify the species involved in the emission of light as well as the possible intracellular sources.

REFERENCES

1. A. Boveris, E. Cadenas, R. Reiter, M. Filipkowski, Y. Nakase and B. Chance, Organ chemiluminescence: noninvasive assay for oxidative radical reactions, Proc. Natl. Acad. Sci. USA 77: 347 (1980).
2. J. F. Turrens, C. Giulivi and A. Boveris, Increased spontaneous chemiluminescence from liver homogenates and isolated hepatocytes upon inhibition of superoxide and hydrogen peroxide utilization, Free Radical Biol. Med. 2: 135 (1986).
3. B. Chance and A. C. Maehly. Assay of catalases and peroxidases. Methods Enzymol. 2: 764 (1955).
4. H. P. Misra and I. Fridovich, The generation of superoxide radical during the autoxidation of ferredoxins, J. Biol. Chem. 246: 6886 (1971).
5. H. U. Bergmeyer, Measurement of catalase activity, Biochem. Z. 327: 255 (1955).
6. P. J. Hissin and A. Hilf, A fluorometric method for determination of oxidized and reduced glutathione in tissues, Anal. Biochem. 74: 214 (1976).
7. P. O. Segeln, Properties of isolated rat liver cells "Methods in Cell Biology," D. M. Prescott , ed. Academic Press, New York (1976).
8. J. M. McCord, Oxygen-derived free radicals in post-ischemic tissue injury, N. Engl. J. Med. 312: 159 (1985).

PREVENTION BY PROPIONYLCARNITINE OF NADPH AND TERT-BUTYL HYDROPEROXIDE-DEPENDENT CHEMILUMINESCENCE IN HEART MICROSOMES

M.L. Schinetti, D. Rossini, and A. Bertelli

Institute of Pharmacology
School of Medicine
University of Pisa, Italy

INTRODUCTION

Carnitine and its esters are essential to the normal functions of various organs playing a central role in fatty acid oxidation. Their acyl buffering action is particularly important in the heart during hypoxic conditions when fatty acids and their metabolites accumulate in the cardiac cell.[1,2] Acylcarnitines, such as propionylcarnitine, have been reported to be more effective than carnitine in improving the recovery of cardiac functions after ischemia and the subsequent reperfusion,[3] when the sudden restoration of blood flow can lead to free radical generation. Oxygen-derived free radicals can be the initiators of a peroxidative chain, leading to lipid degradation and membrane damage.

In the present study propionylcarnitine was evaluated as an inhibitor of the chemiluminescence (CL) arising from cardiac microsomes peroxidized either with NADPH or organic hydroperoxides. The CL is an index of the level of free radicals generated during the oxidative process.[4,5]

MATERIALS AND METHODS

Rat heart microsomal suspensions of 1 ml final volume, containing 0.5 mg protein in 0.1 mM phosphate buffer and the chemiluminigenic probe, were added either with 1 mM NADPH or 2 mM tert-butyl hydroperoxide (t-BOOH) or 2 mM cumene hydroperoxide (cum-OOH) to induce the chemiluminescent response.

The CL was measured in a luminometer at 37°C, after oxygenation of samples for 1 minute, and values were expressed as counts per minute (c.p.m.).

Statistical evaluation of the experimental results was effected by comparing the integrals of the chemiluminescent curve, calculated using Simpson discrete integration, according to Student's t-test for paired samples.

RESULTS AND DISCUSSION

As shown in Figure 1, addition of NADPH to microsomes in the presence of 10^{-5} M lucigenin gives rise to a photoemission that can be prevented in a dose-dependent manner by pretreating the samples for 10 minutes with propionylcarnitine. The CL signal can also be inhibited by scavengers such as SOD as well as by the inhibitors of eicosanoid synthesis eicosatetraynoic acid, nordihydroguaiaretic acid and indomethacin, whereas catalase is ineffective.

The above results and the better sensitivity to lucigenin amplification suggest that superoxide ions play a major role in the CL emission. This is stressed by Allen's demonstration that lucigenin involves, almost exclusively, the superoxide as a reactant.[6]

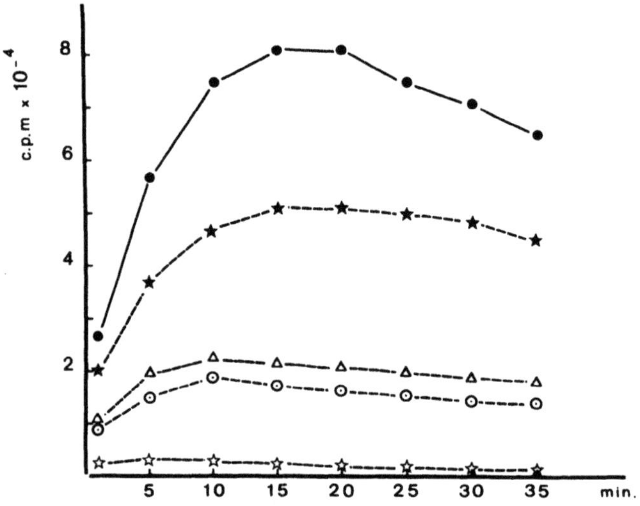

Figure 1. Time-course of NADPH-induced CL (•) and the inhibitory effect of propionylcarnitine: 150 μg/ml (★); 300 μg/ml (Δ); 600 μg/ml (o); 1200 μg/ml (☆).

Moreover, the variations in the shape of the CL curve in the presence of propionylcarnitine seem to suggest that as many free radicals are generated as are destroyed because there is no increase in the rate of free radical production.

Since there is evidence that neutrophil activation during ischemia-reperfusion can contribute to tissue damage,[7] and in order to assess whether propionylcarnitine could directly affect superoxide production, the drug was tested on human neutrophil oxidative metabolism.

The cells were stimulated with either opsonyzed zymosan or phorbol myristate acetate, and free radical production was revealed both by luminol-amplified CL and the cytochrome c reduction test, which specifically detects superoxide generation.

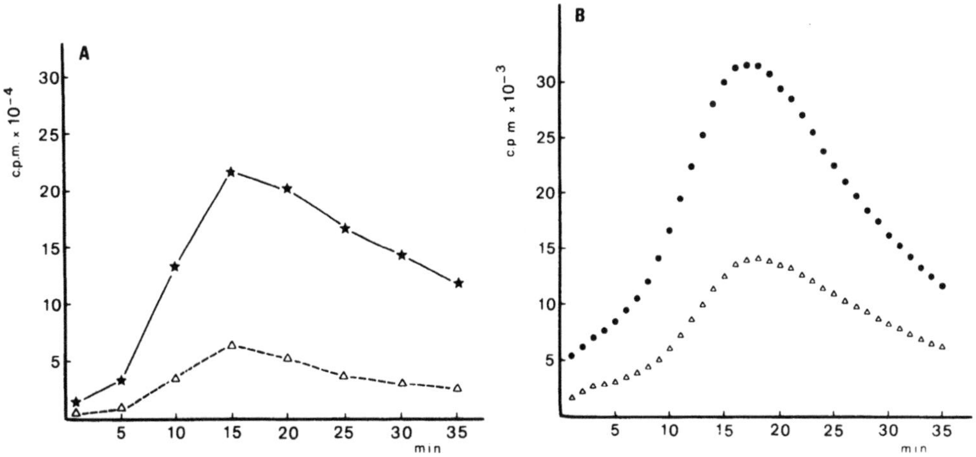

Figure 2. Inhibitory effect of propionylcarnitine on hydroperoxide-induced CL: (a) 2 mM t-BOOH (★) and the effect of 600 μg/ml propionylcarnitine (Δ); (b) 2 mM cum-OOH in the absence (•) or the presence (Δ) of the inhibitor.

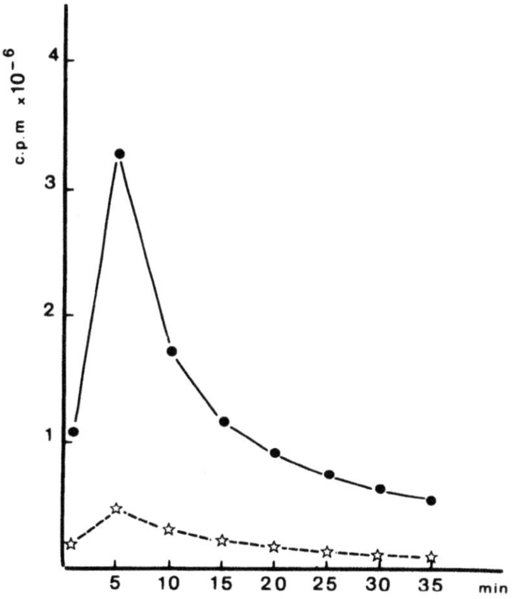

Figure 3. Enhancement of t-BOOH-induced CL by catalase (•) and the inhibitory effect of propionylcarnitine (☆).

Propionylcarnitine did not inhibit neutrophil CL, nor did it have any effect on cytochrome c reduction, demonstrating that it does not behave as a scavenger or have any effect on superoxide generation.

Since these findings made it difficult to establish which level of the peroxidative chain the drug acts on, further tests were carried out on t-BOOH or Cum-OOH-induced chemiluminescence, as the reactive species involved in hydroperoxide-supported oxidation are different from those of NADPH-catalyzed oxidations.

The CL induced in microsomes by 2 mM t-BOOH or 2 mM Cum-OOH, in the presence or the absence of 600 μg/ml propionylcarnitine, was best detected using luminol as a chemiluminigenic probe. In fact, luminol is an indicator of peroxidase activity.

The inhibition exerted by propionylcarnitine on hydroperoxide-induced lipid peroxidation is shown in Figures 2a and 2b. Figure 3 represents the effect of adding 20 μg/ml catalase to the microsomal suspension, just before t-BOOH stimulation, in the presence or the absence of the inhibitor. Catalase, which catalyzes the schism of t-BOOH, dramatically potentiated (1004% enhancement) the extent and the rate of the CL emission, which can be prevented by propionylcarnitine pretreatment.

CONCLUSIONS

As a tentative explanation for the above-mentioned results, it might be hypothesized that the lipophilic drug preserves the microsomes from the peroxidative stress by intercalating into the membrane lipid bilayer, thus preventing structural and functional alterations. This appears to be the most plausible explanation because the substance was effective in a variety of stress conditions induced by various peroxidizing agents and involving different free radical species as well as different microsomal enzyme activities. The inhibitory effects of propionylcarnitine might be due to the drug's action of stabilizing the membrane by preventing the accumulation of acyl-CoA, which is able to deform biological membranes by non-specific intercalation into the lipid bilayer, thus making it more prone to the action of peroxidizing agents.

ACKNOWLEDGMENTS

We thank Sigma Tau Spa. for the gift of propionylcarnitine.

REFERENCES

1. G.J. van der Vusse, and R.S. Reneman, Pharmacological intervention in acute myocardial ischemia and reperfusion, TIPS 6:76 (1985).
2. W.C. Hulsmann, M.L. Dubelaar, and J.M.J. Lamers, Protection by acylcarnitines and phenylmethylsulfonylfluoride of rate heart subjected to ischemia and reperfusion, Biochim. Biophys. Acta 847:62 (1982).
3. D.J. Paulson, J. Traxler, J. Noonan, and A.L. Shug, Effects of carnitine, acetylcarnitine, proprionylcarnitine, and proprionylcarnitin-taurine amide on the ischemic heart, Fed. Proc. 44:abstract 2058 (1982).

4. J.R. Wright, R.C. Rumbaugh, H.D. Colby, and P.R. Miles, The relationship between chemiluminescence and lipid peroxidation in rate hepatic microsomes, Arch. Biochem. Biophys. 192:344 (1979).
5. E. Cadenas and H. Sies, Low level chemiluminescence of liver microsomal fractions initiated by tert-butyl hydroperoxide, Eur. J. Biochem. 124:349 (1982).
6. R. Allen, Phagocytic leukocyte oxygenation activities and chemiluminescence: A kinetic approach to analysis, in: "Meth. Enzymol.," vol. 133, M.A. DeLuca and W.D. McElroy, eds., Academic Press, Inc., Orlando (1986).
7. R. Engler, Granulocytes and oxidative injury in myocardial ischemia and reperfusion, Fed. Proc. 46:2395 (1987).

CHEMILUMINESCENCE FROM VITAMIN E-DEFICIENT ERYTHROCYTE MEMBRANES INDUCED BY XANTHINE OXIDASE REACTION

Hiroshi Yasuda, Masayuki Miki, Yoshito Takenaka,
Hiroshi Tamai, and Makoto Mino

Department of Pediatrics Osaka Medical College
Takatsuki, Osaka 569, Japan

INTRODUCTION

Ultraweak chemiluminescence assay is known to be highly sensitive for oxidative reactions involved in the process of lipid peroxidation.[1-3] The determination of spontaneous chemiluminescence would become a useful noninvasive mean for studying the oxidative damage in biological systems.[4,5] However, the spontaneous chemiluminescence from living tissues may be too weak to be detected and identified, since biological systems are complex structured and contain a lot of substances serving as radical scavengers. Furthermore, the precise mechanism by which chemiluminescence is released still requires detailed study. In the light of the above point, the oxidation of red blood cell (RBC) membranes serves as a good model to study the peroxidative damage in the biological membranes.[6,7] With the hypoxanthine-xanthine oxidase (HX-XOD) reaction, RBC membranes were found to be oxidized and result in hemolysis,[8,9] where vitamin E in the membranes acted as a potent chain-breaking antioxidant and inhibited the peroxidation of membrane lipids as well as hemolysis.[10,11] We have additionally reported that chemiluminescence was observed in the reaction of HX-XOD with vitamin E-deficient RBC.[12]

In this paper, we have studied the relationship between chemiluminescence and membrane lipid peroxidation and tried to determine whether the peroxidation of membrane proteins contributes to chemiluminescence. We also investigated the effect of membrane vitamin E on the intensity of chemiluminescence.

MATERIALS AND METHODS

Preparation of RBC and its ghosts

RBC were obtained from Wistar male rats fed a vitamin E-deficient diet or the control diet for 12 or more weeks.[13] RBC ghosts were prepared by the method of Burton et al.[14]

Reaction systems

The 5% RBC suspensions (v/v) in phosphate-buffered saline containing 0.64 mM hypoxanthine were mixed with the same volume of 0.2 U/ml of xanthine oxidase and incubated at 37°C for 60 min (the final mixture consisted of 2.5% RBC, 0.32 mM HX and 0.1 U/ml XOD) Ghosts were suspended at 4 mg protein/ml in the buffered saline and then the reaction was carried out in the same manner as the above (the final mixture consisted of 2 mg protein/ml ghosts, 0.32 mM HX and 0.1 U/ml XOD).

Assay methods

The assay system for hemolysis, thiobarbituric acid reactive substance (TBARS), and chemiluminescence and the determination of alpha-tocopherol in membranes were described in previous reports.[9,12] The sodium dodecyl sulfate-polyacrylamide gel electrophoresis (SDS-PAGE) of RBC ghosts was performed as reported previously.[7] The content of sulfhydryl (SH) groups of membrane proteins were determined by the method of Fernandes et al.[15]

RESULTS AND DISCUSSION

The results of the oxidation in RBC induced by HX-XOD reaction are shown in Figure 1. With the vitamin E-deficient RBC, an increase in chemiluminescence intensity was observed at first, which was followed by the accumulation of TBARS and eventually by hemolysis. On the other hand, with the vitamin E-sufficient RBC, only a slight increase in chemiluminescence intensity was observed, and neither accumulation of TBARS nor hemolysis was observed throughout the reaction.

Figure 2 shows the effect of catalase added to the reaction medium at the specific incubation time. When catalase was added before the reaction, hemolysis was completely suppressed and chemiluminescence and TBARS formation were not observed. When catalase was added 2 min after the beginning of the reaction, hemolysis was suppressed only a little. At that time, chemiluminescence and TBARS formation were suppressed to half of the control experiment. When catalase was added 5 min after the incubation started, only a minimal suppression of chemiluminescence and TBARS formation were observed with little effect on hemolysis.

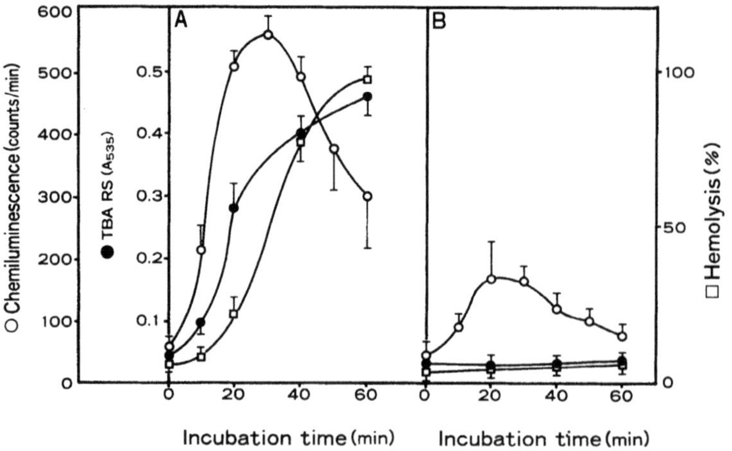

Figure 1. Relationship of chemiluminescence and TBARS formation to hemolysis induced by HX-XOD reaction in vitamin E-deficient (A) and vitamin E-sufficient (B) RBC. Each bar represents M±SD (n=5)

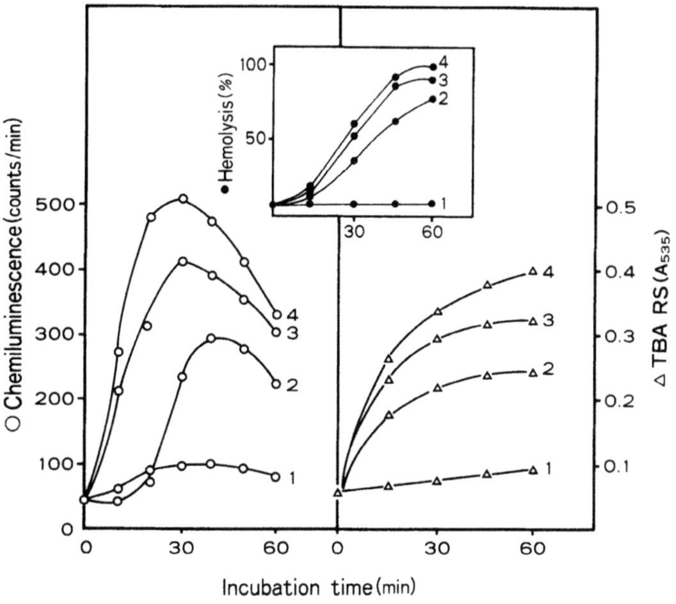

Figure 2. Effect of catalase added to the reaction medium at the specific intervals: 1, catalase was added before the reaction; 2, catalase was added 2 min after the reaction; 3, catalase was added 5 min after the incubation started; and 4, control study without catalase.

The results shown in Figures 1 and 2 suggest that the intensity of chemiluminescence is closely related to the amount of TBARS and to the extent of hemolysis. It has to be noted that the chemiluminescence intensity shown in Figures 1 and 2 is a differential value, while the value for TBARS and hemolysis are integrated ones. Figure 3 shows the plots of integrated, total count of chemiluminescence against TBARS formed, which is one of the most commonly used assay methods for lipid peroxidation. Interestingly, a good correlation exists between chemiluminescence and TBARS.

In the SDS-PAGE of membrane proteins prepared from vitamin E-deficient RBC after the reaction, the most predominant changes were the depletion of higher-molecular-weight proteins, including spectrins, and the appearance of low-molecular-weight (LMW) protein fractions. These changes were already observed in the early stage of the reaction. Even in vitamin E-sufficient RBC, the similar changes were also observed. These findings indicate that membrane vitamin E could not suppress the oxidative damage of proteins efficiently. Moreover, they may support the speculation that RBC membranes would be attacked by radicals from the inside, because the spectrin is localized in the inner side of the membranes.

We also studied changes of the total amount of SH groups in membrane proteins during HX-XOD reaction with RBC ghosts, with relation to chemiluminescence and TBARS formation. As shown in Figure 4, in vitamin E-deficient ghosts, about 20 % of the initial SH groups were consumed within the 20 min incubation. Subsequently, chemiluminescence intensity increased and TBARS were formed with time. In vitamin E-sufficient ghosts, the SH groups decreased similarly in the early stage of the reaction, the

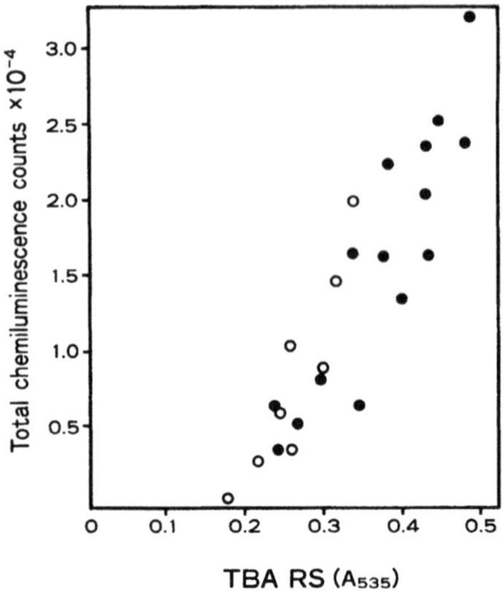

Figure 3. Relationship between total chemiluminescence counts and TBARS formation. (●): data from Figure 1-A, (○): data from curve 2 and 3 in Figure 2.

intensity of chemiluminescence and the formation of TBARS remained in low levels throughout the reaction. These results suggest that the membrane proteins are oxidized irrespective of the presence or absence of vitamin E and that the oxidation of SH groups is not directly correlated with an emission of chemiluminescence.

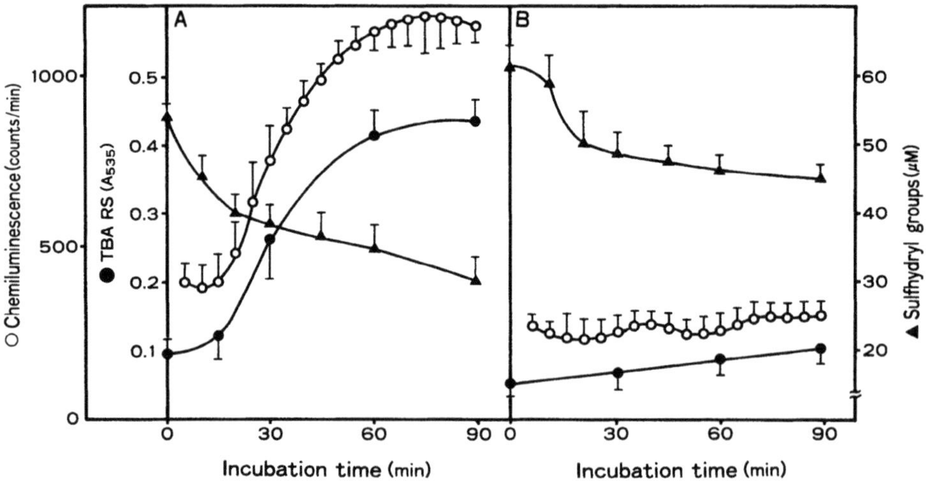

Figure 4. Changes of chemiluminescence, TBARS formation and SH groups in vitamin E-deficient (A) and vitamin E-sufficient (B) RBC ghosts oxidized with HX-XOD reaction. Each bar represents M±SD (n=5)

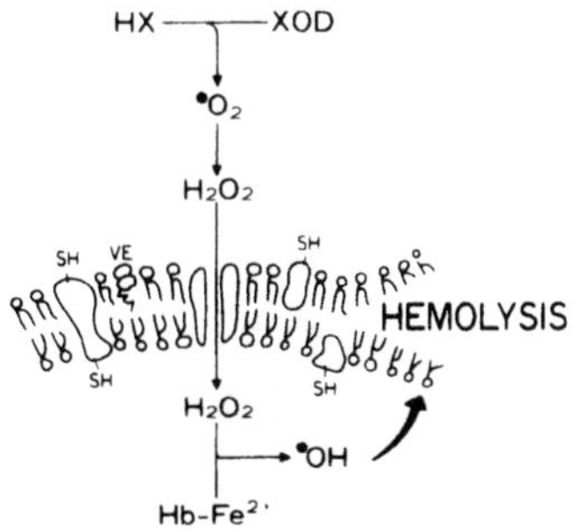

Figure 5. Schematic diagram of the mechanism of hemolysis induced by HX-XOD reaction

The addition of catalase completely suppressed chemiluminescence, accumulation of TBARS, and hemolysis, whereas SOD did not. Even when the RBC were added 5 min after HX was mixed with XOD, hemolysis, TBARS formation and chemiluminescence were also observed. Since $\cdot O_2^-$ and singlet oxygen have too short life spans to participate in the oxidation of membranes, it may be said that the main species in this reaction was H_2O_2. Therefore, the mechanism of hemolysis induced by HX-XOD reaction may be explained as shown in Figure 5. The XOD-catalyzed reaction is thought to produce superoxide anions ($\cdot O_2^-$) initially,[16] which are converted to hydrogen peroxides (H_2O_2) subsequently. The peroxides would penetrate through the RBC membranes and react with hemoglobin iron to form more active species, such as hydroxyl radicals ($\cdot$OH). They then would attack the membranes from the inside.[9] The latter possibility is confirmed by the results from SDS-PAGE as mentioned above and from a experiment in which the iron chelator such as EDTA had the inhibitory effect on the oxidation of ghost membranes.

CONCLUSION

The chemiluminescence arose mainly from the oxidation of membrane lipid and it is a good measure for the extent of oxidative damage of RBC membranes. Chemiluminescence must be emitted from the singlet oxygen (or its oxidation products), and excited triplet carbonyl compounds must be formed in the bimolecular terminating interaction of lipid peroxyl radicals.[17,18] It may be noteworthy thatthe chemiluminescence intensity reflects primarily the rate of chain initiation, while the TBARS value is related to the chain propagation. Therefore, the chemiluminescence and TBARS show the extent of oxidation complementarily.

ACKNOWLEDGMENTS

We are grateful to Prof. E. Niki, Tokyo University, and Prof. M. Nakano, College of Medical Care and Technology, Gunma University, for their helpful advices, discussions, and encouragement and to the Eisai Co. Ltd. for financial assistance and for providing the RRR-alpha-tocopherol and the vitamin E deficient diet.

REFERENCES

1. R. F. Vassilev, and A. A. Vichutinskii, Chemiluminescence and oxidation, Nature 194:1276 (1962).
2. K. Sugioka and M. Nakano, A possible mechanism of the generation of singlet molecular oxygen in NADPH-dependent microsomal lipid peroxidation, Biochim. Biophys. Acta 423:203 (1976).
3. E. Cadenas, H. Wefers, and H. Sies, Low-level chemiluminescence of isolated hepatocytes, Eur. J. Biochem. 119:531 (1981).
4. A. Boveris, E. Cadenas, R. Reiter, M. Filipkowski, Y. Nakase, and B. Chance, Organ chemiluminescence: Noninvasive assay for oxidative radical reactions, Proc. Natl. Acad. Sci. USA 77:347 (1980).
5. E. Cadenas, I.D. Arad, B. Fisher, A. Boveris, and B. Chance, Hydroperoxide-induced chemiluminescence of the perfused lung, Biochem. J. 192:303 (1980).
6. Y. Yamamoto, E. Niki, J. Eguchi, Y. Kamiya, and H. Shimazaki, Oxidation of biological membranes and its inhibition. Free radical chain oxidation of erythrocyte ghost membranes by oxygen, Biochim. Biophys. Acta 819:29 (1985).
7. M. Miki, H. Tamai, M. Mino, Y. Yamamoto, and E. Niki, Free-radical chain oxidation of rat blood cells by molecular oxygen and its inhibition by alpha-tocopherol, Arch. Biochem. Biophys. 258:373 (1987).
8. M. Taniguchi, M. Aikawa, and T. Sakagami, A simple and effective method for hemolysis with a hypoxanthine-xanthine oxidase system and alteration of erythrocyte phospholipid composition during the hemolysis, J. Biochem. 89:795 (1981).
9. H. Tamai, M. Miki, and M. Mino, Hemolysis and lipid changes induced by xanthine oxidase in vitamin E deficient red cells, J. Free Radic. Biol. Med. 2:49 (1986).
10. A. L. Tappel, Measurement of and protection from in vivo lipid peroxidation, Academic Press, New York (1980).
11. M. Mino, S. Nakagawa, H. Tamai, and M. Miki, Clinical evaluation of red blood cell tocopherol, Ann. N. Y. Acad. Sci. 393:175 (1982).
12. H. Yasuda, H. Tamai, M. Miki, and M. Mino, Chemiluminescence in vitamin E-deficient erythrocytes initiated by xanthine oxidase reaction, in relation to the accumulation of thiobarbituric acid substances, J. Nutr. Sci. Vitaminol. (Tokyo) 32:245 (1986).
13. M. Mino, M. Kitagawa, and S. Nakagawa, Changes of alpha-tocopherol levels in red blood cells and plasma with respect to hemolysis induced by dialuric acid in vitamin E-deficient rats, J. Nutr. Sci. Vitaminol. (Tokyo) 27:199 (1981).
14. G. W. Burton, K. U. Ingold, and K. E. Thompson, An improved procedure for the isolation of ghost membranes from human red blood cells, Lipids 16:946 (1981).
15. M. J. Fernandes Dies, D. T. Osuga, and R. E. Feexey, The sulfhydryls of avian ovalbumins, bovine β-lactoglobulin and bovine serum albumin, Arch. Biochem. Biophys. 107:440 (1964)
16. M. J. Thomas, K. S. Mehl, and W. A. Pryor, The role of superoxide in xanthine oxidase-induced autooxidation of linoleic acid, J. Biol. Chem. 250:8812 (1982).
17. E. Cadenas, A. Boveris, and B. Chance, Low-level chemiluminescence of biological systems. Free Radicals in Biology, Academic Press, New York (1984).
18. E. Niki, R. Tanimura, and Y. Kamiya, Oxidation of lipids. Rate of oxidation by alpha-tocopherol and hindered phenols measured by chemiluminescence, Bull. Chem. Soc. Jpn. 55:1551 (1982).

MONITORING THE INTERACTIONS OF ENVIRONMENTAL AGENTS BY CELL-DEPENDENT CHEMILUMINESCENCE: STUDIES WITH POLYCYCLIC AROMATIC HYDROCARBONS

Lorraine E. Twerdok and Michael A. Trush

Department of Environmental Health Sciences
Division of Toxicological Sciences
Johns Hopkins University,
Baltimore, Maryland 21205, U.S.A.

INTRODUCTION

The potential toxic effects of synthetic and naturally occuring substances on human health is of increasing concern. Thus, there is a great deal of interest in the development and application of techniques to monitor the cellular interactions of xenobiotics. The generation of chemiluminescence (CL) represents a potentially powerful probe for examining chemical-cell interactions, especially when it is correlated with other monitors of cellular function or used in conjunction with other analytical techniques.[1]

Neutrophil-derived oxidants have been implicated in both the damage to biomolecules and the metabolic activation of xenobiotics. Studies on the interaction of xenobiotics with polymorphonuclear leukocytes (PMNs) have demonstrated that CL is useful for monitoring the oxidant-dependent oxidation of xenobiotics.[1] For example, the addition of imipramine to human PMNs resulted in MPO-dependent CL.[2] An extension of this observation demonstrated that significant CL was observed when imipramine was incubated with rat liver microsomes and NADPH.[2] Imipramine is known to be metabolized by the microsomal mixed-function oxidase system to a hepatotoxic reactive intermediate, possibly an epoxide which covalently binds to protein.[3]

Like imipramine, the bioactivation of polycyclic aromatic hydrocarbons (PAHs) is essential for their ability to exert a mutagenic and carcinogenic effect.[4] In a series of studies, Seliger and coworkers have demonstrated that CL arises via a dioxetane pathway during the metabolism of PAHs by microsomal P450 enzymes.[3,4] More recently, studies on the interaction of carcinogenic PAHs with human PMNs have shown an MPO-dependent CL from those compounds which are also genotoxic.[5] Similarly, agents which inhibited the CL response from PAHs also interfered with the genotoxic actions of these compounds. The correlation between the ability of a particular PAH to exhibit CL and to exert a genotoxic effect is illustrated in Table 1. Such studies demonstrate that via oxidant-dependent reactions PMNs can metabolically activate carcinogens which may be an important link between the processes of inflammation and the development of neoplasms.

Table 1. Relationship between the Ability Of a PAH To be Metabolized to a Chemiluninescent Derivative with Its Ability to Elicit a Genotoxic Effect

PAH	Exhibits a CL Response in Presence of		Elicits a Genotoxic Effect in Presence of	
	PMNs[a]	Rat Liver Microsomes[b]	PMNs	Cytochrome P-448 (5 pmol)[c]
Benzo[a]pyrene (BP)	-	+[d]	-	-
BP-7,8-dihydrodiol	+	+	+	+
BP-7,8-quinone	-	?	-	-
BP-7,8-dihydrodiol-9,10-dihydro	-	?	-	-
7,8-dihydro BP	+	?	+	+
9,10-dihydro BP	-	?	-	-

[a]Data from ref. 7.
[b]Data from ref. 6.
[c]Data from ref. 4.
[d]Due to its subsequent metabolism to BP-7,8-dihydrodiol.

MATERIALS AND METHODS

Animals and Neutrophil Isolation from Bone Marrow

Male C57Bl/6J and DBA/2J mice (25-30g) were housed in an air conditioned room with a light period from 6 a.m. to 6 p.m. Purina laboratory chow and water were available *ad libitum*.

Bone marrow cells were flushed from the femurs of male mice and enriched for neutrophil content by hypotonic lysis of erythroid cells. Lysed preparations consisted of appoximately 70% ring-form metamyelocytes and neutrophils. The percent of neutrophilic cells was determined for each experiment and the data were corrected for the number of these cells.

Measurement of Chemiluminescence (CL) Responses

CL was monitored in freshly isolated neutrophilic cells from bone marrow in an ambient temperature liquid scintillation spectrometer (Packard model 3320) operated in the out-of-coincidence mode. The counter was set as follows: gain, 100%; window, A to infinity with discriminators set at 0 to 1000; and input selector, 1+2. Experiments were begun by incubating neutrophils in 3 ml phosphate-buffered saline for 10 min at 37°C in dark-adapted polyethylene vials. After the background CL of each vial was determined, BP-7,8-dihydrodiol (C_f 3 uM), luminol (C_f 0.1 uM), lucigenin (C_f 1 uM) or vehicle was added, any response noted, then the reactions were initiated by addition of TPA (C_f 100 ng/ml). CL was monitored for 0.2 min at 100-sec intervals, and vials were maintained at 37°C between counting. All additions to the vials as well as the CL counting were performed in a darkened room. Data are expressed as counts per unit time minus background and area under the curve as approximated by the trapezoid method.

RESULTS AND DISCUSSION

Bone marrow, a relatively neutrophil-rich tissue with low cytochrome P450 activity, is more sensitive to toxicity from orally administered

Table 2. Quantitation of the CL Response from BP-7,8-Dihydrodiol Elicited by TPA-stimulated Preparations of Bone Marrow Neutrophils from C57Bl/6 or DBA/2 Mice

Additions to TPA-stimulated Cells	Source of Bone Marrow Cells: C57Bl/6J	Source of Bone Marrow Cells: DBA/2J
BP-7,8-dihydrodiol (3uM)	$6.18 \pm .95 \times 10^4$ [a]	$1.47 \pm .35 \times 10^5$ [b]
BP-7,8-dihydrodiol + Azide (0.1mM)	$6.68 \pm 1.13 \times 10^3$ (88) [c]	$1.13 \pm .11 \times 10^4$ (86)
BP-7,8-dihydrodiol + SOD (100 ug/ml)	$4.79 \pm .23 \times 10^4$ (46)	$7.45 \pm .08 \times 10^4$ (47)

[a]Data are expressed as total counts of CL (area under curve as estimated by the trapezoid method) per 10^6 neutrophils and represent the mean ± SEM and an N of at least 3.

[b]$p<0.05$ compared with the response elicited by cells from C57Bl/6 mice by Student's _t_-test.

[c]Numbers in parentheses following area under the curve represent percent inhibition of the BP-7,8-dihydrodiol response.

benzo[a]pyrene (BP) in mice with non-inducible P450 systems.[6] Thus, we have compared the oxidant generation of neutrophilic cells isolated from femurs of male DBA/2 and C57Bl/6 mice, strains that are regarded as being susceptible and non-susceptible, respectively, to bone marrow toxicity from BP. Oxidant generation of neutrophilic preparations was assayed by superoxide anion generation and oxidant-dependent CL from luminol or lucigenin. In all assays, cells from DBA mice demonstrated increased oxidant generation. Similarly, a twofold enhancement of oxidant-dependent CL from BP-7,8-dihydrodiol was observed with TPA-stimulated neutrophilic cells from DBA mice as compared with cells from C57 mice (Table 2). This BP-7,8-dihydrodiol-dependent CL was inhibited by the addition of azide and SOD. Similarly, 25% greater covalent binding of BP-7,8-dihydrodiol, the proximate carcinogenic metabolite of BP, to DNA was observed in the presence of TPA-stimulated neutrophils from DBA mice (6.2 pmol/mg DNA) compared with cells from C57 mice (4.7 pmol/mg DNA). These results suggest that the increased risk of DBA mice for BP-induced leukemia may be related to their greater ability to generate oxidants which could participate in the activation of carcinogens as well as promote the development of bone marrow neoplasms.

Based on these observations, we would like to propose a model for the possible participation of neutrophil-derived oxidants in chemically induced bone marrow toxicity (Figure 1). First, there is the appearance of a toxin or its metabolites in the bone marrow where it may or may not require further metabolic activation to biochemically alter a specific target cell population within the bone marrow. These altered or damaged cells could then release a factor which would activate the neutrophilic cells within the bone marrow resulting in extracellular release of oxidants. This oxidant release could result in an amplification of the initial chemical damage in at least two ways. First, the oxidants could metabolically activate residual toxin that may be present in the bone marrow, resulting in damage to non-specific and/or specific target cells, releasing more factor and activating more neutrophils. Secondly, the neutrophil-derived oxidants could themselves non-specifically damage marrow tissue. This damage would then be manifested as cytopenias or anemias or, in the case of severe intoxication, as aplastic anemia or possibly leukemia.

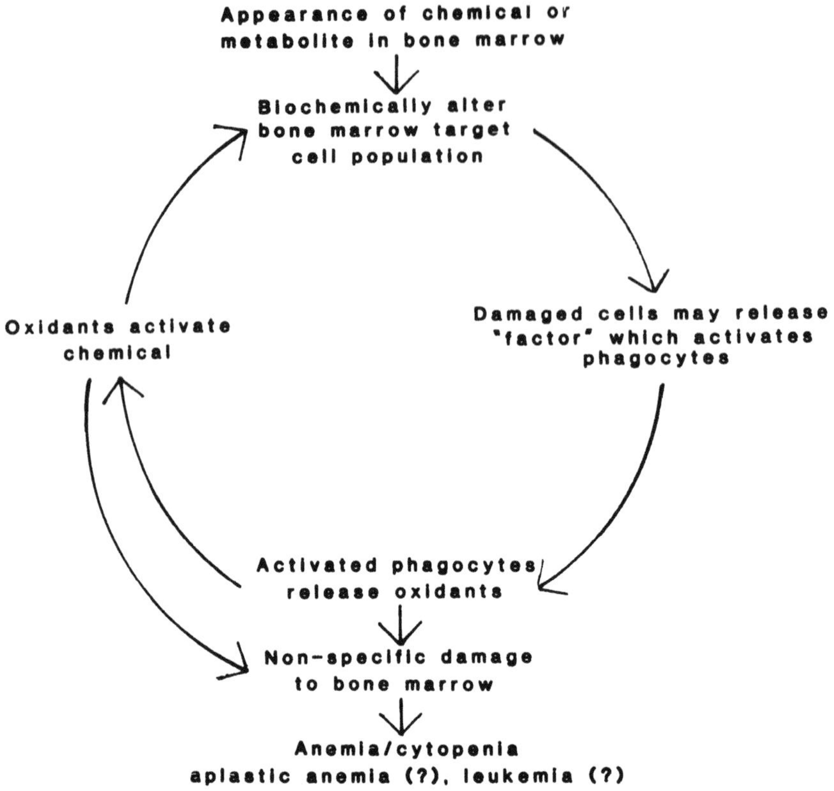

Figure 1. Possible model for investigating the role of neutrophil-derived oxidants in mechanisms of chemically induced bone marrow toxicity

ACKNOWLEDGEMENTS

The authors gratefully acknowledge financial support for this research from the National Institutes of Health ES 03760 and 07141, and the American Cancer Society SIG-3 and #1N-11X.

REFERENCES

1. M. A. Trush, Chemiluminescence as a probe to investigate chemical-cell interactions: a toxicological perspective, in "Cellular Chemiluminescence," K. Van Dyke and V. Castronova, eds., Boca Raton, FL, (1987).
2. M. A. Trush, M. J. Reasor, M. E. Wilson, and K. Van Dyke, Oxidant-mediated electronic excitation of imipramine, Biochem. Pharmacol. 33:1401 (1984).
3. H. Kappus and H. Remmer, Irreversible protein binding of [^{14}C]imipramine with rat human liver microsomes, Biochem. Pharmacol. 24:1079 (1975).

4. A. W. Wood, W. Levin, A. Y. H. Lu, D. Ryan, S. B. West, H. Yagi, H. D. Mah, D. M. Jerina, and A. H. Conney, Structural requirements for the metabolic activation of benzo[a]pyrene to mutagenic products: effects of modifications in the 4,5-, 7,8-, and 9,10-positions, Mol. Pharmacol. 13:1116 (1977).
5. H. H. Seliger, A. Thompson, J. P. Hamman and G.H. Posner, Chemiluminescence of benzo[a]pyrene-7,8-diol, Photochem. Photobiol. 36:359 (1982).
6. J. P. Hamman, H. H. Seliger, and G. H. Posner, Specificity of chemiluminescence in the metabolism of benzo[a]pyrene to its diol epoxide, Proc. Natl. Acad. Sci. USA 78:940 (1981).
7. M. A. Trush, J. L. Seed, and T. W. Kensler, Oxidant-dependent metabolic activation of polycyclic aromatic hydrocarbons by phorbol ester-stimulated human polymorphonuclear leukocytes: Possible link between inflammation and cancer, Proc. Natl. Acad. Sci. USA 82:5194 (1985).
8. D. W. Nebert, N. M. Jensen, R. C. Levitt and J. S. Felton, Toxic chemical depression of the bone marrow and possible aplastic anemia explainable on a genetic basis, Clin. Toxicol. 16:99 (1980).

GEL STATE CHEMILUMINESCENCE: AN ARTIFICIAL ELECTRON TRANSPORT SYSTEM*

Johnathan L. Kiel, David M. Simmons, and David N. Erwin

Radiation Sciences Division (RZP)
USAF School of Aerospace Medicine
Brooks AFB, Texas 78235-5301, U.S.A.

INTRODUCTION

Crosslinked glucose oxidase (GO) and horseradish peroxidase (HRP) can mimic macrophage oxidative function in vivo in respect to tumor cells as targets.[1-3] Both the flavoprotein of NADPH oxidase of macrophages and GO contain flavin adenine dinucleotide as their prosthetic group.[4] HRP has a noncovalently bound hematoporphyrin IX prosthetic group as does the cytochrome b_{559} (green hemoprotein) of NADPH oxidase.[4] Therefore, we anticipated a similarity in biochemical mechanism of action and chemiluminescence (CL) between the semi-synthetic crosslinked GO and HRP and the macrophage NADPH oxidase.

RESULTS AND DISCUSSION

We crosslinked 25 μl of GO (1 mg/ml), HRP (1 mg/ml), and luminol (LH_2)-bovine serum albumin (30 mg/ml) with 0.5% glutaraldehyde in pH 6.9 PBS in a 7-mm diameter filter paper disk to study the CL properties of this macrophage mimicking material (IC-1L).[1,2,5]

Table 1 shows that the CL (recorded with a Turner Designs 20e luminometer) of the disks was driven specifically by glucose (200 μl of 0.5 mg/ml) at pH 6.0, 6.5, 7.0, 7.4, 8.0, and 10.0. The best signal-to-noise ratio, 44,705 to 1, was achieved at pH 7.4. Even without luminol, there was considerable CL from IC-1 (189 to 1 signal-to-noise ratio) when disks treated with glucose were compared with those without glucose at pH 7.4.

The kinetics of the CL also varied with age and handling of the disks. Disks produced maximum CL on the second day after preparation, as did those in Table 1. The peak CL was 6 minutes after activation and persisted for about 3 minutes. On the first day, third day, and subsequent days after synthesis and storage at 4°C, the peak CL of the disks was replaced by a plateau at a lower maximum CL that was achieved slowly

Table 1. Chemiluminescence (relative light units/45 secs ± S.D. after 6 min) of Crosslinked Glucose Oxidase, Horseradish Peroxidase, and Bovine Serum Albumin (IC-1) or Luminol-Bovine Serum Albumin (IC-1L)

pH[a]	IC-1	IC-1L
6.0	0.365 ± 0.0146(n=5)	6.851 ± 0.862(n=6)
6.5	0.370 ± 0.0204(n=5)	16.76 ± 1.778(n=7)
7.0	1.283 ± 0.102(n=5)	112.4 ± 1.329(n=6)
7.4	1.477 ± 0.181(n=4)	348.7 ± 11.57(n=7)
7.4(NG)[b]	---	0.0078 ± 0.0055(n=5)
8.0	1.773 ± 0.337(n=4)	1998.0 ± 228.5(n=6)
8.0(NG)	---	0.4865 ± 0.008(n=6)
10.0	1.045 ± 0.068(n=4)	2397.0 ± 197.0(n=7)
10.0(NG)	---	0.8265 ± 0.011(n=6)

[a]The buffer was phosphate buffered saline (0.01 M sodium phosphate and 0.154 M NaCl) with 0.5 mg/ml D-glucose; at pH 8.0 and 10.0 the PBS pH was adjusted with 0.01 M sodium carbonate.
[b]NG = PBS without glucose.

and persisted for 6 to 10 minutes (Figure 1). Also, the spontaneous CL of IC-1L increased with time as the disk remained at room temperature in pH 7.4 PBS. Luminometry with the Quantitative Luminescence Imaging System (QLIS, USAF) indicated that intense localized long-term (30 minutes or more) CL was produced by IC-1L when compared with standard (1 x 10^{-7} M to 1 x 10^{-6} M) luminol in dimethyl sulfoxide solutions activated by 0.1 N NaOH (1-2 second CL lifetimes).

In Table 2, we report that the increase in bulk buffer pH around the IC-1L gel (a change of 2.6 units) led to an increase of only 1.68 pH

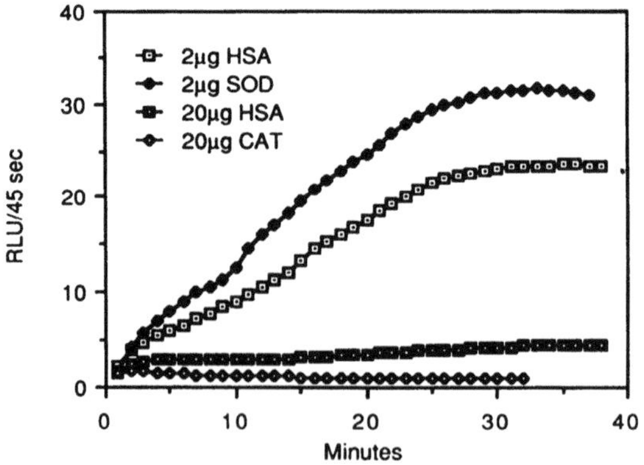

Figure 1. Chemiluminescence of IC-1L disks at room temperature. Each disk was either activated with 200 µl of pH 7.4 PBS containing 100 µg D-glucose or treated with PBS without glucose.

Table 2. Buffering Capacity and Proton Exchange by IC-1L Disks

Components[a]	pH ± S.D.	
Buffer (no gel)	7.4	10.0
Bulk solution		
with gel	7.61 ± 0.04(n=4)	9.09 ± 0.12(n=3)
with gel and glucose	7.66 ± 0.05(n=4)	9.12 ± 0.12(n=3)
Gel surface		
no glucose	6.11 ± 0.07(n=8)	6.98 ± 0.02(n=3)
glucose	5.42 ± 0.09(n=4)	6.90 ± 0.03(n=3)

[a]Each 7-mm paper disk containing 25 μl of gel was treated with 200 μl buffer (0.01 M PBS at pH 7.4 or 0.01 M TRIS/saline at pH 10.0) with or without 0.5 mg/ml D-glucose.

units in the surface of activated gel and 1.76 units in gel without glucose. The gel acted like a buffer with a pKa near 5 or 6 as indicated in Table 2. The addition of glucose consistently decreased the surface pH in either pH 7.4 or pH 10.0 buffer. These data indicate that the polyanionic gel entraps protons (hydrogen ions) even though the bulk solution may be very basic. Activation of CL facilitates the accumulation of protons in the gel.

Figure 2 illustrates the nonspecific and specific antioxidative response of IC-1L to inert human serum albumin (HSA), superoxide dismutase (SOD), and catalase (CAT). Comparing Figures 1 and 2, one can see that at 2 μg, HSA does not affect CL. However, at 2 μg, SOD enhances the CL. At 20 μg, HSA quenches CL by 81%, but CAT suppresses CL by 96%. These data indicate a strong participation of H_2O_2 and a minor but significant abortive role of free $\cdot O_2^-$. Dismutation of released $\cdot O_2^-$ facilitates the CL.

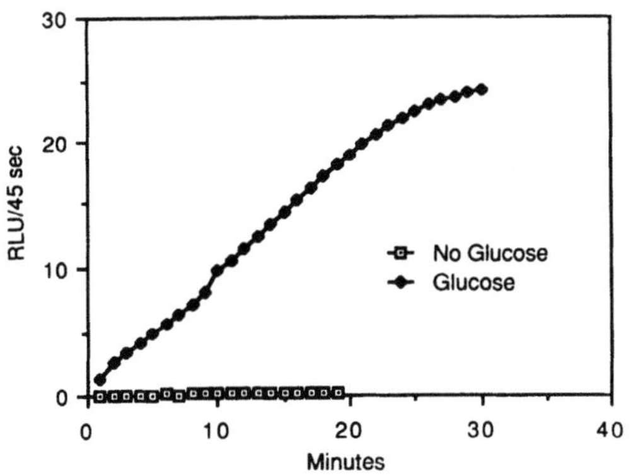

Figure 2. Chemiluminescence of IC-1L disks activated as described in Figure 1, but with human serum albumin, superoxide dismutase, or catalase.

CONCLUSIONS

Based on the data we present here, we conclude that the CL mechanism of IC-1L is as follows (where LO_2 is peroxyluminol or cyclooxyluminol and AP^{-2} is aminophthalate):

$$2\ \text{D-glucose} + 2O_2 \rightarrow 2\ \text{D-gluconic acid} + 2H_2O_2 \quad (1)$$

$$HRP + H_2O_2 + LH_2 \rightarrow HRP(\cdot OH) + H_2O + LH\cdot \quad (2)$$

$$HRP(\cdot OH) + H_2O_2 \leftrightarrow HRP(\cdot O_2^-) + H_2O + H^+ \quad (3)$$

$$HRP(\cdot O_2^-) + LH\cdot \rightarrow HRP + LO_2 + H^+ \quad (4)$$

$$LO_2 \rightarrow AP^{-2} + N_2 + \text{light} \quad (5)$$

ACKNOWLEDGEMENTS

This work was sponsored in part by the U.S. Air Force Office of Scientific Research.

REFERENCES

1. J.L. Kiel, " Insoluble Crosslinked Cytotoxic Oxidase-Peroxidase System," U.S. Patent 4,486,408. Dec 4, 1984 (filed Apr 7, 1981).
2. K.E. Everse, H. Lin, E.L. Stuyt, J.C. Brady, F. Buddingh, and J. Everse, Antitumor activity of peroxidase, Brit. J. Cancer 51:743 (1985).
3. C.F. Nathan and Z.A. Cohn, Antitumor effects of hydrogen peroxide in vivo, J. Exp. Med. 154:1539 (1981).
4. R. Lutter, M.L.J. van Schaik, R. van Zwieten, R. Wever, D. Roos, and M.N. Hamers, Purification and partial characterization of the b-type cytochrome from human polymorphonuclear leukocytes, J. Biol. Chem. 260:2237 (1985).
5. J.L. Kiel, Microwave effects on immobilized peroxidase chemiluminescence, Bioelectromag. 4:193 (1983).

HYDROPEROXIDATION OF UNSATURATED FATTY ESTERS

Edwin N. Frankel

Northern Regional Research Center
U. S. Department of Agriculture
Peoria, Illinois 61604, U.S.A.

INTRODUCTION

Much work has been published on the hydroperoxides of unsaturated lipids. Only recently has significant progress been made, however, to clarify the mechanism and stereochemistry of lipid hydroperoxidation. Although there is considerable information available on the chemistry of lipid oxidation, the biological effects of lipid oxidation products are not well understood. Studies on the source of volatile oxidation products and on their physiological importance are still controversial and difficult to interpret.

This paper reviews structural studies on the primary oxidation products of unsaturated fatty esters with emphasis on mechanistic aspects of the formation and decomposition of lipid hydroperoxides. More recent studies on the biological effects of lipid oxidation products are also presented.

OXIDATIVE PATHWAYS

Free Radical Autoxidation

At normal oxygen pressures and in the presence of initiators such as heat, metals, and light, unsaturated lipids (LH) lose a hydrogen radical, (H•), by reacting with an acceptor radical, R•, to form lipid free radicals, (L•), which rapidly react with oxygen to form peroxy radicals, (LOO•) (Figure 1). The production of free radicals can take place by thermal dissociation, hydroperoxide decomposition, metal catalysis, or by exposure to light. Because the rate of oxidation is independent of oxygen pressure, the overall rate of oxidation is determined by the stability of alkyl radicals. The radical chain is propagated by reacting peroxy radicals with unsaturated lipid to form hydroperoxides as the primary products of autoxidation. Because this reaction step is slow and rate-determining, hydrogen abstraction from unsaturated lipids becomes selective for the most weakly bound hydrogen.

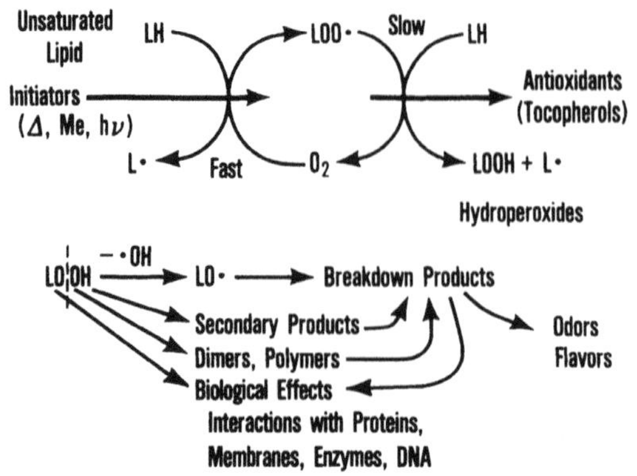

Figure 1. Free radical autoxidation of unsaturated lipids.

Rate studies based on oxygen uptake showed that linoleate (18:2) is 40 to 50 times more reactive with oxygen than oleate (18:1), whereas linolenate (18:3) is only two times more reactive than linoleate.[1,2] The two active methylene groups in methyl linolenate are, therefore, not activated by each other. Recently, Cosgrove et al[3] showed that the oxidizability of 18:2, 18:3, 20:4, and 22:6 are linearly related to the number of doubly allylic positions present in the fatty esters. They predicted an oxidizability of 2.0×10^{-2} $M^{-1/2}$ $sec^{-1/2}$ per activated methylene group at 37°C in chlorobenzene solution in the presence of AIBN [2,2′-azobis(2-methylpropionitrile)] as an initiator.

Reaction sequences leading to lipid autoxidation may be interrupted by phenolic antioxidants (AH) such as tocopherols, which can interfere with either the chain propagation or the initiation processes. By reacting with lipid peroxy radicals, these antioxidants compete with the lipid substrate for the chain-carrying peroxyl radicals to form stable radicals (A-A, LOOA, etc.) that are too unreactive to propagate the chain.[4] Another complementary chain-breaking mechanism involves quinones such as vitamin K, ubiquinone, and alpha-tocopheroquinone, which act as antioxidants by trapping alkyl radicals and by competing with oxygen for the alkyl radicals.

The decomposition of hydroperoxides forms volatile breakdown products by homolytic cleavage to form alkoxy radicals (LO•) and abstraction of a hydroxy radical (•OH). These breakdown products, causing rancidity in foods, include a complex mixture of aldehydes, ketones, alcohols, hydrocarbons, esters, and furans. Secondary products are oxygenated monomeric materials, including cyclic peroxides (hydroperoxy epidioxides), epoxy hydroperoxides, dihydroperoxides, and keto hydroperoxides. These secondary products can also break down to produce volatile materials, contributing to the oxidative deterioration of foods. Dimers and polymers are important condensation products formed from polyunsaturated lipids. They can in turn oxidize and break down into volatile materials causing rancidity.[5]

The interactions of hydroperoxides with proteins and other biological and food components are of most concern to biochemists and food chemists. These reactions can affect vital cell functions by causing damage to enzymes. In foods, these interactions have an impact on stability, flavor, and texture.[6]

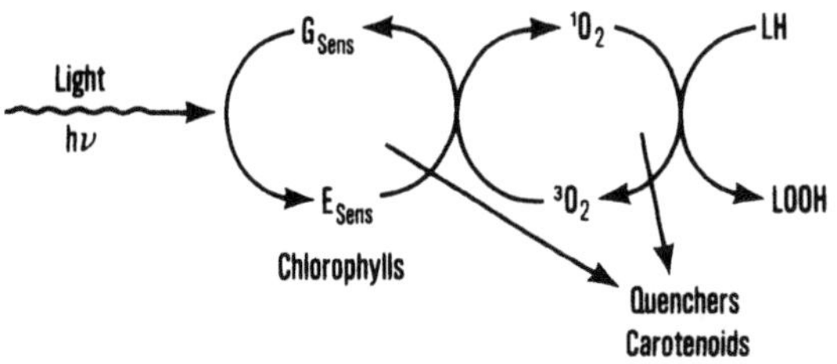

Figure 2. Photosensitized oxidation of unsaturated lipids.

Photosensitized Oxidation

This pathway provides a non-free-radical process of lipid oxidation, which is entirely different than autoxidation. By exposure to light energy, a sensitizer such as chlorophyll and other hemin compounds becomes excited. In the presence of oxygen, the energy is transferred and oxygen from the triplet state becomes excited into the singlet state (Figure 2). The resulting singlet oxygen ($^1O_2$) is a very active species and reacts with linoleate at least 1500 times faster than normal triplet oxygen ($^3O_2$) to form hydroperoxides.[7] For this reason, it has been suggested that hydroperoxides produced by $^1O_2$ break down to initiate normal free radical autoxidation.[8]

Natural quenchers such as carotenoids protect lipids against photosensitized oxidation by interfering with this process. In many foods, carotenoids are bleached out for cosmetic reasons. Recently, we found that β-carotene protects soybean oil against light oxidation when used at concentrations below 20 ppm.[9] At these low concentrations, β-carotene is effectively protected against free radical oxidation by the naturally occuring tocopherols in soybean oil. In earlier studies with distilled methyl esters of soybean oil, we used 1000 ppm of β-carotene to achieve protection against light oxidation.[10] When the soybean esters were treated with carbon black to remove any sensitizers, photosensitized oxidation was inhibited. Our evidence suggested, therefore, that the soybean esters contained sensitizers even after distillation. In soybean oil, β-carotene was not desirable when used at high concentrations because it not only produced off-flavors, but also promoted oxidation, apparently by breaking down into secondary products that can initiate and promote free radical autoxidation.

HYDROPEROXIDE FORMATION

The basic unsaturated fatty acids--oleic, linoleic, and linolenic acids--are present in all foods containing either vegetable or animal fats and are susceptible to oxidation. Arachidonic acid is also present in animal fats and is involved in biological oxidation as a precursor of many physiologically active prostaglandin derivatives.

Oleate Autoxidation

The accepted mechanism of oleate autoxidation involves hydrogen abstraction on allylic carbons-8 and -11 to produce two three-carbon allylic radicals. Oxygen attack at the end-carbon positions of these intermediates produces a mixture of four allylic hydroperoxides with oxygen on carbons-8, -9, -10, and -11 (Figure 3). According to this mechanism, the isomeric allylic hydroperoxides would be expected to form

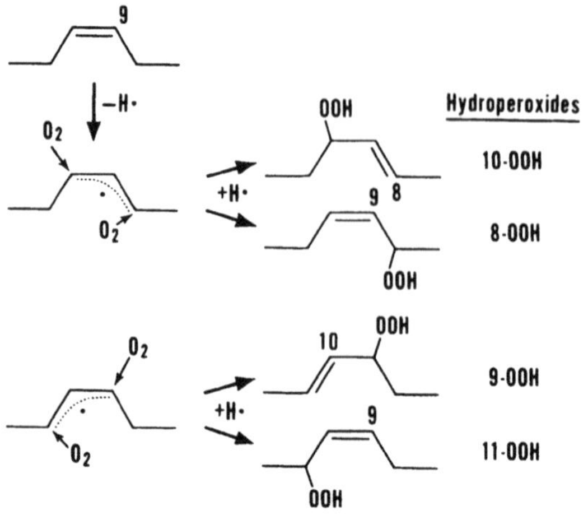

Figure 3. Classical mechanism for the autoxidation of oleate.

in equal amounts. However, our gas chromatographic-mass spectral (GC-MS) and high performance liquid chromatographic (HPLC) studies[11,12] showed a small but consistently higher amount of the 8- and 11-isomers than the 9- and 10-isomers. Stereochemical studies showed that we actually deal with eight cis and trans allylic hydroperoxides. With increasing temperature, we found that the cis-8- and cis-11-isomers decreased, the trans-8- and trans-11-isomers as well as the cis-9- and cis-10-isomers increased, while the trans-9- and trans-10-isomers did not change.[13] To explain these isomeric products, we postulated a mechanism involving conversion of the cisoid allylic radicals (Figure 2) into transoid radicals producing mixtures of trans (8-, 10-, and 11-OOH) and cis (9- and 10-OOH) hydroperoxides (Figure 4).[13] The relatively higher concentrations of the 8- and 11-hydroperoxides observed, especially at low temperatures, may indicate that reaction with oxygen occurs on radical intermediates that are not totally planar and that oxygenation occurs before rotation is complete.

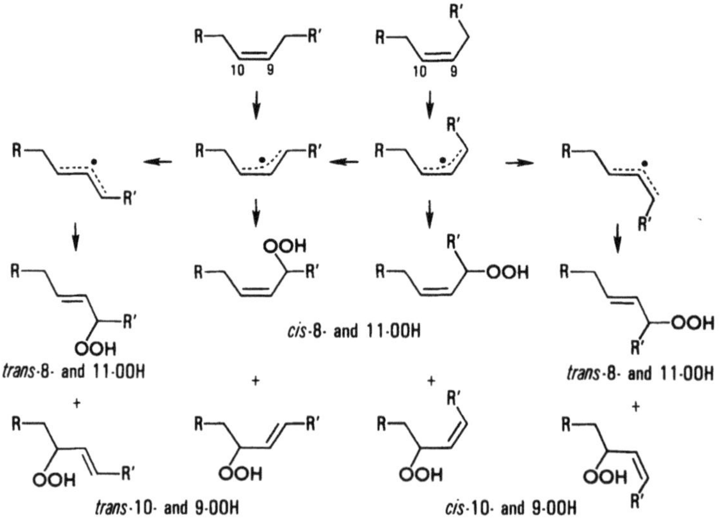

Figure 4. Mechanism II to explain the stereochemistry of oleate autoxidation.[13]

Electron spin resonance (ESR) studies made during photolysis of di-tert-butyl peroxide solutions demonstrated the formation of cisoid allylic radicals in methyl oleate and the corresponding transoid radicals in methyl elaidate.[14] Although methyl elaidate was formed under these photolytic conditions, only unchanged methyl oleate was found under our autoxidation conditions.[14] The cisoid radicals were irreversibly converted to the transoid radicals during photolysis, especially at elevated temperatures. A mechanism was postulated in which the cisoid radicals are interconverted either directly, as suggested in Figure 4,[13] or by reversible oxygen addition via tert-butyl alkoxy radicals.

Linoleate Autoxidation

Hydrogen abstraction at the doubly allylic carbon-11 position of linoleate generates a pentadienyl radical. Reaction of this radical with oxygen at the end carbon positions produces a mixture of two conjugated diene 9- and 13-hydroperoxides (Figure 5). We have found that these isomeric hydroperoxides are formed in equal concentrations under a wide range of autoxidation conditions.[15] Stereochemical studies showed a mixture of four cis,trans and trans,trans conjugated diene hydroperoxides. Initially, the cis,trans isomers are most important. The trans,trans isomers become increasingly important with the degree of oxidation and at elevated temperatures. Chan et al[16,17] showed that the cis,trans-9-and 13-linoleate hydroperoxides are readily interconverted and isomerized into a mixture containing the corresponding trans,trans hydroperoxides and that this process proceeds by exchange of the oxygen atoms in the hydroperoxide groups with atmospheric oxygen. To explain the stereochemistry of linoleate hydroperoxide isomers, a mechanism was postulated that includes the reversible interconversion of cis,cis to trans,cis/cis,trans, and to trans,trans radicals at higher levels of oxidation and elevated temperatures (Figure 6).[18]

In kinetic studies using di-tert-butyl hyponitrite as initiator, Porter et al[19] reported that the trans,cis/trans,trans ratio of conjugated diene hydroperoxides decreased by raising the temperature from 10 to 50°C and increased linearly with the concentration of linoleate.

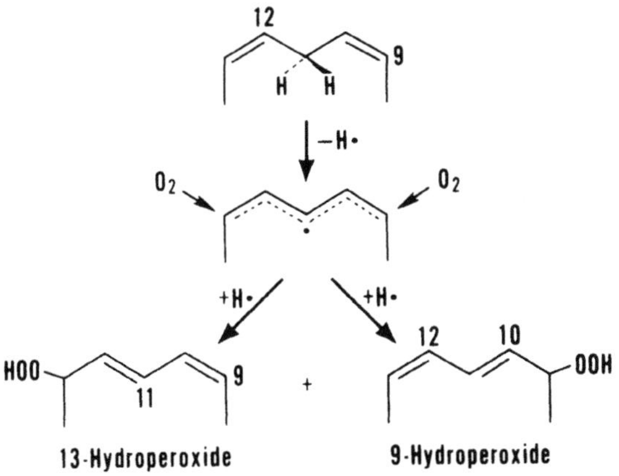

Figure 5. Classical mechanism for the autoxidation of linoleate.

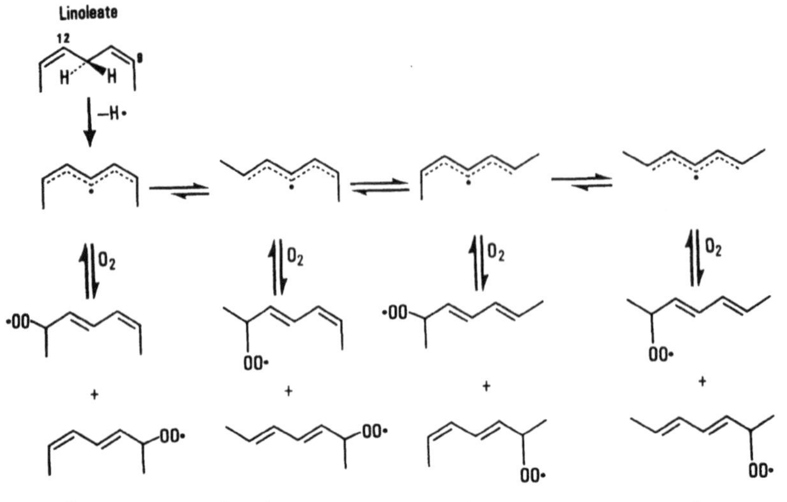

Figure 6. Mechanism II to explain the stereochemistry of linoleate autoxidation.[18]

Kinetic control was achieved in these experiments by limiting the HPLC analyses of hydroperoxide isomers to less than 2% conversion. The mechanism they advanced proceeds by interconversion of carbon and peroxyl radicals, which lose oxygen by -scission (Figure 7). This mechanism is supported by the findings that hydrogen-donor antioxidants such as p-methoxyphenol[19] and alpha-tocopherol[20] greatly increase the trans,cis/trans,trans ratio of linoleate hydroperoxides. ESR spectra taken during photolysis of di-*tert*-butyl peroxide solutions provided direct evidence for the formation of *cis,cis* pentadienyl radicals from methyl linoleate, *cis,trans* and *trans,trans* pentadienyl radicals from the corresponding geometric isomers of linoleate.[14] Because these pentadienyl radicals did not interconvert under their conditions of photolysis, Bascetta et al[14] supported the mechanism of Porter et al[19] involving peroxyl radicals (Figure 7) rather than carbon radicals (Figure 6).

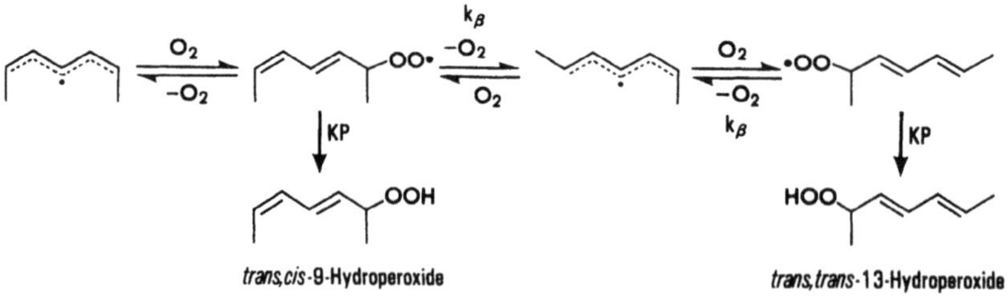

k_β = β-scission of C—O bond
KP = H abstraction

Figure 7. Mechanism proposed by Porter et al[19] to explain the stereochemistry of linoleate autoxidation.

Minor amounts of 8-, 10-, 12- and 14-hydroperoxides (4%) were recently reported by Schiberle et al[21] from the autoxidation of methyl and phenyl linoleate. The minor hydroperoxides are apparently formed by non-selective hydrogen abstraction from the two non-allylic systems between carbon-8 and carbon-10 and between carbon-12 and carbon-14 in the linoleate radicals.

Linolenate Autoxidation

In linolenate we have two doubly allylic carbon-11 and -14, producing two pentadienyl radicals. By the same mechanism as linoleate, reaction with oxygen at the end-carbon positions yields a mixture of four 9-, 12-, 13- and 16-conjugated diene hydroperoxide isomers containing a third isolated double bond (Figure 8). In 1961, we developed a scheme to determine the isomeric composition of linolenate hydroperoxides and showed for the first time that the distribution of isomeric hydroperoxides was uneven in autoxidized methyl linolenate.[22] The hydroxystearate derivatives were dehydrated to a mixture of monoene fatty esters, which were oxidatively cleaved to a mixture of dicarboxylic acids. The isomeric distribution based on the analysis of dicarboxylic acids showed that the proportion of 9- and 16-hydroperoxides was significantly higher than that of the 12- and 13-hydroperoxides. These results on the isomeric composition of linolenate hydroperoxides were later confirmed by two independent analytical methods. In 1977, we used a GC-MS method to analyze the isomeric hydroxystearates.[23] These GC-MS results were confirmed by HPLC analyses[12,24] in showing remarkably good agreement with our earlier work based on chemical degradation. We found the same distribution of linolenate hydroperoxides at different peroxide values and oxidation temperatures.[23] This uneven distribution of isomeric hydroperoxides of linolenate was explained in later studies showing that the internal 12- and 13-hydroperoxides are readily converted into hydroperoxy cyclic peroxides.[25-27] In contrast to linoleate, the cis,trans hydroperoxides of linolenate are not readily isomerized to the trans,trans configuration, apparently because cyclization is much more favorable and competes effectively with geometric isomerization.

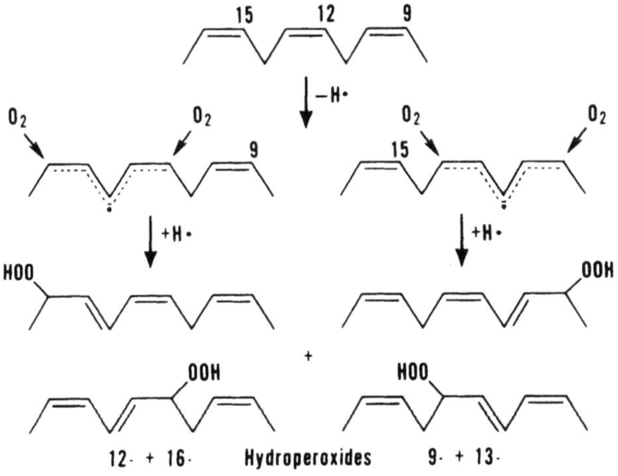

Figure 8. Classical mechanism for the autoxidation of linolenate.

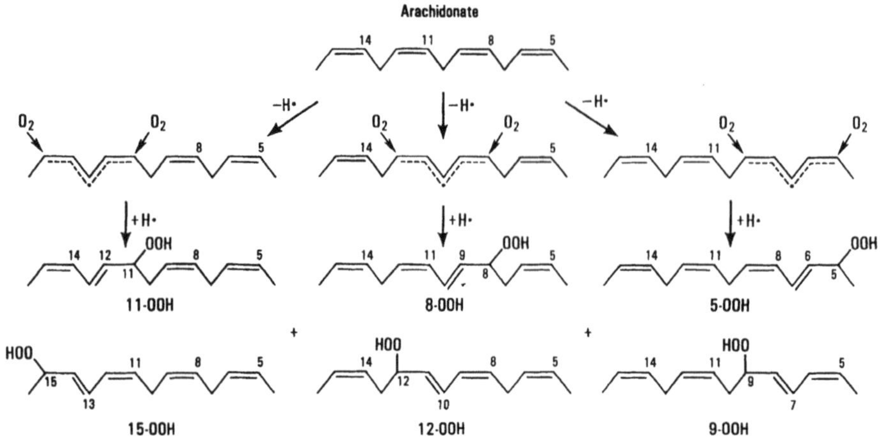

Figure 9. Classical mechanism for the autoxidation of arachidonate.

Arachidonate Autoxidation

For arachidonate the mechanism for autoxidation is the same as that for autoxidation of linoleate and linolenate. Arachidonate has three doubly allylic carbons, 7, 10, and 13. Hydrogen abstraction at these positions produces three pentadienyl radicals. Oxygen attack at either end positions of these radicals produces a mixture of six hydroperoxide isomers with a conjugated diene system and two methylene-interrupted double bonds. Hydroperoxide substitution occurs on carbons-5, 8, 9, 11, 12, and 15 (Figure 9). As with linolenate, the concentrations of the outer 5-hydroperoxide (18.3%) and 15-hydroperoxide (33.8%) are significantly higher than concentrations of the inner 8-, 9-, 11-, and 12-hydroperoxides (9.8-15.6%).[28] This uneven distribution can be attributed, like in linolenate, to the tendency of the inner hydroperoxide isomers to cyclize into hydroperoxy epidioxides. As observed in linolenate, Porter et al[29] found little trans,trans products from arachidonic acid compared to linoleic acid because of competitive 1,3-cyclization of the internal arachidonic acid hydroperoxide isomers.

Photosensitized Oxidation of Unsaturated Fatty Acids

Singlet oxygen reacts with carbon-carbon double bonds directly by a concerted "ene" addition. The hydroperoxides are formed by oxygen addition at each unsaturated carbon.[18] Thus, oleate forms two hydroperoxides, 9 and 10 (with allylic trans unsaturation). Linoleate forms four hydroperoxides, 9 and 13 (conjugated dienes) and 10 and 12 (unconjugated dienes). Linolenate forms six hydroperoxides: 9 + 10 + 12 + 13 + 15 + 16 (Figure 10). According to the ene addition mechanism, one would expect an even distribution of isomeric hydroperoxides in all unsaturated fatty esters. Our results show, however, that only oleate produces the expected even distribution of 9- and 10-hydroperoxide isomers. The hydroperoxide distribution is uneven in linoleate and linolenate.[10] The internal hydroperoxide isomers of linoleate (10 and

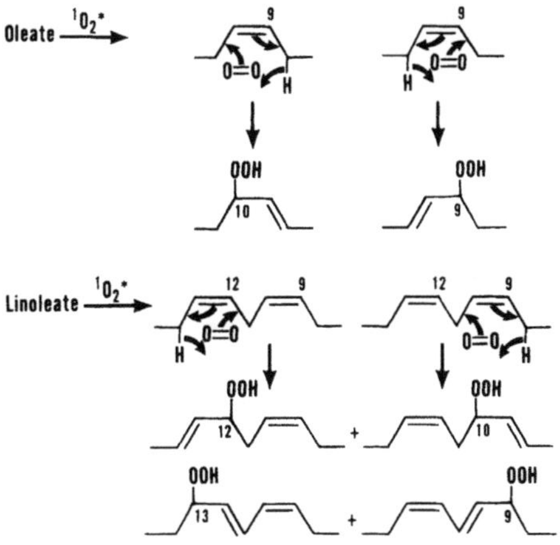

Figure 10. Reaction of singlet oxygen with oleate and linoleate by the "ene" addition mechanism.

12) and linolenate (10, 12, 13 and 15) have a homoallylic structure that allows facile 1,3-cyclization into hydroperoxy epidioxides. This cyclization is very rapid and just as rapid as hydroperoxidation, as indicated by this uneven distribution at very low levels of oxidation in both autoxidized linolenate and photosensitized oxidized linoleate[10] and linolenate.[23]

CYCLIZATION OF HYDROPEROXIDES

Conjugated Diene Hydroperoxides of Linoleate

According to Gardner et al[30] the homolytic decomposition of linoleic hydroperoxides with an Fe(III)-cysteine system forms isomeric mixtures of hydroxy epoxy and keto epoxy monoene compounds. They suggested a free radical mechanism involving the formation of an allylic epoxy radical from the conjugated diene alkoxy radical of linoleic acid. This allylic epoxy radical reacts with either ·OH or with oxygen to form hydroxy or the keto epoxy compounds (Figure 11). By the same mechanism, Gardner et al assumed that the epoxy allylic radical can have either a <u>trans</u> or <u>cis</u> conformation to explain the formation of two <u>trans</u> epoxy hydroperoxy monene and <u>cis</u> and <u>trans</u> epoxy hydroperoxy monoene compounds.

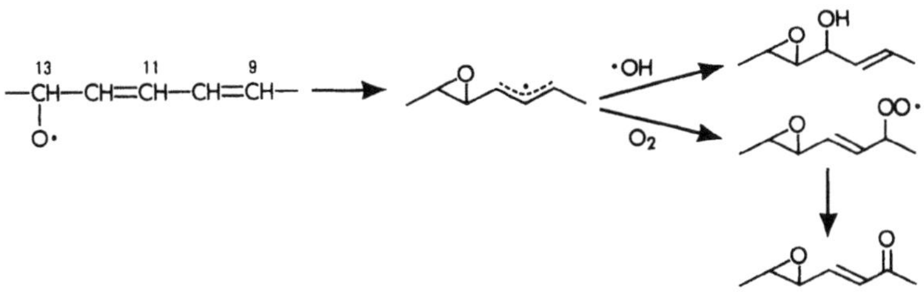

Figure 11. Cyclization of conjugated diene hydroperides.

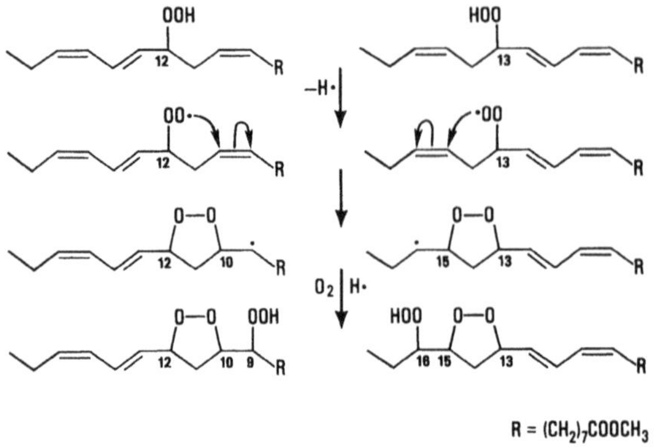

Figure 12. Cyclization of 12- and 13-hydroperoxides of linolenate.

Homoallylic Hydroperoxides of Linolenate

The 12- and 13-hydroperoxides in autoxidized methyl linolenate have a unique homoallylic structure that leads to their rapid cyclization to form hydroperoxy-epidioxides in significant yields[25-27] (Figure 12). This rapid cyclization accounts for the lower concentrations of these isomeric hydroperoxides relative to external 9- and 16-hydroperoxides. The addition of alpha-tocopherol reduced the extent of cyclization during autoxidation of methyl linolenate.[20] In the presence of 5% alpha-tocopherol, cyclization was completely inhibited, and the isomeric hydroperoxides of methyl linolenate were evenly distributed.

Bicycloendoperoxides were first postulated and identified indirectly in autoxidized methyl linolenate by Pryor et al[31], and later identified directly by O'Connor et al.[32] We identified the same compounds in very small amounts in photosensitized oxidized methyl linolenate.[33] These compounds come from the same internal 12- and 13-hydroperoxide isomers that were previously shown to undergo facile 1,3-cyclization to form monocyclic epidioxides. By this mechanism, after 1,3-cyclization occurs, the free radical intermediate undergoes a second cyclization to form a prostaglandin-like endoperoxide structure with an allylic hydroperoxide in either the 9- or 16-carbon position (Figure 13).

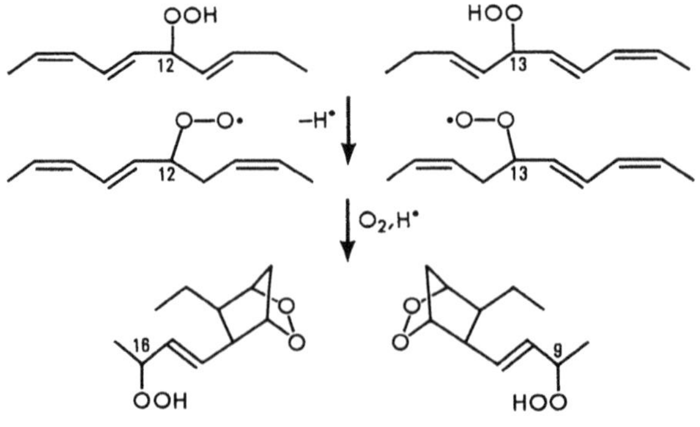

Figure 13. Formation of bicycloendoperoxides from linolenate hydroperoxides.

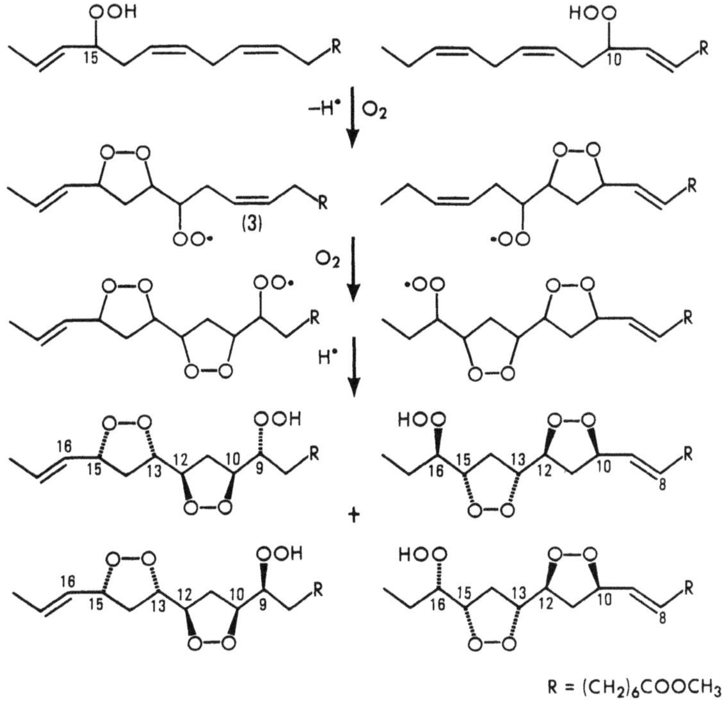

Figure 14. Formation of bis-epidioxides by photosensitized oxidation of methyl linolenate.[33]

The 10- and 15-hydroperoxide isomers formed by photosensitized oxidation of methyl linolenate have a unique structure that allows a second 1,3-cyclization to form bis-epidioxides[33] (Figure 14). The required precursors for these bis-cyclic peroxides are only present in photosensitized oxidized methyl linolenate and not in autoxidized methyl linolenate. Only arachidonate and other fatty acids containing more than three double bonds can produce these types of bis-epidioxides. They were indeed identified in the autoxidation products of the 15-hydroperoxide of arachidonic acid when autoxidation was initiated with di-tert-butyl hyponitrite.[34]

Homoallylic Hydroperoxides of Linoleate

The 10- and 12- homoallylic hydroperoxides from photosensitized oxidized methyl linoleate undergo 1,3-cyclization to produce mono-unsaturated 5-membered epidioxides[35,36] (Figure 15). This cyclization proceeds by the same free radical mechanism as for the 12- and 13-hydroperoxides of methyl linolenate and accounts for the lower relative concentrations of the 10- and 12-hydroperoxides (31-42%) than the 9- and 13-hydroperoxides (58-69%) observed in photooxdized methyl linoleate.[10] The question that arises is how can we have a free radical process under singlet oxidation conditions? Apparently, this is a secondary free radical process occurring as a side reaction during singlet oxidation. This is a very facile independent not photosensitized free radical reaction.[10]

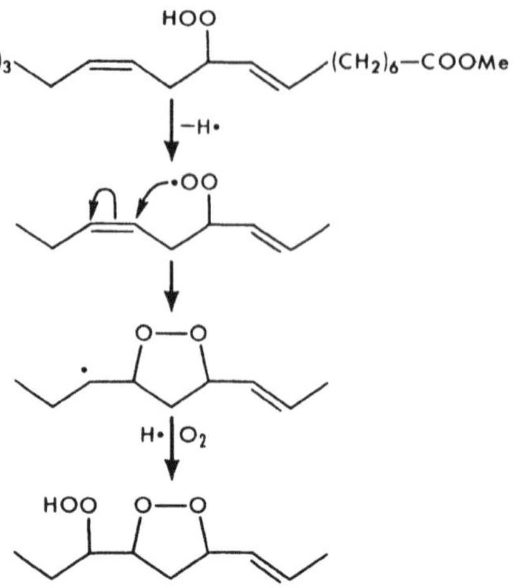

Figure 15. 1,3-Cyclization of 10-hydroperoxide from the photosensitized oxidation of methyl linoleate.

DECOMPOSITION OF HYDROPEROXIDES

Hydroperoxide decomposition involves a very complicated set of reaction pathways. A complex mixture of volatile and nonvolatile products is formed by thermal decomposition involving homolytic cleavage of alkoxy radical intermediates. A more selective heterolytic process is recognized for the acid decomposition of hydroperoxides.

Thermal Decomposition

β-Scission of alkoxy radical intermediate, formed by abstraction of an ·OH radical, proceeds by carbon-carbon cleavage on either side of the alkoxy carbon.[5] A mixture of volatile products is formed, including aldehydes, aldehyde esters, lower esters, olefins, and hydrocarbons. We thermally decomposed pure hydroperoxides directly in the injector port of a gas chromatograph at 180-210°C in an inert gas. The volatiles formed were analyzed quantitatively by GC and qualitatively by GC-MS.[37] More recently we compared the volatile products formed from methyl linolenate hydroperoxides decomposed thermally at 150°C, and catalytically in the presence of $FeCl_3$ and ascorbic acid at room temperature.[38] The thermal decomposition products from the hydroperoxides of autoxidized linolenate included ethane, 2,4-heptadienal, methyl octanoate, decatrienal and methyl 9-oxononanoate. The mechanism advanced to explain the major volatile decomposition products involves homolytic cleavage on either side of the hydroperoxide groups (Figure 16). Significant differences were found in the volatile products obtained by decomposing linolenate hydroperoxides either by heating at 150°C or by treating with ferric chloride and ascorbic acid at room temperature. Although 2,4-heptadienal was the major decomposition product observed with $FeCl_3$-ascorbic acid, it is derived from the 12-hydroperoxide isomer, which constitutes only 12% of the hydroperoxides in autoxidized methyl linolenate. On the other hand, the major thermal decomposition product, methyl octanoate,

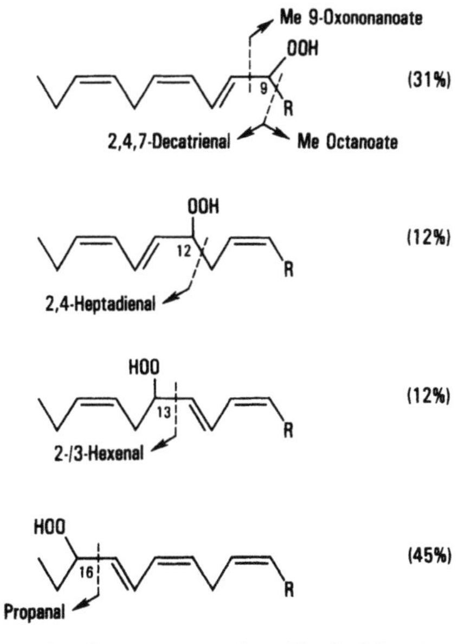

Figure 16. Main thermal cleavages of methyl linolenate hydroperoxides.

originates from the 9-hydroperoxide isomer, which constitutes 31% of the linolenate hydroperoxides. 2,4,7-Decatrienal and methyl 9-oxononanoate, formed in smaller amounts under both thermal and metal-catalyzed decomposition, also originate from the 9-hydroperoxide isomer.

Acid Decomposition

Malonaldehyde has received much attention in the literature because it was reported to be mutagenic and was implicated in reactions causing cell damage by interactions with DNA, enzymes, proteins, and phospholipids.[39-42] Although linolenate hydroperoxides and the derived hydroperoxy epidioxides have been considered to be important sources of malonaldehyde, under our conditions of thermal decomposition, we could not obtain evidence of malonaldehyde or other dialdehydes expected as decomposition products.[5] We developed, therefore, an acetalation-acid decomposition procedure to study the formation of malonaldehyde from monohydroperoxides and secondary oxidation products.[43] Under this milder decomposition procedure, using acid in the presence of methanol, monohydroperoxides are readily cleaved, and produce stable dimethyl and tetramethyl acetals that can be analyzed by GC and GC-MS.

Significant differences were observed between the thermal and acid decomposition products from monohydroperoxides and secondary oxidation products.[44] The differences in product distributions were explained by assuming that thermal decomposition proceeds by homolytic β-scission on both sides of the hydroperoxide group, whereas acid decomposition involves selective heterolytic splitting between the hydroperoxide group and the allylic double bond. A "mixed" homolytic-heterolytic-type mechanism was invoked for the thermal decomposition of hydroperoxides (Figure 17).

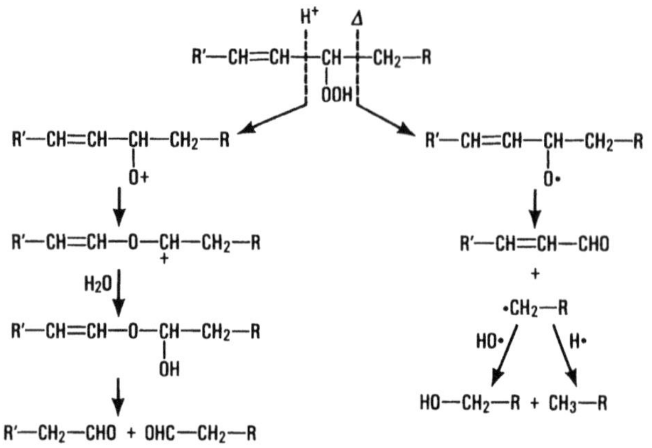

Figure 17. Homolytic-heterolytic mechanism for cleavage of allylic fatty acid hydroperoxides.[44]

According to this mechanism, heterolytic cleavage occurs on the olefinic side to produce two saturated aldehydes. Homolytic cleavage or β-scission on the alkyl side of the hydroperoxide would produce 2-alkenals. The remaining free radical fragments will lead to alcohols, lower esters, and hydrocarbons.

In the presence of methanol containing 5% HCl, cleavage of methyl linoleate hydroperoxides occurs selectively on the allylic sides.[44] The mechanism explaining this selective cleavage involves the formation of carbocation and ether intermediates. The aldehyde products become methoxylated and form dimethyl acetals. The allylic double bond becomes saturated by rearrangement of the vinyl ether (Figure 18).

BIOLOGICAL IMPLICATIONS

Tumor promotion by various free radicals, active oxygen species, and oxidizing agents, including peroxides or hydroperoxides, is curently receiving special attention in the literature.[45] Many organic peroxides and hydroperoxides that are commercially available, such as benzoyl, lauroyl and <u>tert</u>-butyl peroxides, hydroperoxides of vinylcyclohexene, and cumene, have been used in biological studies as

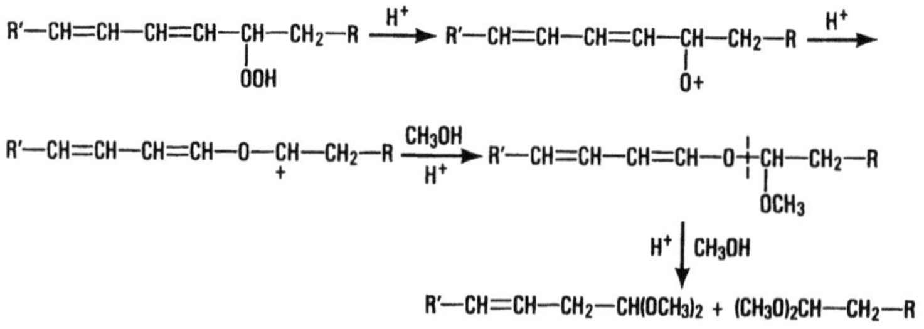

Figure 18. Heterolytic cleavage of methyl linoleate hydroperoxides under acid decomposition conditions.[44]

models of lipid peroxides.[45-47] These compounds are, however, much more stable than lipid peroxides and may not be valid models. Very little work has been reported on the effect of pure fatty acid hydroperoxides, even though they can be readily prepared in pure form by several techniques that have been available for a long time.[48]

The biochemical effect of a series of lipid oxidation products was investigated by determining the fluorescence formed from their interactions with DNA, metals, and reducing agents.[49] We found that at low concentrations of oxidation products, fluorescence formation required the presence of metals and ascorbic acid. In contrast, a positive thiobarbituric acid (TBA) reaction was obtained with the same lipid oxidation products without metals or ascorbic acid. With iron and ascorbic acid, linolenate hydroperoxides gave higher levels of fluorescence than linoleate hydroperoxides. The TBA values showed no correlation with fluorescence. Unoxidized linoleate and linolenate gave high TBA values in the absence or presence of iron and ascorbic acid. In the absence of iron and ascorbate, however, the TBA values were quite low for the linoleate hydroperoxides but relatively high for the linolenate hydroperoxides.

Our results indicate that the TBA test cannot be relied upon to measure either the degree of peroxidation or hydroperoxide decomposition. In future studies a more direct causal relationship needs to be established with pure lipid hydroperoxides to obtain a better understanding of the biological effects of lipid oxidation.

REFERENCES

1. A.J. Stirton, J. Turer, and R.W. Riemenschneider, Oxygen absorption of methyl esters of fatty acids, and the effect of antioxidants, Oil Soap 22:81 (1945).
2. R.T. Holman, and O.C. Elmer, The rates of oxidation of unsaturated fatty acids and esters, J. Am. Oil Chem. Soc. 24:127 (1947).
3. J.P. Cosgrove, D.F. Church, and W.A. Pryor, The kinetics of the autoxidation of polyunsaturated fatty acids, Lipids 22:299 (1987).
4. G. Scott, Antioxidants in vitro and in vivo, Chemistry in Britain 7:648 (1985).
5. E.N. Frankel, Volatile lipid oxidation products, Prog. Lipid Res. 22:1 (1983).
6. E.N. Frankel, Recent advances in the chemistry of rancidity of fats, in: "Recent Advances in the Chemistry of Meat," A.J. Bailey, ed., The Royal Society of Chemistry, London (1984).
7. H.R. Rawls and P.J. Van Santen, A possible source of the original hydroperoxides in fatty acids, Ann. N Y Acad. Sci. 171:135 (1970).
8. H.R. Rawls and P.J. Van Santen, A possible role for singlet oxygen in the initiation of fatty acid autoxidation, J. Am. Oil Chem. Soc. 47:121 (1970).
9. K. Warner and E.N. Frankel, Effects of β-carotetene on light stability of soybean oil, J. Am. Oil Chem. Soc. 64:213 (1987).
10. E.N. Frankel, W.E. Neff, and T.R. Bessler, Analysis of autoxidized fats by gas chromatography-mass spectrometry: V. Photosensitized oxidation, Lipids 14:961 (1979).

11. E.N. Frankel, W.E. Neff, W.K. Rohwedder, B.P.S. Khambay, R.F. Garwood, and B.C.L. Weedon, Analysis of autoxidized fats by gas chromatography-mass spectrometry: I. Methyl oleate, Lipids 2:901 (1977).
12. W.E. Neff and E.N. Frankel, Quantitative analysis of hydroxystearate isomers from hydroperoxides by high-pressure liquid chromatography of autoxidized and photosensitized oxidized fatty esters, Lipids 15:587 (1980).
13. E.N. Frankel, R.F. Garwood, B.P.S. Khambay, G.P. Moss, and B.C.L. Weedon, Stereochemistry of olefin and fatty acid oxidation. III. The allylic hydroperoxides from the autoxidation of methyl oleate, J. Chem. Soc., Perkin Trans. I. 2233 (1984).
14. E. Bascetta, F.D. Gunstone, and J.C. Walton, An electron spin resonance study of fatty acids and esters. Part 1. Hydrogen abstraction from olefinic and acetylenic long-chain esters, J. Chem. Soc., Perkin Trans. II. 603 (1983).
15. E.N. Frankel, W.E. Neff, W.K. Rohwedder, B.P.S. Khambay, R.F. Garwood, and B.C.L. Weedon, Analysis of autoxidized fats by gas chromatography-mass spectrometry: II. Methyl Linoleate, Lipids 12:908 (1977).
16. H.W.-S. Chan, C.T. Costaras, F.A.A. Prescott, and P.A.T. Swoboda, Thermal isomerisations of linoleate hydroperoxides, a phenomenon affecting the determination of isomeric ratios, Biochim. Biophys. Acta 398:347 (1975).
17. H.W.-S. Chan, G. Levett, and J.A. Matthew, The mechanism of the rearrangement of linoleate hydroperoxides, Chem. Phys. Lipids 24:245 (1979).
18. E.N. Frankel, Chemistry of free radical and singlet oxidation of lipids, Prog .Lipid Res. 23:197 (1985).
19. N.A. Porter, B.A. Weber, H. Weenen, and J.A. Khan, Autoxidation of polyunsaturated lipids. Factors controlling the stereochemistry of product hydroperoxides, J. Am. Chem. Soc. 102:5597 (1980).
20. K.E. Peers, D.T. Coxon, and H.W.-S. Chan, Autoxidation of methyl linoleate: The effect of alpha-tocopherol, J. Sci. Food Agric. 32:898 (1981).
21. P. Schieberle and W. Grosch, Detection of monohydroperoxides with unconjugated diene systems as minor products of the autoxidation of methyl linoleate, Z. Lebensm. Unters. Forsch. 173:199 (1981).
22. E.N. Frankel, C.D. Evans, D.G. McConnell, E. Selke, and H.J. Dutton, Autoxidation of methyl linolenate. Isolation and characterization of hydroperoxides, J. Org. Chem. 26:4663 (1961).
23. E.N. Frankel, W.E. Neff, W.K. Rohwedder, B.P.S. Khambay, R.F. Garwood and B.C.L. Weedon, Analysis of autoxidized fats by gas chromatography-mass spectrometry: III. Methyl Linolenate, Lipids 12:1055 (1977).
24. H.W.-S. Chan and G. Levett, Autoxidation of methyl linolenate. Analyses of methyl hydroxylinolenate isomers by high performance liquid chromatography, Lipids 12:837 (1977).
25. W.E. Neff, E.N. Frankel and D. Weisleder, High-pressure liquid chromatography of autoxidized lipids: II. Hydroperoxy-cyclic peroxides and other secondary products from methyl linolenate, Lipids 16:439 (1981).

26. D.T. Coxon, K.R. Price, and H.W.-S. Chan, Formation, isolation and structure determination of methyl linolenate diperoxides, Chem. Phys. Lipids 28:365 (1981).
27. I. Toyoda, J. Terao, and S. Matsushita, Hydroperoxides formed by ferrous ion catalyzed oxidation of methyl linolenate, Lipids 17:84 (1982).
28. J. Terao, and S. Matsushita, Analysis of hemoprotein catalyzed peroxidation products of arachidonic acid by gas chromatography mass spectrometry, Agric. Biol. Chem. 45:595 (1981).
29. N.A. Porter, L.S. Lehman, B.A. Weber, and K.J. Smith, Unified mechanism for polyunsaturated fatty acid autoxidation. Competition of peroxy radical hydrogen atom abstraction, -scission, and cyclization, J. Am. Chem. Soc. 103:6447 (1981).
30. H.W. Gardner, R. Kleiman, and D. Weisleder, Homolytic decomposition of linoleic acid hydroperoxides: Identification of fatty acid esters, Lipids 9:696 (1984).
31. W.A. Pryor, J.P. Stanley, and E. Blair, Autoxidation of polyunsaturated fatty acids: II. Suggested mechanism for the formation of TBA-reactive materials from prostaglandin-like endoperoxides, Lipids 11:370 (1979).
32. D.E. O'Connor, E.D. Mihelich, and M.C. Coleman, Stereochemical course of the autoxidative cyclization of lipid hydroperoxides to prostaglandin-like bicycloendoperoxides, J. Am. Chem. Soc., 106:3577 (1984).
33. W.E. Neff, E.N. Frankel, and D. Weisleder, Photosensitized oxidation of methyl linolenate. Secondary products, Lipids 17:780 (1982).
34. J.A. Khan and N.A. Porter, Serial cyclization of an arachidonic acid hydroperoxide, Angew. Chem. Suppl. 513 (1982).
35. E.D. Mihelich, Structure and stereochemistry of novel endoperoxides isolated from the sensitized photoxidation of methyl linoleate. Implications for prostaglandin biosynthesis, J. Am. Chem. Soc. 102:7141 (1980).
36. E.N. Frankel, W.E. Neff, E. Selke and D. Weisleder, Photosensitized oxidation of methyl linoleate. Secondary and volatile thermal decomposition products, Lipids 17:11 (1982).
37. E.N. Frankel, W.E. Neff, and E. Selke, Analysis of autoxidized fats by gas chromatography-mass spectrometry: VII. Volatile thermal decomposition products of pure hydroperoxides from autoxidized and photosensitized oxidized methyl oleate, linoleate, and linolenate, Lipids 16:279 (1981).
38. E.N. Frankel, W.E. Neff, E. Selke, and D.D. Brooks, Thermal and metal-catalyzed decomposition of methyl linolenate hydroperoxides, Lipids 22:322 (1987).
39. K.S. Chio and A.L. Tappel, Synthesis and characterization of the fluorescent products derived from malonaldehyde and amino acids, Biochemistry 8:2821 (1969).
40. C.J. Dillard and A.L. Tappel, Fluorescent products from reaction of peroxidizing polyunsaturated fatty acids with phosphatidyl ethanolamine and phenylalanine, Lipids 8:183 (1973).
41. W.R. Bidlack and A.L. Tappel, Fluorescent products of phospholipids during lipid peroxidation, Lipids 8:203 (1973).

42. R. Trombly and A. L. Tappel, Fractionation and analysis of fluorescent products of lipid peroxidation, Lipids 10:441 (1975).
43. E.N. Frankel and W.E. Neff, Formation of malonaldehyde from lipid oxidation products, Biochim. Biophys. Acta 754:264 (1983).
44. E.N. Frankel, W.E. Neff, and E. Selke, Analysis of autoxidized fats by gas chromatography-mass spectrometry. IX. Homolytic vs. heterolytic cleavage of primary and secondary oxidation products, Lipids 19:790 (1984).
45. P.A. Cerutti, Prooxidant states and tumor promotion, Science 227:375 (1985).
46. T.J. Slaga, A.J.P. Klein-Szanto, L.L. Triplett, L.P. Yotti, and J.E. Trosko, Skin tumor-promoting activity of benzoyl peroxide, a widely used free radical-generating compound, Science 213:1023 (1981).
47. B.L. Van Duuren, N. Nelson, L. Orris, E.D. Palmes, and F.L. Schmitt, Carcinogenicity of epoxides, lactones, and peroxy compounds, J. Nat. Cancer Inst. 31:41 (1963).
48. E.N. Frankel, Hydroperoxides, in: "Symposium on Foods: Lipids and Their Oxidation," H.W. Schultz, E.A. Day, and R.O. Sinnhuber, eds., The Avi Publishing Co., Inc., Westport, CT (1962).
49. K. Fujimoto, W.E. Neff, and E.N. Frankel, The reaction of DNA with lipid oxidation products, metals and reducing agents, Biochim. Biophys. Acta 795:100 (1984).

LIPID PEROXIDATION AND RETINAL PHOTOTOXIC DEGENERATION

Robert J. Stephens, Dalbir S. Negi,
Sarah M. Short, Frederik J.G.M. van Kuijk,*
Edward A. Dratz,* and David W. Thomas

Cell Biology Program
SRI International
Menlo Park, California 94025, U.S.A.

*Chemistry Department
Montana State University
Bozeman, Montana 59717, U.S.A.

INTRODUCTION

Phototoxic degeneration of the retina was first described by Noell, Walker, Kang, and Berman in 1966 and these authors also advanced the theory that polyunsaturated fatty acid (PUFA) oxidation was the causal mechanism responsible for photoreceptor cell degeneration.[1] Many investigators have contributed to our current understanding of light damage to the retina, and PUFA oxidation continues to be a prominent theory proposed to explain the mechanism of retinal degeneration.[2] PUFA oxidation has also been implicated as an important factor in many other degenerative diseases such as cataract and heart disease.[3] Until recently, however, conclusive evidence that lipid peroxidation takes place _in vivo_ under any condition has been elusive.

Three years ago, we embarked on a program to develop new techniques and procedures with the sensitivity to detect primary peroxidation products (hydroperoxides) in tissues. The goal was to test the concept that PUFA oxidation occurs _in vivo_ in association with retinal phototoxic degeneration. In addition, we sought to identify the toxic molecule(s) that cause photosensitive cells to degenerate when exposed for short periods to intense light (3-6 hrs to ~ 1000 Ft.c.) or to continuous light for extended periods at lower intensities (2-4 days at ~ 150 Ft.c. or 4-6 weeks at ~ 20 Ft.c.).

We have developed procedures to identify primary and secondary products of PUFA oxidation by gas chromatography-mass spectrometry (GC-MS) through the use of pentafluorobenzyl esters and pentafluorobenzyl oximes that provide sensitivity to the picogram level.[4,5] In addition, sample preparation and identification of tissue depot sites have been substantially improved by microdissection of freeze-dried retinas that can be used for GC-MS analysis of both PUFA oxidation products, as well as antioxidants such as vitamin E.[6,7]

EXPERIMENTAL

Our current research centers on the detection, identification, and quantitation of primary (hydroperoxide) and secondary (aldehyde) decomposition products of linoleic (18:2ω6) and docosahexaenoic (22:6ω3) fatty acids. This research makes use of reference oxidized compounds, obtained by photooxidizing phospholipids or methyl esters of the two fatty acids. Rose Bengal is used as a photosensitizer to generate the singlet oxygen (Figures 1 and 2). Reference GC-MS chromatograms of the respective hydroperoxides produced in this manner were compared with those obtained from retinal tissues. Control retinas were obtained from Sprague-Dawley rats reared in cyclical illumination (12 hr light: 12 hr dark) with an intensity of ~ 25 Ft.c. (Figures 3 and 4) and from samples of retina prepared from rats receiving a diet depleted in selenium and vitamin E (Figure 5). Samples of rod outer segments were obtained for analysis by microdissection of freeze-dried retina (Figure 6). Additional GC-MS chromatograms were obtained after the rats were exposed to white fluorescent light with an intensity of ~ 1100 Ft.c. for 6 hr.

The starting material (fatty acid methyl esters) used to obtain the reference chromatogram of 18:2ω6 (Figure 1) contained a small amount of autoxidized product and photooxidation was superimposed on this sample. We obtained a full complex of the possible isomers of the hydroperoxide resulting from oxidation of the two double bonds in 18:2ω6.[8] This chromatogram served to identify the same six isomers in the tissue samples but with differing relative intensities (Figures 3 and 5). The first peak contains two nonconjugated isomers, the second peak contains two conjugated isomers, and the remaining two peaks each contain an additional conjugated isomer as can be determined from specific fragmentation patterns using electron ionization.[4] Chromatograms obtained from retinal tissues match the peaks in the reference traces (cf. 1,3,5). Likewise, the reference chromatographic trace for oxidized 22:6ω3 (Figure 2) is very similar to those obtained from the retina (Figure 4) and outer segments (Figure 6). The number of expected isomers resulting from oxidizing the 6 double bonds in 22:6ω3 is large; therefore, the chromatogram is complex. The complexity is shown by the poorly resolved broad peak to the right of the major peak in the reference chromatographic trace of the conjugated and nonconjugated products (Figure 2). Nevertheless, rather similar traces were obtained from retinal tissues (Figures 4 and 6). Accurate detection and identification of the hydroperoxides of 18:2ω6 and 22:6ω3 has been achieved using the above procedures; however, precise quantitation requires the synthesis of suitable isotopically labeled molecules to be used as internal standards which are not yet available.

Semiquantitation has been achieved by making comparisons between the integrated area of the chromatographic peak due to a saturated fatty acid 20:0 (present in small amounts in retinal tissue) and the hydroperoxides observed in the same samples. Using this procedure, we have found that the amount of hydroperoxide present from both 18:2ω6 and 22:6ω3 in the retinas of rats exposed to 1100 Ft.c. for 6 hrs was sometimes less than, but never more than, those observed in control tissue. These tissues were very heavily damaged by light (Figures 11 and 12) as discussed later. This suggests that hydroperoxides formed in the light may react rapidly to form toxic products. When animals are raised under dim cyclic light, vitamin E protected against degeneration when exposed to an increased light level, while the vitamin E-depleted retinas showed substantial degeneration, suggesting that lipid peroxidation is involved in light damage to the retina, at least under fairly dim (~ 25 Ft.c.), chronic cyclic exposure.

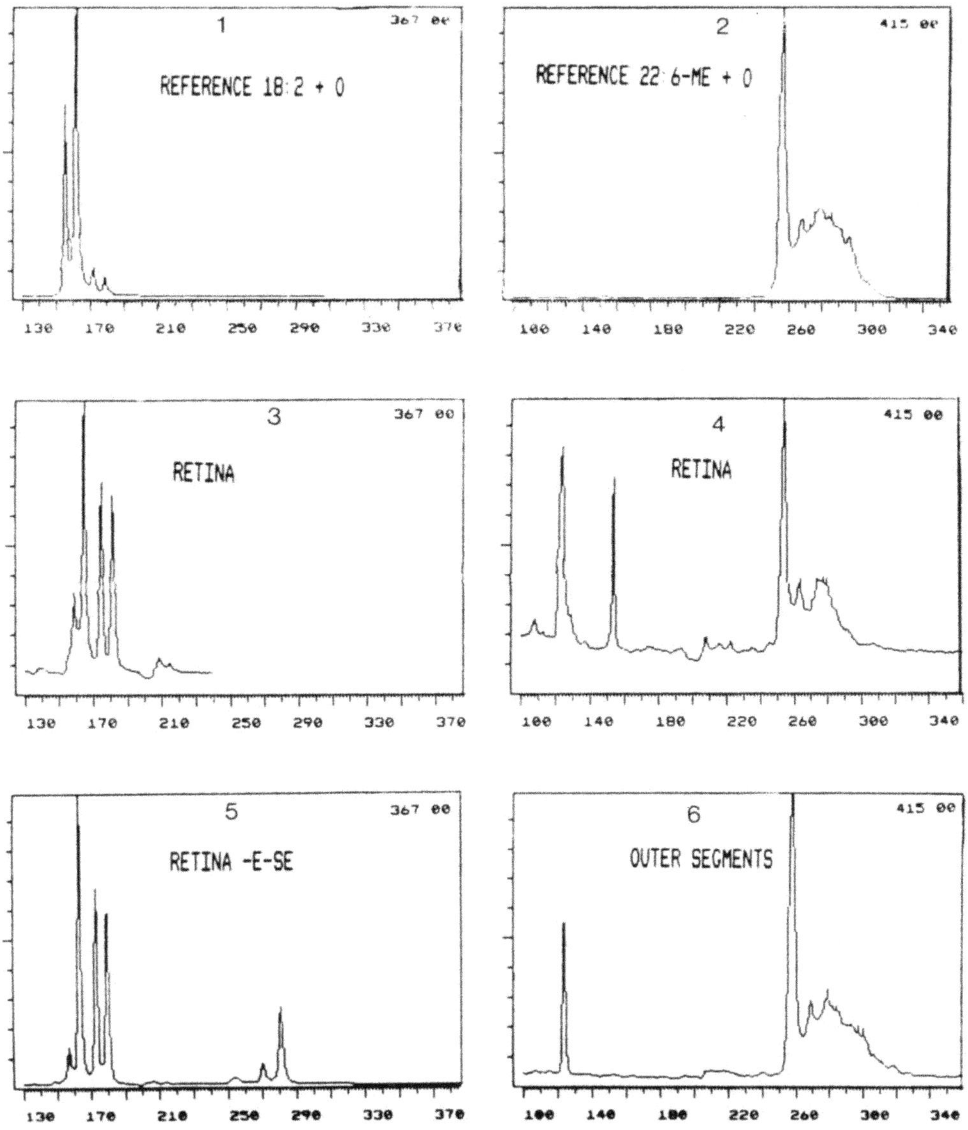

Figures 1-6. These GC-MS chromatograms are of the isomeric hydroperoxides from linoleic (18:2ω6) and docosahexaenoic (22:6ω3) fatty acids. The detection mode used was negative ion chemical ionization (NICI) with selected monitoring mass of 367 and 415 recorded in the upper right corner of each figure, for the identification of the characteristic pentafluorobenzyl ester.

Another product of peroxidation formed from 18:2ω6 is 9-oxononanoate which is a saturated aldehyde derived from the portion of the 18:2ω6 molecule remaining after cleavage of the double bond, resulting in the formation of the hydroperoxide. This aldehyde can also arise from oxidation of other fatty acids in the retina including 16:1, 18:1, and 18:3. The biochemical work-up used to form pentafluorobenzyl esters of the hydroperoxide described above also results in the formation of a pentafluorobenzyl ester of 9-oxononanoic acid. Figure 7 is a GC-MS chromatogram of 9-oxononanoate obtained from retina of vitamin E-, Se-deficient rat exposed for 6 hrs to ~ 1100 Ft.c. of light. By comparing the quantity of 9-oxononanoate in retinal samples, relative to another low-level saturated fatty acid 12:0, we

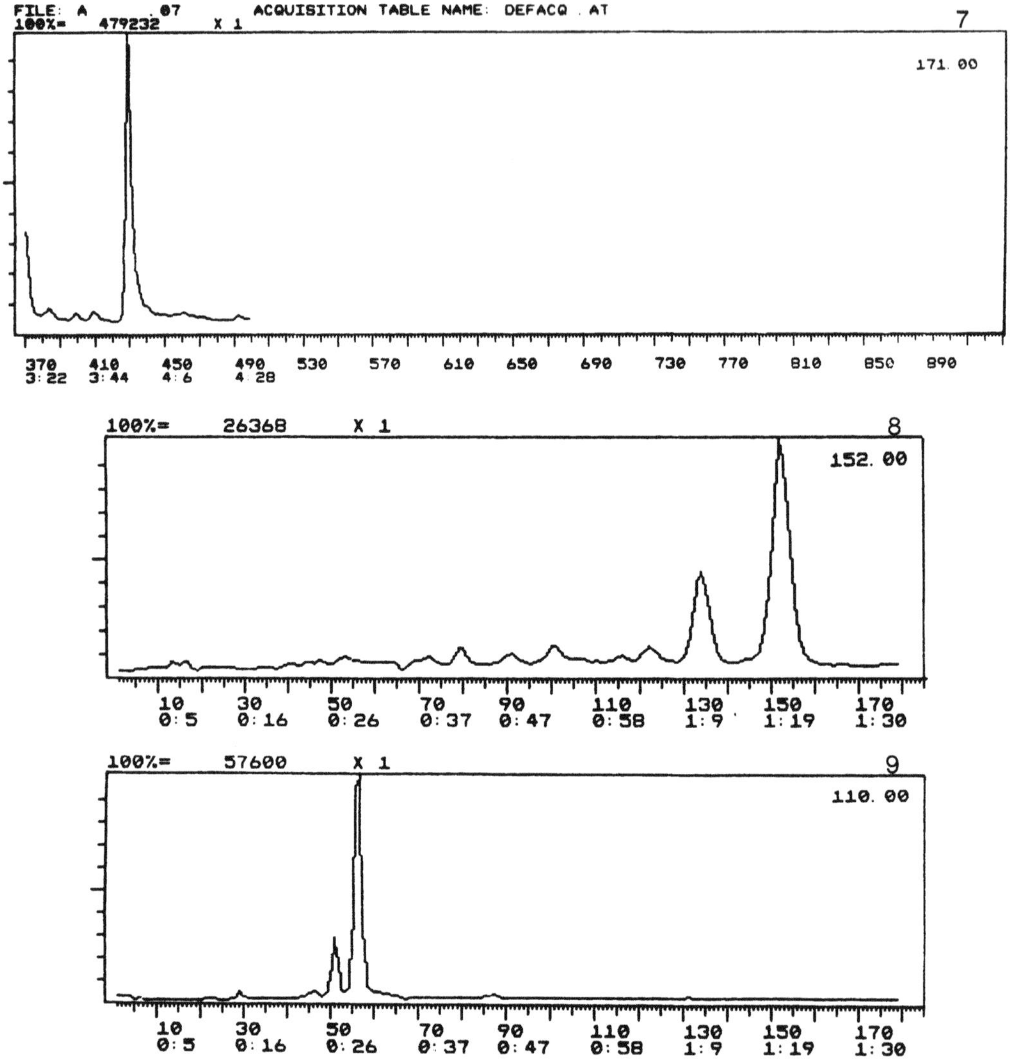

Figures 7–9. The first chromatogram (Figure 7) is of 9-oxononanoate—the fragment of 18:2ω6 remaining after peroxidation of the double bonds form the hydroperoxides. This saturated aldehyde gives a single peak and is relatively stable compared to the hydroperoxides. Figures 8 and 9 are secondary decomposition products of 18:2ω6 and 22:6ω3, respectively. These unsaturated aldehydic compounds, 4-hydroxynonenal and 4-hydroxyhexenal, are highly cytotoxic. The two peaks are the syn- and anti-isomers of the pentafluorobenzyl oximes. Characteristic mass monitoring of 171, 152, and 110 are recorded in the upper right corner of each figure.

observe a consistent elevation of 9-oxononanoate in samples from animals under phototoxic stress compared to their controls, strongly suggesting that peroxidation had increased during exposure.

Two additional and potentially important secondary products of 18:2ω6 and 22:6ω3 peroxidation are their respective hydroxyalkenals, 4-hydroxynonenal (HNE) and 4-hydroxyhexenal (HHE). These reactive unsaturated aldehydes have been shown to be cytotoxic by Esterbauer[9] and Schauenstein,[10] and their chemotactic potency was recently demonstrated by Curzio et al.[11,12] with neutrophil leukocytes in culture. Dr. Esterbauer has provided us with a supply of both HNE and HHE and we have used them to study their toxicity to Chinese hamster ovary (CHO) cells in culture. Both of these reactive 4-hydroxyalkenals were found to be highly toxic to CHO cells, where 50% of the cells were killed in a 24-hr test in which the culture media contained 1.6 μM, HNE or HHE. Both of these compounds, 4-hydroxynonenal and 4-hydroxyhexenal, have been detected and identified in retinal tissues by GC-MS, as a pentafluorobenzyl oxime (Figures 8 and 9). The two peaks in each of these chromatograms are the syn- and anti-isomers of 4-hydroxynonenal (Figure 8) and 4-hydroxyhexenal (Figure 9). Again by using semiquantitative comparisons the amount of HNE in whole retina appears to be from 1-3 ng/mg of tissue with the higher values obtained from exposed rats. More quantitative methods are under development, but it seems clear that there are elevated levels of these aldehydes in vitamin E-deficient retinas.

Parallel structural studies were performed on control and exposed rats (Figures 10-12; all micrographs ×640). For the structural studies, exposed animals were given 4 days to recover after a 6-hr exposure to ~ 1100 Ft.c. of white fluorescent light. This was done to permit the toxicity imposed on the retina during the 6-hr exposure period to be expressed in terms of structural degeneration. In

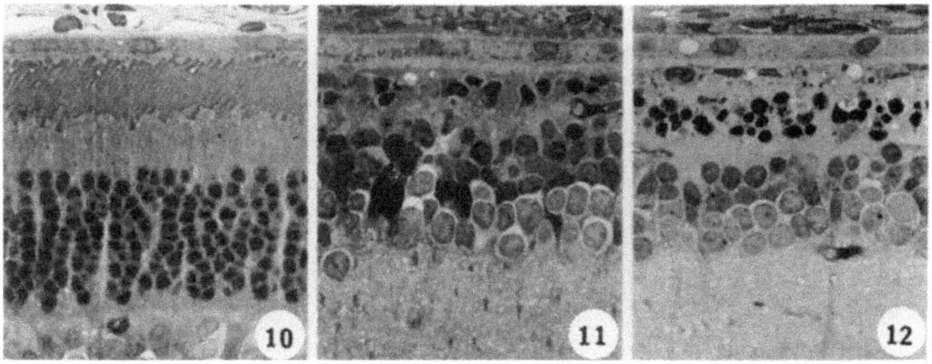

Figures 10-12. Light micrographs of control (Figure 10) and exposed retinas (Figures 11 and 12). Structural characteristics of the photoreceptor cells are normal in the retina from the control animal (Figure 10). In contrast the photoreceptor cells are highly degenerative in retinas from rats exposed to ~ 1100 Ft.c. followed by 4 days of recovery (Figures 11 and 12). Migratory cells are often observed in the subretinal space (Figure 12). ×640.

control animals (Figure 10) the retinal structures, particularly the photoreceptor cells including the outer segments, appear normal in all respects. By comparison, the retinas obtained from exposed animals show extensive loss of photoreceptor cells (Figures 11 and 12). Migratory phagocytic cells were often observed in the subretinal space (Figure 12), sometimes in rather large numbers. Of interest here also are observations made on retinas from animals receiving a vitamin E-depleted diet for significant periods, i.e., up to 40 weeks. One of the first indications of impending retinal degeneration from vitamin E deficiency was a small number of migratory cells observed in the subretinal space and among the outer segments. Migratory cells were not observed in the retinas of rats of the same age receiving chow or vitamin E-supplemented diets. It seems reasonable to propose that the migratory cells observed in the subretinal area have responded to the chemotactic properties of the 4-hydroxyalkenals resulting from lipid peroxidation in the photoreceptor cells.

DISCUSSION

To our knowledge, this is the most direct biochemical evidence to date to demonstrate the presence of lipid hydroperoxides: 4-hydroxynonenal, 4-hydroxyhexenal, and 9-oxononanoate as naturally occurring products of *in vivo* metabolism. One might expect the hydroperoxides to be elevated in the retina at the end of exposure to intense light; however, with semiquantitative procedures, we have found that they frequently appear to be lower than those levels found in control animals. On the other hand, 9-oxononanoate always appears to be elevated in the same retinas as do 4-hydroxynonenal and 4-hydroxyhexenal. We suggest that this data can be explained in the following manner. The primary and secondary products of lipid peroxidation described above are all present in retinas from control animals sacrificed at the end of the dark cycle; thus, as has been predicted earlier by Esterbauer (personal communication), lipid peroxidation occurs at a base rate as part of normal cellular metabolism. Upon elevated photic exposure, peroxidation is increased, as evidenced by elevated 9-oxononanoate. The lack of an observed increase in hydroperoxides, however, suggests that they undergo rapid decomposition to their secondary products. While these products could be numerous, the highly toxic unsaturated 4-hydroxyalkenals are perhaps the most important to retinal degeneration. Photic energy also appears to promote this pathway of *in vivo* decomposition of hydroperoxides since the 4-hydroxyalkenals appear to be elevated after exposure. The 4-hydroxyalkenals appear to be elevated in retinas from vitamin E-deficient animals and, therefore, these products are likely to be formed by lipid peroxidation and not by some other mechanism.

CONCLUSIONS

While it is premature to conclude that lipid peroxidation is responsible for retinal phototoxic degeneration, it has been demonstrated that the elements of a mechanism based on the theory of lipid peroxidation are present in the retina and they indicate that lipid peroxidation may account for the rapid degeneration of the photoreceptor cells when they are exposed to intense or prolonged illumination. Of particular interest is the demonstration of hydroperoxides from 22:6ω3 in the rod outer segments of the photoreceptor cells because this fatty acid represents ~ 50% of all unsaturated fatty acids in the rod outer segments[2] and the first structural indication of phototoxic injury is evident in the rod outer segments. Elevated 9-oxononanoate in exposed tissue indicates that lipid peroxidation increases with increased photic exposure. Thus, the toxicity imposed on the photoreceptor cells may well come from both the reactive hydroperoxides and unsaturated 4-hydroxyalkenals detected, identified and semiquantitated for the first time in retinal samples as a result of normal metabolism.

ACKNOWLEDGMENTS

The authors wish to thank Dr. Hermann Esterbauer for his helpful discussions and his generous gift of 4-hydroxynonenal and 4-hydroxyhexenal. This research was supported by Grant No. EY05910-02 from the National Eye Institute to R. J. Stephens.

REFERENCES

1. W.K. Noell, V.S. Walker, B.S. Kang, and S. Berman, Retinal damage by light in rats, Invest. Ophthalmol. 5:450 (1966).
2. G.J. Handelman and E.A. Dratz, The role of antioxidants in the retina and retinal pigment epithelium and the nature of prooxidant-induced damage, Free Rad. Biol. Med. 2:1 (1986).
3. W.A. Pryor, ed., "Free Radicals in Biology," Vol. IV, Academic Press, New York (1982).
4. F.J.G.M. van Kuijk, D.W. Thomas, R.J. Stephens, and E.A. Dratz, Gas chromatography-mass spectrometry method for determination of phospholipid peroxides. II. Transesterification to form pentafluorobenzyl esters and detection with picogram sensitivity, J. Free Rad. Biol. Med. 1:387 (1985).
5. F.J.G.M. van Kuijk, D.W. Thomas, R.J. Stephens, and E.A. Dratz, Occurrence of 4-hydroxyalkenals in rat tissues determined as pentafluorobenzyl oxime derivatives by gas chromatography-mass spectrometry, Biochem. Biophys. Res. Comm. 139:144 (1986).
6. D.W. Thomas, R.M. Parkhurst, D.S. Negi, K.D. Lunan, A.C. Wen, A.E. Brandt, and R.J. Stephens, Improved assay for α-tocopherol in the picogram range, using gas chromatography-mass spectrometry, J. Chromatogr. 225:433 (1981).
7. D.J. Buntman, D.W. Thomas, R.M. Parkhurst, D.S. Negi, and R.J. Stephens, A practical microbiochemical procedure for analysis of vitamin E in either solid or fluid tissues, Microchemical J. 29:66 (1984).
8. E.N. Frankel, W.E. Neff, and T.R. Bessler, Analysis of autooxidized fats by gas chromatography-mass spectrometry v. photosensitized oxidation, Lipids 14:961 (1979).
9. H. Esterbauer, Aldehydic products of lipid peroxidation, in: "Free Radicals, Lipid Peroxidation and Cancer," D.C.H. McBrien and T. F. Slater, eds., Academic Press, London (1982).
10. E. Schauenstein, Effect of low concentrations of aldehydes on tumor cells and tumor growth, in: "Free Radicals, Lipid Peroxidation and Cancer," D.C.H. McBrien and T.F. Slater, eds., Academic Press, London (1982).
11. M. Curzio, H. Esterbauer, C. Di Mauro, G. Cecchini, and M.U. Dianzani, Chemotactic activity of the lipid peroxidation product 4-hydroxynonenal and homologous hydroxyalkenals, Biological Chemistry Hoppe-Seyler 367:321 (1986).
12. M. Curzio, H. Esterbauer, G. Poli, F. Bissi, G. Cecchini, C. Di Mauro, N. Cappello, and M. Dianzani, Possible role of aldehydic lipid peroxidation products as chemoattractants, Int. J. Tiss. Reac. 9:295 (1987).

PRODUCTION AND METABOLISM OF CYTOTOXIC PHOSPHOLIPIDS GENERATED DURING INCUBATION OF LIPOSOMES WITH OXYHEMOGLOBIN

Keizo Inoue, Hiroyuki Itabe, and Ichiro Kudo

Faculty of Pharmaceutical Sciences
University of Tokyo
Hongo, Tokyo, Japan

INTRODUCTION

It is now widely accepted that liposomes are useful drug delivery systems. The encapsulation of a drug inside liposomes often decreases the toxicity of the drug while maintaining or enhancing its efficacy.[1-4] Liposomal drugs have many applications including anticancer, antiviral, antibacterial, antifungal, and anti-inflammatory therapy. It is therefore increasingly necessary to develop liposomal formulations that satisfy pharmaceutical demands for safety and stability in vivo. Many investigations have been performed to prepare liposomes that are stable in buffer,[5,6] but little information has been available concerning the stability and safety of liposomes after injection into the human body. Since they are used mostly to treat diseases, liposomes must be safe and stable under pathological conditions.

Liposomes composed of egg yolk phosphatidylcholine (egg PC) have been widely used as a drug carrier, and egg PC will continue to be used as a main component of drug-carring liposome because sufficient quantities of this phospholipid can be supplied at a reasonable cost and the membranes formed from it show "proper" flexibility.[2] Liposomes may interact with many factors including oxidants (Fe^{2+}, Cu^{2+}, oxyhemoglobin (HbO_2), etc.) in vivo, being destabilized or even generating toxic products.

The possibility of in vivo peroxidation of liposomal lipid should be carefully considered, particularly when liposomes are applied as liposomal SOD(superoxide dismutase) for therapy of arthritis[7] and hemosomes, which are liposomes containing HbO_2, are used for artificial red blood cells.[8] Liposomes would be exposed to severe oxygen stress under these conditions. In this paper we discuss the possible production, transfer, metabolism, and cytotoxic mechanisms of cytotoxic phospholipids that may be formed in vivo under oxygen stress.

LIPOSOME-BASED PRODUCT DEVELOPMENT

The encapsulation of drugs in liposomes often causes sustained release, protection from breakdown, retention of drugs in injected sites, or change of drug distribution in the body, which results in

Table 1. Medical and Pharmaceutical Applications of Liposomes

In vivo as a drug carrier (for therapy)
1. Sustained release, protection from breakdown, retention of drugs in injected sites, change of drug distribution in body enhanced efficacy, reduced toxicity, reduced frequency of dosage
2. Targeting ("guided missiles")
Natural targeting
Chemical targeting
Physical targeting (temperature, pH, target-sensitive liposomes)
In vitro as a sensor or carrier (for diagnosis, gene or protein transfer)
1. Immune sensor for detection of antigens or antibodies
2. Gene or protein transfer

enhanced efficacy, reduced toxicity, and reduced frequency of dosing (Table 1). Enhanced efficacy accompanied by reduced toxicity is now being developed for several drugs such as amphotericin B, gentamycin, anthracyclines and cisplatin derivatives (Table 2). Sustained release or retention of peptides or enzymes in infected sites will be accomplished by trapping them in liposomes. Trapping drugs in liposome may also reduce their side effects. The application of liposomes as a "guided missiles" is an attractive project. These trials have been successful *in vivo*,[9-11] but not satisfactory *in vivo* because of unstability RES (reticuroendothelial system) may also hinder the distribution of liposomes to target organs. Liposomes can also be used as sensors or carriers *in vitro* for diagnosis[12] or gene and protein transfer into cells.[13]

Table 2. Major Development Projects

Amphotericin B	Antifungal	Enhanced efficacy, reduced toxicity
Gentamycin	Antibacterial	
Doxorubicin and other anthracyclines	Anticancer	
Cisplatin derivatives	Anticancer	
Somatotropin		Sustained release
Interleukin-2		
Pirocarpin	Antiglaucoma	
SOD	Anti-inflammatory	
Indomethacin	Anti-inflammatory	Reduced side effect

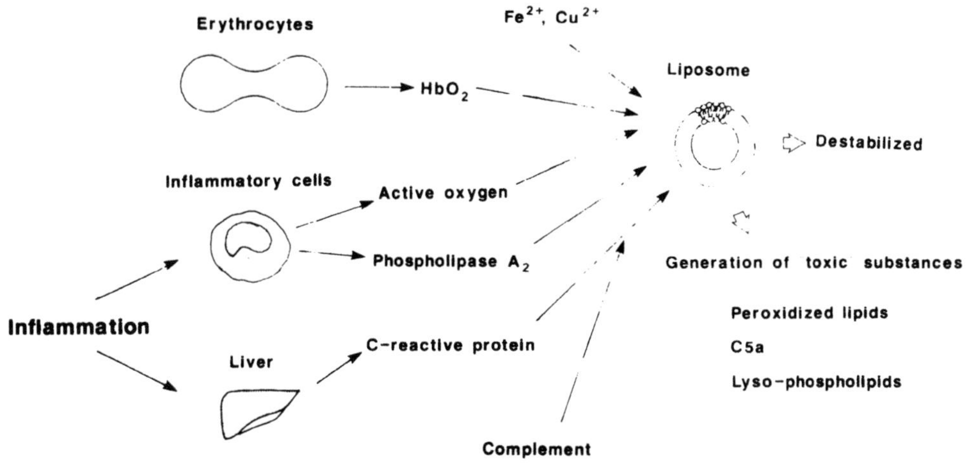

Figure 1. Interaction of liposomes with various extracellular factors under pathological conditions.

STABILITY AND SAFETY OF LIPOSOMES IN VIVO

Liposomes may interact with various serum factors such as fibronectin and various lipoproteins, being destabilized.[14]

Under pathological conditions, liposome injection may be more harmful. Oxidants such as HbO_2, Fe^{2+}, Cu^{2+} ions, active oxygen species generated from inflammatory cells, extracellular phospholipase A_2,[15] and C-reactive protein[16] secreted from hepatocytes may interact with many kinds of liposomes. Such interactions not only destabilize liposomes but also generate cytotoxic products (Figure 1).

DAMAGE OF HUMAN ERYTHROCYTE MEMBRANE UPON INCUBATION WITH EGG PC LIPOSOMES

The membranes of human erythrocytes were damaged when incubated with egg PC liposomes _in vitro_. Various sequential events have been observed[17] before membrane damage of erythrocytes occurred when erythrocytes were incubated with egg PC liposome.

Chemiluminescence was generated shortly after incubation of erythrocytes with egg PC liposomes. The amount of thiobarbituric acid(TBA)-reactive substances increased almost linearly for 30 min, and the conjugated diene concentration in the liposome fraction increased simultaneously. The peroxidation of liposomal lipids was further confirmed by a decrease of polyunsaturated fatty acid content in liposomes. The peroxidation of liposomal lipids during incubation with erythrocytes caused not only liposomal damage but also shedding of acetylcholinesterase(AChE)-rich vesicles, with an average diameter of 150-200 nm, from the erythrocyte surface and hemolysis.

The peroxidation of liposomal lipids might be caused by a minute amount of HbO_2 released from erythrocytes in the medium[18] since peroxidation, which is toxic to red blood cells, can be induced not only by erythrocyte suspension but also by HbO_2. HbO_2-induced peroxidation generates toxic substances, whereas methemoglobin or ferrous ion induced peroxidation without generation of toxic substances,[18] indicating that HbO_2 produces cytotoxic products most efficiently.

The characteristics of peroxidation of liposome by HbO_2 were as follows:

1. It was temperature-dependent and could occur above 30°C.
2. It was inhibited by thiourea, mannitol, or incorporation of α-tocopherol into liposomes but not by SOD, catalase, or EDTA.
3. The peroxidation of liposome, which was monitored by increase of TBA-reactive substances, preceded generation of chemiluminescence, conversion of HbO_2 to methemoglobin, and shedding of AChE-rich vesicles.

Negligible cytotoxic activity was recovered from the medium when peroxidized liposomes were separated from medium by filtration, whereas 60% of TBA-reactive substances were recovered from medium under the same conditions, indicating that the cytotoxic products may stay in membranes.

It can be concluded that when phospholipids in liposomes were peroxidized by HbO_2, the toxic products accumulated in liposome membranes causing not only liposomal damage but also damaging membranes at a distant loci (erythrocytes).[17,18] The transfer of toxic products from liposome to erythrocyte membrane was suggested (Figure 2).

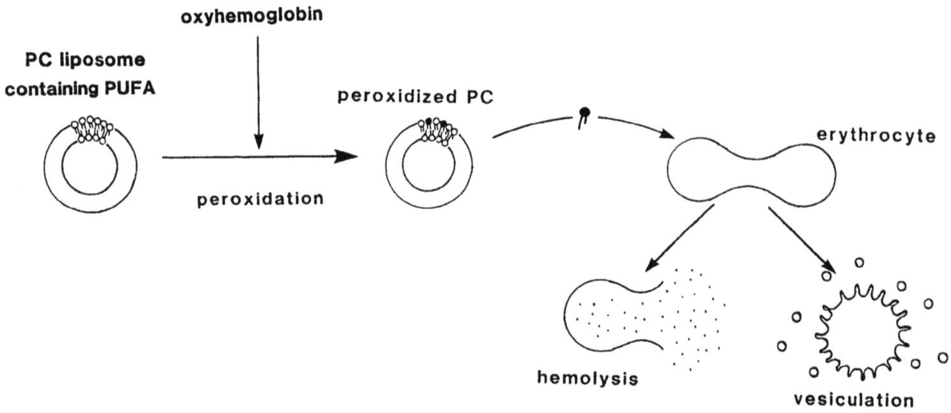

Figure 2. Process of erythrocyte damage induced by liposomes treated with oxyhemoglobin.

IDENTIFICATION OF PEROXIDIZED PRODUCTS THAT ARE TOXIC TO MEMBRANES AT DISTANT LOCI

Among various synthetic PCs, only PCs containing both saturated and polyunsaturated fatty acids gave similar cytotoxic products when incubated with HbO_2.[18] Molecules modified in some way at the polyunsaturated fatty acid residue may be responsible for cytotoxicity.

As for the cytotoxic peroxidized products, 4-OH-alkenal has been already reported[19] to be generated during rat liver microsome peroxidation. This compound or related compounds may not, however, participate in the induction of membrane damage under the present

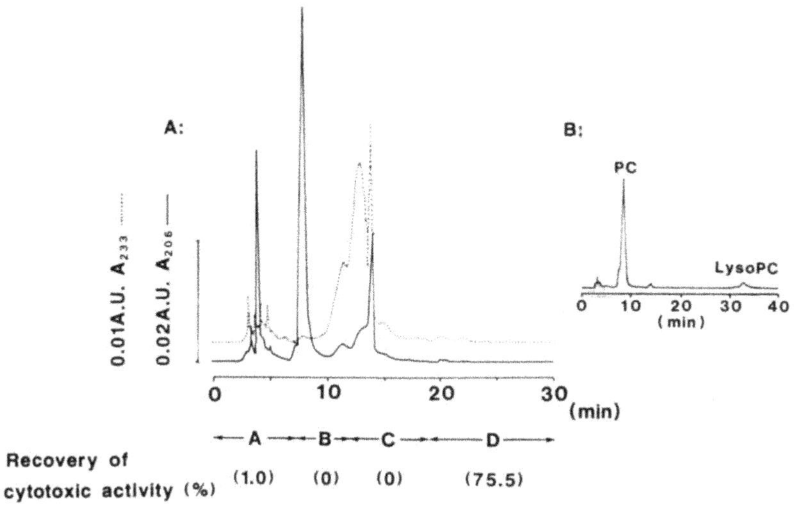

Figure 3. Straight phase HPLC chromatogram of product mixture obtained from oxyhemoglobin-treated 1-O-octadecyl-2-linoleoyl-GPC. 1-O-Octadecyl-2-linoleoyl-GPC (1 mM) was treated with oxyhemoglobin (50μg/ml) at 37°C for 2 hours. Lipids were extracted and analysed (A). Solvent: isopropanol/hexane/ water = 46/46/8; flow rate: 0.7 ml/min; column: Zorbax-SIL (4.6 x 250 mm); UV detection at 206 nm (——) and 233 nm (-----). The eluate was fractionated into four fractions, A-D. The activity of each fraction to induce release of acetylcholinesterase activity was determined as described. The recovery of the activity (% of total activity)was shown in parentheses. (B) HPLC chromatogram of 1-O-octadecyl-2-linoleoyl-GPC and 1-O-octadecyl-2-lyso-GPC under the some conditions.

conditions since the products that can cause both vesiculation and hemolysis cannot be generated from PC having only polyunsaturated fatty acyl residues.

To identify the cytotoxic products, a model phospholipid, 1-O-Octadecyl-2-linoleoyl-glycerophosphocholine, was incubated with erythrocytes or HbO_2. Like egg PC, this phospholipid can cause vesiculation as well as hemolysis when incubated with washed human erythrocytes.

The toxic substances generated during incubation of this phospholipid with HbO_2 was recovered mostly from the organic solvent when fractionated by the method of Bligh and Dyer. The organic layer was further passed through straight and reverse phase HPLC.

The cytotoxic activity monitored by induction of AChE release was mostly recovered from fraction D, which was separated by straight phase HPLC (Figure 3). The hemolytic activity was also recovered from the same fraction. This fraction was eluted between PC and lysoPC, judging from its retention time.

Fraction D was further separated on a reverse phase HPLC, Nucleosil $5C_{18}$ (figure 4). The activity was mostly recovered from fractions 5, 6, and 7. At present, we can identify only the product recovered from fraction 5.

The elution profile of fraction 5, which was observed when rechromatographed on the same column, is also shown in Figure 4B. The structure of the product recovered from fraction 5 was identified as 1-O-octadecyl-2-azelaoyl-glycerophosphocholine (Figure 5A) by secondary ion mass spectrum and NMR spectrum.[20]

A similar product might be formed when 1-palmitoyl-2-linoleoyl-GPC is peroxidized with HbO_2. A peak corresponding to an authentic 1-palmitoyl-2-azelaoyl-GPC (Figure 5B) conjugated with an ADAM reagent (9-anthryldiazomethane) could be detected when 1-palmitoyl-2-linoleoyl--GPC, which was treated with HbO_2 and subsequently with an ADAM reagent, was passed through a Nucleosil column and monitored by the fluorescent intensity at 412 nm.[21]

Synthetic 1-acyl-2-azelaoyl-GPC induced both vesiculation and hemolysis.

The major "intermediate" compounds having conjugated diene structures show rather weak toxic activity towards membranes under these experimental conditions. Other products such as long-chain aldehydes should be also formed,[22,23] but the contribution of such compounds seems rather small since the cytotoxic activity of the whole reaction mixtures was mostly recovered from the fractions containing 1-acyl-2-azelaoyl-GPC. Related compounds having similar structures may also be involved in the cytotoxic effects.

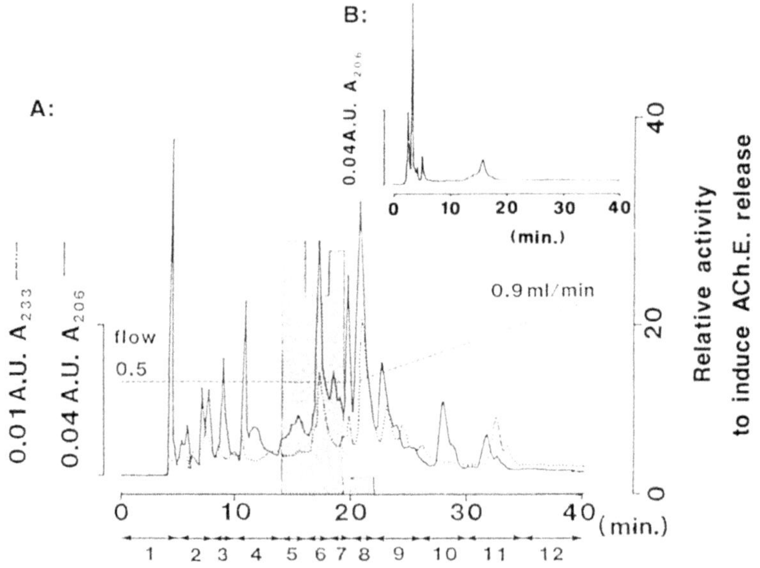

Figure 4. Reverse phase HPLC chromatogram of fraction D (A). Solvent: methanol/20 mM choline chloride= 95/5; flow rate: 0.5 ml/min for 20 min then with a linear gradient to 0.9 ml/min for 20 min; column: Nucleosil $5C_{18}$ (4.6 x 250 mm); UV detection at 206 nm (———), 233 nm (-------). Release of acetylcholinesterase activity was determined (hatched bar). (B) Rechromatogram of fraction 5 under the same conditions.

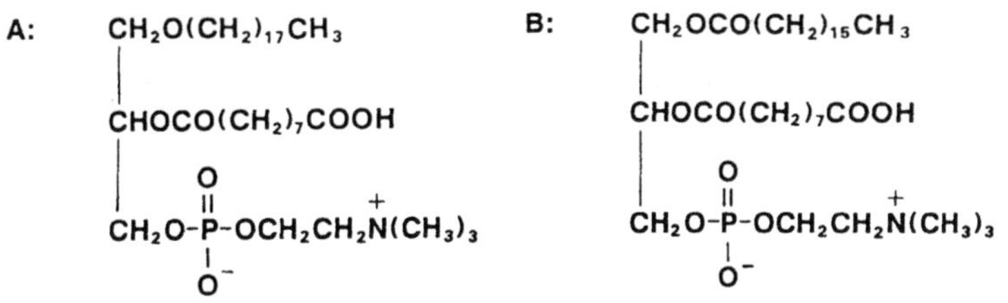

Figure 5. Structure of cytotoxic products formed during lipid peroxidation. (A) 1-O-Octadecyl-2-azelaoyl GPC; (B) 1-palmitoyl-2-azelaoyl GPC.

"OXIDIZED PHOSPHOLIPID-SPECIFIC" PHOSPHOLIPASE A_2 IN INFLAMED SITES

The possible role of phospholipases in protection from lipid peroxidation has been discussed recently.[24,25] Dratz and his group[26] found that lipid peroxidation was effectively inhibited by glutathione peroxidase in the presence of pancreatic phospholipase A_2.

Extracellular phospholipase A_2 activity has been often detected in association with inflammation, for example in experimental models[27,28] and some human diseases.[29] One of these enzymes (13.5 KDa) was purified from peritoneal fluids of rat treated with caseinate.[30] These enzymes cannot, however, differentiate oxidized phospholipids from intact phospholipids.[31] 1-Palmitoyl-2-azelaoyl-GPC was a poor substrate for the enzyme.

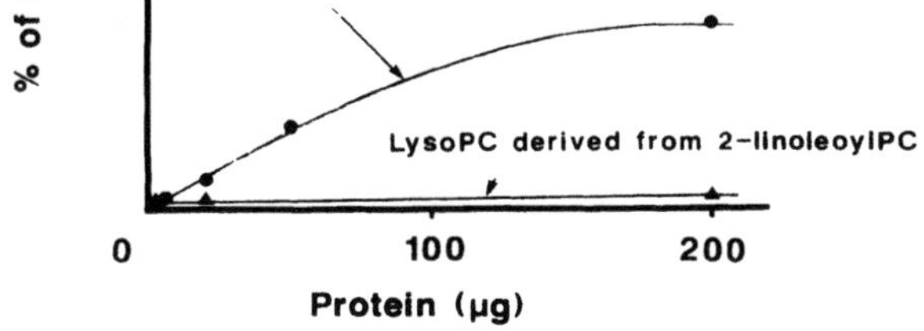

Figure 6. Effect of protein concentration on "oxidized phospholipid specific" phospholipase A_2 activity. The reaction mixture (100 μl) contained 50 mM Tris-HCl(pH 7.0), 1 mM $CaCl_2$ and 20 nmol of either [N-C^3H_3]-1-palmitoyl-2-azelaoyl-GPC (●,o) or [N-C^3H_3]-palmitoyl-2-linoleoyl-GPC (▲,△) both of which were chemically synthesized,[31] and cell-free peritoneal fluids of rats 16 h after casein injecion as an enzyme source. Reaction was carried out at 37°C for 15 min and reaction products were extracted and analysed by TLC.

We have recently found that the peritoneal fluids of rats treated with caseinate contain another phospholipase A_2 that preferentially hydrolyzes 1-acyl-2-azelaoyl-GPC.[31] The formation of lysoPC was observed when radioactive 2-azelaoyl-GPC was incubated with the peritoneal supernatants. 1-Palmitoyl-2-linoleoyl-GPC was not hydrolyzed appreciably under the same conditions (Figure 6).

The characteristics of this enzyme such as substrate specificity, calcium requirement, optimum pH, and heat stability were quite different from 13.5 KDa phospholipase A_2, although both enzymes were secreted into the same extracellular space of inflamed sites.

The enzyme activity might be inducible since it was dramatically increased with an increased number of exudated cells in peritoneal cavity.

In inflamed sites, lipid peroxidation may occur through exposure of membranes to active oxygen species or free metal ions.[32,33] The enzyme that hydrolyzes one of the peroxidized toxic lipids may play an important role as a scavenger of these toxic products (Figure 7).

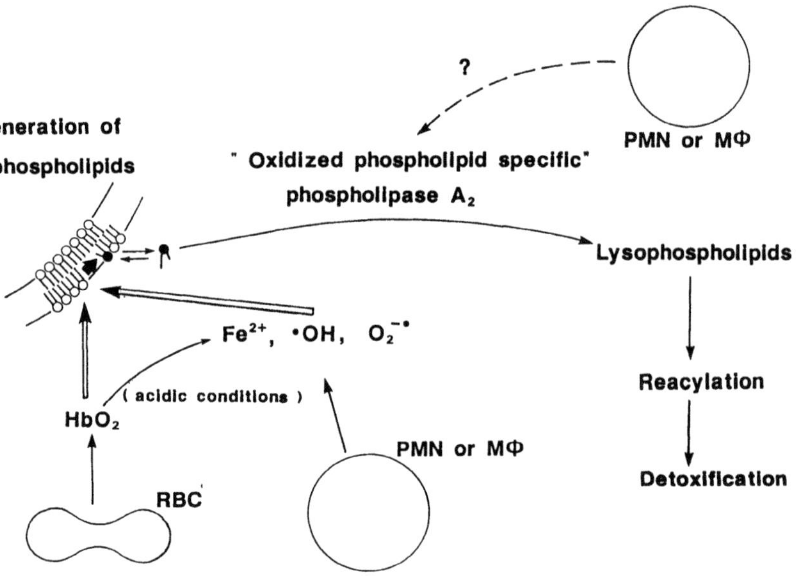

Figure 7. Possible role of "oxidized phospholipid specific" phospholipase A_2 in the detoxification of toxic oxidized phospholipids.

The application of liposomes as drug carriers is highly promising. There is, however, little information about their stability, toxicity, or metabolism after injection into the body. The peroxidation of liposomal lipids may be one possible reaction *in vivo*. Further understanding of fate of liposomes in the body, including peroxidation, is required.

REFERENCES

1. M.J. Ostro, Liposomes, Scientific American 256:90 (1987).
2. G. Poste, R. Kirsh and P. Bugelski, Liposomes as a drug delivery system in cancer therapy, in: "Novel approaches to cancer chemotherapy," P.S. Sunkara, ed., Academic Press, Orlando (1984).
3. G. Gregoriadis, "Liposome technology," vol. 1, Preparation of liposomes, CRC Press, Boca Raton (1984).
4. A.D. Bangham, "Liposome letters," Academic Press, London (1983).
5. S.M. Gruner, R.P. Lenk, A.S. Janoff, and M.J. Ostro, Novel multilayered lipid vesicles: Comparison of physical characteristics of multilamellar liposomes and stable plurilamellar vesicles, Biochemistry 24:2833 (1985).
6. E.M.G. van Bommel, D.J.A. Crommelin, Stability of doxorubicin-liposomes on storage: as an aqueous dispersion, frozen or freeze-dried, Int. J. Pharm. 22:277 (1984).
7. A.M. Michelson, Clinical use of superoxide dismutase and possible pharmacological approaches, in: "Pathology of oxygen," Autor,A.P. ed., Academic Press, New York (1982).
8. J. Szebeni, and K. Toth, Lipid peroxidation in hemoglobin-containing Liposomes. Effects of membrane phospholipid composition and cholesterol content, Biochim. Biophys. Acta 857:139 (1986).
9. M.B. Yatvin, W. Kreutz, B.A. Horwitz, and M. Shinitzky, pH-Sensitive liposomes: Possible clinical implications, Science 210:1253 (1980).
10. J. Connor and L. Huang, Efficient cytoplasmic delivery of a fluorescent dye by pH-sensitive immunoliposomes, J. Cell Biol. 101:582 (1985).
11. R.J.Y. Ho and L. Huang, Target-sensitive immunoliposomes: preparation and characterization, Biochemistry 25;5500 (1986).
12. A.S. Janoff, S. Carpenter-Green, A.L. Weiner, J. Seibold, G. Weissmann, and M.J. Ostro, Novel liposome composition for a rapid colorimetric test for systemic lupus erythematosus, Clin. Chem. 29:1587 (1983).
13. D. Lavelle, M.J. Ostro, W. Paxton, D. Blaustein, and D. Giacomoni, Further studies on liposome-mediated RNA delivery into cells in vitro, Arch. Biochem. Biophys. 215:486 (1982).
14. F. Bonte and R.L. Juliano, Interactions of liposomes with serum proteins, Chem. Phys. Lipids 40:359 (1986).
15. H.W. Chang, I. Kudo, S. Hara, K. Karasawa, and K. Inoue, Extracellular phospholipase A_2 activity in peritoneal cavity of casein-treated rats, J. Biochem. 100:1099 (1986).
16. M. Tsujimoto, K. Inoue, and S. Nojima, Reactivity of human C-reactive protein with positively charged liposomes, J. Biochem. 90:1507 (1981).
17. T. Kobayashi, H. Itabe, K. Inoue, and S. Nojima, Peroxidation of liposomes in the presence of human erythrocytes and induction of membrane damage of erythrocytes by peroxidized liposomes, Biochim. Biophys. Acta 814:170 (1985).
18. H. Itabe, T. Kobayashi, and K. Inoue, Generation of toxic phospholipid(s) during oxyhemoglobin-induced peroxidation of phosphatidylcholines. submitted for publication.

19. A. Benedetti, M. Comporti, and H. Esterbauer, Identification of 4-hydroxynonenal as a cytotoxic product originating from the peroxidation of liver microsomal lipids, Biochim. Biophys. Acta 620:281 (1980).
20. H. Itabe, I. Kudo, Y. Kushi, S. Honda, and K. Inoue, Identification of a toxic phospholipid generated during oxyhemoglobin-induced peroxidation of phosphatidylcholine, in preparation.
21. H. Itabe, I. Kudo, and K. Inoue, Synthesis and charaterization of 2-azelaoyl phosphatidylcholine, one of the toxic products of lipid peroxidation, in preparation.
22. H. Esterbauer, Aldehydic products of lipid peroxidation, in: "Free radicals, lipid peroxidation and cancer", D.C.H. McBrien and T.F. Slater ed., Academic Press, London, (1982).
23. E.N. Frankel, Volatile lipid oxidation products, Prog. Lipid Res. 22:1 (1983).
24. F.J.G.M. van Kujik, A. Sevanian, G.J. Handelman, and E.A. Dratz, A new role for phospholipase A_2: Protection of membranes from lipid peroxidation damage, Trends Biochem. Soc. 12:31 (1987).
25. A. Sevanian and E. Kim, Phosphlipase A_2 dependent release of fatty acids from peroxidized membranes, J. Free Radicals Biol. Med. 1:263 (1985).
26. F.J.G.M. van Kuijk, G.J. Handelman, and E.A. Dratz, Consecutive action of phospholipase A_2 and glutathione peroxidase is required for reduction of phospholipid hydroperoxides and provides a convenient method to determine peroxide value in membrane, J. Free Radicals Biol. Med. 1:421 (1985).
27. R. Franson, R. Dobrow, J. Weiss, P. Elsbach, and W.B. Weglicki, Isolation and characterization of phospholipase A_2 from an inflamatory exudate, J. Lipid Res. 19:18 (1978).
28. T.J. Nevalainen, The role of phospholipase A_2 in acute pancreatitis, Scan. J. Gastroent. 15:641 (1980).
29. E. Stefanski, W. Pruzauski, B. Sternby, and P. Vadas, Purification of a soluble phospholipase A_2 from synovial fluid in rheumatoid arthritis, J. Biochem. 100:1297 (1986).
30. H.W. Chang, I. Kudo, M. Tomita, and K. Inoue, Purification and characterization of extracellular phospholipase A_2 from peritoneal cavity of caseinate-treated rat, J. Biochem. 102:147 (1987).
31. H. Itabe, I. Kudo, and K. Inoue, "Peroxidized phospholipid specific" phospholipase A_2 activity in peritoneal cavity of casein-treated rats, in preparation.
32. R.Y. Youngman, Oxygen activation: Is the hydroxyl radial always biologically relevant?, Trends Biochem. Soc. 9:280 (1984).
33. B. Halliwell and J.M.C. Gutteridge, Iron and free radical reactions: Two aspects of antioxidant protection, Trends Biochem. Soc. 11:372 (1986).

LIPID PEROXIDATION DEPENDENT ON OXYGEN AND CATALYTIC FREE IRON IONS IN SITU STORAGE MUSCLE FOODS

Stella Harel, Joseph Shagalovich, Bella Hazan and Joseph Kanner

Department of Food Science, Volcani Center
Bet Dagan, Israel

INTRODUCTION

Lipid peroxidation is one of the primary mechanisms of quality deterioration in stored foods, and especially in muscle tissues. With regard to lipid peroxidation in muscle tissues, one of the most important questions concerns the source of the primary catalysts that initiate lipid peroxidation in the muscle food in situ. Several authors suggest that such lipid peroxidation is stimulated by enzymic[1] and nonenzymic reactions. Data have been presented suggesting that nonheme iron compounds are the major prooxidants[2]. Some researchers have reported that hemeproteins seem to be the main prooxidants in muscle tissues[3,4], while others concluded that both heme and nonheme iron compounds are important[5,6]. Most of the studies on the mechanisms of lipid peroxidation in foods have been conducted in oils or lipid emulsions[7]. Mechanisms established in those systems may differ from those in organized cells, subcellular organelles and tissues.

Recently, we adopted the utilization of muscle microsomal membranes (sarcosomes) to study enzymic and nonenzymic lipid peroxidation in muscle foods[8,9,10]. Membranal lipid peroxidation initiated by several pathways was inhibited by reducing agents, hydroxyl radical scavengers, chelators, antioxienzymes and food antioxidants[8,9,10].

In this study we have extended our work in order to understand better the mechanism which initiates and stimulates lipid peroxidation in in situ muscle foods.

MATERIALS AND METHODS

All the reagents in our work were of the highest purity commercially available. The hemeproteins and ascorbic acid were treated with Chelex-100. Fresh turkey or chicken muscles were ground in a Moulinex food processor for 60 sec. One part of the minced tissues was incubated in polyethylene bags before or after heating to 68.5°C. Portions (approx. 120 g) of minced muscle were also placed in cans, closed under vacuum, and cooked in 77°C water for 25 min. The samples were stored at 4°C and tested periodically. Isolation of microsomal fraction from turkey tissues was done by a procedure previously described[11]. Thiobarbituric acid reactive compounds (TBA-RC) and proteins were determined as previously described[11]. The accumulation of conjugated dienes was performed by the method of Parr and Swoboda[12]. Free "catalytic" iron ions were determined by a method developed by Gutteridge et al.[13]. The results are means of triplication of two separate experiments, in the figures each error bar (I) denotes the standard deviation.

RESULTS AND DISCUSSION

Lipid peroxidation in muscle tissues could be propagated by breakdown of preformed hydroperoxides. Our results show that a small amount of conjugated dienes is produced even if the minced turkey muscle was canned and heated under vacuum. During storage under vacuum, the amount of conjugated dienes does not change significantly (Figure 1). If, however, the cans have been opened to air and stored at 4°C for several days, the minced muscle starts to peroxidize at a very rapid rate.

The second part of our results shows that lipid peroxidation in minced turkey muscle tissues depends on catalysts, in particular free metal ions. Membranal lipid peroxidation in a model system was adopted for studying the effect of metal ion-chelating agents, especially EDTA, on the oxidative process. Enzymic iron and redox-cycle membranal lipid peroxidation were inhibited by EDTA at the relatively low concentration of 20-25 μM; H_2O_2-activated metmyoglobin (MetMb) membranal lipid peroxidation was also inhibited, but at a concentration of 1 mM and above (Figure 2).

Studying in situ raw and heated minced muscle lipid peroxidation showed that EDTA at a concentration lower than that which could affect activated MetMb inhibits lipid peroxidation (Figure 3). We assume that in turkey minced muscles the most important catalyst for lipid peroxidation is free metal ions and most probably iron ions. We had determined catalytic "free" iron in muscle tissues by the bleomycin method. Bleomycin binds free iron and only this complex is capable of degrading DNA to give base propenals[13].

Our study demonstrated that turkey and chicken muscle contain "free" iron which could act as a catalyzer in oxidative reactions. The degradation of DNA to base propenals was found to be higher by water extracts of dark turkey muscles than by white muscles. DNA decomposition by bleomycin-water muscle extracts was inhibited by desferrioxamine, a known specific chelator of iron ions (Figure 4). The amount of "free" iron in dark muscles was higher than that in white muscles (Table I) and correlates well with the greater instability of dark turkey and chicken muscle toward lipid peroxidation.

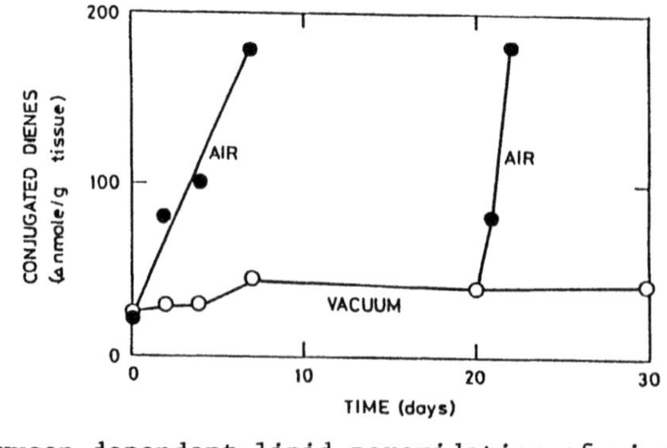

Figure 1. Oxygen-dependent lipid peroxidation of minced dark turkey muscle stored at 4°C.

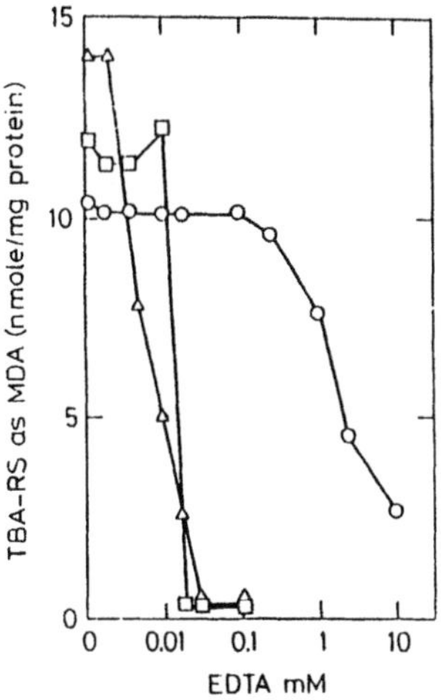

Figure 2. Effect of EDTA concentration on membranal lipid peroxidation at $25^{o}C$. □, enzymic system containing 0.5 mg protein/ml; KCl, 0.12 M; NADPH, 0.4 mM; ADP, 0.4 mM; $FeCl_3$, 0.1 mM; histidine, 5 mM at pH 7.3. Δ, non-enzymic iron-redox system containing 1.0 mg protein/ml; ascorbic acid, 0.2 mM; $FeCl_3$, 0.1 mM; acetate buffer, 50 mM at pH 7.0. O, non-enzymic H_2O_2-activated MetMb containing 1.0 mg protein/ml; 0.03 mM of H_2O_2 and MetMb preincubated for 3 min; acetate buffer, 50 mM at pH 7.0.

During storage of turkey dark muscle at $4^{o}C$ for 7 days, the amount of "free" iron increases almost fourfold. Very similar increases in "free" iron were found with chicken dark and light muscle stored at $4^{o}C$ (results not shown).

Table I. Catalytic "Free" Iron in Muscle Food

Species	Iron (μg/g tissue ± SD)	
	Dark muscle	Light muscle
Chicken	0.55 ± 0.07	0.24 ± 0.09
Turkey	2.50 ± 0.14	0.92 ± 0.16

In situ minced turkey muscle lipid peroxidation was inhibited by ceruloplasmin (Figure 5). This result was obtained with two different sources of turkey muscles, one of high quality and the other of low quality. In both samples, ceruloplasmin prevents between 76% and 56% of the lipid peroxidation process. Ceruloplasmin, which acts as ferroxidase, catalyzed the oxidation of ferrous to ferric ions without the production of oxygen-free radicals. This reaction prevents the breakdown of H_2O_2 and hydroperoxides by ferrous ions to reactive radicals and by this the stimulation of lipid peroxidation. As a great part of the oxidation in situ muscle tissues is prevented by low concentration of EDTA and ceruloplasmin, we conclude that "free" iron ions are the major prooxidant factor in minced turkey muscles. Our results also open a new way for the utilization ceruloplasmin, a natural by-product of animal blood, as a natural antioxidant for muscle foods.

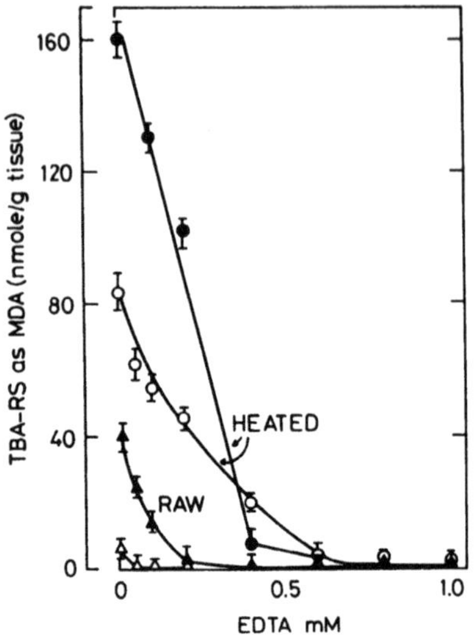

Figure 3. Effect of EDTA concentration on lipid peroxidation of minced turkey dark muscle stored at $4^{o}C$. Δ, raw, zerotime; ▲, raw, 5 days; O, heated, zerotime; ●, heated, 5 days.

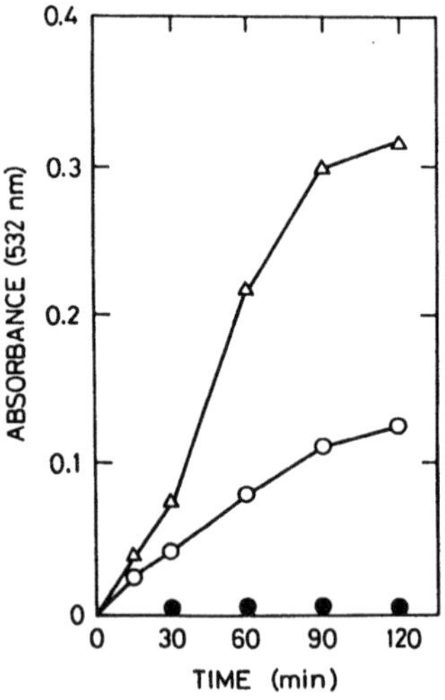

Figure 4. Base propenals derived during the incubation of DNA with bleomycin in mucle tissue extracts. O, water (pyrogen-free, Chelex-100 treated) extracts from light turkey muscle; Δ, water extracts from dark turkey muscles; ●, water extracts from dark turkey muscles without DNA or bleomycin or in the presence of desferrioxamine.

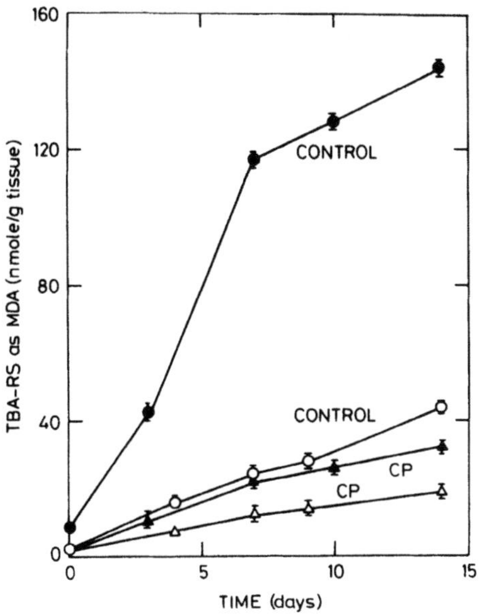

Figure 5. Effect of ceruloplasmin (150 U/gr) on in situ lipid peroxidation in minced turkey muscle tissues. O, control, minced muscle (run 1); Δ, minced muscle (run 1) in the presence of ceruloplasmin; O, control, minced muscle (run 2); Δ, minced muscle (run 2) in the presence of ceruloplasmin.

"Free" iron and other transitional metal ions could affect the quality of muscle foods by changing the stability of the lipid fraction by three main reactions: (a) the free metals interact with reducing agents, such as cysteine, glutathione, ascorbate and α-tocopherol, oxidize them, and thereby reduce the antioxidative tone of the tissue; (b) during the interaction with reducing agents, the metals activate oxygen to active-oxygen species, such as the hydroxyl radical; and (c) reduce metals catalyze the breakdown of hydroperoxides to free radicals, which could initiate a chain reaction of lipid peroxidation.

More work needs to be done to identify the endogenous source of the "free" iron in muscle foods.

REFERENCES

1. H. O. Hultin, Enzyme catalyzed lipid peroxidation in muscle microsomes, in: "Autoxidation in Food and Biological Systems," M. Simic and M. Karel eds, Plenum Press, New York (1980).
2. K. S. Rhee, Factors affecting oxygen uptake in model systems used for investigating lipid peroxidation in meat, J. Food Sci. 43:6 (1978).
3. Y. B. Lee, G. L. Hargus, J. A. Kirkpatrick, D. L. Bemmer, and R. H. Forsyth, Mechanism of lipid peroxidation in mechanically deboned chicken meat, J. Food Sci. 40:964 (1975).
4. J. Fisher and J. C. Deng, Catalysis of lipid peroxidation. A study of mullet (Mugi cehalus) dark flesh and emulsion model systems, J. Food Sci. 42:610 (1977).
5. T. L. Kwoh, Catalysts of lipid peroxidation in meats, J. Am. Oil Chem. 48:550 (1971).

6. H. P. Liu, and B. M. Watts, Catalysts of lipid peroxidation in meats. 1. Linoleate peroxidation catalyzed by MetMb or Fe(II)EDTA, J. Food Sci. 35:950 (1970).
7. M. K. Logani and R. E. Davis, Lipid peroxidation. Review, Lipids 15:485 (1980).
8. B. Hazan, NADPH-dependent membranal lipid peroxidation in muscle microsomes, M.Sc. thesis, Hebrew University of Jerusalem, Israel (1986).
9. S. Harel and J. Kanner, Muscle membranal lipid peroxidation initiated by H_2O_2-activated metmyoglobin, J. Food Chem. 33:1188 (1985).
10. J. Kanner, S. Harel and B. Hazan, Muscle membranal lipid peroxidation by an "Iron Redox Cycle" system: initiation by oxy-radicals and site specific mechanism, J. Food Chem. 34:506 (1986).
11. J. Kanner and S. Harel, Initiation of membranal lipid peroxidation by activated metmyoglobin and methemoglobin, Arch. Biochem. Biophys. 237:314 (1985).
12. L. J. Parr and P. A. T. Swobode, The assay of conjugable oxidation products applied to lipid deterioration in stored foods, J. Food Technol. 11:1 (1976).
13. J. M. C. Gutteridge and B. Halliwell, Bleomycin assay for catalytic iron salts in body fluids, in: "Handbook of Methods for Oxygen Radical Research", R. A. Greenwald, ed., CRC Press, Boca Raton, Florida (1985).

A UNIQUE ROLE OF HISTIDINE IN Fe-CATALYZED LIPID OXIDATION BY FISH SARCOPLASMIC RETICULUM

Marilyn C. Erickson and Herbert O. Hultin

Massachusetts Agricultural Experiment Station
Department of Food Science and Nutrition
Marine Foods Laboratory
University of Massachusetts
Gloucester, Massachusetts 01930, U.S.A.

INTRODUCTION

The sarcoplasmic reticulum (SR) fraction of fish muscle contains an enzymic lipid peroxidation system requiring NADH and Fe that is stimulated by ADP.[1,2] Although in these and other studies, concentrations of metabolites used are representative of the conditions in situ, the amount of Fe is present at a much higher concentration than would normally occur in tissues, e.g., Fe in the low molecular weight fraction of flounder muscle is <3% of total iron.[3] Thus, the amount and state of the Fe is most probably a rate limiting factor in lipid oxidation in situ. A large number of O-, N-, and S-containing compounds are capable of binding Fe in either the Fe^{+2} or Fe^{+3} states, and it is likely, therefore, that Fe is not found free in vivo but is bound to these cellular metabolites. Chelation of Fe is important in lipid oxidation in vitro since it may keep the Fe soluble or change its oxidation-reduction potential. In this study, we demonstrate that the free amino acid histidine can either enhance or inhibit Fe-catalyzed lipid oxidation to a significant degree. The effect of histidine is determined in part by the presence of a nucleotide.

MATERIALS AND METHODS

SR was prepared from the muscle tissue of winter flounder (Pseudopleuronectes americanus).[1] TBA-reactive substances (TBA-RS) were determined by the procedure of McDonald and Hultin[2] and lipid hydroperoxides (LOOH) by the method of Buege and Aust.[4] Protein concentration was determined by Lowry's method.[5]

RESULTS AND DISCUSSION

The order of addition of components in the in vitro assay affected the rate of enzymic lipid peroxidation (Table 1). There was considerable inhibition when histidine was mixed with the Fe^{+3} before the other components, whereas if either ADP or NADH was added prior to the histidine, a higher rate of lipid oxidation was attained. Premixing ADP, NADH and histidine prior to their addition to Fe^{+3} led to an intermediate value of oxidation indicating

Table 1. Effect of Order of Addition of Components in Enzymic Lipid Peroxidation

1.	$FeCl_3$	$FeCl_3$	$FeCl_3$	$FeCl_3$	$FeCl_3$
2.	Histidine	ADP	NADH	ADP-NADH-Hist*	NADH
3.	ADP	Histidine	Histidine	SR	Histidine
4.	NADH	NADH	ADP		SR
5.	SR	SR	SR		
TBA-RS	12.9	35.0	32.3	19.4	83.9

Reaction mixtures contained 0.15 ml 0.5 mM $FeCl_3$, 0.125 ml 4 mM ADP, 0.125 ml 4 mM NADH, 4.5 ml 5.56 mM histidine, pH 6.8, and 0.1 ml of 5 mg/ml SR protein. The numbers in the left column designate the sequence for the addition of assay components. The assay mixture was incubated for 30 min at 6°C. Lipid peroxidative activity was determined by the accumulation of TBA-reactive substances and is expressed as nmoles TBA-RS/mg protein/30 min. *ADP, NADH and histidine were mixed prior to their addition to $FeCl_3$.

the competitive nature of the binding. In the absence of ADP and when NADH was added to the Fe before histidine, greater than twice the lipid peroxidative activity was observed. Lower concentrations of ADP, i.e., equimolar to the amount of Fe^{+3} used, did, however, stimulate the oxidation (data not shown). When the SR membrane was added to the Fe^{+3} before the other components of the assay system, it reduced the activity obtained when ADP or NADH was added before the histidine, and it reduced the inhibition obtained when histidine was added prior to ADP or NADH.

There was an increasing rate of lipid oxidation in the enzymic system (reduction via NADH) up to a concentration of 100 mM histidine (Table 2). Nonenzymic lipid oxidation (Fe^{+3} reduced by ascorbate) also showed increased rates in the presence of histidine with a broad optimum between 2-50 mM.

Table 2. Lipid Peroxidative Activity as a Function of Histidine Concentration

	Peroxidative Activity	
mM Histidine	Enzymic	Nonenzymic
0	8	56
2	---	129
5	40	136
10	45	134
50	111	130
100	142	105

Enzymic assay systems contained 15 μM $FeCl_3$, 100 μM ADP, 100 μM NADH, 0.5 mg SR protein, and varying concentrations of histidine, pH 6.8, in a total volume of 5 ml. The incubation was carried out at 6°C for 30 min. Nonenzymic systems contained 15 μM $FeCl_3$, 15 μM ADP, 60 μM sodium ascorbate, 0.5 mg SR protein, and varying concentrations of histidine, pH 6.8, in a total volume of 5 ml.
The incubation was carried out at 6°C for 3 min. All assay systems exposed the Fe to ADP before the addition of any other components.
Lipid peroxidative activity was determined by the accumulation of TBA-reactive substances and is expressed as nmoles TBA-RS/mg protein/30 min.

Table 3. Peroxidative Activities in the Presence of Various Amino Acids

Amino Acid	Relative Activity TBA-RS	Relative Activity LOOH
None	1.0	1.0
Histidine	6.0	8.7
Asparagine	2.6	2.7
Glutamine	1.3	2.0
Lysine	1.2	1.2
Methionine	0.8	---
Hydroxyproline	0.6	---
Proline	0.5	---
Alanine	0.5	---
Glycine	0.5	---
Arginine	0.2	0.2

Reaction mixtures contained 15 μM $FeCl_3$, 15 μM ADP, 100 μM NADH, 0.5 mg SR protein and 5 mM of the amino acid, pH 6.8. Lipid peroxidation was assessed by quantitating the hydroperoxides and TBA-reactive materials generated during an incubation of 30 min at 6°C. Results were expressed relative to the peroxidative activity obtained in the absence of amino acid (101.5 nmoles LOOH/mg protein/30 min and 9.8 nmoles TBA-RS/mg protein/30 min).

None of the other amino acids tested were seen to have the stimulating effect on the oxidation that histidine had (Table 3). While the presence of asparagine stimulated the reaction somewhat, a number of amino acids were observed to have an inhibitory effect. LOOH and TBA-RS measurements led to similar relative activities among the amino acids indicating that TBA-RS was a reliable indicator of lipid peroxidation and eliminating the possibility that histidine breakdown was generating products that were were TBA-reactive.

To investigate further the inhibitory and stimulatory characteristics of histidine on enzymic lipid peroxidation, several analogs were examined (Table 4). Only histidine (both D- and L-forms) was capable of fully activating the system. The low activity with histidinol, 1-methylhistidine, and urocanate indicated that the carboxyl group, the N-1 group of the imidazole ring and the α-amino group, were necessary for activation. Since the activity with 3-methylhistidine was approximately half of that in the presence of histidine, it is possible that the N-3 group may also contribute to activation. It is likely that histidine is functioning by complexing the Fe since if oxidation of the imidazole ring were a key factor, all of these compounds should behave in a similar manner. In addition, we have observed no measurable loss of histidine during the lipid peroxidation assay (unpublished results).

Less severe structural requirements were exhibited for inhibition of lipid oxidation. All histidine analogs tested led to greater than 50% inhibition implying that a major determinant for inhibition was binding of Fe^{+3} by the imidazole group.

Histidine was also capable of inhibiting the nonenzymic-ascorbate lipid oxidation system but had no effect on the nonenzymic-ferrous system (Table 5), suggesting that histidine interfered with the reduction of Fe^{+3}.

We suggest that a complex is formed between histidine, a nucleotide such as ADP, and Fe. This complex facilitates the reduction of Fe^{+3} to Fe^{+2}

Table 4. Ability of Histidine and Histidine Analogs to Activate or Inhibit Lipid Peroxidation

	Relative Activity	
Compound	Activation System	Inhibition System
None	1	1
L-Histidine	6.5	0.05
D-Histidine	6.1	---
3-Methylhistidine	2.9	0.12
Histidinol	0.7	0.02
1-Methylhistidine	0.4	0.02
Urocanate	0.4	0.41
Imidazole	0.7	0.02

For the activation systems, a 0.5 mM $FeCl_3$ stock solution was prepared in water. An aliquot of this stock solution (0.15 ml) was exposed to ADP (0.125 ml of 4 mM) before the addition of any other components. All compounds tested for activation were at a final concentration of 5 mM, pH 6.8, in the assay system.
For the inhibition systems, the 0.5 mM $FeCl_3$ stock solution was prepared in various solutions of histidine or its analogs at a concentration of 5 mM, pH 6.8. Aliquots of these stock Fe solutions (0.15 ml) were taken and added to an ADP solution (0.125 ml of 4 mM). These Fe-ADP solutions were tested for their ability to peroxidize SR lipids in an assay system containing 5 mM histidine, pH 6.8. Final concentrations of the test components in the assay system were 150 μM.
All assay systems contained 100 μM NADH and 0.5 mg SR protein in a total volume of 5 ml. The incubation was carried out at 6°C for 30 min.
Lipid peroxidative activity was determined by the accumulation of TBA-reactive substances. Where activation was being tested, results were expressed relative to the lipid peroxidative activity obtained from the system containing only $FeCl_3$, ADP, NADH, and SR (6.6 nmoles TBA-RS/mg protein/30 min). Where inhibition was being tested, results were expressed relative to the lipid peroxidative activity measured for the system in which the stock $FeCl_3$ solution was prepared in water (76.0 nmoles TBA-RS/mg protein/30 min).

and requires the carboxylate, the α-amino, and the N-1 group of the imidazole in histidine. It is possible that in the absence of nucleotide, all coordination sites on the Fe^{+3} molecule will be bound by histidine; thus, the Fe^{+3} is not able to participate in lipid oxidation.[6]

Displacement of the nucleotide from the activating complex did not occur even at very high concentrations of histidine (1000:1, histidine:ADP) Since the dissociation of metal ions from ADP is very slow,[7] a kinetic effe may be responsible for the inability of histidine to displace ADP over the time course of our experiments.

Both histidine and nucleotides, like ADP, contain N and O binding sites, the former favoring binding of Fe^{+2} and the latter Fe^{+3}.[2,8] Thus, the effectiveness of the activating complex may be due to the reduction and oxidation of Fe occurring within the same complex.

Table 5. Inhibitory Capacity of Histidine in Enzymic and Nonenzymic Systems

	Peroxidative Activity (TBA-RS)		
		Nonenzymic	
	Enzymic	Ascorbate	Ferrous
Fe Diluted in Water	107.1	102.1	38.8
Fe Diluted in Histidine	5.0	1.5	40.2

A 0.5 mM $FeCl_3$ stock solution was prepared in either water or a 5 mM histidine solution, pH 6.8. Stock solutions of $FeCl_2$ (0.5 mM) were prepared in water or histidine (5 mM) solutions, pH 6.8, previously bubbled with nitrogen.
The $FeCl_2$ stock solutions were capped and used within 30 min. Aliquots of the stock Fe solutions (0.15 ml) were exposed to ADP (0.15 ml of 0.5 mM) before the addition of any other components.
The enzymic system was characterized by the inclusion of 100 uM NADH, while the nonenzymic-ascorbate system contained 60 μM sodium ascorbate. In addition, all assay systems contained 5 mM histidine, pH 6.8, and 0.5 mg SR protein in a total volume of 5 ml. The reaction mixtures were incubated at 6°C for 30 min (enzymic), 3 min (nonenzymic-ascorbate) and 1 min (nonenzymic-ferrous).
Lipid peroxidative activity was determined by the accumulation of TBA-reactive substances and is expressed as nmoles TBA-RS/mg protein/30 min.

CONCLUSIONS

The observations reported here concerning the activity of the complexes between Fe, histidine, and ADP in vitro may have significance to lipid oxidation in stored fish muscle. Free histidine is found in fish muscle at concentrations of approximately 1 to 100 mM.[8] It seems clear that the type of complex formed between the Fe and low molecular weight compounds in the tissue can have very dramatic effects on the ability of Fe to initiate lipid oxidation in the SR. Obvious questions to be answered are, What is the nature of the Fe complexes occurring in fish muscle tissues? and, Is it possible to convert active complexes to inactive complexes in stored fish muscle?

REFERENCES

1. M. Borhan, R.L. Shewfelt, and H.O. Hultin, Sarcoplasmic reticulum from flounder muscle having improved lipid peroxidative activity, Anal. Biochem. 137:58 (1984).
2. R.E. McDonald and H.O. Hultin, Some characteristics of the enzymic lipid peroxidation system in the microsomal fraction of flounder skeletal muscle, J. Food Sci. 52:15 (1987).
3. K.M. Kramer, Form of iron in flounder muscle and its relation to sarcoplasmic reticular lipid oxidation, M.S. Thesis, University of Massachusetts, Amherst (1987).
4. J.A. Buege and S.D. Aust, Microsomal lipid peroxidation, in: "Methods in Enzymology," Vol. 52, S. Fleischer and L. Packer, eds., Academic Press, New York, p. 302 (1978).
5. M.A.K. Markwell, S.M. Hass, L.L. Bieber, and N.E. Tolbert, Modification of the Lowry procedure to simplify protein determination in membrane and lipoprotein samples. Anal. Biochem. 87:206 (1978).

6. E. Graf, J.R. Mahoney, R.G. Bryant, and J.W. Eaton, Iron-catalyzed hydroxyl radical formation. Stringent requirement for free iron coordination site, J. Biol. Chem. 259:3620 (1984).
7. N.H. Kolodny and L.J. Collins, Proton and phosphorus-31 NMR study of the dependence of diadenosine tetraphosphate conformation on metal ions. J. Biol. Chem. 261:14571 (1986).
8. S. Ikeda, Other organic components and inorganic components, in: "Advances in Fish Science and Technology," J.J. Connell, ed., Fishing News Books Ltd., Surrey, England, p. 111 (1980).

ACKNOWLEDGMENTS

This work was supported in part by the Massachusetts Agricultural Experiment Station and by Grant No. I-801-84 from BARD - The United States-Israel Binational Agricultural Research and Development Fund. We thank Marcia M. Stevens for editorial assistance.

CHOLESTEROL EPOXIDATIONS BY DEFINED OXYGEN SPECIES

Leland L. Smith and Krzysztof Jaworski

Department of Human Biological Chemistry and Genetics, University of Texas Medical Branch Galveston, Texas 77550, U.S.A.

INTRODUCTION

Each defined species of active oxygen implicated in oxygen biochemistry, with the exception of superoxide ($\cdot O_2^-$,$HOO\cdot$), is known to oxidize cholesterol under conditions potentially present _in vivo_. Oxysterols found in tissues thus may have at least two origins, enzymic ones related to _in vivo_ metabolism (including lipid peroxidation) and nonenzymic origins in which active oxygen species hydroxyl radical ($HO\cdot$), dioxygen cation (O_2^+), singlet dioxygen ($^1O_2$), ground state dioxygen ($^3O_2$), peroxide (O_2^{2-}, HOO^-,H_2O_2), ozone (O_3), and as yet unrecognized species may be implicated. As the oxysterols formed in all cases have demonstrated biological activities in a variety of bioassays, the question must be posed whether such oxysterols, if formed _in vivo_ and not properly controlled by protective metabolism, are a threat to human health.

Indeed, a considerable body of research is being conducted to test the hypothesis that oxysterols in the diet pose a chronic toxic burden. Other work has established that a rich variety of toxic oxysterols occurs in human blood and tissues. We have demonstrated correlations between levels of 26-hydroxycholesterol in human aortal tissues and severity of aortal atherosclerosis. However, a satisfying proof that endogenous oxysterols have physiological relevance or constitute a chronic health hazard has not been forthcoming.

Our approach to understanding the problem has been to identify the oxidation products of cholesterol formed by defined oxygen species, to identify these oxysterols in tissues and suggest, where possible, oxygen species likely to be implicated in their formation, and to test oxysterol toxicity in additional bioassays. Data of Table 1 summarize in abbreviated fashion results of past efforts that now accord a relatively complete view of the actions of oxygen

Table 1. Action of Oxygen Species on Cholesterol

Site of Attack	Oxysterol Product[a]	Defined Oxygen Species						
		O_3	O_2^+	$^3O_2$	$^1O_2$	$\cdot O_2^-$	O_2^{2-}	$HO\cdot$
C7	7-OOH	-	-	β>α	-	-	-	-
	7-OH	-	+	β>α	-	-	-	β>α
	7-Ketone	-	+	+	-	-	-	+
	5 -OOH	-	-	-	+	-	-	-
Δ^5	5,6-Epoxides	β>α	+	β>α	-	-	β>α	α>β
	3β,5α,6β-Triol	-	-	+	-	-	+	+
	5,6-Secosterols	+	-	-	-	-	-	-
C3	3-Ketones	-	-	+	-	-	-	-
C17-C27	Diverse	-	-	+	-	-	-	-

[a] Abbreviations: 7-OOH, cholesterol 7-hydroperoxides; 7-OH, 7-hydroxycholesterols; 5α-OOH, 3β-hydroxy-5α-cholest-6-ene-5-hydroperoxide.

species on an important compound of the body, cholesterol, which is probably the best known of all tissue components in this respect.

It is clear from these results that several oxysterols uniquely identify the oxygen species involved in their generation. The cholesterol 7-hydroperoxides, with the 7β-epimer predominating, establish $^3O_2$ and autoxidation or lipid peroxidation, however, provoked, as the agents of generation. The role of $^1O_2$ is also uniquely established by the presence of 3β-hydroxy-5α-cholest-6-ene-5-hydroperoxide. However, the commonly encountered 7-hydroxycholesterols and 7-ketocholesterol arise by several means.

The cholesterol 5,6-epoxides are special cases in that they are demonstrated mutagens and suspected carcinogens and are formed from cholesterol by the action of at least five active oxygen species, including O_3 and O_2^+ potentially present in polluted air. Only $^1O_2$ and $\cdot O_2^-$ fail to epoxidize cholesterol.[1,2]

The isomeric cholesterol 5,6-epoxides are formed together in chemical and biochemical system, and both have been detected in human and animal tissues. The ratio of α:β isomers ranges from 1:11 to 8:1 in the chemical and biochemical systems (cf. Table 2); ratios favoring the 5α,6α-epoxide occur in animal tissues. The ratio found may not reflect accurately the relative levels of isomers initially formed or the oxygen species implicated, as more than one oxygen species may be involved in cholesterol expoxidations, and the differential stability of the 5,6-epoxides with respect to hydrolysis to the common product 5α-cholestane-3β,5,6β-triol favors the more stable 5α,6α-epoxide.

Table 2. Epoxidations of Cholesterol or Cholesterol 3β-Acetate

Condition[a]	α:β Ratio
Air, aqueous alkaline dispersion	1:11
Air, aqueous dispersion pH 12	1:5
Air, aqueous dispersion pH 8	1:3.3
Air, aqueous dispersion pH 5.5	1:2
Air, aqueous dispersion pH 3	1.7:1
H_2O_2, aqueous alkaline dispersion	1:8
Sterol hydroperoxides or ROOH, aqueous alkaline dispersion	1:10-11
Ozone, H_2O	1:7-7.5
Ozone, CCl_4	ca. 1:1
Air (solid state)	1:3.6
Soybean lipoxygenase	1:3.7-4
Liver (rat, bovine) microsomes	1:3.3-4.7
Fe(II)-ADP, ROOH or LOOH	1:2-5
Fe(II)-ADP, xanthine oxidase	1:2
Hematin, ROOH	1:4
Fe(III) acetylacetonate, H_2O_2, ROOH or LOOH	1:2.3-5
Fe(III) acetylacetonate, air	1:2.3-4
Pertungstate salts, H_2O_2	1:1.6
MoO_2 acetylacetonate (or $MoCl_5$ or $Mo(CO)_6$, ROOH)	ca.1:1
ROOH, hexamethylphosphotriamide	1:1.2
MoO_5, hexamethylphosphotriamide	1.5-4:1
$Fe(ClO_4)_3$, H_2O_2	2-5:1
$FeSO_4$, H_2O_2	2:1
Perbenzoic acid	2.3:1
2-Hydroxyhexafluoropropyl hydroperoxide	2.8:1
Hydroxyl radical (N_2O, H_2O,γ-radiation)	3.5:1
Methylpyridazine oxide, photolysis	8:1
NO_2-induced _in vivo_ peroxidation	8:1

[a] Abbreviations: ROOH for tert-butyl, tert-amyl, or cumene hydroperoxide; LOOH for fatty acyl hydroperoxide (or methyl ester) or phospholipid hydroperoxides.

RESULTS

The presence of both 5,6-epoxides as products of O_3 attack on cholesterol was previously established for aqueous and alcohol systems, but we have recently demonstrated 5,6-epoxidations of cholesterol in aprotic nonparticipating solvents as well. Solutions of cholesterol in CCl_4 or hexane treated with O_3 yield a complex set of peroxysterols that are resolved with difficulty by high-performance liquid chromatography into at least ten components more mobile than cholesterol. These components are, as determined by proton and carbon nuclear magnetic resonance data, dimeric and oligomeric oxysterols formed by the putativive reaction of the 3β-hydroxyl of cholesterol (or other sterol) with a reactive oxidized sterol, probably the Criegee carbonyl oxide 3β-hydroxy-5-oxo-5,6-secocholestane-6-carbonyl oxide in the same manner as demonstrated for lower alcohol and water.[2]

The cholesterol 5,6-epoxides more polar than cholesterol were not encountered among products, but a dimeric product was recovered, which was characterized by a proton singlet at δ 3.05 ppm, recognized as arising from the 6α-proton of a 5β,6β-epoxide and assigned the structure of 5,6ξ-epidioxy-6ξ-5',6'β-epoxy-5'β-cholestan-3'β-yloxy-5ξ-5,6-secocholestane-3β,5-diol. Acetylation of the unfractionated crude ozonization product mixture further yielded upon high-performance liquid chromatography 3β-acetoxy-5,6ξ-epidioxy-6ξ-5',6'α-epoxy-5'α-cholestan-3'β-yloxy-5ξ-5,6-secocholestan-5-ol and isomeric 3β-acetoxy-5,6ξ-epidioxy-6ξ-5',6'β-epoxy-5'β-cholestan-3'β-yloxy-5ξ-5,6-secocholestan-5-ol in less than 1% yields. Structure assignments as 5,6-epoxides were made on the basis of the characteristic signal of the 6'-proton, δ 2.89 ppm (d, J = 4.1 Hz) and δ 3.06 ppm (s) for the 5'α,6'α- and 5'β,6'β-epoxide acetates, respectively. Corresponding signals for the cholesterol 5,6-epoxides (and acetates) are δ 2.85-2.91 ppm (d, J = 3.8-4.0 Hz) for the 5α,6α-epoxide and δ 3.05-3.07 ppm (d, J = 2-3 Hz) for the 5β,6β-epoxide.[3-7] Other spectral data support the assigned structures in detail.

The dimer 5'α,6'α-epoxide 3β-acetate structure was confirmed by synthesis from another dimer product of cholesterol ozonization, 3β-acetoxy-6ξ-cholest-5'-en-3'β-yloxy-5,6ξ-epidioxy-5ξ-5,6-secocholestan-5-ol. Epoxidation in CCl_4 with m-chloroperbenzoic acid yielded after high-performance liquid chromatography the same dimeric 5'α,6'α-epoxide 3β-acetate previously described. Furthermore, ozonization of cholesterol 3β-acetate in CCl_4 in the presence of tenfold excess of cholesterol 5α,6α-epoxide also gave the same dimer 5'α,6'α-epoxide 3β-acetate and confirmed our formulation how these 5,6-epoxides form in the ozonization of cholesterol.

CONCLUSIONS

These results show that O_3 attacks cholesterol in two ways whether in protic or aprotic systems, the major process being formation of oxidized secosterols and the minor being 5,6-epoxidation. If atmospheric O_3 is capable of in vivo oxidation of cholesterol, whether in an aqueous or an anhydrous (lipid) environment, one may anticipate that not only cholesterol ozonide derivatives but also isomeric cholesterol 5,6-epoxide derivatives form.

ACKNOWLEDGMENTS

The financial support of the Robert A. Welch Foundation of Houston, TX is gratefully acknowledged.

REFERENCES

1. J. Gumulka, J. S. Pyrek, and L. L. Smith, Interception of discrete species of oxygen in aqueous media by cholesterol: Formation of cholesterol epoxides and secosterols, Lipids 17:197 (1982).
2. J. Gumulka and L. L. Smith, Ozonization of cholesterol, J. Am. Chem. Soc. 105:1972 (1983).
3. K. Tori, T. Komeno, and T. Nakagawa, Nuclear magnetic resonance studies on steroids.III. Steroidal epoxides and episulfides, J. Org. Chem. 29:1136 (1964).
4. B. O. Lindgren and C. M. Svahn, Epoxides formed by chlorination of cholesterol and cholest-4-en-3β, 6β-diol, Acta Chem. Scand. 24:2699 (1970).
5. G. Assmann, D. S. Fredrickson, H. R. Sloan, H. M. Fales and R. J. Highet, Accumulation of oxygenated steryl esters in Wollman's disease, J. Lipid Res. 16:28 (1975).
6. G. A. S. Ansari and L. L. Smith, The oxidation of cholesterol by hydroxyl radical, Photochem. Photobiol. 30:147 (1979).
7. L. S. Tsai and C. A. Hudson, Cholesterol oxides in commerical dry egg products: Isolation and identification, J. Food Sci. 49:1245 (1984).

THERMAL DECOMPOSITION OF METHYL LINOLEATE AND METHYL LINOLENATE HYDROPEROXIDES ANALYZED BY CAPILLARY GAS CHROMATOGRAPHY

Edwin N. Frankel, Edward Selke, and William E. Neff

Northern Regional Research Center
U. S. Department of Agriculture
Peoria, Illinois 61604, U.S.A.

INTRODUCTION

Much work has been reported on the volatile oxidation products of unsaturated fats because they cause rancidity in foods and cellular damage in the body.[1] Decomposition of fatty ester hydroperoxides creates a wide range of carbonyl compounds, hydrocarbons, furans, and other materials that contribute to flavor deterioration of foods and that are implicated in biological oxidation. Interaction of some of these degradation products with DNA, proteins, and enzymes may be involved in cell-damaging reactions.[2-4] Although fatty acid hydroperoxides are the recognized precursors of volatile secondary products, relatively little information is available on the mechanism of their decomposition. New information on the volatile lipid oxidation products is needed to better evaluate precursors of oxidative deterioration in lipid-containing foods and to understand the biological significance of lipid peroxidation.

In a previous study, we examined the volatile products from pure methyl linolenate decomposed under different conditions.[5] Analyses by capillary gas chromatography (GC) and mass spectrometry (MS) showed significant differences in the composition of products from linolenate hydroperoxides decomposed thermally at 150°C and catalytically with ferric chloride and ascorbic acid. Although these volatile products represented a small portion of the total decomposition materials (7.43% under thermal decomposition conditions and 2.1% under catalytic conditions), they have an important impact on flavor and biological effects of lipid oxidation.[1] In the present study, we examined volatile products from pure methyl linoleate and linolenate hydroperoxides decomposed at different temperatures. We compared two capillary gas chromatographic techniques to analyze the volatile decomposition products. GC by dynamic headspace permits the analysis of decomposition products at low temperatures (as low as 60°C).[6] GC by direct injection technique is limited to the temperatures used in the injector port (usually above 180°C).[7]

EXPERIMENTAL PROCEDURES

Preparations

Pure hydroperoxides were prepared by silicic acid column chromatography from methyl linoleate and linolenate autoxidized at 40°C.[8] The purity of the hydroperoxides was checked by thin layer chromatography and by peroxide value (6100 meq/kg) determination.

Thermal Decomposition

Hydroperoxides were thermally decomposed either directly in the injector port of a gas chromatograph heated at 180°C,[7] or after purging and trapping in a dynamic headspace system.[6]

Direct Injection Technique. A sample of about 0.2 microliter was introduced into the GC port switched to the splitless mode. After 5 min the GC injector port was switched back to the split mode before temperature programming.

Dynamic Headspace Technique. The sample was purged with helium for 2.5 min at room temperature and heated 7 min between 60 and 180°C. The volatile compounds generated were swept and collected on a trap of porous polymer Tenax (Anspec Co., Ann Arbor, MI). After purging and trapping, the sample was rapidly desorbed at 170°C and back-flushed through a six-port valve to the GC capillary column cooled at -65°C. During desorption the GC injector port was in the splitless mode. After desorption for 5 min the injector port was returned to the split mode.

Gas chromatography (GC)

Volatile decomposition products were analyzed by GC with a Perkin-Elmer instrument (Model Sigma 300, Norwalk, Conn), equipped with a capillary column (Durabond DB-5, J & W Scientific Co., Rancho Cordova, CA; 60 m x 0.32 mm I.D.) cooled to -65°C with liquid nitrogen. Column temperature was programmed to 260°C at 5°C per min with a final hold of 10 min. Volatile compounds were identified by GC-MS.[7,9]

RESULTS AND DISCUSSION

Investigations of the thermal decomposition of methyl linoleate hydroperoxides by capillary GC show significant differences in product distribution according to the volatile collection and analytical technique used. The capillary gas chromatograms show that the main volatile products detected at 180°C by the dynamic headspace technique include pentane and hexanal (Figure 1). In addition to these two compounds, the direct injection technique produces more late peaks, including those attributed to 2,4-decadienal and methyl 9-oxo-nonanoate. Similarly, capillary gas chromatograms of the decomposition products from methyl linolenate hydroperoxides show more early peaks by the dynamic headspace technique and more later peaks by the direct injection technique (Figure 2). The main volatile products detected by dynamic headspace include propenal, propanal, 2,4-heptadienal and methyl octanoate. In addition to these volatile compounds, the direct injection technique shows significant increases in the relative amounts of 2,4,7-decatrienal and methyl 9-oxo-nonanoate.

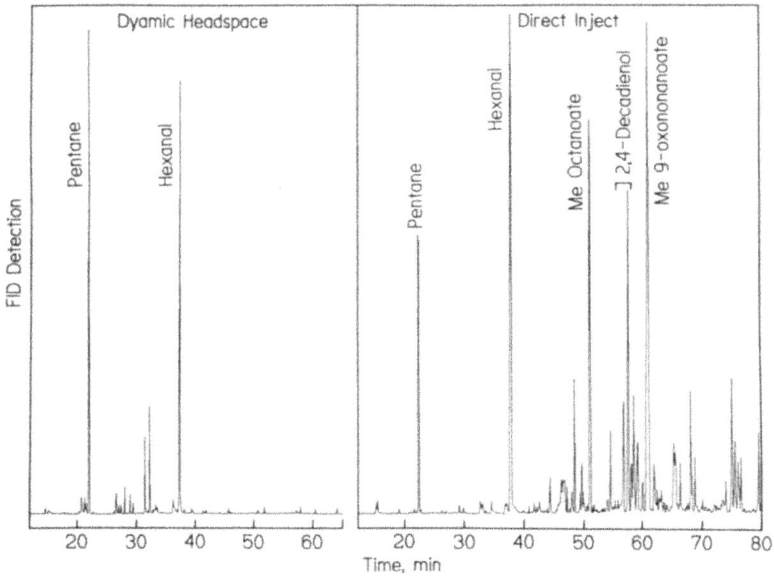

Figure 1. Capillary gas chromatograms of volatile products from the thermal decomposition of methyl linoleate hydroperoxides. Dynamic headspace versus direct injection techniques at 180°C.

Although the absolute concentrations of volatiles products from methyl linoleate and linolenate hydroperoxides increased with the dynamic headspace inlet temperature, dramatic differences were observed in their relative distributions. With methyl linoleate hydroperoxides analyzed between 60° and 180°C, pentane increased

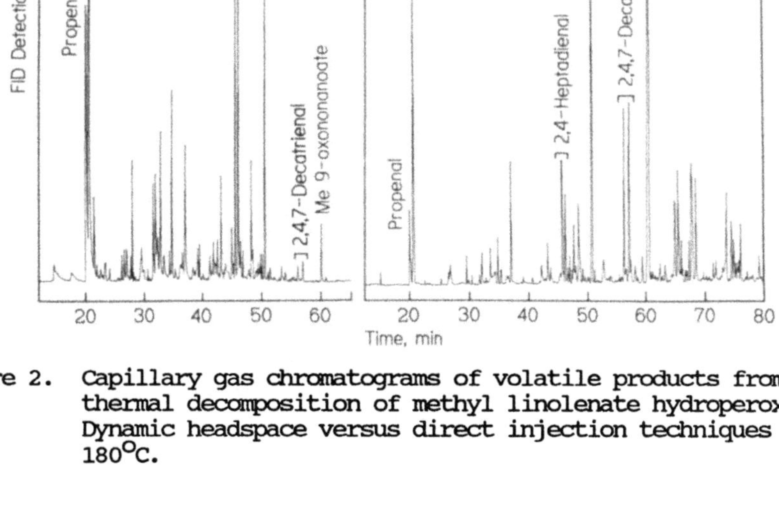

Figure 2. Capillary gas chromatograms of volatile products from the thermal decomposition of methyl linolenate hydroperoxides. Dynamic headspace versus direct injection techniques at 180°C.

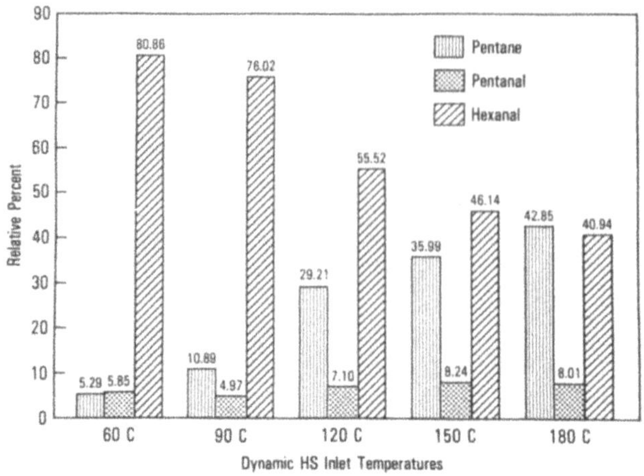

Figure 3. Effect of dynamic headspace temperature on the relative concentrations of pentane, pentanal and hexanal formed by thermal decomposition of methyl linoleate hydroperoxides.

from 5.29 to 42.85%, pentanal did not change much (5.85-8.01%), while hexanal decreased from 80.86 to 40.94% (Figure 3). In the same temperature range, methyl octanoate decreased from 2.15 to 0.34%, 2,4-decadienal increased from 0.02 to 0.38%, and methyl 9-oxo-nonanoate increased from 0.31% to a maximum of 0.49% at 90°C (Figure 4).

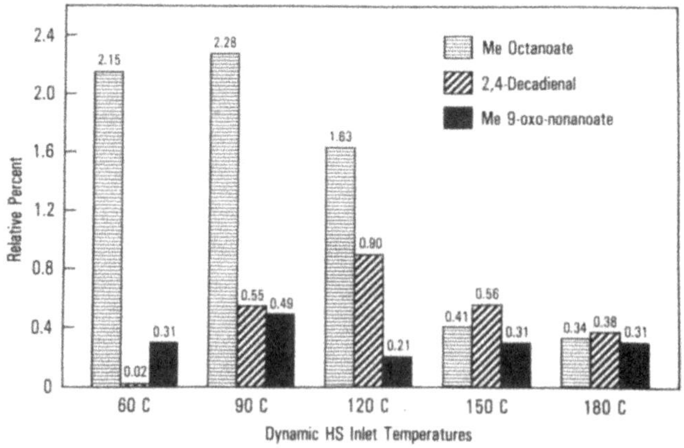

Figure 4. Effect of dynamic headspace temperature on the relative concentrations of methyl octanoate, 2,4-decadienal and methyl 9-oxo-nonanoate formed by thermal decomposition of methyl linoleate hydroperoxides.

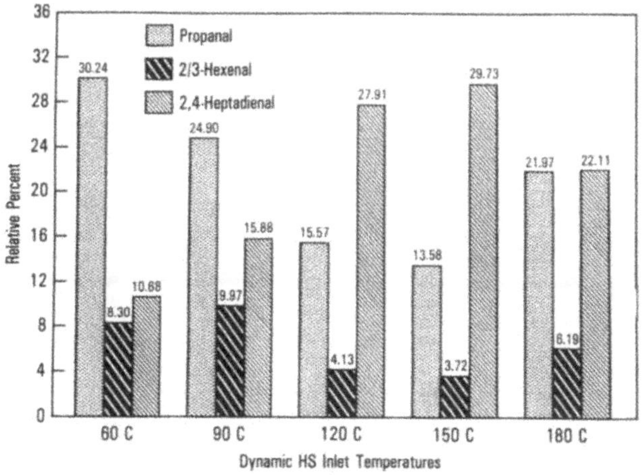

Figure 5. Effect of dynamic headspace temperature on the relative concentrations of propanal, 2/3-hexenal, and 2,4-heptadienal formed by thermal decomposition of methyl linolenate hydroperoxides.

With methyl linolenate hydroperoxides analyzed by dynamic headspace, propanal decreased from 30.24% at 60°C to a minimum of 13.58% at 150°C and then increased to 21.97% at 180° C. 2/3-Hexenal fluctuated from 3.72 at 150°C to 9.97% at 90°C. 2,4-Heptadienal increased from 10.68% at 60°C to 29.73% at 150°C and then decreased at 180°C (Figure 5).

Methyl octanoate increased from 5.05% at 60°C to a maximum of 16.66% at 150°C. 2,4,7-Decatrienal did not change much with temperature (0.44 to 1.11%), while methyl 9-oxo-nonanoate decreased from 3.40 at 60°C to 1.05% at 180°C (Figure 6).

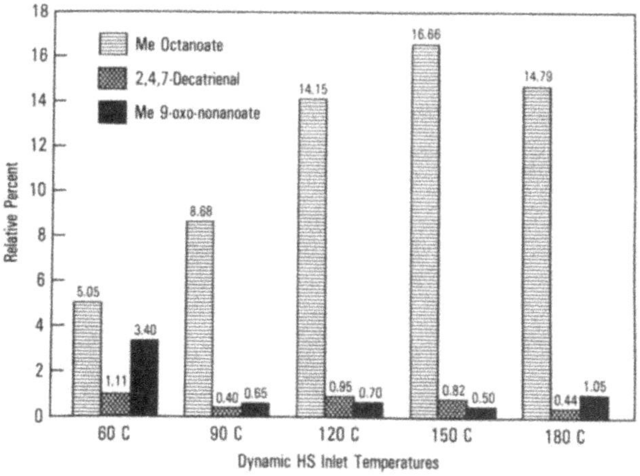

Figure 6. Effect of dynamic headspace temperature on the relative concentrations of methyl octanoate, 2,4,7-decatrienal and methyl 9-oxo-nonanoate formed by thermal decomposition of methyl linolenate hydroperoxides.

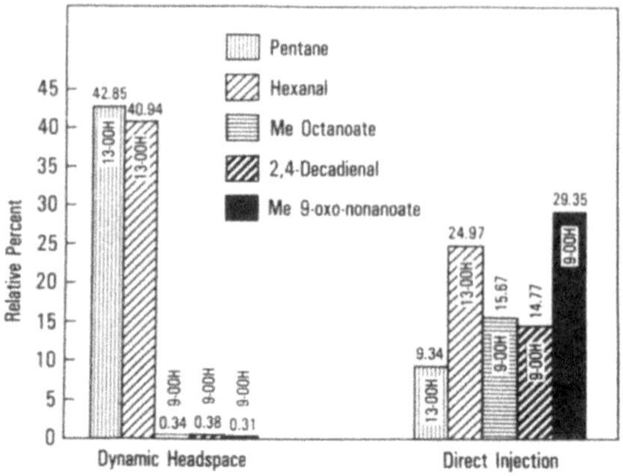

Figure 7. Origins of volatile decomposition products from linoleate hydroperoxides. Dynamic headspace versus direct injection.

The mechanism recognized for the decomposition of methyl linoleate and linolenate hydroperoxides involves homolytic scission on each side of the hydroperoxide group.[1,7,10] Thus, the 9-hydroperoxide of linoleate (representing 50%) produces methyl 9-oxo-nonanoate, 2,4-decadienal, and methyl octanoate, while the 13-isomer (representing the other 50%) forms hexanal and pentane as main decomposition products.[7] If the two capillary GC techniques used in this work are compared at the same temperature of 180°C, GC by dynamic headspace emphasized pentane and hexanal, which originate from 13-linoleate hydroperoxides. However, methyl octanoate, 2,4-decadienal, and methyl 9-oxo-nonanoate, which come from the 9-linoleate hydroperoxide, are only minor products detected by this technique. In contrast, GC analyses by direct injection produced more methyl octanoate, 2,4-decadienal, and methyl 9-oxo-nonanoate, which come from 9-linoleate hydroperoxides (Figure 7). The dynamic headspace technique thus emphasized low boiling volatiles whereas the direct injection technique emphasized the higher-boiling volatiles.

With methyl linolenate hydroperoxides,[5,7] the 9-isomer (31% relative concentration), produces methyl 9-oxo-nonanoate, 2,4,7-decatrienal, and methyl octanoate as main thermal decomposition products. The 12-hydroperoxide isomer (12% relative concentration) forms 2,4-heptadienal; the 13-hydroperoxide isomer (12% relative concentration) produces 3-hexenal, which readily isomerizes into a mixture of 2- and 3-isomers; and the 16-hydroperoxide isomer (45% relative concentration) forms propanal. In this study, the dynamic headspace technique emphasized propanal, which comes from the 16-hydroperoxide, 2,4-heptadienal, which comes from 12-hydroperoxide of linolenate, and methyl octanoate, which comes from the 9-hydroperoxide of linolenate. In contrast, GC analyses by direct injection produced more methyl octanoate (from the 9-hydroperoxide); much more 2,4,7-decatrienal (from the 9-hydroperoxide) and methyl 9-oxo-nonanoate, the most important product, which also comes from the 9-hydroperoxide isomer (Figure 8). As with linoleate hydroperoxides, GC analyses by direct injection exaggerated higher molecular weight compounds and gave different product distributions than were obtained by dynamic headspace GC.

The present study showed that different distribution of volatile products can be obtained according to the GC techniques used to generate and collect these products from the thermal decomposition of linoleate and linolenate hydroperoxides. The inlet temperature used in the dynamic headspace technique was a particularly important factor affecting the distribution of volatile decomposition products of hydroperoxides. At the same sampling temperature of 180°C, the direct injection technique gave a better representation of the volatiles produced from all isomeric hydroperoxides of linoleate and linolenate than did the dynamic headspace technique. GC analyses of volatile lipid oxidation products under different conditions require, therefore, careful interpretation before any conclusion can be made on mechanisms of hydroperoxide decomposition.

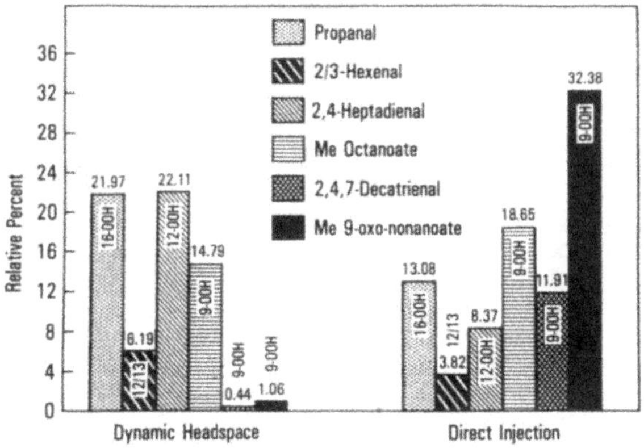

Figure 8. Origins of volatile decomposition products from linolenate hydroperoxides. Dynamic headspace versus direct injection.

The measuring of respiratory ethane and pentane has been considered as a sensitive noninvasive method of following in vivo lipid peroxidation in animals and biological tissues.[3] Ethane is formed by thermolysis of the 16-hydroperoxide isomer of linolenate, while pentane comes from the 13-hydroperoxide isomer of linoleate.[1] The formation of these respiratory hydrocarbons in animals is related to the extent of damage caused by such oxidative agents as ozone, carbon tetrachloride, ethanol, nitrous dioxide, iron compounds, and vitamin E-deficient diets. Dietary vitamin E, other antioxidants and selenium cause, on the other hand, a decrease in these respiratory hydrocarbons.[3] In our in vitro experiments, ethane and pentane are minor products resulting from decomposition of lipid hydroperoxides.[1,7,10] The biological activity of other important volatile decomposition products of lipid hydroperoxides is, therefore, an important research question that remains to be answered.

REFERENCES

1. E.N. Frankel, Volatile lipid oxidation products, Prog. Lipid Res. 22:1 (1982).
2. W.A. Pryor, The formation of free radicals and the consequences of their reactions in vivo, Photochem. Photobiol. 28:787 (1978).
3. A.L. Tappel, Measurement of and protection from in vivo lipid peroxidation, in: "Free Radicals in Biology," Vol. IV, W.A. Pryor, ed., Academic Press, New York (1980).
4. E.N. Frankel, Biological significance of secondary lipid oxidation products, Free Rad. Res. Commun. 3:213 (1987).
5. E.N. Frankel, W.E. Neff, E. Selke, E., and D.D. Brooks, Thermal and metal-catalyzed decomposition of methyl linolenate hydroperoxides, Lipids 22:322 (1987).
6. E. Selke and E.N. Frankel, Dynamic headspace capillary gas chromatographic analysis of soybean volatiles, J. Am. Oil Chem. Soc. 64:749 (1987).
7. E.N. Frankel, W.E. Neff, and E. Selke, Analysis of autoxidized fats by gas chromatography-mass spectrometry: VII. Volatile thermal decomposition products of pure hydroperoxides from autoxidized and photosensitized oxidized methyl oleate, linoleate, and linolenate, Lipids 16:279 (1981).
8. W.E. Neff, E.N. Frankel, and D. Weisleder, High-pressure liquid chromatography of autoxidized lipids: II. Hydroperoxy-cyclic peroxides and other secondary products from methyl linolenate, Lipids 16:439 (1981).
9. E. Selke, W.K. Rohwedder, and H.J. Dutton, Volatile components from triolein heated in air, J. Am. Oil Chem. Soc., 54:62 (1977).
10. E.N. Frankel, W.E. Neff, and E. Selke, Analysis of autoxidized fats by gas chromatography-mass spectrometry. IX. Homolytic vs. heterolytic cleavage of primary and secondary oxidation products, Lipids 19:790 (1984).

EFFECT OF DIETARY LIPID PEROXIDES ON LYMPHOID TISSUES

Motoko Oarada, Teruo Miyazawa, Kenshiro Fujimoto, Takashi Kaneda*, Emiko Ito** and Kiyoshi Terao**

Department of Food Chemistry, Faculty of Agriculture Tohoku University, Sendai 980, Japan

*Kohriyama Women's College, Kohriyama 963, Japan

**Research Institute for Chemobiodynamics, Chiba University, Inohana, Chiba 280, Japan

INTRODUCTION

Several studies have been reported of the toxicity and nutritional problems associated with lipid peroxides produced during the autoxidation of oils. Lipid hydroperoxides, the primary oxidation products, and other oxidation products such as ketones, epoxides and aldehydes are thought to be the substances most responsible for the toxicity. (1) However, it has not been shown whether dietary peroxides affect the functions of immunocompetent cells or not. We have reported that methyl U-^{14}C linoleate hydroperoxide and their radioactive secondary oxidation products (carbonyls such as 9-oxo nonanoate and 4-hydroxy nonenal) were absorbed and incorporated into several organs of rats after the oral administration. (2) To appreciate why oxidized oils that occur in foods show lesser nutritional value or toxicity, it seems important to study the effects of dietary peroxides on lymphoid tissues. We studied on the effects on lymphoid tissues of the thymus of mice induced by orally administered methyl linoleate hydroperoxide (MLHPO).

MATERIALS AND METHODS

MLHPO prepared had a peroxide value of 6100 meq/kg, determined by KI reduction, vs the theoretical value of 6125 meq/kg. Male C57BL/6 mice were allowed to mature to a body weight of 20-23 g with feeding of a commercial diet. Graded amounts (90, 190, 270 and 310 mg) of either MLHPO or 310 mg of unoxidized methyl linoleate (ML) were given to the mice at one time, using a stomach tube. U-^{14}C-MLHPO (200 mg, 654760 dpm) was administered orally to mice. At intervals, thymus, liver, spleen, mesenteric lymph node and bone marrow were excised after collection of blood by heart puncture. The blood and tissues (80-100 mg) were solubilized with NaOH and neutralized with HCl. Radioactivities of the solubilized samples were measured in a toluene scintillation cocktail with an Aloka liquid scintillation system LSC-903. A single photon counting apparatus, Chemiluminescence Analyzer CLA-700 (Tohoku Electronic Ind., Sendai), was used to detect spontaneous photon emission

Table 1. Appearance of Necrosis in Thymus and Adiposis in Liver after Oral Administration of MLHPO to Mice

MLHPO dose	thymus necrosis	adiposis hepatica
[mg/22 g mouse]	[%]*	[%]*
90	0 (0/7)	57 (4/7)
190	29 (2/7)	29 (2/7)
270	40 (3/7)	86 (6/7)
310	50 (2/4)	100 (4/4)

Percentages of mice which showed the thymus necrosis or the hepatic adiposis to the total mice examined. Figures in parenthesis indicate the incident number of mice. MLHPO = methyl linoleate hydroperoxide.

from the tissue homogenates. A 1 ml sample of 10%(w/v) tissue-physiological saline homogenate was placed on a stainless-steel plate and analyzed under an air at 37°C for 10 min. The damages on thymus lymphocytes were also assayed by fluorescence-activated cell sorter analyzer (Becton-Dickinson Co.).

RESULTS AND DISCUSSION

Necrosis of lymphocytes, which are among the epithelioid cells in the thymus, was observed in several of the mice to which MLHPO was orally administered. In the MLHPO-untreated mice, many blast lymphocytes in the cortex near the surface of the thymus could be found. In the case of MLHPO-dosed mice, a considerable degree of lymphocyte necrosis was observed in the thymus. When 190 mg of MLHPO was administered, 29% of the mice showed lymphocyte necrosis in the thymus, and the incidence was increased to 43% and 50% in mice to which 270 mg and 310 mg of MLHPO has been administered, respectively (Table 1).

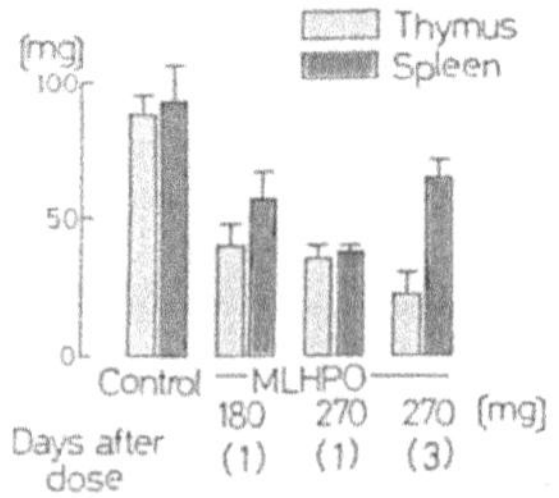

Figure 1. Decrease in thymus and spleen weights after administration of MLHPO to mice. Control = untreated mice. MLHPO = MLHPO-dosed mice.

Table 2. Increase in Tissue Spontaneous Chemiluminescence after Administration of MLHPO to Mice (cpm)

Dose	Days after Dose	Liver*	Thymus*	Spleen*
Untreated	-	648	1182	744
MLHPO 270 mg[a]	1	1518	16116	3048

*Combined tissue of three animals.

[a]mg/22 g mouse.

In the liver, fatty change appeared in 86% and 100% of the mice to which 270 mg and 310 mg of MLHPO had administered, respectively. Lymphocyte necrosis was not observed in the thymi of mice administered unoxidized ML.

With the administration of MLHPO, thymus weights significantly decreased (Figure 1). The decrease was most evident at 3 days after administration of 270 mg of MLHPO. Spleen weights were significantly lower in MLHPO-dosed mice than those of untreated mice.

During of 3-24 hr after the dose of U-^{14}C-MLHPO, the liver contained ^{14}C relatively higher levels, 0.73-1.11% of the total dose. In the blood, the highest ^{14}C level (0.24%) was also detected 12 hr after intubation. A very small amount of ^{14}C was distributed in lymphoid tissues such as the bone marrow, thymus, spleen and mesenteric lymph nodes.

In the liver, the thiobarbituric acid (TBA) value in mice one day after the dose of 270 mg of MLHPO was the highest. In the thymus and spleen, the TBA values were significantly higher than those of the unoxidized ML-dosed mice. As shown in Table 2, oral intubation of MLHPO to mice remarkably increased the spontaneous chemiluminescence intensities of liver, thymus and spleen. Especially after the dose of MLHPO, chemiluminescence of thymus and spleen was increased to 13.6 and 4.1 times of those of untreated mice. No effect was observed on plasma GOT and GPT levels even after the dose of MLHPO.

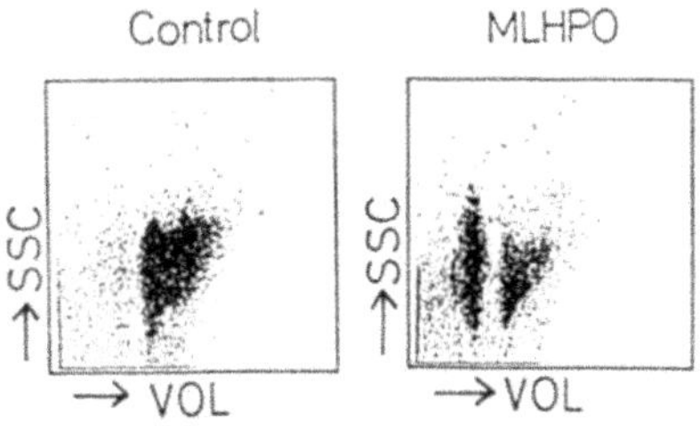

Figure 2. Dot-plot analysis of thymic lymphocytes with fluorescence activated cell sorter. Control = untreated mice. MLHPO = MLHPO (190mg) dosed mice. SSC = light scattering intensity. VOL = lymphocytes volumes.

As shown in Figure 2, abnormal lymphocytes which are smaller volumes and lacking immune responsibility were found with the dot-plot analysis of thymus cells of MLHPO-dosed mice.

CONCLUSIONS

If the significant levels of lymphocyte necrosis are induced by a daily uptake of rancid oils, the immune response of animal body may be gradually compromised. This study showed that oral intake of lipid hydroperoxide caused significant levels of lymphocyte damages in the thymi of mice.

REFERENCES

1. M. Oarada, T. Miyazawa and T. Kaneda, Distribution of ^{14}C after oral administration of U-^{14}C-labeled methyl linoleate hydroperoxides and their secondary oxidation products in rats. Lipids 21:150 (1986).
2. T. Kaneda and T. Miyazawa, Lipid peroxides and nutrition. Wld. Rev. Nutr. Diet. 50:186 (1987).

IMPLICATION OF PROTEIN OXIDATION IN PROTEIN TURNOVER, AGING,

AND OXYGEN TOXICITY

E.R. Stadtman, C.N. Oliver, R.L. Levine, L. Fucci, and A.J. Rivett

Laboratory of Biochemistry
National Heart, Lung, and Blood Institute
National Institutes of Health
Bethesda, Maryland 20892, U.S.A.

INTRODUCTION

A number of enzymatic and nonenzymatic mixed-function oxidation (MFO) systems catalyze the oxidation of enzymes.[1-6] With oxidation, the enzymes are converted to catalytically inactive or less active forms that are more sensitive to heat denaturation[7] and to proteolytic degradation.[8-11] Enzyme oxidation is likely the basis of some normal biological functions, e.g., in the killing of bacteria by neutrophils, in limitation of the inflammatory response, in controlling the switch from anaerobic to aerobic metabolism, and in the regulation of the turnover of some enzymes. Enzyme oxidation also might be implicated in a variety of pathological disorders, including arthritis, aging, carcinogenesis, broncho-pulmonary dysplasia and adult respiratory distress syndrome, retinopathy of prematurity, and reperfusion-mediated ischemic damage.

EXPERIMENTAL RESULTS

Mechanism of Enzyme Oxidation

Results of model studies with *E. coli* glutamine synthetase[1] indicate that oxidation occurs by the mechanism depicted in Figure 1. In the presence of Fe(III), O_2, and a suitable electron donor (e.g., ascorbate, NAD[P]H), MFO systems catalyze the formation of H_2O_2 and Fe(II).

Depending upon the nature of the MFO system and the environmental conditions, the generation of H_2O_2 and the reduction of Fe(III) to Fe(II) may occur directly[1] or indirectly via the intermediate formation of $O_2^{\cdot-}$.[6] The H_2O_2 and Fe(II) thus formed may interact at metal binding sites on the enzyme to generate activated oxygen species (e.g., •OH, perferryl ion, singlet oxygen), which oxidize the side chains of proximal amino acid residues to carbonyl derivatives and perhaps to other as yet unidentified derivatives. This mechanism (Figure 1) is supported by the following observations: (a) catalase inhibits protein oxidation by all MFO systems examined; (b) all MFO systems tested catalyze the reduction of Fe(III) to Fe(II) and also generate H_2O_2; (c) Mn(II), which inhibits the reduction of Fe(III) to Fe(II),[1] inhibits MFO-mediated enzyme inactivation;

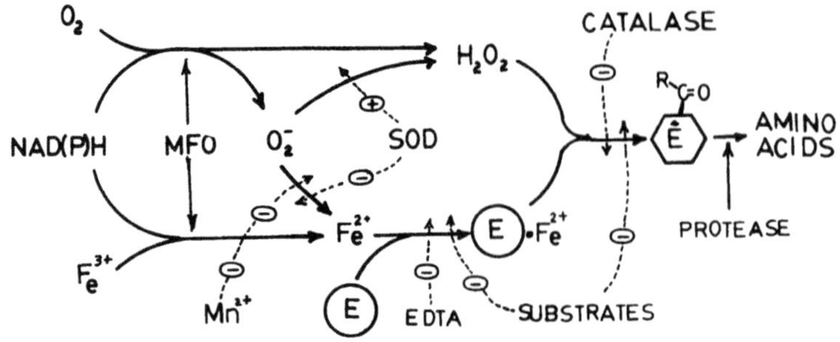

Figure 1. Schematic representation of the role of oxidative modification of enzymes in protein turnover. Abbreviations: MFO, mixed-function oxidation systems; SOD, superoxide dismutase; E_a, native (active) enzyme; E_m-RC=O, oxidized (inactive) enzyme; ⊖, inhibition; ⊕, stimulation.

(d) a mixture of H_2O_2 and Fe(II) catalyzes the inactivation of enzymes in the absence of enzymic MFO systems; (e) strong chelating agents (e.g., EDTA, o-phenanthroline) inhibit the inactivation of most enzymes. However, in a few cases (viz, glucose-6-phosphate dehydrogenase[12] and creatine kinase[4]), EDTA actually stimulates the inactivation of enzymes. To explain this anomaly, it is proposed that for enzymes possessing an Fe(II)-binding site, EDTA will inhibit oxidation by competing for the Fe(II). In this case, •OH generated by peroxidation of the free EDTA•Fe(II) complex will have little opportunity to attack the protein. In cases where EDTA stimulates rather than inhibits the protein oxidation, these proteins may not possess Fe(II)-binding sites but may nevertheless bind EDTA•Fe(II) complexes; peroxidation of the EDTA•Fe(II)•enzyme conjugate would therefore result in a site-directed oxidative attack of the protein.

The composition of some MFO systems that catalyze oxidative inactivation of enzymes is shown in Table 1. Under some experimental conditions, the rates of protein oxidation by most of these systems also is greatly enhanced by addition of a one-electron carrier such as menadione and ferredoxin.[3,6]

Among other changes, the MFO-mediated oxidation of enzymes leads to the conversion of some amino acid residues to carbonyl derivatives. Therefore, to facilitate studies on protein oxidation, several highly sensitive methods have been developed for the assay of protein carbonyl groups. These include procedures in which the carbonyl groups are converted to their 2,4-dinitrophenylhydrazone derivatives,[12,13] or to stable radioactive hydroxyl derivatives by reduction with sodium borotritide.[14] With these techniques, the levels of protein carbonyl groups are readily determined. The data in Table 2 show how the levels of carbonyl groups vary from one protein to another following exposure to two different MFO systems, one comprising ascorbate/Fe(III)/O_2 ± EDTA, the other Fe(II)/Fe(III)/O_2 ± EDTA. To determine which amino acid residues can be oxidized to carbonyl derivatives, the oxidation of various homoamino acid polymers has been studied. The data in Table 2 show that peptides of lysine, histidine, and proline are readily oxidized to carbonyl compounds. These and arginine also (data not shown) are therefore likely targets for MFO attack in protein. In other studies,[15] 2-amino-5-hydroxypentanoic acid was identified among the products obtained following acid hydrolysis of the oxidized prepara-

Table 1. Mixed-Function Oxidation of Systems That Catalyze the Inactivation of Enzymes

A - Enzyme Systems	References
(1) Cytochrome P_{450} reductase/cytochrome P_{450}/NADPH/O_2/Fe(II) or Fe(III)	1-3, 9
(2) Redoxin reductase/putida redoxin/NADH/O_2/Fe(II) or Fe(III)	2
(3) Redoxin reductase/putida redoxin/cytochrome P_{450}/NADH/O_2/Fe(II) or Fe(III)	2
(4) NADH oxidase/NADH/O_2/Fe(II) or Fe(III)	1
(5) NADPH oxidase/NADPH/O_2/Fe(II) or Fe(III)	1
(6) Xanthine oxidase/hypoxanthine or acetaldehyde/O_2/Fe(II) or Fe(III)	6
(7) Nicotinate hydroxylase/NADPH/O_2/Fe(II) or Fe(III)	6
(8) Glucose oxidase/glucose/O_2/ferredoxin or redoxin	6
(9) Peroxidase/H_2O_2/ferredoxin	6
B - Nonenzymic Systems	
(10) Ascorbate/O_2/Fe(III) or Fe(II)	4, 13
(11) Fe(II)/O_2	1, 4
(12) Fe(II)/H_2O_2	1
(13) RSH/Fe(III)/O_2	5

Table 2. Effect of EDTA on MFO-catalyzed Oxidation of Enzymes and Amino Acid Polymers to Carbonyl Derivatives

	MFO System			
Polypeptide or protein	(A) Fe(II)/Fe(III)/O_2 Carbonyl groups, nmol/mg		(B) Ascorbate/Fe(III)/O_2 carbonyl groups, nmol/mg	
	+ EDTA	- EDTA	+ EDTA	- EDTA
Glutamine synthetase	1.2	6.6	2.7	13.9
Ribonuclease A	10.9	3.4	6.5	1.0
Ribonuclease B	6.1	12.7	3.9	1.5
Lysozyme	5.9	15.9	16.3	0.6
Bovine serum albumin	9.5	3.0	26.6	0
Collagen (calf skin)	13.8	3.7	19.0	8.0
Poly-L-proline	8	31.7	67.9	16.5
Poly-L-histidine	12.6	9.0	37.7	14.8
Poly-L-lysine	10.6	7.6	65.8	10.1
Poly-L-lysine: L-alanine 1:1			160.0	10.5
Poly-L-glutamine: L-lysine 1:4	25.1	29.1	128	28

Reaction mixtures: phosphate buffer, 50 mM; KCl, 90 mM; $MgCl_2$, 10 mM. In addition, system (A) contained 2.0 mM Fe(II) and 2.0 mM Fe(III); system (B) contained 25 mM ascorbate, 100 μM Fe(III). After gentle shaking in air for 30 minutes at 30°C, the carbonyl content was measured as previously described.[15]

tions of either poly-L-proline or poly-L-arginine after prior reduction of their carbonyl groups to the corresponding hydroxy derivatives. In addition to as yet unidentified carbonyl derivatives, some histidine residues in polyhistidine are converted to asparagine residues.[15] Asparagine was also identified as a product of the histidine residue which disappears when E. coli glutamine synthetase is inactivated by the ascorbate MFO system.[16]

A number of enzymes which are inactivated by MFO systems are listed in Table 3. It is noteworthy that most of these are either kinases dehydrogenases, or synthetases, i.e., they possess nucleotide binding sites. Most of them require divalent cations for activity, and of those, it has been determined, many contain a histidine residue at the catalytic site. These observations are consistent with the view that the oxidative inactivation of enzymes by MFO systems is a site specific process. It is presumed that Fe(II) either binds directly to intrinsic metal binding sites on the enzyme or is carried to nucleotide binding sites as nucleotide chelate complexes. In either case, it is assumed that the Fe(II) undergoes in situ peroxidation to generate activated oxygen derivatives which attack nearby amino acid residues. This concept explains the inability of radical scavengers (viz, mannitol, formate, dimethylsulfoxide) to protect enzymes

Table 3. Enzymes Inactivated by MFO Systems In Vitro and Enzymes Shown to Accumulate As Altered Forms during Aging[†]

Enzymes	Inactivated by MFO	Altered in aging
Enolase[H]	+	+
Fructose-1,6-bisphosphatase	+	+
Glucose-6-P dehydrogenase	+	+
Lactate dehydrogenase[H]	+	+
Phosphoglycerate kinase[H]	+	+
Pyruvate kinase[H]	+	+
Superoxide dismutase[H]	+	+
Glutamine synthetase[H]	+	+
Acetylcholine esterase	+	?
Alcohol dehydrogenase[H]	+	?
Aspartokinase III	+	?
Creatine kinase	+	?
Glyceraldehyde-3-P-dehydrogenase	+	?
Lysozyme	+	?
Ornithine decarboxylase	+	?
Ribonuclease A & B[H]	+	?
Elongation factor I	?	+
Glutathione reductase	?	+
Hypoxanthine P-ribosyltransferase	?	+
Isocitrate lyase	?	+
Maltase	?	+

[†]See Fucci et al.[1] and Oliver et al.[12] for further details. The "?" indicates that the enzyme has not yet been tested for that characteristic. "H" indicates an enzyme which has been shown to have a histidine at the active site. For enzymes noted to be altered in aging, the specific catalytic activity was shown to decline with age while the amount of protein which cross-reacted with antibody against highly purified preparations of the enzyme remained essentially constant.

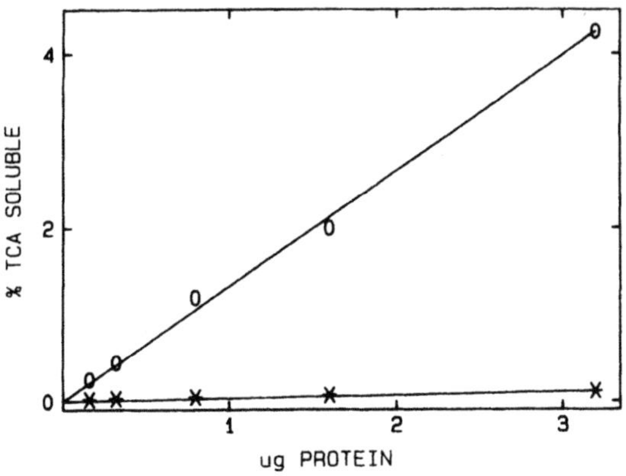

Figure 2. Preferential degradation of oxidized glutamine synthetase by cytosolic mouse liver protease. Various concentrations of the alkaline protease were incubated with 0.8 µM of oxidized ^{14}C glutamine synthetase (-0-0-) or unmodified ^{14}C-labeled glutamine synthetase (-*-*-). Incubation mixtures contained 50 mM Hepes buffer (pH 8.0), 1 mM $MgCl_2$, 2 mM KCl, 0.02 mM EDTA. After 1 hour at 37°C, the TCA-soluble radioactivity was determined. From Oliver et al.[8]

from inactivation by MFO systems. It is consistent also with the fact that histidine residues are particularly susceptible to oxidation by MFO systems. Indeed, in four cases where it has been carefully examined, the inactivation of the enzyme is accompanied by the loss of a single histidine residue.

Protein Turnover

With oxidation, enzymes become highly susceptible to proteolytic degradation, especially by a class of neutral-alkaline cytosolic proteases that have little or no ability to degrade the native (unoxidized) forms[8-11] (Figure 2). The MFO-catalyzed oxidation of enzymes may therefore represent a "marking" step in protein turnover. Since the substrates of enzymes have been shown to protect them from oxidation by MFO systems,[1,6,9] variations in levels of the substrates for a particular enzyme, as might occur in response to changes in metabolism or in the nutritional state of the organism, could be the basis for differential regulation of the turnover of one enzyme with respect to others. In addition, the turnover of enzymes would be controlled by factors that regulate the levels of H_2O_2 and Fe(II) (viz, catalase, superoxide dismutase, peroxidase, Mn(II), radical scavengers, and chelating metabolites), and also by factors that control the activities of MFO systems themselves, as well as the proteases that degrade oxidized proteins.

Implication in Aging

Inactive, less active, and more heat labile forms of some enzymes accumulate during aging.[17] Many of these enzymes (in fact all that have been tested) are also susceptible to oxidation by MFO systems (Table 3). This prompted speculation that the age-related alterations of enzyme might be due to oxidation. This possibility is supported by the demonstration that (a) the level of oxidized protein (i.e., the level of protein carbonyl groups) in human erythrocytes increases with the age of the cell, i.e., with cell density[7] (Figure 3); (b) the level of

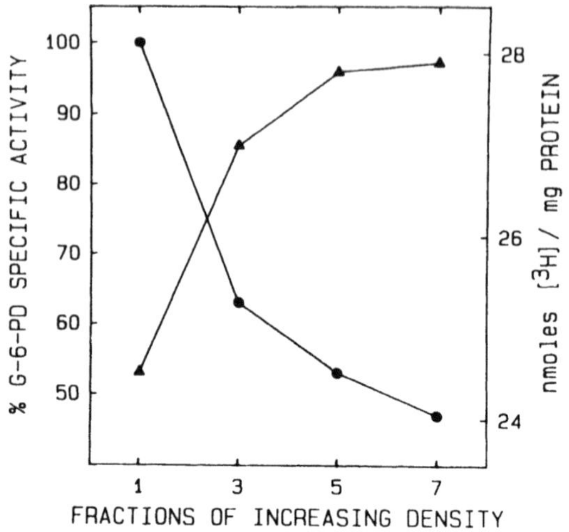

Figure 3. Relationship between cell density (age) of human erythrocytes and the activity of glucose-6-phosphate dehydrogenase (G-6-PD) activity (-●-●-) and the levels of protein carbonyl groups as measured by the ^{3}H-labeling technique (-▲-▲-). From Oliver et al.,[7] where experimental details are described.

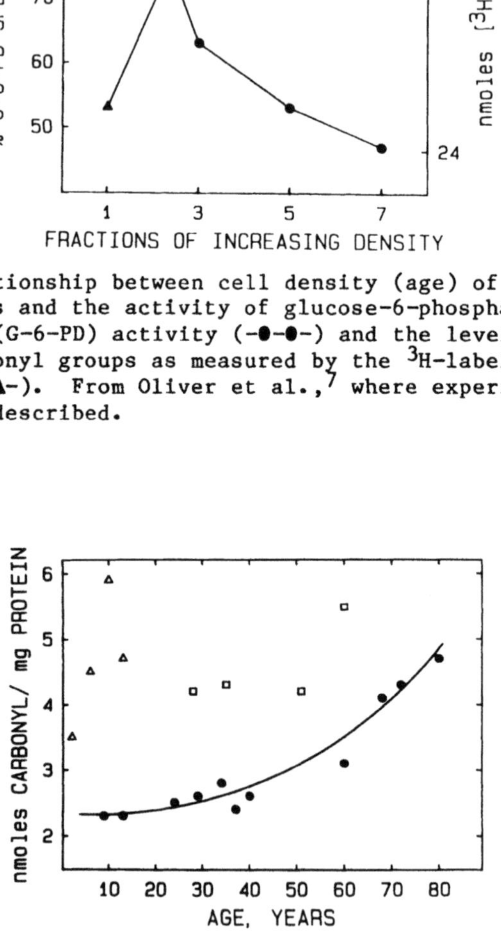

Figure 4. Effect of donor age on the carbonyl content of cultured dermal fibroblasts from normal individuals and individuals with premature aging diseases. Fibroblasts were from normal individuals (-●-●-), individuals with progeria (-Δ-Δ-), and individuals with Werner's syndrome (-□ -□ -) This figure is from Oliver et al.[7] where experimental conditions are described.

protein carbonyl groups in cultured dermal fibroblasts increases with the age of the donor over the range of 9 to 80 years[7] (Figure 4); (c) the levels of protein carbonyl groups in cultured fibroblasts of individuals with premature aging diseases (progeria, Werner's syndrome) are very much higher than those in fibroblasts of normal individuals of the same age[7] (Figure 4) (the levels in fibroblasts from 9-13-year old progeric individuals are similar to that of an 80-year old normal individual); (d) the level of oxidized protein in hepatocytes of rats increases with the age of the animal (see the report of Starke-Reed and Oliver, this volume).

Garland[18] has shown that the level of protein carbonyl groups in extracts of human eye lenses increase with age of the individual over the range of 20 to 95 years, and that the levels in some kinds of cataractous lenses are higher than in normal lenses of individuals of the same age.

That MFO-catalyzed oxidation is responsible for the accumulation of altered forms of enzymes is supported also by the results of immunochemical and heat stability studies showing that the levels of inactive and less active-more heat labile forms of glutamine synthetase and glucose-6-phosphate dehydrogenase (G-6-PD) in rat hepatocytes increase with the time of exposure (0-48 hours) of the rats to high O_2 tensions (100% O_2). Moreover, this increase in the level of altered enzymes is accompanied by a large increase in the levels of protein carbonyl groups;[19] see also the report of Starke-Reed and Oliver, this volume).

Heat Stability

The thermal inactivation of some enzymes from young animals is an apparently first-order process. In contrast, the same enzymes from old

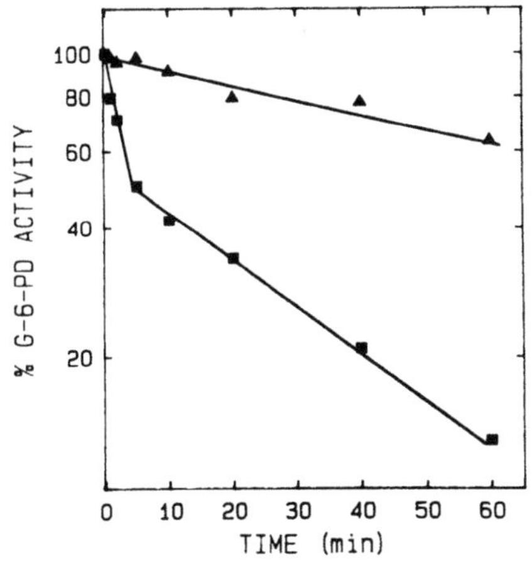

Figure 5. Effect of MFO-catalyzed oxidation of glucose-6-phosphate dehydrogenase (G-6-PD) on heat stability. G-6-PD from bovine adrenals was incubated in air, in a mixture containing 50 mM phosphate buffer (pH 7.2), 10 mM $MgCl_2$, 90 mM KCl, in the presence or absence of 25 mM ascorbate, 2 mM $FeCl_3$ and 2.4 mM EDTA. After 30 minutes at 30°C, samples were gel-filtered and the time-dependent loss of activity at 51°C was determined. Symbols refer to samples incubated in absence (-▲-▲-) or presence (-■-■-) of the ascorbate MFO systems. From Oliver et al.[7] whose experimental conditions are described.

animals often yield biphasic heat inactivation kinetics, indicative of molecular heterogeneity; this has been attributed to the age-dependent accumulation of altered enzyme forms.[17] The possibility that such molecular heterogeneity reflects oxidative damage during aging is supported by the data in Figure 5 showing that preparations of adrenal glucose-6-phosphate dehydrogenase exhibit first-order kinetics of heat inactivation before oxidation by the ascorbate/Fe(III)/O_2 MFO system; however, the activity that remains after limited exposure to the MFO system exhibits a biphasic heat inactivation pattern. The difference in kinetics between oxidized and unoxidized G6PD preparations is therefore similar to that observed between preparations isolated from young and old erythrocytes[19,20] or from cultured fibroblasts of normal young individuals and from individuals with progeria[21] or Werner's syndrome.[22]

Role of Enzyme Oxidation in Neutrophil Function

MFO-catalyzed inactivation of bacterial enzymes may be involved in host defense against bacterial infection. Oliver[23] has shown that when human neutrophils are induced to undergo a period of respiratory burst they catalyze the inactivation of exogenously added *E. coli* glutamine synthetase as well as glutamine synthetase inside *E. coli* cells added to the neutrophil suspension. She showed also that during respiratory burst, a number of endogenous nuetrophil enzymes are inactivated and that there is a substantial increase in the level of carbonyl groups of the neutrophil endogenous proteins.

SUMMARY

It is evident from the results summarized here that a variety of MFO systems catalyze the oxidation inactivation of enzymes. This likely involves site-directed Fenton-chemistry in which Fe(II) bound to metal binding sites on the protein undergoes peroxidation to form active oxygen species that convert proximal amino acid residues to carbonyl derivatives. Such oxidation is likely involved in the accumulation of altered enzymes during aging, in premature aging diseases, in the killing of bacteria by neutrophils and in protein turnover. In view of these results, the possibility that protein oxidation is implicated in various diseases, viz, arthritis, pulmonary dysfunction, and carcinogenesis deserves consideration.

REFERENCES

1. L. Fucci, C. N. Oliver, M. J. Coon, and E. R. Stadtman, Inactivation of key metabolic enzymes by mixed-function oxidation reactions: Possible implication in protein turnover and ageing, *Proc. Natl. Acad. Sci. U.S.A.* 80:1521 (1983).
2. C. N. Oliver, R. L. Levine, and E. R. Stadtman, Regulation of glutamine synthetase degradation, *in*: "Experiences in Biochemical Perception," L. N. Ornston and S. G. Sligar, eds., Academic Press, New York (1982).
3. K. Nakamura, C. N. Oliver, and E. R. Stadtman, Inactivation of glutamine synthetase by a purified rabbit liver microsomal cytochrome P_{450} system, *Arch. Biochem. Biophys.* 240:319 (1985).
4. R. L. Levine, Oxidative modification of glutamine synthetase. II. Characterization of the ascorbate model system, *J. Biol. Chem.* 258:11828 (1983).
5. K. Kim, S. G. Rhee, and E. R. Stadtman, Nonenzymatic cleavage of proteins by reactive oxygen species generated by dithiothreitol and iron, *J. Biol. Chem.* 260:15394 (1985).
6. E. R. Stadtman and M. E. Wittenberger, Inactivation of *Escherichia*

coli glutamine synthetase by xanthine oxidase, nicotinate hydroxylate, putidaredoxin, and menadione, Arch. Biochem. Biophys. 239: 379 (1985).
7. C. N. Oliver, B.-W. Ahn, E. J. Moerman, S. Goldstein, and E. R. Stadtman, Age-related changes in oxidized proteins, J. Biol. Chem. 262:5488 (1987).
8. C. N. Oliver, R. Fulks, R. L. Levine, L. Fucci, A. J. Rivett, J. E. Roseman, and E. R. Stadtman, Oxidative inactivation of key metabolic enzymes during aging, in: "Molecular Basis of Aging," A. K. Roy and B. Chatterjee, eds., Academic Press, New York (1984).
9. R. L. Levine, C. N. Oliver, R. M. Fulks, and E. R. Stadtman, Turnover of bacterial glutamine synthetase: Oxidative inactivation precedes proteolysis, Proc. Natl. Acad. Sci. U.S.A. 78:2120 (1981).
10. A. J. Rivett, J. E. Roseman, C. N. Oliver, R. L. Levine, and E. R. Stadtman, Covalent modification of proteins by mixed-function oxidation: Recognition by intracellular proteases, in: "Intracellular Protein Catabolism," E. A. Khairallah, J. S. Bond, and J. W. C. Bird, eds., Alan R. Liss, New York (1985).
11. A. J. Rivett, Preferential degradation of the oxidatively modified form of glutamine synthetase by intracellular mammalian proteases, J. Biol. Chem. 260:300 (1985).
12. C. N. Oliver, B. Ahn, M. E. Wittenberger, R. L. Levine, and E. R. Stadtman, Age-related alterations of enzymes may involve mixed-function oxidation reactions, in "Modification of Proteins during Aging," R. C. Adelman and E. R. Dekker, eds., Alan R. Liss, New York (1985).
13. R. L. Levine, Oxidative modification of glutamine synthetase: I. Inactivation is due to loss of one histidine residue, J. Biol. Chem. 258:11823 (1983).
14. R. L. Levine, Mixed-function oxidation of histidine residues, Meth. Enzymol. 107:370 (1984).
15. A. Amici, L. Tsai, and E. R. Stadtman, unpublished data.
16. J. M. Farber and R. L. Levine, Sequence of a peptide susceptible to mixed-function oxidation: Probable cation binding site in glutamine synthetase, J. Biol. Chem. 261:4574 (1986).
17. M. Rothstein, Changes in enzymatic proteins during aging, in "Molecular Basis of Aging," A. K. Roy and B. Chatterjee, eds., Academic Press, New York (1984).
18. D. L. Garland, personal communication.
19. P. E. Starke-Reed, C. N. Oliver, and E. R. Stadtman, Modification of hepatic proteins in rats exposed to high oxygen concentration, FASEB J. 1:36 (1987).
20. S. Goldstein and E. J. Moerman, Unstable enzymes in erythrocytes of a family with Hutchinson Gilfar progeria syndrome, in: "The Red Cell," G. J. Brewer, eds., Alan R. Liss, New York (1978).
21. S. Goldstein and E. J. Moerman, Heat labile enzymes in skin fibroblasts from subjects with progeria, New Engl. J. Med. 292:1305 (1975).
22. S. Goldstein and D. P. Singal, Alteration of fibroblast gene products in vitro from a subject with Werner's syndrome, Nature 251:719 (1974).
23. C. N. Oliver, Inactivation of enzymes and oxidative modification of proteins by stimulated neutrophils, Arch. Biochem. Biophys. 253: 62 (1987).

ROLE OF OXIDATIVE MODIFICATION IN THE LABILITY OF CERULOPLASMIN

Paul G. Winyard, Robert C. Hider,* Joseph Lunec,
Alex F. Drake,** and David R. Blake

Dept. of Rheumatology
The Medical School
University of Birmingham
Birmingham, B15, 2TJ, England

*Dept. of Pharmacy
King's College
Manressa Road
London, SW3, 3LX, England

**Dept. of Chemistry
Birkbeck College
20 Gordon Street
London WC1H, OAJ, England

INTRODUCTION

Ceruloplasmin (CP) is a blue copper oxidase found in human serum. It has been suggested that CP functions as an antioxidant by virtue of its ability to scavenge the superoxide anion radical[1] and/or its ability to catalyse the conversion of Fe(II) to Fe(III), thereby inhibiting the Fenton reaction[2]. However, the true physiological role of CP is still not clear. In attempting to define such a role it is important to recognise the relative lability of this protein[3]. Many of the changes in the physicochemical properties of CP_3 which occur during isolation and storage can be ascribed to proteolysis[3]. However, this does not appear to account for all of the modifications which occur during the ageing of CP preparations.

Among its defined physicochemical properties, CP is a single polypeptide chain containing 1046 amino acids, less than 50% of which are thought to participate in beta sheet structure (reviewed in reference 3). In addition, blue oxidases exhibit a visible fluorescence (excitation 330 nm; emission 420 nm), which has been attributed to oxidation products of constituent tryptophan residues, induced by storage, and/or the carbohydrate component[4]. The visible fluorescence previously had been attributed to "type 3" copper (reviewed in reference 4). We demonstrate below that loss of beta structure and induction of visible fluorescence occur as a result of oxidative modification.

EXPERIMENTAL

Low Intensity Ultraviolet Irradiation

CP (Andard-Mount Ltd., Middlesex, U.K: A_{610}/A_{280} =0.044) was dissolved in 40 mM phosphate buffer, 0.15 M NaCl, pH 7.4 at approximately 1.5 mg/ml. Aerobic irradiation was carried out using the 366- plus 254- source previously described[5]. Volumes of 1.2 ml were irradiated in matched quartz cuvettes, 1 cm^2 in cross-section, at a distance of 6 cm from the light source. The light source at 254 nm was a Sylvania G8T5-8W bulb with an average light intensity of 10.5 W/cm^2 at 1m.

High Performance Liquid Chromatography

CP samples were applied to a TSK G3000SW column. Elution conditions and detection by fluorescence (excitation 360 nm; emission 454 nm) and absorption (280 nm) are described elsewhere[6].

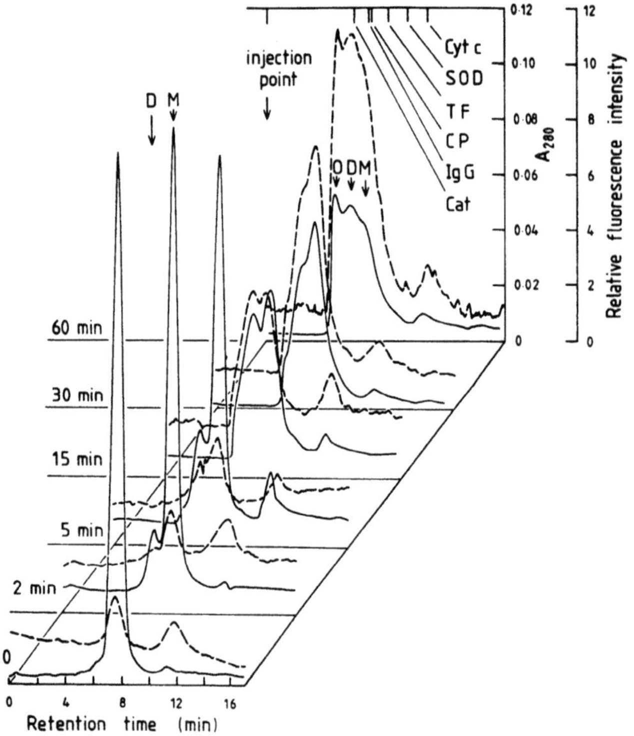

Figure 1. Fluorescent monomeric and oligomeric reaction products of CP generated during timed intervals of exposure of CP to UV irradiation. Elution of CP products was monitored simultaneously for ___, absorbance at 280 nm and ---, visible fluorescence. M_r calibration was performed by monitoring the elution time of pure standards of cytochrome c (cyt c; M_r =12 400), superoxide dismutase (SOD; M_r=32 000), transferrin (TF; M_r=80 000), immunoglobulin G (IgG; M_r=150 000), and catalase (Cat; M_r=230 000).

Optical Spectra

Circular dichroism (CD) spectra were recorded on a JASCO J-40CS CD spectrometer. The protein concentration (using E_{1cm} at 280 nm = 14.9)[7] was kept at approximately 0.7 mg/ml throughout, in 20 mM phosphate buffer, 0.075 M NaCl, pH 7.4, and the near UV and far UV regions required cells of optical pathlength 1 cm and 0.2 mm, respectively. Circular dichroism was expressed as the molar CD absorption coefficient, using a mean residue weight of 113. Absorption spectra were measured prior to CD measurement on a Cary 17 spectrophotometer. The beta sheet content of CP was estimated from the far UV CD spectrum.[8]

RESULTS AND DISCUSSION

Ageing of CP preparations results in a loss of A_{610} and oxidase activity[3]. Similar changes in CP are produced by free radical attack[6], and by low intensity UV irradiation[5] which has been used as a model of protein ageing[4].

The elution profiles of UV irradiated CP, following size exclusion HPLC, are shown in Figure 1. Monitoring was by visible fluorescence (excitation 360 nm; emission 454 nm) and UV absorption (280 nm). The profiles demonstrate the successive production of fluorescence in monomeric CP ("M"), dimers ("D") and oligomers ("O") of $M_r > 10^6$. Oligomers are frequently present in CP preparations[3]. Early oligomerisation appears to involve dimers, but after 60 min of irradiation a substantial proportion of high M_r oligomers had formed.

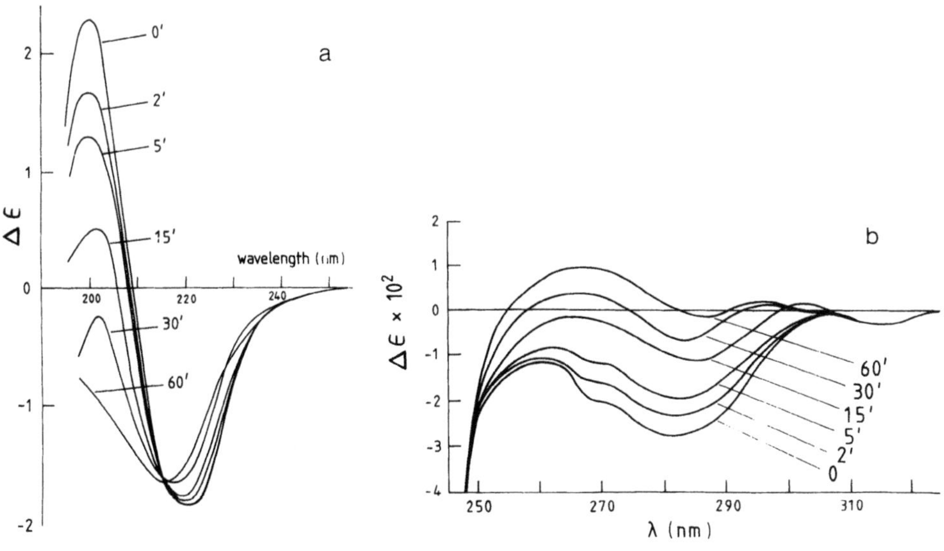

Figure 2. Circular dichroism spectra of CP exposed to timed intervals of UV irradiation.
a, far UV region; b, near UV region. Numbers adjacent to each spectrum represent the number of min or irradiation. Spectra were recorded in 20 mM phosphate buffer, 0.075 M NaCl, pH 7.4.

The far UV CD spectrum of native CP (Figure 2a) shows a strong positive maximum at 200 nm, reflecting an appreciable beta sheet content. The negative band at 220 nm is somewhat less intense than the positive band at 200 nm. When CP is exposed to UV irradiation the trough at 220 nm becomes less intense, broadens and is displaced to lower wavelengths. Simultaneously, the positive extremum at 200 nm is considerably reduced, the peak failing to reach positive values after 30 min of irradiation. There is a blue shift in the 210 nm crossover of the native protein. The isosbestic point at approximately 215 nm indicates that two major conformational states dominate the spectrum. The CD changes in the far UV region are indicative of loss of beta sheet content.

The effect of UV irradiation on the near UV CD spectrum of CP is shown in Figure 2b. The extremum in the region of 260 nm is considerably enlarged and undergoes a red shift. The 280 nm negative band decreases in intensity and shifts to longer wavelengths. After 15 min of irradiation parts of the spectrum reach positive values and after 60 min irradiation a positive peak is clearly visible at 292 nm. The latter band also occurs in the CD spectrum of apo-CP where it was assigned to an "L_b" tryptophan band[9]. The change in sign of many of the near-UV CD bands is likely to result from alterations in tertiary structure, although the chemical modification of some tryptophan side chains in ceruloplasmin may also make a contribution (P.G. Winyard, unpublished data). Difference spectra (not shown) indicate that the partial unfolding of UV-irradiated CP involves the perturbation of tryptophan residues.

Our estimate of the beta sheet content of CP, from the far-UV spectrum is 60-65%. Previous estimates are lower than this, varying between 20 and 46%[3,7,9]. Some of these lower estimates may be due to the loss of beta sheet structure during ageing of CP preparations. The small blue protein plastocyanin, which has sequence homology with CP, contains 60% beta sheet structure based on crystallographic data[10]. The fact that the 190-210 nm region of both CP (Figure 2a) and plastocyanin[11] is virtually identical supports our estimate of 60-65% beta sheet content for CP.

Oxidative stress produces substantial conformational changes in CP. In contrast, limited proteolysis has no significant effect on the far- and near- UV CD spectra of CP[9]. Our estimate of beta sheet content may be higher than those of others because the CP preparation used showed little evidence of oxidative attack, i.e. a high absorption at 610 nm, little sign of aggregation and only a small amount of visible fluorescence. This commercial CP preparation, as supplied, contains the ·OH free radical scavenger mannitol. Furthermore, all experiments were carried out within three days of dissolving the sample and dialysing it to remove the mannitol.

Limited proteolysis does not result in marked changes in the oxidase activity and spectroscopic properties of CP[3]. In contrast, oxidative stress may affect such parameters and could cause a structural rearrangement which alters the nature of the catalytic activity of the copper atoms.

ACKNOWLEDGEMENTS

This work was supported by a project grant to P.G.Winyard and D.R. Blake from The Arthritis and Rheumatism Council.

REFERENCES

1. I.M. Goldstein, H.B. Kaplan, H.S. Edelson, and G. Weissmann, Ceruloplasmin: a scavenger of superoxide anion radicals, J. Biol. Chem. 254:4040 (1979).

2. D.J. Al-Timini and T.L. Dormandy, The inhibition of lipid autoxidation by human ceruloplasmin, Biochem. J. 168:283 (1977).

3. L. Ryden, Ceruloplasmin, in: "Copper proteins and copper enzymes Volume III", R. Lontie, ed., CRC Press, Boca Raton, Florida (1984).

4. L.Avigliano, P. Sirianni, L. Morpurgo, and A Finazzi-Agro, Copper is not responsible for the visible fluorescence of blue oxidases, FEBS Lett. 163;274 (1983).

5. P.G. Winyard, J. Lunec, S. Brailsford, and D.R. Blake, Action of free radical generating systems upon the biological and immunological properties of ceruloplasmin, Int. J. Biochem. 16:-1273 (1984).

6. J. Lunec, D.R. Blake, S.J. McCleary, S. Brailsford, and P.A. Bacon, Self perpetuating mecahnisms of immunoglobulin G aggregation in rheumatoid inflammation, J. Clin. Invest. 76:2084 (1985).

7. S. Freeman and E. Daniel, A spectral study of human ceruloplasmin, Biochim. Biophys. Acta 534:132 (1978).

8. N. Greenfield and G.D. Fasman, Computed circular dichroism spectra for the evaluation of protein conformation, Biochemistry 8:4108 (1969).

9. M. Noyer and F.W. Putnam, A circular dichroism study of undegraded human ceruloplasmin, Biochemistry 20:3536 (1981).

10. P.M. Colman, H.C. Freeman, J.M. Guss, M. Murata, V.A. Norris, J.A.M. Ranshaw, and M.P. Venkatappa, X-ray crystal structure analysis of plastocyanin at 2.7 Å resolution, Nature 272:319 (1978).

11. J.E. Draheim, G.P. Anderson, J.W. Duane, and E.L. Gross, Plastocyanin conformation. An analysis of its near ultraviolet absorption and circular dichroic spectra. Biophys. J. 49:891 (1986).

THE OXIDATIVE MODIFICATION OF LENS PROTEINS

Donita Garland, Paul Russell, and J. Samuel Zigler, Jr.

National Eye Institute
National Institutes of Health
Bethesda, Maryland 20892, U.S.A.

INTRODUCTION

Many post-translational modifications of lens proteins occur during aging.[1] Since the lens continues to grow throughout life and its cells and proteins are not turned over, these modifications accumulate in the older fiber cells.

Oxidative processes have been shown to generate many of the types of changes observed in the lens.[2,3] To further substantiate the idea that these changes are the result of oxidative processes, we have investigated the effect of treating lens crystallins with mixed-function oxidation systems _in vitro_. In this report we will summarize the types of modifications observed.

CARBONYL CONTENT OF HUMAN LENSES

One consequence of the mixed-function oxidation of proteins is the introduction of carbonyl derivatives.[4] The level of carbonyl derivatives in proteins or tissues can be used as a measure of oxidation and can be determined by quantitating the stable hydrazone derivatives formed by reaction with 2,4-dinitrophenylhydrazine.[5]

The level of carbonyl derivatives was measured in apparently normal and cataractous human lenses. For noncataractous lenses, a small but statistically significant increase in the content of carbonyl derivatives was found as a function of age. In over half of the nonbrunescent, cataractous lenses tested, the carbonyl content was higher than the levels found in age-matched control lenses. In brunescent lenses, the carbonyl content in the nuclear region of the lens was consistently several times higher than the age-matched control lenses. For some of these lenses a significant fraction of the hydrazone derivatives was insoluble in 6 M guanidine-HCl. These hydrazone derivatives could be solubilized by pepsin, indicating that carbonyl derivatives were associated with crosslinked protein. These results demonstrate the accumulation of oxidized protein as a function of age and the increased levels of carbonyl derivatives in certain cataractous lenses.

MIXED-FUNCTION OXIDATION OF BOVINE CRYSTALLINS

The oxidation of lens crystallins was further studied by treatment in vitro with a nonenzymic mixed-function oxidation system consisting of ascorbate, metal ion, and atmospheric oxygen. This system was chosen because ascorbate concentration in the lens is approximately millimolar. The oxidation of ascorbate, catalyzed by cupric or ferric ions, produces hydrogen peroxide and reduced metal ion. These react via the Fenton reaction, generating activated oxygen species which have been shown to inactivate enzymes,[2,4,6] induce the site-specific cleavage of proteins,[3] and increase the susceptibility of proteins to proteolysis.[7]

Bovine lens crystallins (α, β, and a low-molecular-weight fraction, LMW, containing γ and βs) were incubated with ascorbate and ferric chloride in the presence of atmospheric oxygen and examined by polyacrylamide electrophoresis and isoelectric focusing.[8] Formation of nondisulfide crosslinks in α and β crystallin fractions (Figure 1) is indicated by a protein band migrating at the position expected for a dimer of crystallin as well as protein migrating at higher molecular weights. These protein crosslinks are not eliminated by heating in the presence of reducing agents and are thus considered to be nondisulfide bonds. In all three fractions there is some nonspecific degradation of the crystallins. In all of the fractions there was a loss of the more basic species and a conversion to new, more acidic species (Figure 1). In the β_h fraction-treatment with ascorbate and ferric ion induced the appearance of non-tryptophan fluorescence. With excitation at 320 nm, fluorescence emission was observed between 400 and 500 nm. The extent of these modifications was dependent on the concentration of ascorbate and metal ion and the time of incubation. Each of these changes has been shown to occur in vivo in mammalian lenses during aging and cataractogenesis.

The incubation of lens homogenates for 10 days under sterile conditions in the presence of ascorbate, ferric ion, and oxygen induced many of the characteristics found in brunescent lenses. The protein became brownish and insoluble. Some of the protein could not be dissociated by denaturants even in the presence of reducing agents indicating the forma-

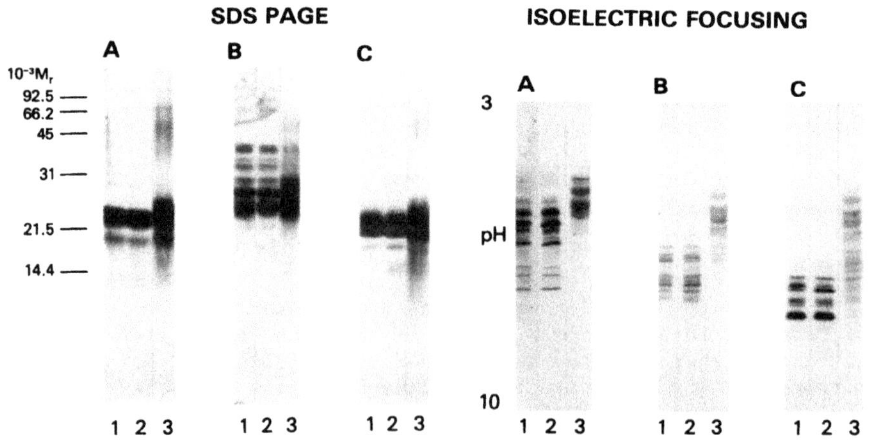

Figure 1. Sodium dodecyl sulfate polyacrylamide gel electrophoresis (PAGE) and isoelectric focusing (IEF) patterns of oxidatively modified (A) α, (B) β, and (C) LMW crystallin fractions. Lane 1, no incubation and no additions; lane 2, no additions; lane 3, 20 mM ascorbate and 100 μM $FeCl_3$.

tion of non-disulfide crosslinks. The level of protein carbonyl derivatives was also increased. There was, however, no direct correlation between carbonyl content and the insolubilization of the protein. At 20 mM ascorbate and trace amounts of iron (none added), the carbonyl level was 30 neq/278 nm O.D. and the protein was soluble. However, at 10 mM ascorbate and 50 μM $FeCl_3$, carbonyl content was 31 neq/278 nm O.D. but 35% of the hydrazone derivatives was associated with 6 M guanidine-HCl insoluble protein. These results suggest two possible mechanisms for the increased protein carbonyl levels upon incubation with ascorbate and ferric ion. One is the oxidation of amino acids in the protein and the second is the association of ascorbate or ascorbate degradation products with the protein, possibly through Schiff bond formation. The latter mechanism is supported by observation of Bensch et al.[9] that radiolabeled ascorbate binds to α- and γ-crystallins.

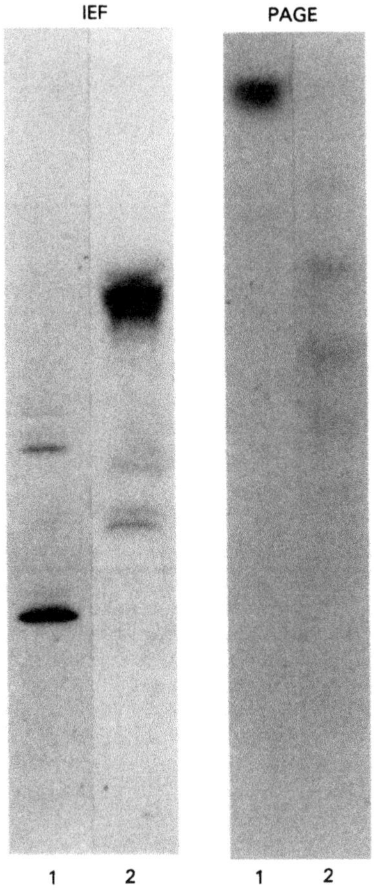

Figure 2. Human 21 kDa γ-crystallin from (lane 1) a 4-month-old lens and (lane 2) a 65-year-old lens. IEF was run with pH range 3.5-9.5 for 1.5 hrs. The anode was at the top of the gel. PAGE was performed with the Laemmli method but without SDS in the buffers. (Reprinted with permission from The FASEB Journal.)

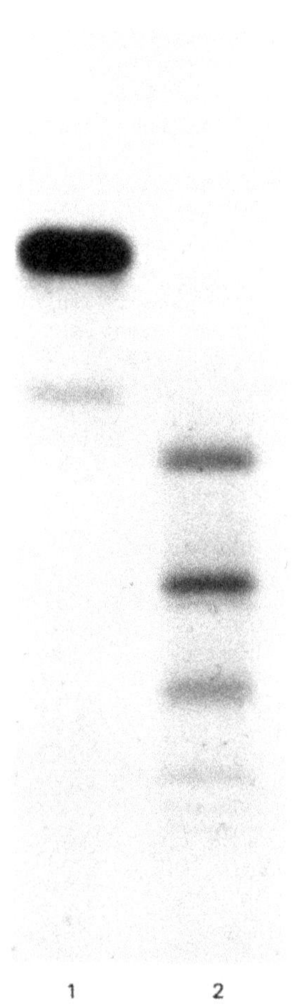

Figure 3. PAGE of the 21 kDa γ-crystallin obtained from mouse L cells and treated with H_2O_2 and ferrous ion. Electrophoresis was performed using the Pharmacia Phast native gel system. Lane 1, no additions; lane 2, H_2O_2 and $FeCl_2$.

HUMAN GAMMA CRYSTALLIN

With increasing age each human γ-crystallin shows an increase in the number of species when electrophoresed under non-denaturing conditions.[10] Figure 2 shows electrophoresis and isoelectrofocusing profiles of the human 21 kDa γ-crystallin proteins isolated from lenses of 4-month- and 65-year-old individuals. The additional more acidic bands in the older sample represent alterations in the charge of the protein since there is no decrease in the molecular weight (data not shown). Unmodified human 21 kDa γ-crystallin was obtained from mouse L cells that had been transfected with the gene for this protein coupled to the metallothionein IIa promoter.[11] After isolation this protein was treated with ascorbate and ferric ion in the presence of oxygen or with hydrogen peroxide and ferrous ion. In Figure 3 are shown the results obtained using hydrogen peroxide and ferrous ion. The new bands observed on electrophoresis represent charge alterations since the molecular weight, as determined by electrophoresis under denaturing conditions, did not decrease (data not shown). Initially upon oxidation, there appeared a single new, more acidic band. These two major bands co-migrated on isoelectric focusing with the two forms of the 21 kDa protein from 4-month-old lenses.[12] With higher reactant concentrations or longer incubation times, additional bands appeared. These bands co-migrated on native polyacrylamide electrophoresis with the bands found in 15-year-old lenses. The data suggest that specific sites are particularly susceptible to alterations in vivo and in vitro. It is not known yet whether the same amino acids are modified in vivo and in vitro.

SUMMARY

In summary, in vitro oxidation of lens crystallins mimics many of the post-translational modifications observed with age and in cataracts (Table 1). These results lend further support to earlier proposals that oxidation is a key factor in cataract formation. The extent to which ascorbate contributes to the oxidation reactions in vivo is not known. In addition to the data presented here that ascorbate can produce these effects in vitro, other observations support the possibility that under certain conditions ascorbate may be involved in the generation of reactive oxygen species in the lens. Using electron spin resonance the ascorbyl radical can be detected early in nuclear cataract formation, and there appears to be a decrease in total ascorbic acid (reduced and oxidized), suggesting further oxidation of ascorbate.[13,14] Iron and copper are both present in mammalian lenses, and there are reports of increased copper levels in the lens with age and in cataracts.[15,16] Increased metal ion concentrations would facilitate these oxidative processes.

Table 1. Summary of Changes Induced in Lens Crystallins by Treatment with Ascorbate, Fe(III) and Oxygen

Non-disulfide covalent crosslinks
Partial degradation
Loss of more basic protein species and increase in more acidic species
Formation of nontryptophan fluorescence
Introduction of carbonyl derivatives
Browning of proteins[a]
Insolubilization of proteins[a]

[a]Incubation for 10 days at 37°C under sterile conditions.

REFERENCES

1. H. J. Hoenders and H. Bloemendal, Aging of Lens Proteins, in: "Molecular and Cellular Biology of the Eye Lens," H. Bloemendal, ed., John Wiley & Sons, Inc., New York (1981).
2. L. Fucci, C. N. Oliver, M. J. Coon, and E. R. Stadtman, Inactivation of key metabolic enzymes by mixed-function oxidation reactions: Possible implication in protein turnover and aging, Proc. Natl. Acad. Sci. U.S.A. 80:1521 (1983).
3. K.-W. Kim, S. G. Rhee, and E. R. Stadtman, Nonenzymatic cleavage of proteins by reactive oxygen species generated by dithiothreitol and iron, J. Biol. Chem. 260:15394 (1985).
4. R. L. Levine, Oxidative modification of glutamine synthetase, J. Biol. Chem. 258:11823 (1983).
5. R. L. Levine, Mixed-function oxidation of histidine residues, in: "Methods of Enzymology," F. Wold and K. Moldave, eds., Academic Press, New York (1984).

6. B. Chance, The reactions of catalase in the presence of the notation system, Biochem. J. 46:387 (1950).
7. A. J. Rivett, Preferential degradation of the oxidatively modified form of glutamine synthetase by intracellular proteases, J. Biol. Chem. 260:300 (1985).
8. D. L. Garland, J. S. Zigler, Jr., and J. Kinoshita, Structural changes in bovine lens crystallins induced by ascorbate, metal, and oxygen, Arch. Biochem. Biophys. 251:771 (1986).
9. K. G. Bensch, J. E. Fleming, and W. Lohmann, The role of ascorbic acid in senile cataract, Proc. Natl. Acad. Sci. U.S.A. 82:7193 (1985).
10. J. S. Zigler, Jr., P. Russell, L. J. Takemoto, S. J. Schwab, J. S. Hansen, J. Horwitz, and J. H. Kinoshita, Partial characterization of three distinct populations of human γ crystallins, Invest. Ophthamol. Vis. Sci. 26:525 (1985).
11. S. O. Meakin, M. L. Breitman, and L.-C. Tsui, Structural and evolutionary relationships among five members of the human - crystallin gene family, Mol. Cell. Biol. 5:1408 (1985).
12. P. Russell, D. L. Garland, J. S. Zigler, Jr., S. O. Meakin, L.-C. Tsui, and M. L. Breitman, Aging effects of vitamin C on a human lens protein produced in vitro, FASEB Journal 1:32 (1987).
13. W. Lohmann, W. Schmehl, and J. Strobel, Nuclear cataract: Oxidative damage to the lens, Exp. Eye Res. 43:859-862 (1986).
14. J. Bellows, Biochemistry of lens: Some studies in vitamin C and the lens, Arch. Ophth. 16:58 (1936).
15. M. Testa, S. Daniels, and M. Bocci, The reactivity of the serum protein-SH groups in the senile cataract, Ophthalmologica 147:485 (1964).
16. R. Nath, S. K. Srivastava, and K. Singh, Accumulation of Cu and inhibition of LDH activity in human senile cataractous lens, Indian J. Exp. Biol. 7:25 (1969).

DEGRADATION OF THE PROTEOGLYCANS OF HUMAN ARTICULAR CARTILAGE BY REACTIVE OXYGEN METABOLITES

Clive R. Roberts, John S. Mort, and Peter J. Roughley

Joint Diseases Laboratory
Shriners Hospital
McGill University
1529 Cedar Avenue
Montreal, Quebec, Canada

INTRODUCTION

The ability of articular cartilage to reversibly resist compression depends on the integrity of the proteoglycan aggregates, which are major components of the tissue. Each proteoglycan aggregate may have a total molecular weight of 200 x 10^6 and consists of a hyaluronic acid filament to which are attached up to 100 proteoglycan monomers in non-covalent interactions which are stabilized by the link glycoproteins. The proteoglycan monomer consists of a core protein of M_r 320,000, whose N-terminal region interacts with the link proteins and hyaluronic acid. To this core protein 130-200 anionic glycosaminoglycan chains are covalently bound. This high density of anionic groups draws water into the tissue until the osmotic swelling pressure is balanced by tension in the inelastic collagen network in which the proteoglycans are entrapped. The retention of intact proteoglycans in the tissue is essential for the high water content of the tissue, and it is the reversible redistribution of this bound water, under mechanical loading, which is responsible for the shock-absorbing properties of the tissue[1].

A number of changes in the structure of articular cartilage proteooglycans have been described in humans; the changes occur with increasing age, and some may be analogous to the degenerative changes which occur in the early stages of osteoarthritis and other joint diseases. These changes include a decrease in the size of the hyaluronic acid molecule, loss of proteoglycan from the tissue, fragmentation of the monomers[2,3] and fragmentation of the link proteins[4]. It is possible to reproduce some of these modifications by treatment of proteoglycans *in vitro* with proteinases derived from cartilage, but some of the changes in the link proteins have not been so reproduced[5]. In order to investigate the possibility that some of the age-related changes in these molecules result from the action of reactive oxygen metabolites, we have studied the effects of hydrogen peroxide on neonatal human articular cartilage proteoglycan aggregates and rat chondrosarcoma proteoglycan aggregates.

METHODS

Articular cartilage was obtained at autopsy from neonatal humans and the proteoglycan aggregates extracted using 4M guanidinium chloride, dialyzed to associative conditions and purified by cesium chloride density gradient centrifugation. Proteoglycans were isolated from Swarm rat chondrosarcoma tissue by the same method. The aggregates (3 mg/ml in 25 mM sodium acetate/acetic acid buffer containing 80 mM NaCl, pH 5.6) were incubated with 0-140 mM hydrogen peroxide for up to 24 hours. Modifications to the proteoglycan aggregates and subunits were studied using Sepharose CL-2B gel chromatography and agarose-polyacrylamide gel electrophoresis. The ability of proteoglycan subunits to interact with hyaluronic acid was studied by gel chromatography in presence or absence of added hyaluronic acid after repurification of treated subunits. Changes in the integrity of the proteins were investigated by SDS-polyacrylamide gel electrophoresis followed by transfer to nitrocellulose and blotting with monoclonal antibodies 9/30/8-A-4, which recognizes two identical epitopes in the link glycoprotein, and 12/20/2-A-5, which recognizes an epitope in the hyaluronic acid-binding region of the proteoglycan core protein.

RESULTS

Treament of proteoglycan aggregates with hydrogen peroxide resulted in a dose-dependent depolymerization of the aggregate as measured by a reduction in viscosity. In addition, the proteoglycan monomers lost their ability to associate with hyaluronic acid in a dose-dependent manner, despite minimal changes in their hydrodynamic size. Modification of the hyaluronic acid-binding region of the core protein could not be ascribed to proteolysis in this region, since treatment with hydrogen peroxide neither generated fragments of core protein nor altered the size of the 60-70 kDa fragment, which subsequent trypsin treatment released from the core protein, and which was recognized by the monoclonal antibody to the hyaluronic acid-binding region. Even at the highest doses of hydrogen peroxide used, there were no changes in the size of the glycosaminoglycan chains after hydrogen peroxide treatment unless transition metal ions were added to the preparation.

In contrast, cleavage of the link proteins did occur. The order in which the major link protein fragments were generated and their electrophoretic mobilities resembled the pattern observed during human aging.[6] The fragments produced from human and rat link proteins by this treatment were of very similar electrophoretic mobilities. The proteolytic changes in the link proteins were dependent on the presence of endogenous trace levels of transition metal ions; the effect was inhibited by the addition of DTPA to 10 μM or more or by the addition of EDTA. The presence of thiourea or tetramethyl urea, but not of urea or sodium formate at 200 mM or more, was also inhibitory. The specific nature of the protein cleavages was lost in the presence of added copper sulphate (0.1-1 mM), suggesting that the sites of chelation by the protein of transition metal ions determine the site of generation of a more reactive molecular species than hydrogen peroxide, which may be the hydroxyl radical, and whose generation results in proximal cleavage of the protein. Separation of the components of hydrogen peroxide-treated proteoglycan aggregate by cesium chloride density gradient centrifugation under associative conditions at starting densities of 1.45 and 1.69 g/cc revealed that fragmented link proteins remained bound to hyaluronic acid filaments after treatment with hydrogen peroxide and retained their ability to reassociate with hyaluronic acid. The elution profile of these hyaluronic acid-bound, fragmented link proteins on

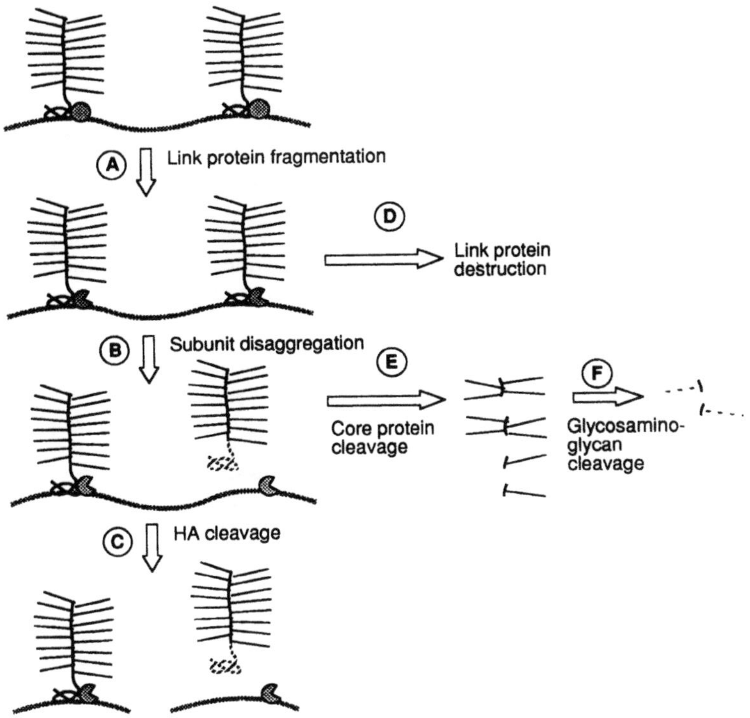

Figure 1. Schematic representation of the action of hydrogen peroxide on cartilage proteoglycan aggregates.

Sepharose CL-2B chromatography demonstrated that hyaluronic acid was cleaved by treatment with high doses of hydrogen peroxide.

We therefore present a scheme (Figure 1) which illustrates the degradation of proteoglycan aggregates by reactive oxygen metabolites generated from hydrogen peroxide; at lower hydrogen peroxide doses (typically 10-30 mM), link glycoproteins are fragmented but remain bound to hyaluronic acid (A) and the hyaluronic acid-binding region of the proteoglycan core protein is modified such that it is no longer functional (B). Hyaluronic acid is fragmented (C). Further cleavage occurs after longer periods of exposure or with higher doses of hydrogen peroxide, with link proteins being destroyed (D) and cleavage of the proteoglycan core protein becoming evident (E). Only after treatment with very high doses of hydrogen peroxide (250 mM or more) or in the presence of added transition metal ions are the glycosaminoglycan chains cleaved (F).

DISCUSSION

The most prevalent human arthritic diseases are strongly age-related, and it is possible that changes in the structure of cartilage with age contribute to the onset of these diseases. The depolymerization of proteoglycan aggregates by hydrogen peroxide, which we have described, is a potential mechanism by which the functional properties of human articular cartilage proteoglycans may change with increasing age; these structural molecules are long-lived and would be expected to accumulate chelated transition metal ions if these were to become available in the joint. It has indeed been shown that transition metal ions which are not protein-bound are released into the synovial fluid during periods of joint inflammation[7], and hydrogen peroxide released by infiltrating phagocytes in addition to that which may be generated as a result of the chondrocytes' oxidative metabolism may participate in the generation of more reactive metabolites of oxygen which damage structural macromolecules. It is noteworthy that synovial fluid from patients with arthritic disease has been whown to have a poor capacity for the detoxification of reactive oxygen metabolites[8]. The mechanism for radical-mediated proteoglycan degradation therefore exists in vivo and is a potentially powerful process by which loss of tissue function may occur as a result of acute inflammatory events, in addition to the more gradual changes which occur with increasing age. The fact that severe cartilage destruction is a well-known consequence of transition metal ion deposition in the joints, for instance in haemophilia, haemochromatosis and Wilson's disease, is consistent with this thesis. In addition, many forms of arthritic disease are autoimmune phenomena, and it is possible that damage to cartilage macromolecules which is mediated by reactive oxygen metabolites may contribute to the development of disease by changing the antigenicity of these structures.

We suggest that some of the age-related changes in cartilage proteoglycan structure in humans, which may be associated with loss of tissue function and predisposition to arthritic disease, are attributable to modification of these molecules by reactive oxygen metabolites.

REFERENCES

1. T. E. Hardingham, M. Beardmore-Gray, D. G. Dunham, and A. Ratcliffe, Cartilage proteoglycans, Ciba Found. Symp. 124:30 (1986).
2. M. T. Bayliss and S. Y. Ali, Age-related changes in the composition and structure of human articular cartilage proteoglycans, Biochem. J. 176:683 (1978).
3. P. J. Roughley and R. J. White, Age-related changes in the struture of of the proteoglycan subunits from human articular cartilage, J. Biol. Chem. 266:217 (1980).
4. J. S. Mort, A. R. Poole, and P. J. Roughley, Age-related changes in the structure of proteoglycan link proteins present in normal human cartilage, Biochem. J. 214:269 (1983).
5. I. K. Campbell, P. J. Roughley, and J. S. Mort, The action of human articular-cartilage metalloproteinase on proteoglycan and link protein, Biochem. J. 237:117 (1986).
6. C. R. Roberts, J. S. Mort and P. J. Roughley, Treatment of cartilage proteoglycan aggregate with hydrogen peroxide: relationship between observed degradation products and those that occur naturally during aging, Biochem. J. 247:349 (1987).
7. J. M. C. Gutteridge, D. A. Rowley, and B. Halliwell, Superoxide-dependent formation of hydroxyl radicals and lipid peroxidation in the presence of iron salts, Biochem. J. 206:605 (1982).
8. D. R. Blake, N.D. Hall, D. A. Treby, B. Halliwell, and J. M. C. Gutteridge, Protection against superoxide and hydrogen peroxide in synovial fluid from rheumatoid patients, Clin. Sci. 61:483 (1981).

FREE RADICAL DENATURATION OF IMMUNOGLOBULIN G IS DUE TO AMINO ACID OXIDATION: IMPLICATIONS FOR RHEUMATOID ARTHRITIS

H.R. Griffiths, J. Lunec, D.R. Blake, and R.L. Willson*

Departments of Rheumatology and Biochemistry
Selly Oak Hospital, Birmingham B29 6JD, U.K.

*Department of Biochemistry
Brunel University, Uxbridge, London, U.K.

INTRODUCTION

Immunoglobulin G (IgG) is a major serum glycoprotein, induced by the presence of foreign proteins. However, in rheumatoid arthritis, it is believed to be chemically altered so that IgG itself is no longer recognised as "self," and an immune response is generated to this apparently foreign protein.[1] This is thought to promote the chronicity of the disease.

It has previously been suggested that oxygen free radicals (ROS) released from activated inflammatory cells, neutrophils, may induce chemical modifications in IgG.[2] Using neutrophils isolated from peripheral blood in the presence of free radical scavengers, •OH has been implicated as the cause of IgG denaturation; denaturation being characterized by protein aggregation (partially mediated via rearrangement of disulphide crosslinks), and a characteristic visible fluorescence (Ex 360 nm, Em 454 nm). It is believed that the fluorescence is, at least in part, a property of oxidised tryptophan metabolites such as kynurenine and N-formyl kynurenine.[3]

In an attempt to define chemical changes associated with denaturation of IgG in vitro and in vivo, we have developed a reverse phase HPLC technique to study oxidised amino acids from protein hydrolysates. Here we describe (a) that the amino acid changes that occur within IgG are caused by chemically generated ROS and (b) that the fluorescence observed in IgG from rheumatoid sera and synovial fluids is due to specific ROS-induced changes.

MATERIALS AND METHODS

Materials

γ-Globulins and pronase were obtained through Sigma. HPLC grade acetonitrile was from M & B. All other reagents were of analytical grade from BDH. Clinical specimens were obtained from patients attending rheumatology clinics at Selly Oak Hospital, Birmingham.

γ-Radiolysis

Aerated solutions of IgG in 40 mM phosphate buffer were exposed to 0, 50 and 100 krads from a cobalt 60 gamma source. 20 mM phenylalanine was added to one incubation to generate a secondary peroxy radical species,[4] whilst 0.2 M formate was added to produce superoxide in solution.

Protein Isolation

IgG was isolated from serum by ammonium sulphate precipitation and purified from any molecular weight contaminants by gel permeation chromatography.

Protein Hydrolysis

Proteins (0.5 mg/ml) in PBS were digested using pronase at 37°C for 40 hours to ensure complete digestion.

Protein hydrolysates were applied to a C18 5 μm spherisorb column and amino acids were eluted at a flow rate of 0.75 ml/min using a gradient of increasing acetonitrile concentration in 0.067 M phosphate buffer, pH 5. Acetonitrile increased by 0.6% per minute over fifteen minutes. The eluent was monitored by absorbance at 240 nm and simultaneous fluorescence measurement (Ex 360 nm, Em 454 nm).

RESULTS

Amino acid chromatography of normal IgG shows that the predominant UV absorbing amino acids in IgG are cysteine, methionine, cystine, phenylalanine, tyrosine and tryptophan. Figure 1 shows a standard chromatogram of these amino acids. The radiolysis of IgG in solution induces chemical modifications to tryptophan and cystine residues. •OH was found to induce fluorescent kynurenine formation and partial oxidation of cystine. The phenylalanine peroxy radical was also found to induce the formation of fluorescent kynurenines and also some fluorescence associated with cystine

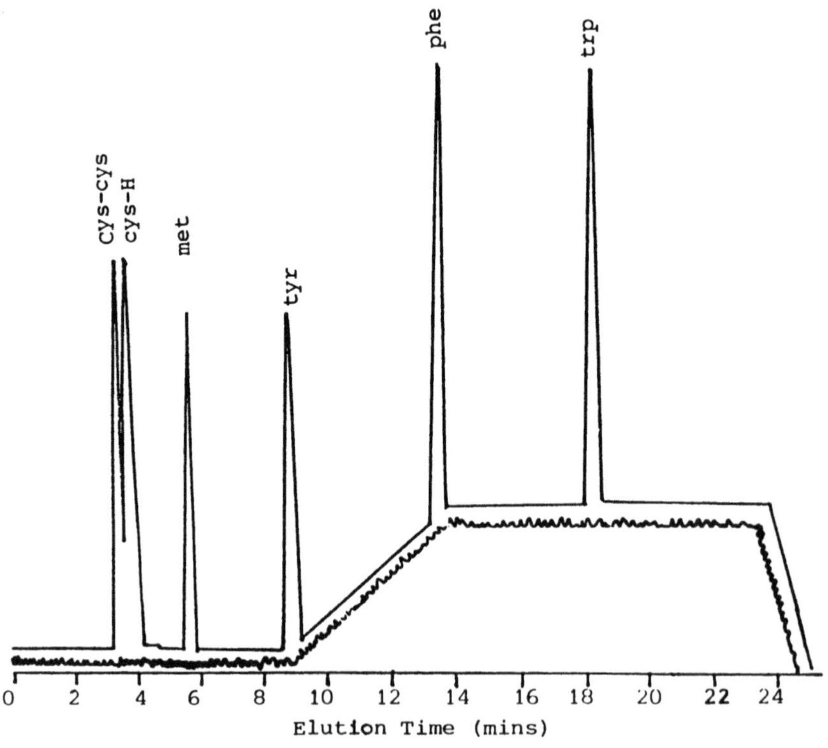

Figure 1. Elution profile of standard amino acids monitoring UV absorbance at 240 nm (-) and visible fluorescence (~~).

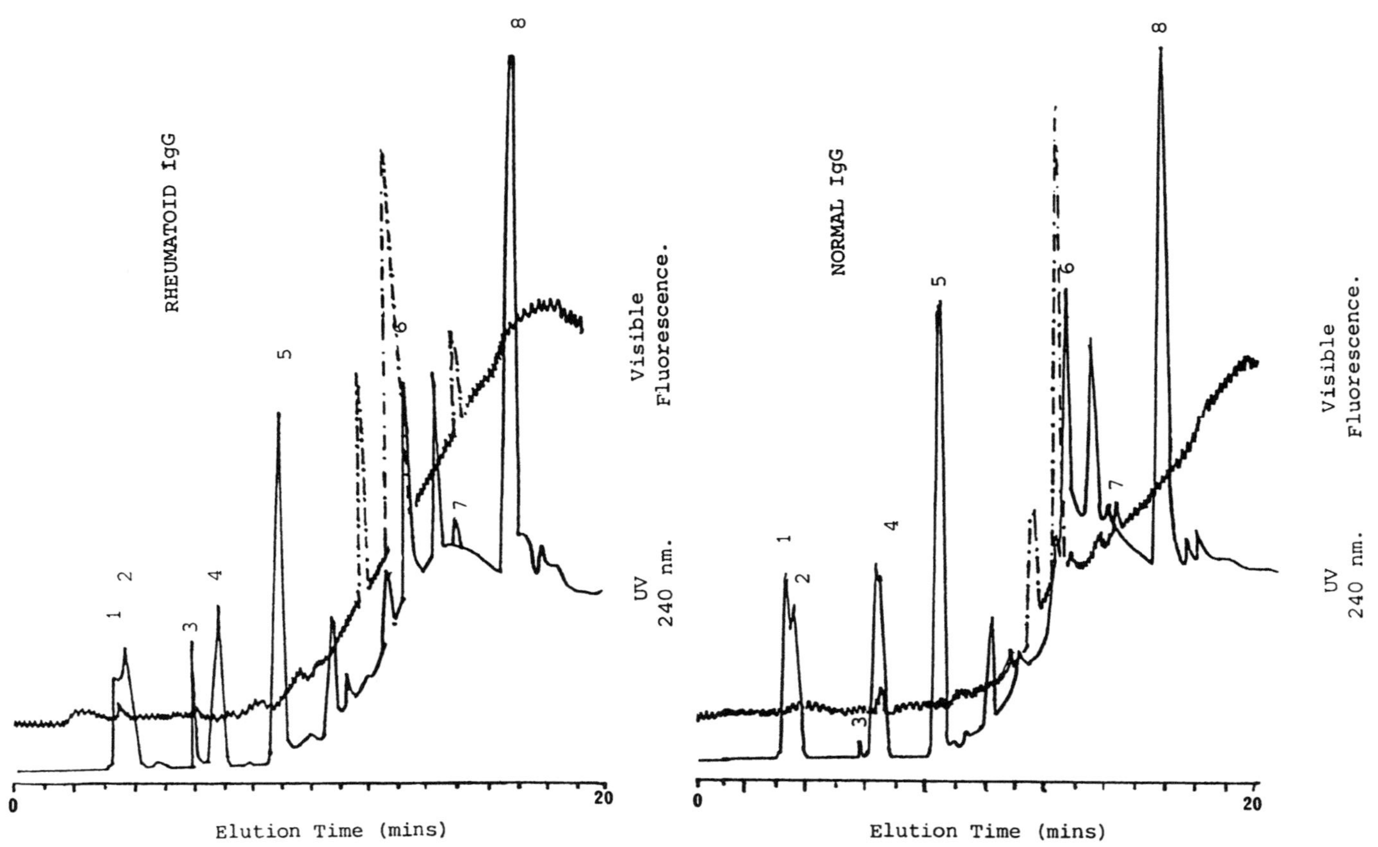

Figure 2. HPLC chromatograms of IgG hydrolysates. Peak identies: 1 cys, 2 cys-H, 3 cysteic acid, 4 met, 5 tyr, 6 phe, 7 N-formyl kynurenine, 8 trp.

metabolites, which included cysteic acid. This was demonstrated by comparison of irradiated cystine products with irradiated protein in terms of their respective absorption spectra and retention times on the Spherisorb reverse-phase column. Superoxide did not induce any changes in fluorescence in the overall protein nor its constituent amino acids. There was however, some disulphide bond splitting, and the formation of cysteic acid.

Figure 2 shows the chromatograms of a serum IgG isolate from a rheumatoid patient compared to that of a normal control IgG. Normal IgG was found to possess fluorescent tryptophan and cystine metabolites. Rheumatoid IgG, however, contained higher levels of the fluorescent tryptophan metabolites kynurenine and N-formyl kynurenine.

DISCUSSION

Much of the literature concerning ROS and proteins has focussed on enzymes which lose their catalytic activity as essential residues are oxidised (reviewed in reference 5). In such studies the sulphydryl and aromatic amino acids prove to be the most susceptible. Within IgG, the cystine disulphide bridges are vital interchain crosslinkers, maintaining the conformation of the four component chains relative to each other. We have shown that tryptophan and cystine are partially oxidised by the attack of radiolytically generated •OH and ROO•, respectively. This oxidation may chemically alter the IgG molecule to induce antigenic changes (i.e., it will be recognised by the immune system as a foreign protein).

IgG from clinical specimens does not contain the fluorescence characteristic of peroxy radical attack, but kynurenines are present. The levels of these oxidised amino acids are higher in rheumatoid patients than in controls. ROS denatured IgG therefore seems to be a more physiologically relevant model for studying IgG as an antigen in rheumatoid arthritis than the established model of heat aggregation (63°C, 20 min.).

From this study, we have provided further evidence to support the role of •OH in inducing chemical changes in IgG in vivo. This knowledge may be useful for the selective use of •OH scavengers therapeutically to reduce ROS damage to IgG in rheumatoid disease.

ACKNOWLEDGMENTS

H.G. and J.L. acknowledge financial support from the Arthritis and Rheumatism Council.

REFERENCES

1. P.M. Johnson and W. Page-Faulk, Rheumatoid factor; Its nature specificity and production in rheumatoid arthritis, Clin. Immunol. and Immunopathol 6:414 (1976).
2. J. Lunec, D.R. Blake, S.J. McCleary, S. Brailsford, and P.A. Bacon, Self-perpetuating mechanisms of Immunoglobulin G aggregation in rheumatoid inflammation, J. Clin. Invest. 76:2084 (1985).
3. M. Sun and S. Zigman, Isolation and identification of tryptophan photoproducts formed from aqueous solutions of tryptophan exposed to near-Q light, Photochem. Photobiol. 29:893 (1979).
4. C.A. Gee, K.J. Kittridge, and R.L. Willson, Peroxy free radicals, enzymes and radiation damage: sensitisation by oxygen and protection by superoxide dismutase and antioxidants, Brit. J. Radiol. 58:251 (1985).
5. S.P. Wolff, A. Garner, and R.T. Dean, Free radicals, lipids and protein degradation, Trends in Biochem. Sci. 11:27 (1986).

A STUDY OF ROS INDUCED DENATURATION OF IgG3 USING MONOCLONAL ANTIBODIES; IMPLICATIONS FOR INFLAMMATORY JOINT DISEASE

H.R. Griffiths, J. Lunec, R. Jefferis*, D.R. Blake, and R.L. Willson**

Departments of Biochemistry and Rheumatology
Shelly Oak Hospital, Birmingham B29 6JD, U.K.

*Department of Immunology
University of Birmingham, Birmingham B15 2TJ, U.K.

**Department of Biochemistry
Brunel University, London, U.K.

INTRODUCTION

Inflammatory joint disease is characterised by an infiltration of phagocytes into articular spaces. When these cells are activated by immune complexes, bacteria or components of the complement cascade, the release proteases and the reactive oxygen species (ROS) $\cdot O_2^-$ and H_2O_2.[1] Whilst the stimulus for phagocytic infiltration and activation remains unknown, it has been shown that the ROS released have the capacity to further damage cartilage,[2] synovial fluid[3] and IgG.[4] The latter is of particular interest since it is believed that production of autoantibodies to a conformationally altered IgG molecule may perpetuate the inflammatory response and promote disease chronicity.[5]

Recently, it has been shown that whilst total IgG is elevated in the synovial fluid, the IgG3 subclass is most significantly elevated.[6] Therefore in this study, we have used a panel of monoclonal antibodies directed against specific epitopes (antigenic sites) within the hinge and Fc portion of the IgG3 molecule in an attempt to demonstrate (a) that ROS can conformationally alter immunoglobulin, and (b) which ROS is responsible for the damage. We propose that ROS reactions can expose new antigenic determinants which may be involved in the autoantigenicity of IgG.

MATERIALS AND METHODS

IgG3 Denaturation

Buffered aerated solutions of IgG3 (0.5 mg/ml) were exposed to 100 krad of continuous irradiation from a cobalt 60 gamma source in the presence or absence of 0.2 M formate.[7] IgG3 was also denaturated by heating at 63°C for 20 min.

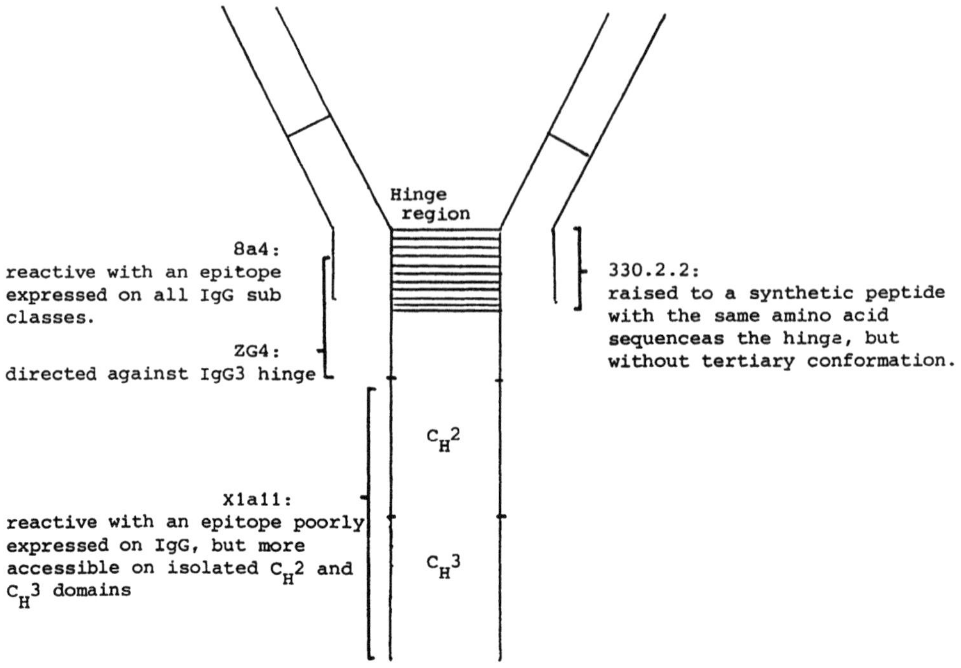

Figure 1. A schematic diagram of the IgG3 molecule which illustrates the sites on the molecule recognised by the McAbs (8a4, ZG4, X1a11 and 330-2.2) used in this study.

Gel Filtration HPLC

Denaturated proteins were applied to a TSK 3000SW column and eluted with 0.067 M phosphate buffer as described elsewhere.[4] The eluent was monitored simultaneously for protein absorbance (280 nm) and fluorescence (Ex 360 nm, Em 454 nm).

Haemagglutination Inhibition (HAI)

Sheep red blood cells were passively sensitised with IgG3 using chromium chloride. The agglutination titre was determined for each McAb by serial twofold dilutions of ascitic fluid. For inhibition assays each McAb was used at a dilution twofold lower than the highest dilution required for agglutination. The inhibitory capacity of the denatured IgG3 was determined following serial twofold dilutions from 0.1 mg/ml incubation with the McAb, followed by addition of the sensitised cells.

RESULTS

Aggregation and Fluorescence

HPLC analysis showed that more than 90% of the IgG3 was present as the monomer. Following heating at 63°C, 20% of IgG3 was aggregated without evidence of autofluorescence. However, radiolysis induced fluorescence and 55% aggregation. In the presence of formate, superoxide is the predominant species generated in aerated solutions; under these conditions no autofluorescence or aggregation was observed.

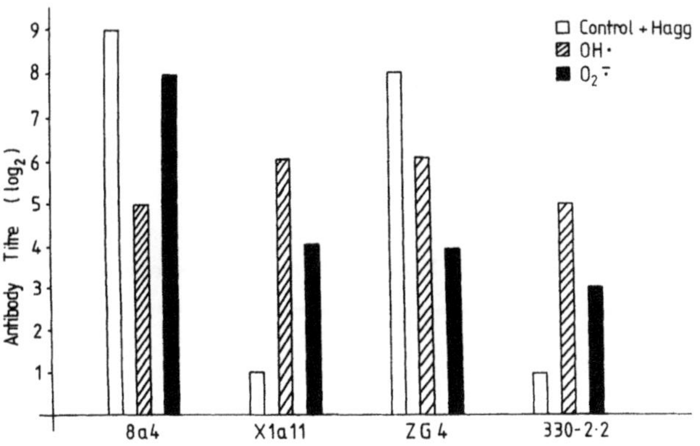

Figure 2. Change in reactivity of McAbs with IgG3 after exposure to ROS. A decrease in observed titre indicates a decrease in epitope expression. Correspondingly, an increase in titre indicates that epitopes are becoming more accessible to the McAb.

Alteration of Epitope Expression

Heat aggregation of IgG3 caused no loss of epitopes (Figure 2). Aggregation induced by radiolysis altered the overall conformation of the molecule (as measured by the McAb, 8a4, in the HAI). Associated with this was the partial destruction of the hinge (as measured by ZG4 reactivity) and increased exposure of the Fc (X1a11). The superoxide ion, on the other hand, had a greater effect on the denaturation of the hinge region, with little effect on the Fc. The use of 330-2.2, a McAb directed against the amino acid sequence of the hinge, has demonstrated that $\cdot OH/\cdot O_2^{-}$ damage to the molecule causes a loss of tertiary conformation.

DISCUSSION

Within normal serum, IgG3 represents only 7% of total IgG, but in the inflamed joint this may rise to 27%. IgG3 apparently mediates most effector, protective functions characteristic of the IgG isotype. This is facilitated by its unusual tertiary structure about the hinge region and, therefore, any conformational change at this site might be expected to influence effector functions.

The primary sequence of the hinge is particularly rich in proline and has 11 interchain disulphide bonds (the other subclasses have four at most). Proline-containing and sulphur-containing residues are particularly susceptible to ROS attack, and this study has demonstrated that the hinge region is conformationally altered; $\cdot O_2^{-}$ is more effective than $\cdot OH$, but the hydroxyl radical causes a greater loss of the tertiary structure. Since epitope expression is commonly associated with the generation of new antigenic determinants, we suggest that ROS may induce antigenicity in IgG at inflammatory sites.

Previous work has shown that serum rheumatoid factor does not react with IgG3.[8] However, more recently it has been shown that IgG3 does bind with high avidity to RFs from synovial fluid (SF).[9] It was also proposed by Robbins et al. that rapid clearance of IgG3-RF immune complexes occurred locally, thus no IgG3-reactive RFs could be detected in serum. This study has demonstrated that the elevated levels of IgG3 in SF can fortuitously scavenge ROS, but this causes epitope loss. We are currently investigating whether the corresponding appearance of new antigenic determinants influences RF binding. Any increase in the avidity or affinity of the subsequent immune complexes formed, would therefore provide a mechanism for promoting the disease chronicity.

ACKNOWLEDGMENTS

H.G. and J.L. acknowledge financial support from the Arthritis and Rheumatism Council.

REFERENCES

1. D.R. Blake, J. Lunec, P.G. Winyard, and S. Brailsford, Oxygen free radicals in inflammatory joint disease, in: "Perspectives in Rheumatology," J.L. Decker and J.T. Scott, eds., Current Medical Literature, London (1985).
2. S.P. Wolff, A. Garner, and R.T. Dean, Free radicals, lipids and protein degradation, Trends in Biochem. Sci. 11:27 (1986).
3. J.M. McCord, Free radicals and inflammation: Protection of synovial fluid by superoxide dismutase, Science 18:529 (1974).
4. J. Lunec, D.R. Blake, S.J McCleary, S. Brailsford, and P.A. Bacon, Self-perpetuating mechanisms of immunoglobulin aggregation in rheumatoid inflammation, J. Clin. Invest. 76:2084 (1985).
5. P.M. Johnson, J. Watkins, and E.J. Holborrow, Antiglobulin production to altered immunoglobulin G in rheumatoid arthritis, Lancet i:611 (1975).
6. P.L. Haber, H. Kubagawa, and N.J. Koopman, Immunoglobulin subclass distribution of synovial plasma cells in rheumatoid arthritis determined by monoclonal antisubclass antibodies, Clin. Immunol. Immunopathol. 35:346 (1985).
7. R.L. Willson, Organic free radicals as ultimate agents in oxygen toxicity, in: "Oxidative Stress," Academic Press Inc., London (1985).
8. D.E. Normansell and C.W. Young, The IgG subclass specificity of anti-IgG immunoglobulin rheumatoid factors, Immunochem. 12:187 (1975).
9. D.L. Robbins, J. Skilling, W.F. Benisek, and R. Wistar, Estimate of the relative avidity of 19S rheumatoid factor secreted by rheumatoid cells for human IgG subclasses, Arthr. and Rheum. 29:722 (1986).

OXY RADICAL-INITIATED PROTEIN S-THIOLATION AND ENZYMIC DETHIOLATION

James A. Thomas and Eun-Mi Park

Department of Biochemistry and Biophysics
Iowa State University
Ames, Iowa 50011, U.S.A.

INTRODUCTION

It has often been suggested that protein S-thiolation/dethiolation (Figure 1) is an important form of metabolic regulation [1,2]. This suggestion usually arises from experiments in which a thiol-disulfide exchange mechanism is considered to be the basis for this process. Since cellular glutathione is normally very reduced, the thiol-disulfide exchange mechanism requires a significant oxidation of this thiol to account for appreciable protein modification in vivo. In addition, both S-thiolation and dethiolation by thiol-disulfide exchange probably require catalysis to occur at significant rates in vivo. These considerations make thiol-disulfide exchange less appealing than a direct oxy radical-initiated mechanism of S-thiolation. This paper describes experiments with electrofocusing technology to assess various mechanisms of protein S-thiolation and dethiolation.

METHODOLOGY

Electrofocusing on thin polyacrylamide gels can detect single charge changes in proteins. This technique, combined with the thiol alkylating agents iodoacetic acid and iodoacetamide, has been developed into an assay for S-thiolation of two enzymes, i.e., glycogen phosphorylase and creatine kinase [3]. Figure 2 shows how this technique can be applied to protein sulfhydryl analysis. Creatine kinase has one reactive sulfhydryl per subunit and phosphorylase has two. In S-thiolation experiments three forms of creatine kinase and five forms of phosphorylase are observed.

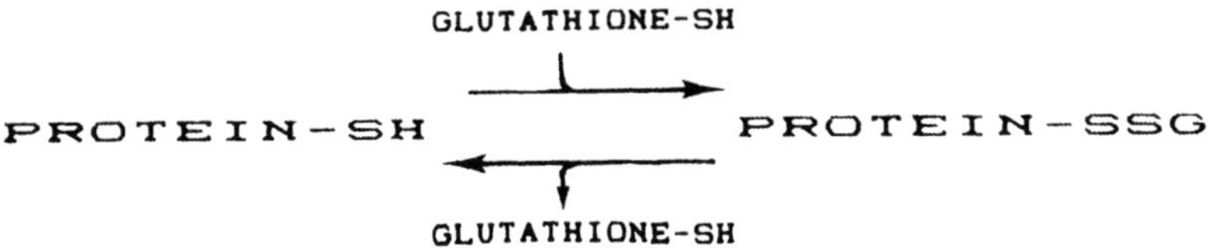

Figure 1. Protein S-thiolation and dethiolation

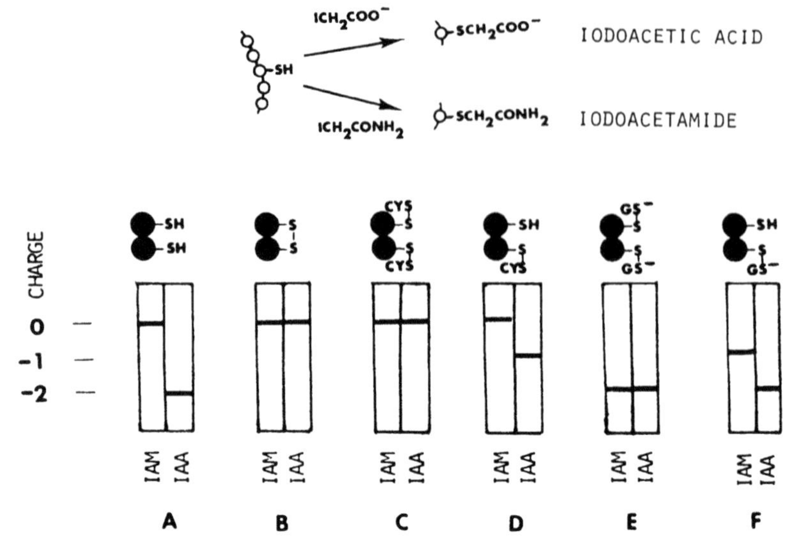

Figure 2. Filled circles show the oxidation status of a protein dimer before alkylating with the agents shown in the top drawing. The electrofocusing analysis is shown with the anode (+) at the bottom. IAA = iodoacetic acid; IAM = iodoacetamide; GS^- = glutathione; CYS = cysteine.

RESULTS AND DISCUSSION

The effect of oxy radicals on glycogen phosphorylase and creatine kinase was assessed in the presence of reduced glutathione to simulate conditions found in a normal cell. Oxy radicals were generated by xanthine oxidase during the entire reaction period. Both S-thiolated proteins and oxidized glutathione were produced, and if reduced glutathione was present at a sufficient concentration, no other protein modifications were evident. Table I shows a comparison of the rate of glycogen phosphorylase S-thiolation and glutathione oxidation. Similar results were obtained with creatine kinase. The oxidized glutathione formed was not sufficient to S-thiolate the proteins by thiol-disulfide exchange. Both proteins could effectively prevent ferricytochrome c reduction in the standard assay for superoxide anion [5]. Thus, these two proteins are extensively S-thiolated without additional modifications under conditions in which oxy radicals are generated by xanthine oxidase. A mechanism of S-thiolation (Figure 3) is suggested in which protein S-thiolation and glutathione oxidation occur independently in the presence of oxy radicals.

The redox state of protein sulfhydryls is a function of both S-thiolation and dethiolation in vivo. Indeed, our experiments with cultured cells [6] showed that dethiolation is a very active process. Figure 4 shows two

Table I Rate of glutathione oxidation and phosphorylase S-thiolation by xanthine oxidase-generated oxy radicals. Glutathione was determined by HPLC [4]. Protein S-thiolation was quantitated by densitometry of gels. Initial reduced glutathione = 100 µM

Minutes	Percent S-thiolation	micromolar oxidized glutathione
10	30	18
20	40	21
40	62	32

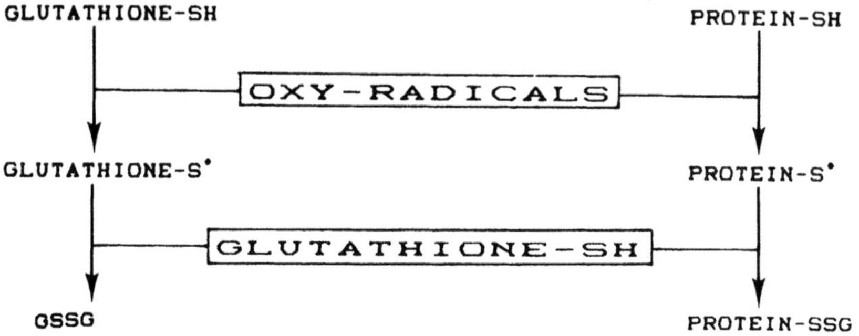

Figure 3. Radical initiated protein S-thiolation and glutathione oxidation.

possible mechanisms of dethiolation [7,8], both requiring enzymic reduction by NADPH. Thiol-disulfide exchange requires glutathione reductase and thioredoxin-exchange requires thioredoxin reductase.

Dethiolation of S-thiolated phosphorylase and creatine kinase was studied by electrofocusing analysis. Significant rates were obtained with reduced glutathione alone, i.e., without enzymic catalysis. Dethiolation of creatine kinase was strongly inhibited by a combination of ATP, creatine, creatine-P, and ADP. The inhibited rate was similar to the uninhibited rate for phosphorylase. Direct reduction with glutathione was too slow to account for rates observed *in vivo* [6].

Two different enzyme activities that stimulated dethiolation were observed in cardiac tissue (Figure 5 - Lanes 3 to 5). One process used reduced glutathione directly to reduce the S-thiolated protein (lane 5) and the other used NADPH for reduction (lane 4). The activity with NADPH was much greater than the activity with reduced glutathione. The NADPH-dependent activity has been resolved into two components with properties similar to thioredoxin reductase and thioredoxin. A survey of other tissues discovered NADPH-dependent activities in rat liver, spleen, lung, and heart.

This study shows that oxy radicals can effectively initiate protein S-thiolation by reactions that are best explained by a preferred attack of the

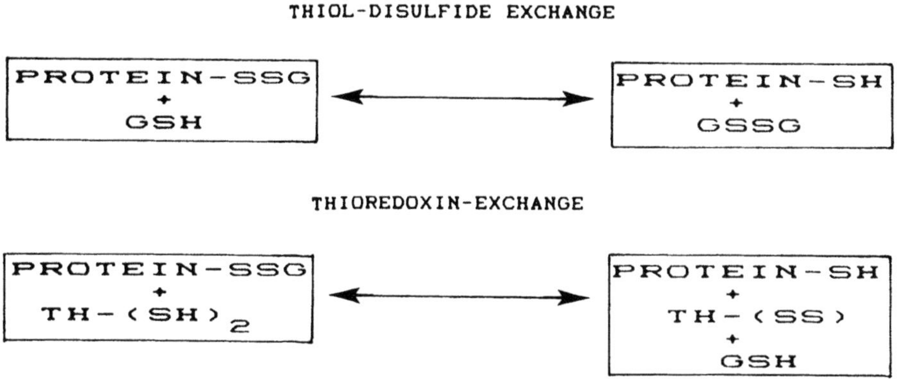

Figure 4. Dethiolation by glutathione (GSH) and thioredoxin ($TH\text{-}(SH)_2$).

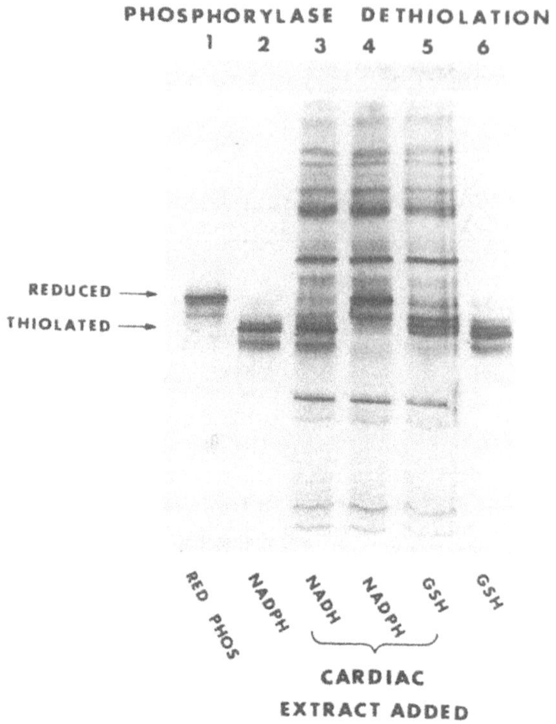

Figure 5. Dethiolation of S-thiolated phosphorylase at 30°, 20 min. Lane 1 - reduced phosphorylase alone; Lane 2 - identical to S-thiolated phosphorylase. One millimolar additions are shown at the bottom.

oxy radicals on protein sulfhydryls. S-thiolated proteins are formed concurrently with the oxidation of glutathione and they can be rapidly dethiolated by enzymic reactions that use either reduced glutathione or NADPH as the direct reductant. The most active dethiolation reaction is catalyzed by an NADPH-dependent enzyme system with properties similar to thioredoxin and thioredoxin reductase.

REFERENCES

1. R. Brigelius, Mixed Disulfides: Biological Functions and Increase in Oxidative Stress, in: "Oxidative Stress", H. Sies, ed., Academic Press, London (1985).
2. D.M. Ziegler, Role of Reversible Oxidation-Reduction of Enzyme Thiol-Disulfides in Metabolic Regulation, Ann. Rev. Biochem. 54:305 (1985).
3. J.A. Thomas and D. Beidler, A Thin-Gel Isoelectric Focusing Method for Quantitation of Protein S-thiolation, Anal. Biochem. 157:32 (1986).
4. M.W. Fariss and D.J. Reed, Glutathione Analysis, Meth. Enzymol. 143:101 (1987).
5. I. Fridovich, Cytochrome c, in: "Handbook of Methods for Oxygen Radical Research", R. Greenwald, ed., CRC Press, Boca Raton (1985).
6. M.W. Collison and J.A. Thomas, S-thiolation of Cytoplasmic Creatine Kinase in Heart Cells Treated with Diamide, Biochim. Biophys. Acta 928:121 (1987).
7. K. Axelsson, S. Eriksson, and B. Mannervik, Purification and Characterization of Cytoplasmic Thioltransferase from Rat Liver, Biochemistry 17:2978 (1978).
8. A. Holmgren, Thioredoxin, Ann. Rev. Biochem. 54:237 (1985).

MODIFICATION OF HUMAN LOW DENSITY LIPOPROTEIN BY LIPID PEROXIDATION

Hermann Esterbauer, Günther Jürgens and Oswald Quehenberger

Institute of Biochemistry and Institute of Medical Biochemistry, University of Graz, A-8010 Graz, Austria

INTRODUCTION

Elevated levels of low density lipoprotein (LDL) have been linked to an increased risk of atherosclerotic vascular diseases, however, the molecular mechanisms by which LDL is involved in atherogenesis are largely unknown. Recently, several reports appeared suggesting that LDL may be oxidatively modified by free radical reactions and that this oxidized LDL exhibits some of the properties which could explain the formation of early atherosclerotic lesions.[1,2] The fatty streaks in plaques originate mainly from lipid laden foam cells, which are derived from smooth muscle cells or macrophages. Cultured macrophages accumulate large amounts of cholesterol-lipids and obtain a foam-cell like appearance if incubated with acetylated LDL,[3] malonaldehyde-treated LDL[4] or oxidized LDL.[5] The uptake of these modified forms of LDL occurs through the so called scavenger receptor, which is not down regulated by the internalized LDL and which obviously recognizes LDL particles which are more negativly charged than native LDL. For acetylated LDL[3] or malonaldehyde-treated LDL,[4] it has been shown with some certainty that the increased negative surface charge is due to the neutralization (acetylation, Schiff´s base formation) of positive charges of lysine amino groups in apo B.[6,7] The commonly used index for the change of the negative charge is the increased relative electrophoretic mobility. With this method it was clearly shown that also oxidized LDL is more negativly charged.[8] Moreover a good correlation exists between the thiobarbituric acid reactive substances (malonaldehyde) in oxidized LDL and the electrophoretic mobility, the loss of lysine amino acid residues and the scavenger receptor mediated uptake by macrophages. LDL is oxidized if incubated with cultured cells such as endothelial cells, smooth muscle cells or monocytes. There is evidence that this modification involves oxygen free radical initiated lipid peroxidation. Antioxidants or chelators for transition metal ions prevent the cell-mediated modification, whereas supplementation with copper or iron greatly accelerates the modification. Moreover, incubation of LDL in oxygen-saturated buffer for long periods (24-48 hours) without cells also results in a modified LDL with properties apparently very similar to the cell-modified LDL. In the cell-free system, copper and iron accelerate the oxidative modification. We report here on some of the chemical and physicochemical properties by which oxidized LDL differs from native LDL.

MATERIALS AND METHODS

Human plasma LDL was prepared by stepwise ultracentrifugation (d = 1.020-1.050 g/ml) from the blood of healthy volunteers under 25 years of age.[8] The antioxidants (EDTA, BHT) added to the blood to prevent LDL oxidation during its isolation were removed by extensive dialysis against oxygen-free 0.01 M phosphate buffer pH 7.4, containing 0.16 M NaCl and 0.1 mg/ml chloramphenicol. These dialyzed LDL samples are designated as "native LDL" (Table 1). For the oxidation experiments in the absence of copper(II)-chloride, the native LDL (1.5 mg/ml) was filled in a dialysis bag which was immersed in the above mentioned buffer.[9] The buffer was continuously gassed with oxygen. At different time points samples were withdrawn and analyzed; given in Table 1 are the values found after 24 hours of incubation. For the oxidation experiments in the presence of copper, the buffer was first gassed with oxygen, then the LDL solution (1.5 mg/ml) was placed in a tube and supplemented with freshly prepared copper(II)chloride in final concentration of 0.1, 0.5, 1.0, 2.0, 5.0 and 10 μM. After 3 hours samples were withdrawn and analyzed; the values found with 10 μM copper are given in Table 1. The measurements of fatty acids, vitamin E, beta-carotene and aldehydes were done as described,[8] the details for the three-dimensional fluorescence measurement of the apo B and the LDL lipids are given elsewhere.[10]

RESULTS

Freshly prepared, EDTA-free LDL was incubated in a cell-free medium with and without supplementation with copper(II)chloride and various parameters indicative for lipid peroxidation were measured at different time points (Table 1, Ref.8, 10).

Oxidation in the Absence of Copper(II)chloride

In the absence of copper all LDL preparations were highly resistant to oxidation, and the polyunsaturated fatty acids 18:2 and 20:4 completely withstood autoxidation for at least three hours. During this time the LDL became progressivly depleted of its antioxidants alpha- and gamma-tocopherol and beta-carotene. Only when the antioxidants had dropped to a certain threshold level, a loss of 18:2 and 20:4 occurred concomitantely with the appearance of lipid oxidation products such as lipid peroxides, malonaldehyde, 4-hydroxynonenal and other aldehydes. After 24 hours the LDL had lost on average 22% of 18:2 and 50% of 20:4 originally present. More or less parallel with the formation of the aldehydes, the fluorescence property of the apo B protein dramatically increased. By the technic of three-dimensional fluorescence spectroscopy,[10,11] it could be proven that the fluorescence increase is primarily due to a newly formed chromophore with an emission maximum at 430 and an excitation maximum of 360 nm. This clearly indicates that the oxidation of the LDL lipids also affects the properties of apo B. This is also supported by increased electrophoretic mobility of the oxidized LDL particle. Between LDL preparations from different donors, a marked variation in the susceptibility to oxidation was observed. Thus, we had occasionally LDL samples which were difficult to oxidize, even when they were depleted of vitamin E and beta-carotene. This suggests, but of course does not prove, that such LDL preparations contained an additional, yet unknown antioxidant factor. The addition of EDTA (1mg/ml) to the LDL solution completely prevented the loss of fatty acids and vitamin E for 24 hours, and no lipid peroxides and aldehydes or fluorescent chromophores were formed. This suggests that traces of bivalent transition metal ions are involved in the initiation of the lipid peroxidation process in LDL. If this is the case, supplementation of the medium with copper or iron ions should greatly enhance the oxidation of

Table 1. Comparison of native and oxidized LDL.

	Native LDL	Oxidized LDL	
		Without $CuCl_2$ 24 h	10 μM $CuCl_2$ 3 h
18:2	345±65	266±63	35
20:4	26±5	12±4	3
α-Tocopherol	2.07±0.4	0	0
τ-Tocopherol	0.16±0.05	0	0
β-Carotene	0.13±0.02	0	0
Peroxides	0.23±0.08	n.e.	7.9
Malonaldehyde	0.46±0.13	2.6±1.9	16
4-Hydroxynonenal	0.29±0.18	0.59±0.35	1.9;5.0
Aldehydes, except MDA	0.92±0.31	4.52±2.9	23.21
Lipid fluorescence	13.5±3.8	18.0±3.4	45
Apo B fluorescence	19.7±4.8	38.0±6.3	140
Electrophoretic mobility	1.00	1.55±0.05	2.5

The values are given in nmol/mg LDL, the fluorescence was measured at Ex 360/Em 430 nm relative to 0.1 μg/ml quinine sulfate. n.e. = not estimated.

LDL. As will be shown below, copper(II)chloride is indeed a powerful prooxidant for LDL, ferrous or ferric ions, on the other hand, are only weak prooxidants for LDL.

Oxidation in the Presence of Copper(II)chloride

In the first studies on the oxidative modification of LDL by cells, incubation were carried out in Ham´s F-10 medium.[12] When in later experiments DME medium was used, no oxidation at all was observed,[2] and it was concluded that the Ham´s medium contains prooxidative metal ions (Fe, Cu), which were absent in the DME medium. And in fact, when this medium was supplemented with 5 to 10 μM Cu(II) or Fe(II), the LDL was again heavily oxidized by the cell cultures.[2,13] Some authors[14] now assume that LDL oxidized by copper or iron in a cell-free medium exhibits most if not all properties of the cell-oxidized LDL.

We have exposed LDL in a cell-free medium to various concentrations of copper(II)chloride (for details see article in this volume by Quehenberger et al.). With a concentration of 0.1 μM, nearly no oxidation occurred within three hours as indicated by the unchanged tocopherol and beta-carotene content. With increasing copper concentration the oxidation was progressivly enhanced and 10 μM led within three hours to a massive oxidation of LDL (Table 1). Under this condition the LDL was completely depleted of vitamin E and beta-carotene and the content of linoleic and arachidonic acid dropped to 10% of the value present in native LDL. Indicative for the heavy oxidation are also the high content of the lipid oxidation product 4-hydroxynonenal and the marked increase of the 430 nm fluorescent chromophores in both the LDL lipid and apo B.

A number of experiments were also performed with free iron and various iron complexes (unpublished). With our LDL preparations the prooxidative effect of iron was much less pronounced than that of copper. A concentration of 20 µM iron(II)sulfate led only in one case to a noteworthy oxidation, whereas in more than five repetitive experiments with other LDL preparations, 20-100 µM Fe(II) failed to induce lipid peroxidation within three hours. An addition of ascorbate (1 mM) to keep the iron in the ferrous form also did not stimulate lipid peroxidation in LDL. On the other hand, a concentration of 5 to 10 µM copper(II)chloride led in all LDL samples so far tested to a strong oxidation.

DISCUSSION

The LDL molecule possesses a molecular weight of about 2.5 million and it can be calculated from Table 1 that one LDL molecule contains about 750 to 1000 molecules of linoleic acid and 50 to 75 molecules of arachidonic acid. These polyunsaturated fatty acids (PUFAs) are bound in phospholipids located at the outer shell of the LDL particles, and in triglycerides and cholesterol representing the core of LDL. We already have emphasized previously that in comparison to other PUFA-containing biological samples (microsomes, tissue homogenates), LDL is highly resistant to oxidative degradation. In copper-free medium LDL had to be exposed for 24 hours to an oxygen-saturated buffer to obtain a substantial oxidation of the PUFAs. This is remarkable since the antioxidant content of LDL is rather low, on average the LDL molecule contains only five molecule alpha-tocopherol and only one out of two LDL contains a molecule of gamma-tocopherol and one out of three contains a molecule of beta-carotene. The latter fact suggests that LDL exists in at least three subpopulations, differing in the number of antioxidant molecules. In both the copper-supplemented medium and in the copper-free medium, we found that lipid peroxidation in LDL is only initiated when it is more or less depleted of the antioxidants. The sequence of the consumption is in the order: alpha-tocopherol, gamma-tocopherol, beta-carotene, PUFAs.

The oxidation of the PUFAs leads to the cleavage of the fatty acid carbon chain resulting in a great number of medium- and short-chain aldehydes.[8] It was assumed by several authors[4,15] that these aldehydes now in turn react with lysine amino groups of the apo B, yielding a modified LDL molecule which is no longer recognized by the LDL receptor, but by the scavenger receptor. If freshly prepared LDL is incubated with reactive aldehydes (malonaldehyde, 4-hydroxynonenal, 2,4-alkadienals), the resulting LDL does in fact show altered chemical and biological properties very similar to oxidized LDL. For example, the 4-hydroxynonenal treated LDL shows exactly the same fluorescent chromophore (Ex 360/Em 430 nm) as does the oxidized LDL.[10] However, the problems with this hypothesis should not be overlooked. Most important is the question of the necessary concentration. For a substantial modification of the apo B by externally added aldehydes, concentrations above 100 µM are necessary. Such aldehyde concentrations, however, are never formed in the medium during oxidation of LDL; the concentration there is in the micromolar range. It has been shown that most of the aldehydes are lipophilic substances and remain therefore mainly associated with the LDL particle, where they can reach local concentrations in the millimolar range. Additional studies are needed to prove the suggestion[8] that the reactive lipophilic aldehydes diffuse within the LDL particle - without being diluted by the surrounding aqueous medium - to the apo B and there attack amino acid side chains, resulting in a fluorescent apo B and an epitope which is no longer recognized by the LDL receptor but, by the scavenger receptor.

ACKNOWLEDGEMENTS

This work was supported by the Österreichischen Fonds zur Förderung der Wissenschaftlichen Forschung Projektnummer P6176B.

REFERENCES

1. J. W. Heinecke, Free radical modification of low-density lipoprotein: Mechanisms and biological consequences, Free Radical Biol. Med. 3:65 (1987).
2. U. P. Steinbrecher, S. Parthasarathy, D. S. Leake, J. L. Witztum, and D. Steinberg, Modification of low density lipoprotein by endothelial cells involves lipid peroxidation and degradation of low density lipoprotein phospholipids, Proc. Natl. Acad. Sci. USA 81:3883 (1984).
3. J. L. Goldstein, Y. K. Ho, S. K. Basu, and M. Brown, Binding site on macrophages that mediates uptake and degradation of acetylated low density lipoprotein, producing massive cholesterol deposition, Proc. Natl. Acad. Sci. USA 76:333 (1979).
4. M. E. Haberland, A. M. Fogelman, and P. A. Edwards, Specificity of receptor-mediated recognition of malonaldehyde modified low density lipoproteins, Proc. Natl. Acad. Sci. USA 79:1712 (1982).
5. D. W. Morel, J. R. Hessler, and G. M. Chisolm, Low density lipoprotein cytotoxicity induced by free radical peroxidation of lipid, J. Lipid Res. 24:1070 (1983).
6. K. H. Weisgraber, T. L. Innerarity, and R. W. Mahley, Role of the lysine residues of plasma lipoproteins in high affinity binding to cell surface receptors on human fibroblasts, J. Biol. Chem. 253:9053 (1978).
7. M. E. Haberland, C. L. Olch, and A. M. Fogelman, Role of lysines in mediating interaction of modified low density lipoproteins with the scavenger receptor of human monocyte macrophages, J. Biol. Chem. 259:11305 (1984).
8. H. Esterbauer, G. Jürgens, O. Quehenberger, and E. Koller, Autoxidation of human low density lipoprotein: Loss of polyunsaturated fatty acids and vitamin E and generation of aldehydes, J. Lipid Res. 28:495 (1987).
9. J. Schuh, G. F. Fairclough, and R. H. Haschemeyer, Oxygen mediated heterogeneity of apo-low-density lipoprotein, Proc. Natl. Acad. Sci. USA 75:3173 (1978).
10. O. Quehenberger, E. Koller, G. Jürgens, and H. Esterbauer, Investigation of lipid peroxidation in human low density lipoprotein, Free Rad. Res. Comms. 3:233 (1987).
11. E. Koller, O. Quehenberger, G. Jürgens, O. S. Wolfbeis, and H. Esterbauer, Investigation of human plasma low density lipoprotein by three-dimensional fluorescence spectroscopy, FEBS Lett. 198: 229 (1986).
12. T. Henriksen, E. M. Mahoney, and D. Steinberg, Enhanced macrophage degradation of low density lipoprotein previously incubated with cultured endothelial cells: Recognition by receptors for acetylated low density lipoproteins, Proc. Natl. Acad. Sci. USA 78:6499 (1981).
13. J. W. Heinecke, L. Baker, H. Rosen, and A. Chait, Superoxide-mediated modification of low density lipoprotein by arterial smooth muscle cells, J. Clin. Invest. 77:757 (1986).
14. U. P. Steinbrecher, J. L. Witztum, S. Parthasarathy, and D. Steinberg, Decrease in reactive amino groups during oxidation or endothelial cell modification of LDL, Arteriosclerosis 7:135 (1987).
15. G. Jürgens, J. Lang, and H. Esterbauer, Modification of human low-density lipoprotein by the lipid peroxidation product 4-hydroxynonenal, Biochim. Biophys. Acta 875:103 (1986).

IS OXIDIZED LOW DENSITY LIPOPROTEIN INVOLVED IN THE RECRUITMENT AND RETENTION OF MONOCYTE/MACROPHAGES IN THE ARTERY WALL DURING THE INITIATION OF ATHEROSCLEROSIS?

Sampath Parthasarathy, Mark T. Quinn, and Daniel Steinberg

Division of Endocrinology and Metabolism
Department of Medicine M-013D
University of California, San Diego
La Jolla, California 92093, U.S.A.

INTRODUCTION

Atheroma or atherosclerosis is a disease affecting major arteries, wherein the intimal layer of the artery is engorged with both intracellular and extracellular lipids, mostly cholesterol esters. Cholesterol, particularly that associated with low density lipoproteins (LDL), has been identified as a major risk factor. The precise mechanism(s) by which LDL cholesterol becomes atherogenic has not been well understood.

Most of the lipid-laden "foam" cells of the early atherosclerotic lesion are macrophage like and are derived from circulating monocytes.[1,2] In fact, one of the earliest changes observed in experimental animals on a cholesterol enriched diet is the adherence of monocytes to the endothelium. In these animals, monocytes can actually be seen penetrating between the endothelial cells to gain entry into the intima.

Paradoxically, macrophages could not be induced to make foam cells with native LDL by way of Goldstein and Brown's B/E receptor.[3] This led to the speculation that some modified form of LDL may be the actual atherogenic lipoprotein. Modified forms of LDL were soon identified, e.g., acetyl LDL[4] that generated foam cells upon incubation with macrophages. However, there has been no evidence thus far that such modified LDLs exist in vivo and are atherogenic.

Henriksen, Mahoney, and Steinberg[5,6] first observed that LDL incubated with cultured rabbit aortic endothelial cells, in addition to being cytotoxic, was modified to a form that was degraded by macrophages, with resultant cholesterol ester accumulation. This degradation was mediated by the same receptor that recognized acetyl LDL,[6] namely the acetyl LDL or scavenger receptor. This cell-induced biological modification of LDL is accompanied by peroxidation of LDL lipids, and this modification can be inhibited by antoxidants and metal chelators.[7] It was soon established that, while lipid peroxidation was essential for modification, the recognition of the modified LDL by the acetyl LDL receptor on the macrophages depended on the generation of specific epitopes on the apoprotein.[8] The apoprotein B-100 of the LDL was oxidatively broken down into several lower molecular weight fragments that carried the recognition sites for the receptor.[9]

Monocyte/macrophages are not present in normal artery, and apparently atherogenic factors must somehow initiate the movement of these cells into the artery. Directional migration of cells (chemotaxis) depends on the presence of chemical gradients of chemotactic factors. Several such factors have been identified for monocyte/macrophages (see ref. 10). It is not, however, known whether any of these factors bear any relation to the initiation of atherosclerosis. Since our findings suggested that oxidized LDL may be involved in the formation of foam cells, we investigated whether oxidized LDL also could play a role in the recruitment and retention of monocyte/macrophages in the artery.[10,11]

EXPERIMENTAL

Using mouse peritoneal macrophages, we observed that native LDL was devoid of any chemotactic activity (Table 1). On the other hand, medium conditioned in the presence of endothelial cells for 24 hrs. showed such activity. We then tested to see whether the medium conditioned as above in the presence of LDL would show increased activity. Our previous studies have shown that LDL is oxidatively modified under these conditions. It was found, however, that the activity under these conditions was very much reduced, to a level even below that of the conditioned medium. Addition of BHT during the conditioning incubation with LDL present restored the activity. This decrease in activity of the conditioned medium was not due to inhibition of release of the chemotactic components; addition of oxidized LDL to medium conditioned in the absence of LDL also resulted in a loss of activity. Based on these findings, we speculated whether oxidized LDL has the property of inhibiting macrophage motility. This was indeed so, as oxidized LDL added to other known chemotactic factors, e.g., zymosan-activated rat serum, also decreased the chemotactic activity of the latter. We also tested the ability of acetyl LDL and several other compounds that are known to interact with the macrophage scavenger receptor to inhibit macrophage chemotaxis. Much to our surprise, acetyl LDL, fucoidin and polyinosinic acid not only failed to have an inhibitory effect, but in fact were themselves chemotactic. The lack of inhibition by these compounds may be due to the fact that the inhibitory component of the oxidized LDL was present in the oxidized lipid fraction.

The endothelial conditioned medium also showed chemotactic activity towards human circulating monocytes (Table 1). However, in contrast to our findings with macrophages, the medium that was conditioned in presence of LDL showed chemotactic activity over and above that of the conditioned medium. In fact, LDL oxidized even in the absence of cells, in presence of copper ions, showed potent chemotactic activity towards human monocytes. This activity was also present in the lipids of the oxidized LDL.

Table 1. Enhancement of Monocyte Chemotaxis and Inhibition of Macrophage Motility by Oxidized LDL[a]

Sample tested	Macrophages	Monocyte
Medium conditioned with EC	4.6	4.1
Medium conditioned with EC and LDL	0.6	9.0
Medium conditioned with EC + oxidized LDL	0.4	9.6
Reisolated EC-modified LDL	0.5	6.4
Lipids from native LDL	0.5	1.1
Lipids from EC-modified LDL	0.3	8.7
Zymosan-activated serum	10.8	N.D.
Zymosan-activated serum + oxidized LDL	3.0	N.D.

[a]Results are express as chemotactic index (CI), the number of cells migrating in response to the test substance divided by the number of cells migrating in controls. Chemotaxis was measured in a 48-well modified Boyden chamber. Seven to eight 400x fields were counted per sample, and averages are shown. All lipoprotein samples were evaluated at 50 μg/ml concentrations.

These results demonstrated that oxidized LDL, while being chemotactic for human monocytes, inhibited the motility of macrophages in vitro. However, does oxidized LDL exist in vivo in the intima of the artery and can it account for the presence of macrophages during lesion development? Oxidized lipids have been isolated from atherosclerotic lesions.[12,13] More direct evidence of the involvement of oxidized LDL has been suggested by recent findings that probucol, an antioxidant that inhibited the modification of LDL in vitro,[14] slowed down the development of atherosclerosis in WHHL rabbits[15] and lowered the degradation of LDL in the lesion areas.[16]

Rabbits, when fed cholesterol, readily develop hypercholesterolemia and foam cell lesions within a few weeks. However, in these animals cholesterol is predominantly associated with β-VLDL.[17] While a specific β-VLDL receptor has been described for macrophages[18] that may account for foam cells, the question still remains as to the mechanism(s) of monocyte recruitment into the artery. Since cholesterol feeding has been reported to increase lipid peroxidation,[19] we tested whether oxidized β-VLDL would have effects similar to oxidized LDL on monocyte/macrophage chemotaxis. The results are summarized in Table 2.

While native β-VLDL showed no chemotactic activity for macrophages, β-VLDL oxidized by copper ions, like oxidized LDL, inhibited the macrophage basal chemotactic response as well as the response to zymosan activated serum. Similarly, oxidized β-VLDL was also strongly chemotactic for human monocytes, severalfold over that of native β-VLDL. The native β-VLDL also contained varying amounts of chemotactic activity for monocytes that could be removed by prior treatment with albumin. These results suggest that also in cholesterol-fed animals, oxidation of the lipoproteins may be necessary for the recruitment of monocytes into the artery.

Table 2. Effect of β-VLDL on Monocyte/Macrophage Chemotaxis

Sample Tested	Macrophages	Monocytes
β-VLDL	1.1	3.5
Oxidized β-VLDL	0	22.0
Zymosan-activated serum	6.3	N.D.
Zymosan-activated serum + β-VLDL	7.0	N.D.
Zymosan-activated serum + Ox. β-VLDL	2.8	N.D.

N.D.: Not determined

CONCLUSIONS

Thus there are four ways by which oxidized LDL may affect atherosclerosis: 1) by favoring monocyte entry in to the endothelium, 2) by preventing the efflux of macrophages, 3) by foam cell formation by way of scavenger receptor and 4) by causing endothelial injury. The following scheme represents the mechanism by which oxidized LDL may favor atherosclerosis:

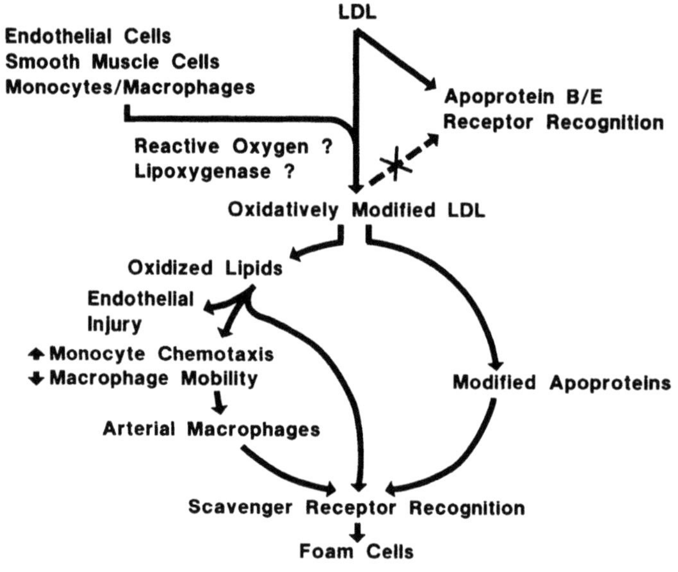

Figure 1

REFERENCES

1. S. Fowler, H. Shio, and W.J. Haley, Characterization of Lipid-laden aortic cells from cholesterol-fed rabbits. Lab. Invest. 41:372 (1979).
2. R.G. Gerrity, The role of monocytes in atherogenesis. I. Transition of blood borne monocytes into foam cells in fatty lesions. Am. J. Pathol. 103:181 (1981).

3. J.L. Goldstein and M.S. Brown, The low-density lipoprotein pathway and its relation to atherosclerosis, Ann. Rev. Biochem. 40:897 (1977).
4. J.L. Goldstein, M.S. Brown and S.K. Basu, A binding site on macrophages that mediates the uptake and degradation of acetylated low density lipoprotein, Proc. Natl. Acad. Sci. U.S.A. 76:333 (1979).
5. T. Henriksen, E.M. Mahoney and D. Steinberg, Enhanced macrophage degradation of low density lipoprotein previously incubated with cultured endothelial cells: recognition by the receptor for acetylated low density lipoproteins, Proc. Natl. Acad. Sci. U.S.A. 78:6499 (1981).
6. T. Henriksen, E.M. Mahoney and D. Steinberg, Enhanced macrophage degradation of biologically modified low density lipoprotein, Arteriosclerosis 3:149 (1983).
7. U.P. Steinbrecher, S. Parthasarathy, D.S. Leake, J.L. Witztum, and D. Steinberg, Modification of low density lipoprotein by endothelial cell involves lipid peroxidation and degradation of low density lipoprotein phospholipids, Proc. Natl. Acad. Sci. U.S.A. 81:3883 (1984).
8. S. Parthasarathy, L.G. Fong, D. Otero, and D. Steinberg, Recognition of solubilized apoproteins from delipidated, oxidized low density lipoprotein (LDL) by the acetyl LDL receptor, Proc. Natl. Acad. Sci. U.S.A. 84:537 (1987).
9. L.G. Fong, S. Parthasarathy, J.L. Witztum, and D. Steinberg, Apoprotein B-100 degradation during the oxidative modification of low density lipoprotein is probably mediated by direct oxidative attack on the polypeptide chain, J. Lipid Res., 28:1466 (1987).
10. M.T. Quinn, S. Parthasarathy, and D.Steinberg, Endothelial cell-derived chemotactic activity for mouse peritoneal macrophages and the effects of modified forms of low density lipoproteins, Proc. Natl. Acad. Sci. U.S.A. 82:5949 (1985).
11. M.T. Quinn, S. Parthasarathy, and D. Steinberg, Oxidatively modified low density lipoproteins: A potential role in recruitment and retention of monocyte/macrophages during atherogenesis, Proc. Natl. Acad. Sci. U.S.A. 84:2995 (1987).
12. W.A. Harland, J.D. Gilbert, and C.J.W. Brooks, Lipids of the human atheroma. VIII. Oxidized derivatives of cholesteryl linoleate, Biochim. Biophys. Acta 316:378 (1973).
13. H. Mowri, K. Chinen, S. Ohkuma, and T. Takano, Peroxidized lipids isolated by HPLC from atherosclerotic aorta, Biochem. Int. 12:347 (1986).
14. S. Parthasarathy, S.G. Young, J.L. Witztum, R.C. Pitman, and D. Steinberg, Probucol inhibits oxidative modification of low density lipoprotein, J. Clin. Invest. 77:641 (1986).
15. T. Kita, Y. Nagano, M. Yokode, K. Ishii, N. Kume, A. Ooshima, H. Yoshida, and C. Kawai, Probucol prevents the progression of atherosclerosis in Watanabe heritable hyperlipidemic rabbit, an animal model for familial hypercholesterolemia, Proc. Natl. Acad. Sci. U.S.A., 84:5928 (1987).
16. T.E. Carew, D.C. Schwenke, and D. Steinberg, An antiatherogenic effect of probucol unrelated to its hypocholesterolemic effect: Evidence that antioxidants in vivo can inhibit low density lipoprotein degradation in macrophage rich fatty streaks and slow the progression of atherosclerosis in the WHHL rabbit, Proc. Natl. Acad. Sci. U.S.A., 84:7725 (1987).

17. R.W. Mahley, Alterations in plasma lipoproteins induced by cholesterol feeding in animals including man, in: "Disturbances in Lipid and Lipoprotein metabolism," Dietschy, et al., eds., Am. Physiol. Soc. Bethesda, MD. (1978).
18. J.L. Goldstein, Y.K. Ho, M.S. Brown, T.L. Innerarity, and R.W. Mahley, Cholesteryl ester accumulation in macrophages resulting from receptor mediated uptake and degradation of hypercholesterolemic canine β-very low density lipoproteins, J. Biol. Chem., 255:1839 (1980).
19. S. Suylok, Bar-Pollak, E. Feher, I. Kemenes, I. Kantor, and J. Feher, Liver lipid peroxidation induced by cholesterol and its treatment with a dihydroquinoline type free radical scavenger in rabbits, Acta Physiol. Hung., 64:437 (1984).

COPPER AND COPPER-NICOTINIC ACID COMPLEXES MEDIATED OXIDATION OF HIGH DENSITY LIPOPROTEIN

Mohamed El-Saadani, Hermann Esterbauer*, Günther Jürgens, Mohamed El-Sayed**, Ahmed Nassar***, Mohamed Goher**, and Anton Holasek

Institute of Medical Biochemistry
University of Graz
Graz, Austria

*Institute of Biochemistry
University of Graz
Graz, Austria

**Faculty of Science
Alexandria University
Alexandria, Egypt

***Department of Biochemistry
Faculty of Medicine Assuit University
Assuit University, Assuit, Egypt

INTRODUCTION

Plasma high density lipoprotein (HDL)-cholesterol and/or apo A-I levels appear to be inversely related to the occurrence of acute myocardial infarctions and have an impact on the onset and progress of atherosclerosis.[1,2] Since the outer surface of the high density lipoprotein particle has a low cholesterol/phosphatidylcholine ratio in comparison to most of the cell membranes,[3] it is suggested to be an effective acceptor for the transport of cholesterol to various tissues. HDL facilitates the removal of the peripheral tissue cholesterol and delivers it within the "reverse cholesterol transport" to the liver.[4] A recent investigation[5] on the scavenger function of sinusoidal liver cells showed that chemically acetylated HDL underwent receptor-mediated endocytosis through a pathway distinct from acetylated low-density lipoprotein (LDL) or formaldehyde-treated serum albumin, which suggests the presence of a scavenger receptor for acetylated HDL.

Since acetylation of HDL does not occur in vivo, the biological function of such a receptor remains obscure. In view of the finding that oxidized LDL is taken up by the same receptor as acetylated LDL,[6] it seems reasonable to assume that also the acetyl-HDL receptor plays a role for the uptake of oxidized HDL. So far no report exists on the oxidative modification of HDL, and therefore the aim of this work was to investigate whether HDL is susceptible to oxidation or not.

In analogy to the studies with LDL, copper(II)chloride was used as prooxidant, and in addition the prooxidant effect of two copper-nicotinic acid (HNA) complexes, which have a pronounced effect on lipid metabolism in vivo[7] were tested.

EXPERIMENTAL PROCEDURES

Chemicals, biochemicals and solvents were of the highest purity grade and purchased from Sigma (Munich, FRG), Merck (Darmstadt, FRG) and Immuno AG (Vienna, Austria). The copper(I)-nicotinic acid complexes were synthesized as described.[8] The $Cu(HNA)_2Cl$ complex contains copper and nicotinic acid in a ratio of 1:2 and forms bright red crystals. The Cu(HNA)Cl complex has a copper to nicotinic acid ratio of 1:1 and exists in yellow crystals. The purity of the two complexes was checked by elementary analysis and infrared.

Human plasma HDL subfractions were prepared by stepwise ultracentrifugation within the following density cut-offs: d = 1.063-1.095 g/ml (HDL_{2b}), d = 1.095-1.125 g/ml (HDL_{2a}), and d = 1.125-1.21 g/ml (HDL_3). The subfraction used in this study was HDL_3. After ultracentrifugation HDL_3 was further purified by gel filtration on a Bio-Gel A 5m (column 1.5 x 90 cm).

Prior to oxidation, the HDL_3 was dialyzed 24 h against vacuum degassed phosphate buffered saline (PBS) pH 7.4 containing 0.1 mg chloramphenicol per ml, to remove EDTA and butylated hydroxytoluene (BHT) which was present throughout all steps of HDL isolation.

Oxidation experiments were carried out essentially as described for LDL.[9] Briefly, 9 ml of the dialyzed HDL_3 (1.5 mg/ml) was placed in a dialysis bag which was immersed in 27 ml PBS and continuously gassed with air. The outside buffer was supplemented as follows: (1) 2.4 μM ascorbic acid (AA); (2) 2.4 μM AA + 5 μM HNA; (3) 2.4 μM AA + 5 μM Cu(II)chloride; (4) 2.4 μM AA + 5 μM Cu $(HNA)_2Cl$; (5) 2.4 μM AA + 5 μM Cu(HNA)Cl. Samples were withdrawn at different time intervals and analyzed for thiobarbituric acid reactive substances (TBARS) according to ref. 10 with some modifications, for fluorescent intensities in the protein and in lipid fractions, and for the electrophoretic mobility on agarose gel.[9]

RESULTS

The kinetics of the oxidation of HDL as measured by the content of TBARS inside the dialysis bag is shown in Figure 1. In the absence of copper ions, no oxidation occurred within the first 12 hours, whereas copper(II)chloride and the two copper complexes strongly enhanced the formation of TBARS. The experiments were highly reproducible with 3 different HDL_3 preparations. Concomitantly with the formation of TBARS, the fluorescence of HDL_3 proteins and HDL_3 lipids remarkably increased. Figure 2 shows as example the fluorescence spectra of the HDL_3 protein

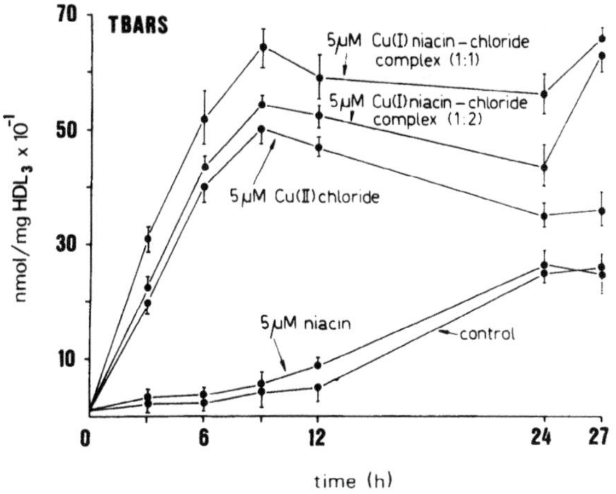

Figure 1. Kinetics of the formation of TBARS during peroxidation of HDL_3. Experimental conditions as in Table 1.

after 6, 12, and 24 hours of oxidation in the presence of Cu(HNA)Cl complex. Similarly to LDL, the formed fluorescent chromophore has an excitation maximum at 360 nm and an emission maximum at 430 nm. A comparison of the five incubation systems in respect to TBARS, fluorescence and relative electrophoretic mobilities of the HDL_3 before and after 24 hours oxidation is represented in Table 1.

DISCUSSION

The investigation clearly shows that HDL_3 is susceptible to oxidation in the presence of copper ions. Since ascorbic acid was present in all incubation systems it should be assumed that the candidate initiating lipid peroxidation is copper in its monovalent form.

The copper niacin 1:1 complex Cu(HNA)Cl acted clearly as a more powerful prooxidant than $Cu(HNA)_2Cl$ or the non complexed copper. This suggests that for copper, as for iron ions, it plays an important role in which form it is added. In comparison to oxidized LDL the most reasonable effect seen in this oxidized HDL is the dramatic increase in the fluorescence intensity. Thus in the oxidation with Cu(HNA)Cl, the 430 nm fluorescence (EX = 360 nm) increased by about tenfold for the lipid and by about 60-fold for the protein moiety. With this strong increase of the fluorescence, we anticipated that also the relative electrophoretic mobility, which is a measure of the charge modification of the protein, also will increase strongly. However, as shown in Table 1, there is only a small increase; furthermore in incubation with AA or AA + HNA, the resulting HDL_3 shows a decreased electrophoretic mobility compared with the native one. Whether such a decrease or increase of the surface charge of HDL_3 will have biological consequences, i.e., binding to macrophages or to the HDL scavenger receptor of the liver, remains for further investigation.

Table 1. Effects of copper and copper(I)nicotinic acid complexes on TBARS, fluorescence and relative electrophoretic mobility (REM) of HDL_3.

	TBARS nmol/mg HDL_3		Fluorescence intensity lipid		apo-protein		REM
Supplement	0 h	24 h	0 h	24 h	0 h	24 h	
no addition	0.052	2.620	17.3	57	54.4	560	0.88
5μM (HNA)	0.047	2.567	17.2	73	53.8	1560	0.87
5μM $Cu(II)CL_2$	0.049	3.608	16.2	158	53.9	2620	1.26
5μM $Cu(HNA)_2Cl$	0.050	6.353	18.1	149	53.8	2785	1.43
5μM Cu(HNA)Cl	0.047	6.643	17.6	170	57.1	3510	1.46

[a] Samples were analyzed at zero-time (0 h) and after 24 h. All incubations contained in addition to the indicated supplements 2.4 μM ascorbic acid.

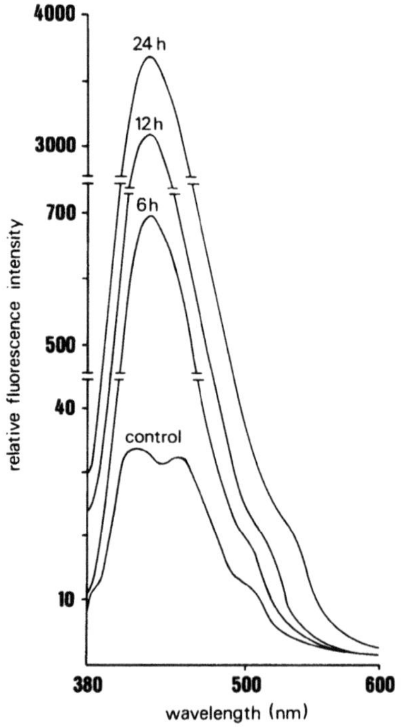

Figure 2. Emission fluorescence spectra at 360 nm excitation, of apo-protein from the native (control) and 6, 12 and 24 hour Cu(HNA)Cl-oxidized HDL_3. Fluorescence intensities are relative to a quinine sulfate standard (0.1 ug/ml 0.1 N H_2SO_4).

ACKNOWLEDGEMENTS

This work was supported in part by the Austrian Science Foundation, project P6176B and by the Egyptian-Austrian Culture Exchange Program.

REFERENCES

1. J. Franzen, B.W. Johansson and A. Gustafson, Reduced high density lipoprotein as a risk factor after acute myocardial infarction, Acta Med. Scand. 221:367 (1987).
2. B. Paigen, P.A. Holmes, D. Mitchell, and D. Albee, Comparison of atherosclerotic lesions and HDL-lipid levels in male, female, and testosterone-treated female mice from strains C57BL/6, BALB/c, and C3H, Atherosclerosis 64:215 (1987).
3. G. Schmitz, H. Robenek, and G. Assmann, Role of the high density lipoprotein-receptor cycle in macrophage-cholesterol metabolism, Klin. Wochenschr. 64:979 (1986).
4. N.E. Miller, A. La Ville and D. Crook, Direct evidence that reverse cholesterol transport is mediated by high-density lipoprotein in rabbit, Nature 314:109 (1985).
5. M. Murakami, S. Horiuchi, K. Takata, and Y. Morino, Distinction in the mode of receptor-mediated endocytosis between high density lipoprotein and acetylated high density lipoprotein: Evidence for high density lipoprotein receptor-mediated cholesterol transfer, J. Biochem. 101:729 (1987).
6. U.B. Steinbrecher, J.L. Witztum, S. Parthasarathy, and D. Steinberg, Decrease in reactive amino groups during oxidation or endothelial cell modification of LDL, Arteriosclerosis 7:135 (1987).
7. M.A. Goher, A.Y. Nassar, N. Monem, and M. El-Saadani, Biological activity of synthetic copper-(I)-nicotinic acid complex on plasma lipids, Proceedings of 3rd Cyprus Conference on New Methods in Drug Research (1987).
8. M.A.S. Goher and T.C.W. Mark, Crystal structure of a polymeric 1:1 complex of copper (I) chloride with isonicotinic acid, Inorg. Chim. Acta 101:L27 (1985).
9. H. Esterbauer, G. Jurgens, O. Quehenberger, and E. Koller, Autoxidation of human low density lipoprotein: Loss of polyunsaturated fatty acids and vitamin E, and generation of aldehydes, J. Lipid Res. 28:495 (1987).
10. J. Schuh, G.F. Fairclough Jr., and R.H. Haschemeyer, Oxygen-mediated heterogenity of apo-low-density lipoprotein, Proc. Natl. Acad. Sci. USA 75:3173 (1978).

OXIDATION OF HUMAN LOW DENSITY LIPOPROTEIN

INITIATED BY COPPER(II)CHLORIDE

Oswald Quehenberger, Günther Jürgens,
Silvia Zadravec and Hermann Esterbauer

Institute of Biochemistry and Institute of
Medical Biochemistry
University of Graz
Schubertstrasse 1, A-8010 Graz, Austria

INTRODUCTION

Human low density lipoprotein (LDL) is a complex of a glycoprotein (apo B), free and esterified cholesterol, phospholipids, triglycerides and lipid soluble vitamins. It has been shown that oxidative modification of LDL mediated by free radical reactions renders LDL recognizable by the scavenger receptors of macrophages, cytotoxic towards fibroblasts and chemotactic towards monocytes.[1-4] These and other properties of oxidized LDL might play an important role in atherogenesis. We have shown that oxidized LDL contains a great variety of aldehydes, such as 4-hydroxynonenal (HNE), derived from the oxidative breakdown of the polyunsaturated fatty acids. Moreover, the apo B of oxidized LDL exhibits a strong fluorescence at 360 nm excitation and 430 nm emission, which we attributed to covalent binding of HNE.[5,6] Traces of metal ions, such as copper or iron, are most probably involved in the initiation of oxidation reactions in LDL. The aim of this investigation was to study the capacity of bivalent copper ions to initiate lipid peroxidation in LDL.

EXPERIMENTAL

All reagents, solvent, buffers and detergents were of analytical grade and purchased from Merck, Darmstadt. Standards required for identification and quantification were the tocopherols (Merck) and beta-carotene (Ega-Chemie Steinheim). 4-Hydroxynonenal was prepared in our laboratory. Human plasma LDL with a density of 1.020 - 1.050 g/ml was prepared by ultra-centrifugation. The two LDL preparations designated LDL-1 and LDL-2 stem from two donors. The initial content of alpha-tocopherol, gamma-tocopherol and beta-carotene in LDL-1 was 695, 76 and 81 ng/mg LDL; the respective values in LDL-2 were 1252, 46, and 49 ng/mg LDL. The oxidation was carried out at room temperature in 30 ml centrifuge tubes. The LDL solutions (1.5 mg LDL/ml 0.01 M phosphate buffer, containing 0.16 M NaCl and 0.1 mg/ml chloramphenicol) were supplemented with various concentrations of copper(II)chloride and incubated up to three hours. The oxidation was terminated by addition of 400 μM EDTA and refrigeration. The free HNE present in the LDL solution after the oxidation was determined by the HPLC method described previously.[7] The peroxides were estimated according to Yagi,[8] and the conjugated dienes by UV-absorption of the LDL lipids dissolved in cyclohexane (0.37 mg lipid/ml cyclohexane) at 234 nm. The tocopherols and beta-carotene were extracted and quantified by HPLC on a Spherisorb C-18 column. The apo B was isolated by extracting 3 ml of a LDL solution (1.5 mg/ml) with chloroform: methanol = 2:1. After washing with water, the precipitated apo B was dissolved with 5% aqueous sodium-dodecylsulfate (SDS). The fluorescence of both the apo B and the lipid moiety was measured with an excitation of 360 nm and an emission wavelength of 430 nm.

RESULTS AND CONCLUSION

Solutions of LDL-1 (1.5 mg/ml) in 0.01 M phosphate-buffered saline were supplemented with 0, 0.1, 0.5, 1.0, 2.0, 5.0 and 10 μM copper(II) chloride. After three hours alpha- and gamma-tocopherol, beta-carotene, lipid peroxides, dienes, HNE and fluorescence were measured (Table 1). With the highest copper concentration tested, no traces of alpha- or gamma-tocopherol or beta-carotene were detectable after the three-hour incubation period. Moreover, the level of HNE increased from 0.04 nmol/mg LDL to 1.90 nmol/mg LDL. Indicative for lipid peroxidation is also the 4.2-fold increase of the fluorescence intensity at an excitation of 360 nm and an emission of 430 nm of the apo B. A chromophore with the same spectral characteristics was also formed in the lipid phase, which could be due to the binding of HNE to distinct phospholipids of LDL.[9] The 10 μM copper-oxidized LDL contained high levels of peroxides, i.e., 7.9 nmol/mg LDL as compared with the control (0.9 nmol/mg LDL). The UV spectrum of the lipid extract showed a clear and sharp maximum at 234 nm, indicative of conjugated dienes, the relative increase of the absorbance at this wavelength being 9.1-fold compared with the control. The lower copper concentrations led to a newly complete depletion of alpha- and gamma-tocopherol and beta-carotene, except the 0.1 μM. The other effects described above for the 10 μM concentration were also pronounced, yet to a lesser extent.

With the lowest copper concentration (0.1 μM), no significant amounts of peroxidation products could be detected after three hours of incubation. The values of alpha- and gamma-tocopherol were comparable with those of native LDL (695 ng/mg LDL and 76 ng/mg LDL), beta-carotene decreased slightly to 80% of its initial content (from 81 to 66 ng/mg LDL). The fluorescence properties of the LDL did not change at all. It can therefore be concluded that the critical copper concentration able to initiate lipid peroxidation lies between 0.1 and 0.5 μM. The fact that the 0.5 μM copper clearly led to formation of lipid peroxides, yet not to lipid degradation products (HNE) or fluorescent materials, together with the finding that this LDL sample still contained some gamma-tocopherol and beta-carotene suggests that these antioxidants were still able to prevent progression of the lipid peroxidation through chain cleavage processes.

The time dependence of the copper-catalyzed lipid peroxidation was tested by incubation of LDL-2 (1.5 mg/ml) in the presence of 2 μM copper up to three hours (Figure 1). After 30 minutes alpha-tocopherol dropped by 59% from 1252 to 512 ng/mg LDL; the peroxide level was approximately doubled (from 0.3 nmol/mg LDL initial to 0.7 nmol/mg LDL). The fluorescence properties of the protein changed slightly and no diene

Table 1. Effect of Various Copper(III)chloride Concentrations on the Vitamin E and β-Carotene Content of LDL-1 and the Formation of HNE and Peroxides

Concentration μM	Tocopherol alpha ng/mg	gamma LDL	β-Carotene ng/mg LDL	HNE nmol/mg	Peroxides LDL
0.0	695	76	81	0.04	0.9
0.1	617	73	66	0.04	1.2
0.5	0	30	50	0.04	2.8
1.0	0	2	18	0.06	3.6
2.0	0	0	7	0.06	n.e.[a]
5.0	0	0	0	0.23	5.8
10.0	0	0	0	1.90	7.9

[a]n.e. = not estimated.

conjugates were detectable. After three hours of incubation the LDL was depleted of alpha- and gamma-tocopherol; beta-carotene dropped from 49 ng/mg LDL initial to 8 ng/mg LDL. The fluorescence intensity of the apo B, the peroxide level and diene conjugates increased to 161%, 571% and 162%, respectively, as compared with the control (= 100%) incubated in the absence of copper.

If the degrees of oxidation under identical conditions (2 μM copper, 3 hours) are compared, it is evident that LDL-1 with the lower initial alpha-tocopherol content is more susceptible to oxidation than LDL-2. This also shows that comparing different LDL preparations must take into account the initial content of their antioxidant activity. On the other hand, with one and the same LDL preparation the kinetics of the oxidation process and its dependence on the copper concentration was highly reproducible.

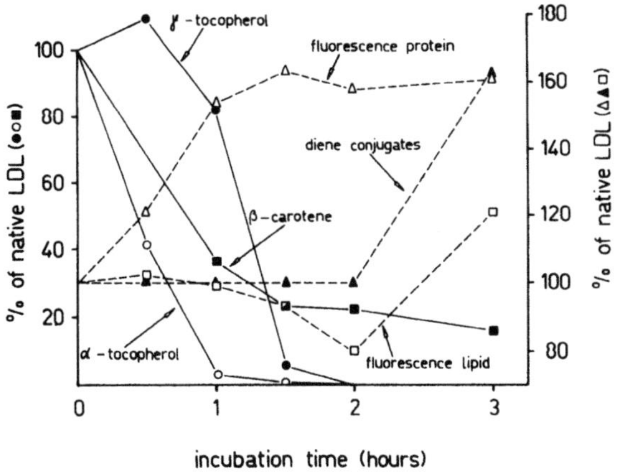

Figure 1. Change of the content of vitamin E, β-carotene conjugated dienes and the fluorescent material during peroxidation of LDL-2 with 2 μM copper(II)chloride.

Copper(II) ion represents a very potent initiator of lipid peroxidation in LDL. The effects seen depend strongly on the copper concentration. A concentration of 0.5 μM copper is sufficient to deplete LDL of alpha-tocopherol, but not of gamma-tocopherol and beta-carotene, and leads to a significant increase of the peroxide level in LDL. However, no lipid degradation products such as HNE or fluorescent compounds are formed. Higher copper concentrations of 2 to 10 μM completely deplete the LDL also of gamma-tocopherol and beta-carotene and with increasing concentration, cause a progressive formation of lipid degradation products.

ACKNOWLEDGMENTS

This work was supported by the Oesterreichischen Fonds zur Foerderung der Wissenschaftlichen Forschung, Projektnummer P6176B.

REFERENCES

1. T. Henriksen, E.M. Mahoney, and D. Steinberg, Enhanced macrophage degradation of low density lipoprotein previously incubated with cultured endothelial cells: Recognition by receptors for acetylated low density lipoproteins, Proc. Natl. Acad. Sci. U.S.A. 78:6499 (1981).
2. U.P. Steinbrecher, S. Parthasarathy, D.S. Leake, J.L. Witztum, and D. Steinberg, Modification of low density lipoprotein by endothelial cells involves lipid peroxidation and degradation of low density lipoprotein phospholipids, Proc. Natl. Acad. Sci. U.S.A. 81:3883 (1984).
3. M.F. Haberland, A.M. Fogelman, and P.A. Edwards, Specificity of receptor-mediated recognition of malondialdehyde modified low density lipoproteins, Proc. Natl. Acad. Sci. U.S.A. 79:1712 (1982).
4. D.W. Morel, J.R. Hessler, and G.M. Chisolm, Low density lipoprotein cytotoxicity induced by free radical peroxidation of lipid, J. Lipid Res. 24:1070 (1983).
5. O. Quehenberger, E. Koller, G. Jurgens, and H. Esterbauer, Investigation of lipid peroxidation in human low density lipoprotein, Free Rad. Res. Commun. 3:233 (1987).
6. H. Esterbauer, G. Jurgens, O. Quehenberger, and E. Koller, Autoxidation of human low density lipoprotein: Loss of polyunsaturated fatty acids and vitamin E and generation of aldehydes, J. Lipid Res. 28:495 (1987).
7. J. Lang, C. Celotto, and H. Esterbauer, Quantitative determination of the lipid peroxidation product 4-hydroxynonenal by high performance liquid chromatography, Anal. Biochem. 150:369 (1985).
8. K. Yagi, Assay for serum lipid peroxide level and its clinical significance, in: "Lipid Peroxides in Biology and Medicine," K. Yagi, ed., Academic Press, New York (1982).
9. Esterbauer, E. Koller, R. Slee, and J.F. Koster, Possible involvement of the lipid peroxidation product 4-hydroxynonenal in the formation of fluorescent chromolipids, Bichem. J. 239:405 (1986).

ISOLATION AND PARTIAL CHARACTERIZATION OF AN OXIDIZED LDL IN HUMANS

Pietro Avogaro, Gabriele Bittolo Bon
and Giuseppe Cazzolato

General Regional Hospital
Unit for Atherosclerosis
Venice, Italy

INTRODUCTION

Foam cells that accumulate in the earliest of atherosclerotic lesions, the so-called fatty streak, arise from two cellular sources: the arterial smooth muscle cells (SMC) and the monocyte-derived macrophage (MM).[2] The latter cell type in culture takes up only a small amount of native low density lipoproteins (LDL) by receptor-mediated endocytosis, but has a distinct receptor system that binds and degrades the more negatively charged LDL.[2,3] Incubation of cultured MM with LDL modified by acetylation (acyl-LDL) results in an accumulation of cholesteryl esters (CE) within the cells thus forming foam cells.[2] However such chemical modification in vivo seems unlikely. Recently, two possible mechanisms by which the more negatively charged LDL can be produced in vivo have been reported. The interaction of LDL with malondialdehyde (MDA) released by aggregating platelets or produced by peroxidation of fatty acids can lead to the formation of MDA-LDL that increase the CE deposition in cultured MM.[4] Moreover, the interaction of LDL with endothelial cells also alters LDL (EC-LDL), thus allowing their uptake by the MM.[5] Furthermore, experimental data have proved that LDL may exert their atherogenic properties by routes other than the interaction between LDL and its specific receptors. Several reports have actually demonstrated that LDL induces cytotoxicity,[6] injury[7] or inhibition of proliferation[7] of cultured endothelial cells. Such toxicity needs the presence of modified LDL and is attributable to the generation of lipid peroxides in the LDL.[6,8,9] Finally, oxidative modification of LDL inhibits the synthesis of prostacyclin by the endothelial cells[10] and inhibits the motility of MM.[11] The "in vivo" correlate of modified LDL has not been yet identified, although negatively charged LDL have been reported in human atherosclerotic lesions.[12]

METHODS

By ion exchange chromatography we have been able to characterize a more electronegative subfraction of LDL in freshly prepared plasma LDL obtained from a series of normal fasting subjects. LDL were prepared by density gradient centrifugation[13] and chromatographed on DEAE Sepharose column using a 0.04% NH_4HCO_3 buffer and a linear gradient from 0 to 0.3 M NaCl. The effluent was monitored at 280 nm for proteins and at 254 nm.

With the latter wavelength the presence of fatty acid conjugated dienes may be detected.[14] The eventual oxidative modification of LDL due to in vitro manipulations was prevented during all the study procedures by the presence in samples of a metal ion chelant (EDTA), an antimicrobial (chloramphenicol), two antioxidants (GSH and d-o-tocopherol), a protease inhibitor (e-aminocaproic acid) and a phospholipase A2 inhibitor (p-bromophenacyl-bromide). Furthermore UV, high temperature and O_2 were stringently avoided.

RESULTS AND DISCUSSION

The elution of LDL by ion exchange chromatography allowed the identification of a LDL subfraction with a more electronegative charge than the LDL bulk. In each case the elution of LDL, as determined by optical density at wavelength of 280 nm, showed that the bulk of LDL proteins was followed by a more or less evident "shoulder" of proteins incompletely resolved from the major peak. With the optical measurement at 254 nm a peak corresponding to the bulk of LDL proteins was observed, followed by a distinct peak that overlapped the protein "shoulder" recorded at 280 nm (Figure 1). Two-milliliter fractions were collected for analysis. Proteins were assayed by the method of Lowry et al.[15] Total cholesterol (TC), free cholesterol (FC), triglycerides (TG) and phospholipids (PL) concentrations were determined enzymatically with specific test kits from Menarini (Firenze, Italy). CE concentrations were calculated as (TC - FC) x 1.68. Diene-conjugated fatty acids esterified as cholesteryl esters (CE-DC), as triglycerides (TG-DC) and diene-conjugated products (DCP) were measured by isocratic high performance liquid chromatography accor-

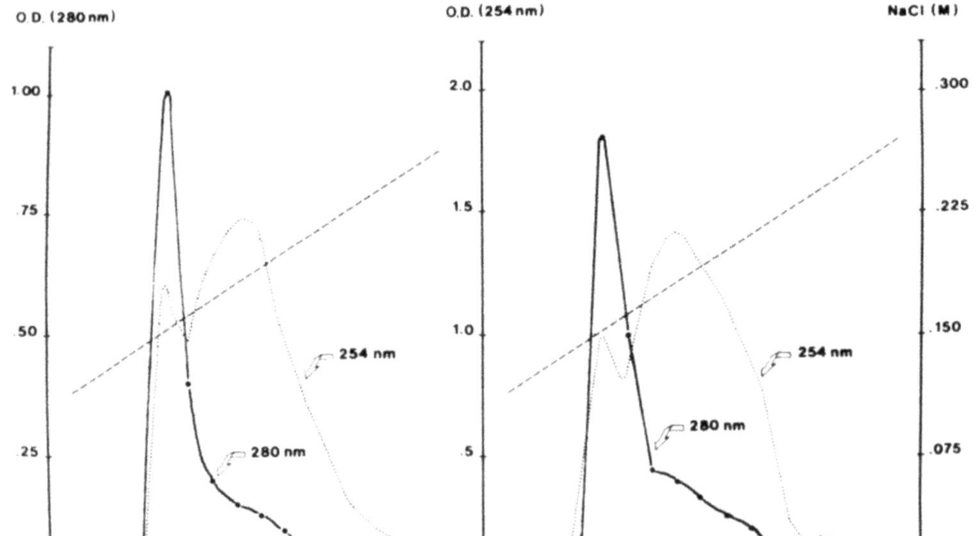

Fig. 1. Representative elution profile by ion exchange chromatography of two different samples of freshly prepared LDL.

ding to Cawood et al.[16] The values of DC and DCP were expressed in integrated area units (i.a.u.). The results were done as CE-DC (i.a.u.)/TC (mmol/l), TG-DC (i.a.u.)/TG (mmol/l) and DCP (i.a.u.)/ (TC + TG) (mmol/l).

The chemical analysis of the fractions eluted by ion exchange chromatography showed a sudden decrease of PL content corresponding to the second peak at 254 nm. Thus we have been able to subdivide LDL in two subfractions, operatively called nLDL (normal LDL) and mLDL (modified LDL). The latter LDL subfraction was present in normal subjects in an amount ranging from 5 to 20% of the total LDL (mean 13%). It contains more proteins and it is characterized by a low CE/FC ratio and expecially by very low levels of PL (Figure 2). This feature may be explained by the finding that oxidized PL are more susceptible to hydrolysis by phospholipases.[17] Recently, it has been demonstrated that phospholipase A2 activity plays an essential role in the endothelial cell induced modification of LDL.[18] Furthermore, the more electronegative LDL subfraction isolated by ion exchange chromatography contains more cholesterol esters and triglycerides conjugated dienes and more lipid peroxides (Figure 2).

Particular attention was paid to the study of apoprotein B (apoB), the main LDL protein, because of its crucial role in the metabolism and in the cellular receptor recognition of LDL. The LDL proteins were separated by SDS polyacrylamide gel electrophoresis (PAGE), as previously described.[19] The bands were identified by Western immunoblotting using appropriate monoclonal antibodies.[20]

The study of LDL proteins by SDS-PAGE showed the presence in nLDL of a band with the appropriate molecular weight of LDL apoB (apoB-100);

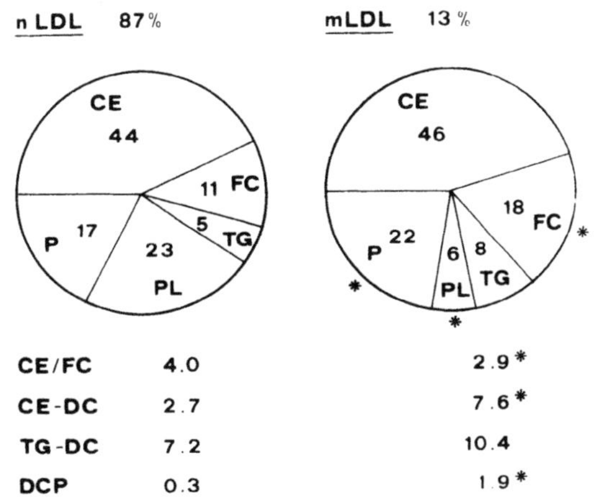

Fig. 2. Mean percentage contribution to total low density lipoproteins (LDL) and mean percentage composition of native LDL (nLDL) and modified LDL (mLDL) obtained from a series of normal subjects. (CE = cholesterol esters; FC = free cholesterol; TG = triglycerides; PL = phospholipids; P = proteins). The values of the CE/FC ratio and of the conjugated dienes of cholesteryl esters (CE-DC), triglycerides (TG-DC) and of conjugated dienes products (DCP) are also given.

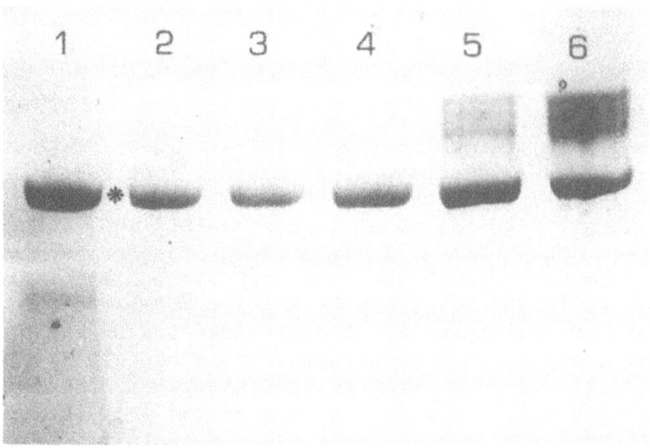

Figure 3. SDS gel electrophoresis pattern of apolipoprotein B in two different samples of LDL subfraction collected by ion exchange chromatography (1,2 = Total LDL; 3,4 = nLDL; 5,6 = mLDL). Asterisk denotes apoB-100 line.

whereas in mLDL, besides apoB-100, the appearance of peptides having a higher molecular weight was observed (Figure 3). By Western blotting these peptides were shown to be reactive against monoclonal anti-apoB antibodies and are probably due to apoB aggregates. The higher molecular weight peptides were not affected by treatment with mercaptoethanol, indicating that S-S bounds are not involved in the aggregate appearance. So, in contrast to endothelial cell-modified LDL, the mLDL we have isolated show the presence of apoB aggregates instead of lower molecular apoB species.[9] Previous studies have shown that the incubation of LDL with MDA[21] and with the lipid peroxidation product 4-hydroxynonenal[22] as well as the treatment with O_2[23] induce a partial conversion of the apoprotein B into higher molecular weight forms. It is also known that lipid peroxidation induces aggregation of various proteins, such as cell membrane proteins, enzymes and immunoglobulins.[24,25] More recently, Davies suggested that aggregation or fragmentation of proteins depends on the species of oxygen radicals involved in protein damage.[26]

The reported results, particularly the chemical composition, the higher presence of fatty acids conjugated dienes, peroxides and the presence of apoB aggregates, support the hypothesis that mLDL are principally produced by LDL peroxidation in vivo. Further evidence for this hypothesis comes from our recent studies on the effect of in vivo administration of antioxidant drugs on mLDL. After a short period of therapy with alpha-tocopherol in a series of patients with coronary atherosclerosis, a significant reduction of free fatty acids conjugated dienes in LDL and of mLDL plasma concentration was observed.[27]

The physiological and/or the pathological role of mLDL is not yet known. Actually it is not known if mLDL and/or an abnormal nLDL/mLDL ratio plays a role in atherogenesis, even when lipids and lipoproteins are kept within "normal" limits. It is noteworthy, however, that an increase of proteins and a decrease of PL were recorded in the LDL derived from myocardial infarction survivors.[28]

REFERENCES

1. R. Ross, Atherosclerosis: A problem of the biology of the arterial wall cells and their interaction with blood components, Arteriosclerosis 1:239 (1981).
2. J.L. Goldstein, Y.K. Ho, S.K. Basu, and M.S. Brown, Binding site on macrophage that mediates uptake and degradation of acetylated low density lipoproteins, producing massive cholesterol ester deposition, Proc. Natl. Acad. Sci. USA 76:333 (1979).
3. O. Stein and Y. Stein, Bovine aortic endothelial cells display macrophage-like properties towards acetylated ^{125}I-labelled low density lipoproteins, Biochim. Biophys. Acta 620:631 (1980).
4. A.M. Fogelman, I. Schechter, I.M. Seager, M. Hokom, J.S. Childs, and P.A. Edwards, Malondialdehyde alteration of low density lipoproteins leads to cholesterol ester accumulation in human monocyte-derived macrophages, Proc. Natl. Acad. Sci. USA 77: 2214 (1980).
5. T. Henriksen, E. Mahoney, and D. Steinberg, Enhanced macrophage degradation of low density lipoprotein previously incubated with cultured endothelial cells; recognition by receptors for acetylated low density lipoproteins, Proc. Nat. Acad. Sci. USA 78: 6499 (1981).
6. J.R. Hessler, D.W. Morel, J.L. Lewis, and G.M. Ghisolm, Lipoprotein oxidation and lipoprotein-induced cytotoxicity, Arteriosclerosis 3: 215 (1983).
7. T. Henriksen, S.A. Evensen, and B. Carlander, Injury to human endothelial cells in culture induced by low density lipoproteins, Scand. J. Clin. Lab. Invest. 39: 361 (1979).
8. J. Schuh, A. Novogradsky, and R.H. Haschemeyer, Inhibition of lymphocyte mitogenesis by autoxidized low density lipoproteins, Biochem. Biophys. Res. Comm. 84: 763 (1978).
9. U.P. Steinbrecher, S. Parthasarathy, D.S. Leake, J.L. Witztum, and D. Steinberg, Modification of low density lipoproteins by endothelial cells involves lipid peroxidation and degradation of low density lipoprotein phospholipids, Proc. Natl. Acad. Sci. USA 81: 3883 (1985).
10. A. Szczeklik and R.J. Gryglewski, Low density lipoproteins (LDL) are carriers for lipid peroxides and inhibit Prostacyclin (PGI2) biosynthesis in arteries, Artery 6: 488 (1980).
11. M.T. Quinn, S. Parthasarathy and D. Steinberg, Endothelial cell derived chemotactic activity for mouse peritoneal macrophages and the effects of modified forms of low density lipoproteins, Proc. Natl. Acad. Sci. USA 82:5949 (1985).
12. H.F. Hoff, LDL with altered surface charge: A new risk factor in atherogenesis? Artery 6: 178 (1979).
13. T.G. Redgrave, D.C.K. Roberts, and C.E. West, Separation of plasma lipoproteins by density gradient ultracentrifugation, Anal. Biochem. 65:42 (1975).
14. W.A. Pryor and L. Castle, Chemical methods for detection of lipid hydroperoxides, in: " Methods in Enzymology: Oxygen Radicals in biological systems, vol. 105", L. Packer, ed., Academic Press, New York (1984).
15. O.H. Lowry, N.J. Rosenburg, A.L. Farr, and R.J. Randall, Protein measurement with the folin phenol reagent, J. Biol. Chem., 193:165 (1951).
16. P. Cawood, D.G. Wickens, S.A. Inversen, J.M. Braganza, and T.L. Dormandy, The nature of diene conjugation in human serum, bile and duodenal fluid, FEBS letters, 162:239 (1983).
17. A. Sevanian, S.F. Muakkassah-Kelly, and S. Montestruque, The influence of phospholipase A2 and glutathione peroxidase on the elimination of membrane lipid peroxides, Arch. Biochem. Biophys. 223: 441 (1983).

18. S. Parthasarathy, U.P. Steinbrecher, J. Barnett, J.L. Witzum, and D. Steinberg, Essential role of phospholipase A2 activity in endothelial cell-induced modification of low density lipoprotein, Proc. Natl. Acad. Sci. USA 82:3000 (1985).
19. G. Cazzolato, G. Bittolo-Bon and P. Avogaro, Apoprotein B-48 is a constant finding in very low density lipoproteins in humans, Arteriosclerosis 5:88 (1985).
20. Y.L. Marcel, M. Hogue, R. Theolis, Jr., and R.W. Milne, Mapping of antigenic determinant of human apolipoprotein B using monoclonal antibodies against low density lipoproteins, J. Biol. Chem. 257:13165 (1982).
21. G. Bittolo-Bon, G. Cazzolato, S. Zago, and P. Avogaro, Effects of pantethine on in vitro-peroxidation of low density lipoproteins, Atherosclerosis 57: 99 (1985).
22. G. Jurgens, J. Lang, and H. Esterbauer, Modification of human low density lipoprotein by the lipid peroxidation product 4-hydroxynonenal, Biochim. Biophys. Acta 875:103 (1986).
23. G. Bittolo-Bon, G. Cazzolato and P. Avogaro, Changes of apolipoprotein B molecular weight and immunoreactivity in malondialdehyde-modified low density lipoproteins, Artery 12: 74 (1983).
24. A.L. Tappel, Measurement and protection from in vivo lipid peroxidation, in: "Free Radicals in Biology," W.A. Pryor, ed, Academic Press, New York (1980).
25. P. Hochstein and C. Rice-Evans, Lipid peroxidation and membrane alterations in erythrocyte survival, in: "Lipid Peroxide in Biology and Medicine", K. Yagi, ed., Academic Press, New York (1982).
26. K.J.A. Davies, Protein damage and degradation by oxygen radicals, J. Biol. Chem. 262:9895 (1987).
27. G. Bittolo-Bon, G. Cazzolato, and P. Avogaro, Presence of a modified LDL in humans: Effect of vitamin E, in: "Clinical and Nutritional Aspects of Vitamin E," O. Hayaishi and M. Mino, eds., Elsevier Science Publishers, Amsterdam (1987).
28. P. Avogaro, Phospholipids in human atherosclerosis, in " Phospholipids and atherosclerosis", P. Avogaro, M. Mancini, G. Ricci and R. Paoletti, eds., Raven Press, Nev York (1985).

CORONARY DISEASE: SPECTROSCOPIC CORRELATION OF STRUCTURE AND SIZE OF PLASMA LIPOPROTEINS WITH PATIENT CLINICAL HISTORY

B. Simic-Glavaski and J. J. Opplt*

Case Center for Electrochemical Sciences
and the Department of Chemistry
Case Western Reserve University
Cleveland, Ohio 44106, USA

INTRODUCTION

The possibility that plasma lipoproteins may be involved in human vascular disease or may act as indicators of such disease states has initiated much interest in the physicochemical properties of these macromolecules [1-6]. Reliable and rapid methods that can be widely used to extend the correlation between disease and the properties of macromolecules could have considerable clinical and research value, particularly if the results of such methods were shown to correlate with a disease state before its actual clinical manifestations.

Optical methods have been a traditional tool in the analysis of the structure and size of macromolecules in fluid suspensions and are widely accessible for making the correlations sought here. Plasma lipoproteins contain α-helical and β-form structural conformations which exhibit different optical and biophysical properties. In the studies reported here circular dichroic (CD) and Fourier transform infrared (FTIR) spectroscopies were used to detect and analyze these structural differences. In addition, the sizes of these lipoproteins were determined by means of self-beating (homodyne) laser spectroscopy. Optical measurements yielded information about the properties of plasma lipoproteins that are compatible with previous data obtained with a variety of techniques [1-6]. It was found that these physicochemical properties correlated with the clinical histories of patients.

METHODS AND PROCEDURE

Human serum lipoproteins were obtained from normal subjects (under 24 years old) and patients with proven coronary artery disease, as indicated by coronary angiography and type IV hyperlipoproteinenmia. These lipoproteins were preisolated as a total moiety by preparative ultracentrifugation under standard conditions so that flotant contained all known classes of plasma lipoproteins. The major portion of the flotant were subjected to molecular filtration on pretreated Sepharose-6B. A

*Department of Pathology, Cleveland Metropolitan General Hospital, Case Western Reserve University, Cleveland, Ohio 44106

neutral bufer at pH = 7.29, composed of 1.5 solution of NaCl and 0.15M solution of Tris-HCl was used as an eluant. In all preparatory procedures precise standard conditions were used, including column length, flow rate, elution volumes of all fractions, pH, temperature, and minimal physical shaking of the fractions; variation in all of these variables is known to affect aggregation sizes of some of the lipoprotein fractions.

Spectroscopic studies of the lipoprotein samples were analyzed by circular dichroism, Fourier transform infrared spectroscopy, and quasi elastic light scattering.

Circular Dichroism (CD)

The analysis of the Cotton effect and CD spectra of lipoproteins was performed in the spectral range of 190-250 nm and Jasco spectropolarimeter at 23°C. CD of lipoproteins was measured in a quartz cell of path length l = 5mm. Absorption of lipoproteins was measured at 280 nm and compared with an albumin standard of known concentration. Assuming that the lipid has negligible U.V. light absorption, it is then possible to calculate the molar ellipticity [θ] of the apoprotein. The mean residue weight of chromophore was taken to be MRW = 112 [7]. From the above elements the molar ellipticity [θ] in cm^2/decimole was obtained

$$[\theta] = \frac{\theta}{10} \frac{MRW}{l.c} \qquad (1)$$

where [Θ] is a recorded ellipticity.

Fourier Transform Infrared Spectroscopy

For infrared investigation, films of diluted lipoproteins were cast onto silver bromide plates. The samples were dried for several hours under lowered pressure at room temperature to remove excess water. The infrared spectra were obtained on a Digilab Model FTS-14 Fourier transform spectrophotometer. Each spectrum was recorded at a resolution of 2 cm^{-1} with total of 200 scans. The investigated frequency range is between 500-2000 cm^{-1}, with a precision of $\pm$ 2 cm^{-1} in the reported frequencies.

Quasi Elastic Light Scattering

The optical homodyne spectrometer was used in our investigation and has been reported earlier [8]. It consists of a Coherent Radiation Model 52B Ar^+ laser, thermoregulated scattering cell, EMI 9656 KR photomultiplier with Keithley 244 high voltage supply, Keithley 104 wide band amplifier and SAICOR SAI-52 real time 400 point spectrum analyzer and digital integrator. It is well established that in a solution of macromolecules, the light scattered from concentration fluctuations for a monodispersed system has a Lorentzian spectral distribution with a half width at half height Γ_d given by

$$\Gamma_d = 2K^2D \qquad (2)$$

where D is the translational diffusion constant and $\vec{K}$ is the scattering wave vector determined by laser wavelength and scattering geometry. By using the Stokes-Einstein equation

$$D = \frac{kT}{6\pi\eta r} \qquad (3)$$

one can calculate molecular radius, r, when the viscosity, η, and temperature, T, of solution are known; k is the Boltsman constant. The accuracy of linewidth measurements was better than 3%.

RESULTS AND DISCUSSION

In this study of the structural forms of lipoproteins, 12 male subjects were selected: 4 normal below 24 years of age, 4 with intermediate clinical indications, and 4 with proven coronary disease with type IV hyperlipoproteinemia.

Circular dichroism spectra obtained from chylomicrons, very low density VLDL, low density LDL, and high density HDL lipoproteins are shown in Figure 1. The spectra obtained from lipoproteins of normal subjects exhibit two distinct troughs at 218 and 208 nm with a crossover at 202 nm. Chylomicron fractions gave a single trough, and the trough intensity indicates their expected lower protein content and therefore smaller molar ellipticity. Troughs at 208 nm are due to $\pi \rightarrow \pi^*$ electron tansitions which are typical of α-helical conformations similar to that of poly-γ-methyl glutamate [9]. The shape of the CD curves and the location of the troughs indicates a considerable α-helical content in these lipoproteins.

The same analysis was performed on spectra obtained from chylomicrons, VLDL, LDL and HDL lipoproteins classes of patients with an intermediate degree and a strong degree of disease, as illustrated in Figure 1. The CD spectra are similar to those reported by Gotto [10]. They exhibit a single through at 218 nm, indicating $n \rightarrow \pi^*$ electron transition and resemble CD curves of poly-L-lysine [11] when it is in the β-form of the pleated sheet. Small molar ellipticity for chylomicrons and VLDL is not surprising since these two classes have a marked predominance of lipids over protein moiety. Although these two classes of lipoproteins are physiologically distinct, common principles govern their overall structure and therefore it is expected that their optical or physical properties are similar. A comparison of the curves from normal and diseased patients shows that coronary artery disease is accompanied by a decrease in the α-helix of plasma lipoproteins. No attempt was made to calculate the ratio of α-helix to β-form in samples obtained from normal subjects because CD spectra are influenced by the alkaline or acidic buffers sodium dodecyl sulfate, urea, temperature and chemical composition of LDL and HDL. However, by taking into account that all parameters were the same for all classes and samples, the most plausible conclusion that may be drawn from above data is that α-helical structure is present in the lipoproteins obtained only from normal subjects and in the HDL class from intermediate patients.

Detailed information about the structural interrelationship between the lipids and protein moiety was obtained with infrared spectroscopy

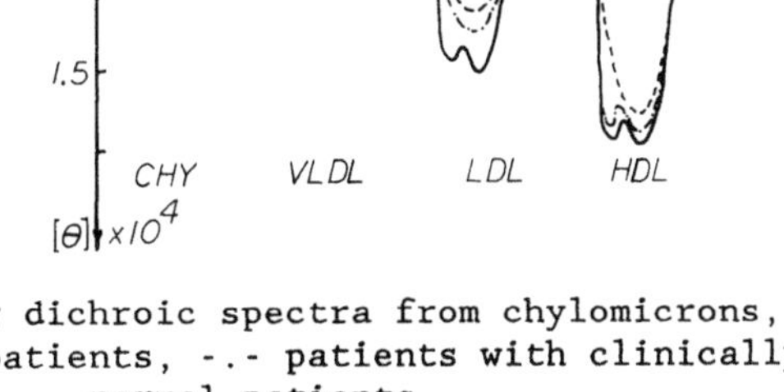

Figure 1. Circular dichroic spectra from chylomicrons, VLDL, LDL and HDL. --- pathological patients, -.- patients with clinically not proven hyperliproproteinemia, ___ normal patients.

(FTIR). The analysis was performed in the frequency range between 500-2000 cm^{-1} for all classes of lipoproteins. Spectra for lipoproteins from normal, intermediate and patients with advanced coronary disease are shown in Figure 2; the main features in these data are extended in Table 1.

The bands at 665 cm^{-1} show that the qualitative differences in the distribution of polypeptide chylomicrons and VLDL is minimal. These molecules appear to present members of the same chemical family, although the different proportion of protein and lipid content may account for their varying size and density. The band at 1745 cm^{-1} indicates the presence of cholesterol ester which occupies the inner core of these particles and is present in higher amounts in samples obtained from patients with coronary disease. A strong band at 1630 cm^{-1}, which

Table 1. Infrared Bands and Their Assignments[a]

Wavelength cm^{-1}	Chylomicrons a	Chylomicrons b	Chylomicrons c	VLDL a	VLDL b	VLDL c	LDL a	LDL b	LDL c	HDL a	HDL b	HDL c	Band Assignment
525	w	w	w	w	w	w	w	w	w	-	-	-	Disulfide stretch
600	s	s	s	s	s	s	s	s	s	s	s	s	Amide VI band
625	sh	sh	sh	sh	sh	sh	sh	sh	sh			sh	Secondary amide VI
665	s	s	s	s	s	s	s	s	s	s	s	s	Amide V disordered polypeptide
725	w	w	w	w	w	w	w.	w	w	-	-	-	N-H Deformation
800	w	w	w	-	-	-	w	w	w	-	-	-	C-H Deformation about c=c
910	M	M	M	M	M	M	M	M	M	M	M	M	C-H Deformation
1030	s	s	s	s	s	s	s	s	s	s	s	s	P-OH in the un-ionized form
1050	s	s	s	s	s	s	s	s	s	s	s	s	Monoglyceride
1135	M	M	M	M	M	M	M	M	M	M	M	M	P-N Stretching
1155	w	w	w	-	-	-	w	w	w	-	-	-	Asymmetric Stretching
1176	w	w	w	w	w	w	w	w	w	w	w	w	CH_2 Wagging
1210	w	w	w	w	w	w	w	w	w	w	w	w	Alcohol-highly symmetrical secondary
1290	s	s	s	s	s	s	s	s	s	s	s	s	Secondary amide III band
1330	ws	ws	ws	-	-	-	ws	ws	ws	-	-	-	C-H Group
1375	Ms	Ms	Ms	-	-	-	Ms	Ms	Ms	-	-	-	$C-CH_3$ Deformation
1389	s	s	s	s	s	s	s	s	s	s	s	s	$C(CH_3)_2$
1392	s	s	s	s	s	s	s	s	s	-	-	-	CH Bending
1399	s	s	s	s	s	s	s	s	s	s	s	s	$C(CH_3)_3$
1460	s	s	s	s	s	s	s	s	s	s	s	s	Amide II band
1525	-	Ms	Ms	-	Ms	Ms	-	-	Ms	Ms	Ms	Ms	β-Form pleated sheet
1555	s	s	s	s	s	s	s	s	s	s	s	s	Amide II band
1630	s	s	s	s	s	s	s	s	s	s	s	s	β-Form pleated sheet
1650	-	-	-	Ss	-	-	Ss	-	-	Ss	ws	-	α-Helix
1745	M	M	M	M	M	M	M	M	M	-	-	-	Triglicerides
1818	w	w	w	w	w	w	-	-	-	-	-	-	Lactones 4 numbered saturated ring

[a]s-strong, M-medium, w-weak, Sh-shoulder, Ss-strong shoulder, Ms-medium shoulder, ws- weak shoulder.

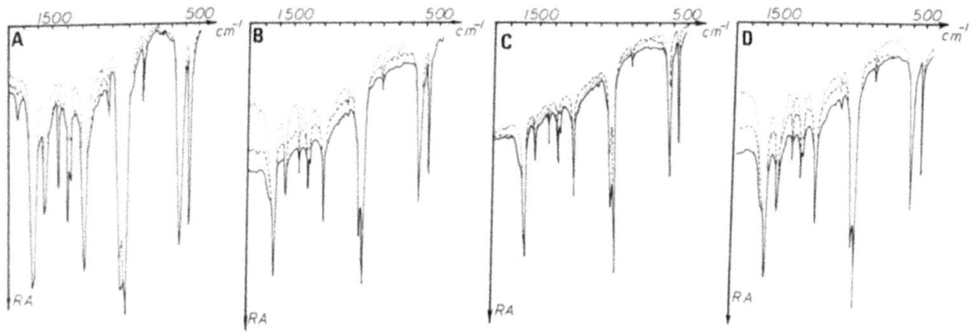

Figure 2. Infrared spectra of lipoproteins obtained from (A) normal subjects, full lines; (B) subjects with intermediate clinical indicators, broken lines; and (C) subjects with proven coronary disease, dotted lines.

represents the β-form of pleated sheets, dominates the infrared region between 1630-1650 cm^{-1}. This band is present for all samples of lipoproteins. A shoulder at 1650 cm^{-1} indicates α-helical structure. It is important to note, however, that the α-helical structure is present only in samples of VLDL, LDL and HDL obtained from normal subjects and is in direct correlation with the CD spectra shown in Figure 1.

The similarity of infrared spectra between VLDL and LDL suggests the role of VLDL molecules as precursors of LDL. Experimental support for this conclusion is also found in NMR studies [12-14].

Measurement of the molecular radii of chylomicrons, VLDL, LDL and HDL were conducted by laser self-beating spectroscopy on 20 normal patients and 130 patients with coronary disease. Spectra of scattered light from all the lipoproteins were a single Lorentzian shape, which sugested that the solution consisted of monodispersed molecules. The mean diameter of the dispersed molecules was calculated from the linewidth of the spectra as reported previously [15]. Figure 3 shows the data on the size of the lipoproteins from normal subjects and from those who presented with advanced coronary disease. It is evident from these results that the sizes of the lipoproteins from the normal subjects are smaller.

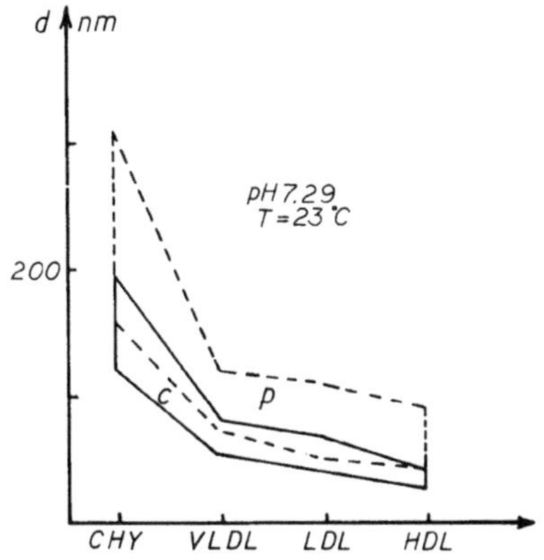

Figure 3. Distribution of mean diameters of lipoproteins obtained from 130 patients with proven hyperlipoproteinemia type IV (P - broken lines) and normal patients (C - full lines). Measurements were made at 23°C.

CONCLUSIONS

The physicochemical properties of the various lipoprotein fractions that we examined are consistent with the reported findings of other workers. Morever, the findings reported here show that these properties of plasma lipoproteins correlate well with the clinical histories of the patients. The optical methods descibed here are reliable, are easily reproducible, and, perhaps most importantly, provide information that may be useful in observing the onset of oxidative processes and in predicting the development of coronary or cardiovascular disease before its actual clinical manifestation.

REFERENCES

1. V. N. Schumaker and G. H. Adams, Circulating lipoproteins, Biochemistry 38:113 (1969).
2. A. M. Gotto, Recent studies on the structure of human serum low- and high-density lipoproteins, Proc. Nat. Acad. Sci. USA 64:1128 (1969)
3. A. M. Gotto, W. V. Brown, R. I. Levey, M E. Birnbaumer, and D. S. Fredrickson, Evidence for identity of the major apoprotein in low density and very low density lipoproteins in normal subjects and patients with familial hyperlipoproteinemia, J. Clin. Investigation 51:1486 (1972).
4. J. D. Morrisett, R. L. Jackson and A. M. Gotto, Lipoproteins: structure and function, Annu. Rev. Biochem. 44:183 (1975).
5. D. W. Morel, P. E. DiCorleto and G. M. Chisolm, Endothelial and smooth muscle cells alter low density lipoprotein in vitro by free radical oxidation, Arteriosclerosis, 4:357 (1984).
6. H. Esterbauer, G. Jurgens, O. Quehenberger and E. Koller, Autoxidation of human low density lipoprotein: loss of polyunsaturated fatty acids and vitamin E and generation of aldehydes, J. Lipid-Res 28:495 (1987).
7. A. Scanu and R. Hirtz, Human serum low-density lipoprotein protein: its conformation studied by circular dichroism, Nature 218:200 (1968)
8. A. M. Jamieson, B. Simic-Glavaski, K. Tansay and A. G. Walton, Studies of elastin coacervation by quazielastic light scattering, Faraday Discussion Chem. Soc. 61:194 (1976).
9. G. Holzwarth and P. Doty, The ultraviolet circular dichroism of polypeptides, J. Am. Chem. Soc. 87:918 (1965).
10. A. M. Gotto, R. I. Levy, A.S. Rosenthal, M E. Birnbaumer, and D. S. Fredrickson, The structure and properties of human beta-lipoprotein and beta-apoprotein, Bioch, Biophys. Res. Comm. 31:(5) 699 (1968).
11. D. G. Fasman and J. Potter, The optical rotatory dispersion of two beta structires, Biochem. Biophy. Res. Comm. 27:209 (1967).
12. R. B. Leslie, D. Chapman and A. M. Scanu, Nuclear magnetic resonance studies of serum low density lipoproteins (LDL_2), Chem. Phys. Lipids 3:152 (1969).
13. J. A. Hamilton, G. Talkawski, E. Williams, E. M. Avila, A. Allerhand, E. Cordes, and G. Comejo, Natural abundance carbon-13 nulear magnetic resonance spectra of human serum lipoproteins, Science 180:193 (1973).
14. J. M. Steim, O. J. Edner and F. G. Bargovt, Structure of human serum lipoproteins: nuclear magnetic resonance supports a micellar model, Science 162:909 (1968).
15. R. W. DeBlois, E. E. Uzgiris, S. K. Davi and A. M. Gotto, Jr., Application of laser self-beat spectroscopic-technique to the study of solutions of human plasma low-density lipoproteins, Biochemistry 12:(14) 2645 (1973).

THE EFFECT OF OXYGEN ON THE OH RADICAL-INDUCED STRAND BREAK FORMATION OF DNA IN VITRO AND IN VIVO

Dietrich Schulte-Frohlinde

Max-Planck-Institut für Strahlenchemie
D-4330 Mülheim/Ruhr, Stiftstr. 34-36, W. Germany

INTRODUCTION

The increased radiosensitivity of oxygenated cells as compared to deoxygenated cells is called "The Oxygen Effect" in radiobiology. Quantitatively the oxygen effect is expressed as the oxygen enhancement factor (OER), which is the ratio of the dose in the absence of oxygen to that in its presence required to obtain the same effect. With survival as the end point, the OER is the dose required to deactivate 63% (37% survival) of the population in the presence of oxygen (D_{37} O_2) divided by that in the absence of oxygen (D_{37} N_2). For E. coli strains the OER values are typically around three. The values depend on the genetic background of the strain, on the medium, on additives (e.g., thiol concentration) and on the status of the cells (e.g., logarithmically growing cells or stationary cells).

Bonura et al.[1] showed for E. coli cells (K-12) that the ratio of the number of double-strand breaks (dsb) in the genome of oxygenated cells divided by the number of deactivated cells is 1 - 1.3. These authors suggested that a dsb is the lethal event in γ-irradiated E. coli cells which is valid as long as the cells carry only one genome.[2] Later on this result was confirmed by various other research groups and it is now generally accepted that an unrepaired dsb is a lethal event in E. coli and yeast cells carrying only one genome per cell.[3-9]

A question under lively debate is that of the pathways leading to strand break (sb) formation. Principally two routes can be distinguished in a cell. One is the formation through chemical reactions, mainly free radical reactions induced by the absorption of radiation energy without participation of enzymes. The other one is the formation of dsb by enzymatic action on DNA damaged sites, e.g., base damages, or single-strand breaks (ssb). Bonura et al.[1] estimate that 70% of the dsb formation in a γ-irradiated E. coli K-12 polA$^-$ strain is due to the activity of nucleases on damaged sites of DNA. Bresler et al.[10] claim that 80% of the observed ssb formation in the DNA of E. coli cells results from the enzymatic pathway. Fielden[11] and his group (Sapora et al.[12]) demonstrated with the help of a rapid-lysis technique that 1.3 endonuclease-sensitive sites were present for each radiation-induced break in the DNA of irradiated E. coli cells. Bryant[13] has shown that with M. luteus endonuclease the number of dsb in Chinese hamster cells after irradiation can be increased by a factor of 1.7. However, Bryant showed also that repair and enzymatic formation of dsb are simultaneous processes in γ-irradiated cells.

A further complication in the determination of the number of dsb produced by the free radical pathway in cells is enzymatic repair. Using pulse irradiation and a cell-lysis technique which allowed determination of dsb within a few minutes before repair of dsb had set in, Tilby and Loverock[14] showed that under these conditions the number of dsb per lethal event also is found to be 1 - 1.3. However, the absolute yield of dsb formation is found to be 2 to 3 times smaller by the fast-lysis technique than by the conventional slow procedure.

If an unrepaired dsb is proposed to be the lethal event, then in the presence of oxygen the number of dsb in the cell should be larger than in its absence. This has indeed been found. The OER values for survival and for dsb formation are the same within the range of error (Table 1).

Table 1. OER values for survival and dsb formation in E. coli cells due to Bonura et al.[1]

	SURVIVAL	**DSB**
OER	**2.8 $\pm$ 0.4**	**2.4 $\pm$ 0.4**

Clearly, one of the effects of oxygen in cells is to increase the number of sb per dose. From pulse irradiation studies of the oxygen effect on cells, it follows that the increase of dsb by oxygen is due to the interaction of oxygen with the radiation-induced free radicals in the cells since addition of oxygen more than 100 ms after the pulse does not result in any oxygen effect.[15] Enzymatic reactions occur not sooner than 2 seconds after the pulse.[16]

In order to study the chemical reactions leading to dsb, it is of advantage to investigate DNA in aqueous solution. If this is done with biologically active DNA (baDNA) which may be plasmid or phage DNA, surprisingly neither for survival[17] nor for dsb formation[18] can an oxygen effect be found. On the contrary, oxygen acts as a protective additive. This result has nothing to do with the fact that the lethal damage of baDNA in aqueous solution is mainly due to single-strand base damage and ssb (for a recent review see reference 19). A dsb is also lethal for baDNA in aqueous solution, but its contribution to the overall lethality is small.[19]

On the basis of the results of Howard-Flanders[20] and of Hutchinson,[17] an oxygen effect on the survival of bacteriophage or plasmid baDNA resp. is observed in aqueous solution when a thiol is present. The necessity of the presence of thiols for the occurrence of an oxygen effect in cells has been demonstrated by showing that cells with a reduced glutathione content exhibit a significantly smaller oxygen effect.[21] The explanation offered is that thiols react in the absence of oxygen with free radicals of the DNA by H transfer, thereby eliminating the free radical state of the DNA. The H transfer is postulated to decrease the biological deactivation (this effect is called repair whatever this may be in detail). However, in the presence of oxygen, because oxygen reacts much faster with the free radicals of DNA than the thiols do, it has to be assumed that the DNA radicals are converted into peroxyl radicals. It is said that "the radiation damage is fixed" by the reaction with oxygen.[22]

In the first part of the present article, the lifetime of the peroxyl radicals of DNA is discussed and the question whether or not peroxyl radicals react with thiols is dealt with. In the second part, recent results on the dose dependence of dsb formation of DNA in aqueous oxygenated solution are shown. In the third part, new results are presented concerning the terminal chemical steps for ssb formation of polyuridylic acid (poly[U]) induced by ·OH radicals in the presence of oxygen. In the last part the present stage of our knowledge of the chemical steps leading to ssb formation of DNA is reviewed.

LIFETIME OF PEROXYL RADICALS OF DNA IN AQUEOUS SOLUTION AND IN CELLS AND THE REACTION OF THE PEROXYL RADICALS WITH THIOLS

As a first step in the investigation of the oxygen effect in more detail, we have studied the lifetime of the postulated peroxyl radicals of DNA in aqueous solution by electron spin resonance (ESR) spectroscopy.[23] First of all, it could be shown that the ESR signals of the DNA radicals which appeared on reacting ·OH radicals with DNA in oxygenated aqueous solution are compatible with those of peroxyl radicals. Secondly, it could be shown that the lifetime of these peroxyl radicals is in the range of seconds, whereas the lifetime of the peroxyl radicals of low molecular weight model compounds is less than milliseconds. The longer lifetime is caused by the polymeric structure of the DNA in combination with the high charge of the polyions. These results show that peroxyl radicals are indeed formed in the presence of oxygen. However, from the work of Willson[24] it follows that not all DNA radicals are converted by oxygen into peroxyl radicals. Certain radicals from the purine bases do not react with oxygen. Up till now we have not seen ESR signals from purine radicals in aqueous solution.

As a next step the reaction of the peroxyl radicals with thiols was studied. First of all it was found that thiols react with the DNA peroxyl radicals. The rate constant is in the range of $k = 10^4$ to 10^5 $M^{-1}s^{-1}$ (Table 3).

The identity of the product of the reaction of a thiol with a peroxyl radical is unknown. Probably it is a hydroperoxide and the faster decay of the peroxyl radicals of DNA in the presence of thiols is due to a peroxyl → hydroperoxide conversion.

Table 2. Lifetime of peroxyl radicals of ss and dsDNA and of some polynucleotides[23]

SUBSTRATE	pH	$[NaClO_4]$ 10^{-3} M	t1/2 (s)
POLY(U)	6.3	0	3.3
	6.3	10	2.0
POLY(A)	7.4	0	0.6
	7.4	10.0	0.2
ssDNA	7.2	1.0	0.65
	7.2	10.0	0.18
dsDNA	7.6	1.0	0.9
	7.5	10.0	0.35

The precise effect of the peroxyl → hydroperoxide conversion on ssb formation and on the biological activity of DNA irradiated in aqueous solution is difficult to determine since the exact number of ·OH radicals that have reacted with DNA or with the thiol has to be known in order to obtain the effect of the peroxyl conversion. It is now clear that more sophisticated mathematical relationships than known hitherto have to be used in order to solve the problem of the reaction of OH radicals with DNA in the presence of scavengers. The new relationship takes into account the polymeric structure of DNA, which leads to a much lower efficiency of scavengers than calculated by ordinary competition kinetics.[25] To my knowledge analyses of appropriate experiments using the new competition kinetics for the determination of the effect of thiols have not yet been performed. In the absence of oxygen the influence of different pH on the activity of biologically active DNA irradiated in aqueous solution in the presence of thiols in comparison with that on sb formation shows that base damage and repair plays a role in the deactivation.[26]

Table 3. Rate constants of peroxyl radicals of ss and dsDNAand of some polynucleotides with various thiols[23]

	POLY(U) pH=6.3	POLY(A) pH=7.4	ssDNA pH=7.2	dsDNA pH=7.5
DITHIOTHREITOL	$5x10^4$	$1.3x10^5$	$3x10^4$	$4x10^4$
CYSTEAMINE	$1.7x10^4$	$1x10^5$	$4x10^4$	$6x10^4$
GLUTATHIONE	$0.8x10^4$	$5x10^4$	$1.5x10^4$	$6x10^3$

The question may be posed whether or not the peroxyl-hydroperoxide conversion occurs in an E. coli cell and whether or not this reaction has an effect on survival and on sb formation. In E. coli cells the main thiol is glutathione in a concentration of up to 7 x 10^{-3} M.[27] This concentration may be just sufficient to convert the peroxyl radical groups of the DNA into hydroperoxyl groups, especially when the lifetimes of the DNA peroxyl radicals are longer due to the high viscosity in the cell.[28] On the other hand, other reducing species, such as ascorbic acid may also reduce the peroxyl groups. However, from the protection which added thiols have on survival of E. coli cells, no conclusion can be drawn whether or not the effect of the thiols is due to ·OH scavenging or to H donation to the peroxyl groups of DNA since the influence of thiols on the efficiency of the enzymatic repair system is so large that it dominates the results. An influence of added cysteamine on the enzymatic repair system of E. coli strains has been demonstrated by showing that strains deficient in the recABC repair system exibit drastically reduced protection in comparison with wild type strains.[28-30] Another argument for an influence of thiols on the enzymatic repair system is the observation that the effect of thiols depends on the growth condition of the cells. Better protection is found in cells in the stationary growth phase than in the logarithmic phase.[31] A further argument is the surprising result that the increase of survival by addition of cysteamine can be removed by the addition of another free radical scavenger, ascorbic acid.[32] These results are not compatible with ·OH radical scavenging as the mechanism of protection by thiols in cells in the presence of oxygen.

An argument in favour of a peroxyl-hydroperoxide conversion in cells is based on a completely different approach. Marklund et al.[33] have compared the radiosensitivity of various human cell lines differing in their content of glutathione peroxidase, catalyse and oxygendismutase (SOD) and found that the content of catalase and SOD had no effect, but an increased peroxidase content decreased the radiosensitivity. This may be taken as a hint that hydroperoxide groups may have a damaging effect on the biological activity of DNA. This result is in line with the assumption that the reactions of oxygen with the free radicals of the DNA in a cell lead to a biological damage not completely repairable by thiols.

Summarizing, it may be said that "the oxygen effect" is basically the result of free radical reactions involving DNA radicals, oxygen and thiols, but the result concerning survival is strongly modified by the influence of the thiols on enzymatic repair. This influence of the thiols depends on concentration, chemical structure of the thiols and the genetic background of the cells. Because of the strength of the effect on enzymatic repair in practice, it is very difficult to distinguish between the influence of a thiol on the free radical contribution to the oxygen effect from that on the enzymatic contribution.

DOSE DEPENDENCE OF SSB AND DSB FORMATION OF DNA INDUCED BY OH RADICALS IN AQUEOUS SOLUTION IN THE PRESENCE OF OXYGEN

Whereas ssb in DNA are formed linearly with dose, dsb is produced linearly as well as quadratically with dose in aqueous solution.[34,18] The quadratic function is easily understood since the dsb formation occurs as the result of the generation of two independently formed ssb within a certain distance (h) along the strand (accumulation process). The distance is found to be h = 43 ± 10 base pairs in vitro (0.1 M $NaClO_4$, neutral pH, calf thymus DNA, O_2 saturation). Others found smaller values (e.g., van Touw et al.[35] found h = 29 ± 6).

The first-order formation of dsb in vitro poses the interesting problem as to how a single ·OH radical in DNA in the presence of oxygen produces a dsb. An attractive hypothesis is that the first step is the formation of a ssb by one ·OH radical. This step involves a sugar radical, which remains attached to the strand. This sugar radical will be a peroxyl radical and due to its long lifetime it should be able to abstract an H atom from the sugar moiety of the complementary strand with a certain probability, initiating a free radical pathway in the complementary strand which eventually leads to ssb formation in that strand thereby generating a dsb.

The efficiency of dsb formation by a single ·OH radical in vitro in the presence of oxygen is ~ 0.6%. For comparison the efficiency of ssb formation is ~ 10 - 14%, based on the data of Siddiqi and Bothe.[18] The efficiency of dsb formation seems to be low. However, the value is large enough to contribute to the first-order formation of dsb in cells up to ~ 150 dsb per genome in yeast[36] and 90 in E. coli[14] (see also the work with thymocytes by Lennartz et al.[37]) under the assumption that mainly ·OH radicals are responsible for dsb formation in cells. Taking the scavenger capacity (the scavenger capacity is defined as the product of the scavenger concentration times the reaction rate constant of ·OH radicals with that scavenger) of _E. coli_ cells as $5 - 8 \times 10^{8} s^{-1}$ Verberne et al.[25] showed that DNA and the scavenger compete for ·OH radicals produced in spurs. The general importance of spur effects to the deactivation of DNA in cells has been already pointed out by Bonura and Smith[38] and by Ward.[39] In spurs it is concievable that dsb are formed with a high probability due to the high concentration of produced radicals. A detailed discussion of the contribution of spur reaction and of the single ·OH radical pathway showed that even for spurs containing several ·OH radicals (e.g., up to 5) the contribution of the single ·OH radical pathway is significant (Bothe and Schulte-Frohlinde, in preparation). The reason why the first order formation of dsb by a single ·OH radical may contribute so considerably to dsb formation in cells is found in the high molecular weight of DNA in cells (up to 2.8×10^9 in E. coli cells) and the fact that the DNA molecules are separated from each other by the cell walls. The number of dsb produced by a single ·OH radical does not depend on the molecular weight of the DNA, however the number of dsb produced by the accumulation pathway decreases with the molecular weight for the same number of ·OH radicals reacting per DNA molecule. For DNA with a molecular weight $\simeq 10^9$ dalton, the probability for the latter process is very small (Bothe and Schulte-Frohlinde, in preparation). From the work of Verberne et al.[25] it can be estimated that the contribution of the direct effect to the deactivation of biologically active DNA may be 25%. It is important to point out that this value was obtained from γ-irradiated aqueous solution in the presence of scavengers at a concentration such that the scavenger capacity of the aqueous solution is equivalent to that of an E. coli cell. The contribution of the direct effect was obtained from the scavenger concentration-independent part of the deactivation which, however, also contains a contribution of deactivation by organic radicals resulting from ·OH reaction with scavengers.

The three postulated pathways nicely explain the formation of dsb by a first-order dose dependence. However, each pathway is accompanied by ssb formation with a higher probability than that for dsb formation. The occurrence of ssb should have the consequence that a quadratic dose dependence for dsb formation should be found. Two causes have to be discussed to account for the absence of the quadratic dsb formation in cells. The first is the fast disappearence of the ssb by repair. The half-life of a ssb in DNA within a cell in the presence of oxygen is less than 15 s.[40,11] The repair of ssb is practically complete in < 6 min.[41] The repair of dsb requires $\simeq 20 - 180$ min. or does practically not occur depending on conditions.[42] The second and main cause is that the probability to find two ssb within a distance of 43 base pairs in a DNA with a molecular weight of 10^9 is too small to lead to a non-linear dsb formation in cells. Another cause for a less efficient dsb formation by accumulation could be found in a smaller h value in vivo (compare Boye[43] for the case of superinfecting phage DNA irradicated in E. coli).

Summarizing, it may be said that the chemically induced first-order formation of dsb in cells is due to three effects: the single ·OH radical dsb formation, the spur effect and the direct effect. The contribution of dsb formation by a ssb accumulation mechanism is not observed due to the fast repair of ssb in the cells and due to the too low density of the ssb along the DNA strand for DNA of $\simeq 10^9$ dalton.

The first-order formation of dsb in cells is thus no reason to exclude ·OH radicals as the damaging intermediate for the irradiation-induced cell death and no reason to postulate direct effects as the only pathway to DNA damage in cells, as has been done various times in the literature.

MECHANISM OF SB FORMATION OF POLY(U)

Strand break formation of polyuridylic acid (poly[U]) in aqueous solution induced by ·OH radicals in the presence of oxygen has been studied in detail.[44-46] The yields of the addition at the various positions where the ·OH adds to the uracil moiety have been determined. Addition occurs at positions 5 (70%) and 6 (23%) of the uracil moiety and H-abstraction from the sugar moiety with 7%.[47] The base radicals contribute considerably to ssb formation since the efficiency of ssb formation initiated by an ·OH radical ($\geq$50%) is much higher than the efficiency of H-abstraction from the sugar (7%). The rate of the major fraction of ssb formation is in accord with the decay of the peroxyl radicals of poly(U).[44] The rate-determining step is most likely H-abstraction from the sugar by peroxyl radicals of the base.[23] Two pathways lead from the sugar radicals to ssb formation, the C-2′ and the C-4′ pathway. Two new results indicate that the C-2′ pathway is the predominant one. The first result is that the efficiency of ssb formation with poly(dU) is only $\leq$ 0.15% per ·OH radical (Adinarayana, Bothe, and Schulte-Frohlinde, in preparation), which is much smaller than that for poly(U) with $\geq$ 50%. In poly(dU) the C-2′ pathway has a low probability due to the absence of the 2′OH group in the sugar. It is therefore proposed that in poly(U) H-abstraction occurs with a high probability at position 2′ followed by a fast cleavage of the phosphoric acid ester bond either via reaction (1) followed by reaction (5) or via reaction (2). Reaction (2) will be the main route.

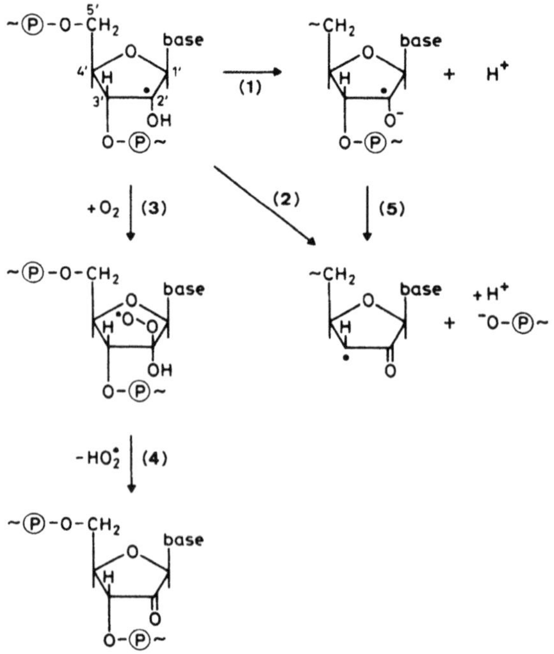

The cleavage of the phosphate bond occurs with a rate faster than that of addition of oxygen (eq. 3). Addition of oxygen would lead to a peroxyl radical which should quickly eliminate $O_2^{\cdot-}$ (eq. 4), and strand break formation would be prevented. If this C-2′ cleavage mechanism is correct, the efficiency of ssb formation should decrease with increasing oxygen concentration due to reaction (3). This has indeed been found. At an oxygen pressure of 100 atm., the efficiency of ssb formation is smaller by a factor of two. A further argument for a fast phosphate cleavage induced by a free spin in β-position and a hydroxy group in α-position to the free spin is the following. The elimination of phosphate in the β-position of a C-radical of glycerol phosphate is so fast that oxygen in air saturated solution does not compete with the phosphate elimination[48] (also M. L. Scholes and C. v. Sonntag, unpublished result, 1987). The fast decay of the C-2′ radical has the consequence that added thiols cannot react with the C-2′ radical. The inhibition of ssb formation of poly(U) by added thiols (e.g., by dithiothreitol) is therefore not due to a reaction with the C-4′ radical but is due to H-donation to the peroxyl radicals of the uracil moiety. A proposal has been made for a pathway leading from the C-4′ sugar peroxyl radical to ssb formation.[45] This pathway contributes to ssb formation in poly(U) with a G value of up to 0.8 on the basis of the isolation of glycolaldehyde which is formed with a G value of 0.8.[49]

The new results indicate that the C-2′ pathway dominates over the C-4′ pathway in poly(U). The C-2′ pathway has a much lower probability in poly(dU) and DNA due to the absence of the hydroxy group at C-2′. The postulated pathway via C-4′ is therefore expected to contribute to a larger extent in DNA and in poly(dU) than in poly(U). The C-4′, mechanism is discussed in the part on ssb formation in DNA. Other reactions of peroxyl radicals in poly(U) lead, for example, to base or nucleoside release or to a consumption of oxygen (up to G $[-O_2]$ values of 20).[49,50] The latter reaction is probably not a chain reaction but a cascade reaction. The observed dsb formation induced by one ·OH radical in the presence of oxygen is an example for a cascade reaction since the postulated radical-transfer step from one strand to the opposite one is characteristic of this type of reaction.

PATHWAYS TO SSB FORMATION IN ssDNA

The efficiency of ·OH radicals for the formation of ssb in single-stranded (ss) DNA in the presence of oxygen is rather low (10 - 14%). Since this percentage is similar to that expected for reaction of ·OH with the sugar (7 - 20%), it is not obvious whether or not base radicals (produced by addition of ·OH to one of the bases of DNA) contribute to ssb formation. Furthermore, it is not clear whether or not H-abstraction from the sugar occurs (in addition to that by ·OH radicals) by peroxyl radicals of the base or by rearranged base radicals which have not reacted with oxygen.

If base radicals contribute to ssb formation in ssDNA, then the interesting question arises whether or not certain base radicals play a greater role than others. Preliminary results show that the efficiency of ssb formation is five times greater for poly(dA) than for all other polydeoxyribohomonucleotides measured (Table 4). This points to a contribution of radicals derived from the adenine moiety to ssb formation of ssDNA.

Table 4. Single-strand break formation of ss and dsDNA and some polydeoxynucleotides induced by ·OH radicals in the presence of oxygen. γ-Radiolysis, N_2O/O_2 (4:1), G values given (Bothe, Adinarayana, and Schulte-Frohlinde, 1987, in preparation)

	G(ssb)	$G(K^+)$
POLY dU	0.15	1.35
POLY dT	0.30	2.71
POLY dC	0.28	2.56
POLY dA	1.59	14.33
POLY dG	0.10	0.96
ssDNA	0.60	4.15

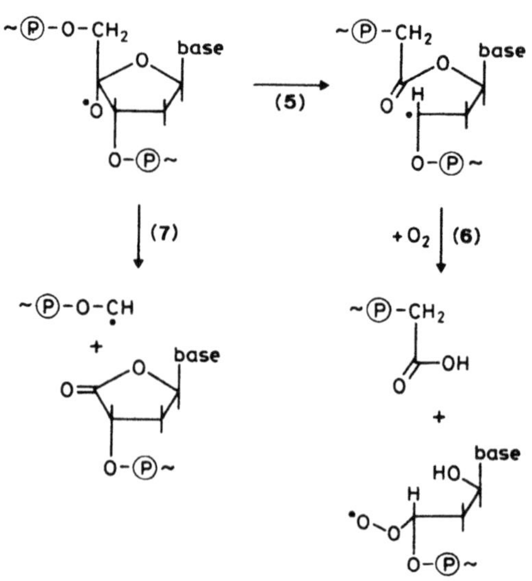

The chemical step leading to strand rupture must involve a sugar radical or a labile sugar derivative. A mechanism has been published in which a pathway is postulated starting from the C-4′ radical in the sugar.[45] This pathway involves addition of oxygen to the C-4′ radical, forming a peroxyl radical which is transformed by a bimolecular reaction with another peroxyl radical into an oxyl radical. In an intramolecular reaction the oxyl radical cleaves the C-3′—C-4′ bond (eq. 5) followed by hydrolysis of the remaining carboxylic acid ester present (eq. 6). This mechanism is attractive because the end group with the structure (P-O-CH_2-COOH) according to this mechanism has been identified by Henner et al.[51] in γ-irradiated DNA in the presence of oxygen. Most probably, however, this pathway is not the only one since the rate of ssb formation of ssDNA induced by ·OH radicals cannot be described by a simple first-order rate constant. In addition, reactions like the following are conceivable: instead of a C′4′—C-3′ bond cleavage, cleavage of the C-4′—C-5′ bond may occur in the oxyl radical (eq. 7), which leads directly to ssb. In addition, cleavage may be initiated by peroxyl radicals at C-5′ followed by cleavage of the C-5′—C-4′ bond by a C-5′ oxyl radical. Cleavage of the C-2′—C-1′ will probably not lead to ssb formation. Another pathway postulated the formation of a hydroperoxide from a peroxyl group at C-4′ followed by two steps leading to the cleavage of the C-3′ - C-4′ bond and hydrolysis of the 3′ phosphate bond[52,53]. However, the conversion of DNA peroxyl radicals into hydroperoxides has been suggested to have protective consequences[54]. The contribution of the various pathways to ssb formation can be obtained from the yield and the assignment of different structures of end groups of the irradiated DNA. However, complete determination of all the end groups has not been carried out as yet.

REFERENCES

1. T. Bonura, C. D. Town, K. C. Smith, and H. S. Kaplan, The influence of oxygen on the yield of DNA double-strand breaks in X-irradiated Escherichia coli K-12, Radiat. Res., 63:567 (1975).
2. F. Krasin and F. Hutchinson, Repair of DNA double strand breaks in Escherichia coli, with requires recA function and the presence of duplicate genome, J. Mol. Biol., 116:81 (1977).
3. H. S. Kaplan, DNA-strand scission and loss of viability after irradiation of normal and sensitized bacterial cells, Proc. Natl. Acad. Sci. USA, 55:1442 (1966).
4. K. S. Y. Ho, Induction of DNA double-strand breaks by X-rays in a radiosensitive strain of yeast Saccharomyces cerevisiae, Mutation Res., 30:327 (1975).
5. R. E. Krisch, F. Krasin, and C. J. Sauri, DNA strand breakage, repair and lethality after ^{125}I decay in rec^+ and recA strains of Escherichia coli, Int. J. Radiat. Biol., 29:37 (1976).

6. M. A. Resnick and P. Martin, The repair of double-strand breaks in the nuclear DNA of Saccharomyces cerevisiae and its genetic control, Molec. Gen. Genetics, 143:119 (1976).
7. M. A. Resnick, Similar responses to ionizing radiation of fungal and vertebrate cells and the importance of DNA double-strand breaks, J. Theoret. Biol., 71:339 (1978).
8. D. Frankenberg, M. Frankenberg-Schwager, D. Blöcher, and R. Harbich, Evidence for DNA double-strand breaks as the critical lesions in yeast cells irradiated with sparsely or densely ionizing radiation under oxic or anoxic conditions, Radiat. Res., 88:524 (1981).
9. J. W. Hülsewede, Doppelstrangbrüche in der DNA in vivo und vitro nach ^{60}Co-γ-Bestrahlung, Dissertation Ruhr-Universität Bochum (1985).
10. S. E. Bresler, L. A. Noskin, and A. V. Suslov, Study of induction and repair of double-strand breaks in DNA in cells of prokaryotes and eukaryotes. I. Use of Zimm elastoviscometer for a study of induction of double-strand breaks in the DNA in γ-irradiated Escherichia coli cells, Mol. Biol., 14:1019 (1980).
11. E. M. Fielden, O. Sapora, and P. S. Loverock, The application of rapid-lysis technique in radiobiology. III. The effect of radiosensitizers on the production of DNA damage and the time course of its repair, Radiat. Res., 75:54 (1978).
12. O. Sapora, E. M. Fielden, and P. S. Loverock, The application of rapid lysis techniques in radiobiology. II. The time course of the repair of DNA fixed damage and single-strand breaks in Escherichia coli mutants, Radiat. Res., 72:308 (1977).
13. P. E. Bryant, Immidiate and repair induced DNA double strand breaks in mammalian cells, in: "Mechanisms of DNA Damage and Repair. Implications for Carcinogenesis and Risk Assessment," M. G. Simic, L. Grossman, and A. C. Upton, Eds., Plenum Press, New York (1986).
14. M. J. Tilby and P. S. Loverock, Measurements of DNA double-strand break yields in E. coli after rapid irradiation and cell inactivation: The effects of inactivation. Technique and anoxic radiosensitizers, Radiat. Res., 96:309 (1983).
15. M. A. Shenoy, J. A. Asquith, G. E. Adams, B. D. Michael, and M. E. Watts, Time-resolved oxygen effects in irradiated bacteria and mammalian cells: A rapid-mix study, Radiat. Res., 62:498 (1975).
16. O. Sapora, E. M. Fielden, and P. S. Loverock, The application of rapid-lysis techniques in radiobiology. I. The effect of oxygen and radiosensitizers on DNA strand break production and repair in E. coli B/r, Radiat. Res., 64:431 (1975).
17. F. Hutchinson, Sulfhydryl groups and the oxygen effect on irradiated dilute solutions of enzymes and nucleic acids, Radiat. Res., 14:721 (1961).

18. M. A. Siddiqi and E. Bothe, Single- and double-strand break formation in DNA irradiated in aqueous solution. Dependence on dose and OH radical scavenger concentration, Radiat. Res. in press (1987).
19. D. Schulte-Frohlinde, Biological consequences of strand breaks in plasmid and vival DNA, Brit. J. Cancer, 55:129 (1987).
20. P. Howard-Flanders, Effect of oxygen on the radiosensitivity of bacteriophage in the presence of sulfhydryl compounds, Nature, 186:485 (1960).
21. M. Edgren, A. Larsson, K. Nilsson, L. Revesz, and O. C. A. Scott, Lack of oxygen effect in glutathione-deficient human cells in culture, Int. J. Radiat. Biol., 37:299 (1980).
22. T. Alper, "Cellular Radiobiology", Cambridge University Press, ... p. 50 (1979).
23. D. Schulte-Frohlinde, G. Behrens, and A. Önal, Lifetime of peroxyl radicals of poly(U), poly(A) and ss and ds DNA and their reaction with thiols. Int. J. Radiat. Biol., 50:103 (1986).
24. R. L. Willson, The reaction of oxygen with radiation-induced free radicals in DNA and related compounds. Int. J. Radiat. Biol., 17:349 (1970).
25. J. B. Verberne, M. V. M. Lafleur, A. Hummel, and H. Loman, Radiation chemistry and biological effects. Non-homogeneous kinetics of reactions of water radicals with biologically active DNA, IAEA Panel Proceedings Series, in press.
26. M. V. M. Lafleur, J. Woldhuis, and H. Loman, Effects of sulphydryl compounds on the radiation damage in biologically active DNA, Int. J. Radiat. Biol., 37:493 (1980).
27. P. Apontoweil and W. Berends, Glutathione biosynthesis in Escherichia coli K_{12} properties of the enzymes and regulation, Biochim. Biophys. Acta, 399:1 (1975).
28. J. W. Hülsewede and D. Schulte-Frohlinde, Radiation protection of E. coli strains by cysteamine in the presence of oxygen, Int. J. Radiat. Biol., 50:861 (1986).
29. S. E. Bresler, L. A. Noskin, I. M. Stepanova, and N. A. Kuzovleva, Mechanism of the radioprotecting action of chemical compounds on Escherichia coli cells, Molec. Gen. Genetics, 163:75 (1978).
30. V. G. Petin and V. L. Matrenina, Radioprotecting action of chemical compounds on γ-irradiated yeast cells of various genotypes, Mol. Gen. Genet., 183:152 (1981).
31. P. H. M. Lohman, O. Vos, C. A. Vansluis, J. A. Cohen, Chemical protection against breaks induced in DNA of human and bacterial cells by X-irradiation, Biochem. Biophys., 224:339 (1970).
32. M. Näslund, L. Ehrenberg, and G. Djalali-Behzad, Antagonism of ascorbate against radioprotective action of cysteamine, Int. J. Radiat. Biol., 30:95 (1976).

33. S. L. Marklund, N. G. Westman, G. Roos, and J. Carlsson, Radiation resistance and the CuZn superoxide dismutase, Mn superoxide dismutase, catalase, and glutathione peroxidase activities of seven human cell lines, Radiat. Res., 100:115 (1984).
34. G. P. van der Schans, J. F. Bleichrodt, and J. Blok, Contribution of various types of damage to inactivation of a biologically-active double-stranded circular DNA by gamma-radiation, Int. J. Radiat. Biol., 23:133 (1973).
35. J. H. van Touw, J. B. Verberne, J. Retel, and H. Loman, Radiation-induced strand breaks in ΦX-174 replicative from DNA: An improved experimental and theoretical approach, Int. J. Radiat. Biol., 48:567 (1985).
36. D. Frankenberg, DNS Doppelstrangbrüche als kritische Schäden für die Zellinaktivierung durch ionisierende Strahlen. Eine Analyse molekularer und zellulärer Strahlenschäden, Habilitationsschrift Universität Frankfurt (1983).
37. M. Lennartz, T. Coquerelle, and U. Hagen, Effect of oxygen on DNA strand breaks in irradiated thymocytes, Int. J. Radiat. Biol., 24:621 (1973).
38. T. Bonura and K. C. Smith, The involvement of indirect effects in cell-killing and DNA double-strand breakage in γ-irradiated Escherichia coli K-12, Int. J. Radiat. Biol., 29:293 (1976).
39. J. F. Ward, Some biochemical consequences of the spatial distribution of ionizing radiation-produced free radicals, Radiat. Res., 86:185 (1981).
40. E. Boye, I. Johansen, and T. Brustad, Timescale for rejoining of bacteriophage Lambda-deoxyribonucleic acid molecules in superinfected POL+ and POLAI strains of Escherichia coli after exposure to 4MEV electrons, J. Bact., 119:522 (1974).
41. N. J. Sargentini and K. C. Smith, Characterization and quantitation of DNA strand breaks requiring recA-dependent repair in X-irradiated Escherichia coli, Radiat. Res., 105:180 (1986).
42. I. Johansen, T. Brustad, and W. D. Rupp, DNA strand breaks measured within 100 milliseconds of irradiation of Escherichia coli by 4 MeV electrons, Proc. Nat. Acad. Sci. USA, 72:167 (1975).
43. E. Boye, Formation and repair of DNA double strand breaks in superinfecting phage λ after ionizing irradiation of Escherichia coli host cells, Radiat. Res., 81:427 (1980).
44. E. Bothe, G. Behrens, E. Böhm, B. Sethuram, and D. Schulte-Frohlinde, Hydroxyl radical-induced strand break formation of poly(U) in the presence of oxygen. Comparison of the rates as determined by conductivity, ESR and rapid-mix experiments with a thiol, Int. J. Radiat. Biol., 49:57 (1986).
45. D. Schulte-Frohlinde and E. Bothe, Identification of a major pathway of strand break formation in poly U induced by OH radicals in presence of oxygen, Z. Naturforsch., 39c:315 (1984).

46. D. Schulte-Frohlinde and C. von Sonntag, Radiolysis of DNA and model system in the presence of oxygen, in: "Oxidative Stress," H. Sies, ed., Academic Press, London, p. 11 (1985).
47. D. J. Deeble, D. Schulz, and C. von Sonntag, Reactions of OH radicals with poly(U) in deoxygenated solutions: sites of OH radical attack and the kinetics of base release, Int. J. Radiat. Biol., 49:915 (1986).
48. S. Steenken, G. Behrens, and D. Schulte-Frohlinde, Radiation chemistry of DNA model compounds. Part IV. Phosphate ester cleavage in radicals derived from glycerol phosphates, Int. J. Radiat. Biol., 25:205 (1974).
49. D. J. Deeble and C. von Sonntag, Radiolysis of poly(U) in oxygenated solutions. Int. J. Radiat. Biol., 49:927 (1986).
50. M. Isildar, M. N. Schuchmann, D. Schulte-Frohlinde, and C. von Sonntag, γ-Radiolysis of DNA in oxygenated aqueous solutions: alterations at the sugar moiety, Int. J. Radiat. Biol., 40:347 (1981).
51. W. D. Henner, S. M. Grunberg, and W. A. Haseltine, Sites and structure of the gamma-radiation-induced DNA strand breaks, J. Biol. Chem., 257:11750 (1982).
52. U. Mahmutoglu, M. E. Scheulen, and H. Kappus, Oxygen radical formation and DNA damage due to enzymatic reduction of bleomycin-Fe(III), Arch. Toxicol. 60:150 (1987).
53. L. Gilomi, M. Takeshita, F. Johnson, C. Iden, and A. P. Grollman, Bleomycin-induced strand-scission of DNA - mechanism of deoxyribose cleavage, J. Biol. Chem., 256:8608 (1981).
54. J. E. Biaglow, M. E. Varnes, E. P. Clark, and E. R. Epp, The Role of Thiols in Cellular Response to Radiation and Drugs, Radiat. Res., 95:437 (1983).

DOUBLE STRAND BREAKAGE IN DNA AND CELLULAR RADIATION SENSITIVITY: LINEAR ENERGY TRANSFER AND THE OXYGEN EFFECT

J. T. Lett

Department of Radiology and Radiation Biology
Colorado State University
Fort Collins, Colorado 80523 U.S.A.

INTRODUCTION

Despite more than three decades of extensive research, identification of the chemical changes responsible for the lethal effects of ionizing radiation has yet to be achieved. Indeed, although the general consensus of opinion associates lethal radiation events with damage, and especially double strand breaks, in DNA, there is still lively debate even about the relative contributions of the direct and indirect effects of sparsely ionizing radiations at the cellular level. Causes for the protraction in this eludication process usually can be traced to the difficulty in determining chemical change in mammalian cells after radiation doses of radiobiological significance, to the inflexibility of classical positions, and to an increase in the reluctance of governmental agencies to underwrite experiments with other than predictable outcomes. Some recent signs are more encouraging, however, because definitive scientific advances now seem possible, but they are more likely to come from laboratories in Europe where the expression of originality has not been blunted by dogma and warranty. Areas where modern ideas must provoke much-needed advances in the understanding of the chemistry of strand breakage induced in cellular DNA by ionizing radiations are discussed here.

In the early 1960's, perceptive experiments with dilute solutions of deoxyribonucleoproteins demonstrated that the constituent proteins provided the DNA with extensive protection against free radicals formed in the surrounding water by radiation (see reference 1). Limitations to the role played by such radicals in the DNA chemistry underlying cellular radiosensitivity, i.e., some of the effects attributed more usually to the indirect action of ionizing radiation, were implied by such behavior. Concurrent and subsequent findings about the efficiencies of strand breakage in DNA, DNA gels of limited water contents, mammalian cells and so forth, also were consistent with that interpretation.[1] Yet, during the last 20 years, the preponderance of chemical studies on radiation-induced change in DNA have been predicated upon the contrary opinion. Justification for that experimental approach often is related to conclusions, which are based upon reaction rates in liquid water, that $\cdot OH$ radicals diffuse 20-30 Å in cells. Nowadays, challenge to the notion of the pre-eminence of such indirect action in determining

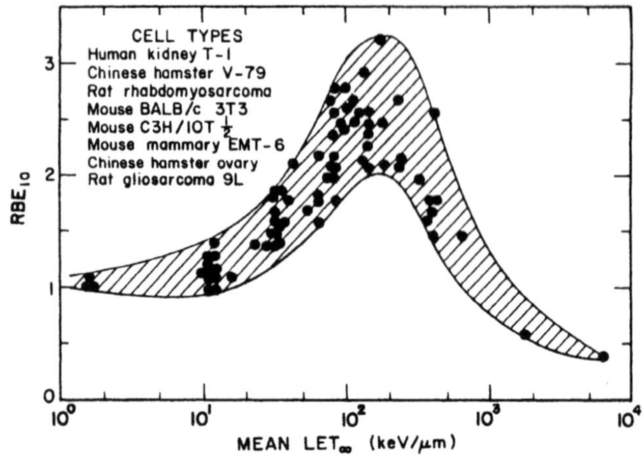

Figure 1. Relationship between aerobic relative biological effectiveness at 10% survival (RBE_{10}) and LET_∞ for a number of different proliferating cell lines *in vitro*. Redrawn from Blakely *et al.*[10] where additional information concerning the symbols can be found.

cellular radiosensitivity is strengthening as understanding broadens about the role played by the structure and hydration (solvation) of deoxyribonucleoprotein in radiation chemical processes involving cellular DNA.

Provocative presentations of these general matters, and lively discussions of them, have appeared recently in the Proceedings of the 13th L. H. Gray Conference. Some of those articles,[1-4] together with selected reviews[5-10] to which the reader is referred for further information, provide the principal sources of reference and the background for this paper. In it, the need to explore new aspects of the radiation chemistry of DNA is examined from the standpoint of providing explanations for the following two radiobiological phenomena, an approach that will be confined here to proliferating systems of mammalian cells. First, cellular radiosensitivities of normally radioresistant cells increase in a specific fashion as the linear energy transfer (LET_∞) of

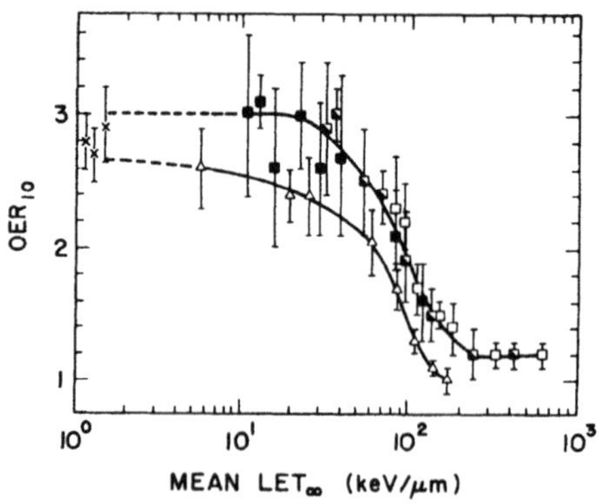

Figure 2. Relationship between the oxygen enhancement ratio at 10% survival (OER_{10}) and LET_∞ for different proliferating cell lines *in vitro*. Redrawn from Blakely *et al.*[10] where additional information concerning the symbols can be found.

ionizing radiations is increased to *circa* 100-200 keV/μm, and then decline progressively as the LET_∞ is increased further (see Figure 1). Second, over that initial range in LET_∞, the propensity for oxygen to increase cellular radiosensitivity is eliminated (see Figure 2).

THE RADIATION TARGET IN THE MAMMALIAN CELL

Although use of general target theory to explain cellular radiosensitivity is unsound, the concept of a cellular radiation target can be retained.[1,11] For ionizing radiations, the target is almost certainly the solvated deoxyribonucleoprotein which contains the nuclear DNA.[1]

Radiant energy deposited directly in the DNA, and the tightly bound water of the primary hydration layer, produces radiation chemical effects by direct action. By such processes, radicals, ions, etc. formed within the DNA can cause DNA damage, or they can react with adjacent, or closely bound water, to cause DNA damage. Reactive species formed in the tightly bound water also can attack the DNA.

Energy deposited in intracellular water will produce reactive species that migrate to, and react with, the DNA. Such processes cause DNA damage by indirect action. Most likely to be involved, however, are water molecules in the immediate proximity of the DNA because radicals produced in water at long range will not reach the DNA.

Other kinds of events can be envisaged as resulting from indirect action. Energy deposited within the proteins of the deoxyribonucleoprotein causes damage in DNA,[12] a finding confirmed recently (J. Hüttermann, private communication), DNA-protein crosslinks, etc. Metallic ions within, or in close proximity to, the deoxyribonucleoprotein structure and polyamines involved its structural geometry can influence not only the expression of radiation chemical events but also their locations. At DNA attachment sites on the nuclear matrix, long-lived radicals formed in lipids may cause damage at specific localizations, perhaps via chain reactions. There are other possibilities too,[1] but for the present purpose a closer examination of the involvement of DNA hydration is needed.

HYDRATION OF DNA

Hydration of DNA[13,14,15] proceeds with a change in structure from the A form to the B form in which DNA usually exists in the natural state. Relative humidities in excess of *circa* 80% are needed to produce the structural transition in DNA fibers, and when it occurs the bound water molecules lie mostly in the major and minor grooves of the twin helix. Indeed, the zig-zag configuration of the water chain in the minor groove is thought to be necessary for the stability of the B form of DNA.[15] Water molecules in this initial layer are bound strongly to the DNA, are in a crystalline state different from the semi-crystalline state of liquid water, and so should be considered as a discrete primary layer of water.[16] Hence, the strongly bound water molecules, which are bound directly to the DNA (10-12 H_2O molecules per nucleotide), can be considered as part of the DNA structure both in basic terms and in terms of the distinction between the direct and indirect deposition of radiant energy.[1] Many of those water molecules, and some of their nearest outlying neighbors, form bonds with the nitrogen and oxygen atoms of the bases in DNA and even form bridges from one DNA strand to the other.[15]

Reactive entities produced in the strongly bound water molecules need move little, if at all, to reach the DNA. Reactive chemical species or excited states formed in DNA itself have ready access to the water molecules, perhaps via the delocalized π-orbital systems in some cases. Since diffusion of radicals through the surrounding medium need not be involved in those interactions, and since the tightly bound water molecules are in a physical state different from liquid water, the chemical reactions involved must surely result from the direct action of ionizing radiation.

Further hydration of the DNA helix introduces water molecules into positions beyond the strongly bound layer, but those molecules have the physical state of liquid water.[15] The extent to which the higher levels of hydration occur in deoxyribonucleoprotein in situ is not known definitively at this time, but constituent proteins, spermines, metallic cations and so forth fulfill specific functions in that complex structure and will be close to, if not in immediate contact with, the DNA. Knowledge of the extent to which such cellular components displace water molecules in the hydration layers of DNA is crucial to the proper understanding of the radiation chemistry of cellular DNA.

Investigations of nucleosomal structures by low angle x-ray scattering, e.g., reference 17, have revealed that the nucleosome is hollow and penetrable by small molecules. One obvious question is, therefore, can •OH radicals formed in the hollows attack the DNA from inside of the nucleosome? Provisional computer simulation of •OH penetrations through the basic nucleosomal structures suggests that they probably cannot (A. Chatterjee, private communication). Much is still to be learned about such facets of deoxyribonucleoprotein structure from the standpoint of the radiation chemistry of cellular DNA. Just consider again, for example, the observation that radiant energy deposited in the proteins of a deoxyribonucleoprotein results in damage to the DNA.[12] Is this caused by direct or indirect action? Are diffusing species involved, or is the energy transferred by electronic pathways?

DAMAGE TO DNA BY •OH RADICALS FROM HYDROGEN PEROXIDE

Hydrogen peroxide can produce many thousands of single strand breaks (SSBs) in the DNA of mammalian cells through the agency of •OH radicals with an average diffusion range of 15 Å and yet be very inefficient at killing them.[2] This interesting result is important for the present discussion because the SSBs do not produce detectable levels of double strand breaks (DSBs) in the cellular DNA.[2] The implication is clear: •OH radicals from H_2O_2 cause effects in DNA different from those of sparsely ionizing radiations, which do produce DSBs in cellular DNA, albeit inefficiently. Sparsely ionizing radiations cause DSBs in cellular DNA through discrete ionizing events, as is the case also with DNA gels of limited water contents,[1] a fact known for 25 years. Formation of DSBs from SSBs in DNA treated in dilute solution with •OH radicals generated from H_2O_2 is not linear but follows the anticipated quadratic function indicative of the random insertion of SSBs in both strands of the twin helix.[18] That result is reminiscent of an even earlier result obtained with deoxyribonuclease by Thomas in 1956,[19] but the metallic ions that catalyse the reaction with H_2O_2 may be located at special sites in cellular nucleoprotein.

FORMATION OF DSBs IN DNA BY THE INDIRECT ACTION OF SPARSELY IONIZING RADIATIONS

Not only are the numbers of DSBs induced by sparsely ionizing radiations in cellular DNA, and DNA gels of limited water content, related linearly to dose but also the processes are similar in efficiency (G ≈0.09). Such findings present significant impediments for the thesis that indirect action plays the dominant role in determining cellular radiosensitivity because DSBs, unrepaired or otherwise, are likely to be the major cause of cellular lethality in normally radio-resistant mammalian cells.

Sparsely ionizing radiations mainly induce DSBs in DNA in dilute aerobic aqueous solution through SSBs introduced at random in both DNA strands by attack from •OH radicals, i.e., a quadratic function of dose, but a linear process of low efficiency (G = 0.035) has been detected at low doses.[3] Two mechanisms involving •OH radicals have been proposed to explain the linear process, and they are discussed in detail by Schulte-Frohlinde.[3] By the first mechanism, one •OH radical is assumed to be responsible for two opposite SSBs in the twin helix through multiple chemical reactions. The second mechanism requires that the DSBs arise from localized concentrations of •OH radicals caused in water[2] by densely ionizing events (clusters, spurs) that are known to occur during the deposition of energy by sparsely ionizing radiation. Neither mechanism could account for the formation of most DSBs in mammalian cells because the efficiency is too low by a factor of circa 3.

Although the range of experimental error for measurements of DSBs in mammalian cells and DNA gels of limited water contents will allow for a contribution by indirect action in the cellular effect,[1] DSBs are produced in the cells mainly by direct action. Ratios of 2:1 or 3:1 are plausible values for the contributions from direct and indirect action in the formation of DSBs in cellular DNA (see the G values given above), and such involvement of the radiation products of water is reasonable in terms of the probable levels of hydration in deoxyribonucleoproteins. An intriguing possibility is that the strongly bound water molecules in the grooves of the twin helix, and especially those bridging the strands, might be responsible for multichemical events involving one •OH radical and for multiple reactions in regions of dense ionization, since they seem especially well situated for the task. If the strongly bound water molecules do contribute to DSBs in this fashion, then on structural grounds the effect should be considered to result from direct action (see also reference 1).

The idea that DSBs are caused by densely ionizing events resulting from the nonuniformity of energy deposition by sparsely ionizing radiations is not new. It was proposed in 1961 to account for the formation of DSBs by high-energy electrons in DNA fibers and DNA gels of limited water content: "The main-chain scission we suggest is due largely to a relatively rare physical event... which leads to a high release of energy confined within a small volume of the DNA molecule. The most probable physical mechanism is a cluster of ionization (i.e., δ rays) and these form a significant fraction of the energy dissipated by sparsely ionizing radiations."[20] That interpretation of the formation of DSBs in DNA by sparsely ionizing radiation prompted an investigation of the damage produced in similar DNA preparations by α-particles,[21] a consideration which, in its turn, brings the development of this discussion to the final phase.

Radiation chemists must begin to address very seriously the question: What changes occur in cellular DNA damage as the LET_{∞} of the incident radiation is increased through the range shown in Figures 1 and 2? Forthcoming investigations must address also the caveat which qualifies the biological relevance of many previous studies with "bulk" DNA.[1] Localized damage at, or near, DNA attachment sites on the nuclear matrix could play the major role in cellular lethality, especially for sparsely ionizing radiations.[1] In that case, a limited amount of specific damage would be important, and some form of indirect action not necessarily determined by studies of DNA damage in dilute solution could predominate over all other types of DNA damage combined, e.g. (see earlier). Other specific variations in deoxyribonucleoprotein structure also could expose the DNA to seemingly insignificant levels of biologically crucial, localized damage. In the cellular and molecular studies performed to date, however, the destructive effects of high LET_∞ radiations seem generally to have outweighed variations in DNA damage introduced by the finer points of deoxyribonucleoprotein structure.

RADIATION CHEMISTRY OF DNA AND THE EFFECT OF LET_∞

In mammalian cells irradiated with sparsely ionizing radiations, DSBs and SSBs are formed in DNA proportionally to dose with efficiencies of circa 1100 eV/break and circa 55 eV/break, respectively. For SSBs, the average dissipation of energy needed to induce a break is essentially the same as the energy deposited per average ionizing event in DNA (60 eV). Although an ionizing event of the size needed to induce a DSB must be rare, there is a common misconception about the process. Deposition of 1100 eV in any particular spot need not be necessary to form a DSB in DNA because dissipation of that amount of energy causes, on average, 1 DSB, 20 SSBs, base damage and so forth. The actual amount of energy involved in the formation of the DSB per se is not known and, indeed, it is unlikely to have a fixed value. Spatial variations in the distribution of energy deposition will cause a spectrum of structurally different types of DSBs to be formed together with a variety of associated chemical damage.

Often another feature of energy absorption from sparsely ionizing radiations also is overlooked. About 5% of the energy dissipation is associated with the ejection of inner shell electrons with energies of several hundred electron volts from carbon, nitrogen, oxygen and phosphorus.[22] Within 10^{-15} sec, emission of Auger electrons follows.[22] Those electrons will travel short distances and could cause ionizing events of the density needed to induce DSBs in DNA. As Ore[22] states: "Clearly, an inner-shell ionization creates an exceptionally high degree of disturbance within and, in condensed media, in the immediate neighborhood of the molecule. The energy deposited... may give rise to some 5-10 ionizations altogether. Even if rare, these kinds of primary processes might give rise to special chemical effects requiring particularly great amounts of energy or high degrees of local disturbance."

As the density of ionization (LET_∞) caused by the incident radiation is raised, the efficiencies of radiation chemical change decrease in most chemical compounds and biological macromolecules. Such is also the case for SSB induction in DNA present in a variety of biological organisms including mammalian cells,[23] (see Figure 3) behavior expected if SSBs are formed with almost maximum efficiency by sparsely ionizing radiations. Quite a different response to LET_∞ occurs for DSB formation,[23] (see Figure 3), and even though LET_∞ is a relatively crude representation of energy deposition and construction of Figure 3 required the approximations necessary to encompass a variety of experimental conditions,[23] the curve in Figure 3 bears a remarkable resemblance to that in Figure 1. Which highlights the question, Are DSBs in DNA responsible for cellular lethality?

To date, there has been no plausible explanation for the DSB response shown in Figure 3 in terms of the free radical processes of indirect action, especially attack by •OH radicals. As the LET_∞ increases, so does the propensity for the recombination of •OH radicals

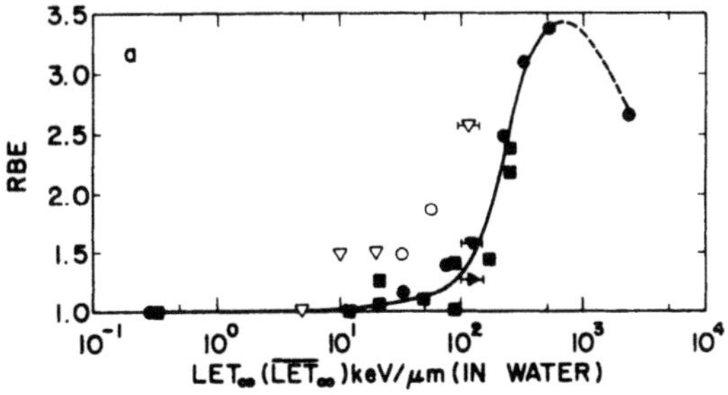

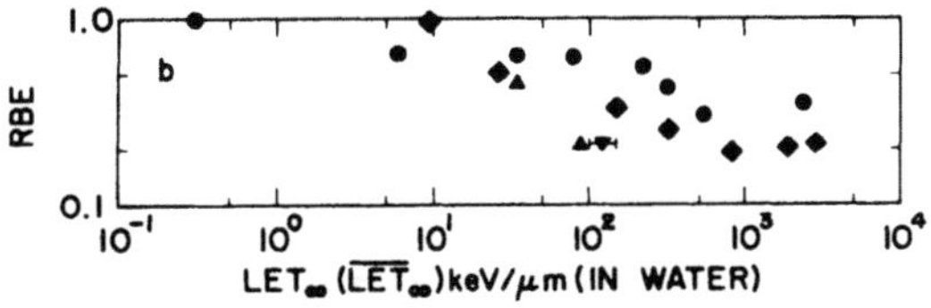

Figure 3. The RBE for DNA strand breakage as a function of LET_∞ under a variety of conditions. Redrawn from Lett et al.[23] where additional information concerning the symbols can be found. a. Double strand breakage; b. Single (or total) strand breakage.

formed close together, which is well known in the case of the Fricke dosimeter, for example. Both on the grounds of chemistry and physics (see remarks by Goodhead in the discussion to reference 1), correlation of the increased efficiency of DSB induction with the free radical processes of indirect action is difficult to justify. Explanation in terms of direct action is simple, however. With increasing LET_∞, the probability of formation of discrete densely ionizing events sufficient in magnitude to form DSBs in the twin helix of DNA also will increase until the point of maximum efficiency is reached. After that, the efficiency of DSB induction will decline. Above circa 100 keV/μm, the efficiency for gross expression of DSBs in DNA exceeds that for SSBs.[23] There are caveats appended to these interpretations, however, and they have been discussed in detail.[1,23] One obvious caveat is that densely ionizing radiations could cause strand breaks in DNA to occur in clusters which cannot be detected by physical techniques.

If the indirect actions of ionizing radiations do contribute to the chemical damage in DNA underlying cellular lethality, the contribution should diminish with LET_∞. Whether such diminution is responsible, at least in part, for the corresponding loss of the oxygen effect is not known, since definitive investigation of the free radical chemistry, and other appropriate radiation chemistry, of cellular deoxyribonucleoproteins has hardly begun. Another possibility is that, as the density of localized ionization increases, the oxygen available to the rapidly reacting species in those localities becomes exhausted, so DNA reactions involving oxygen are reduced progressively in significance, or simply are overwhelmed. Qualitatively analogous responses were described for DNA fibers and DNA gels of limited water contents nearly 30 years ago,[13,20,21,24] but the pertinent chemical processes (free radical, charge transfer, etc.) operating in cellular deoxyribonucleoprotein are virtually unknown even to this day.

EFFECT OF LET_∞ ON THE EFFICIENCY OF CELLULAR REPAIR

Even though the LET_∞ responses in Figures 1 and 3 are similar, it

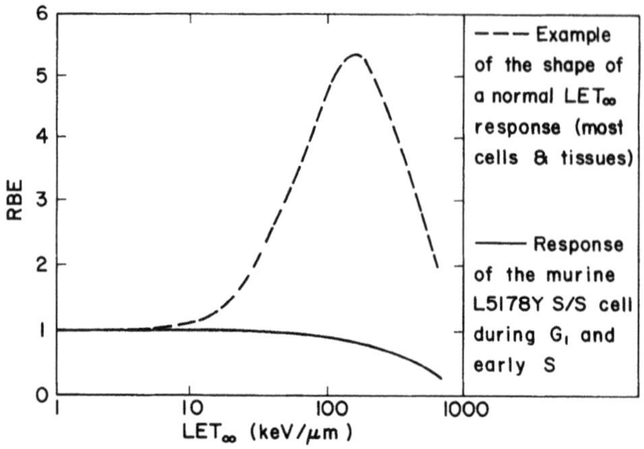

Figure 4. Heuristic comparison of the responses of normally radiosensitive cells and tissues to LET_∞ with that for the L5178Y S/S cells in the G_1- and early S phases of the cell cycle.

is doubtful whether they represent a simple correlation between cause and effect. Another important biological aspect has yet to be considered, namely, the consequences for cellular repair of the changes in DNA damage caused by increasing LET_∞.

One known effect of high LET_∞ is reduction in the efficiency of the rejoining of DNA strand breaks (e.g., reference 25), and recent aerobic experiments with the S/S variant of the L5178Y murine leukemic lymphoblast have demonstrated progressive reduction in both the rate and extent of cellular repair as the LET_∞ is increased until repair is inhibited completely around 450 keV/μm.[11] The S/S variant is a repair-deficient cell with reduced capacity to repair damage caused by sparsely ionizing radiations, especially in the G_1 phase of the cell cycle. At an LET_∞ of 450 keV/μm, the efficiency of DSB formation in cellular DNA should be near its maximum (Figure 3), and yet in the G_1 phase of the cell cycle, the RBE for lethality for the S/S variant decreased to *circa* 0.4. Indeed, in that part of the cycle (see Figure 4), the RBE does not rise above 1 as the LET_∞ is increased from 1 to 450 keV/μm, and the cell "behaves as though it were a chemical or virus" with respect to LET_∞.[26] The heavy ions used in these experiments were selected specifically to give the wide range of strand breaking efficiencies anticipated from the data in Figure 3 because those measurements could not be made at the radiation doses used in the experiments.

Although extended discussion of the causes of lethality in repair deficient cells is not appropriate here, one interpretation of the results obtained with the S/S variant is pertinent. As the LET_∞ is raised, the damage caused in the DNA seems to become increasingly more difficult for the repair systems to rectify, and a point is reached where the damage is so severe that repair enzymes become immobilized by a process akin to substrate inhibition.[1,11] Hence, emphasis is given again to the biochemical roles played in cellular radiosensitivity by the magnitudes, chemical expressions and spatial distributions of the DNA damage in cellular deoxyribonucleoprotein. So sparse is the information currently available on such enzyme-substrate interactions, especially with regard to the involvement of intracellular oxygen, that the situation is in need of prompt remedy.

CONCLUSIONS

Seemingly, the current situation can be summed up better not by how much is known about the radiation chemistry of DNA in dilute solution but how little is known about the radiation chemistry of DNA in macromolecular structures of radiobiological significance. As Wardman and Clarke[4] have noted: "It is perhaps a consequence of the acceptance of the postulate that •OH is the dominant damaging species involved in cellular radiation lethality that the absorption, migration and dissipation of excitation energy have largely been neglected in molecular radiobiology." Even the simple calculations of the diffusion distances of •OH radicals in cells can be challenged. Wardman and Clarke[4] aver with respect to the effects of medium on reaction rates: "These dramatic medium effects must be considered when discussing the likely rate and equilibrium constants for reactions which take place at radical centres which are located in biological macromolecules such as DNA, where the hydration sheath has dielectric properties quite different from bulk water."

For readers interested in the paths that research in the field may take, three recent articles are recommended. One describes a technique for "footprinting" DNA through the reactions of •OH radicals with sections of DNA not screened by specific proteins.[27] Another relates a chemical study of the kind that must be performed if the chemical basis of cellular radiosensitivity is to be resolved and also provides a bibliography of earlier such studies.[28] The third article is entitled "DNA Damage and Oxygen Radical Toxicity"[29]. It contains an interesting statement with regard to SSBs prodcued by H_2O_2 (Fenton reactions, etc.) in dilute aqueous solution: "...about half of the DNA damage generated in vitro is apparently due to a diffusible •OH, while the remaining nicks---are due to a different oxidant. The latter might be an •OH formed directly on the DNA surface, or a distinct radical species".

ACKNOWLEDGEMENTS

Research from the author's laboratories reviewed here was supported in the United Kingdom (1956-1967) by grants to the Chester Beatty Research Institute, Royal Cancer Hospital, London, and in the United States (1968-1987) by the National Cancer Institute, the National Institute on Aging, the National Aeronautics and Space Administration, and the Department of Energy. My thanks are extended to the many colleagues and students who participated in the experiments.

REFERENCES

1. J.T. Lett, Cellular radiation biology in consolidation and transition, Br. J. Cancer (Suppl. VIII) 55:145 (1987).
2. J.F. Ward, J.W. Evans, C.L. Limoli, and P.M. Calabro-Jones, Radiation and hydrogen peroxide damage induced free radical damage to DNA, Br. J. Cancer (Suppl. VIII) 55:105 (1987).
3. D. Schulte-Frohlinde, Biological consequences of strand breaks in plasmid and viral DNA, Br. J. Cancer (Suppl. VIII) 55:129 (1987).
4. P. Wardman and E.D. Clarke, Redox properties and rate constants in free-radical mediated damage, Br. J. Cancer (Suppl. VIII) 55:172 (1987).
5. J.F. Ward, Molecular mechanisms of radiation-induced damage to nucleic acids, Adv. Radiat. Biol. 5:182 (1975).
6. C. von Sonntag, U. Hagen, A. Schön-Bopp and D. Schulte-Frohlinde, Radiation-induced strand breaks in DNA: Chemical and enzymatic analysis of end groups and mechanistic aspects, Adv. Radiat. Biol. 9:110 (1981).
7. W.A. Bernhard, Solid-state radiation chemistry of DNA: The bases, Adv. Radiat. Biol. 9:200 (1981).

8. J.T. Leith, E.J. Ainsworth, and E.L. Alpen, Heavy-ion radiobiology: Normal tissue studies, Adv. Radiat. Biol. 10:192 (1983).
9. C.L. Greenstock, Free-radical processes in radiation and chemical carcinogenesis, Adv. Radiat. Biol. 11:269 (1984).
10. E.A. Blakely, F.Q.H. Ngo, S.B. Curtis, and C.A. Tobias, Heavy-ion radiobiology: Cellular studies, Adv. Radiat. Biol. 11:295 (1984).
11. J.T. Lett, A.B. Cox, R. Okayasu, and M.D. Story, Aspects of DNA damage, DNA repair and radiosensitivity: Responses of mammalian cells to acute doses of radiation, In "Radiation Research," Vol. 2, E.M. Fielden, J.F. Fowler, J.H. Hendry, and D. Scott, eds., Taylor & Francis, New York (1987).
12. P. Alexander, J.T. Lett, and H.G. Ormerod, Energy transfer in irradiated deoxyribonucleic acid and nucleoprotein, Biochim. Biophys. Acta. 51:207 (1961).
13. J.T. Lett and P. Alexander, Crosslinking and degradation of deoxyribonucleic acid gels with varying water contents when irradiated with electrons, Radiat. Res. 15:159 (1961).
14. M. Falk, K.A. Hartman, Jr., and R.C. Lord, Hydration of deoxyribonucleic acid. I. A gravimetric study, J. Am. Chem. Soc. 84:3843 (1962).
15. M.L. Kopka, A.V. Fratini, H.R. Drew, and R.E. Dickerson, Ordered water structure around a β-DNA dodecamer: A quantitative study, J. Molec. Biol. 163:129 (1983).
16. B. Wolf and S. Hanlon, Structural transitions of deoxyribonucleic acid in aqueous electrolyte solutions. II. The role of hydration. Biochem. 14:1661 (1975).
17. K.O. Greulich, J. Ausio, and H. Eisenberg, Nucleosome core particle structure and structural changes in solution, J. Mol. Biol. 186:167 (1985).
18. S.A. Lesko, J-L. Drocourt, and F-U Yang, Deoxyribonucleic acid-protein and deoxyribonucleic acid interstrand cross-links induced in isolated chromatin by hydrogen peroxide and ferrous ethylenediamine tetraacetate chelates. Biochem. 21:5010 (1982).
19. C.A. Thomas, Jr., The enzymatic degradation of deoxyribose nucleic acid, J. Am. Chem. Soc. 78:1861 (1956).
20. J.T. Lett, K.A. Stacey, and P. Alexander, Crosslinking of dry deoxyribonucleic acids by electrons, Radiat. Res. 14:349 (1961).
21. P. Alexander, J.T. Lett, P. Kopp, and R. Itzhaki, Degradation of dry deoxyribonucleic acid by polonium alpha-particles, Radiat. Res. 14:362 (1961).
22. A. Ore, in: "Effects of Ionizing Radiation on DNA", J. Hütterman eds., Springer-Verlag, Berlin (1978).
23. J.T. Lett, P.C. Keng, D.S. Bergtold, and J. Howard, Effect of heavy ions on rabbit tissues: Induction of DNA strand breaks in retinal photoreceptor cells by high doses of radiation, Radiat. Environ. Biophys. 26:23 (1987).
24. P. Alexander and J.T. Lett, Role of oxygen in the crosslinking and degradation of deoxyribonucleic acid by ionizing radiations, Nature 187:933 (1960).
25. M.A. Ritter, J.E. Cleaver, and C.A. Tobias, High-LET radiations induced a large proportion of non-rejoining DNA breaks, Nature 266:653 (1977).
26. J.T. Lett, A.B. Cox, and D.S. Bergtold, Cellular and tissue responses to heavy ions: basic considerations, Radiat. Environ. Biophys. 251 (1986).
27. T.D. Tullius, DNA footprinting with hydroxyl radical, Nature 332:663 (1988).
28. E. Kohfeldt, H. Bertram, and U. Hagen, Action of gamma endonuclease on clustered lesions in irradiated DNA, Radiat. Environ. Biophys. 27:123 (1988).
29. J.A. Imlay, J.A. and S. Linn, DNA damage and oxygen radical toxicity, Science 240:1302 (1988).

EFFECT OF ·OH SCAVENGING ON THE YIELD OF RADIATION-INDUCED DNA DOUBLE-STRAND BREAKS IN ISOLATED NUCLEI

Edward A. Bump, Susan M. Pierce, Ceferino H. Obcemea, Lee Chin, and C. Norman Coleman

Joint Center for Radiation Therapy
Harvard Medical School
Boston, Massachusetts 02115, U.S.A.

INTRODUCTION

Isolated nuclei from Chinese hamster ovary cells were used to study the importance of ·OH scavenging in protection against the formation of locally multiply damaged sites (LMDS) by ionizing radiation. This system was selected because washed nuclei should be relatively free of low molecular weight endogenous ·OH scavengers. LMDS were assayed as DNA double-strand breaks (DSB) by neutral filter elution. The results are compared to the predictions of a model which considers ·OH radical diffusion, recombination, reaction with scavenger, and reaction with DNA.

METHODS

Chinese hamster ovary (CHO) cells were grown in the presence of ^{3}H or ^{14}C thymidine for 48 h and harvested with trypsin. Nuclei were prepared by treatment with 0.3% triton X-100 (in isotonic saline) for 5 min at 4°, with moderate shearing by pipetting gently several times. Nuclei were centrifuged at 2000 g for 10 min and resuspended in an isotonic buffer containing 0.3 mM $MgCl_2$, 1 mM KH_2PO_4, 5 mM $NaHCO_3$, 20 mM HEPES, 5 mM NaCl and 115 mM KCl. Dimethyl sulfoxide (DMSO) was added, and nuclei were incubated unstirred for 30 min under 2% O_2, 5% CO_2 in N_2, at 0° prior to irradiation with 250 KVp x-rays (30 Gy). DSB were assayed by filter elution at pH 9 following lysis and proteinase K digestion[1].

RESULTS AND DISCUSSION

Figure 1 shows representative elution data. Protection factor (PF) was calculated as the ratio of the slope of the control line to that of the treated sample. In the assay, the slopes are proportional to the number of DSB[1]. In the experiment shown, the PF for 120 mM DMSO was 1.4 and the PF for 480 mM DMSO was 2.0. Figure 2 shows the relationship between PF and DMSO concentration. Each point represents a single determination. Although there is considerable scatter, it can easily be seen that the maximum PF is approximately 2 and about 100 mM DMSO is required for half-maximal protection. This pattern is almost identical to that reported for intact cells with cell survival as an endpoint[2]. A very different pattern, with PF spanning a range of several orders of magnitude, has been reported for protection against DNA single-strand breaks (SSB) by ·OH scavengers in dilute solution[3].

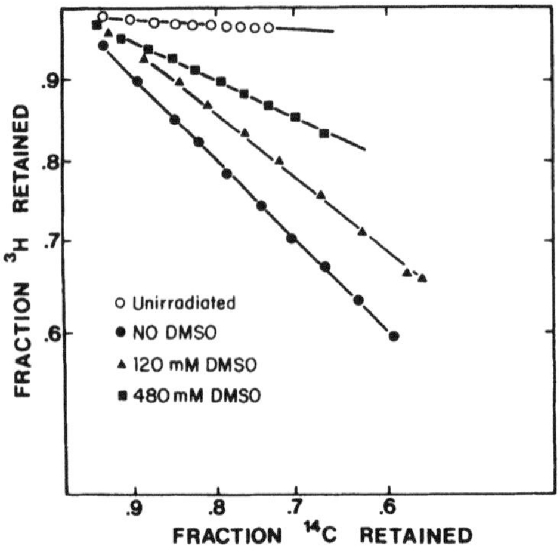

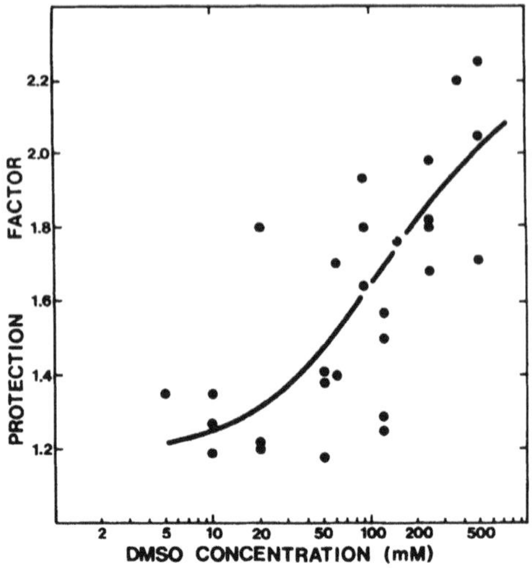

Figure 1 (left). Representative DNA elution rates showing effect of DMSO on radiation-induced DSB in CHO cell nuclei.

Figure 2 (right). Protection factors calculated from neutral elution data. Each point represents a single experiment.

These results can be understood in terms of formation of LMDS by single ionizing events involving multiple radicals[4]. Free radicals are produced in clusters of 3-60 ion pairs by ionizing radiation[5]. The impact of ·OH scavenging on the formation of LMDS is illustrated in Figure 3. A and B represent an ionizing event occurring close to a DNA segment. A double lesion is produced within 10^{-10} s, before scavenging can be effective. C and D represent an event occurring 60 Å away from a segment of DNA. 10^{-9} s are required for radicals to reach the DNA, but radicals are diluted by diffusion and only isolated lesions are formed. Protection by scavengers (D) is irrelevant with regard to LMDS formation.

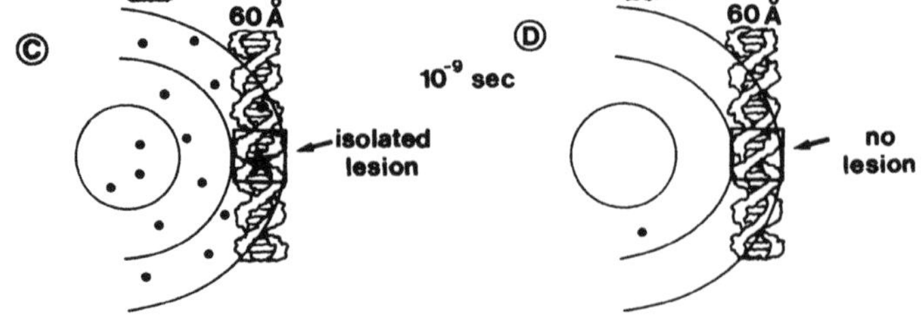

Figure 3. Schematic diagrams of lesion formation in irradiated DNA. Dots indicate the probability density of finding a free radical at a particular location; stars indicate DNA lesions.

Mathematical Model

These considerations point toward the conclusion that only those ionizing events which occur close to the DNA can give rise to LMDS and and that ·OH scavengers should not greatly affect the outcome of this process. We have sought to express this conclusion in a more quantitative manner by constructing a mathematical model. The details of the model will be presented elsewhere[6]. Briefly, we calculated, for various discrete energy depositions[5], the concentration of radicals at 10 Å intervals from the center of the event for various time intervals. This calculation was based on diffusion, recombination and scavenging of radicals, assuming an initial Gaussian distribution of radicals within a volume which depends on the energy of the event, essentially as described by Mozumder and Magee[5] and by Fricke[7]. From these values we determined the probability that single or double lesions will occur in 40 Å segments of DNA arranged randomly about the center of the ionization event. The summation of single and double lesions (SSB and DSB) over time and space for the various event energies (multiplied by their relative frequency) results in SSB and DSB totals for each scavenger concentration (Table 1), expressed per 1.5 Gy per cell. The following assumptions were made in constructing the model:

1. An LMDS, for the purposes of this calculation, is defined as two lesions within a 40 Å length of DNA, occurring on opposite strands.
2. Any ionizations occurring within the volume of the DNA segment (taken to be a cylinder of 10 Å radius), at time = 0, produces a lesion by the direct effect, since the density of DNA in this volume is 1.0.
3. All other ionizations are assumed to produce one ·OH and one H· radical. Any ·OH radicals entering the volume of the DNA segment have a 1/6 chance of producing a lesion by the indirect effect during a 10^{-10} s interval, because the concentration of DNA sugar, base and phosphate moieties in this volume is 7 M; the reaction rate for these groups in DNA with ·OH is approximately $5 \times 10^{8} M^{-1} s^{-1}$; and we assume that only 1/2 of these sites are accessible for reaction. Reactions of H· radicals are ignored, as the reactivity of H· is too low to allow reaction before diffusion away from the site of the LMDS.
4. DNA is assumed to be distributed in space such that there is a 1/10 chance of encountering DNA within "densely packed regions". Radicals produced outside these "densely packed regions" are ignored, as they would not contribute to LMDS formation. DNA content is 7 pg/mammalian cell.
5. The rate constant for ·OH scavenging by DMSO is $6 \times 10^{9} M^{-1} s^{-1}$ [8].
6. An ionization is produced for every 17 eV of energy deposited[5].
7. Radical scavenging by protein can be considered to be a small factor and is not taken into account, since the reaction rate of amino acids with ·OH is low[8], and most of the sites are inaccessible.

Table 1. Theoretical Effect of DMSO on Radiation-Induced DNA Lesions

DMSO Concentration	SSB[a]	DSB[a]	DSB Protection Factor
0	21,600	2,150	
20 mM	17,500	1,800	1.2
50 mM	14,300	1,500	1.4
100 mM	11,400	1,300	1.6
200 mM	8,700	1,050	2.0
500 mM	6,000	830	2.6
1 M	4,800	740	2.9

[a]Per cell per 1.5 Gy dose to "densely packed region" (assumption 4).

The protection factors calculated from the mathematical model agree very well with the actual data presented in Figure 2. Furthermore, if we consider cellular scavenging capacity to be equivalent to 20 mM DMSO, the protection factors calculated relative to the 20 mM protection factor (PF = 2.4 for 1 M DMSO) agree well with the actual cellular protection factors for survival[2]. The implications of the results obtained from the model are summarized below:

1. Even in the absence of scavengers, only those ionizations which occur within 10-30 Å of DNA segment contribute to LMDS. It is therefore not necessary to invoke endogenous scavengers in explaining the difference between protection of cellular DNA by DMSO (maximum PF ≅ 2) and protection against single-strand DNA lesions in dilute solution (maximum PF ≅ 1000). The data would be consistent with the concept that LMDS are the critical lesions for cell survival[4]. Protection factors calculated for single lesions (PF = 3.6 for 1 M DMSO relative to 20 mM DMSO) do not agree as well with cellular protection data, as the PF for DSB.

2. The model predicts about 20 times more lesions than are observed experimentally[4]. This indicates that most lesions are not expressed biologically, and emphasizes the importance of the chemical nature of the lesions[4].

3. Most of the LMDS produced by radiation occur within 10^{-9}s. Diffusion limits the progress of local reactions beyond that time frame, even in the absence of scavengers. Any radicals which are less reactive than ·OH or which are produced by secondary reactions (i.e., amino acid radicals, GSO_2·, H·) will not contribute to LMDS formation because they will diffuse away from the target. Therapeutic doses of radiation do not produce enough radicals to allow a significant probability that the same segment of DNA would be hit twice by independent events (only 1 in 10^6 bases will be damaged by 1.5 Gy).

4. The greatest contribution to LMDS formation is from events involving about four ionizations (0.25 DSB per event). Events involving larger numbers of ionizations are only slightly more effective in producing DSB (0.59 DSB per event involving eight ionizations) since their initial distribution is much more sparce and such events are much less common[5]. Surprisingly, even events involving only two ionizations contribute substantially to LMDS formation (0.07 DSB per event).

5. According to the model, the contribution of the indirect effect to DSB is 70% with no scavenger present and 63% with 20 mM DMSO (to simulate endogenous cellular scavengers). The indirect effect includes lesions which are composed of direct and indirect hits.

REFERENCES

1. M. O. Bradley and K. W. Kohn, X-ray induced double strand break production and repair in mammalian cells as measured by neutral filter elution, Nucleic Acids Res. 7:793 (1979).
2. J. D. Chapman and C. J. Gillespie, Radiation-induced events and their time scale in mammalian cells, Adv. Radiat. Biol. 9:143 (1981).
3. K. D. Held, H. A. Harrop, and B. D. Michael, Effects of oxygen and sulfhydryl-containing compounds on irradiated transforming DNA. I. Actions of dithiothreitol, Int. J. Radiat. Biol. 40:613 (1981).
4. J. F. Ward, Biochemistry of DNA lesions, Radiat. Res. 104:S-103 (1985).
5. A. Mozumder and J. L. Magee, Models of tracks of ionizing radiations for radical reaction mechanisms,Radiat. Res. 28:203 (1966).
6. E. A. Bump, L. Chin, C. H. Obcemea, and C. N. Coleman, in preparation.
7. H. Fricke, Track effect in iochemistry of aqueous solutions, Ann. N. Y. Acad. Sci. 59:567 (1955).
8. G. Czapski, On the use of ·OH scavengers in biological systems, Israel J. Chem. 24:29 (1984).

SONOCHEMISTRY OF NUCLEIC ACID CONSTITUENTS IN AQUEOUS SOLUTION: A SPIN-TRAPPING STUDY

Takashi Kondo*, C. Murali Krishna, and Peter Riesz

Radiation Oncology Branch
National Cancer Institute
National Institutes of Health
Bethesda, Maryland 20892, U.S.A.

INTRODUCTION

Ultrasonic cavitation in aqueous solution is well known to form free radical species produced by the collapse of cavitation bubbles. In this free radical formation, ultrasound is analogous to ionizing radiation.[1,2] The study of the free radical effects of ultrasound on biological systems is of interest because of the widespread application of ultrasound in medical diagnosis and therapy. A recent ESR spin-trapping study has demonstrated that ·OH and H· are formed in water decomposition induced by cavitation.[3] Some reports also have demonstrated stable sonoproducts arising from nucleic acid base destruction induced by ultrasound.[4-6] However, direct evidence for the formation of nucleic acid base radicals was not obtained. In the present work, a water-soluble, non-volatile, aromatic nitroso spin trap, 3,5-dibromo-4-nitrosobenzenesulfonate (DBNBS),[7] has been used for detecting the intermediate radicals arising from sonolysis of nucleic acid constituents, and the results have been compared with those obtained from H_2O_2-photolysis experiments in which only ·OH radicals are the reactive species.

MATERIALS AND METHODS

Pyrimidine bases (thymine, uracil, cytosine) and pyrimidine nucleosides (thymidine, uridine, cytidine, deoxyuridine, deoxycytidine) in aqueous solution were exposed to 50 kHz ultrasound in the presence of DBNBS as the spin trap. Solutions were bubbled with argon gas before and during sonication at 25 °C. In the photolysis experiments, argon-saturated samples were exposed to 275 ± 10 nm UV light from a high-pressure 1000 W Xe lamp coupled with a grating monochromator in the presence of H_2O_2. ESR spectra of sonicated and UV-irradiated samples were measured in a quartz flat cell with a Varian E-9 ESR spectrometer (X-band, 100-kHz field modulation). The observed ESR spectra were compared with computer-simulated ESR spectra.

*Permanent address; Department of Experimental Radiology and Health Physics, Fukui Medical School, Matsuoka, Fukui 910-11, Japan

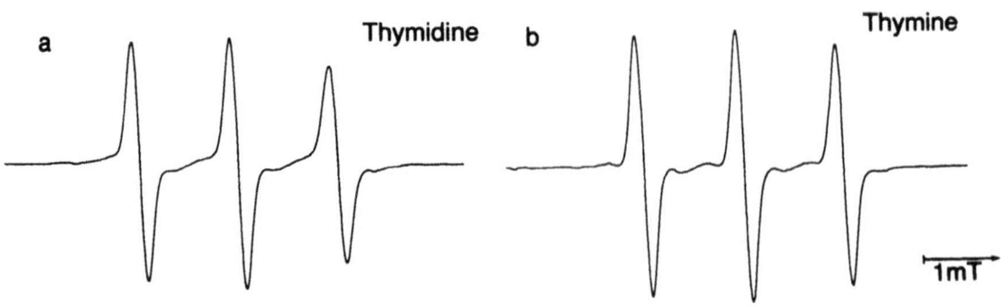

Figure 1. ESR spectra of spin-trapped radicals obtained by sonolysis of argon-saturated aqueous solutions containing DBNBS and (a) thymidine; (b) thymine.

RESULTS AND DISCUSSION

The nitroso spin trap used in the present study reacted with short-lived free radicals of pyrimidine bases and nucleosides induced by sonolysis to produce longer-lived spin adducts. The ESR spectrum of thymidine was composed of a primary nitrogen triplet (a_N = 1.30 mT)(Figure 1a). First, thymidine was used as a model for DNA constituents and the effects of sonication time and of concentration of spin trap, thymidine and sodium formate on the spin adduct yield were determined. The ESR signal intensity of the thymidine spin adduct increased with sonication time and reached a plateau in about 20 min. The ESR signal intensity of this adduct increased with spin trap concentration and remained constant above 5×10^{-4} moldm^{-3}. The ESR signal intensity increased with thymidine concentration in the region between 5×10^{-4} and 2×10^{-1} moldm^{-3}. To elucidate the role of the free radicals induced by sonication, thymidine solutions were sonicated for 20 min in the presence of the various concentrations of sodium formate. The ESR signal intensity of the thymidine spin adducts decreased in

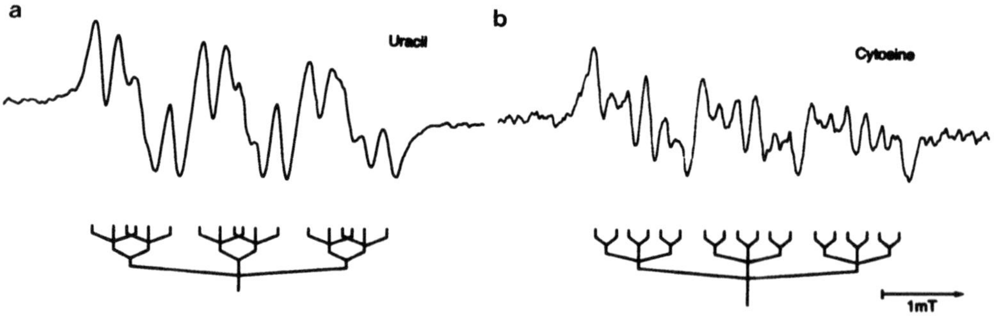

Figure 2. ESR spectra of spin-trapped radicals obtained by sonolysis of argon-saturated solutions containing DBNBS and (a) uracil; (b) cytosine.

Table 1. Primary and Secondary Hyperfine Coupling Constants of the Spin Trapped Radicals of Pyrimidine Derivatives Induced by Ultrasound

Compound	Primary ^{14}N Splitting (mT)	Secondary Splittings (mT) a^{β}_{N}	a^{β}_{H}
Thymine	1.30		
Thymidine	1.30		
Uracil	1.35	0.275	0.44
Uridine	1.30	0.36	0.36
Deoxyuridine	1.275	0.375	0.375
Cytosine	1.35	0.40	0.25
Cytidine	1.30	0.40	0.40
Deoxycytidine	1.30	0.40	0.40
2-Deoxy-D-ribose	1.30		0.55
	1.35		1.25 (2H)

the presence of 2 x 10^{-3} $moldm^{-3}$ sodium formate and reached 10% of the initial ESR signal at 1 $moldm^{-3}$ sodium formate. This result suggests that 90% of the thymidine spin adduct is due to free radical reactions induced by ultrasonic cavitation.

The ESR spectrum of thymine consisted of a triplet (a_N = 1.30 mT) (Figure 1b). The ESR spectra of uracil and cytosine are shown in Figure 2a and b, respectively. Hyperfine coupling constants of other nucleic acid constituents were shown in Table 1. The ESR spectrum of sonicated deoxyribose was mainly composed of six major lines, and these lines were not observed in the ESR spectrum of thymidine, indicating the absence of a radical on the sugar moiety. Sonolysis of ribose solution did not yield an observable ESR spectrum. ESR spectra observed from photolysis were identical to those of the sonolysis experiments. These results suggest that the major-spin trapped radical of thymine and thymidine was the C(5)yl radical and that of cytosine, cytidine, deoxycytidine, uracil, uridine and deoxyuridine was the C(6)yl radical. To compare the radicals induced by sonolysis and photolysis, the decay of the ESR spectra of the thymine and thymidine spin adducts was investigated. The decay curves of thymine and thymidine after sonolysis indicated biphasic decay (half-life of the initial rapid decay was about 10 min and the curve reached a plateau at about 65% of the initial ESR signal). However, after photolysis the spin adducts from both compounds showed very little decay (less than 5% in 2 h). These results indicated that the observed spin adducts in the sonolysis of thymine and thymidine were formed by H· and ·OH addition to the 5,6 double bond; the rapidly decaying spin adduct is assigned to the H· addition. On the other spin adducts of pyrimidine derivatives, H· and ·OH addition to the 5,6 double bond were suggested (Figure 3). At the present, DBNBS appears to be the best spin trap for detecting intermediate radicals in the sonolysis of aqueous solutions. On the other hand, from the experiments using 2-methyl-2-nitrosopropane on the reactions of pyrimidine bases with ·OH, at least three types of radical products were reported for thymine[8-11] and for uracil[12]. The determination of detailed radical structures with DBNBS requires further experiments using other substituted nucleic bases and deuterated DBNBS.

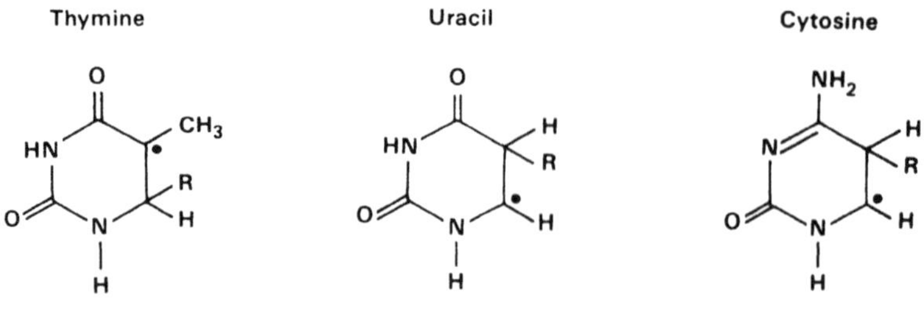

Figure 3. Radicals formed from thymine, uracil and cytosine.

ACKNOWLEDGMENTS

The present research (T. K.) was supported by the Fogarty International Research Fellowship of the NIH (#1 F05 TW0364-01).

REFERENCES

1. P. Riesz, D. Berdahl, and C. L. Christman, Free radical generation by ultrasound in aqueous and nonaqueous solutions, Environ. Health Perspect. 64:233 (1985).
2. A. Henglein, Sonochemistry: historical developments and modern aspects, Ultrasonics 25:6 (1987).
3. K. Makino, M. M. Mossoba, and P. Riesz, Chemical effects of ultrasound on aqueous solutions, Formation of hydroxyl radicals and hydrogen atoms, J. Phys. Chem. 87:1369 (1983).
4. E. L. Mead, R. G. Sutherland, and R.E. Verrall, The ultrasonic degradation of thymine, Can. J. Chem. 53:2394 (1975).
5. J. R. McKee, C. L. Christman, W.D. O'Brien, Jr., and S.Y. Wang, Effect of ultrasound on nucleic bases, Biochemistry 16:4651 (1977).
6. T. -J. Yu, R. G. Sutherland, and R. E. Verall, The sonolysis of uracil, Can. J. Chem. 58:1909 (1980).
7. H. Kaur, K. H. W. Leung, and M. J. Perkins, Water-soluble, nitroso-aromatic spin-trap, J. Chem. Soc. Commun. 142 (1981).
8. A. Joshi, S. Rustgi, and P. Riesz, E.S.R. of spin trapped radicals in gamma-irradiated aqueous solutions of nucleic acids and their constituents, Int. J. Radiat. Biol. 30:151 (1976).
9. S. Kominami, S. Rokushika, and H. Hatano, Studies of short-lived radicals in the gamma-irradiated aqueous solutions of thymidine-5'-monophosphate and cytidine-5'-monophosphate by liquid chromatography and ESR spectroscopy, Radiat. Res. 72:89 (1977).
10. M. Kuwabara, Y. Lion, and P. Riesz, E.S.R. of spin-trapped radicals in gamma-irradiated polycrystalline nucleic acid constituents and their halogenated derivatives, Int. J. Radiat. Biol. 39:465 (1981).
11. M. Kuwabara, O. Inanami, D. Endoh, and F. Sato, Spin-trapping of precursors of thymine damage in X-irradiated DNA, Biochemistry 26:2458 (1987).
12. K. Makino, M. M. Mossoba, and P. Riesz, Electron spin resonance of spin-trapped radicals in gamma-irradiated aqueous uracil solutions. Chromatographic separation of radicals, J. Phys. Chem. 87: 1074 (1983).

GAMMA RADIOLYSIS OF RNA:

AN ESR AND SPIN-TRAPPING STUDY

Alasdair J. Carmichael, Carmen M. Arroyo, and
Thomas L. Walden

Radiation Biochemistry Department
Armed Forces Radiobiology Research Institute
Bethesda, Maryland 20814, U.S.A.

INTRODUCTION

It is well known that the radiolytic cleavage of water produces hydroxyl radicals ($^{\bullet}OH$), hydrogen atoms ($^{\bullet}H$) and hydrated electrons (e^{-}_{aq}).[1] Although these species react rapidly with the individual nucleic acid bases,[2-4] it is generally accepted that when nucleic acid (DNA, RNA) solutions are exposed to ionizing radiation, the principal reaction leading to nucleic acid damage is caused by $^{\bullet}OH$. Besides direct formation, $^{\bullet}OH$ is formed from H_2O_2, also a product of the radiolysis of water, in the presence of trace amounts of divalent metal ions.[5] The study reported here deals with the reaction of RNA with $^{\bullet}OH$, produced in aqueous solutions by gamma radiolysis. Of concern are the processes that occur following this reaction. For instance, are the precursors to the nucleic acid damage localized at specific base sites or do they occur indiscriminately throughout the nucleic acid molecule? Recently Kuwabara et al.[6] using HPLC, spin trapping and ESR identified the precursors of thymine damage in X-irradiated DNA solutions. For the present study, poly A, poly C, poly G and poly U were used as standards for RNA, and RNA was used as a model system for similar reactions that may occur in DNA. The nucleic acid free radical intermediates were detected and identified by ESR and spin trapping following hydrolysis of the spin-trapped nucleic acid molecules.

MATERIALS AND METHODS

To eliminate possible small-molecular-weight contaminants, solutions of RNA, poly A, poly C, poly G and poly U were dialyzed against several changes of doubly deionized distilled water and then lyophilized. The spin trap 2-methyl-2-nitrosopropane (MNP) was prepared by dissolving 20-30 mg of MNP in 10 ml of water and stirring in a water bath at 45°C for approximately 2 h. Solutions of RNA and of the various polynucleotides were prepared by dissolving the lyophilized powders (5×10^{-4} to 1×10^{-3} M) directly in the MNP solution and adjusting the pH to 8. These solutions were ^{60}Co gamma-irradiated at a dose rate of 45 Gy/min to a total dose of 700 Gy. ESR spectra of the spin-trapped RNA and polynucleotides were immediately recorded following irradiation. The spin-trapped RNA samples were hydrolyzed with base (NaOH, pH 12.6) or enzymatically (RNAase). The spin-trapped polynucleotides were hydrolyzed with base. ESR spectra were recorded at various times during hydrolysis until the hydrolysis was complete as judged by the resolution of the ESR spectra. ESR spectra were recorded on a Varian E-109 X-band spectrometer at 100 KHz magnetic field modulation. The spectra were analyzed by comparison with computer-generated ESR spectra.

RESULTS AND DISCUSSION

The ESR spectrum obtained for the spin-trapped RNA is shown in Figure 1a. This spectrum consisting of three broad lines is characteristic of nitroxide groups bound to slowly tumbling large molecules in solution. Figure 1b shows the ESR spectrum obtained following base or enzymatic hydrolysis of the spin-trapped RNA. This spectrum consists of a triplet of sextets indicating that the unpaired nitroxide electron is interacting with the nuclei of a β nitrogen and β hydrogen. The hyperfine coupling constants for this spectrum are a_N=1.48 mT, a_N^β= 0.25 mT and a_H^β= 0.15 mT. Following base hydrolysis of the spin-trapped polynucleotides, only poly C (Figure 2a) and poly U (Figure 2B) yielded ESR spectra consisting of a triplet of sextets. The hyperfine coupling constants for the ESR spectrum corresponding to the hydrolyzed poly C are a_N= 1.49 mT, a_N^β= 0.27 mT and a_H^β= 0.16 mT. For the hydrolyzed poly U, a_N= 1.48 mT, a_N^β= 0.25 mT and a_H^β= 0.15 mT.

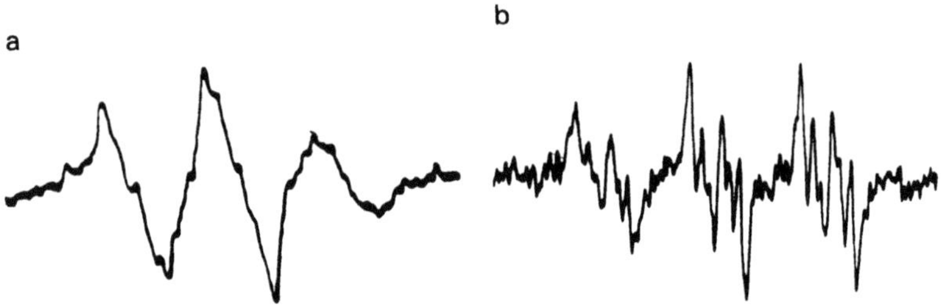

Figure 1.

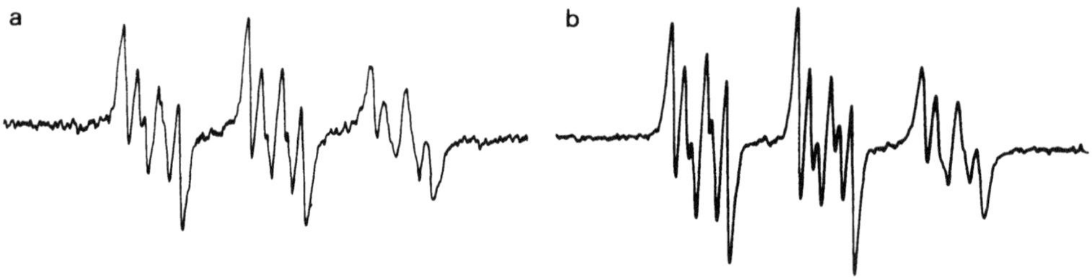

Figure 2.

The results suggest that in RNA only uracil radicals were spin trapped. Furthermore, due to the β nitrogen and β hydrogen interaction with the unpaired nitroxide electron, the results indicate that MNP reacted with uracil-C6-carbon-centered radicals. Because hydroxyl radicals react with all bases at approximately the same rate,[2] two possible explanations for the observed results are immediately obvious: 1) The spin-trapping efficiency is not equal for radicals formed on each base, and 2) the stability of the various spin adducts is not equal. Although still being investigated, the first possibility appears unlikely because [MNP]>[RNA]>[$^{\bullet}$OH] which should permit MNP to react with any base radical formed in RNA. The second possibility is unlikely because the ESR spectra of the polynucleotide spin adducts and of their hydrolyzed products persist for 48 h or longer. Another explanation for the observed results is that the free radical spin density migrates via an intramolecular mechanism, still unclear at this time, to the uracil bases. Radiolysis of poly (A,U) and poly (C,U) in the presence of MNP support this explanation. The results of these experiments also show specific uracil spin trapping.

REFERENCES

1. I. G. Draganic and Z. D. Draganic, "The Radiation Chemistry of Water," Academic Press, New York (1971).
2. L. M. Dorfman and G. E. Adams, "Reactivity of the Hydroxyl Radical in Aqueous Solutions," (NSRDS 46.) U.S. Government Printing Office, Washington, D.C. (1973).
3. M. Anbar, Farhataziz, and A. B. Ross, "Selected Specific Rates of Reactions of Transients from Water in Aqueous Solutions. II . Hydrogen Atoms," (NSRDS-NBS 51.) U.S. Government Printing Office, Washington, D.C. (1975).
4. M. Anbar, M. Bambenek, and A. B. Ross, "Selected Specific Rates of Reactions of Transients from Water in Aqueous Solutions. I . Hydrated Electron," (NSRDS-NBS 43.) U.S. Government Printing Office, Washington, D.C. (1973).
5. W. Bors, M. Saran, and G. Czapski, "Biological and Clinical Aspects of Superoxide and Superoxide Dismutase," (W. Bannister and J. V. Bannister, Eds.) Elsevier, North Holland/New York (1980).
6. M. Kuwabara, O. Inanami, D. Endoh, and F. Sato, Spin Trapping of Precursors of Thymine Damage in X-Irradiated DNA, Biochemistry 26:2458 (1987).

INTERACTIONS OF OXYGEN AND SULFHYDRYLS WITH FREE RADICALS

IN IRRADIATED MAMMALIAN CELLS

Kathryn D. Held

Department of Radiation Medicine
Massachusetts General Hospital Cancer Center
Harvard Medical School
Boston, MA 02114, U.S.A.

INTRODUCTION

Studies on radiosensitization by oxygen and radioprotection by sulfhydryl-containing compounds have often directed attention to the importance of interactions of those two agents -- oxygen and thiols -- in determining radiation response. The so-called repair-fixation theory suggests radiation response is determined by a competition between damage fixation by oxidizing agents (e.g., oxygen) and damage repair by reducing species (e.g., sulfhydryls)[1,2]:

$$T^{\cdot} + O_2 \longrightarrow TOO^{\cdot} \quad \text{(damage fixation - lethal)} \qquad (1)$$

$$T^{\cdot} + RSH \longrightarrow TH + RS^{\cdot} \quad \text{(chemical repair)} \qquad (2)$$

where $T^{\cdot}$ represents an ionizing radiation-induced organic radical on a cellular "target" molecule, TH, such as DNA, and RSH is any thiol. This competition can be clearly demonstrated in simple chemical systems and in nucleic acids such as poly(A) and transforming DNA, and it has been shown that the damage fixation reaction is over 100 times faster than the H donation, chemical repair reaction (reviewed recently[3]). But the occurrence of this competition in mammalian cells is more difficult to demonstrate. This paper will review the current literature indicating the involvement of this chemical competition in determining radiation response in mammalian cells.

DEPLETION OF ENDOGENOUS THIOLS

The major nonprotein sulfhydryl (NPSH) in mammalian cells is glutathione (GSH). The competition theory predicts that depletion of endogenous GSH should lead to radiosensitization of hypoxic cells with minimal effect on oxygenated cells, where the damage fixation reaction is predominant, unless depletion is very severe. Studies to test this theory have been conducted using cells depleted of endogenous thiols. There are numerous chemical methods for depleting intracellular GSH: prevention of synthesis of GSH with subsequent metabolic depletion, as by buthionine sulfoximine (BSO)[4]; NPSH removal by substrates for GSH S-transferase (e.g., diethylmaleate [DEM]); removal of NPSH by oxidation or covalent bond formation (e.g., diamide); removal of NPSH by spontaneous reaction (e.g., CMNI)[5]; and removal of NPSH by metabolic activation under hypoxic conditions (e.g., misonidazole).[5] Of these methods only the first is specific for GSH, although the second, using compounds active with GSH S-transferase, may show some specificity for GSH. Also, there exist human cell strains, from patients with the rare disease 5-oxoprolinuria, which are low in GSH due to a deficiency in the enzyme glutathione synthetase.[6]

When mammalian cells are chemically depleted of thiols, the radiosensitivity of cells irradiated in hypoxia is generally increased, although the sensitivity of cells irradiated in air may or may not be altered. Thus, the oxygen enhancement ratio (ratio of survival curve slopes in the absence and presence of oxygen, OER) may or may not decrease. Some representative data from the literature are summarized in Table 1, and these data will be used here for discussion purposes. Koch et al.[7] reported that when cellular GSH was depleted to 10-20% of control levels using DEM in V79 cells, the enhancement ratio (ER) for both hypoxic and oxygenated cells was the same (1.15-1.35); thus, there was no change in the OER. In contrast, Bump et al.[8] reported that DEM used in CHO cells to reduce GSH to 1% of control increased the radiosensitivity of hypoxic cells (ER=1.7), while having no effect on oxygenated cells; thus, the OER decreased. When cells are depleted of GSH using BSO, the effect on the OER also varies, as seen in Table 1. At least part of this variation results from the fact that sensititization by BSO in air apparently depends on many factors, including GSH depletion protocol, cell line, culture conditions, and medium present during irradiation.[10]

In contrast to the variable results reported when cells are chemically depleted of thiols, glutathione-deficient mutant (GSH^-) cells generally exhibit decreased OERs due to selective sensitization of hypoxic cells (see reviews[17,18]). For example, in early work with these cells, one GSH^- cell line (line VP) with about 5% of normal GSH showed OERs of 1.5 for cell survival,[13] 1.1 for DNA single strand breaks[14] and about 1 for micronuclei formation.[15] Such results are consistent with the competition model for radiosensitization. In more recent studies comparing a number of GSH^- cell strains, no correlation could be found between NPSH or GSH concentrations and radiosensitivity under oxygenated, aerated, or hypoxic conditions.[16] However, those authors did find a correlation between

Table 1. Some Examples of the Effect of Decreased GSH on Radiation Sensitivity

Depletion method	%GSH	Hypoxic ER	OER change*	Reference
DEM	1	1.7	Decreased	8
DEM	10-20	1.15-1.35	No change	7
DEM	<5	1.5	No change	9
BSO	<5	1.1	No change	9
BSO	<10	1.4	Decreased	10
BSO	<5	1.4	3.1--->2.6	11
BSO	10	-	3.1--->2.2	12
GSH^- cells (survival)	5	-	2.9--->1.5	13
GSH^- cells (SSBs)	5	-	3.0--->1.1	14
GSH^- cells (micronuclei)	5	-	2.3--->1	15
GSH^- cells (survival)	varies#	-	2.9--->1.5-2.0#	16

* Several papers do not report OER values; in those cases we indicate only whether there was a decrease or no change.

\# This paper reports on a number of different cell lines with a range of GSH levels and OERs.

hypoxic radiosensitivity and GSH synthetase activity. Other recent studies with GSH^- cells[19] have indicated a role for GSH in repair of radiation-induced damage, but discussion of those reports is beyond the scope of this paper.

When radiosensitivity is plotted as a function of increasing oxygen concentration, the resulting competition plot, often called a "K-curve," can be characterized by two parameters: the K value, the oxygen concentration for half-maximal sensitization and m, the OER at high levels of oxygen. According to the competition theory the K-value should depend on the thiol level in the cells. Generally, this has been confirmed experimentally. Koch et al.[7] showed a 2.5-fold decrease in K in DEM-treated cells when measuring cell survival. Likewise, the K-values for BSO-treated V79 cells[11] and for GSH^- mutant cells[20] were decreased compared to normal controls.

The most convincing evidence that competition does occur between endogenous thiols and oxygen for oxygen-dependent cellular lesions comes from fast mixing experiments. Hodgkiss et al.,[12] using the liquid rapid-mixing technique, have shown that V79 cells depleted of endogenous thiol by BSO pretreatment have increased radiosensitization of cell survival compared to normal, nondepleted cells when oxygen is added milliseconds after irradiation. Using the gas explosion method, the first-order rates of chemical repair of oxygen-dependent lesions in irradiated V79 cells have been determined.[21] For cell lethality (Figure 1) and induction of single and double strand breaks, the reaction rates all increase about twofold when cells are pretreated with BSO to reduce the intracellular GSH to <5%.

INCREASING INTRACELLULAR THIOLS

It would also be predicted from the competition model that increasing intracellular thiols would shift the K-value to higher oxygen levels and provide substantial protection of cells at low oxygen tensions. Thiol levels in cells can be increased by culture conditions, by treatment of cells with certain chemical agents, or by addition of exogenous thiols.

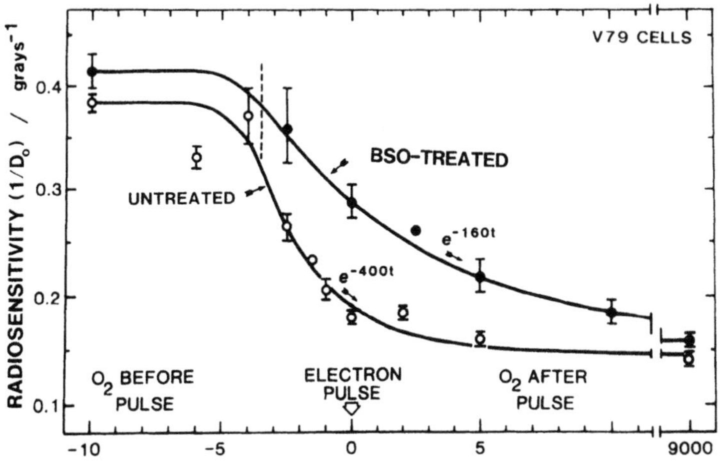

Figure 1. Fast kinetics of the chemical repair of oxygen-dependent lethal damage in V79 cells. From Michael et al.,[21] with permission.

Cullen and co-workers[22-24] have shown that when Ehrlich ascites cells are grown in exponential phase or when plateau-phase V79 cells are stimulated by replating into fresh medium, they have two- to threefold higher NPSH levels than plateau-phase cells. Those cell cultures with the higher NPSH levels have K-values that are 1.4 to 1.7 times larger than the plateau-phase cultures. Russo et al.[25] have used the compound 2-oxothiazolidine-4-carboxylate (OTZ) to increase intracellular GSH levels about twofold. The effect of OTZ was dependent on the oxygen concentration in the V79 cells during irradiation: when irradiated under fully aerated conditions OTZ-pretreated cells had the same sensitivity as untreated cells; at 10 ppm oxygen the cells were protected slightly by OTZ pretreatment (protection factor, PF, of about 1.1), but cells at intermediate oxygen levels, e.g., 0.5%, showed decreased radiosensitivity (PF=1.3) (Figure 2).

Most literature data indicate that aerated cells are protected to a greater degree than hypoxic cells by the addition of SH compounds (see reviews[26,27]). However, the degree of radioprotection by thiol is very dependent on the thiol concentration. For example, our studies[28] on the effect of dithiothreitol (DTT) on cell survival showed that at low concentrations of DTT (0.5-1 mmol dm^{-3}) there is more protection of cells irradiated in hypoxia than in air, but at high concentrations of added DTT the reverse is true: there is more protection of cells in air than in hypoxia (Figure 3). The net effect of this differential protection is that at low concentrations of thiol the OER increases with increasing [DTT], but at higher drug concentrations the OER decreases (Figure 4).

The effect of added thiols on cells is also dependent on the oxygen tension in the cells. Lunec et al.[29] pointed out that Ehrlich cells were protected by dithioerythritol (DTE) only minimally when the cells were fully aerated or hypoxic during irradiation, but were protected to a much greater degree when equilibrated with intermediate oxygen concentrations. Denekamp and colleagues[30-32] have extended such observations into whole

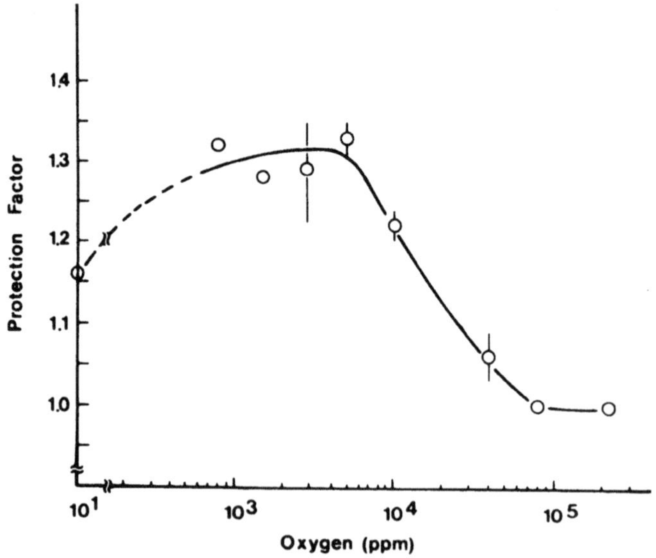

Figure 2. Radiation protection by OTZ in V79 cells depends on oxygen concentration. Protection factor is the ratio of the D_0 from the OTZ-treated survival curve to that for the untreated control. From Russo et al.,[25] with permission.

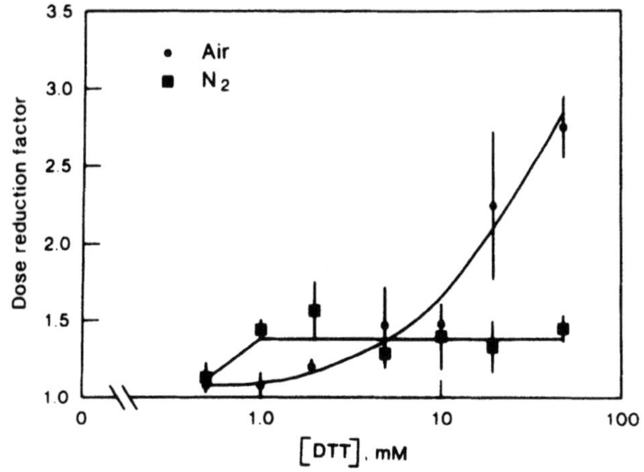

Figure 3. Dependence of radioprotection by V79 cells on DTT concentration when irradiation is in air or hypoxia. Dose reduction factor is the same as protection factor. From Held,[28] with permission.

animals by irrradiating mice breathing gas containing varying oxygen concentrations.[30,31] Under such conditions radioprotection of mouse skin and lung[32] by the cysteamine derivative WR-2721 is greatest at intermediate oxygen concentrations and less at very low or high oxygen levels.

DISCUSSION

Taken as a sum, the data presented above clearly indicate that the competition between thiols and oxygen for radiation-induced cellular

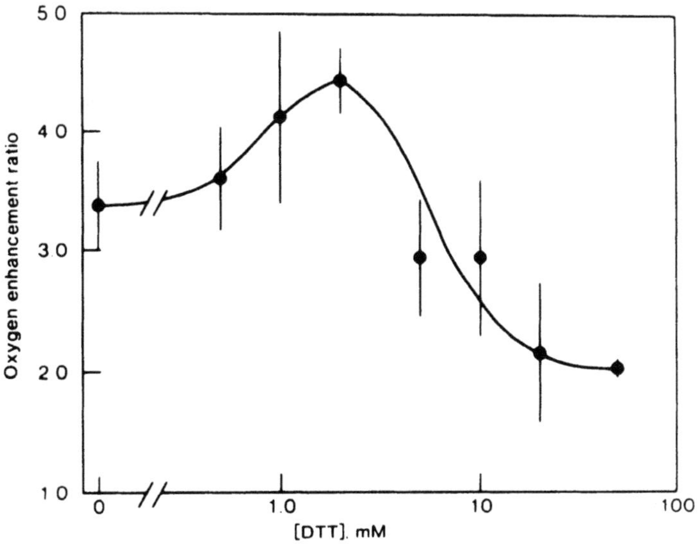

Figure 4. OER depends on the concentration of oxygen present in V79 cells during irradiation. From Held,[28] with permission.

lesions is at least partially involved in determining radiation response in mammalian cells. The fast mix experiments, in particular, clearly show the occurrence of this competition. Yet, the relative importance of this competition, particularly in hypoxic or fully aerated cells, is still uncertain. On the basis of studies with BSO and OTZ, Russo et al.[25] have suggested, for example, that GSH plays a minor role in aerated and hypoxic cells. Yet the data of Edgren and co-workers[13,14] with GSH^- cells indicates large enhancement of hypoxic cell radiosensitivity in GSH-deficient mutants. Perhaps it is not the absolute concentration of GSH per cell that is important for radiosensitivity but the intracellular distribution of that GSH. Edgren[33] has recently shown that BSO does not deplete nuclear GSH as much as cytoplasmic GSH; thus, the concentration of GSH near the DNA in chemically depleted cells may differ substantially from that in GSH^- mutants.

The data of several authors showing that increasing intracellular thiols or adding radioprotectors has the greatest effect at intermediate oxygen concentrations is consistent with the competition hypothesis. For example, if the rate of reaction of oxygen with organic radicals (reaction 1) is 100 times that of thiols with the same radicals (reaction 2), as suggested by studies in simple chemical systems, and if an average cell contains 5 mM NPSH, the oxygen concentration for half maximal effect would be about 0.4%. The work of Russo et al.[25] shows the greatest effect of increasing intracellular GSH at about this oxygen concentration. This observation is consistent with the model: at higher oxygen levels the oxygen reaction would overwhelm the thiol reaction such that little protection from increasing thiol would occur; at lower oxygen levels the normally endogenous thiol should be sufficient for protection so added thiols would have little additional effect.

In its simplest form, as discussed above, the competition model does not include a term for radiation-induced damage which cannot be chemically repaired. Models that do include such a term have been proposed by Koch[35,36] and Scott.[34] Consideration of some of the experimental data presented here indicates this inclusion is needed. For example, as recently pointed out by Scott (personal communication), our data on radioprotection by DTT (Figure 3) are as predicted by the "x-model."[34]

Thiols may have other roles in determining cellular radiation response besides acting via the competion model. They are efficient scavengers of ˙OH radicals, which is a protective reaction, although at low levels of added thiol they will not increase the normal cellular capacity for radical scavenging sufficiently to have an effect. ˙OH scavenging, moreover, should be independent of oxygenation status, and thus cannot account for the additional protection of thiols in oxygenated cells. Some data also indicate that thiols may have a role in enzymatic repair processes,[19] yet further investigation of this is needed.

ACKNOWLEDGMENTS

The author is supported by PHS grant numbers CA18614, CA42167, and CA42151, awarded by the National Cancer Institute, DHHS.

REFERENCES

1. P. Alexander and A. Charlesby, Physiochemical methods of protection against ionizing radiations, in: "Radiobiology Symposium 1954," Z.M. Bacq and P. Alexander, eds., Butterworths, London (1955).
2. T. Alper and P. Howard-Flanders, Role of oxygen in modifying the

radiosensitivity of E. coli B, Nature 178:978 (1956).
3. K.D. Held, Models for thiol protection of DNA in cells, Pharmac. Ther., in press.
4. O.W. Griffith and A. Meister, Potent and specific inhibition of glutathione synthesis by buthionine sulfoximine (S-n-butyl homocysteine sulfoximide), J. Biol. Chem. 254:7558 (1982).
5. J.E. Biaglow, M.E. Varnes, M. Astor, and E.J. Hall, Non-protein thiols and cellular response to drugs and radiation. Int. J. Radiat. Oncol. Biol. Phys. 8:719 (1982).
6. A. Larsson, 5-oxoprolinuria and other inborn errors related to the -glutamyl cycle, in: "Transport and Inherited Disease," N.R. Belton and C. Toothill, ed., MTP Press, Lancaster (1981).
7. C.J. Koch, C.C. Stobbe, and E.A. Bump, The effect on the K_m for radiosensitization at 0° C of thiol depletion by diethylmaleate pretreatment: Quantitative differences found using the radiation sensitizing agent misonidazole or oxygen, Radiat. Res. 98:141 (1984).
8. E.A. Bump, N.Y. Yu, and J.M. Brown, Radiosensitization of hypoxic tumor cells by depletion of intracellular glutathione, Science 217:544 (1982).
9. J.B. Mitchell, A. Russo, J.E. Biaglow, and S. McPherson, Cellular depletion by diethyl maleate or buthionine sulfoximine: No effect of glutathione depletion on the oxygen enhancement ratio, Radiat. Res. 96:422 (1983).
10. E.P. Clark, E.R. Epp, M. Morse-Gaudio, and J.E. Biaglow, The role of glutathione in the aerobic radioresponse. I. Sensitization and recovery in the absence of intracellular glutathione, Radiat. Res. 108:238 (1986).
11. D.C. Shrieve, J. Denekamp, and A. I. Minchinton, Effects of glutathione depletion by buthionine sulfoximine on radiosensitization by oxygen and misonidazole in vitro, Radiat. Res. 102:283 (1985).
12. R.J. Hodgkiss, N.R. Jones, M.E. Watts, and M. Woodcock, Glutathione depletion enhances the lifetime of oxygen-reactive radicals in mammalian cells, Int. J. Radiat. Biol. 46:673 (1984).
13. P.J. Deschavanne, J. Midander, M. Edgren, A. Larsson, E.-P. Malaise, and L. Revesz, Oxygen enhancement of radiation induced lethality is greatly reduced in glutathione deficient human fibroblasts, Biomedicine 35:35 (1981).
14. M. Edgren, A. Larsson, and L. Revesz, Induction and repair of single strand DNA breaks after X-irradiation of human fibroblasts deficient in glutathione. Int. J. Radiat. Biol. 40:355 (1981).
15. J. Midander, Oxygen enhancement ratios of glutathione-deficient fibroblasts determined from the frequency of radiation induced micronuclei. Int. J. Radiat. Biol. 42:195 (1982).
16. D. Debieu, P.J. Deschavanne, J. Midander, A. Larsson, and E.P. Malaise, Survival curves of glutathione synthetase deficient human fibroblasts: correlation between radiosensitivity in hypoxia and glutathione synthetase activity, Int. J. Radiat. Biol. 48:525 (1985).
17. E.-P. Malaise, Reduced oxygen enhancement of the radiosensitivity of glutathione-deficient fibroblasts, Radiat. Res. 95, 486 (1983).
18. L. Revesz, The role of endogenous thiols in intrinsic radioprotection, Int. J. Radiat. Biol. 47:361 (1985).
19. L. Revesz and M. Edgren, Glutathione dependent yield and repair of single-strand DNA breaks in irradiated cells. Br. J. Cancer 49, Suppl. VI:55 (1984).
20. G. Solen, M. Edgren, O.C.A. Scott, and L. Revesz, Cellular glutathione content and K values, Int. J. Radiat. Biol. 51:39 (1987).
21. B.D. Michael, S. Davies, and K.D. Held, Ultrafast chemical repair of DNA single and double strand break precursors in irradiated V79

cells, *in*: "Mechanisms of DNA Damage and Repair," M.G. Simic, L. Grossman, and A.C. Upton, eds., Plenum Press, New York (1986).
22. B.M. Cullen and H.C. Walker, Variation of the radiobiological oxygen constant, K, with the proliferative activity of the cells, *Int. J. Radiat. Biol.* 38:513 (1980).
23. B.M. Cullen, A. Michalowski, H.C. Walker, and L. Revesz, Correlation between the radiobiological oxygen constant, K, and the non-protein sulphydryl content of mammalian cells, *Int. J. Radiat. Biol.* 38:525 (1980).
24. H.C. Walker and B.M. Cullen, The importance of NPSH on the radio-sensitizing effect of oxygen in Chinese hamster V-79 cells, *Int. J. Radiat. Biol.* 51:19 (1987).
25. A. Russo, J.B. Mitchell, E. Finkelstein, W.G. DeGraff, I.J. Spiro, and J. Gamson, The effects of cellular glutathione elevation on the oxygen enhancement ratio, *Radiat. Res.* 103:232 (1985).
26. T. Alper, "Cellular Radiobiology," Cambridge University Press, Cambridge (1979).
27. K.D. Held, Role of free radical scavengers in radiation protection of DNA, *in*: "Free Radicals, Aging, and Degenerative Diseases," J.E. Johnson et al, eds., Alan R. Liss, New York (1986).
28. K.D. Held, Interactions of radioprotectors and oxygen in cultured mammalian cells. I. Dithiothreitol effects on radiation-induced cell killing, *Radiat. Res.* 101:424 (1985).
29. J. Lunec, B. Cullen, H. Walker, and S. Hornsey, A cautionary note on the use of thiol compounds to protect normal tissues in radio-therapy, *Br. J. Radiol.* 54:428 (1981).
30. J. Denekamp, B.D. Michael, A. Rojas, and F.A. Stewart, Thiol radio-protection in vivo: the critical role of tissue oxygen concentra-tion, *Br. J. Radiol.* 54:1112 (1981).
31. J. Denekamp, A. Rojas, and F.A. Stewart, Is radioprotection by WR-2721 restricted to normal tissues?, *in*: "Radioprotectors and Anticarcinogens," O.F. Nygaard and M.G. Simic, eds., Academic Press, New York (1983).
32. C. Parkins, J.F. Fowler, and J. Denekamp, Low radioprotection by thiol in lung: The role of local tissue oxygenation, *Eur. J. Cancer Clin. Oncol.* 19:1169 (1983).
33. M. Edgren, Nuclear glutathione and oxygen enhancement of radiosensi-tivity, *Int. J. Radiat. Biol.* 51:3 (1987).
34. M. Edgren,T. Nishidai, O.C.A. Scott, and L. Revesz, Combined effect of misonidazole and glutathione depletion by buthionine sulphoximine on cellular radiation response, *Int. J. Radiat. Biol.* 47:463 (1985).
35. C.J. Koch and R.L. Howell, Combined radiation-protective and radiation-sensitizing agents. I. Radiosensitivity of hypoxic or aerobic Chinese hamster fibroblasts in the presence of cysteamine and misonidazole: Implications for the "oxygen effect," *Radiat. Res.* 87:265 (1981).
36. C.J. Koch, Competition between radiation portectors and radiation sensitizers in mammalian cells. *in*: "Radioprotectors and Anticarcinogens," O.F. Nygaard and M.G. Simic, eds., Academic Press, New York (1983).

OXYGEN RADICALS, LIPID PEROXIDATION AND DNA DAMAGE IN MITOCHONDRIA

Andrew M. Hruszkewycz and David S. Bergtold*

Department of Pathology, George Washington University Medical Center, Washington, DC
*Center for Radiation Research, National Bureau of Standards, Gaithersburg, MD

INTRODUCTION

Oxygen activation in eukaryotic cells occurs mainly within mitochondria, where the respiratory electron transport chain metabolizes approximately 90% of cellular oxygen. While mitochondrial oxygen metabolism supplies the cell with most of its ATP, it also results in the production of hazardous oxygen radicals.[1,2] Oxygen radical production has been estimated to account for approximately 1 to 2% of mitochondrial oxygen consumption.[3] Although the mitochondrion is protected by an elaborate system of antioxidants and scavengers,[2,4] free radicals may escape their surveillance and cause damage to mitochondrial components. The close spatial relationship between the site of mitochondrial oxygen activation, the peroxidizable lipids of the inner membrane, and mitochondrial DNA (mtDNA) suggests that oxygen radicals and lipid peroxidation may cause mutation of the mitochondrial genome. The condition of mitochondrial DNA structure is of fundamental importance because this genome encodes key enzymes of the respiratory chain.[5] In view of growing evidence indicating that abnormalities of mtDNA may contribute to the etiology of aging[6] and a number of diseases, including cancer,[7,8] the relationship of mitochondrial oxygen radical production and mtDNA integrity merits further consideration.

MITOCHONDRIAL OXYGEN METABOLISM AND LIPID PEROXIDATION

During mitochondrial respiration electrons are added to oxygen one at a time to produce water. This electron flow process is not fully efficient and there is occasional electron leakage which results in the production of activated oxygen intermediates,[1,2,9,10] the first of which is the superoxide anion radical ($O_2^{\cdot-}$). The superoxide is thought to be produced mainly by diversion of electrons to oxygen at the ubiquinone-cytochrome b region of the respiratory chain.[1,11-16] Mitochondrial superoxide dismutase metabolizes superoxide to hydrogen peroxide,[4] and hydrogen peroxide can be scavenged by catalase and mitochondrial selenium dependent glutathione peroxidase.[1,2] However, although enzymes which detoxify active oxygen species are present in high concentrations and are extraordinarily active, their efficiency is not complete. It is estimated that the intramitochondrial steady-state level of $O_2^{\cdot-}$ is approximately 8 x 10^{-12} M,[17] although this will vary with the state of mitochondrial respiration. For example, $O_2^{\cdot-}$ generation will increase as mitochondria approach state 4 respiration, when the components of the respiratory chain are mostly reduced.[3] Also, steady state concentrations of superoxide are higher in mitochondria of older animals.[11]

An inherent danger of the superoxide radical is that it may lead to the formation of oxygen species which are even more reactive, including hydroxyl radicals, singlet oxygen, and hydroperoxy radicals.[18,19,20] The extreme reactivity and short half lives of these latter species suggests that mainly those molecules within a highly circumscribed locus near the site of their production are subject to direct attack. Therefore, the molecules most vulnerable to direct attack by oxygen radicals produced by mitochondria are probably the ubiquinone-cytochrome b complex and the associated membrane lipids.

A potential hazard of oxygen radical attack on membrane lipids is the initiation of lipid peroxidation, which may lead to a highly destructive autocatalytic chain reaction. Lipid peroxidation is initiated by abstraction of a hydrogen atom from a methylene carbon separating double bonds in a polyunsaturated fatty acid present within a membrane.[21] The actual nature of the ultimate oxygen species responsible for initiation of lipid peroxidation is not known, although there is speculation on various possibilities. Neither $O_2^{\cdot-}$ or H_2O_2 themselves have sufficient reactivity to initiate lipid peroxidation directly.[22] Initiation of lipid peroxidation by the hydroxyl radical has been suggested.[23] Aust and coworkers have proposed a ferrous-dioxygen-ferric complex of undetermined structure to be the initiator.[24] Bielski and others have shown that the hydroperoxy radical ($HO_2\cdot$) reacts readily with polyenic fatty acids.[18,25,26]

Whatever the initiation mechanism, hydrogen abstraction results in a lipid radical which reacts with molecular oxygen to form a fatty acid peroxide radical.[27] This lipid peroxide radical propagates the peroxidation reaction by abstracting a hydrogen from another methylene carbon.[28] Ultimately, the unstable fatty acid hydroperoxides decompose to release a large variety of potentially toxic breakdown products, including alkoxyl radicals (LO•), peroxyl radicals (LOO•), and aldehydes.[29,30]

The theoretical strength of the lipid peroxidation hypothesis resides in the dual postulate that spreading peroxidative decomposition of lipids within the inner mitochondrial membrane, coupled with emergence of toxic lipid peroxides or their breakdown products, are processes which provide for action at a distance from the site of initial free radical production. Thus, lipid peroxidation provides a mechanism of amplification by which toxic effects of oxygen radicals can be propagated throughout the mitochondrion. Peroxidative decomposition of mitochondrial membranes has been reported for a variety of situations, including aging,[31] iron overload,[32] irradiation,[33] and xenobiotic toxicity.[33-35] Therefore, membrane oxidation may be invoked as an intermediary mechanism linking free radical production to damage of various mitochondrial components.

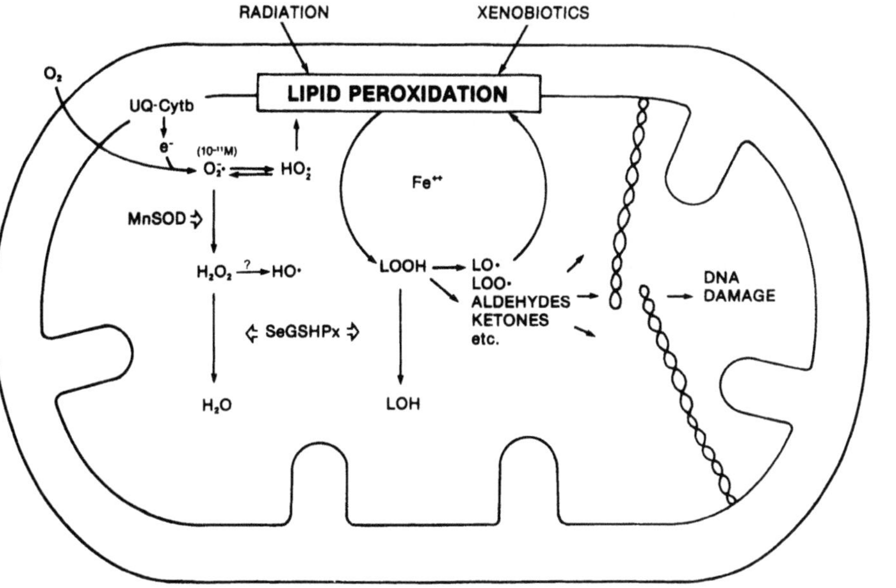

Figure 1. General scheme showing the proposed intermediary role of lipid peroxidation in free radical induced damage to mitochondrial DNA. Lipid peroxidation may be initiated by a variety of agents, including active oxygen species such as the hydroperoxy radical ($HO_2\cdot$), which is in equilibrium with the superoxide radical ($O_2^{\cdot-}$). Lipid peroxidation can spread within the membrane by autocatylytic propagation and produce genotoxic agents. UQ-cytb, ubiquinone-cytochrome b complex; MnSOD, mitochondrial superoxide dismutase; SeGSHPx, selenium-containing glutathione peroxidase; LOOH, lipid hydroperoxide; LO•, alkoxyl radical; LOO•, peroxyl radical; HO•, hydroxyl radical.

MITOCHONDRIAL DNA AS A TARGET OF OXYGEN DERIVED RADICALS

The typical cell contains several hundred mitochondria and each contains several closed circular DNA molecules, which are approximately the size of a bacterial genome. Mitochondrial genes generally represent less than 1% of the total DNA within mammalian cells.[36] Unlike nuclear DNA, mtDNA is not protected by histones.[37] The average rate of nucleotide substitution in the mitochondrial genome is 5 to 10 times greater than that of single-copy nuclear genes.[38] Also, mitochondria lack DNA repair mechanisms comparable to those seen with nuclear DNA.[39] For example, mitochondria fail to repair single strand breaks in DNA,[40] and pyrimidine dimers formed by irradiation are not removed.[41] Many xenobiotic metabolites, such as chemical carcinogens, preferentially react with mtDNA[42-45] and bound agents are not removed from mtDNA as they are with nuclear DNA.[42] Furthermore, aging animals show qualitative alterations in mtDNA[46] and diminution in quantity of mtDNA.[46-49]

There is theoretical and experimental support for the role of oxygen-derived radicals and lipid peroxidation in mtDNA damage. Numerous studies have shown that active oxygen species can cause direct damage to DNA in model experimental systems.[26] However, in the case of biological systems, the presence of many potential radical scavengers would make it more difficult for highly reactive oxygen radicals to survive long enough to traverse the finite distance between the sites of radical production and DNA. This theoretical difficulty is readily overcome by a lipid peroxidation hypothesis, which provides an intervening mechanism which can link localized reactions of oxygen radicals to eventual damage to more distant molecules (Figure 1).

Lipid peroxidation has been shown to cause DNA damage in model systems and prokaryotes. Pietronigro *et al.*[50] reported damage to *Bacillus subtilis* transforming DNA when the DNA was incubated with oxidizing liposomal suspensions. Similarly, Akasaka has observed inactivation of plasmid DNA by lipid peroxidation of liposomes.[51] Others have demonstrated that peroxidizing arachidonic acid causes structural changes in isolated DNA.[52] Also, linoleic acid hydroperoxide has been found to cause site-specific cleavage of double-stranded DNA.[53] Malonic dialdehyde (MDA), a well-known decomposition product of peroxidized polyunsaturated fatty acids, has been shown to covalently bind to DNA *in vitro*,[54,55] and to be mutagenic in bacterial test systems.[55,56] Other aldehydic products of lipid peroxidation induce DNA strand breaks and mutations in bacteria.[57]

While observations of lipid peroxidation-induced damage to DNA in eukaryotic systems are more limited and indirect,[58-62] it has been proposed that lipid peroxidation is a major mode of membrane-mediated chromosomal damage.[63] Recent evidence that mtDNA is attached to the inner mitochondrial membrane[64] suggests that the mtDNA would also be vulnerable to membrane-mediated genotoxic events. Results from our laboratories indicate that the mitochondrial genome is damaged by membrane oxidation.

When mitochondria from rat liver were peroxidized *in vitro*, DNA isolated from these peroxidized mitochondria had completely different electrophoretic mobility than DNA isolated from mitochondria protected from peroxidation.[65] Analysis of the nucleic acids from these peroxidized mitochondria by gas chromatography/mass spectrometry showed a marked increase in 8-hydroxyguanine formation. Alpha-tocopherol, a lipophilic antioxidant that inactivates lipid radicals, concomitantly prevented lipid peroxidation and protected against the genetic damage seen in this *in vitro* system.[66,67] A direct role of oxygen radicals in the mechanism of genotoxicity could not be demonstrated with this system, since superoxide dismutase, catalase, mannitol, or dimethyl sulfoxide (scavengers of $O_2^{\bullet-}$, H_2O_2, and HO•) did not attenuate the electrophoretic mobility change in mtDNA which occurred with lipid peroxidation.[67] Therefore, all our experiments with isolated mitochondria support the hypothesis that peroxidation of membrane lipids is an important intermediary event in free radical-induced mtDNA damage.

Although these findings do not offer a precise mechanism by which lipid peroxidation affects DNA, a direct attack on the DNA by a lipid radical, such as an alkoxyl radical (LO•), is one possibility. This would have formal similarity to postulated mechanisms of hydroxyl radical attack.[68] Lipid free radical interactions with DNA bases have been reported.[69] Also, the interaction of DNA with lipid radicals and other products of lipid peroxidation is the topic of a recent review by Vaca and co-workers.[70]

In view of the large array of potentially genotoxic compounds produced by lipid peroxidation, it is not possible at this time to assert which of these products can penetrate the elaborate cellular antioxidant defense and play an important role in pathophysiologic mechanisms of genetic transformation. Further study of DNA changes in mitochondria undergoing oxidative stress may provide this information and lead to prevention and treatment of disease caused by mtDNA damage.

ACKNOWLEDGMENTS

We gratefully acknowledge the expert assistance of Cassandra Beck and Terry Shuggars during the preparation of the manuscript. This work was supported by NIH grant CA 07955, from the National Cancer Institute.

REFERENCES

1. H. Nohl, Oxygen radical release in mitochondria: Influence of age, in: "Free Radicals, Aging, and Degenerative Diseases," J.E. Johnson, Jr., R. Walford, D. Harman, and J. Miquel, eds., Alan R. Liss, New York (1986).
2. B. Chance, H. Sies, and A. Boveris, Hydroperoxide metabolism in mammalian organs, Physiol. Rev. 59:617 (1979).
3. H.J. Forman and A. Boveris, Superoxide radical and hydrogen peroxide in mitochondria, in: "Free Radicals in Biology," vol. V, W.A. Pryor, ed., Academic Press, New York (1982).
4. I. Fridovich, Superoxide dismutase, Ann. Rev. Biochem. 44:147 (1975).
5. R.W. Yatscoff, S. Goldskin, and K.B. Freedman, Conservation of genes coding for proteins synthesized in human mitochondria, Somatic Cell Genet. 4:633 (1978).
6. J. Miguel and J. Fleming, Theoretical and experimental support for an "oxygen radical-mitochondrial injury," hypothesis of cell aging, in: "Free Radicals, Aging, and Degenerative Diseases," J.E. Johnson, Jr., R. Walford, D. Harman, and J. Miguel, eds., Alan R. Liss, New York (1986).
7. J.W. Shay and H. Werbin, Are mitochondrial DNA mutations involved in the carcinogenic process? Mutat. Res. 186:149 (1987).
8. C.R. Merril and M.G. Harrington, The search for mitochondrial inheritance of human diseases, Trends in Genetics 1:140 (1985).
9. G. Loschen, L. Flohe, and B. Chance, Respiratory chain linked H_2O_2 production in pigeon heart mitochondria, FEBS Lett. 18:261 (1971).
10. A. Boveris, N. Oshino, and B. Chance, The cellular production of hydrogen peroxide, Biochem. J. 128:617 (1972).
11. H. Nohl and D. Hegner, Do mitochondria produce oxygen radicals in vivo? Eur. J. Biochem. 82:563 (1978).
12. G. Loschen, A. Azzi, and L. Flohe, Mitochondrial H_2O_2 formation at site II, Hoppe-Seylers Z. Physiol. Chem. 354:791 (1973).

13. G. Loschen, A. Azzi, and L. Flohe, Mitochondrial hydrogen peroxide formations, in: "Alcohol and Aldehyde Metabolizing Systems," R.G. Thurman, T. Yonetani, J.R. Williamson, and B. Chance, eds., Academic Press, New York (1974).
14. G. Loschen, A. Azzi, and L. Flohe, Mitochondrial H_2O_2 formation: Relationship with energy conservation, FEBS Lett. 33:84 (1973).
15. E. Cadenas, A. Boveris, C.I. Ragan, and A.O.M. Stopani, Production of superoxide radicals and hydrogen peroxide by NADH-ubiquinone reductase and ubiquinol-cytochrome C reductase from beef heart mitochondria, Arch. Biochem. Biophys. 180:248 (1977).
16. H.J. Forman, The role of ubisemiquinone in superoxide production by dihydroorotate dehydrogenase: A proposed mechanism, in: "The Significance of Superoxide and Superoxide Dismutase," vol. I, J.V. Bannister and H.A.O. Hill, eds., Elsevier, Amsterdam (1980).
17. D.D. Tyler, Polarographic assay and intracellular distribution of superoxide dismutase in rat liver, Biochem. J. 147:793 (1975).
18. M.G. Simic and K.A. Taylor, Free radical mechanisms of oxidation reactions, in: "Flavor Quality of Fresh Meat: Warmed-Over Flavor," A.J. St. Angelo and M.E. Bailey, eds., Academic Press, New York (1987).
19. A. Naqai and B. Chance, Reactive oxygen intermediates in biochemistry, Annu. Rev. Biochem. 55:137 (1986).
20. A. Singh, Interconversion of singlet oxygen and related species, Photochem. Photobio. 28:429 (1978).
21. L.K. Dahle, E.G. Hill, and R.T. Holman, The thiobarbituric acid reaction and the autoxidation of polyunsaturated fatty acid methyl esters, Arch. Biochem. Biophys. 98:253 (1962).
22. I. Fridovich, Superoxide radical: an endogenous toxicant, Annu. Rev. Pharmacol. Toxicol. 23:239 (1983).
23. K.L. Fong, P.B. McCary, J.L. Poyer, H.P. Misra, and B.B. Keele, Evidence for superoxide-dependent reduction of Fe^{3+} and its role in enzyme-generated hydroxyl radical formation, Chem. Biol. Interact. 15:77 (1976).
24. S.D. Aust, L.A. Morehouse, and C.E. Thomas, Role of metals in oxygen radical reactions, J. Free Rad. Biol. Med. 1:3 (1985).
25. B.H.J. Bielski, R.L. Arudi, and M.W. Sutherland, A study of the reactivity of HO_2/O_2^- with unsaturated fatty acids, J. Biol. Chem. 258:4759 (1983).
26. C. von Sonntag, "The Chemical Basis of Radiation Biology," Taylor and Francis, London (1987).
27. D.C. Harris and P. Aisen, Facilitation of Fe(II) autoxidation by Fe(III) complexing agents, Biochim. Biophys. Acta 329:156 (1973).
28. G. Schwarzenbach and J. Heller, Komplexone XVIII, Die eisen(II)-und eisen(III)-komplexe des achylenediamintetraessigsaure und ihr redoxgleichgewich, Helv. Chim. Acta 34:576 (1951).
29. H. Esterbauer, Aldehydic Products of Lipid Peroxidation, in: "Free Radicals, Lipid Peroxidation and Cancer," D.C.H. McBrien and T.F. Slater, eds., Academic Press, New York (1982).
30. A.M. Hruszkewycz, E.A. Glende, Jr., and R.O. Recknagel, Destruction of microsomal cytochrome P-450 and glucose-6-phosphatase by lipids extracted from peroxidized microsomes, Toxicol. Appl. Pharmacol. 46:695 (1978).
31. R.G. Hansford, Bioenergetics in aging, Biochim. Biophys. Acta 726:41 (1983).
32. R.B. Bacon, C.H. Park, G.M. Brittenham, R. O'Neill, and A.S. Tavill, Hepatic mitochondrial oxidative metabolism in rats with chronic dietary iron overload, Hepatology 5:789 (1985).

33. Y.A. Vlademirov, V.I. Olenev, T.B. Suslova, and A.P. Chevemisina, Lipid peroxidation in mitochondria membrane, Adv. Lipid Res. 17:173 (1980).
34. R.O. Recknagel and A.K. Ghoshal, Quantitative estimation of peroxidative degeneration of rat liver microsomal and mitochondrial lipids after carbon tetrachloride poisoning, Exp. Mol. Pathol. 5:413 (1966).
35. R. Ogura, Adriamycin-induced lipid peroxidation and its protection, in: "Lipid Peroxides in Biology and Medicine," K. Yagi, ed., Academic Press, London (1982).
36. D.A. Clayton, Replication of animal mitochondrial DNA, Cell 28:693 (1982).
37. I. Salazar, L. Tarrago-Litvak, L. Gil, and S. Litvak, The effect of benzo(a)pyrene on DNA synthesis and DNA polymerase activity of rat liver mitochondria, FEBS Lett. 138:45 (1982).
38. W.M. Brown, M. George, Jr., and A.C. Wilson, Rapid evolution of animal mitochondrial DNA, Proc. Natl. Acad. Sci. U.S.A. 76:1967 (1979).
39. M. Fukanaga and K.L. Yielding, Fate during growth of yeast mitochondrial and nuclear DNA after photolytic attachment of the monoazide analog of ethidium, Biochem. Biophys. Res. Commun. 90:582 (1979).
40. R.A. Lansman and D.A. Clayton, Selective nicking of mammalian mitochondrial DNA in-vivo: photosensitization by incorporation of 5-bromodeoxy uridine, J. Mol. Biol. 99:761 (1975).
41. D.A. Clayton, J.N. Doda, and E.C. Friedberg, The absence of a pyrimidine dimer repair mechanism in mammalian mitochondria, Proc. Natl. Acad. Sci. U.S.A. 71:2777 (1974).
42. B.G. Niranjan, N.K. Bhat, and N.G. Avadhani, Preferential attack of mitochondrial DNA by aflatoxin B_1 during hepatocarcinogenesis, Science 215:73 (1982).
43. J.M. Backer and I.B. Weinstein, Mitochondrial DNA is a major target for a dihydrodiol-epoxide derivative of benzo[a]pyrene, Science 209:297 (1980).
44. V. Wunderlich, I. Tetzlaff, and A. Graffi, Studies on nitrosodimethylamine: Preferential methylation of mitochondrial DNA in rats and hamsters, Chem. Biol. Interact. 4:81 (1971-1972).
45. J.A. Allen and M.M. Coombs, Covalent binding of polycyclic aromatic compounds and nuclear DNA, Nature 287:244 (1980).
46. L. Piko and L. Matsumoto, Complex forms and replicative intermediates of mitochondrial DNA in tissues from adult and senescent mice, Nucl. Acids Res. 4:1301 (1977).
47. H.R. Massie, M.B. Baird, and M.M. McMahon, Loss of mitochondrial DNA with aging, Gerontologia 21:231 (1975).
48. D.M. Stocco and J.C. Hutson, Quantitation of mitochondrial DNA and protein in the liver of Fisher 344 rats during aging, J. Gerontol. 33:802 (1978).
49. R.P. Heumer, K.D. Lee, A.E. Reeves, and C. Bickert, Mitochondrial studies in senescent mice II specific activity, buoyant density and turnover of mitochondrial DNA, Exp. Gerontol. 6:327 (1971).
50. D.D. Pietronigro, W.B.G. Jones, K. Kalty, and H.B. Demopoulos, Interaction of DNA and liposomes as a model for membrane-mediated DNA damage, Nature 267:78 (1977).
51. S. Akasaka, Inactivation of transforming activity of plasmid DNA by lipid peroxidation, Biochim. Biophys. Acta. 867:201 (1986).
52. U. Reiss and A.L. Tappel, Fluorescent product formation and changes in structure of DNA reacted with peroxidizing arachidonic acid, Lipids 8:199 (1973).

53. S. Inouye, Site-specific cleavage of double-stranded DNA by hydroperoxide of linoleic acid, FEBS Lett. 173:231 (1984).
54. B.R. Brooks and O.L. Klamerth, Interaction of DNA with bifunctional aldehydes, Eur. J. Biochem. 5:173 (1968).
55. S. Yonei and H. Furui, Lethal and mutagenic effects of malondialdehyde, a decomposition product of peroxidized lipids on Escherichia coli with different DNA-repair capacities, Mutat. Res. 88:23 (1981).
56. F.H. Mukai and B.D. Goldstein, Mutagenicity of malonaldehyde, a decomposition product of peroxidized polyunsaturated fatty acids, Science 191:868 (1976).
57. L.J. Marnett, H.K. Hurd, M.C. Hollstein, D.E. Levin, H. Esterbauer, and B.N. Ames, Naturally occurring carbonyl compounds are mutagens in Salmonella tester strain TA 104, Mutat. Res. 148:25 (1985).
58. B.N. Ames and R.L. Saul, Oxidative DNA damage as related to cancer and aging, Prog. Clin. Biol. Res. 209A:11 (1986).
59. D.C.H. McBrien and T.F. Slater, eds., "Free Radicals, Lipid Peroxidation and Cancer," Academic Press, New York (1982).
60. I. Emerit, S.H. Khan, and P.A. Cerutti, Treatment of lymphocyte cultures with a hypoxanthine-xanthine oxidase system induces the formation of transferable clastogenic material, J. Free Rad. Biol. Med. 1:51 (1985).
61. T. Ochi and P.A. Cerutti, Clastogenic action of hydroperoxy-5,8,11,13-icosatetraenoic acids on the mouse embryo fibroblast C3H/10T1/2, Proc. Natl. Acad. Sci. U.S.A. 84:991 (1987).
62. U.M. Marinari, M. Ferro, L. Sciaba, R. Finollo, A.M. Bassi, and G. Brambilla, DNA damaging activity of biotic and xenobiotic aldehydes in chinese hamster ovary cells, Cell Biochem. Funct. 2:243 (1984).
63. P.A. Cerutti, I. Emerit, and P. Amstad, Membrane-mediated chromosomal damage, in: "Genes and Proteins in Oncogenesis," I.B. Weinstein and H.J. Vogel, eds., Academic Press, New York (1983).
64. M. Albring, J. Griffith, and G. Attardi, Association of a protein structure of probable membrane derivation with HcLa cell mitochondrial DNA near its origin of replication, Proc. Natl. Acad. Sci. U.S.A. 74:1348 (1977).
65. A.M. Hruszkewycz, Mitochondrial DNA damage during mitochondrial lipid peroxidation, in: "Anticarcinogenesis and Radiation Protection," P.A. Cerutti, O.F. Nygaard, and M.G. Simic, eds., Plenum Press, New York (1988).
66. A.M. Hruszkewycz and D.S. Bergtold, Induction of mitochondrial DNA damage during mitochondrial lipid peroxidation, FASEB J. 2:7494 (1988).
67. A.M. Hruszkewycz, Evidence for mitochondrial DNA damage by lipid peroxidation, Biochem. Biophys. Res. Commun. 153:191 (1988).
68. C. von Sonntag, U. Hagen, A. Schon-Bopp, and D. Schulte-Frohlinde, Radiation-induced strand breaks in DNA: Chemical and enzymatic analysis of end groups and mechanistic aspects, Adv. Radiat. Biol. 9:109 (1981).
69. K.M. Schaich and D.C. Borg, Radiomemetic effects of peroxidizing lipids on nucleic acids and their bases, in: "Oxygen Radicals in Chemistry and Biology," W. Bors, M. Saran, and D. Tait, eds., Walter de Gruyter, Berlin (1984).
70. C.E. Vaca, J. Wilhelm and M. Harms-Raingdahl, Interaction of lipid peroxidation products with DNA. A Review, Mutat. Res. 195:137 (1988).

INVOLVEMENT OF REACTIVE OXYGEN INTERMEDIATES IN THE MUTAGENICITY OF TUMOR PROMOTERS

Neal Wallis Biggart and Edwin C. Murphy, Jr.*

Department of Biology, San Diego State University, San Diego California, 92116-0057, U.S.A.

*M.D. Anderson Hospital and Tumor Institute, University of Texas Cancer System, Houston, Texas 77030, U.S.A.

INTRODUCTION

Reactive oxygen intermediates have been strongly implicated in tumor promotion not only because many tumor promoters generate free radicals either directly or indirectly, but also because many antipromoters are anti-oxidants which inhibit the reactivity of oxygen radicals (for reviews, see references 1-4). Tumor promoting agents can be distinguished from initiating agents on the basis of the DNA damage they induce; whereas true initiators cause irreversible, heritable changes through mutation, tumor promoters are clastogenic, i.e., cause DNA strand breaks, and this damage is reversible to some degree. The clastogenic action of promoters is thought to be due at least in part to the generation of reactive oxygen intermediates and perhaps to the secondary generation of other reactive intermediates from interaction with cellular macromolecules such as membrane lipids. Type II promoters, such as benzoyl peroxide, generate oxygen radicals directly in the target cell without the mediation of other effectors. Type I promoters, such as the phorbol esters, act indirectly via effectors such as protein kinase c to generate superoxide, perhaps as a by-product of the arachidonic cascade. In addition, type I promoters stimulate the respiratory burst of phagocytic cells, and the reactive oxygen (superoxide, hydrogen peroxide) produced during this burst can then act on nearby cells. There is, in fact, a good correlation between the tumor promoting effectiveness of the various phorbol esters and their ability to stimulate superoxide production by phagocytes. Tumor promoters can also be distinguished from initiators in that promoting agents are poorly mutagenic in test systems currently utilized to study mutagenesis, whereas initiating agents are readily studied in these mutagenesis assays.

6M2 MUTAGENESIS ASSAY

It was our interest in the carcinogenic metals which, like tumor promoters, are characterized by poor mutagenicity but strong clastogenicity (5) that led to the development of a mammalian cell assay for both point mutations and clastogenic damage in a single target sequence (6, 7). The cell line utilized (6m2) is a rat kidney fibroblast line infected with a mutant Moloney murine sarcoma virus, MuSVts110. This mutant provirus is defective, stably integrated, and confers temperature-sensitive morphological transformation upon the cells as the result of the temperature-dependent expression of v-mos. The molecular basis for temperature sensitivity lies in a thermodependent splicing event that occurs only at 33°C or lower and that serves to place the v-mos gene in frame with the v-gag gene from which translation begins. Some of the viral primary transcript is spliced at the permissive temperature, and translation of this spliced viral RNA yields a P85gagmos fusion protein encoding a serine kinase known to transform cells whenever it is expressed. The remaining unspliced viral RNA is translated to yield a truncated polypeptide (P58gag) which contains no transforming serine kinase activity.

We have taken advantage of the temperature-dependent morphological transformation and genetic structure of MuSVts110 to study the DNA damaging effects of metals and more recently of the tumor promoting agent 12-O-tetradecanoylphorbol-13-acetate (TPA). The genetic structure of MuSVts110 allows the detection of several types of DNA alterations. v-Mos can only be expressed if it is brought into frame with the gag gene in the viral RNA. There are two ways to accomplish this: by splicing the RNA or by rearranging the DNA sequence to achieve a net frameshift of -2 nucleotides. Because the sequences in the intron (431 nucleotides) are not necessary for production of an active transforming serine kinase, these sequences can be deleted or rearranged without loss of functional coding sequences. In addition, point mutations (base-pair substitutions and frame-shift mutations) in MuSVts110 sequences involved in the splicing process should also be detectable. This is possible because the determinants for defective splicing of MuSVts110 are contained in the viral sequence and because it is well known that single base changes in specific sequences within introns can drastically alter splicing efficiency. Thus, both point mutations and rearrangements in MuSVts110 can alter the expression of v-mos and, consequently, the temperature-dependent morphological transformation of 6m2 cells. By briefly treating 6m2 cells with the test agent and selecting for transformed foci at 39°C (the nonpermissive temperature), cells with mutations in MuSVts110 can be identified, isolated, and established in culture. Of the more than 70 revertant cell lines established from transformed foci, all showed an alteration either in the expression of P85gagmos or in the structure of the gagmos protein. Thus, treatment of 6m2 cells with mutagenic agents (but not with nonmutagenic agents) targets the MuSVts110 DNA for several classes of selectable mutations.

CLASSES OF MUTATIONS INFERRED FROM ALTERATIONS IN VIRAL PROTEINS

We have examined the viral proteins in 6m2-revertants obtained after brief treatment with metal compounds, nitrosomethylurea (NMU, a direct-acting carcinogen that primarily induces base-pair substitution mutations), and TPA (a type I tumor promoter). Four revertant phenotypes have been distinguished at the protein level as shown in Table 1. Nickel chloride and NMU both induced mutations which altered the splicing efficiency of the viral RNA such that P85gagmos was overproduced at all temperatures. However, the extent to which overproduction occurred over a range of temperatures was different, indicating that different mutations had been induced by the two agents. These two phenotypes are believed to have resulted from base-pair substitution mutations in the viral sequences involved in splicing. In contrast, cadmium chloride and sodium chromate induced a class of mutations in which altered transforming proteins were produced in a manner that was independent of splicing. From the size of most of these altered proteins, it appears that they result from the gain or loss of a small number of nucleotides, with a net frameshift of -2 nucleotides. However, one of the cadmium revertants produced a putative transforming protein which would be expected if a deletion of 300 nucleotides had occurred in the MuSVts110 DNA.

Table 1. Representative Protein Phenotypes and Corresponding Mutations in MuSVts110 Thought to Account for the Phenotype

Agent	Protein Phenotype	Proposed Mutation
Nickel chloride	Enhanced production of P85 with partial temperature-sensitive character	Base-pair substitution in specific sequences involved in splicing
NMU	Enhanced production of P85 with no temperature-sensitive character	Base-pair substitution in specific sequences involved in splicing
Sodium chromate	Altered transforming protein (P100gagmos)	Small frameshift mutation
Cadmium chloride	Altered transforming protein (P100gagmos, P90gagmos)	Small frameshift mutation and 300 base deletion
TPA	Altered transforming proteins	Deletions of up to 1000 bases
	Enhanced production of P85	Base-pair substitution

OXYGEN INTERMEDIATES IN CADMIUM CHLORIDE AND TPA MUTAGENESIS

Cadmium chloride is a clastogenic agent, and its ability to induce chromosomal aberrations and DNA strand breaks is inhibited by superoxide dismutase, catalase, and oxygen radical scavengers (8). The mutagenic activity of cadmium chloride in 6m2 cells is also strongly inhibited by these agents, and it therefore appears that the DNA damage induced by cadmium chloride in 6m2 cells may be mediated by reactive oxygen intermediates.

The results of preliminary screening of viral proteins from several TPA revertants are shown in Figure 1. There were several mutations affecting expression (TPA 35, 39, 2, 33) such that P85gagmos was produced at 39°C, a phenotype typical of the nickel-induced revertants in which splicing efficiency was enhanced. There were also several revertant lines (TPA-37, 40, 30) in which proteins of altered sizes were produced at both temperatures. These putative transforming proteins would be expected if TPA treatment had induced deletions of a few to 1,000 nucleotides in MuSVts110 DNA. These preliminary results suggest that TPA induces both point mutations and rearrangements, an observation that is consistent with the induction of mutations by the repair of strand breaks caused by altered bases. The generation of such strand breaks is, in turn, consistent with the type of DNA damage expected to be induced by reactive oxygen intermediates. A further indication that TPA-induced mutation in 6m2 cells may be mediated by reactive oxygen species is the fact that superoxide dismutase, catalase, vitamin E, and mannitol inhibit mutagenesis by TPA in 6m2 cells.

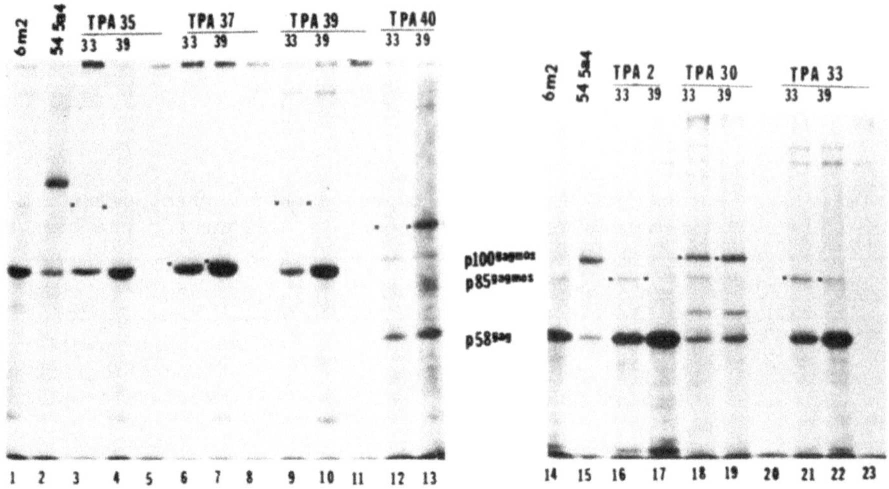

Figure 1. Viral proteins produced by TPA revertants. Viral proteins were immunoprecipitated with gag-specific antisera from cells grown at 33 or 39°C and metabolically radiolabelled with 3-H-leucine. The proteins were separated by SDS-polyacrylamide gel electrophoresis and visualized by fluorography.

Finally, benzoyl peroxide, a type II promoter, increases the yield of transformed colonies in 6m2 cells, and the inhibitors mentioned above all act to diminish this yield. We are examining further the role of reactive oxygen intermediates in the generation of these mutations as well as those induced by TPA and cadmium chloride. We are encouraged in our use of this assay because 6m2 cells appear to detect mutations resulting from strand breaks as well as point mutations, and we believe that the assay will provide another approach to studying the mechanisms by which reactive oxygen intermediates damage DNA.

REFERENCES

1. H.C. Birnboim, Importance of DNA strand break damage in tumor promotion, in: "Radioprotectors and Anticarcinogens", O.F. Nygaard and M.G. Simic, eds. Academic Press, New York (1983).
2. E.S. Copeland, ed., NIH Workshop Report, Free radicals in promotion--a chemical pathology study section workshop, Cancer Res. 43:5631 (1983).
3. T.J. Slaga, ed., "Mechanisms of Tumor Promotion", Vols I-IV, CRC Press, Boca Raton (1984).
4. P.A. Cerutti, Prooxidant states and tumor promotion, Science 227:375 (1985).
5. F.W. Sunderman Jr., Carcinogenic effects of metals, Fed. Proc. 37:40 (1978).
6. N.W. Biggart, G.E. Gallick, and E.C. Murphy, Jr., Nickel-induced heritable alterations in retroviral transforming gene expression, J. Virology 61:2378 (1987).
7. N.W. Biggart and E.C. Murphy, Jr., Analysis of metal-induced alterations in either structure or expression of a stably integrated provirus in a mammalian cell line, Mutation Res., in press (1988).

OXIDATIVE DAMAGE TO DNA IN ADRENOCORTICAL CELLS DURING SENESCENCE IN CULTURE

Kim H. Norris and Peter J. Hornsby

Department of Cell and Molecular Biology
Medical College of Georgia
Augusta, Georgia 30912

INTRODUCTION

Our experiments on the effects of oxidative damage on the functions of adrenocortical cells in culture were initiated by the finding that the loss of activity of cytochrome P-450 enzymes in cultured bovine adrenocortical cells incubated with steroid pseudosubstrates involved oxidative damage.[1] Oxidative damage potentially occurs at a higher rate in cell culture than in vivo. Cultured cells are exposed to several potentially damaging adventitious oxidants under standard culture conditions, as well as oxidants from intracellular metabolic processes.[2] Moreover, we showed that cultured bovine adrenocortical cells suffer from a combined deficiency of vitamin E and selenium. The deficiency of these biological antioxidants in culture was observed as the ability of low concentrations of either nutrient to prevent the toxicity of the aminotransferase inhibitor, aminooxyacetate, and of an organic peroxide, cumene hydroperoxide.[3,4] The two nutrients must be considered together since to a large extent the spectra of their biological activities overlap.[5] Thus, it appeared worthwhile to investigate the possible role of oxidative damage in senescence of bovine adrenocortical cells in culture.

CELLS DEPRIVED OF VITAMIN E AND SELENIUM ARE MORE SUSCEPTIBLE TO PEROXIDE DAMAGE

Cells incubated with cumene hydroperoxide showed a dose-dependent reduction of cloning efficiency (Figure 1).[2,4] The loss of clonogenicity resulting from exposure to 20 and 50 μM cumene hydroperoxide was greatly reduced in cells previously grown in the presence of selenite or α-tocopherol. Some increase in clonogenicity was also observed with selenite or α-tocopherol in control cells (Figure 1). Cumene hydroperoxide also produced extensive breakage of cellular DNA.[2] Strand breaks were assessed by alkaline sucrose gradient centrifugation in DNA from adrenocortical cells that had been incubated with cumene hydroperoxide (Figure 2). Strand breaks were greatly reduced in DNA from peroxide-treated cells that had previously been grown in 20 nM selenite or 1 μM α-tocopherol.

CUMENE HYDROPEROXIDE KILLS CELLS BY NAD^+ DEPLETION RESULTING FROM DNA STRAND BREAKAGE

Single-strand breaks are characteristic of damage to DNA by several oxidants, including superoxide and hydrogen peroxide,[6-10] and breakage of DNA appears to be

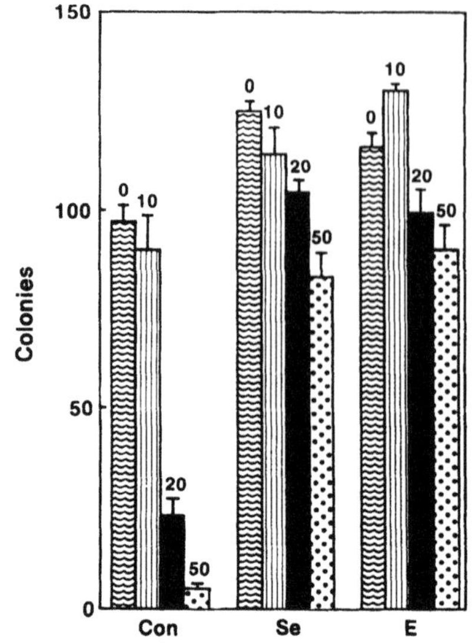

Figure 1 Clonogenicity of adrenocortical cells after exposure to cumene hydroperoxide. Subconfluent monolayers of control cells (Con), cells grown with 20 nM selenite (Se), and cells grown with 1 μM α-tocopherol (E) were exposed to the indicated concentrations of cumene hydroperoxide and were then subcultured for estimation of cloning efficiency. Reprinted, with permission, from reference 2.

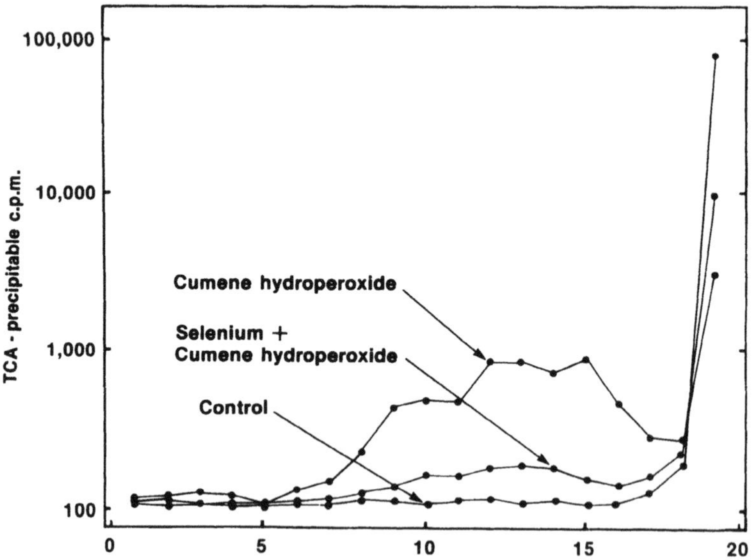

Figure 2 Alkaline sucrose gradient centrifugation of DNA from control cells, cells incubated for 4 hours with 10 μM cumene hydroperoxide, and cells incubated with peroxide that had been grown in the presence of 20 nM selenite. Fraction 1 is the top of the gradient. Reprinted, with permission, from reference 2.

an essential intermediate in peroxide toxicity. The extent of DNA breakage correlates with the incidence of cell death produced by peroxides.[9,11] Cell death appears to result from extensive stimulation of poly(ADP-ribose) synthesis with consequent lethal depletion of NAD^+ in cells with DNA strand breaks,[12] and, at least in some cases, inhibition of poly(ADP-ribose) synthesis with 3-aminobenzamide during peroxide treatment of cells prevents cell death, presumably by preventing the loss of NAD^+.[13-15]

IN BOVINE ADRENOCORTICAL CELLS, 3-AMINOBENZAMIDE PREVENTS CELL DEATH BUT DECREASES TRANSFECTION EFFICIENCY

We confirmed the protective effect of 3-aminobenzamide against oxygen radical-mediated cell death (Table 1). Six mM 3-aminobenzamide partly prevented loss of clonogenicity resulting from incubation with cumene hydroperoxide. In order to search for possible consequences of DNA breakage other than cell death, we also examined effects on transfection of plasmid DNA containing the _neo_ gene, stable integration of which in the genome of mammalian cells is selectable using the antibiotic G418. Cumene hydroperoxide, while decreasing clonogenicity, greatly enhanced the frequency of integration of the plasmid DNA. 3-Aminobenzamide, in contrast, increased cell survival in the presence of cumene hydroperoxide but decreased the transfection rate both in cumene hydroperoxide-treated and control cultures.

Table 1. Effects of Cumene Hydroperoxide and 3-Aminobenzamide on Clonogenicity and Transfection Efficiency in Bovine Adrenocortical Cells

	Surviving colonies/G418-resistant colonies/ Ratio of G418-resistant to surviving colonies	
	Control	With 10 μM cumene hydroperoxide
Control	286/376/1.3	2/139/69
2 mM 3-Aminobenzamide	353/346/1.0	4/71/18
6 mM 3-Aminobenzamide	295/143/0.5	64/0/0

Third-passage bovine adrenocortical cells were transfected by electroporation with 20 μg/ml linearized pSV3neo plasmid. Sixteen hours after transfection, cells were incubated with cumene hydroperoxide and/or 3-aminobenzamide. Ether-extracted serum was used throughout; thus, the medium was completely deficient in vitamin E and selenium.[4] Note that a longer incubation (24 h) with cumene hydroperoxide was used in this experiment than previously.[2] Cells were then subcultured to clonal density for estimation of clonogenicity or were incubated for 10 days in medium with 200 μg/ml G418, followed by subculture to clonal density in the same medium. About 1000 cells/dish were used for cloning efficiency and about 10^5 cells/dish for G418-resistance. Colonies of > 50 cells were counted after 7 days. (S.A. Naseeruddin, D.S.D. Lala, and P.J. Hornsby, unpublished data).

These data are consistent with the following model. In the sequence of events after peroxide exposure, endonuclease/ligase activity, as well as accomplishing the repair of genomic DNA damage, may also integrate exogenously supplied foreign DNA. Thus, as others have shown, DNA strand breakage is associated with increased frequency of integration of transfected DNA.[16,17] (ADP-ribose) synthetase appears to play a role in this phase of repair.[18,19] Inhibition of the enzyme prevents the lethal depletion of NAD^+ but also inhibits endonuclease/ligase activity, thus slowing repair of the peroxide-induced DNA breaks[15] and probably decreasing the accuracy of repair, but also decreasing foreign DNA integration.

VITAMIN E AND SELENIUM PRODUCE SMALL EXTENSIONS OF LIFE SPAN

The susceptibility of antioxidant-deficient cells to oxidative damage prompted us to re-examine the effects of biological antioxidants on replicative life span. Bovine adrenocortical cells were grown continuously, starting at the primary culture, in medium with 10% fetal bovine serum with fibroblast growth factor (FGF), and either 20 nM selenite or 1 μM α-tocopherol. The serum was of the lot used in the experiments on cumene hydroperoxide effects reported above. Selenium increased replicative life span from ~100 doublings to ~110 doublings and α-tocopherol increased the life span to ~130 doublings.[2]

Although there is a modest effect of the antioxidants on replicative life span, the more striking result is that cells are able to proliferate for extended periods in culture in the absence of vitamin E and selenium. Thus, neither vitamin E nor selenium is an essential nutrient for adrenocortical cell growth in culture, at least in the presence of other serum components and at relatively high cell densities. However, there is evidence for effects of vitamin E and selenium on cell growth (i) at clonal density, where, as shown above, vitamin E and selenium produce a small increase in clonogenicity; (ii) under serum-free conditions, where vitamin E and selenium greatly enhance cell survival[20]; (iii) in serum which has been extracted with ether and dialyzed, where vitamin E and selenium had small effects on the growth rate[4]; and (iv) when oxidation of glutamine is blocked with aminooxyacetate, as we have described previously.[3] Under these growth conditions, there may be chronic low-level oxidative damage that is not apparent under standard, higher-density culture conditions, even in late-passage cells.[2] Thus, vitamin E/selenium-preventable damage is not a major factor in adrenocortical cellular senescence. This could be because this kind of damage (i) is not at a level under standard culture conditions to produce large effects; (ii) is repaired efficiently and has no long-term consequences; or (iii) is not involved in senescence.

If the degree of peroxide-induced DNA strand breakage is not sufficient to kill the cell, the damage appears to be rapidly repaired (e.g., H_2O_2 [10]). The capacity for peroxides to kill cells appears much greater than their capacity for yielding permanent damage to DNA. Potentially, however, DNA strand breaks, if not repaired efficiently, could produce permanent DNA damage such as point mutations, chromosome aberrations, or integration of foreign DNA sequences.[21] Peroxides have been shown to be non-mutagenic in some mammalian systems (e.g., H_2O_2 [6,8,22]). In cultured bovine adrenocortical cells, cumene hydroperoxide was not detectably mutagenic as assessed by incidence of resistance to ouabain.[2] However, it should be noted that treatment of cells with ionizing radiation also does not yield ouabain-resistant mutants.[23] Oxidative damage to DNA may be detected only by mutation selection systems that depend on loss of the active gene product rather than those that depend on the production of a dominant product resulting from a specific base change.[24] Alternatively, if oxidative damage of a type that is preventable by biological antioxidants is a feature of normal cellular growth in culture, and also is not fully repaired, then it must be concluded that such damage does not play a role in cellular senescence. Thus, the most likely mechanism for the modest extension of life span seen with biological antioxidant supplementation is prevention of a low level of cell death resulting from minor peroxide damage under standard culture conditions.[2]

CHANGING THE LEVELS OF THE ANTIOXIDANT ENZYMES

The experiments reported above do not rule out oxidative damage as a factor in senescence because oxidative damage could be of a form not preventable by vitamin E or selenium. We are approaching this by assessing the feasibility of altering the level of antioxidant enzymes in the cultured bovine adrenocortical cells. In other work designed to alter the intracellular levels of antioxidant enzymes, liposome-entrapped CuZn-superoxide dismutase (SOD) delivered to cultured endothelial cells prevented oxygen cytotoxicity.[25] Subsequently, transfection was used to create cell

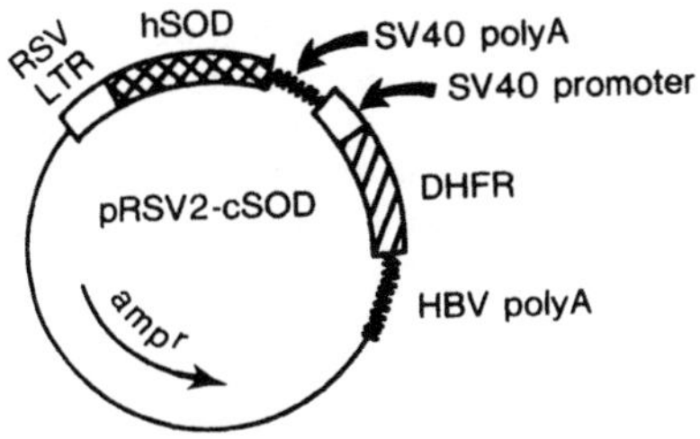

Figure 3 Plasmid pRSV2-cSOD. cDNA for human SOD (hSOD)[27] was generously supplied by Dr. Yoram Groner (Weizmann Institute, Israel) in the pUC18 cloning vector. A PvuII-BamHI fragment containing the hSOD cDNA was inserted in the polylinker of the expression vector pRSV2 (Dr. Cori Gorman, Genentech Inc.) downstream of the long terminal repeat (LTR) promoter from Rous sarcoma virus (RSV). pRSV2 contains the cDNA for DHFR, thus providing for co-amplification of DHFR and of the gene introduced into the vector under the control of the RSV-LTR.

lines overexpressing SOD.[26] These cells were more resistant to paraquat, a herbicide known to increase intracellular production of superoxide.

Overexpression can also be achieved by gene amplification, potentially to yield any desired level of intracellular SOD. Transfected dihydrofolate reductase (DHFR) genes amplify when cells are placed under methotrexate (MTX) selection pressure. Because the unit of amplification includes the selected gene together with flanking DNA, the co-amplification of linked DNA sequences following transfection into mammalian cells is possible through the use of plasmids in which the selected and marker genes have been linked. We have constructed a plasmid containing the cDNA sequence for human SOD and cDNA for DHFR (Figure 3). Bovine adrenocortical cells were co-transfected with pRSV2-cSOD and pSV3neo, thus allowing for initial selection of stable co-transfectants using G418. G418-resistant cells were then exposed to increasing concentrations of MTX in order to select only co-transfectants which have integrated pRSV2-cSOD, and to initiate co-amplification of the DHFR and hSOD sequences. We performed experiments to determine a concentration of MTX that prevents the growth of non-transfected cells. Based on this data, we exposed transfectants initially to 64 nM MTX. Currently, cells resistant to 250 nM MTX are being examined by Southern blotting to determine if transfected DHFR has amplified.

REFERENCES

1. P.J. Hornsby, and J.F. Crivello, The role of lipid peroxidation and biological antioxidants in the function of the adrenal cortex. Part 2, Molec. Cell. Endocrinol. 30:123 (1983).
2. P.J. Hornsby, and S.E. Harris, Oxidative damage to DNA and replicative life span in cultured adrenocortical cells, Exp. Cell Res. 168:203 (1987).
3. P.J. Hornsby, The role of vitamin E in cellular energy metabolism in cultured adrenocortical cells, J. Cell. Physiol. 112:207 (1982).
4. P.J. Hornsby, D.W. Pearson, A.P. Autor, K.A. Aldern, and S.E. Harris, Selenium deficiency in cultured adrenocortical cells: Restoration of glutathione peroxidase and resistance to hydroperoxides on addition of selenium, J. Cell. Physiol. 123:33 (1985).
5. A.L. Tappel, Measurement of and protection from in vivo lipid peroxidation, in: 'Free Radicals in Biology,' Vol. 4, W.A. Pryor, eds., Academic Press, New York, (1980).
6. M.O. Bradley, I.C. Hsu, and C.C. Harris, Relationships between sister chromatid exchange and mutagenicity, toxicity and DNA damage, Nature 282:318 (1979).
7. S.A. Lesko, R.J. Lorentzen, and P.O.P. Ts'o, Role of superoxide in deoxyribonucleic acid strand scission, Biochemistry 19:3023 (1980).

8. M.O. Bradley, and L.C. Erickson, Comparison of the effects of hydrogen peroxide and X-ray irradiation on toxicity, mutation, and DNA damage/repair in mammalian cells (V-79), Biochim. Biophys. Acta 654:135 (1981).
9. M.E. Hoffmann, A.C. Mello-Filho, and R. Meneghini, Correlation between cytotoxic effect of hydrogen peroxide and the yield of DNA strand breaks in cells of different species, Biochim. Biophys. Acta 781:234 (1984).
10. J.F. Ward, W.F. Blakely, and E.I. Joner, Mammalian cells are not killed by DNA single-strand breaks caused by hydroxyl radicals from hydrogen peroxide, Radiat. Res. 103:383 (1985).
11. A.C.M. Filho, M.E. Hoffmann, and R. Meneghini, Cell killing and DNA damage by hydrogen peroxide are mediated by intracellular iron, Biochem. J. 218:273 (1984).
12. D.A. Carson, S. Seto, D.B. Wasson, and C.J. Carrera, DNA strand breaks, NAD metabolism, and programmed cell death, Exp. Cell Res. 164:273 (1986).
13. J.L. Sims, S.J. Berger, and N.A. Berger, Poly(ADP-ribose) polymerase inhibitors preserve nicotinamide adenine dinucleotide and adenosine 5'-triphosphate pools in DNA-damaged cells: Mechanism of stimulation of unscheduled DNA synthesis, Biochemistry 22:5188 (1983).
14. I.U. Schraufstatter, P.A. Hyslop, D.B. Hinshaw, R.G. Spragg, L.A. Sklar, and C.G. Cochrane, Hydrogen peroxide-induced injury of cells and its prevention by inhibitors of poly(ADP-ribose) polymerase, Proc. Natl. Acad. Sci. USA 83:4908 (1986).
15. O. Cantoni, D. Murray, and R.E. Meyn, Effect of 3-aminobenzamide on DNA strand-break rejoining and cytotoxicity in CHO cells treated with hydrogen peroxide, Biochim. Biophys. Acta 867:135 (1986).
16. Y. Nagata, H. Takagi, T. Morita, and M. Oishi, Stimulation of DNA-mediated transformation by UV irradiation of recipient (mouse FM3A) cells, J. Cell Physiol. 121:453 (1984).
17. P.G. Debenham, and M.B. Webb, The effect of X-rays and ultraviolet light on DNA-mediated gene transfer in mammalian cells, Int. J. Radiat. Biol. 46:555 (1984).
18. D. Creissen, and S. Shall, Regulation of DNA ligase activity by poly(ADP-ribose), Nature 296:271 (1982).
19. J.E. Cleaver, W.J. Bodell, C. Borek, W.F. Morgan, and J.L. Schwartz, Poly(ADP-ribose): Spectator or participant in excision repair of DNA damage, Int. Symp. Princess Takamatsu Cancer Res. Fund 13:195 (1983).
20. M.H. Simonian, M.L. White, and G.N. Gill, Growth and function of cultured bovine adrenocortical cells in a serum-free defined medium, Endocrinology 111:919 (1982).
21. P. Cerutti, Prooxidant states and tumor promotion, Science 227:375 (1985).
22. Y. Nishi, M.M. Hasegawa, M. Taketomi, Y. Ohkawa, and N. Inui, Comparison of 6-thioguanine-resistant mutation and sister chromatid exchanges in Chinese hamster V79 cells with forty chemical and physical agents, Cancer Res. 44:3270 (1984).
23. J. Thacker, M.A. Stephens, and A. Stretch, Mutation to ouabain-resistance in Chinese hamster cells: Induction by ethyl methanesulphonate and lack of induction by ionising radiation, Mutat. Res. 51:255 (1978).
24. B. Singh, and R.S. Gupta, Mutagenic response to ultraviolet light and X-rays at five independent genetic loci in Chinese hamster ovary cells, Environ. Mutagen. 4:543 (1982).
25. B.A. Freeman, S.L. Young, and J.D. Crapo, Liposome-mediated augmentation of superoxide dismutase in endothelial cells prevents oxygen injury, J. Biol. Chem. 258:12534 (1983).
26. O. Elroy-Stein, Y. Bernstein, and Y. Groner, Overproduction of human Cu/Zn-superoxide dismutase in transfected cells: Extenuation of paraquat-mediated cytotoxicity and enhancement of lipid peroxidation, EMBO J. 5:615 (1986).
27. Y. Groner, J. Lieman-Hurwitz, N. Dafni, L. Sherman, D. Levanon, Y. Bernstein, E. Danciger, and O. Elroy-Stein, Molecular structure and expression of the gene locus on chromosome 21 encoding the Cu/Zn superoxide dismutase and its relevance to Down syndrome, Ann. N.Y. Acad. Sci. 450:133 (1985).

SUBSTITUTED PHTHALOCYANINES AS PHOTODYNAMIC SENSITIZERS

I. Rosenthal, E. Ben-Hur*, S. Greenberg**,
A. Conception-Lam**, D.M. Drew**, and C.C. Leznoff**

Department of Food Science
A.R.O., The Volcani Center Bet-Dagan, Israel
*Department of Radiobiology
Nuclear Research Center
Negev, Israel
**Department of Chemistry York University
Ontario, Canada

INTRODUCTION

The phthalocyanine dyes have recently been suggested for use in the photodynamic therapy of cancer.[1,2] Their superior attributes, such as the very intense light absorption in the wavelength range preferred for photodynamic therapy (600-700 nm), the lack of toxicity and selective localization of some dyes of this group in experimental tumors, the high efficiency to photosensitize killing of mammalian cells in culture, and finally the ability to induce necroses of transplanted tumors in animals[3,4] have made this suggestion very attractive.

Because of their very short history of use as photobiological sensitizers, it is still premature to define the structural requirements for the ideal photochemotherapeutical phthalocyanine. The only criterion that has emerged for selecting an efficient sensitizer is the requirement of a reasonably long lifetime excited triplet state so that its reactions are facilitated. Thus, phthalocyanines containing paramagnetic metals such as copper or cobalt are not suitable for this application.[5]

It seems that the prevalent view on the insolubility of these compounds has, at least in part, discouraged intensive investigations in the past. Although the unsubstituted phthalocyanines are indeed very insoluble, sulfonation drastically increases their solubility in water. Furthermore, peripheral benzene ring substitution by other chemical groups can extend the solubility range to practically any medium. These chemical modifications also dictate the hydrophilicity or lipophilicity of the sensitizer, which, in turn, might markedly affect the photobiological activity. The optimization of the porphyrin antitumoral activity has been suggested to require hydrophobic sensitizers characterized by high yields in singlet oxygen photoproduction.[6] Thus a final choice for the most useful drug demands phototoxicity screening of various structurally related phthalocyanine dyes.

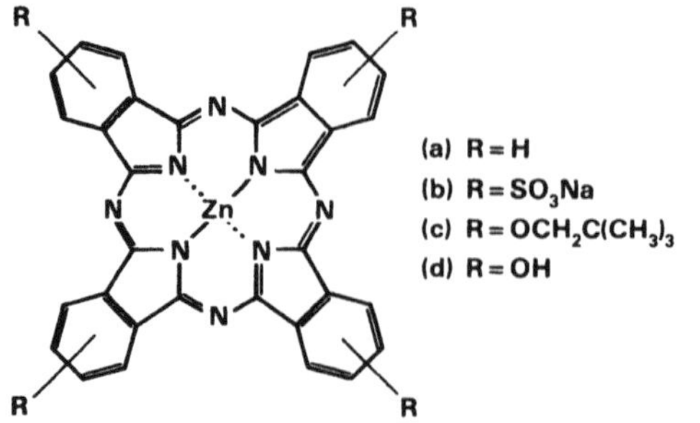

Figure 1. The molecular structure of metallophthalocyanine: (a) zinc phthalocyanine, (b) zinc tetrasulfophthalocyanine, (c) zinc tetraneopentoxyphthalocyanine, and (d) zinc tetrahydroxyphthalocyanine.

MATERIALS AND METHODS

The photoefficiency of zinc phthalocyanines substituted at the benzene rings (Figure 1) has been determined using as a bioassay the colony-forming ability of Chinese hamster cells line V79-B310H as an endpoint.[7]

Among the compounds tested, zinc tetraneopentoxyphthalocyanine[8] is the most lipophilic phthalocyanine compound known; it dissolves in acetone or hexane at concentrations higher than 0.01 M! It was also believed that a tetrahydroxyphthalocyanine would be particularly interesting to test because of its potential solubility and hydrogen-bonding characteristics. Its zinc derivative dissolves in ethanol and acetone (at least 0.01 M), but is hardly soluble in water (less than 10^{-6} M). This compound had not been reported prior to this study, possibly due to the fact that its normal precursor, 4-hydroxyphthalonitrile, does not undergo the expected condensation. The desired zinc tetrahydroxyphthalocyanine could be prepared by ether cleavage of tetraneopentoxyphthalocyanines with boron tribromide in refluxing benzene for 20 h. The tetrahydroxyphthalocyanine could be converted to the tetrahydroxy zinc derivative by treating tetrahydroxy-phthalocyanine with $Zn(OAc)_2$ in dimethylformamide at 150°C. Preferably, tetrahydroxy zinc derivative could be obtained by direct cleavage of the zinc tetraneopentoxyphthalocyanine in 90% yield. The metal is retained under the vigorous conditions of ether cleavage, and the zinc derivative precursor is more easily purified than the metal free by chromatographic methods.[9] Finally, tetrasulfonated and the unsubstituted zinc phthalocyanine were tested. The tetrasulfonated derivative is the most water-soluble derivative (more than 10^{-3} M), while the unsubstituted dye is insoluble in water, alcohol, acetone, and hexane, and only slightly soluble in highly polar solvents such as dimethylformamide.

ROLE OF OXYGEN

A relevant mechanistic question refers to the role of oxygen in the photosensitization process. Because some tumors contain hypoxic regions, complete cell killing is difficult in the partial or total absence of oxygen. The efficiency of photosensitization of cultured Chinese hamster cells by aluminum phthalocyanine is diminished at oxygen concentrations below a threshold of ca. 3.4% oxygen. Very little photoinactivation was observed when the atmosphere equilibrated with cells was oxygen-free nitrogen.[5] While phototoxicity of the traditional photosensitizer for photodynamic therapy, that is, hematoporphyrin derivative, has been mechanistically explained by a singlet oxygen reaction, we could not find yet any confirming evidence that this is true for phthalocyanine dyes as well. However, it seems that only a phthalocyanine dye that can populate the triplet excited state for a sufficiently long time, which is by itself a preliminary requirement for generation of singlet oxygen, can act as a photodynamic sensitizer.[5] The quantum yield for singlet oxygen generation was, therefore, estimated for three out of four phthalocyanine derivatives tested in this study: zinc tetrasulfophthalocyanine, zinc tetrahydroxyphthalocyanine, and zinc tetraneopentoxyphthalocyanine. The low solubility of the unsubstituted zinc phthalocyanine prevented a quantitatively meaningful determination for this compound. The singlet oxygen production was estimated by EPR monitoring of the phthalocyanine-photosensitized conversion of a sterically hindered heterocyclic amine, 2,2,6,6-tetramethyl-4-piperidone, to the corresponding nitroxide radical. This stable radical can be easily detected at ambient temperature as a 1:1:1 triplet of ca. a_N=16.5 G.[10] All phthalocyanines tested were irradiated in the EPR cavity with monochromatic visible light at the lowest energy absorption band in an oxygen-saturated solution of 2,2,6,6-tetramethyl-4-piperidone (0.3 M), and the paramagnetic signal was recorded vs. irradiation time. The dye concentration was high enough to ensure total absorption of the light. The oxidation of the amine was linear in time for the duration of the experiment (10 min) and the relative reaction rates could be estimated from the slopes of the plots and reflect the relative efficiencies for singlet oxygen generation. Because of the different solubilities, the measurements had to be performed in different solvents: water for zinc tetrasulfophthalocyanine, ethanol and acetone for zinc tetrahydroxyphthalocyanine, and acetone for zinc tetraneopentoxyphthalocyanine. All three dyes photosensitized the oxidation of 2,2,6,6-tetramethyl-4-piperidone to the nitroxide radical with the same reaction rate. Since the solvent-induced, radiationless decay constants of singlet oxygen are different for the solvents employed [k_d (water) $=5x10^5$ s^{-1} k_d (ethanol)$=1.4x10^5$, and k_d (acetone)$=3.8x10^4$ s^{-1}], the oxidation rate would be affected if the solvent-induced decay limits the singlet oxygen lifetime in the system. Such an effect could not be observed in the comparative tests with tetrahydroxyphthalocyanine in ethanol and acetone under our experimental conditions. Since the oxidation rates were the same, it was suggested that all three phthalocyanines tested generate singlet oxygen with the same quantum yield, which was estimated as 0.45 ± 0.05. This absolute quantum yield for singlet oxygen production was calculated from the oxidation rates relative to that for a sensitizer of known quantum yield, Rose Bengal, which was subjected to the same technique for detection of singlet oxygen. These results suggests that all these compounds could be efficient photosensitizers.

PHOTOBIOLOGICAL ACTIVITY

In the bioassay, however, the photobiological activity was found to be the highest for the unsubstituted compound, followed in decreasing order by the tetrahydroxy-, tetrasulfonated-, and tetraneopentoxyphthalocyanine derivatives (Figure 2). It was also found that the phototoxicity of tetrahydroxyphthalocyanine was unchanged in the pH range 6 to 8.

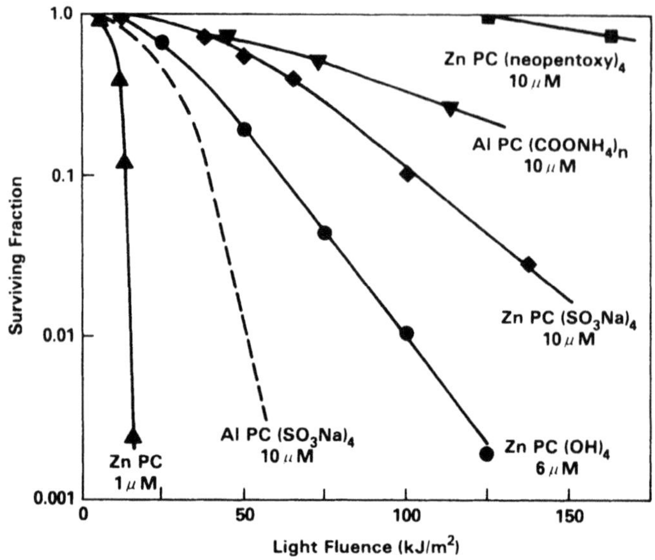

Figure 2. Survival of Chinese hamster cells photosensitized by phthalocyanine derivatives. Cells were incubated in growth medium with the indicated dyes for 16 h and then exposed to light.

The appearance of photosensitization was relatively rapid for zinc phthalocyanine and for the tetrahydroxy derivative during the first 2 h of incubation with the dye. At longer times, further sensitivity developed very slowly. The photosensitivity for zinc tetrasulfophthalo-phthalocyanine develops slowly and continuously over many hours and begins to plateau only after 24 h. These kinetics correlate with the rate of uptake of the various dyes by the cells (Figure 3). Thus, zinc phthalocyanine is taken up at the fastest rate, followed by tetrahydroxy, the tetrasulfonated and tetraneopentoxy derivatives, respectively. Thus, while the cellular uptake, as expected, is a determining factor in the subsequent phototoxicity, it cannot be predicted from the solubility features of the dyes.

Aluminum phthalocyanine and its sulfonated derivative were the only dyes containing aluminum tested so far, and they were found to be among the most efficient sensitizers in this class of compounds.[7]

Another water-soluble aluminum derivative can be prepared by the action of trichloroacetic acid upon unsubstituted phthalocyanine[11] to yield a phthalocyanine carboxylic acid with undefined substitutions. Preliminary tests for photocytotoxic efficiency, using the same bioassay, were run with this compound despite its lack of homogeneity, and it was found to be ca. 3 times less photoactive than the sulfonated counterpart (Figure 2). The cellular uptake of this sensitizer was also lower (Figure 3).

COMPLEXATION WITH BSA

Aluminum phthalocyanines have been shown to complex with BSA.[7] The complexation is visualized by comigration of the dye with BSA when chromatographed on a gel filtration column. Subsequently, all other dyes were tested for binding to serum albumin. Unlike the other three zinc

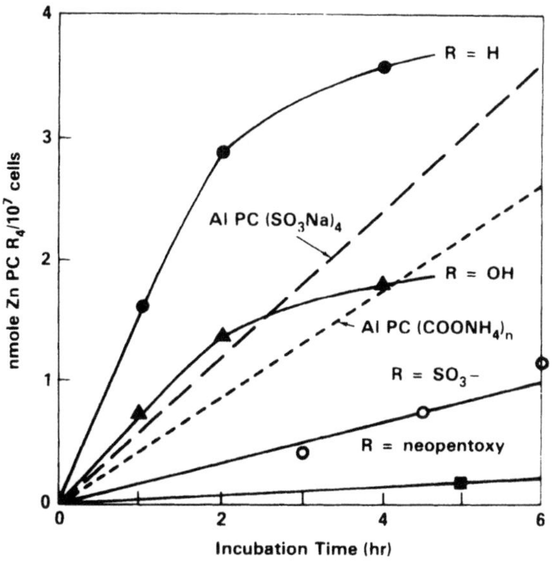

Figure 3. The uptake of phthalocyanine (PC) derivatives into Chinese hamster cells. The dyes were added to log-phase cells at 10 μM ZnPC, 25 μM ZnPC$(OH)_4$, 25 μM ZnPC$(SO_3Na)_4$, 25 μM ZnPC$(neopentoxy)_4$, and 25 μM AlPCs. After various incubation times at 37°C, the cells treated with dyes were rinsed with cold PBS, air dried, and then extracted for 1 h at room temperature with 4 ml extracting solution. The extracting solvents were dimethyl formamide for zinc phthalocyanine, ethanol for zinc tetrahydroxyphthalocyanine, acetone for zinc neopentoxyphthalocyanine, and 0.1 N NaOH containing 1% cetrimide for zinc tetrasulfophthalocyanine and aluminum phthalocyanines. The amount of phthalocyanine in the cell extract was calculated from the absorption at $_{max}$ using standard curves.

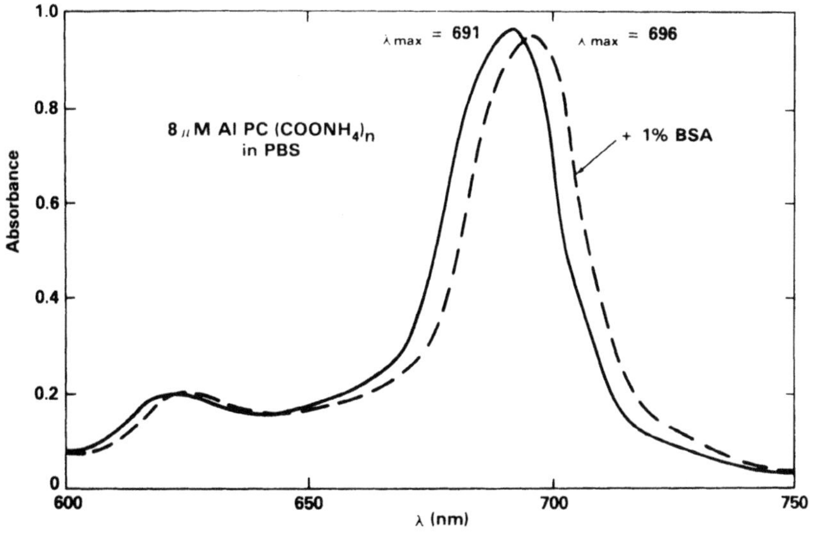

Figure 4. Absorption spectra of free and BSA-complexed aluminum phthalocyanine carboxylic acid.

derivatives, zinc tetraneopentoxyphthalocyanine does not bind to protein at all, and this observation might be related to the absence of a cellular uptake. The lack of binding might be due to the steric hindrance due to the bulky substituents on the phthalocyanine plane. Conversely, aluminum phthalocyanine carboxylic acid binds to BSA (over 90% under experimental conditions employed). The absorption spectrum of the complex was red shifted, as compared to that of the free dye (Figure 4).

CONCLUSION

Peripheral ring substitution can critically affect the photocytotoxicity of phthalocyanine dyes and, subsequently, their ability to act as photodynamic sensitizers. In this context, the phthalocyanines, because of their diversified chemistry, show promise for providing probes for understanding the mechanism of selective retention of the dye in malignant tissue more than any other class of photochemotherapeutical drugs.

REFERENCES

1. E. Ben Hur and I. Rosenthal, The phthalocyanines: Synthetic porphyrin-like dyes with improved potential for cancer phototherapy, abstracts, 3rd Annual Meeting of European Society for Therapeutic Radiology and Oncology, Jerusalem, Israel, 1984.
2. S.G. Bown, J. Wieman, C.J. Tralau, C. Collins, P.R. Salmon, and C.G. Clark, Focal necrosis in liver and a fibrosarcoma in rats produced by photodynamic therapy, Gut 26:563 (1985).
3. C.J. Tralau, A.J. Mac Robert, P.D. Coleridge-Smith, H. Barr, and S.G. Bown, Photodynamic therapy with phthalocyanine sensitization: Quantitative studies in a transplantable rat fibrosarcoma, Br. J. Cancer 55:389 (1987).
4. N. Brasseur, H. Ali, R. Langlois, J.R. Wagner, J. Rousseau, and J.E.van Lier, Biological activities of phthalocyanines. V. Photodynamic therapy of EMT-6 mammary tumors in mice with sulfonated phthalocyanines, Photochem. Photobiol. 45:581 (1987).
5. I. Rosenthal, C. Murali Krishna, P. Riesz, and E. Ben-Hur, The role of molecular oxygen in the photodynamic effect of phthalocyanines, Radiat. Res. 107:136 (1986).
6. C. Emiliani and M. Delmelle, The lipid solubility of porphyrins modulates their phototoxicity in membrane models, Photochem. Photobiol. 37:487 (1983).
7. E. Ben-Hur and I. Rosenthal, Photosensitization of Chinese hamster cells by water soluble phthalocyanine, Photochem. Photobiol. 43:615 (1986).
8. C.C. Leznoff, S.M. Marcuccio, S. Greenberg, A.B.P. Lever, and K.B. Tomer, Metallophthalocyanine dimers incorporating five-atom covalent bridges, Can. J. Chem. 63:623 (1985).
9. I. Rosenthal, E. Ben Hur, S. Greenberg, A. Conception-Lam, D.M. Drew, and C.C. Leznoff, The effect of substituents on phthalocyanine photocytotoxicity, Photochem. Photobiol., in press.
10. S. Cannistraro, A. Van de Vorst and G. Jori, EPR studies on singlet oxygen production by porphyrins, Photochem. Photobiol. 28:257 (1978).
11. I.G. Farbenindustrie, British Patent 487,261.

EFFECTS OF SINGLET OXYGEN ON THE BIOLOGICAL ACTIVITY OF DNA AND ITS INVOLVEMENT IN SINGLE STRAND-BREAK FORMATION

Heribert Wefers[a], Paolo Di Mascio, Hong-Phuc Do-Thi*, Dietrich Schulte-Frohlinde* and Helmut Sies

Institut für Physiologische Chemie I, Universität Düsseldorf
*Max-Planck-Institut für Strahlenchemie, Mülheim a. d. Ruhr
[a]Present address: Hautklinik der Universität Düsseldorf

INTRODUCTION

The formation of singlet molecular oxygen ($^1O_2$) has been shown to occur during photoexcitation of a variety of biological and also clinically applied compounds, such as porphyrins, psoralens and tetracyclines (see, for example references 1-3), but also occurs during enzymatic reactions including cyclooxygenase[4] and chloroperoxidase.[5] Regarding the biological consequences of $^1O_2$ formation, DNA is of particular interest. $^1O_2$ is known to react with guanine [6,7]; however, results on biologically active DNA were conflicting, in particular regarding $^1O_2$-dependent strand-break formation under neutral conditions, i.e. without alkaline treatment.[8-11] Here we emphasize two aspects of $^1O_2$-mediated damage. First, the effect of $^1O_2$ on the transforming activity of pBR322 and M13 DNA in *E. coli* is presented. Second, two different sources of $^1O_2$ are employed for investigation of strand-break formation; these are microwave discharge and the thermodissociable endoperoxide of disodium 3,3'-(1,4-naphthylidene)dipropionate ($NDPO_2$).

MATERIALS AND METHODS

The generation of singlet molecular oxygen ($^1\Delta_g$) by microwave discharge was performed according to Ogryzlo,[12] with slight modifications.[11] The possible reactive contaminants, which might be ozone and O-atoms, were shown to be present only below the detection limit (1.5 nmol per minute).[11] Singlet oxygen generation by photosensitization with Rose Bengal immobilized on a glass plate distant to the DNA solution was performed as described by Midden and Wang,[13] with modifications. The chemical generation of singlet oxygen was performed using the thermal decomposition of the endoperoxide $NDPO_2$.[14]

RESULTS

Singlet Oxygen-Dependent Loss of Transforming Activity

Exposure of plasmid DNA pBR 322 to a gas stream containing singlet

Table 1. Effect of D_2O, NaN_3, Methionine and Methionine Sulfone on DNA Damage by Singlet Oxygen*

	Transforming activity[a]	Supercoiled DNA[b]
	(%)	(%)
Control	= 100	91.9 ± 0.7
$^1O_2$	32.1 ± 4.6	49.1 ± 1.8
$^1O_2$, D_2O	14.1 ± 1.5	30.2 ± 2.2
$^1O_2$, plus NaN_3 (20 mM)	84.3 ± 3.8	86.2 ± 2.6
$^1O_2$, plus methionine (10 mM)	77.5 ± 1.6	69.1 ± 1.7
$^1O_2$, plus methionine sulfone (10 mM)	58.3 ± 6.1	45.8 ± 3.1

*Conditions: DNA solutions were exposed to a gas stream containing $^1O_2$ generated by microwave discharge as described.[11,12]
[a]Data taken from reference 11.
[b]Data from experiments such as in Figure 2.

molecular oxygen generated by microwave discharge led to a pronounced decrease of transforming activity as a function of the exposure time, leading to a transforming activity of 32% after 20 min and less than 10% after 50 min, compared to control samples. Replacing H_2O by D_2O, thereby increasing the lifetime of $^1O_2$ in the buffer, led to a more than two-fold increase of DNA damage regarding transforming activity (Table 1) and to a further loss of supercoiled DNA. Sodium azide (20 mM) as a quencher of $^1O_2$ and methionine (10 mM), known to react with singlet oxygen, lowered the $^1O_2$-dependent loss of transforming activity (Table 1). When methionine sulfone was used as a control, the protective effect on DNA damage assayed by transforming activity was smaller than that observed with methionine (Table 1).

Similar results were found in another series of experiments, using single-stranded and double-stranded bacteriophage M13 DNA. $^1O_2$ was generated by photoexcitation of Rose Bengal, which was immobilized on a glass plate. This was positioned above the DNA-containing solution, and a small gas space (<1mm) was between sensitizer and solution.[13, 14] As shown in Figure 1, the transforming activity decreased depending on the incubation time; this effect was about doubled when H_2O was replaced by D_2O, whereas only minute effects were found in the absence of Rose Bengal or in the presence of methionine as $^1O_2$ scavenger.

Strand-Break Formation upon $^1O_2$ Exposure

Single strand-break formation of pBR322 after $^1O_2$ exposure was detected, as shown in Figure 2, as an increase in the open circular form at the expense of the closed circular (supercoiled) form, measured after electrophoresis with subsequent ethidium-bromide staining. Parallel to the

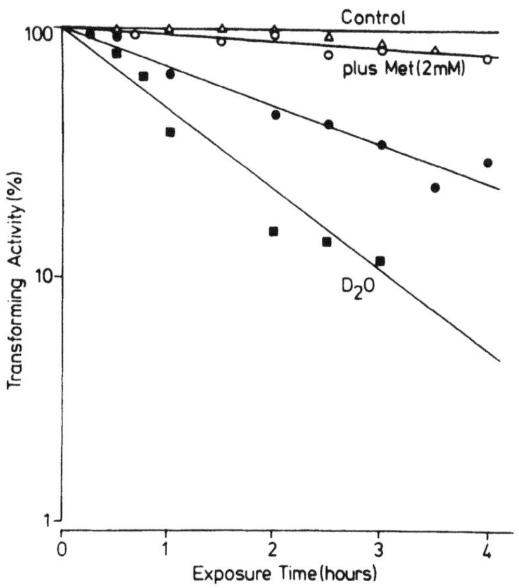

Figure 1. Loss of transforming activity of M13 ss DNA upon $^1O_2$ exposure. $^1O_2$ was generated by photoexcitation of immobilized Rose Bengal. DNA was in 0.1 M $NaClO_4$, pH 7.4 at 16°C. ▲: without Rose Bengal; O: with Rose Bengal, H_2O, methionine 2mM; ●: with Rose Bengal, H_2O; ■: with Rose Bengal, D_2O instead of H_2O.

effects on the transforming activity, D_2O increased strand break-formation, whereas NaN_3 and methionine were protective and methionine sulfone did not show an effect.

Using the thermodissociable endoperoxide $NDPO_2$ in D_2O at concentrations which yielded a total amount of $^1O_2$ of 5 mM, strand-break formation was detected upon exposure of the plasmid, responding to D_2O and the $^1O_2$ quencher NaN_3. A water-soluble anthracene derivative (EAS, see reference 11), which was used as $^1O_2$ scavenger, also decreased strand break formation.

DISCUSSION

Biological effects of $^1O_2$ include DNA damage as evidenced by the loss of transforming activity of pBR322 and ss and ds M13 DNA. Single strand-break formation in pBR322 was detected without alkaline treatment, in agreement with the work of Fiel et al.[9] Strand-break formation seems to be in contrast to work from Nieuwint et al.,[10] who could not detect strand-breaks with the thermodissociable endoperoxide $NDPO_2$ in a bacteriophage and concluded the inability of $^1O_2$ to break the DNA backbone.

The effects of $^1O_2$ quenchers and D_2O and the use of two different sources of $^1O_2$ provides evidence for strand-break formation by $^1O_2$. However, it should be taken into consideration that, according to Cadet et al.[7] attack of dGuo by $^1O_2$ as a primary event of DNA damage is followed by complex mechanisms via unstable intermediates. In one mechanism, the purine ring is left intact, whereas the second pathway leads to a cyanuric acid derivative involving the consecutive addition of two $^1O_2$ molecules.

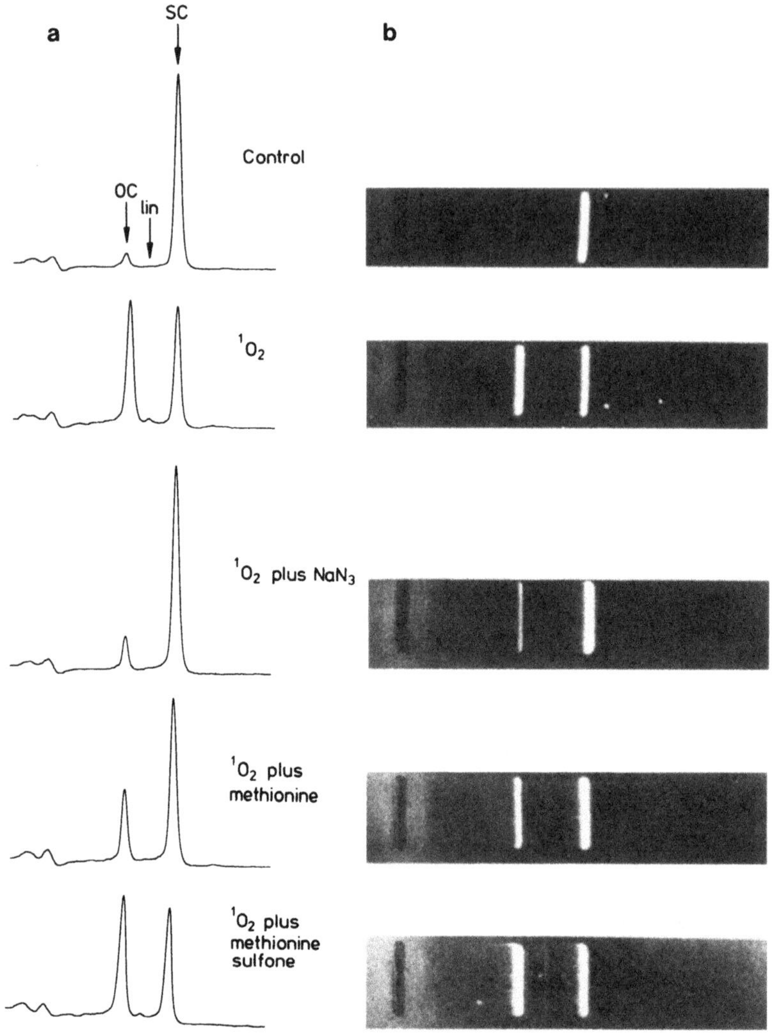

Figure 2. Strand-break formation of pBR322 after exposure to $^1O_2$, generated by microwave discharge (as described in reference 11). Traces are fluorescence scans of agarose gels after electrophoresis and staining with ethidium bromide. SC: supercoiled form (closed circular form); OC: open circular form; lin: linear DNA. Data are given in Table 1.

Reactions leading to strand break formation may be a late event in one of these mechanisms, and thus are not necessarily proportionally related to base damage or loss of transforming activity.

ACKNOWLEDGEMENTS

This work was supported by the Deutsche Forschungsgemeinschaft, the National Foundation for Cancer Research and by Krebsforschung International e.V. (H.-P. Do-Thi).

REFERENCES

1. T.J. Dougherty, Photosensitization of malignant tumors, Seminars in Surgical Oncology 2:24 (1986).
2. W. Poppe, and L.I. Grossweiner, Photodynamic sensitization by 8-methoxypsoralen via the singlet oxygen mechanism, Photochem. and Photobiol. 22:217 (1975).
3. T. Hasan, and A.U. Khan, Phototoxicity of the tetracyclines: Photosensitized emission of singlet delta dioxygen, Proc. Natl. Acad. Sci. USA 83:4604 (1986).
4. E. Cadenas, H. Sies, W. Nastainczyk, and V. Ullrich, Singlet oxygen formation detected by low-level chemiluminescence during enzymatic reduction of prostaglandin G_2 to H_2, Hoppe-Seyler's Z. Physiol. Chemie 364:519 (1983).
5. J.R. Kanofsky, Singlet oxygen production by chloroperoxidase-hydrogen peroxide-halide systems, J.Biol.Chem. 259:5596 (1984).
6. F.R. Hallett, B.P. Hallett, and W. Snipes, Reactions between singlet oxygen and the constituents of nucleic acids, Biophys. J. 10:305 (1970).
7. J. Cadet, C. Decarroz, S.Y. Wang, and W.R. Midden, Mechanisms and products of photosensitized degradation of nucleic acids and related model compounds, Israel Journal of Chemistry 23:420 (1983).
8. E. Boye, and J. Moan, The photodynamic effect of hematoporphyrin on DNA Photochem.Photobiol. 31:223 (1980).
9. R.J. Fiel, N. Datta-Gupta, E.H. Mark, and J.C. Howard, Induction of DNA damage by porphyrin photosensitizers, Cancer Research 41:3543 (1981).
10. A.W.M. Nieuwint, J.M. Aubry, F. Arwert, H. Kortbeek, S. Herzberg, and H. Joenje, Inability of chemically generated singlet oxygen to break the DNA backbone, Free Rad. Res. Comms. 1:1 (1985).
11. H. Wefers, D. Schulte-Frohlinde, and H. Sies, Loss of transforming activity of plasmid DNA (pBR322) in E.coli caused by singlet molecular oxygen, FEBS Lett. 211:49 (1987).
12. E.A. Ogryzlo, in: Singlet Oxygen (H.H. Wassermann and R.W. Murray, eds), pp. 35-56, Academic Press, New York (1979).
13. W.R. Midden, and S. Yi Wang, Singlet oxygen generation for solution kinetics: clean and simple, J. Am. Chem. Soc. 105:4129 (1983).
14. P. Di Mascio and H. Sies, Detection and quantification of singlet oxygen generated by thermolysis of a water-soluble endoperoxide, J. Am. Chem. Soc., submitted (1988).

DETERMINATION OF 8-HYDROXYDEOXYGUANOSINE IN HUMAN URINE:

A POSSIBLE ASSAY FOR *IN VIVO* OXIDATIVE DNA DAMAGE

Kenneth C. Cundy, Ron Kohen, and Bruce N. Ames

Department of Biochemistry
University of California
Berkeley, California 94720, U.S.A.

INTRODUCTION

A sensitive analytical method has been developed for the determination of 8-hydroxydeoxyguanosine (h^8dG) in human urine, based on reversed-phase HPLC combined with electrochemical detection. Previous attempts to use urine as a means to monitor DNA oxidative damage have centered around the detection of thymine glycol, thymidine glycol, and 5-hydroxymethyluracil.[1-3] The present assay offers improvements in detection limits, sample volumes and analysis time.

h^8dG (Figure 1) has been shown to occur in DNA *in vivo* following exposure to oxygen radical-producing agents.[4] The adduct can cause extensive mispairing in *in vitro* DNA replication studies[5] and is removed from DNA *in vivo*.[4] If an excision mechanism exists for h^8dG, then repair of this lesion would lead to excretion of free h^8dG in urine.

The electrochemical detection of h^8dG has previously been demonstrated to be 1000 times more sensitive than detection by UV absorbance,[6] and the present study has utilized this property. In order to reduce the complexity of the EC chromatogram, urine samples are initially purified by a solid-phase extraction (SPE) step, using an aromatic sulfonic acid strong cation-exchange resin (J. T. Baker).

Urine (1 ml, first morning void) was acidified and h^8dG standard (or water) added. The sample was applied to a prewashed 3-ml SPE column and fraction R collected. The column was washed with 1 ml 5% acetic acid (collecting fraction W) and eluted with 1 ml 50 mM KH_2PO_4, pH 7.5

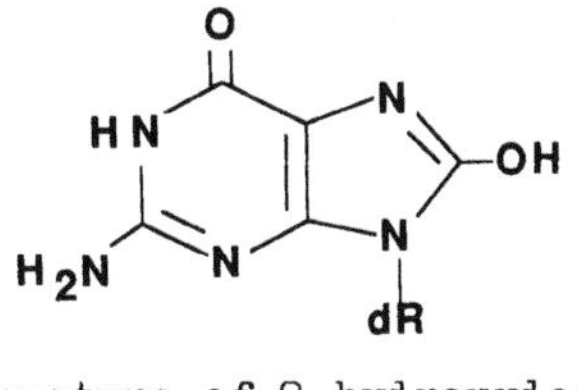

Figure 1. Structure of 8-hydroxydeoxyguanosine.

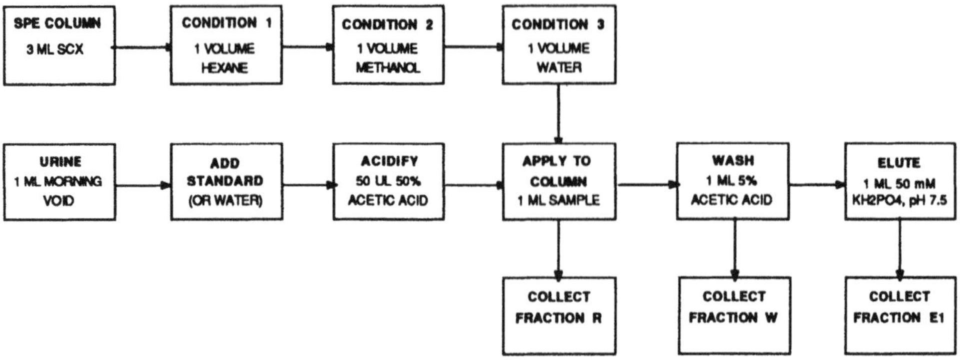

Figure 2. Schematic of the solid phase extraction step.

(collecting fraction E1). The SPE step is summarized in Figure 2. Recovery of h^8dG in this fraction was reproducible based on analysis of urine samples spiked with $[1',2'-^3H]$-h^8dG. Fraction E1 (25-100 µl) was analyzed on the following HPLC system: two Waters Model 6000A pumps; a Waters Model 720 controller; a Waters WISP automatic injector; and two 25-cm Supelco LC-18 analytical columns in series, with a guard column of pellicular ODS material. The mobile phase was 50 mM KH_2PO_4, pH 5.5, with a gradient of methanol (0 to 10% over 80 min). h^8dG was detected by a Model LC-4B electrochemical detector (BAS), using a glassy carbon working electrode at an applied potential of +0.7 volts vs. Ag/AgCl. Data were recorded by a Model 716 Interface (Nelson Analytical) and analyzed by a Model 9816 computer (Hewlett Packard).

Figure 3 shows a standard curve for the determination of h^8dG concentrations between 0 and 250 pmol/ml, based on addition of known amounts of standard to a representative normal urine sample prior to the SPE step. The h^8dG peak was identified by coelution with authentic standard at 68 min (see Figure 4). The curve is linear (R = 0.993) over the range of the experiment.

Table 1 summarizes the results of analysis of four normal human urine samples. The values are normalized to creatinine excretion

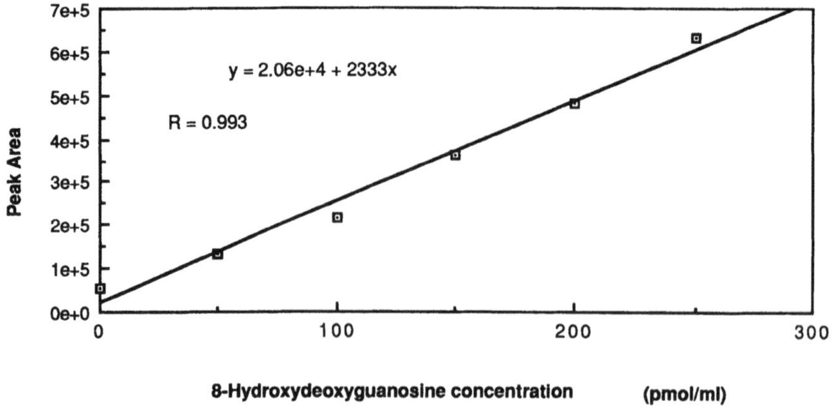

Figure 3. Standard curve for 8-hydroxydeoxyguanosine in human urine.

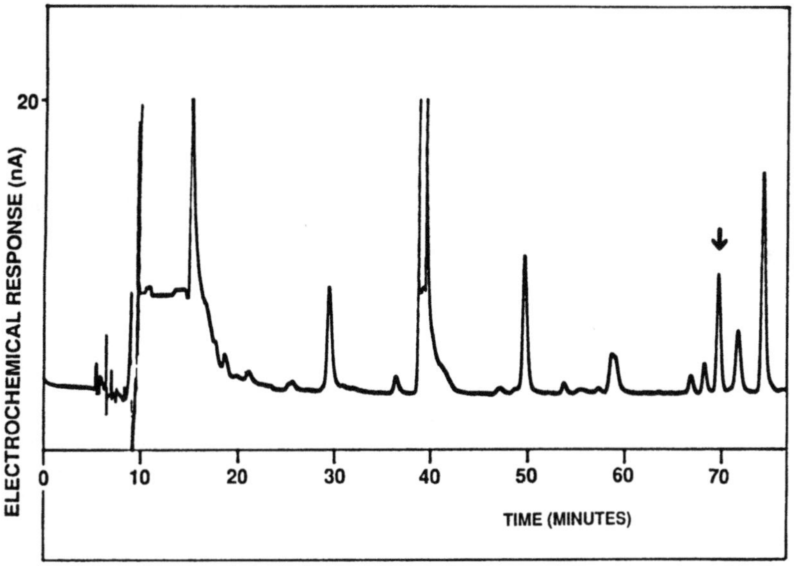

Figure 4. EC chromatogram of 100 l Fraction E1 coinjected with 50 pmol of authentic 8-hydroxydeoxyguanosine standard (indicated by arrow). Chromatographic details are discussed in the text.

Table 1. Urinary excretion of 8-hydroxydeoxyguanosine in normal and CGD subjects

Sample #	Class	h^{8}dG Excretion (pmol/µmol creatinine)	Mean ± SD (n = 2-3)
1	normal	2.31 ± 0.09	
2	normal	1.19 ± 0.05	3.11 ± 1.82
3	normal	3.47 ± 0.35	
4	normal	5.45 ± 0.64	
5	CGD	1.89 ± 0.39	
6	CGD	1.04 ± 0.25	1.34 ± 0.63
7	CGD	1.84 ± 0.19	
8	CGD	0.60 ± 0.04	

(determined by Sigma diagnostic kit). The mean excretion value is 3.11 ± 1.82 pmol/μmol creatinine. Table 1 also shows the results from analysis of four urine samples from patients with chronic granulomatous disease (CGD). This disease is characterized by a complete lack of phagocyte-derived oxidation.[7] There appears to be no significant difference between the normal and CGD values for h^8dG excretion (Student's t test, P = 0.05). This may be interpreted as evidence that the observed levels of h^8dG are the result of oxidation of DNA by species other than phagocyte-derived oxygen radicals. However, these preliminary results are based on a small sample of normal and CGD patients, and no firm conclusions can be drawn until a larger population is examined.

The detection limit for h^8dG by this method is 2-5 pmol/ml (based on injection of 100 μl fraction E1), and the total analysis time for 12 samples is approximately 10 h. We are currently engaged in the isolation of the observed peak for characterization by UV absorption and mass spectrometry. It remains to be demonstrated that the observed levels of h^8dG are not derived from dietary intake of the deoxynucleoside. Additional experiments are now in progress to investigate the urinary excretion of 8-hydroxyguanosine (presumably a product of RNA oxidative damage) and the free base 8-hydroxyguanine.

ACKNOWLEDGMENTS

We are grateful to Drs. John T. Curnette and Bernard M. Babior of the Department of Basic and Clinical Research, Scripps Clinic, and to Dr. Seymour J. Klebanoff of the Department of Medicine, University of Washington, for their gift of CGD urine samples. This work was supported by National Cancer Institute Outstanding Investigator Grant CA39910 to B.N.A. and by NIEHS Center Grant ES01896.

REFERENCES

1. R. Cathcart, E. Schwiers, R. L. Saul, and B. N. Ames, Thymine glycol and thymidine glycol in human and rat urine: A possible assay for oxidative DNA damage, Proc. Natl. Acad. Sci. U.S.A. 81:5633 (1984).
2. R. L. Saul, P. Gee, and B. N. Ames, Free radicals, DNA damage, and aging, in: "Modern Biological Theories of Aging," H. R. Warner et al., eds., Raven Press, New York (1987).
3. B. N. Ames and R. L. Saul, Cancer, aging, and oxidative DNA damage, in: "Theories of Carcinogenesis," O. H. Iversen, ed., Hemisphere Publishing Corporation, Washington, D.C. (1988).
4. H. Kasai, P. F. Crain, Y. Kuchino, S. Nishimura, A. Ootsuyama, and H. Tanooka, Formation of 8-hydroxyguanine moiety in cellular DNA by agents producing oxygen radicals and evidence for its repair, Carcinogenesis 7:1849 (1986).
5. Y. Kuchino, F. Mori, H. Kasai, H. Inoue, S. Iwai, K. Miura, E. Ohtsuka, and S. Nishimura, Misreading of DNA templates containing 8-hydroxydeoxyguanosine at the modified base and at adjacent residues, Nature 327:77 (1987).
6. R. A. Floyd, J. J. Watson, and P. K. Wong, Hydroxyl free radical adduct of deoxyguanosine: Sensitive detection and mechanisms of formation, Free Rad. Res. Commun. 1:163 (1986).
7. J. T. Curnette and B. M. Babior, in: "Advances in Human Genetics, Vol. 16," H. Harris and K. Hirshhorn, eds., Plenum Publishing Corporation, New York (1987).

URINE BIOMARKERS FOR OXIDATIVE DNA DAMAGE

D.S. Bergtold and M.G. Simic

Center for Radiation Research, National Bureau of Standards
Gaithersburg, MD 20899

H. Alessio and R.G. Cutler

Gerontology Research Center, National Institute on Aging
Key Medical Center, Baltimore, MD 21224

INTRODUCTION

The maintenance of macromolecular integrity and function, particularly in DNA, appears to be a major determinant of the longevity and functional capacity of biological systems.[1] DNA damage induced by active oxygen species therefore may be of primary importance in cancer and aging[2-4]. This suggests the need for specific biomarkers (a) to elucidate the biochemical nature of DNA lesions and their repair and (b) to monitor noninvasively the generation and removal of DNA damage in biological systems. Several types of damage are known to result from the interactions of free radicals with chromatin, including single- and double-strand breaks, base and sugar alterations, and DNA-protein crosslinks.[5] Each of these types of damage seems to be amenable to chemical analysis by the techniques of high performance liquid chromatography (HPLC), gas chromatography/mass spectrometry (GC/MS), alkaline and neutral elution, and other chromatographic and filter techniques. Of particular interest have been the measurements of specific products of oxidative damage to DNA and proteins, such as thymine glycol, thymidine glycol, base-amino acid crosslinks, and altered amino acids. Recently, the measurement of thymine glycol (TG), thymidine glycol (dR-TG), and 5-hydroxymethyluracil (HMU) in urine has been suggested to be a suitable approach for the *in situ* assessment of oxidative DNA damage caused by every day metabolic processes. If this is true, such non-invasive assays could be used to study the role oxidative stress may have as a cause of aging, age-related diseases, and other types of pathologic disorders.

In this paper, both the use of GC/MS for the quantitative assessment of TG, dR-TG, 8-hydroxyguanine (8-G-OH) and 8-hydroxy-2'-deoxyguanosine (8-dRG-OH) in the urine and the mechanistic background for the evolution of these products will be presented to evaluate these compounds as specific biomarkers for oxidative damage to DNA.

GC/MS METHODOLOGY

In recent years, GC/MS has proved to be a powerful technique for biochemical trace analyses. Its strengths include the possibility of resolving complex mixtures and providing information in the form of mass spectra for use in accurate chemical identification. Therefore, the identities of many compounds in a biological matrix can be determined simultaneously with great reliability. Furthermore, in the selected ion mode (SIM) GC/MS has the potential to be an extremely sensitive method for the detection, identification and quantitation of minor constituents in a complicated biological system, often without the need for extensive sample manipulation.[6] We have measured as little as 10 fmol of TG by this technique, for example.

GC/MS MEASUREMENTS OF URINE BIOMARKERS

Here we describe an example of the use of GC/MS for the measurement of biomarkers of DNA damage in biological matrices in which thymine glycol, thymidine glycol, 8-hydroxyguanine and 8-hydroxy-2'-deoxyguanosine are quantified in mouse, rat, and human urine. The four compounds were chosen because they are known products of hydroxy radical damage to DNA[1-6] and they have been suggested as markers for in vivo assays of oxidative DNA damage[7-10]. The species examined have maximum life spans of about 3.5, 4, and 110 years, respectively. For analysis of each compound, fractions were isolated from individual urine samples by C_{18} solid phase extraction; $\alpha,\alpha,\alpha,6\text{-}d_4$ thymine glycol (synthesized) and inosine (Sigma) were added as internal standards; then the fractions were freeze-dried, derivatized and analyzed by GC/MS as described in Reference 6.

Direct evidence of the identities of the DNA base damage markers was obtained during the quantitative analysis by monitoring multiple, compound-specific ions at specific GC retention times (Fig. 1 and Ref. 6), which provided information from which to construct abbreviated mass spectra (Fig. 2).

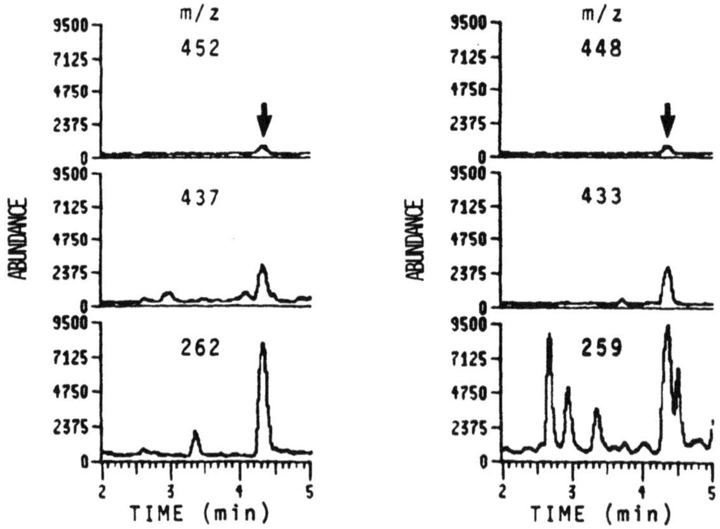

Fig. 1. Representative single ion chromatograms from deuterated thymine glycol internal standard (left panel) and thymine glycol (right panel) as measured in urine. Retention times for each compound are indicated by the arrows.

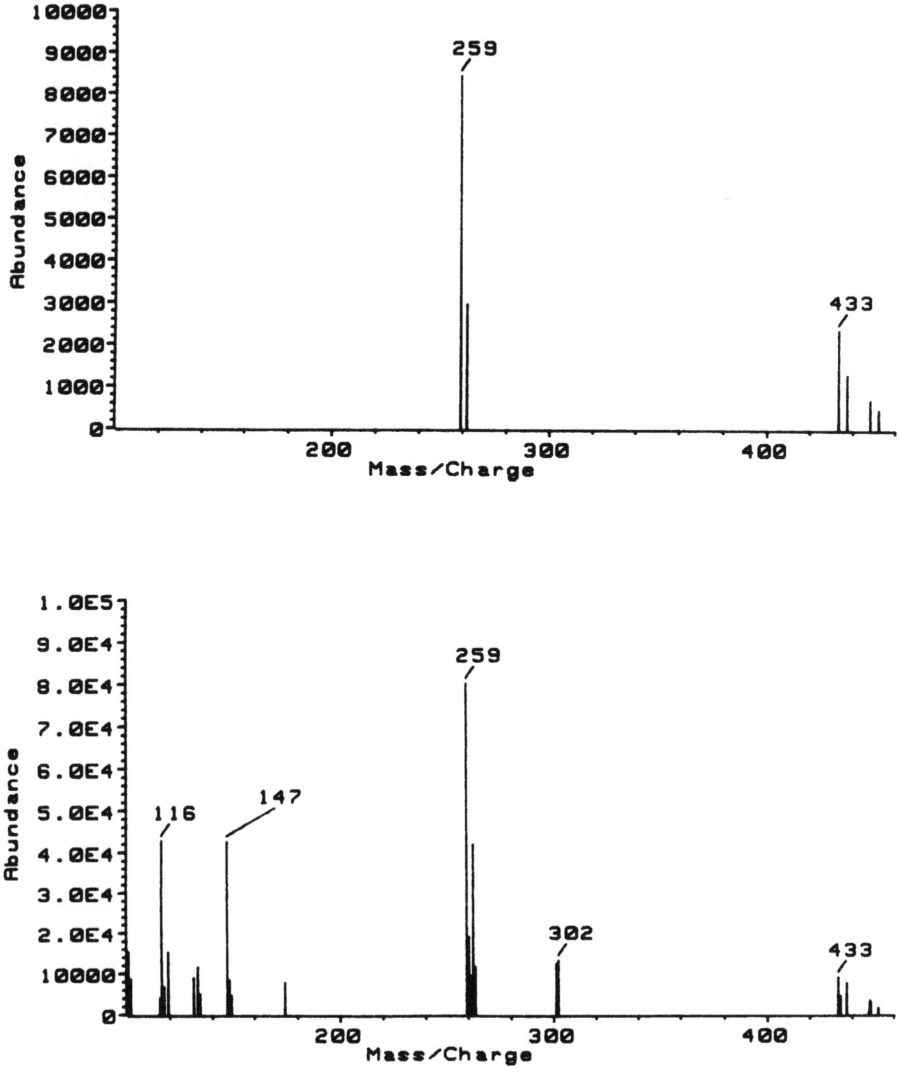

Fig. 2. Comparison of an abbreviated mass spectrum constructed from data such as those of Fig. 1 (top) with the mass spectrum of a mixture of authentic TG and deuterated TG standard (bottom).

The results of our measurements are summarized as Table 1. Comparisons with other results reported in the literature are provided also. The absolute values reported here and elsewhere[7-10] are markedly similar in most cases; calculated values for the relative amounts of thymine and thymidine glycols, 8-hydroxyguanine, and 8-hydroxy-2'-deoxyguanosine are inversely correlated with species longevity.

DISCUSSION

Generation of •OH radicals from hydrogen peroxide by Fe(II) is a well known reaction described first by Haber and Weis in 1932,[11] although the use of hydrogen peroxide and Fe(II) mixtures as an oxidant was described as early as 1894 by Fenton.[12] Ever since, the search for •OH generation in biosystems has been very active because biological cells generate H_2O_2 and also contain iron. More recently, indirect evidence is accumulating for actual generation of •OH in organisms.

Table 1. Comparison of oxidative DNA damage products in urine (nmol/kg/day)

Species	Thymine Glycol TG	Thymidine Glycol dR-TG	8-hydroxy -guanine 8-G-OH	8-hydroxy-2'-deoxyguanosine 8-dRG-OH
Human	0.59 ± 0.26 (0.39 ± 0.36)[a]	0.29 ± 0.11 (0.10 ± 0.05)[a]	3.13 ± 1.58	0.31 ± 0.16 (0.51 ± 0.30)[b]
Rat	9.66 ± 2.93 (5.5 ± 1.7)[a]	4.86 ± 1.46 (1.7 ± 0.8)[a]		
Mouse	27.53 ± 9.03 (~7.8 ± 0.6)[c]	7.23 ± 1.09 (~2.6 ± 0.3)[c]	66.70 ±27.33	10.85 ± 1.43

[a]Reference 7; [b]Reference 8; [c]Figure 5 of Reference 9.

If an •OH radical is generated in the vicinity of DNA it will attack the four DNA bases and deoxyribose. The site of its generation has to be very near to DNA because •OH radical is very reactive and cannot migrate far. Only a few collisions are required before it adds to aromatic rings and double bonds, or abstracts an H atom from C-H bonds.

Since the reactivities of DNA bases with •OH are very similar[13] (within a factor of 2) and the reactivity of deoxribose is lower by a factor of 3 to 5, the distribution of products of •OH attack on DNA should, in the first approximation, reflect the reactivities of the subcomponents. Having multiple reactive sites, each subcomponent also yields multiple products in relative amounts determined by chemical conditions and the microenvironment.

In the case of the thymine reaction with •OH radical there are three reaction sites which correspond to the formation of three free radical intermediates.

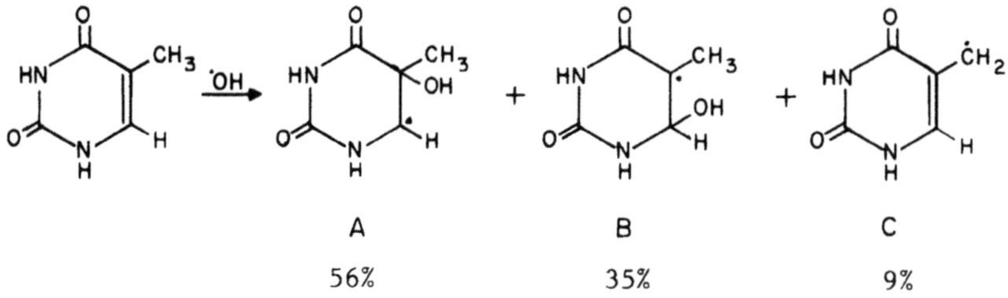

(1)

Intermediate A is the most abundant and is the one that leads to the formation of TG.[14] Since attachment of the TG moeity to DNA would not be affected by the reactions leading to its formation, free TG and dR-TG would appear in the urine only after complete enzymatic digestion (EzD) of damaged DNA, excision of the damage (Ex) or by the repair action of TG glycosylase (TGG)[1]

$$\text{DNA-TG} \begin{cases} \xrightarrow{\text{TGG}} \text{DNA}' + \text{TG} & (2) \\ \xrightarrow{\text{Ex}} \text{DNA}'' + \text{TG} + \text{dR-TG} & (3) \\ \xrightarrow{\text{EzD}} \text{TG} + \text{dR-TG} + \text{Pox} + \text{B} + \text{dRB} & (4) \end{cases}$$

where Pox = oxidative products, B = bases and dRB = nucleotides and nucleosides. Hence, it is apparent that TGG activity alone cannot account for the total TG (TG and dR-TG) observed.

Thymine radical C is a minor product (9%) in the reaction of •OH with thymine — reaction (1). Radical C is a precursor of hydroxymethyluracil[15] (HMU), one of the oxidative products found in urine. The ratio of transients A/C ≈ 6, and the TG/HMU ratio in model systems also should be of that order. The same ratio in human urine is TG/HMU ≈ 0.5. Consequently it must be concluded that if the in vitro data apply in vivo, then either TG is lost to a very large extent or HMU comes from sources other than reaction (1). One possible natural origin for HMU, however, could be 5-hydroxymethyluridine from t-RNA.[16]

8-Hydroxyguanine is only 25% of all G + •OH products, whereas TG is about 50% of the T + •OH reaction.[15] Since, $k(G + \bullet OH) = 1.0 \times 10^{10}\ M^{-1}\ s^{-1}$ is not much higher than $k(T + \bullet OH) = 7.4 \times 10^{9}\ M^{-1}\ s^{-1}$ the total yields of TG and 8-G-OH should be comparable. It is evident from Table 2 that 8-G-OH predominates in human and mouse urine.

Table 2. Total thymine glycol and 8-hydroxyguanine in the urine (nmol/kg/day).

Species	Maximum Life Span (years)	Specific Metabolic Rate (cal/g/day)	$(TG)_t$	$(8\text{-}G\text{-}OH)_t$
human	110 ± 5	25	0.9	3.4
rat	4 ± 0.5	100	15	
mouse	3.5 ± 0.5	180	35	78

The difference between measured $(TG)_t$ = TG + dR-TG and $(8\text{-G-OH})_t$ = 8-G-OH + 8-dG-OH could be attributed to relative instability of TG, existence of site specific •OH generation (predominantly in the G-rich regions), selectivity in the occurrence or repair of damage, or perhaps some other metabolic or unknown processes.

The measured DNA base products excreted in the urine are identical to the •OH radical-induced base products. Assuming that these products come exclusively from DNA via repair enzyme processes, then the results in Tables 1 and 2 indicate a relationship between metabolic rate, longevity and oxidative stress. Species with higher metabolic rates may induce more free radical side reactions. Hence, generation of free radicals at greater rates, in the first approximation, produces more oxidative products, which in turn may lead to a shorter life span.

Oxidative stress in humans is substantially lower than in rodents, as evident from Table 2.[3] This might not be fully explainable by the lower metabolic rate of humans. Instead, a superior defense system, which both prevents and repairs the effects of damaging oxidative processes and thus further reduces the already low, metabolic-rate-related levels of oxidative damage, may have evolved recently and may be responsible for the unusually long human life span.

Preliminary measurements in our laboratories of TG, 8-G-OH and 8-dG-OH in liver DNA from several species, including humans and mice, have demonstrated a correlation between the levels of these products in the DNA and in the urine of individual species.[10] Based on this indication of the validity of our approach, ongoing interspecies correlations between oxidative damage to nucleic acid material, metabolic rate, and longevity can be expected to provide further insight into the mechanisms of aging.

ACKNOWLEDGMENTS

We gratefully acknowledge Gloria Wiersma for her assistance with the preparation of the manuscript. Special thanks are due to Dr. Eric Holwitt of the Armed Forces Radiobiology Research Institute for the synthesis of deuterated thymine glycol and to Edith Cutler for collecting the urine samples (supported by the Paul Glenn Foundation for Medical Research). Part of this work was completed in conjunction with Dr. Robert Young of the Radiation Policy Division of the Defense Nuclear Agency.

REFERENCES

1. E.C. Friedberg, "DNA Repair," W.H. Freeman and Co., New York (1984).
2. P.A. Cerutti, Prooxidant states and tumor production, Science 227:375 (1985).
3. R.G. Cutler, Antioxidants, aging, and longevity, in: "Free Radicals in Biology, Vol VI," W.A. Pryor, ed., pp 371-428 (1984).
4. B.N. Ames, Dietary carcinogens and anticarcinogens: oxygen radicals and degenerative diseases, Science 221:1256 (1983).
5. J.F. Ward, Molecular mechanisms of radiation-induced damage to nucleic acids, Adv. Radiat. Biol. 5:181 (1975).
6. M. Dizdaroglu and D.S. Bergtold, Characterization of free radical-induced base damage in DNA at biologically relevant levels, Anal. Biochem. 156:182 (1986).

7. R. Cathcart, E. Schwiers, R.L. Saul, and B.N. Ames, Thymine glycol and thymidine glycol in human and rat urine: A possible assay for oxidative DNA damage, Proc. Natl. Acad. Sci. USA, 81:5633 (1984).
8. K.C. Cundy, R. Kohen, and B.N. Ames, Determination of 8-hydroxydeoxyguanosine in human urine: A possible assay for in vivo oxidative DNA damage, Proc. 4th Int'l Congress on Oxygen Radicals, June 27-July 3, 1987 (1988).
9. R.L. Saul, P. Gee, and B.N. Ames, Free radicals, DNA damage, and aging, in: "Modern Biological Theories of Aging," H.R. Warner, R.N. Butler, R.L. Sprott, E.L. Schneider, eds., pp. 113-129 (1987).
10. H. Alessio and R.G. Cutler, Average rate of oxy radical oxidation of deoxyguanosine in DNA of mammalian species is proportional to their aging rate, in: The Gerontologist 27:151A (1987).
11. F. Haber and J. Weiss, Uber die Natalyse des Hydroperoxydes, Naturwiss. 20:948 (1932).
12. H.J.H. Fenton, Oxidation of tartaric acid in the presence of iron, J. Chem. Soc. Trans. 65:899 (1894).
13. Farhataziz and A.B. Ross, NSRDS-NBS 59, U.S. Department of Commerce, Washington, D.C. (1977).
14. S.V. Jovanovic and M.G. Simic, Mechanism of OH radical reaction with thymine and uracil derivatives, J. Am. Chem. Soc. 108:5968 (1986).
15. M.G. Simic and S.V. Jovanovic, Free radical mechanisms of DNA base damage, in: "Mechanisms of DNA Damage and Repair," M. G. Simic, L. Grossman, and A.C. Upton, eds., Plenum Press, N.Y. (1986) pp. 39.
16. Y. Iwanami and G. M. Brown, Methylated bases of transfer ribonucleic acid from HeLa and L cells, Arch. Biochem. Biophys. 124:472 (1968).

THE GENERATION OF RADICALS DURING THE NORMAL AND ABNORMAL FUNCTIONING OF CYTOCHROMES P-450

David Dolphin

Department of Chemistry
University of British Columbia
2036 Main Mall, Vancouver, British Columbia
Canada V6T 1Y6

Oxygen exhibits very rich redox chemistry and, as shown in Scheme 1, hemes and hemeproteins play major roles in both the production of and protection from a variety of oxygen and other free radicals. In addition catalase destroys, at a diffusion-controlled rate, hydrogen peroxide by a two-electron process which does not generate hydroxyl radicals. On the other hand hydrogen peroxide, along with a wide variety of transition metal ions, is the principal source of in vivo hydroxyl radicals.

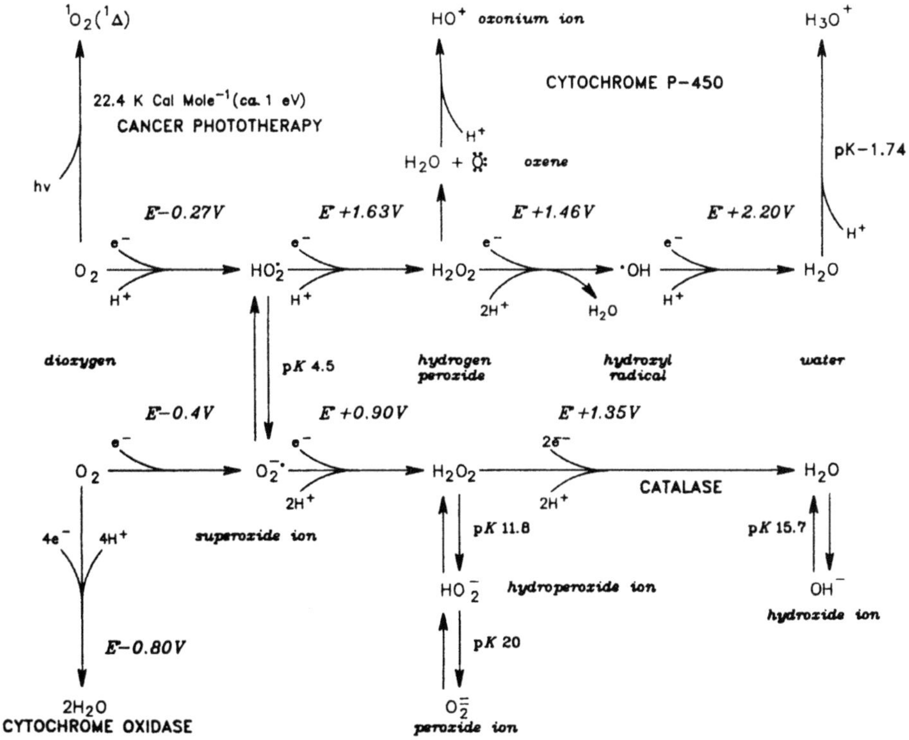

Scheme 1

The catalatic decomposition of peroxide does involve radical production on the heme moiety of the catalase enzyme. Thus the two-electron reduction of peroxide to water involves oxidation of the resting ferric heme to an oxo-ferryl(FeIV) porphyrin π-cation radical (1), eq. (1) where one of the oxidizing equivalents is stored on the porphyrin macrocycle.[1] However, the unpaired electron is sufficiently delocalized and sterically protected by the protein that these porphyrin free radicals are surprisingly stable and consequently exhibit few *typical* radical reactions.[2] The second catalytic step to complete the enzymatic cycle involves the two electron oxidation of peroxide to dioxygen and the reduction of 1 to the resting ferric state, eq. (2).

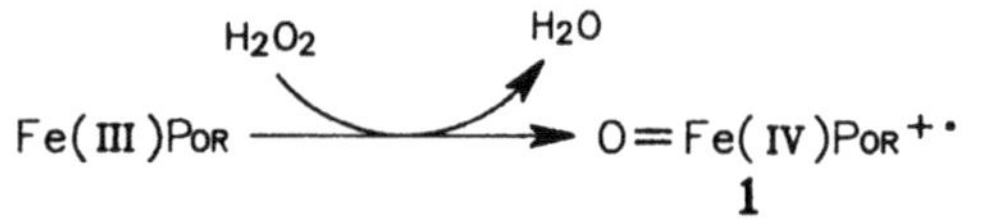

(1)

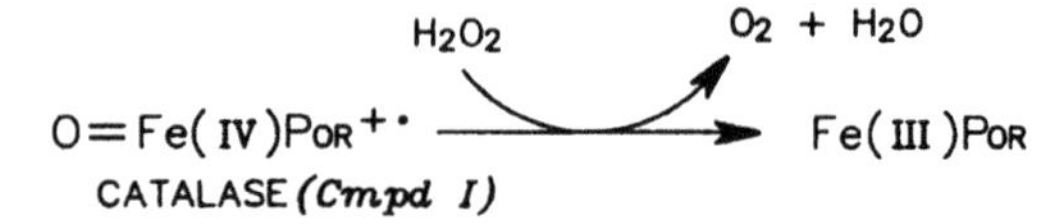

(2)

The peroxidases, typified by horseradish peroxidase (HRP), show the same initial reaction with peroxide to generate 1, but now 1 is used to bring about two single one-electron oxidations of aromatic amines and phenols (eq. 3). The resulting radicals then play important roles in the secondary metabolism of plants.[3]

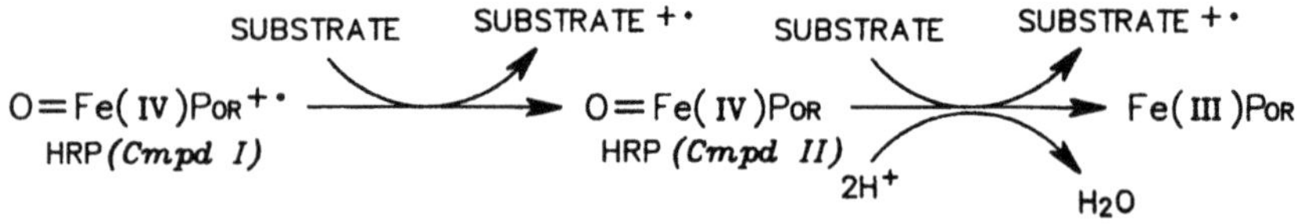

(3)

Cytochrome P-450 is nature's most powerful *in vivo* oxidant as well as being a very powerful reductant. During the normal enzymatic cycle radicals are generated to achieve the final substrate oxidations. During this process a variety of radical (or radical-generating) species may be released into the cell. In order to afford the maximum protection to the heme and the radicals generated in its vicinity, steric protection is provided by the protein, with the heme buried deep in the interior. This can be compared to myoglobin, which does not use highly reactive intermediates during its cycling, where the heme is quite exposed (Figure 1).

Cytochromes P-450 are found in all forms of life and they catalyze a diverse series of oxidations.[4] Nevertheless the basic mechanisms by which they function are similar. Scheme 2 shows the catalytic cycle and it is apparent that radicals, or their precursors, may be generated at each step. Upon binding of substrate (RH) the low-spin resting enzyme (2) is converted into the more easily reduced high-spin complex (3). A one-electron reduction gives the ferrous hemeprotein (4), which may have a redox potential as low as -350 mV![5] Interaction of this powerful reductant with halogenated hydrocarbons may follow several paths. For instance, a one-electron reduction of CCl_4 gives chloride ion, the resting enzyme and the $\cdot CCl_3$ radical. If this radical diffuses from the active site it will, of course, initiate lipid peroxidation. Coupling of an unpaired electron on the ferric heme and the $\cdot CCl_3$ radical generates

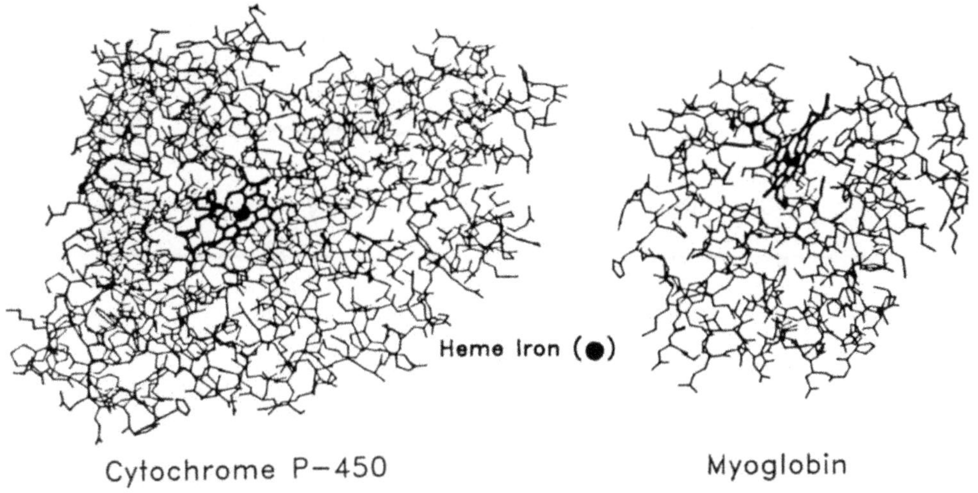

Figure 1. Crystal structures of cytochrome P-450$_{cam}$ and horse heart myoglobin. We thank Drs. T. Poulos and G. Brayer for providing these structures.

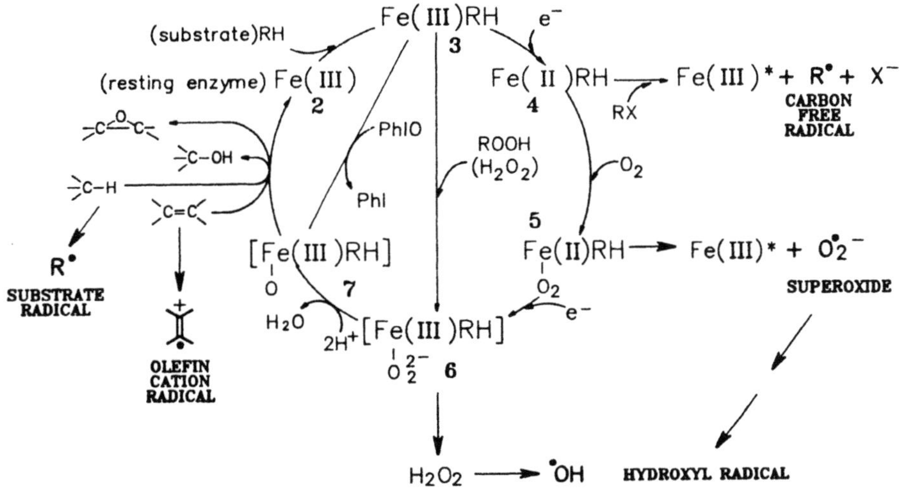

Scheme 2

a σ-alkyliron(III) complex (**8**) and a further one-electron reduction gives the dichlorocarbene iron(II) complex (**9**)[6] (Fig. 2). The σ-alkyl and carbene complexes along with the (N, Fe) bridged methylene iron porphyrin and N-alkyliron porphyrins are all interrelated via redox reactions (Scheme 3), and these reactions of the cytochromes P-450 now represent the most important organometallic chemistry occurring in nature.[7]

Binding of dioxygen to the ferrous P-450 gives the oxygenated intermediate (**5**) (Scheme 2). However, unlike the stable oxyhemoglobin and oxymyoglobin, which are axially coordinated by histidine, oxy P-450 is coordinated by the electron-rich thiolate anion of cysteine.[8] The sulfur ligand places electron density on the iron, which facilitates the release

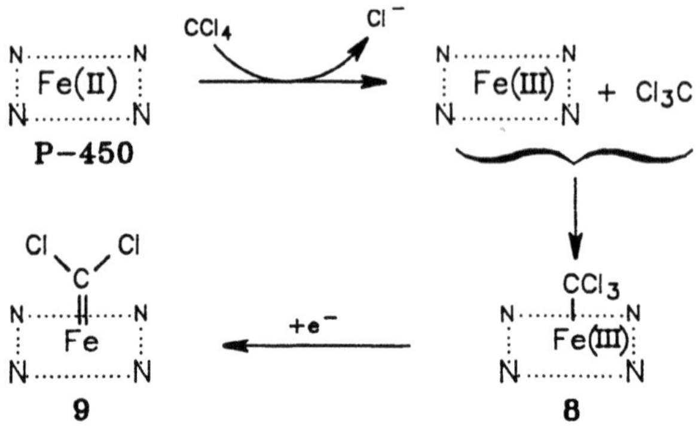

Figure 2. Formation of σ-alkyl- and carbene-iron porphyrin complexes.

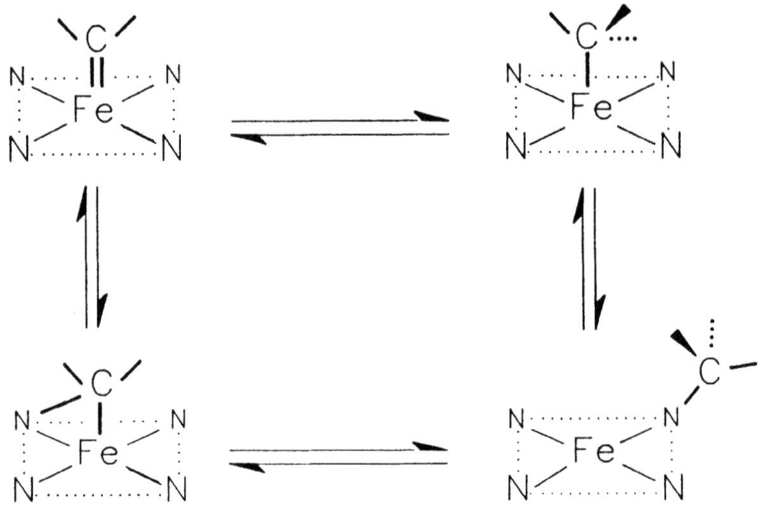

Scheme 3

of superoxide ion, and as discussed elsewhere in this volume, superoxide in its own right, and the peroxide derived from its dismutation, can generate a host of radicals.

Addition of a further electron to the oxygenated enzyme (**5**), which is the rate limiting step of the whole enzymatic cycle, generates a intermediate that is formally a ferric hemeprotein coordinated to the equivalent of peroxide dianion (**6**) (Scheme 2). Supporting this, as shown in scheme 1 (**3**→**6**), the enzyme may be shunted with peroxides replacing dioxygen and two electrons.

After generation of the 'peroxide' intermediate (**6**), hydrogen peroxide can be released from the active site. It is not yet clear how peroxide in fact derives from (**6**) and how much from (**5**) via the loss and dismutation of superoxide.[9]

As noted above (Scheme 2) peroxides, including alkyl hydrogen peroxides, may be used to activate the enzyme in the absence of electron donors. This raises the intriguing possibility that P-450 may in fact remove products of lipid peroxidation by converting them to the corresponding alcohols and indeed there is evidence that this occurs, at least in liver microsomes. This destruction of lipid hydroperoxide must, however, be balanced against the observations that P-450 may also bring about O-O homolytic bond cleavage to generate alkoxyl radicals and thus possibly initiate additional chain reactions.[9,10]

Heterolytic cleavage of the O-O bond of **6** generates, after protonation, water and a species (**7**) that is formally a ferric oxene complex (i.e. Fe(III) bound to an oxygen atom). Structure **7** is but one of several resonance structures that can be written for this very electron deficient species. A more reasonable electronic structure, where the oxygen atom maintains an octet of electrons, is **10** (eq. 4). This is the same electronic structure that we have previously shown for the highest oxidation states (compounds I) of catalase and HRP. The active oxidizing species of P-450 is not well characterized. Its optical spectrum[11] resembles that of compound I of chloroperoxide which is clearly an oxoferryl(FeIV) porphyrin π-cation radical,[3] but the most conclusive evidence as to the electronic structure of **7** (eq. 4) comes from model studies.

$$\underset{\mathbf{7}}{\ddot{\text{O}}\text{:Fe(III)Por}} \longleftrightarrow \cdot\ddot{\text{O}}\text{:Fe(IV)Por} \longleftrightarrow \underset{\mathbf{10}}{:\ddot{\text{O}}\text{:Fe(IV)Por}^{+\cdot}} \qquad (4)$$

Groves and his colleagues[12] have shown that model iron porphyrins and oxygen-transfer agents such as iodosyl benzene or peracids closely mimic the P-450 enzymes (the same oxygen transfer agents can be used to shunt the enzymatic reaction 3→7, Scheme 2). Thus hemins plus iodosyl benzene will hydroxylate aliphatic hydrocarbons showing similar isotope effects to the natural systems as well as showing a preference for tertiary C-H over secondary and then primary.[13] Using the sterically protected meso-tetramesityl hemin **11** (Scheme 4) Groves et. al. have well characterized an oxoferryl (FeIV) porphyrin π-cation radical (**12**) which as would be expected, shows a very short 1.6 A Fe=O bond length (Scheme 4).[14]

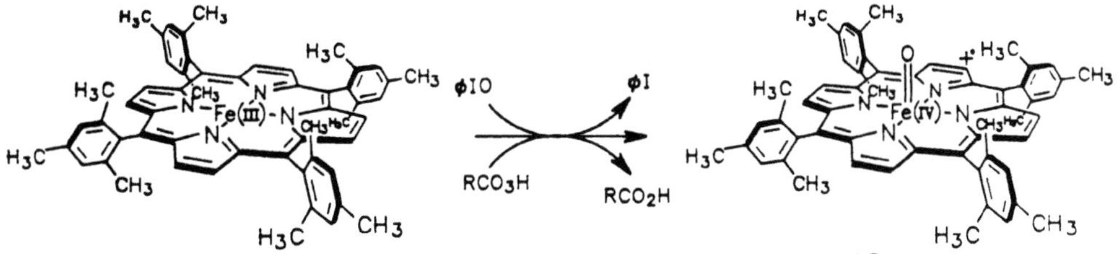

11 **12**

Scheme 4

All of the model studies on hydrocarbon hydroxylation are consistent with an oxygen rebound mechanism[15] as outlined in Scheme 5. An initial abstraction of a hydrogen atom by the oxoferryl π-cation radical generates a substrate radical and the oxoferryl(IV) porphyrin (which is a resonance structure of the ferric porphyrin coordinated by a hydroxyl radical). Recombination of the substrate and hydroxyl radical give the product alcohol and the resting ferric enzyme (Scheme 5). The microsomal metabolism of dieldrin (**13**) to the cyclic ketone **14** (Scheme 6) is a clear example of the radical intermediates.[16]

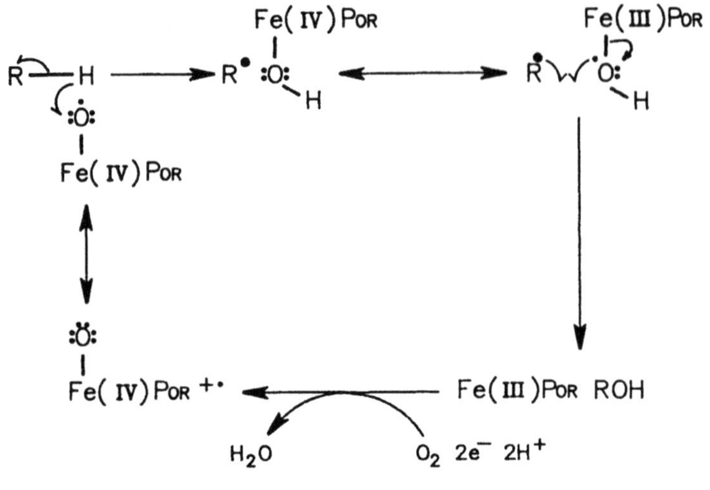

Scheme 5

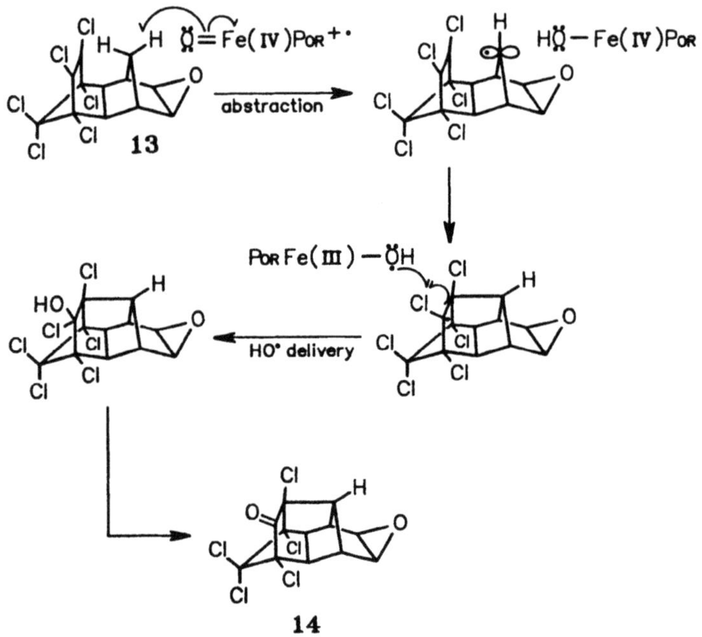

Scheme 6

Escape of either the substrate or hydroxyl radical from the active site seems not to occur during substrate hydroxylation, although oxidation of substrates can generate radicals when mechanisms other than that represented in Scheme 5 occur. A well-documented example is that of the oxidation of dihydropyridines. A one-electron oxidation of the dihydropyridine **15** to the π-cation radical results in alkyl radical extrusion (the driving force being aromatization to the pyridine ring). The alkyl radical **17** can then react with the heme, the protein or be spin trapped away from the active site. Radicals are similarly formed from the oxidation of hydrazines and a wide variety of drugs.[9]

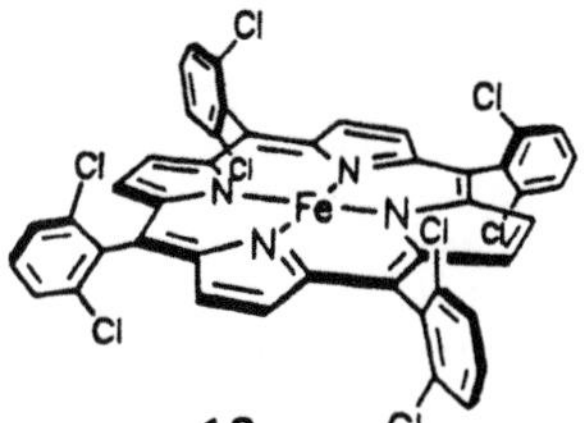

18

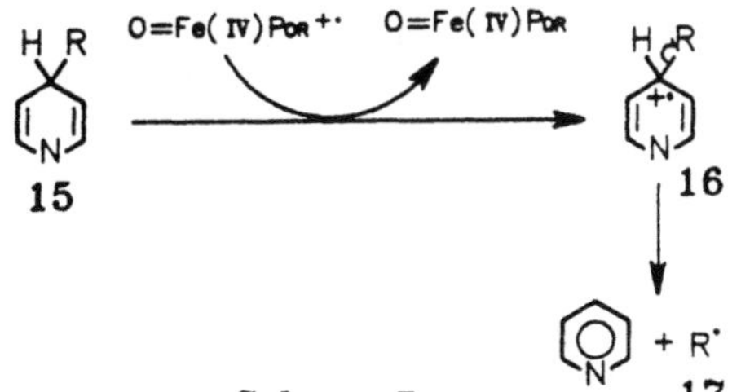

Scheme 7

Radicals are also intermediates in the second major reaction of P-450 enzymes, namely, epoxidation of olefins. Much has been written on this topic, but it can be summarized by some of our recent work. We found that the tetra-(bisdichlorophenyl) heme **18** is, due to the steric and electronic protection afforded by the eight chlorine atoms, a very powerful and robust catalyst that closely mimics P-450 enzymes.[17] Being so robust, high turnover numbers are observed (>10,000) for the epoxidation of norbornene with minimal loss of catalyst. Since large turnovers are observed so is other P-450 chemistry which less stable mimics cannot explore. Thus both the N-alkylation of the heme,[18] a suicide inhibition reaction which occurs during epoxidation with the enzymes themselves, and the rearrangements seen with norbornene[19] (Scheme 8) allow for a common mechanism for epoxidation, N-alkylation and substrate rearrangement, all of which are initiated by an electron transfer from the olefin to the oxoferryl porphyrin π-cation radical to generate the olefin π-cation radical **19** (Scheme 9). Subsequent reactions of this cation radical to give the common intermediate **20** can then explain all of the observed chemistry between olefins and P-450 enzymes, always bearing in mind that unusual substrates may impose unusual reaction mechanisms on the enzymes.

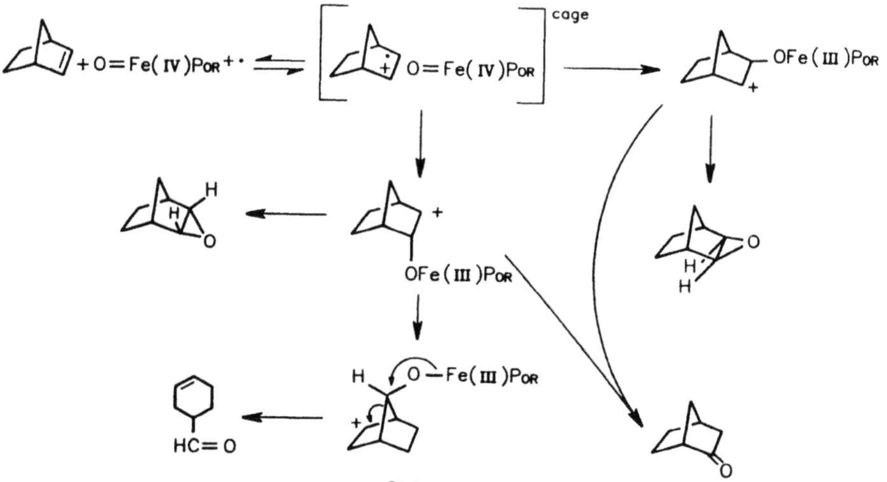

Scheme 8

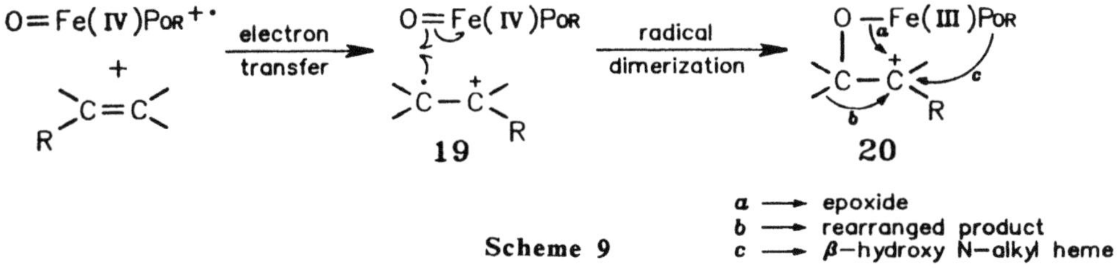

Scheme 9

The enzyme that degrade lignin (the ligninases) are heme proteins and they function, in a manner similar to the peroxidases, to depolymerize lignin.[20,21] The first step in the process is a single electron oxidation of the electron rich aromtic rings of lignin. This is followed by bond cleavage of the propane moiety of the aromatic π-cation radical. The subsequent radicals can then react both aerobically or anaerobically (Scheme 10). The aerobic pathway initiates chain reactions (in a manner similar to lipid peroxidation) and the hydroperoxides so produced could presumably reoxidize the resting enzyme in a manner analogous to the proposed reactions with lipid hydroperoxides discussed above.

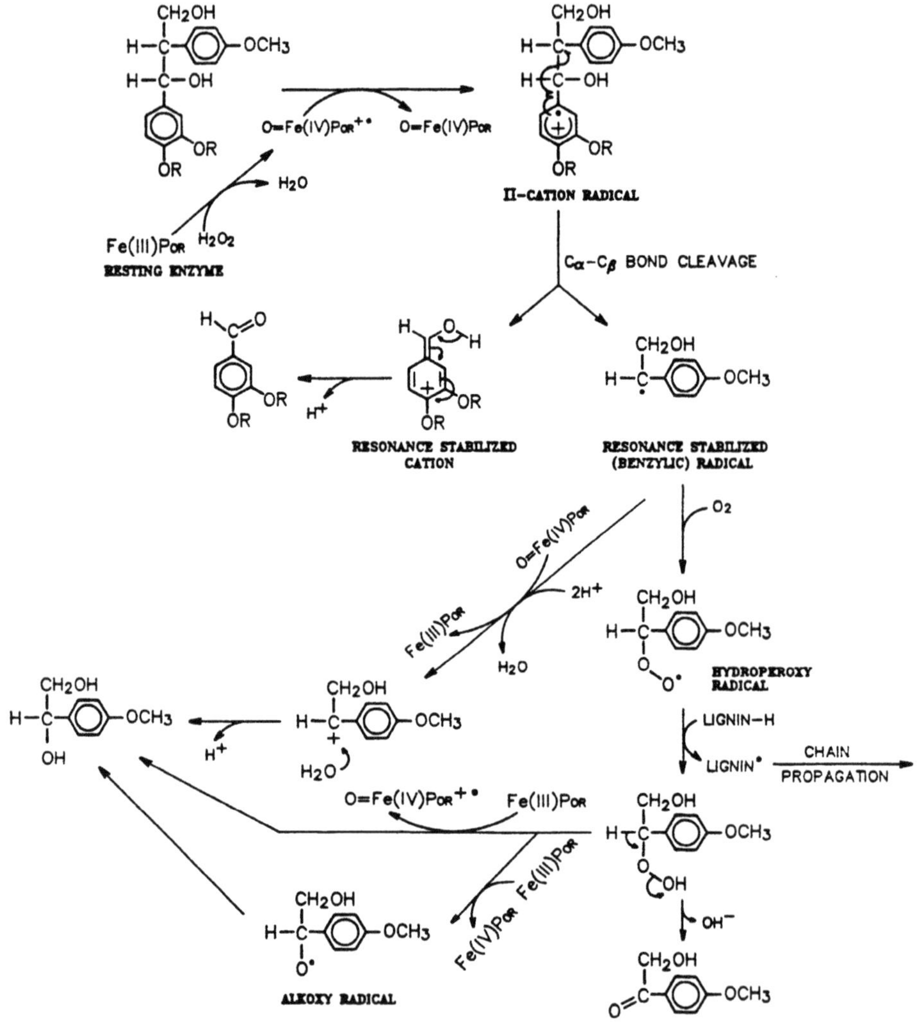

Scheme 10

Since lignin depolymerization requires only a one-electron oxidation for its initiation, it is apparent that a large number of oxidants could fulfill this role. However, if a catalytic oxidant is required, then the tetrasulfonated octachloroheme (21) is a good candidate. The sulfonato groups confer both water solubility and additional electron stabilization over the hemin 18 (Scheme 7). As can be seen from Table 1[22] a wide range of oxidants can prime the catalyst, and oxidation of veratryl alcohol (as a lignin model) occurs over a wide range of pH and temperature.

Table 1. Oxidation of veratryl alcohol

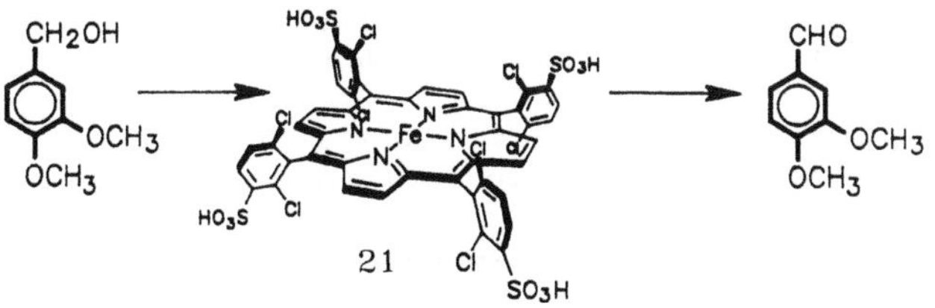

21

Oxidant	(conc M)	Conditions	Yield (based on oxidant)	Comment
NaOCl	(0.067)	–	19%	Moderate rate
NaOCl	(0.067)	pH 10	9%	Fast rate
NaOCl	(0.067)	pH 1.5	37%	Slow rate
t–BuOOH	(0.1)	–	50%	Rapid rate
O-+OOH	(0.1)	–	49%	Rapid rate
KIO_3	(0.5)	pH 7	<0.1%	
KIO_4	(0.043)	pH>1.5	68%	Rapid rate
Peracid	(0.08)		10%	Moderate rate
H_2O_2	(0.1)	pH 1.5	20%	Rapid rate
H_2O_2	(0.1)	pH 7	12%	Rapid rate
H_2O_2	(0.1)	pH 10	5%	Rapid rate
C_6H_5IO	(0.05)	–	30%	Instantaneous
C_6F_5IO	(0.064)	–	28%	Rapid rate

Acknowledgements

This work was supported by the U.S. National Institutes of Health (AM 17989) and the Canadian National Sciences and Engineering Research Council.

References

1. D. Dolphin and R.H. Felton, The biochemical significance of porphyrin cation radicals, Acc. Chem. Res., 7:26 (1974).
2. D. Dolphin, Z. Muljiani, K. Rousseau, D.C. Borg, J. Fajer, and R.H. Felton, The Chemistry of Porphyrin π-Cations, Ann. NY Acad. Sci., 206:177 (1973).
3. W.D. Hewson and L.P. Hager, "Peroxidases, Catalases and Chloroperoxidase" in: "The Porphyrins," D. Dolphin, ed., Academic Press, New York, Vol. VII, pp. 295-332 (1979).
4. P.R. Ortiz de Montellano, Ed., "Cytochrome P-450," Plenum Press, New York, (1986).
5. J.A. Pederson and R.A. Prough, "Cytochrome P-450 Reductase and Cytochrome b_5 in Cytochrome P-450 Catalysis" in: "Cytochrome P-450," P.R. Ortiz de Montellano, Ed., Plenum Press, New York, pp. 89-117 (1986).
6. D. Mansuy, "Cytochrome P-450 and synthetic models", Pure and Appl. Chem., 59:759 (1987).
7. J.-I. Setsune and D. Dolphin, Organometallic aspects of cytochrome P-450 metabolism, Can. J. Chem., 67:459 (1987).

8. T.L. Poulos, "The Crystal Structure of Cytochrome P-450_{cam}", in: "Cytochrome P-450," P.R. Ortiz de Montellano, Ed., Plenum Press, New York, pp. 505-523 (1986).
9. P.R. Ortiz de Montellano, "Oxygen Activation and Transfer", in: "Cytochrome P-450," P.R. Ortiz de Montellano, Ed., Plenum Press, New York, pp. 217-271 (1986).
10. T.Eling, J. Curtis, J. Battista, and L.J. Marnett, Oxidation of (+)-7,8-dihydroxy-7,8-dihydrobenzo[a]pyrene by mouse keratinocyates: evidence for peroxyl radical- and monooxygenase-dependent metabolism, Carcinogenesis (London), 7:1957 (1986).
11. G.C. Wagner, M.M. Palcic, H.B. Dunford, Absorption spectra of cytochrome P-450_{cam} in the reaction with peroxy acids, F.E.B.S. Lett., 156:244 (1983).
12. J.T. Groves, R.C. Haushatter, M. Nakamura, T.E. Nemo, and B.J. Evans, High-valent iron-porphyrin complexes related to peroxidase and cytochrome P-450, J. Am. Chem. Soc., 103:2884 (1981).
13. J.T. Groves, G.A. McClusky, R.E. White, and M.J. Coon, Aliphatic hydroxylation by highly purified liver microsomal cytochrome P-450: Evidence for a carbon radical intermediate, Biochem. Biophys. Res. Commun., 81:154 (1978).
14. J.E. Penner-Hahn, K. Smith Eble, T.J. McMurray, M. Renner, A.L. Balch, J.T. Groves, J.H. Dawson, K.O. Hodgson, Structural characterization of horseradish peroxidase using EXAFS spectroscopy. Evidence for Fe=O ligation in compounds I and II, J. Am. Chem. Soc., 108:7819 (1986).
15. J.T. Groves and G.A. McClusky, Aliphatic hydroxylation via oxygen rebound oxygen transfer catalyzed by iron, J. Am. Chem. Soc., 98:859 (1976).
16. D. Dolphin, A.W. Addison, M. Cairns, R.K. DiNello, N.P. Farrell, B.R. James, D.R. Paulson, and C. Welborn, Heme proteins and metalloporphyrins: redox chemistry and oxygen binding, Int. J. Quantum Chem., 16:311 (1979).
17. P.S. Traylor, D. Dolphin, and T.G. Traylor, Sterically protected hemins with electronegative substituents: efficient catalysts for hydroxylation and epoxidation, J. Chem. Soc., Chem. Commun., 279 (1984).
18. T. Mashiko, D. Dolphin, T. Nakano, and T.G. Traylor, N-Alkylporphyrin formation during the reactions of cytochrome P-450 model systems, J. Am. Chem. Soc., 107:3735 (1985).
19. T.G. Traylor, T. Nakano, B.E. Danlap, P.S. Traylor, and D. Dolphin, Mechanism of hemin-catalyzed alkene epoxidation. The effect of catalyst on the regiochemistry of epoxidation, J. Am. Chem. Soc., 108:2782 (1986).
20. S.K. Glenn, M.A. Morgan, M.B. Mayfield, M. Kuwahara, and M.H. Gold, An extracellular H_2O_2-requiring enzyme preparation involved in lignin biodegradation by the white rot basidiomycete Phanerochaete chrysosporium, Biochem. Biophys. Res. Commun., 114:1077 (1983).
21. M. Tien and T.K. Kirk, Lignin-degrading enzyme from the hymenomycete Phanerochaete chrysosporium burds, Science, 221:601 (1983).
22. D. Dolphin, T. Nakano, T.E. Maione, T.K. Kirk, and R. Farrell, "Synthetic Model Ligninases," in "Lignin enzymic and microbial degradation," Ed. E. Odier, INRA Publications, Paris, pp. 157-162 (1987).

RADICAL INTERMEDIATES IN THE CATALYTIC CYCLES OF CYTOCHROME P-450

Alfin D. N. Vaz, Elizabeth S. Roberts, and Minor J. Coon

Department of Biological Chemistry
Medical School
The University of Michigan
Ann Arbor, Michigan 48109-0606, U.S.A.

INTRODUCTION

Cytochrome P-450 is the term used for an unusual and very large family of heme proteins that have a cysteine thiolate ligated to the fifth coordination site of heme iron, and in the reduced state form unique CO complexes that have a Soret band at around 450 nm.[1] This family of cytochromes consists of constitutive forms, such as the steroid-oxygenating enzymes, and inducible forms that oxygenate a large variety of naturally occurring and foreign compounds.[2] Unlike most enzyme catalysts, this versatile class of heme proteins catalyzes a variety of both oxidative and reductive reactions. Oxidatively, they are capable of effecting both aliphatic and aromatic hydrocarbon hydroxylation, olefin epoxidation, dealkylation at N, O, and S atoms, N-oxidation, sulfoxidation, peroxidation, and carbon-carbon bond cleavage.[1,3] Reductively, they are capable of electron transfer reactions to azo, nitro, N-oxide, epoxide, peroxy, and alkyl halide groups, thus justifying the term "cytochrome" for these catalysts. Substrates for P-450 include endobiotics like steroids, fatty acids, and prostaglandins, and xenobiotics like drugs, pesticides, and petroleum products of aliphatic or aromatic nature.

OXYGENATIVE CATALYTIC CYCLE OF CYTOCHROME P-450

For catalytic activity, the reconstituted microsomal system consists of the flavoprotein, NADPH-cytochrome P-450 reductase, a form of cytochrome P-450, and phospholipid.[4,5] In the oxygenative reaction the cytochrome utilizes electrons from NADPH to activate molecular oxygen to an oxidant that is subsequently inserted into the substrate. The overall equation for the oxygenation of substrates may be represented by the following equation:

$$RH + O_2 + NADPH + H^+ \rightarrow ROH + H_2O + NADP^+ \qquad (1)$$

where RH represents any organic substrate of natural or foreign origin. Work from various laboratories has resulted in the elucidation of distinct steps in the catalytic cycle as shown in Figure 1.[1]

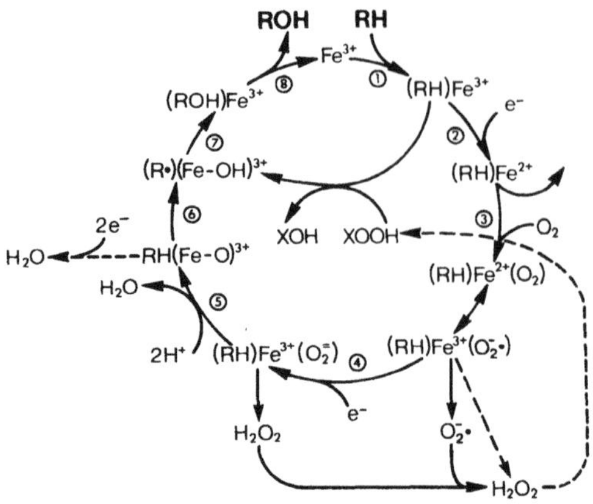

Figure 1. Proposed scheme for mechanism of action of cytochrome P-450, slightly modified from a scheme presented elsewhere.[6]

The individual reactions may be summarized as follows:

1) Substrate binding occurs to the resting Fe^{3+} state of the enzyme, resulting, in some instances, in a low to high spin shift.

2) The first electron transfer from NADPH to the cytochrome is known to be effected only by NADPH-cytochrome P-450 reductase. With some P-450 forms, substrate binding is a prerequisite for electron transfer from the flavoprotein, while with others electron transfer occurs in the absence of substrate, but the rate is enhanced when substrate is bound at the active site.

3) Oxygen binding, as in all molecular oxygen-binding heme proteins, occurs only to the Fe^{2+} form of the enzyme. Thus, reduction of the P-450 by the flavoprotein is a prerequisite for oxygen binding, and in certain instances the reductase may be limiting in amount in the catalytic cycle. EPR and Mössbauer spectroscopy have demonstrated that the oxy-ferrous complex resembles a ferric-superoxide species in the bacterial camphor-metabolizing cytochrome P-450.[7]

4 and 5) The second electron transfer to the iron-oxy complex may occur from either the flavoprotein or from cytochrome $\underline{b}_5$, to yield the iron-peroxy equivalent, which with proton uptake is thought to yield the active oxidant: an "iron oxene."

6-8) The broad variety of substrates oxidized by cytochrome P-450 is thought to involve a common mechanism with stepwise hydrogen atom abstraction from the substrate to produce a transient substrate radical and an iron-bound hydroxyl radical equivalent, which then recombine to yield the oxidized (or hydroxylated) product and the resting enzyme. In some instances, concerted insertion of the iron-bound oxene into the substrate has not been ruled out, as in the epoxidation of olefinic bonds. It should be noted, as shown in Figure 1, that the P-450 system is also capable of 4-electron oxidase activity in which molecular oxygen is reduced to water, presumably by stepwise electron transfer to the iron-oxene with intermediate formation of bound hydroxyl radical.[8]

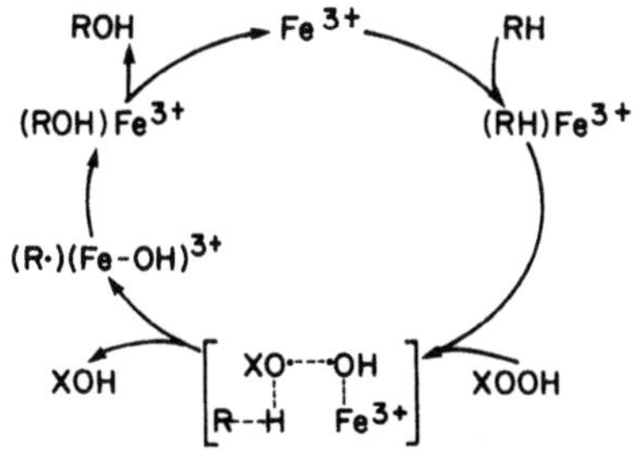

Figure 2. Peroxide shunt, in which a peroxy compound, XOOH, serves as the oxygen donor to the substrate. Taken from a recent paper by Coon and Vaz.[13]

FREE RADICALS IN THE NORMAL OXYGENATIVE CATALYTIC CYCLE OF CYTOCHROME P-450

At several stages in the catalytic cycle of cytochrome P-450, direct and indirect evidence has been obtained for the generation of free radical intermediates of carbon as well as oxygen and other heteroatoms.

Superoxide Radical Anion

Reduced NADPH-cytochrome P-450 reductase has been demonstrated to slowly generate superoxide during its autoxidation[9]; more significant in the cycle is the generation of superoxide by some forms of cytochrome P-450.[1,10]

Hydroxyl Radical

Free hydroxyl radical formation in the P-450 catalytic cycle has not as yet been demonstrated; however, various lines of evidence have implicated a hydroxyl radical equivalent at the heme iron in the hydroxylation of substrates. Evidence from the crystal structure of bacterial cytochrome P-450[11] and ESR-probing techniques with mammalian isozymes[12] indicate the porphyrin to be buried 11-14 A within the protein. Due to the short half-life and the extreme oxidative capacity of hydroxyl radicals, it appears unlikely that their release into bulk solvent occurs.

Carbon and Heteroatom Radicals

Free radicals at carbon and at heteroatom centers are implicated or in some cases identified in the oxidative step because of epimerization of optically active centers during hydroxylation.[14,15] A significant number of P-450-catalyzed hydroxylations occur without epimerization at the hydroxylated center, suggesting a concerted mechanism. These carbon- and heteroatom-centered radicals are presumed to be active site-localized and to be transient species in the hydroxylation reaction.

FREE RADICALS IN THE PEROXYGENATIVE CYCLE OF CYTOCHROME P-450

The ability of some forms of P-450 to utilize alternative oxidants such as hydroperoxides, peracids, iodosobenzenes, and N-oxides as oxidants in place of molecular oxygen and reducing equivalents from NADPH has long been recognized.[1,16,17] This reaction was examined as a possible model for the normal catalytic reaction; however, variations in the products formed in the hydroxylation of substrates and in the stability of the isozyme during catalytic turnover suggest mechanistic differences in the two oxidative reactions. The scheme for the peroxide-supported oxidative reaction[13] is shown at the top of Figure 1 and in greater detail in Figure 2. The decarboxylation of peroxyphenylacetic acid provided evidence for homolytic cleavage of the oxygen-oxygen bond in this reaction.[18] Kinetic studies using hydroperoxides as oxidants further supported the homolytic path for peroxide bond activation and suggested a ternary transition state that involves the peroxy compound coordinated at the heme iron as well as the substrate.[19] Thus, the catalytic events in the peroxygenase reaction of cytochrome P-450 also involve free radical intermediates.

FREE RADICALS IN REDUCTIVE REACTIONS OF CYTOCHROME P-450

As mentioned earlier, cytochrome P-450 may also function as an electron-transfer catalyst in the reduction of various functional groups. Such reactions involve steps 1 and 2 of the cycle shown in Figure 1; molecular oxygen is not involved, but instead the substrate is reduced by ferrous P-450. We have recently shown that cytochrome P-450 catalyzes the reductive cleavage of hydroperoxides by NADPH.[20] This NADPH-dependent reaction is dependent on native cytochrome P-450 and the reductase and is inhibited by CO, indicating that ferrous heme iron is essential for catalysis. Analysis of products from a number of model hydroperoxides revealed that the corresponding alcohol is not formed, but instead products corresponding to fragmentation of the parent hydroperoxide are obtained.[21] For example, cumyl and *t*-butyl hydroperoxides both yield methane and the corresponding ketones, acetophenone and acetone, respectively. From the nature of the products formed in this reductive cleavage, we have proposed a general mechanism involving a stepwise one-electron transfer from NADPH to the hydroperoxide via the reductase and the heme iron of P-450, as outlined below.

$$\text{XRR'C-OOH} + Fe^{2+} \rightarrow [\text{XRR'CO}\cdot] + Fe^{3+} + OH^- \quad (2)$$

$$[\text{XRR'CO}\cdot] \rightarrow \text{XRC=O} + [\text{R'}\cdot] \quad (3)$$

$$H^+ + [\text{R'}\cdot] + Fe^{2+} \rightarrow \text{R'H} + Fe^{3+} \quad (4)$$

$$\text{(Sum)} \quad \text{XRR'C-OOH} + 2e + H^+ \rightarrow \text{R'H} + \text{XRC=O} + OH^- \quad (5)$$

Homolysis of the peroxide bond yields the unstable alkoxy radical, which undergoes β-scission to yield a carbonyl compound and a derived carbon-centered radical. Reduction of this derived radical by the second electron together with proton uptake from solvent yields the hydrocarbon. Two lines of evidence indicate the sequential two-electron reduction of hydroperoxides with the alkoxy radical as an intermediate: (a) the formation of β-scission products (methane, acetophenone, and acetone), a reaction characteristic of alkoxy radicals,[22] and (b) the incorporation of a solvent deuteron into

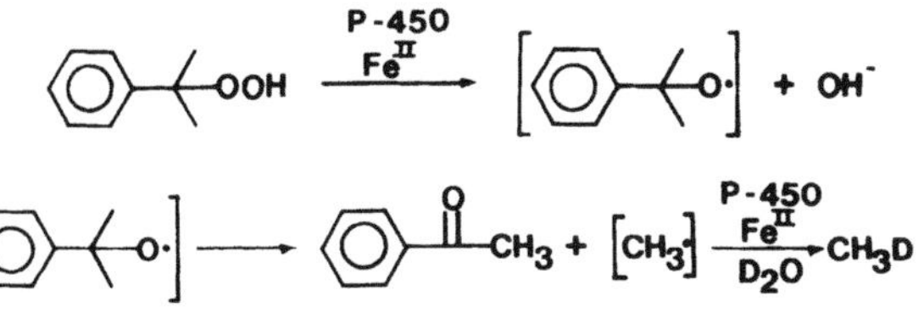

Figure 3. Proposed mechanism of P-450-catalyzed reductive rearrangement of cumyl hydroperoxide that is consistent with the observed products of the reaction.

methane formed from the reaction of cumyl hydroperoxide, which indicates that the derived methyl radical is reduced by one electron prior to proton uptake from solvent.[21]

The scheme in Figure 3 shows in detail the reactions leading to the formation of deuteromethane. This reaction with model hydroperoxides resembles the known physiological process of lipid peroxidation in which lipid hydroperoxides are known to form alkanes and carbonyl compounds. Accordingly, when 13-hydroperoxy-9,11-octadecadienoic acid was examined as a potential substrate, pentane was detected as one of the fragmentation products.[21] More recently we have obtained evidence by correlative HPLC, UV spectroscopy, and reaction with 2,4-dinitrophenylhydrazine that the other fragmentation product is the expected 13-oxo-9,11-tridecadienoic acid.

Six rabbit liver isozymes were examined for their catalytic ability in the reductive reaction with the model hydroperoxides.[23] Alcohol-inducible P-450 form 3a is the most effective in NADPH oxidation, which was used as a measure of catalytic effectiveness in this reaction. However, only P-450 forms 2 and 3a gave good stoichiometry of product formation, indicating different catalytic efficiencies for the cytochromes examined.

SUMMARY

Schemes are presented summarizing current knowledge of the mechanism of action of cytochrome P-450 when it functions either as a monooxygenase with molecular oxygen as the oxygen donor or as a peroxygenase with peroxy compounds as the oxygen donor. In the process, a large variety of physiologically occurring and foreign compounds undergo hydroxylation and oxy and peroxy radicals are generated.

In addition, cytochrome P-450 catalyzes reductive reactions, including a recently discovered reaction in which organic hydroperoxides are cleaved to yield hydrocarbons and aldehydes or ketones. The reaction is believed to involve homolysis of the oxygen-oxygen bond and generation of an alkoxy radical, with β-scission of the latter followed by reduction of the secondary radical to the hydrocarbon. Evidence has been obtained that lipid hydroperoxides are physiological substrates for this reductive cleavage reaction catalyzed by cytochrome P-450.

ACKNOWLEDGEMENTS

This research was supported by Grant DK-10339 from the National Institutes of Health.

REFERENCES

1. R. E. White and M. J. Coon, Oxygen activation by cytochrome P-450, Annu. Rev. Biochem. 49:315-356 (1980).
2. S. D. Black and M. J. Coon, Comparative structures of P-450 cytochromes, in: "Cytochrome P-450," P. R. Ortiz de Montellano, ed., Plenum Press, New York (1986).
3. C. R. Jefcoate, Cytochrome P-450 enzymes in sterol biosynthesis and metabolism, in: "Cytochrome P-450," P.R. Ortiz de Montellano, ed., Plenum Press, New York (1986).
4. A. Y. H. Lu, K. W. Junk, and M. J. Coon, Resolution of the cytochrome P-450-containing ω-hydroxylation system of liver microsomes into three components, J. Biol. Chem. 244:3714-3721 (1969).
5. H. W. Strobel, A. Y. H. Lu, J. Heidema, and M. J. Coon, Phosphatidylcholine requirement in the enzymatic reduction of hemoprotein P-450 and in fatty acid, hydrocarbon, and drug hydroxylation, J. Biol. Chem. 245:4851-4854 (1970).
6. M. J. Coon, A. D. N. Vaz, L. D. Gorsky, and D. Pompon, Alternate pathways for dioxygen and peroxide activation and reduction, in: "Cytochrome P-450, Biochemistry, Biophysics and Induction," L. Vereczkey and K. Magyar, eds., Elsevier, Amsterdam, and Akademiai Kiado, Budapest (1985).
7. M. Sharrock, P. G. Debrunner, C. Schulz, J.D. Lipscomb, V. Marshall, and I. C. Gunsalus, Mössbauer studies of cytochrome $P\text{-}450_{cam}$, Biochemistry 12:258-265 (1973).
8. L. D. Gorsky, D. R. Koop, and M. J. Coon, On the stoichiometry of the oxidase and monooxygenase reactions catalyzed by liver microsomal cytochrome P-450. Products of oxygen reduction, J. Biol. Chem. 259:6812-6817 (1984).
9. B. Bösterling and J. R. Trudell, Spin trap evidence for production of superoxide radical anions by purified NADPH-cytochrome P-450 reductase, Biochem. Biophys. Res. Commun. 98:569-575 (1981).
10. D. D. Oprian, L. D. Gorsky, and M. J. Coon, Properties of the oxygenated form of liver microsomal cytochrome P-450, J. Biol. Chem. 258:8684-8691 (1983).
11. T. L. Poulos, The crystal structure of cytochrome $P\text{-}450_{cam}$, in: "Cytochrome P-450," P. R. Ortiz de Montellano, ed., Plenum Press, New York (1986).
12. A. G. Krainev, L. M. Weiner, I. S. Alferyev, and N. M. Slynko, Bifunctional compound study of the active center location of cytochrome P-450 in a microsomal membrane ('float' molecules method), Biochem. Biophys. Acta 818:96-104 (1985).
13. M. J. Coon and A. D. N. Vaz, Radical intermediates in peroxide-dependent reactions catalyzed by cytochrome P-450, J. Biosci. 11:35-40 (1987).
14. J. T Groves, G. A. McClusky, R. E. White, and M. J. Coon, Aliphatic hydroxylation by highly purified liver microsomal cytochrome P-450. Evidence for a carbon radical intermediate, Biochem. Biophys. Res. Commun. 81:154-160 (1978).
15. O. Augusto, H. S. Beilan, and P. R. Ortiz de Montellano, The catalytic mechanism of cytochrome P-450: Spin trapping evidence for one electron substrate oxidation, J. Biol. Chem. 257:11288-11295 (1982).

16. F. F. Kadlubar, K. C. Morton, and D. M. Ziegler, Microsomal-catalyzed hydroperoxide-dependent C-oxidation of amines, Biochem. Biophys. Res. Commun. 54:1255-1261 (1973).
17. G. D. Nordblom, R. E. White, and M. J. Coon, Studies on hydroperoxide-dependent substrate hydroxylation by purified liver microsomal cytochrome P-450, Arch. Biochem. Biophys. 175:524-533 (1976).
18. R. E. White, S. G. Sligar, and M. J. Coon, Evidence for a homolytic mechanism of peroxide oxygen-oxygen bond cleavage during substrate hydroxylation by cytochrome P-450, J. Biol. Chem. 255:11108-11111 (1980).
19. R. C. Blake, II and M. J. Coon, On the mechanism of action of cytochrome P-450. Evaluation of homolytic and heterolytic mechanisms of oxygen-oxygen bond cleavage during substrate hydroxylation by peroxides, J. Biol. Chem. 256:12127-12133 (1981).
20. A. D. N. Vaz and M. J. Coon, NADPH Oxidation by hydroperoxides: Catalysis by cytochrome P-450 and evidence for homolytic cleavage of the peroxide bond, in: "Cytochrome P-450, Biochemistry, Biophysics and Induction," L. Vereczkey, and K. Magyar, eds., Elsevier, Amsterdam, and Akademiai Kiado, Budapest (1985).
21. A. D. N. Vaz and M. J. Coon, Hydrocarbon formation in the reductive cleavage of hydroperoxides by cytochrome P-450, Proc. Natl. Acad. Sci. USA 84:1172-1176 (1987).
22. P. Neta, M. Dizdaroglu, and M. G. Simic, Radiolytic studies of cumyloxyl radicals in aqueous solutions, Isr. J. Chem. 24:25-28 (1984).
23. A. D. N. Vaz and M. J. Coon, Cytochrome P-450-catalyzed reductive rearrangement of hydroperoxides: Isozyme specificity and mechanistic similarity, Fed. Proc. 46:1956 (1987).

CYTOCHROME P-450 MODEL SYSTEMS: ALKENE OXIDATION

Teddy G. Traylor, Beth E. Dunlap, and Andrew R. Miksztal

Department of Chemistry, D-006
University of California, San Diego
La Jolla, California 92093, U.S.A.

INTRODUCTION

Iron(III) porphyrins efficiently catalyze the epoxidation of alkenes and the hydroxylation of alkanes by iodosylbenzenes and peracids, most likely by the same mechanisms utilized by cytochrome P-450 enzymes. The synthetic catalysts mimic the enzymes in product stereospecificity,[1] the accompanying rearrangements,[2] and the suicide labelling which results in N-alkylhemin formation.[3]

Several mechanisms for epoxidation have been proposed, all of which invoke intermediacy of an iron(IV) porphyrin cation radical oxo species $(Fe^{IV}{=}O)^+$. This intermediate, formed from the iron(III) porphyrin and oxidant, reacts at nearly diffusion-controlled rates with alkenes and is impossible to isolate at ambient temperature. The favored mechanisms are shown in Scheme 1 and Scheme 2. The first involves initial electron transfer from the alkene to form a cation radical caged pair followed by collapse to either a radical or cation species, and the second invokes electrocyclic addition of the alkene to the iron-oxo intermediate to form a metallacyclooxetane. These mechanisms should be distinguishable by product studies.

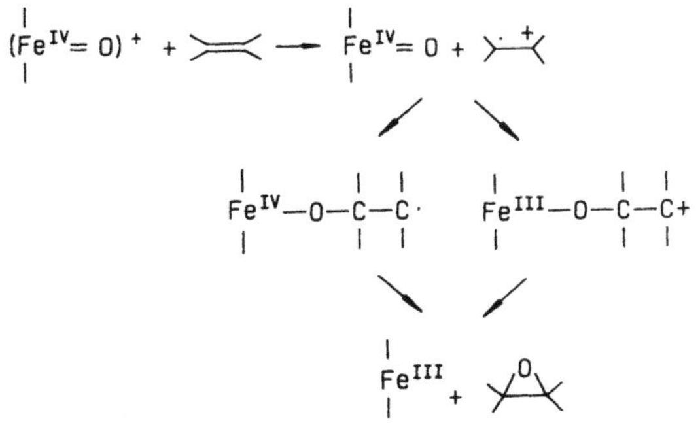

Scheme 1. Proposed electron-transfer mechanism for catalytic alkene oxidation.

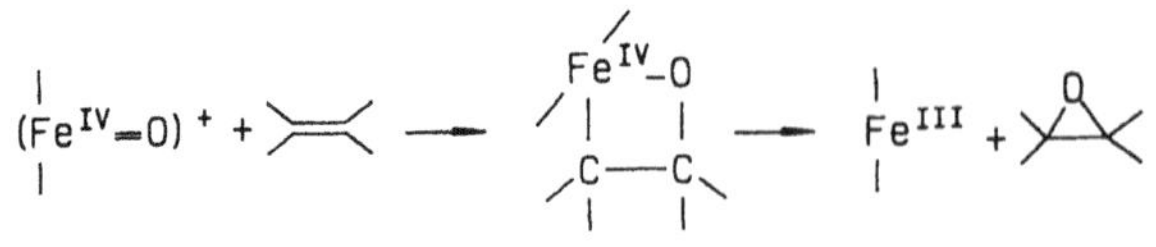

Scheme 2. Proposed electrocyclic addition mechanism for catalytic alkene oxidation.

RESULTS AND DISCUSSION

Catalytic epoxidation of norbornene by iron(III) tetraarylporphyrin chlorides using pentafluoroiodosylbenzene as oxidant yielded substantial amounts of endo-2,3-epoxynorbornane in addition to the usual exo isomer.[4] Formation of endo epoxide is a unique result because electrophilic additions to norbornene occur exclusively on the exo face. Additions to norbornyl radical, however, can be less selective, leading to a mixture of exo and endo adducts. This suggests that collapse to product involves a norbornyl cation radical created by electron transfer. Small amounts of two cation rearrangement products, 2-norbornenone and cyclohexene-4-carboxaldehyde, were also formed, suggesting intermediacy of an alkyl cation. The only mechanism incorporating both sets of results is shown in Scheme 3. Table 1 shows the ratios of exo- to endo-epoxynorbornane produced (in high yields) when 0.32 M equivalent of C_6F_5IO and 1.0 M norbornene in 0.1 ml of CH_2Cl_2 were made 0.001 M in the appropriate catalyst. The decrease in exo/endo product ratio as the catalyst is made more electron deficient most likely results from lower selectivity in the more reactive caged pair.

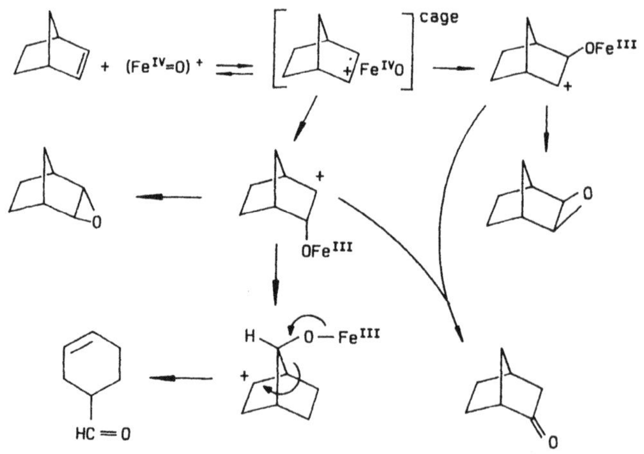

Scheme 3. Catalytic norbornene oxidation pathway.

Table 1. Ratio of Exo- to Endo-epoxynorbornane Obtained from Catalyzed Epoxidation Reactions with Iron(III) Tetraarylporphyrin Chloride

Aryl Group in Catalyst	Ratio of Exo- to Endo-epoxynorbornane
Ph-	35
Mesityl-	23
C_6F_5-	16
2,4-$Cl_2C_6H_3$-	9
C_6Cl_5-	6

Rearrangement products, formed without addition of an oxygen atom, were observed with alkenes known to undergo such rearrangements through intermediacy of a cation radical.[5] The alkenes employed were hexamethyldewarbenzene and diene I, and the catalytic system was iron(III) tetraphenylporphyrin chloride with m-chloroperbenzoic acid as oxidant. These rearrangements accompanied production of unrearranged epoxide, with more polar solvents yielding higher amounts of rearranged products than less polar solvents. This suggests a mechanism involving initial electron transfer to form a caged-pair intermediate. Escape from the cage produces rearranged products, and collapse forms epoxide, as is indicated in Schemes 4 and 5.

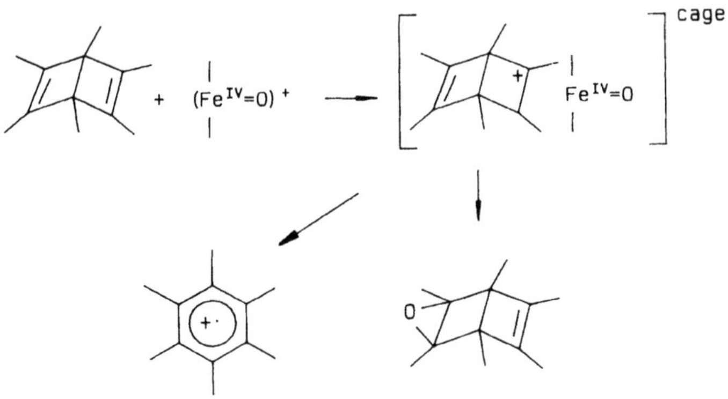

Scheme 4. Catalytic hexamethyldewarbenzene oxidation pathway.

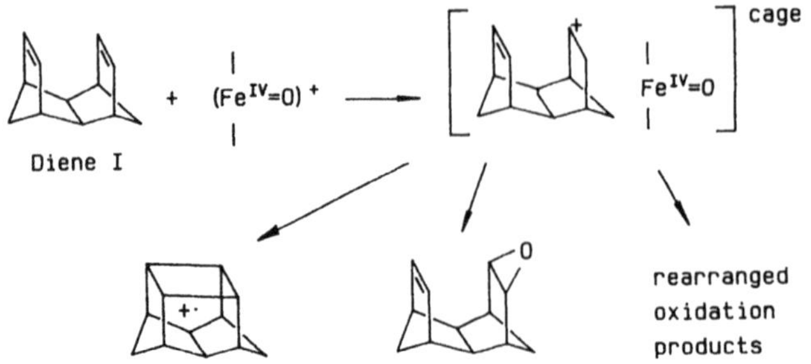

Scheme 5. Catalytic diene I (1,4,4a,5,8,8a-hexahydro-1,4,5,8-endo,endo-dimethanonaphthalene) oxidation pathway.

Although kinetic results appearing to support electrocyclic addition (Scheme 2) have been reported,[6] spectroscopic evidence demonstrates that the species accumulating in those reactions is transiently formed N-alkylhemin, not the suggested metallacyclooxetane.[7] The electron-transfer mechanism (Scheme 1) is consistent with all of the kinetics and product studies published and is the likely mechanism for alkene oxidation by both the model and enzymatic systems.

REFERENCES

1. J. P. Collman, T. Kodadek, and J. I. Brauman, Oxygenation of styrene by cytochrome P-450 model systems: a mechanistic study, J. Am. Chem. Soc. 108:2588 (1986).
2. J. T. Groves and R. S. Myers, Catalytic asymmetric epoxidations with chiral iron porphyrins, J. Am. Chem. Soc. 105:5791 (1983).
3. T. Mashiko, D. Dolphin, T. Nakano, and T. G. Traylor, N-alkylporphyrin formation during the reactions of cytochrome P-450 model systems, J. Am. Chem. Soc. 107:3735 (1985).
4. T. G. Traylor, T. Nakano, B. E. Dunlap, P. S. Traylor, and D. Dolphin, Mechanisms of hemin-catalyzed alkene epoxidation. The effect of catalyst on the regiochemistry of epoxidation, J. Am. Chem. Soc. 108:2782 (1986).
5. T. G. Traylor and A. R. Miksztal, Mechanisms of hemin-catalyzed epoxidations: electron transfer from alkenes, J. Am. Chem. Soc. 109:2770 (1987).
6. J. P. Collman, T. Kodadek, S. A. Raybuck, J. I. Brauman, and L. M. Papazian, Mechanism of oxygen atom transfer from high valent iron to olefins: implications to the biological epoxidation of olefins by cytochrome P-450, J. Am. Chem. Soc. 107:4343 (1985).
7. T. G. Traylor, T. Nakano, A. R. Miksztal, and B. E. Dunlap, Transient formation of N-alkylhemins during hemin-catalyzed epoxidation of norbornene. Evidence concerning the mechanism of epoxidation, J. Am. Chem. Soc. 109:3625 (1987).

CYTOCHROME P-448 INDUCTION RESTORES THE PROPAGATION OF $\cdot O_2^-$-INDUCED LIPID PEROXIDATION IN HEPATOMA MICROSOMES

Tommaso Galeotti and Silvia Borrello

Institute of General Pathology
Catholic University
Rome, Italy

INTRODUCTION

It is well established that intracellular and plasma membranes of cancer cells are resistant to lipid peroxidation and that such resistance is related directly to the growth rate and inversely to the degree of differentiation of the tumor.[1] Lipid peroxidation byproducts are capable of inhibiting the mitotic activity of normal and tumor cells,[2,3] supporting the early hypothesis that the low peroxidizability of tumor membranes may be one of the factors responsible for the increased proliferative activity of malignant cells.[4,5] Consistently, a reasonable approach to the study of lipid peroxidation in tumors is that of establishing the factors which limit the process and possibly finding strategies for restoring it at levels comparable to those of normal cells. The present and previous studies performed in our laboratory are aimed at carrying out this project.

METHODS

In this work we have employed preparations of microsomal membranes from the highly differentiated, slow-growing Morris hepatoma 9618A.[6] Lipid peroxidation was induced by superoxide ($\cdot O_2^-$) radicals, using a xanthine plus xanthine oxidase system, optimized with regard to iron concentration and chelation of iron by ADP.[7] The reaction was measured at 25°C as malondialdehyde (MDA) accumulation during 40 minutes of incubation. Cytochrome P-448 was induced by treatment of the tumor-bearing rats with the carcinogen 3-methylcholanthrene.[8]

RESULTS AND DISCUSSION

It can be seen in Figure 1 that the rate of lipid peroxidation of hepatoma microsomes does not differ from that of rat liver membranes during the first 10 min of incubation. Afterwards, the accumulation of MDA slows down more markedly in hepatoma so that about 60 nmol less end product is formed by tumor microsomes, with respect to control membranes, in the subsequent 30 min of experimental observation. During the first step of microsomal lipid peroxidation ("initiation"), promoted by $\cdot O_2^-$ radicals, lipid hydroperoxides (LOOH) are formed within the membrane, and this process seems to be independent of the availability of cytochrome P-450 which

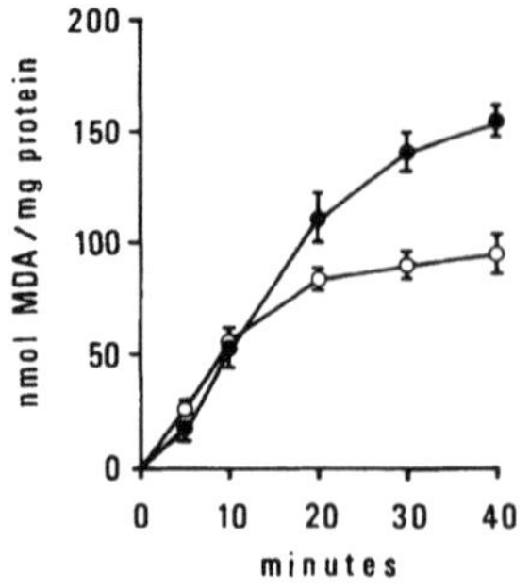

Figure 1. Time course of $\cdot O_2^-$-induced MDA formation in rat liver (●) and Morris hepatoma 9618A (○) microsomes. Microsomes (0.2 mg protein/ml) were incubated at 25°C in an O_2-saturated medium containing 0.15 M KCl, 50 mM Tris-HCl (pH 7.5), 0.33 mM xanthine, 1mM ADP and 50 µM $FeCl_3$. The reaction was started by 50 µg/ml xanthine oxidase. Values are means $\pm$ SE of 4-5 experiments.

is required in the second step ("propagation") for catalyzing the oxidative cleavage of LOOH to lipidoxy radicals (LOO• and LO•).[9] Hepatoma 9618A microsomes have a normal lipid composition, with regard to phospholipid content and degree of fatty acid unsaturation, but lack most of the liver cytochrome P-450 (about 60 to 85% deficiency).[7] Therefore, the long-term inhibition of lipid peroxidation may be reasonably ascribed to such deficiency. This hypothesis is in agreement with previous observations obtained from hydroperoxide (ROOH)-dependent lipid peroxidation experiments carried out with hepatoma 9618A microsomes. In those conditions the cytochrome P-450 deficiency limits the entire process[10] since this hemoprotein is required to catalyze the oxidation of ROOH to initiating radicals (ROO• and RO•) and therefore is involved also in the initiation step of the reaction. Confirmatory evidence that the propagation phase of $\cdot O_2^-$-dependent lipid peroxidation of hepatoma 9618A microsomes is limited by the cytochrome P-450 deficiency and that such inhibition in these tumor microsomes may be overcome is provided by the data presented in Figure 2. These indicate that treatment of the hepatoma-bearing Buffalo rats with the carcinogen 3-methylcholanthrene causes an increment of the tumor cytochrome P-450 content, in the form of cytochrome P-448,[8,10] which is accompanied by a proportional recovery of the lipid peroxidation. Thus induction of 70% of the hemoprotein is already sufficient to re-establish almost normal rates of lipid peroxidation during the propagation phase.

We can conclude that the main rate-limiting factor of lipid peroxidation in microsomes from slow-growing, minimal deviation hepatomas is the cytochrome P-450 availability. The deficiency of this hemoprotein influences, in the $\cdot O_2^-$-dependent system, only the propagation period of lipid peroxidation. Clearly there are means available to restore the process at levels similar to normal. Consequently, if the resistance to membrane lipid peroxidation provides an advantage for cancer cell proliferation,[4,5] such means may become helpful in the control of tumor growth.

ACKNOWLEDGMENTS

The work was supported by a grant of the Italian National Research Council, Special Project "Oncology," contract number 86.00423.44. We thank Mr. Guglielmo Palombini for the skillful technical assistance.

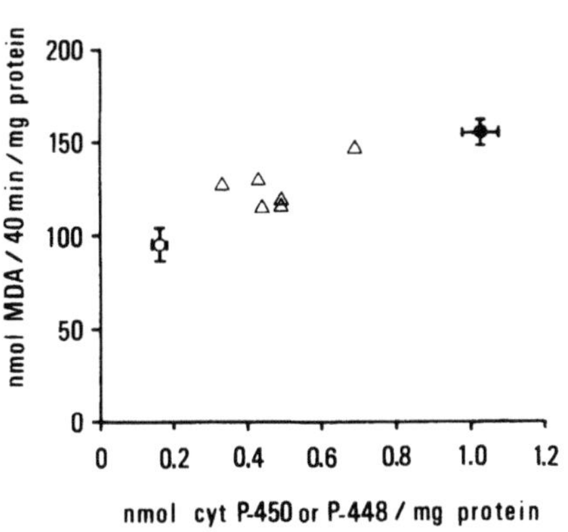

Figure 2. Relationship between MDA accumulation and cytochrome P-450 or P-448 content in microsomes isolated from rat liver (●) and hepatomas 9618A of untreated (○) or 3-methylcholanthrene-treated (△) rats. Cytochrome P-450 and MDA values are means ± SE of 4-5 experiments. The experimental conditions are those of Figure 1.

REFERENCES

1. T. Galeotti, S. Borrello, G. Minotti, and L. Masotti, Membrane alterations in cancer cells: The role of oxy radicals, Ann. N.Y. Acad. Sci. 488:468 (1986)
2. D. G. Cornwell and N. Marisaki, Fatty acids paradoxes in the control of cell proliferation: prostaglandins, lipid peroxides and cooxidation reactions, in: "Free Radicals in Biology," Vol. 6, W. A. Pryor, ed., Academic Press, New York (1984).
3. S. Hauptlorenz, H. Esterbauer, W. Moll, R. Pumpel, E. Schauenstein, and B. Puschendorf, Effects of the lipid peroxidation product 4-hydroxynonenal and related aldehydes on proliferation and viability of cultured Ehrlich Ascites tumor cells, Biochem. Pharmacol. 34:3803 (1985).
4. G. M. Bartoli and T. Galeotti, Growth-related lipid peroxidation in tumour microsomal membranes and mitochondria, Biochim. Biophys. Acta 574:537 (1979).
5. T. Galeotti, G. M. Bartoli, S. Bartoli, and E. Bertoli, Superoxide radicals and lipid peroxidation in tumour microsomal membranes, in: "Biological and Clinical Aspects of Superoxide and Superoxide Dismutase," W. H. Bannister and J. V. Bannister, eds., Elsevier, Amsterdam (1980).
6. H. P. Morris and B. P. Wagner, Induction and transplantation of rat hepatomas with different growth rate (including "minimal deviation" hepatomas), in: "Methods in Cancer Research," Vol. 4, H. Busch, ed., Academic Press, New York (1968).
7. S. Borrello, G. Minotti, G. Palombini, A. Grattagliano, and T. Galeotti, Superoxide-dependent lipid peroxidation and vitamin E content of microsomes from hepatomas with different growth rate, Arch. Biochem. Biophys. 238:588 (1985).
8. T. Ohmachi, I. Sagami, H. Fujii, and M. Watanabe, Microsomal monooxygenase system in Morris hepatoma: Purification and characterization of cytochromes P-450 from Morris hepatoma 5124D of 3-methylcholanthrene-treated rats, Arch. Biochem. Biophys. 236:176 (1985).

9. B. A. Svingen, J. A. Buege, F. O. O'Neal, and S. D. Aust, The mechanism of NADPH-dependent lipid peroxidation: The propagation of lipid peroxidation, J. Biol. Chem. 254:5892 (1979).
10. S. Borrello, T. Galeotti, G. Palombini, and G. Minotti, Restoration of hydroperoxide-dependent lipid peroxidation by 3-methylcholanthrene induction of cytochrome P-448 in hepatoma microsomes, FEBS Lett. 209:305 (1986).

GENERATION OF SUPEROXIDE BY THE MICROSOMAL MIXED-FUNCTION OXIDASE SYSTEM

Lee A. Morehouse and Steven D. Aust

Department of Biochemistry
Michigan State University
East Lansing, Michigan 48824, U.S.A.

INTRODUCTION

The microsomal mixed-function oxidase (MFO) system catalyzes the oxidation of numerous endogenous and exogenous compounds using NADPH as its electron donor. During its oxidation of drug substrates and even in the absence of substrates, this electron transport system has been demonstrated to produce $O_2^{\bar{\cdot}}$, H_2O_2, and excess H_2O. We have investigated the ability of several purified forms of cytochrome P450 to produce $O_2^{\bar{\cdot}}$ when incubated with purified NADPH cytochrome P450 reductase. Addition of cytochrome P450 isozymes to the purified reductase resulted in increased rates of $O_2^{\bar{\cdot}}$ formation, assessed using either EPR spin trapping techniques or acetylated cytochrome c reduction. Conditions necessary for optimal rates of $O_2^{\bar{\cdot}}$ production were similar to those previously shown to be necessary for optimal reconstitution of MFO activity (phospholipid, detergent-solubilized reductase, cytochrome P450:reductase ratios of greater than 3:1). Cytochrome P450d addition to incubations of reductase resulted in greater rates of $O_2^{\bar{\cdot}}$ production than either cytochrome P450c or P450b, especially at equimolar concentrations of cytochrome P450 isozymes and reductase. The contribution of cytochrome P450d to microsomal $O_2^{\bar{\cdot}}$ generation was addressed by investigating $O_2^{\bar{\cdot}}$ production by microsomes isolated from isosafrole-treated rats. Removal of the isosafrole metabolite that binds to cytochrome P450d and prevents dioxygen coordination and MFO activity resulted in a sixfold stimulation of microsomal $O_2^{\bar{\cdot}}$ production and a stoichiometric increase in NADPH oxidation.

METHODS

Male Sprague-Dawley rats (250 - 300 g) were pretreated with phenobarbital, 3,3',4,4',5,5'-hexabromobiphenyl (HBB), or isosafrole (ISF), and liver microsomes were isolated using differential centrifugation. Cytochrome P450 isozymes and reductase were isolated from the microsomes using published procedures.[1-3] Cytochrome c was acetylated by incubating it with a 20-fold excess of acetic anhydride.[4,5] The difference in the rates of acetylated cytochrome c reduction measured in the presence or absence of SOD (SOD-sensitive acetylated cytochrome c reduction) was used as the measure of $O_2^{\bar{\cdot}}$ production. Superoxide production by reconstituted MFO systems was confirmed using EPR spin trapping. A Varian Century-112 EPR spectrometer was used, and 5,5-dimethyl-1-pyrroline-N-oxide (DMPO) was the spin trapping agent.

RESULTS AND DISCUSSION

We had previously shown that the purified reductase generated very little $O_2^{\cdot-}$,[5] confirming the fact that O_2 is an extremely poor electron acceptor for the flavoprotein. However, we found that the addition of purified cytochrome P450 isozymes to the reductase resulted in increased $O_2^{\cdot-}$ production, measured as the SOD-sensitive rate of acetylated cytochrome c reduction. Omission of dilauroylphoshatidylcholine (DLPC), use of the protease-solubilized reductase (that lacks the membrane-binding domain required for its binding to cytochrome P450) instead of the detergent-solubilized enzyme, or heat denaturation of the cytochrome P450 isozyme all resulted in diminution of $O_2^{\cdot-}$ production.

The rate of $O_2^{\cdot-}$ generated by the reconstituted MFO system also depended upon the isozyme of cytochrome P450 added to the system (Figure 1). Cytochrome P450d, the isozyme isolated predominantly in the high spin state from liver microsomes of rats treated with 3-methylcholanthrene-type inducers, generated more $O_2^{\cdot-}$ upon incubation with the purified reductase and DLPC than either cytochrome P450b or cytochrome P450c (the other major cytochrome P450 isozyme induced by 3-methylcholanthrene-type inducers), especially at ratios of cytochrome P450 isozyme:reductase of 1:1 or lower. This result is consistent with the more rapid reduction of the high spin form of cytochrome P450 reported by Tamburini et al.[6] Maximum rates of $O_2^{\cdot-}$ production with any isozyme occurred at ratios of cytochrome P450:reductase greater than 3:1, indicating that the amount of $O_2^{\cdot-}$ generated in the reconstituted system not only was affected by the cytochrome P450 isozyme but also was limited by the concentration of the reductase.

Superoxide production by the reconstituted MFO system was confirmed using EPR spin trapping. As shown in Figure 2, the addition of cytochrome P450b (the major isozyme induced by phenobarbital pretreatment) to purified reductase resulted in a signal resembling that of the DMPO·OOH adduct. The intensity of the adduct signal was dependent upon the concentration of the enzymes (not shown) and substitution of the protease-solubilized reductase resulted in the diminution of the signal. Omission of DLPC reduced the

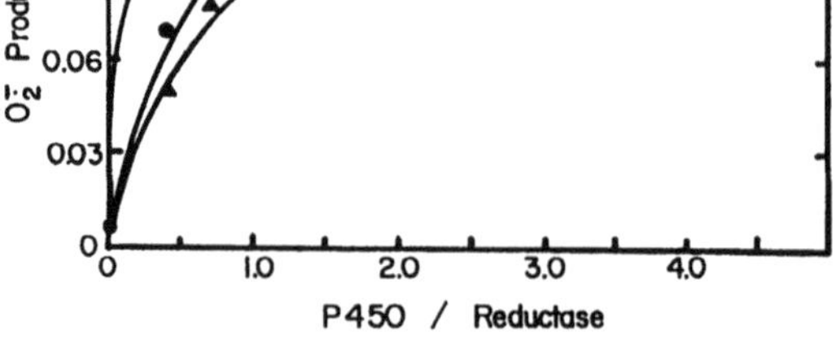

Figure 1. Effect of cytochrome P450 isozymes on superoxide production by the reconstituted mixed-function oxidase system. Incubations contained reductase (0.01 nmol/ml), NADPH (0.1 mM),acetylated cytochrome c (1.6 mg/ml), catalase (500 U/ml), desferrioxamine (0.5 mM), DLPC (20 μg/ml) and varying concentrations of cytochrome P450 isozymes in 50 mM Tris, pH 7.5.

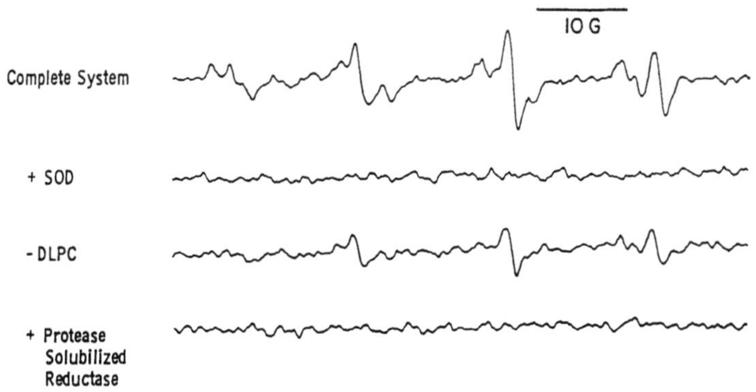

Figure 2. Superoxide production by a reconstituted mixed-function oxidase system containing cytochrome P450b. Incubations contained cytochrome P450b (0.2 nmol/ml), reductase (0.04 nmol/ml), DLPC (30 μg/ml), NADPH (0.1 mM), and DMPO (60 mM). The concentration of SOD was 250 U/ml.

signal intensity by approximately 50%. The addition of SOD also resulted in the loss of a discernable signal, strongly suggesting that the adduct was indeed DMPO·OOH.

Perhaps of greater significance than the tendency of purified isozymes to generate $O_2^{\cdot-}$ in a reconstituted system, however, is their contribution to the generation of $O_2^{\cdot-}$ in intact liver microsomes. Treatment of rats with phenobarbital, HBB, or ISF results in the induction of reductase and/or cytochrome P450 and reductase and a two- to threefold increase in microsomal $O_2^{\cdot-}$ production and concomitant increases in NADPH oxidation (Table 1). However, it is not possible to attribute the increases in $O_2^{\cdot-}$ production to one or even several cytochrome P450 isozymes since the different compounds not only induce different complements of cytochrome P450 isozymes but also differentially induce the reductase. The reductase has been shown to be rate limiting in microsomal MFO reactions and may also limit the amount of microsomal $O_2^{\cdot-}$ generation.

We have used rat liver microsomes isolated from ISF-treated rats to assess the contribution of cytochrome P450d to microsomal $O_2^{\cdot-}$ production. ISF pretreatment of rats results in the induction of cytochrome P450d, but the isolated microsomes contain a metabolite of ISF coordinated to the heme iron of cytochrome P450d[7] that prevents spectral quantitation of

Table 1. Effect of Induction on Microsomal Superoxide Production

Inducer	NADPH Oxidation	Superoxide Production
None	30	3.4
Phenobarbital	40	8.1
HBB	43	9.4
ISF	36	5.2

Incubations contained microsomal protein (50 μg/ml) and NADPH (0.1 mM) in 50 mM Tris pH 7.5. NADPH oxidation was monitored at 340 nm and $O_2^{\cdot-}$ production was monitored using the reduction of acetylated cytochrome c (1.6 mg/ml) at 550 nm in the absence or presence of SOD (250 U/ml). Values are in nmol/min/mg protein.

Table 2. Effect of Butanol Pretreatment on NADPH Oxidation and Superoxide Production by Microsomes from Isosafrole-Treated Rats

	NADPH Oxidation	Superoxide Production	Cytochrome P450 (CO difference)	Total P450
Microsomes	36	5.2	1.45	2.75
+ Butanol	52	31	2.25	2.50

Cytochrome P450 was determined using carbon monoxide difference spectra, and total cytochrome P450 was the sum of that determined with carbon monoxide and that of the difference spectra (reduced vs. oxidized) of the cytochrome P450d-isosafrole metabolite complex having an A_{455}-A_{490} extinction coefficient of 75 mM^{-1}.[8] Rates of NADPH oxidation and $O_2^{\overline{\cdot}}$ production are in nmol/min/mg protein, and cytochrome P450 contents are expressed as nmol/mg protein.

cytochrome P450, oxygen activation by the isozyme, and metabolism of exogenously added drug substrates. The ISF metabolite can be displaced by incubating the microsomes with butanol. Thus, it is possible to assess the contribution of cytochrome P450d (or at least the contribution of cytochrome P450d having metabolite bound to it) to microsomal $O_2^{\overline{\cdot}}$ production by determining the increase of $O_2^{\overline{\cdot}}$ generation upon the displacement of the metabolite (Table 2). The ISF metabolite was removed by incubating the microsomes in butanol (100 mM) and then chromatographing them on Sepharose 2B to remove the metabolite and excess butanol. Butanol treatment of microsomes isolated from ISF-treated rats resulted in a nearly twofold increase in cytochrome P450 detected by CO difference spectra but no conversion to cytochrome P420 and no change in total cytochrome P450 (Table 2). Microsomes freed of the ISF metabolite exhibited a a sixfold increase in $O_2^{\overline{\cdot}}$ production over that generated by microsomes before metabolite displacement. The increase in $O_2^{\overline{\cdot}}$ production was accompanied by a near stoichiometric increase in NADPH oxidation.

CONCLUSIONS

The addition of cytochrome P450 isozymes to a reconstituted MFO system resulted in increased $O_2^{\overline{\cdot}}$ production, assessed using EPR spin trapping or acetylated cytochrome c reduction. Higher rates of $O_2^{\overline{\cdot}}$ production were observed with cytochrome P450d than with either cytochromes P450b or P450c; these differences were accentuated at ratios of cytochrome P450:reductase of approximately 1:1 where cytochrome P450 exhibited rates of $O_2^{\overline{\cdot}}$ production several times those generated by the other isozymes tested. Rates of $O_2^{\overline{\cdot}}$ production plateaued at cytochrome P450:reductase ratios of 3:1, indicating that the reduction of cytochrome P450 is a limiting factor in $O_2^{\overline{\cdot}}$ production, just as it is proposed to be in drug oxidation. The contribution of cytochrome P450d to $O_2^{\overline{\cdot}}$ production by microsomes was assessed using microsomes isolated from ISF-induced rats. Removal of the isosafrole metabolite which is bound to cytochrome P450d resulted in a sixfold increase in $O_2^{\overline{\cdot}}$ production by the microsomes.

REFERENCES

1. A. Astrom and J.W. DePierre, Metabolism of 2-acetylaminofluorene by eight different forms of cytochrome P-450 isolated from rat liver, Carcinogenesis 6:113 (1985).
2. D.J. Waxman and C. Walsh, Phenobarbital-induced rat liver cytochrome P450: Purification and characterization of two closely related isozymic forms, J. Biol. Chem. 257:10446 (1982).

3. Y. Yasukochi and B.S.S. Masters, Some properties of a detergent-solubilized NADPH-cytochrome C (cytochrome P450) reductase purified by biospecific affinity chromatography, J. Biol. Chem. 251:5337 (1976).
4. K. Wada and K. Okunuki, Studies on chemically modified cytochrome c. I. The acetylated cytochrome c, J. Biochem. 64:667 (1968).
5. L.A. Morehouse, C.E. Thomas, and S.D. Aust, Superoxide generation by NADPH-cytochrome P450 reductase: The effect of iron chelators and the role of superoxide in microsomal lipid peroxidation, Arch. Biochem. Biophys. 232:366 (1984).
6. P.P. Tamburini, I. Jansson, L.V. Favreau, W.L. Backes, and J.L. Schenkman, Differences in the spectral interactions between NADPH-cytochrome P450 reductase and a series of cytochrome P450 enzymes, Biochem. Biophys. Res. Comm. 137:437 (1986).
7. M. Dickins, C.R. Elcombe, S.J. Moloney, K.J. Netter, and J.W.Bridges, Further studies on the dissociation of the isosafrole metabolite-cytochrome P450 complex, Biochem. Pharmacol. 28:231 (1979).
8. M. Murray, K. Hetnarski, and C.F. Wilkinson, Selective inhibitory interactions of alkoxymethylene dioxybenzenes towards monooxygenase activity in rat hepatic microsomes, Xenobiotica 15:369 (1985).

MEMBRANE PERMEABILITY AND OXIDANT INDUCED INJURY

Henry Jay Forman, Dianne C. Skelton, George A. Loeb and Raymond J. Dorio

Departments of Pediatrics and Pathology and
Institute of Toxicology
University of Southern California
Childrens Hospital of Los Angeles
Los Angeles, California 90027

INTRODUCTION

This discussion concerns the mechanisms whereby cell functions are compromised by exposure to oxidant stress. One of our principal objectives is to also address the manner in which many of the studies of oxidant stress are performed, that being the use of lactate dehydrogenase (LDH) release as a marker of cell injury and use of millimolar and higher concentrations of peroxides.

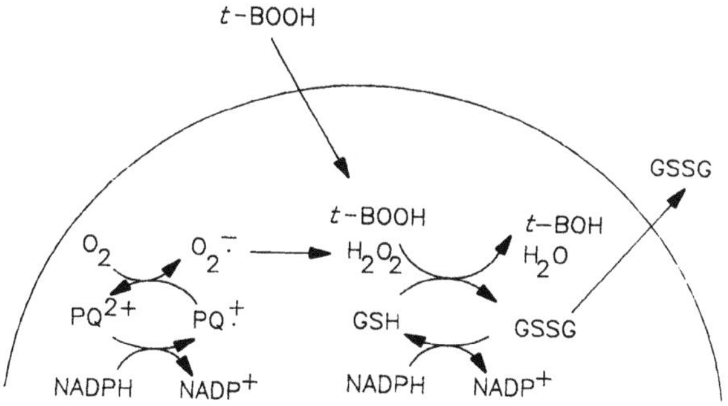

Figure 1. Oxidant stress. Paraquat (PQ) can be reduced within cells by an NADPH-dependent enzyme to a radical, which reacts non-enzymatically with O_2 to form $O_2^{\dot{-}}$. Superoxide can dismute either spontaneously or through the action of superoxide dismutase to H_2O_2. H_2O_2 or added t-butyl hydroperoxide (tBOOH) can be reduced by glutathione (GSH) through the action of the Se-containing enzyme, glutathione peroxidase. GSH is restored through the action of glutathione reductase using NADPH.

Figure 1 shows two ways in which oxidant stress can be imposed upon cells: the intracellular generation of $O_2^{\bar{\cdot}}$ and H_2O_2 by paraquat administration or the introduction of a peroxide such as tertiary butyl hydroperoxide (tBOOH) from the medium. The consequences of such stress can be observed in terms of metabolic alterations in glutathione and NADPH concentrations within the cell. Cells incubated without glucose, which is required to support NADPH production through pentose shunt activity, had increased $NADP^+$ (54% of the total pool versus 27% for plus glucose.[1] As expected, in the absence of glucose, incubation with 1 mM paraquat or 10 μM tBOOH completely oxidized the $NADP^+$ + NADPH pool. When glucose was added after this incubation, the ability to generate NADPH was restored.[1]

Table 1 shows that in the absence of glucose, 10 μM tBOOH caused a very significant increase in the percent oxidized glutathione and in mixed disulfides with a small leak of GSSG into the medium. If glucose was either present or added after incubation with this concentration of the peroxide, these effects were obviated.

On the other hand, dietary deficiencies of selenium, the cofactor for glutathione peroxidase, and sulfur amino acids, precursors of glutathione, have drastic effects upon the survival of rats in hyperoxia. For example, 40% of the rats with either a deficiency in selenium or in cysteine and methionine died within 72 hours of exposure to 80% O_2, a condition which normally does not result in death. When there was a deficiency in both the sulfur amino acids and selenium, 80% of the rats died within 72 hours of exposure to 80% O_2. These dramatic effects correlate with the expected decreases in glutathione peroxidase and glutathione.[2,3] However, correlations do not reveal the underlying biochemical and biophysical mechanisms whereby these alterations in antioxidant defenses result in injury. For that reason we have chosen to investigate the effects of oxidant stress and alteration of the antioxidant system with isolated cells.

Table 1. Effect of 10 μM tBOOH on Glutathione Distribution of Rat Alveolar Macrophages

Incubation	GSH + GSSG	GSSR	$GSSG_{ext}$	%GSSG
	GSH equivalents (nmol/10^6 cells)			
+Glucose -tBOOH	1.02	0.34	0.11	0.57
-Glucose -tBOOH	1.06	0.38	0.10	0.92
-Glucose +tBOOH	0.78*	0.61*	0.18*	7.40*
+Glucose +tBOOH	1.09	0.35	0.12	0.76
-Glucose +tBOOH then 1 min +glucose#	1.14	0.46	0.17*	1.03

Alveolar macrophages were incubated as indicated for 15 min at 37° C. Data for this table is from 1. The glucose concentration was 5 mM. 1 10^6 cells were incubated in 1 ml of Krebs-Ringer phosphate buffer, pH 7.4. GSSR = glutathione conjugates, $GSSG_{ext}$ = GSSG found in medium, %GSSG = % of GSH + GSSG as GSSG. * = $P < 0.05$ versus +glucose -tBOOH. # = incubation was 15 min -glucose +tBOOH followed by 1 min of 5 mM glucose.

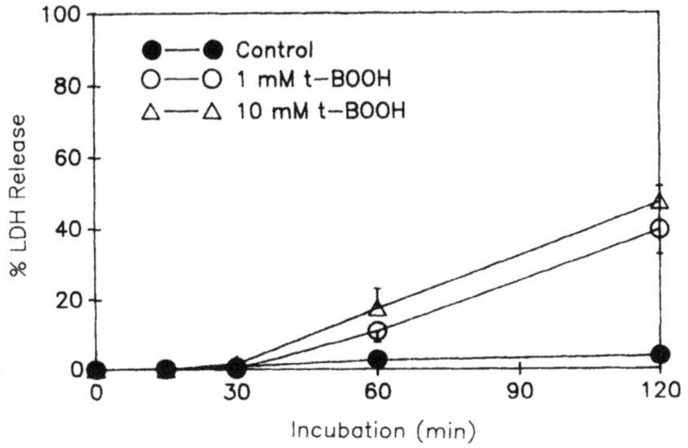

Figure 2. Rat alveolar macrophages were adhered in plastic culture wells and then exposed for the indicated time to indicated concentrations of tBOOH. The release of LDH into the medium was measured at the indicated time of incubation. Note the long delay in LDH release.

In the lung, the alveolar macrophage performs numerous roles including the generation of mediators, such as leukotrienes and prostaglandins, and destruction of bacteria and fungi that invade the lungs. We know that exposure of these cells to hyperoxia either _in vitro_ or _in vivo_ results in the irreversible loss of a membrane functions, such as phagocytosis and stimulated $O_2^{\cdot}$ production[4-6], also called the respiratory burst. This makes the alveolar macrophage not only a convenient but also medically relevant cell to study in terms of oxidant injury. We have shown[7] for example, that prolonged exposure to hyperoxia suppresses the enhancing effect of extracellular Ca^{2+} upon the concanavalin A-stimulated respiratory burst. In control cells concanavalin A stimulation causes an immediate elevation in intracellular Ca^{2+} concentration. The intracellular Ca^{2+} concentration begins to decline very rapidly but is followed by a second transient increase, which lasts for several minutes. This second wave of Ca^{2+} can be blocked by the Ca^{2+}-channel blocker verapamil. Hyperoxic exposure also causes a suppression of this second elevation in intracellular Ca^{2+}. The mechanism appears to involve damage to either the membrane channel responsible for Ca^{2+}-entry or the signal transduction process that results in Ca^{2+}-entry upon concanavalin A stimulation.

MODELS FOR INVESTIGATION OF PEROXIDE TOXICITY

In many studies, LDH release has been used as a measure of cell injury. In order to get results on a time scale of several hours, concentrations of peroxides in excess of 1 mM have commonly been used to get the equivalent of a cell body count (Figure 2). This model can be referred to as "The Thermonuclear Attack Model" and the concentrations of peroxide that have been used as "megatonnage." We admit that this model does provide some useful data but that data must always be considered in terms of possible chemical mechanism rather than of physiologic relevance.

Table 2. Kinetic Parameters for the Glutathione Antioxidant System in Rat Alveolar Macrophages

Enzyme	V_{max} nmol/min / 10^6 cells	Substrate	K_M (μM)	[Steady state] (μM)
GSH Peroxidase	26	tBOOH	54	
		GSH	2018	2045
GSSG Reductase	6.2	NADPH	30	96
		GSSG	27	11
Glucose-6P Dehydrogenase	3.2	$NADP^+$	28	34
		Glucose-6P	134	
Hexokinase	3.5	ATP	1800	2843
		Glucose	1500	

These K_M were measured under conditions in which the alternate substrate was at least 2.5 times the K_M shown for that enzyme. These values are therefore "apparent" K_Ms and should not be considered as reflecting an enzyme mechanism.

Let us explore why we consider this model to be inappropriate to the study of antioxidant defenses. The kinetics of the antioxidant enzymes in the alveolar macrophage indicate that the K_M for the substrates are fairly close to the steady-state concentrations that have been measured in unchallenged cells[1, 8] (Table 2). While we do not know all of the kinetic mechanisms nor the concentration of endogenous inhibitors, it appears that these cells are prepared for antioxidant action. Perhaps this is not surprising for a cell that produces H_2O_2 as part of its physiologic role. However, when 1 x 10^6 cells are exposed to 1 ml of 1 mM tBOOH, it would take hours for the cell to sufficiently lower the tBOOH concentration and restore the steady state of glutathione and NADPH. Under such an overwhelming stress, tests of the effectiveness of any component of the antioxidant system would be pointless.

What happens to alveolar macrophages when lower concentrations that do not produce LDH release are used? Measuring the intrinsic fluorescence in these cells (Figure 3) we can observe that, in the absence of glucose, a concentration of as little as 10 μM tBOOH significantly reduced the level of NADPH in these cells. Addition of glucose overcame the effects of this tBOOH concentration. On the other hand, 100 μM tBOOH decreased NADPH even in the presence of glucose. The apparent mechanism for the decrease in NADPH with 100 μM tBOOH, which is approximately 2 times the K_M for glutathione peroxidase, was that the velocity of glutathione peroxidase under these conditions was fast enough in comparison with the other enzymes to create a drain on the reducing capacity of the cell. Interestingly, there was only a negligible release of LDH caused by 100 μM tBOOH.

ALTERATION OF MEMBRANE FUNCTIONS IN PEROXIDE INJURY

The focus of the rest of this discussion is on the relationship of peroxide injury to loss of cell membrane function. The respiratory burst

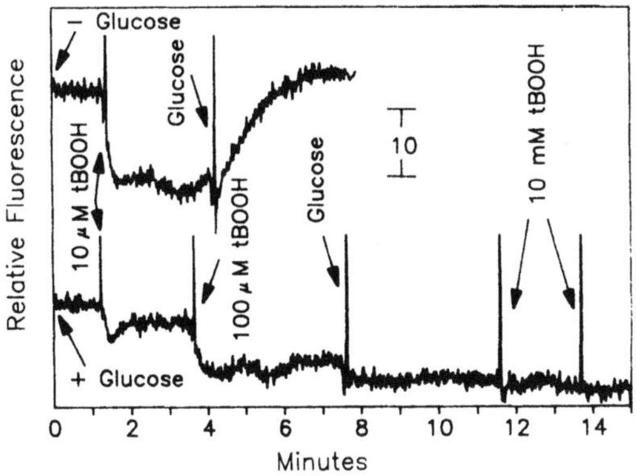

Figure 3. Fluorescence of alveolar macrophages measured with 340 nm excitation and 500 nm emission. The upper trace was for cells without glucose initially. The values are relative with a baseline level of 80. $DiOC_5(3)$ is added to the cuvette first, followed by cells. tBOOH was added after fluorescence reached a steady state. In the absence of tBOOH the fluorescence was constant for at least 30 min. (Adapted from 9).

is an integrated function of alveolar macrophages. We have hypothesized that at least part of the loss of this activity observed in non-lethal oxidant injury could be due to effects on signal transduction. In other recent studies[10] we have observed that an elevation of intracellular Ca^{2+} appears to be involved both in stimulation of $O_2^{\cdot}$ production and in the message the cell requires to end this process. Elevation in intracellular Ca^{2+} has also been implicated by several groups of investigators as crucial in the mechanism of bleb formation and LDH release in hepatocytes and other cells.[11-13] When alveolar macrophages were exposed to a 15 min. incubation with 100 μM tBOOH, which was then removed by washing the cells, subsequent $O_2^{\cdot}$ stimulation by phorbol myristate acetate was inhibited 51%. As described above, 10 μM tBOOH did not inhibit the respiratory burst.

When 100 μM tBOOH was added to alveolar macrophages containing the Ca^{2+}-indicator dye, fura-2, there was a rapid elevation of fluorescence, which is consistent with an increase in intracellular Ca^{2+} concentration (Figure 4). This was not observed with 10 μM tBOOH. When Ca^{2+} was omitted from the external medium, the elevation in intracellular Ca^{2+} concentration occurred as it did in its presence indicating that the source of the increase in intracellular Ca^{2+} concentration must have been an intracellular store. We have scanned the excitation wavelength before and after these apparent changes have occurred to verify that the change is consistent with increased binding of Ca^{2+} to the dye.[9] The sustained change in intracellular Ca^{2+} concentration that occurs with tBOOH is quite different from that observed in the signal transduction involved with concanavalin A stimulation of the respiratory burst. In normal signal transduction the changes in intracellular Ca^{2+} were transient and involved both intracellular and extracellular sources of Ca^{2+} while in

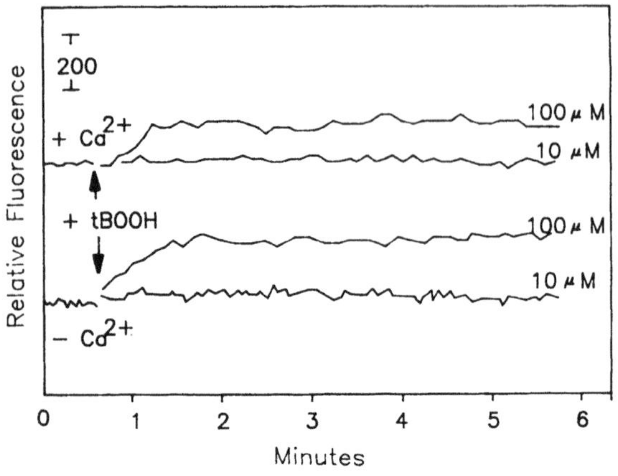

Figure 4. Effect of tBOOH upon intracellular free Ca^{2+} concentration in the presence and absence of extracellular Ca^{2+} (1.3 mM). Cells were incubated with a fura-2 ester, washed and then exposed to the indicated concentration of tBOOH. The fluorescence shown was in relative values but represents an approximate 3 fold increase in intracellular free Ca^{2+} concentration. (Adapted from 17).

tBOOH toxicity only intracellular sources of Ca^{2+} were involved in the rapid responses observed here.

Exposure of alveolar macrophages to tBOOH resulted in a rapid depolarization of the membrane potential as measured by fluorescence of $DiOC_5(3)$. The rate of depolarization was dependent upon the concentration of tBOOH. No change in potential was observed with 10 mM tBOOH in 30 min of incubation, while with 100 µM tBOOH the potential began to slowly decline after approximately 8 min and reached a new steady state after approximately 30 min. When extracellular K^+ was raised from 4 to 50 mM after the new steady state was reached, a second decline in potential was observed, indicating that two separate membrane potentials could be observed (Figure 5). When sucrose replaced all of the ions except for 4 mM KCl in the medium, tBOOH produced a marked decrease in fluorescence that was somewhat smaller in magnitude but at least as rapid as was observed in Na^+ containing buffer. Thus the major source of tBOOH-induced depolarization could not have been Na^+ influx.[14] Mitochondria have a membrane potential of approximately 150 millivolts while the plasma membrane potential of the alveolar macrophage has a potential of 22 millivolts as measured with a microelectrode.[15] Scott and coworkers[16] recently showed that H_2O_2 induced a change in membrane potential and that hypertonic media partially protects against swelling of renal epithelial cells. They proposed that the observed depolarization is due to plasma membrane depolarization preceded by a transient hyperpolarization. Our observations led us to a different conclusion about which membrane was responsible for the major effect; however, the possibility exists that an increase in plasma membrane permeability for either Na^+ or K^+ contributed to depolarization to a minor extent.

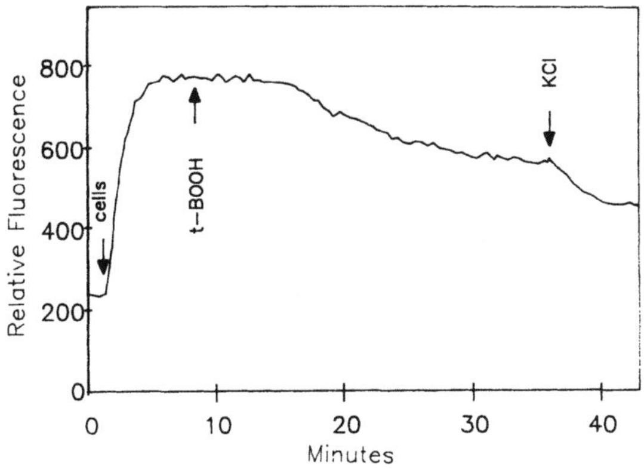

Figure 5. Effect of tBOOH on membrane depolarization of rat alveolar macrophages measured by the fluorescence of $DiOC_5(3)$. Dye was added to the cuvette followed by cells, tBOOH, and finally KCl as described in the text. The fluorescence was approximately 100 times the autofluorescence of cells alone. (Adapted from 9).

We can summarize a proposed sequence of events that occurred in non-lethal but function-altering peroxide-induced injury as follows: The first event was a rapid rise in intracellular Ca^{2+} concentration from an internal source. Beginning shortly thereafter, the mitochondrial membrane depolarized. The elevation in intracellular Ca^{2+} concentration reached an apparent steady state much earlier than did depolarization. The sustained elevation in Ca^{2+} concentration does not resemble the transient changes observed when cells are stimulated. Respiratory burst activity declined during this same period. Both the early elevation of intracellular Ca^{2+} concentration and the timing and concentration dependence of the effect of t-butyl hydroperoxide upon depolarization allow for cause-and-effect relationships for the decline in respiratory burst activity although an exact connection remains to be demonstrated.

We feel that this model potentially represents a more physiologically relevant model for the investigation of oxidant induced damage to cell function than do models involving LDH release and concentrations of oxidants far beyond the capacity of the antioxidant system.

ACKNOWLEDGEMENTS

This work was supported by NIH grant HL37556.

REFERENCES

1. M. W. Sutherland, J. Nelson, G. Harrison, and H. J. Forman, Effects of t-butyl hydroperoxide on NADPH, glutathione, and respiratory burst of rat alveolar macrophages. Arch. Biochem. Biophys. 243:325 (1985).

2. H. J. Forman, E. I. Rotman, and A. B. Fisher, The roles of selenium and sulfur-containing amino acids in protection against oxygen toxicity. Lab. Invest. 49:148 (1983).
3. S. M. Deneke and B. L. Fanburg, Involvement of glutathione enzymes in O_2 tolerance development by diethyldithiocarbamate. Biochem. Pharmacol. 29:1367 (1979).
4. N. Suttorp and M. L. Simon, Decreased bactericidal function and impaired respiratory burst in lung macrophages after sustained in vitro hyperoxia. Am. Rev. Resp. Dis. 128:486 (1983).
5. M. Rister and R. L. Baehner, Effect of hyperoxia on superoxide anion and hydrogen peroxide production of polymorphonuclear leukocytes and alveolar macrophages. Br. J. Haematol. 36:241 (1977).
6. H. J. Forman, J. J. Williams, J. Nelson, R. P. Daniele, and A. B. Fisher, Hyperoxia inhibits stimulated superoxide release by rat alveolar macrophages. J. Appl. Physiol. 53:685 (1982).
7. H. J. Forman, J. Nelson, and G. Harrison, Hyperoxia alters the effect of Ca^{2+} on rat alveolar macrophage superoxide production. J. Appl. Physiol. 60:1300 (1986).
8. M. W. Sutherland, M. Glass, J. Nelson, Y. Lyen, and H. J. Forman, Oxygen toxicity: loss of lung macrophage function without metabolite depletion. J. Free Rad. Biol. Med. 1:209 (1985).
9. H. J. Forman, R. J. Dorio, and D. C. Skelton, Hydroperoxide induced damage to alveolar macrophage function and membrane integrity: alterations in intracellular free Ca^{2+} and membrane potential. Arch. Biochem. Biophys. (in press), 1987.
10. R. J. Dorio, J. Nelson, and H. J. Forman, A dual role for calcium in regulation of superoxide generation by stimulated rat alveolar macrophages. Biochim. Biophys. Acta 928:137 (1987).
11. P. A. Hyslop, D. B. Hinshaw, I. U. Schraufstätter, L. A. Sklar, R. G. Spragg, and C. G. Cochrane, Intracellular calcium homeostasis during hydrogen peroxide injury to cultured $P388D_1$ cells. J. Cell. Physiol. 129:356 (1986).
12. G. F. Rush, J. R. Gorski, M. G. Ripple, J. Sowinski, P. Bugelski, and W. R. Hewitt, Organic hydroperoxide-induced lipid peroxidation and cell death in isolated hepatocytes. Toxicol. Appl. Pharmacol. 78:473 (1983).
13. G. Bellomo, S. Jewell, H. Thor, and S. Orrenius, Regulation of intracellular calcium compartmentation: Studies with isolated hepatocytes and t-butyl hydroperoxide. Proc. Natl. Acad. Sci. USA 79:6842 (1982).
14. P. Mitchell, Chemiosmotic coupling in oxidative and photosynthetic phosphorylation. Biol. Rev. Camb. Phil. Soc. 41:445-502 (1966).
15. A. R. Cameron, J. Nelson, and H. J. Forman, Depolarization and increased conductance precede superoxide release by concanavalin A stimulated rat alveolar macrophages. Proc. Natl. Acad. Sci. USA 80:3726 (1983).
16. J. A. Scott, B.-A. Khaw, C. J. Homcy, and C. A. Rabito, Oxygen radicals alter the cell membrane potential in a renal cell line (LLC-PK1) with differentiated characteristics of proximal tubular cells. Biochim. Biophys. Acta 897:25 (1987).

THE MEASUREMENT OF PROTEIN DEGRADATION IN RESPONSE TO OXIDATIVE STRESS

Robert E. Pacifici, Sharon W. Lin, and Kelvin J. A. Davies

Institute for Toxicology and Department of Biochemistry
University of Southern California
1985 Zonal Avenue, HSC-PSC 614-616
Los Angeles, California 90033, USA

INTRODUCTION

Many elegant techniques have been developed to measure free radical/ oxidant production and toxicity. Such studies can now be conducted with model systems; in cell lysates, extracts, and organelles; and (in some cases) in intact cells, organs, or organisms. One way to find out if a radical/oxidant has altered a cell is to isolate damaged cellular constituents. The most popular index of free radical damage thus far has been the measurement of lipid peroxidation. The effects of radicals on lipids are clearly important; however, mounting evidence suggests that DNA, RNA, carbohydrates, and proteins are also suceptible to oxidative damage. This paper will focus on the measurement of protein degradation following oxidative stress. Recent studies in this laboratory[1-10] and by other groups[11-14] have demonstrated that proteins which have been damaged by radicals or oxidants become excellent substrates for intracellular proteolytic systems and are rapidly degraded.

This laboratory has extensively investigated the effects of oxygen radicals on the primary, secondary, and tertiary structures of various proteins.[2,3,6-10] Invariably, if an oxygen radical was able to alter or abolish protein secondary or tertiary structure (i.e., to cause denaturation) the protein became more susceptible to degradation by a variety of intracellular and purified proteases. These studies support the idea that oxidatively damaged proteins may be selectively recognized and degraded. Primary antioxidant defenses (such as superoxide dismutase, glutathione peroxidase, vitamin E, uric acid, etc.) which react directly with oxidants are ulikeley to be 100% efficient, especially during an episode of oxidative stress. This would allow oxidants to interact with, and damage, vital cellular constituents. Oxidatively damaged proteins have been shown to aggregate, cross link, and fragment to give rise to a host of potentially cytotoxic entities.[1,6-10,13,14] Proteolytic systems which remove such oxidation products, and regenerate useful free amino acids, could therefore be considered to form part of a cell's "secondary antioxidant defenses."[3]

In performing these studies several new methods, as well as modifications of well established techniques, were employed. In each case, protein degradation is quantified by release of free amino acids or small (<5,000 m.w.) peptides. Each method has specific advantages and drawbacks; it is therefore recomended that the investigator scrutinize each of them with his/her question in mind to see which system is most suitable. The methods below have been arranged in three major groupings for ease of presentation; intact cells, cell-free extracts, and purified proteases.

MEASUREMENT OF PROTEIN DEGRADATION IN INTACT CELLS

Erythrocytes and Reticulocytes

Red blood cells (RBC) are easily obtained from humans, rabbits, and cows, and both erythrocytes and reticulocytes may be used for studies of protein degradation.[1-4,10] As reticulocytes mature into erythrocytes, the ability to synthesize proteins is lost. Certain proteolytic activities (e.g. ATP-stimulated proteolysis) decline with red cell age, and this can be used as an interesting experimental variable in studies of oxidative stress.

If erythrocytes are to be studied, whole blood can be collected from humans, rabbits, or cows. Alternatively, reticulocytosis may be induced in rabbits by daily bleedings. The blood is diluted 1:1 with ice-cold 0.9% (w/v) saline containing 10 U/ml heparin, and maintained through the entire isolation procedure at 4°C. RBC (erythrocytes or reticulocytes) are isolated from whole blood by conventional successive centrifugal steps. The washed cells are resuspended to a 10% (v/v) suspension for pre-incubation. At 4°C, RBC membrane permeability is sufficiently low so that intracellular amino acid pools increase significantly during the centrifugal washes. A 20 min pre-incubation at 37°C followed by a centrifugual wash at room temperature is sufficient to deplete amino acid pools. This will prevent a spurious burst of amino acid release during the actual experiment. Red cells are finally resuspended to 5% (v/v) solutions in Krebs-Ringer phosphate buffer (pH 7.4) ± 5 mM glucose, ± 3 mM calcium chloride. Experiments are performed in a shaking water bath at 37°C. Typically, 3 ml of RBC suspension (including all other additions) may be studied in 16mm tubes. Smaller tubes may not allow sufficient exchange of oxygen for radical/oxidant studies.

RBC are unable to synthesize several amino acids whose appearance is, therefore, indicative of protein degradation. The most commonly used amino acid for such studies is alanine, which has the advantage of being relatively resistant to oxidation (as opposed to the aromatic ring of tyrosine for example), and is easily assayed. Alanine formation is measured in perchloric acid extracts of red cells. Perchloric acid (0.4 M final) is added to lyse the cells and precipitate membranes and intact proteins. Next the samples are centrifuged, and aliquots of supernatant are taken to pH 9.0 by addition of KOH and maintained at this pH by the addition of Tris-HCl. Alanine is measured by the reduction of NAD^+ to NADH which occurs during a 1 hr incubation (37°C) with added alanine dehydrogenase, NAD^+, and hydrazine hydrate. NADH is detected fluorometrically at 340 nm excitation and 450 nm emission, in comparison with standards. The assay gives a linear response for alanine contents up to 25 nmol, where 1 mol of NADH is equivalent to 1 mol of alanine.

Two other types of assays may be conducted with the perchlorate extracts: fluorescamine reactivity and complete amino acid analysis. Fluorescamine reacts readily with primary amino groups to give a fluor. Intact proteins have relatively few primary amino groups (N-termini and ε amino groups of lysine). As degradation proceeds, however, more primary amino groups are exposed as amino acids are released. Increased fluorescamine reactivity is, therefore, indicative of protein degradation. Complete amino acid analysis is perhaps the most accurate method of quantitating protein degradation. Since such analyses require expensive equipment and lengthy procedures, however, they may best be reserved for specific problems or confirmatory experiments.

Another technique which has recently been developed takes advantage of the red cell's incredible flexibility. It involves osmotically swelling RBC and "loading in" a potential protein substrate. A radiolabeled protein (± oxidative treatment) is able to enter the cytosol of the red cell via pores in the membrane induced by osmotic swelling. The red cells are then resealed and rejuvenated, and the fate of the exogenous protein(s) can be followed by the formation of trichloroacetic acid-soluble counts. Only small oligopeptides and free amino acids are soluble in this organic acid, and increases in acid-soluble counts are therefore indicative of protein degradation.

Reticulocytes

Reticulocytes can be obtained from rabbits in high yeild by inducing reticulocytosis. Reticulocytosis can be induced by daily bleedings from an ear vein, or by repeated phenylhydrazine injections. The bleeding protocol is preferable for free radical studies, because phenylhydrazine is a red cell oxidant which may produce unwanted effects.

Reticulocyte proteins can be labeled by incubating the cells in a medium which contains a [^{3}H], [^{14}C], or [^{35}S] amino acid. The labeled amino acid is incorporated into newly-synthesized proteins. Following this labeling pulse, the cells are centrifuged, and an excess of "cold" amino acid is added. This pulse-chase technique ensures that unincorporated radiolabeled amino acids are removed. Should the protein be degraded, the "cold chase" ensures that reincorporation of radiolabeled amino acids will be minimal. The labeled reticulocytes are incubated ± oxidants, and protein degradation is measured with high sensitivity by the release of acid-soluble counts.[3,10]

Bacteria

Intact bacteria can be studied by essentially the same labeling procedures as those described (above) for reticulocytes.[3,6,10] Procaryots offer several advantages for studies of protein degradation. Bacteria are readily maintained in culture and divide rapidly. This allows rapid labeling of proteins, with high specific activity. Bacterial studies also provide the versatility of allowing for selection of mutant strains with desirable phenotypes. As with reticulocytes, bacteria are pulse-chased with a radioactive amino acid. The labeled cells are then incubated in the presence or absence of radicals/oxidants. Protein degradation is measured by the release of acid-soluble counts. Interesting comparisons can also be made between bacteria and mammalian mitochondria.[2,3,10] Mitochondria synthesize only a few polypeptides, but can be studied by essentially the same methods as those used with bacteria.

Skeletal Muscle

Intact skeletal muscles from small animals (e.g. rats, mice, rabbits) can be surgically removed and incubated in small flasks or scintillation vials containing a suitable buffer. The muscles (soleus, gastrocnemius, extensor digitorum longus, etc.) should be incubated at 37°C in an O_2/CO_2 mixture, to prevent acidosis via the bicarbonate buffer system. Radical generating systems or oxidants are added to the incubation buffer, and protein degradation is assessed by release of free amino acids into the medium. Alanine is unsuitable (due to metabolic production) but tyrosine has been successfully studied.[2,3,10]

MEASUREMENT OF PROTEIN DEGRADATION IN CELL LYSATES AND CELL-FREE EXTRACTS

Results obtained from in vivo experiments are often obscured by the complexities of cellular metabolism. In an attempt to simplify the understanding of cellular processes, the cell can be fractionated into individual compartments. Cell lysates can be made which retain the cytosolic proteolytic systems. These lysates can be partially purified by dialysis, to produce cell-free extracts. Radicals or oxidants can be added directly to the lysates or extracts for studies of protein degradation. Alternatively, pre-exposed, labeled proteins may be added to the lysates or extracts and studied. Lysates and cell-free extracts may be prepared from erythrocytes and reticulocytes[2,3,4-10] or from bacteria.[2,3,6,10]

Red blood cell lysates are prepared by hypotonic disruption. The preparation of bacterial lysates involves freezing and thawing, followed by lysozyme treatment and incubation in β-mercaptoethanol and KCl. In all cases, cell membranes and other debris are removed by centrifugation. The supernatant (lysate) may then be extensively dialyzed to produce cell-free extracts. Extracts dialyzed against a HEPES-glycerol-DTT buffer can be stored at -80°C for several months without significant loss of proteolytic activity. Lysates and cell-free extracts can be used for several different types of studies. The degradation of endogenous proteins can be measured by alanine formation or by the release of trichloroacetic acid-soluble radioactivity in extracts derived from pre-labeled reticulocytes or bacteria.

Cell-free extracts are particularly useful for studying the degradation of exogenous proteins. Any protein with either a radioactive or fluorometric label can be added to the extracts directly or after exposure to radicals or oxidants. A potential source of error is the release of acid-soluble counts, or fluorometric label, as a result of direct protein fragmentation by radicals/oxidants.[3,6-10,13,14] Fragmentation, particularly when oxygen is present, should be assessed prior to proteolysis assays. Small fragmentation products can usually be removed by dialysis or may themselves be interesting substrates for peptidases. Regardless of whether the protein is endogenous or exogenous, the effects of antioxidants and protease inhibitors or activators can be studied.

MEASUREMENT OF PROTEIN DEGRADATION BY ISOLATED PROTEASES

The proteolytic susceptibility of oxidatively denatured proteins can also be tested with commercially available proteases or with purified or semi-purified proteases from RBC and bacteria.[3,6,9,10] Protein substrates are first radiolabeled by reductive methylation, or a fluorometric label is used. Proteins are then exposed to radicals or oxidants as desired. Protein degradation may be measured by release of acid-soluble counts or fluorometric label after incubation with a given protease. Since both the substrate and protease will be in purified (or semi-purified) form, the overall protein concentration in such experiments will be quite low. Trichloroacetic acid does not precipitate proteins efficiently under these conditons and, to ensure reliable results, an unlabeled carrier protein (e.g., bovine serum albumin) should be added immediatly prior to the acidification.

SUMMARY

This paper briefly outlines several different systems in which radical/oxidant effects on protein degradation can be studied. Erythrocytes, reticulocytes, skeletal muscles, bacteria, mitochondria, cell lysates, cell-free extracts, and isolated proteases are all useful systems, each of which has different advantages and limitations. It is hoped that dissemination of these techniques and approaches will stimulate further study of protein degradation, both as a result and as an index of oxidative stress.

ACKNOWLEDGEMENTS

This work was supported by grant #ES 03598 and by grant #ES 03785 from the National Institutes of Health. Further support was provided by the Faculty Research and Innovation Award of the University of Southern California.

REFERENCES

1. K.J.A. Davies, Free radicals and protein degradation in human red blood cells, in:"Cellular and Metabolic Aspects of Aging: The Red Cell as a Model," J.W. Eaton, D. K. Konzen, and J.G. White, eds., Alan R. Liss, New York (1985).
2. K.J.A. Davies, The role of intracellular proteolytic systems in antioxidant defenses, in: "Superoxide and Superoxide Dismutase in Chemistry, Biology and Medicine," G. Rotilio, ed., Elsevier, Amsterdam (1986).
3. K.J.A. Davies, Intracellular proteolytic systems may function as secondary antioxidant defenses: An hypothesis, J. Free Radicals Biol. Med. 2:155-173 (1986).
4. K.J.A. Davies and A.L. Goldberg, Oxygen radicals stimulate intracellular proteolysis and lipid peroxidation by independent mechanisms in erythrocytes, J. Biol. Chem. 262:8220-8226 (1987).
5. K.J.A. Davies and A.L. Goldberg, Proteins damaged by oxygen radicals are rapidly degraded in extracts of red blood cells, J. Biol. Chem. 262:8227-8234 (1987).
6. K.J.A. Davies, Protein damage and degradation by oxygen radicals: I. General aspects. J. Biol. Chem. 262: 9895-9901 (1987).
7. K.J.A. Davies, M.E. Delsignore, and S.W. Lin, Protein damage and degradation by oxygen radicals: II. Modification of amino acids, J. Biol. Chem. 262:9902-9907 (1987).
8. K.J.A. Davies and M.E. Delsignore, Protein damage and degradation by oxygen radicals: III. Modification of secondary and tertiary structure. J. Biol. Chem. 262:9908-9913 (1987).
9. K.J.A. Davies, S.W. Lin, and R.E. Pacifici, Protein damage and degradation by oxygen radicals: IV. Degradation of denatured protein. J. Biol. Chem. 262:9914-9920 (1987).
10. K.J.A. Davies, Proteolytic systems as secondary antioxidant defenses, in: "Cellular Antioxidant Defense Mechanisms," C.K. Chow, ed., CRC Press, Boca Raton (1988).
11. R.L. Levine, C.N. Oliver, R.M. Fulks, and E.R. Stadtman, Turnover of bacterial glutamine synthetase: Oxidative inactivation precedes proteolysis. Proc. Natl. Acad. Sci. U.S.A. 78:2120-2124 (1987).
12. E.R. Stadtman, Oxidation of proteins by mixed-function oxidation systems: Implication in protein turnover, ageing, and neutrophil function, Trends Biochem. Sci. 11:11-12 (1986).
13. S.P. Wolff, and R.T. Dean, Fragmentation of proteins by free radicals and its effect on their susceptibility to enzymic hydrolysis. Biochem. J. 234:399-403 (1986).
14. S.P. Wolff, A. Garner, and R.T. Dean, Free radicals, lipids, and protein degradation. Trends Biochem. Sci. 11:27-31 (1986).

OXIDATIVE MODIFICATION OF ENZYMES DURING AGING AND ACUTE OXIDATIVE STRESS

Pamela E. Starke-Reed and Cynthia N. Oliver

Laboratory of Biochemistry
National Heart, Lung, and Blood Institute
Bethesda, Maryland 20892, U.S.A.

INTRODUCTION

Previous studies have demonstrated that many cellular enzymes accumulate as catalytically inactive or less active forms during normal aging.[1,2] The mechanism(s) for these changes has (have) not been determined. However, because many of the enzymes which exhibit loss of enzymatic activity during aging are also inactivated by mixed-function oxidation (MFO) systems *in vitro*,[3,4] we have considered the possibility that these changes may be due to oxidative inactivation by MFO systems *in vivo*. In the studies reported here we have examined the levels of oxidatively modified proteins in hepatocytes isolated from young and old rats and from young rats subjected to treatment with 100% oxygen.

METHODS AND RESULTS

MFO-mediated oxidative inactivation of proteins leads to the generation of protein carbonyl derivatives which can be detected by reactivity with 2,4-dinitrophenylhydrazine (DNPH), resulting in the formation of a stable protein hydrazone derivative.[5] Table 1 shows the levels of protein carbonyl groups in the soluble protein fraction of the hepatocytes isolated from rats exposed to 100% oxygen (Experiment 1) or from rats of various ages (Experiment 2). The level of oxidized protein increases with time of exposure to 100% oxygen up to 48 hours and is followed by a sharp decline between 49 and 54 hours. Protein oxidation also increases with age from 3 to 24 months.

Previous studies with purified glutamine synthetase (GS) *in vitro* have indicated that, although MFO-mediated inactivation is associated with the loss of a single histidine in each subunit, further oxidation of the protein leads to increased susceptibility to proteolysis by known proteases as well as a new class of nonlysosomal proteases which exhibit selectivity for the oxidatively modified protein.[7] In order to ascertain whether specific enzymes were inactivated *in vivo* under these conditions, we determined the specific activity and the levels of immunological cross-reacting material of two liver enzymes, glucose-6-phosphate dehydrogenase (G-6-PD) and GS. The results in Figure 1 show that 33% of GS activity and 71% of G-6-PD activity was lost during the 54-hour exposure to 100% oxygen (panels A and B). Similarly, the activity of the two enzymes in 24-month-old rat hepatocytes was 67% and 62% of the initial activity observed at 3 months of age (panels C and D). Antibody titra-

Table 1. Oxidative Modification of Hepatocyte-Soluble Proteins[2]

Experiment 1. Rats Exposed to 100% Oxygen	
Time (hours)	nmoles DNPH/mg protein
0	2.4 ± 0.5
24	5.0 ± 0.8
48	7.1 ± 0.6
54	2.2 ± 0.4
Experiment 2. Rats of Various Ages	
Age (months)	nmoles DNPH/mg protein
3	1.9 ± 0.5
12	2.2 ± 0.4
24	4.1 ± 0.7

[2]Soluble hepatic proteins were isolated from the rats at the times or ages indicated[6]. DNPH-reactive protein was determined according to methods previously described[4].

tion of GS and G-6-PD revealed that the amount of antibody required to precipitate 50% of the activity of either enzyme remained constant or increased with time of exposure to 100% oxygen until 48 hours and then markedly declined (panels A and B), although there was a progressive loss of catalytic activity over the same time period.

During aging the amount of immunologically reactive protein also decreased more slowly than the loss of catalytic activity (panels C and D). These data indicate that inactive protein accumulates under both experimental conditions, and together with the protein oxidation results, suggest that the loss of GS and G-6-PD activity correlates with increased protein carbonyl content and that the loss of enzyme activity precedes the loss of immunological cross reactivity. These results are consistent with the view that loss of GS and G-6-PD activity may be due to oxidative modification and that inactivated enzymes are subsequently degraded.

These and other experiments have suggested that oxidized proteins are readily degraded. However, the accumulation of oxidized proteins during oxygen exposure and aging have suggested that either the rate of oxidation exceeds the proteolytic capacity or the proteases which selectively degrade the oxidized proteins are inactivated or deficient or both. Proteolytic activity of extract preparations from hepatocytes of oxygen-treated rats or rats aged 3, 12, 20, and 26 months was determined at four pH values (2.5, 5.5, 7.4, and 8.8). Following incubation the samples were precipitated with TCA and the content of amino acids and peptides in the soluble fraction was measured fluorometrically after reacting with fluorescamine. The results in Figure 2 show that acid protease activity (pH 2.5 and 5.5) remained relatively constant, but the alkaline protease activity (pH 7.4 and 8.8) increased with time during the first 48 hours of exposure to 100% oxygen. The rate of protein oxidation exceeds the capacity of the cell to degrade the oxidized protein; however, with further exposure (from 48 to 54 hours) there is an

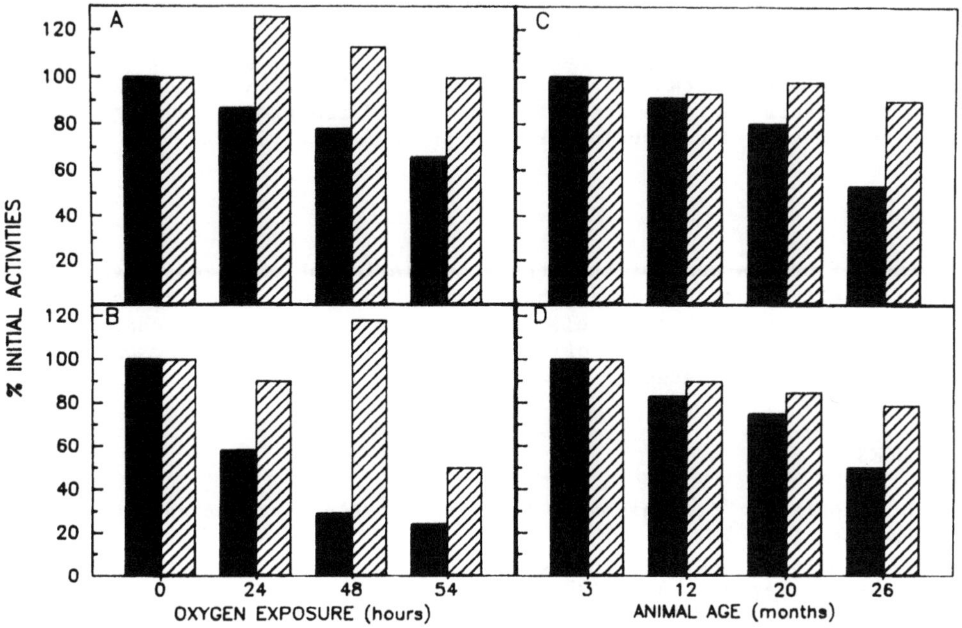

Figure 1. Specific activity and immunological cross reactivity of glutamine synthetase and glucose-6-phosphate dehydrogenase. Left set of panels represents experiments carried out with rats exposed to 100% oxygen (A and B) and right set of panels designates animals of various ages (C and D). GS specific activity (solid) and immunological cross reactivity (hatched) are shown in upper panels (A and C). G-6-PD specific activity (solid) and immunological cross reactivity (hatched) are shown in lower panels (B and D). Assays were carried out according to methods previously described.[8,9]

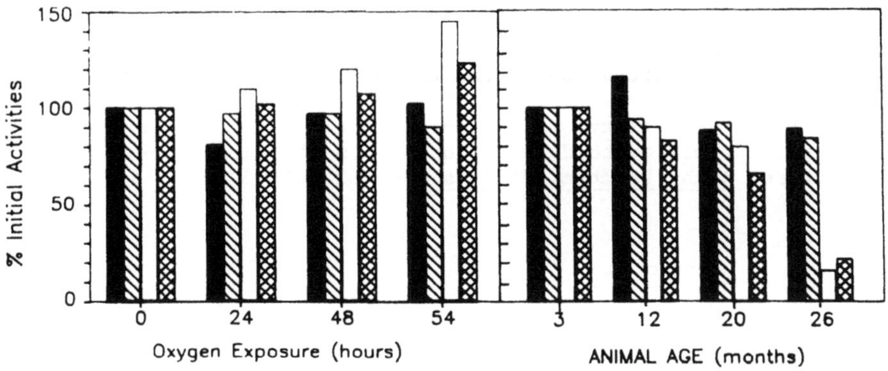

Figure 2. Protease activity in hepatocyte extract from rats treated with 100% oxygen or rats of various ages. Protease activity was determined at four pH values: 2.5 (solid), 5.5 (hatched), 7.4 (blank), and 8.8 (cross-hatched), according to the method of Rivett.[10] The left panel represents data obtained from the oxygen-treated rats and the right panel represents data obtained with animals of various ages.

increase in alkaline protease activity (Figure 2, left panel) leading to a rapid loss of oxidized protein (Table 1, Experiment 1; Figure 1). From a comparison of the data in Table 1, Experiment 2, and in Figure 2, right panel, it appears that the accumulation of oxidized protein during aging may be caused by a decrease in alkaline protease activity.

CONCLUSIONS

These findings suggest that the accumulation of oxidized proteins and the inactivation of enzymes that occur when animals are exposed to 100% oxygen may reflect an acceleration of the same metabolic changes that lead to the accumulation of oxidized proteins during aging and accelerated aging diseases.

REFERENCES

1. M. Rothstein, "Mechanisms of Aging and Development," Academic Press, New York (1977).
2. M. Rothstein, Enzymes, enzyme alteration, and protein turnover, Rev. Biol. Res. Aging 1:305 (1983).
3. L. Fucci, C. N. Oliver, M. J. Coon, and E. R. Stadtman, Inactivation of key metabolic enzymes by mixed-function oxidation reactions: Possible implications in protein turnover and aging, Proc. Natl. Acad. Sci. U.S.A. 80:1521 (1983).
4. C. N. Oliver, B.-W. Ahn, M. E. Wittenberger, and E. R. Stadtman, "Cellular Regulation and Malignant Growth," Scientific Society and Springer-Verlag, Japan (1985).
5. R. L. Levine, Oxidative modification of glutamine synthetase: Modification due to loss of one histidine, J. Biol. Chem. 258:11823 (1985).
6. W. B. Rowe, R. A. Remzio, V. P. Wellner, and A. Meister, Glutamine synthetase (sheep brain), "Methods in Enzymology," Vol. 17a, Academic Press, New York (1970).
7. A. J. Rivett, Preferential degradation of the oxidatively modified form of glutamine synthetase by intracellular mammalian proteases, J. Biol. Chem. 260:300 (1985).
8. R. E. Miller, R. Hadenberg, and H. Gershman, Regulation of glutamine synthetase in cultured 3T3-L1 cells by insulin, hydrocortisone, and dibutyeyl cyclic AMP, Proc. Natl. Acad. Sci. U.S.A. 75:1418 (1978).
9. C. Bishop, Assay of glucose-6-phosphate dehydrogenase and 6-phosphogluconate dehydrogenase in red cells, J. Lab. Clin. Med. 8:149 (1966).
10. A. J. Rivett, The effect of mixed-function oxidation of enzymes on their susceptibility to degradation by a nonlysosomal cysteine proteinase, Arch. Biochem. Biophys. 343:624 (1985).

OXIDATIVE MODIFICATION OF GLUTAMINE SYNTHETASE: COVALENT AND CONFORMATIONAL CHANGES WHICH CONTROL SUSCEPTIBILITY TO PROTEOLYSIS

Rodney L. Levine and A. Jennifer Rivett*

Laboratory of Biochemistry
National Heart, Lung, and Blood Institute
Bethesda, Maryland 20892, U.S.A.

*Department of Biochemistry
University of Leicester
Leicester LE1 7RH, U.K.

INTRODUCTION

Mixed function oxidation of proteins has been implicated in a variety of physiologic and pathologic processes, including intracellular protein turnover, host defense activities, oxygen toxicity, and the aging process. Many key enzymes are susceptible to oxidative modification, usually with loss of catalytic activity.[1] Glutamine synthetase from Escherichia coli is oxidatively modified by a number of enzymic and non-enzymic mixed function oxidation systems.[2,3] To understand the effects of oxidative modification, we have studied the covalent and non-covalent changes in glutamine synthetase which occur upon oxidative modification. Protein of varying extent of modification was prepared by timed exposure to a model system consisting of ascorbate, iron, and molecular oxygen.[1,4] These modified proteins were analyzed for several covalent and conformational changes. We were particularly interested in identifying changes which correlate with increased susceptibility to proteolysis by a novel high molecular weight protease purified from rat liver.[5]

The idealized time courses for modifications in glutamine synthetase are shown in Figure 1. Previous studies demonstrated that oxidative modification caused rapid inactivation of glutamine synthetase which was due to oxidation of a histidine residue.[6] This histidine residue has been identified as His-269,[7] the first of two histidine residues in an interesting pentapeptide (Met-His-Cys-His-Met) which forms a divalent cation binding site.[8] Direct studies of metal binding with radiolabelled Mn^{2+} now confirm that oxidative modification caused loss of a metal-binding site, most likely the high affinity site.[9] Loss of this binding site paralleled inactivation of the enzyme but did not render the protein susceptible to proteolysis. Additional exposure to the mixed function oxidation system had little additional effect on metal binding but did convert the protein to a form susceptible to proteolysis. Covalent modification of the protein continued to increase with the longer incubation, documented by the appearance of carbonyl groups in the protein. Despite the destruction of a metal binding site, there are no major conformational changes in the protein as assessed by lack of change in

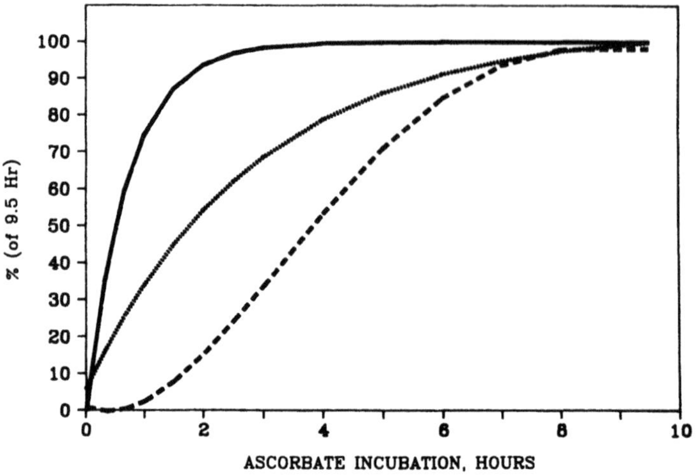

Figure 1. Schematized representation of the time courses observed during exposure of glutamine synthetase to mixed function oxidation. (———), inactivation; loss of one histidine residue; loss of Mn^{2+} binding. (|||||||), introduction of carbonyl groups into the protein. (-- -- --), susceptibility to proteolysis by purified rat liver protease; increase in absorbance at 325 nm; loss of two histidine residues. Results were normalized to the 8-hour value to permit comparison of the various patterns.

the sedimentation velocity. This is in notable contrast to the relaxation induced by removal of all metals upon treatment with a chelator such as EDTA.[9]

As noted, even extensive oxidative modification of glutamine synthetase does not cause a change in sedimentation velocity. Nevertheless, subtle changes in conformation were detected. Specifically, the rate of reaction of sulfhydryl groups with Ellman's reagent increased with increasing extent of oxidative modification. However, the four cysteine residues were not themselves oxidized. Indeed, the only change detected so far in the amino acid composition is loss of histidine residues. The first histidine residue, as noted, is lost during the initial reaction, which causes loss of catalytic activity. This form of the enzyme is not susceptible to proteolysis by the rat liver protease. Further exposure to the oxidizing system leads to loss of a second histidine residue.

The kinetics of loss of the second residue correlated well with induction of susceptibility to proteolysis. Thus, two histidine residues must be lost to "mark" glutamine synthetase for proteolytic degradation by the purified protease. The time course for susceptibility to proteolysis also coincided with the appearance of a UV absorbing band from 300 to 350 nm (data not shown). The chemical basis for this spectral change has not been established, although it obviously could be related to oxidation of the second histidine residue.

These *in vitro* results are matched by *in vivo* results from microinjection of glutamine synthetase into cultured hepatoma cells.[10] Native glutamine synthetase had a half-life of about 60 hours after microinjection. Modified glutamine synthetase which had lost one histidine residue demonstrated the same 60-hour half-life. However, this was not true of

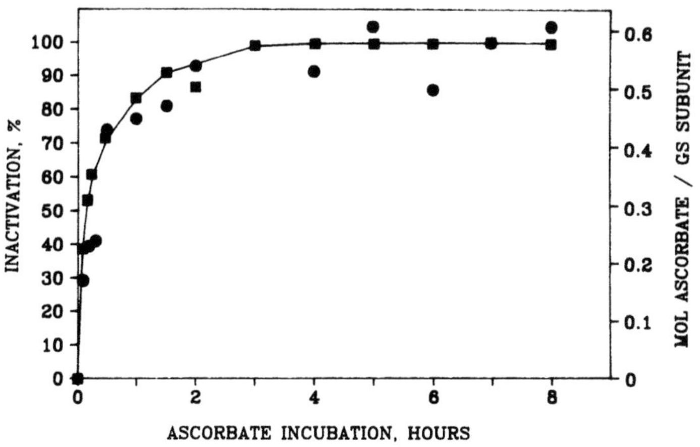

Figure 2. Binding of ascorbate to glutamine synthetase parallels inactivation. Glutamine synthetase was incubated with ^{14}C-labelled ascorbic acid (C1). At intervals, aliquots were withdrawn for determination of enzymatic activity (■ —— ■) and of ascorbate binding (●). Binding was determined after three washes with 10% trichloroacetic acid.

protein which had been further oxidized to modify the second histidine. That species was degraded about 10 times faster than the other two forms.

The detailed studies of glutamine synthetase summarized here all utilized the model mixed function oxidation system based on ascorbate, although many other enzymic and non-enzymic systems also oxidize the protein.[1-3] In addition to oxidative modifications, incubation with the model system could cause covalent binding of ascorbate to glutamine synthetase. Schiff base formation could occur, potentially followed by the Amadori rearrangement to yield a relatively stable derivative. Utilizing ^{14}C-labelled ascorbic acid, we found that ascorbate did form a stable, presumably covalent linkage with glutamine synthetase (Figure 2). The label remained with the protein during repetitive precipitations with trichloroacetic acid or ammonium sulfate and after exhaustive dialysis. As shown in Figure 2, the time course of binding closely followed the course of inactivation and not that of proteolytic susceptibility. Hence, binding of ascorbate itself cannot account for the induction of proteolytic susceptibility. Since many other mixed-function oxidation systems also inactivate glutamine synthetase, binding of ascorbate need not occur for inactivation. The biological significance of protein-ascorbate binding deserves exploration.

Susceptibility to oxidative modification is dramatically altered by substrates and products at physiological concentrations.[4,11] So far, this has proven true for each of the four enzymes in which the phenomenon has been examined[11]. For example, each of the enzymes was protected from oxidative modification by its substrates. Modulation of oxidative modification by small ligands provides a control mechanism by which the cell can regulate the reaction. In the case of protein turnover, such regulation provides overall control of the rate of proteolysis. It may explain, in part, known variations in the rate of enzyme turnover observed in response to alterations in the nutritional or hormonal status of organisms. Intracellular protein turnover can then be viewed as a two-step process:

$$\text{Native Protein} \xrightarrow{\text{Mixed Function Oxidase}} \text{Oxidized Protein} \qquad (1)$$

$$\text{Oxidized Protein} \xrightarrow{\text{Specific Protease}} \text{Proteolytic Products} \quad (2)$$

Regulation of intracellular proteolysis occurs at the first step, mediated in part by substrates and products of the enzyme. Specificity results both from modulation of oxidative modification and from the selectivity of the protease for oxidatively modified proteins.[5,11,12] Other rat liver proteases also discriminate between native and oxidized proteins,[13] and a protease purified from *Escherichia coli* also attacks only the oxidized protein.[12]

ACKNOWLEDGMENTS

We thank Julie Sahakian for valuable technical assistance.

REFERENCES

1. L. Fucci, C. N. Oliver, M. F. Coon, and E. R. Stadtman, Inactivation of key metabolic enzymes by mixed-function oxidation reactions: Possible implication in protein turnover and aging, *Proc. Natl. Acad. Sci. U.S.A.* 80:1521 (1983).
2. R. L. Levine, C. N. Oliver, R. M. Fulks, and E. R. Stadtman, Turnover of bacterial glutamine synthetase: Oxidative inactivation precedes proteolysis, *Proc. Natl. Acad. Sci. U.S.A.* 78:2120 (1981).
3. E. R. Stadtman and M. E. Wittenberger, Inactivation of *Escherichia coli* glutamine synthetase by xanthine oxidase, nicotinate hydroxylase, horseradish peroxidase, or glucose oxidase: Effects of ferredoxin, putidaredoxin, and menadione, *Arch. Biochem. Biophys.* 239:379 (1985).
4. R. L. Levine, Oxidative modification of glutamine synthetase. II. Characterization of the ascorbate model system, *J. Biol. Chem.* 258:11828 (1983).
5. A. J. Rivett, Purification of a liver alkaline protease which degrades oxidatively modified glutamine synthetase. Characterization as a high molecular weight cysteine protease, *J. Biol. Chem.* 260:12600 (1985).
6. R. L. Levine, Oxidative modification of glutamine synthetase. I. Inactivation is due to loss of one histidine residue, *J. Biol. Chem.* 258:11823 (1983).
7. J. M. Farber and R. L. Levine, Sequence of a peptide susceptible to mixed-function oxidation: Probable cation binding site in glutamine synthetase, *J. Biol. Chem.* 261:4574 (1986).
8. R. J. Almassy, C. A. Janson, R. Hamlin, N.-H. Xuong, and D. Eisenberg, Novel subunit-subunit interactions in the structure of glutamine synthetase, *Nature* 323:304 (1986).
9. E. R. Stadtman and A. Ginsburg, The glutamine synthetase of *Escherichia coli*: Structure and control, *in*: "The Enzymes," P. D. Boyer, ed., Academic Press, New York (1974).
10. A. J. Rivett and J. F. Hare, Mixed-function oxidation of glutamine synthetase leads to its rapid degradation *in vitro* and after fusion-mediated injection into hepatoma cells, *Biochem. Soc. Trans.* 14:643 (1986).
11. A. J. Rivett, Regulation of intracellular protein turnover: Covalent modification as a mechanism of marking proteins for degradation, *Curr. Top. Cell Regul.* 28:291 (1986).
12. J. E. Roseman and R. L. Levine, Purification of a protease from *Escherichia coli* with specificity for oxidized glutamine synthetase, *J. Biol. Chem.* 262:2101 (1987).
13. A. J. Rivett, Preferential degradation of the oxidatively modified form of glutamine synthetase by intracellular mammalian proteases, *J. Biol. Chem.* 260:300 (1985).

ROLE OF HEMIN IN OXIDATIVE STRESS INDUCED BY DIVICINE IN VITAMIN E-DEFICIENT RATS

M. D'Aquino and G. Tomassi

Istituto Nazionale della Nutrizione
Via Ardeatina 546, 00179 Rome, Italy

INTRODUCTION

Previous results obtained with erythrocytes of vitamin E + vitamin B_2 deficient rats or vitamin B_2-deficient rats incubated in presence of divicine (the hemolytic factor of the favism) have shown peroxidation of membrane lipid during hemolysis,[1] and decrease in the fluidity of the hydrophobic region of membrane lipids.[2] These modifications appear only in the whole erythrocytes, but not when the ghosts are used (unpublished data).

Recent studies suggest that hemoglobin oxidation has an important role in the initiation of lipid peroxidation.[3] It has also been demonstrated that heme can be released from hemoglobin as hemin (ferriprotoporphyrin IX) and that hemin is a potent hemolytic agent.[4] In normal conditions hemin concentration in red cells is very low because of the very high affinity of the globin for the heme. In the methemoglobin state this linkage becomes very weak, and, therefore, the heme is released as hemin.

The mechanism by which hemin plays its hemolytic action is still unclear. Chau and Fitch showed in mouse erythrocytes that hemin causes hemolysis by a colloid-osmotic mechanism without lipid peroxidation.[5] More recently, using linoleic acid micelles, Gutteridge showed that hemin stimulates lipid peroxidation, which is not inhibited by serum protein haptoglobin.[6]

In the present work, we studied the role of hemin during hemolysis in a condition of elevated susceptibility to lipid peroxidation of erythrocyte membrane (vitamin E-deficiency).

ANIMALS AND DIETS

Male albino Sprague-Dawley rats (60-70 g) were fed a vitamin E-free diet for 4 months. The concentration of alpha-tocopherol in plasma at the end of 4 months was 0.075 ± 0.010 μg/ml in the deficient animals and 1.510 ± 0.060 μg/ml in the control group.

MATERIALS AND METHODS

The Vicine (Serva, Heidelberg) was treated with beta-glycosidase for 20 min at 37°C under nitrogen for the hydrolysis of the glycoside, obtaining the divicine compound as previously described.[1] Blood samples were washed with isotonic phosphate saline buffer pH7.4 and plasma and leucocytes removed. The plasma concentration of alpha-tocopherol was determined by high-pressure liquid chromatography according to the method of Bieri et al.[7] The packed red cells were suspended in the same buffer and incubated with divicine or hemin for 2 hours at 37°C without glucose.

Methemoglobin levels were measured using the standard method of cyanide treatment.[8] The thiobarbituric acid tests were performed by the method of Stocks and Dormandy.[9] The hemolysis at the end of incubation of a 10% suspension of erythrocyte was determined by the amount of hemoglobin in the supernatant at 540 nm according to the method of Chau.[5] The ghosts were prepared according to the method of Dodge.[10] The separation of the lipid core and cytoskeleton proteins of membrane was obtained according to the method of Shaklai et al.[11] and the hemin concentration in the lipid core was determined by the method of Morrison.[12]

RESULTS AND DISCUSSION

The results reported in Table 1 show that divicine induces a very high methemoglobin formation together with a complete hemolysis only in the vitamin E-deficient erythrocytes. In this condition, divicine induces lipid peroxidation in red cells but not in ghosts, as shown by the results of TBA test reported in Table 2, and induces hemin release in the lipid core of membrane (Table 3).

Incubation with hemin at the maximum hemolytic-effect concentration (see Table 4) shows that lipid peroxidation does not occur in control and vitamin E-deficient rats (Table 5).

The obtained results indicate that the presence of hemoglobin is needed to peroxidize lipids in conditions of reduced antioxidative defense capability of the membrane and that hemin is the factor responsible for the hemolytic effect, independently from the peroxidative state of membrane lipids.

Table 1. Methemoglobin formation and hemolysis in red cells of control and vitamin E-deficient rats in presence of divicine

	Control	Control + Div	-Vit.E	-Vit.E + Div.
Methemoglobin %	2.1 ± 0.4	11.04 ± 4.0	3.0 ± 0.3	65.0 ± 8.0
Hemolysis %	3	3	3.4	100%

Table 2. Thiobarbituric acid test in red cells and ghosts of control and vitamin E-deficient rats after incubation with divicine

TBA A532	Control	Control + Div	-Vit.E	-Vit.E + Div.
Red Cells	0.028 ± 0.005	0.030 ± 0.007	0.033 ± 0.004	0.250 ± 0.040
Ghost	0.033 ± 0.004	0.034 ± 0.005	0.050 ± 0.005	0.053 ± 0.003

Table 3. Hemin concentration in lipid core of red cell membranes of control and vitamin E-deficient rats after divicine treatment

Hemin μmol/μ phosphorus			
Control	Control + Div	-Vit.E	-Vit.E + Div.
2.5±0.2	4.3 ± 1.7	2.3 ± 0.2	17.4 ± 1.2

Table 4. Hemolysis in red cells of control and vitamin E-deficient rats at various hemin concentrations

	Hemin concentration (μM)					
	0	4	8	16	24	48
	Hemolysis %					
Control	3	8	12	20	42	100
-Vit.E	3	8	12	22	61	100

Table 5. Thiobarbituric test values in red cells of control and vitamin E-deficient rats after incubation with hemin (48μM for 2 hours)

	Control	Control + hemin	-Vit.E	-Vit.E + hemin
TBA A532	0.027 ± 0.003	0.034 ± 0.008	0.034 ± 0.005	0.035 ± 0.004

ACKNOWLEDGMENTS

This work was supported by a grant from special project I.P.R.A. of C.N.R., Rome, Italy Contract n° 85.02554.55.

REFERENCES

1. M. D'Aquino, S. Gaetani, and M.A. Spadoni, Effect of factors of favism on the protein and lipid components of rat erythrocyte membrane, Biochim. Biophys. Acta 731:161 (1983).
2. M. D'Aquino and C. Rice-Evans, The effects of divicine on erythrocytes under conditions of modified antioxidant protective mechanism:a model for favism, in Free Radicals Cell Damage and Disease, C. Rice-Evans, ed., Richelieu Press, London, 1986.
3. M.A. Sayed Sadrzadah, E. Graf, S. Scoh Panter, P.E. Hallaway, and J.W. Eaton, Hemoglobin, A biologic Fenton reagent, J. Biol. Chem. 259:14354 (1984).
4. A.C. Chau and C.D. Fitch, Hemolysis of mouse erythrocytes by ferriprotorphyrin IX and chloroquine. Chemotherapeutic implications, J. Clin. Invest. 66:856 (1980).
5. A.C. Chau and C.D. Fitch, Mechanism of Hemolysis induced by Ferriprotoporphyrin IX, J. Clin. Invest. 68:672 (1981).
6. J.M.C. Gutteridge, The antioxidant activity of haptoglobin towards haemoglobin-stimulated lipid peroxidation, Biochim. Biophys. Acta 917:219 (1987).
7. J.G. Bieri, J.T. Talliver, and L.G. Catignani, Simultaneous determination of α-tocopherol and retinol in plasma or red cells by high pressure liquid chromatography, Am. J. Clin. Nutr. 32:2143 (1979).
8. J.B. Harley and A.M. Mauer, Studies on the formation of heinz bodies.I° - Methemoglobin production and oxyhemoglobin destruction, Blood 16:1722 (1960).
9. J. Stocks and T.L. Dormandy, The autoxidation of human red cell lipids induced by hydrogen peroxide, Br. J. Haematol. 20:95 (1971).
10. J.T. Dodge, L. Mitchel, and D.J. Hanahran, The preparation and hemical characteristics of hemoglobin-free ghosts of human erythrocytes, Arch. Biochem. Biophys. 100:119 (1963).
11. N. Shaklai, Y. Shviro, E. Rabizadeh, and I. Kirschner-Zilber, Accumulation and drainage of hemin in the red cell membrane, Biochim. Biophys. Acta 821:355 (1985).
12. G.R. Morrison, Fluorimetric Microdetermination of Heme protein, Anal. Chem. 37:1124 (1965).

THE ISOLATION AND PARTIAL CHARACTERIZATION OF STABLE H_2O_2-RESISTANT VARIANTS OF CHINESE HAMSTER FIBROBLASTS

D.R. Spitz, G.C. Li, M.L. McCormick*, Y. Sun*, and L.W. Oberley*

Radiation Oncology Research Laboratory CED-200 Box 0806
University of California
San Francisco, CA 94143, U.S.A.

*Radiation Research Laboratory Rm. 14 Med. Lab.
University of Iowa
Iowa City, Iowa 52242, U.S.A.

INTRODUCTION

Several groups have reported that bacteria exposed to H_2O_2 are capable of adapting to H_2O_2-mediated cytotoxicity.[1-3] The adaptation is accompanied by a shift in relative rates of protein synthesis, increased antioxidant enzymatic activity, and increased resistance to other oxidizing agents and radiation.[1-4] This adaptation phenomenon has been suggested to be a general response to oxidative stress common to all bacteria.[1]

We previously reported that Chinese hamster fibroblasts (HA-1) exposed to a single (one-hour) treatment of H_2O_2 become resistant to a subsequent exposure to H_2O_2.[5] We now report that HA-1 cells exposed chronically (for >200 days) to progressively increasing concentrations (50-800 μM) of H_2O_2 become stably resistant to H_2O_2-mediated killing for at least 240 days following removal from H_2O_2 stress.[6] We also have measured catalase activity and correlated the increases in catalase activity with the increases in H_2O_2-resistance.

PROTOCOL FOR ISOLATION OF H_2O_2-RESISTANT VARIANTS

Exponentially growing HA-1 cells (2.5×10^5) were cultured in complete medium (MEM, 15% fetal calf serum, and 50 μg/ml gentamycin) containing 50 μM H_2O_2. The interaction between H_2O_2 and cells is dynamic; by the end of 4 days of culturing (1 passage) the H_2O_2 was removed from the medium by the cells. Because of this, the cells (2×10^5) were subcultured into fresh medium containing 50 μM H_2O_2. This protocol was followed until the cells were adapted to 50 μM H_2O_2 and capable of forming a confluent monolayer. The original line was then maintained in the presence of 50 μM H_2O_2 and 2×10^5 cells were transferred to 100 μM H_2O_2 and passaged every 4 days until they could form a confluent monolayer. We continued increasing (in twofold steps) the H_2O_2 concentration when the cells could form a monolayer until the cells would no longer grow (1.6 mM H_2O_2). The cells in 50, 100, 200, 400, and 800 μM H_2O_2 were then maintained in H_2O_2 108 days longer, at which time the H_2O_2 was

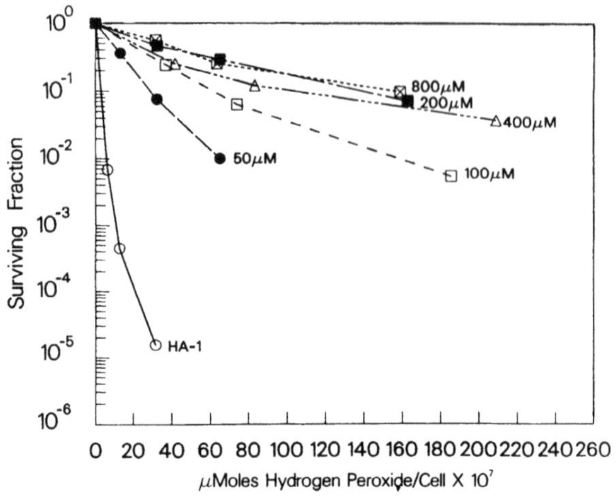

Figure 1. H_2O_2 survival curves obtained following a 1 hr exposure at 37°C to the indicated doses of H_2O_2.

removed from the medium. H_2O_2 resistance and catalase activity were tested periodically following removal from H_2O_2 stress.

Figure 1 shows a representative set of H_2O_2 survival curves obtained for the HA-1 and H_2O_2-resistant cell lines (50-800 μM adapted) 140 days following removal from the H_2O_2 stress.[6] Cell survival is plotted as a function of H_2O_2 concentration. H_2O_2 concentration is expressed as μmoles H_2O_2/cell in order to minimize cell density artifacts.[5] All survival data is normalized to the plating efficiency of each cell line (plating efficiency varied from 60-90% in all cell lines).

Table 1 shows the mean dose modifying factors (DMFs) at 20% isosurvival for each cell line calculated from the complete survival curve data gathered at 56, 140, and 244 days after culturing the H_2O_2-adapted cells in H_2O_2-free medium.[6] Table 1 also shows the levels of catalase activity measured spectrophotometrically in the parental and H_2O_2-resistant cell lines at 172 days following removal from H_2O_2.[7-8]

Table 1. Relationship between H_2O_2 Resistance and Catalase Activity in HA-1 Cells and Their H_2O_2-Resistant Variants

		172 Days Out Catalase Activity[b]			
Cell Line	Mean DMF(range)[a]	k Units/mg Protein x 10^2	Fold Induc.	k Units/Cell x 10^9	Fold Induc.
HA-1	1.0	3.1 ± 0.5	1.0	1.4 ± 0.2	1.0
50μM	6.7 (5.0-8.3)	14.0 ± 2.0	4.5	6.4 ± 0.9	4.5
100μM	11.9 (8.3-14.0)	35.0 ± 4.9	11.3	14.0 ± 2.0	10.0
200μM	25.3 (20.0-29.3)	78.1 ± 7.0	25.2	45.3 ± 4.1	32.3
400μM	21.9 (20.0-23.3)	100.0 ± 34.0	32.3	53.0 ± 18.0	37.6
800μM	30.7 (26.7-36.0)	78.0 ± 16.0	25.2	55.6 ± 12.0	39.4

[a]DMF = $\frac{\text{dose to reach 20\% survival in } H_2O_2\text{-resistant cell line}}{\text{dose to reach 20\% survival in HA-1 cell line}}$

[b]Errors represent the standard deviations of three measurements on one sample.

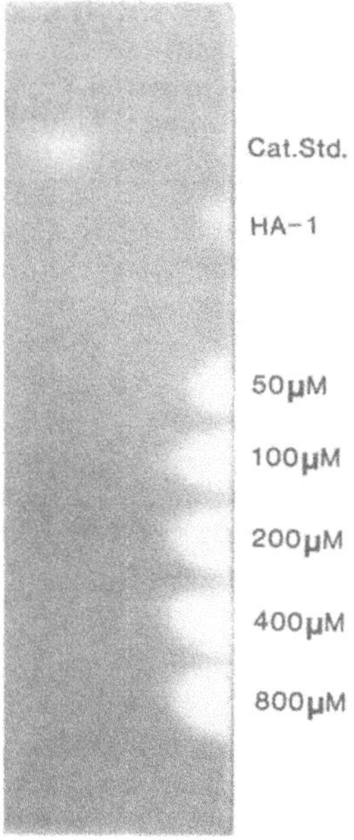

Figure 2. A native polyacrylamide gel[9] (7.5% running gel, 5% stacking gel) stained for catalase activity.[9] This gel was loaded with 0.03 μg of bovine catalase standard, and 20.5 μg of each cell homogenate assayed in Table 1.[6] This data confirms the results obtained in Table 1.

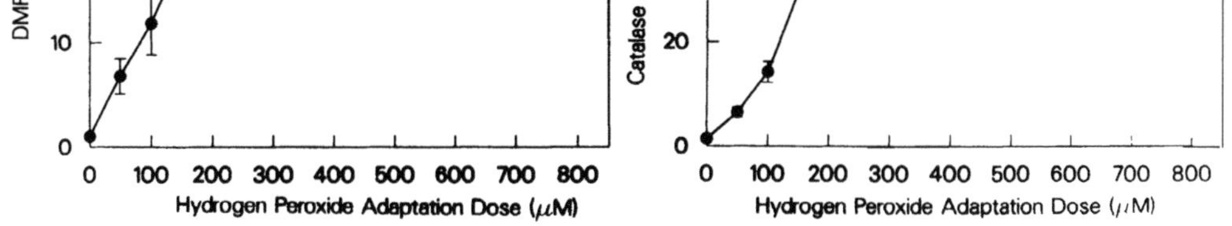

Figure 3. The DMFs and the catalase activities (± 1 std. deviation[6]) from Table 1 replotted as a function of H_2O_2 adaptation dose.

DISCUSSION

Our results show that mammalian cells chronically exposed to progressively increasing concentrations of H_2O_2 (50-800 μM) become stably resistant to H_2O_2 toxicity (DMFs = 5-30) for at least 240 days following removal from H_2O_2. The H_2O_2-resistant variants demonstrate 5- to 40-fold increases in catalase activity. The increases in catalase activity correlate well with the increases in H_2O_2 resistance (Figure 3). These results indicate that a stable change in expression of catalase activity occurs in cultured mammalian cells in response to chronic exposure to H_2O_2. These findings are consistent with the previous reports of H_2O_2 resistance in bacterial systems.[1-2] Therefore, it is very likely that adaptation to H_2O_2-mediated oxidative stress may also be a phenomenon common to mammalian systems.

ACKNOWLEDGMENTS

This investigation was supported by NIH grants CA-31397, CA-41267, CA-09215, and CA-09125. We appreciate Kathleen Wallace for her assistance in the preparation of this manuscript.

REFERENCES

1. M.F. Christman, R.W. Morgan, F.S. Jacobson, and B.N. Ames, Positive control of a regulon for defenses against oxidative stress and some heat shock proteins in Salmonella typhimurium, Cell 41:753 (1985).
2. L. Winquist, U. Rannug, A. Rannug, and C. Ramel, Protection from toxic and mutagenic effects of hydrogen peroxide by catalase induction in Salmonella typhimurium, Mutation Res. 141:145 (1984).
3. B. Demple and J. Halbrook, Inducible repair of oxidative damage in Escherichia coli, Nature 304:466 (1983).
4. R.W. Morgan, M.F. Christman, F.S. Jacobson, G. Storz and B.N. Ames, Hydrogen peroxide-inducible proteins in Salmonella typhimurium overlap with heat shock and other stress proteins, Proc. Natl. Acad. Sci. 83:8059 (1986).
5. D.R. Spitz, W.C. Dewey and G.C. Li, Hydrogen peroxide or heat shock induces resistance to hydrogen peroxide in Chinese hamster fibroblasts, J. Cell. Physiol. 131(3):364 (1987).
6. D.R. Spitz, G.C. Li, M.L. McCormick, Y. Sun and L.W. Oberley, Stable H_2O_2-resistant variants of Chinese hamster fibroblasts demonstrate increases in catalase activity, Radiat. Res. 114:114 (1988).
7. R.F. Beers and I.W. Sizer, A spectrophometric method for measuring the breakdown of H_2O_2 by catalase, J. Biol. Chem. 195:133 (1952).
8. H. Aebi, Catalase in vitro, in: "Methods in Enzymology," Academic Press, Inc. 105:121 (1984).
9. W. Woodbury, A.K. Spencer and M.A. Stahmann, An improved procedure using ferricyanide for detecting catalase isozymes, Anal. Biochem. 44:305 (1971).

REACTIVE OXYGEN METABOLITES AND HYPEROXIC TOXICITY

Dana Jamieson

School of Physiology and Pharmacology
University of New South Wales
Kensington 2033 NSW
Australia

INTRODUCTION

Reactive oxygen species are implicated in the pathology of an ever- increasing number of conditions, including post-ischemic damage,, inflammation, cancer, drug action, radiation and ageing. Hyperoxic damage to cells is also considered to be caused by production of reactive oxygen metabolites at rates which exceed the detoxifying capacities of the cells' antioxidant defense mechanisms. In this article I shall briefly review the current evidence for the role of reactive oxygen species in oxygen toxicity, with particular emphasis on the two main target organs for hyperoxic damage in mammals, namely, lung and brain.

PATHOLOGY OF HYPEROXIA IN MAMMALS

Inspiring pure oxygen above about 500 Torr leads to lung damage[1,2] characterized by such changes as edema, consolidation, atelectasis, fibrin deposits, congestion, inflammation, arteriolar thickening, hypertrophy of alveolar cells and hyperplasia. There are some differences in pulmonary pathology produced during chronic exposure to raised oxygen at normobaric pressure (for example several days at 95% O_2 at 1 atmosphere) compared with the lung damage occurring in acute exposure to higher pressures (such as 4-5 atmospheres absolute (ATA) of oxygen for a few hours). In the former situation there is a large accumulation of fluid in the pleural cavity[3,4] as well as intra-alveolar exudation, while in hyperbaric oxygen toxicity little pleural effusion occurs but there is a large degree of intra-alveolar hemorrhage. It is possible that pleural effusion is associated with neutrophil infiltration, but this has not yet been established. The edema and hemorrhage are due, firstly, to destruction of the capillary endothelial cells, but type I alveolar cells are damaged at a later stage, and surfactant production is also decreased.[5]

Breathing oxygen at pressures above 2-3 atmospheres produces central nervous system (CNS) disturbances with marked hyperexcitability and generalized convulsions, and such CNS effects usually occur before the appearance of gross lung damage.

TISSUE OXYGEN TENSION

In most circumstances lung damage is the cause of death of animals maintained in a hyperoxic environment up to about 6 ATA of O_2. Above this pressure, or if CO_2 is added to the gas mixture, animals may die of CNS damage before lung damage develops fully.[6,7].

Taking into account the actual pO_2 to which brain or lung cells are exposed, it seems that brain tissue is actually more susceptible than lung tissue to the deleterious effect of raised oxygen tension. Alveolar pO_2 is about 100 torr O_2 when animals breathe room air and rises to about 670 torr when 100% O_2 is breathed. However, brain pO_2 in air-breathing animals is about 30 torr or less, and cerebral vasoconstriction induced by hyperoxia may keep cerebral pO_2 values relatively low even in hyperbaric conditions. For example, the highest pO_2 values recorded in brain at 4-5 ATA approximated 600 torr,[8] but most investigators reported considerably lower rises in brain pO_2 values with raised external pO_2.[9-11] Yet convulsions occur within minutes at 4-5 ATA of O_2 breathing, while gross lung damage takes days to develop at 100% normobaric O_2 breathing.

It should also be pointed out that lack of knowledge of tissue pO_2 levels is one of the factors which makes it almost impossible to compare quantitatively the data obtained in vitro (such as measurement of $O_2^{\cdot -}$ and H_2O_2 production in hyperoxia) with results of in vivo experiments. Gassing in vitro preparations with air may lead to a highly hyperoxic environment compared to the situation in situ, especially when dealing with tissues other than the lung.

BACKGROUND BIOCHEMISTRY

Since the first detailed account of oxygen poisoning by Paul Bert[12] and the description of hyperoxic pulmonary damage by Lorraine Smith[13] a plethora of physiological, pharmacological and biochemical data emerged and formed the subject of several extensive reviews[1,2,14,15] but with little illumination of the processes underlying oxygen toxicity. The early biochemical data showed that oxygen consumption was decreased in tissue slices and homogenates exposed to hyperbaric oxygen.[15] Further experiments indicated that the decreased oxygen consumption was due to inhibition of enzymes containing essential sulfhydryl groups. However, the total "dose" of oxygen (taking into account the degree of hyperoxia and the duration of exposure) needed to inactivate such in vitro systems was high compared with the conditions which produced symptoms of oxygen toxicity in vivo. The only early (pre-superoxide dismutase) experiments in vivo in which enzyme activity was shown to be reduced before gross pathological changes occurred were performed by Jamieson et al, [16] who found decreased dehydrogenase activity, accompanied by oxidation of sulfhydryl groups,[17] in lungs of rats exposed to 5 ATA of oxygen. No such changes were found in brain. The overall impression was that enzyme inactivation was a secondary event, rather than the initiating factor, in oxygen toxicity. Further progress on the basic biochemical events in oxygen toxicitywas made in 1965, when Chance et al[18] discovered that pyridine nucleotides (NADH and NADPH) oxidised very rapidly on exposure to raised oxygen tensions. This oxidation occurred at a much faster rate than any other previously measured changes[19] and well before effects were seen in vivo. Oxidation of pyridine nucleotides was found in every system examined, viz. in submitochondrial particles, isolated mitochondria, whole cells, isolated organs and organs in situ in the anaesthetized rat, and the oxidation was not associated with increased cytochrome oxidation.[20] However, it was still not possible at that time to fuse this observation with the known pathology of oxygen poisoning.

OXYGEN RADICALS IN HYPEROXIA

The theory proposing that hyperoxic toxicity may be due to excessive production of oxygen radicals is not new. Gerschman et al[21] first proposed that radical production may be a common factor in irradiation damage and oxygen poisoning in 1954, and this idea was considered with interest in several reviews on the biochemistry of oxygen poisoning.[14,15] However, little more work was done to prove this hypothesis until the now widely acclaimed publication by McCord and Fridovich in 1969 [22] which elucidated the vital role of the enzyme

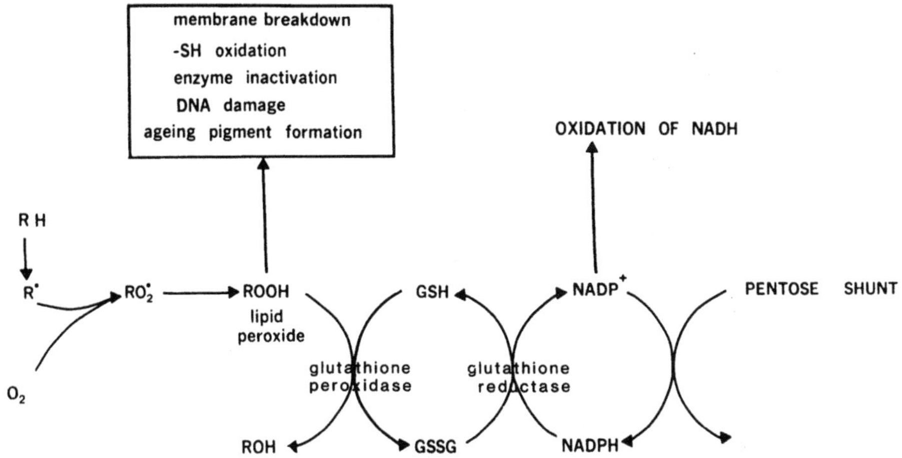

Figure 1. The detoxification of lipid peroxides, and some of the consequences of lipid peroxide formation.

superoxide dismutase in aerobic organisms. Following this publication, techniques were developed to measure superoxide levels and superoxide dismutase activity, and thereafter the importance of the whole antioxidant defense system of the cell became widely recognised.

Methods which could manipulate these antioxidant systems were also devised, such that antioxidant defenses could be raised or lowered. It is now widely accepted that the degree of damage produced by hyperoxia is determined by the balance between the rate at which reactive oxygen species are produced and their rate of removal, which in turn depends on the activity of the total antioxidant defense mechanism. The antioxidant defense mechanism includes superoxide dismutase (SOD), catalase, glutathione peroxidase, glutathione reductase, reduced glutathione, NADPH and the pentose monophosphate shunt pathway which regenerates NADPH. Other known compounds needed for antioxidant activity are vitamin E (tocopherol), vitamin C, and selenium (a cofactor for glutathione peroxidase).

The current hypothesis regarding the role of free radicals in oxygen poisoning is given in bare outline in Figure 1. Reactive oxygen metabolites lead to lipid peroxidation, and the lipid peroxides can be detoxified through the series of reactions shown in Figure 1. Superoxide ($O_2\cdot$) and hydrogen peroxide (H_2O_2) are generated in mitochondria, microsomes and other organelles during normal metabolism. These substances are broken down by superoxide dismutase and catalase or glutathione peroxidase, respectively, but if formed at excessive rates (as during hyperoxia) they can co-react under certain conditions (eg, in the presence of iron) to form the very reactive hydroxyl ($OH\cdot$) radical and/or singlet oxygen ($^1O_2$). Either of these 2 latter substances can abstract hydrogen atoms from lipids (eg membrane lipids) and initiate lipid peroxidation, which is then a self-perpetuating process (inhibited by tocopherol). Lipid peroxidation can lead to the events outlined in Figure 1. It is suggested that excessive formation of reactive oxygen metabolites is the common link between the pathological changes seen in hyperoxia and other conditions such as reperfusion injury after ischemia, radiation damage, toxicity of some antineoplastic drugs, inflammation, and, possibly ageing. Certainly, formation of oxygen radicals with subsequent lipid peroxidation is an attractive explanation for the initiation of the sequelae of events known to occur in oxygen poisoning.

DIRECT MEASUREMENT OF REACTIVE OXYGEN SPECIES IN HYPEROXIA

Direct measurement of oxygen radicals is difficult, owing to the reactivity, and thus short life, of these compounds. However, several such studies have been undertaken in the hyperoxic field and are discussed below.

In Vitro Experiments

Boveris and Chance[23], using spectrophotometric measurement of the cytochrome c peroxidase - H_2O_2 complex, measured H_2O_2 production by pigeon heart and rat liver mitochondria in normoxia and hyperbaric oxygen (up to 20 ATA). A marked increase in H_2O_2 production occurred as the pressure was raised. An increase in H_2O_2 production was seen also in isolated liver cells exposed to hyperbaric O_2 (Chance, personal communication).

A series of papers from Duke University in recent years has shown that hyperoxia increased free radical production in lung tissue and that tissue injury represented by lipid peroxidation was a consequence of the increased $O_2 \cdot^-$ and H_2O_2 production.[24] Both mitochondria and microsomes contributed to increased $O_2 \cdot^-$ and H_2O_2 in these in vitro experiments, but it is impossible at this stage to determine their relative importance in vivo.[25]

Nishiki et al[26] used oxidized glutathione efflux as a measure of lipid peroxidation in perfused isolated whole lungs and livers exposed to hyperbaric oxygen and found increases in lipid peroxidation which began immediately following compression. This effect was greatly increased when tissues from tocopherol-deficient rats were used.

Dirks and Faiman[27] found free radical production in rat brain slices incubated at oxygen pressures above 6 ATA, and increased lipid peroxidation in both rat and mouse brain slices.

In Vivo Experiments

In vivo (or, more correctly ex vivo) experiments have usually measured lipid peroxidation following hyperoxia in hyperbaric oxygen. Most investigators have used malondialdehyde formation to assess lipid peroxidation, and the degree of peroxidation has most commonly been measured in brain tissue from rats and mice which had been exposed to hyperbaric oxygen in the 4-5 ATA range.[28-30] In 1962 Becker et al[28] exposed groups of rats to different levels of hyperoxia, from normobaric 100% O_2 to about 6ATA O_2, for varying time periods depending on the pressure used. Relatively small but significant increases in lipid peroxidation were seen after exposure to hyperbaric oxygen at 3-6 ATA, but lipid peroxide levels did not correlate with the observed convulsions in this series of experiments. However, Jerret et al[30] measured lipid peroxides, and H_2O_2, in brains from normally fed, tocopherol-deficient and tocopherol-supplemented mice and rats. Normal and tocopherol-deficient mice convulsed and had raised lipid peroxide levels after 1 hour's exposure to oxygen at 4ATA, whereas tocopherol-supplemented mice did not convulse, nor was there any increase in brain lipid peroxidation. Noda et al[31] showed that the degree of lipid peroxidation due to hyperbaric oxygen exposure varied in different areas of the brain, and it is likely that changes in specific loci would produce better correlations with CNS activity. However, at present it is not certain which specific loci may initiate oxygen-induced convulsions.

Lipid peroxidation has also been measured in several other tissues following exposure to oxygen at about 5ATA with increases being found in most tissues.[29,32] More recently Freeman et al[24] measured lipid peroxidation in lung tissue of animals which had been maintained in 85% normobaric oxygen for 4 days, and they also found an increase in malondialdehyde production.

In summary, the in vitro experiments have shown that hyperoxia increased levels of superoxide, hydrogen peroxide and lipid peroxides in isolated organelles or perfused organs, while measurement of malondialdehyde formation from animals exposed to hyperoxia indicated that lipid peroxidation had occurred. However, the degree of lipid peroxidation was of relatively small magnitude in many of these experiments unless the animals were tocopherol deficient.

CIRCUMSTANTIAL EVIDENCE FOR THE ROLE OF REACTIVE OXYGEN IN OXYGEN TOXICITY

Production of Tolerance to Hyperoxia

Figure 1 outlines the process by which lipid peroxides are inactivated. Much of the indirect evidence for the role of oxygen radicals in oxygen toxicity has arisen from manipulation of the level of the factors involved in breakdown of toxic oxygen species. For example, some of the most convincing evidence for the idea that oxygen radicals are the primary cause of oxygen poisoning comes from many investigations in which animals have been made tolerant to hyperoxia by induction of the various antioxidant defense mechanisms of the cell.

Various investigators have demonstrated the induction of superoxide dismutase, catalase, glutathione peroxidase and glutathione reductase in animals exposed to sublethal concentrations of oxygen (typically about 85% normobaric oxygen) for several days.[33] Bacterial endotoxin also induced the antioxidant mechanisms, and protected against hyperoxic toxicity. These experiments have been performed in rats, - in contrast, it has also been shown that in species such as mice and guinea pigs, where antioxidant enzymes are not induced by such procedures, tolerance to hyperoxia does not develop.[33]

The overall results of these experiments provide strong evidence that changes in antioxidant enzyme activities correlate well with susceptibility to oxygen poisoning in all systems so far studied.

Increasing the Activity of Antioxidant Enzyme Systems

Another approach has been administration of the enzymes superoxide dismutase and catalase to animals before and/or during exposure to hyperoxic conditions. Initially, the results were negative, which is not surprising as penetration of these enzymes into cells is negligible. Therefore, positive data are much more significant, such as the results reported by McLennan and Autor,[34] who were able to reduce mortality by a continuous delivery system of SOD into rats maintained in 95% O_2. More recently work from Duke University[3,35] has shown that a combination of SOD and catalase, administered in liposomes, caused marked protection against both the convulsions and pulmonary damage caused by hyperbaric oxygen. This excellent series of papers proved that liposomal administration of SOD plus catalase could raise the intracellular levels of these enzymes and was very effective in protecting animals from both normobaric hyperoxia and hyperbaric oxygen. These results confirm those obtained by the measurement of superoxide and H_2O_2 formation in hyperoxia, and the results of the enzyme-induction experiments discussed above.

An increase in the function of glutathione peroxidase, (which implies optimal activity of glutathione reductase, pentose shunt enzymes, sufficient reduced glutathione etc, viz. Figure 1) will detoxify lipid peroxides more rapidly, and administration of substances (such as many sulfhydryl compounds) which maintain reduced glutathione levels has been shown to increase survival times in mice exposed to hyperbaric oxygen.[2,25] Sulfhydryl compounds may also act as radical scavengers.

Depletion of Antioxidant Defense Mechanisms

While it is possible to protect against oxygen toxicity by raising the activity of the antioxidant defenses, it is also possible to potentiate the toxicity of oxygen by lowering the antioxidant defense mechanisms. Several investigators have shown that depletion of glutathione accelerated pulmonary oxygen toxicity and, similarly, a diet deficient in selenium, which is necessary for glutathione peroxidase function, enhanced lung damage.[25]

Diethyldithiocarbamate, which inhibits brain SOD activity, shortened the preconvulsive period in rats exposed to hyperbaric oxygen, as did carmustine, an inhibitor of glutathione reductase.[36] However, it should also be pointed out that Waechter and Faiman[37] were unable to alter preconvulsive times in mice by depleting non-protein sulfhydryls (NPSH) with diethylmaleate, and they concluded that "the mechanisms whereby oxygen produces deleterious effects in the brain may be different from those operative in the lung."

Tocopherol depletion markedly enhances all forms of oxygen poisoning, but tocopherol itself, which inhibits the chain reaction of lipid peroxidation, shows little protection against hyperoxia. This suggests that tocopherol levels may be optimal in the normally fed animals used in these experiments.

ROLE OF HYDROXYL RADICAL OR SINGLET OXYGEN

Owing to difficulties of measurements, the role of these compounds is still speculative. To initiate lipid peroxidation, which can lead to the sequelae of events seen in oxygen poisoning, one needs to postulate the formation of the hydroxyl radical or singlet oxygen (or other highly reactive species) as neither superoxide nor hydrogen peroxide alone can initiate lipid peroxidation. There is evidence from chemical luminescence studies that singlet oxygen is formed during exposure to hyperbaric oxygen, but it is possible that singlet oxygen is a consequence, rather than cause, of lipid peroxidation.[25] There are several scavengers of the hydroxyl radical available, and more experiments with more selective scavengers may help elucidate the crucial role assigned to the hydroxyl radical in oxygen poisoning at this present time. Further studies using chemiluminescence to monitor the production of toxic oxygen metabolites should also assist in the quantitation of the role of singlet oxygen and other oxygen species.

In conclusion, there is much evidence for the hypothesis that hyperoxia increases the formation of reactive oxygen metabolites, and that lipid peroxidation may then result and increase membrane permeability. There is little to contradict this hypothesis as far as mammalian lung is concerned - and lung damage usually causes hyperoxic death. There is evidence that lipid peroxidation occurs also in the brain of animals exposed to hyperbaric O_2. However, the correlation between convulsions, the earliest overt manifestation of oxygen toxicity at pressures above about 3ATA, and reactive oxygen species and lipid peroxidation is not so clear. This may be a reflection of the heterogeneity of the brain, and it may well be that certain areas and certain neurotransmitter pathways, notably those involving GABA, are indeed much more prone to oxidative attack than the brain as a whole.

REFERENCES

1. J.W. Bean, Effects of oxygen at increased pressure, Physiol. Rev. 25:1 (1945).
2. J.M. Clark and C.J. Lambertsen, Pulmonary oxygen toxicity: a review, Pharmacol. Rev 23:37 (1971).
3. J.F. Turrens, J.D. Crapo, and B.A. Freeman, Protection against oxygen toxicity by intravenous injection of liposome-entrapped catalase and superoxide dismutase, J. Clin. Invest. 73:87 (1984).
4. M. Valimaki and J. Niinikoski, Development and reversibility of pulmonary oxygen poisoning in the rat, Aerospace Med. 44:533 (1973).
5. A.B. Fisher, H.J. Forman, and M. Glass, Mechanisms of pulmonary oxygen toxicity, Lung 162:255 (1984).
6. J.M. Clark, Interacting effectsof hypoxia adaptation and acute hypercapnia on oxygen tolerance in rats, J. Appl. Physiol 56:1191 (1984).
7. D. Jamieson and N. Cass, CNS and pulmonary damage in anesthetised rats exposed to hyperbaric oxygen, J. Appl. Physiol. 23:235 (1967).
8. D. Jamieson and H.A.S. van den Brenk, Measurement of oxygen tensions in cerebral tissues of rats exposed to high pressures of oxygen, J. Appl. Physiol. 18:869 (1963).

9. F.G. Hempel, Oxygen tensions measured in cat cerebral cortex under hyperbaric conditions, J. Appl. Physiol. 46:53 (1979).
10. R.C. Hunt, J.G. Blackburn, R.W. Ogilvie, and J.D Balentine, Oxygen tension in the globus pallidus and neostriatum of unanesthetized rats during exposure to hyperbaric oxygen, Exp. Neurol. 62:698 (1978).
11. D. Torbati, D. Parolla, and S. Lavy, Changes in the electrical activity and pO_2 of the rat's brain under high oxygen pressure, Exp. Neurol. 50:439 (1976).
12. P. Bert, Barometric Pressure, translated from the French by M.A. and F.A. Hitchcock. Columbus, Ohio, College Book Co. (1943).
13. J. Lorrain-Smith, The pathological effects due to increase of oxygen tension in the air breathed, J. Physiol. 24:19 (1899).
14. H.C. Davies and R.E. Davies, Biochemical aspects of oxygen poisoning. In: "Handbook of Physiology and Respiration", W.O. Fenn and H. Rahn, Eds Washington D.C,. American Physiological Society, (1965).
15. N. Haugaard, Cellular mechanisms of oxygen toxicity, Physiol. Rev. 48:311 (1968).
16. D. Jamieson and H.A.S. van den Brenk, Pulmonary damage due to high pressure oxygen breathing in rats. 2. Changes in dehydrogenase activity of rat lung, Australian J. Exp. Biol. Med. 40:51 (1962).
17. D. Jamieson, K. Ladner, and H.A.S. van den Brenk, Pulmonary damage due to high pressure oxygen breathing in rats. 4. Quantitative analysis of sulphydryl and disulphide groups in rat lung, Australian J. Exp. Biol. Med. 41:491.
18. B. Chance, D. Jamieson, and H. Coles, Energy-linked pyridine nucleotide reduction: Inhibitory effects of hyperbaric oxygen in vitro and in vivo, Nature 206:257 (1965).
19. B. Chance, D. Jamieson, and J.R.Williamson, Control of the oxidation-reduction state of reduced pyridine nucleotides in vivo and in vitro, in: Proceedings of the 3rd International Conference of Hyperbaric Medicine, I.W. Brown and B.G. Cox, eds. National Academy Sciences, Washington, D.C. (1966).
20. D. Jamieson and B. Chance, Effect of high pressure oxygen on the steady state of cytochromes in rat liver mitochondria, Biochem. J. 100:254 (1966).
21. R. Gerschman, D.L. Gilbert, S.W. Nye, P. Dwyer, and W.O. Fenn, Oxygen poisoning and x-irradiation: A mechanism in common, Science 119:623 (1954).
22. J.M. McCord and I. Fridovich, Superoxide dismutase: An enzymic function for erythrocuprein (hemocuprein), J. Biol. Chem. 244:6049 (1969).
23. A. Boveris and B. Chance,The mitochondrial generation of hydrogen peroxide.Gener al properties and effect of hyperbaric oxygen, Biochem. J. 134:707 (1973).
24. B.A. Freeman, M.K. Topolsky, and J.D. Crapo, Hyperoxia increases oxygen radical production in rat lung homogenates, Arch. Biochem. Biophys. 216:477 (1982).
25. D. Jamieson, B. Chance, E. Cadenas, and A. Boveris, The relation of free radical production to hyperoxia, Ann. Rev. Physiol. 48:703 (1986).
26. K. Nishiki, D. Jamieson, N. Oshino, and B. Chance, Oxygen toxicity in the perfused rat liver and lung under hyperbaric conditions, Biochem. J. 160:343 (1976).
27. C.D. Dirks and M.D. Faiman, Free radical formation and lipid peroxidation in rat and mouse cerebral cortex slices exposed to high oxygen pressure, Brain Res. 248:355 (1982).
28. N.H. Becker and J.F. Galvin, Effect of oxygen-rich atmospheres on cerebral lipid peroxides, Aerospace Med. 33:985 (1962).
29. M.M. Gabivov and K.G. Karagezyan, Study of lipid oxidation in various rat tissues during and after hyperoxia, Bull. Exp. Med. Biol. 91:682 (1981).
30. S.A. Jerrett, D. Jefferson, and C.E. Mengel, Seizures, H_2O_2 formation and lipid peroxides in brain during exposure to oxygen under high pressure, Aerospace Med. 44:40 (1973).
31. Y. Noda, P.L. McGeer, and E.G. McGeer,Lipid peroxide distribution in brain and the effect of hyperbaric oxygen, J. Neurochem. 40:1329 (1983).
32. W.P. Johnson, D. Jefferson, and C.E. Mengel, In vivo formation of H_2O_2 in red cells during exposure to hyperoxia, J. Clin. Invest. 51:2211 (1972).
33. H.J. Forman and A.B. Fisher, Antioxidant defenses. in: "Oxygen and Living Processes. An Interdisciplinary Approach", D.L. Gilbert, ed. Springer-Verlag, New York (1981).

34. G. McLennan and A.P. Autor, Effect of intraperitoneally administered superoxide dismutase on pulmonary damage resulting from hyperoxia, in: "Pathology of Oxygen" A.P. Autor, ed., Academic press New York (1982).
35. T. Yusa, J.D. Crapo, and B.A. Freeman, Liposome-mediated augmentation of brain SOD and catalase inhibits CNS 0_2 toxicity, J. Appl. Physiol. 57:1674 (1984).
36. C.D. Puglia and S.R. Powell, Inhibition of cellular antioxidants: a possible mechanism of toxic cell injury, Environmental Health Perspectives 57:307 (1984).
37. J.M. Waechter and M.D. Faiman, Glutathione and non-protein sulfhydryl in cerebral cortex and lung in mice exposed to high oxygen pressure, Toxicology 23:213 (1982).

INHIBITION OF PEROXIDATION OF MEMBRANES

Etsuo Niki and Erika Komuro

Department of Reaction Chemistry
Faculty of Engineering
University of Tokyo
Hongo, Tokyo 113, Japan

INTRODUCTION

There is an increasing number of observations that suggest the involvement of oxygen radicals in the deterioration of food and in the development of a variety of forms of tissue damage and pathological events such as heart disease, cancer, and aging.[1-5] It is accepted that the non-enzymatic, free radical-mediated chain oxidation of the membranes plays a crucial role. Aerobic organisms are protected from such oxygen toxicity by an array of defense systems,[6-8] which are divided into two categories: the preventive antioxidants and chain-breaking antioxidants. The preventive antioxidants deactivate the active species and possible precursors of free radicals, thereby suppressing the generation of free radicals and reducing the rate of chain initiation.

On the other hand, the chain-breaking antioxidants scavenge free radicals and suppress the free radical-mediated chain oxidation by molecular oxygen. Radicals formed in the aqueous region are scavenged exclusively by hydrophilic antioxidants, such as ascorbic acid and uric acid, before they attack the membranes. The radicals that penetrate the membranes and the radicals generated initially within the membranes cannot be scavenged efficiently by the hydrophilic antioxidants residing in the aqueous region; vitamin E is responsible for scavenging these radicals.[6,8,9]

The rate and mechanism of inhibition of oxidation of lipids in the simple, homogeneous solution are now fairly well understood, whereas those in the membrane systems are not well elucidated yet. In this presentation, I would like to discuss the inhibition of peroxidation of the membranes. The specific question is: Does vitamin E (α-tocopherol) suppress the oxidation of the membranes as efficiently as in the homogeneous solution?

EVALUATION OF ANTIOXIDANT ACTIVITY

The important question is how to evaluate and quantify the antioxidant activity. We have tried to evaluate it by measuring how long the antioxidant suppresses the oxidation and how much it reduces the

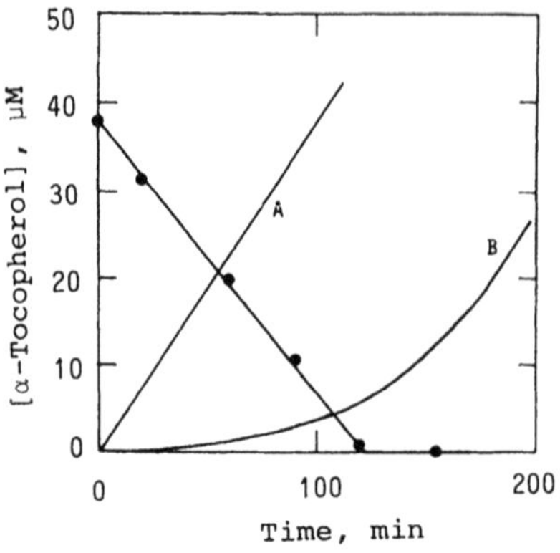

Figure 1. Inhibition of oxidation of 20 mM soybean PC liposomes in 0.1 M NaCl aqueous dispersions at 37°C under air and consumption of α-tocopherol. Lines A and B show oxygen uptake without and with α-tocopherol respectively. [AAPH] = 50 mM

rate of oxidation. Figure 1 illustrates the oxidations of soybean phosphatidylcholine (PC) liposomal membranes, in the absence and presence of α-tocopherol, induced by water-suluble azo compound, 2,2'-azobis(2-amidinopropane) dihydrochloride (AAPH), which generates free radicals at a constant rate in the aqueous phase.[10] As shown, α-tocopherol suppresses the oxidation of soybean PC liposomes and produces a distinct inhibition period. α-Tocopherol is consumed at a constant rate, and when it is depleted, a rapid oxidation takes place at a rate similar to that in the absence of antioxidant. The lower the rate of oxidation during the inhibition period (R_{inh}), the higher the activity of antioxidant.

Figure 2 shows the results of oxidation of methyl linoleate in benzene solution inhibited by α-tocopherol and 2,6-di-tert-butyl-4-methylphenol (BMP), a well-known synthetic antioxidant. α-Tocopherol and BMP suppress the oxidation for the same period under the same reaction conditions. However, the rate of oxidation during the inhibition period in the presence of α-tocopherol is much smaller than that in the presence of BMP, suggesting that α-tocopherol is more active than BMP as an antioxidant. Obviously, the longer it suppresses the oxidation and the more it reduces the rate, the higher the antioxidant activity of the compound.

The rates of oxidations in the absence and presence of antioxidant are given by Eqs. 1 and 2, respectively, where n, R_i, LH, and IH are the stoichiometric number of peroxyl radicals scavenged by each antioxidant, rate of chain initiation, lipid, and antioxidant, respectively, and k_p, k_{inh}, and k_t are the rate constants for chain propagation reaction 3, radical scavenging reaction 4, and chain-terminating reaction 5.[11]

$$R_p = \frac{k_p[LH]R_i^{1/2}}{(2k_t)^{1/2}} \tag{1}$$

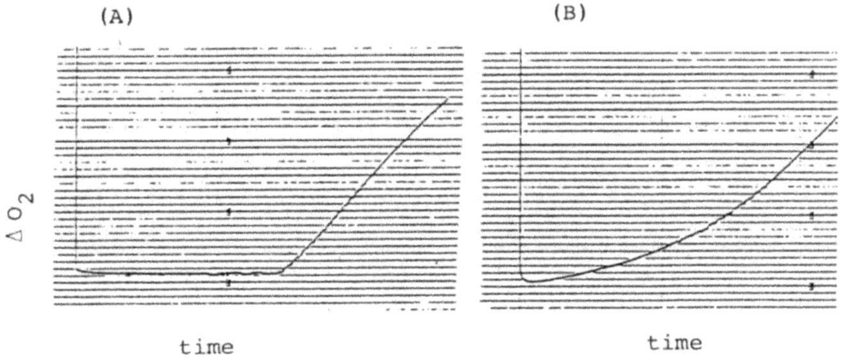

Figure 2. Oxidation of 503 mM methyl linoleate in benzene solution measured as oxygen uptake initiated with 20.4 mM 2,2'-azobis(2,4-dimethylvaleronitrile) at 37°C in the presence of (A) 0.151 mM α-tocopherol and (B) 0.151 mM BHT (2,6-di-tert-butyl-4-methylphenol).
(A) t_{inh} = 73.5 min; R_{inh} = 0
(B) t_{inh} = 71.0 min; R_{inh} = 0.283 μM/s

$$R_{inh} = \frac{k_p[LH]R_i}{nk_{inh}[IH]} \qquad (2)$$

$$LO_2\cdot + LH \xrightarrow{k_p} LOOH + L\cdot \qquad (3)$$

$$LO_2\cdot + IH \xrightarrow{k_{inh}} LOOH + I\cdot \qquad (4)$$

$$LO_2\cdot + LO_2\cdot \xrightarrow{k_t} \text{non-radical products} \qquad (5)$$

The ratio of the rate of inhibited oxidation to that of uninhibited oxidation is given by Eq. 6.

$$\frac{R_{inh}}{R_p} = \frac{(2k_t)^{1/2}R_i^{1/2}}{nk_{inh}[IH]} \qquad (6)$$

The smaller this ratio, the higher the activity of the antioxidant. Figure 3 shows the plots of Eq. 6 for the oxidations of soybean PC liposomal membranes in aqueous dispersions and for the oxidations of methyl linoleate in homogeneous solution inhibited by α-tocopherol. It shows that the ratio R_{inh}/R_p for the membrane system is much higher than that for homogeneous solution, suggesting that the antioxidant activity

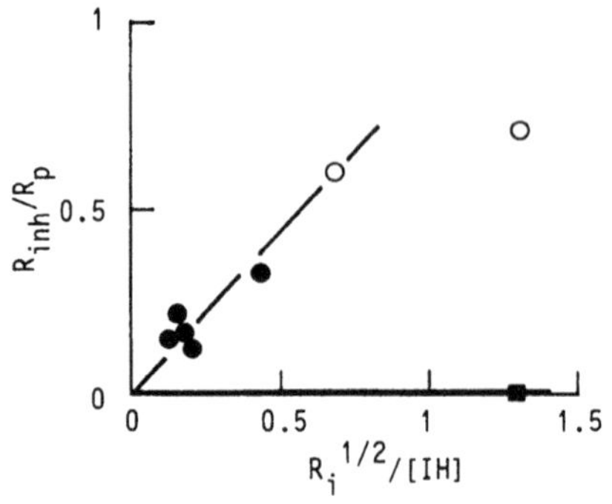

Figure 3. Plot of R_{inh}/R_p as a function of $R_i^{1/2}/[IH]$ in the oxidations of methyl linoleate in benzene solution (■) and of soybean PC liposomal membranes in aqueous dispersions (LUV, ○ ; SUV, ●) inhibited by α-tocopherol at 37°C in air.

of α-tocopherol is much less in the membranes than in the homogeneous solution.

Another important factor is the competition between chain propagation reaction 3 and chain interrupting reaction 4. In order to suppress the oxidation efficiently, the antioxidant should scavenge the chain-carrying peroxyl radicals before they attack another lipid or protein. This competition, the ratio of the rate of reaction 4 to that of reaction 3, is determined by the ratio of the rate constants k_{inh}/k_p.

$$\frac{\text{Rate (4)}}{\text{Rate (3)}} = \frac{k_{inh}[IH]}{k_p[LH]} \qquad (7)$$

This ratio can be determined from the rate of inhibited oxidation and the length of inhibition period (t_{inh}) measured from the rate of oxidation uptake during the oxidation in the presence of antioxidant.[12]

$$\frac{k_{inh}}{k_p} = \frac{[LH]}{t_{inh}R_{inh}} \qquad (8)$$

Table 1 summarizes the results for α-tocopherol and its model compound without phytyl side chain, 2,2,5,7,8-pentamethyl-6-chromanol (PMC). It compares the ratio k_{inh}/k_p for the oxidations in homogeneous solution and in liposomal membranes. In the oxidation of methyl linoleate in homogeneous solution, the ratio of k_{inh}/k_p is 5000 for both α-tocopherol and PMC. It shows that the ratio of reaction 4 to reaction 3 is 50 when the ratio of lipid to antioxidant is 100 to 1, which is compatible to the biological conditions. On the other hand, the ratio k_{inh}/k_p for PC liposomal membranes is approximately 100, which is 50

Table 1. The Ratio k_{inh}/k_p for α-Tocopherol and 2,2,5,7,8-Pentamethyl-6-chromanol (PMC) in the Oxidations of Methyl Linoleate in Benzene Solution and of Soybean PC Liposomal Membranes in Aqueous Dispersions at 37°C

	Homogeneous Solution	Soybean PC Liposomal Membranes MLV	SUV	LUV
α-Tocopherol	$5x10^3$	120	60	60
PMC	$5x10^3$	110		230

times smaller than that in the homogeneous solution. This also implies that the antioxidant activity of α-tocopherol in the membrane is less than that in the homogeneous solution.

CONCLUSION

α-Tocopherol is a potent chain-breaking antioxidant, but its antioxidant activity in the liposomal membranes appears to be less than that in the homogeneous solution. This lower activity may be attributed to a lower mobility of α-tocopherol and to a higher chance of chain propagation in the tightly packed structure of liposomal membranes.

REFERENCES

1. M. G. Simic and M. Karel, eds., "Autoxidation in Food and Biological Systems," Plenum, New York (1980).
2. W. A. Pryor, ed., "Free Radicals in Biology," Vols. I - VI, Academic Press, New York (1976) - (1984).
3. B. Halliwell and J. M. C. Gutteridge, "Free Radicals in Biology and Medicine," Clarendon Press, Oxford (1985).
4. H. Sies, ed., "Oxidative Stress," Academic Press, London (1987).
5. K. Terao and E. Niki, Damage to biological tissues induced by radical initiator 2,2'-azobis(2-amidinopropane) dihydrochloride and its inhibition by chain-breaking antioxidants, J. Free Rad. Biol. Med. 2:193 (1986), and papers cited therein.
6. E. Niki, Antioxidants in relation to lipid peroxidation. Chem. Phys. Lipids, 44:227 (1987).
7. G. W. Burton, D. O. Foster, B. Perly, T. F. Slater, I. C. P. Smith, and K. U. Ingold, Biological antioxidants, Phil. Trans. R. Soc. Lond: B311:565 (1985).
8. E. Niki, Lipid antioxidants: How they may act in biological systems. Br. J. Cancer 55: Suppl. VIII:153 (1987).
9. E. Niki, M. Saito, Y. Yoshikawa, Y. Yamamoto, and Y. Kamiya, Inhibition of oxidation of soybean phosphatidyl-choline and methyl linoleate in aqueous dispersions by uric acid. Bull. Chem. Soc. Jpn. 59:471 (1986).
10. E. Niki, A. Kawakami, Y. Yamamoto, and Y. Kamiya, Synergistic inhibition of oxidation of phosphatidylcholine liposome in aqueous dispersion by vitamin E and vitamin C, Bull. Chem. Soc. Jpn. 58:1971 (1985).

11. E. Niki, Y. Yamamoto, and Y. Kamiya, Oxidation of phosphatidylcholine and its inhibition by vitamin E and vitamin C, in "Oxygen Radicals in Chemistry and Biology," W. Bors, M. Saran, and D. Tait, eds., Walter de Gruyter, Berlin (1984).
12. E. Niki, M. Takahashi, and E. Komuro, Antioxidant activity of vitamin E in liposomal membranes, Chem. Lett. 1573 (1986).

CELLULAR PROTECTION AGAINST DAMAGE BY HYDROPEROXIDES:

ROLE OF GLUTATHIONE

John E. Biaglow, Marie E. Varnes,* Edward R. Epp,**
Edward P. Clark,** Steve Tuttle and Kathryn D. Held

University of Pennsylvania, Philadelphia, PA
*Case Western Reserve University, Cleveland, OH
**Harvard University, Boston, MA

Radicals produced by X-rays may react with oxygen, producing intermediates that are believed to be toxic to cells. We propose a hypothesis that glutathione (GSH)[1] may protect against radiation damage in four ways, as shown in Figure 1.

GSH can react with hydroxyl radicals, organic radicals, and peroxyl radicals, and is a substrate for enzymatic reduction of hydroperoxides and peroxides. Based on this model, we used the GSH-synthesis inhibitor L-buthionine sulfoximine (LBSO) to deplete cellular GSH in order to determine its importance in radiation response.[2] Results showed that GSH depletion to values as low as 0.04-0.1 nmoles/mg protein (Figure 2) increased the aerobic radiation response of A549 cells (dose-modifying factor at 1 Gray is 2). Subsequent inhibition of GSH-transferase and partial inhibition of GSH-peroxidase was observed.[2-4] Both enzymes catalyze hydroperoxide reduction by GSH. Therefore, decreased cellular

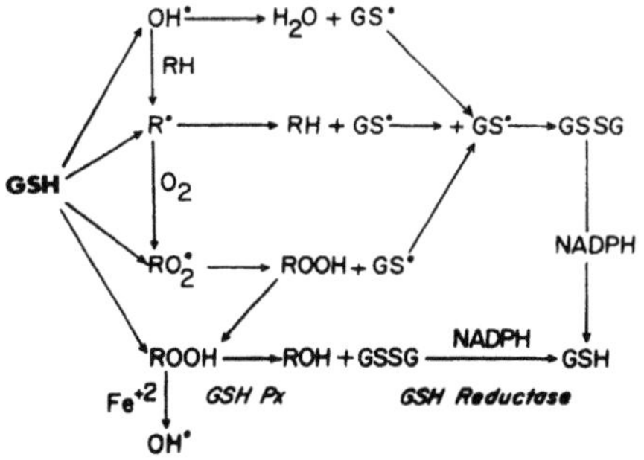

Figure 1. Reaction of glutathione with hydroxyl radical (•OH), organic radicals (R•), peroxyl radicals (ROO•), and hydroperoxides (ROOH). GSH reacts with these radicals to produce thiol radicals that self-associate to produce oxidized glutathione (GSSG). GSH is regenerated by reduction with NAD(P)H generated from the pentose cycle. It is believed that reactions of unbound Fe^{2+} with peroxide can lead to additional generation of hydroxyl radicals and increased production of malondialdehyde (MDA).

capacity to reduce hydroperoxides should result in cellular vulnerability to both peroxide- and hydroperoxide-induced damage.

We have not yet assessed the toxicity of radiation-induced hydroperoxide or subsequent breakdown products. However, we have attempted to understand these toxic mechanisms by investigating the ability of exogenously added hydroperoxides to alter cell biochemistry and ultimately to cause cell death. We have tested the effect of tertiary butylhydroperoxide (TBOOH) on radiation response and cell lethality. We have focused on the effect of glutathione depletion on the inhibition of GSH peroxidase and the non-selenium-containing enzyme GSH-S-transferase, which also can act as a peroxidase. We have also studied the role that the pentose cycle may have in reducing the products of the reactions with GSH (cf. Figure 1).

Intracellular GSH depletion should make cells more vulnerable to hydroperoxides since both transferase and peroxidase activities are altered.[5] Figure 3 represents lethality to A549 cells depleted to less than 0.1 nmoles GSH levels. The top curve shows that some toxicity resulted for control cells; however, the absence of GSH increased lethality by 70-fold with 0.1 mM TBOOH. Also, lesser concentrations of TBOOH increased toxicity in thiol-depleted cells. Lack of significant toxicity in the control cells may reflect the reduction of TBOOH to the more innocuous alcohol tertiary butyl alcohol (TBOH) without appreciable oxidation of GSH. Therefore, to understand the increased TBOOH-induced toxicity toward LBSO-treated cells, we studied its effects on cellular NAD(P)H oxidation and subsequent regeneration.

We have demonstrated that TBOOH added to cells results in oxidation of pyridine nucleotides. This occurs because GSH peroxidase catalyzes TBOOH reduction (2GSH + TBOOH------>GSSG + TBOH + H_2O). GSSG in turn is

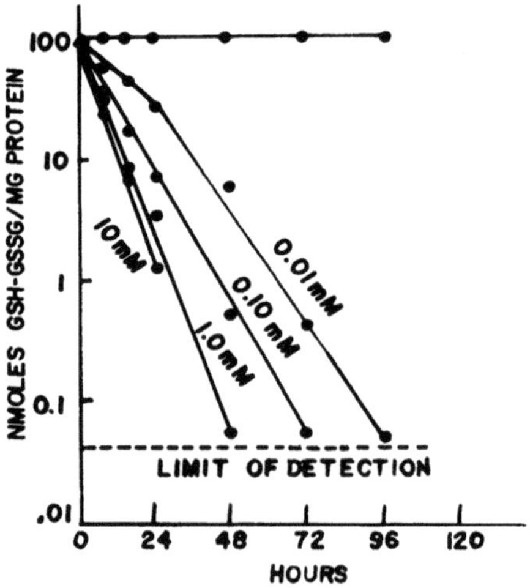

Figure 2. Effect of L-BSO treatment on cellular GSH content. Most rapid depletion of cellular GSH occurs in the presence of 1-5 mM LBSO. The rate of depletion was highest at 0.1 mM LBSO concentration and least with 0.01 mM. The limit of detection is about 2 decades with cell densities of 10^5, and increases to greater than 4 decades with cell densities greater than 10^6. For subsequent experiments we used 0.1 mM LBSO for 72 hr or 0.25 mM for 52 hr to achieve the maximal GSH depletion.

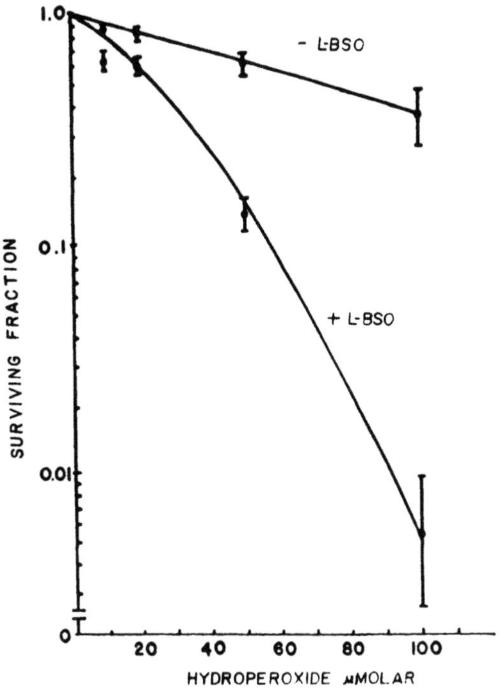

Figure 3. Cytotoxicity of TBOOH. A549 cell cultures were pretreated with 0.1 mM LBSO for 72 hr. Cells were harvested and resuspended in Hank's salt solution (calcium and magnesium free) at a final density of 10^6/ml. TBOOH was added to the cell suspensions at the appropriate concentration for 10 min. Aliquots were removed and plated into fresh growth media to determine clonogenic survival. All incubations were performed at 37°C.

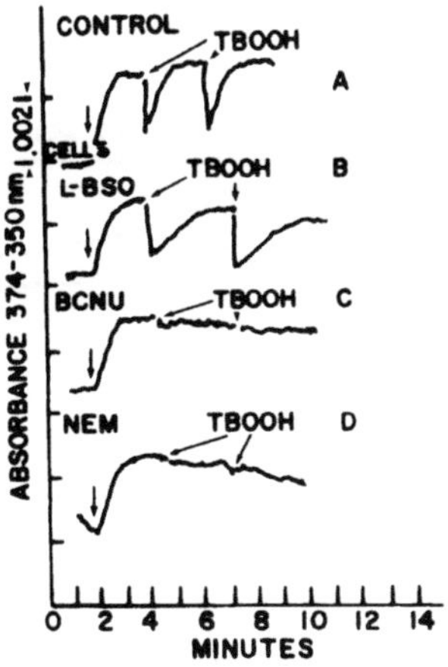

Figure 4. Effect of TBOOH additions on NAD(P)H levels of suspended A549 cells. Cells were suspended after harvesting in 2 ml buffered saline containing 20 mM Hepes, pH 7.4, and 10 mM KCl, 37°C. At the arrow 100 nmoles TBOOH was added. Cyanide conc. was 1 mM and NEM was 0.1 mM. Cells were treated with 0.1 mM LBSO for 72 hr before use and, where indicated, with 0.1 mM BCNU for 1 hr.

reduced by NAD(P)H (GSH reductase) to produce $NADP^+$ and 2GSH. Figure 4 shows pyridine nucleotide changes for A549 cell suspensions following the addition of 100 nmoles of TBOOH. The top curve shows that oxidation of NAD(P)H occurs immediately upon addition of TBOOH, but a return to control levels follows within three minutes. Cells treated with LBSO for 72 hr are shown to regenerate NAD(P)H more slowly, indicating partial inhibition of peroxidase and/or reductase. Bis-chloronitroso urea (BCNU) inhibits glutathione reductase.[6] We then predicted it would inhibit cellular reduction of GSSG produced by TBOOH-linked oxidation of GSH (see Figure 1). Curve C in Figure 4 shows that cyclic oxidation and reduction is inhibited totally by BCNU preincubation. N-ethyl maleimide (NEM) addition (bottom curve) also blocks this process since NEM forms a covalent bond with GSH.[1]

Although we have not yet measured TBOOH disappearance, we have developed methods for measuring the effect of various agents on cellular reduction of peroxide (Figure 5). We found this reduction is not appreciably affected by GSH depletion; however, the amount of NEM required to inhibit peroxide reduction decreased 40-fold after LBSO treatment. We found similar effects for NEM titration of pyridine nucleotide oxidation (data not shown).

Hydroperoxides, if produced following radiation, also may have deleterious effects on cellular macromolecules because they can be a source for hydroxyl radicals (cf. Figure 1). Figure 6 shows that DNA breaks are caused by cells treated with 50 μM TBOOH for 10 min. These breaks are repaired in control cells, but not in GSH-depleted cells (upper curve).

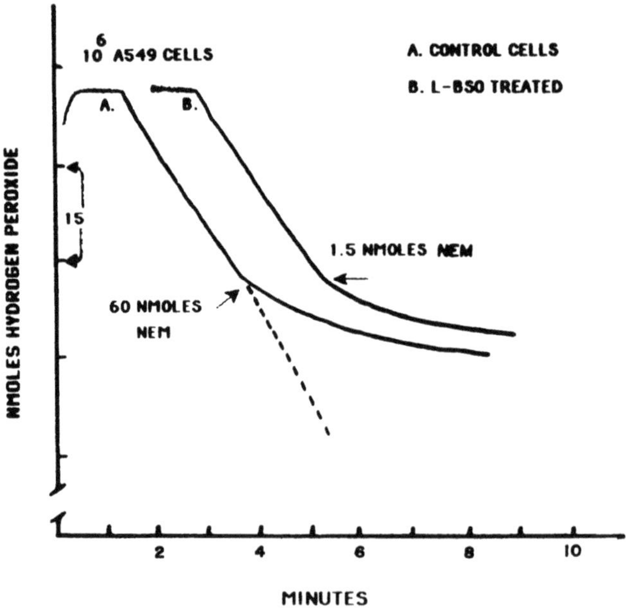

Figure 5. Effect of NEM on peroxide reduction for control cells and for cells treated with 0.25 mM LBSO for 52 hr. Catalase was inhibited by adding sodium azide to a final concentration of 0.02 mM. Cells were incubated in buffered saline (cf. Figure 4). Peroxide was added prior to cell addition. Peroxide reduction is seen as a downward deflection, measured by a peroxide electrode.

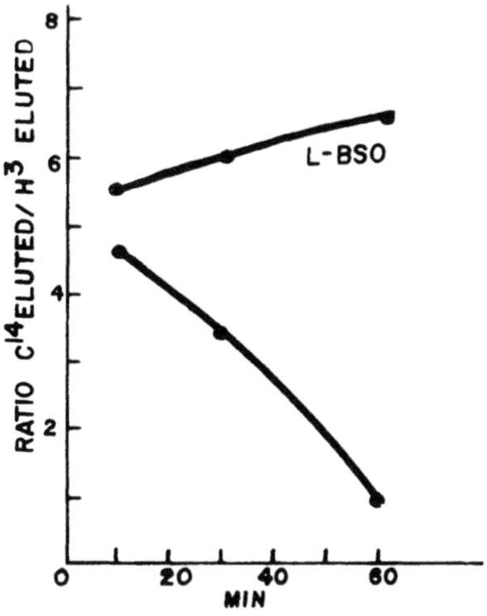

Figure 6. Hydroperoxide-induced lesions in cellular DNA. The alkaline elution technique was used to determine the effect of TBOOH on the integrity of cellular DNA. Cells were pretreated for 52 hr with 0.25 mM LBSO. One million cells were suspended in 10 ml Hank's balanced salts with calcium and magnesium. TBOOH was added to the cell suspension to a final concentration of 50 µM.

According to our model and previously obtained data,[1,7] cellular response to radiation may be altered if hydroperoxides cause depletion of GSH(TBOOH + 2GSH----->TBOH + GSSG + H_2O). Therefore, we have tested the radiosensitizing capacity of TBOOH. Figure 7 shows that TBOOH also can alter the radiation response of control cells (Panel B). Panel A indicates no measurable effect. However, Panel B shows that TBOOH administered post-irradiation enhanced the cell's radiation response. Of major interest is the effect of hydroperoxides on the radiation response of GSH-depleted cells, represented by Panel C. Low thiol levels suggest

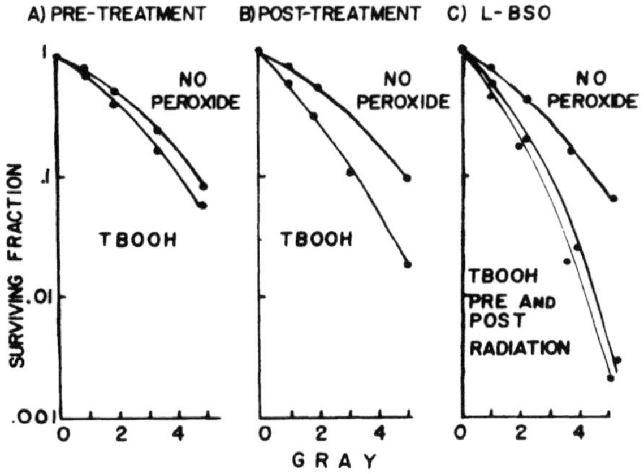

Figure 7. Effect of TBOOH on the radiation response for pretreatment (A), post-treatment (B), and GSH-depleted cells. In the latter case TBOOH was present for 5 min either during or post-irradiation. 200 nmoles of TBOOH was added to 10 ml of 10^6 cells suspended in Hank's solution without magnesium and calcium but containing 20 mM Hepes, pH 7.4, 37°C.

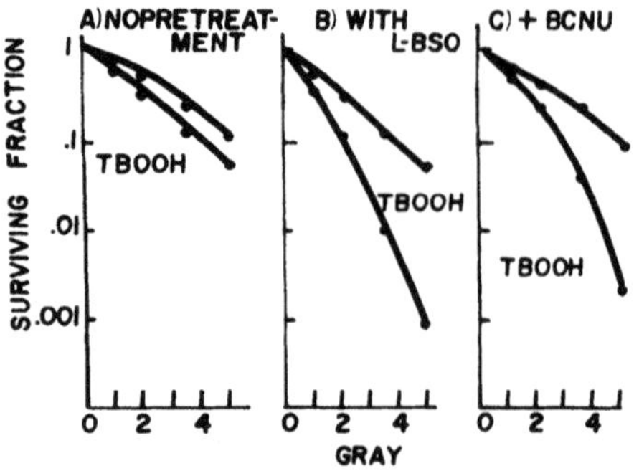

Figure 8. Effects of LBSO and BCNU on the radiation response of cells treated with TBOOH. The incubation conditions are as described for Figure 7. Cells were treated with 0.1 mM BCNU prior to exposure to 20 μM TBOOH.

that TBOOH causes oxidation of "bound" GSH (GSH-PX) resulting in GSSG formation. GSH disulfide will rapidly thiylate protein thiols. GSSG thiylation will result in enzyme inhibition and an inability to repair either X-ray or TBOOH induced damage. It remains to be determined if X-ray or TBOOH damage is at different or similar sites and if the same or different enzymes are involved in the repair process. Figure 8 shows similar effects when cells are pretreated with BCNU, which did not remove GSH but did inhibit GSH reductase, thus affecting the cell's radiation response as much as TBOOH with LBSO-treated cells.

CONCLUSION

In conclusion, our results show that hydroperoxides are toxic to cells and may influence their radiation response. Toxicity is enhanced either by depleting thiols or by blocking the reduction of GSSG by BCNU-linked inhibition of glutathione reductase. Hydroperoxides also increase DNA breaks and alter the cellular radiation response when added post-irradiation. Our results indicate that cellular capacity to reduce peroxides or hydroperoxides is not directly related to effects on cellular lethality. Total inhibition of peroxidase activity does not occur with LBSO treatment, but does occur when residual thiols are depleted.

Cellular redox effects caused by addition of TBOOH to either GSH-depleted cells post-irradiation or to thiol-containing control cells may correlate with inhibition of enzymes (by thiylation as described above) involved in DNA repair. Inhibition of TBOOH reduction by BCNU also suggests that TBOOH affects enzymes involved in DNA breaks and their subsequent repair.

Our results show that LBSO treatment results in GSH depletion to non-detectable values which are still not sufficient to deplete enzyme bound thiols. These thiols may be involved in the radiation response if they are coupled with enzymes to catalyze hydrogen or electron transfer as seen by the lack of inhibition of peroxidase activity.

Figure 9 shows that GSH is located in both the free and bound forms and may be equilibrated freely between the cytosol and the nucleus, but not in the mitochondria. Binding may occur by molecular association to enzymes either in the membrane or in the cytosol. Our data shows that GSH depletion may inhibit GSH-S-transferase to a greater degree than GSH peroxidase.[2] Figure 5 shows that peroxidase inhibition when agents are

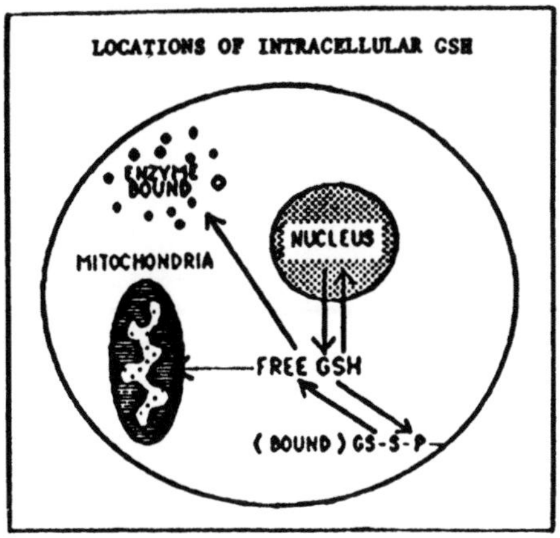

Figure 9. Locations of intracellular GSH. Glutathione is found in the cell cytoplasm, nucleus, and mitochondria, and may exist in the free or bound form. It may be chelated or bound in the disulfide form.

added either depletes thiols or blocks dehydrogenase (GSH reductase) activity (BCNU). It appears that a small pool of thiols may be important in controlling total damage caused by peroxides. This pool of thiols may be enzyme bound and accessible only when covalent binding agents are added to cells. Additional studies are required in this area.

REFERENCES

1. J. E. Biaglow, M. E. Varnes, E. P. Clark, and E. R. Epp, The role of thiols in cellular response to radiation and drugs, Radiat. Res. 95:437 (1983).
2. J. E. Biaglow, M. E. Varnes, E. R. Epp, E. P. Clark, and M. Astor, Factors involved in depletion of glutathione from A549 human lung carcinoma cells: Implications for radiotherapy, Int. J. Radiat. Oncol. Biol. Phys. 10:1221 (1984).
3 J. E. Biaglow, E. P. Clark, E. R. Epp, M. E. Varnes, and J. B. Mitchell, Nonprotein thiols and the radiation response of A549 human lung carcinoma cells, Int. J. Radiat. Biol. 4:489 (1983).
4. J. E. Biaglow, M. E. Varnes, E. R. Epp, E. P. Clark, and M. Astor, Redox enzymes and thiol radicals, in: "Oxygen and Sulfur Radicals in Chemistry and Medicine," M.A.J. Rogers, A. Breccia, and G. Semerano, eds., Lo Scarabeo, Bologna, Italy (1986).
5. J. E. Biaglow, M. E. Varnes, E. P. Clark, and E. R. Epp, The role of cellular glutathione in response of tumor cells to radiation and drugs, in: "Biochemical Modulation of Anticancer Agents: Experimental and Clinical Approaches," F.A. Valeriote and L.H. Baker, eds., Martinus Nijhoff Publishing, Boston (1986).
6. J. E. Biaglow, M. E. Varnes, E. R. Epp, and E. P. Clark, Factors influencing tertiary butylhydroperoxide metabolism, submitted to J. Biol. Chem. (1987).
7. J. E. Biaglow, M. E. Varnes, J. B. Mitchell, and A. Russo, Role of the hexosemonophosphate shunt in the cellular response to radiation and drugs, submitted to Cancer Res. (1987).

ACKNOWLEDGEMENTS

This research was supported by National Cancer Institute Grants #RO1 CA 44982, #RO1 CA 40516, and #RO1 CA 18614.

A SECONDARY ANTIOXIDANT DEFENSE ROLE FOR PROTEOLYTIC SYSTEMS

Kelvin J. A. Davies

Institute for Toxicology and Department of Biochemistry
The University of Southern California
1985 Zonal Avenue, HSC-PSC 614-616
Los Angeles, California 90033, U.S.A.

INTRODUCTION

Free radicals, and related oxidants, have long been studied as agents of tissue damage. Lipid peroxidation is a well-established example of free radical toxicity, as is oxidative damage to RNA, DNA, and even carbohydrates. In contrast, the reactions of proteins with various radicals/oxidants have not been so extensively studied and it is sometimes assumed that proteins are not particularly susceptible to free radical damage. This assumption has been invalidated in recent years, and it is now clear that amino acids, peptides, and proteins are, indeed, vulnerable to attack by a variety of free radicals and related oxidants.[1-19] Although mechanisms of damage may vary greatly, it appears that oxidative processes are the common pathway for toxic reactions between both free radicals and proteins and various oxidants and proteins. I shall, therefore, use the term "oxidatively damaged proteins" to cover a wide variety of possibilities.

Enzymes such as superoxide dismutase, glutathione peroxidase, and DT-diaphorase, and compounds such as vitamin E, β-carotene, and uric acid (this is not an exhaustive list) may be considered to be "primary antioxidant defenses" since they function to prevent initiation and/or propagation reactions of radicals/oxidants with cellular constituents.[20-29] Primary antioxidant enzymes and compounds may protect proteins as efficiently as they guard lipids, nucleic acids, and carbohydrates. Nevertheless, research has shown that protein damage still occurs. Oxidation of particularly sensitive amino acid residues, aggregation or cross-linking, fragmentation, and loss of enzymatic or other functional properties[1-19] are but a few of the documented examples. At best, oxidatively damaged proteins would represent a pool of useless cellular debris. More serious metabolic and immunological problems could be created by large accumulations of protein aggregates, toxic fragments, or deranged enzymes.

Since oxidatively damaged proteins serve no useful roles, one might predict that cells would possess mechanisms for their removal. Proteolytic degradation would be the most obvious means by which this might be achieved, with the obvious advantage that undamaged amino acids might be re-utilized. All cells are thought to contain both soluble[30-38] and membrane-bound[39,40] proteolytic enzymes, and most eucaryotes also possess lysosomes which can injest and degrade proteins.[35] Many proteolytic enzymes are involved in the normal turnover of proteins, while others carry out a wide series of proteolytic activations. In recent years it has become increasingly apparent that several nonsense or missense mutations, as well as products of certain post-translational modifications,[30-40] are also recognized and degraded by proteolytic enzymes. Thus, there is ample precedent for the intracellular degradation of damaged or abnormal proteins.

In this paper I present the hypothesis that proteolytic systems (consisting of proteinases, proteases, and peptidases) may have an important role in overall antioxidant defenses. In this hypothesis proteolytic systems are envisioned as secondary defenses which act to prevent the accumulation of oxidatively damaged proteins and to degrade potentially toxic protein fragments. Proteolytic systems could be of great importance when primary antioxidants (superoxide dismutase, glutathione peroxidase, vitamin E, uric acid, etc.) are insufficient to cope with a major oxidative stress. Even the normal levels of oxidants which cells produce or encounter may necessitate constant protein turnover since no antioxidant is likely to be 100% effective in preventing protein damage.

PROTEIN DEGRADATION DURING "OXIDATIVE STRESS"

"Oxidative stress" may be loosely defined as that state in which exposure to free radicals or other oxidants represents a challenge to normal function, or even to survival. Such situations may arise from an increased exposure to radicals/oxidants, or may be the result of a decreased antioxidant capacity. This paper and its references report multiple cases in which proteins undergo oxidative damage, both in vitro and in vivo. Thus, it is reasonable to state that proteins are susceptible to "oxidative stress".

Recently, data showing that oxidative stress can result in proteolysis have been provided by this laboratory,[16,19,41,46] Dean and co-workers,[47-50] and Stadtman, Levine, Rivett, and co-workers.[51-55] Such studies have included: erythrocytes and reticulocytes from humans, rabbits, and cows; skeletal muscles; cardiac muscle; mitochondria; and bacteria. Protein damage and degradation have been observed following exposure of intact cells to free radicals, related oxidants, and mixed-function oxidases. The proteolytic susceptibility of purified proteins has also been shown to increase with mild exposure to various free radicals and oxidants, but severe damage may actually cause a decrease in proteolytic susceptibility. It now seems clear that oxidatively damaged proteins are, in general, more susceptible to proteolytic degradation than are native proteins. Extensively cross-linked proteins (an extreme form of oxidative damage) appear to be the exception to this rule since they may actually exhibit increased proteolytic resistance.

Free radicals and oxidants (such as •OH and H_2O_2) can alter the primary, secondary, and tertiary structure of proteins (in a variety of ways) to cause denaturation and increased hydrophobicity. With some radicals/oxidants, the presence of oxygen can allow secondary reactions which result in varying degrees of direct protein fragmentation. Both the denatured/hydrophobic products, and the protein fragments are excellent substrates for degradation by intracellular proteolytic systems. In contrast, certain radicals/oxidants under conditions of low or no oxygen, or exposure to extremely high levels of certain radicals/oxidants, can cause extensive protein cross-linking. Such cross-linked products exhibit decreased susceptibility to proteolytic digestion, and would (therefore) be expected to form insoluble aggregates or "clumps" of cellular debris.

In fact, the concept that proteins may undergo increased degradation following exposure to free radicals dates back more than 30 years to the studies of Yallow and Berson.[56] These investigators were intrigued by the finding that [^{131}I]albumin (radio-labeled *in vitro*) was rapidly degraded when injected into human subjects. Yallow and Berson suggested that free radicals generated by decay of the ^{131}I-label subjected the [^{131}I]albumin to damage and denaturation (a form of "self-irradiation"). Yallow and Berson further proposed that such free-radical damage was the cause of increased degradation *in vivo*.

It should also be noted that "oxidative stress" may affect protein degradation in other ways. In red cells, for example, it appears quite clear that agents or situations which increase exposure to O_2^-, H_2O_2, and related radicals or oxidants cause cellular damage.[57-59] Such radicals and oxidants are also known to cause polymerization (or cross-linking) of membrane proteins and lipids, a process which has been associated with increasing age.[60,61] Whether cross-linked aggregates of membrane proteins and lipids would be better or worse substrates for intracellular (and intramembrane) proteolytic enzymes is not yet known.

In other studies of interest, Aft and Mueller[62] have suggested that hemin may cause the fragmentation of proteins in a process involving activated oxygen species. Such fragments would be expected to be rapidly degraded by intracellular proteolytic systems. Several investigators have provided evidence that α_1-antiprotease can be damaged and rendered inactive by free radicals and other oxidants in cigarette smoke and by radicals and oxidants released by neutrophils in the lungs of cigarette smokers.[63-65] Such investigations suggest that the emphysema associated with cigarette smoking may be caused by inability of damaged α_1-antiprotease to inhibit elastase (released by neutrophils), since active elastase will induce emphysema-like changes in laboratory animals. In contrast to this inactivation of a proteolytic inhibitor, Weiss and co-workers[66] have reported that neutrophils can activate their own collagenase by a mechanism involving oxidative modification of the pro-enzyme. These authors propose that such oxidative activation ".....may be important in the pathogenesis of connective tissue damage *in vivo*".

It has been suggested that oxidative damage may also be a mechanism by which cells may intentionally control the degradation of certain proteins.[43,54] Thus, selective oxidation may act as a "marking step" in proteolysis. In addition to other "marking" processes,[54] selective oxidation may play an important role in regulating protein turnover since all cells appear to exert differential control of the synthesis and degradation of individual proteins.[67-70]

Oxidative regulation of protein degradation via transcriptional and translational responses may be suggested by studies which correlate oxidative stress with (aspects of) the phenomenon of "heat-shock". Thus, Ames and his colleagues[71,72] have reported that (in bacteria) some of the heat-shock genes, and their translation products, may also be expressed under conditions of oxidative stress. These authors further propose that (some of) the heat-shock proteins may include antioxidant enzymes (such as superoxide dismutase) which may be important in protecting the cell against oxidative stress. Interestingly, Bond and Schlesinger[73] have reported that ubiquitin is the product of a (mammalian) heat-shock gene. Furthermore, Goff and co-workers[74] have shown that one of the bacterial heat-shock genes codes for protease La, and Ananthan and co-workers[75] propose that denatured proteins may activate (eukaryotic) heat-shock genes in a similar manner to heat shock itself. Such intriguing findings, added to the other evidence presented in this paper, indicate that free radicals and oxidants may have multiple and extensive effects on intracellular and extracellular proteolysis *in vivo*.

RELATIONSHIPS BETWEEN PROTEIN DAMAGE AND PROTEOLYSIS

Denaturation, Increased Hydrophobicity, and Fragmentation Serve to Increase Proteolytic Susceptibility

Studies from this laboratory have indicated that modifications to individual amino acids cannot explain major changes in proteolytic susceptibility.[16-19,41-46] Instead, it appears that alterations to secondary and tertiary structure (which, of course, depend on amino acid modification) underlie proteolytic responses. In particular, protein denaturation and increased hydrophobicity seem to insure rapid proteolysis. We have used measurements of protein solubility in various salt concentrations and at various pH values as a reflection of denaturation and increased hydrophobicity. Such measurements reveal that denaturation and increased hydrophobicity precede the cross-linking reactions induced by exposure to •OH. In fact, denaturation and increased hydrophobicity exhibit essentially the same •OH-dependency as does proteolytic susceptibility. Similar measurements on proteins exposed to •OH + O_2^- (+O_2) indicate that denaturation/hydrophobicity are again associated with increased proteolytic susceptibility, although the results are less clear due to increased degradation of protein fragments. Furthermore, radical-scavenging agents (e.g. mannitol and uric acid) which inhibit denaturation/increased hydrophobicity also prevent increases in proteolytic susceptibility.

Concomitant increases in protein denaturation and hydrophobicity can be caused by the unfolding of a globular protein (such as BSA) into a more open or random form. Such a process may be roughly analogous to heat denaturation, and should affect degradation by many proteases. Our experiments with a wide variety of unrelated animal, plant, yeast, and bacterial (purified) proteases indicate that this is indeed the case. We have found, however, that oxidative damage tends to produce more extreme (and permanent) denaturation than does heat alone. For example, if BSA is heat denatured under O_2, or in air, it is degraded (by extracts and purified proteases) more rapidly than if heat denaturation were carried out under anaerobic conditions. Furthermore, exposure to •OH or heat treatment under O_2 or in air causes greater increases in BSA denaturation/hydrophobicity than does anaerobic heat treatment.

I, therefore, propose that protein denaturation and/or increased hydrophobicity may actually be the major cause(s) of the increased proteolytic susceptibility which accompanies oxygen radical exposure. Limited protein fragmentation by •OH, and more extensive fragmentation by exposure to •OH + O_2^- (+O_2), appears to be a secondary reaction which augments the concentration of readily degradable (modified) protein molecules. Similar conclusions have also been independently reached by Stadtman[54] and by Wolf, Garner, and Dean.[48]

In recent studies (K.J.A. Davies, S.W. Lin, and R.E. Pacifici, unpublished observations) we have detected and partially purified a large proteolytic enzyme (approximately 700,000 molecular weight) in red blood cells which appears to preferentially degrade oxidatively denatured proteins. This protease may recognize hydrophobic sequences of amino acids which are normally shielded in the interior of a folded protein, but which would be exposed upon oxidative denaturation. The protease is inhibited by serine reagents and transition-metal chelators, which also inhibit the degradation of oxidatively denatured proteins in cells and cell extracts. Similar studies with *E. coli* indicate that the degradation of oxidatively denatured proteins is conducted by multiple serine- and metallo-proteases. With oxidatively denatured albumin as substrate, proteases "Do" and "Re" appear to be particularly active. Protease "So", however, is far more effective with oxidatively denatured hemoglobin as a substrate.

Although ATP appears to be involved in the degradation of many proteins (both *in vitro* and *in vivo*), all of our studies indicate that ATP is not required for the degradation of oxidatively denatured proteins. The 700,000 molecular weight protease mentioned above exhibits no ATP stimulation, and the ATP-dependent protease from *E. coli* (protease "La") exhibits little or no activity against oxidatively modified proteins. Furthermore, protease "La" deficient mutants of *E. coli* (engineered by deletion of the *lon* gene) degrade oxidatively damaged proteins just as well as does their wild-type parent strain.

Covalent Cross-Linking Can Decrease Proteolytic Susceptibility

Since modifications to primary structure may not provide a complete mechanistic explanation for increased proteolytic susceptibility, we have also examined secondary and tertiary structural modifications to proteins by •OH, and by •OH + O_2 (and O_2^-).[16-19,41-46] The clearest conclusion of such investigations is that the decrease in proteolytic susceptibility which occurs at •OH/protein molar ratios higher than 20 is due to covalent cross-linking reactions between protein molecules. The formation of covalent cross-links is a minor reaction at •OH/protein molar ratios of less than 5. At exposures of more than 5 •OH radicals per protein molecule, however, cross-linking rapidly becomes a major factor. Such intermolecular cross-links (which include bi-tyrosine) are formed when a protein radical (formed by •OH addition or hydrogen abstraction) reacts with a neighboring protein radical or molecule. Naturally, such covalent bonds are resistant to proteolytic digestion. Furthermore, in the conglomerates of proteins so formed the interior molecules are shielded from protease attack by their exterior (covalent) protein "coats".

Related work by Dizdaroglu, Gajewski, and Simic[76] has shown that cross-links induced in alanine dipeptides by •OH render the peptide bond resistant to proteolysis. In contrast, these authors found that phenylalanine tripeptides exhibit no increase in proteolytic resistance after cross-linking by •OH. Such results indicate that covalent cross-links (induced by •OH) are likely to produce a heterogeneous profile of both resistant and non-resistant peptide bonds within a complete protein.

When oxygen is present during •OH exposure, protein cross-links do not form and no reversal of proteolytic susceptibility is seen. It appears that oxygen reacts with protein radicals (thus inhibiting cross-linking) to form carbon-centered peroxyl radicals. The decomposition of such peroxyl species is thought to result in protein fragmentation (*vide supra*).

PRIMARY AND SECONDARY ANTIOXIDANT DEFENSES

If proteolytic enzymes/systems can preferentially degrade many oxidatively damaged proteins, we might expect that proteolytic inhibitors could prevent this activity. The result of such inhibition should be an accumulation of oxidized proteins which may also be cross-linked due to further oxidation and/or reactions with other cell constituents (e.g. lipids). Evidence for such a scheme has already been provided by Ivy and colleagues[77] who reported that leupeptin and chloroquine cause the accumulation of "lipofuscin like dense bodies in the (rat) brain". Chloroquine decreases the activity of intralysosomal (acidic) proteases by allowing a marked increase in lysosomal pH, whereas leupeptin binds to and inhibits thiol proteases of both cytoplasmic and lysosomal origin. Lipofuscin is a heterogeneous mixture of oxidized proteins and lipids, which are held together in aggregates consisting of both covalent and hydrophobic bonds. Lipofuscin, and lipofuscin-like material, is known to accumulate during ageing in response to several "oxidative stresses", and in a variety of poorly understood diseases.

I propose that proteolytic systems (enzymes) which conduct the degradation of oxidatively damaged proteins should not be considered as part of the damage process but, rather, as a defense mechanism. In this paper I have largely concentrated on soluble or cytoplasmic (non-lysosomal) proteolytic systems, but lysosomal and membrane proteases may also participate. By selectively (and rapidly) degrading oxidatively damaged proteins such proteolytic systems may provide four main benefits to the cell:

1) Degradation of oxidatively damaged, denatured/hydrophobic proteins
2) Inhibition of protein cross-linking or aggregation reactions
3) Degradation of potentially toxic protein fragments (peptides)
4) Proteolytic "release" of undamaged amino acid residues for re-utilization

By selectively degrading oxidatively damaged (denatured/hydrophobic) proteins, proteolytic systems can function directly in the removal of useless cellular debris. Without such proteolytic systems, or when oxidative stress exceeds the capacity of proteolytic systems, oxidatively damaged proteins may accumulate to dangerous levels. It should be noted that only low levels of oxidative damage appear to be necessary for proteolytic recognition (i.e. one or two radical "hits" per protein molecule). In contrast, more extreme oxidative damage (i.e. more than five •OH per protein) may cause covalent cross-linking. By rapidly degrading proteins with low levels of oxidative damage, proteolytic systems may inhibit further reactions in the same protein molecule, thus preventing cross-linking and aggregation. Random peptides produced by oxidative protein fragmentation may induce serious metabolic and immunological toxicities. By degrading such peptides proteolytic systems may prevent a potential plethora of undesired effects. The majority of amino acids released (by proteolysis) from oxidatively damaged proteins and peptide fragments will be undamaged residues. Proteolytic systems, thus, can participate in the return of valuable building blocks to the metabolic pool.

In the scheme below, I have formalized a proposed role for proteolytic systems in overall antioxidant defenses. This scheme differentiates between primary antioxidant defenses such as superoxide dismutase, glutathione peroxidase, DT-diaphorase, catalase, vitamin E, β-carotene, and uric acid,[20-29] and secondary antioxidant defenses such as proteinases, proteases, peptidases, DNA repair systems,[78] and phospholipase A_2. Phospholipase A_2 is included because of recent studies by Sevanian and coworkers[79,80] which indicate that this enzyme can selectively cleave peroxidized lipids from membranes, thus preventing their accumulation and their continued participation in chain reactions.

Proposed Classification of Antioxidant Defenses

Primary Defenses
1. Antioxidant Compound
2. Antioxidant Enzymes

Secondary Defenses
1. Lipolytic Enzymes
2. DNA Repair Systems
3. Proteolytic Systems

In the above scheme primary antioxidant defenses are those enzymes and compounds which prevent the initiation or propagation of radical/oxidant damage, or which act as (radical) chain terminating agents: In other words, the primary antioxidant defenses are the "classical" (or true) antioxidants. In contrast, the secondary antioxidant defenses are enzymes which have no true (or direct) antioxidant role but which are suggested to function, instead, to prevent the accumulation of useless and potentially harmful cellular debris.

As secondary antioxidant defenses, proteinases, proteases, and peptidases would provide a "second line of defense" against an essentially unlimited variety of free radicals and oxidants. While the exact identities of such enzymes have yet to be elucidated it does appear that, in the majority of experiments performed to date, proteolysis is independent of any energy (e.g. ATP) input and may also not require the participation of ubiquitin. We do not yet know if the enzymes of the ATP-stimulated or ubiquitin-dependent proteolytic pathway(s) may be involved in the degradation of oxidatively damaged proteins (without need for ATP or ubiquitin), or if such degradation is conducted by previously unidentified enzymes. It appears probable that cells contain multiple interrelated and independent proteolytic systems, any or all of which may be involved in the degradation of oxidatively damaged proteins. Work is now under way (in my laboratory) to establish the identities and properties of those proteinases, proteases, and peptidases which may be included as elements of secondary antioxidant defense systems.

Proteolytic systems may be required to function as secondary lines of antioxidant defenses at all times, since cells produce and experience radicals and oxidants on a continual basis. The greatest need, however, may arise under conditions of oxidative stress, or when primary antioxidant defenses are compromised. Such conditions arise during exposure to a wide variety of toxic radicals/oxidants, and during various disease states. Interestingly, the ageing process itself has been reported to involve alterations in both primary antioxidant defense capabilities, and proteolytic capacity. Given these thought-provoking potential associations, it appears likely that the role of proteolytic systems in overall antioxidant defense mechanisms will be a fruitful area for further study.

ACKNOWLEDGEMENTS

I should like to thank Dr. Marta E. Delsignore, Ms. Sharon Wu Lin, and Mr. Robert E. Pacifici who have made this paper possible. I am, indeed, fortunate to have such dedicated, able, and energetic colleagues.

This work was supported by grant #ES 03598 and by grant #ES 03785 from the National Institutes of Health. Further support was provided by the Faculty Research and Innovation Award of the University of Southern California.

REFERENCES

1. A.J. Gross and I.W. Sizer, The oxidation of tyramine, tyrosine, and related compounds by peroxidase, J. Biol. Chem. 234:1611-1614 (1959).
2. W.M. Garrison, M.E. Jayko, and W. Bennett, Radiation-induced oxidation of protein in aqueous solution, Radiat. Res. 16:483-502 (1962).
3. G.E. Adams, R.L. Willson, R.H. Bisby, and R.B. Cundall, On the mechanism of the radiation-induced inactivation of ribonuclease in dilute aqueous solution, Int. J. Radiat. Biol. 20:405-415 (1971).
4. K. Baverstock, R.B. Cundall, G.E. Adams, and J.L. Redpath, Selective free radical reactions with proteins and enzymes: the inactivation of α-chymotrypsin, Int. J. Radiat. Biol. 26:39-46 (1974).

5. H. Schuessler and A. Herget, Oxygen effect in the radiolysis of proteins: I. lactate dehydrogenase, Int. J. Radiat. Biol. 37:71-80 (1980).
6. H. Schuessler and K. Schilling, Oxygen effect in the radiolysis of proteins; II. bovine serum albumin, Int. J. Radiat. Biol. 45:267-281 (1984).
7. J. Butler, E.J. Land, W.A. Prutz, and A.J. Swallow, Charge transfer between tryptophan and tyrosine in proteins, Biochim. Biophys. Acta 705:150-162 (1982).
8. G. Boguta and A.M. Dancewicz, Radiolytic and enzymatic dimerization of tyrosyl residues in insulin, ribonuclease, papain, and collagen, Int. J. Radiat. Biol. 43:249-265 (1983).
9. W.H. Garner, M.H. Garner, and A. Spector, H_2O_2-induced uncoupling of bovine lens Na+,K+-ATPase, Proc. Natl. Acad. Sci. USA 80:2044-2048 (1983).
10. E. Shinar, T. Navok, and M. Chevion, The analogous mechanisms of enzymatic inactivation induced by ascorbate and superoxide in the presence of copper, J. Biol. Chem. 258:14778-14783 (1983).
11. D.G. Wickens, A.G. Norden, J. Lunec, and T.L. Dormandy, Fluorescence changes in human gamma-globulin induced by free radical activity, Biochim. Biophys. Acta 742:607-616 (1983).
12. R.L. Levine, Oxidative inactivation of glutamine synthetase: I. inactivation is due to loss of one histidine residue, J. Biol. Chem. 258:11823-11827 (1983).
13. K. Nakamura and E.R. Stadtman, Oxidative inactivation of glutamine synthetase subunits, Proc. Natl. Acad. Sci. USA 81:2011-2015 (1984).
14. E. Gajewski, M. Dizdaroglu, H. Krutzsch, and M.G. Simic, OH radical-induced crosslinks of methionine peptides, Int. J. Radiat. Biol. 46: 47-55 (1984).
15. I. Zs.-Nagy and R.A. Floyd, Hydroxyl free radical reactions with amino acids and proteins studied by electron spin resonance spectroscopy and spin trapping, Biochim. Biophys. Acta 790:238-250 (1984).
16. K.J.A. Davies, Protein damage and degradation by oxygen radicals: I. general aspects, J. Biol. Chem. 262:9895-9901 (1987).
17. K.J.A. Davies, M.E. Delsignore, and S.W. Lin, Protein damage and degradation by oxygen radicals: II. modification of amino acids, J. Biol. Chem. 262:9902-9907 (1987).
18. K.J.A. Davies and M.E. Delsignore, Protein damage and degradation by oxygen radicals: III. modification of secondary and tertiary structure, J. Biol. Chem., 262:9908-9913 (1987).
19. K.J.A. Davies, S.W. Lin, and R.E. Pacifici, Protein damage and degradation by oxygen radicals: IV. degradation of denatured protein, J. Biol. Chem. 262:9914-9920 (1987).
20. J.M. McCord and I. Fridovich, Superoxide dismutase. An enzymatic function for erythrocuprein (hemocuprein), J. Biol. Chem. 244:6049-6055 (1969).
21. I. Fridovich, Oxygen radicals, hydrogen peroxide, and oxygen toxicity. In: Free Radicals in Biology, Volume I (W.A. Pryor, ed.) pp. 239-277, Academic Press, New York (1976).
22. G. Cohen and P. Hochstein, Glutathione peroxidase: The primary agent for the elimination of hydrogen peroxide in erythrocytes, Biochemistry 2:1420-1428 (1963).
23. C. Lind, P. Hochstein, and L. Ernster, DT-diaphorase as a quinone reductase: A cellular control device against semiquinone and superoxide radical formation, Arch. Biochem. Biophys. 216:178-185 (1982).
24. A.T. Quintanilha, L. Packer, J.M.S. Davies, T.L. Racanelli, and K.J.A. Davies, Membrane effects of vitamin E deficiency: Bioenergetic and surface charge density studies of skeletal muscle and liver mitochondria, Ann. NY Acad. Sci. 393:32-47 (1982).
25. A.T. Qunitanilha and K.J.A. Davies, Vitamin E deficiency and photosensitization of electron-transport carriers in microsomes, FEBS

Lett. 139:241-244 (1982).
26. B. Lubin and L.J. Machlin, Vitamin E: biochemical, hematological, and clinical aspects. Ann. NY Acad. Sci. 393:1-506 (1982)
27. B.N. Ames, E. Cathcart, E. Schwiers, and P. Hochstein, Uric acid provides an antioxidant defense in humans against oxidant- and radical caused aging and cancer: A hypothesis, Proc. Natl. Acad. Sci. USA 78: 6858-6862 (1981).
28. A. Sevanian, K.J.A. Davies, and P. Hochstein, Conservation of vitamin C by uric acid in blood, J. Free Radicals Biol. Med. 1:117-124 (1985).
29. K.J.A. Davies, A. Sevanian, S.F. Muakkassah-Kelly, and P. Hochstein, Uric acid-iron complexes: A new aspect of the antioxidant functions of uric acid, Biochem. J. 235:747-754 (1986).
30. A.L. Goldberg and F.J. Dice, Intracellular protein degradation in mammalian and bacterial cells, Ann. Rev. Biochem. 43:835-869 (1974).
31. A.L. Goldberg and A.C. St.John, Intracellular protein degradation in mammalian and bacterial cells: Part 2, Ann. Rev. Biochem. 45:747-803 (1976).
32. A.L. Goldberg, R. Voellmy, and K.H. Sreedhara Swamy, Studies of the pathway for protein degradation in Escherichia coli and mammalian cells, in: "Biological Functions of Proteinases," H. Holzer and H. Tschesche, eds., pp. 35-48, Springer-Verlag, Berlin (1979).
33. A.L. Goldberg, N.P. Strnad, and K.H. Sreehara Swamy, Studies of the ATP dependence of protein degradation in cells and cell extracts, in: "Protein Degradation in Health and Disease," Ciba Foundation Symposium 75, pp. 227-251, Exerpta Medica/Elsevier, North Holland (1980).
34. A. Hershko, A. Ciechanover, H. Heller, A.L. Haas, and I.A. Rose, Proposed role of ATP in protein breakdown: Conjugation of proteins with multiple chains of the polypeptide of ATP-dependent proteolysis, Proc. Natl. Acad. Sci. USA 77:1783-1786 (1980).
35. A. Hershko and A. Ciechanover, Mechanisms of intracellular protein breakdown, Ann. Rev. Biochem. 51:335-364 (1982).
36. S. Pontremoli, E. Melloni, F. Salamino, B. Sparatore, M. Michetti, U. Benatti, A. Morelli, and A. De Flora, Identification of proteolytic activities in the cytosolic compartment of mature human erythrocytes, Eur. J. Biochem.. 110:421-430 (1980).
37. M.J. Mc Kay, R.S. Daniels, and A.R. Hipkiss, Breakdown of aberrant protein in rabbit reticulocytes decreases with cell age, Biochem. J. 188:279-283 (1980).
38. M.J. McKay and A.R. Hipkiss, ATP-independent proteolysis of globin cyanogen bromide peptides in rabbit reticulocyte cell-free extracts. Eur. J. Biochem. 125:567-573 (1982).
39. S. Pontremoli, F. Salamino, B. Sparatore, E. Melloni, A. Morelli, U. Benatti, and A. De Flora, Isolation and partial characterization of three acidic proteinases in erythrocyte membranes, Biochem. J.. 181:559-568 (1979).
40. E. Melloni, B. Sparatore, F. Salamino, M. Michetti, and S. Pontremoli, Proteolysis of human reticulocyte membrane proteins: Evidence for a physiological role of the acid endopeptidases, Arch. Biochem. Biophys 218:579-584 (1982).
41. K.J.A. Davies, Free radicals and protein degradation in human red blood cells, in: "Cellular and Metabolic Aspects of Aging: The Red Cell as a Model," J.W. Eaton, D.K. Konzen, and J.G. White, eds., pp. 15-27, Alan R. Liss, New York (1985).
42. K.J.A. Davies, The role of intracellular proteolytic systems in antioxidant defenses, in: "Superoxide and Superoxide Dismutase in Chemistry, Biology, and Medicine," G. Rotilio, ed., pp. 443-450, Elsevier, Amsterdam (1986).
43. K.J.A. Davies, Intracellular proteolytic systems may function as secondary antioxidant defenses: an hypothesis, J. Free Radicals Biol. Med. 2:155-173
44. K.J.A. Davies and A.L. Goldberg, Oxygen radicals stimulate intracellular

proteolysis and lipid peroxidation by independent mechanisms in erythrocytes, J. Biol. Chem. 262:8220-8226 (1987).
45. K.J.A. Davies and A.L. Goldberg, Proteins damaged by oxygen radicals are rapidly degraded in extracts of red blood cells, J. Biol. Chem. 262:8227-8234 (1987).
46. S.W. Lin and K.J.A. Davies, E. coli proteases selectively degrade oxidized proteins, Fed. Proc. 45:1597 (1986).
47. R.T. Dean and J.K. Pollak, Endogenous free radical generation may influence proteolysis in mitochondria, Biochem. Biophys. Res. Commun. 126:1082-1089 (1985).
48. S.P. Wolf, A. Garner, and R.T. Dean, Free radicals, lipids, and protein degradation, Trends Biochem. Sci. 11:27-31 (1986).
49. S.P. Wolff and R.T. Dean, Fragmentation of proteins by free radicals and its effect on their susceptibility to enzymic hydrolysis, Biochem. J. 234:399-403 (1986).
50. R.T. Dean, S.M. Thomas, and A. Garner, Free-radical-mediated fragmentation of monoamine oxidase in the mitochondrial membrane, Biochem. J. 240:489-494 (1986).
51. R.L. Levine, C.N. Oliver, R.M. Fulks, and E.R. Stadtman, Turnover of bacterial glutamine synthetase: Oxidative inactivation precedes proteolysis, Proc. Natl. Acad. Sci. USA 78:2120-2124 (1981).
52. L. Fucci, C.N. Oliver, M.J. Coon, and E.R. Stadtman, Inactivation of key metabolic enzymes by mixed-function oxidation reactions: Possible implication in protein turnover and ageing, Proc. Natl. Acad. Sci. USA 80:1521-1525 (1983).
53. A.J. Rivett, Preferential degradation of the oxidatively modified form of glutamine synthetase by intracellular mammalian proteases, J. Biol. Chem. 260:300-305 (1985).
54. E.R. Stadtman, Oxidation of proteins by mixed-function oxidation systems: Implication in protein turnover, ageing and neutrophil function, Trends Biochem. Sci. 11:11-12 (1986).
55. Roseman, J.E. and Levine, R.L., Purification of a protease from Escherichia coli with specificity for oxidized glutamine synthetase, J. Biol. Chem. 262:2101-2110 (1987)
56. R.S. Yallow and S.A. Berson, Chemical and biological alterations induced by irradiation of I^{131} labeled human serum albumin, J. Clin. Invest. 36:44-50 (1957).
57. E.W. Kellogg, III, and I. Fridovich, Liposome oxidation and erythrocyte lysis by enzymatically generated superoxide and hydrogen peroxide, J. Biol. Chem. 252:6721-6728 (1977).
58. B. Goldberg and A. Stern, The role of the superoxide anion as a toxic species in the erythrocyte, Arch. Biochem. Biophys. 178:218-225 (1977).
59. H.C. Sutton, P.B. Roberts, and C.C. Winterbourn, The rate of reaction of superoxide radical ion with oxyhaemoglobin and methaemoglobin, Biochem. J. 155:503-510 (1976).
60. S.K. Jain and P. Hochstein, Polymerization of membrane components in aging red blood cells, Biochem. Biophys. Res. Commun. 92:242-254 (1980).
61. P. Hochstein and S.K. Jain, Association of lipid peroxidation and polymerization of membrane proteins with erythrocyte aging, Fed. Proc. 40:183-188 (1981).
62. R.L. Aft and G.C. Mueller, Hemin-mediated Oxidative degradation of Proteins, J. Biol. Chem. 259:301-305 (1984).
63. J.C. Fantone and P.A. Ward, Role of oxygen-derived free radicals and metabolites in leukocyte-dependent inflammatory reactions, Am. J. Pathol. 107:395-418 (1982).
64. S.J. Weiss and A.F. LoBuglio, Phagocyte-generated oxygen metabolites and cellular injury, Lab. Invest. 47:5-18 (1982).
65. W.A. Pryor, M.M. Dooley, and D.F. Church, Inactivation of human α-1-

proteinase inhibitor by gas-phase cigarette smoke, Biochem. Biophys. Res. Commun. 122:676-681 (1984).

66. S.J. Weiss, G. Peppin, X. Ortiz, C. Ragsdale, and S.T. Test, Oxidative autoactivation of latent collagenase by human neutrophils, Science 227:747-749 (1985).

67. J. Mandelstam, Turnover of proteins in growing and non-growing populations of Escherichia coli, Biochem. J. 69:110-119 (1958).

68. D. Schlessinger and F. Ben-Hamida, Turnover of protein in Escherichia coli, Biochim. Biophys. Acta 119:171-182 (1966).

69. R.D. Mosteller, R.V. Goldstein, and K.R. Nishimoto, Metabolism of individual proteins in exponentially growing Escherichia coli, J. Biol. Chem. 255:2524-2532 (1980).

70. R.D. Mosteller and B.E. Goldstein, A mathematical model that applies to protein degradation and post-translational processing of proteins and to analogous processes for other molecules in non-growing and exponentially growing cells, J. Theor. Biol. 108:597-621 (1984).

71. P.C. Lee, B.R. Bochner, and B.N. Ames, AppaA, heat-shock stress, and cell oxidation, Proc. Natl. Acad. Sci. USA 80:7496-7500 (1983).

72. M.F. Christman, R.W. Morgan, F.S. Jacobson, and B.N. Ames, Positive control of a regulon for defenses against oxidative stress and some heat shock proteins in Salmonella typhimurium, Cell 41:753-762 (1985).

73. U. Bond and M.J. Schlesinger, Ubiquitin is a heat shock protein in chicken embryo fibroblasts, Mol. Cell Biol. 5:949-956 (1985).

74. S.A. Goff, L.P. Casson, and A.L. Goldberg, Heat shock regulatory gene htpR influences rates of protein degradation and expression of the lon gene in Escherichia coli, Proc. Natl. Acad. Sci. USA 81:6647-6651 (1984).

75. J. Ananthan, A.L. Goldberg, and R. Voellmy, Abnormal proteins serve as eukaryotic stress signals and trigger the activation of heat shock genes, Science 232:522-524 (1986).

76. M. Dizdaroglu, E. Gajewski, and M.G. Simic, Enzymatic digestibility of peptides cross-linked by ionizing radiation, Int. J. Radiat. Biol. 45:283-295 (1984).

77. G.O. Ivy, F. Schottler, J. Wenzel, M. Baudry, and G. Lynch, Inhibitors of lysosomal enzymes: Accumulation of lipofuscin-like dense bodies in the brain, Science 226:985-987 (1984).

78. B. Demple and J. Halbrook, Inducible repair of oxidative DNA damage in Escherichia coli, Nature 304:466-468 (1983).

79. A. Sevanian, S.F. Muakkassah-Kelly, and S. Montestruque, Influence of phospholipase A_2 and glutathione peroxidase on the elimination of membrane lipid peroxides, Arch. Biochem. Biophys. 223:441-452 (1983).

80. A. Sevanian and E. Kim, Phospholipase A_2 dependent release of fatty acids from peroxidized membranes, J. Free Radicals Biol. Med. 1:263-271 (1985).

RADIOPROTECTION OF MICE WITH ASCORBIC ACID, DESFERAL AND MERCAPTOETHYLAMINE

Ajit Singh, Harwant Singh, James S. Henderson*, Robert D. Migliore*, Jacques Rousseau**, and Johan E. Van Lier**

Radiation Applications Research Branch
Atomic Energy of Canada Limited Research Company
Whiteshell Nuclear Research Establishment
Pinawa, Manitoba, Canada ROE 1L0

* Pathology Department
Faculty of Medicine
University of Manitoba
Winnipeg, Manitoba, Canada R3E 0W3

** Department of Nuclear Medicine and Radiobiology
M.R.C. Group on Radiation Sciences
Faculté de médecine, Centre hospitalier universitaire de Sherbrooke
Université de Sherbroke
Sherbrooke, Québec, Canada J1H 5N4

INTRODUCTION

Radiation damage in biological systems is initiated by the ionic, excited and free radical species formed during the energy deposition events. It progresses with time through a variety of mechanisms.[1] Because of the variety of these mechanisms, any single radioprotective agent can be only partially effective. Simultaneous use of several radioprotective agents is warranted.[1]

We have used three modifiers of radiation biological damage in our work on radiation protection of Balb/c mice: vitamin C (ascorbic acid, AA), Desferal (desferrioxamine, DF) and mercaptoethylamine (MEA). AA has been used in several radiation protection studies in cells, mice and rats. It is a radioprotector for *E. coli*, Chinese hamster ovary cells and rats.[2-4] In the case of mice, however, it is a radiosensitizer rather than a radioprotector.[5,6] MEA is a widely used and very effective radioprotector for mice.[5,7] However, administration of AA together with MEA reduces radioprotection by the latter.[5] Some chelating agents, e.g., penicillamine, are known to act as radioprotectors.[5,8] Based on the observed protection of lipid peroxidation by DF *in vitro*, Gutteridge et al.[9] suggested its use as a radioprotector. DF is an approved drug sold under the trade name Desferal. We report the results of our preliminary work on radioprotection of Balb/c mice by AA, MEA and DF.

EXPERIMENTAL METHODS

Each treatment group consisted of ten 8- to 10-week-old Balb/c mice[10] along with a control group of five or ten mice. The amounts of the modifiers administered intraperitoneally in physiological saline solutions 15 min before the irradiations were AA, 200 mg/kg; DF, 50 mg/kg and MEA, 200 mg/kg. The controls were similarly injected with the saline only solutions. The ^{60}Co irradiations were done with a Theratron (Winnipeg) or an Eldorado (Sherbrooke) irradiator. An equivalent treatment and control group was irradiated at each centre.

RESULTS AND DISCUSSION

The survival of mice was followed for 30 days following irradiation. The results are summarized in Table 1. Some of the key results are also shown in Figure 1.

The protection seen with MEA is similar to that reported by several workers[5,11] and is attributable to the ability of the sulfhydryl groups to repair free radical damage and induce favourable biochemical effects.[1,5,11]

The sensitization seen with AA is consistent with that reported by Forsberg et al.[5] and Tewfik et al.[6], although we see sensitization at doses 7 - 10 Gy, whereas Tewfik et al.[6] reported sensitization below 7 Gy. This difference is attributable to different strains of mice and/or the longer duration of AA administration by Tewfik et al.[6] (one week before and four weeks after) in drinking water. The reduction of the protection by MEA caused by simultaneous administration of AA, seen by us and by Forsberg et al.,[5] can be attributed to several factors: (i) antagonism between sulfhydryl compounds and AA[2,5]; (ii) adverse reaction of the monodehydroascorbate anion with MEA,[2] and (iii) the effects of transition metal[5] ions that are normally present, e.g., Fe^{++}. The third factor appears to be very important. AA reduces Fe^{+++} to Fe^{++},[13] thereby facilitating damaging Fenton-type reactions [1,9,11,13] in irradiated mice. This view is strongly supported by the protection seen when AA and DF are administered together, compared to the effects of either alone, at 7 Gy. Chelation of available Fe^{+++} with DF[9,13] seems to prevent its reduction by AA, thus preventing the

Table 1. Effect[a,b] of Ascorbic Acid, Desferal and Mercaptoethylamine on Radiation Survival of Balb/c Mice

	Modifiers					
Dose (Gy)	AA	DF	MEA	AA & DF	MEA & AA	MEA & DF
7	--[d]	--	+++	++	+++	+++
8	--	nd	+++	nd	nd	nd
10	-	+	+++	ne	++	++++
13	+		+++	+	++	++

[a] Compared to the effect on mice given physiological saline solution, intraperitoneally, 15 min before irradiation.
[b] Modifiers were given intraperitoneally, 15 min before irradiation, in physiological saline solution. AA, ascorbic acid; DF, Desferal; MEA, mercaptoethylanine.
[d] - and + refer to sensitization and protection, respectively; their numbers indicate relative magnitude of the effect; nd, not determined; ne, no effect.

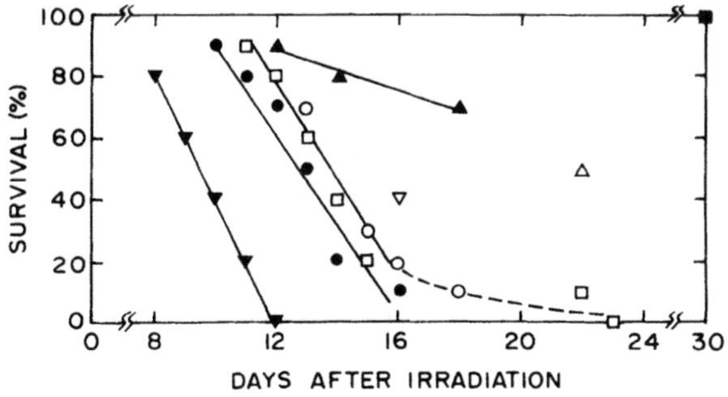

Figure 1. Effects of radiation modifiers on the 30-day survival of Balb/c mice: 7 Gy- ▿ saline (no death after day 16), ○ DF, □ AA, △ AA & DF (no death after day 22); 10 Gy- ▾ saline, ▲ MEA (no death after day 18), ● MEA & AA (no death after day 16), ■ MEA & DF (all survived).

toxicity of the Fenton-type reactions. Since DF along shows slight sensitization at 7 Gy, the large protection seen with AA & DF cannot be ascribed solely to chelation of iron and prevention of the Fenton-type reactions. It appears that AA has at least two effects: (i) the radiosensitizing effect via the reduction of Fe^{+++}, and (ii) a radioprotective effect, as an antioxidant and free radical scavenger.[14,15] In the presence of DF, the protective effect of AA seems to predominate. The slight protection by AA at 13 Gy indicates a relative increase in its protective effect, as compared to its sensitizing effect, with increasing dose.

The main reason for the reduction in the protective action of MEA by AA appears to be the antagonism between the two.[2,5] Both MEA and AA can reduce Fe^{+++} to Fe^{++}.[13,16] Therefore, the extent of the Fenton-type reactions is likely to be unaffected when they are given together, and they are not likely to contribute to the reduction in protection of MEA by AA.

The protection by DF at 10 and 13 Gy is consistent with its chelation of Fe^{+++}. The slight sensitization at 7 Gy requires further investigation. The improved protection when MEA and DF are given together is also attributable to the chelating properties of DF. Thus the Fenton-type reactions taking place in the presence of MEA would be prevented by DF.

We hope to pursue various aspects of this work, particularly the enhanced protection by AA & DF and MEA & DF. These results suggest that the three component mixture of MEA & DF & AA may provide even better protection by combining the protective effects of MEA and AA and by eliminating the toxic effects of free iron (Fenton-type reactions) with DF. The increased protection seen with the use of the combinations of different types of protectors supports the case made earlier[1] for such studies.

REFERENCES

1. A. Singh and H. Singh, Time-scale and nature of radiation-biological damage: Approaches to radiation protection and post-irradiation therapy, Prog. Biophys. Molec. Biol. 39:69 (1982).
2. M. Naslund, L. Ehrenberg, and G. Djalali-Behzad, Antagonism of Ascorbate against the radioprotective action of cysteamine, Int. J. Radiat. Biol. 30:95 (1976).

3. M.K. O'Connor, J.F. Malone, M. Moriarty, and S. Mulgrew, A radioprotective effect of vitamin C observed in chinese hamster ovary cells, Brit. J. Radiol. 50:587 (1977).
4. L. Ala-Ketola, R. Varis, and K. Kiviniitty, Effect of ascorbic acid on the survival of rats after whole body irradiation, Strahlentherapie 148:643 (1974).
5. J. Forsberg, M. Harms-Ringdahl, and L. Ehrenberg, Interaction of ascorbate with the radioprotective effect of mercaptoethylamin. An exploratory study in mice, whole animals and cell cultures, Int. J. Radiat. Biol. 34:245 (1978).
6. H.H. Tewfik, F.A. Tewfik, and E.F. Riley, The Influence of ascorbic acid on survivial of mice following whole body X-irradiation, in: "Vitamin C: New Clinical Applications in Immunology, Lipid Metabolism and Cancer," A. Hanck, ed., Huber Publishers, Bern (1982)
7. J.R. Maisin, G. Mattelin, and M. Lambiet-Collier, Chemical protection against the long-term effects of a single whole-body exposure of mice to ionizing radiation. I. Life shortening, Radiat. Res. 71:119 (1977).
8. W.F. Ward, A. Shih-Hoellwarth, and P.M. Johnson, Survival of Penicillamine-treated mice following whole-body irradiation, Radiat. Res., 81:131 (1980).
9. J.M.C. Gutteridge, R. Richmond, and B. Halliwell, Inhibition of the iron-catalysed formation of hydroxyl radicals from superoxide and of lipid peroxidation by desferrioxamine, Biochem. J. 184:469 (1979).
10. J.S. Henderson and J.L. Weeks, A study of the carcinogenicity for skin of a polyphenyl coolant, Ind. Med. Surg. 42(2):10 (1973).
11. J.B. Storer, Chemical protection of the mouse against radiation-induced life shortening, Radiat. Res. 47:537 (1971).
12. T.L. Dormandy, Free-radical oxidation and anti-oxidants, Lancet 1:647 (1978).
13. S.D. Aust, L.A. Morehouse and C.E. Thomas, Role of metals in oxygen radical reactions, J. Free Rad. Biol. Med. 1:3 (1985)
14. B.H.J. Bielski, Chemistry of ascorbic acid radicals, in: "Ascorbic Acid: Chemistry, Metabolism, and Uses," P.A. Seib and B.M. Tolbert, eds., Adv. Chem. Ser., 200, American Chemical Society, Washington, D.C. (1982).
15. R.L. Willson, Free radical mechanisms and the interactions of glutathione and vitamins C and E, in: "Radioprotectors and Anticarcinogens," O.F. Nygaard and M.G. Simic, eds., Academic Press, New York (1983).
16. G. Capozzi and G. Modena, Oxidation of thiols, in: "The Chemistry of the Thiol Group," Part 2, S. Patai, Ed., John Wiley and Sons, London (1974).

ANTIINFLAMMATORY, ANALGESIC, AND ANTIULCER ACTIVITIES OF COPPER COMPLEXES SUGGEST THEIR USE IN A PHYSIOLOGIC APPROACH TO TREATMENT OF ARTHRITIC DISEASES

John R.J. Sorenson

Division of Medicinal Chemistry
College of Pharmacy
University of Arkansas for Medical Sciences
4301 W. Markham Street
Little Rock, Arkansas 72205, U.S.A.

Copper is an essential metalloelement, and therefore is required by all human cells for normal metabolism. Amounts of copper found in various body tissues and fluids have been tabulated elsewhere.[1] Calculated amounts of ionic copper suggested to be present in biological systems (10^{-18} M in plasma) are too small to be measured using the most sensitive instrumentation available. As a result, measurable tissue copper content reflects content of copper complexes, and these complexes account for absorption, distribution, and biologically active forms, including copper-dependent enzymes.[1,2]

Ingested copper complexes representing the recommended 3 mg (47 μmol) daily intake of copper follow a pathway leading to gastric and instestinal absorption. These copper complexes then undergo systemic circulation to all tissues and are 1) utilized by tissues following ligand exchange with apoenzymes and apoproteins to form metalloenzymes and metalloproteins, 2) stored in the liver following ligand exchange with thioneine to form copper-thioneine, or 3) excreted if tissue needs have been met and stores replenished. Since excessive copper storage has never been reported in any normal population and copper complex absorption must vary in these populations, efficient homeostatic mechanisms regulate retention and excretion of varying amounts of absorbed copper complexes in all normal individuals without Wilson's disease. Homeostatic release of copper complexes from the liver meets normal copper-dependent physiologic requirements of body tissues, including <u>de novo</u> synthesis of copper-dependent enzymes, provided dietary intake is adequate. Unfortunately, many or nearly all modern diets studied do not supply required amounts of this and other essential metalloelements.[3] It is then reasonable to suggest that marginal or deficient intakes or impaired absorption may lead to decreased enzyme activity and manifestations of acute inflammatory disease in the short term as well as manifestations of chronic inflammatory disease in the long term.

During the active phases of inflammatory diseases, elevations in plasma copper complexes are generally two to three times the normal concentration. With disease remission, plasma copper complex concentrations return to normal. The central question concerning this observation is whether the elevation is a cause of these diseases or a physiological response to them. Since there is no evidence that elevation of normal blood copper-containing components causes any of these diseases, the hypothesis that elevation of plasma copper-containing components is a general physiologic response to inflammatory diseases merits serious consideration. The fact that low molecular weight copper complexes have antiinflammatory activity in humans and animal models of inflammation is offered as evidence that associated increases in blood copper-containing components is a physiologic response which has a role in mediating remission, when this component of the response to inflammation is adequate and remission occurs.

To date, over 140 copper complexes have been studied as antiinflammatory agents. Results of these studies, which have been reviewed, confirm as well as extend original observations that copper complexes of inactive ligands and active antiinflammatory drugs are more active than the parent ligand (drugs) or inorganic copper. Table 1 is a brief presentation of data from the original report suggesting that copper complexes are active metabolites of antiarthritic drugs.[4] These data show that copper complexes are more effective than their parent ligands and $Cu(II)_2(acetate)_4$. Since amounts of copper in these complexes do not correlate with activity, it is suggested that pharmacologic activity may be better correlated with physicochemical properties of these complexes.

Acute and chronic toxicity studies were done early in the course of this work, and it was found that the copper complexes studied were less toxic than their parent ligands and $Cu(II)_2(acetate)_4$. The projected human dose of $Cu(II)_2(aspirinate)_4$ is 3 to 6 μmol/kg daily, with a decrease to some maintenance dose with remission. This regimen seems to be safe enough to correct what may be, in part, a copper deficiency or inadequacy disease since the recommended safe daily intake of copper is 47 μmol/day, and increasing this intake can be monitored to assure safety.

Antiinflammatory activities of copper complexes partially explain the earlier observations by Fenz, Forestier, Kuzell, and Hangarter that copper complexes are effective in treating arthritic and other human degenerative disease.[5] Another partial explanation comes from the observation that copper complexes have potent analgesic activity.

It has very recently been reported that copper complexes of nonsteroidal antiinflammatory agents are more effective analgesics than their parent drugs.[6] As shown in Table 2, copper complexes of 3,5-DIPS, aspirin, niflumic acid, and indomethacin were more effective analgesics than their parent compounds and copper chloride or copper acetate. In addition, the

Table 1. Antiinflammatory Activities of Some Copper Complexes[a]

Compound	Carrageenan Paw Edema	Adjuvant Arthritis	Copper (%)
$Cu(II)_2(acetate)_4(H_2O)_2$	A at 0.02	I at 0.089	31.8
3,5-DIPS acid	I at 0.90	I at 0.135	
$Cu(II)_2(3,5\text{-DIPS})_4$	A at 0.01	A at 0.001	12.5
Aspirin	A at 0.36	A at 0.033	
$Cu(II)_2(aspirinate)_4$	A at 0.01	A at 0.001	15.0
D-Penicillamine	I at 1.34	I at 0.201	
$Cu(I)(D\text{-pen})(H_2O)1.5$	A at 0.03	Not Tested	26.7
$Cu(II)_2(D\text{-pen disulfide})_2(H_2O)_2$	A at 0.01	A at 0.040[b]	15.4

[a]All compounds were given by subcutaneous injection and the lowest active (A) dose or the inactive (I) initial screening dose expressed as millimole per kilogram of body weight; [b]only dose tested.

Table 2. Analgesic Effects of Parent Ligands and Copper Complexes in Acetic Acid-Induced Writhing Pain in Mice and Adjuvant-Induced Arthritis Pain in Rats

Compound	Writhing Pain: ED_{50} (mmol/kg)[a] (95% C.L.)	Arthritic Pain: ED_{50} (mmol/kg)[a] (95% C.L.)
3,5-DIPS	0.4 (0.29-0.67)	>0.90
$Cu(II)_2(3,5\text{-DIPS})_4$	0.22 (0.10-0.45)	>0.20
Aspirin	0.48 (0.22-1.04)	1.41 (0.50-3.96)
$Cu(II)_2(aspirinate)_4$	0.14 (0.09-0.25)	0.09 (0.04-0.21)
Niflumic acid	0.21 (0.11-0.40)	0.48 (0.19-1.20)
$Cu(II)_2(niflumate)_4$	0.10 (0.05-0.20)	>0.16
Indomethacin	0.01 (0.01-0.03)	0.02 (0.01-0.03)
$Cu(II)_2(indomethacin)_4$	0.002 (0.001-0.003)	0.002 (0.001-0.003)
$Cu(II)(chloride)_2$		>2.24
$Cu(II)_2(acetate)_4$		>0.83
Morphine Hydrochloride[b]	0.002 (0.001-0.003)	0.002 (0.001-0.005)

[a]administered orally in 5% propylene glycol and 1.4% polyvinyl alcohol in water. [b]administered subcutaneously in saline.

copper complex of indomethacin was found to be as effective as morphine in both pain models, and the amount of copper in the 1 μmol dose of $Cu(II)_2$-$(indomethacin)_4$ is one-fiftieth the recommended safe daily intake of copper. These observations also support the hypothesis that copper complexes are active forms, formed in vivo, of the non-steroidal antiinflammatory agents and the hypothesis that these complexes activate opioid receptors.[6]

Based upon observations that copper complexes appeared to facilitate the synthesis of superior granuloma around cotton pellets in the cotton-wad granuloma model of inflammation, it was hypothesized that these com-complexes promoted connective tissue synthesis, as in wound repair, in response to inflammatory insults. They were subsequently found to be potent antiulcer agents in the Shay and corticoid-induced models of ulcer.[4] To date, over 70 copper compounds have been shown to have anti-ulcer activity using six different models of gastric ulcer. Representative data from the original report[4] are presented in Table 3. It is most likely that antiulcer activity is related to the physiochemical properties of these complexes. It is also worth noting that the amount of copper in these ED_{50} doses represents one-fourth to one-fifteenth of the recommended daily intake of copper and these are the projected adult doses.

The original report of antiulcer activity and reduced gastric acid secretion for copper complexes has been confirmed and extended by many others.[7] The report by West[8] showing that $Cu(II)(salicylate)_2$ is more effective than cimetidine in protecting against aspirin-, indomethacin-, and cold-stress-induced ulcers (Table 4) is most definitive.

Table 3. Oral Antiulcer Activity (ED_{50}) of Non-Steroidal Antiinflammatory Agent Copper Complexes

Compound	Antiulcer Activity (μmol/kg)	% Copper
$Cu(II)_2(acetate)_4(H_2O)_2$	53	32
$Cu(II)_2(aspirinate)_4$	13	15
$Cu(II)(salicylate)_2 \cdot 8H_2O$	11	15
$Cu(II)(butazolidine)_2$	7	9
$Cu(II)_2$(D-pen disulfide)$_2 \cdot 3H_2O$	6	16
$Cu(II)_2(niflumate)_4$	4	10
$Cu(II)_2(indomethacin)_4$	3	8
$Cu(II)_2$(3,5-DIPS)$_4$	2.5	13

Table 4. Comparison of $Cu(II)(Salicylate)_2$ and Cimetidine, Given Orally, in Protecting Against Gastric Lesions Produced by Oral Ulcerogenic Doses of Aspirin (2.22 mmol/kg), Indomethacin (56 μmol/kg), and Cold-Stress in Rats

Ulcerogen	Antiulcer Agent	Dose (μmol/kg)	% inhibition Mean(+/-SE)
Aspirin	Cimetidine	148	19(5)
		297	16(4)
	$Cu(II)(Salicylate)_2$	22	21(6)
		44	61(10)[a]
Indomethacin	Cimetidine	148	10(3)
		297	24(6)
	$Cu(II)(Salicylate)_2$	22	37(7)[a]
		44	50(5)[a]
Cold-Stress	Cimetidine	148	38(8)[a]
		297	61(9)[a]
	$Cu(II)(Salicylate)_2$	22	58(8)[a]
		44	79(7)[a]

[a]$p<0.05$ for comparison with values obtained for non-treated rats

REFERENCES

1. J.R.J. Sorenson, ed., "Biology of Copper Complexes," Humana Press, Clifton, New Jersey (1987).
2. J.R.J. Sorenson, ed., "Inflammatory Diseases and Copper," Humana Press, Clifton, New Jersey (1982).
3. J.R.J. Sorenson, V. Kishore, A. Pezeshk, L.W. Oberley, S.W.C. Leuthauser, and T.D. Oberley, "Copper complexes: A physiological approach to the treatment of "inflammatory diseases." Inorg. Chim. Acta 91:285 (1984).
4. J.R.J. Sorenson, Copper complexes as possible active forms of the antiarthritic agents, J. Med. Chem. 19:135 (1976).
5. J.R.J. Sorenson and W. Hangarter, Treatment of rheumatoid and degenerative diseases with copper complexes, Inflammation 2:217 (1977).
6. S. Okuyana, S. Hashimoto, H. Aihara, W.M. Willingham, and J.R.J. Sorenson, Copper complexes of non-steroidal antiinflammatory agents: Analgesic activity and possible opioid receptor activation, Agents and Actions 21:130 (1987).
7. J.R.J. Sorenson, The antiinflammatory activities of copper complexes, in: "Metal Ions in Biological Systems-14," H. Siegel, ed., Marcel Dekker, New York (1982).
8. G.B. West, Testing for drugs inhibiting the formation of gastric ulcers, J. Pharmacol. Methods 8:33 (1982).

TOCOPHEROL BEHAVIOR AND MEMBRANE CONSTITUENTS IN ERYTHROCYTES WITH OXIDANT STRESS

Masayuki Miki, Tadashi Motoyama, Yoshito Takenaka, and Makoto Mino

Department of Pediatrics, Osaka Medical College
Takatsuki, Osaka 569, Japan

INTRODUCTION

Recently, toxicity caused by oxygen and its active species has received much attention in connection with a variety of pathological events.[1] However, the mechanism responsible for the peroxidation of biomembranes is still unclear. Tocopherols (vitamin E) are located in the membranes and act there as an only chain-breaking antioxidant.[2,3] When biomembranes are damaged mainly by a free radical chain oxidation,[4] the effects of tocopherols will be spotlighted. Nevertheless, the mechanisms by which membrane tocopherols inhibit the peroxidation still require detailed study, partly because the active species and the sites of their generation have not usually been known or controlled.

The role of tocopherols in the biomembranes as a radical-scavenger can be most clearly understood by the use of azo compounds (A-N=N-A), which generate initiating radicals at a known and constant rate and at a specific site.[5,6] For example, water-soluble 2,2'-azobis(2-amidinopropane) dihydrochloride (AAPH) generates free radicals in the aqueous region, whereas lipid-soluble 2,2'-azobis(2,4-dimethylvaleronitrile) (AMVN) generates ones within the lipid region. The geminate radicals formed by the thermal decomposition of the azo compound may recombine to give a stable radical or diffuse apart to become free radicals (reaction 1). The rate of generation of free radicals can be easily controlled and measured by adjusting the concentration of the compounds. They rapidly react with oxygen to give peroxyl radicals (reaction 2), which attack the membrane constituents such as lipids (LH in reaction 3), and abstract hydrogen atoms to give lipid radicals. The lipid peroxyl radicals then induce a sequence of propagation reactions to give lipid hydroperoxides (reactions 4 and 5, repeatedly).

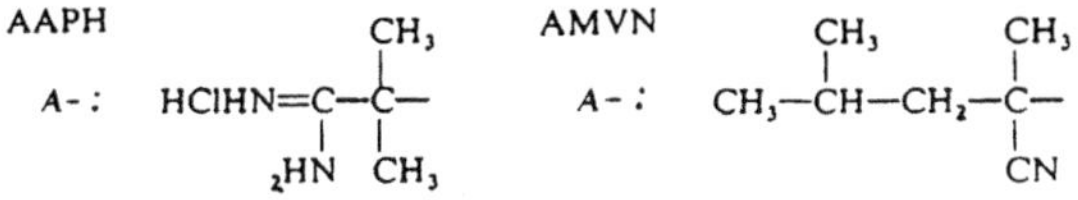

$$\text{A-N=N-A} \longrightarrow (1-e)\text{A-A} + 2e\text{A}^{\cdot} + \text{N}_2 \quad (1)$$
$$\text{A}^{\cdot} + \text{O}_2 \longrightarrow \text{AOO}^{\cdot} \quad (2)$$
$$\text{AOO}^{\cdot} + \text{LH} + \text{O}_2 \longrightarrow \text{AOOH} + \text{LOO}^{\cdot} \quad (3)$$
$$\text{LOO}^{\cdot} + \text{LH} \longrightarrow \text{LOOH} + \text{L}^{\cdot} \quad (4)$$
$$\text{L}^{\cdot} + \text{O}_2 \longrightarrow \text{LOO}^{\cdot} \quad (5)$$

In this review, we will show some of our results from experiments using rat-erythrocyte ghosts as a model of biomembranes with azo compounds, with the aim of understanding tocopherol behavior in membranes.

THE MOBILITY OF TOCOPHEROLS IN THE MEMBRANES

One of the important factors in the activity of tocopherols is their mobility in the membranes. The ratio of α-tocopherol to PUFA in biomembranes is small, usually below one to a thousand. Therefore, tocopherols and/or peroxyl radicals must move around rapidly to meet each other in order to suppress the peroxidation efficiently.

It has been found that among tocopherol series, α-tocopherol is the most reactive toward peroxyl radicals and α-tocopheroxyl radical is the most stable.[7,8] Therefore, as shown in Figure 1-A, when both α- and γ-tocopherols are incorporated into soybean-phosphatidylcholine (PC) liposomes, where they can move easily, and are subjected to the oxidation with AAPH, only α-tocopherol is initially consumed and γ-tocopherol remains almost constant, and then γ-tocopherol decreases after all of α-tocopherol is depleted. This phenomenon is not interpreted solely by the differences of tocopherols toward peroxyl radicals, but it suggests that rapid interaction of α-tocopherol and γ-tocopheroxyl radical takes place.[8] On the other hand, as shown in Figure 1-B, when α- and γ-tocopherols are incorporated into the individual multilamellar liposomes, where it is hard for them to contact with each other, they decrease simultaneously during the oxidation.

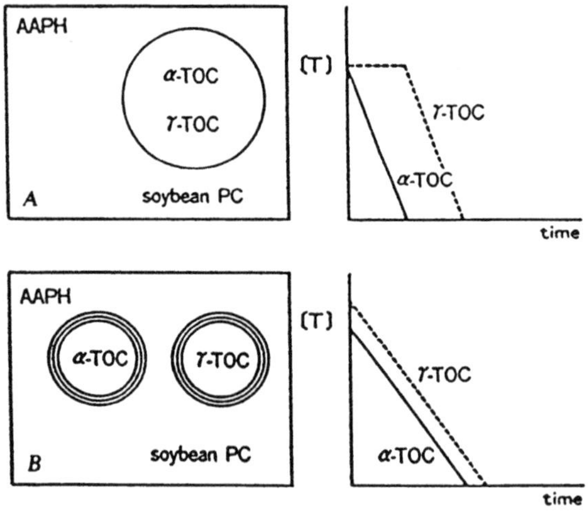

Figure 1. The sparing actions between α- and γ-tocopherols during the oxidation of liposomes.

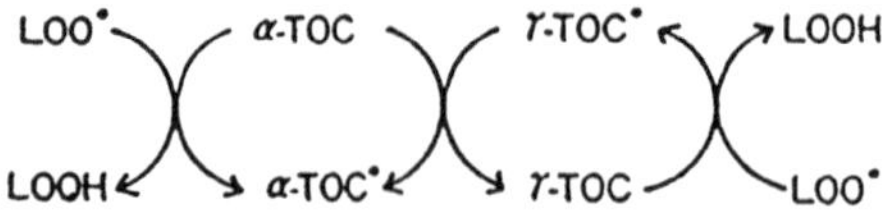

Figure 2 shows changes in α- and γ-tocopherols in RBC ghost membranes during the oxidation with AAPH. The above sparing actions between these two tocopherols were not observed, that is, they were consumed linearly with time, similarly as shown in Figure 1-B. It suggests that

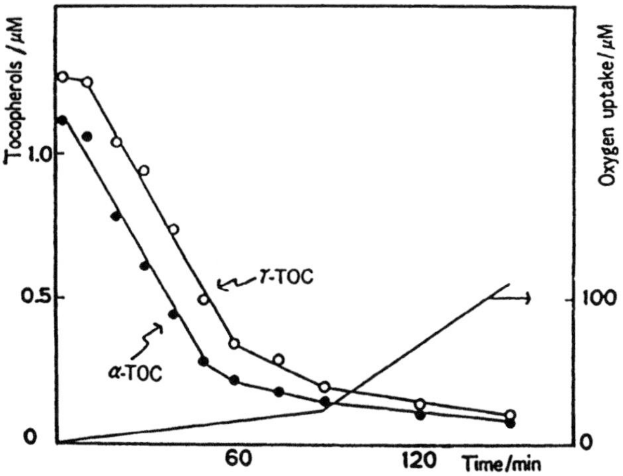

Figure 2. Rates of α- and γ-tocopherols disappearance and the oxygen uptake in the oxidation of RBC ghosts induced by AAPH.

α-tocopherol and γ-tocopheroxyl radical cannot interact efficiently in the biomembranes. When an appreciable amount of tocopherols was present, the oxygen uptake was suppressed to form an induction period. From the rate of oxidation during inhibition period and the length of the period, i.e., from the extent to which the rate of oxidation is reduced and for how long, the antioxidant activity of tocopherols in the biomembranes has been evaluated by Niki et al. and reported to be lower than for the oxidation of methyl linoleate in homogeneous solution.[9] These results may be attributed to the lower mobility of tocopherols in the biomembranes and the higher chance of chain propagation in the tightly packed structure of the membranes.

CONSUMPTION AND REGENERATION OF α-TOCOPHEROL IN THE MEMBRANES

One of the key factors that determine the antioxidant activity is how many peroxyl radicals can be trapped by the antioxidant. From the kinetic study in the homogeneous solution using azo compounds, it is known that α-tocopherol traps one peroxyl radical rapidly to generate α-chromanoxyl radical, which can trap another peroxyl radical.[5] Therefore, one molecule of α-tocopherol can scavenge two molecules of peroxyl radicals and interrupt the chain propagation efficiently. In the presence of reducing agents, such as ascorbic acid and glutathione, α-chromanoxyl radical is reduced to regenerate α-tocopherol.[10,11] Accordingly, α-tocopherol can scavenge more peroxyl radicals than expected until the reducing agents are consumed. Thus, the synergistic inhibition of peroxidation by the combination of α-tocopherol and ascorbic acid has been known and proved to contribute even in the liposomal membrane system.[12,13] Moreover, it must contribute to maintaining the vitamin E level in tissues.

It is shown in Figure 3 how many α-tocopherols are consumed as a function of the amount of free radicals generated from AAPH. The amount of radicals generated in aqueous phase at 37 °C was calculated from the equation 6.[14]

$$[\text{radicals}] = 1.30 \times 10^{-6}\ [\text{AAPH}]\ [\text{sec}] \qquad (6)$$

The solid line shows the relationship with micelled α-tocopherol, and the broken line shows the relationship with ghost-membrane α-tocopherol. α-

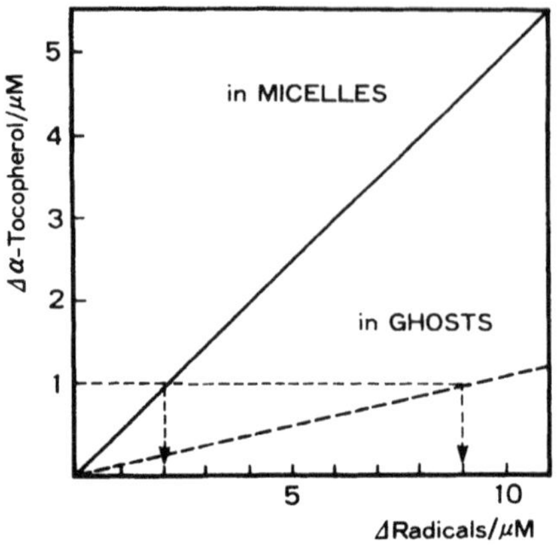

Figure 3. Consumption of α-tocopherol as a function of the amount of free radicals from AAPH.

Tocopherol in ghosts appears to trap more radicals than in micelles. It may be because other membrane constituents trap the initiating radicals before tocopherol does, or because the sulfhydryl (SH) groups in the membrane proteins may reduce tocopherol radicals to regenerate tocopherols.

In order to investigate the behavior of membrane SH groups, we studied the effect of L-cysteine in aqueous phase on the oxidation of α-tocopherol. It is found that cysteine reacts easily with the alkylperoxyl radical from AAPH and that it is depleted linearly with time. When the micelled α-tocopherol was oxidized with AAPH, as the dotted line in Figure 4 shows, it also disappeared linearly with time. On the

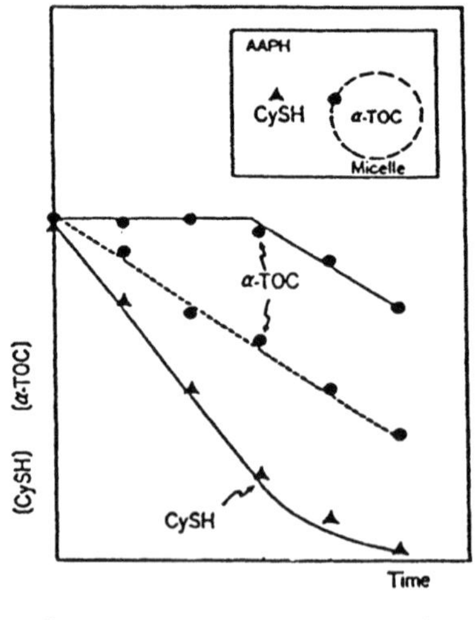

Figure 4. Effects of cysteine in aqueous phase on the oxidation of the micelled α-tocopherol induced by AAPH.

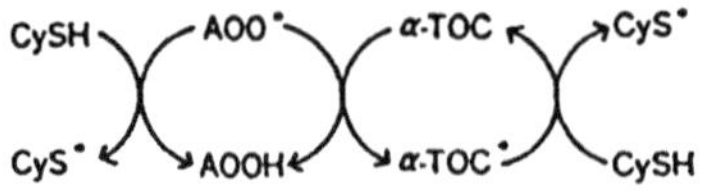

other hand, the solid lines show the result of oxidation of the micelled α-tocopherol in the presence of cysteine in aqueous phase. Cysteine was consumed first and tocopherol remained almost constant, and then tocopherol was consumed after cysteine was exhausted. The finding that the rate of disappearance of tocopherol was suppressed by cysteine may be due to the trapping by cysteine of radical from AAPH and/or the reduction of tocopherol radical by cysteine to regenerate tocopherol.

It was examined whether cysteine could interact with tocopherol radical and inhibit the oxidation synergistically. Figure 5 illustrates the profile of oxygen uptake during the oxidations of soybean-PC liposomes initiated with azo compounds. In the absence of antioxidants the soybean-PC liposomes were smoothly oxidized at a constant rate without any noticeable induction period with long kinetic chain length (lines 1). When the radicals were generated initially in the aqueous region by using a water-soluble initiator (AAPH), as shown in Figure 5-A, both cysteine and α-tocopherol suppressed the oxidation and produced a clear induction period individually (lines 2 and 3, respectively). However, cysteine suppressed oxygen uptake less significantly during the induction period than α-tocopherol. In the presence of both α-tocopherol and cysteine (line 4), the length of the induction period was close to the sum of the individual induction periods observed, that is, the effect of cysteine is additive and not synergistic. In this case, therefore, cysteine was thought to readily trap the radicals in the aqueous phase before they reached the membranes and reacted with α-tocopherol. On the other hand, when the radicals were generated initially within the membranes by using a lipid-soluble initiator (AMVN) incorporated into membranes, as shown in Figure 5-B, cysteine was not an efficient antioxidant, that is, it retarded the oxidation but did not produce a clear induction period (line 2). Nevertheless, cysteine increased the induction period when α-tocopherol was present in the membranes (line 4). In this case, probably, the radicals derived from the initiation and/or lipid peroxyl radicals were trapped by α-tocopherol within the membranes and the resulting α-tocopherol radical

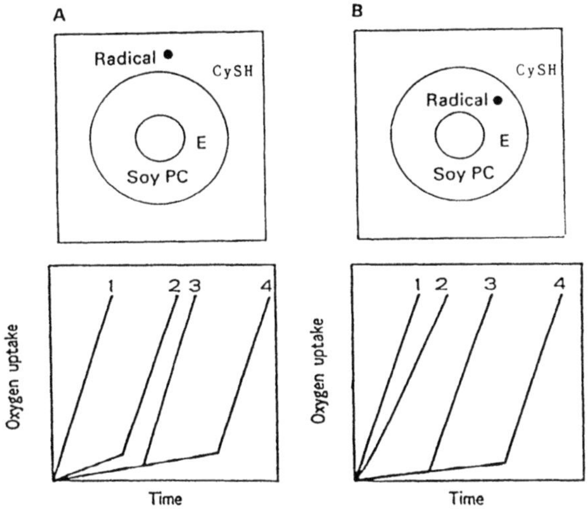

Figure 5. Inhibition of the oxidation of soybean-PC liposomes in aqueous dispersions by cysteine and/or α-tocopherol. The oxidation was initiated with AAPH (A) or with AMVN (B), 1, without antioxidants; 2, with cysteine dispersed in water; 3, with α-tocopherol incorporated into membranes; and 4, in the presence of both α-tocopherol and cysteine.

reacted with cysteine to regenerate α-tocopherol, presumably at the surface of the membranes. The results given above strongly suggest that cysteine can regenerate α-tocopherol by reacting with α-chromanoxyl radical in the membranes and can contribute to the synergistic inhibition of the oxidation.

A number of investigations have been performed recently to explore the possibility that reducing agents such as vitamin C might reduce the tocopheroxyl radical and permit a single molecule of tocopherol to scavenge many radicals in vivo. The tocopherol radical may undergo a few competing reactions: it may react with another peroxyl radical to give a stable adduct, react with another tocopherol radical to give a dimer, or react with the reducing agent to regenerate tocopherol. The relative importance of the above reactions is determined by the reactivities, concentrations, and accessibilities of the reducing agents. For example, if the concentrations of the lipid peroxyl radical and tocopherol radical are low, the tocopherol radical has more chance to react with the reducing agent to regenerate tocopherol. It should be well understood that whether or not the antioxidative synergism takes place and its extent are dependent upon several factors. Although cysteine has lower reactivity and accessibility toward tocopheroxyl radical than vitamin C,[15] the clearest experimental evidence for a synergistic effect of cysteine has been obtained in this study.

This conclusion is supported by the findings in electron spin resonance spectroscopy. As shown in Figure 6, chromanoxyl radical disappeared quite rapidly when mixed with cysteine, and the spin adduct of cysteine radical was observed when a spin trap, α-(4-pyridyl-N-oxide)N-t-butyl nitrone (POBN), was added simultaneously.

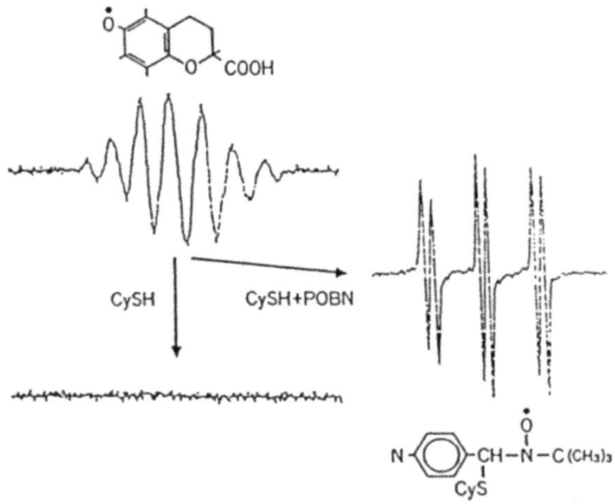

Figure 6. Changes of chromanoxyl radical ESR spectrum when mixed with cysteine and in the presence of cysteine and POBN.

THE ANTIOXIDANT EFFICIENCY OF MEMBRANE α-TOCOPHEROL FOR THE OXIDATION OF MEMBRANE PROTEINS

In the case of β-thalassemia major, active oxygen species were suspected to generate massively within erythrocytes from the abnormal hemoglobins. When patients with this disease were treated with vitamin E, as reported by Rachmilewitz et al.,[16] the increased formation of malondialdehyde in erythrocytes was corrected. Nevertheless, vitamin E could not prevent the decrease in SH groups in RBC membrane proteins and could not correct anemia efficiently. Therefore, the investigation as to whether membrane α-tocopherol can inhibit the oxidation of membrane proteins as well as lipids seems to be attractive.

Here, we studied the change of SH groups in hemoglobin-free ghosts in relation to membrane α-tocopherol consumption during the oxidation with AAPH. In the early stage of oxidation, the membrane α-tocopherol decreased linearly with time. Interestingly, SH groups were depleted concurrently with tocopherol within 60 min-incubation, and then about 40% of the initial SH groups remained unchanged as shown in Figure 7. After almost all of tocopherol was consumed, the oxygen uptake proceeded rapidly and the arachidonic acids in membrane lipids decreased. This indicates that membrane tocopherol could not efficiently prevent the oxidation of SH groups and that more than 60% of SH groups in membrane proteins could react easily with free radicals in aqueous phase generated from AAPH.

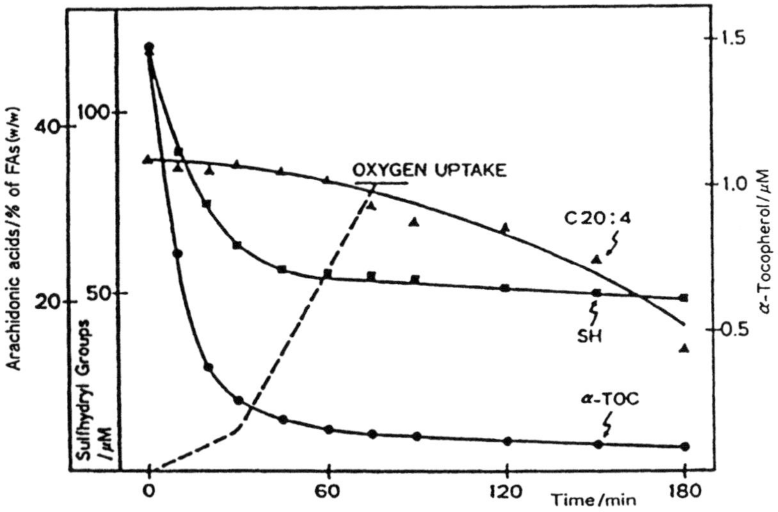

Figure 7. Changes in ghost membrane constituents during the oxidation induced by AAPH.

Figure 8 shows that the decrease of arachidonic acids, the formation of TBA reactants, and the detection of fluorescence in membrane lipids were suppressed until membrane α-tocopherol fell below a critically low level. The finding indicates clearly that tocopherol could inhibit the oxidation of membrane lipids completely as far as it existed.

Figure 9 shows the changes of SDS-polyacrylamide gel electrophoresis of ghosts during the oxidation with AAPH. This indicates that the membrane proteins were damaged by AAPH-induced oxidation and that the high-molecular-weight (HMW) proteins were formed even in the early stage of the oxidation, concurrently with the depletion of SH groups. At that time, the appreciable amount of membrane α-tocopherol was present and the peroxidation of membrane lipids was inhibited, measured as no formation of TBARS. Hochstein and Jain have stated that the formation of fluorescent chromolipids and HMW proteins are subsequent to peroxidation of membrane lipids.[17] However, Koster et al. cautioned that other polymerization reactions are possible, e.g., a direct radical attack on the proteins.[18] This latter possibility is substantiated somewhat by our finding that under oxidative stress, the process of HMW protein genesis occurred independently of the lipid peroxidation process. That is, the SH groups at the outer surface of the membranes must react easily with radicals from the aqueous region before membrane tocopherol scavenges the radicals, and the polymerization of membrane proteins may be formed due to the direct crosslinking of oxidized proteins, and not due to the formation of Schiff-base adduct with malondialdehydes.

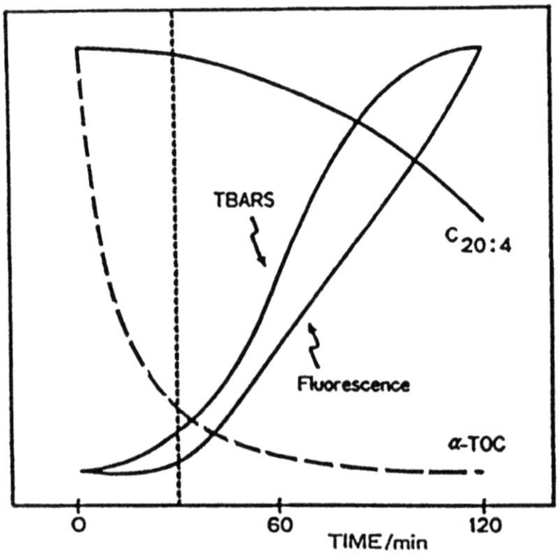

Figure 8. Membrane lipid peroxidation in the ghosts oxidized with AAPH in relation to the change in α-tocopherol.

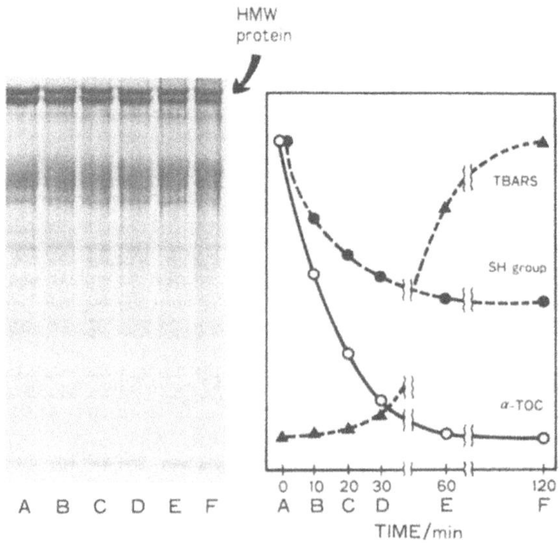

Figure 9. SDS-gel electrophoresis of the ghosts oxidized with AAPH in relation to the changes of α-TOC, SH group, and TBARS.

We compared the effects of the sites of initial radical generation on the oxidations of ghost membranes. In the case of AAPH, the free radicals were generated initially in aqueous region and they would attack the membranes from the outside. On the other hand, in order to generate the radicals within membrane-lipid region, the ghosts were incubated with dimyristoyl-PC liposomes containing AMVN. The membrane α-tocopherol can efficiently scavenge the radicals induced by AMVN and the formation of TBARS was inhibited completely; the membrane SH groups remained almost constant. After tocopherol was exhausted, membrane lipids were oxidized and, concomitantly, SH groups were consumed. Thus, membrane SH groups seem to be oxidized by lipid peroxyl radicals and consumed. Therefore, this oxidation of SH groups is suppressed by tocopherol more significantly than in the case of AAPH, as shown in Figure 10.

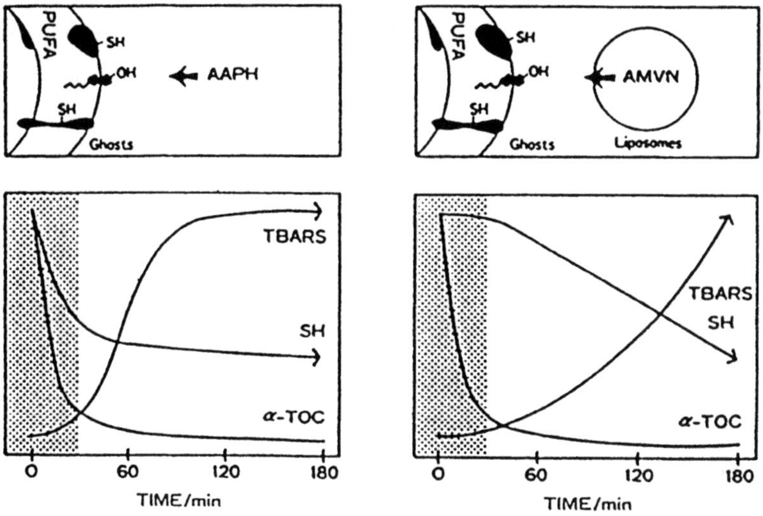

Figure 10. Effects of the sites of initial radical generation on the antioxidant efficiency of membrane α-tocopherol.

From a physiological point, the depletion of membrane SH groups is very interesting with respect to the mulfunction of the receptors and the inactivation of membrane enzymes induced by the oxidation. In conclusion, membrane α-tocopherol can suppress the oxidation of SH groups in the ghosts to some extent, but its efficacy depends on the sites of initial radical generation. Furthermore, it may be said that the SH groups in membranes could not regenerate α-tocopherol since they did not interact with tocopherol radicals induced by AMVN and remained unchanged.

ACKNOWLEDGMENTS

We are very grateful to Professor Etsuo Niki, University of Tokyo for his numerous suggestions and encouragement, and also to the Eisai Company and the Ministry of Education, Science and Culture, Japan.

REFERENCES

1. W. Bors, M. Saran, and D. Tait, "Oxygen Radicals in Chemistry and Biology," Walter de Gruyter, Berlin (1984).

2. M. Mino, S. Nakagawa, H. Tamai, and M. Miki, Clinical evaluation of red blood cell tocopherol, Ann. NY Acad. Sci. 393:175 (1982).
3. G.W. Burton, A. Joyce, and K.U. Ingold, Is vitamin E the only lipid-soluble, chain-breaking antioxidant in human blood plasma and erythrocyte membrane?, Arch. Biochem. Biophys. 221:281 (1983).
4. M. Miki, H. Tamai, M. Mino, Y. Yamamoto, and E. Niki, Free-radical chain oxidation of rat red blood cells by molecular oxygen and its inhibition by α-tocopherol, Arch. Biochem. Biophys. 258:373 (1987).
5. Y. Yamamoto, S. Haga, E. Niki, and Y. Kamiya, Oxidation of lipids. V. Oxidation of methyl linoleate in aqueous dispersion, Bull. Chem. Soc. Jpn. 57:1260 (1984).
6. L.R.C. Barclay, S.J. Lccke, J.M. MacWeil, J. VanKessel, G.W. Burton, and K.U. Ingold, Autoxidation of micelles and model membranes. Quantitative kinetic measurements can be made by using either water-soluble or lipid-soluble initiators with water-soluble or lipid-soluble chain-breaking antioxidants, J. Am. Chem. Soc. 106:2479 (1984).
7. G.W. Burton and K.U. Ingold, Autoxidation of biological molecules. 1. The antioxidant activity of vitamin E and related chain-breaking phenolic antioxidants in vitro, J. Am. Chem. Soc. 102:6472 (1981).
8. E. Niki, J. Tsuchiya, Y. Yoshikawa, Y. Yamamoto, and Y. Kamiya, Oxidation of lipids. XIII. Antioxidant activities of α-, β-, γ-, and δ-tocopherols, Bull. Chem. Soc. Jpn. 59:497 (1986).
9. E. Niki, Lipid antioxidants: How they may act in biological systems, Br. J. Cancer 55:153 (1987).
10. E. Niki, T. Saito, A. Kawakami, and Y. Kamiya, Inhibition of oxidation of methyl linoleate in solution by vitamin E and vitamin C, J. Biol. Chem. 259:4177 (1984).
11. E. Niki, J. Tsuchiya, R. Tanimura, and Y. Kamiya, Regeneration of vitamin E from α-chromanoxyl radical by glutathione and vitamin C, Chem. Lett. 789 (1982).
12. E. Niki, A. Kawakami, Y. Yamamoto, and Y. Kamiya, Oxidation of lipids. VIII. Synergistic inhibition of oxidation of phosphatidylcholine liposome in aqueous dispersion by vitamin E and vitamin C, Bull. Chem. Soc. Jpn. 58:1971 (1985).
13. T. Doba, G.W. Burton, and K.U. Ingold, Antioxidant and co-antioxidant activity of vitamin C. The effect of vitamin C, either alone or in the presence of vitamin E or a water-soluble vitamin E analogue, upon the peroxidation of aqueous multilamellar phospholipid liposomes, Biochim. Biophys. Acta 835:298 (1985).
14. E. Niki, M. Saito, Y. Yoshikawa, Y. Yamamoto, and Y. Kamiya, Oxidation of lipids. XII. Inhibition of oxidation of soybean phosphatidylcholine and methyl linoleate in aqueous dispersions by uric acid, Bull. Chem. Soc. Jpn. 59:471 (1986).
15. J. Tsuchiya, T. Yamada, E. Niki, and Y. Kamiya, Interaction of galvinoxyl radical with ascorbic acid, cysteine, and glutathione in homogeneous solution and in aqueous dispersions, Bull. Chem. Soc. Jpn. 58:326 (1985).
16. E.A. Rachmilewitz, A. Shifter, and I. Kahane, Vitamin E deficiency in β-thalassemia major: Changes in hematological and biochemical parameters after a therapeutic trial with α-tocopherol, Am. J. Clin. Nutr. 32:1850 (1979).
17. P. Hochstein, and S.K. Jain, Association of lipid peroxidation and polymerization of membrane proteins with erythrocyte aging, Fed. Proc. 40:183 (1981).
18. J.F. Koster and R.G. Slee, Lipid peroxidation of human erythrocyte ghosts induced by organic hydroperoxides, Biochim. Biophys. Acta 752:233 (1983).

PORCINE STRESS SYNDROME, FREE RADICALS AND VITAMIN E

Garry Duthie, John Arthur and *Peter Hoppe

Rowett Research Institute
Aberdeen AB2 9SB
Scotland, U.K.

*BASF - Tierernahrungsstation
D-6745 Offenbach
West Germany

INTRODUCTION

Widespread among certain breeds of pigs is an inherited tendency to develop a fatal malignant hyperthermia when subjected to stresses such as transportation and exercise. The malignant hyperthermia response can also be pharmacologically triggered in these pigs, and in predisposed humans, by exposure to halothane anesthesia. Characteristically, prior to the rapid increase in body temperature, the limbs become rigid, and at post-mortem, muscle tissue is severely disrupted. Although abnormalities of calcium transport consistent with an inherent membrane defect have been implicated in the etiology of the syndrome, the initial causative factors which trigger the response are unknown.[1]

The following evidence[2] suggests that the malignant hyperthermia response is consistent with an antioxidant disorder leading to oxidative damage to cell membranes: (1) Both vitamin E deficient animals and stress susceptible pigs have elevated activities in plasma of the muscle isozymes of pyruvate kinase and creatine kinase and an increased tendency for erythrocyte lysis. These could be ascribed to free radical-mediated damage to cell membranes. (2) Halothane produces potent free radicals which could exacerbate inadequate antioxidant defense mechanisms. (3) The rapidity of development of the MH crisis is reminiscent of rapid free radical-mediated peroxidation that can arise in the polyunsaturated fatty acid components of cell membranes. (4) Ad hoc observations by farmers suggest that feeding large amounts of vitamin E to pigs reduces stress-related mortalities.

This paper reviews some attempts to assess whether stress-susceptible pigs have abnormalities in antioxidant status and whether dietary supplementation with vitamin E and vitamin C is of therapeutic value.

MATERIALS AND METHODS

Animals

The British Landrace pigs used in this study were homozygous for the presence or absence of the gene which confers sensitivity to halothane anesthesia. They were maintained on standard ration containing 10 I.U. vitamin E/kg or supplemented with 235 I.U. vitamin E/kg or 235 I.U. vitamin E/kg + 1000 ppm vitamin C. Prior to the introduction of these experimental diets and 4 weeks later, blood was removed via heparinised vacutainer from the jugular vein of each pig. Subsequently, one of the following stresses was imposed (a) 30 min transportation[4], (b) halothane anesthesia[3], (c) subcutaneous injection of 'myostress' which contains neostigmine bromide and induces shivering of the muscles for approximately 1 h.

Biochemical analysis

Saline-washed erythrocytes were suspended in 0.9% NaCl and incubated with shaking for 90 min at 37°C. Incubations also contained final concentrations of hydrogen peroxide of 0, 0.3, 0.9 or 1.5%. Reactions were stopped with 10% trichloroacetic acid and thiobarbituric acid reactive substances (TBARS) were measured at 535 nM. The antioxidant-related parameters given in Table 1 were measured as previously described.[2,5]

RESULTS

Antioxidant-related parameters

Prior to dietary supplementation with vitamin E or vitamins E and C, plasma TBARS and conjugated dienes were significantly elevated in the reactor pigs and muscle damage was apparent from the increased plasma activities of pyruvate kinase and creatine kinase (Table 1). However, there were no differences in vitamin E concentrations or in antioxidant enzyme activities between reactors and non-reactors (Table 1).

Table 1. Antioxidant-Related Parameters of Halothane Reactor and Non-Reactor Pigs. (Means ± SEMs)

Parameter	Reactor (n=10)	Non-Reactor (n=10)	P
Plasma:			
Pyruvate kinase (mU/ml)	1028 ± 182	256 ± 51	<0.001
Creatine kinase (mU/ml)	1431 ± 296	563 ± 156	<0.001
Vit. E (μg/ml)	0.89 ± 0.08	0.98 ± 0.10	NS
Vit. C (μmoles/l)	13 ± 1	19 ± 3	NS
GSHPx (U/ml) 3.4 ± 0.2	3.6 ± 0.2	NS	
Caeruloplasmin (mU/ml)	203 ± 12	210 ± 16	NS
TBARS (nmoles MDA/ml)	1.95 ± 0.29	1.16 ± 0.12	<0.025
Conj. dienes (mm/ml)	157 ± 9	130 ± 7	<0.050
Erythrocyte:			
Vit. E (μg/g Hb)	1.38 ± 0.32	1.00 ± 0.35	NS
GSHPx (U/g Hb)	279 ± 29	276 ± 35	NS
Catalase (k/g Hb)	160 ± 8	162 ± 14	NS
SOD (μg/g Hb)	639 ± 55	620 ± 22	NS

Erythrocyte lipid peroxidation

Erythrocytes of those reactors which had been maintained on 10 I.U. vitamin E/kg diet produced significantly more TBARS than the non-reactors at hydrogen peroxide concentrations of 0.9 and 1.5% in the incubations (Figure 1). Dietary supplementation with vitamin E or vitamins E and C reduced erythrocyte TBARS production to that of non-reactors.

Effects of stress

Following the imposition of either transport stress, halothane or 'myostress', those reactors consuming diets containing 10 I.U. vitamin E/kg had high concentrations of TBARS in plasma. Consumption of the diets supplemented with the vitamins in all cases appeared to suppress plasma TBARS concentrations (Figure 2).

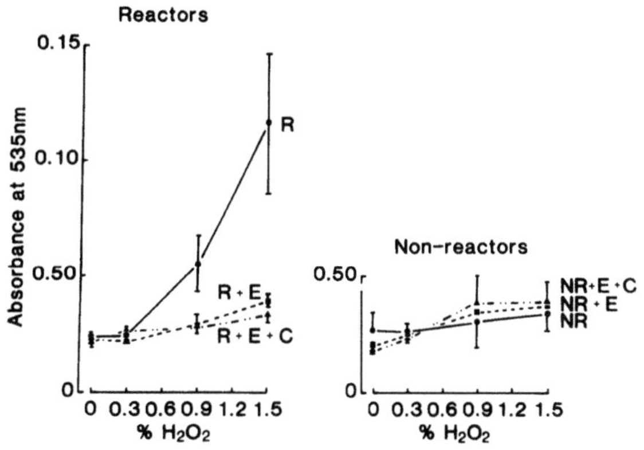

Figure 1. TBARS formation by erythrocytes of reactor (R) and non-reactor (NR) pigs after 4 weeks on experimental diets containing either 10 I.U. vitamin E/kg, 235 I.U. vitamin E/kg (+E) or 235 I.U. vitamin E/kg and 1000 ppm vitamin C (+E+C).

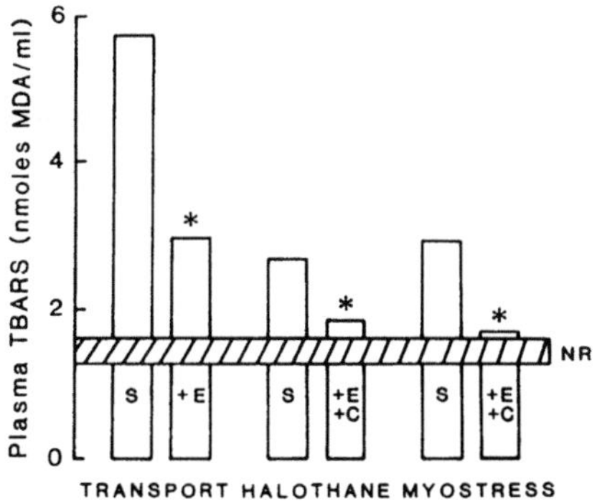

Figure 2. Effects of transport, halothane or 'myostress' on plasma TBARS concentrations of reactors. S denotes standard ration containing 10 I.U. vitamin E/kg. +E denotes 235 I.U. vitamin E/kg. +E+C denotes 235 I.U. vitamin E + 1000 ppm vitamin C. * indicates significant difference (P<0.001) between pigs subjected to the same stress on standard ration or with vitamin supplementation. The shaded bar indicates the range of values obtained for non-reactors (NR).

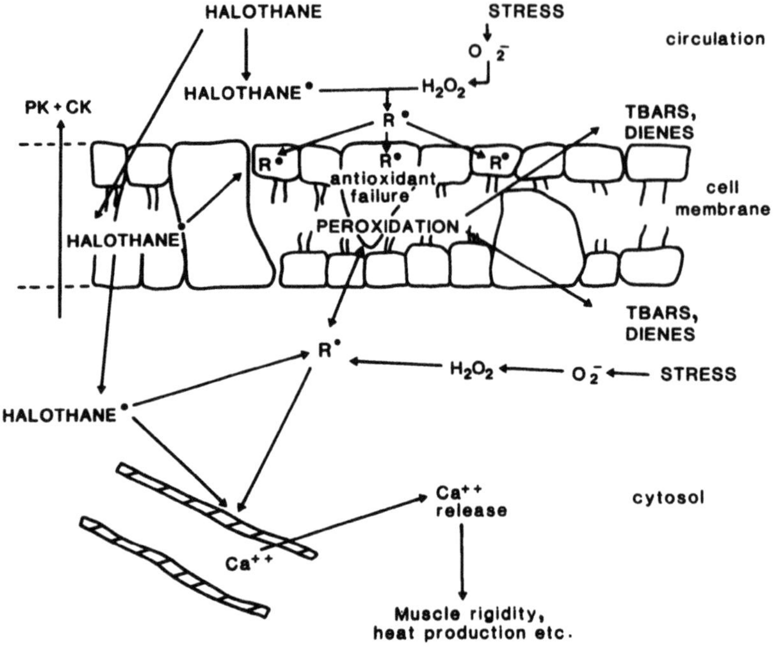

Figure 3. Proposed antioxidant abnormality in halothane reactor pig which promotes peroxidation of cell membranes with subsequent leakage of creatine kinase and pyruvate kinase (PK and CK) to plasma and increases plasma TBARS and conjugated dienes.

DISCUSSION

The increased concentrations of TBARS and conjugated dienes in plasma (Table 1) and the enhanced tendency of erythrocyte peroxidation (Figure 1) strongly suggest that the halothane-reactor pig has an antioxidant abnormality. Such an abnormality appears to be exacerbated by imposed stressors and is ameliorated by dietary supplementation with antioxidants. Although the site and nature of the defect remains unclear, we tentatively suggest that the fulminant stress response is related to an inability to accommodate a sudden increase in free radical production from either endogenous or exogenous sources. The resulting damage to cell membranes would lead to the observed increase in plasma of TBARS and creatine kinase and pyruvate kinase activities. Such peroxidation of membranes could cause release of calcium from organelles, inducing the characteristic development of limb rigidity, and also could promote the uncoupling of metabolic processes, resulting in excess heat production with consequent rapid rise in body temperature (Figure 3).

ACKNOWLEDGEMENTS

This work was funded by BASF, W. Germany. The Animal Breeding Research Organisation, Edinburgh, Scotland, U.K. kindly supplied the pigs.

REFERENCES

1. G. Mitchell and J. J. A. Heffron, Porcine stress syndromes, Adv. Food Res. 28:167 (1982).
2. G. G. Duthie, J. R. Arthur, C. F. Mills, P. Morrice and F. Nicol, Anomalous tissue vitamin E distribution in stress-susceptible pigs after dietary vitamin E supplementation and effects on plasma pyruvate kinase and creatine kinase activities, Livestock Prod. Sci. in press.
3. A. J. Webb, The halothane test: A practical method of eliminating porcine stress syndrome, Vet. Rec. 106:410 (1980).
4. K. Bickhardt and L. Richter, Methodische aspetite des creatin-kinase-tests beim schwein, Dtsch. Tieraerztl. Wochenschr. 87:296 (1980).
5. G. G. Duthie and J. R. Arthur, Blood antioxidant status and plasma pyruvate kinase activity of halothane reactor pigs. Am. J. Vet. Res. 48:309 (1987).

INCREASED OXIDATION OF TOCOPHEROLS IN CHICKENS WITH INHERITED MUSCULAR DYSTROPHY

Michael E. Murphy and James P. Kehrer

Division of Pharmacology and Toxicology
College of Pharmacy
The University of Texas at Austin
Austin, Texas 78712-1074, U.S.A.

INTRODUCTION

Inherited muscular dystrophy in chickens shares many characteristics with inherited muscular dystrophy in humans. Both are conditions caused by the defect of a single recessive gene, both lead to the degeneration of muscle tissue over time, and the underlying biochemical basis for cell death is unknown in both disorders. Numerous hypotheses have seen suggested to explain muscular dystrophy.[1] One of these postulates the basic biochemical defect of inherited muscular dystrophy is one which causes oxidative stress in muscle tissue, and that this is the underlying cause of cell death.[2] Table 1 summarizes the changes found in dystrophic muscle from both chickens and humans that closely resemble those caused in muscle by oxidative processes.[1] Oxidative stress in muscle cells could be caused by two basic biochemical defects. First, the defect could be one which leads to the excessive production of free radicals in muscle tissue. Second, the defect could be in one of the antioxidant systems important to normal muscle integrity, although apparently not in any of the antioxidant systems analyzed to date, all of which show enhanced activity in dystrophic muscle.[1]

It has been known for many years that vitamin E, including both α- and γ-tocopherol (α-TH and γ-TH), are critical to normal muscle function. Researchers have generally rejected the theory, first formulated 40 years ago, that a defect in vitamin E metabolism in muscle tissue is the underlying basis of muscular dystrophy.[3] However, previous investigations into the levels and, more importantly, the antioxidant activity of vitamin E in muscle were limited

Table 1. Similarities between Oxidative Stress and Muscular Dystrophy

Evidence	Species: Human	Mouse	Chicken
1. ⇑ products of lipid peroxidation	√	√	√
2. ⇓ proportion of PUFAs	√	√	√
3. ⇑ activities of antioxidant enzymes	√	√	√
4. ⇓ activity of SR calcium-ATPase	√	√	√
5. similarity to vitamin E deficiency	√	√	√
6. ⇓ content of tocopherols	inconclusive	inconclusive	√
7. ⇑ content of tocopheryl quinones	not done	not done	√

by the relatively insensitive techniques used to measure the tocopherols and the tocopheryl quinones, such as α- and γ-tocopheryl quinones (α-TQ and γ-TQ), which can be formed upon the oxidation of vitamin E.

RESULTS AND DISCUSSION

Tissue Tocopherol Content

We have developed a new technique, using high performance liquid chromatography with redox-cycling electrochemical detection, to measure the endogenous levels of tocopherols and tocopheryl quinones at the levels normally found in tissue.[4] Although only the pectoralis major muscle is affected by inherited muscular dystrophy in chickens, it is possible that any underlying change in the level or activity of vitamin E might be expressed in unaffected tissues, such as the soleus muscle and liver. We have, therefore, analyzed three tissues of normal (strain S11) and muscular dystrophic (strain S10) chickens for their content of tocopherol and tocopheryl quinone using this new, more sensitive assay.

γ-TQ was found at significantly higher levels in dystrophic tissues, particularly dystrophic pectoralis major muscle, compared to normal (Table 2). α-TQ showed a similar trend, although statistical significance was not achieved because of the greater variability of these data. The variation in the measurement of α-TQ was due in part to the effects of an interfering compound which made quantification difficult . The level of γ-TQ in dystrophic pectoralis major muscle, not subject to this interference, was nearly 10 times the level seen in normal muscle. Although not statistically significant, a trend toward a decreased tocopherol content in dystrophic tissues was apparent in all tissues examined. Normal and dystrophic tissues differed significantly in the ratio of α- to γ-TH. Dystrophic tissues had a lower proportion than normal of α-TH. Because of its chemical structure, α-TH is more easily oxidized than γ-TH and may be depleted more quickly *in vivo* during oxidative stress. The diet contained a preponderance of γ-TH but because all birds received the same diet, the proportion of α- and γ-TH reaching the tissues should be initially similar in normal and dystrophic animals, assuming the two strains do not differ in their relative absorption of α- and γ-TH (which favors α-TH). Thus, a greater depletion of α-TH in dystrophic animals is consistent with the hypothesis that an oxidative stress is occurring in these tissues.

Table 2. Content of α-TQ, γ-TQ, Total Tocopherols and the Ratio of α-TH to γ-TH in Tissues of 4-Week-Old Normal and Dystrophic Chickens

Tissue Source	α-Tocopheryl quinone[a]	γ-Tocopheryl quinone[a]	Total tocopherols[b]	Ratio of α-TH to γ-TH
Pectoralis major muscle				
Normal	2.55 ± 1.58[c]	0.17 ± 0.12	1.93 ± 0.38	4.33 ± 0.09
Dystrophic	4.64 ± 1.89	1.71 ± 0.43 *	1.39 ± 0.17	3.40 ± 0.26 *
Soleus muscle				
Normal	8.48 ± 3.57	0.46 ± 0.06	1.16 ± 0.23	3.95 ± 0.13
Dystrophic	11.39 ± 4.06	1.05 ± 0.32	0.79 ± 0.06	2.80 ± 0.19 *
Liver				
Normal	4.63 ± 1.58	0.38 ± 0.09	0.54 ± 0.06	4.08 ± 0.15
Dystrophic	6.72 ± 1.20	0.95 ± 0.16 *	0.47 ± 0.07	2.44 ± 0.31 *

[a]Tocopheryl quinone contents are expressed as pmol/mg lipid.
[b]Total tocopherol (α-TH & γ-TH) contents are expressed as nmol/mg lipid.
[c]Data are expressed as mean ± S.E. n=4 for all groups.
*Significantly different from normal tissue ($p<0.05$) using Student's t test.

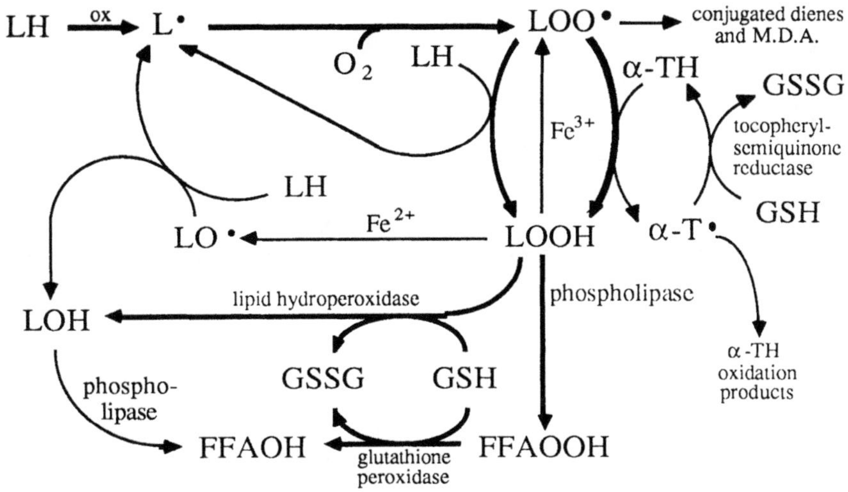

Figure 1. Possible pathways of autoxidative and antioxidative reactions of lipids in muscle which lead to formation of free hydroxy-fatty acids (FFAOH) and free fatty acid hydroperoxides (FFAOOH)

Tocopherol Oxidation in Microsomes

The results presented in Table 2 provide evidence that α- and γ-TH are being oxidized to a greater extent than normal in tissues of dystrophic birds. However, these data do not provide an indication of whether this is a secondary result of pathological changes in the muscle or is directly related to the underlying biochemical basis of cell death. The increased oxidation of tocopherols in cellular membranes could have several causes: 1) an increased production of free radicals which initiate lipid peroxidation, 2) an increased propagation of lipid peroxidation following a single initiation event, or 3) an increased likelihood of tocopherol oxidation following its initial scavenging of a free radical. Investigators have recently characterized enzymatic antioxidants, some of which depend on glutathione, that protect cellular membranes from free radical damage.[5-7] Figure 1 illustrates how some of these may limit the initiation and propagation of free radical damage through the membrane, while others may be able to regenerate tocopherols from the tocopheryl semiquinones produced initially by free radical scavenging. The role of these systems in muscle has not been investigated, but could be important in the increased tocopherol oxidation seen in dystrophic muscle.

We have analyzed the content and changes in tocopherols and tocopheryl quinones in microsomal preparations from both normal and dystrophic chicken liver subjected to oxidative stress *in vitro*. Figure 2 shows the time course of α-TH disappearance in liver microsomes after the addition of Fe^{++}/ADP/ascorbate. The rate of α-TH oxidation was the same in normal (Figure 2a) and dystrophic (Figure 2b) liver microsomes, and was similar to the oxidation of γ-TH in these preparations (data not shown). Oxidation of both tocopherols was slightly slowed in the presence of glutathione (GSH). Heating microsomes before the experiment reduced the initial α-TH content and eliminated the protective effect of GSH. The protective effect of GSH was minor, however, compared to that seen in microsomes of rat liver where this effect was first reported.[5-7] In addition, we could not demonstrate any protective effect of GSH in muscle microsomes obtained from either rat or chicken (data not shown). Two consistent observations concerning the oxidation of liver and muscle microsomes of both chickens and rat were: 1) a small amount of α-tocopherol (≈10%) always resisted oxidation even after long incubation periods, and 2) tocopheryl quinones were never produced during or after the incubation.

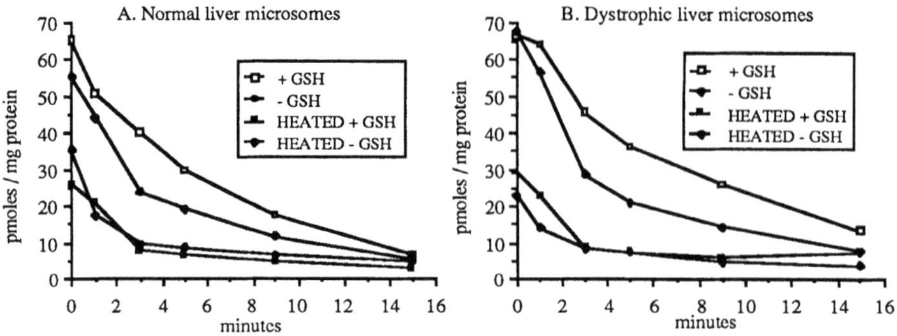

Figure 2. Time course of α-tocopherol depletion in liver microsomes of normal and dystrophic chickens. Microsomes were heated under N_2 at 100°C for 15 minutes. Glutathione was added to a final concentration of 1mM. Microsomal lipids were extracted and analyzed for tocopheryl compounds as noted earlier.

The role of oxidative stress in the pathogenesis of muscular dystrophy is still unclear. Our data continue to support the theory that oxidative stress occurs in tissues of dystrophic birds, although the source of this stress is unknown. These results do not indicate GSH-dependent factors that protect vitamin E from oxidation are defective in dystrophic birds, or even that they are normally important to muscle tissue in general. However, further research is necessary to determine which antioxidant systems are actually critical to normal muscle, and what the main source of oxidative stress may be in this tissue.

ACKNOWLEDGEMENTS

JPK is the recipient of Research Career Development Award HL 01435 and is the Gustavus Pfeiffer Centennial Fellow of Pharmacology. This work was supported by Biomedical Research Support Grants RR07091 and RR05849.

REFERENCES

1. M.E. Murphy and J.P. Kehrer, Free radicals: A potential pathogenic mechanism in inherited muscular dystrophy, Life Sci., 39: 2271 (1986).
2. S.T. Omaye and A.L. Tappel, Glutathione peroxidase, glutathione reductase, and thiobarbituric acid-reactive products in muscles of chickens and mice with genetic muscular dystrophy, Life Sci., 15: 137 (1974).
3. A.T. Milhorat, J.B. Mackenzie, S. Ulick, H. Rosenkrantz and W.E. Bartels, Observations on a biologically active vitamin E derivative present in hog gastric mucin and in hog stomach lining. The biologic activity of dl,alpha-tocopheryl-quinone, Ann. NY Acad. Sci., 52: 334 (1949).
4. M.E. Murphy and J.P. Kehrer, Simultaneous measurement of tocopherols and tocopheryl quinones in tissue fractions using high-performance liquid chromatography with redox- cycling electrochemical detection, J. Chromatogr. Biomed. Appl., 421: 71 (1987).
5. F. Ursini, M. Maiorino and C. Gregolin, The selenoenzyme phospholipid glutathione peroxidase, Biochim. Biophys. Acta, 839: 62 (1985).
6. D.D. Gibson, J. Hawrylko and P.B. McCay, GSH-Dependent inhibition of lipid peroxidation: Properties of a potent cytosolic system which protects cell membranes, Lipids, 20: 704 (1985).
7. G.R.M.M. Haenen and A. Bast, Protection against lipid peroxidation by a microsomal glutathione-dependent labile factor, FEBS Lett., 159: 24 (1983).

VITAMIN E AND IMMUNE FUNCTIONS

Adrianne Bendich

Clinical Nutrition
Hoffmann La Roche Inc.
Nutley, NJ 07110, U.S.A.

INTRODUCTION

All circulating cells that are derived from the common hemopoietic stem cell precursor demonstrate membrane changes when vitamin E is absent from the diet. Red blood cells from vitamin E deficient animals and humans lyse in the presence of hydrogen peroxide, whereas red blood cells from vitamin E supplemented animals are not lysed. Platelets from vitamin E-deficient animals are more adhesive and produce more thromboxane than platelets from vitamin E-replete animals. Neutrophils from vitamin E-deficient animals have an increased level of peroxidized lipid in their membranes, produce more hydrogen peroxide, and have depressed chemotaxis and phagocytosis.[1]

It should not be suprising that cells involved in the generation of specific immune responses, macrophages, T lymphocytes and B lymphocytes are also affected by the level of vitamin E in the diet. The objective of this report is to examine the effects of vitamin E on immune responses. In addition, the immunomodulating effects of other nutritional components associated with vitamin E will be discussed.

VITAMIN E AND IMMUNE CELL FUNCTION

The initiation of the immune response is considered to occur at the level of the cell membrane.[2] Vitamin E, the major fat-soluble antioxidant, is an essential constituent of all cellular membranes, including the outer cellular membrane and the internal mitochondrial and nuclear membranes.[3] In addition, vitamin E is important in maintaining the lymphocyte membrane fluidity necessary for proliferative responses.[4]

The vitamin E content of lymphocytes and mononuclear cells is usually ten times greater than that found in platelets and red blood cells.[5] The importance of vitamin E was clearly demonstrated when lymphocyte membranes from vitamin E-deficient animals were compared with those from animals given the vitamin. Electron micrographs of mitochondrial membranes from lymphocytes from vitamin E-deficient animals showed swollen and disrupted areas.[6]

When laboratory animals were placed on vitamin E deficient diets, many immune responses were severely depressed.[7-10] Macrophage membrane receptors were altered, interleukin 2 production was lowered (Bendich, in press) and production of prostaglandins was altered.[11-13] Low levels of dietary vitamin E, sufficient to protect against many of the adverse effects of vitamin E deficiency, were insufficient to enhance lymphocyte proliferative responses even though the diet contained all other nutrients at recommended levels.[14]

In certain instances, the immune system can develop aberrant responses which can result in harmful self-destructive effects. An example would be the chronic inflammatory response seen in the joints of arthritics. The damage caused in these joints has been associated with the increase in free radicals generated in the inflamed tissues. A recent study by Tappel's group has directly linked the level of dietary vitamin E with the degree of lipid peroxidation and inflammatory responses seen in an animal model of arthritis. Vitamin E deficient animals given an injection of vitamin E had significantly less inflammation and produced less of the peroxidative intermediates. Furthermore, vitamin E protected these animals against chronic inflammation and bone distortion.[15]

In the initial studies to determine the effect of vitamin E on immunity, higher than normal dietary levels were fed to laboratory and farm animals. These animals were then challenged with pathogenic bacteria. Vitamin E supplementation resulted in increased resistance to infection and decreased morbidity and mortality.[16-17] In agreement with these findings of a correlation of high vitamin E status with enhanced lymphocyte responses to infection in animals, Chavance et al.,[18] in an epidemiological study, found a significant association between high plasma vitamin E levels and a lower number of infections in healthy adults over the age of 60.

VITAMIN E AND PUFA: EFFECTS ON IMMUNITY

It is well recognized that the requirement for vitamin E increases as the level of polyunsaturated fats in the diet increases. If the fat source in animal diets was corn oil (high in polyunsaturated fatty acids [PUFA]), rather than lard or coconut oil (high in saturated fatty acids), immune responses of T and B lymphocytes to mitogens and in vitro antigen dependent antibody production were depressed.

The degree of immunodepression was directly related to the concentration of dietary PUFA. Shapiro et al.[19] found no depression of mouse T and B lymphocyte mitogen responses when 5% PUFA replaced 5% saturated fat in the diet. Corwin and Schloss [20] found depression when 8% PUFA replaced saturated fat in mouse diets. Bendich et al. [21] found significant depression in these responses when rats were fed 10% PUFA instead of lard. Vitamin E deficiency as well as low vitamin E (15 IU/kg diet) exacerbated this immunosuppression. Higher than normal dietary levels of vitamin E could partially overcome the immunosuppressive effects of these high PUFA diets (Fig 1).

VITAMIN E AND VITAMIN C

The antioxidant capacity of vitamin E has been associated with the donation of an electron from vitamin C and the subsequent regeneration of tocopherol from the tocopheryl free radical.[22] These cell-free experiments have prompted us to determine the interaction of vitamin E and vitamin C on T and B lymphocyte responses to mitogens. Guinea pigs were fed diets with either a low but anti-scorbutic level of vitamin C or a high level of vitamin C and either 0, 30, or 200 IU/kg vitamin E (six

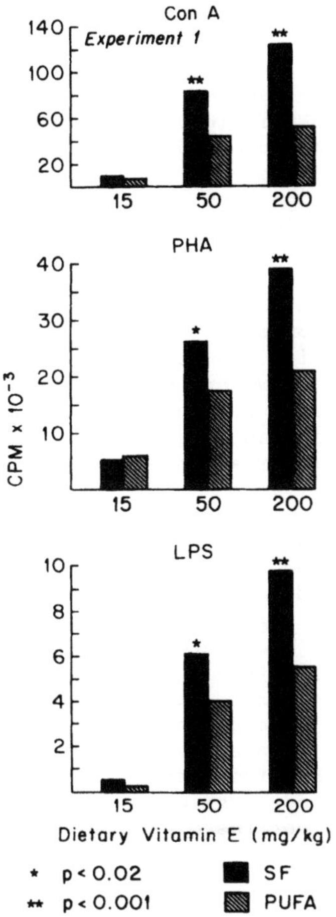

Fig 1. Groups of rats were fed diets containing 10% saturated fat (SF) or 10% polyunsaturated fatty acids (PUFA) and either 15, 50 or 200 mg/kg vitamin E for 10 wk. Splenic lymphocyte proliferative responses were determined using the T lymphocyte mitogens, Con A and PHA, and the B lymphocyte mitogen, LPS.

dietary groups). As in mice and rats, vitamin E-deficient guinea pigs had depressed responses to mitogens compared to guinea pigs fed diets containing vitamin E. The two levels of vitamin C did not affect these responses. However, guinea pigs fed the high vitamin C diet had a significantly higher concentration of vitamin E in the lung and plasma than the guinea pigs fed the low level of vitamin C, even though both groups were fed diets containing the same level of vitamin E.[23] These data could be interpreted as an in vivo verification of the cell-free experiments showing that vitamin C can help regenerate vitamin E.

VITAMIN E AND SELENIUM

Along with the antioxidant vitamins, there are several enzymes which can protect against damage which can be caused by free radicals. While vitamin E can protect against lipid peroxidation in cell membranes, the enzyme glutathione peroxidase can reduce the level of lipid peroxidation within the cell. Selenium is part of the metalloenzyme glutathione peroxidase, which can scavenge free radicals formed during lipid peroxidation and is also one of the catalysts for the destruction of hydrogen peroxide.[24]

When rats were fed a vitamin E and selenium-deficient diet for 4-8 wk, the T and B lymphocyte responses to mitogens were severely depressed. When 0.2 mg/kg sodium selenite was added to the diet, the responses were not significantly enhanced; however, when 200 IU/kg vitamin E was added to the basal diet, both T cell responses to the mitogen PHA and B cell responses to LPS were significantly enhanced. When both nutrients were included in the diet, a further enhancement of T cell responses to Con A was noted.

In experiments reported by Mulhern et al.[25, 26] the lymphocyte mitogen responses of first generation selenium-deficient, vitamin E-adequate mice were the same as animals fed the diet containing selenium; however, specific antibody responses (IgG) were reduced following challenge with sheep red blood cells. When mice were kept selenium deficient through two generations, mitogenesis, specific and nonspecific antibody titers as well as thymic size were depressed. It may be that selenium is required during the critical period of embryonic development when immune functions are iniated. This would be a possible explanation for the immunodepression seen in the mice born to selenium-deficient dams. If this is the case, the immunodepression seen in the second generation mice should be found whether or not the diet contains selenium.

SUMMARY

Vitamin E, the major lipid-soluble antioxidant present in all cellular membranes, is an important nutrient for optimal immune function. When animals are fed nutritionally complete diets lacking vitamin E, immune responses are adversely affected. Supplementation of these diets with higher than nutritionally adequate levels of vitamin E enhances immune responses.

High levels of PUFA are immunosuppressive, and vitamin E can partially overcome this immunosuppression. High levels of vitamin C can protect tissue levels of vitamin E and may indirectly contribute to the immunoenhancement by vitamin E. Severe selenium deficiency is immunosuppressive. Vitamin E can protect some aspects of immune responses from the adverse effects of selenium deficiency.

These data clearly indicate that nutrients that affect the overall antioxidant status have important effects on immune functions. In addition, antioxidant nutrient interactions can synergize to overcome the adverse effects of polyunsaturated fatty acids on immune functions (Fig 2).

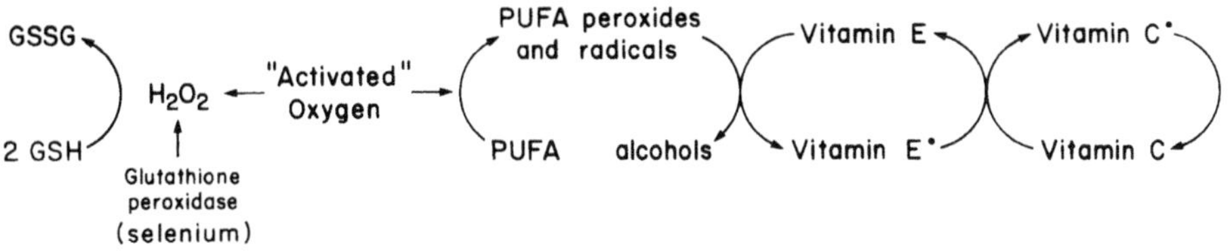

Fig 2. Antioxidant nutrient interactions.

REFERENCES

1. L. J. Machlin, Vitamin E, in: "Handbook of Vitamins," L.J. Machlin, ed., Marcel Dekker, Inc., New York, pp. 99-145, (1984).
2. I. M. Roitt, "Essential Immunology," Blackwell Scientific Publications, London (1984).
3. I. Molennaar, C. E. Hulstaert, and M. J. Hardonk, Role in function and ultrastructure of cellular membranes, in: "Vitamin E: A Comprehensive Treatise," L. J. Machlin, ed., Marcel Dekker, New York, pp. 372-389 (1984).
4. M. W. Fountain, C. Dees, J. D. Weete, and R. D. Schultz, Mol. Immunology 19:1491-1498 (1982).
5. L. J. Hatam and H. J. Kayden, J. Lipid Res. 20:639-645 (1979).
6. J. Lehmann and M. McGill, J. Lipid Res. 23:299-306 (1982).
7. A. Bendich, E. Gabriel, and L. J. Machlin, J. Nutr. 113:1920-1926 1983).
8. L. M. Corwin and R. K. Gordon, Ann. N Y Acad. Sci. 393:427-451 (1982).
9. L. M. Corwin and J. Shloss, J. Nutr. 110:2497-2505 (1980).
10. A. Gebremichael, E. M. Levy, and L. M. Corwin, J. Nutr. 114:1297-1205 (1984).
11. R. O. Likoff, D. R. Guptill, L. M. Lawrence, C. C. McKay, M. M. Mathias, C. F. Nockels, and R. P. Tengerdy, Am. J. Clin. Nutr. 34:245-251 (1982).
12. S. N. Meydani, M. Meydani, C. P. Verdon, J. B. Blumberg, and K. C. Hayes, Fed. Proc. 43:478 (1984).
13. L. M. Lawrence, M. M. Mathias, C. F. Nockels, and R. P. Tengerdy, Nutr. Res. 5:497-509 (1985).
14. A. Bendich, E. Gabriel, and L. J. Machlin, J. Nutr. 116:675-681 (1986).
15. C. J. Dillard, K. J. Kunert and A. L. Tappel, Res. Commun. Chem. Pathol. Pharmacol. 37:143-146 (1982).
16. C. K. Chow, Vitamin E and blood, in: "World Nutritional Determinants," G. H. Bourne, ed., Karger, Basel, pp. 133-166 (1985).

17. R. S. Panush and J. C. Delafuente, Vitamins and immunocompetence, in: "World Nutritional Determinants," G. H. Bourne, ed., Karger, Basel, pp. 97-133 (1985).
18. M. Chavance, G. Brubacher, B. Herbeth, G. Vernhes, T. Mikstacki, F. Dete, C. Fournier, and C. Janot, Immunological and nutritional status among the elderly, in: "Lymphoid Cell Functions in Aging," A. L. de Weck, ed., Eurage, Rijswijk, pp. 231-237 (1984).
19. A. Shapiro, S. N. Meydani, J. B. Macauley, M. Meydani, J. B. Blumberg, Fed. Proc. 44:1150 (1984).
20. L. M. Corwin and J. Shloss, J. Nutr. 110:916-923 (1980).
21. A. Bendich, E. Gabriel, and L. J. Machlin, Fed. Proc. 44:934 (1985).
22. A. Bendich, L. J. Machlin, O. Scandurra, G. W. Burton, and D. D. M. Wayner, Adv. in Free Radical Biology and Medicine 2:419-444 (1986).
23. A. Bendich, P. D'Apolito, E. Gabriel, and L. J. Machlin, J. Nutr. 114:1588-1593 (1984).
24. R. J. Shamberger, "Biochemistry of Selenium," Plenum Press, New York, pp. 239-243 (1983).
25. S. A. Mulhern, V. C. Morris, A. R. Vessey, and O. A. Levander, Fed. Proc. 40:935 (1981).
26. S. A. Mulhern, G. L. Taylor, L. E. Magruder, and A. R. Vessey, Nutr. Res. 5:201-210 (1985).

BENEFICIAL EFFECT OF DIETARY ANTIOXIDANTS ON THE AGING IMMUNE SYSTEM

S. Nikbin Meydani, T. Furukawa, M. Meydani, and J.B. Blumberg

Nutritional Immunology and Toxicology Laboratory
USDA Human Nutrition Research Center on Aging
Tufts University
Boston, Massachusetts, 02111, U.S.A.

INTRODUCTION

Aging is associated with depressed immune function. Decline in immunological vigor contributes to many age-associated diseases and the prolonged post-illness recovery often observed in the elderly. Well-documented, age-related functional changes have been defined for both humoral and cell-mediated responses.[1] Although all four major cell types of the immune system, i.e., stem cells, macrophages, T-cells and B-cells, show age-related changes, the major alterations have been demonstrated in the T-cell.[2] In vivo T-cell-dependent, cell-mediated functions such as primary delayed hypersensitivity, graft vs. host reaction, and resistance to challenge with syngeneic and allogeneic tumors and parasites are depressed with age.[3] In vitro, the proliferative response of human and rodent lymphocytes to phytohemagglutinin (PHA) and concanavalin A (ConA) as well as natural killer cell activity become depressed with age.[4]

The immune response is orchrestrated by macrophage-lymphocyte interactions. Arachidonic acid metabolites, e.g., prostaglandins (PG), hydroxyeicosatetranoic acid (HETE), leukotrienes (LT) and other oxidative products of activated macrophages, such as hydrogen peroxide play an important role in maintaining these interactions under physiological an pathological conditions. Numerous biological factors impinge upon arachidonic acid metabolism in these cells including PG-LT substrates in phospholipids (20:4ω6, 20:5ω3) and changes in peroxide tone which are partially diet-dependent. Dietary antioxidants such as vitamin E and prooxidants such as polyunsaturated fatty acids also influence the macrophage-lymphocyte relationship by altering production of peroxides and the activity of cyclooxygenase and lipoxygenase.

Dietary manipulation of the immune system has been proposed as a practical approach to intervening in its age-related functional decline. Although several studies report encouraging effects of calorie restriction on life span and immune functions of long- and short-lived rodents,[5] the effect of individual dietary components with known or potentially important roles in the immune system has been largely overlooked. Nonetheless, early evidence suggests an important role of dietary antioxidants in regulating immune function.[6] The apparent beneficial effect of these nutrients may derive from their ability to dampen peroxidative events resulting from free radical reactions.

VITAMIN E AND THE AGING IMMUNE SYSTEM

Tengerdy[7] first reported that chickens given 100 mg/kg vitamin E had significantly increased generation of anti-sheep red blood cell (SRBC) plaque-forming cells (PFC). Mice fed diets containing 60 mg vitamin E/kg had significantly increased humoral immune responses as measured by PFC and antibody responses to SRBC and tetanus toxoid.[7] Vitamin E deficiency decreases the PFC response to SRBC in mice, an effect restored to normal by vitamin E but not by the antioxidant N-N-diphenyl-p-phenylenediamine.[8] Pigs supplemented with vitamin E show enhanced proliferation of peripheral blood lymphocytes to PHA.[9]

Since numerous studies have indicated increased lipid peroxidation with aging, and antioxidants like vitamin E have been demonstrated to be immunostimulatory in young animals, we examined the effect of vitamin E on immune responsiveness of aged mice and the relationship of this effect to changes in arachidonic acid metabolism.[10] Three- and 24-month-old specific pathogen free (SPF) male C57BL/6Nia mice were obtained from the National Institute of Aging colonies. Groups (N = 10) of young and old mice were fed a casein-based semipurified diet which contained 30 ppm dl-α-tocopheryl acetate. A group of 24-month-old mice was fed the same diet supplemented with 470 ppm dl-α-tocopheryl acetate. Water and food was provided ad libitum. Mice were maintained on the diets for 6 weeks under cage level barrier condition. In vivo cellular immunity was measured by delayed hypersensitivity skin test (DHST) as ear swelling with 2,4-dinitrofluorobenzene. Blastogenic responses to various doses of the mitogens ConA, PHA and _E. coli_ lipopolysaccharide (LPS) were measured by thymidine uptake after a 68-hour incubation period. Ex-vivo PGE_2 synthesis was measured by radioimmunoassay of spleen homogenates incubated for 10 minutes at 37°C. Serum tocopherol level was measured by high performance liquid chromatography (HPLC). Supernatants harvested from ConA-stimulated cultured spleen cells were assayed for interleukin 2 (IL-2) activity by a quantitative microassay of thymidine incorporation into CTLL cells. Kidney amyloid deposits were quantified histologically.

Alpha-tocopherol concentration in serum was noted to decrease with age and was elevated almost threefold with dietary supplementation at 500 ppm. Vitamin E supplementation of aged mice enhanced percent ear swelling to DNFB as well as the mitogenic response of splenocytes to ConA and LPS. Increases in PHA-stimulated lymphocyte proliferation in vitamin E-supplemented mice were not statistically significant. Spleen homogenates from old mice fed 30 ppm vitamin E had a significantly higher PGE_2 level than young mice fed 30 ppm vitamin E and old mice fed 500 ppm vitamin E. Old mice fed 30 ppm vitamin E had a lower activity of the T-cell growth factor IL-2 than young mice ($P \leq 0.01$); vitamin E supplementation induced a twofold increase in IL-2 activity ($P \leq 0.07$). Results are summarized in Table 1. Furthermore, while 40% of the old mice had histological evidence of kidney amyloidosis, the young mice and the old mice supplemented with 500 ppm vitamin E showed no signs of amyloid deposition.[10]

These results show that splenocytes from aged mice synthesize more PGE_2 and less IL-2 than those from young mice. Since PGE_2 has an inhibitory effect on lymphocyte proliferation, the decline in lymphocyte proliferation and DHST in aged mice might be due, in part, to increased synthesis of PGE_2. Vitamin E has been shown to decrease PG synthesis and, in this study, aged mice supplemented with vitamin E had significantly less PGE_2 than non-supplemented old mice. Thus, the immunostimulatory effect of tocopherol in the aged mice might reflect inhibition of PGE_2 synthesis. Further, since changes in the immune system have been implicated in the pathogenesis of amyloidosis, the beneficial effects of vitamin E on kidney amyloidosis might be mediated by its immunostimulatory effect.

Table 1. Effects of Vitamin E on Immune Responsiveness of 24-Month-Old Mice[a]

Parameters	Control	Dietary Vitamin E[b] 30 ppm	500 ppm
Serum alpha-tocopherol	236 ± 39 μg/dl	71	194*
Delayed Hypersensitivity	59.9 ± 13.8%	36*	75
T-cell lymphocyte proliferation	119,611 ± 35,491 cpm	5*	38
B-cell lymphocyte proliferation	25,074 ± 5,149 cpm	24*	85
Ex-vivo splenic PGE_2 synthesis	2.60 ± 0.08 μg/g	123*	89
Interleukin 2	27 ± 4 unit/ml	44*	85

[a]Data adapted from Meydani et al.[11]
[b]All values expressed as percent of 3-month-old control group fed 30 ppm vitamin E
*Significantly different from control and other experimental group, $P \leq 0.05$

GLUTATHIONE AND THE AGING IMMUNE SYSTEM

Age-related decreases in glutathione (GSH) have been observed in the liver, kidney, heart and blood of mice and may be responsible for some aspects of the aging process.[12] The depletion of GSH has been reported to lower mitogenic responsiveness and addition of this tripeptide in culture medium reverses the effect.[13] Lymphocytes exposed to sulfhydryl oxidizing agents also show a reduced proliferative response to mitogens.[14] Dietary precursors for GSH can be provided as cysteine or methionine, but diets high in these amino acids are hepatotoxic.

Oral supplementation with GSH appears nontoxic and effectively raises tissue GSH levels.[15] Thus, we have examined the potential of dietary GSH to alter immune response in aging mice.[15] Four-, 17, and 24-month old SPF male C57BL/6Nia mice were obtained from the National Institute of Aging colonies and fed casein-based semipurified diets containing 0 (control) to 1.0% reduced GSH for 4 weeks (N = 7-12/group). Tests were conducted as described above except DHST was assessed by uptake of [^{125}]iodo-deoxyuridine and expressed as a stimulation index (SI). Splenocyte sulfhydryl [-SH] content was measured by a N-ethyl-[^{14}C] maleimide-binding (NEM) assay expressed as μmol NEM/10^6 viable cells counted by the trypan blue exclusion method. Reduced and oxidized GSH were determined by HPLC.

Age-related declines in splenic GSH, DHST, and the mitogenic response to ConA in the 17-and 24-month-old mice were partially but significantly reversed by GSH at levels of 0.1, 0.5 and/or 1.0% of diet. This change was associated with an increase in splenocyte cell viability. Partial results are summarized in Table 2. Splenic levels of oxidized GSH (0.45-0.55 μmol/g) and splenocyte capacity to synthesize PGE_2 were unchanged by diet. After incubation, splenocyte thiol content and ConA responsiveness were positively correlated (r = 0.679).

Table 2. Effects of GSH on Immune Responsiveness of 24-Month Old Mice[a]

Parameter	Control	Dietary GSH 0%	Dietary GSH 1.0%
Splenic GSH	2.65 ± 0.12 μmol/g	82*	99
Delayed hypersensitivity	2.81 ± 0.910 SI	80*	100[c]
T-cell lymphocyte proliferation	14,900 ± 5,000 cpm	27*	49
Splenocyte viability[d]	---	45.8 ± 0.4*	55.9 ± 1.5

[a]Data adapted from Furukawa et al[15]
[b]All values expressed as percent of 4-month-old control group fed 0% GSH
[c]Response to 0.5% dietary GSH
[d]Actual percent viable cells
*Signficantly different from control and/or other experimental group, $P \leq 0.05$

These results show GSH supplementation of a semipurified, nutritionally adequate diet enhances in vitro T-cell lymphocyte proliferation and in vivo T-cell-mediated response as measured by DHST. GSH, however, appears to promote immune function through a different mechanism than vitamin E, as no effect on PGE_2 was observed. The targets for GSH and thiol oxidizing agents could include membrane and cytoplasmic proteins which regulate lymphocyte proliferation. Thus, splenocytes from GSH-fed aged mice appear to maintain a more vigorous cellular antioxidant status. This antioxidant ability to inhibit free radical formation may underlie its immunostimulatory action.

DISCUSSION

The age-related accumulation of deleterious free radical reactions and increased rate of lipid peroxidation may contribute to some of the chronic degenerative conditions associated with aging. These peroxidative changes influence arachidonic acid metabolism and autocoid formation, which may in turn contribute to the alterations in immune responsiveness observed in the elderly. The immunostimulatory effect of GSH and vitamin E provides further support that these events play an important role in the aging immune system. However, the exact nature of the immunostimulatory effect of vitamin E and GSH and the role of arachidonic acid metabolites and other oxidative products of activated macrophages in the age-associated changes in immune responsiveness is not well understood. Further studies in which changes in free radical formation are measured would yield more direct evidence for the involvement of reactive oxygen metabolites in alterations of immune function.

Dietary antioxidants may act to increase immune responsiveness by altering macrophage events mediated by cyclooxygenase (generation of PG) and lipoxygenase (generation of HETE and LT) or IL-1 production. The mechanism(s) of such action could be based upon antioxidant dampening of enzyme activity, quenching of lipid perocidation, and/or altering fatty acid percursor pools of splenocyte phospholipids which dictate PG and LT synthesis.

With the exception of calorie restriction, efforts to enhance immune responsiveness in aged organisms has not been successful and/or practical. Intervention in the aging process through increased intake of dietary antioxidants might be beneficial, but further research is necessary to substantiate this hypothesis.

REFERENCES

1. M. M. B. Kay, An overview of immune aging, Mech. Ageing Dev. 10:1979 (1975).
2. T. Makinodan, Cellular basis of immunological aging, in: "Biological Mechanism's in Aging," R. T. Shimke, ed., USDA NIH, Bethesda (1981).
3. J. S. Goodwin, R.P. Searles, and K.S.K. Tung, Immunological responses of a healthy elderly population, Clin. Exp. Immunol. 48:4023 (1982).
4. M. M. B. Kay, An overview of immune ageing, Mech. Ageing Dev. 9:35 (1979).
5. E. J. Masoro, Nutrition and aging - a current assessment, J. Nutr. 115:842 (1985).
6. J. B. Blumberg and S. N. Meydani, Role of dietary antioxidants in aging, in: "Nutrition and Aging," M. Hutchinson, H. Munro, eds., Academic Press, NY (1986).
7. P. Tengerdy and J. C. Brown, Effect of vitamin E and A on humoral immunity and phagocytosis in E. coli infected chickens, Poultry Sci. 56:957 (1977).
8. P. Tengerdy, Effect of vitamin E on immune responses, Basic Clin. Nutr. 1:429 (1980).
9. H. J. Larsen and S. Tollersrud, Effect of dietary vitamin E and selenium on the phytohaemagglutin response of pig lymphocytes, Res. Vet. Sci. 31:301 (1981).
10. S. N., Meydani, E. S. Cathcart, R. E. Hopkins, M. Meydani, K. C. Hayes, and J. B. Blumberg, Antioxidants in Experimental Amyloidosis of Young and Old Mice, in: "Amyloidosis," (Proceedings of the IV International Congress), G. G. Glenner, E. F. Osserman, E. Benditt, E. Calkins, A. S. Cohen, D. Zucker-Franklin, eds., Plenum Press, New York (1986).
11. S. N. Meydani, M. Meydani, C. P. Verdon, A. C. Shapiro, J. B. Blumberg, and K. C. Hayes, Vitamin E supplementation suppresses prostaglandin E_2 synthesis and enhances the immune response of aged mice, Mech. Ageing Dev. 34:191 (1986).
12. G. A. Hazelton and C. A. Lang, Glutathione content of tissues in the aging mouse, Biochem. J. 188:25 (1980).
13. M. A. Gougerst-Poidals, M. Fay, Y. Roche, P. Lacombe, and C. Marquetty, Immune oxidative injury induced in mice exposed to normobaric O_2: Effects of thiol componds on the splenic cell sulfhydryl content and ConA proliferative response, J. Immunol. 135:2045 (1985).
14. R. J. Noelle and D. A. Lawrence, Determination of glutathione in lymphocyte and possible association of redox state and proliferative capacity of lymphocytes. Biochem. J. 198:571 (1981).
15. T. Higashi and K. Yamaguchi, in: "Glutathione: Storage, Transport and Turnover in Mammals.," Y. Sakamoto, T. Higashi, and N. Tateishi, eds., Japan Sci. Soc. Press Tokyo VNU Science Press, Utrecht (1983).
16. T. Furukawa, S. N. Meydani, and J. B. Blumberg, Reversal of age-associated decline in immune responsiveness by dietary glutathione supplementation in mice, Mech. Ageing Dev. 38:107 (1987).

INCLUSION OF ANTIOXIDANTS IN RESUSCITATION FLUIDS

Bo E. Hedlund

University of Minnesota
Minneapolis, MN 55455, U.S.A.

INTRODUCTION

Hemorrhagic shock is, whenever possible, treated with rapid volume replacement. Such treatment is often initiated by paramedical personnel as soon as possible following blood loss, e.g., at the site of an accident and during transport to an emergency room. The fluids used for this type of volume replacement include crystalloids, i.e., solutions of sodium chloride and other inorganic salts. Other volume expanders include biocompatible polysaccharides, so-called colloids, such as hydroxyethyl starch, dextran and human albumin. A considerable amount of research is aimed at developing resuscitation fluids based on solutions of human and bovine hemoglobin. Such stroma-free hemoglobin solutions are intended to deliver oxygen to critical organs where severe hypoxia may lead to irreversible damage.

RELEASE OF IRON

The decreased microvascular blood flow associated with hemorrhagic shock is particularly pronounced in the renal cortex and the intestine.[1] During ischemic insults substantial quantities of ferrous iron are released from ferritin, a process which is well documented in liver and probably occurs in other organs as well.[2,3] This process is mediated by xanthine oxidase and involves the superoxide anion as a reducing agent. During resuscitation, which can be considered a global reperfusion, the cascade of events leading to radical-mediated tissue injury begins in tissues exposed to transient ischemia with associated anoxia or hypoxia.[4] Iron is a critical participant in catalyzing the reaction between superoxide and hydrogen peroxide to yield the highly reactive hydroxyl radical. Of even greater importance is the ability of "free" iron to act as a catalyst in promoting non-enzymatic lipid peroxidation. Based on the critical importance of iron participation in these reactions, iron sequestration represents an effective means of minimizing radical-mediated tissue injury occuring secondary to hemorrhagic shock. None of the volume replacement solutions presently used during the initial resuscitation provide any antioxidant potential. In the case of stroma-free hemoglobin, oxidant injury may actually be further aggravated by release of iron from the protein, particularly following oxidation to methemoglobin.

The addition of biocompatible iron chelators to resuscitation fluids will enhance the iron sequestration potential in organs at risk for reperfusion injury. This approach is quite similar to the enhancement of antioxidant capacity accomplished by including superoxide dismutase (SOD) and other oxygen radical scavengers in these solutions. A great deal of additional research remains to be carried out before the relative therapeutic efficacy of these two approaches can be determined. Possibly the best results are obtained when both enzymatic oxygen radical scavengers and biocompatable iron chelators are included in the same solution. It is likely that the two mechanisms invoked for prevention of oxygen-radical-mediated tissue injury, i.e., the direct radical scavenging by enzymes and the indirect effect of removing catalytically active iron by chelators, act in synergy.

PROTECTION BY TRANSFERRIN AND CERULOPLASMIN

The two most prominent proteins in human plasma acting as indirect antioxidants are transferrin and ceruloplasmin.[5] Iron bound to transferrin cannot act as a catalyst for hydroxyl radical formation and lipid peroxidation. Ceruloplasmin acts by effectively oxidizing ferrous to ferric iron without the formation of oxygen radical intermediates. Thus, ceruloplasmin very effectively prevents lipid peroxidation initiated by ferrous iron. Both transferrin and ceruloplasmin can, in principle, be incorporated into resuscitation fluids. However, these proteins may have limitations similar to those of enzymatic scavengers in terms of storage and long-term stability in solutions intended for emergency use. Synthetic iron chelators may provide equal therapeutic potential without being susceptible to denaturation and contamination problems associated with proteins obtained from human blood.

PROTECTION BY DFO

Recent research involving ischemia and reperfusion has established that deferrioxamine (DFO), an iron chelator used clinically, does indeed protect against oxygen- and/or lipid-radical-mediated injury. Thus, in the perfused ischemic rabbit heart, two groups have shown that DFO does afford protection against reperfusion injury comparable to that of SOD.[6-8] Similar conclusions can be drawn from studies using oxygen radical scavengers and the iron chelator in a model of renal ischemia. Paller and collaborators showed that DFO protects the kidney as well as SOD against reperfusion injury.[9,10] However, the toxicity associated with intravenous infusion of DFO, combined with the short vascular half-life of this drug, may limit its potential as a therapeutic ingredient in resuscitation fluids.

Isovolemic models involving cardiopulmonary and cerebral resuscitation in animals have also provided some evidence that DFO may afford protection against radical-mediated tissue injury. Thus, Babbs and his group[11,12] have shown that DFO decreased mortality in a rat model of cardiopulmonary resuscitation. Similarly, White et al.[13] have obtained results suggesting improved neurological recovery following DFO treatment at the point of resuscitation in a canine model of cardiopulmonary and cerebral resuscitation.

CONCLUSIONS

Morbidity and mortality resulting from shock and trauma can be decreased if antioxidant therapy is initiated as early as possible during resuscitation. Research aimed at determining which ingredients provide the best protection against oxygen-radical-mediated tissue injury has yet to be carried out in a systematic fashion. In addition to "conventional" enzymatic radical scavengers, both transferrin and ceruloplasmin should be considered in this context. Other iron-binding compounds, including synthetic iron chelators, may have an advantage over protein based antioxidants in terms of stability and storage properties.

REFERENCES

1. H. Hirasawa, M. Odaka, Y. Tabata, H. Kobayashi, and H. Sato, Tissue blood flow in brain, liver, renal cortex and renal medulla in experimental hemorrhagic shock, Crit Care Med. 5;141 (1977).
2. A. Mazur, S. Baez, and E. Shorr, The mechanism of release from ferritin as related to its biological properities, J. Biol. Chem. 213:147 (1955).
3. S. Green and A. Mazur, Relation of uric acid metabolism to release of iron from hepatic ferritin, J. Biol. Chem. 219:653 (1957).
4. J.M. McCord, Oxygen-derived free radicals in post-ischemic tissue injury, N. Engl. J. Med. 312:159 (1985).
5. J. Stocks, J.M.C. Gutteridge, R.J. Sharp, and T.L. Dormandy, Assay using brain homogenate for measuring the antioxidant activity of biological fluids, Clin. Sci. Mol. Med. 47:215 (1974).
6. C.L. Meyers, S.J. Weiss, M.M. Kirsch, B.M. Shepard, and M. Shlafer, Effect of supplementing hypothermic crystalloid cardioplegic solutions with catalase, superoxide dismutase, allopurinol and deferoxamine on functional recovery of globally ischemic and reperfused isolated hearts, J. Thoracic Cardiovasc. Surg. 91:281 (1986).
7. G. Ambrosio, M.L. Weisfeldt, W.E. Jacobus, and J.T. Flaherty, Evidence for a reversible oxygen radical mediated component of reperfusion injury: Reduction by recombinant human superoxide dismutase administered at time of reflow, Circulation 75:282 (1987).
8. G. Ambrosio, J.L. Zweier, W.E. Jacobus, M.L. Weisfeldt, and J.T. Flaherty, Improvement of postischemic myocardial function and metabolism induced by administration of deferoxamine at the time of reflow: the role of iron in the pathogenesis of reperfusion injury, Circulation 76:906 (1987).
9. M.S. Paller, J.R. Hoidal, and T.F. Ferris, Oxygen free radicals in ischemic acute renal failure in the rat, J. Clin. Invest. 74:1156 (1984).
10. M.S. Paller, Role of iron in ischemic and hemoglobinuric acute renal failure, abstract, Kidney International 31:373 (1986).
11. C.F. Babbs, Role of iron in the genesis of reperfusion injury following successful cardiopulmonary resuscitation: Preliminary data and a biochemical hypothesis, Ann. Emerg. Med. 10:501 (1985).
12. S.D. Kompala, C.F. Babbs, and K.E. Blaho, Effect of deferoxamine on late deaths following CPR in rats, Ann Emerg. Med. 15:405 (1986).
13. B.C. White, G.S. Krause, S.D. Aust, and G.E. Eyster, Postischemic tissue injury by iron-mediated free radical lipid peroxidation, Ann. Emerg. Med. 14:804 (1985).

SERUM ANTIOXIDANT ACTIVITY AS A DETERMINANT

OF PULMONARY DYSFUNCTION IN CIGARETTE SMOKERS

Raymond B. Bridges and Stanley R. Rehm

Departments of Oral Biology and Medicine
University of Kentucky and VA Medical Centers
Lexington, Kentucky 40536-0084, U.S.A.

INTRODUCTION

Cigarette smoking is a significant risk factor for the development of chronic obstructive pulmonary disease (COPD), and the primary etiology of COPD in smokers is presumably a pulmonary protease-antiprotease imbalance.[1] Oxidants in cigarette smoke[2,3] and those generated by activated phagocytes[4] have both been shown to inactivate the pulmonary antiproteases. This cross-sectional study examined the effects of smoking on total serum antioxidant activity (AOA), the components of peripheral blood which contributed to this AOA, and the relationships between AOA and smoking history or pulmonary dysfunction.

METHODS AND RESULTS

The Caucasian populations studied consisted of 170 male smokers and 170 age- and sex-matched nonsmokers who were relatively young (mean age 37 years). The smokers consumed a mean of 25 cigarettes per day, with a mean cumulative smoking history of 24 pack-years and 20 years smoked. Pulmonary function was assessed by forced expiratory spirometry and the single breath nitrogen test.[5] Data for smokers were adjusted for age and height from prediction equations derived from the nonsmoking population using analysis of covariance and expressed as percent predicted values. Consistent with their relatively young age and brief smoking history, the smokers had modest but significant impairment of pulmonary function.[5]

Serum concentrations of ceruloplasmin were determined by radial immunodiffusion (RID) (Behring Diagnostics, La Jolla, CA) and by p-phenylenediamine (PPD) oxidase activity.[6] Serum iron was determined by blood chemistries (SMA-24), while antioxidant activity (AOA) was determined by the ability of serum to inhibit the autoxidation of lipids in bovine brain homogenate.[7] Smokers had modest but significant increases in ceruloplasmin, as measured by either RID or PPD oxidase activity (Table 1). However, serum iron and AOA were not significantly different in smokers and nonsmokers.

Determining the linear correlations between smoking history and serum ceruloplasmin, iron, or AOA (Table 2), ceruloplasmin correlated significantly (or marginally) with all of the measures of present or

Table 1. Serum Ceruloplasmin, Iron, and Antioxidant Activitya

	n	Nonsmokers	Smokers	% change	P
Ceruloplasmin (mg/dl)	154	25.9 ± 0.4	27.9 ± 0.4	+ 7.7%	0.0006
PPD Oxidase Activity[b]	125	216.7 ± 4.1	241.2 ± 4.7	+11.3%	0.0001
Serum Iron (μg/dl)	153	117.4 ± 3.3	119.9 ± 3.6	+ 2.1%	0.61
Antioxidant Activity[c]	119	30.8 ± 0.8	31.8 ± 0.9	+ 3.2%	0.41

[a]Results are expressed as means (± S.E.M.).
[b]Millimoles PPD oxidized/minute/liter of serum.
[c]Percentage inhibition of lipid peroxidation by 25 μl of sera.

cumulative smoking history, whereas serum iron correlated only marginally with pack-years smoking history. Although AOA was not significantly different in smokers and nonsmokers, it correlated significantly with both present cigarette consumption and pack-years smoking history, but not years smoked.

Table 2. Relationships of Smoking History with Serum Ceruloplasmin, Iron, or Antioxidant Activity[a]

	Cigarettes/day		Years Smoked		Pack-Years	
	r	p	r	p	r	p
Ceruloplasmin	0.157	0.05	0.139	0.08	0.125	0.11
PPD Oxidase Activity	0.158	0.07	0.250	0.004	0.290	0.0009
Serum Iron	-0.090	0.26	-0.056	0.48	-0.129	0.10
Antioxidant Activity	0.178	0.05	0.113	0.22	0.212	0.02

[a]Pearson's correlates for 122 smokers.

Serum AOA correlated significantly with iron ($r = -0.386$, $p = 0.0001$) and ceruloplasmin (RID, $r = 0.202$, $p = 0.02$; PPD oxidase, $r = 0.177$, $p = 0.05$). Multiple regression analysis revealed that iron was the single most important determinant of AOA, accounting for 16.03% of its variation ($p \leq 0.0001$). Although ceruloplasmin also added significantly ($p \leq 0.02$) to the further prediction of AOA, iron and ceruloplasmin together accounted for only 19.82% of the variation in AOA.

Decrements in FEV1 or FVC were significantly associated with increased AOA but not with ceruloplasmin (Table 3). Significant (or marginally significant) correlations were also observed between decrements in FEV1 or FVC and decreased serum iron.

Table 3. Relationships between FEV1 or FVC and Serum Ceruloplasmin, Iron, or Antioxidant Activity[a]

	FEV1		FVC	
	r	p	r	p
Ceruloplasmin	-0.112	0.17	-0.112	0.17
PPD Oxidase	-0.086	0.35	-0.139	0.13
Serum Iron	0.144	0.08	0.166	0.04
Antioxidant Activity	-0.222	0.02	-0.195	0.04

[a]Pearson's correlates for 122 smokers.

Cumulative smoking history was previously shown to be the best predictor of decrements in FEV1 or FVC in smokers.[5,8] To assess the role of AOA, multiple regression analyses were performed with FEV1 or FVC as the dependent variable and present daily cigarette consumption, years smoked, pack-years, AOA, total leukocyte counts, serum α1-proteinase inhibitor (α1-PI) concentration, and neutrophil lysosomal enzyme activities (i.e., elastase or myeloperoxidase) as the independent variables (Table 4). Although AOA correlated significantly with cumulative smoking history, it contributed significantly to the further prediction of both FEV1 and FVC over that obtained with present or cumulative smoking history. The independent variables not listed in Table 4 did not add significantly to the prediction of the dependent variables.

Table 4. Regression Analysis of Pulmonary Function

Independent Variable	FEV1 (% predicted)	FVC (% predicted)
Years Smoked	$p < 0.0002^a$	$p < 0.0001^b$
Cigarettes/day	$p = 0.03$	$p = 0.03$
Antioxidant Activity	$p = 0.01$	$p = 0.04$
Total R^2	0.2394	0.2728

[a]Years smoked alone, R^2=0.147
[b]Years smoked alone, R^2=0.2063

DISCUSSION

An increased oxidant burden either from cigarette smoke itself or generated from a smoking-induced inflammatory process may contribute to the development of COPD in smokers. Ceruloplasmin potentially provides antioxidant protection by preventing the oxidative inactivation of α1-PI.[9] Transferrin, due to its iron-dependent effects and its presence in alveolar epithelial lining fluid, may also be an important pulmonary antioxidant.[10] These observations provided the impetus to study serum ceruloplasmin and AOA and to correlate their levels with pulmonary dysfunction.

Like others,[11] we found that serum ceruloplasmin is elevated in smokers. However, in contrast, AOA was not significantly different in smokers and nonsmokers, and elevated AOA was significantly associated with increased smoking history. Consistent with the previous report that AOA is more directly related to available iron-binding capacity of transferrin than to ceruloplasmin,[11] we found that AOA was more dependent upon iron than ceruloplasmin.

Possibly most notable is our observation that elevated AOA is significantly correlated with pulmonary dysfunction in smokers and adds to the prediction of both FEV1 and FVC over that obtained by the smoking history variables. Increased AOA likely does not bear a causal relationship to pulmonary dysfunction in smokers. Rather, a smoking-induced inflammatory process associated with pulmonary dysfunction elevates serum AOA by increasing ceruloplasmin, decreasing iron, and possibly affecting other as yet unknown determinants of AOA.

We previously demonstrated that this inflammatory reaction in smokers is also characterized by elevated neutrophil counts, α1-PI, and the ninth component of complement (C9) and that the levels of these inflammatory mediators are significantly correlated with both smoking history and decrements in FEV1 and FVC.[5] Elevations of neutrophil counts, α1-PI, and ceruloplasmin and depressed serum iron associated with pulmonary dysfunction are all consistent with an inflammatory reaction mediated by the monokine, interleukin-1 (IL-1).[12] This hypothesis appears plausible since alveolar macrophages treated with lipopolysaccharide produce IL-1.[13] Thus, our hypothesis is that increased AOA associated with pulmonary dysfunction in smokers is a result of an inflammatory reaction, possibly mediated by IL-1.

ACKNOWLEDGMENTS

We thank Ms. May Fu and Ms. Judy Combs for expert technical assistance, Ms. Donna Reed for data entry, and Mr. John Turbek for statistical analysis of the data. This study was supported by Grant No. 5-41091 from the Tobacco and Health Research Institute, University of Kentucky, Lexington, Kentucky.

REFERENCES

1. A. Janoff, State of the Art: Elastases and emphysema. Current assessment of the protease-antiprotease hypothesis, Am. Rev. Respir. Dis. 132:417 (1985).
2. W.A. Pryor, M.M. Dooley, and D.F. Church, Inactivation of human alpha-1-proteinase inhibitor by gas-phase cigarette smoke, Biochem. Biophys. Res. Commun. 122:676 (1984).
3. K. Ohlsson, U. Fryksmark, and H. Tegner, The effect of cigarette smoke condensate on α_1-antitrypsin, antileukoprotease and granulocyte elastase, Eur. J. Clin. Invest. 10:373 (1980).
4. M.C. Zaslow, R.A. Clark, P.J. Stone, J.D. Calore, G.L. Snider, and C. Franzblau, Human neutrophil elastase does not bind to alpha-1-protease inhibitor that has been exposed to activated human neutrophils, Am. Rev. Respir. Dis. 128:434 (1983).
5. R.B. Bridges, R.J. Wyatt, and S.R. Rehm, Effects of smoking on inflammatory mediators and their relationship to pulmonary dysfunction, Eur. J. Respir. Dis. 69 (suppl 146):145 (1986).
6. P.L. Wolf, Ceruloplasmin: Methods and clinical use, CRC Critical Rev. Clin. Lab. Sci. 17:229 (1982).
7. J. Stocks, J.M.C. Gutteridge, R.J. Sharp, and T.L. Dormandy, Assay using brain homogenate for measuring the antioxidant activity of biological fluids, Clin. Sci. Mol. Med. 47:215 (1974).
8. G.J. Beck, C.A. Doyle, and E.N. Schachter, Smoking and lung function, Am. Rev. Respir. Dis. 123:149 (1981).
9. J. Taylor, and L. Oey, Ceruloplasmin: plasma inhibitor of the oxidative inactivation of $alpha_1$-protease inhibitor, Am. Rev. Respir. Dis. 126:476 (1982).
10. E.R. Pacht, and W.B. Davis, Transferrin: an important antioxidant protein in alveolar epithelial lining fluid, Am. Rev. Respir. Dis. 135: A265 (1987).
11. M. Galdston, J.G. Feldman, V. Levytska, and B. Magnusson, Antioxidant activity of serum ceruloplasmin and transferrin available iron-binding capacity in smokers and nonsmokers, Am. Rev. Respir. Dis. 135:783 (1987).
12. C.A. Dinarello, Interleukin-1, Rev. Infect. Dis. 6:51 (1984).
13. E.A. Rich, D.J. Tweardy, H. Fujiwara, and J.J. Ellner, Spectrum of immunoregulatory functions and properties of human alveolar macrophages, Am. Rev. Respir. Dis. 136:258 (1987).

PROBING ENZYME-SUBSTRATE RECOGNITION AND CATALYTIC MECHANISM IN CU,ZN SUPEROXIDE DISMUTASE

John A. Tainer, Robert A. Hallewell, Victoria A. Roberts,
Hans E. Parge, and Elizabeth D. Getzoff

Dept. of Molecular Biology
Research Institute of Scripps Clinic
La Jolla, California 92037, U.S.A.

INTRODUCTION

Superoxide dismutase (SOD) is a ubiquitous enzyme of aerobic organisms that protects against the toxic effects of dioxygen metabolism. SOD catalyzes the dismutation of the superoxide radical, $\cdot O_2^-$, to molecular oxygen and hydrogen peroxide through the alternate reduction and oxidation of the active site metal ion (Cu, Mn, or Fe)[1]:

$$\cdot O_2^- + \cdot O_2^- + 2H^+ \rightarrow H_2O_2 + O_2$$

Cu,Zn SOD occurs almost entirely in eukaryotic cells, while Mn or Fe SODs occur predominantly in prokaryotes and mitochondria. The catalytic rate of SOD is very rapid ($2 \times 10^9\ M^{-1}sec^{-1}$),[2] suggesting the evolution of an optimal active site for the recognition and chemical catalysis of the substrate by the enzyme. This remarkably high catalytic rate makes the enzyme a simple model system in which to study the role of electrostatic forces in molecular recognition.

STRUCTURE OF SOD

In order to understand the detailed molecular events that lead to enzymatic catalysis, it is important to have an accurate definition of the atomic positions in the active site and to interpret these positions in the light of the known biochemistry. The x-ray crystal structure of Cu,Zn bovine SOD has been determined.[3-5] Figure 1 is a schematic of the active site residues of bovine Cu,Zn SOD, including the metal-liganding residues, the activity-important Arg 141,[6] and the Cu and Zn ions. Twenty-one residues contribute to the active site channel. Residues Glu 131, Lys 134, Thr 135, Gly 136, Asn 137, Ala 138, Gly 139, and Arg 141 form one rim of the channel and residues Thr 56, Ala 58, Gly 59, Pro 60, and Asn 63 form the other rim. The metals and their seven liganding residues form the floor of the channel with the buried Asp 122 lying beneath. Compared with the rest of the molecule, the residues that make up the active site channel are highly conserved in the published sequences.[7-23]

In cross section, the channel funnels downward stepwise from a large shallow depression about 24Å across to a deeper well about 10Å wide and 5Å deep, then narrows to less than 4Å forming a complementary pocket for the superoxide anion directly above the Cu ion. Docking experiments of a superoxide molecule into the active site channel show that the superoxide can be positioned in the channel with an excellent fit to the surface. One oxygen of the substrate lies 2Å away from the copper, while the other oxygen forms a hydrogen bond with the Arg 141 side chain. Thus, the crystal structure indicates that the channel, made up of conserved residues, is optimal for the fit and positioning of the superoxide molecule.

ACTIVE SITE

A unique feature of the active site of this enzyme is the bridge formed between the Zn and Cu ions by the His 61 side chain. Several experimental studies, including EPR[24] and NMR,[25] indicate that this bridge is broken when superoxide reduces the Cu ion, relaxing the geometry around the Cu(I) ion to a more tetrahedral conformation. His 61 then picks up a proton from solvent. This proton is now positioned for protonation of a second molecule of superoxide, which binds to the Cu(I) ion, resulting in the production of hydrogen peroxide. With the formation of hydrogen peroxide and oxidation of the Cu ion, the His 61 bridge is reformed. Protonation by this histidine provides an excellent explanation for the pH independence of the catalytic rate between pH 5 and pH 9.5 for Cu,Zn SOD.[2]

The dismutation of superoxide ($\cdot O_2^-$) by Cu,Zn SOD is almost as fast as that by aqueous Cu ion.[26] This enzymatic rate is surprising, since Cu,Zn SOD must overcome a large handicap to compete with free copper; $\cdot O_2^-$ can approach aqueous copper from any coordinate position, but must find the tiny fraction (5Å^2) of the molecular surface of Cu,Zn SOD (6500Å^2) that forms the one free Cu coordination site.[27] In addition, Cu,Zn SOD has a net negative charge at physiological pH; the pI value for bovine enzyme is 4.9, while other species exhibit pI values ranging from 4.6 to 6.8.[28] This net negative charge would be expected to create an electrostatic barrier for incoming $\cdot O_2^-$ that should be diminished with increasing ionic strength, which would accelerate the reaction. Instead, raising the ionic strength decreases the reaction rate, implying that electrostatic forces facilitate the reaction.[29]

From the x-ray structure of bovine SOD, we have calculated the electrostatic potential and electrostatic fields around the enzyme,[27] and have identified electrostatic forces involved in both long-range[30] and short-range recognition and orientation of the negatively charged superoxide substrate. The electrostatic potential of bovine Cu,Zn SOD shows no organized pattern over most of the surface, but reveals a positive region that extends over the long, deep channel above the Cu ion (Figure 2). The positive potential near the Cu and Zn ions is very high, forming an electrostatically complementary binding site for the negatively charged $\cdot O_2^-$ substrate. In addition, two residues at the top of the channel, Lys 120 and Lys 134, appear to guide the substrate into the active site channel. Both Lys 134 and Lys 120 form salt bridges with negatively charged residues (Glu 131 and Glu 119, respectively), which may prevent non-productive binding of superoxide with these two lysine residues.

The positive potential surrounding the active site channel, composed almost exclusively of sequence-conserved amino acid residues, represents approximately 1/10th of the surface area of the subunit. The resulting positive field allows the superoxide to be attracted to a much larger area of the protein than just the available surface area of the copper alone. Guidance of superoxide into the active site pocket by the electrostatic field appears to be significant at distances of 14Å, thus coding for active site recognition prior to actual collision at the molecular surface (Figure 3).

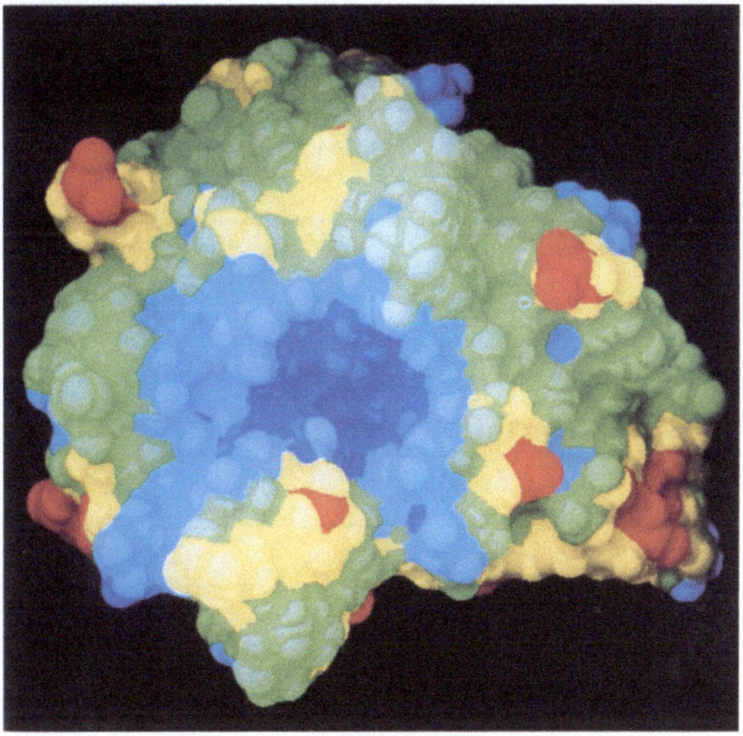

Figure 2. The electrostatic potential color-coded onto the surface of a solid, space-filling model of bovine Cu,Zn SOD. The active site channel is characterized by highly positive (dark blue) potential over the metal ions surrounded by positive (blue) potential. The rest of the surface is neutral (green) with areas of negative (yellow) and highly negative (red) potential. For details of the electrostatics calculations, see Getzoff et al., 1983.[27]

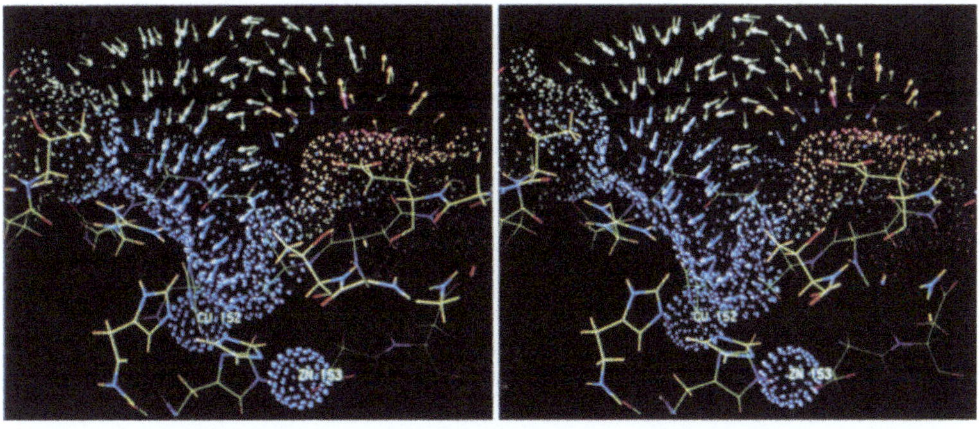

Figure 3. A stereo pair showing the deep active site channel of SOD with electrostatically color-coded molecular surface and electrostatic field vectors.[27] Both the molecular surface (dots) and the field vectors are color-coded by electrostatic potential as in Figure 2. Electrostatic field vectors indicate the direction a negative charge and hence the superoxide molecule, would move in the field. The skeletal model includes hydrogens and is colored by atom type (red = O, blue = N, green = C, and yellow = H).

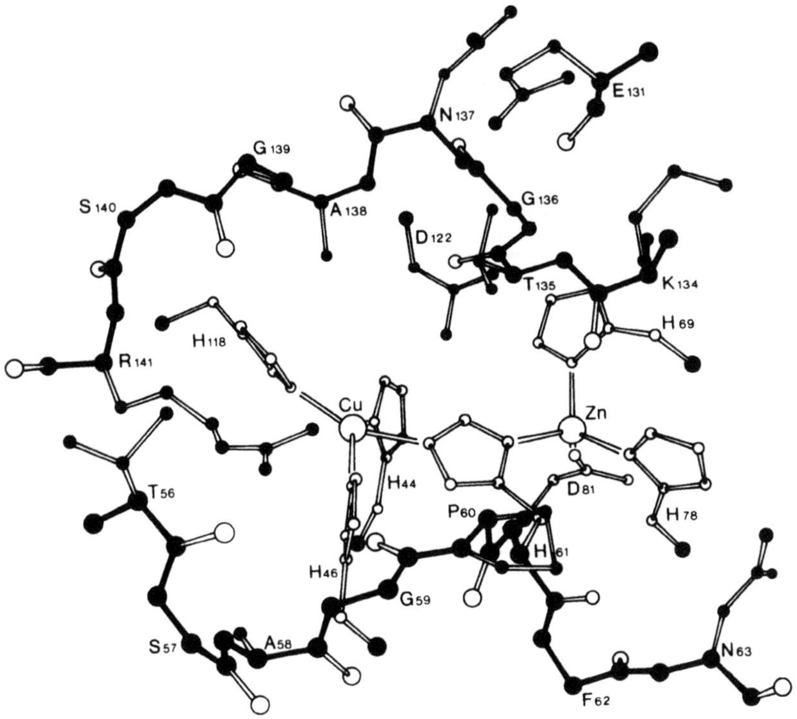

Figure 1. The critical active-site channel residues of bovine SOD as viewed from the solvent. The main chain is shown in solid black, ligand side chains are open, and other side chains are shown by solid atoms and open bonds. Residues are labeled with the one letter amino acid code and residue number. Residues 131, 134-139, and 141 form one rim of the channel, residues 56, 58-60, and 63 form the other rim, and the metal ions and their ligands form the floor. Main-chain atoms of residues 57, 62, and 140 have also been included in the drawing for continuity. The Cu(II) ion (left center) is liganded by histidines 118, 44, 61, and 46 (clockwise from upper left) with an uneven tetrahedral distortion from a square planar geometry. The geometry of Zn ligands (right center) 61, 69, 78, and Asp 81 (clockwise from left) is tetrahedral, with the largest distortion involving His 69. Asp 81 is behind the Zn and is buried. His 61 bridges the Cu and Zn ions, with its ring almost planar to the metals but tilted towards the solvent by ≈ 0.3Å on the Cu ion side. In the metal-liganding residues, all five of the non-liganding imidazole nitrogens and the non-liganding Asp 81 carboxyl oxygen are intramolecularly hydrogen-bonded.[4] Reduction of the Cu ion during catalysis should weaken the hydrogen bonds of the histidine rings bound to the Cu ion, allowing small shifts in ligand geometry that could give a more tetrahedral Cu site. The carboxyl group of solvent-inaccessible Asp 122, which forms a second, less direct bridge between the metals by hydrogen-bonding to both Zn ligand His 69 and Cu ligand His 44, may couple the Cu oxidation state to the Zn geometry; coupled movement of Asp 122 and His 69 away from His 44 and towards His 78 in the Cu(I) enzyme would decrease the distortion in the Zn tetrahedral geometry. Based upon the interpretation of the active site atomic structure, the techniques of site-directed mutagenesis allow the specific role of each side chain to be tested. (Adapted from Tainer et al.[3] with permission.)

The chances of productive collision are also improved by dimer formation, which buries about 1000Å^2 of the molecular surface of each monomer (that would lead to unproductive collision) in the dimer contact. In addition, the two active sites are oriented in opposite directions in the dimer, so that superoxide molecules at the different sites do not affect each other.

Inside the channel, Arg 141 affects the field direction near the Cu ion. The *catalytic* importance of Arg 141 in the complementary fit of the $\cdot O_2^-$ into the active channel is supported by the large decrease in activity that results from either covalent modification[6,31] or replacement using site-directed mutagenesis[32] of this single residue. Covalent modification of Arg 141,[6] however, leaves the ionic strength response for the enzyme's residual activity unchanged. These results suggest that the lysine residues on the surface of the channel are important in providing long-range electrostatic guidance for $\cdot O_2^-$, while Arg 141 facilitates short-range orientation of superoxide in the active site.

STRUCTURE-FUNCTION

Some of the questions concerning the catalytic mechanism of SOD that can now be addressed include the influence of the metal geometry on electron transfer between the enzyme and substrate; the possible importance of specific hydrogen-bonding patterns to the proposed mechanism; whether or not movements of atoms or groups are probable and how much movement might be involved; the source and geometry for the protons transferred to the superoxide; the structural and catalytic importance of the Zn ion; and the implications of the active site structure for inner-sphere versus outer-sphere catalysis. These questions cannot be readily addressed from the structure alone, but our crystallographic and computational results have provided a basis for the design of specific mutants of human SOD to be used to probe the roles of specific residues in the active site of SOD.

Human Cu,Zn SOD is particularly well suited for structure-function studies of proteins produced by in vitro mutagenesis because the human SOD gene has been sequenced,[33] cloned, and expressed at high levels from both *E. coli*[33] and yeast,[34] allowing site-directed mutagenesis experiments. Both recombinant human SODs have normal specific activity when compared to the enzyme purified from human erythrocytes, but recombinant human SOD from *E. coli* lacks the N-acetylation found in the eukaryotic enzyme. X-ray crystallographic studies on human SOD are in progress.[35] Over 30 specific site-directed mutants have already been designed, cloned, and expressed, as well as 10^6 more mutants with random amino acids at selected sites. The functions of several active channel residues are already being examined using site-directed mutants of Thr 56, Asp 122, Glu 131, Lys 134, Thr 135, and Arg 141 (numbering from the bovine sequence, as in Figure 1).

Site-directed mutagenesis experiments guided by crystallographic and computational information show great promise for the exploration of the role of specific residues in the mechanism of enzyme reactions. Since enzymes have evolved to optimize a number of conflicting requirements, it may be possible to design improved catalytic sites. In addition, a detailed understanding of the reaction mechanism may be applied to the redesigning of enzyme active sites to perform different reactions.

REFERENCES

1. I. Fridovich, Superoxide and superoxide dismutase, *in:* ''Advances in Inorganic Biochemistry,'' G. L. Eichhorn and L. G. Marzilli, eds., Elsevier/North Holland, New York (1979).

2. D. Klug, J. Rabani, and I. Fridovich, A direct demonstration of the catalytic action of superoxide dismutase through the use of pulse radiolysis, *J. Biol. Chem.* 247:4839 (1972).
3. J. A. Tainer, E. D. Getzoff, J. S. Richardson, and D. C. Richardson, Structure and mechanism of copper, zinc superoxide dismutase, *Nature* 306:284 (1983).
4. J. A. Tainer, E. D. Getzoff, K. M. Beem, J. S. Richardson, and D. C. Richardson, Determination and analysis of the 2Å structure of copper, zinc superoxide dismutase, *J. Mol. Biol.* 160:181 (1982).
5. J. S. Richardson, K. A. Thomas, B. H. Rubin, and D. C. Richardson, Crystal structure of bovine Cu, Zn superoxide dismutase at 3Å resolution: Chain tracing and metal ligands, *Proc. Natl. Acad. Sci. U.S.A.* 72:1349 (1975).
6. D. P. Malinowski and I. Fridovich, Chemical modification of arginine at the active site of the bovine erythrocyte superoxide dismutase, *Biochemistry* 18:5909 (1979).
7. D. Barra, F. Martini, J. V. Bannister, M. E. Schininà, G. Rotilio, W. H. Bannister, and F. Bossa, The complete amino acid sequence of human Cu/Zn superoxide dismutase, *FEBS Lett.* 120:53 (1980).
8. H. M. Steinman, V. R. Naik, J. L. Abernethy, and R. L. Hill, Bovine erythrocyte superoxide dismutase. Complete amino acid sequence, *J. Biol. Chem.* 249:7326 (1974).
9. H. M. Steinman, The amino acid sequence of copper-zinc superoxide dismutase from bakers' yeast, *J. Biol. Chem.* 255:6758 (1980).
10. J. R. Jabusch, D. L. Farb, D. A. Kerschensteiner, and H. F. Deutsch, Some sulfhydryl properties and primary structure of human erythrocyte superoxide dismutase, *Biochemistry* 19:2310 (1980).
11. K. Lerch and D. Ammer, Amino acid sequence of copper-zinc superoxide dismutase from horse liver, *J. Biol. Chem.* 256:11545 (1981).
12. H. M. Steinman, Copper-zinc superoxide dismutase from *Caulobacter crescentus* CB15. A novel bacteriocuprein form of the enzyme, *J. Biol. Chem.* 257:10283 (1982).
13. G. J. Steffens, A. M. Michelson, K. Puget, and L. Flohé, The amino-acid sequence of rat Cu-Zn superoxide dismutase, *Biol. Chem. Hoppe-Seyler* 367:1017 (1986).
14. G. J. Steffens, A. M. Michelson, F. Ötting, K. Puget, W. Strassburger, and L. Flohé, Primary structure of Cu-Zn superoxide dismutase of *Brassica oleracea* proves homology with corresponding enzymes of animals, fungi and prokaryotes, *Biol. Chem. Hoppe-Seyler* 367:1007 (1986).
15. G.-J. Steffens, J. V. Bannister, W. H. Bannister, L. Flohé, W. A. Günzler, S.-M. A. Kim, and F. Ötting, The primary structure of Cu-Zn superoxide dismutase from *Photobacterium leiognathi*: Evidence for a separate evolution of Cu-Zn superoxide dismutase in bacteria, *Hoppe-Seyler's Z. Physiol. Chem.* 364:676 (1983).
16. K. Hering, S.-M. A. Kim, A. M. Michelson, F. Ötting, K. Puget, G. J. Steffens, and L. Flohé, The primary structure of porcine Cu-Zn superoxide dismutase. Evidence for allotypes of superoxide dismutase in pigs, *Biol. Chem. Hoppe-Seyler* 366:435 (1985).
17. M. E. Schininà, D. Barra, M. Simmaco, F. Bossa, and G. Rotilio, Primary structure of porcine Cu,Zn superoxide dismutase, *FEBS Lett.* 186:267 (1985).
18. Y. M. Lee, D. J. Friedman, and F. J. Ayala, Superoxide dismutase: An evolutionary puzzle, *Proc. Natl. Acad. Sci. U.S.A.* 82:824 (1985).
19. M. E. Schininà, D. Barra, S. Gentilomo, F. Bossa, C. Capo, G. Rotilio, and L. Calabrese, Primary structure of a cationic Cu,Zn superoxide dismutase. The sheep enzyme, *FEBS Lett.* 207:7 (1986).
20. K. Lerch and E. Schenk, Primary structure of copper-zinc superoxide dismutase from *Neurospora crassa*, *J. Biol. Chem.* 260:9559 (1985).

21. Y. Kitagawa, S. Tsunasawa, N. Tanaka, Y. Katsube, F. Sakiyama, and K. Asada, Amino acid sequence of copper,zinc-superoxide dismutase from spinach leaves, *J. Biochem.* 99:1289 (1986).
22. R. E. Cannon, J. A. White, and J. G. Scandalios, Cloning of cDNA for maize superoxide dismutase 2 (SOD2), *Proc. Natl. Acad. Sci. U.S.A.* 84:179 (1987).
23. H. A. Rocha, W. H. Bannister, and J. V. Bannister, The amino-acid sequence of copper/zinc superoxide dismutase from swordfish liver. Comparison of copper/zinc superoxide dismutase sequences, *Eur. J. Biochem.* 145:477 (1984).
24. M. E. McAdam, E. M. Fielden, F. Lavelle, L. Calabrese, D. Cocco, and G. Rotilio, The involvement of the bridging imidazolate in the catalytic mechanism of action of bovine superoxide dismutase, *Biochem. J.* 167:271 (1977).
25. I. Bertini, C. Luchinat, and R. Monnanni, Evidence of the breaking of the copper--imidazole bridge in copper/cobalt-substituted superoxide dismutase upon reduction of the copper(II) centers, *J. Am. Chem. Soc.* 107:2178 (1985).
26. W. H. Koppenol, The physiological role of the charge distribution on superoxide dismutase, *in:* ''Oxygen and Oxy-Radicals in Chemistry and Biology,'' E. L. Powers and M. A. J. Rodgers, eds., Academic, New York (1981).
27. E. D. Getzoff, J. A. Tainer, P. K. Weiner, P. A. Kollman, J. S. Richardson, and D. C. Richardson, Electrostatic recognition between superoxide and copper, zinc superoxide dismutase, *Nature* 306:287 (1983).
28. M. L. Salin and W. W. Wilson, Porcine superoxide dismutase, *Mol. Cell. Biochem.* 36:157 (1981).
29. A. Cudd and I. Fridovich, Electrostatic interactions in the reaction mechanism of bovine erythrocyte superoxide dismutase, *J. Biol. Chem.* 257:11443 (1982).
30. W. H. Koppenol, On the reactivity of the superoxide anion and the biological function of superoxide dismutase, *in:* ''Oxidases and Related Redox Systems,'' T. E. King, M. S. Mason, and M. Morrison, eds., Pergamon, Oxford (1982).
31. C. L. Borders, Jr. and J. T. Johansen, Identification of Arg-143 as the essential arginyl residue in yeast Cu,Zn superoxide dismutase by use of a chromophoric arginine reagent, *Biochem. Biophys. Res. Commun.* 96:1071 (1980).
32. W. F. Beyer, Jr., I. Fridovich, G. T. Mullenbach, and R. Hallewell, Examination of the role of arginine-143 in the human copper and zinc superoxide dismutase by site-specific mutagenesis, *J. Biol. Chem.* 262:11182 (1987).
33. R. A. Hallewell, F. R. Masiarz, R. C. Najarian, J. P. Puma, M. R. Quiroga, A. Randolph, R. Sanchez-Pescador, C. J. Scandella, B. Smith, K. S. Steimer, and G. T. Mullenbach, Human Cu/Zn superoxide dismutase cDNA: Isolation of clones synthesising high levels of active or inactive enzyme from an expression library, *Nucl. Acids Res.* 13:2017 (1985).
34. R. A. Hallewell, R. Mills, P. Tekamp-Olson, R. Blacher, S. Rosenberg, F. Otting, F. R. Masiarz, and C. J. Scandella, Amino terminal acetylation of authentic human Cu,Zn superoxide dismutase produced in yeast, *Biotechnology* 5:363 (1987).
35. H. E. Parge, E. D. Getzoff, C. S. Scandella, R. A. Hallewell, and J. A. Tainer, Crystallographic characterization of recombinant human CuZn superoxide dismutase, *J. Biol. Chem.* 261:16215 (1986).

BACTERIAL SUPEROXIDE DISMUTASES

Howard M. Steinman

Department of Biochemistry
Albert Einstein College of Medicine
Bronx, N.Y.

INTRODUCTION

Superoxide dismutase (SOD) has played a major role in establishing the biological relevance of oxyradicals. To many oxyradical researchers, "superoxide dismutase" connotes a copper-zinc SOD (CuZnSOD), the first dismutase to be purified and the form of prime interest clinically. However, superoxide dismutases containing manganese or iron as cofactor are also known and are widespread among bacteria.[1] The manganese and iron SODs (MnSOD and FeSOD) are homologous in primary and three-dimensional structure and have clearly evolved from a common ancestor.[2,3] Although structurally homologous, the Mn- and FeSODs are functionally distinct. Only in rare instances can the endogeneous Mn (or Fe) be replaced by Fe (or Mn) with retention of catalytic activity. The Mn- and FeSODs are further distinguished in their distribution among bacterial species. Strict anaerobes contain one SOD, an FeSOD. Bacterial aerobes usually contain an MnSOD or both Mn- and FeSODs. (The MnSOD is also widely found in eukaryotes. The FeSOD is also found in primitive eukaryotes and some green plants. Distributions of the Mn- and FeSODs among eukaryotes have been discussed elsewhere.[1,4])

The subjects of this review are three case studies of bacterial superoxide dismutases. What benefits could possibly arise from study of superoxide dismutases found only in bacteria? First, comparisons between species, e.g., bacterial and human, has been a vital source of information about protein structure-function relationships. Secondly, bacteria often reveal ingenious solutions to biochemical problems. Tolerance of oxidative stress is a problem faced by both eukaryotes and prokaryotes. Bacterial SODs, as described below, do show novelties not found in eukaryotic dismutases. These novelties may suggest ways to engineer human dismutases or suggest new ways to formulate long-standing questions of clinical relevance.

THE PHYSIOLOGICAL ROLE OF SUPEROXIDE DISMUTASE

Mutation is a classical approach for establishing the physiological role of a bacterial gene product. First, specific mutants and revertants in a gene are obtained. Then, the physiological role of the gene product is inferred by comparing wild type, mutant, and revertant phenotypes. Shortly after the identification of bacterial SODs, the isolation of SOD mutants was attempted. However, no stable mutants were isolated by classical approaches.

In the absence of evidence from classical mutation, an alternative approach was used for demonstrating the physiological role of SOD. This approach relied upon induction of SOD activity.[5] SOD (usually MnSOD) can often be induced by exposing bacterial cultures to oxidative stress in the form of hyperoxia or certain redox active drugs. Induced strains are more resistant than wild type to still higher levels of the oxygen toxicity agent. This increased tolerance is evidence that bacterial SOD is important in the defense against oxygen toxicity. However, induction data are not definitive. Other enzymes may be induced by the oxidative stresses. From this approach, SOD could not be implicated as the sole agent responsible for the increased aerotolerance.

Definitive evidence for superoxide dismutase as a physiological antioxidant has been obtained from making E. coli mutants in SOD using recombinant DNA technology.[6] E. coli contains both Mn- and FeSOD. Both genes had been cloned then inactivated by insertion of a Mu transposon. The insertionally inactivated genes were exchanged with the wild type allele. The net result is replacement of the two active wild-type Mn- and FeSOD genes by defective, insertionally inactivated genes. Mutants were thus made which specifically lacked either MnSOD or lacked FeSOD or contained no SOD, but were otherwise genetically identical to wild type. The wild type was compared with these E. coli superoxide dismutase mutants in several types of studies.[6,7]

One study evaluated the rate of spontaneous mutation to rifampicin resistance in aerobic cultures. The mutant lacking FeSOD is similar to wild type in its rate of mutation. However, when MnSOD is absent, the mutation rate is 9 times greater than wild type. The mutation rate in the mutant lacking both SODs is 40 times wild type. A similar pattern is found on studying sensitivity towards hydrogen peroxide. The wild-type and the mutant lacking FeSOD are killed at comparable rates. The mutant without MnSOD is more sensitive to peroxide, and the double mutant with no SOD is exquisitively sensitive. In these or similar studies, wild type phenotype is regained by transformation of the appropriate mutant with a plasmid harboring a cloned SOD gene. This reversion to wild-type phenotype clearly indicates that the sensitivity to oxidative stress is attributable to the absence of SOD. These studies firmly establish the importance of superoxide dismutase as a major defense against oxygen toxicity.

What about the role of SOD for normal aerobic physiology in the absence of any added oxidative stress? In rich medium, SOD is not required for aerobic growth of E. coli. The double SOD mutant survives. However, in minimal medium the mutant lacking both MnSOD and FeSOD will not grow aerobically. Restriction of nutrients places mutants without SOD at a competitive disadvantage. Clearly, SOD is required for optimal aerobic survival.

The availability of cloned SOD genes and specific SOD mutants will now allow more insightful studies of SOD function. These tools will permit definitive studies of the regulation of expression of E. coli dismutases and analysis of specific oxygen-toxicity targets protected by superoxide dismutase.

STRUCTURAL COMPARISONS OF Mn AND Fe SODS

E. coli is ordinarily parsimonius with respect to enzyme activity. Isoenzymes are present only rarely. The superoxide dismutases are one such rare occurrence of isoenzymes. The presence of an MnSOD and a FeSOD in E. coli (and other bacteria as well) has perplexed SOD enzymologists. Multiple SOD forms in the same eukaryotic cell can be related to their different intracellular locations.[1,4] However, the Mn and FeSODs in E. coli are both cytoplasmic,[8] so that explanation is not suitable. The aerobic inducibility of MnSOD and not E. coli FeSOD has been proposed as an explanation for the presence of two SODs.[6] That explanation implies that the critical difference between the two isoenzymes is quantitative. However, the inability to exchange Mn for Fe with retention of catalytic activity[1,9] suggests that qualitative differences exist as well.

The structural basis of isoenzymic bacterial superoxide dismutases has been investigated by protein crystallographers. In 1984, homologies were identified between the polypeptide backbone tracings of E. coli FeSOD and MnSOD from Thermus thermophilus.[2] The only major difference was an extra loop in the Mn enzyme, important in its association to a tetramer. The E. coli FeSOD lacks the loop and forms a dimer. This general backbone homology was not unexpected because Mn- and FeSODs showed amino acid sequence homologies.[1,3] A major issue addressed by the crystallographers was the nature of their metal-binding sites. Given the general inability to replace one divalent metal ion by another with retention of activity, how similar are Mn and FeSOD? Only recently have amino acid sequence data become available for E. coli FeSOD[10,11] and T. thermophilus MnSOD,[12] allowing interpretation of the X-ray data at higher resolution.[11] The answer is that the immediate ligands are identical, 3 histidine and 1 glutamic acid residues, as is their geometry. Furthermore, there is a striking similarity in the general organization in the hydrophobic environment surrounding the metal cofactor.

These results have created a quandary. How can the marked similarity of Mn- and Fe-binding sites be reckoned with the inability to exchange metals with retention of catalysis? Answers may come from the next round of research. Data at higher resolution may indicate differences not presently seen. In addition, the diversity of bacterial SODs may reveal insights. There are a small number of SODs which can exchange Mn for Fe, and vice versa, with retention of catalysis.[13,14,15] This exchange is accomplished by manipulating the metal content of the medium[13,14] or by culturing anaerobically vs. aerobically.[15] In either case, an apparently identical protein core complexes either manganese or iron. Crystallographic three dimensional analysis of such "less specific" superoxide dismutases may provide clues to the structural basis for the strict specificity found in most other bacterial dismutases.

BACTERIOCUPREINS, BACTERIAL CuZnSODs

As a rule, bacteria contain SODs based on either manganese or iron. As a rule, the copper-zinc SOD is characteristic of eukaryotes. Bacteriocupreins are exceptions to this rule. Bacteriocupreins are bacterial SODs that contain copper and zinc. The first bacteriocuprein was discovered in Photobacterium leiognathi, a luminescent marine bacterium that can live symbiotically in Leiognathid fish.[16] The sequence of this protein clearly established that P. leiognathi bacteriocuprein is in the same family as eukaryotic CuZnSODs. Residues known to be critical for metal chelation or structural stability appear at homologous positions in bacterial and eukaryotic CuZnSODs.[17]

The presence of copper-zinc SODs in bacteria has raised several questions. What physiological role is filled by the CuZnSOD? Is it a role that cannot be filled by the Mn and FeSODs ordinarily found in bacteria? What is the evolutionary connection, if any, between bacterial and eukaryotic CuZnSODs? Some information about these questions has come from the recent cloning and sequencing of the gene encoding *P. leiognathi* bacteriocuprein.[18]

The sequence of the cloned gene encoded the entire primary structure previously determined for the protein.[17] However, between the ATG translation start and the codon for the amino terminal glutamine, 63 nucleotides were found. The implicated leader peptide was directly demonstrated by amino acid sequence analysis of the cell-free translation product of the cloned gene. The sequence of this leader peptide is homologous to leader sequences in proteins known to be exported from the cytoplasm. This homology suggests that *P. leiognathi* bacteriocuprein is extracytoplasmic. As a rule, bacterial SODs are cytoplasmic. (Exceptions have been suggested for SODs of certain pathogens.[18])

The presence of a leader sequence in *P. leiognathi* bacteriocuprein could be a clue to its function. Perhaps bacteriocupreins appear in bacterial species experiencing oxygen toxicity threats outside the cytoplasm and necessitating an extracytoplasmic SOD. Two avenues of investigation may reveal information about bacteriocuprein function. The first is comparison of *P. leiognathi* bacteriocuprein with the bacteriocuprein of *Caulobacter crescentus*. A bacteriocuprein has been purified from *C. crescentus*, a non-luminescent, non-symbiotic, fresh-water bacterium.[19] Comparison of its amino acid sequence with that of *P. leiognathi* and determination if the recently cloned *C. crescentus* bacteriocuprein gene contains a leader sequence will be important.[20] This information will identify which portions of the amino acid sequence are characteristic of bacterial CuZnSODs and which are characteristic of eukaryotic CuZnSODs, and will determine if the extracytoplasmic hypothesis for bacteriocuprein presence remains feasible. Secondly, crystallographic determination of the three-dimensional structure of *P. leiognathi* bacteriocuprein may identify structures that are different from the well-characterized structure of the eukaryotic CuZnSOD. Efforts to crystallize *P. leiognathi* bacteriocuprein are underway.[21]

The presence of an ordinarily eukaryotic CuZnSOD in *P. leiognathi* was originally attributed to transfer of the CuZnSOD gene from the eukaryotic (fish) host to the bacterial symbiont.[22,23] Subsequently, alternative hypotheses of bacteriocuprein origin were proposed.[4,24] Recently, the gene encoding human extracellular SOD (ecSOD, the major SOD in extracellular fluids) has been cloned and sequenced.[25,26] It has been shown that ecSOD is a CuZnSOD, showing sequence homology with the cytoplasmic CuZnSOD widespread in eukaryotes, but containing a leader sequence, consistent with its extracellular location. The demonstration that human ecSOD is a copper-zinc dismutase and the discovery of a leader sequence in *P. leiognathi* bacteriocuprein suggest that the bacterial CuZnSOD may be the ancestor of the eukaryotic cytoplasmic and ecSODs. According to this scenario, the CuZnSOD arose in bacteria. The eukaryotic cytoplasmic CuZnSODs have lost the leader sequence, and the ecSODs retained it. This model of bacteriocuprein evolution contrasts with the original model.[22,23] In the original model, bacterial CuZnSOD is viewed as a branch arising from host-to-symbiont transfer of DNA from an established line of eukaryotic descent.

REFERENCES

1. H. M. Steinman, Superoxide dismutases: Protein chemistry and structure-function relationships, in "Superoxide Dismutase, Volume I", L. W. Oberley, ed., CRC Press, Boca Raton (1982).
2. W. C. Stallings, K. A. Pattridge, and M. L. Ludwig, Manganese and iron superoxide dismutases are structural homologs, J. Biol. Chem. 259:10695 (1984).
3. D. Barra, M. E. Schinina, W. H. Bannister, J. V. Bannister, and F. Bossa, The primary structure of iron-superoxide dismutase from Photobacterium leiognathi, J. Biol. Chem. 262:1001 (1987).
4. J. Kwiatowski, Comparison of chloroplast and cytosolic Cu/Zn superoxide dismutase isozymes from tomato in relation to superoxide dismutase evolution, Isozymes: Curr. Top. Biol. Med. Res. 15:21 (1987).
5. I. Fridovich, Superoxide radical: An endogenous toxicant, Ann. Rev. Pharmacol. Toxicol. 23:239 (1983).
6. A. Carlioz and D. Touati, Isolation of superoxide dismutase mutants in Escherichia coli: Is superoxide dismutase necessary for aerobic life?, EMBO J. 5:623 (1986).
7. S. B. Farr, R. D'Ari, and D. Touati, Oxygen-dependent mutagenesis in Escherichia coli lacking superoxide dismutase, Proc. Natl. Acad. Sci. (USA) 83:8268 (1986).
8. L. Britton and I. Fridovich, Intracellular localization of the superoxide dismutases of Escherichia coli: A reevaluation, J. Bacteriol. 131:815 (1977).
9. D. E. Ose and I. Fridovich, Superoxide dismutase. Reversible removal of manganese and its substitution by cobalt, nickel, or zinc, J. Biol. Chem. 251:1217 (1976).
10. M. E. Schinina, L. Maffey, D. Barra, F. Bossa, K. Puget, and A. M. Michelson, The primary structure of iron superoxide dismutase from Escherichia coli, FEBS Letters 221:87 (1987).
11. A. Carlioz, M. L. Ludwig, W. C. Stallings, J. A. Fee, H. M. Steinman, and D. Touati, Iron superoxide dismutase: Nucleotide sequence of the gene from E. coli K12 and correlations with crystal structure, J. Biol. Chem., in press.
12. S. Sato, Y. Nakada, and K. Nakazawa-Tomizawa, Amino acid sequence of a tetrameric, manganese superoxide dismutase from Thermus thermophilus HB8, Biochim. Biophys. Acta 912:178 (1987).
13. B. Meier, D. Barra, I. F. Bossa, L. Calabrese, and G. Rotilio, Synthesis of either Fe- or Mn-superoxide dismutase with an apparently identical protein moiety by an anaerobic bacterium dependent on the metal supplied, J. Biol. Chem. 257:13977 (1982).
14. M. E. Martin, B.R. Byers, M. O. J. Olson, M. L. Salin, J. E. L. Arceneauz, and C. Tolbert, A Streptococcus mutans superoxide dismutase that is active with either manganese or iron as a cofactor, J. Biol. Chem. 261:9361 (1986).
15. C. D. Pennington and E. M. Gregory, Isolation and reconstitution of iron- and manganese-containing superoxide dismutase from Bacteroides thetaiotaomicron, J. Bacteriol. 166:528 (1986).
16. K. Puget and A. M. Michelson, Isolation of a new copper-containing superoxide dismutase, bacteriocuprein, Biochem. Biophys. Res. Commun. 58:830 (1974).
17. G.-J. Steffens, J. V. Bannister, W. H. Bannister, L. Flohe, W. A. Gunzler, S.-M. A. Kim, and F. Otting, The primary structure of Cu-Zn superoxide dismutase from Photobacterium leiognathi: Evidence for a separate evolution of Cu-Zn superoxide dismutase in bacteria, Hoppe-Seyler's Z. Physiol. Chem. 364:675 (1983).

18. H. M. Steinman, Bacteriocuprein superoxide dismutase of Photobacterium leiognathi. Isolation and sequence of the gene and evidence for a precursor form, J. Biol. Chem. 262:1882 (1987).
19. H. M. Steinman, Copper-zinc superoxide dismutase from Caulobacter crescentus CB15. A novel bacteriocuprein form of the enzyme, J. Biol. Chem. 257:10283 (1982).
20. H. M. Steinman, unpublished data.
21. J. A. Tainer and E. D. Getzoff, unpublished data.
22. J. P. Martin, Jr. and I. Fridovich, Evidence for a natural gene transfer from the ponyfish to its bioluminescent bacterial symbiont Photobacterium leiognathi, J. Biol. Chem. 256:6080 (1981).
23. J. V. Bannister, W. H. Bannister, and G. Rotilio, Aspects of the structure, function, and applications of superoxide dismutase, CRC Critical Rev. Biochem. 22:11 (1987).
24. H. M. Steinman, Chemical aspects of structure, function, and evolution among superoxide dismutases: The general scenario and the bacteriocuprein exceptions, in "Oxy Radicals and Their Scavenger Systems. Volume I, Molecular Aspects," G. Cohen and R. A. Greenwald, eds., Elsevier Biomedical, New York (1983).
25. S. L. Marklund, L. Tibell, K. Hjalmarsson, G. Skogman, A. Engstrom, and T. Edlund, Sequence of complementary DNA encoding human extracellular-superoxide dismutase and production of recombinant enzyme, in present volume.
26. K. Hjalmarsson, S. L. Marklund, A. Engstrom, and T. Edlund, Isolation and sequence of complementary DNA encoding human extracellular superoxide dismutase, Proc. Natl. Acad. Sci. (USA), 84:6340 (1987).

EXTRACELLULAR-SUPEROXIDE DISMUTASE ASSOCIATION WITH CELL SURFACE-BOUND SULFATED GLUCOSAMINOGLYCANS

Kurt Karlsson and Stefan L. Marklund

Department of Clinical Chemistry
Umeå University Hospital
S-901 85 Umeå, Sweden

INTRODUCTION

Extracellular-superoxide dismutase (EC 1.15.1.1., abbreviation EC-SOD) is a tetrameric glycoprotein with an apparent subunit molecular weight of around 30 kDa.[1] Like the CuZn SODs, EC-SOD contains one Cu and one Zn atom per subunit[1] and is inhibited by cyanide, azide, and diethyldithiocarbamate,[2] and H_2O_2 is the product of the catalysed reaction.[3] Still, despite the similarities, the amino acid compositions of human CuZn SOD and EC-SOD are clearly different,[1] and no cross-reaction between polyclonal rabbit antibodies directed towards the enzymes have been observed.[2] EC-SOD is the major SOD isoenzyme in extracellular fluids like plasma, lymph,[4] and synovial fluid,[5] but also occurs in tissues.[6,7] The total plasma EC-SOD concentration varies greatly between species, whereas the tissue concentrations are more equal.[7]

EC-SOD isolated from human lungs was heterogenous with regard to binding to Heparin-Sepharose and could be separated into three fractions: A, without affinity; B, with weak affinity; and C, with relatively high affinity.[1] Many other proteins with affinity for heparin are known. For some of these, such as lipoprotein lipase, hepatic lipase,[8] diamine oxidase,[9] and platelet factor 4,[10] it has been found that injection of heparin leads to vast increases in the plasma content of these proteins. The release of these factors into plasma is apparently due to displacement from heparin analogues/heparan sulfate on endothelialcell surfaces.[11,12] The heparin affinity of EC-SOD C thus lead to the suspicion that the enzyme in the body might be partially bound to sulfated glucosaminoglycans (S-GAG) such as heparan sulfate. To test this hypothesis a number of experiments have been performed.

INTRAVENOUS INJECTION OF HEPARIN INTO DIFFERENT SPECIES

Plasma from man,[13] pig, cat, rabbit, guinea pig, and mouse was separated on Heparin-Sepharose, and in all cases the EC-SOD was found to be divided into A, B, and C fractions. Intravenous injection of heparin into these species was found to result in a prompt rise in plasma EC-SOD activity. Addition of heparin to whole blood had no effect on plasma EC-SOD activity. The released form of EC-SOD was, as expected, type C. Rat plasma contained only fractions A and B, and i.v. heparin had no effect on the plasma EC-SOD activity. The likely mechanism of the heparin effect is displacement from S-GAG located on endothelial cell surfaces. EC-SOD C apparently forms an equilibrium between the plasma phase and endothelial cell surfaces, with most of the enzyme surface associated. A and B might be intended for a protective function in the fluid phase.

INTRAVENOUS INJECTION OF DIFFERENT S-GAG AND PROTAMIN INTO PIGS

To test for the specificity of the effect of intravenous heparin, heparin fractionated with regard to antithrombin III affinity was also tested in the pig. Low-affinity (LA) and (HA) high-affinity heparin were as active as unfractionated heparin. Dextran sulfate 5000 was even more effective than heparin. Since many different negatively charged polyelectrolytes were able to bring about release, apparently no specific configuration is necessary for binding to EC-SOD C. Intravenous injection of the heparin antidote protamine was also found to release EC-SOD C to plasma. Protamine is a strongly positively charged protein, and we speculate that the effect is due to binding to the negatively charged heparan sulfate, leading to EC-SOD C displacement.

BINDING OF HUMAN EC-SOD C TO PIG AORTA ENDOTHELIUM

Binding to endothelium was further studied on pig aortas. They were cut open and mounted between perspex blocks with holes drilled so as to form wells with endothelium in the bottom. EC-SOD C was found to bind to the endothelium with a very high capacity. The enzyme could be released from endothelium by dextran sulfate 5000, heparin, heparan sulfate, and dextran sulfate 500000 (potencies in deareasing order). LA- and HA heparin were equally active and did not differ from unfractionated heparin. Chondroitin sulfate and hyaluronic acid had no significant releasing potency. Pretreatment of the endothelium with heparitinase (low conc.) and trypsin (high conc.) abolished subsequent EC-SOD C binding. Chondroitinase ABC had no effect. In conclusion, the binding substance on endothelium appears to be heparan sulfate.

BINDING OF HUMAN EC-SOD C FROM S-GAG´S - SUBSTITUTED SEPHAROSE

Binding of EC-SOD C to sulfated glucosaminoglycans coupled to Sepharose 4B was also studied. The binding strength was assessed by measuring the displacement of EC-SOD C caused by increasing concentrations of NaCl. Heparin was found to bind EC-SOD stronger than heparan sulfate did.

Chondroitin sulfate bound only weakly, and EC-SOD C was almost completely released at physiological NaCl concentrations.

TURN OVER OF ^{125}I-LABELLED HUMAN EC-SOD IN RABBITS

^{125}I-labelled human EC-SOD C injected i.v. into rabbits was rapidly sequestered from the blood. Injection of heparin within one hour resulted in immediate, complete return of the ^{125}I-EC-SOD C to the plasma. Later heparin injections resulted in less release, but a significant effect could still be demonstrated after 72 h. The half-life of heparin-releasable EC-SOD C was found to be between 10 and 15 hours. The organ distribution of sequestered ^{125}I-EC-SOD C was determined. Most enzyme was bound to liver, followed by kidney, spleen, heart and lung.

CONCLUSION

Most EC-SOD C in the vasculature is bound to cellular surfaces on endothelium and possibly also to hepatocytes (which have a close connection with plasma). The binding substance is in all probability heparan sulfate. This surface association seems to be a new principle for protection of cells against superoxide radicals.

REFERENCES

1. S. L. Marklund, Human copper-containing superoxide dismutase of high molecular weight, Proc. Natl. Acad. Sci. USA 79;7634 (1982).
2. S. L. Marklund, Properties of extracellular-superoxide dismutase from human lung, Biochem. J. 220;269 (1984).
3. S. L. Marklund, Product of extracellular-superoxide dismutase catalysis, FEBS Lett. 184;237 (1985).
4. S. L. Marklund, E. Holme, and L. Hellner, Superoxide dismutase in extracellular fluids, Clin. Chim. Acta. 126;41 (1982).
5. S. L. Marklund, A. Bjelle, and L. G. Elmqvist, Superoxide dismutase isoenzymes of the synovial fluid in rheumatoid arthritis and in reactive arthritides, Ann. Rheum. Dis. 45;847 (1986).
6. S. L. Marklund, Extracellular-superoxide dismutase in human tissues and human cell lines, J. Clin. Invest. 74;1398 (1984).
7. S. L. Marklund, Extracellular-superoxide dismutase and other superoxide dismutase isoenzymes in tissues from nine mammalian species, Biochem. J. 222;649 (1984).
8. R. M. Krauss, H. C. Windmüeler, and D. S. Fredrickson, Selective measurement of two different triglyceride lipase activities in rat post heparin plasma, J. Lipid. Res. 14;286 (1973).
9. R. Hansson, C. G. Holmberg, G. Tibbling, N. Tryding, A. Westling, and H. Wetterquist, Heparin-induced diamine oxidase increase in human blood plasma, Acta. Med. Scand. 180;533 (1966).

10. J. Dawes, R. C. Smith and, D. S. Pepper, The release, distribution, and clearance of human ß-thromboglobulin and platelet factor 4, Tromb. Res. 12;851 (1978).
11. A. Robinson-White, S. B. Baylin, T. Olivecrona, and M. A. Beaven, Binding of diamine oxidase activity to rat and guinea pig microvascular endothelial cells, J. Clin. Invest. 76;93 (1985).
12. C. Busch, J. Dawes, D. S. Pepper and, Å. Wasteson, Binding of platelet factor 4 to cultured human umbilical vein endothelial cells, Tromb. Res. 19;129 (1980).
13. K. Karlsson and S. L. Marklund, Heparin-induced release of extracellular-superoxide dismutase to human blood plasma, Biochem. J. 242;55 (1987).

CATALASES--WITH AND WITHOUT HEME

Wayne F. Beyer, Jr., and Irwin Fridovich

Department of Biochemistry
Duke University Medical Center
Durham, North Carolina 27710, U.S.A.

Catalases ($H_2O_2:H_2O_2$ oxidoreductase; EC 1.11.1.6) are metalloenzymes that catalyze the elimination of H_2O_2 according to the equation:

$$2H_2O_2 \xrightarrow{\text{catalase}} 2H_2O + O_2 \qquad (1)$$

Catalases, together with superoxide dismutases and peroxidases, provide aerobic cells with a defense system for removal of superoxide radical and of hydroperoxides. H_2O_2 is a cellular toxicant in its own right. Considerable interest in the production of hydroxyl radical by the reduction of H_2O_2 (equation 2)

$$H_2O_2 + e^- \longrightarrow OH\cdot + OH^- \qquad (2)$$

mediated by transition metals or other cellular reductants has arisen.

Until recently, all known catalases contained heme-iron as the prosthetic group. We will now briefly review the properties of heme catalases with emphasis on recent findings. The remainder of this chapter will be devoted to the recently purified non-heme catalases.

HEME CATALASES

The majority of heme catalases are homotetramers with a molecular weight of 240 Kda, although bacterial catalases have been reported with higher molecular masses.[1-4] Each subunit has a molecular weight of about 60,000 daltons and contains a single ferriprotoporphyrin group. The presence of carbohydrate moieties in mammalian catalases has been reported.[5] The integrity of the tetramer does not depend upon disulfide bonds[6] nor is there any evidence that association of subunits depends upon other covalent bonding.[7] Subunit assembly appears to depend primarily upon hydrophobic interactions.[8] The oligomeric protein dissociates easily during storage, dilution, freeze drying or exposure of the enzyme to extremes of pH. The subunits lack catalase activity but have substantial peroxidase activity. It has been noted that some forms of the human disease acatalasemia are attributable to the formation of a catalase of normal specific activity but a greater tendency to dissociate into subunits.[9]

The accepted reaction mechanism for the enzyme involves the alternate two-electron oxidation-reduction of the heme center by H_2O_2 as follows:

$$Fe^{+3}\text{-cat} + H_2O_2 \longrightarrow \underset{\text{Compound I}}{Fe^{+5}\text{-cat}} \quad (3)$$

$$\underset{\text{Compound I}}{Fe^{+5}\text{-cat}} + H_2O_2 \longrightarrow Fe^{+3}\text{-cat} + 2H_2O + O_2 \quad (4)$$

Compound I, is actually an Fe(IV) π cation radical, i.e. during the divalent oxidation of the resting enzyme, one electron is removed from the iron and one from the porphyrin.[10] At low levels of H_2O_2, a steady-state concentration of Compound I is present, and in the presence of small electron donors (e.g., methanol, ethanol, formic acid, phenols) the oligomeric enzyme exhibits peroxidase activity:

$$\text{Compound I} + RH_2 \longrightarrow \underset{\text{(Compound II)}}{Fe^{+4}\text{-cat}} + RH\cdot + H^+ \quad (5)$$

$$\text{Compound II} + RH_2 \longrightarrow Fe^{+3}\text{-cat} + RH\cdot + H^+ \quad (6)$$

The reaction (1) is rapid ($k \sim 10^7 M^{-1} sec^{-1}$); whereas the peroxidative reactions occur much more slowly with $k \sim 10^2 - 10^3$.[11] Heme catalases are very efficient enzymes and are difficult to saturate with H_2O_2.

Heme catalases are reversibly inhibited by CN^-, N_3^- and S^{-2} due to ligation of these anions at the heme iron. Irreversible inactivation of catalases <u>in vivo</u> or <u>in vitro</u> is achieved by incubation of the enzyme with 3-amino-1,2,4 triazole (3-AT)[12] or cyanamide[13] in the presence of H_2O_2. No inactivation is observed in the absence of H_2O_2, and Margoliash and Novogrodsky have demonstrated the modification of histidine in compound I by 3-AT.[12] Diaminobenzidine (DAB) is a potent inhibitor of catalase,[14] and this inhibition may be reversible or irreversible depending upon the presence of H_2O_2. Superoxide radical rapidly inhibits catalase[15,16] by formation of compound III.

$$Fe^{+3}\text{-cat} + O_2^{\bar{\cdot}} \longrightarrow Fe(III)\text{-}O_2^{\bar{\cdot}} \longleftrightarrow \underset{\text{(Compound III)}}{Fe(II)\text{-}O_2} \quad (7)$$

This rapid inhibition can be prevented and reversed by superoxide dismutase.[15] A slower inactivation of catalase by superoxide could be prevented and reversed by ethanol but not SOD. This inactivation was proposed to involve formation of compound II.[15]

$$\underset{\text{(Compound I)}}{Fe^{+5}\text{-cat}} + O_2^{\bar{\cdot}} \longrightarrow O_2 + \underset{\text{(Compound II)}}{Fe^{+4}\text{-cat}} \quad (8)$$

Catalase was rapidly inactivated at H_2O_2 concentrations exceeding 0.1 M or by exposure of catalase to a continuous source of H_2O_2 supplied by microdiffusion.[17] The mechanism of substrate inactivation is not clear, but conversion of the active enzyme-H_2O_2 Compound I to the inactive Compounds II or III has been proposed.[18,19]

The human erythrocyte and bovine liver catalases contain 4 molecules of tightly bound NADPH per tetramer.[17,19] The dissociation constant for NADPH was less than 10 nM. The order of affinity was NADPH>NADH>NADP$^+$> NAD$^+$. The NADPH was not essential for catalase activity during brief assays, but extended exposure to a flux of H_2O_2 resulted in oxidation of NADPH and concomitant loss of catalase activity.[17,19] Recent experiments indicate that the bound NADPH reduces the rate of formation of the inactive compound II.[19]

The crystal structures of catalases from bovine liver (2.5Å resolution[20]) and Penicillium vitale (2.0Å resolution[21]) have been reported. The structures of these two enzymes are remarkably similar. The bound NADPH in the bovine liver catalase is located near the C-terminus, about 14Å from the buried heme moiety. The cofactor is present in a folded conformation, in contrast to the extended conformation found in other NADPH-binding proteins. The fungus P. vitale catalase contains a flavodoxin-like (α/β structure) domain near the C-terminus, suggesting a similar NADPH binding site. The presence of NAD(P)H in bacterial catalases has not yet been reported.

NONHEME OR PSEUDOCATALASE

Many bacteria that are aerotolerant yet cannot synthesize heme produce an azide and cyanide-insensitive catalase. This nonheme or "pseudo" catalase was first observed in pediococci [22] and soon thereafter in lactobacilli, leuconostocs, and streptococci.[23] Pseudocatalase was partially purified from Streptococcus faecalis[24] and was found to be stable to dialysis against EDTA at pH 7.0. Addition of Mn and of Fe to the growth medium increased the level of pseudocatalase in Lactobacillus plantarum, suggesting that it might be an iron or manganese enzyme. The enzyme was partially purified, but assays for Mn or Fe were negative.[25] Attempts to demonstrate the presence of heme and flavin were also fruitless.[26] Kono and Fridovich[27] isolated this pseudocatalase from Lactobacillus plantarum and found it to be an oligomeric manganienzyme. Initial studies of its structure and metal content suggested a homohexameric protein containing one atom of manganese per subunit.[27] Recent investigations have demonstrated the presence of a Mn-containing catalase in the thermophilic bacteria Thermoleophilum album[28] and Thermus thermophilus HB-8.[29]

We have extended our investigation of the Mn-catalase from L. plantarum and reported data suggesting that the protein is a homopentameric enzyme with two atoms of Mn per subunit.[30] Because of the uniqueness of the Mn-catalase, its stability to freezing and thawing and its potential use for removal of H_2O_2 in reaction mixtures containing CN^- or N_3^-, the similarities and differences of the Mn-catalase from L. plantarum and the thermophiles will be reviewed.

FUNCTIONAL SIGNIFICANCE OF MN-CATALASE

A strain of L. plantarum which was unable to produce Mn-catalase (ATCC 8014) grew more rapidly and to a higher plateau density than the Mn-catalase-positive stain (ATCC 14431), and this was the case during aerobic or anaerobic growth.[31] But, under aerobic conditions, the Mn-catalase-positive strain maintained viability during the stationary phase whereas the Mn-catalase-negative strain suffered a pronounced loss of viability. Earlier studies[27] indicated that the catalase-negative strain of L. plantarum accumulated H_2O_2 during the late log and stationary phases, whereas the Mn-catalase-positive strain did not. These results suggested the importance of the Mn-catalase in protection against the cellular toxicant H_2O_2. This was confirmed by demonstrating that the Mn-catalase positive strain, when grown to stationary phase and exposed to NH_2OH, suffered a significant loss of viability.[27] NH_2OH is a powerful inhibitor of the Mn-catalase but only in the presence of H_2O_2.[32] Indeed, in cells where the Mn-catalase had been inactivated by treatment with NH_2OH, the cells exhibited intolerance towards added H_2O_2 when exposure was performed in the presence of puromycin to prevent resynthesis of active Mn-catalase.[31] Catalase-negative strains of L. plantarum could be protected against loss in viability from H_2O_2 if bovine liver catalase was added to the growth medium, or if these cells were grown in the presence of hemin, which allowed synthesis of a CN^--sensitive catalase.[31]

Paraquat increased cyanide-resistant respiration in several strains of thermophilic bacteria and induced the synthesis of a Mn-catalase.[33] This induction was attributed to the $O_2^{\cdot-}$ dependent formation of H_2O_2. Further studies demonstrated that catalase could be induced by adding H_2O_2 to the growth medium of the thermophiles. Paraquat at 2.5 μM induced the Mn-catalase approximately 9-fold in Thermoleophilum album,[28] and this procedure was used to facilitate purification and characterization of Mn-catalase from this thermophile. Addition of H_2O_2 to the growth medium of Thermus thermophilus likewise resulted in 15-fold induction in catalase activity.[28] Induction of Mn-catalase by H_2O_2 or by paraquat was not detected in catalase-positive L. plantarum, and paraquat at 100 μM did not impart any measurable lethality to this strain.[34] It has been suggested that L. plantarum lacks a NADPH:paraquat oxidoreductase necessary for intracellular redox cycling of this herbicide.[35,36]

In the examples cited, the data offers strong evidence that the Mn-catalase, like its heme analog, functions to minimize the toxic effects of H_2O_2. Clearly this is an important factor in the aerobic viability of L. plantarum and Thermophilic bacteria.

PROPERTIES OF THE MN-CATALASE

To date, the most complete characterization has been performed on the Mn-catalase from L. plantarum.[27,30] The purified enzyme is orange-pink in concentrated solutions. Mn-catalase has been purified 180-fold from cell-free extracts of L. plantarum, with greater than 50 percent recovery of the initial activity.[30] The specific activity of the enzyme is typically about 7000 U/mg. Under similar assay conditions, the specific activity of the bovine liver catalase is about 40,000 U/mg. It follows that the heme enzyme is about six times more active than the Mn-catalase on a per mg protein basis.

Unlike the heme catalase, the Mn-enzyme was not inhibited at high concentrations of H_2O_2. From the kinetic plot we calculate K_m $H_2O_2 \sim 220$ mM, and from V_m a turnover number of 3.7×10^5 mole of H_2O_2 per mole of enzyme per second. For comparison, the turnover number of bovine liver catalase was reported to be 3.5×10^6 per second.[37] Because saturation with substrate is difficult, the kinetic parameters given serve merely as a comparison rather than suggesting a rigorous quantitative difference between the two proteins.

Sedimentation equilibrium was used to establish the molecular masses of the oligomeric protein (169,000 $\pm$ 4,200) and of the subunits (33,500 $\pm$ 500).[30] The metal content, determined using graphite atomic absorption or neutron activation analysis, indicated 9-12 g-atoms of Mn per 172,000 g purified protein. Collectively, the data indicated a homopentameric protein with two Mn atoms per subunit. The possibility of a binuclear cluster was suggested.[30]

The native protein was exposed to the crosslinking agent dimethyl-suberimidate (DMS) and the crosslinked products were analyzed by SDS-PAGE.[38,39] After 30 minutes of reaction, five bands became evident. At progressively longer reaction times, additional minor bands appeared. The amino acid composition of the Mn-catalase indicated that the sum of the basic amino acids Lys + Arg = 150.[30] We anticipated 31 tryptic peptides from a homopentamer and 26 from a homohexamer. Thirty-two unique peptides

Table 1. Effect of Some Inhibitors on the Activity of Pseudocatalase[a]

Additions	Relative Rate
mM	
None	100
NaCN (1-8)	100-93
NaN_3[b] (4-10)	100-98
PCMPS (0.1-1.0)	98-93
Iodoacetate (0.1-0.2)	100-81
3-Aminotriazole (100-400)	88
Hydroxylamine (0.001)	54
Hydroxylamine (0.008)	0

[a] Catalase was assayed by following the decrease of H_2O_2 at 240 nm. The reaction mixture (3 ml) contained 50 mM potassium phosphate, pH 7, 10 mM H_2O_2 and 1.7 µM catalase. PCMPS, p-chloromercuriphenyl sulfonate.

[b] Catalase was assayed with an oxygen electrode using a reaction mixture (2 ml) containing 50 mM potassium phosphate, pH 7, 10 mM H_2O_2, and 2.55 µM catalase.

were resolved from the tryptic map, supporting the proposed homopentameric structures of the protein.

The hydrodynamic properties of the native protein were measured at two rotor speeds. The Stokes radius was calculated to be 44Å, and the calculated frictional coefficient of 1.2 suggests that the Mn-catalase is a globular protein.[30]

Table 1 lists the effect of some inhibitors on the activity of Mn-catalase. The heme inhibitors CN^- or N_3^- had little or no effect on the activity of the Mn-catalase. Hydroxylamine in the presence of H_2O_2 proved to be a potent inhibitor of the Mn-catalase, and this inhibition was not reversed by dialysis. Inactivation was associated with a bleaching of the visible spectrum of the enzyme, suggesting a change in oxidation state at the Mn active site.[32] The metal chelators EDTA, α,α'-dipyridyl, 8-hydroxyquinoline, diethyldithiocarbamate, thioglycolate, and o-phenanthroline had no inhibitory effect on enzyme activity when present at 2.0 mM at pH 7.0.[32] Reducing agents such as sodium bisulfite, sodium sulfide and sodium nitrite also failed to inhibit the protein. The $NH_2OH + H_2O_2$-inactivated enzyme could be partially reactivated by exposure to dithionite or $O_2^{\cdot-}$.[32] Exposure of the native enzyme to an enzymatic or photochemical source of $O_2^{\cdot-}$ resulted in an inactivation of the enzyme to an apparent limit of 50%. This inactivation was prevented but not reversed by superoxide dismutase. The limit of inactivation was explained in terms of the ability of $O_2^{\cdot-}$ to both inactivate and reactivate the Mn-catalase.[32] Other oxidants and reductants such as ferricyanide, persulfate or ascorbate were not efficacious in restoring activity to Mn-catalase inactivated by $NH_2OH + H_2O_2$.[32] Inactivation of the Mn-catalase by carbonyl reagents has also been reported.[40] Ethanol did not restore catalase activity to the $NH_2OH + H_2O_2$-inactivated Mn-catalase, suggesting that the Mn-catalase, unlike the heme catalase, would not act as a

Table 2. Comparison of Properties of the Known Mn-Catalases

1) Lactobacillus plantarum (ATCC 14431): (incapable of heme synthesis)

   M_r~172 Kda; 5 equiv. subunits M_r~34 Kda
   2 Mn atoms/subunit (EPR silent)
   $S_{20,w}$ 9.6 S, R_S=44 Å
   Inhibited by $O_2^{\cdot-}$, $NH_2OH + H_2O_2$
   Not inhibited by CN^-, N_3^-, 3-AT, EDTA
   Lacks peroxidase activity
   Not inactivated by freezing and thawing
   Stable between 25-50°C

2) Thermus thermophilus HB8: (capable of heme synthesis)

   M_r~210 Kda; 6 equiv. subunits M_r~35 Kda
   Mn content not reported
   $S_{20,w}$ 10.0 S
   Inhibited by $NH_2OH + H_2O_2$, N_3^- ($I_{50\%}=10^{-4}M$)
   Not inhibited by 3-AT, EDTA
   Stable between 25-95°C

3) Thermoleophilum album NM:

   M_r~141 Kda; 4 equiv. subunits M_r~34 Kda
   1.4 ± 0.4 Mn atoms/subunit
   Inhibited by $NH_2OH + H_2O_2$, N_3^- ($I_{50\%}=10^{-2}M$), H_2O_2
   Inactivated by freezing and thawing
   Not inhibited by EDTA, 3-AT
   Exhibits sigificant peroxidase activity
   Stable between 25-80°C
   Induced 10-fold by paraquat

peroxidase. This enzyme also failed to act as a peroxidase towards nitrite, dianisidine, pyrogallol, NAD(P)H, guaiacol, ascorbate or ferrocytochrome c.[32] We were unable to demonstrate that the native or $NH_2OH + H_2O_2$-inactivated enzyme would act as a superoxide dismutase.

The visible absorption spectrum of the Mn-catalase has a maximum at 470 nm (ϵ~147 $M^{-1}cm^{-1}$ per Mn atom) with significant fine structure. This absorption is similar but not identical to the Mn-superoxide dismutase which has been demonstrated to contain Mn^{+3} at its active site.[41,42] The resting state of the Mn-catalase is EPR silent.[7] The EPR of the $NH_2OH + H_2O_2$-inactivated enzyme yields a complex 16-line pattern at 100 K. This pattern suggested a Mn(IV) mixed-valence system and is consistent with the proposal of a binuclear Mn center at the active site of the inactivated enzyme.[43] Denaturation of the enzyme with HCl generates a 6-line EPR signal of Mn^{+2}.

The Mn-catalases from the thermophilic bacteria Thermus thermophilus[29] and Thermoleophilum album NM[28] have been purified and characterized. Table 2 suggests some remarkable differences between the enzymes from the two thermophiles and the L. plantarum protein. T. thermophilus, unlike L. plantarum, is capable of heme biosynthesis and makes several types of cytochromes, yet it produces non-heme catalase.[44] Mn-catalase

Table 3. Effect of Inhibitors on the Activity of the Catalase from *T. album*[27]

Inhibitor	Final Concn (mM)	% Inhibition
Hydroxylamine hydrochloride	0.01	56
	0.1	100
$HgCl_2$	1.0	100
Histidine	5.0	40
	50.0	100
Sodium azide	10.0	45
	25.0	84
KCN	10.0	19
	100.0	37
3-Amino-1,2,4-triazole	10.0	0
	150.0	13
EDTA	5.0	0
	10.0	18
H_2O_2	40.0	43
p-Phenylenediamine	3.0	0

Catalase was assayed with an oxygen electrode in a reaction mixture (3 ml) containing 50 mM potassium phosphate (pH 8.0), 20 mM H_2O_2, and 5 μg of catalase.

was purified 14-fold from *T. album* NM and found to be a homotetramer (M_r~ 141 K daltons) with 1.4 ± 0.4 Mn atoms per subunit. The *T. thermophilus* enzyme was determined to be a homohexamer (M_r~210 K daltons), the Mn content was not reported. It is interesting to note that the subunit molecular weights of the Mn-catalase from the thermophiles and the *L. plantarum* enzyme are identical. Both thermophilic enzymes exhibit some sensitivity to N_3^-, but inhibition by 3-AT, a potent heme-catalase inhibitor, was not detected. $NH_2OH + H_2O_2$ was a powerful inhibitor of the thermophilic Mn-catalases. Not surprisingly, the thermophilic Mn-catalases were much less sensitive to thermal inactivation than the *L. plantarum* enzyme. The enzyme from *T. album* NM exhibited substantial peroxidase activity towards p-phenylenediamine but could not use diaminobenzidine, o-dianisidine, guaiacol, glutathione or ferrocytochrome *c* as substrates.[28] H_2O_2 at concentrations above 20 mM inhibited *T. album* NM catalase activity. Table 3 summarizes the effects of other inhibitors examined for this protein. The EPR spectrum of the *T. thermophilus* enzyme suggested that the enzyme active center was formed by binuclear complexes of Mn^{+2}.[45] The spectra consisted of a superposition of a 22-component and a 16-component signal. The latter signal was shown to result from a Mn^{+3} (s=2)-Mn^{+4}(s=3/2) complex with a strong antiferromagnetic interaction.[45] The three-dimensional structure of the Mn-catalase from *T. thermophilus* was determined to 3Å resolution.[46] Crystals of the protein were cubic with space group $P2_13$ and unit cell dimension a=133.4Å. Isomorphic replacement was used to solve the structure. A model of the peptide backbone of the enzyme subunit was presented. Unique features included the presence of 4 highly ordered α-helices each incorporating 25-30 amino acid residues arranged in parallel. The binuclear manganese site is located in the middle of the 4 helices.

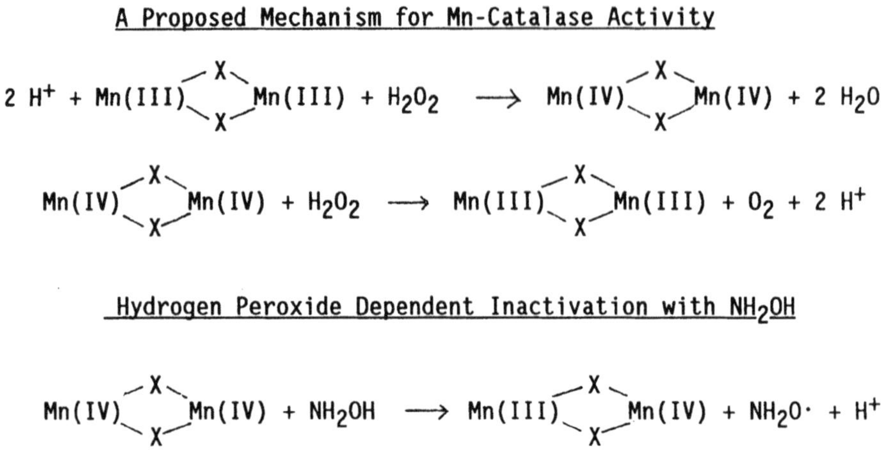

Scheme 1.

The presence of the Mn binuclear cluster in enzymes has stimulated an interest in the synthesis of small molecular weight inorganic complexes that have electronic absorption characteristics similar to the Mn-catalase.[47,48] Crystallographic studies of the models indicate Mn-Mn distances of about 3.2Å and the presence of bridging ligands between the two Mn atoms. The electrochemistry of these systems indicate that it is possible to vary the oxidation states of the manganese centers between +2 and +4 by reversible one-electron transfers without destroying the coupling of the binuclear complexes.

Scheme 1 presents a proposed mechanism for catalase activity in the Mn-catalase. The generation of a mixed-valence species upon inactivation with $NH_2OH + H_2O_2$ is consistent with the experimentally observed EPR signal of this species. Presumably, the mixed-valence species is catalytically inactive due to the inability to undergo the 2-electron exchange required for decomposition of each H_2O_2 molecule. A real test for the small inorganic models prepared will be the demonstration of catalytic dismutation of H_2O_2 in aqueous solvents.

ACKNOWLEDGEMENTS

We would like to thank Dr. Jacqueline Reynolds for the hydrodynamic measurements and Dr. Yasuhiko Nozaki for the dry weight measurements. In addition we are grateful to Dr. Charles Dismukes for many useful discussions.

This work was supported by research grants from the Council for Tobacco Research-U.S.A., Inc., the United States Army Research Office, the National Science Foundation, and the National Institutes of Health.

REFERENCES

1. R. K. Poole, B. S. Baines, and C. A. Appleby, Haemoprotein b-590 (Escherichia coli), a reducible catalase and peroxidase: Evidence for its close relationship to hydroperoxidase I and a cytochrome alb preparation, J. Gen. Microbiol. 132:1525 (1986).

2. A. Claiborne, D. Malinowski, and I. Fridovich, Purification and characterization of hydroperoxidase II of Escherichia coli B, J. Biol. Chem. 254:11664 (1979).
3. A. Claiborne, and I. Fridovich, Purification of the o-dianisidine peroxidase from Escherichia coli B. Physicochemical characteri zation and analysis of its dual catalatic and peroxidatic activities, J. Biol. Chem. 254:4245 (1979).
4. P. C. Loewen and J. Switala, Purification and characterization of catalase HPII from Escherichia coli K12, Biochem. Cell. Biol. 64:638 (1986).
5. M. Pegg, D. Crane, and C. Masters, Confirmation that catalase is a glycoprotein, Biochem. Int. 12:831 (1986).
6. S. Morikofer-Zwey, M. Cantz, H. Kaufman, J. P. von Wartburg, and H. Aebi, Heterogeneity of erythrocyte catalase. Correlations between sulfhydryl group content, chromatographic and electrophoretic properties. Eur. J. Biochem. 11:49 (1969).
7. G. R. Schonbaum and B. Chance, Catalase, in: "The Enzymes," vol. 13, P. Boyer, ed., Academic Press, New York (1976).
8. H. Furuta, A. Hachimori, Y. Ohta, and T. Samejima, Dissociation of bovine liver catalase into subunits on acetylation, J. Biochem. 76:481 (1974).
9. H. Aebi, S. R. Wyss, B. Schorz, and F. Skvaril, Heterogeneity of erythrocyte catalase. II. Isolation and characterization of normal and variant erythrocyte catalase and their subunits, Eur. J. Biochem. 48:137 (1974).
10. D. Dolphin, A. Forman, D. C. Borg, J. Fajer, and R. H. Felton, Compounds I of catalase and horseradish peroxidase: π-cation radicals, Proc. Natl. Acad. Sci. USA 68:614 (1971).
11. B. Chance, An intermediate compound in the catalase-hydrogen peroxide reaction, Acta Chem. Scand. 1:236 (1947).
12. E. Margoliash and A. Novogrodsky. A study of the inhibition of catalase by 3-amino-1:2:4-triazole, Biochem. J. 68:468 (1958).
13. E. G. DeMaster, B. Redfern, F. N. Shirota, and H. T. Nagasana, Differential inhibition of rat tissue catalase by cyanamide, Biochem. Pharmacol. 35:2081 (1986).
14. D. Darr and I. Fridovich, Inhibition of catalase by 3,3′-diaminobenzidine, Biochem. J. 226:781 (1985).
15. Y. Kono and I. Fridovich, Superoxide radical inhibits catalase, J. Biol. Chem. 257:5751 (1982).
16. N. Shimizu, K. Kobayashi, and K. Hayashi, The reaction of superoxide radical with catalase. Mechanism of the inhibition of catalase by superoxide radical, J. Biol. Chem. 259:4414 (1984).
17. H. N. Kirkman and G. F. Gaetani, Catalase: a tetrameric enzyme with four tightly bound molecules of NADPH, Proc. Natl. Acad. Sci. USA 81:4343 (1984).
18. H. Aebi, Catalase in vitro, Meth. Enzymol. 105:121 (1984).
19. H. N. Kirkman, S. Galiano, and G. F. Gaetani, The function of catalase-bound NADH. J. Biol. Chem. 262:660 (1987).
20. W. R. Melik-Adamyan, V. V. Barynin, A. A. Vagin, V. V. Borisov, B. K. Vainshtein, I. Fita, M. R. Murthy, and M. G. Rossman, Comparison of beef liver and Penicillium vitale catalases. J. Mol. Biol. 188:63 (1986).
21. B. K. Vainshtein, W. R. Melik-Adamyan, V. V. Barynin, A. A. Vagin, A. I. Grebenko, V. V. Borisov, K. S. Bartels, I. Fita, and M. G. Rossman, Three-dimensional structure of catalase from Penicillium vitale at 2.0A resolution, J. Mol. Biol. 188:49 (1986).
22. E. A. Delwiche, Catalase of Pediococcus cerevisiae, J. Bacteriol. 81:416 (1961).
23. M. A. Johnston and E. A. Delwiche, Catalase of the Lactobacillaceae, J. Bacteriol. 83: 936 (1962).

24. D. Jones, D. H. Diebel, and C. F. Niven, Jr., Catalase activity of two Streptococcus feacalis strains and its enhancement by aerobiosis and added cations, J. Bacteriol. 88:602 (1964).
25. M. A. Johnston and E. A. Delwiche, Distribution and characteristics of the catalases of Lactobacillaceae, J. Bacteriol. 90:347 (1965).
26. M. A. Johnston and E. A. Delwiche, Isolation and characterization of the cyanide resistant and azide-resistant catalase of Lactobacillus plantarum, J. Bacteriol. 90:352 (1965).
27. Y. Kono and I. Fridovich, Isolation and characterization of the pseudocatalase of Lactobacillus plantarum. A new manganese-containing enzyme, J. Biol. Chem. 258:6015 (1983).
28. G. S. Allgood and J. J. Perry, Characterization of a manganese-containing catalase from the obligate Thermophile thermoleophilum album, J. Bacteriol. 168:563 (1986).
29. V. V. Barynin and A. I. Grebenko, T-catalase-a nonheme catalase of the extremely thermophilic bacterium Thermus thermophilus HB8, Dokl. Acad. Nauk. SSSR 286:461 (1986).
30. W. F. Beyer, Jr., and I. Fridovich, Pseudocatalase from Lactobacillus plantarum: evidence for a homopentameric structure containing two atoms of manganese per subunit, Biochem. 24:6460 (1985).
31. Y. Kono and I. Fridovich, Functional significance of manganese catalase in Lactobacillus plantarum, J. Bacteriol. 155:742 (1983).
32. Y. Kono and I. Fridovich, Inhibition and reactivation of Mn-catalase. Implications for valence changes at the active site manganese, J. Biol. Chem. 258:13646 (1983).
33. G. S. Allgood and J. J. Perry, Paraquat toxicity and effect of hydrogen peroxide on Thermophilic bacteria. J. Free Rad. Biol. Med. 1:233 (1985).
34. W. F. Beyer, Jr., unpublished results.
35. F. S. Archibald and I. Fridovich, Manganese and defenses against oxygen toxicity in Lactobacillus plantarum. J. Bacteriol. 145:442 (1981).
36. F. S. Archibald and I. Fridovich, Manganese, superoxide dismutase, and oxygen tolerance in some lactic acid bacteria, J. Bacteriol. 146: 928 (1981).
37. B. Chance, The composition of catalase-peroxide complexes. J. Biol. Chem. 179:1311 (1949).
38. G. E. Means and R. E. Feeney, "Chemical Modification of Proteins" Holden-Day, San Francisco, CA (1971).
39. G. E. Davies and G. R. Stark, Use of dimethyl suberimidate, a cross-linking reagent in studying the subunit structure of oligomeric proteins, Proc. Natl. Acad. Sci. USA 66:651 (1970).
40. Y. Kono, Mn-catalase of Lactobacillus plantarum: inhibition by carbonyl reagents, in "Superoxide and Superoxide Dismutase in Chemistry, Biology and Medicine," G. Rotilio, ed., Elsevier (1986), p. 231.
41. B. B. Keele, Jr., J. M. McCord, and I. Fridovich, Superoxide dismutase from Escherichia coli B. A new manganese containing enzyme. J. Biol. Chem. 245:6176 (1970).
42. J. J. Villafranca, F. J. Yost, Jr., and I. Fridovich, Magnetic resonance studies studies of manganese(III) and iron(III) superoxide dismutase. Temperature and frequency dependence of proton relaxation rates of water, J. Biol. Chem. 249:3532 (1974).
43. W. F. Beyer, Jr., J. J. Villafranca, and I. Fridovich, unpublished results.
44. T. A. Kent, E. Munck, R. W. Dunham, W. F. Filter, K. L. Findling, T. Yoshida, and J. A. Fee, Mossbauer study of a bacterial cytochrome oxidase: cytochrome c_1aa_3 from Thermus thermophilus, J. Biol. Chem. 257:12489 (1982).
45. S. V. Khangulov, V. V. Barynin, V. R. Melik-Adamyan, A. I. Grebenko, N. V. Voevodskaya, D. L. A. Blynmenfel, S. N. Dobryakov, and V. B.

Il-Yasova, EPR study of T-catalase from Thermus thermophilus, Bioorg. Khim. 12:741 (1986).
46. V. V. Barynin, A. A. Vagin, V. R. Melik-Adamyan, A. I. Grebenko, S. V. Khangulov, A. N. Popov, M. E. Andrianonva, and B. K. Vainshtein, Three-dimensional structure of the T-catalase with a 3A resolution, Dokl. Acad. Nauk. SSSR 288:877 (1986).
47. J. E. Sheats, R. S. Czernuszewicz, G. C. Dismukes, A. L. Rheingold, V. Petroleas, J. Stubbe, W. H. Armstrong, R. H. Beer, and S. J. Lippard, Binuclear manganese(III) complexes of potential biological significance. J. Amer. Chem. Soc. 109:1435 (1987).
48. K. Wrighardt, V. Bossek, J. Bonvoisin, P. Beauvillain, J. Girerd, B. Nuber, J. Weiss, and J. Heinze, Dinuclear manganese (II,III,IV) model complexes for the active center of the metalloprotein photosystemII: synthesis, magnetism, and crystal structure of [L Mn^{III} (μ-O)(μ-CH_3CO_2)$_2$ Mn^{IV}L][ClO_4], Angew. Chem. Int. Ed. 25:1030 (1986).

GLUTATHIONE PEROXIDASE

L. Flohé

Gruenenthal GmbH
Center of Research, Zieglerstr. 6
D-5100 Aachen, FRG

DEFINITION

Glutathione peroxidase (E.C. 1.11.1.9)[1] shall mean the selenoenzyme catalyzing the general reaction shown in eg. 1.

$$2GSH + ROOH \longrightarrow GSSG + ROH + H_2O \quad (1)$$

The term glutathione peroxidase here is not used for the so-called "non-Se-glutathione peroxidase"[2] which, in fact is a transferase catalysing *inter alia* the transfer of GSH to H_2O_2 according to eq. 2.

$$GSH + ROOH \longrightarrow GSOH + ROH \quad (2)$$

which is followed by the presumably non-enzymatic reaction

$$GSOH + GSH \longrightarrow GSSG + H_2O. \quad (3)$$

Glutathione peroxidase is also distinct from a similar selenoenzyme acting on hydroperoxo groups of complex lipids integrated in membranes.[3]

THE BIOLOGICAL CONTEXT

Glutathione peroxidase reduces a large variety of hydroperoxides. It is therefore considered to contribute to the defense system protecting cells against oxidative damage.[1] It appears to be particularly important for maintaining the structural and functional integrity of the cytosolic compartment and the mitochondria or, more generally, of cells or cellular compartments poor in catalase and other substituting activities.[4]

The redundance of enzymes acting on hydroperoxides makes it difficult to define exactly the relative contribution of glutathione peroxidase under *in vivo* conditions. However, persuasive evidence for its importance is supplied by selenium deficiency syndromes.[5] These are, of course, associated with low activities of glutathione peroxidase and, in most examples, characterized by an impaired protection against oxidative damage.[1,5] In fact, the predominant selenium deficiency syndrome of birds, i.e., exudative diathesis, can be mimicked by *in vivo* inhibition of glutathione peroxidase with D-penicillamine.[6] In rats, glutathione peroxidase deficiency due to experimental selenium depletion results in lysis of H_2O_2-exposed erythrocytes, a finding leading to the identification of selenium as an integral constituent of the enzyme.[7] Similarly, glutathione peroxidase-deficient human red blood cells are prone to hemolysis when exposed to pro-oxidant drugs.[8]

The potential participation of glutathione peroxidase in oxidative arachidonic acid metabolism has attracted much interest but remains controversial.[1] It is obviously irrelevant to the metabolization of prostaglandin G_2 and 5-hydroperoxyarachidonic acid, the key intermediate of leukotriene biosynthesis,[9] but it does appear to reduce 12-hydroperoxy-arachidonic acids in platelets.[10] A regulatory role of glutathione peroxidase for various lipoxygenases has been demonstrated in vitro,[11] but its physiological relevance remains to be established.

The broad range of hydroperoxides accepted by glutathione peroxidase provides ample opportunity to speculate about potential roles of the enzyme in balancing specific events such as inflammation, proliferation, mutation, cancerogenesis and atherogenesis.[1] What can be stated with safety so far is that it is a housekeeping enzyme preventing tissue damage when overproduction of hydroperoxides occurs.[1]

STRUCTURE AND FUNCTION

The amino acid sequence of bovine GSH peroxidase[12] and its three-dimensional structure[13] have been determined. These sets of data in combination with functional studies on kinetics, substrate specificity and inhibitors allow a fairly reliable description of the mechanism of action.[1]

The catalytic cycle starts with the enzyme in a reduced form. A selenolate residue of a selenocysteine which is integrated in the protein backbone is sterically exposed at a flat lipophilic impression of the protein. Thus, it is ideally suited to be oxidized by any kind of hydroperoxide which can approach it. This explains the close to diffusion-limited reaction rates and the lack of specificity for hydroperoxides. The active site selenocysteine is thereby presumably oxidized to a selenenic acid derivative. Glutathione is then bound electrostatically via its carboxylic functions to two arginine residues adjacent to the catalytic center, causing its SH group to be directed towards the oxidized selenium. The SH group of glutathione reacts with the selenenic acid residue to form a selenadisulfide bridge, which is then split by the second GSH molecule under regeneration of the reduced enzyme. Briefly,

$$\text{E-Se}^- + \text{H}^+ + \text{ROOH} \xrightarrow{k_1} \text{E-SeOH} + \text{ROH} \qquad (4)$$

$$\text{E-SeOH} + \text{GSH} \xrightarrow{k_2} [\text{E-SeOH GSH}] \qquad (5)$$

$$[\text{E-SeOH GSH}] \xrightarrow{k_3} \text{E-Se-SG} + \text{H}_2\text{O} \qquad (6)$$

$$\text{E-Se-SG} + \text{GSH} \xrightarrow{k_4} \text{E-Se}^- + \text{GSSG} + \text{H}^+ \qquad (7)$$

The dependency of inhibitor efficiency on the functional state of the enzyme has contributed to the elucidation of this catalytic sequence.

Only the reduced form can be trapped by alkylation[14] or superoxide anions.[15] Elimination of Se by CN at alkaline pH requires previous oxidation.[16] Excess hydroperoxide results in overoxidized species of the enzyme (seleninic acid derivative?) which is only slowly regenerated by thiols. Some alternate substrate inhibitors (mercaptosuccinate, penicillamine) are supposed to yield dead-end intermediates E-Se-SR which sterically resist the approach of the second GSH molecule.[17]

The ping-pong kinetics of glutathione peroxidase are in excellent agreement with the reaction scheme proposed (egs. 4 - 7). The fastest step is the oxidation of $E\text{-}Se^-$ by the hydroperoxide. The regeneration by GSH is two orders of magnitude slower than, for example, the reaction of $E\text{-}Se^-$ by H_2O_2 (eq. 4). However, with excess GSH, as present physiologically, E-SeOH does not accumulate. Neither do any other complexes or intermediates accumulate since the electrostatic orientation of the first GSH molecule (eq. 5) is slow compared to the subsequent reactions, as is indicated by lack of limiting maximum velocities. The kinetic pattern would thus suggest a simple sequence of bimolecular reactions if not for the specificity of the enzyme for GSH reinforcing the assumption of a (short-lived) enzyme/GSH complex.

IMPLICATIONS OF THE MECHANISM FOR THE INTERPRETATION OF ACTIVITY TESTS AND IN VIVO RATES

Due to the ping-pong kinetics with infinite maximum velocities, glutathione peroxidase cannot be determined at "limiting substrate concentrations" and correspondingly "international units" are meaningless without proper definition of test conditions since the turnover rate under conventional test procedures strictly depends on the GSH concentration.[1,18] This further implies that the k_2 which is determined may vary between species. Comparisons of enzyme molarities between species based on activity measurements are therefore risky.

The high physiological GSH concentrations guarantee that the enzyme is usually present in its reduced state, i.e., ready to scavenge instantly hydroperoxides at the presumed below micromolar steady-state levels. Under these `normal conditions', variations of GSH concentrations within the millimolar range do not affect the turnover rate. This will depend exclusively on the enzyme molarity, the concentration of hydroperoxide and the kind of hydroperoxide which, of course, determines k_1. However, once the hydroperoxide generation exceeds the regeneration of reduced glutathione by glutathione reductase, the cellular GSH content drops sharply, and the defense system collapses.

BIOSYNTHESIS OF GLUTATHIONE PEROXIDASE

The recent elucidation of cDNA encoding glutathione peroxidases of various species has partially unravelled the problem of how selenocysteine is incorporated into the enzyme. Surprisingly, TGA - formerly known as a termination codon - codes for the incorporation of the selenocysteine of glutathione peroxidase in all species investigated[19,20] and also in a bacterial formate dehydrogenase.[16] How the presumed selenocysteinyl-t-RNA is formed remains to be elucidated. Also, it is still unclear why TGA is sometimes obligatory read as coding for selenocysteine and in other cases as `end of message.'

Several structural elements of the gene encoding glutathione peroxidase are reminiscent of elements regulating gene expression and, in fact, the enzyme activity appears to respond to environmental and endogenous factors (for review see reference 1). But until now no definite conclusions appear justified as to what ultimately determines its expression.

RESEARCH NEEDS

Although the knowledge of the mechanism of action of glutathione peroxidase appears advanced, it might be rewarding to investigate enzymes from remote species in respect to structure-activity relationships in order to unravel the still unresolved details of its mechanism of action.

It also appears extremely interesting to delineate when nature learned to use selenium catalysis to cope with the risks of aerobic life. A more detailed analysis of the molecular background of reported "glutathione peroxidase activities" might be revealing in this respect.

More importantly, we should try to distinguish which selenium deficiency syndromes are related to glutathione peroxidase deficiency and which to other metabolic defects. Since it has become obvious from the Chinese studies on Keshan[22,23] and Kashin-Beck[24] disease that selenium deficiency is not only a veterinary problem, the potential implications of selenium biochemistry on human health care can no longer be ignored.

REFERENCES

1. L. Flohe, The selenoprotein glutathione peroxidase, in: "Coenzymes and Cofactors," D. Dolphin, R. Poulson, and O. Avramovic, eds., John Wiley & Sons, Inc., New York, in press.

2. R.F. Burk and R.A. Lawrence, Non-selenium-dependent glutathione peroxidase, in: "Functions of Glutathione in Liver and Kidney," H. Sies and A. Wendel, eds., Springer-Verlag, Berlin (1978).

3. F. Ursini, M. Maiorino, and C. Gregolin, The selenoenzyme phospholipid hydroperoxide glutathione peroxidase, Biochim. Biophys. Acta 839:62 (1985).

4. L. Flohé and W. Schlegel, Glutathion-Peroxidase, IV. Intrazelluläre Verteilung des Glutathion-Peroxidase-Systems in der Rattenleber, Hoppe-Seyler's Z. Physiol. Chem. 352:1401 (1971).

5. Subcommittee on Selenium; Committee on Animal Nutrition; Board on Agriculture; National Research Council. "Selenium in Nutrition," National Academy Press, Washington (1983).

6. S.J. Mercurio and G.F. Combs, Jr., Drug-induced changes in selenium-dependent glutathione peroxidase activity in the chick, J. Nutr. 115:1459 (1985).

7. J.T. Rotruck, W.G. Hoekstra, A.L. Pope, H. Ganther, A. Swanson, and D. Hafeman, Relationship of selenium to GSH peroxidase, Fed. Proc. 31:691 (1972).

8. T.F. Necheles, The clinical spectrum of glutathione-peroxidase deficiency, in: "Glutathione," L. Flohé, H.Ch. Benöhr, H. Sies, H.D. Waller, and A. Wendel, eds., Georg Thieme Publishers, Stuttgart (1974).

9. W.A. Günzler, M. Lemke, G. Loschen, W. Martin, and W.H. Söhngen, Investigations on the hypothetical role of glutathione peroxidase in leukotriene formation by rat white blood cells, in: "Superoxide and Superoxide Dismutase in Chemistry, Biology and Medicine," G. Rotilio, ed., Elsevier Science Publishers B.V., Elsevier (1986).

10. R.W. Bryant and J.M. Bailey, Altered lipoxygenase metabolism and decreased glutathione peroxidase activity in platelets from selenium-deficient rats, Biochem. Biophys. Res. Commun. 92:268 (1980).

11. M.A. Warso and W.E. Lands, Presence of lipid hydroperoxide in human plasma, J. Clin. Invest. 75:667 (1985).

12. W.A. Günzler, G.J. Steffens, A. Grossmann, S.-M.A. Kim, F. Ötting, A. Wendel, and L. Flohé, The amino-acid sequence of bovine glutathione peroxidase, Hoppe-Seyler's Z. Physiol. Chem. 365:195 (1984).

13. R. Ladenstein, O. Epp., W.A. Günzler, and L. Flohé, Glutathione peroxidase on approval, Life Chem. Rep. 4:37 (1986).

14. L. Flohé and W.A. Günzler, Glutathione peroxidase, in: "Glutathione," L. Flohé, H.Ch. Benöhr, H. Sies, H.D. Waller, and A. Wendel, eds., Georg Thieme Publishers, Stuttgart (1974)

15. J. Blum and I. Fridovich, Inactivation of glutathione peroxidase by superoxide radical, Arch. Biochem. Biophys. 240:500 (1985).

16. R.J. Kraus, J.R. Prohaska, and H.E. Ganther, Oxidized forms of ovine erythrocyte glutathione peroxidase. Cyanide inhibition of a 4-glutathione:4-selenoenzyme, Biochim. Biophys. Acta 615:19 (1980).

17. J. Chaudiere, E.C. Wilhelmsen, and A.L. Tappel, Mechanism of selenium-glutathione peroxidase and its inhibition by mercaptocarboxylic acids and other mercaptans, J. Biol. Chem. 259:1043 (1984).

18. L. Flohé and W.A. Günzler, Assays of glutathione peroxidase, in: "Methods in Enzymology," Vol. 105, L. Packer, ed., Academic Press, Inc., Orlando (1984).

19. I. Chambers, J. Framptom, P. Goldfarb, N. Affara, W. McBain, and P.R. Harrison, The structure of mouse glutathione peroxidase gene: the selenocysteine in the active site is encoded by the "termination" codon, TGA, EMBO J. 5:1221 (1986).

20. G.T. Mullenbach, A. Tabrizi, B.D. Irvine, G.I. Bell, J.A. Tainer, and R.A. Hallewell, Analysis of bases flanking "stop" codons encoding selenocysteine residues: three glutathione peroxidase cDNA sequences, in press.

21. F. Zinoni, A. Birkmann, W. Leinfelder, and A. Böck, Cotranslational insertion of selenocysteine into formate dehydrogenase from Escherichia coli directed by a UGA codon, Proc. Nat. Acad. Sci. 84:3156 (1987).

22. Keshan Disease Research Group of The Chinese Academy of Medical Sciences, Observations on effect of sodium selenite in prevention of Keshan disease, Chin. Med. J. 92:471 (1979).

23. Keshan Disease Research Group of The Chinese Academy of Medical Sciences, Epidemiologic studies on the etiologic relationship of selenium and Keshan Disease, Chin. Med. J. 92:477 (1979).

24. Mo Dong-xu, Pathology and selenium deficiency in Kaschin-Beck disease, Acta Acad. Med. Xi'an 1:32 (1985).

THE ROLE OF GLUTATHIONE TRANSFERASE IN THE DETOXICATION AND REPAIR OF LIPID AND DNA HYDROPEROXIDES

Brian Ketterer, David J. Meyer, and Kia Hong Tan

Department of Biochemistry
University College and Middlesex School of Medicine
London W1P 6DB, U.K.

INTRODUCTION

It is commonly accepted that cumulative effects of free radical damage to membranes and macromolecules are part of the aging process.[1] Accumulation of damage takes place despite the presence in the cell of a number of protective mechanisms, including those which inhibit oxygen-centred free radical formation, detoxify active oxygen species and repair free radical damage. Peroxidation of membrane lipids and of DNA are two toxic consequences of the action of oxygen-centred free radicals,[2] and in the present communication we discuss the significance of GSH transferase activity for their detoxication and repair. Attention will be drawn to a rarely studied isoenzyme from the rat, GSH transferase 5-5,[3] which is very active in reducing DNA hydroperoxides and appears to be a major form in the nucleus.

THE DETOXICATION OF LIPID PEROXIDATION BY GSH-DEPENDENT SYSTEMS

GSH peroxidase activity inhibits lipid peroxidation. This activity is associated with GSH transferases and Se-dependent GSH peroxidase. The importance of GSH transferases relative to Se-dependent GSH peroxidase for the reduction of organic peroxides varies with the species, the tissue and the cell compartment. For example, in the the rat liver cytosol, GSH transferases account for 35% of the total peroxidase activity. This proportion is much higher in the testis cytosol (95%) and within the hepatocyte much higher in the nucleus (80%). In other species the proportion in the liver cytosol may also be higher; thus, in man it is 80% and in the guinea pig 100%.[4] These enzymes do not use esterified fatty acid hydroperoxides as substrates but rather free fatty acid hydroperoxides (see Table 1); consequently, they can only detoxify phospholipid hydroperoxides when assisted by the concomitant action of phospholipase A (see Figure 1 reactions I and II).[5-9]

If phospolipid hydroperoxide is not hydrolysed by phospholipase A, or fatty acid hydroperoxide is not reduced by GSH transferase or Se-dependent GSH peroxidase, redox metal catalysed decomposition of both phospholipid hydroperoxide or the free fatty acid hydroperoxide may

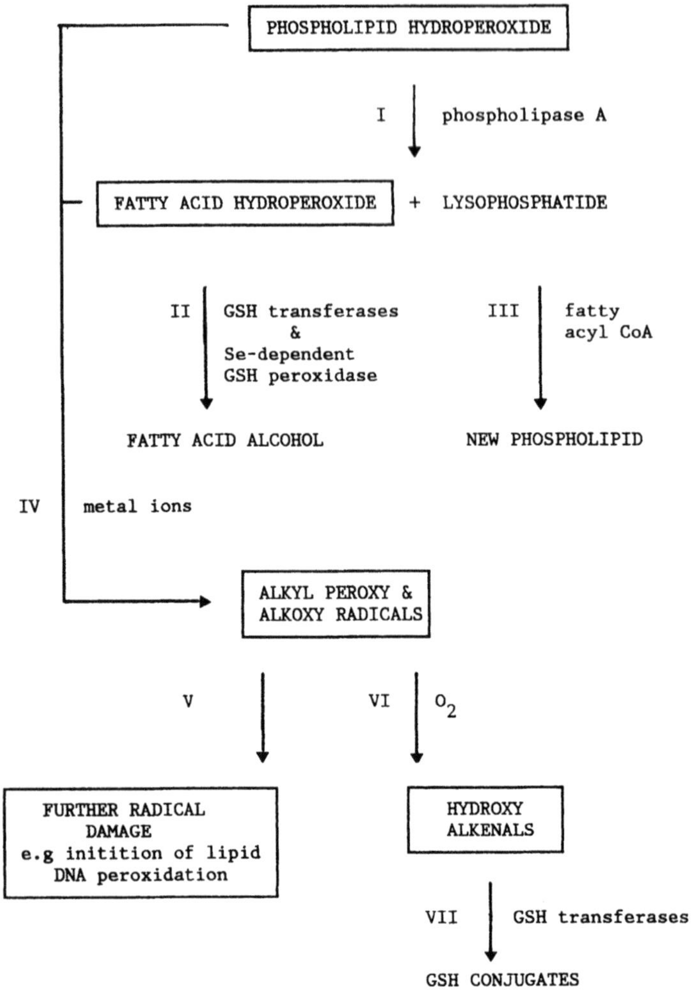

Figure 1. The toxicity of phospholipid hydroperoxides and enzymic reactions involved in their excision, detoxication and repair.

produce alkoxy and alkylperoxy radicals[10] capable of doing further damage either in situ or, after diffusion, at a distance (see Figure 1 reactions IV and V). Alternatively, these radicals may also undergo oxidative decomposition to a range of products, some of which are nontoxic, such as alkanes (the expiration of which can be used to monitor lipid peroxidation), and some genotoxic or cytotoxic, such as hydroxy-alkenals, which are electrophilic substrates for GSH transferases (see Table 1 and Figure 1 reactions VI and VII).[11,12]

Thus, phospholipase A excises the peroxidized fatty acyl group, the GSH transferases and Se-dependent GSH peroxidase detoxify the hydroperoxide and the damaged phospholipid is finally replaced after the action of lysophosphatide fatty acyl CoA transferase (see Figure 1 reaction III). These three enzymes act together to detoxify and repair radical damage to membranes resulting from the attack of radicals in the presence of oxygen.

Table 1. Peroxidation Products Which Are Substrates For GSH Transferases[a]

GSH transferase	µmol/min/mg enzyme				
	CHE	HNE	LinOOH	HOOMU	DNAOOH
1-1	0.14	2.6	2.8	0.13	<.002
2-2	nd	0.67	1.8	1.01	<.002
3-3	<.001	2.7	0.2	0.5	0.018
4-4	<.001	6.9	0.2	0.3	0.032
5-5	<.001	nd	5.3	nd	0.500
6-6	nd	nd	0.06	1.3	0.01
7-7	nd	nd	1.5	1.3	0.005
8-8	nd	170.0	nd	nd	nd

[a]CHE = 5α,6α-epoxy cholesterol; HNE = 4-hydroxy non-2-enal; LinOOH = linoleic acid hydroperoxide; HOOMU = 5-hydroperoxymethyl uracil; DNAOOH = DNA peroxidized by X-irradiation; nd = not determined

THE DETOXICATION OF DNA HYDROPEROXIDES

DNA hydroperoxides may arise in vivo either from exposure to high-energy irradiation or the attack of alkyl peroxy and alkoxy radicals generated during unrestrained lipid peroxidation (see Figure 1). Once formed, DNA hydroperoxides are potential mutagens[13] and may be one of the

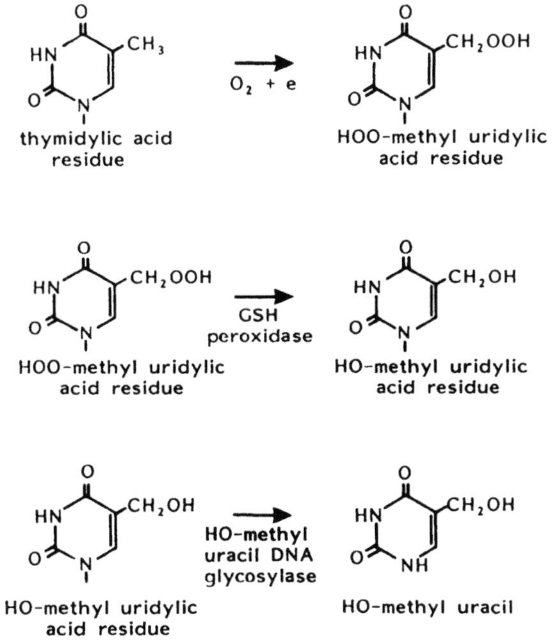

Figure 2. This shows first a radical attack, in the presence of oxygen, on the 5-methyl group of a thymine residue in DNA to give a 5-hydroperoxy methyl uracil residue and, subsequently, its reduction by GSH peroxidase and excision by 5-hydroxymethyl uracil DNA glycolsylase.

factors responsible for mutagenicity of high-energy irradiation or the carcinogenicity in the rat of a choline- and methionine-deficient diet which is associated with lipid peroxidation in the nucleus[14,15] and strand breaks in nuclear DNA.[16] The reduction of these hydroperoxide residues by GSH transferases or Se-dependent GSH peroxidase should be an important step in the limitation of their potential for further DNA damage by decomposition to radicals. The resulting reduction products, 5-hydroxymethyl uracil or thymine glycol, are potentially mutagenic because they inhibit DNA polymerase; however, they may be removed by the appropriate DNA glycosylases.[17,18] The further action of endonucleases, DNA polymerase and DNA ligase should allow complete repair of DNA to occur (see Figure 2).

If such a repair mechanism were to operate, it is essential that peroxidized DNA be accessible to GSH transferases. In the case of nuclear DNA, GSH transferases (or Se-dependent GSH peroxidase) would of necessity be nuclear proteins similar to other repair enyzmes. We have separated two GSH transferase fractions from the nucleus. One is soluble in 0.075 M NaCl/0.05 M EDTA (the free fraction) and the other is extracted from the residue with 8.5 M urea (the bound fraction). Analysis of the GSH transferase content of these two fractions shows that, while the free fraction contains all the GSH transferases present in the cytosol, the bound fraction is more specific, consisting of 95% of a GSH transferase similar in pI, molecular weight and substrate specificity to GSH transferase 5-5 and 5% consisting of GSH transferase 6-6 (see Figure 3).

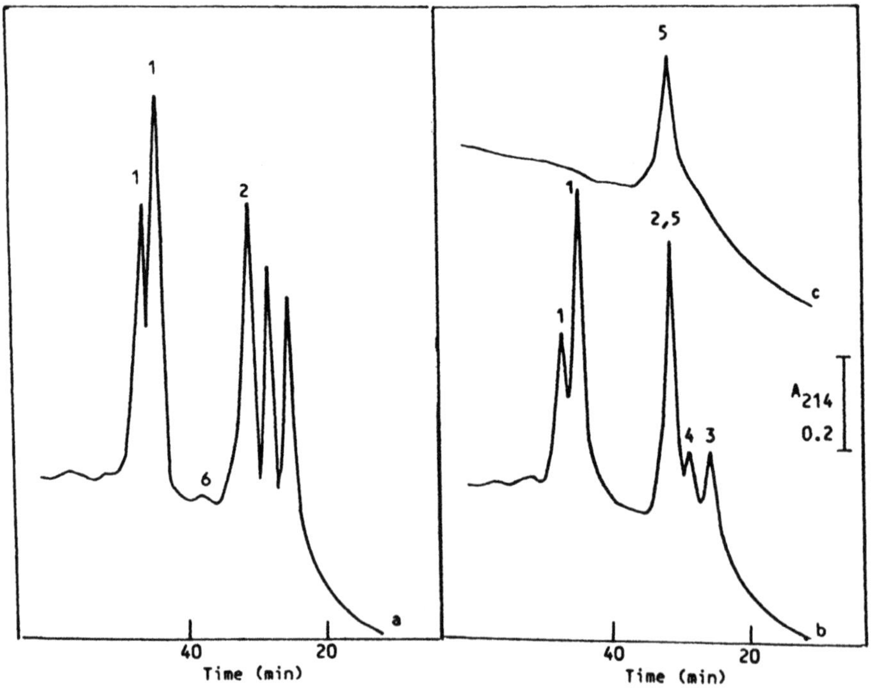

Figure 3. Reverse phase HPLC separation of GSH transferase subunits present in the following subcellular fractions of the rat liver: a) soluble supernatant, b) saline-EDTA extract of nucleus and c) 8.5 M urea extract of the resultant nuclear residue (see method in ref. 19).

As stated above, GSH transferases account for 80% of the nuclear GSH peroxidase activity. Should DNA glycosylases release both hydroperoxy- and hydroxy-thymines, the hydroperoxythymine might be detoxified by nuclear GSH transferase 6-6 (see 5-hydroperoxymethyl uracil in Table 1).

CONCLUSION

In this paper we have focused attention on the GSH peroxidase activity of GSH transferases and shown that the GSH transferases may often be important alternatives to Se-dependent GSH peroxidase where reduction of organic hydroperoxides is required. We have shown that this enzymic activity may be important in dealing with peroxidized lipids and peroxidized DNA. In the proposed mechanism for the inhibition of lipid peroxidation which involves excision of fatty acid hydroperoxide by phospholipase A, its reduction by a GSH peroxidase and the resynthesis of phospholipids by lysophosphatide fatty acyl CoA transferase the relative importance of the cytosolic GSH transferases varies with the species and, in any one species, the tissue involved. In the rat liver the GSH transferases are probably a little less important than the Se-dependent GSH peroxidase. In the case of detoxication and repair of peroxidized DNA in the rat liver nucleus, which is proposed to involve GSH peroxidase, DNA glycosylase, endonuclease, DNA polymerase and DNA ligase, GSH transferases (notably GSH transferase 5-5) are likely to play the major role since the Se-dependent GSH peroxidase has a low abundance there.

Direct evidence for the proposed reduction of DNA hydroperoxides is not yet forthcoming. No means of analysis of thymine hydroperoxide in DNA is as yet available, but thymine glycol and 5-hydroxymethyl uracil residues have been detected in DNA from cells exposed either to the "oxygen burst" of the cytotoxic cellular immune system[20] or to ionizing radiation.[21,22] These hydroxy derivatives may result from the reaction of carbon-centred radicals in thymine with O_2 to give peroxy radicals, which abstract H to give hydroperoxides.[23] The known rate of reaction of O_2 with carbon radicals and the relative abundance of O_2 intracellularly suggests that a large proportion of the 5-hydroxymethyl uracil and thymine glycol detected arises from thymine hydroperoxides and requires GSH transferase activity.

Our investigation of this system arose out of observations by Ames et al.[24] that thymine glycol and 5-hydroxymethyl uracil are present in urine and that when one species is compared with another the amounts found are proportional to the metabolic rate or oxygen consumption of the species.

REFERENCES

1. B.N. Ames, Dietary carcinogens and anticarcinogens: Oxygen radicals and degenerative diseases, Science 221:1256 (1983).
2. K.H. Tan, D.J. Meyer, B. Coles, N. Gillies, and B. Ketterer, Detoxication of peroxidized DNA by GSH transferases, Biochem. Soc. Trans. 15:628 (1987).
3. D.J. Meyer, L.G. Christodoulides, K.H. Tan, and B. Ketterer, Isolation, properties and tissue distribution of rat GSH transferase E, FEBS Lett. 173:327 (1984).
4. R.A. Lawrence and R.F. Burk, Species, tissue and subcellular distribution of non Se-dependent GSH peroxidase activity, J. Nutr. 108:211 (1978).
5. K.H. Tan, D.J. Meyer, and B. Ketterer, Inhibition of lipid peroxidation in rat liver microsomes by GSH S-transferase B, Biochem. Soc. Trans. 11:308 (1983).
6. K.H. Tan, D.J. Meyer, J. Belin, and B.Ketterer, Inhibition of

microsomal lipid peroxidation by GSH transferases B and AA: role of phospholipase A_2, Biochem. J. 220:243 (1984).
7. A. Sevanian, S. F. Muakkassah-Kelly, and S. Montestruque, The influence of phospholipase A_2 and glutathione peroxidase on the elimination of membrane lipid peroxides, Arch. Biochem. Biophys. 233:441 (1983).
8. A. Grossman and A. Wendel, Activity of the selenoenzyme glutathione peroxidase with enzymatically hydroperoxidized phospholipids, Eur. J. Biochem. 135:549 (1983).
9. F.J.G.M. van Kuijk, A. Sevanian, G.J. Handelman, and E.A. Dratz, A new role for phospholipase A_2: protection of membranes from lipid peroxidation damage, Trends in Biochem. Sci. 12:31 (1987).
10. W.A. Pryor, The role of free radical reactions in biological systems in: "Free Radicals in Biology," W.A. Pryor, ed., Academic Press, New York (1976).
11. B. Mannervik, P. Alin, C. Guthenberg, H. Jensson, and M. Warholm, Glutathione transferases and the detoxication of products of oxidative metabolism in: "Microsomes and Drug Oxidations," A.R. Boobis, J. Caldwell, F. DeMatteis and C.R. Elcombe, eds., Taylor & Francis, London, (1985).
12. H. Jensson. C. Guthenberg, P.Alin, and B. Mannervik, Rat glutathione transferase 8-8, an enzyme efficiently detoxifying 4-hydroxyalk-2-enals, FEBS Letts. 203:207 (1986).
13. H.F. Thomas, R.M. Harriott, B. Hahn, and S.Y. Wang, Thymine hydroperoxide as a mediator in ionising radiation mutagenesis, Nature 259:341 (1976).
14. T.H. Rushmore, Y.P. Lin, and A.K. Ghoshal, Rapid lipid peroxidation in the nuclear fraction of rat liver induced by a diet deficient in choline and methionine, Cancer Lett. 24:251 (1984).
15. K.H. Tan, D.J. Meyer, and B. Ketterer, Lipid peroxidation in choline-methionine deficiency, Free Rad. Res. Comms. 3:273 (1986).
16. T.H. Rushmore, E. Farber, A.K. Ghoshal, S. Parodi, M. Pela, and M. Taningher, A choline-devoid diet, carcinogenic in the rat, induces DNA damage and repair, Carcinogenesis 7:1677 (1986).
17. M.C. Hollstein, P. Brooks, S. Linn, and B.N. Ames, Hydroxymethyl-uracil DNA glycosylase in mammalian cells, Proc. Natl. Acad. Sci. USA. 81:4003 (1984).
18. B.N. Ames, Carcinogens and anticarcinogens in: "Antimutagenesis and Anticarcinogenesis Mechanisms," D.M. Shankel, P.E. Hartman, T. Kada, and A. Hollaender, eds., Plenum Press, New York (1985).
19. A-K Ostlund Farrants, D.J. Meyer, B. Coles, C. Southan, A. Aitkin, P.J. Johnson, and B. Ketterer, The separation of glutathione transferase subunits by using reverse-phase high-pressure liquid chromatography. Biochem. J. 245:423 (1987).
20. J.L. Swinehart, W.S. Lin, and P.A. Cerutti, Gamma-ray induced damage in thymine in mononucleotide mixtures and in single and double stranded DNA, Radiat. Res. 58:166 (1974).
21. K. Frankel, M.S. Goldstein, and G.W. Teebor, Identification of the cis-thymine glycol moiety in chemically oxidized and gamma-irradiated deoxy-ribonucleic acid by high-pressure liquid chromatography, Biochemistry 20:7566 (1981).
22. K. Frankel, A. Cummings, J. Solomon, J. Cadet, J.J. Steinberg, and G.W. Teebor, Quantitative determination of the 5-(hydroxymethyl) uracil moiety in the DNA of gamma-irradiated cells, Biochemistry 24:4527 (1985).
23. C. von Sonntag, "The Chemical Basis of Radiation Biology", Taylor and Francis, London, (1987).
24. R. Cathcart, E. Schwiers, R.L. Saul, and B.N. Ames, Thymine glycol and thymidine glycol in human and rat urine: A possible assay for DNA damage, Proc. Natl. Acad. Sci. USA. 81:5633 (1984).

n-3 FATTY ACIDS AND HYDROPEROXIDE ACTIVATION OF FATTY ACID OXYGENASES

William E.M. Lands and Robert B. Pendleton

Department of Biological Chemistry
University of Illinois at Chicago
Chicago, Illinois, 60612, U.S.A.

EICOSANOIDS AND DISEASE

The eicosanoids are extremely potent regulators of cellular and tissue function which are derived from the twenty-carbon fatty acid, arachidonate (20:4n-6). There are two large classes of eicosanoids, the prostaglandins and the leukotrienes (Figure 1). The first committed step in the biosynthesis of each class is catalyzed by a fatty acid oxygenase that produces a hydroperoxide product and that requires a hydroperoxide activator. Thus each oxygenase constitutes an amplifier of hydroperoxide concentration within the cell in which it functions. The amplification process may represent an important molecular tactic for intercellular signaling; an explosive increase in the rate of eicosanoid synthesis is probably necessary to overcome the prevailing rapid rate of eicosanoid inactivation found in most cells. However, a continued excessive rate of eicosanoid synthesis is associated with a variety of pathophysiological situations that are often treated with pharmacological agents that can decrease the biosynthetic rate, such as the steroidal and non-steroidal anti-inflammatory drugs. As long as the rate of metabolic inactivation tends to exceed the rate of biosynthesis, the eicosanoids that are formed do not accumulate to a concentration sufficient to interfere with the behavior of the surrounding cells.

A hypothesis that is currently being examined in increasing detail is that dietary n-3 fatty acids may antagonize pathological formation and function of n-6 eicosanoids.[1] Imbalanced overproduction of n-6 eicosanoids is thought to occur in immune-inflammatory disorders like allergy and arthritis, in metastasis of cancer, and in thrombotic disorders such as heart attacks and strokes, and every institute of the National Institutes of Health has invited biomedical researchers to explore further the potential health benefits of n-3 fats. Clearly dietary n-3 fats can diminish the thrombotic tendencies that depend upon a fast rate of production of the n-6 eicosanoid, thromboxane A2 (ref 2; see TXA_2 in Figure 1), and they may diminish excessive rates of eicosanoid

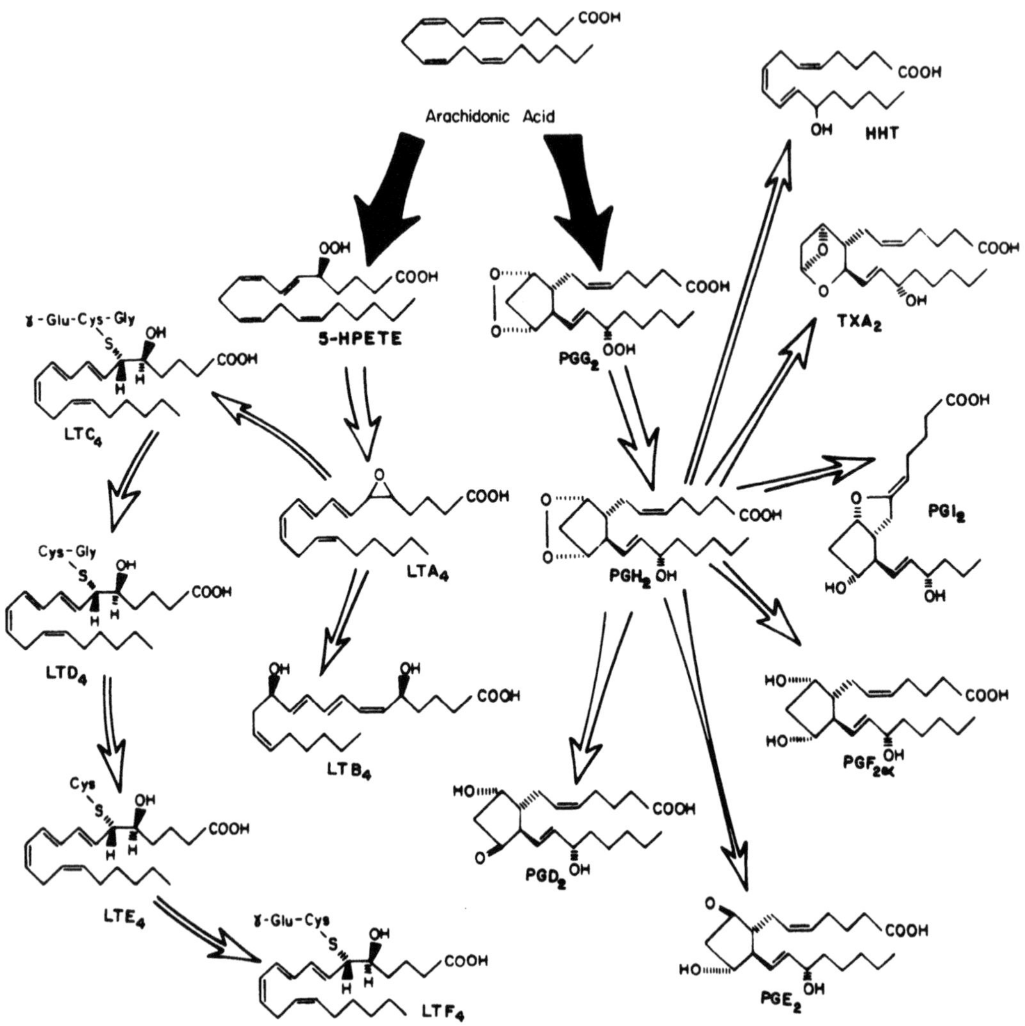

Figure 1. Fatty acid oxygenases control eicosanoid biosynthesis. Arachidonic acid is converted to leukotrienes by reaction with lipoxygenase (left bold arrow) and to prostaglandins by prostaglandin synthase (right bold arrow). Reprinted from Free Rad. Biol. Med. 1:97 (1985) with permission from Pergamon Press.

synthesis in tissues other than platelets. Three means by which the n-3 acids can decrease the n-6 eicosanoid synthetic rate are competition for the n-6 substrate site, increased self-inactivation of the fatty acid oxygenase, and impaired amplification of activating hydroperoxides. In addition to decreasing the rate of synthesis of n-6 eicosanoids, the n-3 acids may form eicosanoid products that antagonize the function of the n-6 eicosanoids at their cellular receptor sites. Thus there are four different ways by which the antagonism may occur, and different cells and tissues may exhibit these to different degrees. This report examines the action of the n-3 acids in impairing the enzyme-catalyzed amplification of the activating hydroperoxides.

HYDROPEROXIDES CONTROL FATTY ACID OXYGENASES

The requirement of all fatty acid oxygenases for a hydroperoxide activator makes the amount of intracellular hydroperoxide an important regulator of the rate of conversion of polyunsaturated fatty acids into prostaglandins and leukotrienes. The lipoxygenase activity that leads to leukotriene formation and the cyclooxygenase activity of prostaglandin H synthase both require hydroperoxides for maximal rates of oxygenation to occur. Thus intracellular hydroperoxides are physiological mediators that may be needed in amounts of 10 to 100 nM to promote intercellular signaling by eicosanoids. The explosive kinetics of the rapid burst of oxygenase activity reflects the amount of oxygenation activity present, with greater amounts of enzyme activity giving a rapid rise in reaction velocity as the hydroperoxide concentration is rapidly amplified (e.g. 560 - 1000 units in Figure 2), whereas smaller amounts of enzyme give a less dramatic, slow rate of biosynthesis (e.g. 110 - 220 units in Figure 2). The removal of hydroperoxides by peroxidases antagonizes the amplification of hydroperoxide and therefore creates a constraint upon the explosively amplified rate of oxygenation.[3]

To illustrate the importance of hydroperoxides to the oxygenation of n-3 and n-6 acids, we treated the reaction systems with increased amounts of glutathione peroxidase. In reaction mixtures with 40 units (0.3 μg) of PGH synthase, the optimal velocity of oxygenation with 50 μM 20:4n-6 was decreased to 39% by 830 units of added glutathione peroxidase, whereas the optimal velocity attained with 50 μM 20:5n-3 was decreased to 47% by only 33 units and to 3% by 330 units.

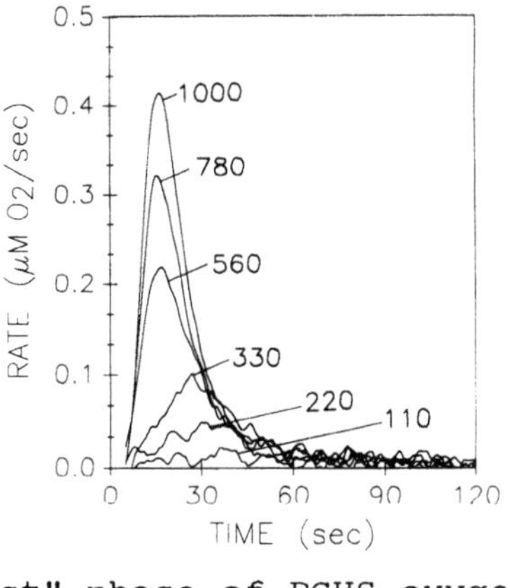

Figure 2. The "burst" phase of PGHS oxygenation is enzyme dependent. Reaction conditions: 25 μM peroxide free 20:5n-3, 1 mM phenol, and 100 mM potassium phosphate, pH 7.2, 25 ^{o}C. Reaction was initiated by adding the indicated units of PGH synthase at time zero.

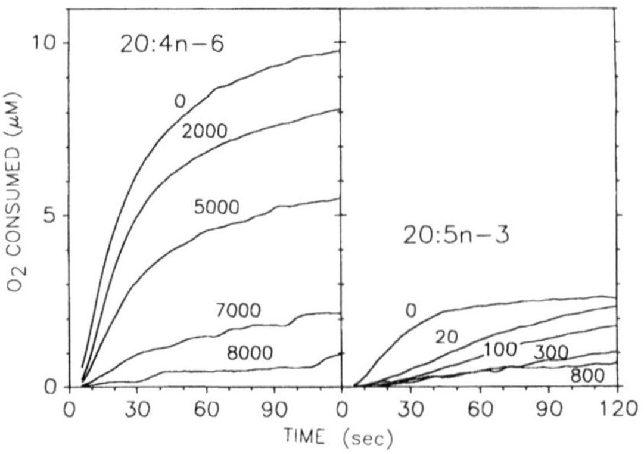

Figure 3. Oxygenase suppression by glutathione peroxidase. Reaction conditions are given in Figure 2; 1 mM glutathione (GSH) was added as cosubstrate for the indicated units of peroxidase. One unit PGHS is 1 nmol O_2/min. with 100 μM 20:4n-6 at 30°C, pH 7.2; one unit peroxidase is 1 nmol reduced GSH/min. at 25°C, pH 7.0.

Only the higher ambient levels of hydroperoxides associated with low levels of added peroxidase permitted an appreciable velocity of prostaglandin biosynthesis with 20:5n-3. A comparison of the two acids is shown in Figure 3. Using 20:4n-6 and 110 units PGHS (1.1 μg), the "burst" phase of the reaction was almost completely suppressed when the amount of glutathione peroxidase was increased from 0 to 8000 units of activity, leaving only a low level of synthetic activity that resembled baseline drift. Suppression of oxygenation was much more easily achieved with 20:5n-3 than with 20:4n-6 as the substrate, requiring only 300 to 800 units peroxidase to achieve over 95% suppression (Figure 3). This result confirms an earlier report[4] that under conditions in which n-6 acids reacted rapidly, the n-3 fatty acids reacted at rates that were comparable to baseline drift. That earlier study led us to demonstrate the effect of dietary n-3 fats in decreasing the injury during cerebral ischemia[5] and in myocardial infarction secondary to coronary artery thrombosis.[6] Those experimental models were designed to provide conditions of eicosanoid overproduction that might be moderated by the antagonistic action of the n-3 fatty acids.

PEROXIDASES CONTROL HYDROPEROXIDE AVAILABILITY

We believe that the high sensitivity of 20:5n-3 to modest rates of hydroperoxide removal makes it an effective antagonist of high rates of prostaglandin biosynthesis from 20:4n-6 under physiological conditions. By decreasing the rate of prostaglandin biosynthesis, 20:5n-3 may allow the catabolism of prostaglandins by degradative enzymes to prevent the local

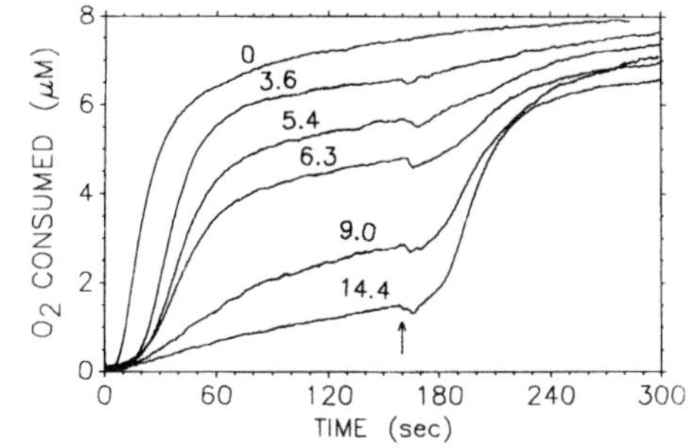

Figure 4. Suppression and recovery of oxygenase activity with 20:5n-3. The ratio of peroxidase to oxygenase is indicated for each incubation. 3 mM n-ethyl-maleimide (NEM) was added after 160 seconds reaction (arrow) to quench peroxidase action. The upper curve is "unsuppressed"; NEM was added 15 seconds before the oxygenase to a mixture containing no peroxidase.

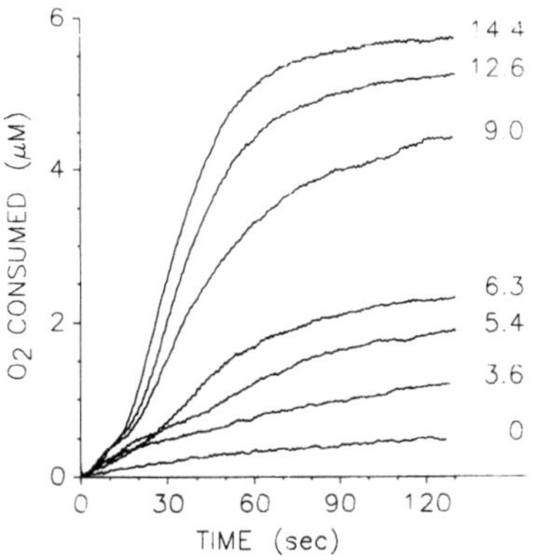

Figure 5. Recovery of latent oxygenase activity. Results are from Figure 4 with zero time being the point at which NEM was added.

accumulation of eicosanoids that is associated with certain pathophysiologic events.

An unusual property of fatty acid oxygenases is that they undergo self-catalyzed inactivation so that under our experimental conditions, enzyme rather than substrate limits the extent of reaction.[4] This feature of the activated enzyme (e.g curve 3.6 in Figure 4) was convenient in demonstrating that significant amounts of enzyme remained in an apparently unactivated state when the amplification of hydroperoxides was suppressed by peroxidase (Figure 4). When n-ethylmaleimide was added to reaction systems after the rates of oxygenation had approached baseline drift, we found that the previously

Table 1. Removal and generation of hydroperoxides. Values are reported as units for tissues obtained from rat; data from reference 3.

Tissue	GSH Peroxidase	PGH Synthase	Ratio
Liver	33,000	18	1,800
Heart	9,200	10	920
Kidney	8,100	11	740
Spleen	7,900	36	220
Lung	4,700	45	105
Macrophage	700	23	30

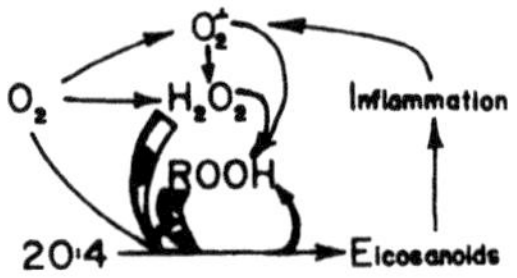

Figure 6. Cell to cell signaling: the interplay of eicosanoids and inflammation. Oxygen free radicals and their metabolites may act as intercellular signals. Reprinted from "Rheumatology-85" with permission of Elsevier Science Publishers.

unactivated enzyme proceeded to react to give a final overall extent of reaction similar to the unsuppressed system.

The activity recovered by releasing the suppression of the peroxidase is presented in a rearranged form in Figure 5. These data confirm that the predominant mechanism by which peroxidase action suppresses prostaglandin biosynthesis is to decrease hydroperoxide availability. Comparison of the results in Figures 3 and 4 with those in Table 1 emphasizes the high degree of suppression of oxygenases activity that must occur in normal tissues by endogenous peroxidase. Many tissues had a ratio of peroxidase to oxygenase greater than a hundred, and even the macrophage, a cell which is known to participate in eicosanoid signaling, had a peroxidase to oxygenase ratio of 30 (Table 1). As a result, we expect that 20:5n-3 will not easily activate oxygenase molecules _in vivo_ to participate in the rapidly amplified response that is characteristic of enzyme behavior _in vitro_. The highly suppressed conditions _in vivo_ makes more important the probable role of hydrogen peroxide in triggering intracellular oxygenase activity. Although hydrogen peroxide is a much less effective activator of fatty acid oxygenases than lipid hydroperoxides[7], it can be formed in large amounts by neutrophils and mononuclear leukocytes. Since hydrogen peroxide can diffuse readily into and out of cells, it may serve a significant role in cell-cell signaling. Thus a combination of intracellular and extracellular amplification of hydroperoxides (Figure 6) probably participates in the cell-cell signaling that is characteristic of immune-inflammatory diseases. We must check further to see how the vulnerability of n-3 oxygenation to peroxidase action may function to decrease inflammation in vivo.

ACKNOWLEDGEMENTS

This research was supported in part by a USPHS grant (HL-34045) and a Pfizer Biomedical Research Award (WEML).

REFERENCES

1. W.E.M. Lands, "Fish and Human Health," Academic Press, Orlando, FL, pp. 1-166 (1986).
2. W.E.M. Lands, B.R. Culp, A. Hirai, and R. Gorman, Relationship of thromboxane generation to the aggregation of platelets in humans: Effect of eicosapentaenoic acid, Prostaglandins 30:819 (1985).
3. P.J. Marshall, R.J. Kulmacz, and W.E.M. Lands, Constraints on prostaglandin biosynthesis in tissues, J. Biol. Chem. 262:3510 (1987).
4. W.E.M. Lands, P.R. LeTellier, L.H. Rome, and J.V. Vanderhoek, Inhibition of prostaglandin biosynthesis, Adv. Biosci. 9:15 (1973).
5. K.L. Black, B. Culp, D. Madison, O.S. Randall, and W.E.M. Lands, The protective effects of dietary fish oil on focal cerebral infarction, Prostaglandins and Med. 3:257 (1979).
6. B.R. Culp, W.E.M. Lands, B.R. Lucchesi, B. Pitt, and J. Romson, The effect of dietary supplementation of fish oil on experimental myocardial infarction, Prostaglandins 20:1021 (1980).
7. P.J. Marshall, R. Kulmacz and W.E.M. Lands, Lipid hydroperoxides and prostaglandin synthesis, in: "Oxygen Radicals in Chemistry and Biology", de Gruyter, Berlin, p. 299 (1984).

SEQUENCE OF COMPLEMENTARY DNA ENCODING HUMAN EXTRACELLULAR-SUPEROXIDE DISMUTASE AND PRODUCTION OF RECOMBINANT ENZYME

Stefan L. Marklund, Lena Tibell,* Karin Hjalmarsson,* Gunnar Skogman,* Åke Engström,** and Thomas Edlund+

Department of Clinical Chemistry
Umeå University Hospital
S-901 85, Umeå, Sweden

*Department of Molecular Genetics and Cell Biology
SYMBICOM AB, Box 1451
S-901 24 Umeå, Sweden

**Department of Immunology
Biomedical Center, Box 582
S-751 23 Uppsala, Sweden

+Department of Microbiology
University of Umeå
S-901 87 Umeå, Sweden

INTRODUCTION

Extracellular-superoxide dismutase (EC-SOD; 1.15.1.1,) is the major SOD isoenzyme in extra- cellular fluids such as plasma, lymph[1] and synovial fluid[2] but also occurs in tissues.[3,4] EC-SOD is heterogenous with regard to binding to Heparin-Sepharose and can be separated into three fractions: A, without affinity; B with weak affinity; and C, with relatively high affinity.[5] Most EC-SOD in the vascular system apparently is bound to endothelial cell surfaces.[6] Membrane-bound heparan sulfate is the likely receptor for the enzyme, and EC-SOD fraction C is the form that binds.[6] EC-SOD is a tetrameric glycoprotein with an apparent subunit molecular weight of around 30 kDa.[5] Like the CuZn SOD's, EC-SOD contains one Cu and one Zn atom per subunit.[5] Still, despite the similarities, the amino acid compositions of human CuZn SOD and EC-SOD are clearly different,[5] and no cross-reaction between polyclonal rabbit antibodies directed towards the enzymes have been observed.[7]

ISOLATION AND SEQUENCE OF cDNA ENCODING HUMAN EC-SOD

cDNA's encoding EC-SOD were isolated from a human placenta cDNA library using a 48-meric deoxyoligonucleotide

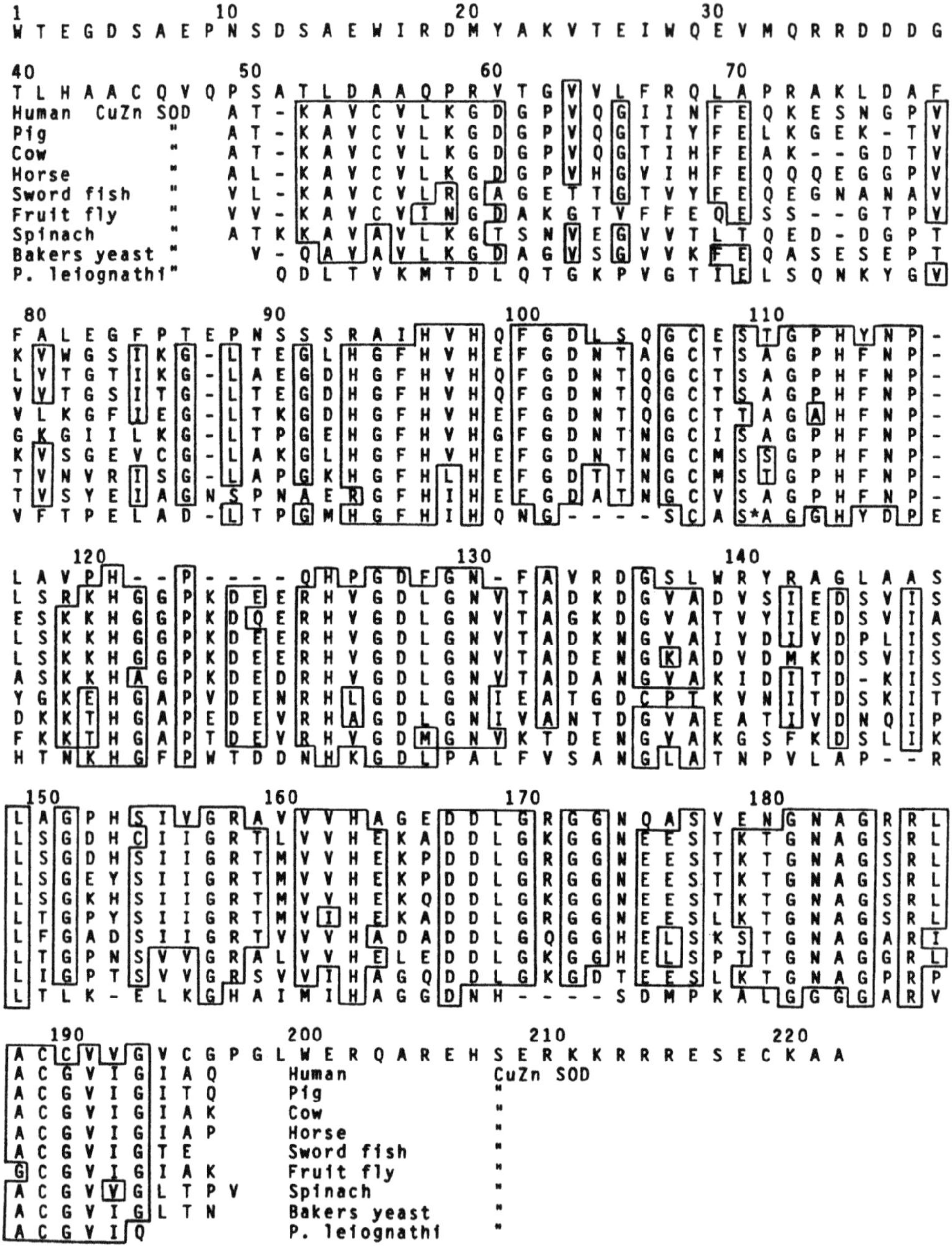

Figure.1. Comparison of the deduced amino acid sequence of mature human EC-SOD with those of CuZn SODs from man, pig, cow, horse, swordfish, fruit fly, spinach, bakers yeast, and P. leiognathi (references for the sequences are found in ref 8). Homologous amino acids are enclosed in boxes in positions where 6 or more are shared by the 9 CuZn SODs. *; sequence (SEKDGKVVLGGA) in P. leiognathi not presented.

probe synthesized on the basis of the amino-terminal amino acid sequence. The largest of the clones found was subcloned and sequenced,[8] according to standard techniques. The cDNA was 1396 bp long and contained an open reading frame of 240 amino acids surrounded by 69 bp and 607 bp untranslated regions of the 5´- and 3´ ends respectively. An 18-amino-acid-long signal peptide preceded the mature EC-SOD. The deduced amino acid sequence of EC-SOD is presented in Figure 1 together with the sequences of CuZn SOD´s.

The first 95 amino acids contain a N-glycosylation site (Asn 89), but show no significant sequence homologies with other sequenced proteins. The carboxyl-terminal end, residues 194-222, is very hydrophilic and contains 9 positively charged amino acids. This sequence probably confers the affinity of EC-SOD for heparin and heparan sulfate. The amino acid sequence from residue 96 through 193 shows strong nearly equal homologies (around 50%) with the final two-thirds of the sequences of all known eukaryotic CuZn SOD´s, whereas the homology with the Photobacterium leiognathi CuZn SOD is clearly lower (Figure 1).

The finding indicates that the EC-SODs may have evolved from the CuZn SODs or that they have a common ancestor. The analysis of the sequence homologies (Figure 1) indicates that the divergence has occurred before the evolution of fungi and plants. The EC-SODs may be widely distributed among higher phyla. The CuZn SOD (or bacterio- cuprein) from P. leiognathi was recently shown to be synthesized with a signal peptide,[9] indicating that it is a secretory protein. Maybe the EC-SODs are the direct descendants of the bacteriocupreins. The non-secretory eukaryotic CuZn SODs may then have evolved from the bacteriocupreins/EC-SODs to finally replace the Mn SODs and Fe SODs in the cytosol.

If EC-SOD is compared with bovine CuZn SOD, the three-dimensional structure of which is known,[10] it is found that all ligands to Cu (His-96, His-98, His-113 and His-163) and Zn (His-113, His-121, His-124 and Asp-127) can be identified in EC-SOD. The cysteines forming the intrasubunit disulfide bridge in the CuZn SODs also have their counterparts in EC-SOD, Cys-107 and Cys-189. The Arg in the active site entrance of the CuZn SODs,[11] the chemical modification of which leads to inactivation of the enzyme, is also found in EC-SOD, Arg-186. The active site of EC-SOD is thus similar to that of the CuZn SODs. This would explain the similarities of EC-SOD and CuZn SODs with respect to specific activity,[5] and sensitivity to inhibitors such as cyanide, azide, diethyldithiocarbamate and hydrogen peroxide.[7]

RECOMBINANT HUMAN EC-SOD PRODUCED BY CHO CELLS

An SV 40 expression vector containing the human EC-SOD cDNA was constructed and transfected into Chinese hamster ovary (CHO) cells.[12] The medium of a three-day culture of the transfected cells was found to contain 97.5% and the cells 2.5% of the total EC-SOD activity of the culture. The cells thus secreted human EC-SOD into the culture medium. The secreted recombinant (r-) EC-SOD was isolated in high yield with a three-step procedure beginning with immobilized monoclonal anti-EC-SOD antibodies. The properties of the r-EC-SOD were compared with native (n-) EC-SOD isolated from human umbilical cords. The specific activities and amino-terminal amino acid sequences were identical. The amino acid compositions were virtually identical and very similar to the composition deduced from the complementary DNA sequence, indicating that the latter is correct.

Both r-EC-SOD and n-EC-SOD contained 4 Cu and 4 Zn atoms per molecule, and the presence of Zn in EC-SOD is thus now established. The r-EC-SOD produced is of type C since its affinity for Heparin-Sepharose was identical to that of n-EC-SOD type C. Both enzymes bound to concanavalin A, lentil lectin and wheat germ lectin and are thus glycoproteins. r-EC-SOD and n-EC-SOD seem to have the same subunit structure and composition as analysed by polyacrylamide gel electrophoresis and gel chromatography.

REFERENCES

1. S. L. Marklund, E. Holme, and L. Hellner, Superoxide dismutase in extracellular fluids, Clin. Chim. Acta. 126:41 (1982).
2. S. L. Marklund, A. Bjelle, and L. G. Elmqvist, Superoxide dismutase isoenzymes of the synovial fluid in rheumatoid arthritis and in reactive arthritides, Ann. Rheum. Dis. 45:847 (1986).
3. S. L. Marklund, Extracellular-superoxide dismutase in human tissues and human cell lines, J. Clin. Invest. 74:1398 (1984).
4. S. L. Marklund, Extracellular-superoxide dismutase and other superoxide dismutase isoenzymes in tissues from nine mammalian species, Biochem. J. 222:649 (1984).
5. S. L. Marklund, Human copper-containing superoxide dismutase of high molecular weight, Proc. Natl. Acad. Sci. USA:79, 7634 (1982).
6. K. Karlsson and S. L. Marklund, Heparin-induced release of extracellular-superoxide dismutase to human blood plasma, Biochem. J. 242:55 (1987).
7. S. L. Marklund, Properties of extracellular-superoxide dismutase from human lung, Biochem. J. 220:269 (1984).
8. K. Hjalmarsson, S. L. Marklund, Å. Engström, and T. Edlund, Isolation and sequence of complementary DNA encoding human extracellular-superoxide dismutase, Proc. Natl. Acad. Sci. USA 84:6340 (1987).
9. H. M. Steinman, Bacteriocuprein superoxide dismutase of Photobacterium leiognathi; isolation and sequence of the gene and evidence for a precursor form, J. Biol. Chem. 262:1882 (1987).
10. J. A. Tainer, E. D. Getzoff, K. M. Beem, J. S. Richardson, and D. C. Richardson, Determination and analysis of the 2 Å structure of copper, zinc superoxide dismutase, J. Mol. Biol. 160:181 (1982).
11. C. L. Borders, J. E. Saunders, D. M. Blech, and I. Fridovich, Essentiality of the active-site arginine residue for the normal catalytic activity of Cu, Zn superoxide dismutase, Biochem. J. 230:771 (1985).

12. L. Tibell, K. Hjalmarsson, T. Edlund, G. Skogman, Å. Engström, and S. L. Marklund, Expression of human extracellular-superoxide dismutase in Chinese hamster ovary cells and characterization of the product, Proc. Natl. Acad. Sci. USA 84:6634 (1987).

A cSOD-NULL MUTANT OF DROSOPHILA CONFERS SENSITIVITY TO PARAQUAT AND REDUCED LONGEVITY

J.P. Phillips

Department of Molecular Biology and Genetics
University of Guelph, Guelph, Ontario, Canada N1G 2W1

INTRODUCTION

Many widely accepted models of biological defense against toxic oxygen species invoke the active participation of superoxide dismutase as a front-line enzymatic defense (1). Such models rest primarily upon a variety of indirect observation and *in vitro* experiments; very few definitive *in vivo* experiments have been done. Consequently, we still do not have a clear understanding of the function of these enzymes *in vivo*, particularly as they relate to such complex biological phenomena as inflammation, mutagenesis, carcinogenesis and aging in higher eukaryotes.

A powerful but little used approach to this problem is that of genetic analysis wherein mutations are utilized to selectively disrupt specific components of the oxygen defense system. We have undertaken a genetic and molecular analysis of the enzymatic components of the oxygen defense system in *Drosophila melanogaster*. In so doing, we are capitalizing on the genetic system of Drosophila through which we can isolate mutants and perform genetic analyses. In addition, the well-defined reproductive, developmental and behavioural systems of Drosophila are amenable to a thorough evaluation of the biological consequences of such mutations. In this report, I describe some of our initial results on the isolation and biological characterization of a mutant of Drosophila deficient in cSOD.

RESULTS

Isolation of a cSOD-Null Mutation

Previous work on cSOD allozyme variants localized the cSOD locus to 68A-C on chromosome 3 (2, 3). We reasoned that mutants devoid of cSOD activity would be recessive lethal; EMS-induced mutations were, therefore, screened for recessive lethality over the deficiency *lxd-9*, which covers the cSOD locus. Recessive lethals were then scored for cSOD activity in *cSOD*^{F}/lethal heterozygotes. cSOD-null mutants were then balanced over TM3 for further investigation. The remainder of this report will deal with a single such cSOD-null mutant, 1-108 (3). The 1-108/TM3 balanced stock provides a convenient genetic system for evaluating the biological consequences of the 1-108 mutation. This stock will produce offspring of two types, 1-108/TM3 and 1-108/1-108, in a ratio of 2:1, respectively (TM3 carries a recessive embryonic lethal which precludes the appearance of TM3 homozygotes). The ratio (K) of the two progeny types eclosing as adults after larval exposure to various conditions provides an unambiguous biological index of the selective effects of cSOD deprivation. Furthermore, because the two types of progeny differ only at discreet genetic loci, the question of variable genetic background is virtually eliminated.

1-108 Confers Hypersensitivity to Paraquat

Figure 1 demonstrates that 1-108 homozygous larvae are selectively killed at increasing concentrations of dietary paraquat. In this experiment, eggs from 1-108/TM3 parents are raised on paraquat-containing media and the ratio (K) of the two progeny types eclosing as adults is taken as an index of the selective effect of paraquat on the development of cSOD-null homozygotes. The results are precisely what would be predicted on the assumption that paraquat toxicity in *Drosophila* is mediated by superoxide radicals generated by the redox cycling of paraquat.[4,5] This experiment demonstrates clearly that cSOD alone can provide physiological protection in larvae against a xenobiotic agent known to generate superoxide radicals and that protection against the type of oxygen radical-induced toxicity conferred by paraquat is related to cSOD level in these genotypes.

Figure 2 shows that cSOD-null (1-108 homozygous) adults are likewise hypersensitive to dietary paraquat. In this experiment, the viability of newly eclosed adults (0-48 hr old) is assessed after a 48 hr exposure to paraquat-containing media. The results parallel the results with larvae and reinforce the conclusion that cSOD plays a vital role in defense against paraquat-generated oxy radical toxicity.

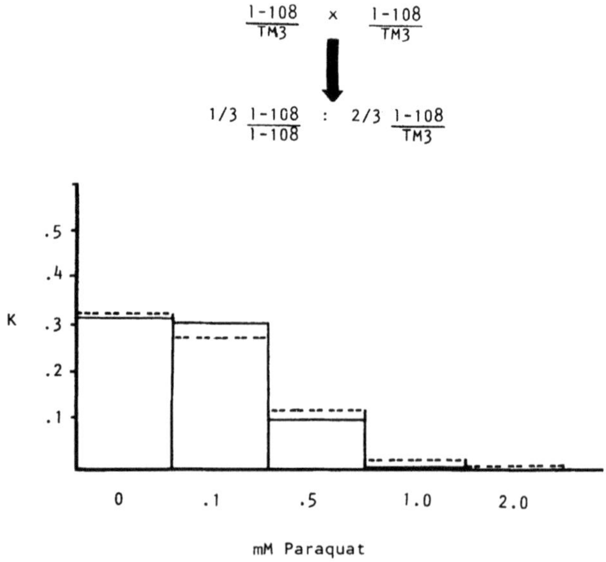

Figure 1. Toxicity of paraquat in <u>1-108</u> larvae relative to <u>1-108</u>/TM3 sibs (K). Solid and dotted lines represent duplicate independent experiments.

cSOD Activity Is Not an Absolute Requirement for Viability

cSOD-null embryos develop into larvae which successfully metamorphose into viable (although short-lived) adults. Although the time of pre-adult development is extended by about 25%, the viability *per se* of cSOD-null genotypes stands as a serious challenge to the belief that cSOD activity is essential to eukaryote cell viability in aerobic conditions. Whether the survival of cSOD-deficient cells and whole organisms relies upon compensatory activity by mSOD or upon a generous repair capacity provided by glutathione peroxidose, glutathione transferase, etc. remains to be investigated.

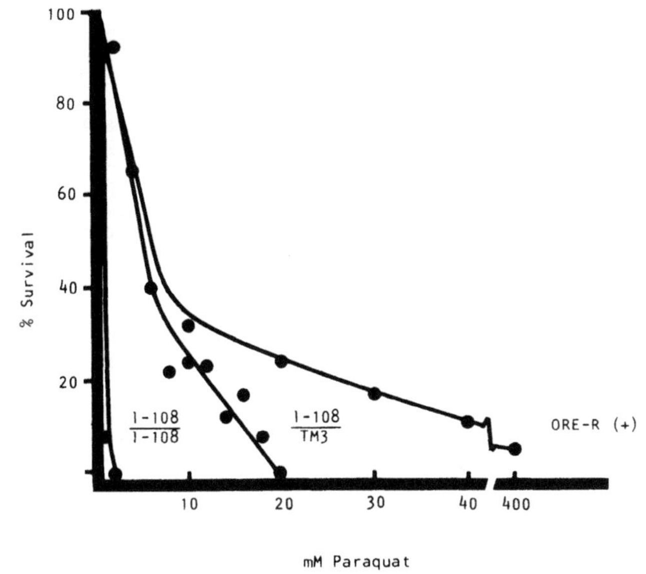

Figure 2. Effect of 1-108 on the toxicity of paraquat in adults. Newly eclosed adults are assessed for viability after a 48 hr exposure to paraquat. ORE-R is a wild-type control.

1-108 Reduces Adult Longevity

The 1-108 mutation has a dramatic foreshortening effect on adult lifespan (Figure 3). The mean adult lifespan of 1-108 homozygotes is only 11.2 days (16%) compared to 70.2 days for wild-type (ORE-R) controls. The intermediate life span of 1-108/TM3 heterozygous siblings as well as their intermediate hypersensitivity to paraquat (Figure 1 and 2) probably result from the additive effects of the TM3 balancer chromosome. This chromosome carries several recessive genes as well as multiple inversions which in sum contribute to a decreased viability of all TM3-balanced heterozygotes. The reduced adult lifespan of the 1-108 homozygotes, however, demonstrates for the first time a clear correspondence between cSOD activity and longevity in strains of a higher eukaryote which differ only at defined genetic loci.

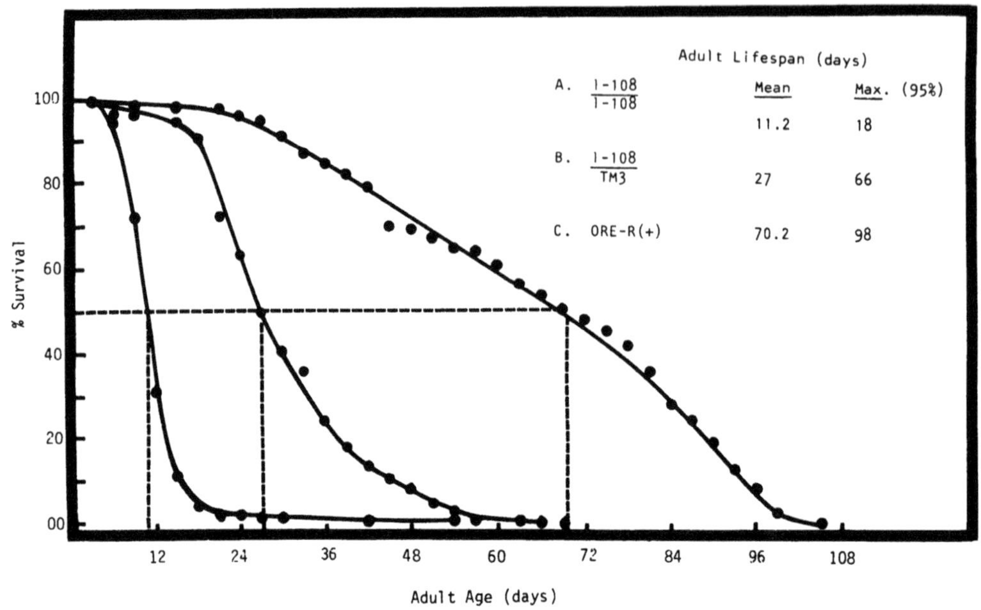

Figure 3. Effect of 1-108 on longevity in Drosophila adults. ORE-R is a wild-type control.

These results confirm the utility of genetic analysis in Drosophila for investigating the in vivo status of putative components of the proposed oxygen defense system. Current work on 1-108 and other mutants focuses on the biological effects of other dietary oxidants, hyperbaric O_2 and ionizing radiation, and on phenotypic rescue by dietary antioxidants. Genotoxic consequences of oxy radical damage to DNA in cSOD-null mutants are being examined by assaying both somatic and germ line mutational events in 1-108 and other cSOD deficient mutants. Finally, we have cloned the Drosophila cSOD gene[6] and will examine the biological consequences of hyper-cSOD levels in p element-mediated cSOD transformants.

ACKNOWLEDGEMENTS

The participation of S. Campbell, D. Michaud and M. Charbonneau in this work is gratefully acknowledged. This work was supported by the Natural Science and Engineering Research Council of Canada.

REFERENCES

1. L.W. Oberley, "Superoxide Dismutase," Vols. I-III, CRC Press, Boca Raton (1985).
2. J.E. Jelnes, Identification of hexokinases and localization of a fructokinase and a tetrazolium oxidase locus in Drosophila melanogaster, Hereditas 67:291 (1972).
3. S.E. Campbell, A.J. Hilliker, and J.P. Phillips, Cytogenetic analysis of the cSOD microregion in Drosophila melanogaster, Genetics 112:205 (1986).

4. L.L. Smith, Paraquat toxicity, Phil. Trans. R. Soc. Lond. B 311:647 (1985).
5. J.A. Farrington, M. Ebert, E.J. Land, and K. Fletcher, Bipyridylium quaternary salts and related compounds: Pulse radiolysis studies on the reaction of paraquat radical with oxygen, Biochim. Biophys. Acta 314:372 (1973).
6. K.C. Kirkland and J.P. Phillips, Isolation and chromosomal localization of genomic DNA sequences coding for cytoplasmic superoxide dismutase from *Drosophila melanogaster*, Gene 61:415 (1987).

INDUCTION OF SUPEROXIDE DISMUTASE ACTIVITY BY PARAQUAT OR EDU IN HUMAN GINGIVAL FIBROBLASTS

Janet S. Kerr, George A. Boswell, Jr., Neil R. Ackerman, and Theresa M. Stevens

Medical Products Department
E.I. du Pont de Nemours & Co., Inc.
Wilmington, Delaware, 19898, U.S.A.

INTRODUCTION

Oxygen free radicals mediate cell injury. These radicals are formed as the reduction of molecular oxygen occurs, initially resulting in the formation of superoxide anions ($\bullet O_2^-$), then H_2O_2 and finally H_2O. The first line of defense against such radicals is the enzyme superoxide dismutase (SOD), which dismutes $\bullet O_2^-$ to H_2O_2. Subsequent reactions may involve the antioxidant enzymes catalase (CAT) and glutathione peroxidase (GSHPx). The production of $\bullet O_2^-$ by the redox cycling of paraquat may be responsible for the induction or derepression of SOD activity. The purpose of this study was to develop a model using human cells to investigate the effects of pharmacologic agents which might act as inducers of the antioxidant enzymes SOD, CAT and GSHPx. We developed this model in human gingival fibroblasts using paraquat (PQ) as the positive control. One pharmacologic agent, EDU, N-[2-(2-oxo-1-imidazolidinyl) ethyl]-N-phenylurea (Figure 1), was an inducer of SOD and CAT activities in plants. It protected plants against acute and chronic ozone-induced injury,[1] and this tolerance was correlated with increased antioxidant enzyme activities.[2] We hypothesized that this heterocyclic urea might induce these enzymes in mammalian cells.

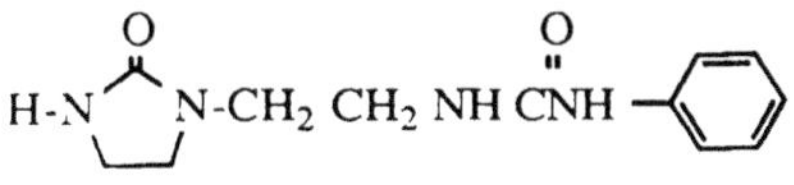

Figure 1. Structure of EDU.

METHODS

Human gingival fibroblasts (Gin-1) were obtained from the American Type Culture Collection (Bethesda, MD). Cells were cultured in Dulbecco's modified Eagles medium supplemented with 2 mM glutamine and 10% heat-inactivated fetal calf serum at 37°C with 5% CO_2 in air. Cells were passaged every third day after reaching confluency. PQ was dissolved in phosphate-buffered saline (PBS) to 100 mM and then diluted in Dulbecco's medium. EDU was dissolved in DMSO and then diluted to 100 mM with deionized H_2O and further diluted with Dulbecco's medium. The final concentration of DMSO in cell cultures was < 0.1%.

Except for the time-course experiments, the incubation period with PQ or EDU was 24 hr. Lactate dehydrogenase (LDH) activity released into the medium was determined from cell preparations as an index of cell viability (Sigma Diagnostic kit #500). LDH release was expressed as a percent of total LDH activity released from cells treated with 0.4% Triton X.[3] Less than 10% of the total LDH activity in the media was considered normal.

Prior to enzymatic assays, Gin-1 cells were removed from the culture flasks using 0.25% trypsin, centrifuged and washed. The pellet was suspended in 1.0 ml PBS at 2-3 X 10^{-6} cells/ml. The cells were disrupted using a Branson Sonifier, then the sonicates were centrifuged at 100,000 x g for 10 min. On the same day as harvesting the cells, SOD and GSHPx activities were measured. The remaining supernatant was stored at -70°C until CAT activity was measured. Total protein was determined in each sonicate using the Bradford dye-binding assay with bovine serum albumin as the standard.[4]

Total SOD (EC 1.15.11) was determined by the method of McCord and Fridovich.[5] This assay depends upon SOD activity to inhibit cytochrome c reduction mediated by the $\bullet O_2^-$ generated from the xanthine oxidase system. One unit of total SOD activity was defined as the amount of SOD required to reduce the maximum rate of the reaction by 50%. CAT (EC 1.11.1.6) was determined by the method of Beers and Sizer.[6] One unit of CAT activity was defined as 1 µmol H_2O_2 removed per minute. GSHPx (EC 1.11.1.1.9) activity was determined by the method of Gunzler et al.,[7] using t-butylhydroperoxide as substrate. One unit of GSHPx was defined as the activity of glutathione peroxidase which oxidized 1 nmol NADPH per minute. All enzyme activities were expressed as units of activity per mg protein. Results were compared by Student's t-test.[8] The level of significance was taken as $p < 0.05$.

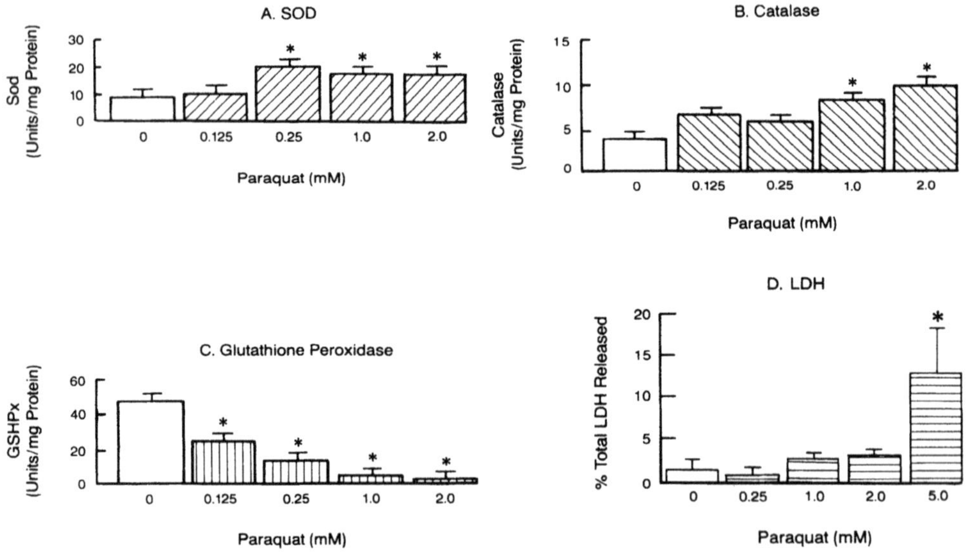

Figure 2. Activities of antioxidant enzymes from PQ-treated Gin-1 cells. Cells were incubated with no PQ (controls), or 0.125 - 2.0 mM PQ for 24 hrs. Fig. 2A is total SOD activity (N=8). *$p < 0.05$ compared with controls or 0.125 mM PQ. Fig. 2B is catalase activity (N=7). *$p < 0.05$ compared with controls, 0.125 or 0.25 mM PQ. Fig. 2C is GSHPx activity (N=6). *$p < 0.05$ compared with controls. Fig. 2D is LDH released from PQ-treated Gin-1 cells (N=5). *$p < 0.05$ compared with controls or 0.25 - 2.5 mM PQ.

RESULTS

Human gingival fibroblasts (1 x 10^6 cells/ml) were incubated without PQ (controls) or 1 mM PQ for various times. Total SOD activity in controls remained constant at 18, 24 and 48 hr, ranging from 6.6 ± 1.3 to 10.3 ± 0.8 Units/mg Protein (n=5 per time point). In the presence of 1.0 mM PQ, SOD activity ranged from 13.6 ± 1.9 to 18.1 ± 1.2 Units/mg Protein (n=5 per time point) and was increased significantly compared with untreated controls ($p < 0.05$). There were no significant differences between controls or between PQ-treated cells at 18, 24 or 48 hrs.

The Gin-1 cells were incubated with 0.125 - 2.0 mM PQ for 24 hrs and the activities of total SOD, CAT and GSHPx were measured. With 0.25 mM PQ, SOD activity increased from 10.3 ± 0.8 Units/mg Protein to 20.9 ± 2.9 Units/mg Protein (N=8; $p < 0.05$). This increase remained constant over the range of 0.25 - 2.0 mM PQ (Figure 2). CAT activity did not significantly increase above 4.0 ± 0.08 Units/mg Protein until 1.0 mM PQ, when CAT activity increased to 7.2 ± 0.06 Units/mg Protein (N=8; $p < 0.05$). In contrast, GSHPx activity decreased in a concentration-dependent manner, from 47.9 ± 3.9 Units/mg Protein in controls to 2.7 ± 0.8 Units/mg Protein with 2.0 mM PQ, indicating PQ inactivated GSHPx. LDH activity was not significantly increased until 5 mM PQ was present.

In the presence of 0.125 - 2.0 mM EDU, total SOD activity increased from 7.8 ± 0.7 Units/mg Protein in controls to an average of 13.9 ± 1.5 Units/mg Protein in EDU-treated cells (range = 11.6 ± 1.4 to 20.7 ± 3.3 Units/mg Protein; $p < 0.05$; N=5 per concentration) (Figure 3). Catalase activity increased in a concentration-dependent manner from 5.6 ± 0.09 Units/mg Protein to 13.9 ± 0.16 Units/mg Protein, with CAT activity almost 200% above controls with 2.0 mM EDU. EDU had no effect on glutathione peroxidase activity, with the activity remaining constant at control levels of 44 ± 4 Units/mg Protein (range = 44 ± 4 to 52 ± 5; N=3). In contrast to PQ, EDU did not inactivate GSHPx. EDU was not cytotoxic at concentrations up to 5.0 mM.

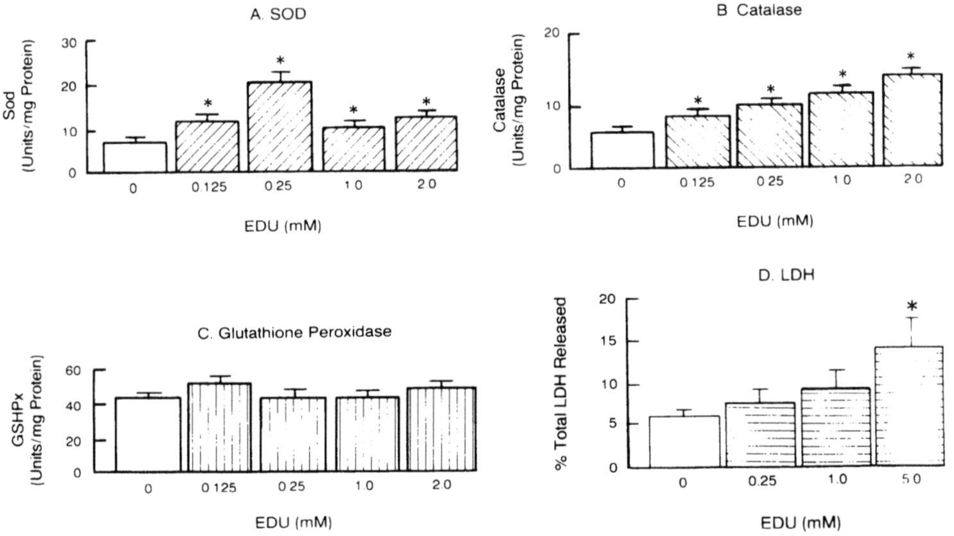

Figure 3. Activities of antioxidant enzymes from EDU-treated Gin-1 cells. Gin-1 cells were incubated with no EDU (controls) or 0.125 - 2.0 mM EDU for 24 hrs. Fig. 3A is total SOD activity (N=5). *$p < 0.05$ compared with controls. Fig. 3B is catalase activity (N=4). *$p < 0.05$ compared with controls. Fig. 3C is GSHPx activity (N=6). Fig. 3D is LDH released from EDU-treated Gin-1 cells (N=3). *$p < 0.05$ compared with controls or 0.25 - 1.0 mM EDU.

DISCUSSION

These data indicate that human Gin-1 cells are a useful model in which to investigate the effects of pharmacologic agents on the activities of the antioxidant enzymes superoxide dismutase, catalase and glutathione peroxidase. Paraquat serves as the positive control for studying the mechanisms by which the activities of these enzymes may be affected. Induction of SOD actiivty by paraquat has been previously reported in Chinese hamster fibroblasts[9] and cultured pneumocytes.[10] The heterocyclic urea EDU, reported to induce SOD and catalase activities in plants, also induces these enzyme activities in human gingival fibroblasts.

The mechansim by which EDU does induce SOD and catalase activities is unclear. EDU does not act as a scavenger of $\cdot O_2^-$ in human neutrophils.[11] Nor does it activate these cells to produce $\cdot O_2^-$ or H_2O_2[11] (unpublished observations). It is interesting to speculate that although SOD and catalase are on different chromosomes, there could be coordinate regulation between them. This is based on the observation that the activities of both enzymes increased in the presence of both paraquat and EDU. Perhaps these inducers derepress or induce a common transduction pathway operating on both genes. By understanding the molecular events through which paraquat and EDU can function, novel pharmacologic agents will be discovered.

REFERENCES

1. J.E. Carnahan, E.L. Jenner, and E.K.W. Wat, Prevention of ozone injury to plants by a new protectant chemical, Phytopathology 68:1225 (1978).
2. E.H. Lee, and J.H. Bennett, Superoxide dismutase, A possible protective enzyme against ozone injury in snap beans (Phaseolus Vulgaris L.), Plant Physiol. 69:1444 (1982).
3. P.G. Cabaud, and Wroblewski, Colorimetric measurement of LDH activity of body fluids, Am. J. Clin. Pathol. 30:234 (1958).
4. M.M. Bradford, A rapid and sensitive method for the quantitation of microgram quantities of protein utilizing the principle of protein-dye binding, Anal. Biochem. 72:248 (1976).
5. J.M. McCord, and I. Fridovich, Superoxide dismutase: An enzymatic function for erythrocuprein, J. Biol. Chem. 244:6049 (1969).
6. R.F. Beers, and I.W. Sizer, A spectrophotometric method for measuring the breakdown of hydrogen peroxide by catalase, J. Biol. Chem. 195:133 (1952).
7. W.A. Gunzler, H. Kramers, and L. Flohe, An improved coupled test procedure for glutathione peroxidase in blood, Z. Klin. Chem. Biochem. 12:444 (1974).
8. W.G. Snedecor, Sampling from a normally distributed population. In: Statistical Methods. 5th ed. The Iowa State Univ. Press p. 45-47 (1964).
9. T.M. Nicotera, A.W. Block, Z. Gibas, and A.A. Sandberg, Induction of superoxide dismutase, chromosomal aberrations and sister-chromatoid exchanges by paraquat in Chinese hamster fibroblasts, Mutation Res. 151::263 (1985).
10. K. Saito, Effects of paraquat on macromolecule synthesis in cultured pneumocytes, Tohoku J. Exp. Med. 148:303 (1986).
11. M.A. Amoruso, G. Witz, and E. Brennan, Inhibiton of superoxide anion radical production in stimulated phagocytic cells by N-[2-(2-oxo-1-imidazolidinyl)ethyl]-N-phenylurea, The Toxicologist, 7:148 (1987).

SUPEROXIDE DISMUTASE IN CULTURED BENIGN AND MALIGNANT TUMORS OF THE COLON

Richard L. Nelson

University of Illinois College of Medicine at Chicago
Chicago, Illinois 60612, U.S.A.

INTRODUCTION

A relative deficiency of the mitochondrial allele of superoxide dismutase (SODb) has been found in a variety of cancers both in vivo and in vitro. A decrease in the activity of SODb also has been described in cells from patients with two conditions that often are associated with cancer development, Down's syndrome and Dubin-Johnson-Sprinz syndrome. Because of these associations it has been hypothesized that SODb deficiency not only is a good biochemical marker for the malignant phenotype, but also may play a central role in oncogenesis.[1] More specifically, cells that are unable to induce or turn on SODb either prior to or as a result of initiation are more susceptible to the oxidative damage induced by tumor promoters.[2,3]

If this model accurately described a condition that is common to all cancers, then initiated premalignant cells, i.e., those cells that have a very high probability of becoming cancers also should demonstrate a deficiency in SODb. The best example of such an initiated precancer is the colonic adenoma[4] because the adenoma-carcinoma sequence is widely accepted as the mechanism by which colon cancers arise.[5] This hypothesis was investigated in this report by comparing the activity of cytosolic SODa and SODb in colon cancers, a colonic adenoma, normal colonic tissue, dermal fibroblasts, and red blood cells. Cell culture was used to obtain pure cell types for analysis. Activities were analyzed by comparing the intensity of the histochemical staining of isozyme bands after separation of the SOD isozymes by starch gel electrophoresis. Other enzymes with mitochondrial alleles were analyzed similarly to determine whether the pattern of activity was seen in SODb was due to an overall alteration in mitochondrial enzyme concentration or was specific to that enzyme.

MATERIALS AND METHODS

Two colon cancer lines were obtained from the American Type Culture Collection (WiDr & COLO 205). The colonic adenoma was obtained from a surgical specimen of a patient with familial polyposis coli. A colon cancer, normal colonic mucosa, and normal skin were also obtained from surgical specimens from individuals with and without hereditary colon cancer. These cultures were initiated in RPMI 1640 or in modified medium 199 and maintained in RPMI 1640 with 15% fetal calf serum.[6] The epithelial origin of these cultures was confirmed by gross morphology of rounded cells densely adherent to the flask and positive staining for mucin using the periodic acid Schiff (PAS) reaction.

Isozyme analysis was performed on lysates of the above cells using horizontal starch gel electrophoresis. Results of enzyme-specific histochemical analysis of the following enzymes are contained in this report: superoxide dismutase (E.C. 1.15.1.1.) (SODa and SODb), malate dehydrogenase (E.C. 1.1.1.37) (MDHs and MDHm), and isocitrate dehydrogenase (E.C.1.1.1.42) (ICDs and ICDm).[7]

RESULTS

Electrophorectic separation of cell lysates revealed two zones of SOD activity, the more anodal corresponding to cytosolic SODa and the row closer to the origin corresponding to SODb (Figure 1 and Table 1). Comparison of the intensity of the histochemical staining for SOD revealed roughly equivalent activity of SODa in all tissues tested, including red blood cells (RBC). SODb activity was not seen in RBCs (as there are no mitochondria in RBCs) and only very weak bands of activity were seen in the colon cancer lines. All other tissues tested, including dermal fibroblasts, colonic epithelium, colonic fibroblasts, and most importantly, colonic adenoma cells demonstrated normal SODb activity as measured by the intensity of the stain.

The opposite or mirror-image slice of the gel was stained for MDH as a mitochondrial control. Two zones of activity were once again seen, the more anodal corresponding the MDHs and the cathodal, to MDHm, the mitochondrial allele of MDH. The intensity of the MDHs stain was equivalent in all cell types. There was marked variation in the staining of MDHm, as strong activity was seen in the cancer cells, but weak or absent bands of activity seen in all other cells. The findings with ICD were the same with marked activation of the mitochondrial allele in the cancers only.

DISCUSSION

It is shown in this report that a colonic adenoma grown in tissue culture has normal activity of SODb when compared to the deficiency seen in

Table 1. Staining Intensity of Isozymes in Colon Cancers,Polyp, Mucosa, and Red Blood Cells

	SOD		MDH		ICD		N
Tissue	a	b	s	m	s	m	
Cancer	++	<+	++	+++	+++	++	3
Adenoma	++	++	++	+	++	0	1
Mucosa	++	++	++	+	+++	0	3
Skin	++	++	++	0	+++	0	3
RBC	++	0	ND	ND	ND	ND	2

Staining intensity was read from 0 to +++.
Subscripts a and s denote cytosolic alleles.
Subscripts b and m denote mitochondrial alleles.
ND - not done.

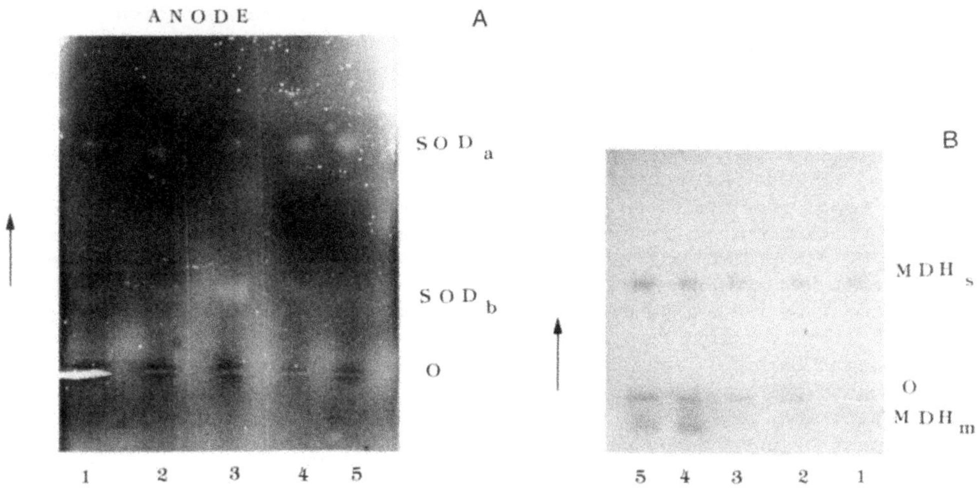

Fig.1 (A) A gel stained for SOD. Bands close to the anode represent SODa and those more cathodal, SODb. 0 represents the origin. (B) The opposite of mirror-image slice of the gel stained for MDH. The more anodal band of activity is MDHs and the cathodal bands of activity seen in slots 4 and 5 are the mitochondrial allele of MDH, MDHm. Slot 1, dermal fibroblasts; slot 2, colonic adenoma; slot 3, colonic epithelium, slots 4 and 5, colon cancers.

colon cancers also grown in tissue culture. It also is shown that two other mitochondrial enzymes, specifically two tricarboxilic acid cycle enzymes, malate dehydrogenase (MDHm) and isocitrate dehydrogenase (ICDm), have increased activity in the colon cancers WiDr and COLO 205 (ATCC), as well as in a colon cancer line initiated in this laboratory when compared to benign cells. This demonstrates that the SODb deficiency observed in colon cancers is not due to an overall deficiency in mitochondrial enzymes, but to a specific decrease in SODb activity.

Previous studies of SOD in colon cancer have shown either higher[8] or lower[9] levels of activity of SOD when whole tumor biopsies were compared to adjacent normal epithelium. Other enzyme activities described in those studies tended to parallel those seen for total SOD, including SODb. The importance of isolating epithelial cells from tumors has been stressed,[10] as only a small minority of the tissue in a colon cancer is of epithelial origin. After epithelial cell isolation, total SOD, and SODb activities after cyanide inhibition of SODa, were compared in dimethylhydrazine-induced colon cancers of rats to adjacent epithelium and normal colonic epithelium. SODa was increased in cancers over all other cell types except for those cells isolated from normal colonic crypts, where cell division normally takes place. This is consistent with the finding that one of the earliest detectable neoplastic changes in the colonic mucosa is a change in the proliferative activity of the cells at all levels of the colonic villus to the proliferative rate normally seen only in the proliferative zone of the crypt.[11] SODb was decreased in the epithelium adjacent to the tumors and formed a similarly lower overall percentage of SOD activity in the colon tumors, although an absolute deficiency of SODb was not seen in the cancers.[10] This may have been due to incomplete inhibition of the activated SODa by cyanide.

Abnormalities in mitochondrial activity have long been thought to play a significant role in the malignant phenotype. Warburg first described cancer as a disease of respiration in which fermentation (anaerobic glycolysis) predominated.[12] Others have thought that all the events that lead to malignancy could be explained exclusively by events occurring within mitochondria.[13] However, the discovery of cellular oncogenes has demonstrated the role of mutation and altered expression of nuclear genes in the onset of cancer.[14]

The most widely held theory of oncogenesis states that the first event is initiation, an irreversible mutational event of a nuclear gene, followed by promotion, which is a reversible modification of gene expression that is thought to be mediated through oxygen radicals.[3,4] Oberly has hypothesized that a deficiency of SODb may greatly effect a cell's susceptibility to promotion,[2] and therefore may be a necessary precondition for malignant change. Yet, it is shown in this study that SODb is not deficient in a colonic adenoma when compared to the deficiency seen in colon cancer cells and, therefore, that SODb deficiency is more likely a result of malignant change in the colon rather than a cause of it.

REFERENCES

1. L. W. Oberly and G. R. Buettner, Role of superoxide dismutase in cancer: A review, Cancer Res. 39:1141 (1979).
2. L. W. Oberly and T. D. Oberly, The role of superoxide dismutase and gene amplification in carcinogenesis, J. Theor. Biol. 106:403 (1984).
3. P. A. Cerutti, Prooxidant states and tumor promotion, Science, 227:375 (1985).
4. E. Farber, The multistep nature of cancer development, Cancer Res. 44:4217 (1984).
5. G. C. Vitale and J. S. Spratt, The adenoma carcinoma sequence, an update. in "Controversies in Colon Cancer," R. L. Nelson, ed. Lippincott, Philadelphia (1987).
6. E. A. Friedman, P. J. Higgins, M. Lipkin, H. Shinya and A. M. Gelb, Tissue culture of human epithelial cells from benign colonic tumors, In Vitro, 17:632 (1981).
7. H. Harris and D. A. Hopkinson, "Handbook of Enzyme Electrophoresis in Human Genetics," North Holland, Amsterdam (1976).
8. G. Baur and A. Wendel, The activity of the peroxide-metabolizing system in the human colon, J. Cancer Res. Clin. Oncol. 97:267 (1980).
9. C. E. J. Hoffman, N. R. Webster, P. A. Wiggins, E. M. Chisolm, G. R. Giles, and S. H. Leveson, Free radical detoxifying systems in human colorectal cancer, Br. J. Cancer, 51:127 (1985).
10. D. P. Loven, L. W. Oberly, F. M. Rousseau, and R. H. Stevens, Superoxide dismutase activity in 1,2-dimethylhydrazine-induced rat colon adenocarcinoma, J. Nat. Cancer Inst. 65:377 (1980).
11. M. Lipkin and E. E. Deschner, Early proliferative changes in intestinal cells, Cancer Res. 36:2665 (1976).
12. O. Warburg, On the origin of cancer cells, Science, 123:309 (1956).
13. J. Hartung, Might cancer be a failed response to renegade mitochondria? J. Theor. Biol. 94:173 (1982).
14. J. M. Bishop, The molecular genetics of cancer, Science, 235:305 (1987).

MICROINJECTION OF ANTIOXIDANT ENZYMES TO PROTECT CELLS FROM OXYGEN DERIVED FREE RADICALS

Carine Michiels and José Remacle

Unité de Biochimie Cellulaire
Facultés Universitaires ND de la Paix
rue de Bruxelles, 61
B-5000 Namur Belgium

INTRODUCTION

In recent years, there have been many reports on the protective effect of superoxide dismutase (SOD), catalase and glutathione peroxidase (Gpx) in preventing oxidative damages induced by a variety of toxic agents or occuring in pathological conditions[1]. Despite all these works, the relative importance of each of these antioxidant enzymes is not yet well established. We undertook the present study in order to determine the respective efficiency of these three enzymes in protecting human fibroblasts against oxygen-derived free radicals generated either by oxygen or by nitrofurantoin.

RESULTS AND DISCUSSION

We developed an experimental model by incubating human diploid fibroblasts WI-38 under 2 atm of oxygen and took advantage of the microinjectiontechnique[2] to introduce increasing concentrations of each of the three above-mentioned enzymes into these cells and to observe their protective capacity[3]. The survival curves for enzyme-injected cells were compared with the curves for buffer-injected cells; the results showed that the survival curves for enzyme-injected cells was higher than for buffer-injected ones and, therefore, indicated that the enzymes protected the cells against oxygen toxicity (Figure 1). The fact that heat-denaturated enzymes did not protect cells indicates that this protection was due to the enzyme catalytic activity and was not the result of the injection of protein.

We tested different concentrations of SOD, catalase and Gpx in protecting cells against 2 atm of O_2, and we observed that this protection was enzyme-concentration dependent. The enzyme subunit concentration in the injected solutions which gave 20% protection was 6250×10^{-9}, 20×10^{-9} and 3.1×10^{-9} mol/ml respectively for SOD, catalase and Gpx. The glutathione peroxidase was thus more efficient than catalase and much more than SOD[4].

The same study was performed for cells incubated in the presence of 5×10^{-4} M nitrofurantoin instead of oxygen. This nitrofurantoin concentration induced a cellular mortality comparable to the mortality

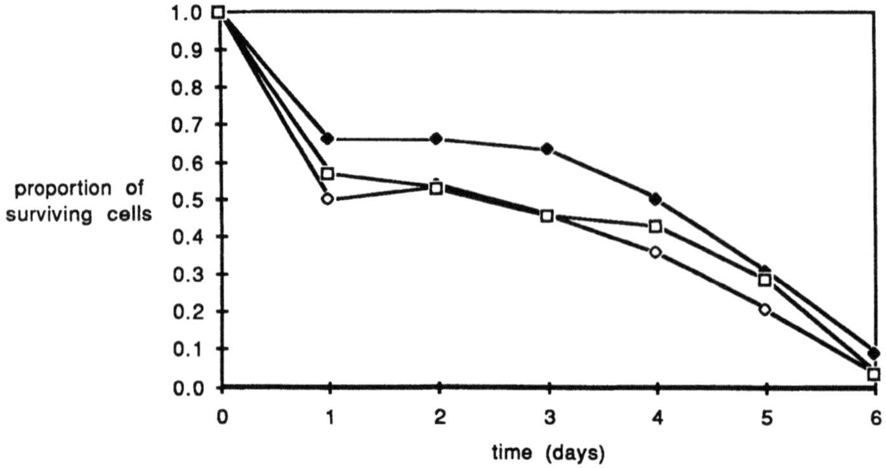

Figure 1 : Evolution of the number of surviving cells exposed to 2 atm of O_2. Cells were injected with a solution of 55 U/ml of catalase (●), with heat-denaturated catalase (○) or with buffer (□). Thirty cells were considered for each curve.

of cells exposed to 2 atm of oxygen. The three enzymes were injected at different concentrations. Here also, the enzymes protected the cells against the nitrofurantoin-derived free radicals, and the protection was related to the enzyme concentration. The enzyme subunit concentration in the injected solutions which gave 20% protection was respectively 2400×10^{-9}, 37×10^{-9} and 1.9×10^{-9} mol/ml for SOD, catalase and Gpx. Here also, Gpx is the most efficient enzyme.

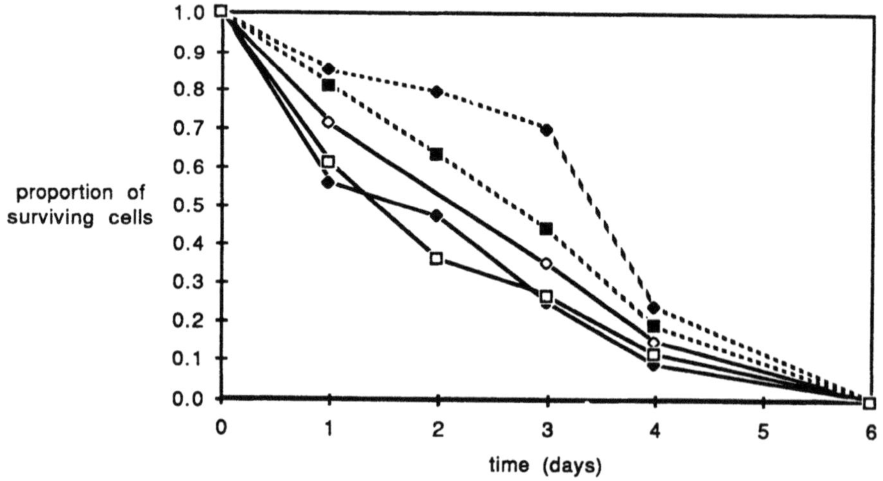

Figure 2 : Evolution of the number of surviving cells exposed to 2 atm of O_2 when cells are cultivated with (-----) or without (——) α-tocopherol (5×10^{-4} M). Cells were injected with a solution of 12 U/ml Gpx (○), 6 U/ml Gpx (●), 3 U/ml Gpx (■) with buffer (□).

We found, as in the case of the oxygen-induced toxicity, the same protective effects within the same range of concentrations. This suggests that, in both cases, the cellular death is provoked by identical mechanisms of free radical attacks. Inasmuch as glutathione peroxidase and catalase are much more efficient than superoxide dismutase, the oxygen species involved are probably hydrogen peroxide derivatives.

In the second part of this work, we tested combinations of the enzymatic systems together with a chemical antioxidant when cells were exposed to 2 atm of oxygen. All three enzymes were injected in the presence of 5×10^{-4} M of α-tocopherol. This molecule alone at 5×10^{-4} M slightly protected the cells. The results showed that the protection of each enzyme was greatly increased in the presence of α-tocopherol. The addition of α-tocopherol to Gpx was even synergistic (Figure 2).

We also tested combinations of two enzymes. Different concentrations of catalase or SOD were injected with or without Gpx into cells exposed to 2 atm of O_2: The results showed that the protection of two enzymes was greater than the protection by each enzyme alone. The protection by catalase or SOD was actually additive to the protection by Gpx. These observation could be explained by complementary roles of membrane protection by Gpx and of cytosol protection by catalase or SOD, in keeping the cellular integrity against the free radical attacks.

On the other hand, when different concentrations of SOD were added to catalase, we observed a slight protection at high SOD concentrations but a toxic effect at low SOD concentrations (Figure 3). We can explain this toxicity by the fact that, when detoxifying superoxide anion, SOD generates hydrogen peroxide which could lead to an overwhelming of the defense system and could be toxic for the cells.

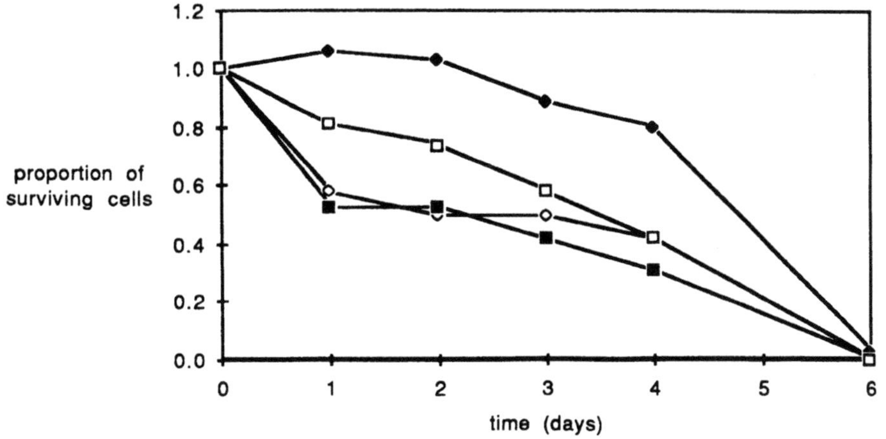

Figure 3 : Evolution of the number of surviving cells exposed to 2 atm of O_2. Cells were injected with a solution of 0.6×10^6 U/ml SOD + 27.5 U/ml catalase (●), 0.1×10^6 U/ml SOD + 27.5 U/ml catalase (○), 0.01×10^6 U/ml SOD + 27.5 U/ml catalase (■) or with buffer (□).

In this work, we were able to test and quantify the intracellular enzymatic protection against free radical attacks. We found an important protective effect of catalase and glutathione peroxidase, which suggests that the protection against hydrogen peroxide or its derivatives is of critical importance to maintain cellular integrity. The fact that Gpx is the most efficient enzyme might indicate that it act on more sensitive targets in the cells by removing hydroperoxides from cellular membranes.

ACKNOWLEDGMENTS

C. Michiels was supported by the "Institut pour l'encouragement de la Recherche Scientifique dans l'Industrie et l'Agriculture".

REFERENCES

1. B. Halliwell and J.M.C. Gutteridge, Oxygen toxicity, oxygen radicals, transition metals and disease, Biochem. J. 219:1 (1984).
2. A. Graessmann, M. Graessmann, C. Mueller, Microinjection of early SV40 DNA fragments and T antigen, Meth. Enzymol. 65:816 (1980).
3. C. Michiels and J. Remacle, A new experimental model to study oxygen toxicity, Arch. Int. Physiol. Biochem. 94:S13 (1986).
4. M. Raes, C. Michiels, J. Remacle, Comparative study of the enzymatic defense systems against oxygen-derived free radicals : the key role of glutathione peroxidase, Free Rad. Biol. Med. 3:3 (1987).

HUMAN FIBROBLAST ANTIOXIDANT DEFENSE RESPONSE TO ALTERATION IN OXYGEN TENSION

Arthur K. Balin, Robert G. Allen, and Richard Reimer

Laboratory for Investigative Dermatology
The Rockefeller University
1230 York Avenue
New York, NY 10021

INTRODUCTION

Our previous studies have revealed that partial pressures of oxygen (PO_2) ranging from 6 through 720 mm Hg modulate growth of human diploid fibroblasts. We have demonstrated that even physiological oxygen concentrations (PO_2 6 - 134 mm Hg) modulate the proliferative life span of these cells. Furthermore, proliferatively aged cells that exhibit a decreased growth rate are more sensitive to oxidative stress than are actively growing cells at lower population doubling levels. Inhibition of cell growth by partial pressures of oxygen above PO_2 300 mm Hg and the limited proliferative capacity of cells in culture under physiological oxygen concentrations are probably distinct phenomena resulting from different mechanisms. Our studies indicate that fibroblasts contain multiple oxygen-sensitive sites which exhibit differential sensitivity to oxygen; however, total cellular metabolism is not generally inhibited by hyperoxic exposure. Thus, oxygen-induced inhibition of cell growth may result from a relatively selective process.[1]

In the present study, we probed the regulation of the antioxidant defense system of human cells by quantifying cellular responses to a gradient consisting of varied levels of oxidative stress. Specifically we examined the intracellular concentration of glutathione (oxidized and reduced), and the activities of total and cyanide-insensitive superoxide dismutase, glutathione peroxidase, glutathione-S-transferase, glucose-6-phosphate dehydrogenase and catalase as a function of the ambient oxygen tension. The results demonstrate that cells increase their concentrations of reducing equivilants as a primary response and that enzymatic changes constitute a secondary response to oxidative stress.

METHODS

Actively growing cultures of human fetal lung fibroblasts (WI-38) were subcultured and inoculated into polystryene growth vessels at 10,000 cells/cm^2, equilibrated with a PO_2 of 32 mm Hg and grown at 37°C for either 4 days (exponential) or 7 days (confluent) before the oxygen tension was altered as described below. The cells were grown in Dulbecco's Modified Eagles Medium (DMEM) supplemented with 10% fetal calf serum (FCS). After 7 days the confluent cells were refed DMEM containing 0.5 % FCS (confluent, low serum) to insure that the cells were in a non-dividing state. After 4 days of exponential growth or after 10 days the flasks were refed with fresh medium containing 10% FCS or 0.5% FCS, respectively, and equilibrated with one of 6 oxygen tensions: PO_2 8.5, 32, 122, 218, 308, or 570 mm Hg. Replicate flasks were harvested by trypsinization (after 24 and 48 hours for the exponential cells and after 24, 48, 72, 96, 168 hours for the confluent cells), combined to yield triplicate samples at each oxygen level and analyzed for all the above mentioned parameters except catalase. Catalase was measured on confluent cells exposed to the various oxygen tensions for 48 and 168 hours.

Assay Methods

Total glutathione was determined by the cycling method of Tietze[2] and oxidized glutathione by the method of Griffith.[3] Superoxide dismutase activity was determined by the positive assay of Misra and Fridovich[4]; catalase by the method described by Luck[5]. Glutathione peroxidase was quantified by a modification of the procedure of Tappel[6]. Glutathione -S-transferases were determined with 1-chloro-2,4 dinitrobenzene as a substrate by the procedure of Warholm, et al.[7] Glucose-6-phosphate dehydrogenase was measured by the method of Deutsch.[8]

Cell number was electronically determined after trypsinization with a coulter counter. Protein determination was performed using a Bradford assay kit obtained from Bio-Rad (Richman, CA). Other cell culture techniques have been previously described.[9]

RESULTS AND DISCUSSION

Glutathione

Glutathione concentration was higher in cells during exponential growth than in cells that were quiescent. Total glutathione (GSH + GSSG) increased dramatically as a function of oxygen tension. This response took place within 24 hours of transfer. In exponentially growing cells exposed to the various oxygen tensions for 24 hours total glutathione concentration (micrograms per mg total protein) was PO_2 8.5, 2.1 micrograms; PO_2 32, 5.75 micrograms; PO_2 122, 8.75 micrograms; PO_2 218, 10.5 micrograms; PO_2 308, 12.25 micrograms; and PO_2 570, 16.25 micrograms. Confluent cells exposed to the various oxygen tensions for 168 hours exhibited GSH concentrations ranging from 3.3 micrograms/mg protein at PO_2 8.5 mm Hg to 9.2 micrograms/mg protein at PO_2 570 mm Hg. A parallel increase in GSSG concentration was observed; it ranged from 0.4 micrograms/mg total protein at PO_2 8.5 mm Hg to 1.2 micrograms/mg protein at PO_2 570 mm Hg at 168 hr. The cellular redox balance, defined as the ratio of GSSG/GSH, remained virtually constant at approximately 0.09.

Superoxide Dismutase

There was no induction of superoxide dismutase activity by any of the oxygen tensions employed after 24 or 48 hours exposure of cells in exponential growth or after 24, 48, or 72 hours of exposure during the confluent nonproliferative state. However, total and cyanide-insensitive SOD activity progressively increased at all oxygen tensions with time. For example, cyanide-insensitive SOD at a PO_2 of 122 mm Hg was 1.4 units/mg dialyzed protein at 24 hours of exponential growth, 2.0 units at 48 hours of exponential growth, 4 units after 24 hours at confluence and 5.2 units after 72 hours of confluence. Induction of SOD occurred with prolonged exposure of confluent cells to the higher oxygen tensions; the most dramatic increase occurred in the cyanide insensitive SOD. After 96 hours cyanide-insensitive SOD activity ranged from 5.4 units/mg dialyzed protein at PO_2 8.5 mm Hg to 10.8 units/mg dialyzed protein at PO_2 570 mm Hg.

Glucose-6-Phospate Dehydrogenase

In general G6PD increased as a function of oxygen tension up to a PO_2 308 mm Hg; it decreased slightly at a PO_2 570 but its activity was still greater than that of cells incubated at a PO_2 122 or less. Incubation of confluent cells for 168 hours at the various oxygen tensions resulted in activities of: PO_2 8.5, 230 U/mg protein; PO_2 32, 265 U/mg protein; PO_2 122, 320 U/mg protein; PO_2 218, 362 U/mg protein;PO_2 308, 420 U/mg protein;and PO_2 570, 360 U/mg protein G6PD. One unit was defined as the amount activity that led to the formation of 1 nmol NADPH/min.

Glutathione Peroxidase

Glutathione peroxidase was stimulated by oxygen in cells maintained under oxygen tensions ranging from PO_2 8.5 to PO_2 32 mm Hg. Tensions greater than PO_2 32 mm Hg progressively depressed the enzyme up to PO_2 308 mm Hg; the activity was similar at PO_2 308 and 570 mm Hg. After 48 hours of exponential growth the glutathione peroxidase activity in U/mg soluble protein was PO_2 8.5, 19 U/mg protein; PO_2 32, 23.5 U/mg protein; PO_2 122, 21.5 U/mg protein; PO_2 218, 19 U/mg protein; PO_2 308, 15 U/mg protein; and PO_2 570, 16 U/mg protein. One unit was equated with the amount of activity that promoted oxidation of 1 nmol NADPH/min. The relative decline in glutathione peroxidase activity at elevated oxygen tensions was greater with more prolonged incubation of confluent cells.

Glutathione-S-Transferase

Oxygen did not induce glutathione-S-transferase activity regardless of the duration of exposure. Confluent cells incubated at the various oxygen tensions for 96 or 168 hours had about 260 U/mg soluble protein activity at at all oxygen tensions tested. A unit was defined as the amount of enzyme that catalysed the formation of 1 nmol of product/min.

Catalase

In general, fibroblasts have low levels of catalase. The catalase content of the fibroblasts decreased with exposure to elevated oxygen tensions. For example, the catalase content of confluent cells exposed to the various oxygen tensions for 168 hours was (units/mg soluble protein) PO_2 8.5, 1.07 units; PO_2 32, 1.14 units; PO_2 122, 1.14 units; PO_2 218, 0.82 units;PO_2 308, 0.56 units;and PO_2 570, 0.11 units.

CONCLUSIONS

Early cellular responses to oxidative stress involve increased glucose consumption[10] and glutathione synthesis. Cells maintain a constant ratio of GSSG/GSH, suggesting that conservation of redox balance is a cellular priority. Many of the antioxidant protective enzymes are not rapidly induced by alterations in oxygen tension. G6PD is among the most readily induced; it takes 24-48 hours to show an increase in activity. Overall, our results indicate that the primary response of human fibroblasts to oxygen-induced stress is a compensatory increase in reducing equivalents. The alteration of enzymatic activity of the antioxidant protective enzymes constitutes a secondary response to alterations in oxygen tension, and occurs only after prolonged exposure.

ACKNOWLEDGMENTS

The expert technical assistance of Ina Leong, Wende Reenstra, and Steven Lilie is gratefully acknowledged. Sincere appreciation is extended to Professor D. Martin Carter who provided the laboratory environment that made this research possible. Particular thanks to Susan Haas for typing the manuscript. This work was supported by PHS grants AG00282, AG04993 and AG05346. This research was also supported in part by a General Clinical Research Center Grant (RR00102) from The National Institutes of Health to The Rockefeller University Hospital and with general support from the Pew Trusts.

REFERENCES

1. A. K. Balin, D. Goodman, H. Rasmussen, and V. J. Cristofalo, Oxygen-sensitive stages of the cell cycle of human diploid cell, J. Cell Biol. 78:390 (1978).

2. F. Tietze, Enzymatic method for quantitative determination of nanogram aounts of total and oxidized glutathione: application to mammalian blood and other tissues, Anal. Biochem. 27:502 (1969).

3. O. W. Griffith, Determination of glutathione and glutathione disulfide using glutathione reductase and 2-vinyl pyridine, Anal. Biochem. 1016:207 (1980).

4. H. Misra and I. Fridovich, Superoxide dismutase: positive spectrophotometric assays, Arch. Biochem. Biophys. 183:511 (1977).

5. H. Luck, Catalase, in: "Methods of Enzymatic Analysis," H. Bergmeyer, ed., Academic Press, New York (1965).

6. A. L. Tappel, Glutathione peroxidase and hydroperoxides, in: "Methods in Enzymology," S. Fleischer and L. Packer, eds., Academic Press, New York (1978).

7. M. Warholm, C. Guthenbert, C. von Bahr and B. Mannervik, Glutathione transferases from human liver, in: "Methods in Enzymology," A. Meister, ed., Academic Press, New York (1985).

8. J. Deutsch, Glucose-6-phosphate dehydrogenase, in: "Methods of Enzymatic Anaylsis," H. Bergmeyer, ed., Verlag Chemie, Deerfield Beach, FL (1983).

9. A. K. Balin, A. J. Fisher, and D. M. Carter, Oxygen modulates growth of human cells at physiologic partial pressures, J. Exp. Med. 160:152 (1985).

10. A. K. Balin, D. Goodman, H. Rasmussen, and V. J. Cristofalo, The effect of oxygen tension on the growth and metabolism of WI-38 cells, J. Cell Physiol. 89:235 (1976).

RELATIONSHIP BETWEEN THE ANTIOXIDANT ENZYME DT-DIAPHORASE AND TUMOR RESPONSE TO MITOMYCIN C TREATMENT

Chris A. Pritsos, Lani L. Pardini, Andrew J. Elliott, and Ronald S. Pardini

Department of Biochemistry
University of Nevada
Reno, Nevada 89557, USA

INTRODUCTION

The flavoprotein DT-diaphorase (NAD(P)H: quinone oxidoreductase, EC 1.6.99.2) is an enzyme which has been described as catalyzing the two-electron reduction of quinoid compounds to their representative hydroquinone forms.[1] The enzymatic reduction of quinones to their respective hydroquinones may represent part of the cellular defense mechanism against quinone-induced toxicity, because fully reduced quinones are capable of being conjugated by sulfate, sugars and amino acids and subsequently excreted.[2] Both chemically and in biological systems, however, it is easier to reduce quinones by one electron to their semiquinones than by two electrons to their hydroquinone forms. The formation of these semiquinone radicals can lead to alkylation reactions and/or reactions with molecular oxygen to form oxygen radicals, resulting in oxidative stress and quinone toxicity. This mechanism of toxicity has been suggested to be responsible for at least part of the mechanism of antineoplastic activity for some of the quinoid anticancer drugs such as Adriamycin (Adr) and mitomycin C (MMC).[3,4] Because DT-diaphorase reduces quinones by two electrons and thereby circumvents the potentially toxic one-electron reduction reactions of these compounds, it is possible that tumor response to treatment by these compounds may be greatly influenced by the endogeneous DT-diaphorase activity in the tumors. We tested the response of three solid tumors to MMC treatment as well as their endogenous DT-diaphorase levels to see whether a relationship exists between the DT-diaphorase activity in these tumors and their response to MMC treatment. MX-1, human mammary carcinoma; LT 583, equine carcinoma[5]; and BLU 284, canine sarcoma,[6] were the three previously characterized solid tumors tested in these studies.

RESULTS AND DISCUSSION

Tumors were grown in athymic nude mice and treated with MMC and monitored for their response to the treatment. The tumor response data for all three tumors at three doses of MMC are shown in Figure 1. The data compares the calculated tumor weights of the treated tumors and the untreated tumors. The data show that all three tumor lines responded to MMC treatment at the three doses tested. MX-1, however, had the greatest response, with a T/C ratio (the final weight of the treated tumor divided by the final weight of the control or untreated tumor) of less than 0.05 at the lowest dosage of 1 mg/kg body weight, while LT 583 and BLU 284 gave T/C ratios of approximately 0.6 and 0.65, respectively. Greater tumor response was observed at the higher dosages of MMC of 2 mg/kg and 3 mg/kg body weight. From these studies it is apparent that while all three tumors respond to MMC treatment, MX-1 responds the best, followed by LT 583 and BLU 284. With a differential response to MMC treatment established for the three tumors, we then measured the DT-diaphorase enzyme activity for each.

Total diaphorase and DT-diaphorase levels for the three tumor lines are shown in Table 1. BLU 284 and LT 583 have similar total diaphorase and DT-diaphorase enzyme activities, while MX-1 is much lower for both of these enzyme activities. The total diaphorase and DT-diaphorase activities for MX-1 were significantly different from those for BLU 284 and LT 583. The lower total diaphorase activity of MX-1 appears to be due to the lower DT-diaphorase levels and not due to lower activities of non DT-diaphorases. These results show that the tumor with the greatest sensitivity to MMC treatment had significantly less DT-diaphorase activity than the other two tumors.

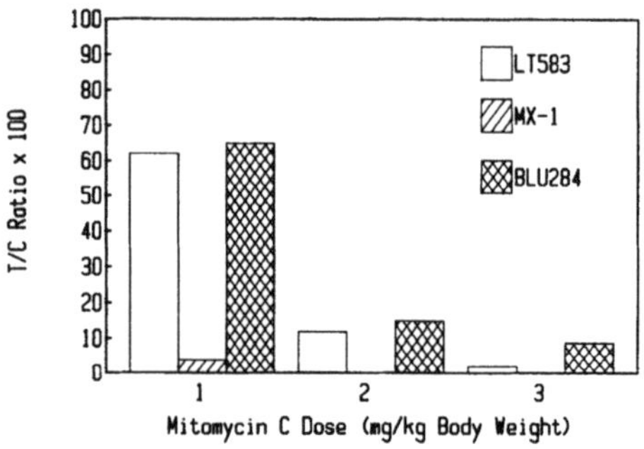

Figure 1. Antineoplastic effect of mitomycin C on three tumors. Tumor trochars were implanted subcutaneously into athymic nude mice and allowed to grow to approximately 25 g prior to treatment. When tumor weights were approximately 25 g, mitomycin C treatment was begun. Mitomycin C was given i.p. once a week for four weeks at one of three doses, 1.0, 2.0 or 3.0 mg/kg body weight. Tumor weights were recorded weekly with the final measurement taken 5 days following the final mitomycin C treatment and were compared to control tumor weights from untreated mice. The final measurement was used to calculate the T/C ratios shown in the chart. The values obtained represent the average value from five animals per group.

The enzyme activities shown in Table 1 were obtained from untreated tumors. The MMC treatment protocol for these tumors in nude mice included drug treatment once a week for four weeks. Tissues often respond to insult by a xenobiotic challenge by enhancing their detoxification mechanisms. We wanted to test whether any of these tumors would increase their DT-diaphorase activity as a result of the MMC treatments. Nude mice were inoculated with the respective tumors (Figure 1) and MMC treatment was begun as previously described. Two days following the second MMC treatment, the tumors were excised and tested for their diaphorase activities (Table 2). The data in Table 2 show that LT 583 and BLU 284 did not change their total diaphorase or DT-diaphorase levels during the treatment. MX-1, however, has a significant increase in DT-diaphorase activity, which was also reflected as an increase in its total diaphorase activity. The observation that the MX-1 tumor responded to the MMC treatment by increasing its DT-diaphorase levels, whereas LT 583 and BLU 284 did not is an interesting finding since MX-1 had significantly lower DT-diaphorase activities in the untreated tumors.

If the tumor response to MMC treatment (Figure 1) is related to the DT-diaphorase levels of the tumors (Table 1) and if the increase in DT-diaphorase activity for MX-1 following partial MMC treatment (Table 2) is a response to the MMC, then MMC must be a substrate for the diaphorase enzymes. Using partially purified beef liver DT-diaphorase, MMC was shown to be a substrate for the enzyme with a K_m value of approximately 1.5 μM. That MMC is a potential in vivo substrate for DT-diaphorase is consistent with the hypothesis that DT-diaphorase may react with MMC to inhibit its ability to generate oxygen radicals and subsequently induce oxidative damage.

Table 1. Diaphorase Activity in Three Solid Tumors

Tumor	Diaphorase Activity μmole cytochrome c reduced/min/mg protein	
	Total Diaphorase Activity ± SD	DT-Diaphorase Activity ± SD
BLU 284	0.29 ± 0.08	0.25 ± 0.09
LT 583	0.33 ± 0.09	0.29 ± 0.09
MX-1	0.10 ± 0.04	0.06 ± 0.02

Diaphorase activity was measured for three solid tumors. Crude homogenates were first sonicated and then centrifuged at 60,000g for 30 minutes. The supernatant was removed and used for the enzyme assays. The enzymatic reaction mixture consisted of 50 mM Tris Buffer, pH 7.5, 0.5 mM NADH, 75 μM cytochrome c and 35 μM menadione. 5 μM dicoumarol was added to the reaction mixture in order to separate DT-diaphorase activity from total diaphorase activity. Each value represents an average of at least five determinations ± S.D.

In summary, these studies measured the DT-diaphorase levels and the response to MMC treatment of three solid tumor xenographs in nude mice. The results suggest that for these three tumors an inverse relationship exists between DT-diaphorase levels and their response to MMC treatment.

Table 2. Diaphorase Activity in Three Solid Tumors Following Mitomycin C Treatment

Tumor	Diaphorase Activity µmole cytochrome c reduced/min/mg protein	
	Total Diaphorase Activity ± S.D.	DT-Diaphorase Activity ± S.D.
BLU 284	0.21 ± 0.05	0.16 ± 0.05
LT 583	0.33 ± 0.10	0.29 ± 0.10
MX-1	0.24 ± 0.07	0.20 ± 0.07

Tumors were grown in athymic nude mice as previously discussed, and mitomycin C treatment was begun at 1 mg/kg body weight as described in Figure 1. Two days following the second injection the tumors were excised and diaphorase activity was measured as described in Table 1. Each value represents an average of at least four determinations ± S.D.

REFERENCES

1. L. Ernster, in: "Methods in Enzymology," vol. 10, R.W. Estabrook and M.E. Pullman, eds., Academic Press, New York (1967).
2. C. Lind, H. Vadi, and L. Ernster, Metabolism of benzo (a) pyrene-3,6-quinone and 3-hydroxybenzo (a) pyrene in liver microsomes from 3-methylcholanthrene-treated rats, Arch. Biochem. Biophys. 190:97 (1978).
3. N.R. Bachur, S.L. Gordon, M.V. Gee, and H. Kon, NADPH-cytochrome P-450 reductase activiation of quinone anticancer agents to free radicals, Proc. Natl. Acad. Sci. USA 76:954 (1979).
4. C.A. Pritsos and A.C. Sartorelli, Generation of reactive oxygen radicals through bioactivation of mitomycin antibiotics, Cancer Res. 46:3528 (1986).
5. L.L. Pardini, L.M. Aaronson, P.L. Hering, and R.S. Pardini, Heterotransplantation of equine carcinoma cells in athymic (nude) mice, Am. J. Vet. Res. 47:610 (1986).
6. L.M. Aaronson, L.L. Pardini, and R.S. Pardini, Heterotransplantation of canine sarcoma cells (Blu 284) in athymic (nude) mice, Am. J. Vet. Res. 48:816 (1987).

MICROSOMAL FATTY ACID COMPOSITION AND Ca^{2+} UPTAKE MODIFICATIONS INDUCED BY SOD DEPLETION IN RAT LIVER

G.M. Bartoli, P. Palozza, E. Piccioni, B. Giannattasio, G. Calviello, A.M. Guerrera, D. Bossi, G. Maccauro, and A. Cittadini*

Institute of General Pathology
Catholic University
Rome, Italy

*Department of Cell Biology and Physiology
University of L'Aquila
L'Aquila, Italy

INTRODUCTION

The hepatic metabolism of certain substances may result in the release of reactive free radicals capable, among other damaging effects, of initiating lipid peroxidation in biological membranes. This effect, in turn, can modify different cell functions.[1,2] Animal cells have developed several defense mechanisms against oxygen toxicity, both enzymatic and non-enzymatic. Superoxide dismutase, catalase and glutathione peroxidase scavenge initiating reactive species of oxygen; glutathione, α-tocopherol, and ascorbic acid represent species of endogenous antioxidants. The impairment of defense systems should result in the increase of endogenous lipid peroxidation, as in the case of vitamin E and selenium deficiency.[3,4] Moreover elevation of intracellular superoxide dismutase correlates with increased resistence toward oxygen toxicity,[5] while SOD diminution[6] favours the increase of intracellular $\cdot O_2^-$ production. The aim of this work is to investigate if a diminution of SOD induces an oxidative stress and influences lipid composition and function of rat liver microsomal membranes. For this purpose, we have treated rats with a copper-deficient diet that inhibits Cu-Zn-SOD but does not significantly affect other important enzymes, such as the cytochrome oxidase.[7] During the treatment we have measured hepatic SOD level, fatty acid composition, in vitro lipid peroxidation and Ca^{2+} uptake of endoplasmic reticulum membranes.

RESULTS AND DISCUSSION

Male weanling Wistar rats were fed for 8 weeks on a copper deficient diet and deionized water ad libitum.[7] SOD activity was determined in the liver cytosol as adrenochrome formation.[8] The in vitro lipid peroxidation induced by either NADPH or xanthine/xanthine oxidase system, was measured after membrane incubation at 37°C as malondialdehyde (MDA) production revealed by thiobarbituric acid assay.[9] Phospholipid fatty acid composition was analyzed by reversed phase HPLC with precolumn derivatization.[10] Microsomal

Table 1. Liver SOD Content in Cu-Deficient Rats

	Weeks of Treatment				
	2	4	6	8	8+2*
			μg SOD/g wet weight		
Control	0.45 (2)	1.06 (2)	1.20 (2)	1.20 ± 0.05 (8)	1.19 ± 0.10 (6)
Cu-deficient	0.15 ± 0.005 (4)	0.22 ± 0.007 (4)	0.32 ± 0.007 (4)	0.31 ± 0.053 (4)	1.29 ± 0.11 (6)

*Animals fed for 2 weeks on Cu-supplemented diet.
Values are mean ± SEM (number of experiments).

ATP-dependent calcium uptake was measured at pH 6.8 at 37°C with 100 uM external Ca^{2+}, according to Moore [11] by the $^{45}Ca^{2+}$ technique.

Table 1 shows cytosolic SOD content of liver from rats on copper-deficient diet. It can be seen that SOD increases with age, but copper-deficient animals show significantly lower SOD activity; in fact, it is already inhibited by almost 70% at 2 weeks and reaches the minimum at 4 weeks of treatment. Cu refeeding recovers SOD content, showing the reversibility of this phenomenon. We have previously shown that microsomal membranes after 8 weeks of Cu-deficiency[12] exhibit a more saturated pattern of phospho lipid fatty acids. In these experiments (Table 2) the double-bond index, which is an indication of fatty acid saturation, decreases starting from the sixth week of treatment. Cu-supplementation completely reverses this effect.

We believe that the SOD depletion causes an oxidative stress by increasing the possibility of endogenous lipid peroxidation after defense mechanism inhibition. The cell would react to this situation by adapting its membrane lipid composition and synthesizing more saturated fatty acids.

Table 2. Double Bond Index of Fatty Acid Microsomal Membranes from Cu-Deficient Rats

	Weeks of Treatment				
	2	4	6	8	8+2*
Control	126	110	134	183	186
Cu-deficient	129	113	127	94	169

*Animals fed for 2 weeks on Cu-supplemented diet.

Table 3. MDA Production in Microsomal Membranes from Cu-Deficient Rats

	Weeks of Treatment					
	0	2	4	6	8	8+2*
	nmoles/mg protein/30 min					
NADPH	37.4 ± 3.6 (7)	45.8 ± 7.8 (4)	40.7 ± 5.1 (4)	39.5 ± 9.4 (4)	29.1 ± 2.9 (3)	39.1 ± 2.1 (6)
Xanthine/ Xanthine oxidase	48.6 ± 4.0 (8)	58.9 ±4.2 (4)	53.7 ± 5.5 (4)	52.2 ± 8.3 (4)	27.3 ± 7.4 (3)	50.2 ± 4.2 (6)

*Animals fed for 2 weeks on Cu-supplemented diet.
Values are mean ± SEM (number of experiments).

This hypothesis is supported by the fact that membranes isolated from copper-deficient rats become more resistant to in vitro peroxidation than control membranes after 8 weeks of treatment (Table 3). This resistance is evident later than SOD depletion since until 6 weeks of treatment no change in MDA production can be detected. SOD restoration through Cu-supplementation reverses this effect. To investigate if the changes observed in the Cu-deficient membranes may influence other cellular functions, we have studied microsomal Ca^{2+} uptake. It has been shown that microsomal Ca^{2+} uptake involved in the modulation of cytosolic free Ca^{2+} concentration is sensitive to variation of membrane lipid composition.[13] As observed for SOD activity,

Table 4. Rate of Calcium Uptake of Microsomal Membranes from Cu-Deficient Rats

	Weeks of Treatment				
	2	4	6	8	8+2*
	nmoles $^{45}Ca^{2+}$/mg protein/2 min				
Control	6.42 (2)	7.70 (2)	11.76 (2)	13.07 ± 0.43 (3)	16.06 ± 1.50 (4)
Cu-deficient	5.81 ± 0.49 (4)	6.79 ± 1.20 (4)	6.63 ± 0.10 (3)	5.95 ± 0.27 (6)	16.93 ± 1.30 (4)

*Animals fed for 2 weeks on Cu-supplemented diet.
Values are mean ± SEM (number of experiments).

the rate of microsomal Ca^{2+} uptake increases with age, while during copper deficiency it diminishes progressively to less than 50% of the control. Also in this case Cu-supplementation restores completely Ca^{2+} uptake, indicating a close relationship with membrane fatty acid composition.

The data presented above suggest that SOD deficiency causes an endogenous oxidative stress that modifies membrane fatty acid composition. Such modification could lead to serious functional impairment, with consequent cell damage.

ACKNOWLEDGMENTS

The work was supported by a grant of the Italian National Research Council, Special Project "Oncology", contract number 86.00423.44. and funds from Associazione Italiana per la Ricerca sul Cancro.

REFERENCES

1. R. A. Floyd, Free radicals in arylamine carcinogenesis, in: " Free Radicals in Biology", Vol. 4, W. A. Pryor, ed., Academic Press, New York (1980).
2. R. P. Mason, Free-Radical intermediates in the Metabolism of Toxic Chemicals, in: "Free Radicals in Biology", Vol. 5 W. A. Pryor, ed., Academic Press, New York (1982).
3. D. G. Hafeman and W. G. Hoekstra, Lipid peroxidation in vivo during vitamin E and selenium deficiency in the rat as monitored by ethane evolution, J. Nutr. 107:666 (1977).
4. L. A. Herschberge and A. L. Tappel, Effect of vitamin E on pentane exhaled by rats treated with methyl ketone peroxide, Lipids 17:686 (1982).
5. B. Freeman, S. L. Young, and J. Crapo, Liposome-mediated augmentation of superoxide dismutase in endothelial cells prevents oxygen injury, J. Biol. Chem. 258:12534 (1983).
6. B. D. Goldstein, M. G. Rozen, J. C. Quintavalla, and M. A. Amoruso, Decrease in mouse lung and liver glutathione peroxidase activity and potentiation of the lethal effects of ozone and paraquat by the superoxide dismutase inhibitor diethyldithiocarbamate, Biochem. Pharmacol. 28:27 (1979).
7. D. I. Paynter, R. J. Moir, and E. J. Underwood, Changes in activity of the Cu-Zn superoxide dismutase enzyme in tissues of rat with changes in dietary copper, J. Nutr. 109:1570 (1979).
8. J. M. McCord and I. Fridovich, Superoxide dismutase, J. Biol. Chem. 244:6049 (1969).
9. F. E. Hunter, J. M. Gebicki, P. E. Hoffsten, J. Weinstein, and A. Scott, Swelling and lysis of rat liver mitochondria induced by ferrous ions, J. Biol. Chem. 238:828 (1963).
10. R. F. Borch, Separation of long chain fatty acids as phenacyl esters by high pressure liquid chromatography, Anal. Chem. 47:2437 (1975).
11. L. Moore, Inhibition of liver microsomes calcium pump by in vivo administration of CCl_4, $CHCl_3$ and 1,1-dichloroethylene, Biochem. Pharmacol. 29:2505 (1980).
12. G. M. Bartoli, B. Giannattasio, and A. Cittadini, A., The loss of superoxyde dismutase mimics the alteration observed in membranes of tumour cells, in: "Cell Membranes and Cancer", T. Galeotti, A. Cittadini, G. Neri, S. Papa, and L. A. Smeths, eds., Elsevier Science Publishers B.V., Amsterdam (1985).
13. L. Moore, T. Chen, H. R. J. Knapp, and E. J. Landon, Energy-dependent calcium sequestration activity in rat liver microsomes, J. Biol. Chem. 250:4562 (1975).

THE USE OF OXYGEN FREE RADICAL SCAVENGERS IN THE RAT PERITONITIS MODEL

Whitney M. Limm, Susan N. Ishikawa, Thomas Grover,*
J. Judson McNamara, and Lawrence H. Piette *

Cardiovascular Research Laboratory
Queen's Medical Center
Honolulu, Hawaii 96813, U.S.A.

*Department of Chemistry and Biochemistry
Utah State University
Logan, Utah 84322, U.S.A.

INTRODUCTION

In the 1960s, Davis and Yull noted synergistic effect on mortality between *E. coli* and red blood cells (RBCs) instilled into the peritoneum of rats.[1] Specifically, when *E. coli* at a level of 10^9 bacteria/cc and hemoglobin (Hb) in a concentration of 4 gm% were injected intraperitoneally, the 24-hour mortality rate was 90%. When *E. coli* or hemoglobin was injected separately, there was minimal mortality. This lethal synergistic action has been repeatedly demonstrated as investigators have noted mortality rates of 70 to 90% with this combination.[2-5] However, except for the implication of iron as the critical agent,[6] little progress has been made in elucidating the adjuvant effect of hemoglobin. Furthermore, no effective treatment has been found.

A hypothesis that the synergistic effect of hemoglobin may be mediated by oxygen free radicals was tested by treating infected rats with free radical scavengers-superoxide dismutase (SOD), catalase and mannitol. Sadrzadeh et al. have demonstrated that a biologic Fenton reaction can occur in the presence of hemoglobin.[7] In this reaction the iron in the hemoglobin molecule can convert the relatively inert superoxide radical ($\cdot O_2^-$) to the more reactive hydroxyl radical ($\cdot OH$). Conceivably, the lethal effect of hemoglobin may be due to its ability to drive the Fenton reaction.

METHODS

Sprague Dawley rats weighing 180 to 250 grams were used. The rats were housed under standard conditions and had access to food and water continuously. Intraperitoneal injections of unanesthetized rats were performed using 25 gauge needles.

E. coli (strain ATCC:25922) was incubated at 37°C in blood agar plate for 18 hours, and single colonies were then incubated in trypticase soy broth (TSB, American Products) for an additional 18 hours to achieve a suspension of 10^9 organisms/cc. Bacterial counts were performed manually using the hemocytometer and confirmed by the serial dilution plate method.

Baboon RBCs were lysed by freezing to -20°C. On the morning of the experiment the RBCs were thawed and added to the broth to achieve a hemoglobin concentration of 4 gm%.

Superoxide dismutase (Sigma) and catalase (Sigma) were added to the E. coli/Hb mixture at a concentration of 2,600 units/kg (body weight) and 11,000 units/kg, respectively. Mannitol (Merck 0.7 gm/kg) was given to the second treatment group.

The control group comprised 10 animals that were injected with the E. coli/Hb mixture. There were 10 rats in the SOD/catalase group and 12 rats in the mannitol group. The blood adjuvant and the test substances were added to the E. coli broth immediately prior to innoculation. An additional 6 rats received E. coli without hemoglobin. For all groups the volume of injection was 5 cc/kg. Mortality rates were calculated at 24 hours.

RESULTS

The mortality rate for the E. coli/Hb group was 80%. Mortality rates were significantly lower in the treatment groups. Addition of SOD/catalase to the E. coli/Hb mixture lowered the mortality rate to 30% ($p<0.025$). Addition of mannitol lowered the mortality to 42% ($p<0.05$). There was one death among the rats that received E. coli alone.

To determine if the free radical scavengers inhibited bacterial growth, SOD/catalase and mannitol were added to separate E. coli broths and incubated at 37°C for 24 hours. Bacterial counts were then performed by the serial dilution plate method. The counts were 1.2×10^9 cells/cc and 0.9×10^9 cells/cc, respectively. They did not differ markedly from the control value of 1.1×10^9 cells/cc.

DISCUSSION

The adjuvant effect of hemoglobin in E. coli peritonitis has obvious practical significance and has been known for many years. However, it is not clear how hemoglobin exerts this effect. Simmons, Diggs and Sleeman demonstrated that hemoglobin did not increase the growth rate nor the virulence of E. coli.[4] The suspicion that hemoglobin exerted a local effect was confirmed by the loss of synergism noted when hemoglobin was injected subcutaneously or intramuscularly. Furthermore, the effect of hemoglobin was minimized if injected intraperitoneally more than one hour before or after the E. coli injection.

A large number of investigations have focused on the hypothesis that host defenses are inhibited by hemoglobin. Filler and Sleeman demonstrated that hemoglobin decreased absorption of dye-tagged proteins from the peritoneal cavity.[3] Hemoglobin was thought to exert its effect by mechanically blocking absorption of bacteria. However, using nonviable ^{125}I-labeled bacteria, Dunn et al. found no influence on bacterial clearance by hemoglobin.[8]

Hemoglobin may exert its effect by influencing polymorphonuclear leukocyte (PMN) function. The data on this, however, is not entirely clear. In vitro studies by Hau et al. showed that hemoglobin impairs chemotaxis at low concentrations.[9] By using serial bacterial colony counts and WBC counts Hau et al. also noted an influx of PMNs into the peritoneal cavity after intraperitoneal injection of E. coli.[10] This resulted in a rapid decline of bacteria colony counts. However, when hemoglobin was included the PMN influx was blunted and bacterial growth was unchecked.

More recently, using sroma-free hemoglobin, Dunn et al. confirmed the lethal adjuvant effect of hemoglobin but found no inhibition of PMN influx during bacterial peritonitis.[11] Furthermore, in vitro human PMN functions were not altered by the addition of stroma-free hemoglobin. The authors concluded that the interference of PMNs noted by other investigators may be due to insoluble material including stromal elements in the hemoglobin preparations used. The functional importance of the iron component in hemoglobin led us to postulate the possible role of oxygen free radicals in this model. In the presence of $\cdot O_2^-$ and H_2O_2, iron can catalyze the formation of the hydroxyl radical:

$$2\cdot O_2^- + 2H^+ \rightarrow H_2O_2 + O_2$$

$$\cdot O_2^- + Fe^{3+} \rightarrow Fe^{2+} + O_2$$

$$H_2O_2 + Fe^{2+} \rightarrow \cdot OH + OH^- + Fe^{3+} \text{ (Fenton Reaction)}$$

The hydroxyl radical is an extremely reactive oxidizing species that can react with biomembranes and induce injuries by lipid peroxidation. In the presence of a superoxide-generating system, hemoglobin can promote hydroxyl formation in a dose-dependent fashion.[7]

Such a system exists when PMNs are stimulated and undergo the "respiratory burst."[12] During this period, there is a large increase in PMN oxygen consumption. The oxygen is then converted to $\cdot O_2^-$ and H_2O_2. It is not clear if $\cdot O_2^-$ has bactericidal properties, but H_2O_2 is capable of destroying bacteria by itself by oxidizing halide ions to hypohalites.[13,14]

Perhaps a "respiratory burst" follows the peritoneal leukocytosis that is seen after the E. coli injection, and in the presence of hemoglobin the Fenton reaction, driven by Fe^{2+}, takes place and hydroxyl radicals are formed. This may result in the local compromise of the host defense system, perhaps by autodegradation of PMNs and local cellular injury.

The use of SOD/catalase is based on the following reactions:

$$\cdot O_2^- + \cdot O_2^- + 2H^+ \xrightarrow{SOD} H_2O_2 + O_2$$

$$2H_2O_2 \xrightarrow{catalase} O_2 + H_2O$$

Essentially, the enzymes are called antioxyenzymes because they eliminate the superoxide radical and hydrogen peroxide by converting them into O_2^- and H_2O. Mannitol, on the other hand, is a hydroxyl radical scavenger.

CONCLUSION

SOD/catalase and mannitol were able to significantly modify the lethal combination of E. coli and hemoglobin in the rat peritonitis model. This may be due to their ability to scavenge oxygen free radicals. Bacterial growth in TSB was not altered by the addition of the scavengers.

REFERENCES

1. J.H. Davis and A.B. Yull, A possible toxic factor in abdominal injury, J. Trauma 2:291 (1962).
2. J.H. Davis and A.B. Yull, A toxic factor in abdominal injury. II. The role of the red cell component, J. Trauma 4:84 (1964).
3. R.M. Filler and H.K. Sleeman, Pathogenesis of peritonitis. I. The effect of Escherichia coli and hemoglobin on peritoneal absorption, Surgery 61:385 (1967).
4. R.L. Simmons, J.W. Diggs, and H.K. Sleeman, Pathogenesis of peritonitis. III. Local adjuvant action of hemoglobin in experimental E. coli peritonitis, Surgery 63:810 (1968).
5. H.K. Sleeman, J.W. Diggs, D.K. Hayes, and H.F. Hamit, Value of antibiotics, corticosteroids and peritoneal lavage in the treatment of experimental peritonitis, Surgery 66:1060 (1969).
6. J.T. Lee, D.H. Ahrenholz, R.D. Nelson, and R.L. Simmons, Mechanisms of the adjuvant effect of hemoglobin in experimental peritonitis. V. The significance of the coordinated iron component, Surgery 86:41 (1979).
7. S.M. Sadrzadeh, E. Graf, S.S. Panter, P.E. Hallaway, and J.W. Eaton, Hemoglobin, a biologic Fenton reagent, J. Biol. Chem 259:14354 (1984).
8. D.L. Dunn, R.A. Barke, J.T. Lee, R.M. Condie, E.W. Humphrey, and R.L. Simmons, Mechanisms of the adjuvant effect of hemoglobin in experimental peritonitis. VII. Hemoglobin does not inhibit clearance of Escherichia coli from the peritoneal cavity, Surgery 94:487 (1983).
9. T. Hau, R.D. Nelson, V.D. Fiegel, R. Levenson, and R.L. Simmons, Mechanisms of the adjuvant action of hemoglobin in experimental peritonitis. II. Influence of hemoglobin on human leukocyte chemotaxis in vitro, J. Surg. Res. 22:174 (1977).
10. T. Hau, R. Hoffman, and R.L. Simmons, Mechanisms of the adjuvant effect of hemoglobin in experimental peritonits. I. In vivo inhibition of peritoneal leukocytosis, Surgery 83:223 (1978).
11. D.L. Dunn, R.D. Nelson, R.M. Condie, and R.L. Simmons, Mechanisms of the adjuvant effect of hemoglobin in experimental peritonitis. VI. Effects of stroma-free hemoglobin and red blood cell stroma on mortality and neutrophil function, Surgery 93:653 (1983).
12. B.M. Babior, Oxygen dependent microbial killings by phagocytes. Part I, New Engl. J. Med. 298:659 (1978).
13. S.J. Klebanoff, Iodination of bacteria: A bactericidal mechanism, J. Exp. Med. 126:1063 (1967).
14. R.J. McRipley and A.J. Sbarra, Role of phagocyte in host-parasite interactions. XII. Hydrogen peroxide-myeloperoxidase bactericidal system in the phagocyte, J. Bacteriol. 94:1425 (1967).

ADAPTATIONS TO PLANT PRO-OXIDANTS IN A PHYTOPHAGOUS INSECT MODEL: ENZYMATIC PROTECTION FROM OXIDATIVE STRESS

Ronald S. Pardini, Chris A. Pritsos, Susan M. Bowen, Sami Ahmad, and Gary J. Blomquist

Department of Biochemistry
University of Nevada Reno
Reno, Nevada 89557-0014, U.S.A.

INTRODUCTION

Adaptations in antioxidant enzyme activities of superoxide dismutase (SOD), catalase (CAT), glutathione peroxidase (GPOX) and glutathione reductase (GR) against superoxide free radicals ($\cdot O_2^-$) and prevention of their cascade into cytotoxic H_2O_2 and hydroxyl radical ($\cdot OH$) have been demonstrated in prokaryotes and vertebrates,[1-3] but not yet in insects.[4]

Some plant pro-oxidants like the furanocoumarins, e.g., xanthotoxin, are light activated and generate both singlet oxygen (${}^1O_2$) and $\cdot O_2^-$,[5] while others like quinones or flavonoids, e.g., quercetin, are metabolically activated and produce $\cdot O_2^-$, H_2O_2 and $\cdot OH$[6,7] (Figure 1).

The primary emphasis on insect adaptation towards plant pro-oxidants has been focused on enzymes that metabolize the xenobiotics directly to metabolites that are innocuous and readily excretable. Much of the detoxification is attributed to initial attack by the phase-I enzyme cytochrome P-450 polysubstrate monooxygenase (PSMO), followed by conjugations to highly polar metabolites by phase-II enzymes.[8]

In our phytophagous insect model, the cabbage looper (Trichoplusia ni) naturally feeds on many plants but mostly on crucifers, which are devoid of xanthotoxin and contain only trace amounts of quercetin. The

Quercetin (3,5,7,3',4'-Pentahydroxyflavone)

Xanthotoxin (8-Methoxypsoralen)

Figure 1. Quercetin is metabolically activated to produce ($\cdot O_2^-$), H_2O_2, $\cdot OH$. Xanthotoxin is photochemically activated to produce ${}^1O_2$, $\cdot O_2^-$, $\cdot OH$.

southern armyworm (*Spodoptera eridania*) because of polyphagy often encounters and feeds on plant pro-oxidants of both types. On the other hand, the black swallowtail butterfly (*Papilio polyxenes*) feeds with impunity on many species of *Umbelliferae* and *Rutaceae* known to contain the photoactive pro-oxidant, xanthotoxin. These three insect species comprise our insect model of low, moderate and high dietary exposure to plant pro-oxidants, as shown: *T. ni* << *S. eridania* <<< *P. polyxenes*.

We now report on the activities of the antioxidant enzymes SOD, CAT, GPOX and GR in these three insects in relation to the acute toxicity of xanthotoxin and quercetin.

RESULTS AND DISCUSSION

The data in Table 1 show that *T. ni*, which is naturally exposed to low levels of pro-oxidants, possesses low endogeneous levels of SOD and GR. These findings are consistent with the observations that the fifth instar *T. ni* are highly sensitive to both xanthotoxin and quercetin with LC_{50} values of 0.0004 and 0.0045% (w/w in the diet), respectively.[9] The extremely high endogeneous activities of CAT observed during all stages of larval development remain unexplained but apparently are unrelated to their low intake of plant pro-oxidants.

The *S. eridania* possessed high endogeneous activity of SOD and moderate activity of CAT. GR activity is low to moderate depending on larval stage. These findings are consistent with the highly polyphagous nature of this insect which is reflected in its occasional exposure to plant pro-oxidants. Delayed toxicity of xanthotoxin towards the first

Table 1. Antioxidant Enzyme Activities of *Trichoplusia ni*, *Spodoptera eridania* and *Papilio polyxenes*

Enzyme	Larval Instar	Insect Species[a]					
		T. ni		*S. eridania*		*P. polyxenes*	
Superoxide dismutase[b]	Third	1.0	0.08	5.5	1.20	7.4	0.70
	Fourth	1.3	0.09	8.4	1.67	5.5	0.25
	Fifth	1.5	0.04	2.8	0.15	2.1	0.98
Catalase[c]	Third	215.8	10.25	27.1	1.57	144.2	3.97
	Fourth	217.0	7.50	20.3	2.47	171.8	1.44
	Fifth	238.7	13.83	77.2	13.60	321.8	8.14
Glutathione reductase[d]	Third	0.6	0.09	0.6	0.15	6.8	2.05
	Fourth	0.7	0.09	1.1	0.22	5.1	0.70
	Fifth	1.2	0.10	3.7	0.58	5.9	1.26

[a] Data are from Ahmad et al.[9] for *T. ni*, from Pritsos et al.[10] for *S. eridania*, and unpublished for *P. polyxenes*.
[b] Activities are means ± S.D.s and are expressed as units$\cdot mg^{-1}$ protein$\cdot min^{-1}$ as defined by McCord and Fridovich.
[c] Activities are means ± S.D.s and are expressed as µmole H_2O_2 decomposed mg^{-1} as defined by Aebi.[12]
[d] Activities are means ± S.D.s and are expressed as units$\cdot mg^{-1}$ protein$\cdot min^{-1}$ as defined by Racker.[13]

Table 2. Relative Toxicity of Pro-oxidant Allelochemicals Towards Insect Model

Xanthotoxin	T. ni	>	S. eridania	>>>	P. polyxenes
	(0.0004%)[a]		(0.1%)[b]		(> 1%)[c]
Quercetin	T. ni	>	S. eridania		N.D.
	(0.0045%)[a]		(> 1%)[b]		

[a]The LC_{50} values are shown in parentheses as percent of diet.
[b]No acute toxicity observed; insects did not mature past second instar.
[c]LC_{50} from Berenbaum.[15]

instar S. eridania[14], as well as the fifth instar,[10] has been reported, whereas the larvae exhibited no 24-h acute toxicity to quercetin up to a dietary concentration of 1% (w/w) (Table 2). This differential toxicity is at present unexplained. It is tempting to speculate that xanthotoxin is toxic because it generates $^1O_2$ and $\cdot O_2^-$ whereas quercetin has only been shown to generate $\cdot O_2^-$ and H_2O_2. S. eridania appears to have little defense against $^1O_2$ but its complement of SOD and CAT may protect it from $\cdot O_2^-$ generating xenobiotics like quercetin.

The preliminary data shown in Table 1 indicate that P. polyxenes is well endowed with SOD, CAT and GR activities at all stages of larval development. Berenbaum and Feeney[15] reported that xanthotoxin did not adversely affect the growth rate or survival of P. polyxenes larvae but acted as a feeding stimulant. The insensitivity towards xanthotoxin has been attributed to a highly active xanthotoxin-specific PSMO system present in the butterfly larvae.[8] The findings reported here suggest that a second line of defense against pro-oxidant allelochemicals is their compliment of antioxidant enzymes.

A noteworthy observation throughout the investigation was the negligible activity observed for GPOX in all three insect species tested. These observations are consistent with those of Allen et al.[16-18] who demonstrated the absence of GPOX and presence of GR in the housefly, Musca domestica. These findings raise questions regarding the role of GPOX as an antioxidant enzyme in lepidopteran species. The inducibility of GPOX in the three insects is currently under investigation.

ACKNOWLEDGMENTS

This work was supported by USDA Science and Education competitive grant 86-CRCR-1-2038.

REFERENCES

1. J. M. McCord, B. B. Keele, and I. Fridovich, An enzyme-based theory of obligate anaerobiosis, Proc. Natl. Acad. Sci. USA 68:1024 (1971).
2. L. Frank, Prolonged survival after paraquat: Role of the lung antioxidant enzyme systems, Biochem. Pharmacol. 30:2319 (1981).

3. C. A. Pritsos, L. M. Aaronson, and R. S. Pardini, Metabolic consequences of dietary 2,3-dichloro-1,4-napthoquinone (CNQ) in the rat: Alteration in anti-oxidant enzyme activities, Biochem. Pharmacol. 35:1131 (1986).
4. R. A. Larson, Insect defenses against photoxic plant chemicals, J. Chem. Ecol. 12:859 (1986).
5. P. C. Joshi and M. A. Pathak, Production of singlet oxygen and superoxide radicals by psoralens and their biological significance, Biochem. Biophys. Res. Commun. 112:638 (1983).
6. H. M. Hassan and I. Fridovich, Intracellular production of superoxide radical and hydrogen peroxide by redox active compounds, Arch. Biochem. Biophys. 196:385 (1979).
7. W. F. Hodnick, F. S. Kung, W. J. Roettger, C. W. Bohmont, and R. S. Pardini, Inhibition of mitochondrial respiration and production of toxic oxygen radicals by flavonoids: A structure-activity study, Biochem. Pharmacol. 35:2345 (1986).
8. S. Ahmad, L. B. Brattsten, C. A. Mullin, and S. J. Yu, Enzymes involved in the metabolism of plant allelochemicals, in: "Molecular Aspects of Insect-Plant Associations," L. B. Brattsten and S. Ahmad, eds., Plenum Press, New York (1986).
9. S. Ahmad, C. A. Pritsos, S. M. Bowen, K. E. Kirkland, G. F. Blomquist, and R. S. Pardini, Activities of enzymes that detoxify superoxide anion and related toxic oxyradicals in Trichoplusia ni, Arch. Insect Biochem. Physiol. 6:85 (1987).
10. C. A. Pritsos, S. Ahmad, S. Bowen, G. J. Blomquist, and R. S. Pardini, Antioxidant enzyme activities in the southern armyworm, Spodoptera eridania, Comp. Biochem. Physiol. C (in press).
11. J. M. McCord and I. Fridovich, Superoxide dismutase, J. Biol. Chem. 244:6049 (1969).
12. H. Aebi, Catalase in vitro, Methods Enzymol. 105:121 (1984).
13. E. Racker, Glutathione reductase (liver and yeast), Methods Enzymol. 2:722 (1955).
14. M. Berenbaum, Toxicity of a furanocoumarin to armyworms: A case of biosynthetic escape from insect herbivores, Science 201:532 (1978).
15. M. Berenbaum and P. Feeny, Toxicity of angular furanocourmarins to swallowtail butterflies: Escalation in a coevolutionary arms race? Science 212:927 (1981).
16. R. G. Allen, K. J. Farmer, and R. S. Sohal, Effect of catalase inactivation on levels of inorganic peroxides, superoxide dismutase, oxygen consumption and life span in adult houseflies (Musca domestica), Biochem. J. 216:503 (1983).
17. R. G. Allen, K. J. Farmer, and R. S. Sohal, Effect of diamide administration on longevity, oxygen consumption, superoxide dismutase, catalase, inorganic peroxides and glutathione in the adult housefly, Musca domestica, Comp. Biochem. Physiol. 78C:31 (1984).
18. R. G. Allen, K. J. Farmer, R. K. Newton, and R. S. Sohal, Effects of paraquat administration on longevity, oxygen consumption, lipid peroxidation, superoxide dismutase, catalase, glutathione reductase, inorganic peroxides and glutathione in the adult housefly, Comp. Biochem. Physiol. 78C:283 (1984).

MECHANISMS FOR THE GENERATION OF OXYGEN RADICALS BY DRUGS

Daniel M. Ziegler

Clayton Foundation Biochemical Institute and Department of Chemistry, The University of Texas at Austin
Austin, Texas 78712, USA

INTRODUCTION

In this report I will give a brief overview of drug-dependent metabolic reactions that can generate reactive oxygen species in tissues. The term drug is used here in its widest sense and includes any physiologically useless compound, either synthetic or naturally occurring, that can transfer one or two electrons to dioxygen via enzyme-catalyzed reactions. At the molecular level it is an enzyme that recognizes a specific structure and sustains the reaction. Therefore, a discussion of biochemical mechanisms is primarily concerned with the nature of the reaction and the identity of enzymes that catalyze reactions. The following review follows this general plan and focuses on enzymes that catalyze oxidation or reduction of functional groups in xenobiotics that often lead to the generation of oxygen radicals.

In addition, some of the problems in evaluating drug-induced oxidative stress in tissues or whole animals by measuring the oxidation of GSH to GSSG will be addressed. While a substantial number of xenobiotics can stimulate oxidation of GSH, this is not, as frequently assumed, always due to the generation of hydrogen peroxide or superoxide anions.

OXIDASES

Oxidases catalyze the transfer of electrons from substrate to oxygen, generating $O_2^{\overline{\cdot}}$, H_2O_2, or H_2O. The ones of interest here are those that catalyze one- or two-electron oxidation of xenobiotics by oxygen, yielding either $O_2^{\overline{\cdot}}$ or H_2O_2. While a number of oxidases could be listed in this category, the monoamine and aldehyde oxidases are the only oxidases implicated in the metabolism of a substantial number of drugs. The specificities of others (D-amino acid oxidase, diamine oxidase, etc.) are quite restricted and their contribution to oxidative stress is limited to relatively few xenobiotics.

Monoamine Oxidase

Monoamine oxidase is an integral membrane flavoprotein tightly bound to the outer mitochondrial membrane. Two different forms, designated A and B, have been identified in liver and other tissues. The two forms, although similar in composition and physical properties, are immunologically

Table 1. Typical Substrates for Monoamine Oxidases[a]

Amines
Benzylamine
Dopamine
N-Methylbenzylamine
N,N-Dimethylbenzylamine
MPTP
Hydrazines
Procarbazine
1,2-Dimethylhydrazine

[a]Examples taken from references 3 and 4.

distinct proteins.[1,2] However, the two forms are similar in catalytic properties, and only some features common to both forms will be considered here.

The mitochondrial monoamine oxidases catalyze the oxidation of a variety of amines and hydrazines, as shown by examples listed in Table 1. In general, alkylamines containing an unsubstituted methylene adjacent to nitrogen show substrate activity, and activity has been demonstrated with primary, secondary and tertiary amines[3] and some hydrazines.[4] While alicyclic primary alkylamines are the preferred substrates, monoamine oxidases also catalyze the oxidation of some cyclic tertiary amines,[5] and the substrate specificity of these enzymes may be much broader than generally assumed.

Studies on mechanism[6] suggest that substrate binds to the oxidized flavoprotein and is oxidized by sequential transfer of two electrons from substrate to flavin. Reoxidation of the reduced flavin generates hydrogen peroxide and completes the catalytic cycle, as illustrated in Figure 1. Oxygen is the only physiological electron acceptor and hydrogen peroxide is formed stoichiometrically with substrate oxidation products. As a result the oxidation of xenobiotic substrates by monoamine oxidases can generate H_2O_2 within cells and contribute to drug-induced oxidative stress.

Aldehyde Oxidase

Aldehyde oxidase present in the cytosol fraction of liver and in other tissues is a complex metalloflavoprotein that catalyzes hydroxylation of a variety of xenobiotics as illustrated in equation 1.

$$AH + H_2O + 2\ O_2 \longrightarrow AOH + 2\ O_2 + 2H^+ \quad (1)$$

The oxygen incorporated into the hydroxylated product (AOH) is derived from H_2O rather than O_2, which distinguishes the enzyme from the oxygen-

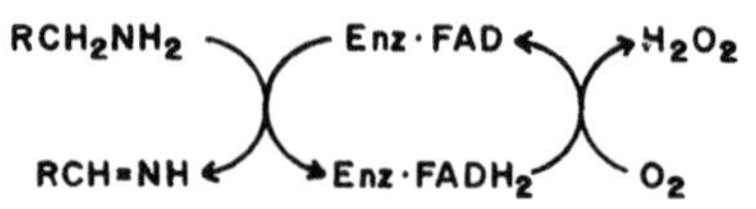

Figure 1. Catalytic cycle of monoamine oxidase.

ases. The enzyme appears to be a true oxidase since physiological electron acceptors other than oxygen have not been described.[7]

As the name implies the aldehyde oxidase catalyzes the oxidation of aldehydes to carboxylic acids, but an exceptional variety of substituted pyridines, pyrimidines, purines and pteridines also show substrate activity.[7] However, substrate specificity varies with species,[8,9] and a good substrate for aldehyde oxidase in one species may show little or no activity with the enzyme from another source. For instance, methotrexate is a good substrate for the rabbit enzyme[9] but not for the human aldehyde oxidase.[7]

REDUCTASES

The one-electron reduction of xenobiotics to products that are rapidly reoxidized by oxygen is the most widely used procedure for generating $O_2^{\overline{\cdot}}$ in the intact tissue. The reductive step is catalyzed by reductases, or in a few cases, the xenobiotic is reduced non-enzymically by reductants so abundant in the strong reducing environment within cells. While a variety of different functional groups present in drugs and other xenobiotics are susceptible to metabolic reduction, only a limited number are capable of sustained reduction and reoxidation by oxygen. In a few instances the contribution of a specific enzyme, or enzymes, to the reduction of a functional group is known, but for others the reactions are known only as activities in crude preparations. For a more complete description of reductive reactions in the biotransformation of xenobiotics the reader is referred to the recent excellent review by McLane et al.[10]

Quinone Reductases

Quinones, both synthetic and naturally occurring, are susceptible to metabolic one- or two-electron reduction. One-electron reduction produces the semiquinone radical, which is rapidly reoxidized by dioxygen forming the superoxide anion radical and the parent quinone, as illustrated in Table 2. Because of its availability and simplicity, menadione has been a favorite agent for studying the effects of drug-induced oxidative stress on the structure and function of hepatocytes.[17,18] However, metabolic redox cycling of the other quinones in Table 2 has also been demonstrated, and the cardiotoxicity of adriamycin has been attributed to the generation of oxygen radicals as a result of metabolic redox cycling.[13,14]

The velocity of quinone-dependent oxygen radical formation depends largely on the velocity of the reductive step. While not all reductants have been identified, a number of flavoproteins are usually implicated of which the NADPH-cytochrome c reductase is the major catalyst in liver.[19] In other tissues, reduction catalyzed by pyridine nucleotide-dependent ferrodoxin reductase, lipoamide dehydrogenase, cytochrome b_5 reductase, and ubiquinone reductase may be significant.[10] While xanthine oxidase is also capable of reducing quinones,[7] the contribution of this flavoprotein to the reduction of quinones *in vivo* is uncertain. In addition to these flavoproteins, mammalian tissues also contain a flavoprotein, the NAD(P)H: acceptor oxidoreductase (DT-diaphorase), that catalyzes the direct two-electron reduction of quinones to hydroquinones without formation of intermediate semiquinones. The action of this enzyme effectively stops redox cycling maintained by one-electron reduction. Manipulation of this enzyme by selective induction or inhibition can alter the toxicity of menadione in isolated hepatocytes.[12]

While the enzyme-catalyzed reductions are usually the more efficient, alloxan is also readily reduced by thiols,[20] and rapid reduction by

Table 2. Metabolic Redox Cycling of Quinones

REACTIONS

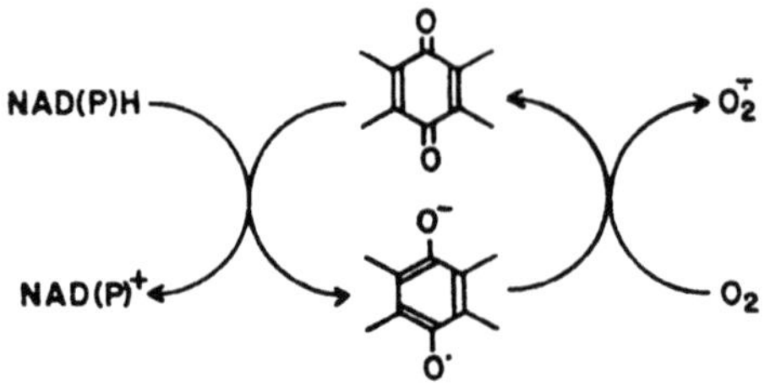

EXAMPLES	REFERENCE
Menadione	11,12
Alloxan	11
Plumbagin	11
Adriamycin	13,14
Bleomycin	15
Daunomycin	10
Mitomycin	16

thioredoxin has been reported.[21] Reoxidation of the reduced product (5-hydroxybarbituric acid) by oxygen produces superoxide anion radicals which may be responsible for destruction of pancreatic β-cells in susceptible animals.[21]

Viologen Reductases

The salts of quaternary derivatives of 4,4-dipyridyl (viologens) and

Table 3. Metabolic Redox Cycling of Dipyridyls

REACTIONS

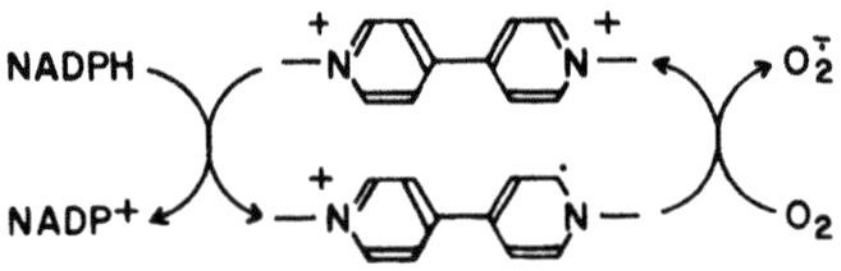

EXAMPLES	REFERENCES
4,4-Dipyridyl	11
Methylviologen (paraquat) Ethylviologen	11,22,23
Diquat	26,27

of the closely related 2,2-dipyridyl are known to generate $O_2^{\bar{\cdot}}$ in aerobic organisms.[11] Of the various derivatives available, the herbicides paraquat (N,N-dimethyl-4,4-bipyridinum dichloride) and diquat (1,1-ethylene-2,2- dipyridylium dichloride) have been used extensively to study the effects of oxidative stress in animals. Both compounds produce liver necrosis and lipid peroxidation in rats[22] and Smith et al.[23] suggest that, based on biliary efflux of GSSG, diquat is one of the more potent agents tested. Changes in the thiol status of lung tissue due to redox cycling has also been reported in rats treated with paraquat.[24]

Both paraquat and diquat are reduced to radicals by homogenates of liver, kidney, or lung,[25] and the bipyridyl radicals are reoxidized by dioxygen as illustrated in Table 3. The reductive step is apparently catalyzed most rapidly by a NADPH-dependent cytosolic enzyme,[26] although other pyridine nucleotide dehydrogenases undoubtedly contribute to the one-electron reduction of these compounds.

Nitro- and Azo-Reductases

Reduction of the nitro group to the amine is a significant route for the metabolism of a variety of xenobiotics.[10] The six electrons required for reduction to the amine go through several intermediates, but the first step, reduction of the nitro to the nitro anion radical (Table 4), is primarily responsible for generation of superoxide. Aerobically, reoxidation of the nitro anion radical is favored over disproportionation, which leads to futile redox cycling of nitroaromatic compounds, with the generation of superoxide.

The reduction of aromatic nitro compounds is catalyzed by a number of flavoproteins including xanthine oxidase, lipamide dehydrogenase, DT-diaphorase and NADPH-cytochrome _c_ reductase.[10,28] The latter enzyme along with cytochrome P-450 is largely responsible for the nitro reductase activity of liver microsomes,[27] and may be the principal catalyst for one-electron reduction of aromatic nitro compounds in this tissue.

Nitroso derivatives formed upon disproportionation or reduction of the nitro anion radical are readily reduced to hydroxylamines[29] as illustrated in equation 2.

$$ArN{=}O + 2GSH \longrightarrow ArNHOH + GSSG \qquad (2)$$

Table 4. Metabolic Redox Cycling of Aromatic Nitro Compounds

REACTIONS

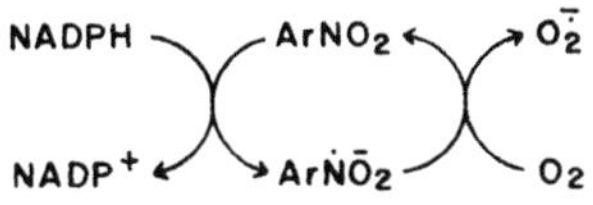

EXAMPLES[a]

Nitrobenzene	Nitrofurantoin
Chloroamphenicol	4-Nitropyridine N-oxide
Niridazole	Nitrazepam
Metronidazole	

[a]Examples taken from reviews by McLane et al.[10] and Hewick.[28]

Table 5. Metabolic Redox Cycling of Azo Compounds

REACTIONS

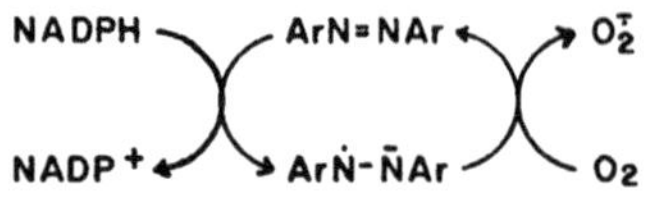

EXAMPLES[a]

Amarath	Methyl red
Azobenzene	Neoprontosil
Butter yellow	Sulfasalazine

[a]Examples taken from the review by Hewick.[28]

The arylhydroxylamines formed are capable of reducing dioxygen to $O_2^{\cdot-}$ and/or H_2O_2 in the presence of metals, but the contribution of this reaction to the generation of oxygen radicals in animals treated with aromatic nitro compounds is largely unknown.

Azo compounds are also capable of metabolic redox cycling as illustrated in Table 5. The azo nitrogen, reduced to the azo anion radical is reoxidized to the parent compound by dioxygen.[30] In isolated microsomes the reaction is catalyzed by NADPH-cytochrome c reductase,[34] and this enzyme along with P-450 may be the major catalyst for maintaining futile redox cycling of azo compounds in mammalian tissues.

MONOOXYGENASES

Monooxygenases also catalyze the oxidation of drugs by dioxygen, but they differ from the oxidases in that they usually require a reductant in addition to substrate and one atom of oxygen is incorporated into substrate while the other is reduced to water (equation 3).

$$NADPH + H^+ + O_2 + S \longrightarrow NADP^+ + SO + H_2O \qquad (3)$$

Although the monooxygenase-catalyzed reaction does not produce oxygen radicals, a few xenobiotics are oxidized to products, usually quinones, that can serve as electron acceptors for one or more of the reductases described in the preceeding section. For example, the oxidation of benzo[a]pyrene catalyzed by P-450 yields quinones in addition to phenols and diols.[31] The phenols are probably intermediates in the formation of hydroquinones as shown schematically in equation 4.

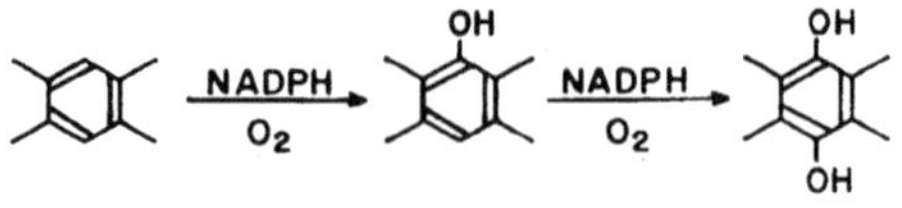

(4)

In addition, O-Dealkylation of butylated hydroxyanisole (BHA) yields t-butylhydroquinone (equation 5) which is the metabolite responsible for futile redox cycling[32] in microsomes incubated with BHA.

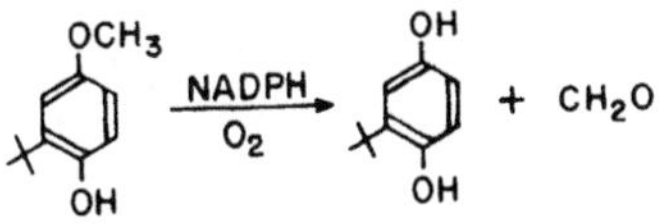

(5)

Thus, it is evident that cytochrome P-450 catalyzed hydroxylations of some aromatic hydrocarbons and ethers yield products susceptible to auto-oxidation and futile cycling *in vivo*, but drug-dependent uncoupling of cytochrome P-450 observed *in vitro*[33] apparently does not occur in the intact tissue. The efflux of glutathione from liver[34] and hepatocytes[35] treated with aminopyrine and a few other drugs is apparently due to formation of a labile conjugate (S-hydroxymethylglutathione) and not to drug-induced hydrogen peroxide formation.[36]

DRUG-DEPENDENT GSH OXIDATION

Since liver is the primary site for drug metabolism, and excess GSSG is actively transported into bile, the biliary efflux of GSSG is frequently used to measure intrahepatic oxidative stress. While there is no question that drug-induced generation of $O_2^{\bar{\cdot}}$ and/or H_2O_2 will stimulate biliary GSSG efflux, drug-dependent efflux is not always due to oxidation of GSH by peroxide. For instance, a variety of compounds bearing a thiol function are rapidly oxidized by the microsomal flavin-containing monooxygenase to sulfenates. The sulfenates react non-enzymically with GSH, regenerating the parent drug and GSSG, as shown in Table 6. The rapid biliary efflux of GSSG by this mechanism in live animals treated with methimazole and in liver perfused with thiocarbamides has been described.[37] Other metabolic peroxide-independent pathways for oxidation of GSH to GSSG by drugs have not been as fully documented, but some may exist. For example, the reduction of nitrosoarenes by GSH generates GSSG.[29] Since nitroarenes are intermediates in the metabolic reduction of aromatic nitro compounds, GSSG efflux may be a poor index of oxygen radicals generated *in vivo* by aromatic nitro compounds. While GSH oxidation is a useful index for estimating oxidative stress, the results must be interpreted with considerable

Table 6. Metabolic cycling of Xenobiotic Thiols

REACTIONS

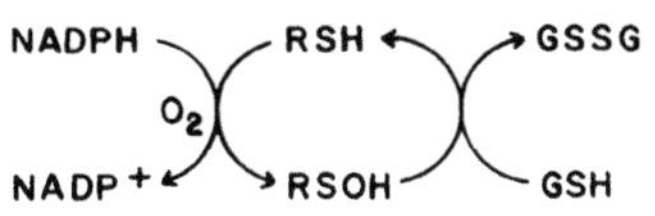

EXAMPLES[a]

Methimazole	Propylthiouracil
Thiourea	2-Mercaptoethanol
Dithiocarb	

[a]Examples taken from references 37 and 38.

caution. Devising methods for accurate estimation of drug-induced oxidative stress *in vivo* is one of the more challenging tasks facing molecular toxicologists.

REFERENCES

1. R.M. Denny, R.R. Fritz, N.T. Patel, and C.W. Abell, Human liver MAO-A and MAO-B separated by immunoaffinity, Science 215:1400-1403 (1982).
2. L.M. Kochersperger, A. Waguespack, J.C. Patterson, L.C. Hsieh, W. Weyler, J.I. Salach, and R.M. Denny, Immunological uniqueness of human monoamine oxidases A and B: New evidence from studies with monoclonal antibodies to human monoamine oxidase A, J. Neurosci. 5:2874-2881 (1985).
3. K.F. Tipton, Monoamine Oxidase, in: "Handbook of Physiology," H.Blaschko and A.D. Smith, eds., American Physiological Society, Washington D.C. (1975).
4. M.W. Coomes and R.A. Prough, The mitochondrial metabolism of 1,2-disubstituted hydrazines, procarbazine and 1,2-dimethylhydrazine, Drug Metab. Dispos. 11:550-555 (1983).
5. J.I. Salach, T.P. Singer, N. Castagnoli, and A.J. Trevor, Oxidation of the neurotoxic amine 1-methyl-4-phenyl-1,2,3,6-tetrahydropyridine (MPTP) by monoamine oxidases A and B and suicide inactivation of the enzymes by MPTP, Biochem. Biophys. Res. Comm. 125:831-835 (1984).
6. R.B. Silverman, C.K. Hiebert, and M.L. Vagquez, Inactivation of monoamine oxidase by allylamine does not result in flavin attachment, J. Biol. Chem. 260:14648-14652 (1985).
7. K.V. Rajagopalan, Xanthine oxidase and aldehyde oxidase, in: "Enymatic Basis of Detoxication" Vol. 1, W.B. Jakoby ed., Academic Press, New York, New York (1980).
8. T.A. Krenitsky, M.N. Shannon, G.B. Elion, and G.H. Hitchings, A comparison of the specificities of xanthine oxidase and aldehyde oxidase, Arch. Biochem. Biophys. 150:585-599 (1972).
9. D.G. Johns, A.T. Jannotti, A.C. Sartorelli, B.A. Booth, and J.R. Bertino, The identity of rabbit liver methotrexate oxidase, Biochem. Biophys. Acta 105:380-382 (1965).
10. K.F. McLane, J. Fisher, and Ramakrishnan, Reductive drug metabolism, Drug Meta. Revs. 14:741-799 (1983).
11. H.M. Hassan and L. Fridovich, Intracellular production of superoxide radical and of hydrogen peroxide by redox active compounds, Arch. Biochem. Biophys. 196:385-395 (1979).
12. H. Thor, M.T. Smith, P. Hartzell, G. Bellomo, S.A. Jewell, and S. Orrenius, The metabolism of menadione (2-Methyl-1,4-Napthoquinone) by isolated hepatocytes: A study of the implications of oxidative stress in intact cells, J. Biol. Chem. 257:12419-12425 (1982).
13. W. Thayer, Adriamycin stimulates superoxide formation in submitochondrial particles, Chem. Biol. Interact. 19:265-278 (1977).
14. J.H. Doroshow and J. Reeves, Daunorubicin-stimulated reactive oxygen metabolism in cardiac sacrosomes, Biochem. Pharmacol. 30:259-262 (1981).
15. N. Yamoaka, T. Kato, K. Nishida, S. Shimizu, M. Fukushima, and K. Ota, Increase of antitumor effect of bleomycin by reduced nicotinamide adenine dinucleotide phosphate and microsomes in vitro and in vivo, Cancer Res. 40:2051-2053 (1980).
16. M. Tomasz and R. Lipman, Reductive metabolism and alkylating activity of mitomycin c induced by rat liver microsomes, Biochemistry. 20:5056-5061 (1981).
17. D. DiMonte, G. Bellomo, H. Thor, P. Nicotera, and S. Orrenius, Menadione-induced cytotoxicity is associated with protein thiol

oxidation and alteration in intracellular Ca^{++} homeostasis, Arch. Biochem. Biophys. 235:343-350 (1984).
18. S. Orrenius, H. Thor, and D. DiMonte, Mechanisms and implications of oxidative stress studied with intact cells, in: "Oxidative Damage and Related Enzymes," G. Rolillo and J.V. Bannister eds., Harwood Academic Publishers, London and New York (1987).
19. B.S.S. Masters, Bilimoria, H. Kamin, and Q.H. Gibson, The mechanism of 1- and 2-electron transfers catalyzed by reduced triphosphopyridine nucleotide-cytochrome c reductase, J. Biol. Chem. 240:4081-4088 (1965).
20. J.W. Patterson, A. Lazarow, and S. Levey, Reactions of alloxan and dialuric acid with the sulfhydryl group, J. Biol. Chem. 177:197-204 (1949).
21. A. Holmgren and C. Lyckeborg, Enzymatic reduction of alloxan by thioredoxin and NADPH-thioredoxin reductase, Proc. Natl. Acad. Sci. USA 77:5149-5152 (1980)
22. R.J. Burk, R.A. Lawrence, and J.M. Lave, Liver necrosis and lipid preroxidation in the rat as a result of paraquat and diquat administration, J. Clin. Invest. 65:1024-1031 (1980).
23. C.V. Smith, H. Hughes, B.H. Lauterburg, and J.R. Mitchell, Oxidant stress and hepatic necrosis in rats treated with diquat, J. Pharmacol. Exp. Ther. 235:172-177 (1985).
24. R.L. Keeling, L.L. Smith, and W.N. Aldridge, The formation of mixed disulfides in rat lung following paraquat administration, Biochem. Biophys. Acta 716:249-257 (1982).
25. J.C. Gage, The action of paraquat and diquat on the respiration of liver cell fractions, Biochem. J. 109:757-767 (1968).
26. R.C. Baldwin, A. Pasi, J.T. MacGregor, and C.H. Hine, Rates of radical formation from the dipyridylium herbicides paraquat, diquat, and morfamquat in homogenates of rat lung, kidney, and liver. inhibitory effect of carbon monoxide, Toxicol. Appl. Pharmac. 32:298-304 (1975).
27. R.P. Mason and J.L. Holtzman, The role of catalytic superoxide formation in the O_2 inhibition of nitroreductase, Biochem. Biophys. Res. Comm. 67:1267-1274 (1975).
28. D.S. Hewick, Reductive metabolism of nitrogen-containing functional groups, in: "Metabolic Basis of Detoxication," W.B. Jakoby, J.R. Bend and J. Caldwell eds., Academic Press, New York, New York (1982).
29. P. Eyer, Reactions of nitroarenes with sulfhydryl groups: Reaction mechanism and biological significance, in: "Biological Oxidation of Nitrogen in Organic Molecules," J.W. Gorrod and L.A. Damani eds., Ellis Horwood, Chichester (1985).
30. R.P. Mason, F.J. Peterson, and J.L. Holtzman, Inhibition of azoreductase by oxygen, Mol. Pharmacol. 14:665-671 (1978).
31. G. Holder, H. Yagi, P. Dansette, D.M. Jerina, W.Levin, A.Y.H. Lu, and A.H. Conney, Effects of inducers and epoxide hydrase on the metabolism of benzo[a]pyrene by liver microsomes and a reconstituted system: Analysis by high pressure liquid chromatography, Proc. Nat. Acad. Sci. USA 71:4356-4360 (1974).
32. S.W. Cummings and R.A. Prough, Butylated hydroxyanisole-stimulated NADPH oxidase activity in rat liver microsomal fractions, J. Biol. Chem. 258:12315-12319 (1983).
33. A.G. Hildebrandt and Roots, Reduced nicotinamide adenine dinucleotide phosphate (NADPH)-dependent formation and breakdown of hydrogen peroxide during mixed function oxidation reactions in liver microsomes, Arch. Biochem. Biophys 171:385-397 (1975).
34. H. Sies, A. Wahllander, and C. Waydas, Properties of glutathione disulfide (GSSG) and glutathione-S-conjugate release from perfused rat liver, in: "Functions of Glutathione in Liver and Kidney" H. Sies and A. Wendel eds., Springer, Berlin (1978).

35. D.P. Jones, L. Eklow, H. Thor, and S. Orrenius, Metabolism of hydrogen peroxide in isolated hepatocytes: Relative contribution of catalase and glutathione peroxidase in decomposition of endogenous generated H_2O_2, Arch. Biochem. Biophys. 210:505-516 (1981).
36. P.A. Krieter, D.M. Ziegler, K.E. Hill, and R.F. Burk, Studies on the biliary efflux of GSH from rat liver due to metabolism of aminopyrine, Biochem. Pharmacol. 34:955-960 (1985).
37. P.A. Krieter, D.M. Ziegler, K.E. Hill, and R.F. Burk, Increased biliary GSSG efflux from rat livers perfused with thiocarbamide substrates for the flavin-containing monooxygenage, Mol. Pharmacol. 26:122-127 (1984).
38. D.M. Ziegler, Microsomal flavin-containing monooxygenase: Oxygenation of nucleophilic nitrogen and sulfur compounds, in: "Enzymatic Basis of Detoxication," W.B. Jakoby ed., Academic Press, New York (1980).

ENHANCEMENT OF CHEMICAL ACTIVATION VIA RADICAL-DEPENDENT MECHANISMS: AN EMERGING CONCEPT IN CHEMICAL-CHEMICAL INTERACTIONS

Michael A. Trush and David C. Thompson

Department of Environmental Health Sciences
Division of Toxicological Sciences
Johns Hopkins University
Baltimore, Maryland 21205, U.S.A.

INTRODUCTION

The biological effects of exogenous chemicals including drugs, are initiated through physiochemical interactions between the chemical and specific tissue macromolecules, such as enzymes, structural and receptor proteins, and nucleic acids. In most cases, such interactions provide the molecular basis for both the pharmacologic and toxic effects of xenobiotics. One of the possible results of the interaction of a chemical with cellular enzymes is its metabolism to a bioreactive intermediate.[1] The reaction of either this initial reactive metabolite or a secondary reactive product with critical target molecules can then bring about changes in cellular function. Reactions which result in the enhanced activation of a chemical can then result in an increased pharmacologic or toxicologic effect. Chemical-mediated induction of enzymes involved in the metabolism of xenobiotics is a well-established mechanism whereby a chemical can indirectly enhance the activation and hence the potential toxic actions of another chemical.[2]

Recently, another mechanism has emerged which also results in the increased formation of biological reactive intermediates but in this case is mediated directly through the actions of either reactive oxygen species or radical intermediates. For example, superoxide from xanthine oxidase has been shown to convert 2-hydroxyestrogens and alpha-methyldopa to intermediates capable of binding covalently to protein.[3,4] Likewise, McManus and Davies[5] demonstrated that paraquat, by virtue of its enzyme-catalyzed redox cycling, enhanced the superoxide-mediated activation of DOPA to a species which covalently binds to protein. Reed et al.[6] have since demonstrated that the direct interaction of a peroxy radical of phenyl-butazone with benzo(a)pyrene-7,8-dihydrodiol, the proximate carcinogenic metabolite of benzo(a)pyrene, results in the increased epoxidation of benzo(a)pyrene-7,8-dihydrodiol which was accompanied by increased covalent binding and formation of tetraols. In this manuscript, we will illustrate several additional examples of chemical activation via radical-dependent mechanisms and discuss their pharmacologic and toxicologic implications.

METHODS

Bleomycin-mediated DNA Deoxyribose Cleavage

Cleavage of DNA deoxyribose yields a product which reacts with thiobarbituric acid, yielding a chromophore which can be measured at 533 nm.[7] Accordingly, microsomes (final concentration, 0.5 mg microsomal protein per ml) prepared from mouse lung, Lewis lung carcinoma and B-16 melanoma were incubated in the presence or absence of bleomycin A_2 (100 μM), calf thymus DNA (0.25 mg/ml), mitomycin C (500 μM), and an NADPH-generating system and analyzed for thiobarbituric acid reactive products (TBAR) as previously described.[8]

BHT-induced Mouse Lung Toxicity

Male CD-1 mice (4-5 weeks old) were given subcutaneous injections of BHA (250 mg/kg) 30 min prior to intraperitoneal injection of either BHT (175 mg/kg) or deuterated BHT (175 mg/kg). Lung (wet) and body weights were recorded on day 4 after injection. Toxicity was measured as an increase in lung/body weight ratio (%).

RESULTS AND DISCUSSION

Bleomycin A_2 (Blm), a glycopeptide antibiotic currently used in chemotherapy, is activated in the presence of iron and reactive oxygen species (ROS) to an intermediate which can cleave DNA deoxyribose,[9] yielding toxic base propenals[10] by the following reaction:

$$\text{Blm-Fe}^{3+} \xrightarrow{\text{ROS}} \text{Blm-Fe}^{2+}\text{-O}_2 \longrightarrow \begin{matrix}\text{Activated}\\ \text{Bleomycin}\\ \text{Intermediate}\end{matrix} \longrightarrow \begin{matrix}\text{DNA}\\ \text{Deoxyribose}\\ \text{Cleavage}\end{matrix}$$

The above reaction has been shown to be catalyzed by enzymes associated with the endoplasmic reticulum or nuclear membrane, to be dependent on aerobic conditions, and to be inhibited by superoxide dismutase (SOD) and metal chelators.[11,12] As shown in Table 1, the addition of mitomycin C, an antibiotic with redox cycling ability, to microsomes prepared from mouse lung or tumors enhanced the bleomycin-mediated cleavage of DNA deoxyribose. Mitomycin C alone did not cleave deoxyribose. Addition of SOD inhibited the cleavage of DNA deoxyribose by bleomycin either in the absence or presence of mitomycin C. The enhanced generation of "activated" bleomycin, as reflected by DNA deoxyribose cleavage, within tumor cells could be considered beneficial, while, on the other hand, occurrence of such a process in host cells would not. The above interaction with mitomycin C is of interest in light of the demonstration that the simultaneous administration of bleomycin and mitomycin C to rats with AH66 hepatoma resulted in an increased life span of the rats but also resulted in increased lung toxicity.[13] In this regard, pulmonary toxicity is the dose-limiting toxicity of bleomycin clinically.[13]

Like bleomycin, the administration of butylated hydroxytoluene (BHT) to mice results in pulmonary toxicity, as reflected in damage to type I alveolar cells.[14] This activity of BHT has been attributed to its metabolism to a quinone methide derivative.[15] Recently, we have observed

Table 1. Enhancing Effect of the Reactive Oxygen-Generating Compound Mitomycin C on Bleomycin A_2-Mediated DNA Dexyribose Cleavage Catalyzed by Lung and Tumor Microsomes[a]

Additions to Microsomes	Source of Microsomes		
	Lung	B16 Melanoma	Lewis lung Carcinoma
BLM A_2, DNA	43.7[b]	13.7	7.2
BLM A_2, DNA, Mitomycin C	82.7	35.9	21.3

[a]Reactions were conducted under O_2 atmosphere in the presence of an NADPH-generating system.[8]
[b]nmoles TBAR/mg protein/60 min; Mean of N=3-6.

that the horseradish peroxidase or prostaglandin synthetase-catalyzed metabolism of BHT to quinone methide and its covalent binding to protein can be enhanced significantly in the presence of chemicals which form phenoxy, but not peroxy, radicals and include agents such as BHA, diethylstilbestrol, and phenol.[16] In addition, this enhancing activity of BHA can be catalyzed by microsomes from various animal tissues, including the lung.[17] The mechanism for the enhanced activation of BHT by various compounds is illustrated in Figure 1 and is based on a scheme described by Kurechi and Kato.[18] As illustrated in Figure 1, a substrate compound is preferentially oxidized by a peroxidase enzyme to a phenoxy radical (X·), which is recycled to the parent compound by accepting an electron from BHT. During this process, BHT is converted first to its radical intermediate and then to its two-electron reduced electrophilic species, BHT-quinone methide.

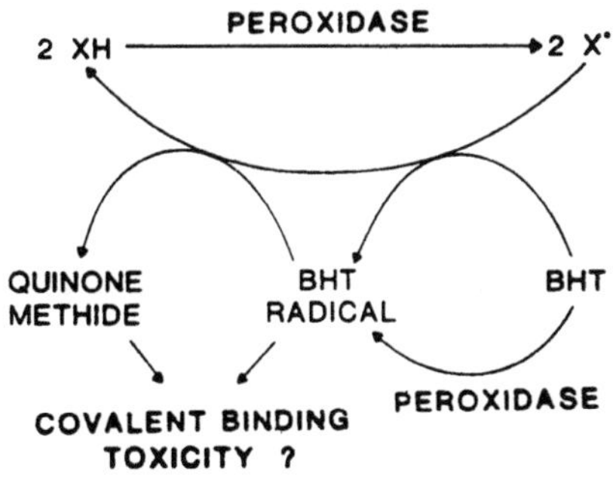

Figure 1. Proposed mechanism for the increased formation of BHT-quinone methide in the presence of enhancing compounds such as BHA, phenol and estradiol.

TABLE 2. Ability of BHT and d_3-BHT to Elicit Mouse Lung Toxicity in the Presence of BHA

Group	Lung/Body Weight (%)
Control-corn oil	0.59
BHT (175 mg/kg)	0.62
d_3-BHT (175 mg/kg)	0.60
BHT (175)/BHA (250 mg/kg)	1.06[a]
d_3-BHT (175)/BHA (250 mg/kg)	0.70

[a]Significantly different ($p < 0.05$) from control.

We have extended these *in vitro* observations and have shown that the subcutaneous administration of BHA (50-250 mg/kg) 30 min prior to subthreshold toxic doses of BHT (0-250 mg/kg i.p.) resulted in an enhanced lung/body weight ratio indicative of BHT-induced lung damage.[17] The lung/body weight ratio elicited by this combination is nearly equivalent to that seen with a 400 mg/kg dose of BHT. BHA alone had no effect on the lung/body weight ratio up to a dose of 500 mg/kg (s.c.). Subsequent studies were then undertaken to determine if the mechanism by which BHA enhanced the *in vivo* pulmonary toxicity of BHT is similar to what we have observed *in vitro* with peroxidase enzymes. One approach we undertook was to use deuterated BHT. The deuterated BHT had a significantly reduced ability to both form quinone methide *in vitro* (34.8 nmoles versus 13.3 for d_3-BHT) and to elicit pulmonary toxicity in the presence of BHA (Table 2). These observations support the concept that BHA is increasing the activation of BHT *in vivo*. Likewise, we have observed that BHA increased the covalent binding of BHT to mouse lung slices and that the addition of H_2O_2 or hydroperoxides, including BHT hydroperoxide, to mouse lung microsomes resulted in increased BHA-mediated covalent binding of BHT to protein.[19] Addition of BHA alone stimulated the formation of H_2O_2 by 4.7-fold in mouse lung microsomes probably as a result of its P-450 catalyzed metabolism to TBHQ. Interestingly, the NADPH-cytochrome *c* (P-450) reductase catalyzed redox cycling of TBHQ supported the horseradish peroxidase catalyzed BHA-enhanced formation of BHT-quinone methide. Collectively, the above results suggest that BHA facilitates the activation of BHT to a reactive intermediate in the lung as a result of both its metabolism to TBHQ, resulting in increased H_2O_2 formation, and the subsequent utilization of this H_2O_2 for the peroxidase-dependent conversion of BHA to a phenoxy radical, which then mediates the activation of BHT in a manner shown in Figure 1.

In summary, the results described here with bleomycin A_2 and BHT extend our knowledge regarding the involvement of reactive oxygen or chemical-derived radical intermediates, particularly phenoxy radicals, in enhancing the activation of xenobiotics. Previous studies have established a role for peroxy radical intermediates in mediating the epoxidation of retinoic acid and benzo(a)pyrene-7,8-dihydrodiol.[6,20] The stimulation of catecholamine oxidation by a chlopromazine radical and the enhancing effect of phenol on the peroxidase-catalyzed oxidation of 0,0-biphenol are additional examples of this concept.[21,22] Similarly, Eastmond *et al*.[23] have shown that phenol also enhances the peroxidase-catalyzed activation of hydroquinone. Phenol and hydroquinone are both metabolites of benzene and interestingly their simultaneous, but not separate, administration to mice results in myelotoxicity. Thus, it would appear that the enhancement of chemical activation via radical-dependent mechanisms provides a novel mechanism of chemical-chemical interactions both in vitro and in vivo.

ACKNOWLEDGMENTS

The authors gratefully acknowledge financial support for this research from the National Institutes of Health ES 03760 and 07141, National Institute of Occupational Safety and Health OH 01833, American Cancer Society SIG-3, and Johns Hopkins CAAT. The assistance of Mrs. Beverly Taylor in the preparation of this manuscript also is gratefully acknowledged.

REFERENCES

1. M.R. Boyd, Biochemical-mechanisms in chemical-induced lung injury: Roles of metabolic activation. CRC Crit. Rev. Toxicol. 7:103 (1980).
2. R.W. Estabrook and E. Lindenlaub, eds, "The Induction of Drug Metabolism," F.K. SchattauerVerlag, Stuttgart-New York (1979).
3. E. Dybing, S.D. Nelson, J.R. Mitchell, H.A. Sasame, and J. R. Gillette, Oxidation of alpha-methydopa and other catechols by cytochrome P-450-generated superoxide anion: Possible mechanism of methyldopa hepatitis, Mol. Pharmacol. 12:911 (1976).
4. S.D. Nelson, J.R. Mitchell, E. Dybing, and H.A. Sasame, Cytochrome P-450-mediated oxidation of 2-hydroxyestrogens to reactive intermediates, Biochem. Biophys. Res. Commun. 70:1157 (1976).
5. M.E. McManus and D.S. Davies, Paraquat stimulated binding of dopa to liver and lung microsomal protein, Xenobiotica 10:745 (1980).
6. G.A. Reed, E.A. Brooke, and T.A. Eling, Phenylbutazone-dependent epoxidation of 7,8-dihydroxy-7,8-dihydrobenzo(a)pyrene. J. Biol. Chem. 259:5591 (1984).
7. M.T. Kuo and C.W. Hardle, Characterization of chain breakage in DNA by bleomycin, Biochim. Biophys. Acta 335:109 (1973).
8. M.A. Trush, E.G. Mimnaugh, E.Ginsburg, and T.E. Gram, Studies on the interaction of bleomycin A_2 with rat lung microsomes I. Characterization of factors which influence bleomycin-mediated DNA chain breakage, J. Pharmacol. Exp. Ther. 221:152 (1982).
9. R.M. Burger, J. Peisach and S.B. Horwitz, Activated bleomycin: a transient complex of drug, iron and oxygen that degrades DNA, J. Biol. Chem. 256:11634 (1984).
10. A.P. Grollman, M. Takeshita, K.M.R. Pillar, and F. Johnson, Origin and cytotoxic properties of base propenals from DNA, Cancer Res. 45:1127 (1985).
11. M.A. Trush, E.G. Mimnaugh, E. Ginsburg, and T.E. Gram, Studies on the interaction of bleomycin A_2 with rat lung microsomes. II. Involvement of adventitious iron and reactive oxygen in bleomycin-mediated DNA chain breakage, J. Pharmacol. Exp Ther. 221:159 (1982).
12. M.A. Trush, K.A. Kennedy, B.K. Sinha, and E.G. Mimnaugh, Bleomycin-mediated deoxyribrose cleavage in rabbit lung nuclei, The Pharmacologist 25:225 (1983).
13. N. Yamanaka, M. Fukushima, T. Koto, K. Koizumi, and K. Ota, Experimental results with the combination of bleomycin plus mitomycin C, in: "Recent Results in Cancer Research," S.K. Carter, H. Umezawa, J. Douros and Y. Sakurai, eds., Springer-Verlag, Berlin, (1978).
14. H. Witschi and S. Lock, Toxicity of butylated hydroxytoluene in mouse following oral administration, Toxicology 9:137 (1978).

15. T. Mizutani, I. Ishida, K. Yamamoto, and K. Tajima, Pulmonary toxicity of butylated hydroxytoluene and related alkylphenols: Structural requirements for toxicity potency in mice, Toxicol. Appl. Pharmacol. 62:273 (1982).
16. D.C. Thompson, Y.N. Cha, and M.A. Trush, The peroxidative activation of butylated hydroxytoluene to BHT-quinone methhide and stilbenequinone, in "Biological Reactive Intermediates, III.," J.J. Kocsis, D.J. Jollow, C.M. Witmer, J.O. Nelson and R. Snyder, eds., Plenum Press, New York, (1986).
17. D.C. Thompson and M.A. Trush, The toxicological implications of the interaction of butylated hydroxytoluene with other antioxidants and phenolic chemicals, Fd. Chem. Toxicol. 24:1189 (1986).
18. T. Kurechi and T. Kato, Studies on the antioxidants. XVI. Synergistic reactions between butylated hydroxyanisole and butylated hydroxytoluene in hydrogen donation to 2,2-diphenyl-1-picrylhydrazyl, Chem. Pharm. Bull, Tokyo 30:2964 (1982).
19. D.C. Thompson and M.A. Trush; Studies on the mechanism of BHT-induced mouse lung toxicity by BHA, The Pharmacologist 28:216 (1986).
20. V.M. Samokyszyn and L.J. Marnett, Hydroperoxide-dependent cooxidation of 13-cis-retinoic by prostaglandin H synthase, J. Biol. Chem. 262:14119 (1987).
21. R.A. Lovstad, Activating effect of chlopromazine on the peroxidase-catalyzed oxidation of catecholamines, Gen. Pharmacol. 10:437 (1979).
22. V.V. Subrahmanyam and P.J. O'Brien, Phenol oxidation product(s) formed by a peroxidase reaction that bind to DNA, Xenobiotica 15:873 (1985).
23. D.A. Eastmond, M.T. Smith, and R.D. Irons, An interaction of benzene metabolites reproduces the myelotoxicity observed with benzene exposure, Toxicol. Appl. Pharmacol. 91:85 (1987)

ROLES OF OXYGEN AND OXYGEN SUBSTITUTES IN DNA SUGAR DAMAGE

BY ANTITUMOR ANTIBIOTICS

Irving H. Goldberg, Lizzy Kappen, and Der-Hang Chin

Department of Biological Chemistry and Molecular Pharmacology
Harvard Medical School
Boston, Massachusetts 02115, U.S.A.

INTRODUCTION

Oxidative damage to the deoxyribose backbone of DNA produced by ionizing radiation and antitumor antibiotics (e.g., bleomycin) generally involves oxygen at two stages in the damage process: 1) "reactive oxygen" (OH radical or its equivalent) as the hydrogen atom abstracting species, generating a carbon-centered radical on the deoxyribose, and 2) addition of dioxygen to the latter to "fix" the lesion in a form not readily repaired. By contrast, neocarzinostatin (NCS) is a member of a class of antitumor antibiotics that binds specifically to DNA and is itself converted to a radical species that directly attacks the DNA sugar. Dioxygen or its substitute is involved only after generation of the DNA damage intermediate. This review will focus on the roles of dioxygen and oxygen substitutes, the nitroaromatic radiation sensitizers, in the formation of novel types of oxidative DNA sugar damage by NCS.

NEOCARZINOSTATIN STRUCTURE AND REVERSIBLE BINDING TO DNA

NCS is a member of a family of macromolecular (M_r >10,000) antitumor antibiotics obtained from culture filtrates of _Streptomyces_. All biological activity resides in a methanol-extractable nonprotein chromophore (M_r = 659) that is tightly ($K_D \sim 10^{-10}$ M) and specifically bound to an apoprotein (M_r = 11,000) (see references 1 and 2 for references within). The latter acts as a carrier and protects the labile chromophore (NCS-Chrom) from degradation.

NCS-Chrom has been purified by high-pressure liquid chromatography, and a structure has been proposed for its major form (Figure 1).[3-5] It consists of three subunits: a 5-methyl-7-methoxy naphthoate, a 2,6-dideoxy-2-methylamino-galactose moiety and an interconnecting highly unsaturated C_{12}-subunit bearing a cyclic carbonate and an epoxide. The C_{12}-subunit consists of a novel, highly strained bicyclo(7,3,0)dodecadiyne system. Mercaptan treatment results in the addition of one molecule of mercaptan as well as in the reduction of the C_{12}-substructure, consuming a total of three sulfhydryls (or four hydrogens from $NaBH_4$).[3]

NCS-Chrom binds to DNA in a two-step process.[6] The first involves external binding and the second, intercalation of the chromophore between adjacent DNA base pairs. NCS-Chrom binds tightly to DNA with a $K_D = \sim 10^{-6}$M

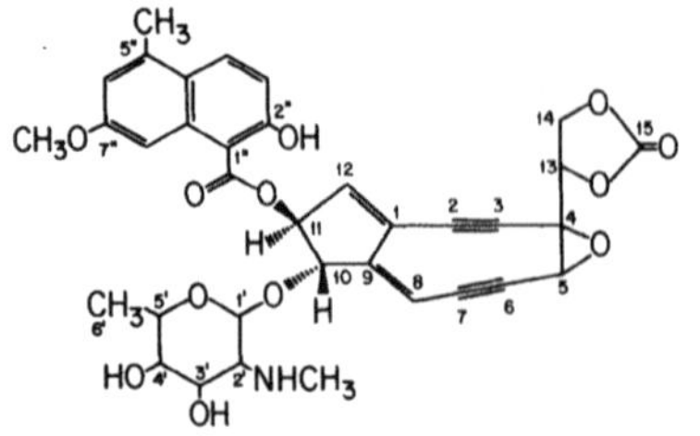

Figure 1. Structure of NCS-Chrom.

and has a preference for DNA high in T and A residues. A relatively slow off rate (0.75 s^{-1}) probably places the chromophore on the DNA for a sufficient length of time to be activated by thiol. Thiol activates NCS to a labile (presumably radical) species, which, in the absence of DNA, converts rapidly to an inactive form.

Accumulated data have led us to propose a model for NCS-Chrom binding to DNA in which the naphthoic acid part of the chromophore intercalates between the DNA base pairs via the minor groove of DNA, and the positively charged amino sugar binds electrostatically to either the negatively charged oxygen of the DNA sugar-phosphate backbone or to the deep electrostatic potential well of the minor groove.[7,8] These two attachment sites are viewed as serving as anchors for the proper placement in the minor groove of the highly strained diyne-containing 9-membered ring close to C-5' of deoxyribose (see below).

TYPES OF NCS-INDUCED DNA DAMAGE

NCS or NCS-Chrom treatment of mammalian cells induces single-strand breaks in the DNA, which can be repaired to a considerable extent. Experiments using the nucleoid sedimentation technique show a quantitative correlation between NCS-induced inhibition of replicon initiation and generation of sufficient strand breakage to relax domains of supercoiling in DNA (target size of about 10^9 daltons) of mammalian cells, implying that a single event, presumably a strand break, is adequate to disrupt DNA structure sufficiently to interrupt replicon initiation. DNA damage occurs *in vitro*, provided that a reducing agent, such as a thiol, DNA and drug are simultaneously present. Some DNA cleavage is found at very high levels of NCS in the absence of thiol, but the reaction is stimulated at least 1000-fold by thiol. High levels of thiol block DNA damage, presumably by scavenging radical forms of the drug or DNA (see below). With NCS-Chrom the reaction is fast even at 0°C, but higher temperatures are required in the reaction with the holoantibiotic since the apoprotein must first unfold in order to release the chromophore. Not only does NCS degrade naked DNA, it also solubilizes chromatin from mammalian cells and isolated nuclei by attacking primarily the linker DNA between nucleosomes.

At least four types of lesions are formed in DNA *in vitro* by NCS-Chrom in the presence of sulfhydryl cofactors. The predominant lesion is a strand break (T>A>>C>G), resulting from selective, oxygen-dependent oxidation of deoxyribose to a 5'-aldehyde.[9,10] The next most frequent lesion is release of free base (T>A>>C>G), presumably associated with some as yet undetermined form of sugar oxidation. In addition, at least two types of adducts between DNA and NCS-Chrom have been detected.[11-13] In the presence of 2-mercapto-ethanol and oxygen, a labile adduct containing an oxidized deoxyribose is formed at a very low yield. In the presence of dithiothreitol, a much more stable adduct species is produced at a somewhat higher yield. Under anaerobic conditions, the yield of this type of adduct is further increased (while strand breakage and base release are suppressed) so that stable adducts become the predominant DNA lesion.[13] This stable adduct species, and probably the labile adduct species as well, involves covalent linkage of the chromophore to deoxyribose in DNA.

Strand Breakage, Base Release, and Deoxyribose Damage

Studies using the DNA sequencing technique of Maxam and Gilbert, in which $^{32}PO_4$ is attached to the 5'-end of the DNA, show that NCS cleaves double-stranded DNA restriction fragments almost exclusively at T and A residues. Overall, T residues are attacked much more frequently than are A residues, although there is variability in the attack rate for both at different locations in the DNA molecule, presumably because of tertiary structural constraints (see above). While all T residues are sites of scission, not all A residues are cleavage sites. These studies also confirmed earlier studies showing that there is a phosphate moiety at the 3'-end of the break. In addition to spontaneous breaks, there are alkali-labile breaks the exact chemistry of which is yet to be elucidated, although they are presumably mainly at sites of base release (see below).

The base sequence-dependent alteration in the binding affinity, as recorded by polynucleotide-chromophore binding studies, is clearly reflected in the observation that not all T-residues in DNA restriction fragments are attacked with equal efficiency by NCS. Furthermore, an occasional A-residue will be attacked much more frequently than adjoining T-residues. Similarly, there is considerable heterogeneity in the formation of alkali-labile apurinic/apyrimidinic (AP) sites in NCS-treated DNA. This is most clearly seen in the increased frequency of AP site formation at the C-residue in AGC, which is a frequent site of GC to AT transition mutation.[14,15] Although there is no clear-cut base sequence specificity for NCS-induced strand breakage, adenine is more frequently found on the 5'-side and guanine on the 3'-side of the thymine attacked by NCS-Chrom.

Earlier studies with DNAs labeled with different forms of radioactive thymidine indicated that DNA strand breakage is accompanied by damage to the deoxyribose moiety. These findings were confirmed by DNA sequencing experiments, using DNA restriction fragments with $^{32}PO_4$ at the 3'-end, that showed that the 5'-end of the strand break possessed a modified nucleoside, not an exposed phosphate group.[9,10] In these studies the radioactive band at the point of a break moved about two nucleotides more slowly on the electrophoresis gel than the marker produced by chemical treatment in the Maxam-Gilbert procedure, which has a 5'-phosphate end. The modified nucleoside could be removed by alkali so as to generate a phosphate-ended fragment. The modified nucleoside in thymidine-labeled DNA was liberated by nuclease digestion and was characterized as thymidine 5'-aldehyde (see 3 in Scheme 1). Since nucleoside 5'-aldehyde would be expected to release its base (and sugar) via β-elimination reactions upon alkaline treatment, the finding that over 80% of base-release is alkali-dependent indicates that most of the breaks result from oxidation at the C-5' of deoxyribose in DNA. Similarly, DNA sequencing experiments show that strand breaks (gaps) with phosphate moieties at both ends account for less than 20% of the spontaneous breakage.

Evidence has been presented that similar DNA damage specificity and mechanisms exist in intact HeLa cells treated with NCS.[16] Further, in cells depleted of glutathione, DNA strand breakage is greatly decreased, in agreement with the findings of DeGraff et al., who showed that NCS is much less toxic and mutagenic in glutathione-depleted mammalian cells.[17]

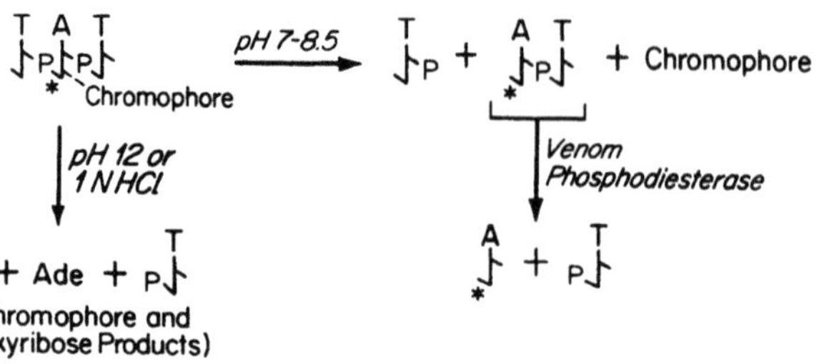

Figure 2. Labile adduct hydrolysis patterns. Incubation of intact adduct at pH 12 or in 1 M HCl releases adenine as well as both 3'dTMP and 5'dTMP. Incubation at pH 7 or pH 8.5 releases, simultaneously, chromophore, 3'-dTMP, and a modified (*) d(ApT) that is a substrate for venom phosphodiesterase.

NCS-Chrom-Deoxyribose Adducts

In addition to causing DNA strand breaks with nucleoside 5'-aldehyde and 5'-phosphate termini, NCS-Chrom forms novel labile covalent adducts with DNA.[11] The predominant adduct recovered from nuclease digests of NCS-Chrom-treated poly(dA-dT)·poly(dA-dT) in the presence of 2-mercaptoethanol and dioxygen is a compound with the structure chromophore-d(TpApT) in which the attached chromophore rendered both phosphodiester bonds refractory to endonuclease S1. This adduct fragment was completely hydrolyzed at pH 12, releasing adenine, 3'-dTMP, and 5'-dTMP (Figure 2). At pH 7 the adduct fragment slowly released chromophore and 3'-dTMP with parallel kinetics, leaving a modified d(ApT), which was cleaved by snake venom phosphodiesterase to yield 5'-dTMP and a modified deoxyadenosine. These hydrolyses patterns are unlike those of any previously characterized base or phosphotriester DNA adduct but indicate the presence of an altered deoxyadenosine sugar with chemical properties similar to those of thymidine 5'-aldehyde. The modified deoxyadenosine was identified as deoxyadenosine 5'-aldehyde by a series of chemical derivatizations. The formation of adducts containing a modified deoxyribose suggests that deoxyribose is the site of covalent chromophore attachment. Support for such a possibility comes from experiments in which this compound was subjected to mild acid hydrolysis to release free adenine and to snake venom exonuclease (pH 6.5) treatment to release 5'-TMP so as to leave, in both cases, adducts of slightly altered chromatographic mobility. These results eliminate adenine and 5'-dTMP as possible sites of covalent chromophore attachment.

A similar but much more stable adduct species has also been isolated, again with structure chromophore-d(TpApT).[12,13] Acid hydrolysis of this adduct released free adenine, as well as both 3'-dTMP and 5'-dTMP, leaving a compound which contained only chromophore and the deoxyadenosine sugar moiety. These results clearly indicate the formation of a covalent linkage between chromophore and deoxyribose in DNA. Formation of this adduct shows an oxygen dependence that is the inverse of that seen for nucleoside 5'-aldehyde formation, i.e., the yield of adduct is maximal under anaerobic conditions. This apparent competition between DNA sugar oxidation and adduct formation suggests that the two lesions share a common precursor. A chromophore-induced carbon-centered radical on C-5' of deoxyribose that could react by addition with the same chromophore molecule to form the stable covalent adduct or with O_2 to form the 5'-aldehyde would be the most logical candidate for such a precursor (see below).

The resistance of both types of chromophore-oligonucleotide adducts to further nuclease digestion suggested that the adducts block exonucleolytic digestion of the DNA and that the sequence specificity of adduct formation might then be determined by mapping the adducts as exonuclease termination sites in defined-sequence DNA. Such a study revealed that the stable adducts occurred with the same base specificity as for strand-break formation.[18] The similarity in sequence specificity of adducts and strand breaks suggests that a common form of nascent DNA damage may be a precursor to both lesions. Again, a chromophore-induced free radical on C-5' of deoxyribose, subject to competitive fixation by addition reactions with either oxygen or chromophore, is the most likely candidate for such a precursor.

A reaction entirely analogous to NCS-deoxyribose adduct formation occurs when DNA is γ-irradiated.[19,20] The cycloaddition products 8,5'-cyclo-2'-deoxyguanosine and 8,5'-cyclo-2'-deoxyguanosine have recently been identified in γ-irradiated DNA. These products result from the intramolecular cyclization between C-5' of deoxyribose and C-8 of the purine base following hydrogen atom abstraction from C-5' by radiation-generated hydroxyl radicals. Instead of the carbon-centered radical at C-5' reacting with a well-positioned drug, as in the case of NCS, it reacts with its own base to form the cycloaddition product.

Involvement of Thiol and Dioxygen

DNA strand breakage, base release and the formation of nucleoside 5'-aldehyde by NCS require O_2. Kinetic studies have indicated that in the presence of DNA, O_2 and sulfhydryls, the chromophore undergoes a single oxidation-reduction cycle, which results in efficient DNA damage and leaves an inactive product devoid of further redox activity.[21] The process can be resolved into three steps. The first step, which is accompanied by marked changes in chromophore spectral properties but which does not require O_2, is a reaction between the chromophore and a single sulfhydryl, probably an addition reaction. ESR studies show that free radical signals are generated during the thiol reaction with NCS, but their relationship to "activated" drug remains unclear. Formation of this activated chromophore species is rapidly followed (within a few seconds) by the uptake of one mole of O_2 per mole of chromophore. O_2 uptake is then followed by uptake of at least an additional sulfhydryl group. Since consumption of the second sulfhydryl does not occur in the absence of O_2, it may represent a one-electron reduction of a reaction intermediate, which includes some form of reduced oxygen (see below). The oxygen that ends up in the 5'-aldehyde at the strand break is derived entirely from O_2 and not H_2O, as shown by ^{18}O incorporation experiments.[22]

There is no evidence that a reduced form of oxygen, such as superoxide, peroxide or hydroxyl free radical, or transition metals, is involved in NCS-induced DNA damage, although NCS-Chrom can undergo autoxidation with the production of superoxide free radical.[23] Superoxide free-radical formation, however, is not involved in the production of DNA damage, whether or not thiol is present.

Hydrogen Atom Abstraction From C-5' of Deoxyribose

The results of ^{18}O incorporation studies, which show that dioxygen, not water, is the source of the oxygen in the 5'-aldehyde generated at the 5'-end of an NCS-induced DNA break, are compatible with a mechanism in which dioxygen adds to a carbon-centered radical at C-5' to form a peroxyl radical adduct (see below for modified versions). The carbon-centered radical on the deoxyribose forms when the thiol-activated NCS-Chrom abstracts a hydrogen atom from the C-5' position. Using DNA in which the hydrogens of the deoxyribose were replaced by tritium in the 1', 2' or 5' positions, it

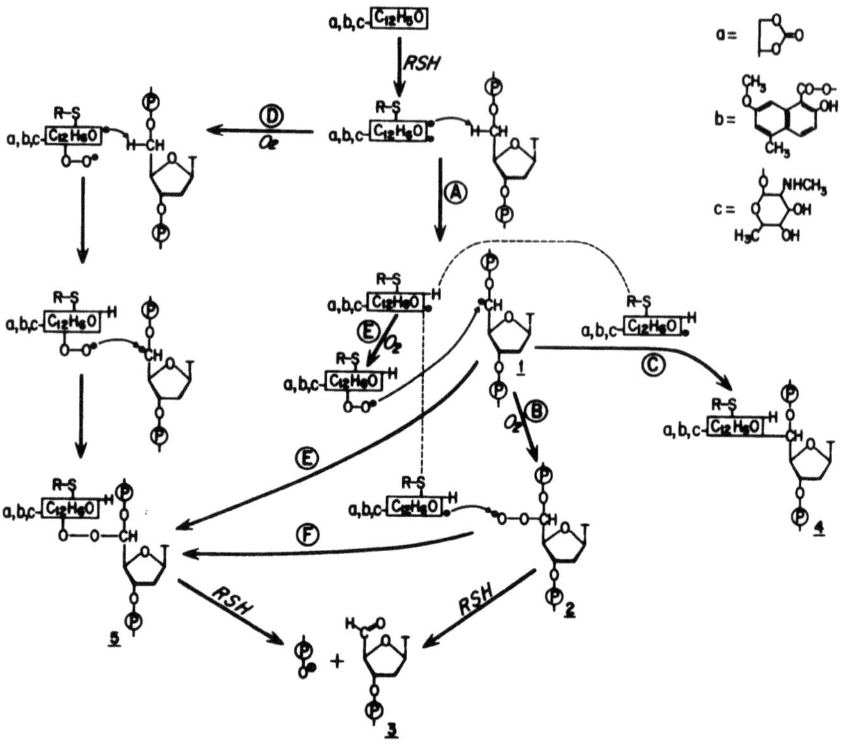

Scheme 1

was found that the activated NCS-Chrom selectively abstracted tritium only from the 5'-position and incorporated it into a stable, non-exchangeable form of the chromophore.[24,25] Drug activation and the subsequent hydrogen abstraction reaction do not require molecular oxygen. DNA damage is dependent on thiol, as is 5'[^{3}H] abstraction into the chromophore. Further, in the aerobic reaction there is a good correlation between the amount of 5'[^{3}H] abstracted into the chromophore and the extent of DNA damage over a wide range of thiol concentrations; at low levels of thiol there is a nearly linear increase both in 5'[^{3}H] abstraction and DNA damage. At the highest level of thiol used, there is a small but significant reduction in both DNA damage and ^{3}H-abstraction, probably due to a faster rate of inactivation of the drug prior to its interaction with DNA.

MOLECULAR MECHANISM OF NCS-INDUCED DNA DAMAGE

Based on the accumulated data, we have formulated a reaction scheme in which a carbon-centered radical at C-5' is an intermediate in the formation of the various DNA damage products formed under either aerobic or anaerobic conditions (Scheme 1).[25] The first step involves the generation of an active species of the drug (shown as a free radical) upon thiol addition to the highly unsaturated C_{12} subunit of the chromophore. The activated drug abstracts a hydrogen atom from the 5'-carbon of deoxyribose (pathway A) to generate a carbon-centered radical (1) on the DNA, the fate of which depends on whether or not O_2 is available. Addition of O_2 to the 5'-carbon-centered radical (pathway B) gives rise to a peroxyl radical (2) to produce DNA strand breaks with 5'-terminal nucleoside aldehyde (3). The involvement of thiol in the degradation of a peroxyl structure to an aldehyde is compatible with our experiments in which at least a second molecule of thiol is consumed in the reaction beyond the one used in the initial activation of the drug. In the absence of O_2, the carbon-centered radical on the deoxyribose (1) may instead react with the chromophore containing the abstracted hydrogen (pathway C) to form a stable covalent chromophore-DNA adduct (4), possibly by addition of the C-5' radical across an unsaturated bond in the chromophore, or may be repaired by abstraction of a hydrogen from thiol. This analysis, involving a common DNA intermediate, is consistent with studies showing that there is a competition between stable chromophore-DNA adduct formation under anoxia and DNA strand breakage and base release in O_2.

Although drug activation and hydrogen-atom abstraction from the 5'-position of deoxyribose in DNA can proceed in the absence of O_2, they do not rule out the possibility that in the presence of O_2 the activated chromophore can add O_2 to form a peroxyl derivative of the activated chromophore (pathway D). The peroxyl radical on the drug cannot itself be responsible for hydrogen-atom abstraction since the latter is incorporated into the chromophore in a non-exchangeable form. Nevertheless, a radical elsewhere in the peroxyl form of the drug could abstract the hydrogen atom and then react (possibly as part of a concerted reaction) with the carbon-centered radical on the DNA by way of its peroxyl radical to form a labile adduct, possibly with the structure shown in (5), as an intermediate in strand break formation. Again, this peroxyl adduct may be cleaved by thiol to form (3). Consistent with such a second role for thiol in the formation of the 5'-aldehyde are experiments showing that 2-mercaptoethanol accelerates the decomposition of the labile adduct (L.F. Povirk and I.H. Goldberg, unpublished results) and that high thiol concentrations increase both the absolute yield of 5'-aldehyde and the ratio of aldehyde to released free base (L.S. Kappen and I.H. Goldberg, unpublished results).

It is also possible that the activated chromophore adds O_2 after it abstracts the hydrogen atom from DNA (pathway E) and then reacts with 1 to form the labile adduct (5), or it may instead react with the peroxyl form of the deoxyribose (2) to form (5) via pathway F. Since the chromophore is anchored on the DNA by intercalation and electrostatic interactions, it would not be surprising if it participated in such reactions after the initial hydrogen abstraction. Reaction pathway E, involving a peroxyl form of the drug and a labile chromophore-DNA adduct intermediate, best fits data showing that one molecule of O_2 is consumed for each molecule of chromophore, whereas DNA damage is somewhat less than stoichiometric. Furthermore, pathway E is more attractive than D, since it does not require postulating a different hydrogen-abstracting form of the chromophore under aerobic and anaerobic conditions. Finally, it should be noted that it is possible to combine pathways A and E in a concerted mechanism involving a ternary complex of activated chromophore, O_2 and C-5' of deoxyribose.

The proposed scheme focuses on the 5'-carbon of deoxyribose as the predominant site of attack. It is consistent with the findings that over 80% of the DNA strand breaks (major lesion) have nucleoside 5'-aldehyde at their 5'-termini and that C-5' oxidation also occurs in the formation of the labile chromophore-DNA adducts. The absence of any ^3H abstraction from the C-1' and C-2' of thymidine in DNA eliminates these positions as targets. Further, virtually no ^3H could be found in the solvent H_2O. It should be noted, however, that C-1' appears to be the primary site of hydrogen abstraction when radiolytically-activated NCS reacts with DNA to form alkali-labile, abasic lesions.[26] In these experiments the anaerobic reaction of poly(dA-dT) with NCS activated by the carboxyl radical $CO_2^{\cdot -}$, an electron donor generated from γ-ray radiolysis of nitrous oxide-saturated formate buffer, was characterized. The activated NCS-Chrom intermediate was shown to be a free radical. A similar mechanism may occur in the formation of an abasic site at the C-residue in AGC (see below).

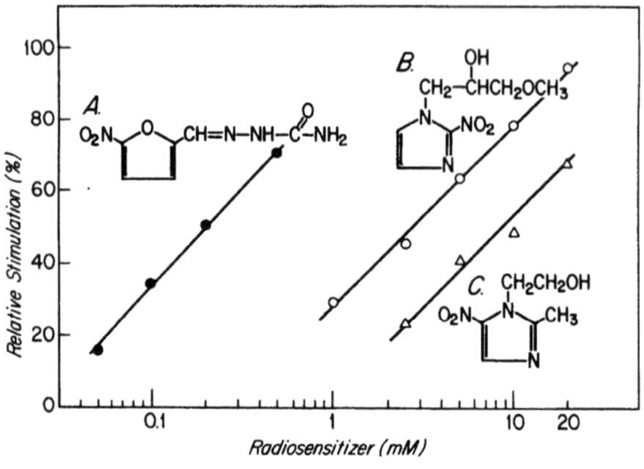

Figure 3. Comparison of the ability of different nitroaromatic radiation sensitizers to substitute for oxygen in the stimulation of NCS-dependent DNA damage in nitrogen. (A) Nitrofurazone, E_7^1/mv = -257; (B) Misonidazole, E_7^1/mv = -389; and (C) Metronidazole, E_7^1 = -486.

Nucleoside 5'-aldehyde is the predominant product at the 5'-end of a DNA strand break under the aerobic conditions ordinarily used to study NCS-induced DNA damage in vitro. On DNA sequencing gels, however, a band of ^{32}P radioactivity (from a DNA restriction fragment labeled with $^{32}PO_4$ at its 3'-end) can also be seen with a mobility somewhat faster than that of the nucleoside 5'-aldehyde ended DNA fragment (but not as fast as the Maxam-Gilbert procedure produced marker that has a phosphate at its end and is also one nucleotide shorter). The electrophoretic and chemical properties of this DNA fragment suggest that it has a 5'-carboxylic acid at its 5'end, raising the possibility discussed in detail in reference 2, that it was formed by the bimolecular decay of the peroxyl radical via a short-lived tetroxide intermediate in a disproportionation reaction (Russell mechanism).

Mechanism of Misonidazole Action

When misonidazole substitutes for O_2 in the NCS-DNA reaction, the attack site specificity is the same as in O_2, but in the presence of misonidazole, base release and gap formation (with 3'- and 5'-phosphoryl termini) predominate.[27] The amount of DNA damage produced varies with the one-electron redox potential of the nitroaromatic compound (Figure 3), with the more electron-affinic compounds being more effective. Despite the difference in final DNA products, abstraction of 5'-hydrogen by thiol-activated drug to form a carbon-centered radical on C-5' of deoxyribose appears to be a common initial step in both the O_2- and misonidazole-dependent reactions. In fact, there appears to be a competition between the misonidazole-dependent and the oxygen-dependent pathways, as revealed by cleavage patterns on DNA sequencing gels. Since misonidazole is much less effective than O_2 as a radiation sensitizer, it is not surprising that the nitroaromatic compound competes less efficiently with scavenging thiol, as indicated by the finding that the ability to produce DNA damage under anaerobic conditions is inhibited by lower thiol levels than in the aerobic reaction without misonidazole. It seems likely that after the induction of the same nascent lesion on the DNA, participation of misonidazole in the subsequent steps, by a mechanism different from that of O_2, leads to different products.

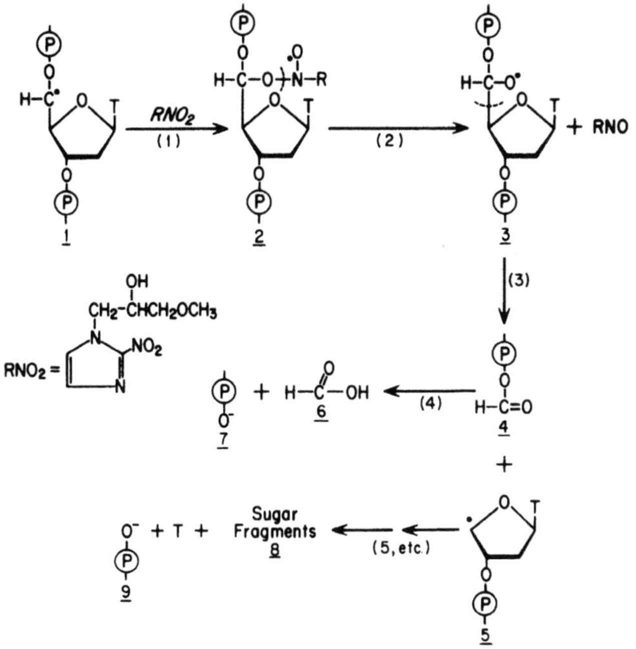

Scheme 2

We have obtained evidence that ^{14}C-labeled misonidazole undergoes reduction in the reaction, probably to form the glutathione adduct.[28] Further, we have found that under anaerobic conditions with misonidazole, NCS causes, in addition to the release of thymine, the release of substantial amounts of a 5'[^{3}H]-labeled sugar fragment (from 5'[^{3}H]thymidine-labeled DNA) that elutes from HPLC at 4 min.[25] This substance is uncharged,[28] but when treated with either acid or alkali is converted into formic acid. We have identified the formic acid precursor as formyl-Tris (Tris being an amino group-containing molecule used as the buffer in the reaction). Similarly, when hydroxylamine substitutes for Tris, formyl hydroxamate is formed. The amount of the latter formed is comparable to the spontaneously released thymine.

It appears that under anaerobic conditions in the presence of misonidazole a reactive form of formate from C-5' of deoxyribose of thymidylate is generated, and the formate can be transferred to available nucleophiles. These results, plus the evidence that[28] misonidazole reduction is DNA-dependent, suggest a mechanism (Scheme 2)[28] in which the carbon-centered radical formed at C-5' (1) by hydrogen atom abstraction by thiol-activated NCS reacts anaerobically with misonidazole to form a nitroxide radical intermediate (2) (reaction 1), which fragments (reaction 2) to produce an oxyradical at C-5' (3). β-Scission (reaction 3) results in cleavage between C-5' and C-4', with the generation of 3'-formyl phosphate-ended DNA, a high-energy form of formate. The latter spontaneously hydrolyzes, releasing formate and creating a 3'-phosphate-end, or transfers the formyl moiety to available nucleophiles. A similar mechanism, involving dioxygen addition, is probably responsible for the 10-15% DNA gap formation in the aerobic reaction. In the cell nucleus the activated formate intermediate can transfer the formate moiety to neighboring nucleophiles, which include histones and other proteins in addition to the DNA itself, generating lesions that are difficult if not impossible to repair.

Support for the proposed mechanism comes from earlier reports (see reference 28 for references) on the addition of carbon-centered radicals to the oxygen of the nitro group of tetranitromethane and nitrobenzenes to give intermediate nitroxide radicals, as revealed by electron spin resonance spectroscopy and pulse radiolysis. The ease of formation of such adducts depends on the one-electron redox potential of the nitro compound and is in agreement with our results showing a similar relationship for NCS-induced DNA damage. While these adducts can act as intermediates in one-electron oxidations, as has been described in radiation sensitization reactions with pyrimidines and their derivatives, they can also undergo fragmentation (reaction 2) to form oxy radicals (3) and the nitroso reduction product of misonidazole, a two-electron process. Oxy radicals undergo β-fragmentation resulting in cleavage between C-5' and C-4' (reaction 3), with the formation of 3'-formyl phosphate-ended DNA (4) and other fragments from the rest of the sugar (8). Spontaneous hydrolysis of the labile formyl phosphate DNA or transfer of the activated formyl moiety to a nucleophile results in the formation of a gap with phosphates at both ends (7 and 9). It appears that the nitroxide radical adduct does not undergo OH^--catalyzed heterolysis (a one-electron oxidation) with hemi-acetal formation to a significant extent since a strand break with a nucleoside 5'-aldehyde at the 5'-end is not an important product of the reaction. Under aerobic conditions the peroxyl radical formed at C-5' results mainly (>80%) in oxidation of C-5' to form a nucleoside 5'-aldehyde at the 5'-end of the break. A small fraction (10-15%), however, appears to be converted to the oxy radical form (perhaps via a concerted reaction involving a tetroxide intermediate), which fragments to form 3'-formyl phosphate-ended DNA.

Formate has been identified as a product in the irradiation of DNA by X-rays under aerobic conditions,[29] but evidence for its generation via a formyl phosphate derivative has been lacking. On the other hand, X-ray treatment of monoethyl phosphoric acid leads to the formation of acetyl phosphate, as shown by its conversion to the corresponding hydroxamic acid derivative.[30] Ionizing radiation-induced cleavage between C-5' and C4' dependent on misonidazole has not been reported before, although the formation of N^1-formyl N^2-pyruvylurea due to cleavage between C-5 and C-6 of thymine, increases in the presence of nitroaromatic compounds upon irradiation of thymine in the absence of dioxygen.[31] No mechanism has been previously offered to account for this effect, but it would appear to be analogous to that proposed here. Similarly, the increase in inorganic phosphate released by -irradiation of nucleotides in the presence of misonidazole[32] may also be due to the newly formulated mechanism. Further, since it appears that the formation of thymine oxidation products of the 5,6-dihydroxydihydrothymine-type due to one-electron oxidation are not responsible for the radiosensitization by misonidazole,[33] it is possible that the proposed two-electron process is important in this process as well.

ABASIC SITE FORMATION

The abasic sites at C residues in the sequence AGC[14] are mutagenic lesions, accounting for most of the GC to AT transitions in the cI gene of lambda phage treated with NCS.[15] Preliminary results show that hydrogen atom abstraction occurs from C-1' not C-5' of the C residue to produce a carbonyl group at C-1' (or the carboxylic acid in the open form). These lesions are much more susceptible to phosphodiester bond cleavage than the ordinary apurinic site produced by acid treatment of DNA, which has an aldehyde at C-1' (in the open form). It is of considerable interest that the AGC sequence generates a structure that modifies the interaction with NCS in such a way as to alter its attack-site specificity. The basis for this change in specificity of attack is under study.

CONCLUSIONS

The antitumor antibiotic, NCS, is one of a small number of agents that selectively attacks the deoxyribose moiety of DNA. Although ionizing radiation also causes oxidative DNA sugar damage, DNA bases appear to be its main target. Nevertheless, many of the biochemical and biological effects of the two agents are similar (see reference 2). Because antibiotics interact with DNA by forming specific complexes, the precise attack site on the deoxyribose may vary. Thus, bleomycin mainly abstracts a hydrogen atom from C-4', whereas NCS acts mainly at C-5'. Further, NCS forms adducts at the attack site on the DNA, whereas bleomycin appears not to. Finally, oxygen plays different roles with the different agents. "Reactive oxygen" is the hydrogen-atom-abstracting species for ionizing radiation and bleomycin, whereas a radical form of NCS itself is the reactive species; in both types of reactions, however, dioxygen (and presumably nitroaromatic radiation sensitizers) add to the generated carbon-centered radical on the deoxyribose to produce irreversible DNA damage.

NCS has served as a mechanistic model for agents that are activated by cellular thiol to a DNA-attacking radical species. Recently, reports have appeared on the identification of exceptionally potent antitumor antibiotics that appear to be related in structure and mechanism to NCS. The calichemicins[34] and esperamicins[35] contain ene-diyne systems that can be triggered by thiol to aromatize via a free radical intermediate (aryl diradical) capable of hydrogen atom abstraction from the deoxyribose of DNA. These agents are the most potent antitumor drugs presently known; against murine tumors calichemicin is about 4000 fold more potent than adriamycin.

The term "radiomimetic" has been misused in the past to describe drugs that had some biological effects similar to those of X-rays, but whose mechanisms were clearly unrelated to X-rays. Based on data such as described in this review, it now appears justified to apply this term to agents belonging to the neocarzinostatin-bleomycin class of antitumor antibiotics.

REFERENCES

1. I. H. Goldberg, T. Hatayama, L.S. Kappen, M.A. Napier, and L.F. Povirk, Protein antibiotics as DNA damaging agents, in: "Molecular Actions and Targets for Cancer Chemotherapeutic Agents," A.C. Sartorelli, J.R. Bertino and J.S. Lazo, eds., Second Annual Bristol-Myers Symposium in Cancer Research, Academic Press, New York, p. 163 (1981).
2. I.H. Goldberg, Free radical mechanisms in neocarzinostatin-induced DNA damage, Free Radical Biol. Med. 3:41 (1987).
3. O.D. Hensens, R.S. Dewey, J.M. Liesch, M.A. Napier, R.A. Reamer, J.L. Smith, G. Albers-Schönberg, and I.H. Goldberg, Neorcarzinostatin chromophore: Presence of a highly strained ether ring and its reaction with mercaptan and sodium borohydride, Biochem. Biophys. Res. Commun. 113:538 (1983).
4. M. Shibuya, K. Toyooka, and S. Kubota, Synthesis of the naphthalene-carboxylic acid derivative obtained from neocarzinostatin (NCS): a structure revision, Tetradedron Lett. 25:1171 (1984).
5. K. Edo, M. Mizugaki, Y. Koide, H. Seto, K. Furikata, N. Otake, and N. Ishida, The structure of neocarzinostatin chromophore possessing a novel bicyclo[7,3,0]dodecadiyne system, Tetradedron Lett. 26:331 (1985).
6. D. Dasgupta, D.S. Auld and, I.H. Goldberg, Cryospectrokinetic evidence for the mode of reversible binding of neocarzinostatin chromophore

to poly(deoxyadenylic-deoxythymidylic acid), Biochemistry 24:7049 (1985).

7. D. Dasgupta and I.H. Goldberg, Mode of reversible binding of neocarzinostatin chromophore to DNA: Evidence for binding via the minor groove, Biochemistry 24:6913 (1985).

8. D. Dasgupta and I.H. Goldberg, Base sequence dependency of binding, Nucleic Acid Res. 14:1089 (1986).

9. L.S. Kappen, I.H. Goldberg, and J.M. Liesch, Identification of thymidine 5'-aldehyde at DNA strand breaks induced by neocarzinostatin chromophore, Proc. Natl. Acad. Sci. USA 79:744 (1982).

10. L.S. Kappen and I.H. Goldberg, DNA damage by neocarzinostatin chromophore: Strand breaks generated by selective oxidation of C-5' of deoxyribose, Biochemistry 22:4872 (1983).

11. L.F. Povirk and I.H. Goldberg, Covalent adducts between DNA and the nonprotein chromophore of neocarzinostatin contain a modified deoxyribose, Proc. Natl. Acad. Sci. USA 79L369 (1982).

12. L.F. Povirk and I.H. Goldberg, Neocarzinostatin chromophore-DNA adducts: Evidence for a covalent linkage to the oxidized C-5' of deoxyribose, Nucleic Acids Res. 10:6255 (1982).

13. L.F. Povirk and I.H. Goldberg, Competition between anaerobic covalent linkage of neocarzinostatin chromophore to deoxyribose in DNA and oxygen-dependent strand breakage and base release, Biochemistry 23:6295 (1984).

14. L.F. Povirk and I.H. Goldberg, Endonuclease-resistant apyrimidinic sites formed by neocarzinostatin at cytosine residues in DNA: Evidence for a possible role in mutagenesis, Proc. Natl. Acad. Sci. USA 82:3182 (1985).

15. L.F. Povirk and I.H. Goldberg, Base substitution mutations induced in the cI gene of lambda phage by neocarzinostatin chromphore: Correlation with depyrimidination hotspots at the sequence AGC, Nucleic Acids Res. 14:1417 (1986).

16. L.S. Kappen, T.E. Ellenberger, and I.H. Goldberg, Mechanism and base specificity of DNA breakage in intact cells by neocarzinostatin, Biochemistry 26:384 (1987).

17. W.G. DeGraff and J.B. Mitchell, Glutathione dependence of neocarzinostatin cytotoxicity and mutagenicity in Chinese hamster V-79 cells, Cancer Res. 45:4760 (1985).

18. L.F. Povirk and I.H. Goldberg, Detection of neocarzinostatin chromophore-deoxyribose adducts as exonuclease-resistant site in defined-sequence DNA, Biochemistry 24:4035 (1985).

19. A.F. Fuciarelli, G.G. Miller, and J.A. Raleigh, An immunochemical probe for 8,5'-cycloadenosine-5'-monophosphate and its deoxy analog in irradiated nucleic acids, Radiat. Res. 104:272 (1985).

20. M. Dizdaroglu, M.L. Dirksen, M.G. Simic, and J.H. Robbins, Identification of the novel lesion 8,5'-cyclo-2'-deoxyguanosine in DNA isolated from γ-irradiated human cells, Fed. Proc. 45:1626 (1986).

21. L.F. Povirk and I.H. Goldberg, Stoichiometric uptake of molecular oxygen and consumption of sulfhydryls by neocarzinostatin chromophore bound to DNA, J. Biol. Chem. 258:11767 (1983).

22. D.-H. Chin, S.A. Carr, and I.H. Goldberg, Incorporation of $^{18}O_2$ into thymidine 5'-aldehyde in neocarzinostatin chromophore-damaged DNA, J. Biol. Chem. 259:9975 (1984).

23. D.-H. Chin and I.H. Goldberg, Generation of superoxide free radical by neocarzinostatin and its possible role in DNA damage, Biochemistry 25:1009 (1986).

24. R.L. Charnas and I.H. Goldberg, Neocarzinostatin abstracts a hydrogen during formation of nucleotide 5'-aldehyde on DNA, Biochem. Biophys. Res. Commun. 122:642 (1984).

25. L.S. Kappen and I.H. Goldberg, Activation of neocarzinostatin chromophore and formation of nascent DNA damage do not require molecular oxygen, Nucleic Acids Res. 13:1637 (1985).

26. V. Favaudon, R.L. Charnas, and I.H. Goldberg, Polydeoxyadenylic-deoxythymidylic acid damage by radiolytically activated neocarzinostatin, Biochemistry 24:250 (1985).
27. L.S. Kappen and I.H. Goldberg, Nitroaromatic radiation sensitizers substitute for oxygen in neorcarzinostatin-induced DNA damage, Proc. Natl. Acad. Sci. USA 81:3312 (1984).
28. D.-H. Chin, L.S. Kappen, and I.H. Goldberg, 3'-formylphosphate-ended DNA: a high-energy intermediate in antibiotic-induced DNA sugar damage, Proc. Natl. Acad. Sci. USA, in press.
29. C. von Sonntag, U. Hagen, A. Schon-Böpp, and D. Schulte-Frohlinde, Radiation-induced strand breaks in DNA: Chemical and enzymatic analysis of end groups and mechanistic aspects, Adv. Radiat. Bol. 9:109 (1981).
30. G. Scholes and J. Weiss, Formation of acetylphosphate in the action of X-rays on monoethyl phosphoric acid in aqueous solutiion, Nature 173:267 (1954).
31. S. Nishimoto, H. Ide, T. Wada, and T. Kagiya, Radiation-induced hydroxylation of thymine promoted by electro-affinic compounds, Int. J. Radiat. Biol. 44:585 (1983).
32. T. Honna, M. Kuwabara, and G. Yoshii, Effect of misonidazole on free radical formation and inorganic phosphate release in gamma-irradiated nucleotides, Int. J. Radiat. Biol. 39:457 (1981).
33. J.F. Remsen, Effect of misonidazole on formation of thymine damage by gamma rays, Radiat. Res. 101:306 (1985).
34. M.D. Lee, T.S. Dunne, C.C. Chang, G.A. Ellestad, M.M. Siegel, G.O. Morton, W.J. McGahren, and D.B. Borders, Calichemicins, a novel family of antitumor antibiotics. 2. Chemistry and structure of calichemicinγ_1^I, J. Am. Chem. Soc. 109:3466 (1987).
35. J. Golik, G. Dubay, G. Groenewold, H. Kawaguchi, M. Konishi, B. Krishnan, H. Ohkuma, K. Saitoh, and T.W. Doyle, Esperamicins, a novel class of potent antitumor antibiotics. 3. Structures of esperamicins A_1,A_2, and A_{1B}, J. Am. Chem. Soc. 109:3462 (1987).

REDUCTION OF 5-NITROIMIDAZOLES, NITROFURAZONE, AND 2,4-DINITROPHENOL TO THEIR FREE RADICAL METABOLITES BY *TRITRICHOMONAS FOETUS* HYDROGENOSOMES

Silvia N.J. Moreno and Roberto Docampo

Instituto de Microbiologia
Universidade Federal do Rio de Janeiro
Cidade Universitária, Bloco I
Rio de Janeiro 21944, RJ, Brazil

INTRODUCTION

Oxidative decarboxylation of pyruvate is catalyzed in trichomonads by pyruvate:ferredoxin oxidoreductase (PFO). This enzyme is localized in characteristic membrane-bound organelles that are termed hydrogenosomes on the basis of their biochemical characteristics. The main metabolic function of hydrogenosomes is the conversion of pyruvate to acetate via acetyl-CoA, accompanied by substrate level phosphorylation coupled to production of molecular hydrogen (Figure 1).

We have recently demonstrated the formation of two substrate free radicals as intermediates in the reaction of PFO in *Tritrichomonas foetus* hydrogenosomes.[1] The first step in this reaction consists in the binding of pyruvate to thiamine diphosphate (PPT), followed by decarboxylation to yield hydroxyethyl-thiamine diphosphate (Figure 2). This intermediate subsequently donates one electron to the iron-sulfur cluster of the enzyme to form a carbon-centered radical at C-3 of the parent pyruvate molecule (C-2 of the hydroxyethyl group).

The iron-sulfur cluster is then reoxidized by ferredoxin to give an enzyme-free radical intermediate complex. This complex is stable until coenzyme A reduces the iron-sulfur cluster, converting itself into a thiyl radical derivative. The thiyl radical then accepts the acetyl group from thiamine diphosphate generating acetyl-CoA. Reoxidation of the iron-sulfur cluster by ferredoxin completes the cycle (Figure 2).

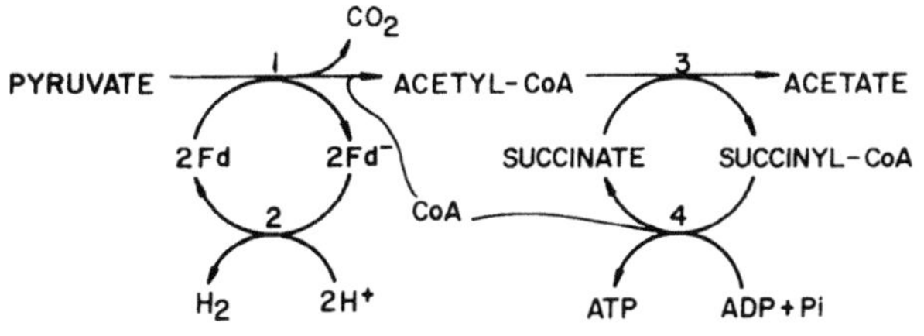

Fig.1. Map of pyruvate metabolism in *Tritrichomonas foetus* hydrogenosomes. Pyruvate:ferredoxin oxidoreductase (step 1); hydrogenase (step 2); acetate:succinate-CoA transferase (step 3); succinate thiokinase (step 4); Fd, ferredoxin.

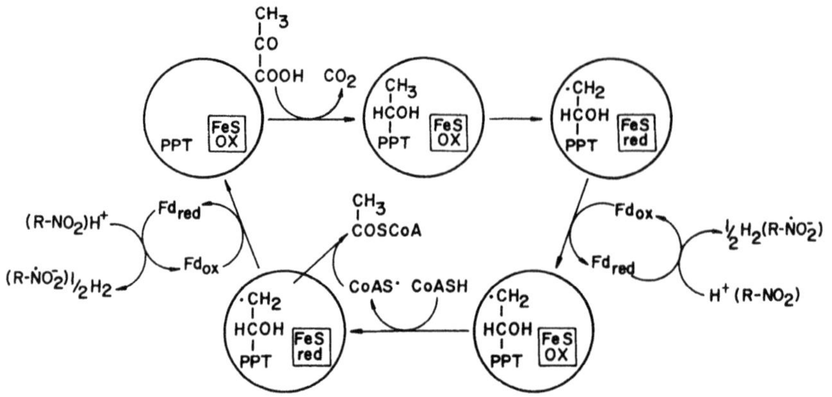

Fig. 2. Catalytic cycle of pyruvate:ferredoxin oxidoreductase from T. foetus. Fd, ferredoxin, R-NO_2, nitro-compound, R-NO_2^-, nitro anion radical.

In addition to its physiological role in the formation of acetyl-CoA and H_2, this system is also involved in the reduction of the nitro-compounds that are used in the chemotherapy of the diseases caused by these organisms.[2-4]

2,4-Dinitrophenol (2,4-DNP), 5-nitroimidazoles, and nitrofurazone exert significant effects on trichomonad flagellates and parasitic amoebae.[2-5] These observations are not easy to interpret in the case of 2,4-DNP because these aero-tolerant anaerobic protozoa do not carry out oxidative phosphorylation.[6] Thus, the effect of 2,4-DNP is not likely to be due to its uncoupling properties.[5] In this communication we show that T.foetus hydrogenosomal fractions are able to reduce 2,4-DNP to its free radical metabolite as occurs with 5-nitroimidazoles and nitrofurazone. Redox cycling of these nitro-compounds was also demonstrated.

RESULTS AND DISCUSSION

The incubation of nitrofurazone (Figure 3A), dimetridazole (Figure 4A), ornidazole (Figure 4B), and bamnidazole (Figure 4C) with T. foetus hydrogenosomal fractions in the presence of pyruvate and CoA generated multi-line ESR spectra identified as their nitro anion radicals. Omission of CoA led to a total loss of activity. No ESR signals could be detected when using either heat-denatured hydrogenosomes (70°C, 30 min) or when the drugs were omitted (Figure 3B and not shown).

The spectra of the anion radicals could not be observed under aerobic conditions. This is in agreement with the redox cycling of nitro-compounds as shown in reaction 5.

$$ArNO_2^- + O_2 \longrightarrow ArNO_2 + O_2^- \quad (5)$$

Accordingly, aerobic incubation of the hydrogenosomal fraction containing nitrofurazone (Figure 5A) or 5-nitroimidazoles (not shown), pyruvate, CoA, and the spin trap DMPO had spectra containing contributions from both the superoxide adduct and the hydroxyl adduct, although the superoxide spin adduct predominated in incubations with nitrofurazone (Figure 5A) and the hydroxyl spin adduct predominated in incubations with 5-nitroimidazoles (not shown). Both signals were greatly reduced by superoxide dismutase. Catalase had no effect on the observed spectra.

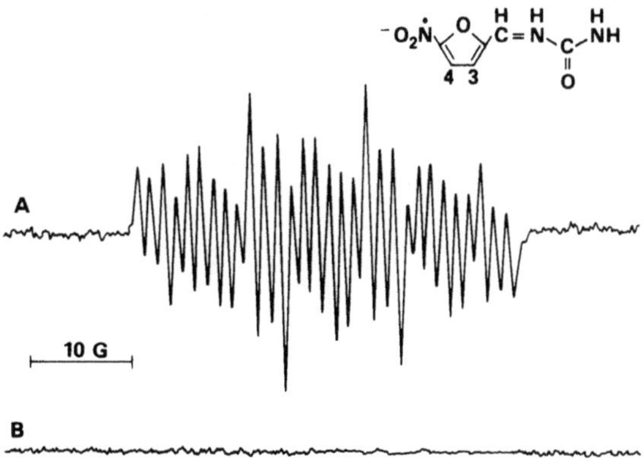

Fig. 3. The ESR spectra of T. foetus hydrogenosomes in the presence and absence of nitrofurazone. A, the ESR spectrum of nitrofurazone anion radical. The spectrum was observed after anaerobic incubation of 1 mM nitrofurazone with 5 mM pyruvate, 1 mM CoA, and T. foetus hydrogenosomal fraction (0.6 mg/ml) in the buffer described in reference 4. The spectrum was obtained with a nominal microwave power of 20 mW and a modulation amplitude of 0.5 G. B, identical with A, but without nitrofurazone. Other experimental conditions were given in reference 4.

These results show that the DMPO-hydroxyl adduct is formed by decomposition of the DMPO-superoxide adduct and/or by the trapping of the hydroxyl radical generated by the decomposition of the DMPO-superoxide radical adduct.[7,8] The DMPO adduct signals were not observed in the absence of pyruvate or CoA. Weak superoxide dismutase-sensitive DMPO-adduct signals were observed in the presence of pyruvate, CoA, and T. foetus hydrogenosomal fraction as a result of pyruvate plus CoA-supported superoxide anion formation by the hydrogenosomes.[4]

The incubation of 2,4-DNP (Figure 6) with T. foetus hydrogenosomal fraction in the presence of pyruvate and CoA generated a free radical metabolite with hyperfine splittings similar to those observed by Borg[9] after reduction of 2,4-DNP in a chemical system, and which were attributed to the anion radical derivatives of 2,4-DNP.[9] No ESR spectrum could be observed when using heat-denatured hydrogenosomes or when pyruvate or CoA were omitted. The spectrum of this free radical metabolite of 2,4-DNP was only observed under anaerobic conditions. Aerobic incubations of the hydrogenosomal fraction containing 2,4-DNP, pyruvate, and CoA gave weak superoxide dismutase-sensitive DMPO-adduct signals, thus indicating redox cycling of 2,4-DNP.

Since the toxic effects of 2,4-DNP on T. foetus are observed at substrate level concentrations[5] that are higher than those which uncouple mitochondrial respiration, these effects can be explained by the role of this compound as an electron acceptor. In addition, redox cycling of 5-nitroimidazoles and 2,4-DNP under aerobic conditions might be considered as a detoxification reaction by inhibiting net reduction of the drugs,[4] thereby inhibiting their uptake. The resulting low steady-state concentrations of O_2^{-} are easily detoxified by the superoxide dismutase present in these parisites[10]. On the other hand, redox cycling of nitrofurazone or other compounds with more positive reduction potential will result in formation of high steady state concentrations of oxygen-derived metabolites that might be of toxicological significance.

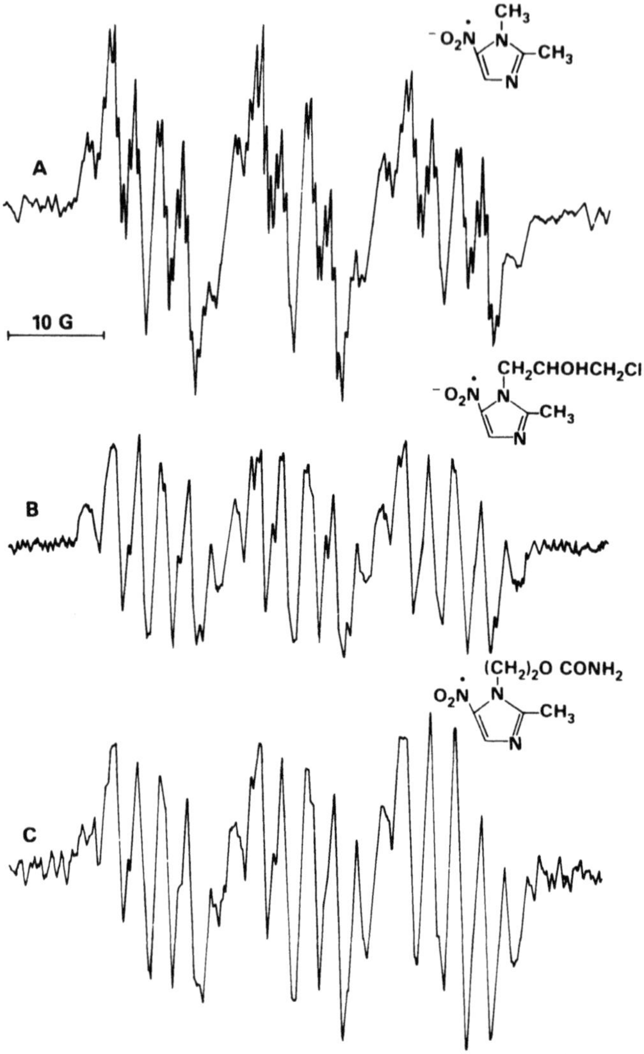

Fig. 4. The ESR spectra of T. foetus hydrogenosomes in the presence of 5-nitroimidazoles. A, the ESR spectrum of dimetridazole anion radical. The spectrum was observed after anaerobic incubation of 20 mM dimetridazole with 5 mM pyruvate, 1 mM CoA, and T. foetus hydrogenosomal fraction (9.5 mg/ml) in the buffer described in Figure 3. The spectrum was obtained with a nominal microwave power of 20 mW and a modulation amplitude of 1.6 G. B, identical with A, but with 20 mM ornidazole instead of dimetridazole. C, identical with A, but with 20 mM bamnidazole instead of dimetridazole.

It is interested to note that redox cycling of 2,4-DNP has been suggested to explain the enhanced cyanide-resistant respiration of rat lung mitochondria induced by this compound.[11] However, no evidence of the formation of the 2,4-DNP anion free radical was provided in that work.[11]

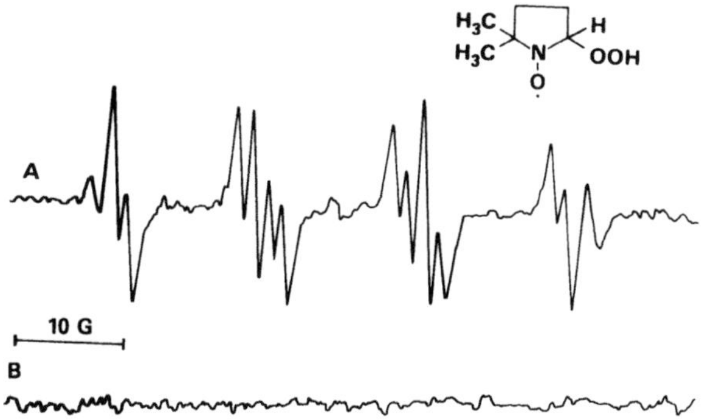

Fig. 5. ESR spectra of the DMPO-superoxide and DMPO-hydroxyl adducts. A, the incubation contained 0.6 mg/ml of hydrogenosomal protein in the same buffer as in Figure 3, 1 mM nitrofurazone, 100 mM DMPO, 5 mM pyruvate, and 1 mM CoA. The nominal microwave power was 20 mW, and the modulation amplitude was 0.3 G. B, same as A but in the presence of superoxide dismutase (20 μg/ml).

Fig. 6. The ESR spectrum of T. foetus hydrogenosomes in the presence of 2,4-DNP. The ESR spectrum was observed after incubation of 3 mM 2,4-DNP with 5 mM pyruvate, 1 mM CoA, and T. foetus hydrogenosomal fraction (9.0 mg/ml) in the buffer described in Figure 3. The spectrum was obtained with a nominal microwave power of 20 mW and a modulation amplitude of 2.5 G. The incubation was anaerobic.

ACKNOWLEDGMENTS

This work received financial support from the National Institute of Environmental Health Sciences, and from CNPq and FINEP, Brazil.

REFERENCES

1. R. Docampo, S.N.J. Moreno and R.P. Mason, Free radical intermediates in the reaction of pyruvate ferredoxin oxireductase in Tritrichomonas foetus hydrogenosomes, J. Biol. Chem. 262:12417 (1987).
2. S.N.J. Moreno and R. Docampo, Mechanism of toxicity of nitro compounds used in the chemotherapy of trichomoniasis, Environm. Hlth. Perspect. 64:199 (1985).
3. S.N.J. Moreno, R.P. Mason, R.P.A. Muniz, F.S. Cruz and R. Docampo, Generation of free radicals from metronidazole and other nitroimidazoles by Tritrichomonas foetus, J. Biol. Chem. 258:4051 (1983).

4. S.N.J. Moreno, R.P. Mason and R. Docampo, Distinc reduction of nitrofurans and metronidazole to free radical metabolites by Tritrichomonas foetus hydrogenosomal and cytosolic enzymes, J. Biol. Chem. 259: 8252 (1984).
5. M. Müller, V. Nseka, S.R. Mack and D.G. Lindmark, Effect of 2,4-dinitrophenol on trichomonas and Entamoeba invadens, Comp. Biochem. Physiol. 64B:97 (1979).
6. M. Müller and D.G. Lindmark, Respiration of hydrogenosomes of Tritrichomonas foetus. II. Effect of CoA on pyruvate oxidation, J. Biol. Chem. 253:1215 (1978).
7. E. Finkelstein, G.M. Rosen, E.J. Rauckman and J. Paxton, Spin trapping of superoxide, Mol. Pharmacol. 16:676 (1979).
8. E. Finkelstein, G.M. Rosen and E.J. Rauckman, Production of hydroxyl radical by decomposition of superoxide spin-trapped adducts, Mol. Pharmacol. 21:262 (1982).
9. D.C. Borg, Electron spin resonance applied to free radicals of physiological, pharmacological or biochemical interest, in: "Biological Applications of Electron Spin Resonance," H.M. Swartz, J.R. Bolton and D.C. Borg, eds., Wiley Interscience, New York, (1964).
10. D.G. Lindmark and M. Müller, Superoxide dismutase in the anerobic flagellates Tritrichomonas foetus and Monocercomonas sp., J. Biol. Chem. 249:4634 (1974).
11. B.A. Freeman and J.D. Crapo, Hyperoxia increases oxygen radical production in rat lungs and lung mitochondria. J. Biol. Chem. 256, 10986-10992 (1981).

ROLE OF FREE RADICALS IN ETOPOSIDE (VP-16,213) ACTION

Birandra K. Sinha

Clinical Pharmacology Branch National Cancer Institute, National Institutes of Health Bldg.10 Rm.6N-119, Bethesda, Maryland 20892, USA

INTRODUCTION

Etoposide (VP-16, Figure 1), a semisynthetic derivative of phodophyllotoxin, is clinically active as a single agent and in combination with other antitumor drugs, e.g. adriamycin and *cis*-platinum. VP-16 has shown promising activity in the treatment of small cell lung carcinoma, testicular tumors and malignant lymphomas[1,2]. VP-16 has been shown to induce DNA strand breaks in tumor cells and it is believed that topoisomerase II is the likely intracellular target for this DNA damage[3,4]. Although the DNA strand breaking activity of VP-16 has been implicated in its cytotoxicty, the molecular mechanism remains to be defined. For DNA damage and the biological activity, the presence of cellular components and the presence of free hydroxyl group in the C-4' are essential[5] suggesting other factors may also be important in the biochemical mechanism of the drug.

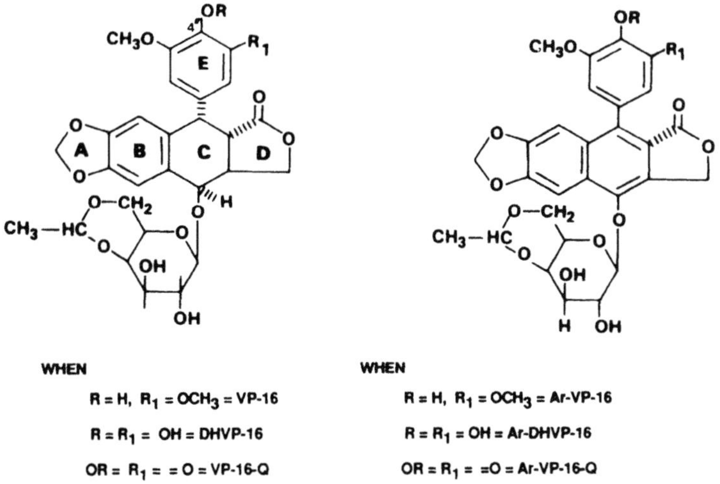

WHEN

R = H, R_1 = OCH_3 = VP-16

R = R_1 = OH = DHVP-16

OR = R_1 = =O = VP-16-Q

WHEN

R = H, R_1 = OCH_3 = Ar-VP-16

R = R_1 = OH = Ar-DHVP-16

OR = R_1 = =O = Ar-VP-16-Q

Fig 1: Structures of VP-16 and its metabolites formed during enzymatic activation.

Studies reported from our laboratory have shown that VP-16 is O-demethylated in the presence of liver microsomal enzymes and NADPH to an o-dihydroxy compound (DHVP). Using stimulators and inhibitors, it was shown that cytochrome P-450 was responsible for this O-demethylation[6]. DHVP was rapidly oxidized either chemically or enzymatically to the corresponding semiquinone free radical and subsequently to a reactive o-quinone derivative (VP-16-Q) resulting in P-450-dependent covalent binding of the drug/ metabolites to proteins and exogeneously added DNA[6-8]. In addition to cytochrome P-450, peroxidases (horseradish or prostaglandin synthase) also catalyzed the activation of VP-16 and the related antitumor drug, VM-26[9,10]. The peroxidase-catalyzed activation resulted in the formation of a free radical intermediate (Figure 2), as detected by electron spin resonance (ESR) spectrometry.

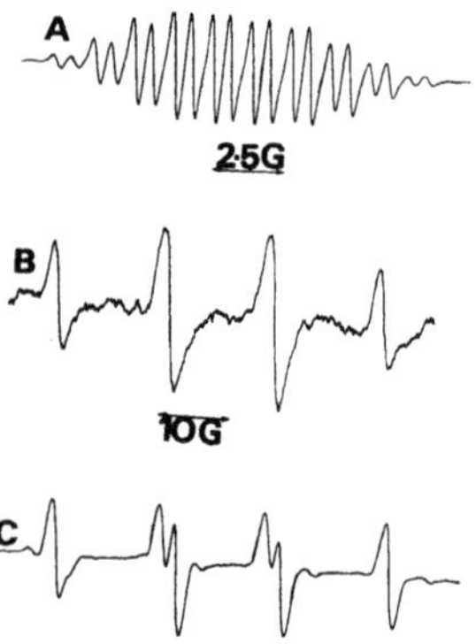

Fig 2: The ESR spectrum obtained from (A) VP-16 in the presence of peroxidase (HRP or prostaglandin synthase) in the presence of hydrogen peroxide or arachidonic acid. The radical was extracted into organic solvent, concentrated under nitrogen and treated with (B) GSH in the presence of DMPO in aqueous buffer and (C) identical to (B) except cysteine was used. The hyperfine splitting constants for DMPO-SG and DMPO-Scyst. are $a^N = 15.0$ G and $a^H = 16.3$ G: $a^N = 15.3$ G and $a^H = 17.0$ G, respectively.

This radical was shown to be the phenoxy radical of VP-16, VP-16·, formed from the one-electron oxidation of the free OH group in the C-4' position. The peroxidase-catalyzed oxidation of VP-16 also resulted in O-demethylation with the formation of VP-16-Q[10]. Moreover, this peroxidase-dependent oxidation of VP-16 also formed reactive intermediates that bound covalently to proteins and DNA. The peroxidase-catalyzed O-demethylation required both the presence of the free OH group and the formation of the phenoxy radical, since the 4'-o-methyl derivative of VP-16 neither formed the phenoxy radical nor the o-quinone compound when incubated with the peroxidase systems.

Recently, we have found that VP-16 is extremely toxic to human promyelocytic leukemia cells, HL-60, while the o-methyl derivative of VP-16 is devoid of any antitumor activity. Interestingly, VP-16 rapidly formed the phenoxy radical when incubated with HL-60 cells in the presence of hydrogen peroxide while no radical was detected from the o-methyl derivative under identical conditions. The half life of the phenoxy radical in HL-60 cells was less than 5 minutes, in contrast to that obtained by the peroxidase activation, indicating a rapid reduction of the radical by cellular reductants such as GSH. This is consistent with our earlier studies and we have found that the phenoxy radical is readily reduced by soluble thiols to the parent drug and with the oxidation of thiols (11). Using spin-trapping ESR techniques, we have shown that this reduction of the VP-16 radical by thiols resulted in the formation of thiyl radicals (RS·, Figure 2), which were intermediates for the formation of the oxidized thiols. Furthermore, in vivo administration of VP-16 to mice decreased total thiol pools in liver and concomitantly increased the oxidation of thiols, detected as GSSG, suggesting that the phenoxy radical may also be formed in_vivo.

$$\text{VP-16}^{\cdot} + \text{RSH} \longrightarrow \text{VP-16} + [\ \text{RS}^{\cdot}\ ] \longrightarrow \text{RSSR}$$

Since GSH is known to play a significant role in the cellular defenses against oxidant and DNA alkylation, VP-16-induced depletion of GSH may lead to cellular sensitization to VP-16 by enhancing bioalkylation of critical cellular proteins and DNA by VP-16-derived reactive species causing cellular dysfunction and ultimately cell death.

Thus it may be concluded that:1) VP-16 undergoes both cytochrome P-450 and peroxidase-dependent metabolism with the formation of the VP-16 phenoxy radical;2) the phenoxy radical is the pivotal intermediate for the the formation of reactive species that alkylate proteins and DNA;3) while the phenoxy radical does not react with molecular oxygen to generate either superoxide or hydroxyl radicals, it is rapidly reduced by glutathione with regeneration of the parent drug and the oxidation of GSH to GSSG; and 4) administration of VP-16 in mice results in a decrease in the soluble thiols, presumably through the formation of the phenoxy radical; VP-16-induced GSH depletion may enhance tumor cell sensitization to VP-16 and to other antitumor drugs.

REFERENCES

1. M. Rozencweig, D. D. Von Hoff, J. E. Henney, and F. M. Muggia, VM-26 and VP-16-213: a comparative analysis, Cancer 40: 334 (1977).
2. P. J. Dwyer, B. Leyland-Jones, M. T. Alonso, S. Marsoni, and R.E. Wittes, Etoposide (VP-16-213): current status of an active anticancer drug, N. Engl. J. Med., 312: 692 (1985).
3. A. J. Wojniac, and W. E. Ross, DNA damage as a basis for 4'-demethylepipodophyllotoxin-9-(4,6-O-ethylene-ß-D-glucopyranoside (etoposide) cytotoxicity, Cancer Res., 43: 120 (1983).

4. B. H. Long, S. T. Musial, and M. G. Brattain, Comparison of cytotoxicity and DNA breakage activity of congeners of podophyllotoxin including VP-16-213 and VM-26: a quantative structure-activity relationship, Biochemistry, 23: 1183 (1984).
5. J. D. Loike, and S. B. Horwitz, Effects of VP-16-213 on the intracellular degradation of DNA in HeLa cells, Biochemistry 15: 5443 (1976).
6. N. Haim, J. Nemec, J. Roman, and B. K. Sinha, In vitro metabolism of etoposide (VP-16-213) by liver microsomes and irreversible binding of reactive intermediates to microsomal proteins, Biochem. Pharmacol., 36: 527 (1987).
7. B. K. Sinha, and C. E. Myers, Irreversible binding of etoposide (VP-16-213) to deoxyribonucleic acid and proteins, Biochem. Pharmacol., 33 : 3725 (1984).
8. B. K. Sinha, M. A. Trush, and B. Kalayanaraman, Microsomal interactions and inhibition of lipid peroxidation by etoposide (VP-16-213): Implications for mode of action, Biochem.Pharmacol., 34: 2036 (1985).
9. B. K. Sinha, M. A. Trush, and B. Kalayanaraman, Free radical metabolism of VP-16 and inhibition of anthracycline-induced lipid peroxidation,Biochem. Pharmacol., 32: 3495 (1983).
10. N. Haim, J. Roman, J. Nemec, and B. K. Sinha, Peroxidative free radical formation and O-demethylation of etoposide (VP-16) and teniposide (VM-26), Biochem. Biophys. Res. Commun., 135: 215 (1986).
11. A. G. Katki, B. Kalayanaraman, and B. K. Sinha, Interactions of the antitumor drug, etoposide, with reduced thiols in vitro and in vivo, Chem. Biol. Interact., 62, 237 (1987).

THE MECHANISM OF QUINONE FORMATION FROM CARCINOGENIC BENZO[a]PYRENE MEDIATED BY LIPID PEROXIDATION IN LIPOSOMAL SUSPENSION

Junji Terao and Setsuro Matsushita

Research Institute for Food Science
Kyoto University, Uji, Kyoto 611, Japan

INTRODUCTION

Polycyclic aromatic hydrocarbons, including benzo[a]pyrene (B[a]P), are widely distributed in the environment and some of them are known to possess potent carcinogenicity. B[a]P is metabolized to ultimate or proximate carcinogenic form by subcellular mixed function oxidase system. Lipid peroxides and their derived radicals are suggested to play a role in the metabolic activation of B[a]P. Earlier works showed that B[a]P quinones were formed as main products when B[a]P was cooxidized by nonenzymatic lipid peroxidation of polyunsaturated fatty acids[1] or of liver microsomal membrane[2]. The contribution of quinone formation to the carcinogenic activation of B[a]P is suggested by the following observations: (i) Intermediates in quinone formation can bind to DNA and cause DNA strand scission. (ii) B[a]P quinone possesses mutagenicity. (iii) B[a]P quinone generates active oxygen species by the reduction and oxidation cycle involving quinone, hydroquinone and molecular oxygen. Such active oxygens are recently recognized to act as potent promoters in carcinogenesis. The purpose of this work is to know the mechanism of lipid peroxidation-mediated quinone formation from B[a]P in biomembranes by using a water-soluble azo compounds as the free radical initiator in a liposomal suspension[3].

COOXIDATION OF B[a]P BY LIPID PEROXIDATION IN LIPOSOMES

Our experimental procedure is illustrated in Figure 1. Multilamellar liposomes containing B[a]P were obtained from rat liver phosphatidylcholine (PC), and oxidation was initiated by the addition of aqueous solution of 2-2'-azobis(2-amidinopropane)hydrochloride (AAPH). After the incubation was completed, the reaction mixture was extracted by chloroform and methanol , and then B[a]P fraction was subjected to reverse-phase HPLC analysis. B[a]P oxidation products collected by HPLC were then analyzed by GC-MS. The pathway of lipid peroxidation in liposomal membrane induced by AAPH has been well clarified as follows (Figure 2). Peroxy radical (AOO•) is produced in aqueous phase by thermal decomposition of the azo compound and the following reaction with molecular oxygen. Lipid radical (L•) is generated by the attack of the radical and subsequently molecular oxygen reacts rapidly with lipid radical, resulting in lipid peroxy radical (LOO•). A radical chain reaction proceeds via LOO• as an obligatory intermediate. As a result of these reactions, hydroperoxides (LOOH) accumulated exclusively.

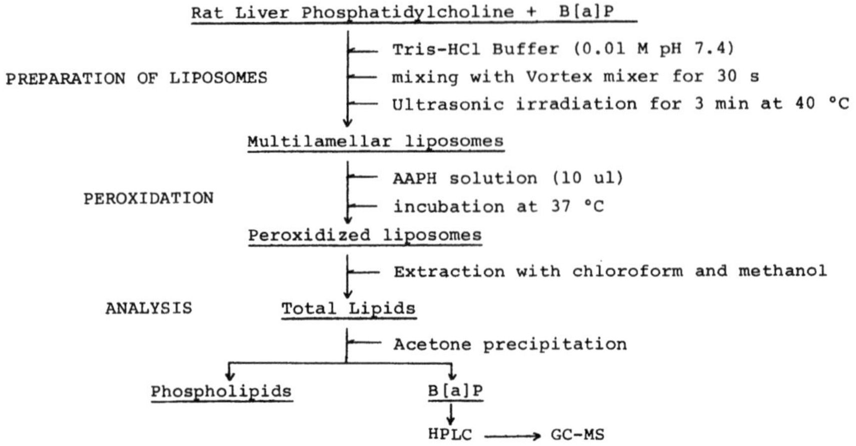

Figure 1. Experimental procedure for cooxidation of B[a]P by lipid peroxidation in liposomal suspension.

In this reaction system, B[a]P was found to be cooxidized by AAPH-induced lipid peroxidation. Main oxidation products which accumulated in peroxidized liposomes were B[a]P 1,6-, 3,6-, and 6,12-quinone. Antioxidants acting as scavengers of chain propagating LOO•, α-tocopherol, butyl hydroxy-toluene, and butyl hydroxyanisole effectively prevented B[a]P oxidation as well as lipid peroxidation in the liposomal suspension. The finding that PC-hydroperoxides, the primary products of PC oxidation, did not react with B[a]P in the absence of the azo compounds suggests that LOO•, not LOOH, is responsible for the formation of these quinones. On the other hand, B[a]P quinones were formed to some extent when B[a]P was incubated with AAPH in the liposomes of saturated PC (dimyristoyl PC). This means that peroxy radical generated from the thermal decomposition of AAPH (AOO•) can serve as an initiator of the formation of B[a]P quinones. Therefore, it can be concluded that peroxy radicals derived from both lipids and AAPH are responsible for the attack to B[a]P when lipid peroxidation is initiated by the azo compounds.

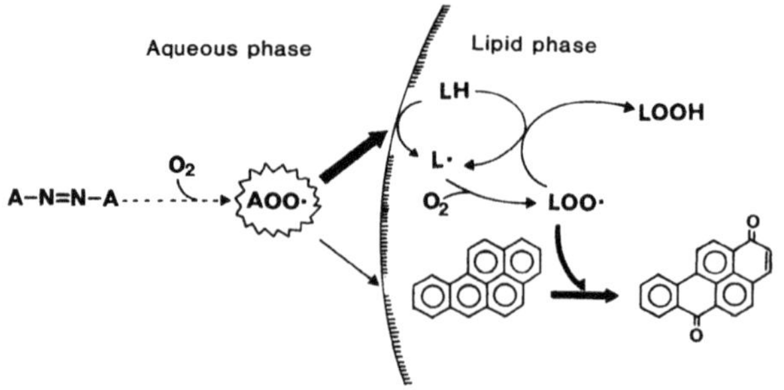

Figure 2. AAPH-induced lipid peroxidation in liposomal suspension and the formation of B[a]P.

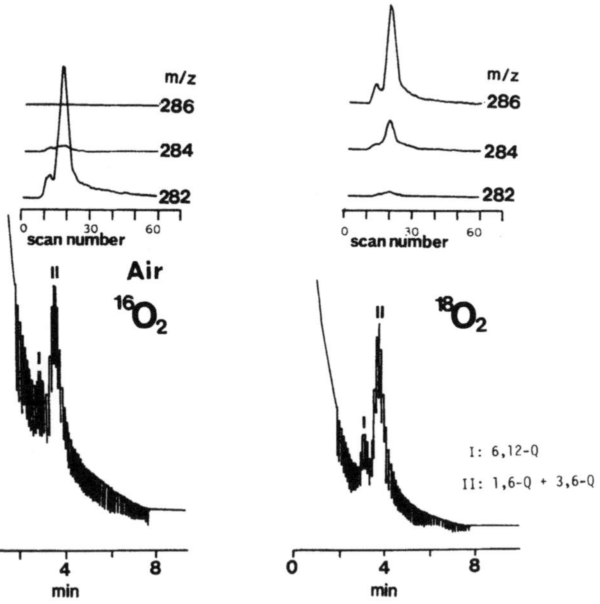

Figure 3. Mass chromatography of B[a]P Quinones obtained from peroxidized rat liver PC liposomes under air or $^{18}O_2$-rich atmosphere[3]. The liposomal suspension (1.0 ml) containing rat liver PC (1.25 µmol) and B[a]P (0.25 µmol) was peroxidized by the addition of AAPH (10 mM) for 24 hr at 37°C.

INCORPORATION OF $^{18}O_2$ INTO B[a]P QUINONES

To clarify the origin of the oxygen in B[a]P quinone, we first carried out the experiment using $^{18}O_2$ gas. A liposomal suspension of rat liver PC containing B[a]P was peroxidized by the addition of AAPH in the air or under $^{18}O_2$-rich atmosphere. The mass chromatography of B[a]P quinone formed under $^{18}O_2$ gas showed the prominent peak at m/z 286, which was 4 mass unit higher than the molecular ion of B[a]P quinones (Figure 3). Therefore, the origin of the two oxygens in B[a]P quinone was confirmed to be molecular oxygen.

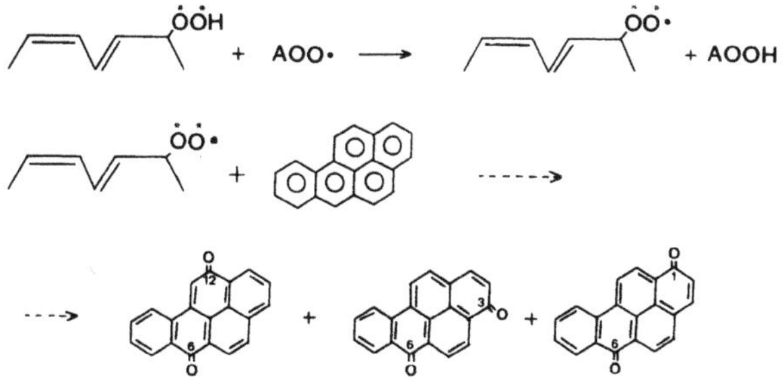

Figure 4. Possible reaction of B[a]P with ^{18}O-labelled ML-OOH.

Table 1. Relative Intensities of the Molecular Ions of B[a]P Obtained from the Reaction with ^{18}O-Labeled ML-OOH[3].

	Reaction with ^{18}O-labeled ML-OOH		Lipid Peroxidation in the air		
m/z	6,12-quinone	1,6-quinone + 3,6-quinone	1,6-quinone	3,6-quinone	6,12-quinone
282	100	100	100	100	100
284	41	26	4	5	4
286	nd	2	nd	nd	nd

Dimyristoryl-PC (1.25 μmol), B[a]P (0.15 μmol), and ML-OOH (1.25 μmol) were dispersed in 0.5 ml of Tris-HCl buffer (0.01M, pH 7.4). AAPH (20 mM) was added to the suspension and incubated at 37°C for 4 hr.

REACTION OF B[a]P WITH ^{18}O-LABELED LIPID HYDROPEROXIDES

To know whether or not the oxygen of LOO• is transfered into B[a]P, an experiment using ^{18}O-labeled LOOH was designed as shown in Figure 4. Dimyristoyl-PC liposomes containing B[a]P and methyl linoleate hydroperoxides (ML-OOH), whose hydroperoxy oxygen was labeled by ^{18}O (O*) at 54% excess, were mixed and the reaction was initiated by the addition of AAPH. Labeled LOO• was assumed to be produced by the action of AOO•, and GC-MS analysis of the reaction products, B[a]P quinones, was supposed to clarify the transfer of oxygen from LOO• to B[a]P. Table 1 demonstrates that ion at m/z 284, 2 mass unit higher than the molecular ion, was accompanied by the molecular ion at significant level, whereas the ion intensity at m/z 286, 4 mass unit higher than the molecular ion, was negligible. Thus, it can be concluded that LOO•binds to B[a]P directly, and the origin of oxygen in B[a]P quinone is partly due to LOO . The source of the other oxygen may be the molecular oxygen itself or peroxy radical derived from AAPH.

From the above results, we propose the mechanism of B[a]P quinone formation in which LOO• generated in polyunsaturated fatty acid moiety binds with the reactive 6-position of B[a]P. The subsequent attack of molecular oxygen at 1, 3, or 12-position of the intermediate may result in the formation of B[a]P quinones. However, the mass spectrometric analysis of B[a]P quinone could not distinguish between the position of labeled oxygen and that of unlabeled oxygen. A study clarifying the intermediate of this reaction would confirm the mechanism of quinone formation by the peroxidative attack on B[a]P in a membrane system.

REFERENCES

1. J. D. Gower and E. D. Wills, The generation of oxidation products of benzo[a]pyrene by lipid peroxidation: A study using γ-irradiation, Carcinogenesis 5:1183 (1984).
2. R. Morgenstern, J. W. DePierre, C. Lind, C. Guthenberg, B. Mannervik, and L. Ernster, Benzo[a]pyrene quinones can be generated by lipid peroxidation and are conjugated with glutathione by glutathione S-transferase B from rat liver, Biochem. Biophys. Res. Comm. 99:682 (1981).
3. J. Terao, P. L. Boey, H. Murakami, and S. Matsushita, Quinone formation from carcinogenic benzo[a]pyrene mediated by lipid peroxidation in phosphatidylcholine liposomes, Arch. Biochem. Biophys. 254:472 (1987).

LIPID PEROXIDATION, PROTEIN THIOLS AND CALCIUM HOMEOSTASIS IN BROMOBENZENE-INDUCED LIVER DAMAGE

A.F. Casini, E. Maellaro, A. Pompella, M. Ferrali, and M. Comporti

Istituto di Patologia Generale, Università di Siena
Via del Laterino 8, 53100 Siena, Italy

INTRODUCTION

Previous studies from our laboratory[1] have shown that bromobenzene or iodobenzene administration to mice results in the development of lipid peroxidation when the hepatic glutathione (GSH) depletion reaches critical values. Liver necrosis behaves similarly and strictly correlates with the extent of lipid peroxidation. The treatment of the intoxicated animals with Trolox C (6-hydroxy-2,5,7,8-tetramethylchroman-2-carboxylic acid), a lower homolog of vitamin E, completely prevented both lipid peroxidation and necrosis, while not changing at all the extent of the covalent binding of bromobenzene metabolites to liver protein. This would suggest that lipid peroxidation is an important factor in the pathogenesis of the bromobenzene-induced liver cell death.

Evidence has been accumulated[2,3] according to which perturbation of calcium homeostasis is the crucial and irreversible event leading to cell death in the liver injury produced by a number of toxins. As far as GSH-depleting agents, a great deal of experimental work carried out with isolated hepatocytes undergoing an oxidative stress suggests that calcium homeostasis is altered by the loss of protein -SH groups secondary to GSH depletion.[3-5] The loss of protein sulphydryl groups would be mainly the result of oxidative processes occurring during the metabolic redox cycling of some chemicals.[3-5]

Therefore, in the present study the relationships among GSH depletion, lipid peroxidation, loss of protein thiols, calcium homeostasis and liver necrosis were studied in bromobenzene-intoxicated animals.

MATERIALS AND METHODS

Overnight-fasted, male NMRI albino mice were given bromobenzene intragastrically (13 mmol/Kg body wt). Desferrioxamine (DFO) was given i.p.

7 and 13 hr after the intoxication (250 mg/Kg body wt). Liver damage was assessed as serum glutamate-pyruvate transaminase (SGPT) activity. GSH was evaluated according to Sedlak and Lindsay.[6] Protein thiols were measured essentially according to Di Monte et al.[4] Lipid peroxidation was measured in the whole liver as tissue content of malonic dialdehyde (MDA) according to Casini et al.[7]; in microsomal and mitochondrial fractions as the amount of carbonyl functions in phospholipids according to Benedetti et al.[8] The energy-dependent calcium sequestration activity of the microsomal fraction was measured in the presence of oxalate.[9] The calcium uptake by the mitochondrial fraction was evaluated in an ATP-supported system.[10]

RESULTS AND DISCUSSION

Liver necrosis (SGPT level) appeared between 9 and 12 hr and increased considerably at 18 hr. Lipid peroxidation (hepatic MDA content), already detectable at 6 hr, increased thereafter, showing a good correlation ($r = 0.837$, $P < 0.001$) with the severity of liver necrosis. Despite a quite fast depletion of hepatic GSH, nearly maximal at 6 hr, a significant decrease in protein thiols could be observed at 12-18 hr only. The decrease in protein thiols was correlated with the lipid peroxidation values ($r = 0.875$, $P < 0.001$).

In both the microsomal and mitochondrial fractions from intoxicated animals, a behaviour similar to that observed for the whole liver was evident. Lipid peroxidation was already detectable at 6-9 hr in some animals and a good correlation was observed between protein thiol decrease and lipid peroxidation ($r = 0.849$, $P < 0.001$ for microsomes; $r = 0.692$, $P < 0.02$ for mitochondria).

As far as Ca^{2+} homeostasis, in vivo models do not allow the sophisticated approach as that which is possible in vitro (e.g., isolated hepatocytes). However good information can be obtained by evaluating the Ca^{2+} sequestration activity of the microsomal and mitochondrial fractions. This activity is impaired in both fractions after bromobenzene intoxication. However, the microsomal calcium pump was decreased only at 18 hr, while the mitochondrial calcium uptake was already lowered significantly at 6 hr. A good inverse correlation was observed between the calcium sequestration activity of both microsomes and mitochondria and the corresponding lipid peroxidation values (Figire 1). A significant correlation was also observed between the impairment of calcium pump in microsomes and mitochondria and the loss of protein thiols ($r = 0.894$, $P < 0.001$ for microsomes; $r = 0.861$, $P < 0.001$ for mitochondria).

In order to further evaluate the role of lipid peroxidation, DFO, a powerful iron chelator and inhibitor of lipid peroxidation, was administered 7 and 13 hr after bromobenzene intoxication. Fifteen hours after intoxication, lipid peroxidation, liver necrosis and protein thiol decrease were extremely reduced in the DFO-treated animals as compared to untreated ones, while the GSH-depletion was only minimally affected. This would suggest a major role of lipid peroxidation in the bromobenzene-induced protein thiol decrease and liver cell death.

At a later time (24 hr), notwithstanding that the results were essentially similar to those obtained at 15 hr, in some animals elevations in SGPT were found which were not matched by decrease in protein -SH groups and lipid peroxidation.

The results of the present work clearly show that during the first phase of bromobenzene intoxication of mice (up to 15-18 hr), a strict correlation exists among lipid peroxidation, loss of protein thiols, impairment of the mitochondrial and microsomal Ca^{2+} uptake, and liver necrosis. It is therefore possible that, as suggested by others,[2-5] the loss of protein -SH groups is an important factor in the bromobenzene-induced impairment of the calcium sequestration activity. However, the decrease in protein thiols does not appear strictly linked to the hepatic GSH-depletion. In fact, first, the kinetics of the two phenomena show a different behaviour, and second, the DFO treatment of the intoxicated animals almost completely prevents the loss in protein thiols, while not affecting the hepatic GSH-depletion.

On the contrary, the above results are consistent with the hypothesis that lipid peroxidation is responsible for the chain of events leading to cell death. It is however possible that in some animals, in which for some reason lipid peroxidation does not develop, another mechanism of liver damage unrelated to lipid peroxidation occurs at later times.

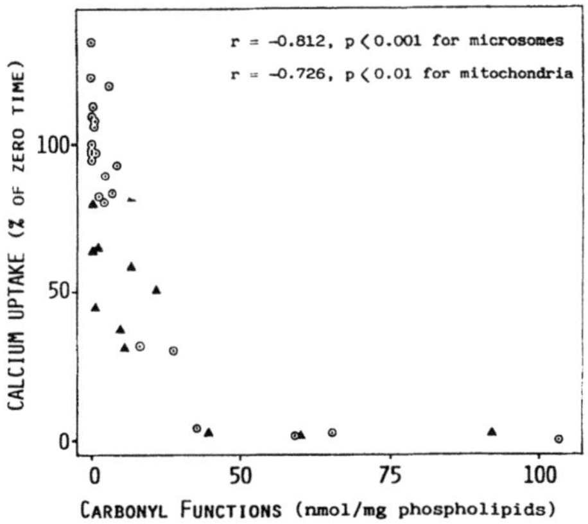

Figure 1. Plot of the values for calcium uptake against the corresponding lipid peroxidation values (amount of carbonyl functions in phospholipids) in both microsomes (○) and mitochondria (▲) from livers of bromobenzene intoxicated mice. The values obtained 6-18 hr after poisoning were used. Calcium uptake was given as per cent values with respect to the zero time, i.e., control values.

ACKNOWLEDGMENTS

This research was supported by a grant (no. 85.02107.44) from CNR, Italy, Special Project "Oncology". Additional funds were derived from CNR, Group of Gastroenterology, and from the Association for International Cancer Research, Great Britain.

REFERENCES

1. A. F. Casini, A. Pompella, and M. Comporti, Liver glutathione induced by bromobenzene, iodobenzene, and diethylmaleate poisoning and its relation to lipid peroxidation and necrosis, Am.J.Path. 118:225 (1985).
2. G. Bellomo, S. A. Jewell, M. T. Smith, H. Thor, and S. Orrenius, Perturbation of Ca^{2+} homeostasis during hepatocyte injury, in: "Mechanisms of Hepatocyte Injury and Death," D. Keppler, H. Popper, L. Bianchi, and W. Reutter, eds., MTP Press Limited, Lancaster (1984).
3. G. Bellomo and S. Orrenius, Altered thiol and calcium homeostasis in oxidative hepatocellular injury, Hepatology 5:876 (1985).
4. D. Di Monte, G. Bellomo, H. Thor, P. Nicotera, and S. Orrenius, Menadione-induced cytotoxicity is aassociated with protein thiol oxidation and alteration in intracellular Ca^{2+} homeostasis, Archs.Biochem.Biophys. 235:343 (1984).
5. M. Moore, H. Thor, G. Moore, S. Nelson, P. Moldeus, and S. Orrenius, The toxicity of acetaminophen and N-acetyl-p-benzoquinone imine in isolated hepatocytes is associated with thiol depletion and increased cytosolic Ca^{2+}, J.Biol.Chem. 260:13035 (1985).
6. J. Sedlak and R. H. Lindsay, Estimation of total, protein-bound, and nonprotein sulfhydryl groups in tissue with Ellman's reagent, Analyt.Biochem. 25:192 (1968).
7. A. F. Casini, M. Ferrali, A. Pompella, E. Maellaro, and M. Comporti, Lipid peroxidation and cellular damage in extrahepatic tissues of bromobenzene-intoxicated mice, Am.J.Path. 123:520 (1986).
8. A. Benedetti, R. Fulceri, M. Ferrali, L. Ciccoli, H. Esterbauer, and M. Comporti, Detection of carbonyl functions in phospholipids of liver microsomes in CCl_4- and $BrCCl_3$-poisoned rats, Biochim. Biophys.Acta 712:628 (1982).
9. L. Moore, T. Chen, H. R. Knapp, Jr., and E. J. Landon, Energy-dependent calcium sequestration activity in rat liver microsomes, J.Biol.Chem. 250:4562 (1975).
10. J. Bielawski and A. L. Lehninger, Stoichiometric relationships in mitochondrial accumulation of calcium and phosphate supported by hydrolysis of adenosine triphosphate, J.Biol.Chem. 241:4316 (1966).

VANADATE STIMULATION OF PYRIDINE NUCLEOTIDE OXIDATION IN MAMMALIAN LIVER MICROSOMAL MEMBRANES

Roger A. Coulombe, Jr., David W. Reif, Randal J. Keller, Donald P. Briskin,* Steven D. Aust, and Raghubir P. Sharma

Graduate Programs in Toxicology and Molecular Biology and Biochemistry, Utah State University, Logan, Utah 84322-4620 U·S.A. *Department of Agronomy, University of Illinois, Urbana Illinois, 61801, U.S.A.

INTRODUCTION

Vanadate has been shown in numerous experiments to markedly stimulate the oxidation of NAD(P)H in biological membranes. However, explanations of the nature of this stimulation often have been contradictory. Vanadate-catalyzed oxidations have been postulated to be a result of a specific membrane-associated NADH oxidase.[1] Liochev and Fridovich[2] have reported that this phenomenon reflects a vanadate stimulation of NAD(P)H oxidation by superoxide rather than by vanadate-stimulated NAD(P)H oxidases or dehydrogenases. The purpose of the present study was to characterize the nature of the vanadate stimulation of pyridine nucleotide oxidation in membranes from rat liver and to determine if vanadate is capable of stimulating NADPH oxidation in systems containing purified NADPH cytochrome P-450 reductase.

METHODS

Vanadate-dependent oxidation of NAD(P)H was typically measured in a 2.5 ml reaction volume containing 0.1 mM NAD(P)H, 0.4 mM sodium orthovanadate (Na_3VO_4), 30 mM $NaHPO_4$ buffer, pH 6.5, and ca. 1.8 mg of microsomal protein. In experiments with the purified reductase, a 1 ml reaction volume contained 0.3 M chelexed NaCl, pH 7.2, 0.95 μg of protease-solubilized reductase and 0.30 mg of antibody, in addition to the above amounts of NADPH and vanadate, where indicated. In these experiments, the antibody was allowed to incubate for at least 10 min with either the purified reductase or microsomes before being added to the reaction mixture.

RESULTS AND DISCUSSION

The addition of vanadate to microsomes caused a several-fold increase in NADPH or NADH oxidation. When boiled microsomes were used, the oxidation of pyridine nucleotides decreased dramatically (Table 1). When the salt of a similar transition element, sodium molybdate, was used, no such stimulation was seen (Table 1). At pH 6.5 in the presence of membranes, vanadate stimulated the oxidation of NADPH at rate over two times that of NADH.

Table 1. Properties of the Vanadate-Dependent NAD(P)H Oxidase Activity Associated with Rat Liver Microsomes[3]

Assay	Oxidase Activity (μmol min^{-1} mg^{-1}) NADH	%	NADPH	%
Microsomes $+V^{+5}$	0.021	100	0.054	100
Microsomes $-V^{+5}$	0.004	20	0.005	9
Heat-denatured microsomes $+V^{+5}$	0.007	33	0.006	11
Microsomes $+Na_2MoO_4$	0.003	14	0.003	5
-Microsomes $+V^{+5}$	0.008	38	0.009	17

The vanadate-stimulated oxidation of NADPH had an optimum of pH 6.0-6.5, whereas the NADH component had a pH optimum of 5.0 (Figure 1). In the absence of microsomes, non-enzymatic vanadate-stimulation of pyridine nucleotides was favored at low pH (Figure 1). Previous studies in our laboratory have shown that the NADPH component was insensitive to mitochondrial inhibitors and exhibited saturation kinetics with respect to vanadate concentration (0.0-4.0 mM) and oxidase activity, whereas the NADH component exhibited a sigmoidal relationship with respect to vanadate concentration and NADH oxidase activity, and was sensitive to NaCN and rotenone.[3] The substrate specificities, as well as kinetic differences associated with the vanadate-stimulated oxidation of NADPH and NADH, provide evidence for specific interactions between vanadate and microsomal membrane enzymes.

Table 2. Vanadate-Stimulated NADPH Oxidation under Aerobic and Anaerobic Conditions

Assay	NADPH oxidation (mol min^{-1} mg^{-1}) $+O_2$	$-O_2$
Microsomes	0.007	0.007
Microsomes $+V^{+5}$	0.046	0.063

The vanadate-mediated oxidation of NADPH and NADH in microsomes occurred with a concomitant consumption of oxygen, indicating that oxygen may be an electron acceptor in this process.[3] However, when oxygen was eliminated from the system by gassing all components of the reaction with argon and adding glucose/glucose oxidase to remove trace oxygen, similar rates of NADPH oxidation in the presence of vanadate were observed (Table 2). Thus, oxygen species may not be obligate intermediates in vanadate-stimulated NADPH oxidation.

It was of interest to investigate the possible effect of vanadate on the NADPH oxidase activity of purified NADPH cytochrome P-450 reductase since this enzyme is a likely candidate for specific interactions with vanadate in membranes. Indeed, earlier experiments in our laboratory showed that the vanadate-stimulated activity was located in isolated endoplasmic reticulum.[3] A polyclonal antibody raised to the NADPH cytochrome P-450 reductase was used to probe possible interactions between vanadate and the reductase.

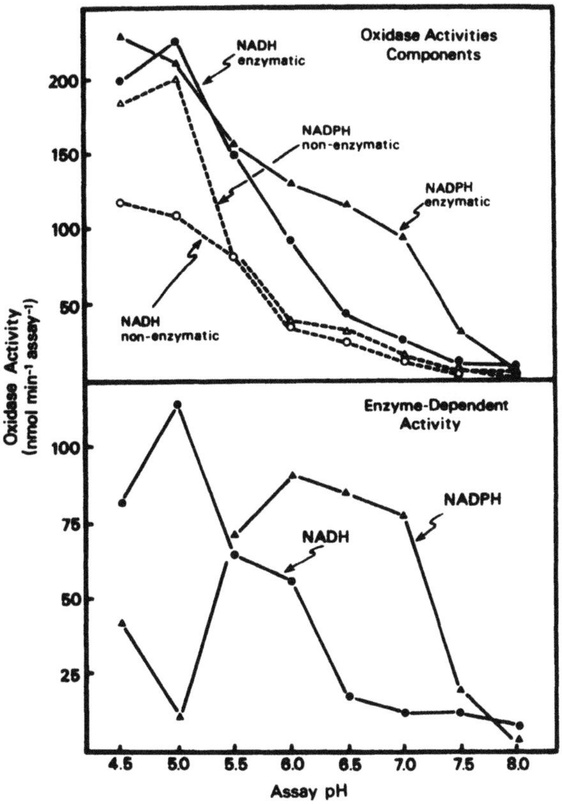

Figure 1. Enzymatic and non-enzymatic Na_3VO_4-dependent oxidation as a function of assay pH. The top frame shows the total activity of each component; the enzyme-dependent NAD(P)H oxidase activities are the sum of enzymatic and non-enzymatic (no microsomes) activity. Enzyme-dependent activity (bottom frame) has non-enzymatic activity subtracted. From Coulombe et al. (1987).[3]

As a confirmation of the specificity of the anti-cytochrome P-450 reductase antibody, the addition of the antibody caused an inhibition of cytochrome c reduction catalyzed by either the NADPH cytochrome P-450 reductase or by microsomes (Table 3).

Table 3. Effect of Anti-Cytochrome P-450 Reductase Antibody on Cytochrome c Reduction Associated with Purified Reductase or Rat Liver Microsomes.

Assay	Cytochrome c reduction (μmol min^{-1} mg^{-1})
Reductase	32.5
Reductase + Ab	12.0
Microsomes	0.19
Microsomes + Ab	0.003

The oxidation of NADPH was low in reactions containing purified reductase or microsomes without vanadate (Table 4). However, in the presence of vanadate, a marked stimulation of NADPH oxidation by purified reductase occurred. When the catalytic activity of the reductase was inhibited by the addition of the antibody, the ability of vanadate to stimulate NADPH oxidation was likewise inhibited. A similar inhibition was seen when the antibody was added to reactions containing microsomes, vanadate, and NADPH (Table 4). These results indicate that vanadate stimulates oxidation of NADPH in membranes via specific interactions with NADPH cytochrome P-450 reductase.

Table 4. Vanadate Stimulation of NADPH Oxidation Associated with Purified P-450 Reductase or Rat Liver Microsomes: the Effect of Anti-Cytochrome P-450 Reductase Antibody.

Assay	NADPH oxidation (μmol min^{-1} mg^{-1})
Reductase	0.3
Reductase + V^{+5}	4.0
Reductase + V^{+5} + Ab	2.5
Microsomes	0.009
Microsomes + V^{+5}	0.067
Microsomes + V^{+5} + Ab	0.012

These observations are consistent with the hypothesis that vanadate stimulates the oxidation of pyridine nucleotides by serving as an electron acceptor for at least one membrane-bound enzyme system, such as the NADPH cytochrome P-450 reductase. However, since vanadate stimulates the oxidation of pyridine nucleotides in the absence of membranes, it is likely that other mechanisms are involved.

REFERENCES

1. S. Vijaya, F.L. Crane, and T. Ramasarma, A vanadate-stimulated NADH oxidase in erythrocyte membranes generates hydrogen peroxide, Mol. Cell. Biochem. 62:175 (1984).
2. S. Liochev and I. Fridovich, The vanadate-stimulated oxidation of NAD(P)H by biomembranes is a superoxide-initiated free radical chain reaction, Arch. Biochem. Biophys. 250:139 (1986).
3. R. A. Coulombe, D.P. Briskin, R.J. Keller, W.R. Thornley, and R.P. Sharma, Vanadate-dependent oxidation of pyridine nucleotides in rat liver microsomal membranes, Arch. Biochem. Biophys. 255:267 (1987).

ACKNOWLEDGMENTS

This research was supported in part by a University Faculty Research Grant from Utah State University (R.A.C.). Technical paper No. 3474 of the Utah State Agricultural Experiment Station.

MECHANISM OF OXIDATION OF 1-METHYL-4-PHENYL-3,3-DIHYDROPYRIDINIUM ($MPDP^+$)

B. Kalyanaraman, W. Korytowski, and C. C. Felix

National Biomedical ESR Center, Department of Radiology
8701 Watertown Plank Road, Milwaukee, WI 53226, U.S.A.

INTRODUCTION

The compound 1-methyl-4-phenyl-1,2,3,6-tetrahydropyridine (MPTP) is a neurotoxin found as a contaminant in certain street drugs. Individuals using these drugs can develop symptoms similar to those of Parkinson's disease.[1-3] The symptoms have been attributed to MPTP, which is known to produce Parkinsonian symptoms in some animals[4,5] and which selectively destroys neuronal cells in the melanin-containing substantia nigra of humans and primates. The conversion of MPTP to reactive metabolites by monoamine oxidase in brain tissue is crucial for MPTP neurotoxicity.[6-7] Production of oxygen radicals from metabolites of MPTP has been proposed to account for its selective neuronal damage.[8-13]

The first isolatable metabolite of MPTP is $MPDP^+$, formed from two-electron oxidation of MPTP by MAO-B (Scheme).[14] The fate of this intermediate has not been fully determined.[15,16] Whether $MPDP^+$ undergoes enzymatic or non-enzymatic conversion to MPP^+ has not yet been unambiguously established,[16] and no information is available concerning oxygen activation during conversion of $MPDP^+$ to MPP^+.

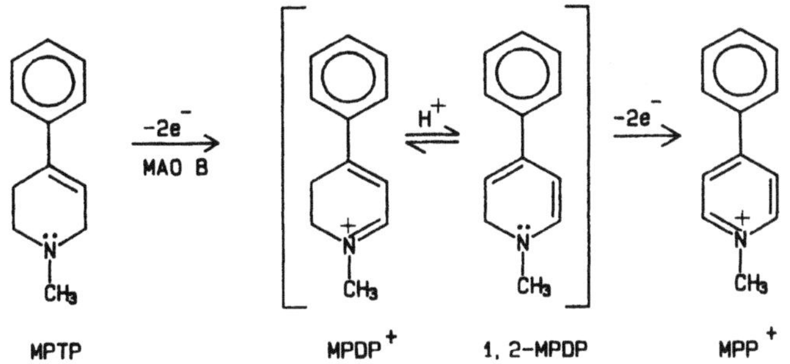

Scheme. Proposed intermediates during monamine oxidase-catalyzed oxidation of MPTP.

In this paper, we report (a) the production of hydrogen peroxide during oxidation of $MPDP^+$ to MPP^+ and (b) the effect of iron chelates on the rate of oxidation of $MPDP^+$.

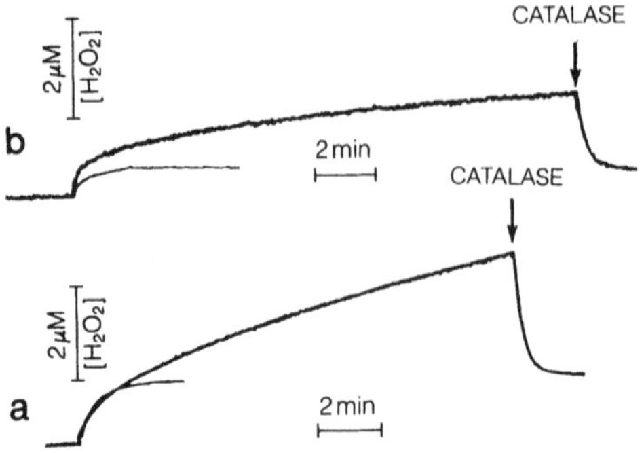

Figure 1. Production of hydrogen peroxide from $MPDP^+$ (30 μM) in a phosphate buffer, pH 8.0, during autoxidation at 37°C (a) in the presence of 100 μM EDTA and (b) in the presence of 30 μM desferrioxamine. Arrows denote points of addition of catalase (10 μg/ml). Control denotes the response of the electrode in N_2-saturated buffer containing $MPDP^+$ (30 μM). The control signal was insensitive to added catalase.

RESULTS

The addition of $MPDP^+$ to an aerated phosphate buffer (pH 8.0) containing EDTA produced a slow release of hydrogen peroxide (0.25 ± 0.02 μM/min), which was continuously monitored using the oxidase electrode (Figure 1a). The addition of catalase rapidly decomposed the hydrogen peroxide to the control level obtained when $MPDP^+$ was added to a nitrogen-saturated buffer (Figure 1). The addition of superoxide dismutase did not affect significantly the rate of production of hydrogen peroxide. In the presence of desferrioxamine the initial rate of production of hydrogen peroxide decreased to 0.15 ± 0.02 μM/min (Figure 1b). At higher concentrations of desferrioxamine (~100 μM) a further decrease in the rate of production of hydrogen peroxide was observed (data not shown). Inhibition by desferrioxamine implies a catalytic role for metal ions, iron for example. The addition of MPTP or MPP^+ under conditions of Figure 1 did not produce measurable amounts of hydrogen peroxide.

During autoxidation of $MPDP^+$ its spectral maximum was gradually replaced by that of MPP^+ (Figure 2). The isosbestic point at 300 nm is consistent with the existence of the two species, MPP^+ and $MPDP^+$.[17]

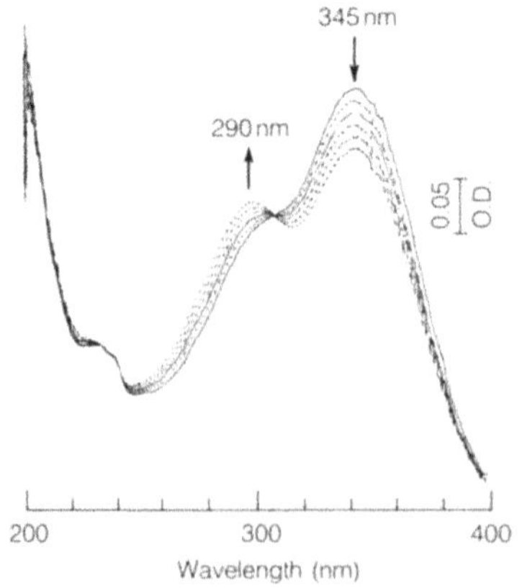

Figure 2. Repeated optical scans following autoxidation of $MPDP^+$ under conditions identical to Figure 1a. The maxima at 345 and 290 nm belong to $MPDP^+$ and MPP^+, respectively.

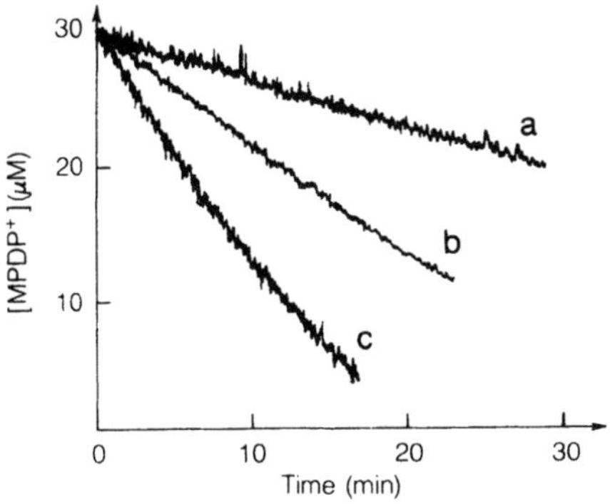

Figure 3. The rate of depletion of $MPDP^+$ in a phosphate buffer (pH 8.0) at 37°C (a) $MPDP^+$ (30 μM) and 100 μM EDTA (b) $MPDP^+$ (30 μM) and 10 μM Fe(III)-EDTA and (c) $MPDP^+$ (30 μM) and 70 μM Fe(III)-EDTA. The rate of depletion of $MPDP^+$ was monitored optically at 345 nm.

The rate of decomposition of $MPDP^+$ (0.33 μM/min) was monitored using the decrease in absorbance at 345 nm (Figure 3a). The rate was found to increase to 0.91 μM/min, and 1.95 μM/min respectively, in the presence of 10 and 70 μM Fe(III)-EDTA (Figures 3b and 3c). Conversely, the addition of desferrioxamine or Fe(III)-desferrioxamine chelates inhibited the rate of decomposition of $MPDP^+$. These experiments suggest that metal ions are involved during dismutation of $MPDP^+$.

From the data (Figures 1a and 3a), the stoichiometry between $MPDP^+$ disappearance and H_2O_2 production is estimated to be ca. 1.0:0.8.

DISCUSSION

The present results show that hydrogen peroxide is also formed during oxidation of $MPDP^+$ to MPP^+, and that the rate of production of hydrogen peroxide is sensitive to added metal-ion chelators.

Based on the following rate equation

$$\frac{d[H_2O_2]}{dt} = k[MPDP^+][O_2] = k'[MPDP^+] \qquad (1)$$

where $k' = k[O_2]$, a pseudo-first-order rate constant ($k' = 2.3 \pm 0.2 \times 10^{-3}\ s^{-1}$) for oxidation of $MPDP^+$ by molecular oxygen is obtained from Figure 1a.

At high concentrations of $MPDP^+$, it has been suggested that $MPDP^+$ decomposes to MPP^+ via a second-order dismutation reaction[18] (equation 2).

$$2\ MPDP^+ \xrightarrow{2\ H^+} MPP^+ + MPTP \qquad (2)$$

Although conflicting results have previously been reported with regard to the stability of MPTP in the presence of metal ions,[11,17] the present optical data (Figure 4) show that the decomposition of $MPDP^+$ to MPP^+ in part involves a reaction or reactions (for example one-electron oxidation/ reduction) catalyzed by metal chelates. The substantia nigra reportedly contains a high concentration of iron deposits.[19] This, combined with the selective destruction by MPTP of nigral cells, may make the following reactions particularly relevant to an understanding of MPTP toxicity.

It is conceivable that potentially toxic free radical intermediates are involved during the iron-catalyzed decomposition of $MPDP^+$ to MPP^+ as shown:

$$Fe^{3+}\text{-EDTA} + MPDP^+ \rightarrow Fe^{2+}\text{-EDTA} + MPP\bullet \quad (3)$$

$$Fe^{2+}\text{-EDTA} + O_2 \rightarrow Fe^{3+}\text{-EDTA} + \bullet O_2^- \quad (4)$$

$$2\ \bullet O_2^- + 2H^+ \rightarrow H_2O_2 + O_2 \quad (5)$$

In the presence of air MPP· probably undergoes rapid oxidation to form MPP^+ and superoxide (eq. 6), to give an overall reaction stoichiometry (eq. 7) almost consistent with that observed.

$$MPP\bullet + O_2 \rightarrow MPP^+ + \bullet O_2^- \quad (6)$$

$$MPDP^+ + O_2 \xrightarrow{Fe^{3+},\ 2H^+} MPP^+ + H_2O_2 \quad (7)$$

ACKNOWLEDGMENTS

This work was supported by National Institutes of Health Grant GM-29035 and RR-01008.

REFERENCES

1. G. C. Davis, A. C. Williams, S. P. Markey, M. H. Ebert, E. D. Caine, C. M. Reichert, and I. J. Kopin, Chronic Parkinsonism secondary to intravenous injection of meperidine analogues, Psychiatr. Res. 1:249 (1979).
2. J. W. Langston, P. Ballard, J. W. Tetrvd, and I. Irwin, Chronic Parkinsonism in humans due to a product of meperidine-analog synthesis. Science 219:979 (1983).
3. J. M. Wright, R. A. Wall, T. L. Perry, and D. W. Paty, Chronic Parkinsonsim secondary to intranasol administration of a product of meperidine-analog synthesis, N. Eng. J. Med. 310:325 (1984).
4. R. S. Burns, C. C. Chiueh, S. P. Markey, M. H. E. Ebert, D. Jacobowitz, and I. J. Kopin, A primate model of Parkinsonism: Selective destruction of dopaminergic neurons in the pars compacta of the substantia nigra by N-methyl-4-phenyl-1,2,3,6-tetrahydropyridine, Proc. Natl. Acad. Sci. USA 80:4546 (1984).
5. J. W. Langston, L. S. Forno, C. S. Rebert, and I. Irwin, Selective nigral toxicity after systemic administration of 1-methyl-4-phenyl-1,2,3,6-tetrahydropyridine (MPTP) in the squirrel monkey, Brain Res. 292:390 (1984).
6. R. E. Heikkila, L. Manzino, F. S. Cabbat, and R. S. Duvoisin, Protection against the dopaminergic neurotoxicity of 1-methyl-4-phenyl-1,2,3,6-tetrahydropyridine of monamine oxidase inhibitors, Nature (Lond.) 311:467 (1984).
7. J. W. Langston, I. Irwin, E. B. Langston, and L. S. Forno, Pargyline prevents MPTP induced Parkinsonism in primates, Science 225:1480 (1984).
8. R. Lewin, Brain enzyme is the target of drug toxin - a chemical known as MPTP causes a Parkinson-like state in humans and monkeys; Biochemical and autoradiographical studies are closing in on the mechanism, Science 225:1460 (1984).
9. J. M. Johannesson, C. C. Chiueh, R. S. Burns, and S. P. Markey, Differences in the metabolism of MPTP in the rodent and primate parallel differences in sensitivity to its neurotoxic effects, Life Sci. 36:219 (1985).

10. V. W. Young, T. L. Perry, and A. A. Krisman, Depletion of glutathione in brainstem of mice caused by N-methyl-4-phenyl-1,2,3,6-tetrahydropyridine is prevented by antioxidant pre-treatment, Neurosci. Lett. 63:56 (1986).
11. J. Poirier and A. Barbeau, A catalyst function for MPTP on superoxide formation. Biochem. Biophys. Res. Commun. 131:1284 (1985).
12. J. Poirier, J. Donaldson, and A. Barbeau, The specific vulnerability of substantia nigra to MPTP is related to the presence of transition metals, Biochem. Biophys. Res. Commun. 128:25 (1985).
13. G. V. Corsini, S. Pintus, C. C. Chiueh, J. F. Weiss, and I. J. Kopin, 1-Methyl-4-phenyl-1,2,3,6-tetrahydropyridine (MPTP) neurotoxicity in mice is enhanced by pretreatment with diethyldithiocarbamate, Eur. J. Pharmacol. 119:127 (1985).
14. K. Chiba, A. Trevor, and N. Castagnoli, Jr., Metabolism of the neurotoxic tertiary amine, MPTP, by brain monamine oxidase, Biochem. Biophys. Res. Commun. 120:574 (1984).
15. N. Castagnoli, Jr., K. Chiba, and A. J. Trevor, Potential bioactivation pathways for the neurotoxin 1-methyl-4-phenyl-1,2,3,6-tetrahydropyridine (MPTP), Life Sci. 36:225 (1985).
16. R. R. Fritz, C. W. Abell, N. T. Patel, W. Gessner, and A. Brossi, Metabolism of the neurotoxin in MPTP by human liver monamine oxidase B, FEBS Lett. 186:224 (1985).
17. E. Y. Wu, K. Chiba, A. J. Trevor, and N. Castagnoli, Jr., Interactions of the 1-methyl-4-phenyl-2,3-dihydropyridinium species with synthetic dopamine-melanin, Life Sci. 39:1695 (1986).
18. K. Chiba, L. A. Peterson, K. P. Castgnoli, A. J. Trevor, and N. Castagnoli, Jr., Studies on the molecular mechanism of biotransformation of the selective nigrostriatal toxin 1-methyl-4-phenyl-1,2,3,6-tetrahydropyridine, Drug. Metab. Dispos. 13:342 (1985).
19. J. M. Hill and R. C. Switzer, The regional distribution and cellular localization of iron in the rat brain, Neurosci. 11:595 (1984).

REDUCTION OF NITROHETEROCYCLIC DRUGS BY ASCORBATE AND CATECHOLAMINES: A POSSIBLE MECHANISM FOR THE NEUROTOXICITY OF NITROHETEROCYCLIC DRUGS

D.N. Ramakrishna Rao and Ronald P. Mason

Laboratory of Molecular Biophysics
National Institute of Environmental Health Sciences
PO Box 12233
Research Triangle Park, North Carolina 27709, U.S.A.

INTRODUCTION

Nitroheterocyclic drugs such as 5-nitrofuran and 2- and 5-nitroimidazole derivatives are used as antimicrobial agents, some of which are selective for hypoxic cells.[1] These compounds are also used as radiation sensitizers,[2] and the 2-nitroimidazole derivative misonidazole was tested clinically. Unfortunately, a dose of 14 g/m^2 needed to obtain sufficient tissue levels of this drug for sensitization produced a high incidence of peripheral neuropathy,[3] hence this drug is of limited use.[3] Nitrofurantoin and metronidazole also produce neuropathy.[4,5] At lower doses nitrofurantoin is used as a urinary antiseptic,[6] while metronidazole is used for amebic dysentery and for microbial infections of the vagina.[7]

The mechanism of action of nitroheterocyclic drugs as cytotoxic agents for hypoxic cells has been widely studied.[8,9] In general, the first step in the mechanism is the reduction of the nitro group of the drug to the corresponding nitro anion radical.[10] Nitro anion radicals can be further reduced under hypoxic conditions to form their corresponding nitroso, hydronitroxide and amine compounds.[1] In the presence of oxygen, the nitro anion radicals reduce oxygen to the superoxide anion radical and hydrogen peroxide.

Recently we showed that ascorbate and catecholamine neurotransmitters such as norepinephrine, dopamine and related catechol derivatives can reduce nitro compounds to their respective nitro anion radicals.[11,12] In this paper we present evidence for the ascorbate and catecholamine reduction of nitro drugs and discuss its relevance to the neurotoxicity of these drugs.

MATERIALS AND METHODS

Hepatocytes isolated from CD male rats by the procedure of Moldéus et al.[13,14] were heated (2-3 x 10^7 cells/mL) at 70°C for 10 min under a stream of nitrogen gas. The heat-denatured hepatocytes were cooled and centrifuged at 27,000 x g for 10 min. The supernatant was assayed for the generation of nitro anion radicals. Chromaffin granules were isolated and purified by the method of Trifaro and Dworkind.[15]

Electron spin resonance measurements were performed at room temperature (23°C) with a Varian E-109 spectrometer equipped with a TM_{110} cavity as previously described.[16] A 5-mL solution of the nitro compound was bubbled with a stream of nitrogen gas. Then a biochemical reducing agent, either sodium ascorbate, glutathione (reduced), norepinephrine, dopamine, serotonin or any catechol derivative, was dissolved to obtain the desired concentration. The pH of this solution was then adjusted to the desired value, and the electron spin resonance spectrum was recorded. Spectral measurements on chromaffin granules were performed in a 1.5-mL solution at a protein concentration of 44 mg/mL. Computer simulation of the experimental spectra and assignment of the hyperfine coupling constants were carried out on a Hewlett Packard 9835B desk-top computer using the correlation technique described previously.[17] Oxygen consumption was measured at 30°C using a Clark oxygen electrode (Yellow Spring Instruments). Assays of oxygen consumption were performed in 0.1 M Tris-HCl buffer, pH 8.2 or 9.0, and additions were made as stated in the figures. In the case of chromaffin granules (protein, 2.9 mg/mL), the pH was adjusted to 7.8 or 8.0 just before oxygen consumption measurements. The ascorbic acid content of chromaffin granules and hepatocytes were determined by the method of Roe and Kuether.[18] Protein was assayed by the method of Lowry et al.[19]

RESULTS

Ascorbate

All three nitro compounds, i.e., nitrofurantoin, misonidazole and metronidazole, were reduced to their respective nitro radical anions by ascorbate at alkaline pH. The formation of both nitro anions and semidehydro ascorbate radicals were observed by ESR spectroscopy. Nitrofurantoin and misonidazole can be detected at *ca* pH 8.5 and pH 10.0, respectively, but metronidazole could only be detected in 0.025 M NaOH. Nitro radical anions were not detected in control samples that did not contain ascorbate. The nitro anion radicals formed were identified by the computer simulation of the experimental spectra.[11]

The ascorbate-supported oxygen consumption by nitro compounds was observed in all three nitro drugs. This oxygen consumption is a measure of the rate of formation by the respective nitro radical anions. The reaction of a nitro drug such as misonidazole with ascorbate at physiological pH is shown by the oxygen-consumption experiments (Figure 1). The rate of oxygen consumption decreased when either superoxide dismutase or catalase was added. These results can be explained on the basis of reactions shown in Figure 2.

Catecholamines

Catechols such as epinephrine, norepinephrine, dopamine and L-dihydroxyphenylalanine (L-DOPA) readily reduce 5-nitrofurans such as nitrofurantoin, nifurtimox, nifuroxime and nitrofurazone to their corresponding nitro anion radicals. In our studies with 5-nitrofurans, a wide range of norepinephrine concentrations and pH values were used (data not shown). The steady-state concentration of the nitro anion radicals in all these studies was lower when the pH was below 9.0. Nitro anion radicals of nitrofurantoin, nifurtimox and nitrofurazone were observed in the presence of dopamine or L-DOPA at about pH 9.5. Semiquinone radicals of catecholamines formed by the oxidation were also observed at a higher pH (*ca* pH 11.0). Serotonin reduced nitrofurantoin to its anion radical at pH 10.1. In general, reduction of drugs containing 5-nitrofurans by dopamine, L-DOPA or serotonin were not observed below pH 9.5. Drugs containing nitroimidazoles were reduced at a higher pH. Thus, drugs such as ronidazole, secnidazole and benznidazole were reduced by norepinephrine in 0.1 M NaOH (*ca* pH 12.4).

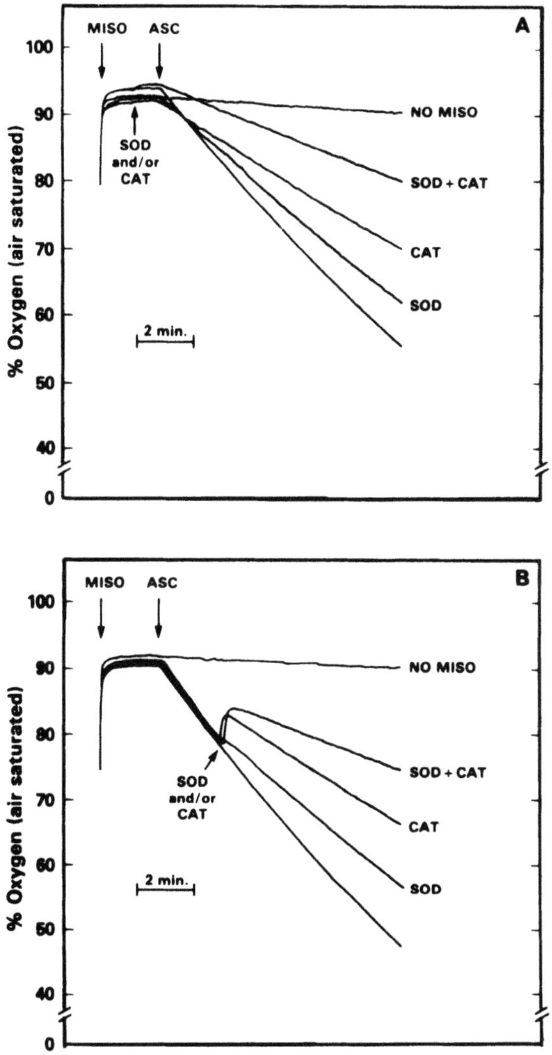

Figure 1. The oxygen uptake curves for the misonidazole (50 mM)/ascorbate (5 mM) system showing the effects of superoxide dismutase (SOD; 330 µg/mL) and catalase (CAT; 5000 units/ml) in Tris/HCl buffer, pH 7.5 (with 1.0 mM DTPA).
A) Addition of ascorbate was preceded by the addition of SOD and CAT. B) Addition of ascorbate was followed by SOD and CAT (from reference 11, with permission).

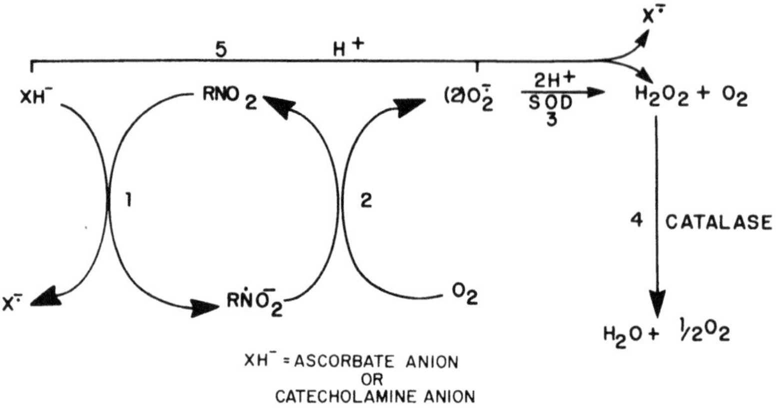

Figure 2. Scheme of reactions.

The hyperfine coupling constants for radicals of nifurtimox, nifuroxime and nitrofurazone obtained by catecholamine reduction have been reported.[12] The values reported in this paper were obtained by the correlation technique,[17] and gave coupling constants much improved over those reported in the literature.[12] The results presented above clearly show the formation of nitro anion and semiquinone radicals by the catecholamine reduction of nitroheterocyclic drugs.

Administration of nitro drugs such as misonidazole and metronidazole leads to their distribution into different tissues and organs. Hence, it was interesting to study the effect of these drugs on the adrenal medullae, which contain chromaffin cells that synthesize and store catecholamines in chromaffin granules.[20]

Figure 3 shows the electron spin resonance spectrum of nitrofurantoin anion radical formed in chromaffin granules at pH 10.2. At pH 7.4, nitrofurantoin anion radicals were not observed. However, a small increase in the pH of the solution to pH 7.8 resulted in the detection of the nitrofurantoin anion radical.[12] Strong electron spin resonance signals of this nitro anion radical were observed as the pH was increased. The measured ascorbate content of the chromaffin granules (0.8 mM) may also contribute to the reduction of nitrofurantoin.

In neutral solutions of norepinephrine, i.e., at pH 7.4, nitrofurantoin-dependent oxygen consumption was very slow; however, at a slightly alkaline pH, the oxygen uptake was significant.[12] Addition of superoxide dismutase and/or catalase reduced the rate of oxygen consumption. When both the enzymes were present, the effect was additive. Similar results were obtained with other neurotransmitters such as dopamine and serotonin and also with L-DOPA. The oxygen consumption in solutions containing norepinephrine and misonidazole or metronidazole (data not shown) had similar characteristics of oxygen consumption by nitrofurantoin. These results are explained by the equations of Figure 2. Catechol derivatives such as 3,4-dihydroxy norephedrine, 3,4-dihydroxybenzylamine, isoproterenol, 3,4-dihydroxy mandelic acid and 6-hydroxy dopamine have been used in the solutions containing nitrofurantoin at pH 9.0. Oxygen consumption in all these systems had the same characteristics of Figure 1.

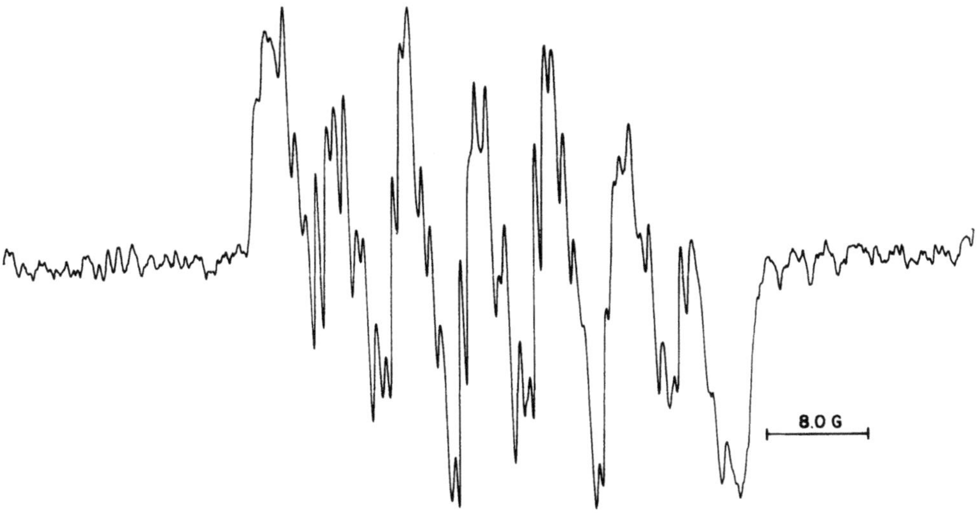

Figure 3. The ESR spectrum of nitrofurantoin radical anion produced in a system of nitrofurantoin (3 mM) and chromaffin granules (protein, 44 mg/mL) in 0.1 M HEPES, pH 10.2. Instrumental conditions were 20 mw microwave power, 0.53 G modulation amplitude, 2 sec time constant and 6.25 G/min scan rate.

Figure 4 shows the ESR spectrum of the nitrofurantoin radical formed by the supernatant of heat-treated hepatocytes (ascorbate, ca 0.5 mM) at pH 8.6. The ascorbyl radical was observed immediately after the addition of nitrofurantoin. After a few minutes the ESR signal from the nitrofurantoin radical was observed.

DISCUSSION

Ascorbate concentration in mammalian brain is generally very high. Rat brain homogenates contain ca 2.6 mM ascorbate, and there is a very marked gradient of ascorbate throughout the brain.[21] In general, the ascorbate concentration tends to be high in regions rich in catecholamines, and synaptosomal cytoplasm contains the highest specific content of ascorbate.[21] One of the important functions of ascorbate is related to the release and storage of neurotransmitters in the nerve terminals. It has been shown that Ca^{2+}-dependent acetylcholine release from the synaptic vesicles is enhanced by ascorbate.[22]

Our results presented above clearly demonstrate that nitro drugs can be reduced by ascorbate to nitro anion, thus forming the ascorbyl radical. Loss of ascorbate due to oxidation may reduce the acetylcholine release from synapse.[22] This process may also weaken muscle cells since ascorbate acts as a releasable neurotrophic factor which promotes proline hydroxylation and collagen formation, thereby facilitating assembly of the basement membrane around muscle cells.[23] Loss of ascorbate may also affect dopaminergic transmission by specifically inhibiting dopamine-sensitive adenyl cyclase to levels below that which is normally present in the brain.[24] In peripheral nerves such as the sciatic nerve a decrease in ascorbate level has been shown to cause increased lipid peroxidation due to this decreased anti-

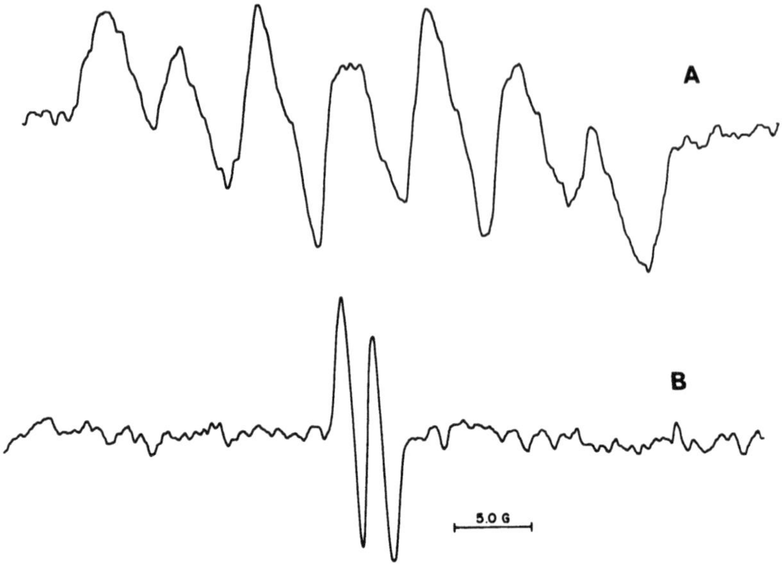

Figure 4. A) The ESR spectrum of nitrofurantoin radical anion generated by heat-treated hepatocytes and supernatant and nitrofurantoin (5 mM) at pH 8.6 after ca 10 min. B) The ESR spectrum of ascorbate radical anion generated in the above system immediately after the addition of nitrofurantoin (ca 5 min).

oxidant defense.[25] In the presence of oxygen-derived radicals formed by the reaction with nitro radical anions, an enhanced lipid peroxidation may ensue resulting in the degeneration of peripheral nerves.

Loss of ascorbate and catecholamines due to oxidation by a nitro drug may produce a cooperative effect in the degeneration of the nervous system. Generally higher concentrations of ascorbate in the brain parallel higher concentrations of catecholamines. Secondly, ascorbate facilitates catecholamine storage.[26] Catechol derivatives such as 6-hydroxydopamine, 6-aminodopamine and 5,6-dihydroxytryptamine are known to accumulate in the catecholamine neurons, causing nerve degeneration and neuropathy.[27] It has been shown that the auto-oxidation of these compounds is related to their neurotoxicity.[28] In the presence of oxygen, these catechol derivatives undergo facile oxidation to produce semiquinone (or semiquinoneimine) free radicals, the corresponding quinone (or quinoneimine) and oxygen-derived radicals. The quinones can bind covalently to the nucleophilic groups of proteins and DNA and cause the degeneration of neurons.[28,29] A similar explanation has been offered for manganese neuropathy.[30] It has been observed that the manganese neurotoxication in neonatal rats resulted in the induction of lipid peroxidation in several regions of rat brain. This has been attributed to neurotransmitters such as dopamine, apparently oxidized by Mn^{3+} (formed by the oxidation of Mn^{+2}) to form semiquinone radicals which act as cytotoxic agents.[30]

Our results presented here clearly establish that catechol derivatives reduce nitro drugs such as 5-nitrofurans and 2- and 5-nitroimidazoles with the concomitant oxidation of catecholamines to produce semiquinone or quinoneimine radicals. Catecholamines, which undergo facile oxidation, reduced nitro drugs more easily; this was observed in the oxygen consumption experiments. Solutions containing dopamine and 5-nitrofurans had significant oxygen consumption, but this effect is much more pronounced in the solutions of 6-hydroxydopamine or 6-aminodopamine and nitrofurantoin. Thus, oxygen consumption measurements show the rate of formation of nitro anion radicals, superoxide and hydrogen peroxide. Optical measurements show the formation of oxidation products (quinones) of catechol derivatives, while electron spin resonance measurements show the formation of both semiquinone radicals of catecholamines and nitro anion radicals of nitroheterocyclic drugs.

Heat-treated hepatocytes also reduce nitrofurantoin, suggesting that ascorbate or some other biochemical reducing agent may be partly responsible for the reduction of nitro drugs in intact cells. Recently we showed that glutathione can reduce nitro drugs (D.N.R. Rao and R.P. Mason, unpublished observations). This result may be very significant, since rat neurons and neurofilaments are known to contain large concentrations of reactive thiol groups which are supposed to be involved in the axonal transport of proteins by peripheral nerves of frog.[31,32]

In conclusion, reaction of a nitro drug with ascorbate, catecholamine neurotransmitters or glutathione may lead to the impairment of some neuronal functions, which may eventually lead to the degeneration of neurons.

REFERENCES

1. R. P. Mason and P. D. Josephy, Free radical mechanism of nitroreductase, in: "Toxicity of Nitroaromatic Compounds," D.E. Rickert, ed., Hemisphere Publishing Corp., Washington (1985).
2. A. Breccia and J. F. Fowler, "New Chemo and Radiosensitizing Drugs," Edizioni Scientifiche (Lo Scarabeo), Bologna, Italy (1985).
3. T. Girinski, M. H. Pejovic, C. Haie, M. Bonnay, A. Gerbaulet, J. J. Mazeron, E. Malaise, and D. Chassagne, Radical irradiation and misonidazole in the treatment of advanced cervical carcinoma: Results of a phase II trial, Int. J. Radiat. Oncol. Biol. Phys. 11:1783 (1985).
4. G. P. Rose, A. J. Dewar, and I. J. Stratford, A biochemical neurotoxicity study relating the neurotoxic potential of metronidazole and nitrofurantoin with misonidazole, Int. J. Radiat. Oncol. Biol. Phys. 8:781 (1982).
5. C. Yiannikas, J. D. Pollard, and J. G. McLeod, Nitrofurantoin neuropathy, Aust. N. Z. J. Med. 11:400 (1981).
6. W. M. Brutinel and W. J. Martin, II, Chronic nitrofurantoin reaction associated with T-lympocyte alveolitis, Chest 89:150 (1986).
7. S. N. J. Moreno and R. Docampo, Mechanism of toxicity of nitro compounds used in the chemotherapy of trichomoniasis, Environ. Health Perspect. 64:199 (1985).
8. P. D. Josephy, B. Palcic, and L. D. Skarsgard, Ascorbate-enhanced cytotoxicity of misonidazole, Nature 271:370 (1978).
9. C. J. Koch, R. L. Howell, and J. E. Biaglow, Ascorbate anion potentiates cytotoxicity of nitro-aromatic compounds under hypoxic and anoxic conditions, Br. J. Cancer 39:321 (1979).
10. R. P. Mason and J. L. Holtzman, The mechanism of microsomal and mitochondrial nitroreductase. Electron spin resonance evidence for nitroaromatic free radical intermediates, Biochemistry 14:1626 (1975).
11. D. N. R. Rao, L. Harman, A. Motten, J. Schreiber, and R. P. Mason, Generation of radical anions of nitrofurantoin, misonidazole, and metronidazole by ascorbate, Arch. Biochem. Biophys. 255:419 (1987).

12. D. N. R. Rao and R. P. Mason, Generation of nitro radical anions of some 5-nitrofurans, 2- and 5-nitroimidazoles by norepinephrine, dopamine, and serotonin, J. Biol. Chem. 262:11731 (1987).
13. P. Moldéus, J. Högberg, and S. Orrenius, Isolation and use of liver cells, Methods Enzym. 52:60 (1978).
14. J. R. Fry, Preparation of mammalian hepatocytes, Methods Enzym. 77:130 (1981).
15. J. M. Trifaró and J. Dworkind, A new and simple method for isolation of adrenal chromaffin granules by means of an isotonic density gradient, Anal. Biochem. 34:403 (1970).
16. R. P. Mason, Assay of in situ radicals by electron spin resonance, Methods Enzymol. 105:416 (1984).
17. A. Motten and J. Schreiber, Correlation analysis of ESR spectra on a small computer, J. Magn. Reson. 67:42 (1986).
18. S. T. Omaye, J. D. Turnbull, and H. E. Sauberlich, Selected methods for the determination of ascorbic acid in animal cells, tissues, and fluids, Methods Enzymol. 62:3 (1979).
19. O. H. Lowry, N. J. Rosebrough, A. L. Farr, and R. J. Randall, Protein measurement with the folin phenol reagent, J. Biol. Chem. 193:265 (1951).
20. H. Winkler and E. Westhead, The molecular organization of adrenal chromaffin granules, Neuroscience 5:1803 (1980).
21. C- H. Kuo, N. Yonehara, F. Hata, and H. Yoshida, Subcellular distribution of ascorbic acid in rat brain, Jpn. J. Pharmacol. 28:789 (1978).
22. C- H. Kuo and H. Yoshida, Ascorbic acid, an endogenous factor required for acetylcholine release from the synaptic vesicles, Jpn. J. Pharmacol. 30:481 (1980).
23. C. Kalcheim, E. Bachar, D. Duksin, and Z. Vogel, Ciliary ganglia and spinal cord explants release an ascorbate-like compound which stimulates proline hydroxylation and collagen formation in muscle cultures, Neurosci. Lett. 58:219 (1985).
24. L. C. Tolbert, T. N. Thomas, L. D. Middaugh, and J. W. Zemp, Effect of ascorbic acid on neurochemical, behavioral, and physiological systems mediated by catecholamines, Life Sci. 25:2189 (1979).
25. M. Chvapil, J. M. Kern, R. Misiorowski, and P. R. Weinstein, Endogenous antioxidants and rate of malondialdehyde formation in central and peripheral nervous systems, Exp. Neurol. 78:765 (1982).
26. M. I. Sears, Vitamin C as a requirement for the storage of norepinephrine by the iris, Biochem. Pharmacol. 18:253 (1969).
27. R. M. Kostrzewa and D. M. Jacobowitz, Pharmacological actions of 6-hydroxydopamine, Pharmacol. Rev. 26:199 (1974).
28. D. G. Graham, S. M. Tiffany, W. R. Bell, Jr., and W. F. Gutknecht, Autoxidation versus covalent binding of quinones as the mechanism of toxicity of dopamine, 6-hydroxydopamine, and related compounds toward C1300 neuroblastoma cells in vitro, Mol. Pharmacol. 14:644 (1978).
29. A. Saner and H. Thoenen, Model experiments on the molecular mechanism of action of 6-hydroxydopamine, Mol. Pharmacol. 7:147 (1971).
30. J. Donaldson, D. McGregor, and F. LaBella, Manganese neurotoxicity: A model for free radical mediated neurodegeneration?, Can. J. Physiol. Pharmacol. 60:1398 (1982).
31. A. Mäkinen, H. Savolainen, E. Lehtonen, and H. Vainio, Reduced sulfhydryl groups of rat neurons, glial cells and neurofilaments, Res. Commun. Chem. Pathol. Pharmacol 16:577 (1977).
32. A. Edström and H. Mattsson, Inhibition and stimulation of rapid axonal transport in vitro by sulfhydryl blockers, Brain Res. 108:381 (1976).

ROLE OF ACTIVE OXYGEN IN PARAQUAT AND 1-METHYL-4-PHENYL-1,2,3,6-TETRAHYDROPYRIDINE (MPTP) CYTOTOXICITY

Martha S. Sandy, Donato Di Monte, Phyllis Cohen, and Martyn T. Smith

Department of Biomedical and Environmental Health Sciences, School of Public Health, University of California, Berkeley, California 94720, U.S.A.

INTRODUCTION

The cyclic tertiary amine 1-methyl-4-phenyl-1,2,3,6-tetrahydropyridine (MPTP) is a contaminant of the illicitly synthesized meperidine analog MPPP and is able to induce symptoms which closely resemble idiopathic Parkinson's disease (PD) in humans.[1] This appears to be caused by a selective destruction of dopaminergic neurons in the zona compacta of the substantia nigra, as seen following the administration of MPTP to primates.[2] Metabolic activation of the compound seems to be essential for its cytotoxic action.[3] Mitochondrial monoamine oxidase type B converts MPTP to $MPDP^+$, which is further oxidized to MPP^+, the ultimate toxic metabolite.[4]

It has been noted that the structure of MPP^+ is similar to that of the bipyridylium herbicide paraquat (PQ^{++})(Figure 1), leading to the suggestion that the two compounds may share a similar mechanism of toxicity. Moreover, it has been proposed[5] that agricultural chemicals present as environmental contaminants may be involved in the etiology of idiopathic PD. A retrospective epidemiological study[5] has indeed shown a correlation between the use of pesticides and the prevalence of PD within the province of Quebec in Canada.

PQ^{++} toxicity has been related to the generation of oxygen radicals and to the consequent perturbation of cellular redox status known as "oxidative stress."[6] The involvement of oxidative stress in MPP^+ toxicity was consequently suggested, and needed to be tested experimentally. Isolated rat hepatocytes have been widely used as a model for studying the biochemical effects induced by numerous xenobiotics. Pretreatment with an inhibitor of glutathione reductase, 1,3-bis(2-chloroethyl)-1-nitrosourea (BCNU), predisposes hepatocytes to the toxic effects of oxygen radicals by impairing the ability of glutathione peroxidase to metabolize hydrogen peroxide (H_2O_2). We had previously[7] studied the effects of three bipyridyl compounds, including PQ^{++}, in BCNU-pretreated hepatocytes[9], and therefore decided to perform a comparative study of the effects of MPTP and MPP^{+} in this model system.

MECHANISM OF PARAQUAT CYTOTOXICITY

A comparison of cell toxicity in non-treated vs. BCNU-treated hepatocytes revealed that PQ^{++} was much more potent in BCNU-treated hepatocytes in which glutathione reductase activity was inhibited Table 1). Reduced glutathione (GSH) was rapidly decreased by PQ^{++} in BCNU-treated cells and was mostly recovered in the media in its oxidized form, GSSG.[7,8] Lipid peroxidation and α-tocopherol depletion accompanied the expression of PQ^{++} cytotoxicity (Table 2). Addition of the antioxidant DPPD (N,N'-diphenyl-p-phenylenediamine) completely inhibited this lipid peroxidation, and delayed but did not prevent cell death (Table 1). The toxicity of PQ^{++} in BCNU-treated hepatocytes could be potentiated by the addition of transition metals to the incubations and prevented by the ferric ion chelator desferrioxamine (Table 1), suggesting the involvement of a Fenton-type reaction between H_2O_2 and transition metals.[8,9]

Table 1. Effect of Antioxidants and BCNU-Pretreatment On the Toxicity of Paraquat, MPP^{+} and MPTP to Isolated Hepatocytes

	% TRYPAN BLUE UPTAKE AT 2 OR 3 HRS			
Incubation Conditions	Control 3 hrs	PQ^{++} 3 hrs	MPP^{+} 3 hrs	MPTP 2 hrs
No pretreatment	16±5	14±7	78±6	91±4
BCNU pretreatment	16±7	74±8	95±4	93±2
BCNU + DPPD (10 μM)	19±6	26±5	76±9	91±5
BCNU + Desferrioxamine (100 μM)	19±4	27±8	92±7	90±4

PQ^{++}, MPP^{+} and MPTP were all added to a final concentration of 1.5 mM. DPPD = diphenyl-p-phenylenediamine, a potent antioxidant. Values represent the mean ± S.D. of 3 experiments.

These results clearly point to the generation of oxygen radicals via a mechanism of redox cycling as a critical event in the toxicity of PQ^{++}. The relative importance of the individual reactive oxygen species (e.g., $O_2^{\cdot-}$, H_2O_2, ·OH) remains to be established, however. For example, the role played by superoxide anion radical remains particularly controversial. Although intraveneously administered superoxide dismutase (SOD) delayed the toxicity of PQ^{++} to rats,[10] the addition of exogenous SOD to BCNU-treated hepatocytes provided no protection against bipyridyl herbicide cytotoxicity.[9] Studies[13] with transformed Escherichia coli expressing high SOD activity have also shown that these bacteria are more susceptible to PQ^{++} toxicity than wild type E. coli.[11] On the other hand, PQ^{++}-resistance has been observed in human and mouse cell lines expressing elevated SOD activity[12] and in cells "scrape-loaded" with SOD.[13] The importance of the role played by superoxide anion radical in

PQ^{++} cytotoxicity thus appears to be dependent upon the model system used. Its precise role, if any, in the <u>in vivo</u> toxicity of PQ^{++} still remains to be elucidated, however.

MECHANISM OF MPTP AND MPP^+ TOXICITY TO HEPATOCYTES

Initial studies[8,14] indicated that both MPTP and MPP^+ were toxic to isolated rat hepatocytes (Table 1) and depleted intracellular GSH. In contrast to the results with PQ^{++}, a comparison of MPTP and MPP^+-induced cytotoxicity in non-treated vs. BCNU-treated hepatocytes revealed no significant potentiation resulting from the inhibition of glutathione reductase (Table 1). Furthermore, the rate of GSH depletion induced by either MPTP or MPP^+ was not significantly increased by pretreatment with BCNU.[8,15] In fact, the mechanism of glutathione depletion by the two compounds was found[15] to be unrelated to oxidation and due to an increased efflux of GSH. GSSG was subsequently formed in the extracellular space and its measurement was not, therefore, a true indicator of intracellular H_2O_2 generation. Another difference between the action of PQ^{++} and MPP^+ in isolated hepatocytes was the lack of protection afforded by the ferric ion chelator desferrioxamine against MPP^+-induced cytotoxicity (Table 1).[8] Desferrioxamine was similarly ineffective when added to cell incubations exposed to MPTP. These results suggest that the generation of oxygen radicals is unlikely to be an important factor in the cytotoxicity of either MPTP or MPP^+ to hepatocytes.

Table 2. Effect of Paraquat, MPTP and MPP^+ on Lipid Peroxidation, α-Tocopherol and ATP Content in Hepatocytes

	<u>Lipid Peroxidation</u> (nmol TBA-reactive products/10^6 cells)	<u>α-Tocopherol</u> (% of zero time values)	<u>ATP</u> (% of zero time values)
	<u>150 min</u>	<u>60 min</u>	<u>150 min</u>
<u>Treatment</u>			
No addition	1.8 ± 0.5	62 ± 9	97 ± 3
PQ^{++} (BCNU -pretreated hepatocytes)	11.5 ± 1.1	48 ± 5	85 ± 4
MPTP	1.0 ± 0.3	72 ± 6	N.D.
MPP^+	2.6 ± 0.4	72 ± 5	22 ± 8

PQ^{++} and MPP^+ were added to 1.5 mM and MPTP to 1 mM final concentration. ND = none detected. Values represent the mean ± S.D. of 3 experiments.

A similar conclusion arises from studies on the effects of MPTP and MPP^+ on the peroxidation of membrane lipids[16] and on the cellular content of α-tocopherol. The levels of lipid peroxidation products measured in incubations of either BCNU-treated or untreated cells exposed to MPP^+ was only slightly above control levels (Table 2). Furthermore, DPPD addition did not protect against MPP^+-induced cell death (Table 1),[8] although it did completely inhibit lipid peroxidation. MPTP was unexpectedly found[16] to possess antioxidant activity in isolated hepatocytes, and also protected rat liver microsomes against both chemically and non-chemically induced lipid peroxidation. Measurement of the α-tocopherol content of hepatocytes exposed to either MPTP or MPP^+ did not reveal any increased loss during the course of the incubation, as compared to control cells (Table 2).

If oxidative stress is not involved in the toxic effects of MPTP and MPP^+ in hepatocytes, what alternative mechanism of cytotoxicity might play a more essential role? The fact that MPP^+ is able to cause the inhibition of NADH-linked substrate oxidation in mitochondria[17] prompted us to investigate the effects of this toxicant on cellular energy supplies in the form of ATP. Exposure of isolated hepatocytes to MPP^+ indeed led to a rapid dose-dependent depletion of intracellular ATP which preceded the onset of cytotoxicity (Table 2).[18] Similarly, a complete loss of ATP was caused by the parent compound MPTP, while PQ^{++} only slightly decreased intracellular ATP (Table 2). Thus, mitochondrial damage rather than oxidative stress seems to be a critical step in the biochemical events leading to cell death after MPTP and MPP^+ exposure. Furthermore, the depletion of ATP might be involved in the increased efflux of GSH from hepatocytes via perturbations of membrane potentials.[15]

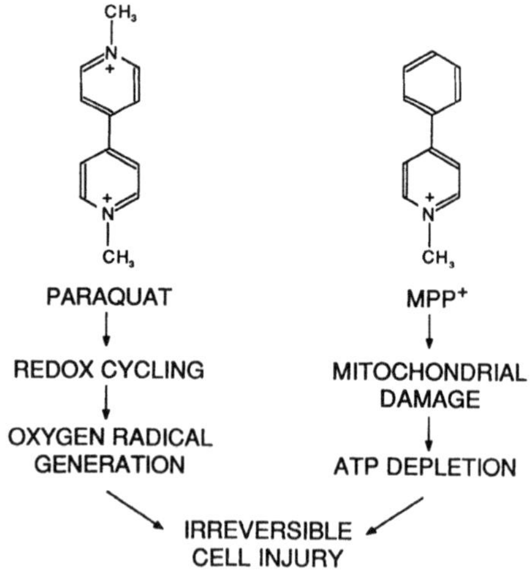

Figure 1. Proposed Mechanisms of Paraquat and MPP^+ Cytotoxicity.

CONCLUSIONS

Our studies with isolated rat hepatocytes have shown that the general mechanisms of cytotoxicity of MPP^+ and PQ^{++} are different (Figure 1). PQ^{++} toxicity is likely to be related to a process of one-electron redox cycling which generates oxygen radicals and ultimately leads to oxidative stress. In contrast, we found no evidence for a critical involvement of oxidative stress in the cytotoxicity of MPP^+. The different abilities of PQ^{++} and MPP^+ to generate active oxygen species is probably the consequence of their different redox potentials which reflect the free energy necessary for their reduction. The redox potential of PQ^{++} is quite low (-0.446 V),[19] and its ability to redox cycle is, therefore, relatively poor as compared to its congener diquat or to quinonoid drugs.[7] Electrochemical measurement of the redox potential of MPP^+ has yielded the value of -1.17 V, well below the limits of activity of known cellular reducing systems.[19] This result further supports the conclusion that the redox cycling of MPP^+ with molecular oxygen is unlikely to occur to any significant extent or to be involved in its cytotoxic effects.

As mentioned previously, it has been hypothesized that exposure to agricultural chemicals, such as PQ^{++}, might increase the risk for parkinsonism. Even if PQ^{++} and MPP^+ are toxic via different mechanisms, it might be possible that the two compounds share the property of a selective neurotoxicity *in vivo*. In contrast to MPP^+, however, PQ^{++} is not able to deplete striatal dopamine when either administered to the mouse[20] or infused directly into the striatum of rats.[21] Furthermore, because of its charged structure, the PQ^{++} molecule is unlikely to penetrate the blood-brain barrier or damage selectively the neurons of the substantia nigra. The selective action of MPP^+ (which is also charged) seems to be due to its generation within the CNS from its lipophilic parent compound MPTP and its accumulation within neurons of the substantia nigra via the dopamine uptake system.[22] There is as yet no direct evidence, therefore, linking PQ^{++} exposure to PD and further studies are needed to clarify the possible role played by oxygen radicals in the selective damage of dopaminergic neurons during both MPTP-induced and idiopathic parkinsonism.

ACKNOWLEDGMENTS

We thank the Health Effects Component of the University of California Toxic Substances Program, the Northern California Occupational Health Center and the National Institute for Environmental Health Sciences (ES01896) for supporting this work.

REFERENCES

1. J. W. Langston, P. Ballard, J. W. Tetrud, and I. Irwin, Chronic parkinsonism in humans due to a product of meperidine-analog synthesis, Science 219:979 (1983).
2. R. S. Burns, C. C. Chiueh, S. P. Markey, M. H. Ebert, D. M. Jacobowitz, and I. J. Kopin, A primate model of parkinsonism: Selective destruction of dopaminergic neurons in the pars compacta of the substantia nigra by N-methyl-4-phenyl- 1,2,3,6-tetrahydropyridine, Proc. Natl. Acad. Sci. USA 80:4545 (1983).

3. R. E. Heikkila, L. Manzino, F. S. Cabbat, and R. C. Duvoisin, Protection against the dopaminergic neurotoxicity of 1-methyl-4-phenyl-1,2,3,6-tetrahydropyridine by monoamine oxidase inhibitors, Nature (London) 311:467 (1984).
4. K. Chiba, A.J. Trevor, and N. Castagnoli, Jr. Metabolism of the neurotoxic tertiary amine, MPTP, by brain monoamine oxidase, Biochem. Biophys. Res. Commun. 120:547 (1984).
5. A. Barbeau, M. Roy, T. Cloutier, L. Plasse, and S. Paris, Environmental and genetic factors in the etiology of Parkinson's disease, in "Advances in Neurology," vol. 45, M. D. Yahr and K. J. Bergman, eds., Raven Press, New York (1986).
6. J. S. Bus and J. E. Gibson, Paraquat: Model for oxidant- initiated toxicity, Environ. Health Perspec. 55:37 (1984).
7. M. S. Sandy, P. Moldeus, D. Ross, and M. T. Smith, Role of redox cycling and lipid peroxidation in bipyridyl herbicide cytotoxicity. Studies with a compromised isolated hepatocyte model system, Biochem. Pharmacol. 35:3095 (1986).
8. D. Di Monte, M. S. Sandy, G. Ekstrom, and M. T. Smith, Comparative studies on the mechanisms of paraquat and 1-methyl-4-phenylpyridine (MPP^+) cytotoxicity, Biochem. Biophys. Res. Commun. 137:303 (1986).
9. M. S. Sandy, P. Moldeus, D. Ross, and M. T. Smith, Cytotoxicity of the redox cycling compound diquat in isolated hepatocytes: Involvement of hydrogen peroxide and transition metals, Arch. Biochem. Biophys. 259:29-37 (1987).
10. A. P. Autor, Reduction of paraquat toxicity by superoxide dismutase, Life Sci. 14:1309 (1974).
11. M. D. Scott, S. R. Meshnick, and J. W. Eaton, Superoxide dismutase-rich bacteria. Paradoxical increase in oxidant toxicity, J. Biol. Chem. 262:3640 (1987).
12. O. Elroy-Stein, Y. Bernstein, and Y. Groner, Overproduction of human Cu/Zn-superoxide dismutase in transfected cells: extenuation of paraquat-mediated cytotoxicity and enhancement of lipid peroxidation, EMBO J. 5:615 (1986).
13. A. C. Bagley, J. Krall, and R. E. Lynch, Superoxide mediates the toxicity of paraquat for Chinese hamster ovary cells, Proc. Natl. Acad. Sci. USA 83:3189 (1986).
14. M. T. Smith, G. Ekstrom, M. S. Sandy, and D. Di Monte, Studies on the mechanism of 1-methyl-4-phenyl-1,2,3,6-tetrahydropyridine cytotoxicity in isolated hepatocytes, Life Sci. 40:741 (1987).
15. D. Di Monte, M. S. Sandy, and M. T. Smith, Increased efflux rather than oxidation is the mechanism of glutathione depletion by 1-methyl-4-phenyl-1,2,3,6-tetrahydropyridine (MPTP), Biochem. Biophys. Res. Commun. 148:153 (1987).
16. G. Ekstrom, D. Di Monte, M. S. Sandy, and M. T. Smith, Comparative toxicity and antioxidant activity of 1-methyl-4-phenyl-1,2,3,6-tetrahydropyridine and its monoamine oxidase B-generated metabolites in isolated hepatocytes and liver microsomes, Arch. Biochem. Biophys. 255:14 (1987).
17. W. J. Nicklas, I. Vyas, and R. E. Heikkila, Inhibition of NADH-linked oxidation in brain mitochondria by 1-methyl-4-phenylpyridine, a metabolite of the neurotoxin, 1-methyl-4-phenyl-1,2,3,6-tetrahydropyridine, Life Sci. 36:2503 (1985).
18. D. Di Monte, S. A. Jewell, G. Ekstrom, M. S. Sandy, and M. T. Smith, 1-methyl-4-phenyl-1,2,3,6-tetrahydropyridine (MPTP) and 1-methyl-4-phenylpyridine (MPP^+) cause rapid ATP depletion in isolated hepatocytes, Biochem. Biophys. Res. Commun. 137:310 (1986).
19. D. M. Frank, P. K. Arora, J. L. Blumer, and I. M. Sayre, Model study on the bioreduction of paraquat, MPP+, and analogs. Evidence against a "redox cycling" mechanism in MPTP neurotoxicity,

Biochem. Biophys. Res. Commun. 147:1095 (1987).

20. T. L. Perry, V. W. Yong, R. A. Wall, and K. Jones, Paraquat and two endogenous analogs of the neurotoxic substance N-methyl-4-phenyl-1,2,3,6-tetrahydropyridine do not damage dopaminergic neurostriatal neurons in the mouse, Neurosci. Lett. 69:285 (1986).

21. H. Rollema, G. Damsma, A. S. Horn, J. B. De Vries and B. H. C. Westerink, Brain dialysis in conscious rats reveals an instantaneous massive release of striatal dopamine in response to $MPP.^{+}$ Eur. J. Pharmacol. 126:345 (1986).

22. S. H. Snyder and R. J. D'Amato, MPTP: A neurotoxin relevant to the pathophysiology of Parkinson's disease, Neurology 36:250 (1986).

FREE RADICAL FORMATION BY CYTOTOXIC

ALKYLAMINOANTHRAQUINONES IN LIVER MICROSOMES

Laurence H. Patterson, Jaspal Basra,
and Jeffrey R. Brown*

Department of Pharmaceutical Chemistry
School of Pharmacy
Leicester Polytechnic
Leicester LE1 9BH, UK
*Department of Pharmaceutical Chemistry
Faculty of Pharmaceutical Sciences
Sunderland Polytechnic
Sunderland SR2 7EE, UK

INTRODUCTION

The anthracycline antibiotics as exemplified by doxorubicin and daunorubicin are widely used antitumour agents. The mechanism of toxicity of these agents probably involves damage to multiple sites within cell systems since DNA binding, DNA strand breakage, enzyme inactivation, and membrane effects have all been described.[1] The planar chromophore and cationic nature of the clinically important anthracyclines are essential to many of their interactions with biological systems especially intercalative binding to DNA[2] and free radical generation.[3] Mitozantrone is an alkylaminoanthraquinone antitumour agent which has been developed from a consideration of the modes of action of the anthracyclines and although a more simplified derivative it has a similar activity with respect to DNA binding[4] and also generates free radicals.[5] To explore further the structural factors influencing DNA binding and free radical generation by alkylaminoanthraquinones we have prepared derivatives based on the anthraquinone chromophore.[6] In this study metabolic free radical generation by four alkylaminoanthraquinones (compounds 1-4) that differ in the substitution pattern is described (Figure 1).

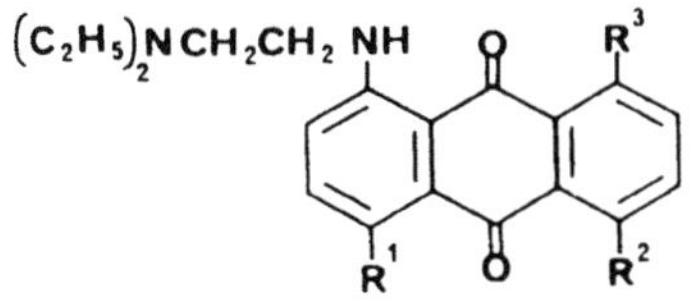

compound	R^1	R^2	R^3
1	H	H	H
2	$NH(Me)_2NEt_2$	H	H
3	H	$NH(Me)_2NEt_2$	H
4	H	H	$NH(Me)_2NEt_2$

Figure 1. The four alkylaminoanthraquinones studied here.

METHODS

Alkylaminoanthraquinones 1-4 were synthesised and identified as described previously.[6] Electron spin resonance spectrometry (ESR) was used to directly monitor alkylaminoanthraquinone free radical formation in NADPH supplemented mouse liver microsomes as previously described[5]. The effect of alkylaminoanthraquinones on the rate of microsomal NADPH oxidation and superoxide anion formation was carried out as previously described[5].

RESULTS

ESR spectra were obtained with g-values between 2.0040-2.0045 and line widths of 3.8-5.5G when compounds 1-4 (100μM) were incubated anaerobically with NADPH-supplemented mouse liver microsomes. These spectra were attributed to metabolically formed alkylaminoanthraquinone free radicals since none were obtained when NADPH, native microsomes, or compounds were absent. The broad line widths and lack of hyperfine structure of the ESR spectra are probably a consequence of hindered rotation of the free radicals when generated by the membrane-bound enzymes of the microsomal system. Figure 1 shows a typical ESR spectrum for compound 3. These alkylaminoanthraquinones stimulated basal rate microsomal NADPH oxidation which was typically 12.1 + 1.5 nmole /min /mg microsomal protein (see Figure 3a). This was accompanied by a dose-dependent increase in microsomal superoxide anion formation for all compounds except 1 as shown in Figure 3b. The basal rate of microsomal superoxide anion formation was 8.0 + 0.2 nmole/ min /mg microsomal protein. The apparent lack of stimulation of superoxide anion formation by compound 1 is not consistent with its observed ESR spectrum and requires further investigation.

DISCUSSION

The results of this study show that in biological systems alkylaminoanthraquinones generate free radicals that in the absence of oxygen accumulate to a steady state equilibrium observable by ESR. The loss of these alkylaminoanthraquinone free radicals in the presence of air (results not shown) and the stimulation of superoxide anion formation in the presence of alkylaminoanthraquinones suggest that these compounds, via a free radical intermediate, reduce dioxygen to the superoxide anion. The results also show that NADPH is implicated as a source of reducing equivalents in this alkylaminoanthraquinone-mediated superoxide anion

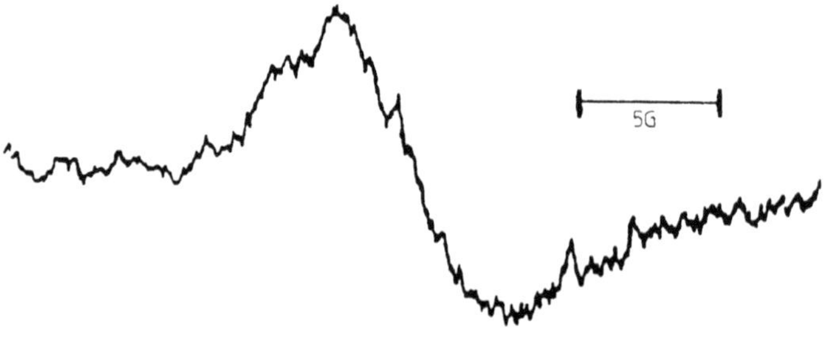

Figure 2. ESR spectrum generated when compound 3 (100μM) was incubated anaerobically at 21-22°C in the presence of NADPH-supplemented mouse liver microsomes. ESR operating conditions were microwave power 5mW, modulation amplitude 4G, time constant 3s, scan time 4 min, receiver gain $2x10^6$ and microwave frequency 9.5 MHz.

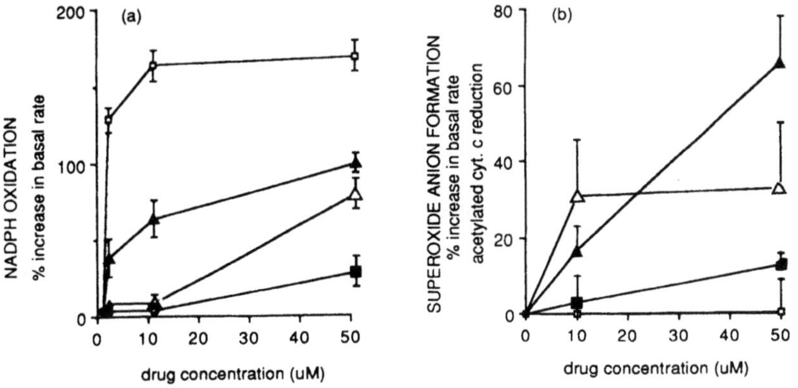

Figure 3. Effect of alkylaminoanthraquinones on mouse liver (a) NADPH oxidation and (b) superoxide anion formation measured as superoxide dismutase (300μg/ml) inhibitable acetylated cytochrome c reduction. Basal rates of these processes, as given in the text, have been deducted from the present data which are the mean + SD of at least three determinations. Compounds are -□- 1; -▲- 2; -■- 3; -△- 4.

since NADPH is required for generation of alkylaminoanthraquinone free radicals and NADPH oxidation is stimulated markedly in the presence of these compounds. Hence alkylaminoanthraquinone free radicals generated metabolically, presumably by one-electron reduction, shuttle electrons from NADPH to dioxygen resulting in superoxide anion formation. This process is directly analagous to that described for free radical formation by doxorubicin and hence would be expected to be involved in the toxicity associated with these compounds. The redox cycling scheme involving alkylaminoanthraquinone free radical formation is shown in Figure 4.

The four compounds used in this study differ in the position of the alkylamino substituent. Such differences appear to have influenced the extent of NADPH utilisation with 1 > 2 > 4 > 3 in order of increasing activity. This may indicate the extent of interaction of these compounds with NADPH-dependent microsomal reductases responsible for one electron reduction of quinones.[7] However these compounds may also be substrates for microsomal oxidative routes of metabolism which would be expected to

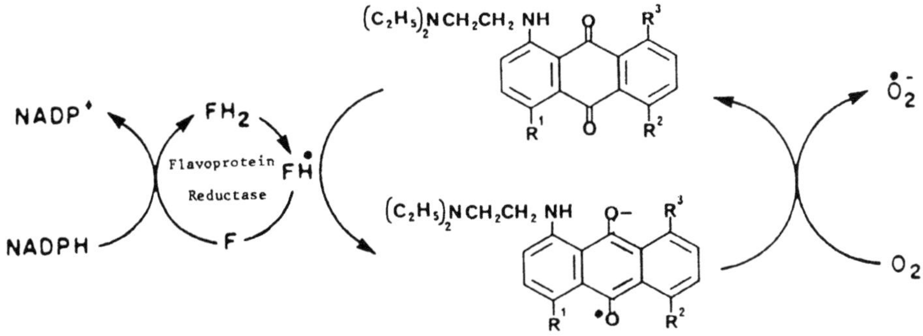

Figure 4. Redox cycling by alkylaminoanthraquinones.

contribute to the overall rates of NADPH oxidation observed. This may help explain the lack of correlation between NADPH oxidation and stimulation of superoxide anion formation which is in the order 4 > 2 > 3 > 1. Previously it was shown that there is a correlation in the cytotoxicity of these compounds with their relative DNA binding affinities.[6] Since these compounds have now also been shown to generate superoxide anions it is possible that DNA-binding acts in concert with reactive oxygen generation to produce the cytotoxicity observed.

ACKNOWLEDGEMENTS

This work was supported by The Cancer Reseach Campaign UK (grant SP1669 to LHP and SP1423 to JRB).

REFERENCES

1. L. Gianni, B. L. Gorden, and C.E. Meyers, The biochemical basis of anthracycline toxicity and antitumour activity, in "Reviews in Biochemical Toxicity, Vol. 5", E. Philpot and E. Bend, eds, Elsevier, New York (1984).
2. J. R. Brown and S. H. Imam, Recent studies on doxorubicin and its analogues, Prog. Med. Chem. 21:169 (1984).
3. B. Kalyanaraman, E. Perez-Reyes, and R. P. Mason, Spin-trapping and direct electron spin resonance investigations of the redox metabolism of quinone anticancer drugs, Biochim. Biophys. Acta 630:119 (1980).
4. F. Traganos, Dihydroxyanthraquinone and related Bis (substituted)-aminoanthraquinones: a novel class of antitumour agent. Pharmacol. Ther. 22:199 (1983).
5. J. Basra, C. R. Wolf, J. R. Brown, and L. H. Patterson, Evidence for human liver mediated free-radical formation by doxorubicin and mitozantrone, Anti-Cancer Drug Design 1:45 (1985).
6. S. A. Islam, S. Neidle, B. M. Gandecha, M. B. Partridge, and L.H. Patterson, Comparative computer graphics and solution studies on the DNA interaction of substituted anthraquinones based on doxorubicin and mitoxantrone, J. Med. Chem. 28:857 (1985).
7. R. P. Mason and C. F. Chignell, Free radicals in pharmacology and toxicology: selected topics, Pharmacol. Rev. 31:189 (1982).

LIPID PEROXIDATION AND HALOALKYLATION IN CCl_4-INDUCED LIVER FATTY DEGENERATION AND NECROSIS

Giuseppe Poli, Elena Chiarpotto, Fiorella Biasi, Emanuele Albano, Rita Carini, Umberto M. Marinari,* and Mario U. Dianzani

Department of Experimental Medicine and Oncology of the University, C. Raffaello 30, 10125, Torino, Italy

*Institute of General Pathology of the University Via L.B. Alberti 2, 16132, Genova, Italy

INTRODUCTION

Liver fatty accumulation following acute CCl_4 poisoning is triggered by a block of lipoprotein secretion which occurs very soon after treatment.[1,2] Various steps of the secretory pathway are probably involved in the expression of such damage. Recent studies with isolated rat hepatocytes have shown a marked impairment of the maturation and transport of lipoprotein micelles at the level of the Golgi apparatus soon after poisoning with CCl_4.[3]. In the present paper the demonstration of early disturbances of liver Golgi function occurring in the whole animal dosed with CCl_4 is reported, together with studies on the pathogenesis of this cellular perturbation. In addition, the relative role played by CCl_4-dependent alkylation of macromolecules and lipid peroxidation in the necrogenic effect of the toxin has been evaluated in in vivo experiments.

EXPERIMENTAL PROCEDURES

In order to characterize the relative implication of the two recognized CCl_4 damaging mechanisms in fatty liver and necrosis, the pretreatment of animals with a single large amount of alpha-tocopherol was adopted. The vitamin, in fact, has been proved in the whole rat to completely inhibit the prooxidant effect of the haloalkane without significantly modifying the haloalkylation of cell molecules by its free radical metabolites (data not shown).

RESULTS AND DISCUSSION

The vitamin E-loading treatment protected the CCl_4-dosed animals against the inhibition of Golgi glycosylation reaction, but only slightly protected (as shown in Table 1) against the lipid accumulation in the Golgi, consequent to a block of the efflux of lipoprotein micelles from the apparatus. This finding does not support a major role of lipid peroxidation, at least in the early block of lipoprotein Golgi transport due to CCl_4, indirectly suggesting the main implication of the covalent binding of CCl_4 metabolites.

Table 1. ^3H-Palmitate Incorporation into the Lipids of Golgi Fractions Obtained from Rats Poisoned with CCl_4, Pretreated or Not with Vitamin E

Exp. groups	Golgi fractions F_1	F_2	F_3
Control	46,700±2,400	30,100±2,300	15,300± 900
Vitamin E	45,500±3,600	32,200±1,100	14,700±1,300
CCl_4	66,300±3,000(+42%)[a]	48,500±4,000(+61%)[a]	18,000±1,200(+18%)
Vitamin E + CCl_4	58,700±2,600(+29%)[b]	48,400±3,600(+50%)	17,400±1,000(+18%)

Values obtained 1 hour after dosing rats with CCl_4 (2.5 ml/kg) or mineral oil alone are expressed as cpm/mg protein and represent the means ± S.D. of six to eight experiments.
Values in parentheses are the percent increase compared with the corresponding control: a, $P < 0.05$ with respect to the control group and b, $P < 0.05$ with respect to CCl_4 group.

In relation to the necrogenic effect of the hepatotoxin, the monitoring of serum transaminases levels up to 24 hour poisoning showed their significant and progressive increase after the fourth hour treatment (Table 2). When rats undergoing intoxication with CCl_4 (2.5 ml/kg) received i.p. 100 mg/kg vitamin E 15-16 hours in advance, the extent of cell necrosis (mainly but not only localized in the liver), estimated in terms of serum transaminases, was markedly reduced by half (Table 2).

Table 2. Serum Aspartate Amino Transferase(U/l) Levels of Control and CCl_4-Poisoned Rats Pretreated or Not with Vitamin E[a]

Exp. groups	1 h	4 h	12 h	24 h
Control	55 ± 5	51 ± 8	87 ± 6	88 ± 20
Vitamin E	51 ± 9	55 ± 6	72 ± 4	72 ± 5
CCl_4	55 ± 11	195 ± 52[b]	825 ± 66[c]	4252 ± 462[c]
CCl_4 + Vit.E	53 ± 8	96 ± 7[d]	583 ± 71[d]	1952 ± 292[e]

a, All values are means ± S.D. of four experiments; for each experimental group six animals were used.
b, $P < 0.05$ with respect to the control group
c, $P < 0.001$ with respect to the control group
d, $P < 0.05$ with respect to the CCl_4 group
e, $P < 0.005$ with respect to the CCl_4 group

Moreover, the same membrane enrichment in alpha-tocopherol afforded a complete protection against necrosis when a concentration of CCl_4 five times lower was given to the rats. Of interest is the observation in the latter experimental condition of partial prevention of fatty liver due to the poison.

In relation to the cholestatic effect of CCl_4 reported by other authors in the past,[4] a new finding is represented by the protection achieved against it by pretreatment of animals with vitamin E.

Table 3. Serum Alkaline Phosphatase (U/l) Levels of Control and CCl_4-Poisoned Rats Pretreated or Not with Vitamin E[a]

Exp. groups	1 h	4 h	12 h	24 h
Control	90 ± 43	90 ± 11	107 ± 11	107 ± 15
Vitamin E	140 ± 21	117 ± 20	117 ± 50	100 ± 20
CCl_4	147 ± 59[d]	110 ± 36	225 ± 47[b]	255 ± 50[b]
CCl_4 + Vit. E	120 ± 21[e]	83 ± 25	140 ± 52	160 ± 21[c]

a, All values are means ± S.D. of four experiments; for each experimental group six animals were used.
b, $P < 0.01$ with respect to the control group
c, $P < 0.05$ with respect to the CCl_4 group;
d, Not significant as to the control group;
e, Not significant as to the CCl_4 group.

Actually, the cholestasis induced by the toxin appears mainly a consequence of the irreversible hepatocyte damage due to the intoxication. In fact, the level of serum alkaline phosphatase, the proper marker of cholestasis, is significantly increased in CCl_4-treated animals as to untreated only later than four hour poisoning when, on the contrary, transaminase levels of the experimental group are about four time as higher as the control. In addition, the vitamin E-induced prevention of cholestasis is quantitatively consistent with that of necrosis (Table 3).

Combined with identical results previously obtained in isolated liver cells,[5] the present data support the hypothesis that lipid peroxidation is the main cause of hepatocyte irreversible damage, at least in acute poisoning with CCl_4.

REFERENCES

1. T.F. Slater, Hepatotoxicity of carbon tetrachloride: Fatty degeneration, in "Free Radical Mechanisms in Tissue Injury," J.R. Lagnado, ed., Pion Ltd, London (1972).
2. W.J. Brattin, E.A. Glende, Jr., and R.O. Recknagel, Pathological mechanisms in carbon tetrachloride hepatotoxicity, J. Free Rad. Biol. Med. 1:27 (1985).

3. G. Poli, E. Chiarpotto, E. Albano, D. Cottalasso, G. Nanni, U.M. Marinari, A.M. Bassi, and M.U. Dianzani, Carbon tetrachloride-induced inhibition of hepatocyte lipoprotein secretion: Functional impairment of Golgi apparatus in the early phases of such injury, Life Sciences 36:533 (1985).
4. E.A. Rodriguez Garay, and J. Pahissa Campa, Bilirubin metabolism in experimental liver injury, J. Lab. Clin. Med. 72:582 (1968).
5. G. Poli, E. Albano, F. Biasi, G. Cecchini, R. Carini, G. Bellomo, and M.U. Dianzani, Lipid peroxidation stimulated by carbon tetrachloride or iron in hepatocyte death: Protective effect of vitamin E, in: "Free Radicals in Liver Injury," G. Poli, K.H. Cheeseman, M.U. Dianzani, and T.F. Slater, eds., IRL Press, Oxford (1985).

ACKNOWLEDGMENTS

The authors wish to thank the Consiglio Nazionale delle Ricerche and the Ministero della Pubblica Istruzione, Roma, Italy.

PARAQUAT-RESISTANT HeLa CELLS

J. Krall, A.C. Bagley, G.M. Mullenbach*, R.A. Hallewell*, and R.E. Lynch

Research Service, VA Medical Center, Salt Lake City, Utah 84148 and Department of Pathology, University of Utah Medical School, Salt Lake City, Utah 84132, U.S.A.

*Chiron Corporation, 4560 Horton St., Emeryville California 94608, U.S.A.

INTRODUCTION

The cytotoxic action of chemicals which transfer electrons from biological reductants to molecular oxygen is believed to result from the formation of $\cdot O_2^-$, H_2O_2, and $\cdot OH$. Paraquat is one such agent which appears to act in Escherichia coli primarily by generating $\cdot O_2^-$, but necessarily H_2O_2 as well. To have its cytotoxic action by promoting the intracellular formation of $\cdot O_2^-$ paraquat must first be taken up by the cell, it must then be reduced by one electron to form the paraquat radical, and finally it must auto-oxidize to yield $\cdot O_2^-$. If this scheme is correct for mammalian cells as it appears to be for prokaryotes,[1] cells which adapt to growth in the presence of paraquat should modify one or more of these steps, probably by mutation or other genetic adaptation, such as amplification of genes for the superoxide dismutases or for catalase, as a means of increasing the cellular content of enzymes which scavenge $\cdot O_2^-$ or H_2O_2, respectively.

It appeared important to determine how paraquat acted because if its toxicity is due mainly to the intracellular generation of $\cdot O_2^-$, paraquat would be a convenient means to explore the range of the effects of $\cdot O2^-$ in cultured mammalian cells.

RESULTS

HeLa cells Resistant to 40 μM Paraquat

To determine some of the necessary steps in the toxicity of paraquat in mammalian cells we isolated paraquat-resistant HeLa cells after lethal increments in the concentration of paraquat in the medium. Lethal increments were chosen in an attempt to avoid adaptations resulting from the exclusion of paraquat. The rare colonies which formed when 5,000 to 500,000 cells were seeded per dish were cloned with glass cylinders and expanded in tissue culture for further characterization. Cells selected in the first increment in concentration of paraquat were found to have increased the cellular content of both superoxide dismutases. The cellular content of the Mn superoxide dismutase increased from 1.36 units/mg (± 0.10, SEM) of cellular protein to 3.0 ± 0.34 units/mg and the cellular content of the CuZn superoxide dismutase increased from 4.48 (± 0.34) units/mg of cellular protein to 6.9 (± 0.47) in the cells resistant to 40 μM paraquat (PQRHM40S). Cells selected on other occasions also exhibited increased cellular content of both superoxide dismutases of similar magnitudes. The cellular content of catalase was not significantly different in the wild-type and the paraquat-resistant cells.

Because the increased cellular content of superoxide dismutases might represent a non-specific response to growth under disadvantageous conditions and not a protective adaptation, the cells resistant to 40 μM paraquat were cultivated for five months in the absence of paraquat. If the increased cellular content of superoxide dismutases dissociated from resistance to paraquat, one could not be the cause of the other. At intervals cells were reassayed for resistance to paraquat and for the cellular content of superoxide dismutases and catalase. After five months of growth in the absence of paraquat the increased cellular content of superoxide dismutases was retained and the cells remained resistant to paraquat. The concentration of paraquat which killed 50% of the paraquat-resistant cells grown continuously in the presence of paraquat, 47 μM, was similar to the concentration which killed 50% of the cells which were grown for five months in the absence of paraquat, 52 μM. Both concentrations were greater than that which killed 50% of the wild-type cells, 18 μM. The cellular contents of the Mn superoxide dismutase, 2.9 units/mg, and of the CuZn superoxide dismutase, 8.7 units/mg, were not significantly less than the values for cells grown continuously in the presence of 40 μM paraquat. These data suggest that the increased content of superoxide dismutases was not a non-specific response to growth under stressful conditions and that it could have caused resistance to the toxicity of paraquat.

Resistance to Paraquat of Cells Enriched for Superoxide Dismutase

NIH/3T3 cells which were enriched with the human CuZn superoxide dismutase by transcription of transfected genes were also resistant to the toxicity of paraquat. The cellular content of superoxide dismutase was increased in NIH/3T3 cells by transcription of a modified cDNA for the human CuZn superoxide dismutase[2] driven by the human metallothionein promoter in the same plasmid in which the SV40 early promoter drives transcription of the gene for aminoglycoside phosphotransferase.[3] The gene for aminoglycoside phosphotransferase permitted us to select cells in which this plasmid was stably integrated into the chromosome of the host and expressed by exposure of cells to an aminoglycoside which is toxic for mammalian cells, G-418. Cells expressing the transfected cDNA could metabolically transform G-418 to a non-toxic form and, thereby, survive. NIH/3T3 cells cotransfected by the calcium phosphate precipitation technique[4,5] with this plasmid and with another, which encodes a dihydrofolate reductase with a high Km,[6] were selected first for resistance to the aminoglycoside G-418 and then for resistance to methotrexate, first 0.25 μM and then 2 μM. The first concentration of methotrexate was chosen to permit survival only of the cells which expressed the transfected cDNA for the dihydrofolate reductase with the aberrantly high Km because amplification of the resident chromosomal gene as a means of resistance is extremely unlikely in a single step at such a great initial concentration. The second selection in 2 μM methotrexate was done to isolate cells which had amplified the gene for the transfected dihydrofolate reductase and, we hoped, the cotransfected cDNA for the human CuZn superoxide dismutase as well. By this means cells were obtained with a four- to fivefold increase in the cellular content of superoxide dismutase; the specific activity in these cells was 29 units/mg of cellular protein compared with 3-7 units/mg in untransfected controls and with 7 units/mg in cells (controls) transfected with the same two plasmids and treated similarly except that no cDNA had been inserted downstream from the metallothionein promoter. Expression of the transfected cDNA for the human CuZn superoxide dismutase could be further documented by the appearance of new bands of activity in acrylamide gels after electrophoresis of cellular extracts[7] which, by mobility, appeared to correspond to human-human and to human-murine dimers. The concentration of paraquat which killed 50% of these cells, 15 μM in both of two experiments, was approximately twice that required for controls which had been treated similarly except that no cDNA had inserted into the transfected expression vector, 7.5 and 9.5 μM.

When combined with the data showing increased resistance to paraquat's toxicity by cells enriched for superoxide dismutase by scrape-loading[8] and by other mammalian cells enriched for the human CuZn superoxide dismutase by transcription of the transfected cDNA,[9] these data strongly suggest that in mammalian cells, as in prokaryotes,[1,10] superoxide mediates the toxicity of paraquat.

Cyanide-resistant Respiration of Paraquat-resistant HeLa Cells

To verify that paraquat entered the PQRHM40S cells and underwent reduction, we compared the respiration of wild-type and PQRHM40S cells in the presence and in the absence of cyanide. The respiration in the two types of cells was very similar. In both cases cyanide inhibited respiration by about 97% in the absence of paraquat. When paraquat was added to the chamber of the oxygraph little effect was observed on overall respiration or on the cyanide-resistant component. However, when either type of cell was grown in the presence of 40 μM paraquat for 24 hours, a substantial fraction of cellular respiration (12.3-14.1%) was not inhibited by cyanide. These data suggest that paraquat enters both types of cells, undergoes reduction, and auto-oxidizes. The difference is that the wild-type cells are killed (after several days) but the PQRHMR40S cells are not. The adaptation which is able to account for survival in the presence of continued diversion of electrons to O_2 appears to be the augmented cellular content of both superoxide dismutases.

CONCLUSIONS

These paraquat-resistant mammalian cells suggest that in some cultured cells, as in prokaryotes, the toxicity of paraquat is dependent on its intracellular reduction and on the subsequent auto-oxidation of the paraquat radical to yield superoxide. The means by which a stable increase in the cellular content of superoxide dismutases in the paraquat-resistant HeLa cells has occurred is not yet clear. Preliminary data from Southern blots of genomic DNA digested with Pst I or Hind III indicate that the increase in superoxide dismutases has not resulted from amplification of genes for these enzymes, which reside on separate chromosomes.

REFERENCES

1. H. M. Hassan and I. Fridovich, Superoxide radical and the oxygen enhancement of the toxicity of paraquat in Escherichia coli, J. Biol. Chem. 253:8143 (1978).

2. R. A. Hallewell, F. R. Masiarz, R. C. Najarian, J. P. Puma, M. R. Quiroga, A. Randolph, R. Sanchez-Pescador, C. J. Scandella, B. Smith, K. S. Steimer, and G. T. Mullenbach, Human Cu/Zn superoxide dismutase cDNA: isolation of clones synthesizing high levels of active or inactive enzyme from an expression library, Nucleic Acids Res. 13:2017 (1985).
3. C. Colbere-Garapin, F. Horodniceanu, P. Kourilsky, and A-C. Garapin, A new dominant hybrid selective marker for higher eukaryotic cells, J. Mol. Biol. 150:1 (1981).
4. C. Gorman, High efficiency gene transfer into mammalian cells, in: "DNA Cloning, Volume II," D. M. Glover, ed., IRL Press, Oxford & Washington, D. C. (1985).
5. F. Graham and A. van der Eb, A new technique for the assay of infectivity of human adenovirus 5 DNA, Virology 52:456 (1973).
6. C. C. Simonsen and A. D. Levinson, Isolation and expression of an altered mouse dihydrofolate reductase cDNA, Proc. Natl. Acad. Sci. U. S. A. 80:2495 (1983).
7. C. Beauchamp and I. Fridovich, Superoxide dismutase: improved assays and an assay applicable to acrylamide gels, Analyt. Biochem. 44:276 (1971).
8. A. C. Bagley, J. Krall and R. E. Lynch, Superoxide mediates the toxicity of paraquat for Chinese hamster ovary cells, Proc. Natl. Acad. Sci. U. S. A. 83:3189 (1986).
9. O. Elroy-Stein, Y. Bernstein, and Y. Groner, Overproduction of human Cu/Zn-superoxide dismutase in transfected cells: extenuation of paraquat-mediated cytotoxicity and enhancement of lipid peroxidation, EMBO J. 5:615 (1986).
10. A. Carlioz and A. Touati, Isolation of superoxide dismutase mutants in Escherichia coli: is superoxide dismutase necessary for aerobic life?, EMBO J. 5:623 (1986).

THE RESPIRATORY BURST OXIDASE

Bernard M. Babior

Research Institute of Scripps Clinic
La Jolla, CA 92037, U.S.A.

INTRODUCTION

Neutrophils are phagocytic cells whose major function is the ingestion and destruction of invading microorganisms. On exposure to certain stimuli, these cells begin to manufacture large quantities of microbicidal oxidants through a metabolic event known as the "respiratory burst" (reviewed in reference 1). The key reaction in the respiratory burst is the one-electron reduction of oxygen to $\cdot O_2^-$ at the expense of a reduced pyridine nucleotide (eq. 1).

$$2\ \cdot O_2 \ + \ NAD(P)H \ \longrightarrow \ 2\ \cdot O_2^- \ + \ NAD(P)^+ \ + \ H^+ \qquad (1)$$

Subsequent reactions include the formation of H_2O_2 by the dismutation of $\cdot O_2^-$, and finally the production of the following microbicidal oxidants: hypochlorite and chloramines through the myeloperoxidase-catalyzed oxidation of Cl^- by H_2O_2 and highly reactive oxidizing radicals (e.g., $OH\cdot$) through metal-catalyzed reactions involving inter alia $\cdot O_2^-$ and H_2O_2.

The enzyme responsible for $\cdot O_2^-$ production is the respiratory burst oxidase (more formally known as NADPH oxidase). To date this enzyme has been found exclusively in "professional" phagocytes--neutrophils, eosinophils and mononuclear phagocytes[2]. In resting phagocytes the enzyme is inactive, but when the cells are stimulated it acquires catalytic activity. In this chapter, work by our laboratory on the properties of the $\cdot O_2^-$-forming activity from human phagocytes is reviewed.

PROPERTIES OF THE OXIDASE

The respiratory burst oxidase was first described as an oxygen-consuming activity that could be measured when phagocyte homogenates were incubated with NADH or NADPH in the presence of Mn^{++}. In early days this activity was inconsistently defined: it was reported to occur in homogenates of both resting and activated phagocytes[3] (resting < activated, was variably detected in particles[3,4] or cytosol[5], and in various hands showed a preference for NADH[5] or NADPH[3,6]. Starting with the discovery that $\cdot O_2^-$ was the true product of the respiratory burst oxidase[7,8], and the contingent finding that the apparent requirement for Mn^{++} was artifactual[9,10], the picture gradually came into focus, and it is now universally agreed that the respiratory burst oxidase is a membrane-bound enzyme that catalyzes $\cdot O_2^-$ production at the expense of NADPH, as shown in eq. 1.

The properties of the oxidase were first studied in particulate preparations from neutrophil homogenates. $\cdot O_2^-$-forming particles could be obtained from neutrophils activated with opsonized zymosan, but not from resting neutrophils or from zymosan-treated neutrophils obtained from patients with chronic granulomatous disease (CGD), a disorder (more accurately, a set of disorders) in which phagocytes are unable to express a respiratory burst because of a defect in the respiratory burst oxidase or its activating apparatus (reviewed in reference 11). Fractionation of these $\cdot O_2^-$-forming particles showed that the oxidase was located in the plasma membrane of the cells[12,13], where it was oriented vectorially with a portion projecting into the cytosol[14]. Measurements of activity revealed a pH optimum near neutrality[14], a strong preference for NADPH ($K_m \approx$ 30-40 μM) over NADH ($K_m \approx$ 500-1000 μM)[14], and a relatively high oxygen affinity ($K_m \approx$ 10 μM, equivalent to the concentration of dissolved oxygen in water under an atmosphere of 1% $\cdot O_2$)[15,16]. The high oxygen affinity of the respiratory burst oxidase, demonstrated with both neutrophil particles and whole neutrophils, implies that phagocytes are able to manufacture microbicidal oxidants even at the relatively low oxygen tensions prevailing at sites of infection. Finally, a requirement for FAD was demonstrated by particles assayed in the presence of Triton X-100, but not by particles assayed in the absence of detergent[17].

After a great deal of effort, solubilization of the respiratory burst oxidase was finally achieved. Solubilization was initially accomplished by extraction of $\cdot O_2^-$-forming particles from zymosan-activated neutrophils with a low ionic strength buffer that contained Triton X-100 as the solubilizing detergent and ethylene glycol as stabilizer[18,19]. Studies of the Triton-solubilized oxidase revealed a pH optimum and K_m values for NADPH and NADH similar to those previously found with the particulate enzyme[20]. The K_m for FAD was measured at ≈ 0.06 μM[20]. This enzyme preparation also demonstrated an absolute requirement for phospholipid. This requirement could be fulfilled by phosphatidyl-ethanolamine, but not phosphatidylserine or phosphatidylcholine[19]. The oxidase did not show a metal requirement, nor was it inhibited by EDTA or o-phenanthroline[20]. Its activity was, however, inhibited by sulfhydryl reagents (e.g. p-chloromercuribenzene sulfonic acid)[21], but not by heme ligands (e.g., CN^-)[22]; the latter finding is consistent with numerous observations with whole phagocytes documenting the insensitivity of the respiratory burst to CN^-[23].

Starting with material solubilized under somewhat different conditions (deoxycholate or deoxycholate plus Lubrol PX as detergent and glycerol as stabilizer), the $\cdot O_2^-$-forming activity was purified by a one-step procedure involving "dye affinity" chromatography over red agarose[24,25]. (A rapid purification procedure was essential because of the instability of the enzyme once it was solubilized; under the best conditions, the oxidase lost over half its activity after only 24 hours at 4°.) This procedure yielded a preparation that showed a single major band on non-denaturing gel electrophoresis. When the non-denaturing gel was sliced and subjected to fluorescence spectroscopy, a flavin was found to co-migrate with the major band; when the same procedure was applied to a non-denaturing gel of material obtained by "dye affinity" chromatography of a solubilized preparation from resting neutrophil particles, no flavin was found anywhere in the gel. Analysis of the purified $\cdot O_2^-$-forming activity by SDS-PAGE revealed peptides at 67K, 48K and 32K, plus minor contaminants. The same 3 major peptides were seen when the major band from the non-denaturing gel was electroeluted and subjected to SDS-PAGE. These results suggested that the respiratory burst oxidase was composed of 3 polypeptides of unequal size and that its minimum molecular weight was ≈147K. On the other hand, the occurrence of the 32K and 48K peptides in various "dye affinity"-purified $\cdot O_2^-$-forming preparations was not absolutely consistent, suggesting as an alternative that the association of one or both of the 2 smaller peptides with the $\cdot O_2^-$-forming activity might be coincidental. The case for the 67K peptide, however, seems relatively secure: a 67K peptide has appeared in every preparation of purified $\cdot O_2^-$-forming activity obtained to date, including a preparation isolated in Kakinuma's laboratory by a new procedure[26]; and affinity labelling of the respiratory burst oxidase by NaCN[^{3}H]BH_3 plus periodate-oxidized NADPH yielded a radioactive 67K peptide[27,28].

Certain of the properties of the "dye affinity"-purified oxidase have been examined[25]. Kinetic studies revealed 1) a specific activity of 9.2 ± 3.2 μmoles $\cdot O_2^-$/min/mg protein at 25°, corresponding to a turnover number of ≈1,200/min (assuming a MW of 147K); and 2) K_m values for both NADPH and NADH that were similar to those obtained with activated neutrophil particles and crude solubilized extract. Determinations were also made of the quantities of certain redox cofactors associated with the purified oxidase. These were of particular interest because of studies with cruder systems that had suggested the participation not only of flavin, but also of ubiquinone and a heme protein as electron carriers in the $\cdot O_2^-$-forming reaction. The results of these measurements are given in Table 1.

Table 1. Redox Carrier Content of Purified Oxidase

Redox carrier	Moles/mole enzyme
FAD	3.0 ± 0.2
CoQ_{10}	<0.06
Heme	0.015 ± 0.004

Of the carriers measured, only FAD was found in the purified oxidase. Similar results were obtained with Kakinuma's preparation[26]. The absence of heme from the purified material is of special interest, because it implies that the purified oxidase is devoid of cytochrome b_{558}, a heme protein that is closely connected with the respiratory burst apparatus[29,30] and has been invoked as an obligatory electron carrier in the $\cdot O_2^-$-forming reaction. The absence of cytochrome b_{558} from these $\cdot O_2^-$forming preparations seems to suggest that this heme protein is not an obligatory component of the electron transport path from NADPH to $\cdot O_2^-$. Alternatively, it may be that the oxidase purified in my laboratory is actually only the NADPH-binding component of a complex electron transport system that normally includes cytochrome b_{558} as another of its components, and that the ability of this preparation to catalyze $\cdot O_2^-$ production is artificial, the result of its separation from the rest of the system.

A FULLY SOLUBLE OXIDASE ACTIVATING SYSTEM

A characteristic feature of the respiratory burst oxidase is its ability to be turned on and off in response to various stimuli[1]. In the past 3 years, several laboratories have reported the development of cell-free systems in which oxidase activation can be studied[31-34]. These systems consist of membranes and cytosol from resting neutrophils or macrophages, Mg^{++} ion, and a suitable anionic detergent such as an unsaturated long-chain fatty acid, SDS[35] or sodium dodecylsulfonate. Other metal ions (in particular, Ca^{++}) are not required for oxidase activation in these systems, and activation is independent of ATP or other sources of energy.

We have recently been able to bring the cell-free oxidase activating system into a fully soluble form using the solubilizing cocktail employed previously for extracting activated oxidase from particles obtained from zymosan-stimulated neutrophils[36]. The time course of $\cdot O_2^-$ production by this fully soluble system was consistent with the simple model shown in eqs. 2 and 3.

$$E_d \xrightarrow{k} E_a \qquad (2)$$

$$\cdot O_2 \xrightarrow[E_a]{V} \cdot O_2^- \qquad (3)$$

where E_d is the dormant oxidase, E_a is the active enzyme, k is the first-order rate constant for the activation of the oxidase, and V is the V_{max} for the reduction of oxygen to $\cdot O_2^-$ by the active oxidase. To obtain further information regarding the properties of the soluble activating system, we examined k and V as a function of the concentrations of the various components of the system[37]. When the concentration of detergent was varied, k rose monotonically, but V passed through a maximum that probably represented the offsetting effects of an increase in k and a decrease in oxidase stability with rising detergent concentration. Increases in the concentrations of other components caused changes in k and V as shown in Table 2.

Table 2. Kinetics of the Soluble Oxidase Activating System

Component	Effect on k	Effect on V
Membrane	Slight	Saturation behavior
Cytosol	None	Saturation behavior, but quadratic at low concentrations
Mg^{++}	Saturation behavior ($K_m \approx 1$ mM)	Linear increase through 10 mM Mg^{++}

Those results suggested that the activation of the oxidase in the fully soluble system was the outcome of a complex set of interactions between the components of the system. To look for simpler interactions underlying this complex process, preincubation experiments were carried out to see if a subset of components could be activated to a species that catalyzed $\cdot O_2^-$ production when the rest of the components were added. The results (Table 3) suggested the reversible formation of a complex between membranes, Mg^{++} and cytosol that was converted to active oxidase upon addition of detergent.

Table 3. Search for a Preactivation Complex

Component omitted from preincubation mixture	k (min^{-1})	V (nmoles $\cdot O_2^-$/min)
No preincubation (control)	0.78	25.9
Cytosol	0.46	5.1
Membrane	0.67	17.3
SDS	1.19	19.4
Mg^{++}	2.17	3.4
NADPH	19.5	16.2

Assay mixtures were prepared containing all but the indicated component. After 5 minutes, the missing component was added to start the reaction.

The formation of a complex is indicated by the reduction in V resulting from the omission of cytosol or Mg^{++}, a finding consistent with the idea that cytosol plus Mg^{++} form a complex with the membrane-associated component that stabilizes the latter against denaturation by SDS. The lack of an effect of preincubation on k indicates that complex formation precedes the activation step whose rate determines the value of k.

The foregoing kinetic results are explained by the following scheme, where M refers to the membrane-associated component(s) of the oxidase, S to the cytosolic factor that stabilizes the membrane against denaturation by SDS, and C to a cytosolic factor that associates with the activated complex MS^* to form the activated oxidase:

$$M + S \underset{Mg^{++}}{\rightleftharpoons} MS \xrightarrow[Mg^{++},\,SDS]{k} MS^* + 2C \rightleftharpoons MS^*C_2, \text{ the active oxidase} \quad (4)$$

(For a full rationalization of this scheme, see reference 37.) The activation of the oxidase thus appears to be a complex process involving at least 2 different cytosolic components, with many sites at which oxidase activity can be regulated up or down as circumstances require.

REFERENCES

1. J.T. Curnutte and B.M. Babior, Chronic granulomatous disease, in: "Advances in Human Genetics", Volume 16, H. Harris and K. Hirschhorn, ed., Plenum Publishing Corporation, New York (1987).
2. A.W. Segal, R. Garcia, A.H. Goldstone, A.R. Cross and O.T.G. Jones, Cytochrome b-245 of neutrophils is also present in human

monocytes, macrophages and eosinophils, Biochem. J. 196:363 (1981).
3. P. Patriarca, R. Cramer, S. Moncalvo, F. Rossi and D. Romeo, Enzymatic basis of metabolic stimulation in leucocytes during phagocytosis: The role of activated NADPH oxidase, Arch. Biochem. Biophys. 145:255 (1971).
4. F. Rossi and M. Zatti, Biochemical aspects of phagocytosis in polymorphonuclear leucocytes. NADPH oxidation by the granules of resting and phagocytizing cells, Experientia 20:21 (1964).
5. R.H. Cagan and M.L. Karnovsky, Enzymatic basis of the respiratory stimulation during phagocytosis, Nature 204:255 (1965).
6. D.C. Hohn and R.I. Lehrer, NADPH oxidase deficiency in X-linked chronic granulomatous disease, J. Clin. Invest. 55:707 (1975).
7. B.M. Babior, R.S. Kipnes and J.T. Curnutte, Biological defense mechanisms: The production by leukocytes of superoxide, a potential bactericidal agent, J. Clin. Invest. 52:741 (1973).
8. J.T. Curnutte, D.M. Whitten and B.M. Babior, Defective superoxide production by granulocytes from patients with chronic granulomatous disease, N. Engl. J. Med. 290:593 (1974).
9. J.T. Curnutte, M.L. Karnovsky and B.M. Babior, Manganese-dependent NADPH oxidation by granulocyte particles, J. Clin. Invest. 57:1059 (1976).
10. P. Patriarca, P. Dri, K. Kakinuma, F. Tedesco and F. Rossi, Studies on the mechanism of metabolic stimulation in polymorphonuclear leukocytes during phagocytosis. I. Evidence for superoxide anion involvement in the oxidation of NADPH, Biochim. Biophys. Acta 385:380 (1975).
11. J.T. Curnutte, ed., Clinics in Haematology, W.B. Saunders, Philadelphia, in press.
12. B. Dewald, M. Baggiolini, J.T. Curnutte and B.M. Babior, Subcellular localization of the superoxide-forming enzyme in human neutrophils, J. Clin. Invest. 63:21 (1979).
13. T. Yamaguchi, K. Sato, K. Shimada and K. Kakinuma, Subcellular localization of $O2^-$ generating enzyme in guinea pig polymorphonuclear leukocytes; fractionation of subcellular particles by using a Percoll density gradient, J. Biochem. (Tokyo) 91:31 (1982).
14. B.M. Babior, J.T. Curnutte and B.J. McMurrich, The particulate superoxide-forming system from human neutrophils: Properties of the system and further evidence supporting its participation in the respiratory burst, J. Clin. Invest. 58:989 (1976).
15. T.G. Gabig, S.I. Bearman and B.M. Babior, Effects of oxygen tension and pH on the respiratory burst of human neutrophils, Blood 53:1133 (1979).
16. K. Kakinuma and M. Kaneda, Apparent Km of leukocyte $O2^-$ and H2O2 forming enzyme for oxygen, in: "Biochemistry and Function of Phagocytes", F. Rossi and P. Patriarca, ed., Plenum Press, London (1982).
17. B.M. Babior and R.S. Kipnes, Superoxide-forming enzyme from neutrophils. Evidence for a flavin requirement, Blood 50:517 (1977).
18. T.G. Gabig, R.S. Kipnes and B.M. Babior, Solubilization of the $O2^-$-forming activity responsible for the respiratory burst in human neutrophils, J. Biol. Chem. 253:6663 (1978).
19. T.G. Gabig and B.M. Babior, The $O2^-$-forming oxidase responsible for the respiratory burst in human neutrophils: Properties of the solubilized enzyme, J. Biol. Chem. 254:9070 (1979).
20. B.M. Babior and W.A. Peters, The $O2^-$-producing enzyme of human neutrophils: Further properties, J. Biol. Chem. 256:2321 (1981).

21. G.L. Babior, R.E. Rosin, B.J. McMurrich, W.A. Peters and B.M. Babior, Arrangement of the respiratory burst oxidase in the plasma membrane of the neutrophil, J. Clin. Invest. 67:1724 (1981).
22. P. Bellavite, M.C. Serra, A. Davoli, J.V. Bannister and F. Rossi, To be entered, Mol. Cell. Biochem. 52:17 (1983).
23. A.J. Sbarra and M.L. Karnovsky, The biochemical basis of phagocytosis. I. Metabolic changes during the ingestion of particles by polymorphonuclear leukocytes, J. Biol. Chem. 234:1355 (1959).
24. M. Markert, G.A. Glass and B.M. Babior, Respiratory burst oxidase from human neutrophils: Purification and some properties, Proc. Natl. Acad. Sci. USA 82:3144 (1985).
25. G.A. Glass, D.M. DeLisle, P. DeTogni, T.G. Gabig, B.H. Magee, M. Markert and B.M. Babior, The respiratory burst oxidase of human neutrophils, J. Biol. Chem. 261:13247 (1986).
26. K. Kakinuma, Y. Fukuhara and M. Kaneda, The respiratory burst oxidase of neutrophils, J. Biol. Chem. 262:12316 (1987).
27. T. Ume, K. Takeshige and S. Minakami, NADPH binding component of neutrophil superoxide-generating oxidase, J. Biol. Chem. 261:5229 (1986).
28. T. Umei, K. Takeshige and S. Minakami, NADPH-binding component of the superoxide-generating oxidase in unstimulate, Biochem. J. 243:467 (1987).
29. A.W. Segal, A.R. Cross, R.C. Garcia, N. Borregaard, N. Valerius, J.F. Soothill and O.T.G. Jones, Absence of cytochrome b-245 in chronic granulomatous disease: A multicenter European evaluation of its incidence and relevance, N. Engl. J. Med. 308:245 (1978).
30. J.A. Badwey, J.T. Curnutte, C.B. Berde and M.L. Karnovsky, Cytochalasin E diminishes the lag phase in the release of superoxide by human neutrophils, Biochem. Biophys. Res. Commun. 106:170 (1982).
31. Y. Bromberg and E. Pick, Unsaturated fatty acids stimulate NADPH-dependent superoxide production by cell-free system derived from macrophages, Cell. Immunol. 88:213 (1984).
32. J.T. Curnutte, Activation of human neutrophil nicotinamide adenine dinucleotide phosphate, reduced (triphosphopyridine nucleotide, reduced) oxidase by arachidonic acid in a cell-free system, J. Clin. Invest. 75:1740 (1985).
33. L.C. McPhail, P.S. Shirley, C.C. Clayton and R. Snyderman, Activation of the respiratory burst enzyme from human neutrophils in a cell-free system: Evidence for a soluble cofactor, J. Clin. Invest. 75:1735 (1985).
34. R.A. Heyneman and R.E. Vercauteren, Activation of a NADPH oxidase from horse polymorphonuclear leukocytes in a cell-free system, J. Leukocyte Biol. 36:751 (1984).
35. Y. Bromberg and E. Pick, Activation of NADPH-dependent superoxide production in a cell-free system by sodium dodecyl sulfate, J. Biol. Chem. 260:13539 (1985).
36. J.T. Curnutte, R. Kuver and B.M. Babior, Activation of the respiratory burst oxidase in a fully soluble system from human neutrophils, J. Biol. Chem. 262:6450 (1987).
37. B.M. Babior, R. Kuver and J.T. Curnutte, Kinetics of activation of the respiratory burst oxidase in a fully soluble system from human neutrophils, J. Biol. Chem., in press.

REACTIONS OF MYELOPEROXIDASE WITH SUPEROXIDE AND HYDROGEN PEROXIDE: SIGNFICANCE FOR ITS FUNCTION IN THE NEUTROPHIL

Christine C. Winterbourn and Anthony J. Kettle

Pathology Department
Christchurch School of Medicine
Christchurch, New Zealand

INTRODUCTION

Myeloperoxidase (MPO) in its purified form acts as a conventional peroxidase and, in addition, catalyses the reaction

$$H_2O_2 + Cl^- \longrightarrow HOCl + OH^- \quad (1)$$

It has a pH optimum of 5-6, but shows some activity at neutral pH. HOCl and the MPO-H_2O_2-Cl^- system are very reactive with most biological molecules, and both are potently microbicidal. Thus the MPO system seems the ideal candidate to provide the neutrophil with antimicrobial activity.

Yet there are a number of paradoxes in the relationship between MPO and neutrophil function: 1) MPO deficiency is common and rarely associated clinically with infection. Killing of some micro-organisms by MPO-deficient cells is impaired, but many others are killed normally. Hence MPO is either unimportant in the killing process, or other back-up mechanisms take over when it is absent. 2) MPO is released either into phagosomes, or to the exterior of stimulated neutrophils. Exocytosed enzyme will function largely at neutral pH, and the pH within the phagosome during the early stages of ingestion is also neutral or slightly alkaline.[1] This is when killing takes place. The pH gradually drops, but so then does NADPH-oxidase dependent production of MPO's substrate, H_2O_2. Thus most MPO activity is likely to occur under neutral conditions, far removed from its pH optimum. 3) MPO released into phagosomes is likely to encounter high H_2O_2 concentrations. With NADPH-oxidase activity directed into the narrow space between the phagosome membrane and ingested particle, millimolar H_2O_2 concentrations could easily be achieved.[2] 4) MPO released during neutrophil stimulation reacts with co-released $\cdot O_2^-$ and is converted to the superoxide adduct or compound III (see Figure 1 for definition of MPO intermediates). This can be seen in difference spectra of normal vs. stimulated cells.[2]

We were intrigued by the $\cdot O_2^-$ reaction, and its significance to neutrophil function, as well as by the apparent non-ideal conditions under which MPO acts. We therefore undertook to examine the reactions of purified MPO under these conditions.

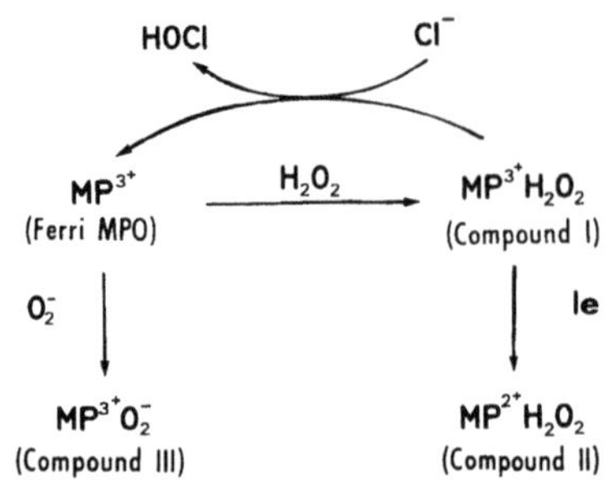

Figure 1. Transformations of the intermediates of myeloperoxidase.

CATALASE ACTIVITY OF MYELOPEROXIDASE

We have found that MPO can act as a catalase, breaking down H_2O_2 (to form O_2) but not HOCl. Its chlorinating activity is inhibited by H_2O_2 so that under neutral conditions, the rate of HOCl formation reaches a maximum between 50-100 μM and then declines. The rate of H_2O_2 consumption increases slightly so that the efficiency of conversion to HOCl falls, and the enzyme functions almost exclusively as a catalase (Figure 2). Thus, high H_2O_2 concentrations in the neutrophil might favour catalase rather than peroxidase activity.

FORMATION AND REACTIONS OF MYELOPEROXIDASE SUPEROXIDE ADDUCT (COMPOUND III)

Compound III can be formed either by reduction of ferric MPO with dithionite and reoxygenation[3] or by its reaction with a $\cdot O_2^-$-generating system such as xanthine plus xanthine oxidase or stimulated neutrophils. It slowly autoxidizes, with a half life of approximately 12 min at 20 °C, to the native enzyme plus $\cdot O_2^-$.

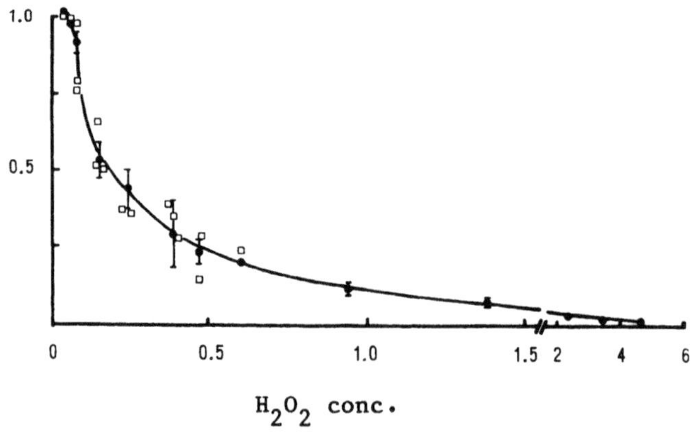

Figure 2. Effect of H_2O_2 on the efficiency of conversion of H_2O_2 to HOCl by MPO. Reactions were at pH 7.6 with 1.8 μM MPO and 0.1 M NaCl. HOCl was measured as chlorination of monochlorodimedon.

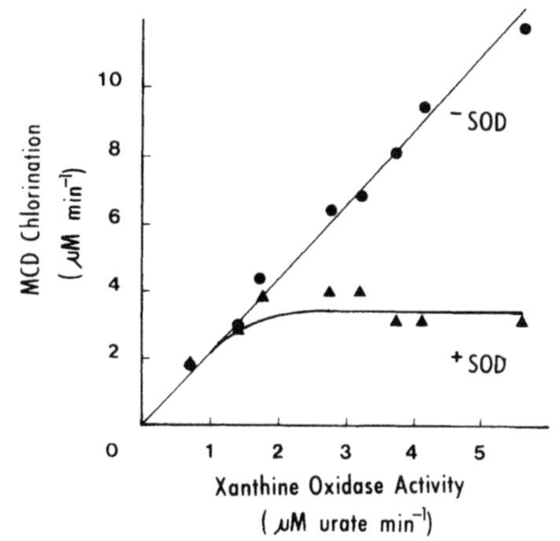

Figure 3. Effect of superoxide dismutase (SOD) on rate of HOCl formation from 0.1 M NaCl, 0.2 μM MPO, 50 μM hypoxanthine (HX) and varying concentrations of xanthine oxidase, pH 7.8.

The accepted dogma is that compound III of peroxidase is an inactive side product. However, we observed that at pH 7.6, chlorination by preformed compound III and H_2O_2 occurs at the same rate as with native enzyme. Furthermore, chlorination by xanthine oxidase/xanthine and MPO is no faster in the presence of superoxide dismutase, even though it prevents conversion of most of the MPO to compound III. In fact, with high xanthine oxidase to MPO ratios, superoxide dismutase inhibits (Figure

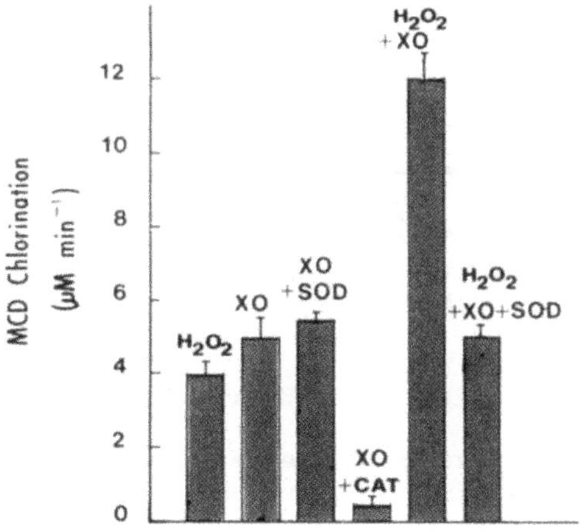

Figure 4. Effect of xanthine oxidase (XO) on HOCl formation by 0.6 μM MPO, 500 μM H_2O_2 and 0.1 M NaCl. Concentrations of hypoxanthine SOD and catalase were 50 μM, 20 and 1 μM respectively.

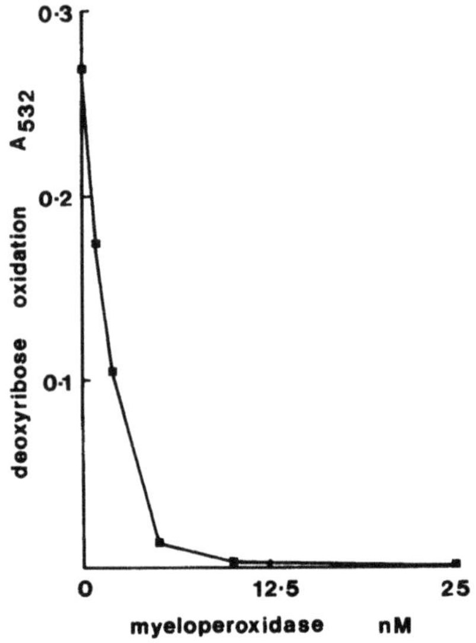

Figure 5. Inhibition of $\cdot$OH production by myeloperoxidase. $\cdot$OH was generated with HX, XO and 2 μM Fe(EDTA), and detected with deoxyribose.

3).[5] Formation of catalytically inactive compound II (see Figure 1), which occurs in the presence of superoxide dismutase, appears to be responsible for this inhibition. Hence, $\cdot O_2^-$ prevents build-up of compound II, which is formed in progressively greater amounts as the H_2O_2 concentration increases, and allows MPO to remain in its catalytic cycle. In addition, $\cdot O_2^-$ enhanced the rate of MPO dependent chlorination by H_2O_2 (Figure 4). Compound III is not a dead-end complex but is turned over by H_2O_2 to form compound II. Thus $\cdot O_2^-$ produced by the neutrophil would enable MPO to remain active at H_2O_2 concentrations that would otherwise be inhibitory.

MPO can break down $\cdot O_2^-$ catalytically. Although true superoxide dismutase activity has been implicated,[6] this may not be the case since catalase prevents MPO-dependent chlorination by the xanthine oxidase system. However, catalytic concentrations of MPO compete with nitroblue tetrazolium for $\cdot O_2^-$.[5] This suggests a mechanism involving breakdown of both $\cdot O_2^-$ and H_2O_2, i.e., MPO acting as a combined superoxide dismutase and catalase. Maybe it should have the alternative name of superoxide-hydrogen peroxide oxidoreductase!

MYELOPEROXIDASE AS AN INHIBITOR OF HYDROXYL RADICAL PRODUCTION

A still unresolved question is whether neutrophils produce $\cdot$OH as part of their microbicidal armory. The conventional mechanism for $\cdot$OH production involves the reaction between $\cdot O_2^-$ and H_2O_2 catalysed by an iron or other transition metal catalyst. The question arises how MPO released by the cells would affect such a mechanism. Either through HOCl formation or catalase/ superoxide dismutase activity, it would consume H_2O_2 and $\cdot O_2^-$ and, therefore, $\cdot$OH production would have to compete with these reactions. As seen in Figure 5, MPO added to a xanthine oxidase/Fe(EDTA) system does, this very effectively. So, too, do neutrophils stimulated with fMet-Leu-Phe and cytochalasin B. PMA-stimulated neutrophils release very little myeloperoxidase and are much less inhibitory, and myeloperoxidase-deficient cells do not inhibit at all.[7] Thus the release of myeloperoxidase would not favour endogenous $^{\bullet}OH$ production.

In summary, the reactions of MPO in the neutrophil are probably not those seen with the purified enzyme acting at its pH optimum. Its catalase/superoxide dismutase activity, prevention of compound II build-up and enhancement of chlorination by $\cdot O_2^-$, formation of chlorinating species via compound III, and its ability to remove precursors of $^{\bullet}OH$ must all be accommodated.

REFERENCES

1. A.W. Segal, M. Geisow, R. Garcia, A.M. Harper and R. Miller, The respiratory burst of phagocytic cells is associated with a rise in vacuolar pH. Nature 290:406 (1981).
2. C.C. Winterbourn, R. Garcia and A.W. Segal, Production of the superoxide adduct of myeloperoxidase (compound III) by stimulated human neutrophils and its reactivity with hydrogen peroxide and chloride. Biochem. J. 228:583 (1985).
3. T. Odajima and I. Yamazaki, Myeloperoxidase of the leukocyte of normal blood. III. The reaction of ferric myeloperoxidase with superoxide anion. Biochim. Biophys. Acta 284:355 (1972).
4. I. Yamazaki and K. Yokota, Oxidation states of peroxidase. Mol. Cell Biochem. 2:39 (1973).
5. A.J. Kettle and C.C. Winterbourn, Superoxide modulates the activity of myeloperoxidase and optimizes the production of hypochlorous acid. Biochem. J. in press (1988).
6. R.A. Cuperus, A.D. Muijers and R. Wever, The superoxide dismutase activity of myeloperoxidase: formation of compound III. Biochim. Biophys. Acta 871:78 (1986).
7. C.C. Winterbourn, Myeloperoxidase as an effective inhibitor of hydroxyl radical production: Implications for the oxidative reactions of neutrophils. J. Clin. Invest. 78:545 (1986).

BACTERIAL KILLING BY PHAGOCYTES: POTENTIAL ROLE(S) OF HYPOCHLOROUS ACID AND HYDROGEN PEROXIDE IN PROTEIN TURNOVER, DNA SYNTHESIS, AND RNA SYNTHESIS

Susan M. McKenna and Kelvin J.A. Davies

Institute for Toxicology, Department of Biochemistry
and Department of Hematology and Oncology, The University of
Southern California, and Childrens Hospital of Los Angeles
1985 Zonal Avenue, HSC-PSC 614-616
Los Angeles, California 90033, U.S.A.

INTRODUCTION

Although phagocytic cells clearly play a primary role in host defense against bacterial infection, the exact mechanisms of bacterial killing are largely unknown. Stimulation of an NADPH oxidase system during the "respiratory burst" in phagocytes generates a series of oxidants. Molecular oxygen is reduced to superoxide which then rapidly dismutates to form hydrogen peroxide (H_2O_2). Myeloperoxidase from phagocyte granules then uses the H_2O_2 to oxidize chloride ions to form hypochlorous acid (HOCl). The importance of these oxidants is underscored by studies of patients with chronic granulomatous disease (CGD). CGD patients suffer severe, recurrent bacterial infections (with catalase postive bacteria) because their neutrophils are unable to reduce oxygen to superoxide. The exact role(s) of oxygen metabolites (thought to act synergistically with cytotoxic proteins) in bacterial killing remain(s) undetermined.

Substantial data support the hypothesis that oxidatively damaged proteins are rapidly degraded by intracellular proteolytic systems in red cells, skeletal and cardiac muscle, mitochondria, and bacteria.[1-11] Work in our laboratory[1,2,5,9] has shown that _Escherichia coli_ undergo a generalized increase in protein degradation after exposure to H_2O_2. We hypothesized that H_2O_2 and possibly HOCl, produced by phagocytes, might exert a bactericidal effect in a similar manner, i.e., by oxidizing bacterial proteins and thus rendering them more susceptable to proteolysis. Arguing against this hypothesis, however, is the observation that neutrophil ingested _E. coli_ rapidly lose the ability to divide, but retain their gross structural integrity for prolonged periods.[12]

RESULTS AND DISCUSSION

We undertook a series of experiments to compare the bactericidal activity of HOCl with that of H_2O_2. We also attempted to test our hypothesis that oxidative damage to bacterial proteins might lead to augmented degradation of these proteins by intracellular proteases. The role(s) of protein synthesis, DNA synthesis, and RNA synthesis in the bactericidal activities of HOCl and H_2O_2 were also examined.

HOCl exerted a dramatic effect on *E. coli* growth at physiologically relevant concentrations. *E. coli* strain RM312, a derivative of the W3110 strain of *E. coli* K-12, was used in all but one experiment (noted below). Cells grown by overnight culture were subcultured to the exponential phase of growth in M-9 minimal medium plus glucose for each experiment. HOCl (1.0 µM - 0.5 mM) or H_2O_2 (1.0 µM - 0.5 mM) were then added directly to suspensions of *E. coli* (5×10^8 cells/ml) for incubation at 37°C. *E. coli* survival was assessed both by colony formation on solid medium and by light scattering measurements of cell suspensions at 595 nm. Only 20 µM HOCl was required for 50% inhibition of colony formation, and 50 µM HOCl inhibited growth by 94%. These effects occurred rapidly (within 5 min), and were not reversed by extensive washings. The efficiency of H_2O_2 was only 6% of that obtained with HOCl, exerting 50% inhibition of colony formation at a concentration of 300 µM.

HOCl, particularly at high concentrations, reacts avidly with many proteins, causing deamination, decarboxylation, and cleavage of peptide bonds. Preliminary experiments with tritiated bovine serum albumin ([^{3}H]BSA) suggested that direct fragmentation by HOCl occurred only at HOCl concentrations greater than 10 mM. *E. coli* protein degradation was measured in cells whose proteins were pre-labeled by incorporation of [^{3}H]leucine (0.09 µCi/ml) in a pulse-chase procedure. The *E. coli* were labeled at a concentration of 5×10^8 cells/ml, and then incubated with HOCl for up to 1 hr at 37°C. Protein degradation (and fragmentation) was measured by counting the radioactivity released from cellular proteins in acid-precipitable forms after cell lysis by trichloroacetic acid. Analogous to our experiments with [^{3}H]BSA, little radioactivity was released from cells treated with less than 10 mM HOCl, i.e., less than 5% protein degradation or fragmentation occurred at concentrations of HOCl below 10 mM. Hydrogen peroxide induced 5-10% degradation of endogenous proteins at concentrations of 0.3-0.5 mM. Thus, although both HOCl and H_2O_2 can cause protein degradation/fragmentation in *E. coli*, the relative amounts of protein actually digested are small. Furthermore, the incubation time required for protein degradation/fragmentation with HOCl as well as the actual concentrations of HOCl required were found to be significantly higher than those needed for inhibition of growth; although this did not appear to be true for H_2O_2. These results may indicate either that essential enzymes are damaged and degraded in response to HOCl and/or H_2O_2 or that proteolysis is only a secondary effect of these oxidants.

A series of experiments were designed then to test for other effects of HOCl and H_2O_2 which might directly result in the inhibition of cell division. Protein synthesis, measured by incorporation of [3H]leucine (at 5×10^8 cells/ml), was effectively inhibited by 50 µM HOCl. The time course of this reaction, however, suggested a secondary effect. Protein synthesis in cells treated with 50 µM HOCl was only 5% lower than in controls at 5 min, and maximal inhibition (80%) was not attained until 20 min of incubation. When *E. coli* were incubated for 1 hr with as much as 1.0 mM H_2O_2, only a 17% inhibition of protein synthesis was observed.

Both DNA synthesis (assessed by incorporation of [^{3}H]thymidine at 5.0 µCi/ml) and and RNA synthesis (assessed by incorporation of [^{3}H]uridine at 0.05 µCi/ml) were rapidly inhibited by low concentrations of HOCl. DNA synthesis in both a wild-type derivative (RM312) and a thymidylate synthetase-deficient mutant (RM712) decreased to only 42% of control levels after a 1 min exposure to 50 µM HOCl, and fell to only 4% of control values by 5 min. Impairment of [^{3}H]thymidine uptake could not explain the inhibition of incorporation. RNA synthesis was inhibited by HOCl with similar reaction kinetics and HOCl requirements to those observed for DNA synthesis. H_2O_2 also rapidly inhibited both DNA and RNA synthesis, but at least 2.5 mM H_2O_2 was required for 50% inhibition.

A summary of results for cell growth, protein degradation or fragmentation, protein synthesis, DNA synthesis, and RNA synthesis in the presence of HOCl is shown in Table 1. A summary of the effects of H_2O_2 on these cellular functions is given in Table 2.

Table 1. The Effects of HOCl on *E. coli* Function

Function Assessed	EC_{50} for HOCl	Reaction Time
Cell division	≃ 20 μM	< 5 min
Protein degradation	> 20 mM	> 20 min
Protein synthesis	≃ 25 μM	> 20 min
DNA synthesis	≃ 20 μM	< 1 min
RNA synthesis	≃ 20 μM	< 1 min

Cell division was assessed by the growth of colonies on solid medium. Protein degradation was measured by release of acid-soluble counts from [^{3}H]leucine-labeled cell proteins. Protein synthesis reflects the incorporation of [^{3}H]leucine into acid-precipitable forms. DNA and RNA synthesis values report the incorporation of [^{3}H]thymidine and [^{3}H]uridine (respectively) into new DNA or RNA. EC_{50} values refer to the concentration of HOCl at which 50% of the maximum inhibition of a given function occurred.

Table 2. The Effects of H_2O_2 on *E. coli* Function

Function Assessed	EC_{50} for H_2O_2	Reaction Time
Cell division	≃ 0.3 mM	< 5 min
Protein degradation	≃ 0.1 mM	> 20 min
Protein synthesis	> 1.0 mM	> 30 min
DNA synthesis	> 2.5 mM	< 1 min
RNA synthesis	> 2.5 mM	< 1 min

EC_{50} values refer to the concentration of H_2O_2 at which 50% of the maximum inhibition of a given function occurred. All measurements were performed as described in the legend to Table 1.

The data of Table 1 suggest that low concentrations of HOCl exert a selective effect on *E. coli* division. This effect involves increased protein degradation and decreased protein synthesis, but may be more directly related to inhibition of DNA and RNA synthesis. We suggest that HOCl may react with membrane components which are crucial for cell division and may alter membrane/DNA interaction or elements of the replication machinery. Alternatively, HOCl may inhibit synthesis of critical proteins required for cell division. Our experiments with purified HOCl may provide a model which closely resembles the early events in bacterial killing by phagocytes, and thus may allow further definition of the molecular events involved.

The pattern of H_2O_2 effects shown in Table 2 is clearly different from results observed with HOCl (Table 1). Although H_2O_2 rapidly inhibits both DNA and RNA synthesis, the concentrations of H_2O_2 required are more than twice those required to inhibit cell division. Interestingly, significantly lower concentrations of H_2O_2 are needed to stimulate protein degradation and inhibit protein synthesis. Comparisons between Tables 1 and 2 also clearly point out the increase in bactericidal activity which occurs on conversion of H_2O_2 to HOCl (catalyzed by myeloperoxidase in neutrophils).

ACKNOWLEDGEMENTS

This work was supported by grant #ES 03598 and by grant #ES 03785 from the National Institutes of Health. Further support was provided by the Faculty Research and Innovation Award of the University of Southern California.

REFERENCES

1. K.J.A. Davies. The role of intracellular proteolytic systems in antioxidant defenses. in: "Superoxide and Superoxide Dismutase in Chemistry, Biology, and Medicine," G. Rotilio, ed., pp. 443, Elsevier, Amsterdam (1986).
2. K.J.A. Davies, Intracellular proteolytic systems may function as secondary antioxidant defenses: An hypothesis, J. Free Radicals Biol. Med. 2:155 (1986).
3. K.J.A. Davies and A.L. Goldberg, Oxygen radicals stimulate intracellular proteolysis and lipid peroxidation by independent mechanisms in erythrocytes, J. Biol. Chem. 262:8220 (1987).
4. K.J.A. Davies and A.L. Goldberg, Proteins damaged by oxygen radicals are rapidly degraded in extracts of red blood cells, J. Biol. Chem. 262:8227 (1987).
5. K.J.A. Davies, Protein damage and degradation by oxygen radicals: I. General aspects, J. Biol. Chem. 262:9895 (1987).
6. K.J.A. Davies, M.E. Delsignore, and S.W. Lin, Protein damage and degradation by oxygen radicals: II. Modification of amino acids, J. Biol. Chem. 262:9902 (1987).
7. K.J.A. Davies and M.E. Delsignore, Protein damage and degradation by oxygen radicals: III. Modification of secondary and tertiary structure, J. Biol. Chem., 262:9908 (1987).
8. K.J.A. Davies, S.W. Lin, and R.E. Pacifici, Protein damage and degradation by oxygen radicals: IV. Degradation of denatured protein, J. Biol. Chem. 262:9914 (1987).
9. S.W. Lin and K.J.A. Davies, E. coli proteases selectively degrade oxidized proteins, Fed. Proc. 45:1597 (1986).
10. E.R. Stadtman, Oxidation of proteins by mixed-function oxidation systems: Implication in protein turnover, ageing and neutrophil function, Trends Biochem. Sci. 11:11 (1986).
11. S.P. Wolff, A. Garner, and R.T. Dean, Free radicals, lipids, and protein degradation, Trends Biochem. Sci. 11:27 (1986).
12. P. Elsbach, P. Pettis, S. Beckerdite, and R. Franson, Effects of phagocytosis by rabbit granulocytes on macromolecular synthesis and degradation in different species of bacteria, J. Bacteriol. 115:490 (1973).

CELLULAR TARGETS OF H_2O_2 AND HOCl INDUCED INJURY

I.U. Schraufstätter, P.A. Hyslop, J. Jackson,
and C.G. Cochrane

Research Institute of Scripps Clinic
10666 North Torrey Pines Road
La Jolla, California 92037, U.S.A.

INTRODUCTION

During the respiratory burst stimulated neutrophils or macrophages release cell damaging concentrations of oxidants in any area of inflammation.[1-3] In order to evaluate the role of various oxidants in target cell injury, we have studied the primary targets of oxidants in cells and the biochemical pathways responsible for cellular damage. The target cells used were P388 D1 murine macrophage-like tumor cells, GM 1380 human embryo lung fibroblasts, human peripheral lymphocytes and rabbit alveolar macrophages. The oxidants used were H_2O_2, HOCl and phorbol myristate acetate (PMA)-stimulated PMNs or neutroplasts - enucleated PMNs that still undergo the respiratory burst.

CELLULAR INJURY INDUCED BY H_2O_2

Seconds after the addition of H_2O_2 to P388 D1 cells, DNA strand breaks were detected[4] as determined by an alkaline unwinding technique.[5] DNA damage resulted in the activation of poly-ADP-ribose polymerase within minutes after the addition of H_2O_2. At the same time, intracellular levels of NAD--the substrate of poly-ADP-ribose polymerase[4]--and ATP[6] dropped, and nicotinamide accumulated extracellularly.[4] Glyceraldehyde-3-phosphate dehydrogenase of the glycolytic pathway became inhibited at the same time. Ten to 15 min after the addition of H_2O_2, intracellular Ca^{++} concentrations[7] determined by Quin 2 fluorescence[8] increased. After 1 hr the cells became permeable to extracellular Na^+ and Ca^{++} (determined by atomic absorption),[7] actin-rich cytoskeletal elements polymerized and aggregated,[9] and blebbing of the plasma membrane was observed.[9] Finally, after 3-10 hr the cells lysed.[10,11] Figure 1 summarizes the sequence of these events.

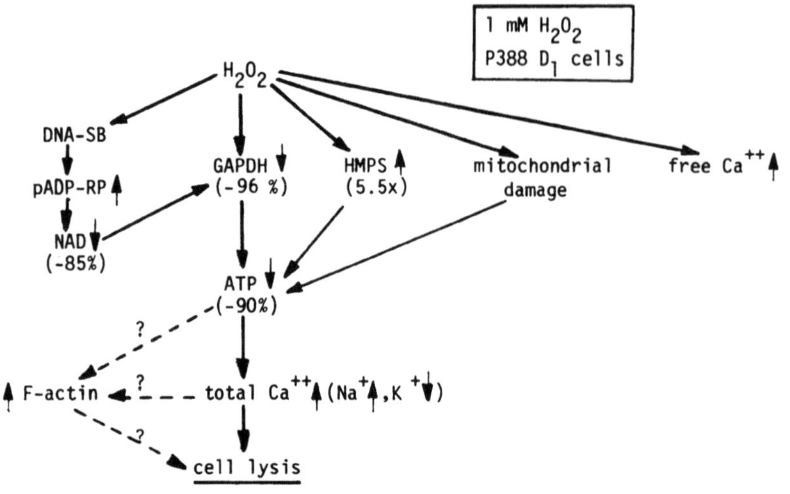

Figure 1 - Effect of H_2O_2 on P388D1 cells

Inhibition of poly-ADP-ribose polymerase with 2.5 mM 3-aminobenzamide or 5 mM nicotinamide not only prevented the fall in NAD and, in part, ATP, but also prevented Ca^{++} influx, actin polymerization and cell lysis.[12] However, DNA strand breaks could not be repaired, and the cells were unable to replicate. A similar sequence of events following the formation of DNA strand breaks has been described for a variety of DNA damaging agents.[13,14] Sublethal doses of H_2O_2 (100 uM in GM 1380 fibroblasts) led to the formation of DNA strand breaks, inhibition of DNA and RNA synthesis and prolonged (35 hr) inhibition of cell replication.

CELLULAR INJURY INDUCED BY HOCl

HOCl proved considerably more cytotoxic (LD_{50} for P388 D1 cells of about 70 µM compared to 350 µM for H_2O_2). It reacted very efficiently with sulfhydryls: 50 nmoles HOCl oxidized 5-6 nmoles of cellular sulfhydryls. Amino acid transport (uptake of ^{3}H-aminoisobutyric acid) and glucose transport (uptake of ^{3}H-glucose +/- cytochalasin B) were half-maximally inhibited with 25 µM HOCl. Higher concentrations of HOCl inhibited metabolic functions, e.g., ATP synthesis in a nonspecific way. No DNA strand breaks were, however, observed with up to 200 µM HOCl.

DNA DAMAGE INDUCED BY VARIOUS OXIDANTS

Studies were performed to determine if H_2O_2 was responsible for damage to intracellular DNA. Susceptibility to H_2O_2-induced DNA strand breaks varied by an order of magnitude in various target cells (P388 D1 cells, GM 1380 fibroblasts, lymphocytes, PMNs and macrophages) and was inversely correlated with cellular catalase content (Table 1). Catalase inhibitors (azide, aminotriazole) enhanced the susceptibility to DNA strand breakage. SOD and GSH peroxidase were unaffected by the inhibitors. In the presence of these inhibitors the divergent dose-response of H_2O_2-induced DNA damage in the various target cells became the same: 10 µM H_2O_2 would induce a 30% decrease of alkali stable double-stranded DNA in all cells under these conditions. This suggested that the susceptibility to DNA strand breaks was dependent on the amount of H_2O_2 that was able to diffuse into the nucleus in any particular cell.

While catalase obviously played a major role in protecting from H_2O_2-induced DNA strand breaks, stimulated neutrophils release other species of oxidants which could contribute to the DNA damage. When mononuclear cells composed of 75% lymphocytes and 25% monocytes were exposed to a H_2O_2-producing system (glucose oxidase/glucose) or $\cdot O_2^-$ producing system (xanthine oxidase/purine) or stimulated to produce $\cdot O_2^-$ by the addition of PMA, DNA strand breaks developed in proportion to the concentrations of H_2O_2 produced. In all cases strand breaks were inhibited by addition of extracellular catalase but not by superoxide dismutase. These data indicated that the extracellular concentration of H_2O_2 determined the formation of DNA strand breaks.

Table 1. Correlation Between Cellular Catalase Content and Susceptibility to DNA Strand Breaks Induced by H_2O_2 in Various Target Cells

Cell	D_{50}[a] $\mu M\ H_2O_2$	Catalase Content ng/10^6 cells)
Macrophage	170	850
PMN	94	503
GM1380	58	213
P388D1	35	20
Lymphocyte	20	4
Macrophage + 12 mM AT	13.5	< 2

[a] D_{50} represents the concentration of H_2O_2 that led to unwinding in alkali of 50% of the DNA as extrapolated from dose-response curves. Catalase content in supernatants of cell sonicates from 5-20 x 10^6 cells was determined by following the disappearance of H_2O_2 at 240 nm spectrophotometrically.[20]

Further studies suggested that $\cdot OH$, generated within the cell after exposure to H_2O_2, was responsible for the DNA damage. In related studies with isolated PM2 phage DNA, H_2O_2-induced DNA strand breaks, and this was dependent on the presence of $FeSO_4$. Only when $FeSO_4$ was added could $\cdot OH$ formation be detected by electron spin resonance, using DMPO as a spin trap.[15-17] Iron chelators (phenanthroline, desferrithiocin) as well as $\cdot OH$ scavengers (DMSO, mannitol, benzoate, ethanol) prevented the formation of DNA strand breaks in the isolated DNA. In whole cells, H_2O_2-induced strand breaks could be prevented by the addition of cell permeable Fe-chelators (phenanthroline,[18] desferrithiocin). $\cdot OH$ scavengers were ineffective in protecting whole cells, presumably due to the insufficient concentration at the site of injury on the DNA. Table 2 shows the effect of various antioxidants on inhibition of H_2O_2-induced DNA strand breaks.

Preliminary results indicate the formation of 8-hydroxydeoxyguanosine[19] which is formed by $\cdot OH$ attack in P388 D1 cells exposed to 250 uM H_2O_2. The data suggest that of the various oxidants generated in the inflammatory process, H_2O_2 has the greatest capability of penetrating cells to gain access to the nucleus. In the presence of transition metal, probably through the generation of $\cdot OH$, DNA damage occurs as revealed by DNA strand breaks.

Table 2. Influence of Oxidant Scavengers on the Formation of H_2O_2 Induced DNA Strand Breaks in P388D1 Cells [a]

Scavenger	Scavenged Species	Concentration	% Inhibition of DNA Strand Breaks
SOD	O_2^-	1 mg/ml	0%
Catalase	H_2O_2	5000 U	100%
DMTU		800 mM	0%
DMSO		1 M	36%
DMSO		0.1 M	3%
Ethanol	$OH^\cdot$	0.33 M	- 10%
Na-benzoate		0.1 M	8%
Mannitol		0.5 M	0%
DMPO		0.1 M	25%
DFO (overnight inc.)		100 µM	86%
DFT	Fe chelator	100 µM	90%
DTPA		100 µM	- 5%
Phenanthroline		100 µM	74%
Vitamin E (overnight preinc.)	lipid peroxides	20 µM	4%

[a] $2x10^6$ P388D1 cells/ml were incubated with the respective scavenger for 15 min at 37°C. A dose range of H_2O_2 (20,35,50,100 uM) was added to parallel samples in the presence and absence of the scavenger. After a 5 min incubation at 37°C the samples were centrifuged for 10 sec, placed on ice and prepared for the determination of DNA strand breaks. Inhibition of strand break formation was calculated for the concentration of H_2O_2 which induced 50% DNA unwinding in cells in the absence of scavenger.

REFERENCES

1. G.O. Till, K.J. Johnson, R. Kunkel, and P.A. Ward, Intravascular activation of compliment and acute lung injury: Dependency on neutrophils and toxic oxygen metabolites, J. Clin. Invest. 69:1126 (1982).
2. D.M. Shasby, K.M. Vanbenthysen, R.M. Tate, S.S. Shasby, I. McMurtry, and J.E. Repine, Granulocytes mediate acute edematous lung injury in rabbits and in isolated lungs perfused with phorbol myristate acetate: Role of oxygen radicals, Am. Rev. Resp. Dis. 125:443 (1982).
3. C.G. Cochrane, R. Spragg, and S.D. Revak, Studies on the pathogenesis of the adult respiratory distress syndrome: Evidence of oxidant activity in bronchoalveolar lavage fluid, J. Clin. Invest. 71:754 (1983).
4. I.U. Schraufstatter, D.B. Hinshaw, P.A. Hyslop, R.G. Spragg and C.G. Cochrane, Oxidant injury of cells: DNA strand breaks activate poly-ADP-ribose-polymerase and lead to depletion of nicotinamide adenine dinucleotide, J. Clin. Invest. 77:1312 (1986).
5. H.C. Birnboim and J.J. Jevcak, Fluorometric method for rapid detection of DNA strand breaks in human white blood cells reproduced by low doses of radiation, Cancer Res. 41:1889 (1981).
6. R.G. Spragg, D.B. Hinshaw, P.A. Hyslop, I.U. Schraufstatter, and C.G. Cochrane, Alterations in ATP and energy charge in cultured endothelial and P388D1 cells following oxidant injury, J. Clin. Invest. 76:1471 (1985).

7. P.A. Hyslop, D.B. Hinshaw, I.U. Schraufstatter, L.A. Sklar, R.G. Spragg, and C.G. Cochrane, Intracellular calcium homeostasis during hydrogen peroxide injury of cultured P388D1 cells, J. Cell. Physiol. 129:356 (1986).
8. R.Y. Tsien, T. Pozzan, and T.J. Rink, Calcium homeostasis in intact lymphocytes: Cytoplasmic free calcium monitored with a new intracellularly trapped fluorescent indicator, J. Cell Biol. 94:325 (1982).
9. D.B. Hinshaw, L.A. Sklar, B. Bohl, I.U. Schraufstatter, P.A. Hyslop, M.W. Rossi, R.G. Spragg, and C.G. Cochrane, Cytoskeletal and morphologic impact of cellular injury, Am. J. Pathol. 123:454 (1986).
10. I.U. Schraufstatter, D.B. Hinshaw, P.A. Hyslop, R.G. Spragg, and C.G. Cochrane, Glutathione cycle activity and pyridine nucleotide levels in oxidant induced injury of cells, J. Clin. Invest. 76:1131 (1985).
11. C.F. Nathan, S.C. Silverstein, L.H. Brukner, and Z.A. Cohn, Extracellular cytolysis by activated macrophages and granulocytes. II. Hydrogen peroxide as a mediator of cytolysis, J. Exp. Med. 149:100 (1979).
12. I.U. Schraufstatter, P.A. Hyslop, D.B. Hinshaw, R.G. Spragg, L.A. Sklar, and C.G. Cochrane, Hydrogen peroxide-induced injury of cells and its prevention by inhibitors of poly-ADP-ribose polymerase, Proc. Natl. Acad. Sci. 83:4908 (1986).
13. J.L. Sims, S.L. Berger, and N.A. Berger, Poly(ADP-ribose)polymerase inhibitors preserve nicotinamide adenine dinucleotide and adenosine 5'-triphosphate pools in DNA damaged cells, Biochemistry 22:5188 (1983).
14. K. Wielckens, A. Schmidt, E. George, R. Bredehorst, and H. Hilz, DNA fragmentation and NAD depletion: Their relation to the turnover of endogenous mono(ADP-ribosyl) and poly(ADP-ribosyl) proteins, J. Biol. Chem. 257:12872 (1982).
15. E. Finkelstein, G.M. Rosen, and E.J. Rauchman, Spin trapping of superoxide and hydroxyl radical: Practical aspects, Arch. Biochem. Biophys. 200:1 (1980).
16. J.H. Jackson, I.U. Schraufstatter, P.A. Hyslop, K. Vosbeck, R. Sauerheber, S.A. Weitzman, and C.G. Cochrane, Role of oxidants in DNA damage: Hydroxyl radical mediates the synergistic DNA damaging effects of asbestos and cigarette smoke, J. Clin. Invest. 80:1090-1095 (1987).
17. R.A. Floyd and C.A. Lewis, Hydroxyl free radical formation from hydrogen peroxide by ferrous iron-nucleotide complexes, Biochemistry 22:2645 (1983).
18. A.C. Mello Filho and R. Meneghini, In vivo formation of single strand breaks in DNA by hydrogen peroxide is mediated by Haber-Weiss-reaction, Biochem. Biophys. Acta 781:56 (1984).
19. H. Kasai, P.F. Crain, Y. Kuchino, S. Nishimura, A. Ootsuyama, and H. Tanooka, Formation of 8-hydroxyguanine moiety in cellular DNA by agents producing oxygen radicals and evidence for its repair, Carcinogenesis, 7:1849 (1986).
20. H. Aebi, Catalase: In: Methods of Enzymatic Analysis. (H. Bergmeyer, editor), Verlag Chemie, Weinheim 2:673-684 (1974).

List of Abbreviations: PMN - polymorphonuclear leukocyte; PMA - phorbol myristate acetate; pADP-RP-poly-ADP-ribose polymerase; DNA-SB - DNA strand breaks; GAPDH - glyceraldehyde-3-phosphate dehydrogenase; HMPS - hexose monophosphate shunt; DMSO - dimethylsulfoxide; DMPO - 5,5-dimethyl-1-pyroline-N-oxide; AT - aminotriazole; DMTU - dimethylthiourea; DFO - desferrioxamine; DFT - desferrithiocine; DTPA - diethylenetriaminepentacetic acid

OXIDATIVE MODIFICATION OF ENZYMES BY STIMULATED NEUTROPHILS

Cynthia N. Oliver

Laboratory of Biochemistry
National Heart, Lung, and Blood Institute
National Institutes of Health
Bethesda, Maryland 20892, U.S.A.

INTRODUCTION

We have previously demonstrated that a variety of enzymic and non-enzymic mixed-function oxidation (MFO) systems catalyze the oxidative inactivation of *E. coli* glutamine synthetase (GS) and other key metabolic enzymes.[1,2] Considerable evidence has been presented that the inactivation of GS correlates with the loss of a single histidine residue and is accompanied by the generation of a protein carbonyl derivative.[3] Further oxidation renders the enzyme highly susceptible to known proteases[4] as well as to a new class of nonlysomal proteases which exhibit selectivity for the oxidatively modified enzyme.[5] These *in vitro* studies have suggested that oxidative inactivation of enzymes is a marking step for selective degradation of enzymes and is under metabolic control. We have subsequently undertaken studies to examine the possible physiological role of this process in biological systems, and one such system is activation of polymorphonuclear leukocytes (PMN, neutrophils).

Neutrophils can be stimulated by a variety of chemical or phagocytic agents to undergo a respiratory burst which is characterized by increased oxygen consumption, a change in cellular energy metabolism, and the generation of activated oxygen species.[6] This process is required for efficient microbicidal activity of these cells. Two neutrophil enzymes are thought to play a major role in microbial cell killing, namely, myeloperoxidase and NADPH oxidase. Although individuals with myeloperoxidase deficiency are not as seriously impaired as those with a defect in the NADPH oxidase pathway, both types of deficiences lead to a decreased capacity to fight microbial infections.[7] Because both microbial and mammalian NADH oxidases catalyze the oxidative inactivation of several enzymes *in vitro*,[2] we have considered the possibility that the activated NADPH oxidase in stimulated neutrophils catalyzes the oxidative inactivation of enzymes as a mechanism of neutrophil microbicidal activity.

EXPERIMENTAL RESULTS

GS Inactivation by Stimulated (Differentiated) HL-60 Cells

HL-60 cells are a promyelocytic cell line which can be differentiated to a neutrophil-like cell by treatment with agents such as dimethyl-

Table 1. Inactivation of Free GS and GS in Intact E. coli[a]

Addition	% GS Inactivation
GS	< 1.0
GS + PMA	< 1.0
GS + PMN	< 1.0
GS + PMN + PMA	80.0
E. coli	< 1.0
E. coli + PMA	< 1.0
E. coli + PMN	5.3
E. coli + PMN + PMA	20.7

[a]All inactivation assays were carried out as previously described.[10] E. coli cells remaining in suspension after light centrifugation were permeabilized by cetyltrimethylammonium bromide (CTAB)in order to assay residual GS activity.

sulfoxide (DMSO) or retinoic acid.[8] When differentiated HL-60 cells were incubated with GS in the presence of phorbol myristate acetate (PMA) or formyl L-methionyl-L-leucyl-L-phenylalanine (fMLP) inactivation of GS occurred. Undifferentiated or unstimulated HL-60 cells did not inactivate the enzyme. The data in Figure 1 show that the capacity of these cells to inactivate GS correlates closely with the development of respiratory burst competence during differentiation with DMSO. Similar GS inactivation was observed when HL-60 cells were stimulated with fMLP although both the extent and the rate of inactivation were diminished compared to PMA-stimulated cells.

GS Inactivation by Stimulated Neutrophils

Experiments were also carried out with freshly isolated human neutrophils. The data in Table 1 show that when either free GS or unopsonized E. coli containing GS was incubated with PMA-activated neutrophils inactivation occurred. Control experiments demonstrated that PMA had no effect on GS activity or on E. coli, and neutrophils have less than 2% of the total GS activity in these experiments. Moreover, because GS inactivation occurred readily in bacterial cells which remained suspended after light centrifugation, direct association with neutrophils or phagocytes is not required for inactivation. These results are consistent with earlier in vitro experiments[11] which indicated that inactivation can be mediated by diffusable products (viz, H_2O_2 and Fe(II)) produced by MFO systems and that complex formation or contact of the MFO system with the target enzyme is not required for inactivation.

The properties of GS inactivation catalyzed by activated neutrophils are similar to the properties of the in vitro MFO-mediated inactivation. The reaction (Figure 2) requires molecular oxygen and is stimulated by the addition of Fe(III) although inactivation occurs without Fe(III) addition. GS inactivation was inhibited by Mn(II), catalase, EDTA, and o-phenanthroline. The neutrophil reaction was also slightly inhibited by SOD. This result could be explained if $\cdot O_2^-$ produced by activated neutrophils contributed directly to the reduction of Fe(III).

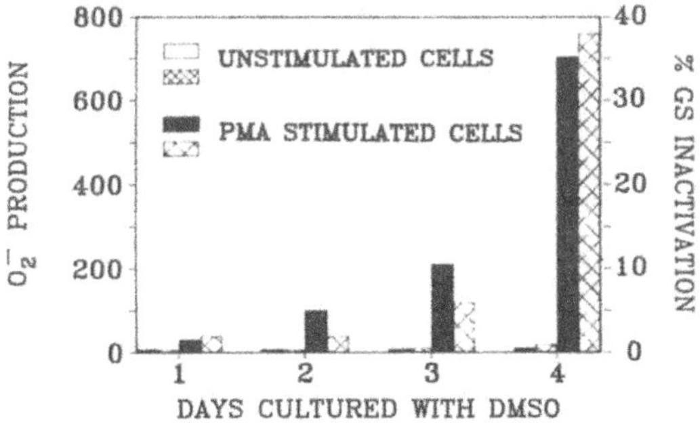

Figure 1. $\cdot O_2^-$ generation and GS inactivation by stimulated HL-60 cells during culture with DMSO. Growth and differentiation of HL-60 cells with dimethylsulfoxide (DMSO) was carried out according to previously described procedures.[9] $\cdot O_2^-$ generation and GS inactivation were determined by standard methods.[10] $\cdot O_2^-$ generation is depicted by open and filled bars in unstimulated and PMA-stimulated cells, respectively. GS inactivation is represented by small hatched and large hatched bars in unstimulated and PMA-stimulated cells.

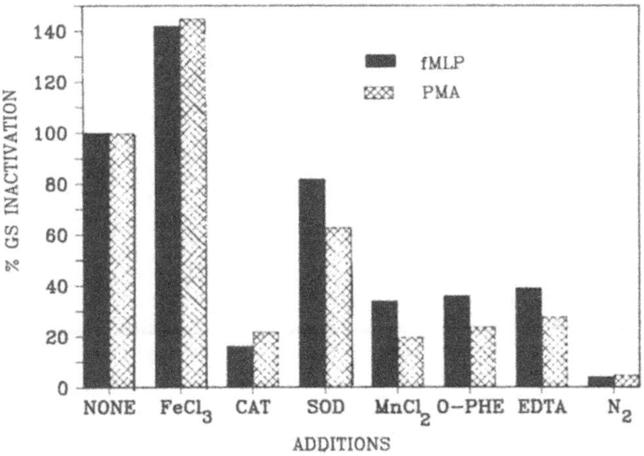

Figure 2. Properties of GS inactivation by freshly isolated neutrophils stimulated with PMA or fMLP. GS inactivation assays were carried out as previously described.[10] For purpose of comparison, 100% was 50% inactivation for fMLP-stimulated cells and 82% inactivation for PMA-stimulated cells. The concentrations of effectors are as follows: 0.1 mM $FeCl_3$, 0.42 mM catalase (CAT), 60 units/ml superoxide dismutase (SOD), 0.1 mM $MnCl_2$, 0.1 mM o-phenanthroline (o-phe), 1 mM ethylenediaminetetraacetic acid (EDTA).

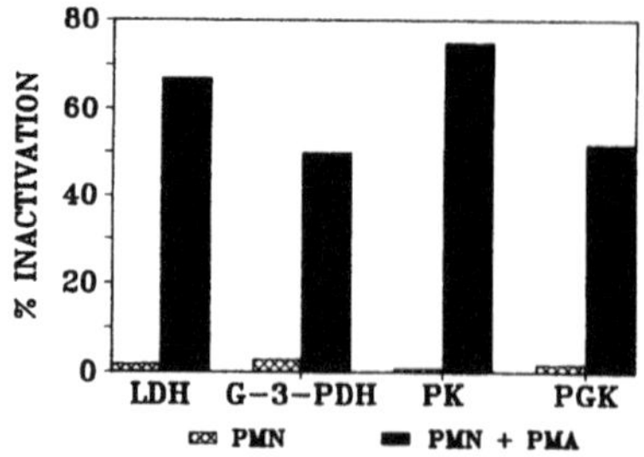

Figure 3. Inactivation of endogenous neutrophil enzymes following stimulation with PMA. Freshly isolated neutrophils were incubated in the presence or absence of PMA and then collected by centrifugation. The cells were broken by sonication and insoluble debris was removed by centrifugation. The specific activities of four soluble enzymes, lactate dehydrogenase (LDH), glyceraldehyde-3-phosphate dehydrogenase (G-3-PDH), pyruvate kinase (PK), and phosphoglycerate kinase (PGK) were determined according to standard procedures.[2]

Generation of Protein Carbonyl Derivatives in Stimulated Neutrophils

Oxidative inactivation of GS is accompanied by the generation of a protein carbonyl derivative which can be detected by reactivity with 2,4-dinitrophenylhydrazine (DNPH) resulting in the formation of a stable protein hydrazone derivative.[3] DNPH reactivity can be determined in crude extract preparations and protein hydrazones can be separated from other cellular material by extraction with organic solvents. The carbonyl content was determined in PMA-stimulated neutrophils and control cells. Stimulated neutrophils exhibit a time-dependent increase in carbonyl content and the isolated protein hydrazone derivatives possess spectral properties similar to those obtained with purified enzymes and MFO systems *in vitro*.[11,12] Reverse phase HPLC analysis of the hydrazone mixtures revealed that protein recovery was essentially the same in samples from PMA-stimulated neutrophils and control cells, but the samples from the stimulated cells exhibited a 10-30% increase in the carbonyl content of several peaks. It is well known that neutrophil myeloperoxidase reactions give rise to chlorinated amines which decompose to form protein aldehydes and it might be argued that these reactions are the primary source of protein carbonyl derivatives. However, considerable *in vitro* evidence indicates that GS is not inactivated by this mechanism.[3] In addition, the enzyme is remarkably stable to treatment with hydrogen peroxide, sodium hypochlorite or hydrogen peroxide and sodium hypochlorite.[3] Moreover, although carbonyl derivatives are abundant in stimulated karyogranuloplast preparations, they are also generated readily in cytoplast preparations which are largely depleted of myeloperoxidase granules.

These results also suggested that activated neutrophils oxidatively modify endogenous proteins and perhaps inactivate endogenous enzymes. When the activities of four enzymes, lactate dehydrogenase (LDH), pyruvate kinase (PK), phosphoglycerate kinase (PGK), and glyceraldehyde-3-phosphate dehydrogenase (G-3-PDH) were determined before and after PMA stimulation, 50-75% inactivation occurred (Figure 3). These same enzymes are highly susceptible to inactivation by MFO systems *in vitro*.

These studies indicate that stimulated neutrophils and neutrophil-like (differentiated) HL-60 cells catalyze the inactivation of free GS, GS in *E. coli* as well as endogenous enzymes. This inactivation is associated with neutrophil respiratory burst and possesses properties similar to oxidative inactivation of GS by isolated MFO systems *in vitro*.[1,2,3,11] The reaction is accompanied by the generation of protein carbonyl derivatives which can be derivatized by treatment with DNPH. The resulting protein hydrazones exhibit spectral characteristics similar to GS inactivated by MFO systems *in vitro*. Failure of sodium hypochloride acid and hydrogen peroxide to inactivate GS *in vitro*[3] and failure to incorporate ^{131}I into GS inactivated *in vitro*[3] argue against a direct role of myeloperoxidase chlorination mechanism in the GS inactivation catalyzed by stimulated neutrophils.

It is likely that the inactivation of key metabolic enzymes is an important mechanism of neutrophil microbicidal activity. In addition, the oxidative modification of endogenous neutrophil enzymes suggests that neutrophils fail to discriminate host tissues from invading organisms and this process may contribute to tissue damage during chronic inflammation.

REFERENCES

1. R. L. Levine, C. N. Oliver, R. M. Fulks, and E. R. Stadtman, Turnover of bacterial glutamine synthetase: Oxidative inactivation precedes proteolysis, *Proc. Natl. Acad. Sci. U.S.A.* 78:2120 (1981).
2. L. Fucci, C. N. Oliver, M. J. Coon, and E. R. Stadtman, Inactivation of key metabolic enzymes by mixed-function oxidation reactions: Possible implication in protein turnover and aging, *Proc. Natl. Acad. Sci. U.S.A.* 80:1521 (1983).
3. R. L. Levine, Oxidative modification of glutamine synthetase. II. Characterization of the ascorbate model system, *J. Biol. Chem.* 258:11828 (1983).
4. J. M. Farber and R. L. Levine, Oxidative modification of the glutamine synthetase of *E. coli* enhances its susceptibility to proteolysis, *Fed. Proc.* 41:865 (abstr.) (1982).
5. A. J. Rivett, R. L. Levine, and E. R. Stadtman, Mixed-function oxidation of proteins can modify multiple amino acids leading to various effects on the protein, *Fed. Proc.* 44:1399 (abstr.) (1985).
6. S. J. Klebanoff and R. A. Clark, "The Neutrophil: Function and Clinical Disorders," North Holland, New York (1985).
7. B. L. Babior and C. A. Crawley, Chronic granulomatous disease and other disorders of oxidative killing by phagocytes, *in*: "The Metabolic Basis of Inherited Disease," J. B. Stanbury, J. B. Wyngaarden, D. S. Frederickson, J. L. Goldstein, and M. S. Brown, eds., McGraw-Hill, New York (1983).
8. S. J. Collins, F. W. Ruscetti, R. E. Gallagher, and R. C. Gallo, Normal functional characteristics of cultured human promyelocytic leukemia cells (HL-60) after induction of differentiation, *J. Exp. Med.* 149:969 (1979).
9. T. R. Breitman, B. R. Keene, H. Hemmi, Studies of growth and differentiation of human myelomonocytic leukemia cell lines in serum-free medium, *in*: "Methods for Serum-Free Culture of Neuronal and Lymphoid Cells," D. A. Barnes, D. A. Sirbasku, G. H. Sato, eds., Alan R. Liss, New York (1984).

10. C. N. Oliver, Inactivation of enzymes and oxidative modification of proteins by stimulated neutrophils, Arch. Biochem. Biophys. 253: 62 (1987).
11. K. Nakamura, C. N. Oliver, and E. R. Stadtman, Inactivation of glutamine synthetase by a purified rabbit liver microsomal cytochrome P_{450} system, Arch. Biochem. Biophys. 240:319 (1985).
12. R. L. Levine, Mixed-function oxidation of histidine residues, "Methods in Enzymology,", Vol. 107, Academic Press, New York (1984).

OXIDATIVE INACTIVATION OF NEUTROPHIL GRANULE PROTEINASES: IMPLICATIONS FOR NEUTROPHIL-MEDIATED PROTEOLYSIS

Margret C.M. Vissers and Christine C. Winterbourn

Pathology Department
Christchurch School of Medicine
Christchurch, New Zealand

INTRODUCTION

Oxidants released by stimulated neutrophils are likely contributors to tissue damage at sites of inflammation. One way they could act would be to modify structural components of the tissue to enhance their susceptibility to proteolytic digestion. In that neutrophils concurrently release oxidants and granule proteinases, this would be a mechanism where the two could act synergistically.

OXIDANT ENHANCEMENT OF BASEMENT MEMBRANE COLLAGEN PROTEOLYSIS

Several authors have reported that pretreatment with oxidants, particularly .OH, enhances subsequent proteolysis of various soluble proteins. We have examined digestion of insoluble glomerular basement membrane collagen (Type IV) by elastase, the enzyme responsible for most of the digestion of this collagen by intact neutrophils.[1] Pretreatment with H_2O_2 alone had no effect, but H_2O_2 in combination with Cl^- and neutrophil myeloperoxidase caused a threefold increase in the rate of proteolysis.

On this basis, it might be expected that when basement membrane was exposed to stimulated neutrophils, scavengers of H_2O_2 or myeloperoxidase-derived oxidants would decrease the rate of collagen degradation.

OXIDANT INACTIVATION OF NEUTROPHIL GRANULE ENZYMES

Oxidative inactivation of some neutrophil granule enzymes, including lysozyme, β-glucuronidase and vitamin B_{12} binding protein, has been reported.[2,3] To see whether released proteinases are also susceptible to oxidants, a cell-free medium containing exocytosed neutrophil granule constituents was prepared.[4] Serial additions of H_2O_2 were made and the activities of various constituent enzymes measured. As shown in Figure 1, all enzymes were progressively inactivated. Of the neutral proteinases, gelatinase and collagenase (not shown) were readily inactivated, cathepsin G less so, and elastase was relatively more resistant. Inactivation was entirely dependent on myeloperoxidase since it was prevented by azide or methionine.

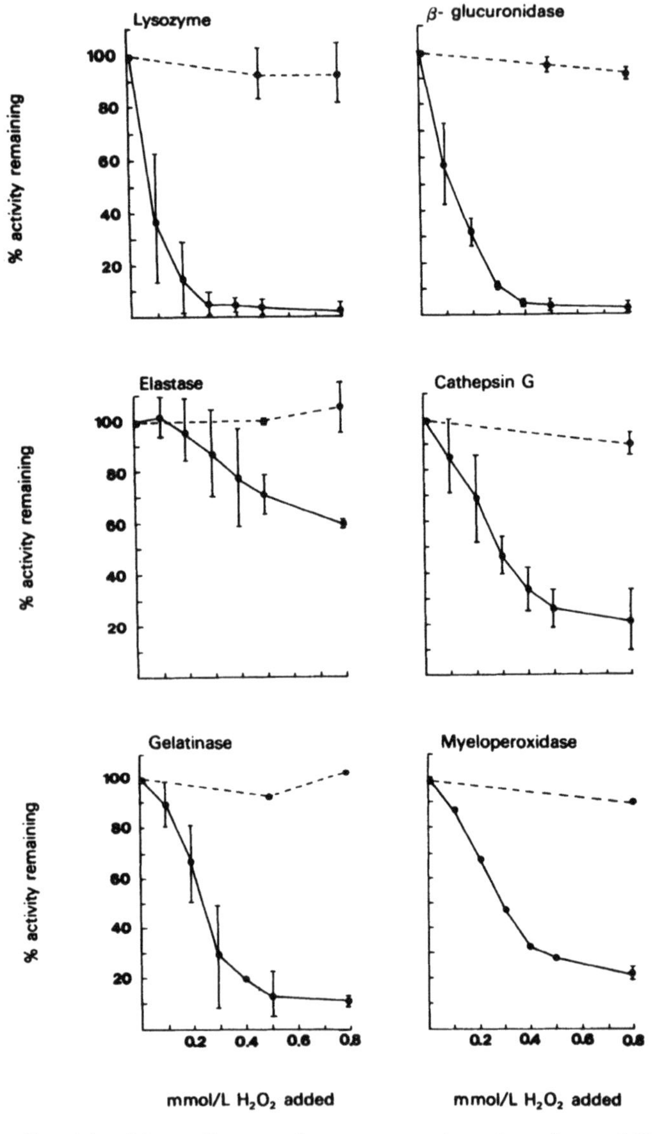

Figure 1. Inactivation of granule enzymes by stepwise additions of 100 µmol H_2O_2 to the supernatant of stimulated neutrophils. Dotted lines, with azide present, except with myeloperoxidase to which catalase was added.[2]

Our findings with gelatinase and collagenase contrast with those of others[5,6] who have shown oxidative activation of both enzymes. It appears that there is a fine dividing line between activation and inactivation, with activation occurring only within a narrow range of myeloperoxidase activities. In our hands, both latent enzymes are proteolytically activated when released from fully stimulated neutrophils,[7] and inactivation by oxidants is most likely.

On the basis that myeloperoxidase-derived oxidants inactivate neutrophil proteinases, it might be expected that oxidants released by stimulated neutrophils would decrease the digestive capacity of the cells and oxidant scavengers would enhance digestion.

NEUTROPHIL-MEDIATED DEGRADATION OF BASEMENT MEMBRANE COLLAGEN

To determine the outcome of the two opposing effects described above, we investigated the role of oxidants in degradation of isolated glomerular basement membrane by intact neutrophils. In the model system employed, the basement membrane contains immune complexes which induce the cells to adhere, release oxidants, degranulate, and digest the membrane proteins.[8] We found that with catalase present to scavenge H_2O_2, or with myeloperoxidase-deficient or chronic granulomatous disease neutrophils, the release of active granule enzymes, including elastase, was 1.5 to 3 times higher than from untreated normal cells. Despite the increase in enzyme actvity, however, these treatments had only a small effect on the rate of protein degradation per cell (Table 1). Degradation per unit of enzyme activity, however, was low when either H_2O_2 or myeloperoxidase was not available.

We interpret these findings as resulting from a combination of two effects of myeloperoxidase-derived oxidants causing (a) inactivation of released neutrophil proteinases and (b) modification of basement membrane protein to increase its susceptibility to digestion. These two reactions largely cancel each other out so that the ultimate outcome is a situation in which the contribution of oxidants to neutrophil-mediated collagen degradation is scarcely apparent.

Table 1. Basement Membrane Collagen Degradation and Elastase Release by Normal, Myeloperoxidase-Deficient (MPO-Def) and Chronic Granulomatous Disease (CGD) Neutrophils.

Cells	Elastase release (% normal control)	Hydroxyproline solubilized (% normal control)	
		per 10^7 cells	per unit elastase
Normal	100	100	100
Normal + catalase	155	93	66
MPO-Def	200	105	57
CGD	320	134	42

REFERENCES

1. M.C.M. Vissers and C.C. Winterbourn, The effect of oxidants on neutrophil-mediated degradation of glomerular basement membrane collagen. Biochim. Biophys. Acta, 889:277 (1986).

2. A.A. Voetman, R.S. Weening, M.N. Hamers, L.J. Meerhof, A.A.M. Bot and D. Roos, Phagocytosing human neutrophils inactivate their own granular enzymes. J. Clin. Invest., 67:1541 (1981).
3. R.A. Clark and N. Borregaard, Neutrophils autoinactivate secretory products by myeloperoxidase-catalysed oxidation. Blood, 65:375 (1985).
4. M.C.M. Vissers and C.C. Winterbourn, Myeloperoxidase-dependent oxidative inactivation of neutrophil neutral proteinases and microbicidal enzymes. Biochem. J., 245:277 (1987).
5. S.J. Weiss, G. Peppin, X. Ortiz, C. Ragsdale and S.T. Test, Oxidative autoactivation of latent collagenase by human neutrophils. Science, 227:747 (1985).
6. G.J. Peppin and S.J. Weiss, Activation of the endogenous metalloproteinase, gelatinase, by triggered human neutrophils. Proc. Natl. Acad. Sci. U.S.A., 83:4322 (1986).
7. M.C.M. Vissers and C.C. Winterbourn, Activation of human neutrophil gelatinase by endogenous serine proteinases. Biochem. J. 249:327 (1988).
8. M.C.M. Vissers, C.C. Winterbourn and J.S. Hunt, Degradation of glomerular basement membrane by human neutrophils in vitro. Biochim. Biophys. Acta, 804:154 (1984).

EFFECTS OF ALPHA,BETA-UNSATURATED ALDEHYDES ON MACROPHAGE AND NEUTROPHIL MEMBRANE FUNCTION, FLUIDITY AND SULFHYDRYL STATUS

Gisela Witz, Nancy J. Lawrie, Bernard D. Goldstein,
Jill Ryer-Powder and Marie A. Amoruso

UMDNJ-Robert Wood Johnson Medical School, Department
of Environmental and Community Medicine, 675 Hoes Lane
Piscataway, New Jersey 08854-5635, U.S.A.

We have postulated that oxidative damage in the lung may be mediated by alpha,beta-unsaturated aldehydes formed during oxidant stress by ozone. As such, we are investigating the toxic effects of alpha,beta-unsaturated aldehydes on alveolar macrophages since these are key cells in the lung, important for the detoxification and removal of inhaled particles and for their interaction with lymphoid cells for immune functions. Since many of these functions are critically dependent on the integrity of the macrophage plasma membrane, we have concentrated our efforts on the effects of alpha,beta-unsaturated aldehydes on macrophage plasma membrane function and dynamic properties. The alpha,beta-unsaturated aldehydes inhibit the plasma membrane NADPH-dependent oxidase which metabolizes oxygen to superoxide anion radical, a precursor species to bactericidal active oxygen species such as ·OH and H_2O_2. The observed IC_{50}'s for inhibition of the macrophage oxidase are 24, 77 and 240 uM for acrolein, trans-4-hydroxynonenal and crotonaldehyde, respectively.[1] Similar IC_{50}'s were obtained for the inhibition by these aldehydes of the corresponding oxidase in human neutrophils.[2]

Alpha,beta-unsaturated aldehydes are capable of directly alkylating cellular sulfhydryls, and it is possible that alkylation of plasma membrane SH may be involved in the inhibition of the oxidase and in changes of the other membrane functions and dynamic properties described below. In polymorphonuclear leukocytes, decreases in available surface sulfhydryls are greater than decreases in intracellular soluble SH (mainly GSH) at low aldehyde concentrations (Table 1). Thus, at 10uM acrolein there is a 42% decrease in available surface SH compared with a 28% decrease in intracellular soluble SH. For crotonaldehyde at 100uM, the decreases are 28 and 9% for available surface SH and soluble SH, respectively. If one assumes that the decreases in sulfhydryls are due to a Michael addition reaction of thiols across the reactive alpha,beta-unsaturated double bond, the data suggest that the plasma membrane is acting as a partial barrier against

Table 1. Decrease in Plasma Membrane Surface Sulfhydryl Groups and Soluble Intracellular Sulfhydryl Groups of Polymorphonuclear Leukocytes Treated With Alpha,Beta-Unsaturated Aldehydes

Aldehyde	Concentration (uM)	% Decrease Surface SH*	Soluble SH*
Acrolein	10	42 (3)	28 (3)
	100	53 (3)	41 (3)
Crotonaldehyde	100	28 (3)	9 (3)
	1000	46 (3)	45 (2)

*Control values for surface SH ranged from 2.0 to 6.5 nmoles/10^6 cells and for soluble SH ranged from 2.0 to 6.7 nmoles / 10^6 cells. (Adapted from Reference 1)

intracellular entry of reactive aldehydes at low concentrations of these compounds. This capability appears to be saturated at higher aldehyde concentrations, i.e. 100uM acrolein and 1mM crotonaldehyde, at which available surface SH and intracellular soluble SH are decreased to almost equal extents, i.e. about 50%.

A comparison of acrolein with crotonaldehyde reveals that at equimolar concentrations (100 uM), crotonaldehyde is approximately 50% less effective in decreasing available surface SH compared with acrolein and about 25% as effective as acrolein in decreasing intracellular soluble SH. Since acrolein has been reported to be about 150 times more reactive than crotonaldehyde toward alkylation by glutathione,[3] and if one again assumes that the observed decreases in sulfhydryls are due to alkylation by these aldehydes, these data suggest that biological interactions of reactive aldehydes in cellular systems cannot be predicted strictly on the basis of their chemical reactivity in homogeneous aqueous systems. Additionally, rates of detoxification via metabolism may play an important role in modifying structure-activity relationships based on chemical reactivity.

In vitro exposure of rat pulmonary macrophages to acrolein, crotonaldehyde and trans-4-hydroxynonenal decreases membrane lipid fluidity as seen by the increases in the fluorescence polarization (Table 2) of the lipid probe all trans-1,6-diphenyl-1,3,5-hexatriene (DPH). The reactivity differences observed for acrolein and crotonaldehyde with respect to their effects on cellular sulfhydryls are also evident with respect to their relative ability to decrease membrance lipid fluidity and phagocytosis (data not shown).

Since the binding of the fluorescent lipid probe DPH is probably not restricted to the plasma membrane, the polarization values are not strictly indicative of fluidity changes in the plasma membrane. Rather, they most likely reflect decreases in the lipid fluidity of an array of cellular membranes, including the plasma membrane, after treatment of macrophages with aldehydes. The mechanisms by which reactive aldehydes bring about changes in membrane lipid fluidity are not known.

Table 2. Effect of Alpha,Beta-Unsaturated Aldehydes on DPH* Fluorescence Polarization in Macrophages

Aldehyde	Concentration (uM)	DPH Polarization Value at 37°**
Control	0	0.241
Acrolein	10	0.250
	100	0.259
Crotonaldehyde	10	0.245
	100	0.248
	240	0.252
Trans-4-hydroxy-nonenal	10	0.244
	50	0.257
	100	0.261

*DPH, all trans-1,6-diphenyl-1,3,5-hexatriene.
**Average of 2-3 determinations.

The effects of alpha,beta-unsaturated aldehydes described above mirror those observed in alveolar macrophages obtained from rats after acute exposure to ozone.[4,5] Whether reactive aldehydes of this chemical nature are indeed formed in the lung as a result of oxidant stress remains yet to be elucidated.

ACKNOWLEDGMENTS. This work was supported by NIH grant ES02510. The authors thank Mrs. Elaine Randolph and Mrs. Cecile Stapleton for their expert secretarial assistance in the preparation of this manuscript.

REFERENCES

1. G. Witz, N.J. Lawrie, M.A. Amoruso, and B.D. Goldstein, Inhibition by reactive aldehydes of superoxide anion radical production from stimulated polymorphonuclear leukocytes and pulmonary alveolar macrophages. Effects on cellular sulfhydryl groups and NADPH oxidase activity, Biochem. Pharmacol. 36:721 (1987).
2. G. Witz, N. J. Lawrie, M.A.Amoruso, and B. D. Goldstein, Inhibition by reactive aldehydes of superoxide anion radical production in stimulated human neutrophils, Chem.-Biol. Interactions 53:13 (1985).
3. E. Schauenstein, H. Esterbauer and H. Zollner, ,ß-Unsaturated aldehydes. In: "Aldehydes in Biological Systems. Their Natural Occurrence and Biological Activities," Methuen, Inc., New York (1977).
4. M.A. Amoruso, G. Witz, and B.D. Goldstein, Decreased superoxide anion radical production by rat alveolar macrophages following inhalation of ozone or nitrogen dioxide, Life Sci. 28:2215 (1981).
5. J.E. Ryer, M.A. Amoruso, B. D. Goldstein, and G. Witz, Ozone inhalation affects rat alveolar macrophage superoxide anion radical production, phagocytosis, membrane lipid fluidity and membrane sulfhydryl group status, Toxicologist 7:127 (1987).

LATEX BEADS AS VEHICLES FOR THE STUDY OF FREE RADICAL PROCESSES IN PHAGOCYTIC CELLS

Michael J. Thomas and Catherine C. Hedrick

Department of Biochemistry, Bowman Gray School of Medicine
Winston-Salem, North Carolina, 27103, U.S.A.

INTRODUCTION

The phagocytosing human polymorphonuclear leukocyte (PMN) generates large quantities of oxidants: superoxide ($HO_2^\cdot$ and $\cdot O_2^-$), H_2O_2, and a chlorine-containing oxidant reputed to be hypochlorite. These oxidants are believed to aid in the killing of microorganisms since patients whose PMNs cannot undergo this "respiratory burst" suffer severe, recurrent infections. There appears to be good,[1] but not unequivocable evidence for the participation of superoxide in bacterial killing by the human PMN. Extensive[2] research has shown that $\cdot O_2^-$ is neither a strong oxidant nor a strong reductant and, therefore, it is probably not a potent bactericide. The conjugate acid, $HO_2^\cdot$, should be sufficiently reactive[3] to attack many components of the bacterial cell. Since the human PMN phagosome generally is assumed to be acidic, the acid-base equilibrium will be shifted toward $HO_2^\cdot$. However, the disproportionation reaction rate increases rapidly as the pH approaches the pK_a of $HO_2^\cdot$, 4.8.[3] It is not clear that attack on the susceptible bacterial cell components would be competitive with disproportionation.

Participation of the hydroxyl radical ($HO\cdot$) in bacterial killing has been suggested by several authors.[4] However, most of the methods used to test for the participation of oxy radicals have been questioned.[5] The source of $HO\cdot$ in the phagosome has been attributed to a Fenton-type reaction, the H_2O_2 being generated by disproportionation of superoxide. However, much of the H_2O_2 produced in the phagosome could be consumed by myeloperoxidase, thereby reducing the significance of Fenton chemistry.[6] Two recent attempts to detect $HO\cdot$ formation by PMNs have been unsuccessful.[7,8]

Since there appeared to be significant ambiguity concerning the participation of oxy radicals in the bactericidal process, we set out to determine whether oxy radicals might be involved. Linoleyl alcohol adsorbed onto latex beads[8] was used as the phagocytosable probe for oxy radicals and singlet oxygen ($^1O_2$). Linoleyl alcohol is oxidized readily by $^1O_2$ and by free radical chain autooxidation initiated by a wide variety of oxy radicals, e.g. $HO_2^\cdot$.[9] Although $^1O_2$ is not generally considered to be a major oxidant in the phagosome, the use of a 1,4-diene such as linoleyl alcohol also allows us to test for its participation. We have determined the relative sensitivity of the linoleyl alcohol-latex bead system to oxidation by hydroxyl radical ($HO\cdot$) and by superoxide. In an earlier study in which the linoleyl alcohol-latex bead system was used to probe the phagosome, we showed that the phagocytosing PMN does not oxidize linoleyl alcohol on latex beads unless iron salts are included in the medium.[8] The preliminary work presented here now makes it possible to estimate a lower limit for the flux of radicals in the phagosome.

EXPERIMENTAL

Studies on the autooxidation of linoleyl alcohol by HO· and superoxide were carried out in ^{60}Co sources located in the Department of Chemistry of Brookhaven National Laboratory. Superoxide was prepared by scavenging HO· with 1 M sodium formate in oxygen and N_2O-saturated solutions. For these studies, 10 mL of solution containing 5 x 10^9 beads/mL in 10 mM phosphate buffer was irradiated in quartz tubes. Sodium formate was recrystallized from Milli-Q water containing ethylenediaminetetraacetic acid, then twice from pure Milli-Q water. Sodium phosphate was "puratronic grade." Sodium hydroxide was "Baker Grade" and sulfuric acid was "nitronic grade." Standards were prepared by dye sensitized photooxidation in oxygen sparged methanol. $NaBH_4$ reduced samples of oxidized linoleyl alcohol were analyzed by GC-mass spectrometry.[8]

RESULTS AND DISCUSSION

The rates of formation of HO· and superoxide were approximately equal in the experiments reported here (Table 1). By way of comparison, if we assumed that the rate of superoxide formation by the human PMN was equal to the rate of oxygen consumption by 1.3 x 10^8 cells,[8] then approximately 50 times more superoxide was being formed by the PMNs (Table 1). However, because of the differences in the volumes of the two systems, the steady-state concentration of superoxide in the phagosome was very much higher than that in the irradiated solution. Assuming a phagosomal pH of 4.5, the superoxide concentration is calculated to be approximately 6.34 x 10^{-5} M, as compared to a calculated value of 2.3 x 10^{-9} M in the irradiated solution. The following assumptions facilitated the calculation: (1) The phagosome diameter was that of the latex bead (0.732 μmeter) and (2) There was an average of three beads per PMN.[8] These assumptions are supported by several experiments, including electron microscopy (data not shown). Even with these high steady-state concentrations of superoxide, the PMN does not oxidize linoleyl alcohol.[8] In the PMN studies no effort was made to remove transition metals from the phosphate-buffered saline medium used with the PMNs. Oxidized products were detected only when iron salts were added to the medium. However, both HO· and $HO_2^{\cdot}$ formed by irradiation gave significant yields of oxidized products (dienediols), >100 μg per sample. This leads us to conclude that the rate of formation of HO· and the steady-state concentration of $HO_2^{\cdot}$ in the phagosome are significantly lower than that in the radiation experiments. These results imply that much less than 0.004% of the oxygen utilized during the respiratory burst is converted to a radical that is situated to initiate the free radical chain autooxidation of linoleyl alcohol. Our results are consistent with a recent study that reports HO· adducts with spin traps could not be detected when human PMNs were treated with zymosan.[7]

Solutions of linoleyl alcohol-coated latex beads exposed to $HO_2^{\cdot}$ gave only conjugated dienediols after methanolic $NaBH_4$ reduction (Table 2). No oxidation products were detected when the coated beads were treated with $\cdot O_2^-$. When HO· was the oxidant, large amounts of the conjugated dienediols were obtained. Small amounts of nonconjugated dienediols were also detected. From the types of products that were isolated, we conclude

Table 1. The Rate of Formation of Oxidants from Several Sources

Oxidant	Source	Rate (moles/sec)
$HO\cdot$ [a]	^{60}Co	2.36×10^{-10}
$HO_2^{\cdot}/O_2^{-}$ [b]	^{60}Co	2.36×10^{-10}
X [c]	PMN	1.29×10^{-8}

[a] pH = 7.0; [b] pH = 1.6-8.0; [c] The first oxidant is superoxide from the reduction of oxygen by a neutrophil NADPH oxidase. The rate of oxidant formation was assumed to equal the rate of oxygen consumption measured with a Clark electrode. The pH of the medium was 7.4.

that both $HO_2^{\cdot}$ and $HO\cdot$ can initiate free radical chain autooxidation of linoleyl alcohol coated on latex beads. By way of comparison, the human PMN gave roughly equal amounts of both conjugated and nonconjugated dienediols when iron salts were added to the medium.[8]

In these preliminary experiments relatively small amounts of nonconjugated dienediols were obtained only in the presence of $HO\cdot$. Therefore, the formation of these nonconjugated dienediols in the PMN phagosome cannot be explained by processes attributed to the direct reaction of $HO_2^{\cdot}$ or $HO\cdot$ with linoleyl alcohol. In homogeneous solution the formation of roughly equal amounts of nonconjugated and conjugated dienediols is characteristic of the $^1O_2$ oxidation of 1,4-dienes.[8,10] Potential sources of $^1O_2$ in the phagosome are the reaction of hypochlorite with H_2O_2 and the disproportionation of superoxide. However, the reaction between hypochlorite and H_2O_2 should not take place in an acidic phagosome. Careful studies of the pH dependence of $^1O_2$ formation by this reaction have shown that it does not proceed efficiently below pH 8.[11,12] Attempts to detect $^1O_2$ from disproportionation of superoxide in aqueous solutions have been unsuccessful.[13-16] Explanation of the product distribution obtained with iron salts awaits further experimentation.

In summary, low concentrations of $HO_2^{\cdot}$ and $HO\cdot$ effectively initiate the free radical chain autooxidation of linoleyl alcohol on latex beads. However, it would appear that the levels of oxy radicals in the phagosome

Table 2. The Distribution of Oxidized Products Resulting from Reaction of Linoleyl Alcohol with Various Oxidizing Systems

Oxidant	10-OH [a] t8,c12	12-OH c9,t12	9-OH t10,c12	13-OH c9,t11	9-OH t10,t12	13-OH t9,t11
$^1O_2$ [b]	++	++	++	++		
$HO\cdot$ [b]	+	+	++	++	+	+
$HO_2\cdot$ [c]			++	++	+	+
PMNLs,Fe [b]	++	++	++	++	+	+

++ Major product; + minor product; [a] All products are octadecadienediols having a hydroxy moiety at carbon 1. The first line delineates the position of the second hydroxy moiety. The second line describes the position and <u>cis,trans</u>-stereochemistry of the double bonds. [b] pH = 7-7.4; [c] pH = 1.6.

are much lower than the concentrations required to initiate detectable autooxidation. No special precautions were taken to remove iron salts in the PMN studies. Therefore, the small quantities of iron normally found in buffer systems were not sufficient to catalyze autooxidation. Our observations do not exclude the possibility that site-specific damage to a bacterium could be catalyzed by iron in the presence of PMN oxidants.

ACKNOWLEDGMENTS

We would like to thank Dr. B. H. J. Bielski of Brookhaven National Laboratory for his advice and for the use of his facilities. This work was supported in part by PHS grants GM 29661 and HL 33532.

REFERENCES

1. R. B. Johnston, B. B. Keele, H. P. Misra, J. E. Lehmyer, L. S. Webb, R. L. Baehner, and K. V. Rajagopalan, The role of superoxide anion generation in phagocytic bactericidal activity, J. Clin. Invest. 55:1357 (1975).
2. D. T. Sawyer and J. S. Valentine, How super is superoxide? Acc. Chem. Res. 14:393 (1981).
3. B. H. J. Bielski, D. E. Cabelli, R. L. Arudi, and A. B. Ross, Reactivity of HO_2/O_2 radicals in aqueous solution, J. Phys. Chem. Ref. Data 14:1041 (1985).
4. J. A. Badwey and M. L. Karnovsky, Active oxygen species and the functions of phagocytic leukocytes, Annu. Rev. Biochem. 49:695 (1980).
5. H. J. Forman and M. J. Thomas, Oxidant production and bactericidal activity of phagocytes, Annu. Rev. Physiol. 48:669 (1986).
6. C. C. Winterbourn, Myeloperoxidase as an effective inhibitor of hydroxyl radical production, J. Clin. Invest. 78:545 (1986).
7. B. E. Britigan, G. M. Rosen, Y. Chai, and M. S. Cohen, Do human neutrophils make hydroxyl radical? J. Biol. Chem. 261:4426 (1986).
8. M. J. Thomas, P. S. Shirley, C. C. Hedrick, and L. R. DeChatelet, Role of free radical processes in stimulated human polymorphonuclear leukocytes, Biochemistry 25:8042 (1986).
9. J. M. Gebicki and B. H. J. Bielski, Comparison of the capacities of the perhydroxyl and the superoxide radicals to initiate chain oxidation of linoleic acid, J. Am. Chem. Soc. 103:7020 (1980).
10. M. J. Thomas and W. A. Pryor, Singlet oxygen oxidation of methyl linoleate: isolation and characteriation of the $NaBH_4$-reduced products, Lipids 15:544 (1980).
11. A. M. Held, D. J. Halko, and J. K. Hurst, Mechanisms of chlorine oxidation of hydrogen peroxide, J. Am. Chem. Soc. 100:5732 (1978).
12. T. Kajiwara and D. R. Kearns, Direct spectroscopic evidence for a deuterium solvent effect on the lifetime of singlet oxygen in water, J. Am. Chem. Soc. 95:5886 (1973).
13. C. S. Foote, F. C. Shook, and R. A. Abakerli, Chemistry of superoxide ion: singlet oxygen is not a major product of dismutation. J. Am. Chem. Soc. 102:2503 (1980).
14. M. J. Thomas, K. Mehl, and W. A. Pryor, The role of superoxide in xanthine oxidase-induced autooxidation of linoleic acid, J. Biol. Chem. 257:8343 (1982).
15. R. L. Arudi, B. H. J. Bielski, and A. O. Allen, Search for singlet oxygen luminescence in the disproportionation of HO_2/O_2^-, Photochem. Photobiol. 39:703 (1984).
16. T. Nagano and I. Fridovich, Does the aerobic xanthine oxidase reaction generate singlet oxygen? Photochem. Photobiol. 41:33 (1985).

SUPEROXIDE GENERATION OF PHAGOCYTIZING POLYMORPHONUCLEAR

LEUKOCYTES FROM PATIENTS WITH CHRONIC RENAL FAILURE

Toru Tanigawa, Toshikazu Yoshikawa, Norimasa Yoshida, Shigeru Sugino, and Motoharu Kondo

First Department of Medicine
Kyoto Prefectual University of Medicine, Kyoto, Japan

INTRODUCTION

Patients with chronic renal failure (CRF) are susceptible to microorganisms and often suffer from severe infection. Whether this is caused in part by decreased generation of active oxygen by polymorphonuclear leukocytes (PMN) is unclear. Active oxygen generation by PMN in CRF has been reported to be augumented (NBT reduction assay)[1], normal (NBT reduction assay, cytochrome C reduction assay)[2,3] decreased (luminol-dependent chemiluminescence)[3].

We investigated the generation of superoxide ($\cdot O_2^-$) by PMN in CRF patients by chemiluminescence (CL) employing a newly introduced CL probe, 2-methyl-6-phenyl-3,7-dihydroimidazo[1,2-a]pyrazine-3-one, a Cypridina luciferin analog (CLA), which is sensitive to and specific for $\cdot O_2^-$.[3,4]

MATERIALS AND METHODS

Peripheral venous blood was obtained from eight CRF patients on chronic hemodialysis (CHD), three CRF patients under conservative therapy, two patients with mild renal impairment (serum creatinine 2.1-5.0 mg/dl) and 23 normal healthy adults. CRF was defined as serum creatinine greater than 5 mg/dl. The profile of the patients are shown in Table 1.

PMN were isolated from venous blood by dextran sedimentation followed by metrizoate-ficoll (Lymphoprep) separation and hypotonic lysis of contaminating erythrocytes. PMN were suspended in Hanks' balanced salt solution (HBSS) at pH 7.4 and stored on ice until used.

CLA-dependent CL was measured with a Luminescence Reader (Aloka Co., Tokyo). The standard assay mixture contained 4×10^5 PMN, 2 mg of opsonized zymosan (OZ) and 1 μM CLA in HBSS at a total volume of 2 ml. The experiment with the same mixture from which OZ was omitted was used as a control. PMN in HBSS were preincubated for 3 min and the reaction was initiated by the addition of OZ. During the luminescence measurement, the incubation mixture was agitated by rotation at 37°C in the Luminescence Reader.

In some cases, urea, creatinine (Cr), uric acid (UA) or methylguanidine (MG) was added prior to the incubation in order to know the effects of these

Table 1 The Profile of Patients

	Age	Sex	Renal Disease[a]	Serum BUN (mg/dl)	Serum Creatinine (mg/dl)	Serum Uric Acid (mg/dl)	Duration of Dialysis (y)
Patients on chronic hemodialysis							
G.Y.	76	M	CGN	34.5	6.3	5.9	5
Y.M.	46	M	DM	57.5	7.7	7.7	2
T.H.	58	F	CGN	84.3	9.5	7.8	10
I.O.	52	M	CGN	63.0	12.2	8.4	2
K.I.	42	M	CGN	59.9	11.3	6.1	2
S.I.	60	M	CGN	60.5	10.9	7.0	3
S.M.	47	F	CGN	100.2	9.3	7.0	6
T.T.	60	M	CGN	71.9	8.4	3.2	19
(Mean ± S.D.)				(66.5 ± 18.3)	(9.5 ± 1.8)	(6.6 ± 1.5)	
Patients under conservative therapy							
S.I.	66	M	DM	83.0	5.1	6.8	-
S.M.	69	F	CGN	80.0	5.6	6.2	-
F.H.	72	F	CGN	107.0	8.8	6.8	-
(Mean)				(90.0)	(6.5)	(6.6)	
Patients with minimal renal impairment							
H.M.	25	M	CGN	30.0	2.9	8.8	-
M.U.	57	F	CGN	59.0	4.3	6.8	-
(Mean)				(44.5)	(3.6)	(7.8)	

[a]DM: diabetic nephropathy, CGN: chronic glomerulonephritis

substance on peak CL. To determine the scavenging or promoting effects of these substance on $\cdot O_2^-$, CL was measured using hypoxanthine (100 μM) and xanthine oxidase (Sigma, 10 mU/ml) system as another source of $\cdot O_2^-$ in the presence and in the absence of the substance.

RESULTS

Peak CL by phagocytizing PMN from CRF patients on CHD (1137 ± 341 X 10^3 cpm) and that of CRF patients under conservative therapy (1055 ± 252 X 10^3 cpm) were higher than those of normal subjects (549 ± 142 X 10^3 cpm). Peak CL by phagocytizing PMN from patients with mild renal impairment was higher than normal subjects and lower than CRF patients (828 ± 10^3 cpm) (Figure 1). Peak CL by phagocytizing PMN had no correlation with serum BUN or Cr but was positively correlated with serum UA (Figure 2).

Urea had no effect on peak CL by PMN at concentrations wich can be seen under normal and uremic conditions (10^{-3} M, 10^{-2} M) but suppressed it at a higher concentration (10^{-1} M). Cr also had no effects at pathophysiological concentrations. UA increased peak CL at the concentrations which are comparable to serum concentration (Figure 3). MG had no effect on peak CL at the concentrations in uremia, but increased peak CL at higher concentrations (10^{-3} M, 10^{-2} M).

In hypoxanthine and xanthine oxidase system, UA decreased peak CL slightly (Figure 3), but this may be the result of inhibition of xanthine oxidase. Cr had no effect at pathophysiological concentrations but increased CL at a higher concentration. Urea and MG neither decreased nor increased the peak CL. Thus the increase of CL by phagocytizing PMN induced by UA was probably due to its enhancement of $\cdot O_2^-$ generation by PMN.

DISCUSSION

Phagocytizing PMN generate active oxygen species which play a major role in destruction of invading microorganisms. Once these active oxygen species are liberated, they may be harmfull to surrounding cells and result in tissue damage.

We found that the CLA-dependent CL of phagocytizing PMN was increased in uremic patients both under conservative therapy and on CHD. This indicates that $\cdot O_2^-$ generation by phagocytizing PMN in uremic patients is higher than in normal subjects and, consequently, the increased susceptibility to infection in uremic patients is not caused by decreased $\cdot O_2^-$ generation by PMN.

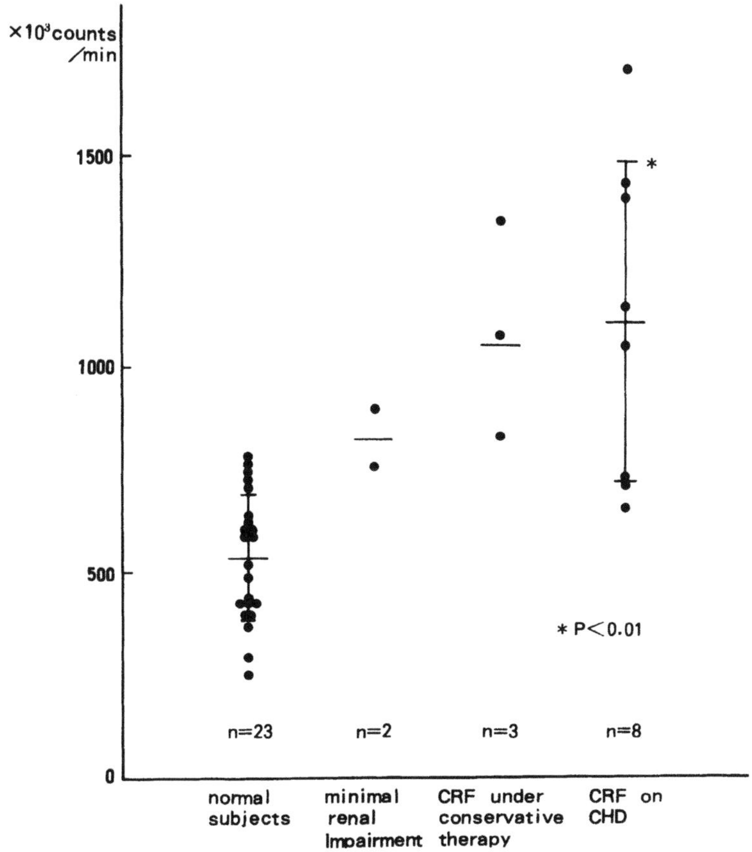

Figure 1 CLA-dependent chemiluminescence of PMN from patients and normal healthy adults.

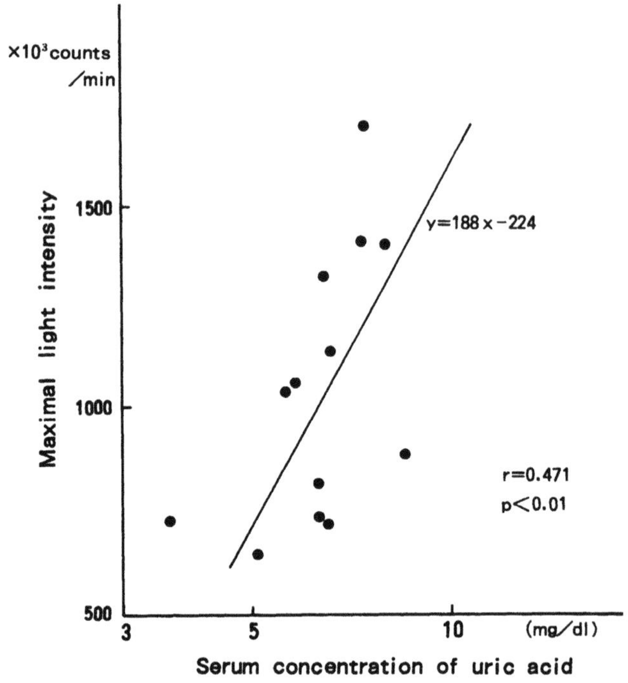

Figure 2 Correlation between serum uric acid and CLA-dependent CL.

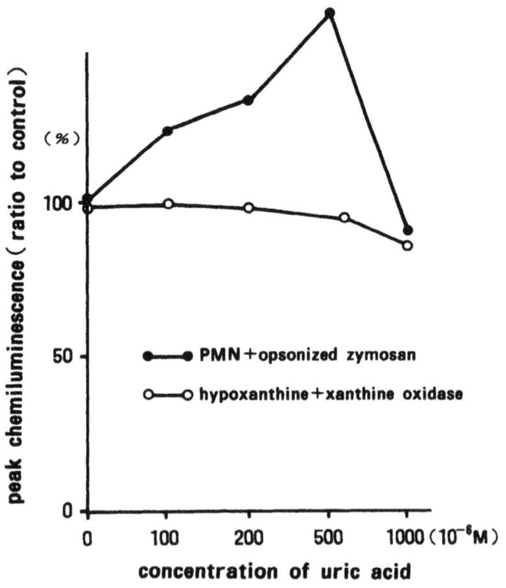

Figure 3 Effect of uric acid on CLA-dependent chemiluminescence

We examined the effects of several substances which increase in the sera of uremic patients on the generation of $\cdot O_2^-$ by phagocytizing PMN in vitro, and UA increased peak CL at comparable serum concentrations in uremic patients. This suggests that UA is one of the causative factors that enhance $\cdot O_2^-$ generation by phagocytizing PMN.

Although the effects of the increased $\cdot O_2^-$ generation by phagocytizing PMN is unclear, the renal damage by infiltrated PMN might be promoted, and some uremic toxins like methylguanidine, which is reported to be a oxidative product by active oxygen,[6] might be increased.

ACKNOWLEDGEMENT

We are grateful to Professor Minoru Nakano of Gunma University for providing CLA.

REFFERENCES

1. W. Sutowicz, Z. Komrowska, Z. Hanicki, T. Cichocki and O. Smolenski, The enzymatic activities and NBT reduction test of granulocytes in untreated and dialysed uremic patients, Acta Haematol.(Basel) 61:144 (1979).
2. J.H. Chretien and V.F. Garagusi, Phagocytosis and nitroblue tetrazolium reduction in uremia, Experimentia 29:612 (1973).
3. E.R. Ritchey, J.D. Wallin and S.V. Shah, Chemiluminescence and superoxide anion production by leukocytes from chronic hemodialysis patients, Kidney Int. 19:349 (1981).
4. A. Nishida, K. Sugioka, M. Nakano and K. Yagi, Use of 2-methyl-6-phenyl-3,7-dihydroimidazo[1,2-a]pyrazin-3-one for assay of superoxide anions in phagocytizing granulocytes, J. Clin. Biochem. Nutr. 1:5 (1986).
5. K. Sugioka, M. Nakano, S. Kurashige, Y. Akuzawa and T. Goto, A chemiluminescent probe with a Cypridina luciferin analog, 2-methyl-6-phenyl-3,7-dihydroimidazo[1,2-a]pyrazine-3-one, specific and sensitive for $\cdot O_2^-$ production in phagocytizing macrophages, FEBS Lett. 197:27 (1986).
6. S. Nagase, K. Aoyagi, M. Narita and S. Tojo, Synthesis of methylguanidine by active oxygen, with special refference to the significant role of hydroxyl radical, Jpn. J. Nephrol. 27:1149 (1985).

CLASTOGENIC FACTOR IN ISCHEMIA-REPERFUSION INJURY:

PROTECTIVE EFFECT OF ALLOPURINOL

Ingrid Emerit and Jean-Noel Fabiani*

Laboratoire de Genetique
Institute Biochemical des Corderliers
CNRS-Universite Paris VI
15 Rue de l'Ecole de Medicine
75006, Paris, France

*Clinique de Chirurgie Cardiovasculaire
Hopital Broussais 96
rue Didot, 75014
Paris, France

INTRODUCTION

Many studies have established a major role for toxic metabolites of oxygen as mediators of tissue injury following ischemia-reperfusion in a wide variety of species and organs.[1-3] The source of the oxygen free radicals is still the subject of debate. Tissue damage may be due in part to direct radical attack, in part to longer-lived secondary compounds derived from lipid peroxidation of membranes.

We have previously described the formation of DNA-damaging material under the term "chromosome breakage factor" or "clastogenic factor" (CF), released by cells under conditions of oxidative stress.[4,5] It is observed not only after irradiation and in chronic inflammatory diseases, but may be induced also *in vitro* by generation of a respiratory burst or by addition of hypoxanthine plus xanthine oxidase to the medium of cell cultures.[6,7] Oxy radical formation via the xanthine oxidase reaction was proposed as a pathogenic mechanism responsible for tissue damage after reperfusion.[1] We wondered whether CF was detectable in the plasma of patients submitted to ischemia-reperfusion during open heart surgery and whether allopurinol, an inhibitive competitor of xanthine oxidase, was protective.

MATERIAL AND METHODS

A total of 14 randomized patients submitted to coronary bypass surgery for stable angina was studied. Seven patients received 100 mg/l allopurinol in a cold antegrade cardioplagia solution injected immediately after cross-clamping of the aorta. Two or three bypasses were performed with an ischemia time in the range of 50 ± 10 min.

Before umclamping the aorta, a second cardioplegia was done. This so-called "reperfusion cardioplegia" was a warm blood cardioplegia containing again 100 mg allopurinol and injected into the bypass graft. The other seven patients were operated without allopurinol. Anesthesiologic and cardio-pulmonary bypass management were the same for all patients.

The blood samples were immediately transferred to the Cytogenetics Laboratory, where the origin of the samples, whether from treated or from untreated patients, remained unknown. Part of each blood sample was used for the study of the patient's chromosomes. The blood cultures were set up according to conventional cytogenetic procedures. Each culture flask contained 5 ml of tissue culture medium (TCM) 199 (Flow Lab., Paris), 1.5 ml heat-inactivated fetal calf serum and 0.5 ml of whole blood. The cell cultures were stimulated to divide with phytohemagglutinin P (Difco Lab., Detroit). After 68-72 hr of incubation at 37°C, the mitoses were arrested in metaphase by addition of colchicine. Slides were prepared for chromosome analysis, and 50 mitoses were studied for each sample. The results are given as the number of aberrations observed in 100 mitoses.

The majority of the blood was used for isolation of CF according to the procedure developed in this laboratory for CF from the plasma of irradiated persons or patients with chronic inflammatory diseases.[4,5] The plasma was filtered through Diaflo ultrafiltration filters (YM 10) retaining substances with a molecular weight of more than 10,000 Da. The ultrafiltrates were then concentrated 10 times by filtration through a second filter (YM 2) having a cut off at 1000 Da. The concentrates were tested for their chromosome damaging (clastogenic) effect in test cultures set up with whole blood from healthy volunteers. Aliquots of 0.5 ml were added to each culture flask containing 5 ml TCM 199, 1.5 ml fetal calf serum and 0.5 ml blood, as described for the cultures of patients' blood. Test cultures were set up in duplicate. The ultrafiltrates from samples a and b were studied simultaneously on the chromosomes from the same blood donor.

RESULTS

The frequency of structural chromosome aberrations (gaps, breaks, fragments, complex rearrangements) was in the normal range in all 14 patients before the operation (samples a) and corresponded to the observed in the cultures of the healthy donors used as the test system for ultrafiltrates (Table 1, Ia and II).

The concentrated ultrafiltrates prepared from these samples did not induce chromosome damage in the blood cultures of healthy donors (Table 1, IIa). An increased aberration rate was induced in 9 of the 14 patients set up with the samples collected after 20 min reperfusion (samples b). The same samples induced chromosome damage after addition to the test cultures (Table 1, IIb).

After identification of the origin of the blood samples, it could be concluded that 7 of the 9 patients with chromosome damage belonged to the group of untreated patients, while the 5 patients without chromosome damage had received allopurinol. The mean aberration rate of the 7 untreated patients was increased from 6 to 26 per cent in the patients' blood cultures and from 6 to 26 per cent in the test cultures to which the concentrates of patients' plasma had been added.

Table 1. Chromosome Analysis of Blood Cultures from Patients and of Control Cultures Exposed to CF From Patients's Plasma

Untreated patients					Patients with allopurinol				
Ia	Ib	II	IIa	IIb	Ia	Ib	II	IIa	IIb
6	32	4	2	16	0	6	2	6	0
3	36	16	18	40	6	8	4	4	4
3	20	6	6	24	0	18	2	2	12
4	18	2	2	18	0	6	2	2	2
8	20	2	4	22	4	4	6	6	10
8	14	2	2	32	10	8	10	10	10
8	28	10	8	18	2	18	8	14	24

The data represent numbers of chromosome aberrations observed in 100 mitoses studied. I = blood cultures of patients, set up with blood taken in the coronary sinus before ischemia (a) or 20 min after reperfusion (b). II = blood cultures of control persons to which concentrated ultrafiltrates of patients plasma were added. Prepared from blood taken before ischemia (a), or taken 20 min. after reperfusion (b).

Two of 7 patients treated with allopurinol had increased chromosome damage (18 percent aberrations). Their plasma induced an aberration rate of 12 and 24 percent respectively in the test cultures. In the other 5 patients no increase was observed either in patients' blood cultures or in the test cultures.

DISCUSSION

In agreement with our working hypothesis, 9 of 14 patients had chromosome damage 20 min after unclamping of the aorta, and their plasma was clastogenic for chromosomes of controls. No diagnostic X-rays were applied during the operations, and except for allopurinol, no drugs were given. Therefore, the chromosome damage cannot be explained by the well-known chromosome damaging agents, such as ionizing radiation or clastogenic chemicals. We may conclude that CF were formed as a consequence of the oxidative stress situation created upon reperfusion.

The pathogenic mechanisms responsible for tissue damage in the heart following temporary coronary artery occlusion are still incompletely understood and the source of oxy radicals is a matter of debate. Morphologic cellular damage is most evident after reperfusion, when oxygen is reintroduced. The respiratory burst of neutrophils invading the ischemic tissue has been considered as one possible source of $\cdot O_2^-$, and subsequent studies in neutrophil depleted dogs support this theory.[8] The xanthine oxidase reaction was another source of oxyradicals supposed to be of importance in ischemia-reperfusion injury.[1] In healthy tissue xanthine oxidase exists as a dehydrogenase using NAD as an electron acceptor and does not produce superoxide. It has been shown in the rat heart that a short period of ischemia is sufficient to convert the enzyme to an oxidase which uses molecular oxygen as electron acceptor and produces $\cdot O_2^-$. Furthermore an accumulation of oxypurine is the expected consequence of hypoxia induced dephosphorylation and degradation of intracellular ATP. Hence during reperfusior of ischemic heart, xanthine oxidase, hypoxanthine and molecular oxygen are all present to produce $\cdot O_2^-$. Blockade of xanthine oxidase with allopurinol afforded (in the rat heart) a degree of protection in the infarction process as observed with the $\cdot O_2^-$ scavenging enzyme superoxide dismutase.[9] In a large series of patients, operated on at Broussais Hospital, allopurinol seemed to be beneficial.[10] The protection given by allopurinol against CF production in the present study might be an argument in favor of xanthine oxidase related oxy radical production in the human heart, where the biochemical confirmation of this pathogenic mechanism is still lacking. Other explanations for the protective effect of allopurinol have been proposed.[11]

Besides the indirect evidence for a role of free radicals in ischemia-reperfusion injury coming from the protective effects of free radical scavengers, direct measurement of free radical generation has been demonstrated in the rabbit heart by electron paramagnetic resonance spectroscopy.[12] These experiments demonstrated that reactive oxygen-centered radicals are generated during ischemia and that a burst of oxy radical generation occurs within moments after reperfusion. In this context, it is noteworthy that the clastogenic material described as CF in this study is detectable after 20 min. Samples collected in the coronary sinus of untreated patients immediately after unclamping of the aorta did not contain clastogenic activity (data not shown). The exact time of appearance of detectable CF levels is under study.

Since the clastogenic activity is transferable from one culture system to another, more long lived chemical species than free radicals have to be considered as components of CF. Toxic effects of secondary products have also been evoked by others. Conjugated dienes derived from lipid peroxidation were found to be at a fivefold higher activity on reperfusion than at the end of ischemia.[13] Substances with conjugated diene structure as well as thiobarbituric acid reactive material were greater in CF preparations from hypoxanthine plus xanthine oxidase treated cell cultures.[7] Similar studies have not yet been done on CF preparations from patients undergoing open heart surgery.

All CF observed either <u>in vivo</u> or <u>in vitro</u> after exposure of cells to a source of $\cdot O_2^-$ pass through Diaflo filter YM 10 indicating that their components have a molecular weight under 10,000 Da. The fact that the ultrafiltrates can be concentrated by a second filtration through filter YM 2 does not preclude compounds with a molecular weight under 1000 Da, since certain molecules may be retained by specific binding to the filter membrane.

The chromosome-breaking material present in the plasma of patients after reperfusion was prepared according to the usual ultrafiltration protocol and is therefore in the molecular weight range of CF. Further investigation will show whether the formation of this CF can be prevented by superoxide dismutase, as is the case for all CFs described until now. This would be proof for the implication of the superoxide radical, $\cdot O_2^-$.

The study of CF formation may be helpful for the elucidation of the pathogenic mechanisms involved in reperfusion injury and also may be of interest for screening of therapeutic agents using easily applicable cytogenetic methods.

REFERENCES

1. J.M. McCord, Oxygen-derived free radicals in post-ischemic tissue injury, New Eng. J. Med. 312:159 (1985).
2. D.A. Peterson, R.W. Asinger, K.J. Elsperger, D.C. Homans and J.W. Eaton, Reactive oxygen species may cause myocardial reperfusion injury, Biochem. Biophys. Res. Comm. 127:87 (1985).
3. D.N. Granger, M.E. Hollwarth and D.A. Parks, Ischemia-reperfusion injury: role of oxygen-derived free radicals, Acta Physiol. Scand. 548:47 (1986).
4. I. Emerit, Chromosome breakage factors: origins and possible significance, Progr. Mutat. Res. 4:61 (1982).
5. I. Emerit, Properties and action mechanism of clastogenic factors, in: "Lyxmphokines," Edgar Pick, ed., Academic Press, (1983).
6. I. Emerit and P. Cerutti, Tumor promoter phorbol-myristate-acetate induces a clastogenic factor in human lymphocytes, Proc. Natl. Acad. Sci. USA 79:7509 (1982).
7. I. Emerit, S.H. Khan and P. Cerutti, Treatment of lymphocyte cultures with a hypoxanthine-xanthine oxidase system induces the formation of transferable clastogenic material, J. Free Radicals Bio. Med. 1:51 (1985).
8. R. Jolly, W.J. Kane, M.B. Bailie, G.D. Abrams and B.R. Lucchesi, Canine myocardial reperfusion injury. Its reduction by the combined administration of superoxide dismutase and catalase, Circ. Res. 54:277 (1984).
9. D.E. Chambers, D.A. Parks, G. Patterson, S. Yoshida, K. Burton, L.F. Parmley, J.M. McCord and J. M. Downey, Role of oxygen derived radicals in myocardial ischemia, Fed. Proc. 47:1093 (1983).
10. J.N. Fabiani, J. Relland and A. Carpentier, Myocardial protection via the coronary sinus in cardiac surgery: comparative evaluation of two techniques, in: "The Coronary Sinus," W. Mohl, ed., Steinkopff Verlag Darmstadt (1984).
11. D.V. Godin and S. Bhimji, Effects of allopurinol on myocardial ischemic injury induced by coronary artery ligation and reperfusion, Biochem. Pharmacol. 36:2101 (1987).
12. J.L. Zweier, J.T. Flaherty and M.L. Weisfeldt, Direct measurement of free radical generation following reperfusion of ischemic myocardium. Proc. Natl. Acad. Sci. USA 84:1404 (1987).
13. A.D. Romaschin, G.J. Wilson and D.A.G. Mickle, Hydroxy-conjugated dienes, products of free radical attack on polyunsaturated phospholipids in reversibly and irreversibly ischemic myocardium, Symposium on "Free Radicals in Biology and Medicine: Ischemia/Reperfusion Injury." Point Clear, Alabama, March 17-19 (1986).

INACTIVATION OF CREATINE PHOSPHOKINASE BY SUPEROXIDE DURING REPERFUSION INJURY

Joe M. McCord and William J. Russell

Department of Biochemistry
College of Medicine
University of South Alabama
Mobile, AL 36688, U.S.A.

INTRODUCTION

Creatine phosphokinase (CK) release is commonly used to assess myocardial damage in laboratory models as well as clinically. At least part of the myocardial damage observed during reoxygenation of the heart appears to be mediated by superoxide produced by xanthine oxidase.[1] While many reports have shown protection of reoxygenated myocardium by superoxide dismutase (SOD), occasional reports fail to observe protection by SOD.[2] When we examined CK release by the continuously perfused anoxic/reoxygenated rat heart we saw marked protection by SOD;[1] when we varied the model and examined the rat heart subjected to global no-flow ischemia/reperfusion, we saw an apparent paradoxical exacerbation of CK release in the presence of SOD, as described below. This led us to examine the possibility that SOD might, in fact, protect CK from partial inactivation by $\cdot O_2^-$.

MATERIALS AND METHODS

Creatine phosphokinase (type III, Sigma) from bovine heart was further purified by chromatography on Sephadex G-100 to remove traces of SOD. Xanthine oxidase (XO) was purified from bovine cream by the method of Waud et al.[3] Superoxide production by XO was assayed by monitoring cytochrome c reduction at 550 nm.[4] Yeast Cu/Zn SOD was from Pharmacia; catalase was from Boehringer; desferrioxamine (Desferal Mesylate) was from Ciba-Geigy.

Male Wistar rats were decapitated and hearts with lungs attached were excised and cannulated on a nonrecirculating perfusion apparatus within 1 min of excision. The perfusion apparatus represented a modification of the working heart model as described by Neely.[5]

Hearts subjected to hypoxia/reoxygenation were perfused for 40 min with hypoxic (95% N_2/5% CO_2) glucose-deficient buffer, then reoxygenated for 10 min with oxygenated buffer containing glucose. One-minute samples of coronary effluent were collected. Other hearts were subjected to 80 min of no-flow ischemia by clamping off the aortic perfusion line. The hearts were reperfused for 10 min with oxygenated glucose-containing buffer by removing the clamp. One-minute samples of coronary effluent were again collected. SOD, when present, was added to the perfusate solutions at 20 μg/ml. Collected coronary effluent was chilled and assayed for CK activity using Sigma procedure #520.

To determine total CK in the myocardium, hearts were homogenized in 10 ml of chilled buffer containing 50 mM potassium phosphate, 0.25 mM EGTA, 1 mM allopurinol, 0.04% deoxycholate, and 1 mM phenylmethylsulfonylfluoride at pH 7.5. After centrifugation, the supernates were assayed for CK activity.

RESULTS AND DISCUSSION

The ability of SOD to attenuate the release of CK from continuously perfused, post-hypoxic hearts is shown in Figure 1. The control hearts released an average of 9,610 units of CK during 10 min of reoxygenation. The presence of SOD in the perfusion buffers caused a 75% reduction in CK release (2,420 units/10 min). Also shown in Figure 1. is the pattern of CK release from hearts subjected to 80 min of no-flow ischemia, followed by reperfusion with oxygenated buffer, in the absence and presence of SOD. Here, the control hearts released 14,330 units/10 min. Surprisingly, the addition of SOD to the perfusate resulted in the release of 28,820 units/10 min, an increase of 100%. Two possible explanations for this apparent paradox were considered: either 1) SOD was truly increasing the amount of damage to the organ, perhaps by scavenging a reducing radical to tilt the balance in favor of strongly oxidizing radicals such as the hydroxyl radical, or 2) that SOD allowed the release of more CK activity by protecting the CK against inactivation by $\cdot O_2^-$. The latter case was easily tested by a "bookkeeping" experiment, the results of which are shown in Figure 2. Hearts subjected to 80 min of no-flow ischemia contained as much CK activity per gram as non-ischemic hearts (i.e., ischemia per se caused no loss of activity.) After 10 min of oxygenated reperfusion, however, only 80% of the activity could be accounted for, summing the activity released in the coronary perfusate with that remaining in the hearts. If SOD were present in the buffers, no such loss of CK occurred. The sum of that remaining and that released equaled that originally present. SOD prevented the reoxygenation-induced inactivation of CK.

A more direct demonstration of the ability of superoxide to inactivate CK is presented in Figures 3 and 4. Figure 3 shows that incubation of CK with increasing concentrations of XO in the presence of saturating xanthine for 30 min leads to progressive loss of CK activity. It should be noted that rat myocardium

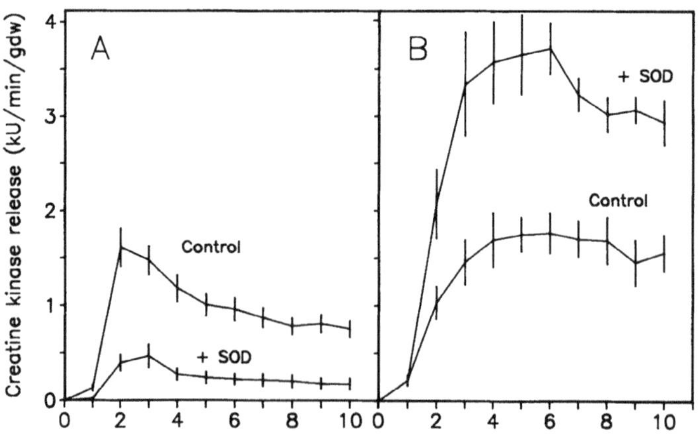

Figure 1. The effect of SOD (20 ug/ml) on CK release during reoxygenation following (A) 40 min of hypoxic glucose-free perfusion or (B) 80 min of complete no-flow ischemia. Hypoxia: n = 12 for controls; n = 7 for SOD treated. Ischemia: n = 8 for controls; n = 4 for SOD treated.

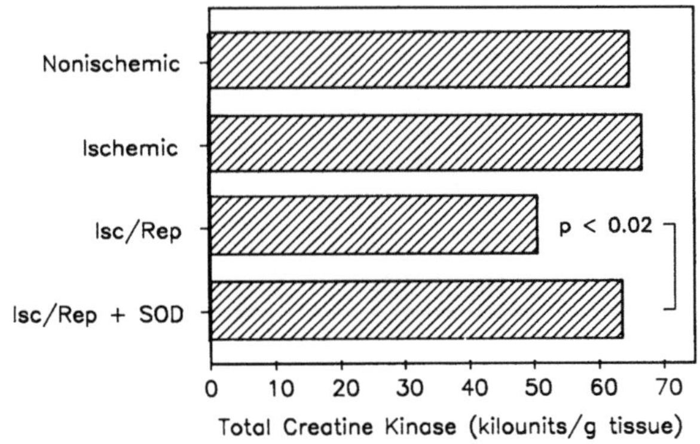

Figure 2. Superoxide dismutase prevents inactivation of myocardial creatine kinase during reperfusion. Nonischemic isolated hearts were perfused with oxygenated, glucose-containing buffer for 90 min. Ischemic hearts were subjected to 80 min of complete no-flow ischemia, but were not reperfused. Isc/Rep were subjected to 80 of ischemia followed by 10 of reperfusion. In the case of Isc/Rep + SOD, the buffer contained, in addition, 20 ug/ml SOD. Total CK reflects the activity present in the homogenized organ plus that released into the coronary perfusate.

contains approximately 30 times the concentration of XO required for half-maximal inactivation under these conditions. Figure 4 shows the ability of SOD to prevent the inactivation of CK by XO. Half-maximal protection was seen at less than 0.4 ug SOD/ml, with maximal protection being >90%. These data were obtained at pH 7, but similar results were seen at pH 6 and pH 8 (data not shown). Catalase also provided substantial protection of CK at pH 8 (70%), but the extent of protection dropped to 50% at pH 7 and to 25% at pH 6. Half-maximal protection was seen between 0.1 and 0.2 ug/ml of catalase. These results are shown in Figure 5.

The near-complete protection of CK provided by SOD at all pH values is consistent with the actual inactivating species being $\cdot O_2^-$ or a species derived secondarily from $\cdot O_2^-$ such as $HO\cdot$. It also rules out H_2O_2 per se as being the species responsible, as SOD does nothing to eliminate H_2O_2. The low degree of

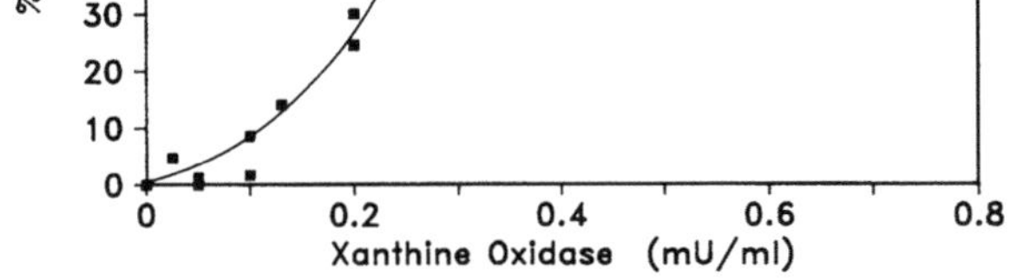

Figure 3. Creatine kinase inactivation by xanthine oxidase/xanthine. CK (0.5 U/ml) was incubated for 30 min at 37°C in 50 mM potassium phosphate, 0.1 mM EDTA, and 0.1 mM xanthine at pH 7 with the indicated concentration of XO. Aliquots were then assayed for remaining CK activity.

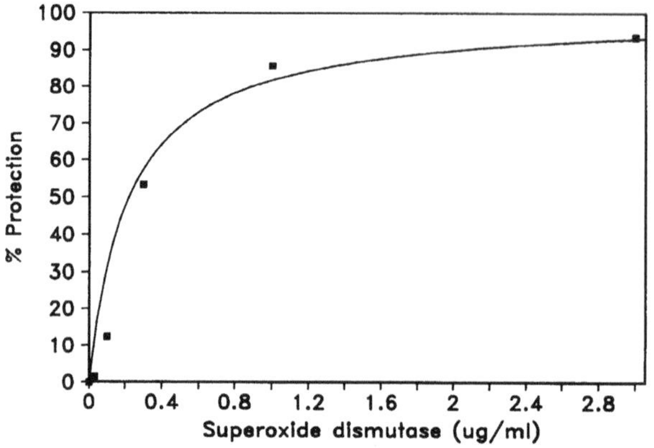

Figure 4. Superoxide dismutase protects creatine kinase against xanthine oxidase-dependent inactivation. CK (0.26 U/ml) was incubated with XO (0.39 mU/ml) as described in Figure 3., with the indicated concentrations of SOD. After 30 min, aliquots were assayed for CK activity. In the absence of SOD, the CK was >90% inactivated.

protection afforded by catalase at pH 6 indicates that H_2O_2 is not a major participant in the process at this pH. The substantial protec-tion afforded by catalase at pH 8, however, suggests that the presence of both O_2^- and H_2O_2 is necessary for the major mechanism of inactivation at this pH. The iron-catalyzed Haber-Weiss reaction is such a mechanism. Experiments with the potent iron chelator desferrioxamine support this hypothesis, as shown in Figure 6. Again, greatest protection is seen at pH 8; little if any is seen at pH 6. The addition of 0.1 mM ethylenediaminetetraacetic acid (EDTA) in place of desferrioxamine had no protective effect on CK, in keeping with earlier studies showing that the iron-EDTA complex is catalytically active in the production of hydroxyl radical.[6]

An approximate rate constant for the inactivation of CK by superoxide was calculated from data obtained by inactivating CK in the presence of catalase. Catalase eliminated hydroxyl radical production and allowed measurement of CK inactivation due only to superoxide. Superoxide production at 3.5×10^{-9} M/sec

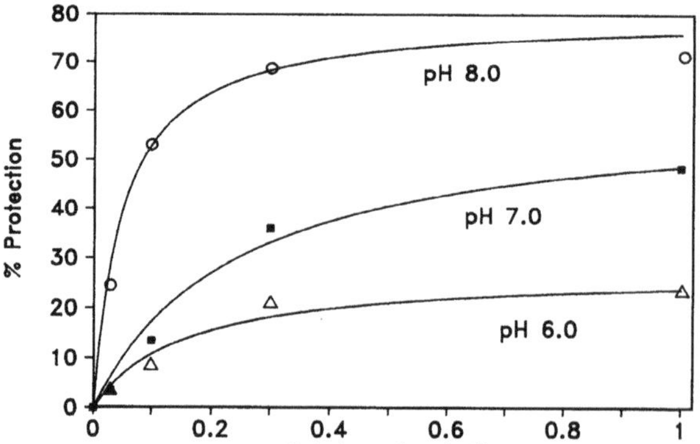

Figure 5. Catalase protects creatine kinase against xanthine oxidase-dependent inactivation. CK was incubated with XO as described in Figure 3., with the indicated concentrations of catalase and at the indicated pH. After 30 min, aliquots were assayed for CK activity. In the absence of catalase the CK was >90% inactivated.

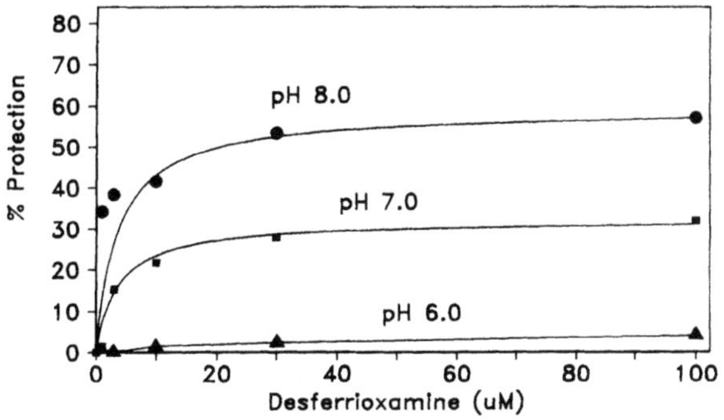

Figure 6. Desferrioxamine protects creatine kinase against xanthine oxidase-dependent inactivation. CK was incubated with XO as described in Figure 3., with the indicated concentrations of desferrioxamine and at the indicated pH. After 30 min, aliquots were assayed for CK activity. In the absence of desferrioxamine the CK was >90% inactivated.

inactivated 6.1 nM CK at a rate of 1.3%/min. The rate constant was calculated by making several reasonable assumptions, viz., that the steady state concentration of $\cdot O_2^-$ was determined by its rates of spontaneous and SOD-catalyzed dismutation. (The dismutation of superoxide by SOD is included because CK was found to contain SOD as a contaminant at 9.3 ug SOD/mg CK.) The calculated rate constant for the reaction of $\cdot O_2^-$ with CK is $1.2 \times 10^5\ M^{-1}\ sec^{-1}$. This value, it should be noted, is approximately equal to the rate constant for the reaction of $\cdot O_2^-$ with ferricytochrome c, considered by most to be quite an avid reaction.[4]

This work adds CK to the list of enzymes which are inactivated by the superoxide radical. It thereby delineates a molecular mechanism for free radical-mediated injury to the reperfused myocardium. It is unclear why the phenomenon seems to occur more in hearts subjected to no-flow ischemia than to continuously perfused hypoxia. The finding also suggests that under certain conditions the release of CK activity may be a poor indicator of the extent of myocardial injury, whether in the clinic or in the laboratory.

REFERENCES

1. M. B. Grisham, W. J. Russell, R. S. Roy, and J. M. McCord, Reoxygenation injury in the isolated perfused working rat heart: roles of xanthine oxidase and transferrin, in: "Superoxide and Superoxide Dismutase in Chemistry, Biology and Medicine," G. Rotilio, ed., Elsevier, Amsterdam (1986).
2. C. L. Myers, S. J. Weiss, M. M. Kirsh, and M. Shlafer, Involvement of hydrogen peroxide and hydroxyl radical in the 'oxygen paradox': reduction of creatine kinase release by catalase, allopurinol or deferoxamine, but not by superoxide dismutase, J. Mol. Cell. Cardiol. 17:675 (1985).
3. W. R. Waud, F. O. Brady, R. D. Wiley, and K. V. Rajagopalan, A new purification procedure for bovine milk xanthine oxidase: effect of proteolysis on the subunit structure, Arch. Biochem. Biophys. 169:695 (1975).
4. J. M. McCord and I. Fridovich, Superoxide dismutase: an enzymic function for erythrocuprein (hemocuprein), J. Biol. Chem. 244:6049 (1969).
5. J. R. Neely, H. Liebermeister, E. J. Battersby, and H. E. Morgan, Effect of pressure development on oxygen consumption by isolated rat heart, Am. J. Physiol. 212:804 (1967).
6. J. M. McCord and E. D. Day, Jr., Superoxide-dependent production of hydroxyl radical catalyzed by iron-EDTA complex, FEBS Lett. 86:139 (1978).

THE POST TREATMENT EXPERIMENT: AN OPERATIONAL DEFINITION OF REPERFUSION INJURY

Charles F. Babbs

Biomedical Engineering Center, Purdue University
West Lafayette, Indiana 47907, U.S.A.

INTRODUCTION

During the past few years evidence for a remarkable hypothesis about the pathophysiology of ischemia and reperfusion has begun to accumulate.[1-5] The hypothesis states that significant tissue damage resulting from a period of tissue ischemia, followed by reperfusion, may actually occur during the reperfusion phase, rather than during the period of ischemia, and may be caused by oxyen-derived free radicals. This concept has been described as the oxygen paradox,[6,7] namely, that although restoration of oxygen supply to ischemic tissue is necessary for return of normal function, oxygen may simultaneously participate in certain deleterious chemical reactions. Such paradoxical oxygen toxicity occurring during reperfusion involves, not molecular oxygen, but rather the partially reduced oxygen species: superoxide anions, hydrogen peroxide, and hydroxyl radicals. These species are capable of initiating chain reactions that chemically modify cellular proteins, lipids, and amino acids to produce what is referred to as reperfusion injury.

A simple experimental means to distinguish reperfusion injury from ischemic injury involves what may be called a post-treatment experiment, in which a drug is given after the period of ischemia or hypoxia at the onset of reperfusion. Such experiments are based upon the assumption, that if post-treatment with drug X prevents death, damage, functional impairment after ischemia and reperfusion, then X must have prevented some form of reperfusion injury, since X could have acted only during the reperfusion phase. A sufficient number of such studies have now been published[8-12] to establish with some confidence that reperfusion injury, thus defined, is both real and preventable by treatment with existing and relatively safe drugs, such as allopurinol and deferoxamine, given at the beginning of the reperfusion phase.

As early as 1973 Toledo-Pereyra[8] reported that preservation of post-ischemic gut by perfusion with allopurinol containing solution resulted in prolonged graft survival. More recently, Demetriou and coworkers[9] reported that post-treatment with intraperitoneal dimethylsulfoxide decreased ischemic bowel necrosis in rats. Ambrosio and coworkers have recently[13] described improved ventricular function and recovery of high energy phosphates following post-treatment with superoxide dismutase in isolated hearts after 30 min. of ischemia.

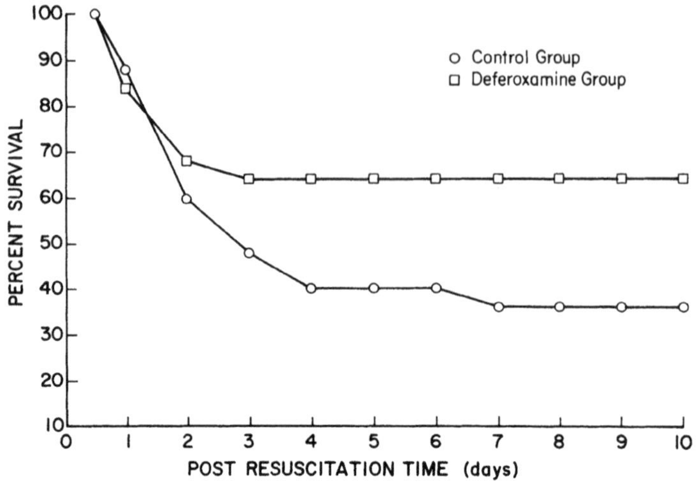

Figure 1. Sequential incidence of survival of deferoxamine (50 mg/kg, i.v. over 3 min) post-treated versus untreated control rats for 10 days following resuscitation from 7 min. of cardiopulmonary arrest (n = 25 for each gloup). Data from reference 10.

In post-treatment studies of cardiac arrest and resuscitation, Kompala and coworkers[10] administered the iron chelator, deferoxamine, at the onset of reperfusion, following 7 min. of total circulatory arrest and subsequent cardiopulmonary resuscitation in rats. In this animal model, total circulatory arrest was induced in anesthetized rats by intracardiac injection of cold, isotonic KCl solution. After the 7 min. arrest time, chest massage and ventilation were performed until sinus rhythm and a palpable apex heartbeat returned. Then saline vehicle or deferoxamine (50 mg/kg) was injected intravenously as a post-treatment. Analysis of a population of 25 rats that were post-treated with deferoxamine vs. the 25 controls that were not (Figure 1), shows a significant difference in survival at 10 days following reperfusion. The effectiveness of deferoxamine can be attributed to blockade of the superoxide driven, iron-catalyzed Haber-Weiss reaction,[3,14] a source of toxic hydroxyl radicals.

In a similar study Badylak and coworkers[11] investigated the potential of the xanthine oxidase inhibitor, allopurinol, as post-resuscitative therapy, thinking that xanthine oxidase might be a source of superoxide ions and hydrogen peroxide that could react with iron to initiate cellular injury, as has been proposed by McCord.[1] Allopurinol sodium salt was given intravenously in a single dose of 25 mg/kg immediately after restoration of the circulation. Survival at 10 days was more than doubled in the allopurinol post-treated group (Figure 2).

Notably, the results in Figures 1 and 2 were obtained with only a single dose of either of two drugs previously thought to have absolutely nothing to do with resuscitation. Results of such post-treatment experiments with deferoxamine or allopurinol support the assertion that preventable reperfusion injury happens after the circulation is restored and that such injury is of serious consequence to intact animals.

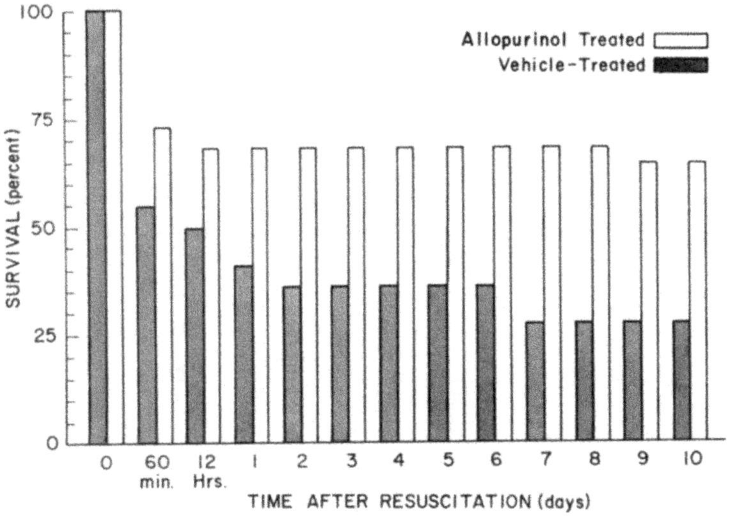

Figure 2. Sequential incidence of survival of allopurinol (25 mg/kg, i.v. over 3 min) post-treated vs. untreated control rats for 10 days following resuscitation from 7 min. of cardiopulmonary arrest (n = 22 for each group). Data from reference 11.

Post-treatment experiments can also be applied to isolated organs. Using the isolated rat heart preparation, Badylak et al.[12] observed the effects of oxypurinol, a xanthine oxidase inhibitor, and deferoxamine upon myocardial reperfusion injury following 60 min. of warm ischemia. To date we have done 24 hearts, divided into 3 groups of 8. First, the isolated hearts were perfused for 15 min. with a modified Krebs-Henseleit perfusate solution to permit stabilization of the preparation. Each heart was then subjected to 60 min. of total ischemia at 37°C followed by 60 minutes of reperfusion with either control perfusate, oxyprinol-containing perfusate (200 mcg/ml), or deferoxamine-containing perfusate (8 mcg/ml). Reperfusion injury was evaluated by the total amount of creatine phosphokinase (CPK) released into the perfusate during the first 60 min. of reperfusion, by changes in myocardial vascular resistance, and by morphologic examination. The placebo-treated group released an average of 3,492 ± 1420 Sigma units of CPK into the perfusate vs. 1073 ± 426 units for the oxypurinol-treated group ($t = 2.61$, $df = 14$, $p < 0.05$) and 1108 ± 497 S.U. for the deferoxamine-treated group ($t = 2.31$, $df = 14$, $p < 0.05$). The time course of CPK release in the three treatment groups is presented in Figure 3. The myocardial vascular resistance increased an average of 1,006 ± 321 peripheral resistance units (PRU) in the placebo-treated group during the reperfusion phase vs. 135 ± 47 PRU for the oxypurinol-treated group ($t = 2.87$, $df = 14$, $p < 0.02$), and 441 ± 193 PRU for the deferoxamine-treated group d ($t = 2.15$, $df = 14$, $p < 0.05$).

Direct visual examination of treated hearts typically revealed vigorous beating compared to the feeble ventricular activity of untreated, ischemitized, and reperfused hearts. Ultrastructural examination of one randomly selected heart from each group after 60 min. of reperfusion showed marked attenuation of mitochondrial and endoplasmic reticulum swelling, increased maintenance of membrane integrity, and diminished separation of myocardial myofilaments in the oxypurinol-treated and deferoxamine-treated hearts.

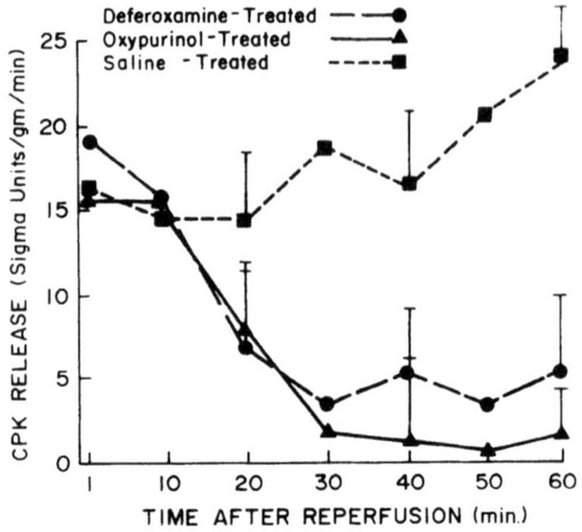

Figure 3. Sequential mean (± SEM) values for creatine phosphokinase release for deferoxamine post-treated, oxypurinol post-treated, and saline post-treated isolated rat hearts after 60 min. of complete, normothermic ischemia. Data from reference 12.

CONCLUSIONS

It is convenient to define reperfusion injury as tissue damage that occurs during restoration of circulation to transiently ischemic tissues. Such injury is to be distinguished from primary ischemic injury that occurs during the ischemic phase and ceases upon restoration of the circulation. Post-treatment experiments suggest that a substantial proportion of the total injury seen after 5 to 60 minute periods of circulatory arrest may actually develop during the reperfusion phase. Current theories of the genesis of reperfusion injury invoke oxygen centered free radicals, which are generated only after previously ischemic tissue has been resupplied with oxygen. Most evidence in support of the free-radical hypothesis has been derived from pre-treatment studies, in which putative blockers of free radical generation or putative free radical trapping agents are given before the onset of ischemia. We propose that such drugs can be equally effective when given after ischemia, at the onset of reperfusion. To the extent that such post-treatment are protective, the evidence for physiologically significant reperfusion injury is substantially strengthened.

ACKNOWLEDGMENTS

Supported by Grant HL-29398 from the National Heart, Lung, and Blood Institute, U.S. Public Health Service, Bethesda, Maryland.

REFERENCES

1. J.M. McCord, Oxygen-derived free radicals in postischemic tissue injury, N. Eng. J. Med. 312:159 (1985).
2. B.C. White, S.D. Aust, K.E. Arfors, and L.D. Aronson, Brain injury by ischemic anoxia: Hypothesis extension--a tale of two ions?, Ann. Emerg. Med. 13:862 (1984).
3. C.F. Babbs, Role of iron ions in the genesis of reperfusion injury following successful cardioipulmonary resuscitation: Preliminary data and a biochemical hypothesis, Ann. Emerg. Med. 14:777 (1985).

4. C. Guarnieri, F. Flamigni, and C.M. Caldarera, Role of oxygen in the cellular damage induced by re-oxygenation of hypoxic heart, J. Mol. Cell. Cardiol. 12:797 (1980).
5. F.Z. Meerson, V.E. Kagan, Y.P. Kozlov, L.M. Belkina, and Y.V. Arkhipenko, The role of lipid peroxidation in pathogenesis of ischemic damage and the antioxidant protection of the heart, Basic Res. Cardiol. 77:465 (1982).
6. M.L. Hess and N.H. Manson, Molecular oxygen: Friend and foe. The role of the oxygen free radical system in the calcium paradox, the oxygen paradox, and ischemia/reperfusion injury, J. Mol. Cell. Cardiol. 16:969 (1984).
7. C.L. Myers, S.J. Weiss, M.M. Kirsh, and M. Shlafer, Involvement of hydrogen peroxide and hydroxyl radical in the oxygen paradox: reduction of creatinine kinase release by catalase, allopurinol, or deferoxamine, but not by superoxide dismutase, J. Mol. Cell. Cardiol. 17:675 (1985).
8. L.H. Toledo-Pereyra, R.L. Simmons, and J.S. Najarian, Comparative effects of chlorpromazine, methylprednisolone and allopurinol during small bowel preservation, Am. J. Surg. 126:631 (1973).
9. A.A. Demetriou, P.K. Kagoma, S. Kaiser, E. Seifter, X.T. Niu, and S.M. Levenson, Effect of dimethylsulfoxide and glycerol on acute bowel ischemia in the rat, Am. J. Surg. 149:91 (1985).
10. S.D. Kompala, C.F. Babbs, and K.E. Blaho, Effect of deferoxamine on late deaths following cardiopulmonary resuscitation in rats, Ann. Emerg. Med. 15:405 (1986).
11. S.F. Badylak, C.F. Babbs, C. Kougias, and K. Blaho, Effect of allopurinol and dimethylsulfoxide on long-term survival in rats after cardiorespiratory arrest and resuscitation, Amer. J. Emerg. Med. 4:313 (1986).
12. S.F. Badylak, A. Simmons, J. Turek, and C.F. Babbs, Protection from reperfusion injury in the isolated rat heart by postischemic deferoxamine and oxypurinol administration, Cardiovasc. Res. 21:500 (1987).
13. G. Ambrosio, M.L. Weisfeldt, W.E. Jacobus, and J.T. Flaherty, Evidence for a reversible oxygen-radical mediated component of reperfusion injury: Reduction by recombinant human superoxide dismutase administered at the time of reflow, Circulation 75:282 (1987).
14. J.M.C. Gutteridge, R. Richmond, and B. Halliwell, Inhibition of the iron-catalyzed formation of hydroxyl radicals from superoxide and lipid peroxidation by desferrioxamine, Biochem J. 184:469 (1979).

MECHANISMS OF OXIDANT-MEDIATED MICROVASCULAR INJURY FOLLOWING REPERFUSION OF THE ISCHEMIC INTESTINE

Barbara J. Zimmerman, Matthew B. Grisham, and
D. Neil Granger

Department of Physiology and Biophysics
Louisiana State University Medical Center
Shreveport, Louisiana 71130-3932, U.S.A.

INTRODUCTION

It is now well recognized that ischemia/reperfusion-induced tissue injury is a major factor in the pathogenesis of certain life threatening diseases such as stroke and coronary artery disease. Studies indicate that this injury occurs at the time of reperfusion rather than during the period of ischemia[1]. There is now a large body of experimental data which suggests that oxy radicals mediate the microvascular and parenchymal cell damage observed during reperfusion of ischemic tissues. Reactive oxygen metabolites have been implicated in ischemia/reperfusion injury in the small intestine, heart, liver, pancreas, and brain.

ROLE OF XANTHINE OXIDASE

Reperfusion of ischemic small intestine results in substantial tissue damage that is largely mediated by oxygen-derived free radicals. This contention is based on observations that ischemia/reperfusion injury in the small intestine is attenuated by pretreatment with either superoxide dismutase, catalase, or dimethylsulfoxide[2]. There are several potential sources of oxy radicals in the small intestine. Much of the emphasis has been placed on the enzyme xanthine oxidase. Several lines of evidence support the importance of xanthine oxidase as a major source of oxidants associated with reperfusion injury of the ischemic bowel: 1) The intestinal mucosa is a rich source of the enzyme. 2) Xanthine dehydrogenase is rapidly converted to xanthine oxidase during intestinal ischemia (a 6- to 8-fold increase in oxidase during 1 hour of ischemia). 3) Tissue hypoxanthine concentration increases 10- to 20-fold during intestinal ischemia. 4) Xanthine oxidase inhibitors (allopurinol, pterin aldehyde) provide significant protection against the ischemia/reperfusion-induced increase in microvascular permeability and mucosal lesion formation. 5) Protease inhibitors, which prevent xanthine dehydrogenase-to-oxidase conversion, also protect against ischemia/reperfusion injury. 6) Local intra-arterial infusion of hypoxanthine and xanthine oxidase (that produces blood levels comparable to those measured in the ischemic bowel) increases vascular permeability in the normal (nonischemic) bowel to the same extent observed in the postischemic intestine; the hypoxanthine-xanthine oxidase infusion-induced increase in vascular permeability is largely prevented by superoxide dismutase or dimethyl sulfoxide. 7) The increased spontaneous chemiluminescence (an index of oxidant production) observed after reperfusion of the ischemic bowel is prevented by pretreatment with allopurinol[3].

POLYMORPHONUCLEAR LEUKOCYTES

Another potential source of reactive oxygen metabolites is the inflammatory polymorphonuclear leukocyte (PMN). A role for PMNs (predominately neutrophils) in postischemic myocardial damage was first proposed by Romson et al[4]. They demonstrated that neutrophil depletion prior to occlusion of the left anterior descending coronary artery resulted in a significant reduction in infarct size. Based on these findings, Romson et al proposed that the accumulation of PMNs following reperfusion may exacerbate tissue injury presumably through the release of reactive oxygen metabolites. Mullane et al [5] have shown that BW755C, an inhibitor of both the lipoxygenase and cyclooxygenase pathways of arachidonic acid metabolism, attenutes leukocyte infiltration into the infarcted myocardium and significantly reduces infarct size. Others have also noted an attenuation of ischemia/reperfusion injury with PMN depletion. However, this effect has been attributed to improved blood flow during the ischemic period as a result of limited microvascular plugging by PMNs [6-8]. Some investigators[9] have been unable to demonstrate protection against ischemia/reperfusion injury with PMN depletion. Therefore, the role of PMN as a source of reactive oxygen metabolites following reperfusion of ischemic tissues remains controversial.

PMN-XANTHINE OXIDASE INTERACTION

Recently Grisham et al[10] have proposed a mechanism that interrelates xanthine oxidase-derived free radicals, neutrophil infiltration, and tissue injury. Using tissue-associated myeloperoxidase activity as an index of PMN infiltration, they demonstrated that reperfusion of ischemic mucosa resulted in extensive PMN infiltration (Figure 1). Pretreatment with either allopurinol or administration of superoxide dismutase significantly decreased the extent of PMN infiltration, suggesting that xanthine oxidase-derived oxy radicals may play a role in the recruitment of PMNs into the post-ischemic tissue. There are two possible explanations for these findings: 1) Superoxide or $O_2^{\cdot-}$-derived oxidants such as hydrogen peroxide and/or hydroxyl radicals may damage the intestinal mucosa and cause the release of certain tissue proteases capable of interaction with the complement system to yield substances that attract and activate PMNs. This would explain why superoxide dismutase or allopurinol not only protects the microvasculature but also decreases neutrophil influx. 2) Superoxide may interact with certain lipoproteins to yield chemoattractants. In this situation PMNs would directly mediate the microvascular damage. Prevention of $O_2^{\cdot-}$ production by allopurinol or superoxide dismutase administration would eliminate the chemoattractant and the subsequent PMN influx. Based on the results of their study Grisham and coworkers have expanded the xanthine oxidase hypothesis to include a role for PMNs in postischemic tissue injury.

CHEMOATTRACTANTS

The existance of a superoxide-dependent chemoattractant for PMNs was first proposed by Petrone et al[11] and Perez et al[12]. Using human PMNs, Petrone et al demonstrated that plasma exposed to $O_2^{\cdot-}$ generated by the interaction between hypoxanthine and xanthine oxidase becomes potently chemotactic. They also demonstrated that intradermal injection of $O_2^{\cdot-}$-treated plasma or hypoxanthine and xanthine oxidase causes heavy infiltration of PMNs at the injection site. These observations, coupled to the ability of superoxide dismutase to inhibit PMN migration to sites of inflammation, led Petrone et al to propose that $O_2^{\cdot-}$ reacts with certain

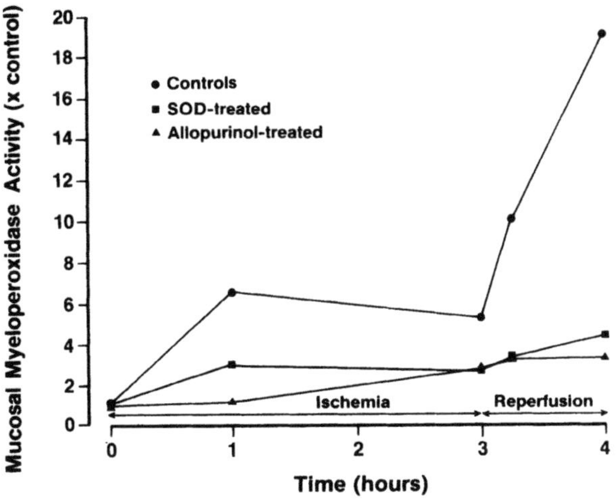

Figure 1. The effects of ischemia/reperfusion on neutrophil infiltration in the cat intestine[10].

substances in extracellular fluid to produce a potent chemoattractant. This $O_2^{\cdot-}$-dependent chemoattractant consists of an as yet undefined lipid component that is bound to serum albumin. According to the mechanism proposed by Grisham et al the $O_2^{\cdot-}$-dependent chemoattractant described by Petrone et al may be responsible for the neutrophil infiltration associated with ischemia/reperfusion in the intestinal mucosa. We[13] have recently examined the chemotactic potential of xanthine oxidase-treated feline plasma using in vitro and in vivo methods. Chemotactic performance of purified feline PMNs was assessed in vitro by the Boyden method using the procedure of Petrone et al. Autologous feline plasma (20%) was incubated for 30 minutes at 37°C with 500 μM hypoxanthine and 20 mU/ml xanthine oxidase. The reaction was terminated by the addition of 100 μg/mL each of superoxide dismutase and catalase. The chemotactic potential of $O_2^{\cdot-}$-treated plasma was compared to results obtained using either untreated plasma, hypoxanthine-xanthine oxidase in buffer, or LTB_4. An increased PMN migration in response to $O_2^{\cdot-}$-treated plasma was considered as evidence for the production of a $O_2^{\cdot-}$-dependent chemoattractant in the cat.

The chemotactic potential of xanthine oxidase-treated plasma was tested in vivo by measuring MPO activity of mucosal samples obtained before and for 3-30 minute intervals during local intra-arterial infusion of hypoxanthine-xanthine oxidase-treated plasma. A high interstitium to blood gradient of a protein-bound chemoattractant would initiate the movement of PMNs out of the marginated pool and into the blood stream. Therefore, mucosal MPO activity should fall with local intra-arterial infusion of the O_2^--dependent chemoattractant.

The results obtained from both in vitro and in vivo studies indicate that the chemotactic potential of feline extracellular fluid (plasma) is not enhanced by exposure to hypoxanthine-xanthine oxidase. Therefore, these studies do not support the hypothesis that O_2^--dependent chemoattractants play an important role in ischemia/reperfusion-induced neutrophil infiltration in the feline small intestine. Although we are unable to obtain evidence implicating O_2^--dependent chemoattractants in ischemia-reperfusion injury, there are several naturally occurring substances which could mediate the neutrophil infiltration associated with reperfusion of ischemic tissue. These include bacterial products (N-formylated peptides, endotoxin), leukotriene B_4, and activated complement components (C5a). Feline neutrophils are chemotactic to leukotriene B_4 and activated complement components. However, recent reports[14] demonstrate that feline neutrophils do not respond chemotactically to N-formylated peptides. The identity of the chemoattractants that are responsible for the ischemia/reperfusion-induced PMN infiltration remains unknown, and thus further work on this area is warranted.

PMN INFILTRATION: A CAUSE OR EFFECT OF ISCHEMIA/REPERFUSION INJURY?

An important question that arises from the data relating neutrophil infiltration to xanthine oxidase-derived reactive oxygen metabolites is whether neutrophils are a cause or an effect of ischemia/reperfusion injury. Hernandez et al[15] have recently addressed this question. Two approaches were used to assess the importance of PMNs in mediating the microvascular injury associated with ischemia/reperfusion in the small intestine, i.e., neutrophil depletion and prevention of neutrophil adherence with a monoclonal antibody (Mo 60.3) directed against a specific membrane-associated glycoprotein that modulates neutrophil adherence to endothelium. Their results demonstrated that PMN depletion afforded significant protection against ischemia/reperfusion-induced microvascular injury and prevention of PMN adherence was equally effective in attenuating ischemia/reperfusion-induced microvascular injury. The observation that PMN depletion and prevention of neutrophil adherence are equally effective in attenuating the microvascular injury suggests that circulating PMNs must adhere to microvascular endothelium in order to mediate ischemia/reperfusion injury. Although other investigators[6-8] have attributed the protective effect of PMN depletion to an increased perfusion during the ischemic phase, Hernandez et al eliminated this possibility by precise control of total intestinal blood flow during the ischemic period in all experimental groups.

Although the results of Hernandez et al provided strong support for a role of neutrophils in ischemia/reperfusion-induced microvascular injury, the chemical mediators of this injury process remains undefined. It is tempting to attribute the injury process entirely to reactive oxygen metabolites since oxy radical scavengers also protect against ischemia/reperfusion injury. However, this observation alone does not constitute strong support for oxy radicals as final mediators of injury since superoxide dismutase also prevents neutrophil infiltration, indicating that oxy radicals may function primarily to recruit neutrophils into postischemic tissue. In addition to reactive oxygen metabolites, activated PMNs release a variety of non-oxidative toxins that are capable of damaging the microvasculature. These include lactoferrin, elastase, collagenase, cationic proteins, and arachidonate metabolites. There is also evidence indicating that oxy radical production is required in order for some of these non-oxidative toxins to produce tissue injury (e.g., collagenase activation requires hypochlorous acid)[16,17]. The relative contribution of PMN-derived oxy radicals and non-oxidative toxins to ischemia/reperfusion microvascular injury remains to be defined.

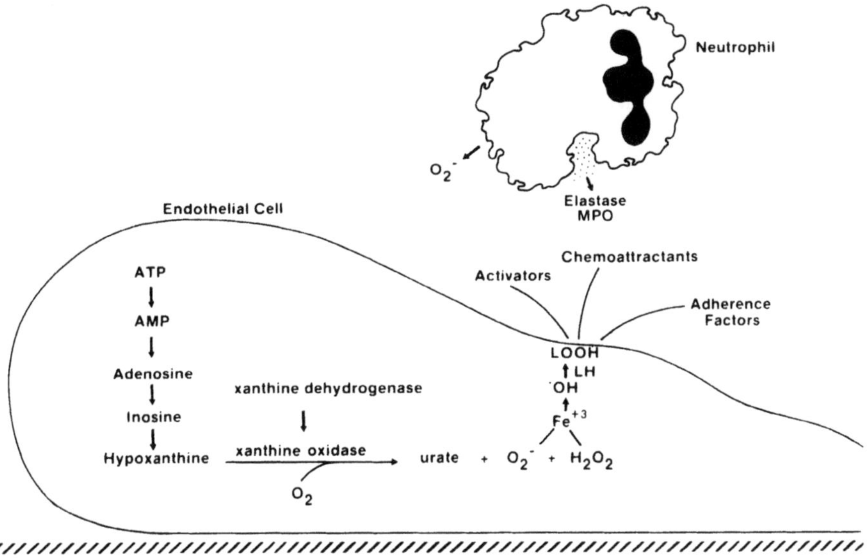

Figure 2. Proposed interaction between xanthine oxidase-derived oxidants, PMN infiltration, and microvascular injury following reperfusion of the ischemic intestine.

SUMMARY

Based on work from our laboratory and studies by others, we propose the following hypothesis to explain the interaction among xanthine oxidase, PMNs, and tissue injury in the postischemic small intestine (Figure 2). During the ischemic period, ATP is catabolized to yield hypoxanthine. The hypoxic stress also triggers the conversion of NAD-reducing xanthine dehydrogenase to the oxygen radical-producing xanthine oxidase via a protease. When the intestine is reperfused, molecular oxygen is reintroduced into the tissue where it reacts with hypoxanthine and xanthine oxidase to produce a burst of superoxide anion and hydrogen peroxide. In the presence of ferric iron, superoxide anion and hydrogen peroxide react via the Haber-Weiss reaction to form hydroxyl radicals. This highly reactive and cytoxic free radical then initiates lipid peroxidation of cell membrane components and the subsequent release of substances that activate, attract, and promote the adherence of PMN to microvascular endothelium. The adherent PMN then causes further endothelial cell injury via the release of superoxide and various proteases.

REFERENCES

1. D. A. Parks and D. N. Granger, Contributions of ischemia and reperfusion to mucosal lesion formation, Am. J. Physiol. (Gastrointest. Liver Physiol. 13) 250:G749 (1986).
2. D. N. Granger, M. E. Hollwarth, and D. A. Parks, Ischemia-reperfusion injury: Role of oxygen-derived free radicals, Acta Physiol. Scan. Suppl. 548:47 (1986).
3. D. N. Granger, L. A. Hernandez, and M. B. Grisham, Reactive oxygen metabolites: Mediators of cell injury in the digestive system, Viewpoints on Digestive Disease, 18(4):13 (1986).
4. J. L. Romson, B. G. Hood, S. L. Kunkel, G. D. Abrams, M. A. Schork, and B. R. Lucchesi, Reduction of the extent of ischemic myocardial injury by neutrophil depletion in the dog, Circulation 67(5):1016 (1983).
5. K. M. Mullane, N. Read, J. A. Salmon, and S. Moncada, Role of leukocytes in acute myocardial infarction in anesthetized dogs: Relationship to myocardial salvage by anti-inflammatory drugs, J. Pharm. Exp. Ther. 228(2):510 (1984).
6. R. L. Engler, M. D. Dahlgren, M. A. Peterson, A. Dobbs, and G. W. Schmid-Schonbein, Accumulation of polymorphonuclear leukocytes during 3-Hr experimental myocardial ischemia, Am. J. Physiol. 251:H93 (1987).
7. G. W. Schmid-Schonbein and R. L. Engler, Granulocytes as active participants in acute myocardial ischemia and infarction, Cardiovasc. Path. 1:15 (1987).
8. S. M. Smith, L. Holm-Rutili, M. A. Perry, M. B. Grisham, K. E. Arfors, D. N. Granger, and P. R. Kvietys, Role of neutrophils in hemorrhagic shock-induced gastric mucosal injury in the rat, Gastroenterology, 93(2):466 (1987).
9. M. G. Sarr, G. B. Bulkley, and J. L. Cameron, The role of leukocytes in the production of oxygen-derived free radicals in acute experimental pancreatitis, Surgery 101:292 (1987).
10. M. B. Grisham, L. A. Hernandez, and D. N. Granger, Xanthine oxidase and neutrophil infiltration in intestinal ischemia, Am. J. Physiol. (Gastrointest. Liver Physiol. 14) 251:G567 (1986).
11. W. F. Petrone, D. K. English, K. Wong, and J.M. McCord, Free radicals and inflammation: Superoxide-dependent activation of a neutrophil chemotactic factor in plasma, Proc. Natl. Acad. Sci. 77(2):1159 (1980).
12. H. D. Perez, B. B. Weksler, and I. M. Goldstein, Generation of a chemotactic lipid from arachidonic acid by exposure to a superoxide-generating system, Inflammation 4(3):313 (1980).
13. B. J. Zimmerman, M. B. Grisham, and D. N. Granger, Role of superoxide-dependent chemoattractants in ischemia-reperfusion induced neutrophil infiltration, Fed. Proc. 46(4):1124 (1987).
14. G. D. Gray, G. M. Ohlmann, D. R. Morton, and R. G. Schaub, Feline polymorphonuclear leukocytes respond chemotactically to leukotriene B_4 and activated serum but not to F-Met-Leu-Phe, Agents and Actions 18, 3/4:401 (1986).
15. L. A. Hernandez, M. B. Grisham, B. Twohig, K. E. Arfors, J. M. Harlan, and D. N. Granger, Role of neutrophils in ischemia/reperfusion-induced microvascular injury, Am. J. Physiol. (Heart Circ. Physiol. 22) 253:H699 (1987).
16. G. J. Peppin and S. J. Weiss, Activation of the endogenous metalloproteinase, gelatinase, by triggered human neutrophils, Proc. Natl. Acad. Sci. USA 83:4322 (1986).
17. S. J. Weiss, G. Peppin, X. Ortiz, C. Ragsdale, and S. T. Test, Oxidative autoactivation of latent collagenase by human neutrophils, Science 227:747 (1985).

THE AMPLIFICATIVE ROLE OF PAF-ACETHER IN THE OXIDATIVE STRESS FOLLOWING REPERFUSION OF ISCHEMIC STOMACH

Marie-Therese Droy-Lefaix, Yvette Drouet, Gerard Geraud, and Pierre Braquet

Institut Henri Beaufour
17 avenue Descartes
92350 Le Plessis Robinson, France

INTRODUCTION

It is now well demonstrated that neutrophil infiltration during ischemia-reperfusion may be responsible for tissue injury.[1] In conjunction with an oxidative burst, neutrophils can generate oxygen free radicals such as superoxide anion or hydrogen peroxide. These oxygen metabolites can produce the hydroxyl radical which is particularly toxic to cells.[2,3]

Platelet-activating factor (PAF-acether) is an endogenous mediator synthesized from polymorphonuclear leucocytes.[4] It is known to be involved in gastrointestinal mucosal ulceration.[5,6] Recent experiments show that PAF-acether can serve to amplify neutrophil oxidative response at sites of inflammation.[7] We have investigated the potential significances of PAF-acether upon ischemia-reperfusion, using a PAF-acether antagonist, BN 52021 (ginkgolide B).[8] Its effects were compared with those of superoxide dismutase (SOD) and allopurinol.

METHODS

Experimental Protocol

Female Sprague-Dawley rats (190-200 g) were randomly separated into six groups of 8 animals each and anesthetized with 50 mg/kg of sodium pentobarbital. After abdominal incision, coeliac artery occlusion was performed for one hour, followed by a 24-hour reperfusion period. After 18 h fasting, rats were sacrificed. The stomachs were removed, opened along the greater curvature and washed at body temperature.

Gross mucosal necrosis was quantitated using an arbitrary scale between 0 and 5. Tissue samples were fixed with Bouin liquid for mucosal histology and with Trump liquid for scanning electron microscopy. 10 µl of adherent mucous gel was collected by scraping the gastric mucosa. Spinability, or thread-forming properties of mucus, a rheological parameter reflecting the glycoprotein polymerisation,[9] was measured automatically on a Sefam apparatus. Statistical analysis was evaluated by Student's t test.

Drug Administration

Superoxide dismutase was intravenously injected at a dose of 15,000 U/kg 45 minutes before arterial occlusion. Allopurinol was delivered by gavage at a dose of 50 mg/kg for two days before ischemia induction, and BN 52021 was administered by gavage at doses of 2.5 and 5 mg/kg, 45 minutes before clamping. Controls and ischemia rats received no pretreatment.

RESULTS

Macroscopic Mucosal Changes

Ischemia-reperfusion produced severe and extensive hemorrhagic necrotic lesions. Lesion score showed a significant increase from 0.25 $\pm$ 0.25 in the control group to 3.6 $\pm$ 0.44 in the ischemia group ($p < 0.001$). Drug pretreatments significantly reduced the gastric mucosa damage ($p < 0.001$) according to the following lesion score : SOD (0.68 $\pm$ 0.25), allopurinol (1.31 $\pm$ 0.34), BN 52021 2.5 mg/kg (1.06 $\pm$ 0.39), 5 mg/kg (0.81 $\pm$ 0.42).

Mucus Spinability

With ischemia-reperfusion, the spinability significantly decreased from 7.93 $\pm$ 0.30 mm in the control group to 3.62 $\pm$ 0.44 mm ($p < 0.001$). Pretreatment with drugs significantly reduced this drop in spinability ($p < 0.001$). The spinability values are SOD (7.52 $\pm$ 0.41 mm) ; allopurinol (10.36 $\pm$ 1.59 mm) ; BN 52021 2.5 mg/kg (7.06 $\pm$ 0.61 mm) and 5 mg/kg (7.92 $\pm$ 0.40 mm) (Figure 1).

Mucosal Histology

Major pathologic changes were caused by ischemia-reperfusion of gastric mucosa. Large deep and hemorrhagic necrotic lesions with significant edema were seen with substantial infiltration of neutrophils (Figure 2). Cell exfoliation into the lumen was observed. Pretreatment with SOD, allopurinol and BN 52021 protected gastric mucosa from ischemia-reperfusion prominent changes.

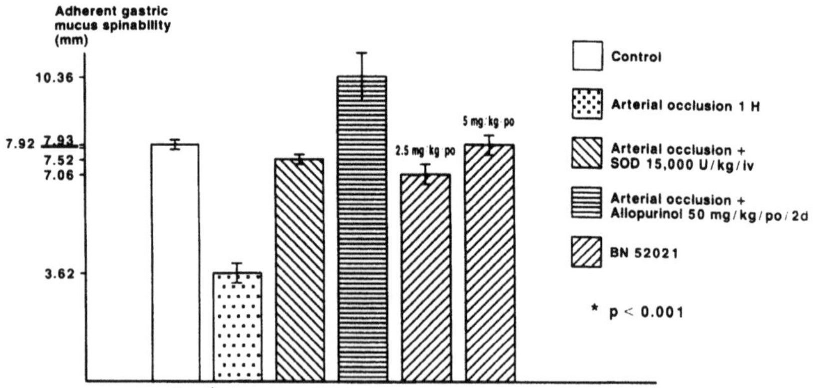

Figure 1. Effect of PAF-acether antagonist (BN 52021), superoxide dismutase (SOD) and allopurinol on adherent mucus spinability drop induced by 1 h arterial occlusion and 24 h reperfusion.

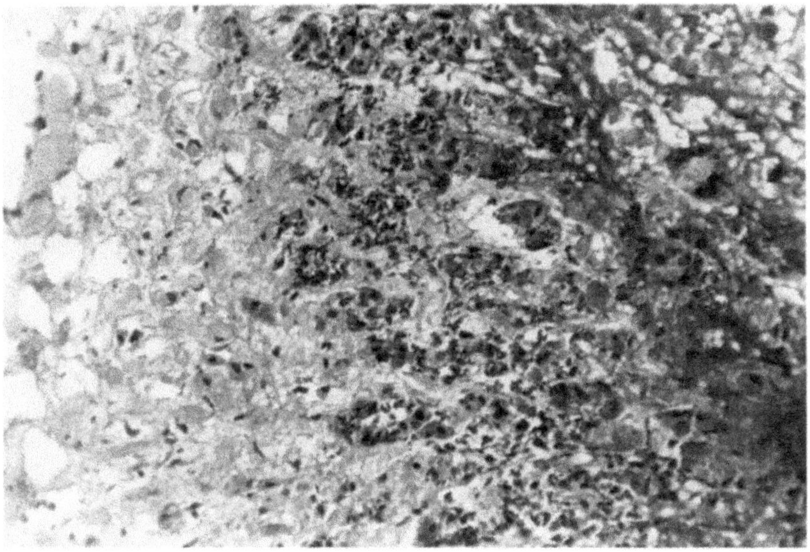

Figure 2. Histology of rat gastric mucosa after ischemia-reperfusion showing severe necrotic lesions and significant neutrophil infiltration (X 250).

Scanning Electron Microscopy

Severe ultrastructural injury was observed after ischemia-reperfusion. The loss of the epithelial cells was marked with extrusion of gastric gland cells. In large areas the lamina propria was totally denuded and necrotic. Pretreatment with SOD, allopurinol and BN 52021 reduced this cellular abrasion considerably.

DISCUSSION

This study directly implicates oxygen free radicals and platelet-activating factor in ischemia-reperfusion gastric lesions. All the parameters studied demonstrated severe gastric mucosal alterations with a drop in adherent mucous gel spinability, deep necrosis and marked influx of neutrophils.

Pretreatment with allopurinol, a xanthine oxidase inhibitor, superoxide dismutase, a superoxide scavenger and BN 52021, a platelet-activating factor antagonist, significantly reduced morphological changes induced by ischemia-reperfusion.

It is now well established that oxygen radical production by xanthine oxidase[10] and probably by neutrophils[7] induces tissue damaging effects.[11-14] Platelet-activating factor is known to be an inflammatory and ulcerogenic mediator[15] and enhances the rate of superoxide anion production.[7]

Since with ischemia-reperfusion there is evidence of neutrophil infiltration, it would appear rational to presume that the neutrophil oxygen free radical system contributes to mucosal ulceration.[16]

BN 52021, which is not a free radical scavenger, inhibits the enhancement of superoxide radical production by human polymorphonuclear leukocytes induced by addition of very low concentrations of PAF-acether to C5a or FMLP. These data suggest that BN 52021 alone or in association with free radical scavengers may inhibit ischemia-reperfusion damage.[17]

REFERENCES

1. M. B. Grisham, L. A. Hernandez, and D. N. Granger, Xanthine oxidase and neutrophil infiltration in intestinal ischemia, Am. J. Physiol. 251:G567 (1986).
2. J. C. Fantone and P. A. Ward, Role of oxygen-derived free radicals and metabolites in leukocyte - dependent inflammatory reactions, Am. J. Pathol. 107:397 (1982).
3. R. T. Briggs, J. M. Robinson, M. L. Karnovsky, and M. J. Karnovsky, Superoxide production by polymorphonuclear leukocytes, Histochemistry 84:371 (1986).
4. M. Sanchez-Crespo, F. Alonso, G. M. Garcia, J. Gomez-Cambronero, and M.L. Nieto, Synthesis of plateletactivating factor from human polymorphonuclear leukocytes:Regulation and pharmacological approaches, Int. J. Tiss. Reac. VII:345 (1985).
5. P. Braquet, A. Etienne, J. M. Mencia-Huerta, and F. Clostre, Possible role of Platelet-activating factor in gastrointestinal ulcerations as evidenced by the use of the specific antagonists, BN 52021 and BN 52063, Pharmacol. Research Commun, in press.
6. J. L. Wallace and B. J. R. Whittle, Inhibition of endotoxin-induced gastrointestinal plasma leakage by PGE2 and a PAF-antagonist, Gastroenterology 92:1685 (1987).
7. J. C. Gay, J. K. Beckman, K. A. Zaboy, and J. N. Lukens, Modulation of neutrophil oxidative responses of soluble stimuli by platelet-activating factor, Blood 67:931 (1986).
8. P. Braquet, B. Spinnewyn, M. Braquet, R. H. Bourgain, J. F. Taylor, A. Etienne, and K. Drieu, BN 52021 and related compounds:A new series of hightly specific PAF-acether receptor antagonists isolated from Ginkgo biloba, Blood Vessels 6:559 (1985).
9. J. P. Arnould, J. M. Zahm, G. Pottier, C. Duvivier, and E. Puchelle, A new apparatus for measuring the thread-forming property (spinability) of biological fluids, Biorheology 1:123 (1984).
10. J. M. McCord and R. S. Roy, The pathophysiology of superoxide:Roles in inflammation and ischemia, Can. J. Physiol. 60:1346 (1982).
11. D. N. Granger, G. Rutili, and J. M. McCord, Superoxide radicals in feline intestinal ischemia, Gastroenterology 81:22 (1981).
12. D. A. Parks, A. K. Shah, and D. N. Granger, Oxygen radicals:Effects on intestinal vascular permeability, Am. J. Physiol. 247:G167 (1984).
13. M. Itoh and P. H. Guth, Role of oxygen-derived free radicals in hemorrhagic shock induced gastric lesions in the rat, Gastroenterology 88:1162 (1985).
14. D. N. Granger, M. E. Hollwarth, and D. A. Parks, Ischemia-reperfusion injury:Role of oxygen-derived free radicals, Acta Physiol. Scand. 548:47 (1986).
15. A. Rosam, J. L. Wallace, and B. J. R. Whittle, Potent ulcerogenic actions of platelet-activating factor on the stomach. Nature 319:54 (1986).
16. N. W. R. Wickham, G. M. Vercelotti, H. Q. Yin, C. F. Moldow, and H. S. Jacob, Neutrophils are printed to release toxic oxidants by contact with thrombinstimulated endothelium:Role of endothelial cell generated platelet activating factor, Fed. Proc, in press.
17. B. Spinnewyn, N. Blavet, F. Clostre, N. Bazan, and P. Braquet, Involvement of platelet-activating factor (PAF) in cerebral post-ischemic phase in mongolian gerbils, Prostaglandins, in press.

EQUIEFFECTIVE PROTECTION BY HUMAN AND BOVINE SOD AGAINST RENAL REPERFUSION DAMAGE IN RATS

Johannes Schneider, Elmar Friderichs, and Hubert Giertz

Grünenthal GmbH
Center of Research
Zieglerstr. 6
5100 Aachen, FRG

INTRODUCTION

Oxygen-derived free radicals are considered to cause postischemic tissue injury in many tissues.[1,2] In ischemic and reoxygenated kidneys of different animal species comparable protective effects have been obtained by allopurinol pretreatment or superoxide dismutase infused concomitantly with reoxygenation.[3,4] These observations indicate that the free radical damage is initiated by superoxide ions possibly generated by xanthine oxidase,[1] although the precise mechanism of superoxide formation and the cascade of pathogenic events resulting therefrom is still debated.

With regard to the therapeutic implications of these findings, an authentic human superoxide dismutase (hum-SOD) has been prepared from recombinant yeast[5] and is being developed for clinical use. We here describe the effects of hum-SOD on renal function after ischemia/reperfusion in rats in comparison with a proper set of controls, i.e., rats treated with bovine SOD (bov-SOD) or the corresponding copper-free apoenzyme (Apo-SOD) and untreated sham-operated animals.

MATERIALS AND METHODS

Male Wistar SPF rats (200-250 g) were anesthetized with pentobarbitone (50 mg/kg i.p.). After median laparatomy renal ischemia was produced by clamping both renal arteries. Five minutes before occlusion of the renal arteries the animals received heparin (5 mg/kg i.v.). In a sham-operated group of rats only a laparatomy, but no renal artery occlusion, was performed. Kidney reperfusion was started by releasing the clamps after an ischemic period of 30 or 60 min. The different SOD forms and the vehicle (saline) were infused intravenously for 15 min, starting at the moment of kidney reperfusion. After 30 min of reperfusion, the inulin and p-aminohippuric acid (PAH) clearance test was performed by a single injection method: 40 mg inulin and 5 mg PAH were injected intravenously. At 10, 12.5, 15, 17.5, and 20 min after this injection, venous blood samples were taken to determine the plasma concentrations of inulin and PAH that are excreted exclusively by the kidneys. The clearance values were calculated from the first-order decrease of the plasma inulin and PAH concentrations, concordant to the formula:

Cl = $k_2 \cdot V$ where k_2 = elimination rate (min^{-1}) and V = distribution volume (ml). The specific activity of hum-SOD and bov-SOD yielded 3,500 U/mg protein. The residual enzymatic activity of Apo-SOD, produced by serial dialysis of bov-SOD against Na_2-EDTA (20 mN), NaCl (100 mM) and H_2O, was found to be 0.1 % of the activity of bov-SOD and hum-SOD.

RESULTS

The clearance rates in sham-operated (non-ischemic) control animals were 5.1 ± 0.6 $ml \cdot min^{-1} \cdot kg^{-1}$ for inulin and 14.1 ± 1.1 $ml \cdot min^{-1} \cdot kg^{-1}$ for PAH. Renal ischemia of 30 (60) min, followed by reperfusion, reduced the inulin clearance to 3.0 ± 0.7 (2.0 ± 0.2) $ml \cdot min^{-1} \cdot kg^{-1}$ and the PAH clearance to 6.2 ± 0.8 (5.5 ± 0.5) $ml \cdot min^{-1} \cdot kg^{-1}$. The depressed inulin and PAH clearance values after the 60-min ischemic period were significantly improved by hum-SOD and bov-SOD in the dosage range between 4.64 and 21.5 $mg \cdot kg^{-1} \cdot 15\ min^{-1}$. Apo-SOD (10.0 $mg \cdot kg^{-1} \cdot 15\ min^{-1}$) did not improve the reduced clearance rates (Figures 1 and 2). In the experiments with 30 min of renal occlusion followed by reperfusion only the 10.0 $mg \cdot kg^{-1} \cdot 15\ min^{-1}$ doses of bov-SOD and hum-SOD were investigated (Figures 3 and 4). Again, both drugs clearly improved the diminished inulin and PAH clearance values.

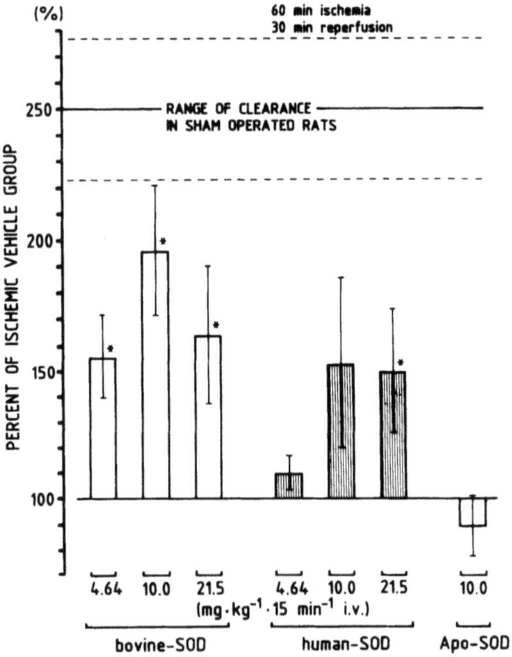

Figure 1. Effects of bov-SOD, hum-SOD, and Apo-SOD (n = 8 per group) on inulin clearance after 60 min of renal ischemia and 30 min of reperfusion. The results are given as percent of clearance value in the vehicle treated ischemic control group. The lined area signifies the range of clearance in sham-operated rats, relative to the clearance in the ischemic vehicle group. Data are presented as the mean ± SEM. Asterisks indicate statistical difference between active drug- and vehicle-treated animals: $p < 0.05$.

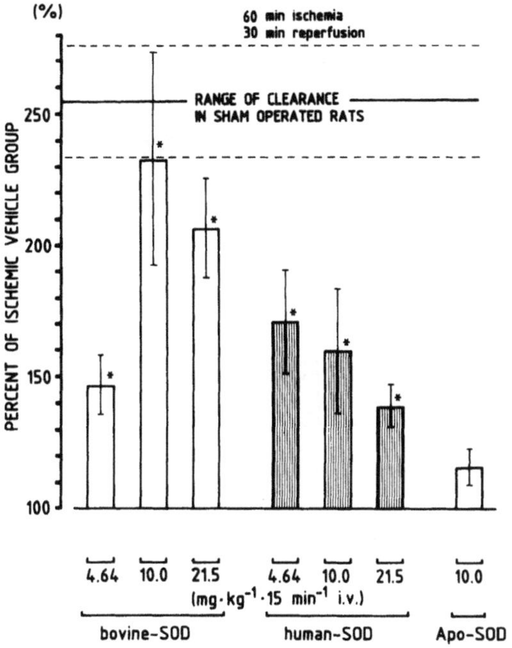

Figure 2. Effects of bov-SOD, hum-SOD, and Apo-SOD (n = 8) per group) on PAH clearance after 60 min of renal ischemia and 30 min of reperfusion. For further explanation, see Figure 1.

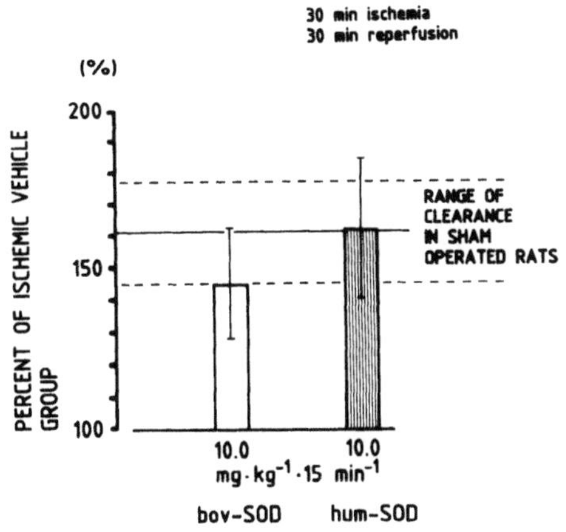

Figure 3. Effects of bov-SOD and hum-SOD (n = 8 per group) on inulin clearance after 30 min of renal ischemia and 30 min of reperfusion. For further explanation, see Figure 1.

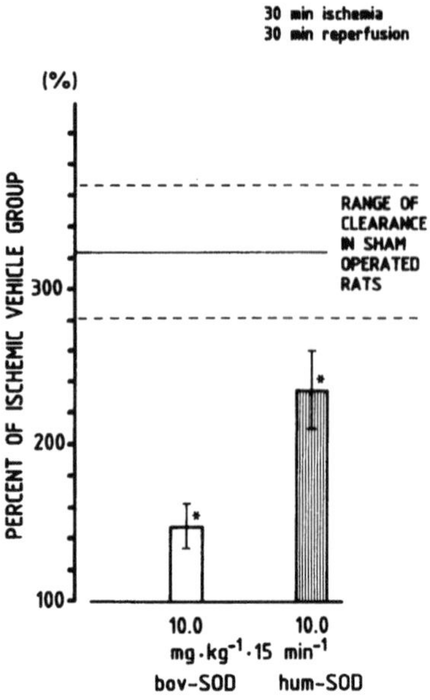

Figure 4. Effects of bov-SOD and hum-SOD (n = 8 per group) on PAH clearance after 30 min of renal ischemia and 30 min of reperfusion. For further explanation, see Figure 1.

SUMMARY

Both hum-SOD and bov-SOD improved the impaired renal function during kidney reperfusion. In the experiments with a 60-min occlusion period the effects of bov-SOD tended to be more pronounced than the effects of hum-SOD, whereas after a 30-min occlusion period the improvement of renal function by hum-SOD was more striking. In essence, therefore, hum-SOD proved to be equipotent to bov-SOD. The apoenzyme (Apo-SOD), lacking the active Cu moiety, virtually did not improve the renal function impairment. Thus, the protective effects of hum-SOD and bov-SOD must be ascribed to the enzymatic function of these proteins.

REFERENCES

1. J. M. McCord, Oxygen-derived free radicals in postischemic tissue injury, N. Engl. J. Med. 312: 159 (1985).
2. R. A. Weisiger, Oxygen radicals and ischemic tissue injury, Gastroenterology 90: 494 (1986)
3. M. S. Paller, J. R. Hoidal, and Th. F. Ferris, Oxygen free radicals in ischemic acute renal failure in the rat, J. Clin. Invest. 74: 1156 (1984).
4. K. Ouriel, N.G. Smedira, and J. J. Ricotta, Protection of the kidney after temporary ischemia: Free radical scavengers, J. Vasc. Surg. 2: 49 (1985).
5. R. A. Hallewell, F. R. Masiarz, R. C. Najarian, J. P. Puma, M. R. Quiroga, A. Randolph, R. Sanchez-Pescador, C. J. Scandella, B. Smith, K. S. Steimer, and G. T. Müllenbach, Human Cu/Zn superoxide dismutase CDNA: Isolation of clones synthetizing high levels of active or inactive enzyme from an expression library, Nucleic Acids Res. 13: 2017 (1985).

POTENTIAL OF METHYLENE BLUE TO BLOCK OXYGEN RADICAL GENERATION IN REPERFUSION INJURY

Michael J. Kelner, Richard Bagnell, Braden Hale, and Nicholas M. Alexander

Division of Laboratory Medicine, Department of Pathology
University of California, San Diego Medical Center
225 Dickinson Street
San Diego, California 92103 U.S.A.

INTRODUCTION

Superoxide and hydroxyl radical generation have been implicated in a variety of pathological conditions, including ischemic injury in myocardial, renal, and skin-flap tissue, as reperfusion of the ischemic tissue presumably leads to a burst of free radical generation [1-4,14]. Tissue hypoxia during ischemia results in the conversion of xanthine dehydrogenase to xanthine oxidase and concomitant breakdown of ATP to hypoxanthine [14]. Upon reperfusion the xanthine oxidase oxidizes the hypoxanthine, and large quantities of superoxide, hydrogen peroxide, and possibly hydroxyl radicals are produced. Previous attempts to inhibit oxygen radical toxicity have focused on removing the superoxide or hydroxyl radicals after their production.

A more efficacious approach may be to prevent oxygen radical formation. Allopurinol, a specific inhibitor of xanthine oxidase, effectively blocks the production of superoxide by inhibiting xanthine oxidase. It does not, however, allow for removal of the purine oxidation products that accumulate during ischemia. An ideal drug would prevent the production of superoxide radicals while allowing conversion of hypoxanthine to uric acid. The drug should be water soluble, be able to permeate cell membranes, and be approved for human drug use.

While reviewing literature on xanthine oxidase, we noted that methylene blue "converted" the oxidase enzyme back to the dehydrogenase form in a way similar to that of dithiothreitol [15]. It is unlikely that methylene blue acts by reducing disulfide bonds, as dithiothreitol does, because methylene blue actually oxidized sulfhydryl groups in a redox reaction [6]. Futhermore, this conversion occurred with proteolytically generated xanthine oxidase, whereas dithiothreitol was ineffective. This suggested that methylene blue was interacting at a common point in the electron pathways of the two forms of the enzyme, producing the impression that the oxidase was converted back to the dehydrogenase form. If this were true, then the superoxide radical production by xanthine oxidase could be eliminated in the presence of methylene blue. Indeed, McCord and Fridovich noted that superoxide radical production was decreased in the presence of methylene blue [7]. However, at that time, the role of oxygen radicals in reperfusion injury was not appreciated and the finding had no apparent clinical significance. With the current knowledge of oxygen radicals and reperfusion injury, we decided to characterize the ability of methylene blue to eliminate superoxide and hydroxyl radical production by xanthine oxidase and to determine the classical type of inhibition, the potency, and the mechanism of the inhibition.

RESULTS

The initial experiments of McCord and Fridovich [7] were repeated quantitatively. Methylene blue was found to be a potent inhibitor of superoxide production because greater than 50% inhibition of superoxide production could be achieved with molar ratios of oxygen to methylene blue of 170:1, indicating a marked preference by the enzyme for the dye.

The desulfo and deflavinated forms of xanthine oxidase were prepared from fresh preparations of rat hepatic tissue and milk [7-10]. Addition of methylene blue to the desulfo form (inactivation of the molybdenum disulfide site) failed to reactivate the enzyme (Table 1). Others have demonstrated that molybdo free enzyme still interacts with methylene blue [16]. Together these results indicate that methylene blue interacts with the enzyme at the FAD site (where molecular oxygen interacts) or at the iron-sulfide site. Deflavo xanthine oxidase possessed no oxidase activity during aerobic incubation with xanthine, as measured by uric acid production at 292 nM. The deflavo preparation retained the ability to directly reduce cytochrome C at the iron-sulfide site, as previously reported [11]. When FAD was added to the deflavo enzyme preparation, the oxidase function was slowly regained (5 minutes). The addition of methylene blue very quickly (15 seconds) restored the ability of deflavinated enzyme to oxidize xanthine and to produce uric acid and also accelerated the reduction of cytochrome C. The addition of MB and FAD together was no different than when only MB was added to the deflavinated enzyme. In other experiments, when methemoglobin was substituted for cytochrome C, no reduction of the hemeprotein by superoxide radicals was observed (630 nM). However, the addition of MB caused a slow reduction of methemoglobin to oxyhemoglobin (data not shown).

Because the autoxidation of methylene blue can generate hydrogen peroxide in the absence of an appropriate electron acceptor, there was the possibility that the dye could actually contribute to tissue damage by increasing hydroxyl radical production. However, the production of hydroxyl radicals, as measured by conversion of methional to ethylene or dimethylsulfoxide to formaldehyde, actually decreased to a small extent in the presence of methylene blue (Table 2). When an appropriate electron acceptor was available, such as cytochrome C or hemoglobin, an even larger decrease occurred.

Table 1 Reactivation of Deflavo Xanthine Oxidase by Methylene Blue

Additions	Uric acid Production[a] (nanomoles/min)	Cytochrome C reduction (nanomoles/min)
None	< 0.4	9.0
FAD (5 minutes)	2.8	2.3
MB, 2.8 uM (15 seconds)	9.5	9.7
FAD + MB	8.8	9.2

[a] Uric acid production assayed in the absence of cytochrome C.

Table 2 Methylene Blue Inhibition of Hydroxyl Radical Formation

Dimethylsulfoxide/xanthine oxidase assay

Additions	Formaldehyde (nmoles)
Methylene blue + cytochrome C (no xanthine oxidase)	< 1
Cytochrome C	264
Methylene blue, 3 uM	217
Methylene blue + cytochrome C	63

Methional/xanthine oxidase assay

Additions	Ethylene (nmoles)
Enzyme only	9.5
Cytochrome C	6.2
Cytochrome C + methylene blue (1 uM)	4.0
Methemoglobin	8.9
Methemoglobin + methylene blue (1 uM)	6.3

SUMMARY

Methylene blue interacts with xanthine oxidase at the iron-sulfide site in the electron pathway (Scheme I) that is known to serve as an electron-sink connecting the reductive and oxidative sites in both the oxidase and dehydrogenase forms [12]. Thus, shunting of electrons to methylene blue at this site effectively diverts their flow away from the FAD site where molecular oxygen is converted to superoxide radicals [12]. Since the electron affinity constants of xanthine oxidase for electron acceptors are FAD > iron/sulfide > molybdenum [13], methylene blue falls between the FAD and iron-sulfide site. Thus, methylene blue effectively inhibits superoxide and hydroxyl radical production while accelerating the conversion of xanthine to uric acid. As methylene blue is already approved for medicinal use in humans and is relatively nontoxic, the drug may have a role in reducing tissue injury associated with reperfusion. We are currently investigating this possibility in animal models.

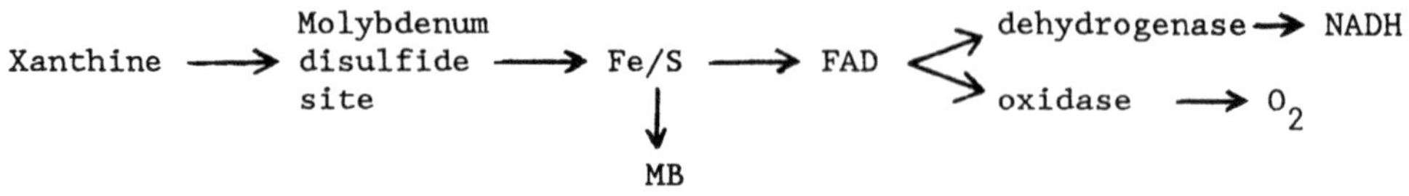

Scheme I

ACKNOWLEDGMENTS

Michael J. Kelner was supported by a Clinical Investigator Award from the National Institute of Environmental Health Sciences, ES 00142-02 and by an award from the University of California Academic Senate.

REFERENCES

1. D.A. Parks, G.B. Bulkley, D.N. Granger, S.R. Hamilton, and J.M. McCord, Ischemic injury in the cat small intestine: Role of superoxide radicals, Gastroenterology. 82:9 (1982).
2. R.A. DeWall, K.A. Vasko, E.L. Stanley, and P. Kezdi, Responses of ischemic myocardium to allopurinol, Am. Heart J. 82:362 (1971).
3. M.J. Im, W. Shen, C.J. Pak, P.N. Manson, G.B. Bulkley, and J.E. Hoopes, Effect of allopurinol on the survival of hyperemic island skin flaps, Plast. Reconstr. Surg. 73:276 (1983).
4. P.N. Manson, R.M. Anthenell, M.J. Im, G.B. Bulkely, and J.E. Hoopes, The role of oxygen-free radicals in ischemic tissue injury, Ann. Surg. 198:87 (1983).
5. I. Fridovich, Quantitative aspects of the production of superoxide radicals by milk xanthine oxidase, J. Biol. Chem. 245:4053 (1972).
6. M.J. Kelner and N.M. Alexander, Methylene blue directly oxidizes glutathione without the intermediate formation of hydrogen peroxide, J. Biol. Chem. 261:1636 (1986).
7. J.M. McCord and I. Fridovich, The utility of superoxide dismutase in studying free radical reactions, J. Biol. Chem. 245:1374 (1970).
8. K.V. Rajogopalan and P. Handler, Purification and properties of chicken liver xanthine dehydrogenase, J. Biol. Chem. 242:4097 (1967).
9. V. Massey and D. Edmundson, On the mechanism of inactivation of xanthine oxidase by cyanide, J. Biol. Chem. 245:6595 (1970).
10. H. Komai, V. Massey and G. Palmer, The preparation and properties of deflavo xanthine oxidase, J. Biol. Chem. 244:1692 (1969).
11. M. Kanda, O. Brady, R.V. Rajagopalan and P. Handler, Studies on the dissociation of flavin adenine dinucleotide from metalloflavoproteins, J. Biol. Chem. 247:765 (1971).
12. H. Komai and V. Massey, in: "Flavins and Flavoproteins", H. Kamin, ed., pp 399-415, University Park Press, Baltimore (1971).
13. JS Olson, DP Bellow, G Palmer, and V Massey, The mechanism of action of xanthine oxidase, J. Biol. Chem. 249:4363 (1974).
14. J.M. McCord, Oxygen-derived free radicals in post ischemic tissue injury, New Eng. J. Med. 312:159 (1985).
15. F. Stirpe and E. Della Corte, The regulation of rat liver xanthine oxidase, J. Biol. Chem. 244:3855 (1969).
16. S. Gardlik, M.J. Barber, and K.V. Rajagopalan, A molybdopterin-free form of xanthine oxidase, Arch. Biochem. Biophys. 259: 363 (1987).

SUPEROXIDE RADICAL RELEASE INTO THE CYTOPLASM OF HEART CELLS BY AN NADH-DRIVEN OXYGEN ACTIVATOR

Hans Nohl

Institute of Pharmacology and Toxicology
Veterinary University of Vienna
A-1030 Vienna, Linke Bahngasse 11, Austria

INTRODUCTION

It is now unequivocally accepted, mainly on the basis of indirect evidence[1-4] that heart mitochondria (RHM) release small amounts of superoxide radicals as a physiological byproduct of respiration. The existence of this process indicates not only that the oxygen radical generator is able to establish an adequate redox couple with molecular oxygen but also its interaction with the physiological electron acceptor exhibits kinetic restraints, causing electron deviation from the normal pathway to oxygen at a nonphysiological site of the respiratory chain. In isolated, normally functioning mitochondria, the latter condition was found to be closely related to energy coupling since superoxide radical formation was only observed under conditions in which, due to a lack of ADP, mitochondria were hindered from ATP-synthesis (= state 4 conditions of respiration). Although such conditions are not likely to exist in highly active energy-consuming organs, there is convincing evidence that O_2^- formation occurs in the living heart[2,3].

An answer to these obviously conflicting observations may be provided by our recent findings regarding the existence of an as yet unknown NADH dehydrogenase, an enzyme associated with the inner membrane of heart mitochondria[5]. This enzyme was shown to directly introduce reducing equivalents from the cytosol into the respiratory chain. Thus, inhibition of NADH consumption was possible with classical respiratory inhibitors. Apart from "normally respiring" mitochondria, inhibitor-induced barriers of NADH-dependent respiration could be partially overcome by nonphysiological cytosolic electron acceptors[5,6]. Electron transfer along this cytosolic shunt was sensitive to SOD, indicating the involvement of superoxide radicals. More detailed investigations on the oxygen activation capacity of the "exogenous NADH dehydrogenase" revealed that this organ-specific enzyme is likely to be the most potent superoxide generator in heart tissue, also operating under <u>in vivo</u> conditions.

RESULTS

We recently have shown the involvement of the exogenous NADH-dehydrogenase in the reduction of cytochrome (cyt) b from external NADH[5]. Since

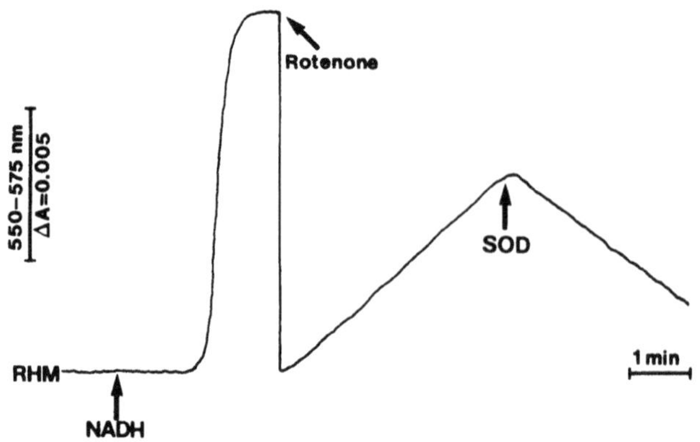

Figure 1. Demonstration of cyt c reduction of rat heart mitochondria (RHM) by external NADH. 0.5 mg of mitochondrial protein was suspended in 1 ml of the reaction medium (300 mM sucrose, 2 mM EGTA, 20 mM triethanolamine-HCl pH 7.4) at room temperature. Final concentrations of NADH, 250 μM, rotenone, 20 μM, SOD, 2 μM. Redox difference spectrum was followed at 550 against 575 nm using a Shimadzu UV 3000 spectrophotometer, dual wavelength mode.

the novel enzyme was found to introduce reducing equivalents into the respiratory chain from cytosolic NADH at the level of complex I, inhibiton of cyt b reduction observed in the presence of rotenone was to be expected[5]. Figure 1 shows that also cyt c, which is on the oxidising side of cyt b, can be reduced by external NADH after a relatively long lag phase. However, in contrast to cyt b, cyt c became slowly reduced again after rotenone induced reoxidation. SOD was found to prevent cyt c reduction in the presence of rotenone, indicating the involvement of $O_2^{\cdot-}$ radicals.

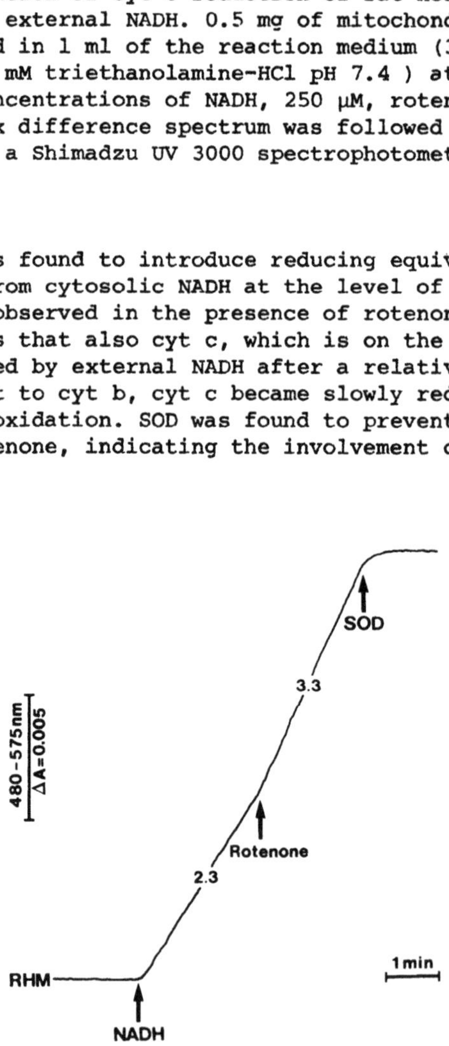

Figure 2. Initiation of $O_2^{\cdot-}$ generation in RHM following the addition of NADH. 1 mg of RHM was suspended in the reaction medium of Figure 1 at 25° C; $O_2^{\cdot-}$ generation was followed by measuring (continued)

spectrophotometrically SOD-sensitive adrenochrome formation from epinephrine (1 mM final conc) at 480-575 nm. The values associated with the curve indicate nmol $O_2^{\cdot -}$ formed x min^{-1} x mg^{-1}. The curve represent typical results of five experiments. Final conc of rotenone, 40 µM.

Figure 2 shows that RHM are, in fact, an active $O_2^{\cdot -}$ source in the presence of exogenous NADH and, in addition, that rotenone increases the formation rate of this oxygen species. Since the "classical $O_2^{\cdot -}$ generator" in RHM was found to depend on a high reduction state of ubiquinone[3] and cyt b_{566}[7], the stimulatory effect of rotenone clearly excludes the involvement of this source in $O_2^{\cdot -}$ radical release observed. Furthermore, in contrast to "classical $O_2^{\cdot -}$ sources", $O_2^{\cdot -}$ radicals from the NADH-related generator could be detected in the reaction medium despite the presence of SOD in the mitochondrial matrix. Because of spontaneous dismutation of $O_2^{\cdot -}$ in the medium, small amounts of H_2O_2 could also be observed (Figure 3). However, addition of external SOD initiated catalytically H_2O_2 formation from NADH-induced $O_2^{\cdot -}$ release at a stoichiometric ratio of 1/2. This proves that $O_2^{\cdot -}$ radicals from the novel source were almost totally released into the extramitochondrial space.

Bioenergetic properties of the NADH-dependent $O_2^{\cdot -}$ generator with respect to oxygen concentration were inferred from the Lineweaver-Burk plot and compared with the respective plot obtained for $O_2^{\cdot -}$ release of succinate-respiring submitochondrial particles (to remove matrix bound SOD) in the presence of antimycin ("classical conditions for $O_2^{\cdot -}$ formation) (Figure 4). The greater V_{max} value and the higher affinity to oxygen suggest that the novel $O_2^{\cdot -}$ generator is more active under physiological conditions, e.g., based on physiological oxygen concentrations in the heart cell, the NADH-driven $O_2^{\cdot -}$ generator is calculated to be 25% more active than the classical $O_2^{\cdot -}$ source.

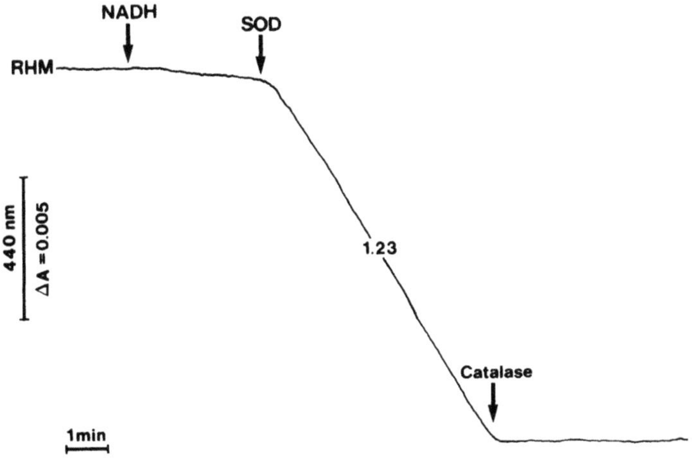

Figure 3. SOD-induced H_2O_2 formation in heart mitochondria (2 mg/ml) after the addition of NADH (20 µM final conc). H_2O_2 formation rate (nmol x min^{-1} x mg^{-1}), indicated by the inserted number, was inferred from changes of o-dianisidine absorption at 440 nm[8]. The trace represents a typical curve similar to those obtained in four other experiments. Catalase (5 µM final conc) was added to demonstrate the existence of H_2O_2.

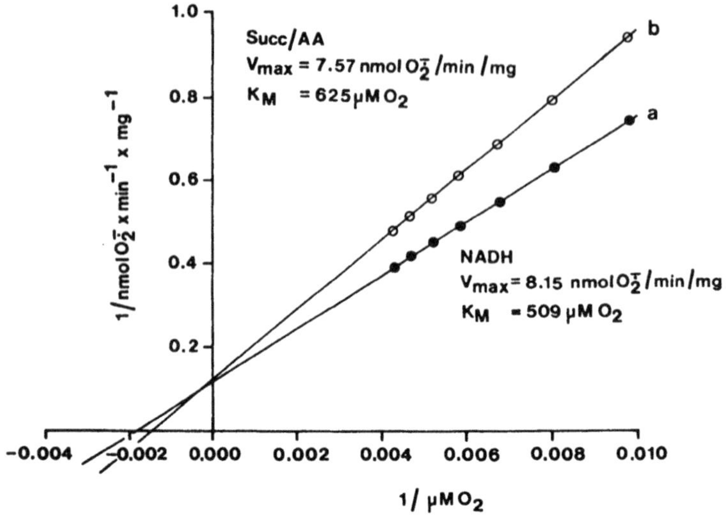

Figure 4. Determination of the Michaelis constant and V_{max} values for the monovalent reduction of molecular oxygen (a) by the exogenous NADH dehydrogenase of RHM and (b) by antimycin A (AA) inhibited succinate respiration of submitochondrial particles. For experimental details see reference[5].

Solubilization of the exogenous NADH-dehydrogenase from the inner mitochondrial membrane[5] permitted the demonstration of the identity of this active NADH-dependent O_2^{-} generator (Figure 5) since the isolated enzyme also generated O_2^{-} radicals at the expense of NADH. Rotenone had no effect on the activation rates of oxygen by the isolated enzyme. Thus, the rotenone-induced increase of O_2^{-} release, observed with the membrane-associated exogenous NADH-dehydrogenase (see Figure 2) elicited a deviation of the normal e^{-} transfer to oxygen at the level of the novel enzyme, as soon as the e^{-} flux in mitochondria was inhibited.

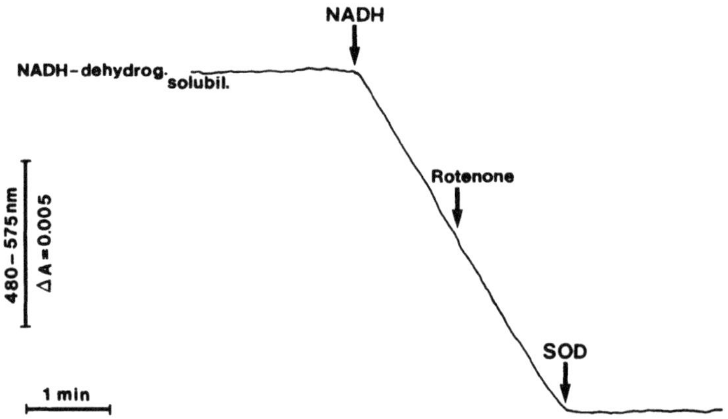

Figure 5. Detection of O_2^{-} formation by exogenous NADH-dehydrogenase solubilized from NADH[5]. Experimental conditions were similar to those described in Figure 2.

DISCUSSION

Our investigations demonstrate the existence of an active O_2^- generator in RHM which operates independently of energy-linked respiration. The requirement for external (cytosolic) NADH to run this source and the formation of O_2^- radicals by the solubilized exogenous NADH-dehydrogenase makes it likely that this enzyme is identical with the novel oxygen activator. The poor substrate selectivity at the oxidising side and its one electron redox potential[5,6] suggest that the exogenous NADH-dehydrogenase can form an active redox couple with molecular oxygen, yielding univalently reduced dioxygen molecules. The total recovery of O_2^- radicals in the cytosol indicates that in contrast to "classical O_2^- sources," O_2^- from the above enzyme is not released into the mitochondrial matrix, thus escaping dismutation to H_2O_2 by SOD, the enzyme which is localised in this compartment. The novel superoxide generator was shown to have a higher affinity to oxygen than "classical O_2^- sources" earlier reported and also exhibits higher rates of O_2^- formation under physiological oxygen concentrations. Since, in contrast to classical O_2^- sources, O_2^- release from the novel generator is not linked to the energy-conserving process of respiration and starts already in the presence of cytosolic NADH, the latter oxygen activator is likely also to operate *in vivo* and to be responsible for O_2^- generation in the living cell.

REFERENCES

1. G. Loschen, A. Azzi, C. Richter and L. Flohe, Superoxide radicals as precursors of mitochondrial hydrogen peroxide. *FEBS Lett.* 42: 68 (1974).
2. H. Nohl and D. Hegner, Do mitochondria produce oxygen radicals *in vivo*? *Eur. J. Biochem.* 82: 563 (1978).
3. E. Cadenas, A. Boveris, C.I. Ragan and A.O.M. Stoppani, Production of superoxide radicals and hydrogene peroxide by NADH-ubiquinone reductase and ubiquinol-cytochrome c reductase from beaf heart mitochondria. *Arch. Biochem. Biophys.* 180: 248 (1977).
4. H. Nohl, The biochemical mechanism of the formation of reactive oxygen species in heart mitochondria, in: "Advances in Studies on Heart Metabolism", C. M. Caldarera and P. Harris, eds, CLUEB, Bologna (1982).
5. H. Nohl, Demonstration of the existence of an organo-specific NADH dehydrogenase in heart mitochondria. *Eur. J. Biochem.* 169: 585 (1987).
6. H. Nohl, Identification of the site of adriamycin activation in the heart cell. *Biochem. Pharmacol.* in press (1988).
7. H. Nohl and W. Jordan, The mitochondrial site of superoxide formation. *Biochem. Biophys. Res. Commun.* 138: 533 (1986).
8. E. Bernt and H.U. Bergmeyer, Anorganische Peroxide, in: "Methoden der enzymatischen Analyse", H.U. Bergmeyer, ed, Verlag Chemie, Weinheim (1974).

DETECTION AND MEASUREMENT OF FREE RADICAL

GENERATION DURING CARDIOVASCULAR INJURY

Carmen M. Arroyo[†], Benjamin F. Dickens[†], Jay H. Kramer[†], Roy H. Leiboff[†], Gertrud W. Mergner*, I. Tong Mak[†], and William B. Weglicki[†‡]

Departments of [†]Medicine, [‡]Physiology, and *Anesthesiology
The George Washington University, Washington, D.C. 20037

INTRODUCTION

It has been suggested that during myocardial ischemia short-lived free radicals such as superoxide ($O_2^{\bar{\bullet}}$), hydroxyl (•OH), or lipid peroxide (RO•) are produced. The evidence, however, of the participation of such radicals in cardiovascular injury has been, for the most part, based upon activation of lipid peroxidation and on myocardial protection provided by scavengers.[1,2] Direct detection of such radicals, as might be possible by electron spin resonance spectroscopy (ESR) has not provided conclusive identification of radicals produced during these injuries. However, spin trapping and ESR offer an indirect technique that provides the evidence necessary to confirm the involvement of these labile intermediates. Spin trapping has been applied very successfully to studies of free radical mechanisms in a variety of chemical and photochemical reactions; however, it has only recently been applied to biological systems.[3] The majority of the spin traps being used are molecules containing a nitroso or nitrone group which, upon addition of a labile free radical, form a stable nitroxide that is easily detected by ESR.

The general mechanism of spin trapping can be illustrated with the following scheme:

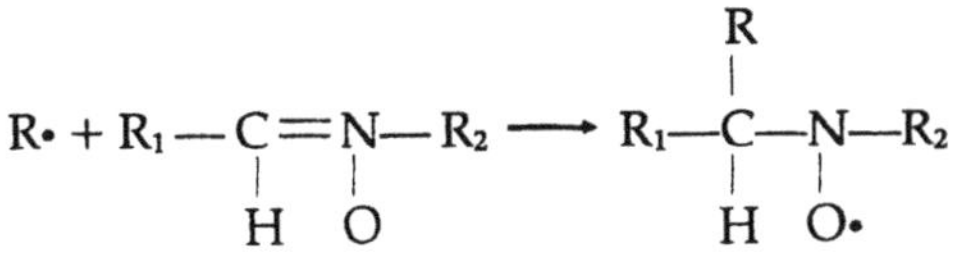

Such a reaction scheme involves the trapping of a reactive free radical by an addition reaction to produce a more stable radical, detectable by ESR with a discernable hyperfine pattern that permits identification of the initial radical trapped. We have utilized spin trapping techniques to study biologically generated free radicals in different systems: in an in vivo regional ischemic canine heart and in an in vitro working rat heart model of ischemic and reperfusion injury.

MATERIAL AND METHODS

Phenyl-tert-butyl nitrone (PBN), 2-methyl-2-nitroso propane (MNP), and 5,5'-dimethyl-1-pyrroline-N-oxide (DMPO) were obtained from Aldrich Chemical Company (Milwaukee, WI). DMPO was purified according the method of Buettner and Oberley[4] by repeatedly treating the solution with activated charcoal until all the free radical impurities were eliminated. The experimental room was kept in darkness throughout the study to prevent any photolytic degradation of PBN, MNP or DMPO. The experimental procedures utilized in in vivo regional ischemic canine heart have been described previously by Arroyo et. al.[5] In an in vitro working rat heart models the methods used have been given in references 6 and 7. Unless otherwise stated, spectra were taken with a Bruker ER100 series, X-band spectrometer (9.5 GHz) with a field modulation frequency of 100 KHz, using microwave power of 10 mW and modulation amplitude of 1 G.

RESULTS AND DISCUSSION

In Vivo Regional Ischemic Canine Heart

Two different spin-trap agents were used, phenyl-tert-butyl nitrone (PBN) and 2-methyl-2-nitroso propane (MNP), since they are known to form relatively long-lived radical adducts with carbon and oxygen-centered radicals.[8] Perfusion buffers containing either 50mM of PBN or 50mM of MNP were infused in a

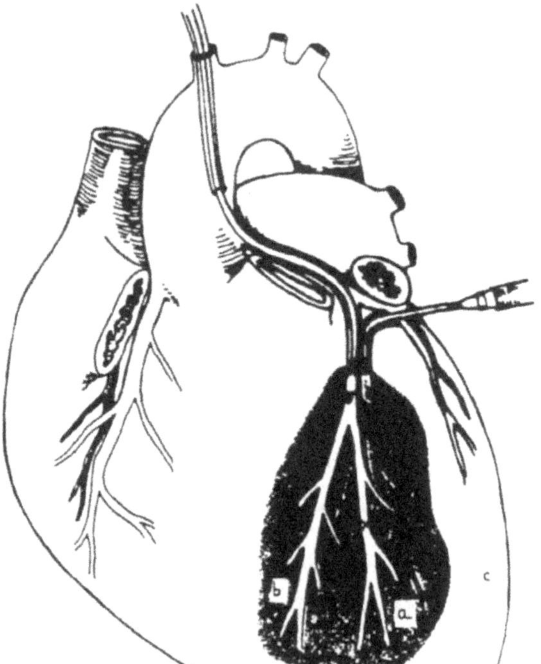

Figure 1. Diagram of the left anterior descending coronary artery (LAD) in the open chest dog. A diagonal branch of the LAD was ligated for 30 min. Myocardial samples were taken from a) ischemic, b) non-ischemic tissue of the LAD, and c) from non-perfused tissue of the circumflex coronary artery.

blood-free solution for approximately 30 min as described by Arroyo et al.[5] After the loading procedure, the heart was subjected to different intervals of regional ischemia, by ligating a diagonal branch of the left anterior descending (LAD) coronary artery.

Myocardial tissue samples, taken from perfused ischemic and non-ischemic tissue regions of the LAD, and from non-perfused tissue region of the circumflex coronary artery (Figure 1), were place in a flat quartz tissue cell and fitted into the cavity of the Bruker ER 100E X-band spectrometer.

The ESR spectrum of the spin adduct observed in ischemic canine heart tissue containing the spin trap (PBN) is illustrated in Figure 2A. The detection of this complex PBN ESR signal indicates that a mixture of two adducts is present. One adduct is identified as a carbon-centered free radical adduct (a_N =

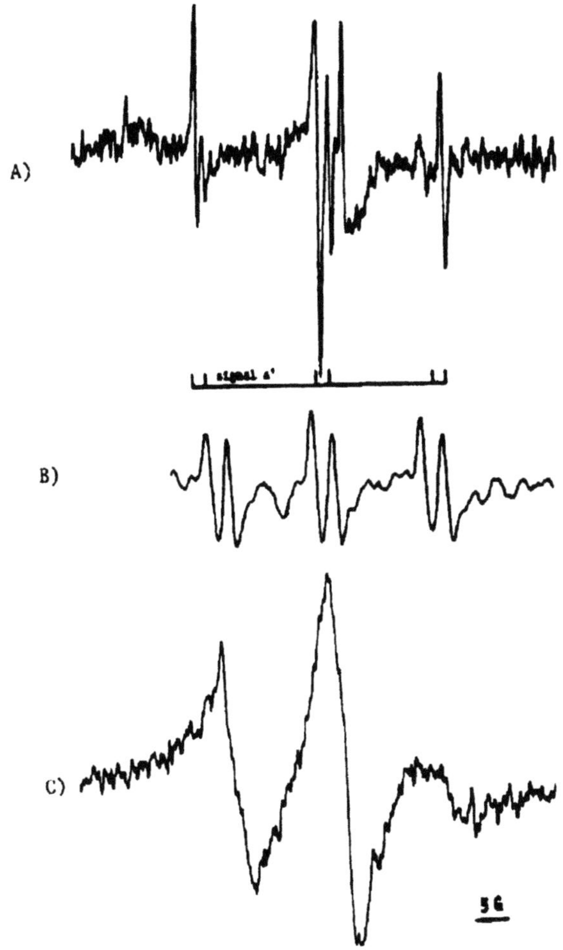

Figure 2. ESR spectra from ischemic heart tissue. Canine hearts were subjected to 90 min of ischemia a) in the presence of 50 mM PBN; the stick diagram shows the hyperfine splitting for RO• (a_N = 13.60G, a_H^β= 1.56 G), b) ESR spectrum of lipid extracted from tissue sample a, and c) in the presence of 50 mM MNP.

15.20 G; a^{β}_{H}= 3.85 G) and since the signal is present in the washed chloroform phase of a Folch extract (Figure 2B), these adducts must be derived from membrane lipid radicals. By using only the spacing between the low field doublet and the center lines, hyperfine splitting constants (a': a_N = 13.60 G, a^{β}_{H} = 1.56 G) are obtained for the second PBN adduct. These values are consistent with previously published primary and secondary oxy-adducts of PBN.[9,10] Unlike the PBN-R adduct, the PBN-OR adduct signal was not observed in the washed lipid extract (Figure 2B). Apparently, the RO• adduct decays during lipid extraction and concentration, while the carbon-centered R• adduct is more stable to such manipulation.

When ischemic canine heart tissue was loaded with MNP (50mM) and analyzed by ESR, nitroxy radical adduct (with a characteristic ESR triplet hyperfine splitting of 14.75 G) was observed (Figure 2C). As the duration of ischemia increased, the intensity of this triplet signal increased. The extraction of the tissue samples in chloroform/methanol (2:1) causes a broadening of the triplet; the resulting spectra are similar to those in Figure 2C. The trapped species can be considered to be a secondary species derived from the interaction of a strong oxidant, possibly oxygen-centered free radical, with cardiac membrane lipids. The high stability of the nitroxide radical adduct precludes the possibility that MNP may have trapped oxygen-centered radicals. Therefore, we conclude that the MNP adduct is a lipid-derived radical adduct probably formed by the interaction of oxygen-centered radical with cardiac membrane lipids.

In Vitro Working Rat Heart Model

Using 5,5' dimethyl-1-pyrroline-N-oxide (DMPO), carbon-centered radicals were identified during ischemia and oxy-radical adducts in post-ischemia rat heart.[6] The formation of these spin adducts was inhibited by superoxide dismutase (SOD), suggesting that superoxide plays a role in the adduct formation.

When global ischemia was carried out in the presence of 100mM DMPO, the ESR spectra of the coronary effluent indicated that two free radical adducts were formed (Figure 3A). One of the radicals appears to be •OH and its generation is dependent on the presence of O_2 (DMPO, a_N = 14.9G; a^{β}_{H}= 14.9 G). The other spin-trapped radical is a derivative of an unknown carbon compound and its formation is decreased by the presence of oxygen. The hyperfine constants are a_N = 15.5 G and a^{β}_{H}= 22.6 G for the DMPO-carbon radical adduct. Besides the possibility that DMPO itself could lead to a similar signal,[11] our results support the assumption that the carbon-centered radicals might be derived from lipid peroxidation.

Post-ischemic, reperfused rat hearts in the presence of DMPO resulted in the formation of different paramagnetic species. ESR spectra obtained can be attributed essentially to the formation of DMPO-OH species, but characteristics of DMPO-OOH and DMPO-OOR adducts were also detected (Figure 3B). Superoxide dismutase (SOD) was found to interfere with this process by inhibiting the formation of the DMPO adducts. This argues in favor of decomposition of DMPO-OOH to DMPO-OH.

The absence of a detectable DMPO-OOH ESR signal during global ischemia does not negate the presence of a small amount of superoxide in the process, but merely defines the stability of the spin adduct under the experimental

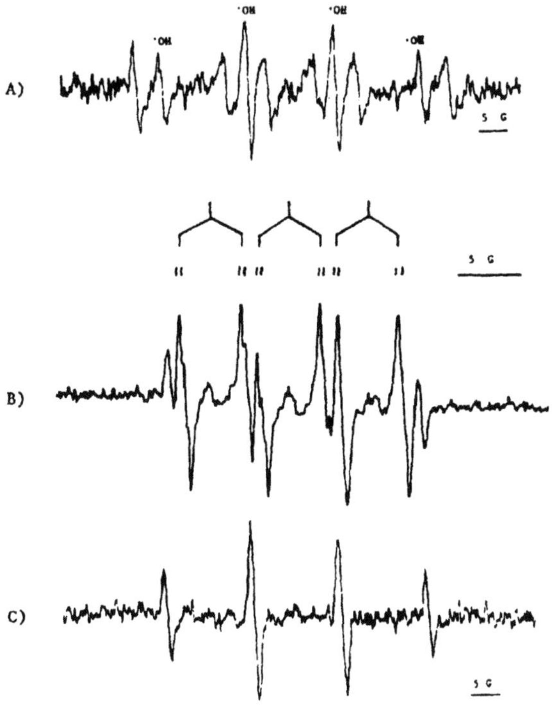

Figure 3. ESR spectra of DMPO adducts in ischemic and post-ischemic rat hearts. Hearts were subjected to 40 min of global ischemia in the presences of 100 mM DMPO. Scan time 4 min; receiver gain 1.25 x 10^5; modulation amplitude 0.5G. a) DMPO-OH adduct and DMPO-carbon radical adduct. b) DMPO-OOH and DMPO-OOR adduct observed 20 seconds after reperfusion was initiated. Stick diagram shows the hyperfine couplings for DMPO-OOH (a_N = 14.25G; a_H^β= 11.3G; a_H^γ = 1.4G), c) DMPO-OH adduct observed 1 min after reperfusion was initiated.

conditions used. This was supported by the observation that SOD completely inhibited the spin adducts formation during ischemia and reperfusion.

CONCLUSIONS

Our results demonstrate that short-lived free radicals — superoxide anion, hydroxyl, and carbon-centered radicals — are formed during myocardial ischemia and post-ischemic reperfusion. Although spin-adduct formation is considered to be the most direct technique for the detection of free radicals, some disadvantages are encountered. The spin-trapping agents frequently have been claimed to have toxic effects on myocardial tissue; a high concentration (60-200 mM) of spin trap should be used to allow effective competition with spontaneous dismutation. In preliminary studies of spin trapping we found no release of regional creatine phosphate kinase (CPK) from myocardium perfused for 30 min with 50 mM PBN containing 8.5% DMSO. We did observe some systemic cardiovascular changes. There was a mild decrease in blood pressure and an increase in heart rate. Peripheral resistance was decreased, cardiac output increased, and systolic pulmonary artery pressure was elevated.[12]

In addition, spin trapping should be accompanied by the appropriate controls. It should be subjected to all the added inhibitors in separate experiments so that the lifetime of the spin adduct in the presence of each reagent is known. In spite of these drawbacks, the spin-trapping technique allows us to confirm the nature of the free radicals produced during myocardial injury, and it offers a tool to investigate the effectiveness of myocardial intervention studies on free radical production.

ACKNOWLEDGMENT

This study was supported by research grants from The George Washington University (NIH-2-S07-RR05359) and the National Institutes of Health (RO1-HL36418 and PO1-HL38079).

REFERENCES

1. M. Bernier, D. J. Hearse, and A. S. Manning, Reperfusion-induced arrhythmias and oxygen-derived free radicals. Studies with "anti-free radical" interventions and a free radical-generating system in the isolated perfused rat heart. Circ. Res. 58:331 (1986).
2. J. M. McCord, Oxygen-derived free radicals in post-ischemic tissue injury, N. Engl. J. Med. 312:159 (1985).
3. E. G. Janzen, H. J. Stronks, C. M. Dubose, J. L. Poyer, and P. B. McCay, Chemistry and biology of spin-trapping radicals associated with halocarbon metabolism in vitro and in vivo. Environ. Health Perspect. 64:151 (1985).
4. G. R. Buettner and L. W. Oberley, Considerations in spin trapping of superoxide and hydroxyl radical in aqueous systems using 5,5-dimethyl-1-pyrroline-N-oxide. Biochem Biophys. Res. Commun. 83:69 (1978).
5. C. M. Arroyo, J. H. Kramer, R. H. Leiboff, G. W. Mergner, B. F. Dickens, and W. B. Weglicki, Spin trapping of oxygen and carbon-centered free radicals in ischemic canine myocardium. Free Radical Biol. and Med. 3(5):313 (1987).
6. C. M. Arroyo, J. H. Kramer, B. F. Dickens, and W. B. Weglicki, Identification of free radicals in myocardial ischemia/reperfusion by spin trapping with nitrone DMPO. FEBS Lett. 221:101 (1987).
7. J. H. Kramer, C. M. Arroyo, B. F. Dickens, and W. B. Weglicki, Spin-trapping evidence that graded myocardial ischemia alters post-ischemic superoxide production. Free Radical Biol. and Med. 3:153 (1987).
8. E. G. Janzen, A critical review of spin trapping in biological systems. In: "Free Radicals in Biology", W. A. Pryor, ed., Academic Press, New Yok (1980).
9. M. V. Merritt and R. A. Johnson, Spin trapping, alkyl-peroxy radicals and superoxide akyl halide reaction, J. Am. Chem. Soc. 99: 3713 (1977).
10. P. B. McCay, E. K. Lai, J. L. Poyer, C. M. Dubose, and E. G. Janzen, Oxygen and carbon-centered free radical formation during carbon tetrachloride metabolism. Observation of lipid radical in vivo and in vitro. J. Biol. Chem. 259:2135 (1984).
11. G. M. Rosen and E. J. Rauckman, Spin trapping of free radicals during hepatic microsomal lipid peroxidation, Proc. Natl. Acad. Sci. U.S.A. 78:7346 (1981).
12. D. J. Hearse and A. Tosaki, Free radicals and reperfusion-induced arrhythmias: Protection by spin trap agent PBN in the rat heart, Circ. Res. 60:375 (1987).

EVIDENCE THAT FREE RADICALS RESULT IN REPERFUSION INJURY IN HEART MUSCLE

Myron L. Weisfeldt, Jay Zweier, Giuseppe Ambrosio, Lewis C. Becker and John T. Flaherty

Johns Hopkins Medical Institutions
Cardiology Division
600 N. Wolfe Street
Baltimore, Maryland 21205, U.S.A.

INTRODUCTION

Recently, with the advent of thrombolytic therapy and emergency angioplasty for acute infarction interest in reperfusion injury has intensified. The question is whether all myocardial necrosis is reflective of the duration and severity of ischemia or whether there are cells which are alive at the end of the period of ischemia and undergo necrosis as a result of events that occur during or after reperfusion. These comments are entirely limited to three specific issues free-radical-induced reperfusion injury. Our group came into this area of free radical injury to heart muscle subsequent to the early studies of Drs. McCord[1], Luchessi[2] and Hearse[3] and their colleagues. Therefore, we had the opportunity to focus our attention to three specific issues that we viewed as pivotal: (1) Are mechanisms that have been proposed to produce free radicals at the moment of reperfusion supported by direct measured oxygen based free radicals? (2) Does superoxide dismutase (SOD) administered as a therapeutic agent into the perfusate have any directly measurable effect on the presumed burst of oxygen free radicals on reperfusion. There has been considerable doubt as to whether SOD in the perfusate can reduce injurious, possibly intracellularly generated radicals? (3) Does a drug given only at the end of the period of ischemia result in reduced injury since the benefit of a drug given after ischemia can only effect reperfusion injury? A positive study would establish unambiguously the existence of reperfusion injury.

ESR SPECTRA OF HEART RADICALS

The methods that were used to measure oxygen free radicals directly were developed by Jay Zweier[4] using electron paramagnetic resonance (EPR) spectroscopy. The observed signals are quite distinctive for individual radical species (Figure 1). A carbon-centered semiquinone radical exhibits a symmetrical first-derivative gaussian line with a central g value (Figure 1). An oxygen radical for which there is most interest possess axial symmetry, which is defined by two individual g values (g II and g I). Finally, a nitrogen centered radical exhibits a triplet signal due to hyperfine coupling of the electron to the nitrogen nucleus.

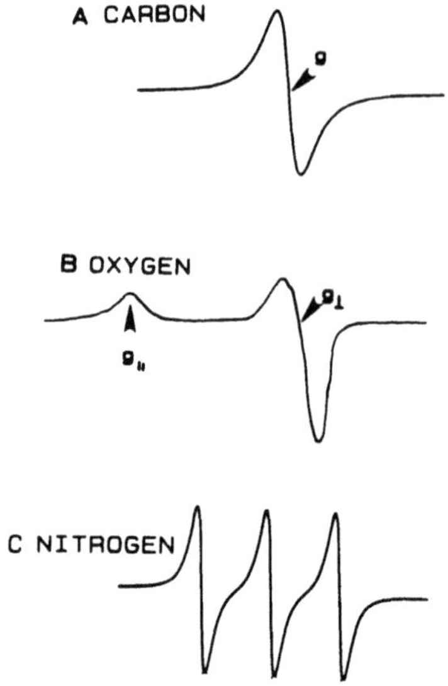

Figure 1. Typical EPR spectra of A - carbon, B - oxygen and C - nitrogen centered free radicals.

We used rabbit hearts, which like those of humans have low levels of xanthine oxidase, and perfused them with a Krebs-Ringers bicarbonate buffer solution. The hearts were freeze-clamped and maintained in liquid nitrogen at 77°K, ground to a fine powder, and then inserted into 3 mm diameter tubes for EPR measurement. Figure 2 shows the signal which appears after freeze-clamping a perfused non-ischemic heart. The only signal present has spherical symmetry and is similar appearance to the signal obtained from a carbon-centered semiquinone radical. In freeze-clamped samples obtained after 10 minutes of global ischemia at 37°, there is considerable change in not only the configuration but also the magnitude of the signal (Figure 2). There appears to be an increase in the complexity of the signal, with other non-spherically symmetric components appearing. Finally, if we add 10 minutes of ischemia and 10 seconds of reperfusion with an oxygenated solution, the overall magnitude of the signal increases markedly as does the complexity of the signal. The amount of radical can be determined by comparison with a known standard.

The next question that we asked was whether this increase in magnitude and change in character of the free radicals signal was related to generation of oxygen based radicals. Comparison was made between heart reperfused with nitrogen vs. oxygen containing perfusate. Figure 3 shows EPR signals obtained from three other hearts: freeze-clamped pre-ischemia, control, end ischemia, and ischemia followed by 10 seconds of reperfusion. There was less signal intensity noted during ischemia, and again with 10 seconds of reperfusion with oxygenated solution there was an increase in the complexity and the magnitude of the signal. Figure 3 is the EPR signal obtained from another heart subjected to 10 minutes of ischemia. Reperfusion with a nitrogen-containing solution rather than oxygen resulted in a radical signal which was similar that seen for the non-reperfused ischemic heart. This appears to answer the first of the three questions: there does appear to be a burst of free radicals generated at the end of ischemia, which is related oxygen in the reperfusion solution.

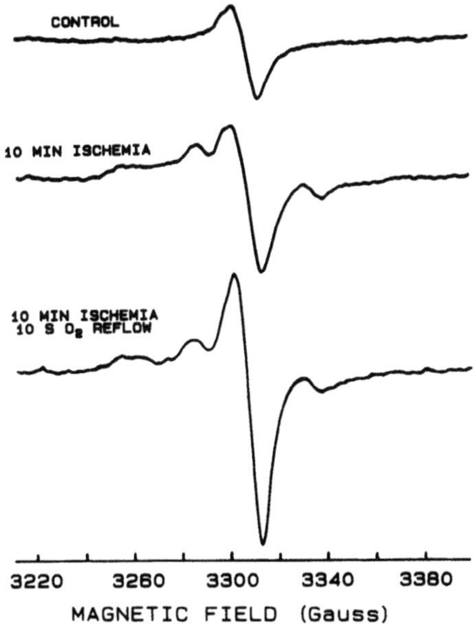

Figure 2. EPR spectra of control, ischemic, and postischemic reperfused heart samples at 77°K. Microwave frequency, 9.278 GHz; microwave power, 1.0 mW, modulation amplitude, 2.5 G.

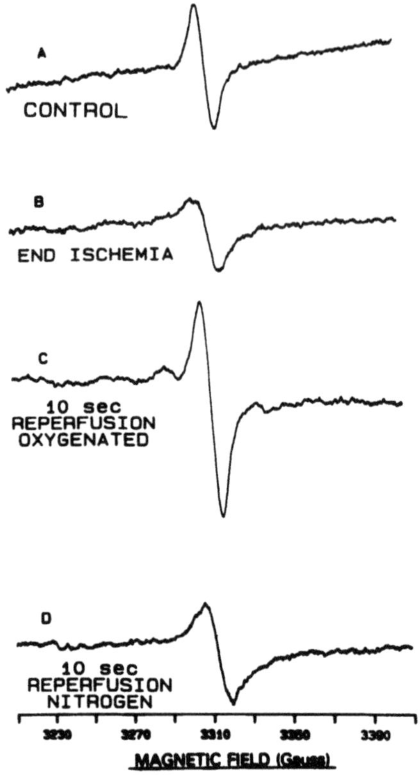

Figure 3. EPR spectra of heart samples after A - control perfusion, B - 10 minutes ischemia, C - 10 minutes ischemia followed by 10 seconds of reperfusion with oxygenated perfusate and D - 10 minutes ischemia followed by 10 seconds of reperfusion with nitrogen equilibrated perfusate. EPR spectra were recorded as described in Figure 2.

The question was taken one step further. Individual radicals are differentially temperature sensitive. The frozen sample is rewarmed to -80°C. The radicals are differentially temperature labile. The nitrogen-based signal is the most temperature labile and the oxygen next most. After prolonged warming only a carbon centered radical remains. By rewarming the sample, then cooling it again to 77°K and measuring the difference spectrum, nitrogen-centered and oxygen-centered radical signals were distinguished. Again, the quantity of each radical species was then determined by comparison with a known standard. Figure 4 summarizes the data for the three individual radicals over time. Each time point represents EPR data obtained from a single rabbit heart. At the end of the 10 minute period of ischemia there is some rise in the oxygen-centered radical and the nitrogen centered radical concentration and a fall in quantity of carbon-centered radicals. On reperfusion there is a burst of oxygen radicals generated, which results in a sevenfold increase in concentration after 10 seconds of reflow that can be distinguished from the other two radicals. Thus, these data would confirm that there is a burst of oxygen radical generation occurring at the moment of reperfusion.

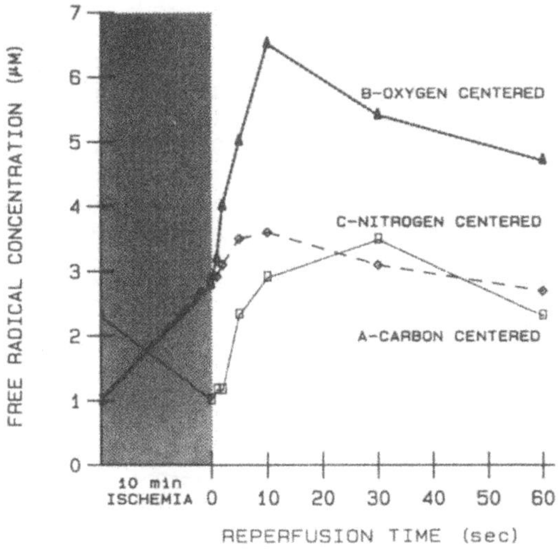

Figure 4. Time course of free radical generation; each point corresponds to measurements performed on hearts freeze-clamped at different points in time. Signal A, Signal B, , Signal C. Quantitation was performed by comparison of the double integral of the signals to that of the known concentration of peroxylamine standard.

PROTECTION BY SOD

The second question was whether or not SOD given at the moment of reperfusion would be successful at reducing the burst of oxygen free radicals[5]. Human recombinant SOD from Biotechnology General Corp. was infused at the moment of reperfusion. There was obliteration of the burst of oxygen radicals generated during the early seconds of reperfusion by virtue of the presence of superoxide dismutase. Thus, the answer to the second question is also yes, the administration of exogenous SOD can eliminate the burst of oxygen radicals generated on reperfusion.

Now what is the value of administering superoxide dismutase only at the moment of reperfusion? What does it do to recovery of metabolism and function after ischemia? Perfused rabbit hearts were placed in the bore of a Bruker 180 NMR spectrometer to measure P^{31} spectra and a balloon filled with fluid was placed in the left ventricle was used to monitor ventricular pressure as an index of function[6]. Two groups of hearts were subjected to 30 minutes of 37°C global ischemia, a very significant period of ischemia for the rabbit heart. There was a time-related fall NMR detectable phosphocreatine (Figure 5), which was similar in both groups during 30 minutes of ischemia. At the end of 30 minutes, one group of 12 hearts received a bolus and a constant infusion of SOD for 15 minutes and the other group of 12 hearts did not. Even at 5 minutes, phosphocreatine concentration is higher in the SOD-treated hearts. The balloon in the left ventricle remained deflated during the first 15 minutes of reperfusion. The balloon was reinflated after 15 min of reperfusion and remained inflated during the remainder of reperfusion. ATP content also declined similarly over the 30 minute of ischemia as first 30 minutes of reperfusion. However, there was a small but statistically significantly difference in ATP content by the end of 45 minutes of reperfusion. Beginning at 15 minutes of reperfusion the mean left ventricular developed pressure was higher in treated hearts than in controls. Treated hearts recovered to ~70% of pre-ischemic levels compared to ~40% in the control hearts. Thus, at the end of the experiment there was higher ATP, higher phosphocreatine and greater function in hearts given only SOD at the moment of reperfusion.

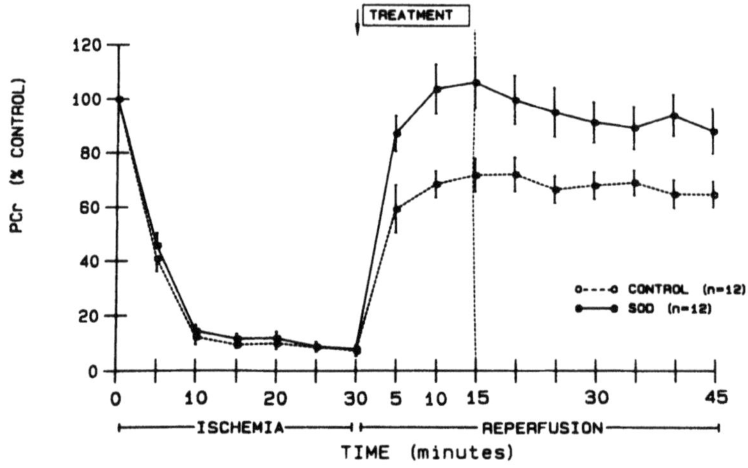

Figure 5. Time course of myocardial PCr content, expressed as a percentage of the preischemic control value (% control), during 30 min of normothermic (37°C) ischemia and 45 min of reperfusion. Control measurements were obtained immediately before cross-clamping of the perfusion line (time 0). The arrow (treatment) indicates the time of bolus administration (i.e., reflow), the vertical dotted line marks the resumption of cardiac isovolumic work (i.e., balloon reinflation) and the end of treatment after 15 min of reperfusion. Two-way ANOVA showed that recovery of PCr during reperfusion is significantly different in SOD-treated vs. control hearts ($p < .01$).

Finally, to approach the issue of the clinical relevance of reperfusion injury due to oxygen radicals in a regionally ischemic blood perfused situation, we performed a study in anesthetized open chest mongrel dogs in which 90 minutes of circumflex coronary artery occlusion followed by reperfusion was performed[7]. In one group of dogs at the moment of reperfusion 400,000 units of SOD was given in a bolus in the left atrium followed by 300,000 units in an intravenous infusion over a 60 minute period. The control group received saline. After completion of this acute phase of the study the chest of the animal was closed and the animals were placed back in their cages. At 48 hours infarct size was examined. At this time we were able to examine infarct size directly. Risk region was quantified by a post-mortem angiographic method. Figure 6 presents infarct size as a percent of left ventricle and percent of risk region. Risk region was the same in the SOD and control groups but infarct size was nearly 40% smaller with SOD. There was another aspect of this particular study. Regression lines relating infarct size as a percent of risk region to collateral epicardial flow were determined for control and SOD-treated dogs. In the control dogs (Figure 7) shown by the upper regression line, the usual relationship

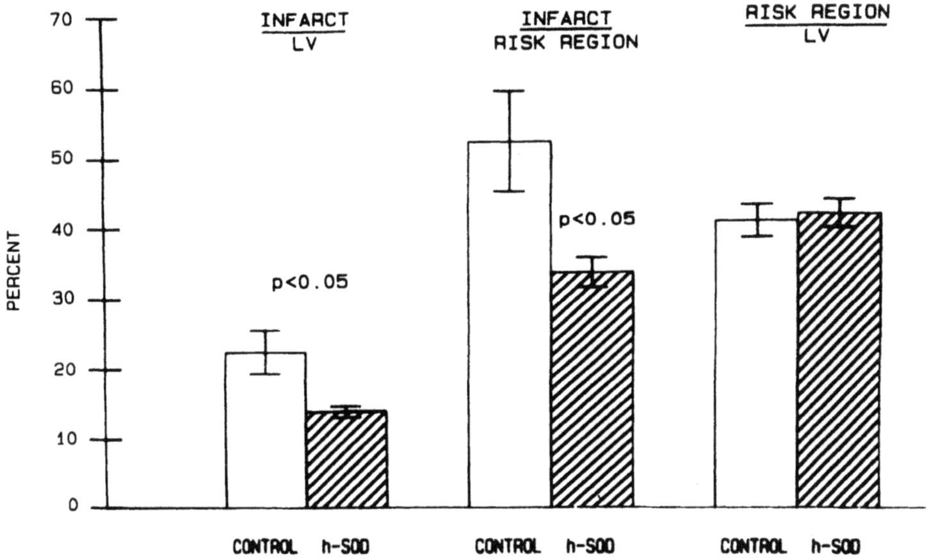

Figure 6. Infarct size (expressed as percent of the left ventricle or of the risk region) and average risk region (expressed as percent of the left ventricle) in control (open bars) and h-SOD-treated dogs (hatched bars).

exists between collateral flow during ischemia and infarct size: that is, dogs with low collateral flow had big infarcts, dogs with high collateral flow had small infarcts. In the SOD-treated dogs the slope of the regression line was quite different. The animals that appear to have the greatest benefit by virtue of administering SOD at the moment of reperfusion were the animals that had the lowest collateral blood flow during the period of ischemia. Thus, animals with the most severe ischemia had the most reduction of infarct size by SOD. Also, the rabbit hearts with the lowest level of ATP at the end of the ischemic period also appeared to obtain the greatest benefit from SOD.

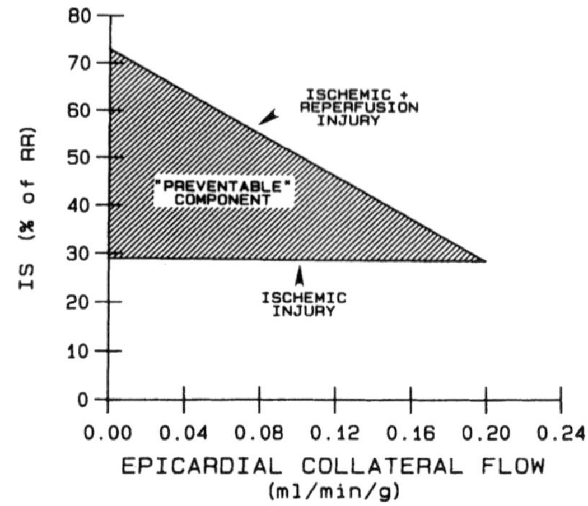

Figure 7. Superimposition of the two regression lines from the data in Figure 6 relative to the epicardial flows. The upper line represents the correlation of blood flow and infarct size in control animals. Necrosis in this group was the result of ischemic injury combined with reperfusion injury. The lower flat line refers to the h-SOD-treated group, which might be considered representative of "pure" ischemic injury, since in these dogs the treatment significantly reduced reperfusion injury. The hatched area between the two lines should therefore represent the extent of h-SOD-preventable reperfusion injury as a function of collateral flow during ischemia.

CONCLUSION

In summary, we tried to address three questions in the area of oxygen free radical injury during reperfusion. The first question was whether we could document a burst of oxygen free radicals at the moment of reperfusion. The EPR studies show this answer to be yes. The second question was whether SOD administered as a pharmacologic agent at the time of reflow could in fact reduce that burst of oxygen radical generation. SOD eliminated the burst. The third question we tried to answer was whether the administration of only SOD at reperfusion after a period of ischemia could result in better functional and metabolic improvement. In both the rabbit and the dog model it did. In the dog study the period of reperfusion was sufficiently long to allow direct gross pathological evaluation of infarct size. Finally, we noted a most interesting observation in the two models: the more severe the ischemia, the more beneficial this type of therapy. This observation has not been made with any other therapy that causes a reduction of infarct size.

ACKNOWLEDGMENTS

Supported by Ischemic Heart Disease Specialized Center of Research Grant, HL17655-13, National Heart, Lung and Blood Institute, National Institutes of Health, Bethesda, Maryland.

REFERENCES

1. JM McCord: Oxygen-derived free radicals in postischemic tissue injury. N. Engl. J. Med. 312:159, 1985.
2. JR Jolly, WJ Kane, Bailie MB, Abrams GD, Lucchesi BR: Canine myocardial reperfusion injury. Its reduction by the combined administration of superoxide dismutase and catalase. Circ. Res.54:277, 1984.
3. AS Manning, DJ Coltart and Hearse DJ: Ischemia and reperfusion-induced arrhythmias in the rat. Effects of xanthine oxidase inhibition with allopurinol. Circ. Res. 55:545-548, 1984.
4. JL Zweier, JT Flaherty, Weisfeldt ML: Direct measurement of free radical generation following reperfusion of ischemic myocardium. Proc. Nat'l Acad. Sci. 84:1404-1407, 1987.
5. JL Zweier, BK Rayburn, Flaherty JT, Weisfeldt ML: Recombinant superoxide dismutase reduces oxygen free radical concentrations in reperfused mycoardium. J. Clin. Invest. 80:1728-1734, 1987.
6. G Ambrosio, ML Weisfeldt, Jacobus WE, Flaherty JT: Evidence of a reversible oxygen radical mediated component of reperfusion injury: Reduction by recombinant human superoxide dismutase administered at the time of reflow. Circulation, 75:282-291, 1987.
7. G Ambrosio, LC Becker, Hutchins GM, Weisman HF, Weisfeldt ML: Reduction in experimental infarct size by recombinant human superoxide dismutase: Insights into the pathophysiology of reperfusion injury. Circulation 74:1424-1433, 1986.

IDENTIFICATION OF FREE RADICAL SPECIES IN THE MYOCARDIUM

Hiroe Nakazawa, Kazunobu Ban, Kohji Ichimori,
Yoshiro Sinozaki, Haruka Okino, Takashi Masuda, Nobuo Aoki,
Shingo Hori,* and Masahiro Kono**

Department of Physiology
Tokai University
Bohseidai, Isehara, Japan
* Keio University
Tokyo, Japan
** Japan Electronic Optic Laboratory
Tokyo, Japan

INTRODUCTION

With the increasing interest in the role of oxygen-derived free radical generation in various diseases, several attempts to directly evaluate free radicals in tissue have been performed using EPR.[1-8] However, none of the previous studies has fully determined the species of the radicals and none has deliberated on the possible existence of artifactual radicals which are likely to be generated during sample handling. The purposes of this study were 1) to establish an artifact-free sampling method for the quantification of freeze-trapped free radicals in myocardium by EPR, 2) to identify artifact radicals and the radicals native to the myocardium and 3) to demonstrate the relationship of these radicals to reperfusion injury.

MATERIALS AND METHODS

Rapid freezing of tissue was selected as the means of trapping radicals in the myocardium. Myocardial specimens were obtained from open-chest dogs and from isolated perfused rat and rabbit hearts. For the canine hearts, a guillotine device[9] was used. For rat and rabbit hearts, a freeze-clamping method with precooled tongs was used. The guillotine device was specifically built for rapid sampling which enables cross-sectioning and freezing of canine beating heart within 120 msec in situ. For each experiment, mitochondria were isolated from the non-frozen myocardium and immediately frozen in liquid N_2.

EPR Measurement Conditions and In Vitro Generation of Free Radicals

Frozen samples were cut into sections of approximately 3×5×3 mm with a chisel under liquid N_2 and were inserted into a quartz EPR tube (3.5 mm i.d.). EPR measurements were made with a JEOL JES-2XG, X-band spectrometer using a magnetic field modulation frequency of 100 KHz at 77 K and/or at 123 K. After measurements were completed, the protein concentration was determined in each sample by the Lowry method and radical concentration was expressed as spins/mg protein. Plausible radicals reported to exist in various tissues,[1-5,9] such

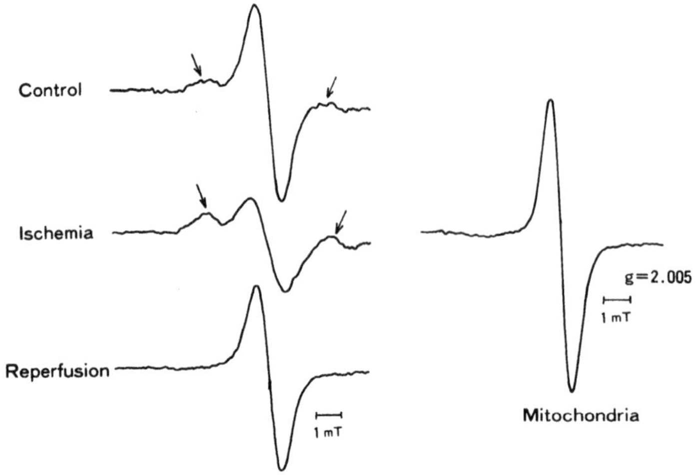

Figure 1. EPR spectra of myocardium and mitochondria.

as that of coenzyme Q_{10}, flavoprotein, ascorbic acid and lipid peroxide, were generated by electrolysis and by excitation with UV light.

RESULTS AND DISCUSSION

EPR Spectra

EPR spectra obtained from control, ischemic and reperfused myocardium and mitochondria are shown in Figure 1. The control spectrum disclosed a sharp isotropic pattern which was identical to the spectrum of mitochondria, whereas that of ischemic myocardium showed a marked decrease in peak and broadening of line width of the mid-signal, with a g-factor of 2.005. These changes were consistently observed and did not differ among canine, rat and rabbit hearts. However unlike these radicals, the small humps bilateral to the mid-signal (arrows) did not show any consistent correlation between the signal magnitude and experimental conditions or between animal species. These radicals were not observed in the mitochondria signal. Besides, we found that these humps could be minimized by gently cutting the sample with a chisel. Additionally the axial anisotropy with g =2.033 signal which was reported to be an oxygen-derived radical by Zweier et al.,[7] using a method identical to ours, was not demonstrated in our sample. The only difference between their method and ours was that they pulverized the samples for EPR

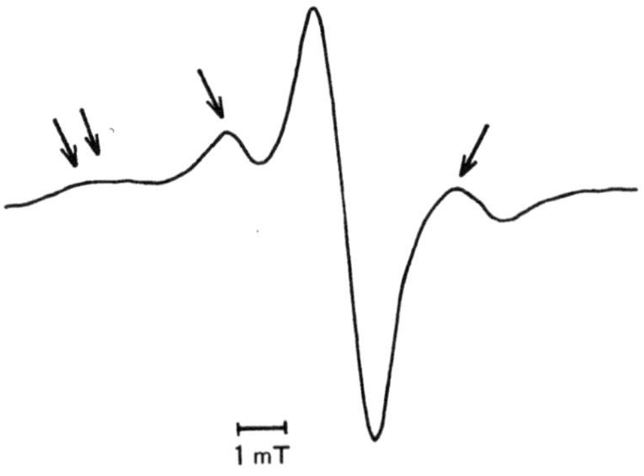

Figure 2. Spectrum of pulverized sample.

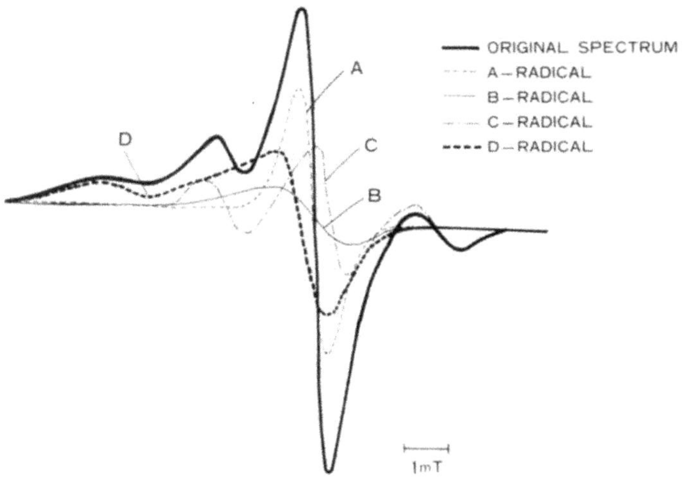

Figure 3. Isolation of four radicals. A-radical: g=2.005 (line width: 0.8mT), B-radical: g=2.005 (line width: 1.2mT), C-radical: g=2.002, coupling constant: 2.7mT, D-radical: $g_{//}$=2.033, $g_{\perp}$=2.006

tube transfer. Therefore, in order to test the generation of artifactual radicals by mechanically breaking covalent bonds, the control sample shown in Figure 1 was taken out of the EPR tube, pulverized to a fine powder, and then transferred once again into an EPR tube and EPR measurement was repeated. After this manipulation, the spectrum was identical to that of Zweier et al. (Figure 2). A marked increase in bilateral humps was demonstrated with a new small peak at the lower field, indicating that both the radical bilateral to the mid-signal (arrows) and the radical with an axial symmetry (double arrow) were artifactually generated radicals due to mechanically damaged frozen tissue.

The Isolation and Identification of Several Components in Spectra

The spectrum of each component in myocardium including artifactual radicals was isolated by the subtraction method and the temperature annealing procedure (Figure 3). The mitochondria spectrum was designated as the A-radical and assumed to be the basic spectrum in control myocardium. This spectrum was consistently isotropic and by the power saturation procedure it behaved as a single component. The radical which emerged in ischemic myocardium was delineated by subtracting the spectrum of control from that of ischemic myocardium. This was designated as the B-radical. The signal of bilateral humps was separated by temperature annealing because it disappeared when the sample was warmed to 173 K for 5 min or by subtracting the spectrum of non-pulverized sample from that of the pulverized sample. This is shown

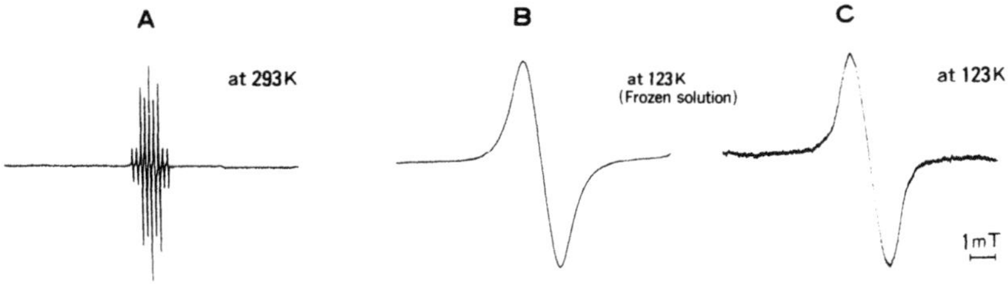

Figure 4. EPR spectra of coenzyme $\cdot Q_{10}^{-}$ generated in vitro and myocardium. A and B: Electrochemically generated coenzyme $\cdot Q_{10}^{-}$, C: Control non-ischemic myocardium.

Table 1. Comparison of Bioassayed Coenzyme $Q_9 + Q_{10}$ and A-radical[a]

	Brain	Heart	Kidney	Spleen	
Coenzyme Q	1.13±0.11	3.14±0.58	2.39±0.13	0.36±0.10	(μg/mg prot)
A-radical	2.4 ±1.6	8.2 ±2.8	5.5 ±2.5	0.9 ±0.3	(10^{12} spins/mg prot)

[a] The bioassayed coenzyme Q values agreed with the A-radical values.

as the triplet spectrum and designated as the C-radical. The anisotropic signal with axial symmetry, which appeared only after sample pulverization, was designated as the D-radical. This could be eliminated after warming the sample to 193 K for 1 h and this temperature stability was also in agreement with that previously reported to be an oxygen-derived free radical.[7]

Identification of Radical Species

Regarding the A-radical, among radicals generated in vitro, only spectrum of coenzyme $\cdot Q_{10}^-$ disclosed identical EPR parameters such as g-factor, line width and power saturation behavior (Figure 4). The spectrum of myocardium was also in agreement with that of coenzyme $\cdot Q_{10}^-$ in the ubiquinone cytochrome bc_1 complex.[10] Additionally, the spin concentration of A-radical in mitochondria was 2.5 times higher than in myocardium, which supports the identification of the A-radical as coenzyme $\cdot Q_{10}^-$ since coenzyme Q_{10} is predominantly found in mitochondria. In order to confirm further this identification, the concentration of coenzyme Q in various rat organs was analyzed by HPLC and compared with that of A-radical (Table 1). Since rats have coenzyme Q_9 as the major homolog with Q_{10} and both will give identical EPR patterns, the total coenzyme $Q_9 + Q_{10}$ was used as a comparison. The samples were obtained by freeze-clamping method while blood pressure was monitored and maintained above 120 mmHg.

The spectrum of B-radical agreed with that of UV light-excited arachidonic acid, however this identification was not confirmed and remains uncertain. The C-radical was considered to be a peptide sequence with nitrogenous dangling bonds and the D-radical, an oxygen-centered radical such as superoxide or peroxyl radical, but a confirmation study was not performed. We found that tissue from ischemic or reperfused heart was more susceptable to the generation of these artifactual radicals, which could lead to the reason why the presence of the C-radical and D-radical has been linked to reperfusion injury in the previous report.[7]

Artifactual Changes in Coenzyme $\cdot Q_{10}^-$ Values and Generation of B-radical with Freezing Delay in the Myocardium

Since radicals are extremely reactive and the concentration may be distorted by even a minute sampling delay, the ischemic artifact should be

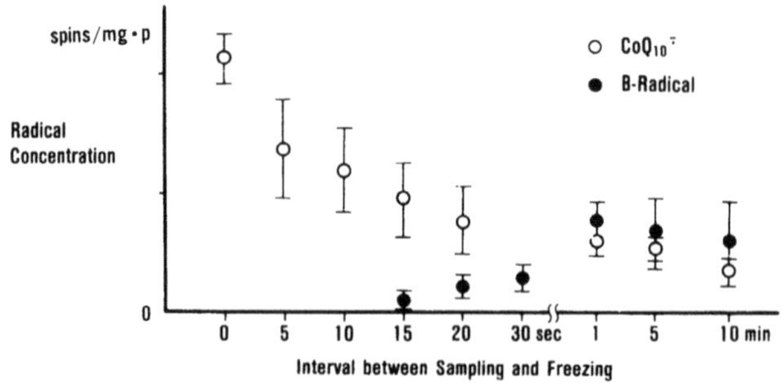

Figure 5. Changes in coenzyme $\cdot Q_{10}^-$ value and generation of B-radical with freezing delay.

clarified in a systematic manner for the future applications of radical measurements. This is particularly true in experiments which concern the changes associated with ischemia or hypoxia, because ischemic artifacts caused by delayed sampling can be misleading in the quantification of radicals. The artifactual changes by delayed sampling were systematically examined and are shown in Figure 5.

Coenzyme $\cdot Q_{10}^-$ concentration decreased abruptly with only a 5-sec freezing delay and the B-radical appeared after 30 sec. The most rapidly sampled specimens demonstrated a stable value, verifying that coenzyme $\cdot Q_{10}^-$ and B-radical can be quantified as indices of ischemia or reperfusion injury if samples are frozen without delay.

We also examined the artifactual generation of radicals at the time of in vivo sampling by comparing needle biopsy, drill biopsy, guillotine device and freeze-clamping methods. None of these methods produced the C- or D-radical and we assume that unlike frozen tissue, the mechanical cleavage of covalent bonds does not readily occur when the tissue is soft. The importance of avoiding mechanical damage in tissue is also suggested by the study in which a signal like non-heme iron protein was shown to be related to reperfusion injury in pulverized myocardium.[8] This signal was not observed in our sample. We feel that various kinds of radicals could be produced by altering the speed or the length of pulverization.

It is possible that spectra distortion and artifactual generation of radicals occur when samples are lyophilized, though we did not examine this. Therefore it will be necessary to reevaluate the previous studies[3,6] which reported oxygen-derived radical generation in lyophilized and homogenized tissue.

Quantification of Radicals in Myocardium Subjected to Ischemia/Reperfusion Injury

By use of artifact-free samples without freezing delay, coenzyme $\cdot Q_{10}^-$ and B-radical concentrations were quantified in anesthetized and open-chest dogs (12 dogs) which were subjected to ischemia and reperfusion. The anterior descending branch of left coronary artery was occluded for 20 min or 90 min and reperfused after occlusion. Then a heart was cross-sectioned using a guillotine device and EPR measurements were performed in samples taken from non-ischemic and ischemic areas. Concentration of coenzyme $\cdot Q_{10}^-$ was high in control but lower in ischemic area during occlusion period (Figure 6). The B-radical concentration was trivial in non-ischemic myocardium but was sig-

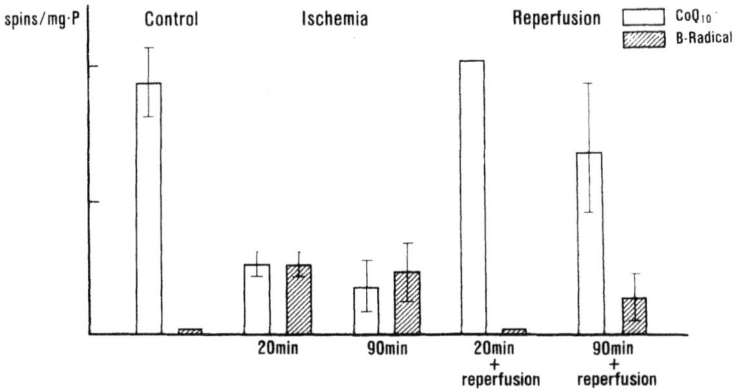

Figure 6. Concentrations of coenzyme $\cdot Q_{10}^-$ and B-radical in myocardium subjected to ischemia/reperfusion.

nificantly high in ischemic myocardium. With reperfusion, the concentration of both radicals returned to the non-ischemic radical pattern; however, the concentration of coenzyme $\cdot Q_{10}^-$ was lower and B-radical was higher in 90-min ischemia compared with that of 20-min ischemia. This indicates that an impaired recovery of coenzyme $\cdot Q_{10}^-$ may imply the disturbance of electron transport in mitochondria and the presence of severe tissue damage.

Since the concept of a burst of superoxide production upon reperfusion has been given some credence by numerous indirect studies and we failed to demonstrate the oxygen-derived radical in the artifact-free sample, we evaluated perfused rat and rabbit hearts pre-treated with superoxide dismutase inhibitor (sodium diethyldithiocarbamate)[11] in order to increase the possibility of generating superoxide. The samples were freeze-clamped at 10 sec of reperfusion after 10-min ischemia, the likely burst time of superoxide. Despite this treatment there was still no evidence of the presence of superoxide in the artifact-free samples. This failure to demonstrate superoxide however, does not disprove its existence in myocardium since the concentration of superoxide may be below the detection range of the EPR.

CONCLUSIONS

It was shown that for radical evaluation, it is crucial to avoid freezing delay and the generation of artifactual radicals. Superoxide radical was not detected in artifact-free frozen myocardium with this method even during the reperfusion period. However, direct measurements of radicals using EPR demonstrated coenzyme $\cdot Q_{10}^-$ and B-radical as radicals native to the myocardium.

REFERENCES

1. B. Commoner, J. Townsend, and G.E. Pake, Free radicals in biological materials, Nature 4432:689 (1954).
2. B. Commoner and J.L. Ternberg, Free radicals in surviving tissues, Proc. Natl. Acad. Sci. U.S.A. 47: 1374 (1961).
3. Y. Suzuki, Studies on the free radicals in myocardial mitochondria by electron spin resonance (ERS) spectrometry (studies on experimental infarction dog), Jpn. Circ. J. 39:683 (1975).
4. N. Yamazaki, Y. Suzuki, N. Kamikawa, and T. Kamikawa, "Studies on myocardial mitochondria in ischemic dog hearts by electron spin resonance (ESR) spectrometry," University Park Press, Baltimore (1980).
5. N. J. F. Dodd, and H. M. Swarts, ESR signals of lyophilized tissue, Br. J. Cancer 42:349 (1980).
6. P. S. Rao, M. V. Cohen, and H. S. Mueller, Production of free radicals and lipid peroxides in early experimental myocardial ischemia, J. Mol. Cell. Cardiol. 15:713 (1983).
7. J. L. Zweier, J. T. Flaherty, and M. L. Weisfeldt, Direct measurement or free radical generation following reperfusion ischemic myocardium, Proc. Natl. Acad. Sci. U.S.A. 84:1404 (1987).
8. W. Limm, M. Mugiishi, L. H. Piette, and J.J McNamara, Quantitative assessment of free radical generation during ischemia and reperfusion in the isolated rabbit heart, 4th International Congress on Oxygen Radicals, La Jolla, California, bstract, Thurs., p. 123, (1987).
9. S. Hori, H. Nakazawa, A. Murayama, Guillotine freeze-clamping of canine beating heart in short axis plane, Circulation 68 (suppl):323 (1983).
10. J. R. Bowyer, and T. Ohnishi, "EPR spectroscopy in the study of ubisemiquinones in redox chains," Wiley, New York (1985).
11. I. Kerckaert, and F. Roels, Myocardial H_2O_2 production in the unanaesthetized rat - influence of fasting, myocardial load and inhibition of SOD and monoamine oxidase, Basic Res Cardiol 81:83 (1986).

ABNORMAL ELECTRICAL ACTIVITY INDUCED BY H_2O_2 IN ISOLATED CANINE MYOCYTES

Peggy L. Barrington[1], Charles F. Meier, Jr.,* and William B. Weglicki[2]

Cardiovascular Research
Oklahoma Medical Research Foundation
Oklahoma City, Oklahoma 73104, U.S.A.

*Department of Pharmacology
University of Oklahoma Health Sciences Center
Oklahoma City, Oklahoma 73190, U.S.A.

INTRODUCTION

Free oxygen radicals may participate in a variety of pathological cardiac conditions which are associated with an increased incidence of arrhythmias. For example, adriamycin, an anti-cancer agent that undergoes redox-cycling and produces radicals, causes arrhythmias that restrict its therapeutic usefulness.[1] The arrhythmias that accompany reperfusion of ischemic hearts are reduced by the presence of allopurinol[2] or free radical scavengers[3,4] and enhanced by Fe-ADP.[4] Such reports imply that radicals have a direct effect on the electrical activity of myocardial cells. Yet this indirect evidence is weak because it relies on the production of radicals during pathological conditions which simultaneously produce hypoxia, acidosis, hyperkalemia or other biochemical changes. Previously, we have described a distinct sequence of electrophysiological changes that occur in the presence of free-radical-generating systems consisting of 3 mM dyhydroxyfumaric acid or xanthine:xanthine oxidase.[5,6] The changes induced by exogenous generating systems follow a reproducible pattern. They are not readily reversed by removal of the generating system but can be partially prevented by the addition of the free radical scavengers superoxide dismutase and catalase.[6] The generating systems used previously should produce a spectrum of free radicals in our superfusion solutions, but they also require the addition of one or more compounds to the superfusate. Hydrogen peroxide (H_2O_2), which occurs naturally at very low levels in aqueous solutions, can produce free radicals in the presence of oxygen. We now report the effects of a simplified H_2O_2 free-radical-generating system on the isolated canine myocyte.

METHODS

Canine myocytes were isolated using collagenase solutions. Cells were placed in a bath and superfused with a HEPES buffered Tyrode's solution (in mM: NaCl, 115; KCl, 5; $CaCl_2$, 5; HEPES, 10; dextrose, 10; sucrose, 70; pH

[1]Current address: Department of Pharmacology, Northwestern Universtity, Chicago, Illinois 60611.
[2]Current address: Department of Medicine, George Washington University Medical School, Washington, D.C. 20037.

7.35 at 37°C and bubbled with 100% O_2). Transmembrane action potentials were recorded following impalement of a cell with a conventional microelectrode and an equilibration period of 10 to 20 minutes. Cells were then superfused with a solution equilibrated with 100% O_2 and containing different concentrations of H_2O_2 (0.01 to 0.000001%). During superfusion the action potential of the cells were recorded at 1-minute intervals for the first 5 minutes and subsequently at 5-minute intervals or when unusual electrical activity occurred. In experiments using catalase, the scavenger was superfused a minimum of ten minutes prior to the addition of H_2O_2 to the solution containing the enzyme. Data were recorded by a Digital LSI 11/23 computer for subsequent off-line analysis of the following action potential parameters: duration at full repolarization, peak plateau and pre-plateau notch voltages, upstroke velocity and resting membrane potential.

RESULTS AND DISCUSSION

Superfusion of myocytes with solutions containing H_2O_2 produced a time dependent sequence of changes in the cellular action potential which resembled those observed with other free-radical-generating systems.[5,6] The sequence of changes consisted of three stages when cells were exposed to high concentrations of H_2O_2 (0.01 to 0.0001%). First, the plateau potential became progressively more positive, while the pre-plateau notch, typical of isolated myocytes, was reduced in depth and the action potential duration increased (Figure 1A). The second stage consisted of early and late after-depolarizations (1B), which could cause spontaneous activity (1C:#2).

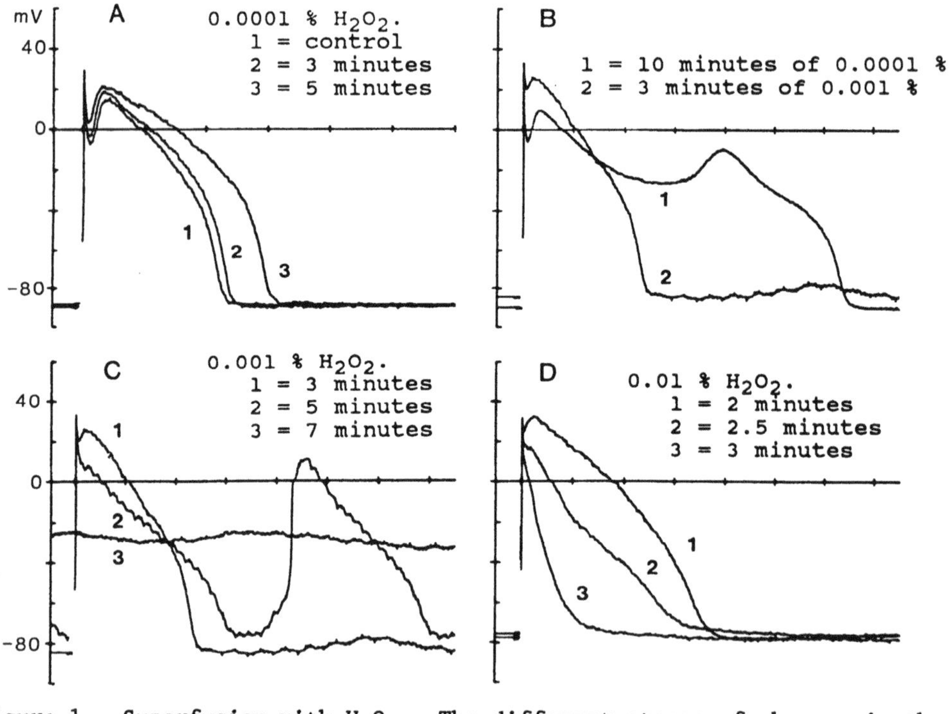

Figure 1. Superfusion with H_2O_2. The different stages of changes in the myocyte action potential during exposure to superfusion solutions containing H_2O_2. A) Stage 1: increase in the action potential plateau peak and duration and decrease in the pre-plateau notch. B) Stage 2: afterdepolarizations. C and D) Stage 3: partial depolarization and development of inexcitability. Timing marks indicate 100 msec intervals.

Table 1. Mean Time to Inexcitability with Different Concentrations of H_2O_2

Treatment	Inexcitable (minutes)	Excitable (30 minutes)	n
H_2O_2 0.01%	9 ± 2	0/6	6
H_2O_2 0.001%	10 ± 3	0/6	6
H_2O_2 0.0001%	19 ± 3	0/6	6
H_2O_2 0.00001%	27(1 cell)	4/5	5
H_2O_2 0.000001%	> 30	6/6	6
Catalase 100 μg/ml	> 30	4/4	4

Values are means ± SE.

The third stage was partial depolarization of the cell and development of inexcitability. Cells exposed to 0.001 and 0.0001% H_2O_2 failed to fully repolarize during stage two, and exhibited slow, continuous, oscillatory potentials (1C:#3). This resembles the development of stage 3 in cells exposed to the xanthine:xanthine oxidase-generating system[6] and 7 out of 10 cells exposed to dihydroxyfumaric acid.[5] Cells exposed to the highest concentration of H_2O_2 (0.01%) progressed quickly through stages 1 and 2. As the afterdepolarizations of stage 2 ceased, the plateau voltages became more negative and the action potential duration shortened as the plateau potential merged with the resting potential (1D). A similar pattern preceded the partial depolarization in a 3 out of 10 cells exposed to dihydroxyfumaric acid.[5] All cells exposed to 0.0001% H_2O_2 or higher concentrations became partially depolarized and inexcitable with the exposure time required to achieve inexcitability decreasing with higher concentrations of H_2O_2 (Table 1). Cells exposed to lower concentrations of H_2O_2 maintained a normal resting membrane potential and remained excitable for more than 35 minutes. The presence of catalase in the H_2O_2 solution eliminated the effects completely. Thus, exposure of cells to the H_2O_2 free-radical-generating system resulted in a three stage sequence of changes in the electrical activity of the cell. Many aspects of the observed changes, including the increased action potential duration, afterdepolarizations and inexcitability, could contribute to arrhythmias in whole hearts. The observed results suggest that if endogenously produced free radicals have access to the myocyte membrane then they could produce changes in electrical function that might cause arrhythmias.

The time dependence of the different electrophsyiological parameters are shown in Figure 2. The biphasic increase (stage 1) and then decrease (transition to stage 3) in the action potential duration occurred within 5 minutes in cells exposed to 0.01% H_2O_2. The 0.001 and 0.0001% H_2O_2 solutions also caused increases in the action potential duration, but at these lower concentrations the cells required a longer exposure to the solutions (3 (minutes and 5 minutes, respectively, $p < 0.05$). Shortening of the action potential duration was occasionally observed at these concentrations before the development of inexcitability but the tremendous variability in the action potential shape and duration caused by the afterdepolarizations (stage 2) prevented the assessment of action potential parameters. No significant effects on the action potential duration were observed during superfusion with 0.00001% or 0.000001% H_2O_2 for 35 minutes, although a slight prolongation in the duration of the action potential and afterdepolarizations were observed in cells exposed to the 0.00001% solution. The peak plateau voltage had a similar rapid onset of stage 1 changes with the highest H_2O_2 concentration having the fastest and greatest effect. Again no effects were observed with the 0.00001 and 0.000001% H_2O_2 concentrations. The pre-plateau notch voltage (data not shown) had a time course similar to the peak plateau voltage and action potential duration. Changes in the resting membrane potential did not occur as rapidly during H_2O_2 superfusion as the action potential duration or plateau voltage changes. The reduction of the resting membrane potential was seen only during superfusion with the highest H_2O_2 concentration. Similar rapid depolarizations occurred with 0.001 and 0.0001% H_2O_2 prior to the development of inexcitability but are not shown as they occurred during stage 2 afterdepolarizations. Changes in the upstroke velocity (data not shown) did not occur without an accompanying reduction of the resting membrane potential. Thus, superfusion of myocytes with H_2O_2 produced significant time-dependent changes that were first observed in the action potential duration and voltages of the plateau peak and pre-plateau notch.

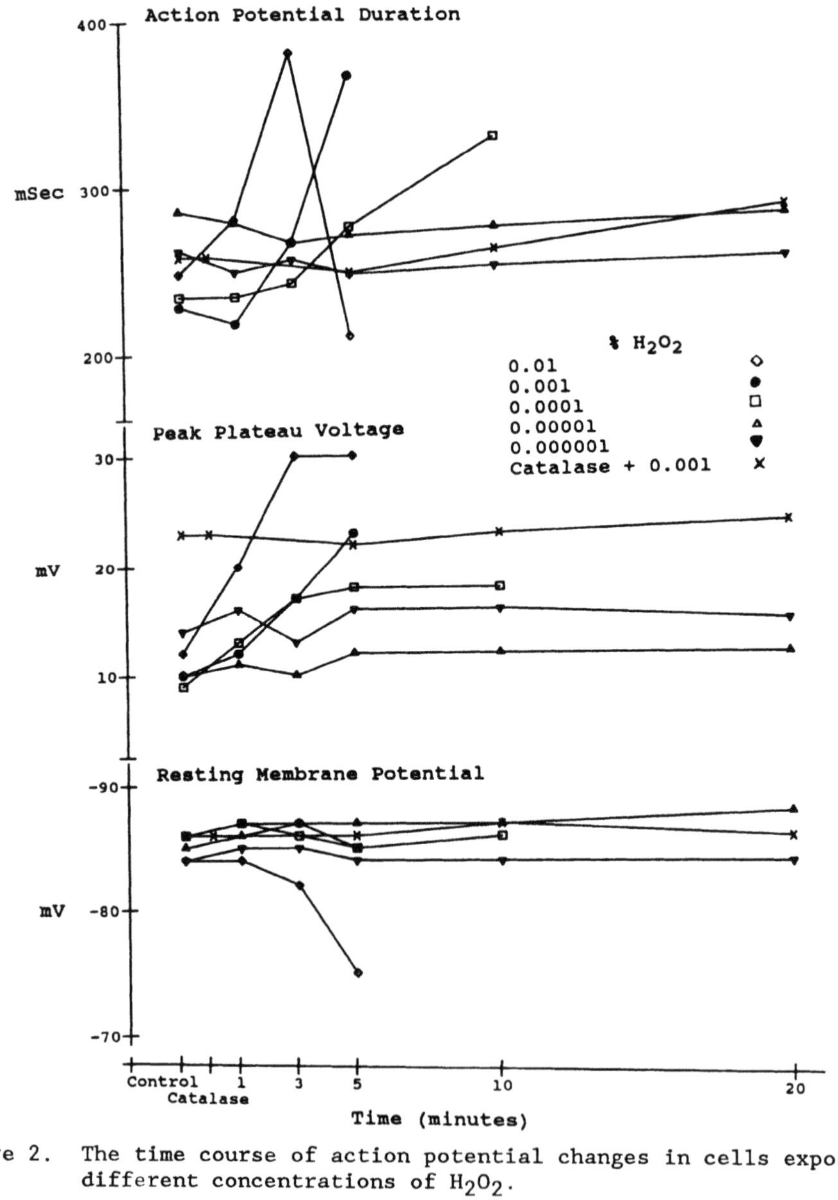

Figure 2. The time course of action potential changes in cells exposed to different concentrations of H_2O_2.

These initial changes were followed by depolarization of the resting membrane potential and inexcitability (Table 1).

The concentration dependence of the H_2O_2 effects on the myocyte action potential can be seen as the difference between pre-treatment control action potentials and the effects after 5 minutes of exposure to different concentrations of H_2O_2 (Figure 3). The parameters which changed earliest during exposure to H_2O_2 (peak plateau voltage, pre-plateau notch voltage and action potential duration) are effected after 5 minutes of exposure to 0.0001% H_2O_2. Only the highest concentration of H_2O_2 (0.01%) caused changes in the resting membrane potential after just 5 minutes of exposure. The effects of H_2O_2 on the action potential duration and peak plateau voltage were reversed after 5 minutes at this high concentration. These results suggest that the ionic currents responsible for the plateau and the duration of the action potential are more sensitive to the effects of H_2O_2 solutions.

The ionic currents that contribute to the action potential can be altered by changing the ionic distribution across the membrane. Increases in intracellular calcium can alter the height and shape of the action potential plateau and cause afterdepolarizations similar to those seen during exposure to free radical generating systems. Free radicals have been reported to alter the permeability of membranes to calcium. Thus, we investigated whether

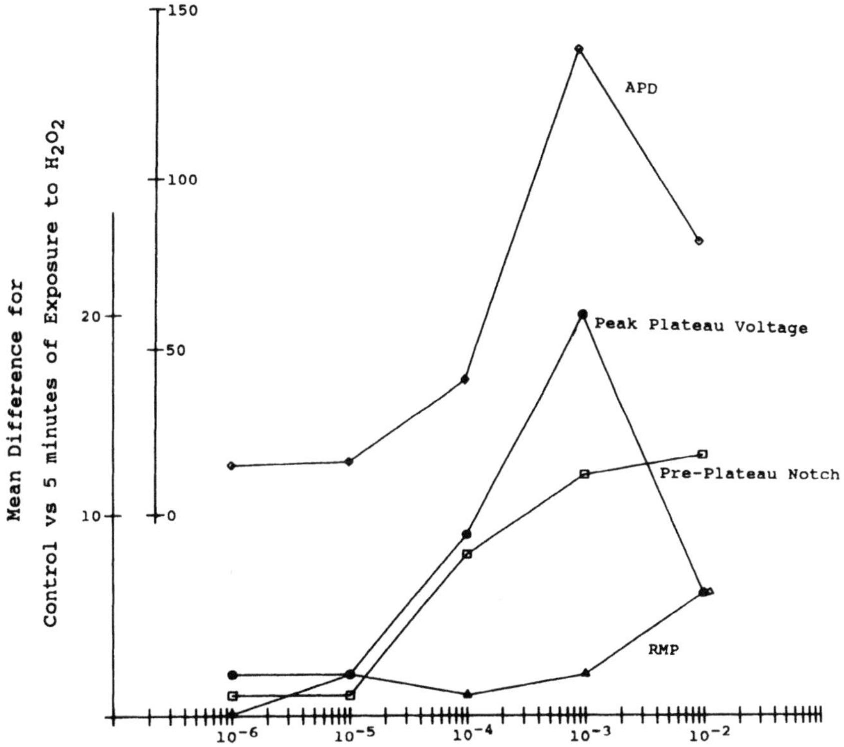

Figure 3. The concentration dependence of the different action potential parameters after 5 minutes of exposure to H_2O_2. The differences for the peak plateau voltage, pre-plateau notch and resting membrane potential (RMP) are reported in mVolts. The difference for the action potential duration is reported in msec.

Table 2. Mean Time to Inexcitability with Reduced Calcium or Different Stimulus Rates

Treatment	Inexcitable (minutes)	Excitable (30 minutes)	n
H_2O_2 0.001% at 1 Hz	10 ± 3	0/6	6
1.25 mM Ca^{2+}	12 ± 3	0/7	7
Quiescence 4.5 minutes	11 ± 4	0/6	6
0.5 Hz	11 ± 3	0/5	5
2 Hz	11 ± 2	0/6	6

Values are means ± SE.

procedures that modify calcium gradients across the cell (reduced extracellular calcium) or alter the influx of calcium into the cell (stimulus rate) would change the response of a cell to the free-radical-generating system. The reduction of extracellular calcium from 5 to 1.25 mM decreased the plateau voltage and action potential duration during control superfusion but did not alter the three stages of free radical effects nor the time course for the development of inexcitability during exposure to 0.001% H_2O_2 (Table 2). Thus the effects of H_2O_2 do not appear to be dependent on the extracellular calciur concentration. Similarly, the effects of the H_2O_2 are not dependent on the stimulation rate since the sequence and time course of effects during exposure to 0.001% H_2O_2 was unaltered by quiescence during the first 4.5 minutes of H_2O_2 exposure or stimulation rates from 0.5 to 2 Hz (Table 2).

We have demonstrated that our H_2O_2 free-radical-generating system produces the same three stages of effects on the myocyte action potential as other extracellular free radical generating systems.[5,6] The initial specificity of the changes for the plateau potential and action potential duration indicate that free radicals probably alter specific currents that contribute to the myocyte action potential. This observation suggests that certain functions of membranes in living cells may be more susceptible and, therefore, preferentially altered during exposure to free radicals.

REFERENCES

1. R. Kehoe, D.H. Singer, A. Trapani, M. Billingham, R. Levandowski, and J. Elson, Adriamycin-induced cardiac dysrhythmias in an experimental dog model, Cancer Treat. Rep. 62:963 (1978).
2. A.S. Manning, D.J. Coltart, and D.J. Hearse, Ischemia and reperfusion-induced arrhythmias in the rat. Effects of xanthine oxidase inhibition with allopurinol, Circ. Res. 55:545 (1984).
3. B. Woodward, and M.N.M. Zakaria, Effects of some free radical scavengers on reperfusion induced arrhythmias in the isolated rat heart, J. Mol. Cell. Cardiol. 17:485 (1985).
4. M. Bernier, D.J. Hearse, and A.S. Manning, Reperfusion-induced arrhythmias and oxygen-derived free radicals. Studies with "anti-free radical" interventions and a free radical-generating system in the isolated perfused rat heart, Circ. Res. 58:331 (1986).
5. P.L. Barrington, C.F. Meier, Jr., and W.B. Weglicki, Effect of free radicals on the canine myocyte action potential, Fed. Proc. 44:1578 (1985).
6. P.L. Barrington, C.F. Meier, Jr., B.F. Dickens, and W.B. Weglicki, Free radical scavengers protect canine myocytes from radical-induced changes in the action potential, Circ. 72 (suppl III):228 (1985).

ROLE OF OXYGEN RADICALS IN ENDOTOXIN SHOCK AND DISSEMINATED INTRAVASCULAR COAGULATION

T. Yoshikawa, N. Yoshida, H. Miyagawa, T. Tanigawa,
T. Takemura, Y. Morita, S. Sugino, and M. Kondo

First Department of Medicine
Kyoto Prefectural University of Medicine
Kamigyo-ku, Kyoto, 602, Japan

INTRODUCTION

The generation of oxygen free radicals seems to be implicated in the pathogenesis of shock and disseminated intravascular coagulation (DIC) caused by endotoxin[1,2]. Radical generation is increased in some critical states directly related to shock and DIC, such as tissue hypoxia[3], incomplete ischemia and tissue reperfusion[4], activation of arachidonic acid cascade[5], and complement-induced granulocyte aggregation[6].

It was reported that exposure to endotoxin can prime neutrophils for increasing their release of superoxide[7] and that it may ultimately result in death[8]. Endotoxin activates the complement system, producing C5a. Stimulated by the complement components, polymorphonuclear leukocytes (PMNs) generate significant amounts of superoxide anion[9].

Lipid peroxidation is the reaction of oxidative deterioration of polyunsaturated lipids. Peroxidation involves the direct reaction of oxygen and lipid to form free radical intermediates and to produce toxic peroxides.

In the following communication, we will briefly review recently obtained experimental data which demonstrate the role of oxygen radicals, complement system, and PMNs in the pathogenesis of shock and DIC induced by endotoxin.

MATERIALS AND METHODS

The animals used in the study were female rats (190-220 g) of the Wistar strain obtained from Keari Co., Ltd., Osaka. Endotoxin [Escherichia coli 055:B5 lipopolysaccharide (LPS) B; Difco Lab., Detroit, Michgan] was dissolved in pyrogen-free physiological saline before every experiment.

Experimental shock was induced by a single injection of 100 mg/kg of endotoxin, diluted in 1.0 ml of physiological saline, into the femoral vein. Experimental DIC was induced by sustained infusion of 100 mg/kg of endotoxin, diluted in 11.4 ml of physiological saline, into the femoral vein for 4 h using a syringe.

For examination of the effects of superoxide dismutase (SOD) and catalase on the experimental shock and DIC, the rats were injected subcutaneously with SOD from bovine blood (3,050 U/mg protein. Sigma Chemical Co., St. Louis, MO) at 50.0 mg/kg or catalase from bovine liver (40,000 U/mg protein, Sigma Chemical Co., St. Louis, MO) at 1.0 mg/kg 12 and 1 h before the injection of endotoxin (100 mg/kg).

Rats were intraperitoneally injected with 200 U/kg body weight (in 0.5 ml sterile saline) of purified cobra venom factor (CVF) 6 h before the endotoxin treatment to induce complement depletion. Rats were depleted of leukocytes by subcutaneous injection of 120 mg/kg of cyclophosphamide monohydrate (CPM) 4 days prior to endotoxin treatment. K76-COOH (provided by Otsuka Pharmaceutical Co., Tokushima, Japan), an oxidized derivative of a sesquiterpene compound which inhibits complement activity, especially the C5 step[10] (Figure 1), was injected intraperitoneally at a dose of 3.0 mg/kg, 30 min before the endotoxin treatment.

The serum α-tocopherol content and thiobarbituric acid (TBA) reactive substances were determined by the method of Abe et al.[11] and by the method of Yagi[12], respectively. The severity of shock was determined by systolic blood pressure[13], and lysosomal enzymes, such as acid phosphatase[14] and β-glucuronidase[15]. The severity of DIC was determined by various parameters, such as fibrinogen and fibrin degradation products (FDP)[16], fibrinogen[17], prothrombin time (PT)[18], partial thromboplastin time (PTT), platelet count, and percent glomerular fibrin deposites (GFD); 100 glomeruli were counted and those having fibrin thrombi were expressed as a percentage.

RESULTS

Endotoxin shock

Immediately after the single injection of endotoxin (100 mg/kg), systolic blood pressure was reduced and heart rate was increased. Serum acid phosphatase and β-glucuronidase activities were increased. These changes were most remarkable 45 min after the injection of endotoxin[1]. After the treatment with endotoxin, serum α-tocopherol/cholesterol ratio was significantly decreased (Figure 2), and serum TBA reactive substances were significantly increased (Figure 3). These results suggest that peroxidation may occur in several organs, and that lipid peroxides produced by radicals may aggravate these pathological conditions.

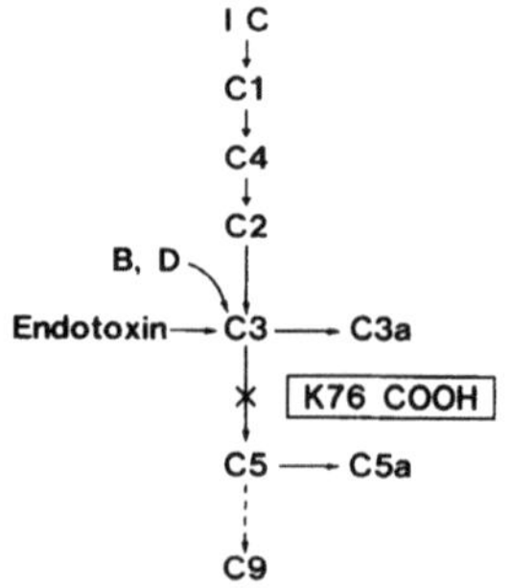

Figure 1. Inhibition of complement activation by K76-COOH.

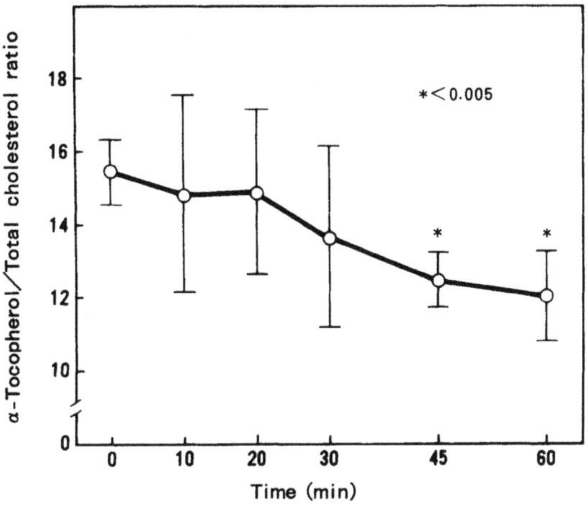

Figure 2. Changes in serum α-tocopherol/cholesterol ratio (10^{-3}) after the i.v. injection of endotoxin (100 mg/kg). Results are expressed as mean value ± SD of 8 rats. $p < 0.005$ for difference from the value before the endotoxin injection.

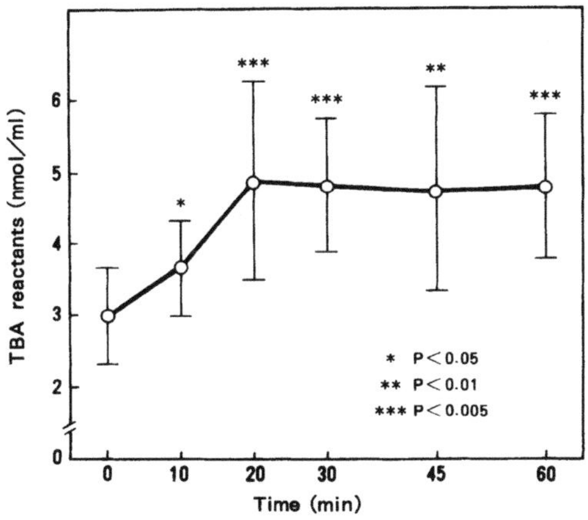

Figure 3. Changes in serum TBA reactive substances after the i.v. injection of endotoxin (100 mg/kg). Results are expressed as mean value ± SD of 8 rats.

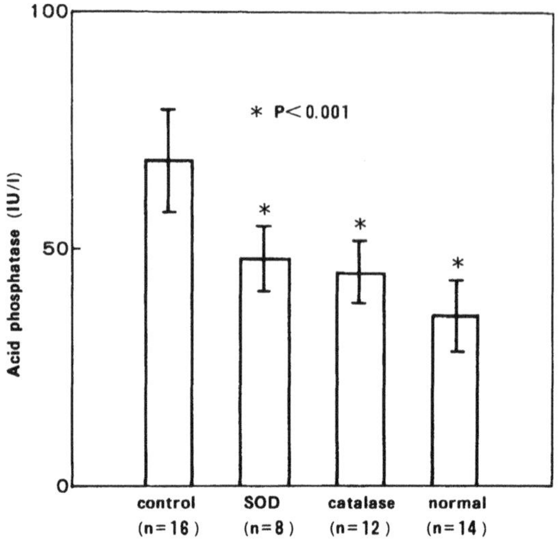

Figure 4. Effects of SOD (50 mg/kg) and catalase (1.0 mg/kg) on serum acid phosphatase activity 45 min after the endotoxin injection. Results are expressed as mean value ± SD.

The reduction of systolic blood pressure after the endotoxin injection was significantly inhibited by the treatment with 50 mg/kg of SOD, or by 1.0 mg/kg of catalase. The increase in serum lysosomal enzymes was also significantly inhibited by the administration of SOD or catalase (Figure 4). The increase in serum TBA reactants was inhibited by the administration of SOD, but not by catalase. SOD was partially inactivated by heating. The heat denatured SOD did not show any protective effect on endotoxin-induced shock states.

When rats were depleted of either complement (by i.p. injection of CVF) or leukocytes (by s.c. injection of CPM) prior to endotoxin injection, shock states were remarkably prevented. Moreover, K76-COOH, an anticomplementary agent prevents the aggravation of endotoxin shock (Figure 5).

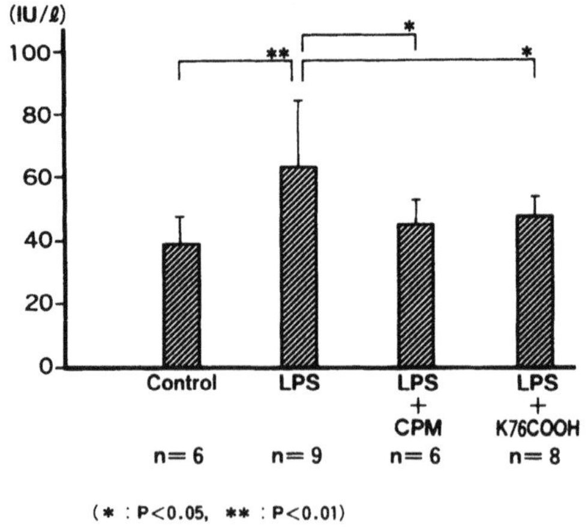

Figure 5. Effect of CPM (120 mg/kg) and K76-COOH (3 mg/kg) on serum acid phosphatase activity 45 min after the endotoxin injection.

Table 1. Effects of SOD (50 mg/kg) and Catalase (1.0 mg/kg) on Experimental DIC in Rats

	FDP (µg/ml)	PT (s)	Platelet count ($x10^9/1$)	GFD (%)	N
Control	56.1 ± 12.1	32.4 ± 6.4	66 ± 9	87.6 ± 12.4	12
SOD	12.4 ± 6.2*	19.1 ± 4.2*	221 ± 71*	28.6 ± 5.4*	16
Catalase	13.9 ± 6.1*	19.2 ± 5.7*	227 ± 35*	44.0 ± 8.0*	19
Normal	3.8 ± 0.8*	12.7 ± 0.4*	603 ± 81*	0*	12

Results are expressed as mean value ± SD. *p<0.001 for difference to the value of control rats by Student's *t*-test.

These observations provide striking evidence that both complement activation products and blood leukocyte are important factors in the pathogenesis of endotoxin shock.

Endotoxin DIC

Experimental DIC could be induced by 4 h sustained infusion of endotoxin (100 mg/kg), diluted in 11.4 ml of physiological saline, into the femoral vein[20]. FDP and %GFD increased. PT and PTT were prolonged, fibrinogen and platelet count were depressed, and TBA reactive substances in serum increased gradually during the infusion of endotoxin[20-22]. These changes were remarkable in vitamin E deficient rats and only slight in those supplemented with vitamin E[23]. By treatment with SOD, catalase, CPM, K76-COOH or CVF, the aggravation of DIC states was prevented (Table 1, Figures 6, 7).

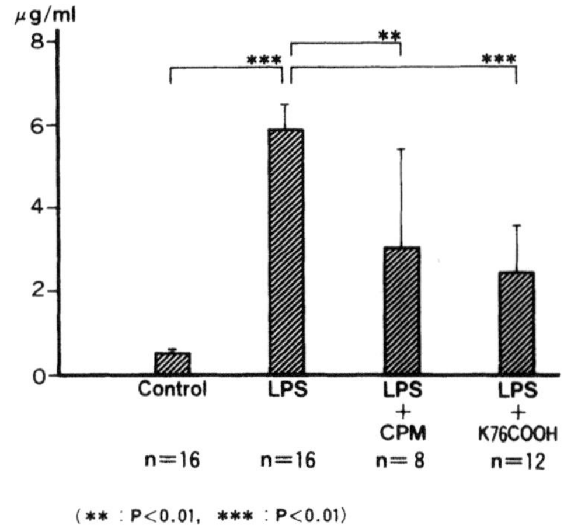

Figure 6. Effects of CPM (120 mg/kg) and K76-COOH (3 mg/kg) on FDP levels 4 h after the endotoxin infusion. Results are expressed as mean value ± SD.

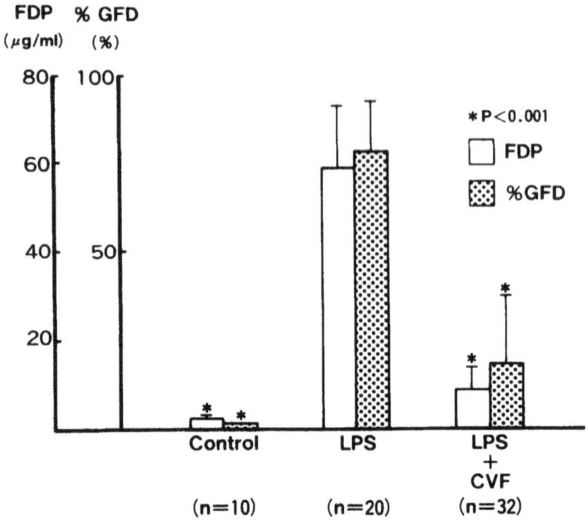

Figure 7. Effects of CVF (200 U/kg) on FDP levels and %GFD 4 h after the endotoxin infusion. Results are expressed as mean value ± SD.

DISCUSSION

Endotoxin has many biological effects. Hydrogen peroxide is released from macrophages activated by endotoxin[24]. In addition, endotoxin activates the complement system. In shock and DIC states, systolic blood pressure was decreased and ischemia could be induced in several organs. Increasing evidence indicates that ischemia-associated organ injury is related to the production of oxygen radicals either within affected tissue or by PMNs. The mechanism for the production of oxygen radicals in ischemic tissues appears to be affected by changes in purine metabolism within ischemic cells.

Hydrogen peroxide and superoxide radicals have been reported to cause aggregation of blood platelets[25,26]. Platelet aggregation has been shown to be inhibited by a variety of antioxidants and by scavengers of hydroxyl radicals and singlet oxygen[27,28]. The injury to the endothelial cells, platelet or any kind of tissues may activate the coagulation system. These findings suggest that reactive oxygen species may play some role in the thrombus formation in the experimental DIC induced by endotoxin. Treatment of rats with SOD or catalase results in significant protection from endotoxin-induced shock and DIC.

K76-COOH, a monocarboxylic acid derivative of K-76, purified from the culture filtrate of Stachybotrys complementi nov. sp. K-76, inhibits complement activity. Its inhibitory action is mainly on C5 step[10]. The anticomplementary agent inhibits the aggravation of endotoxin-induced shock and DIC. Depletion experiments have also suggested the essential role played by complement-activated PMNs in the pathogenesis of shock and DIC.

A number of scientists observed that oxygen radical-generating systems can stimulate the peroxidation of fatty acids or membranes, and peroxidation may be initiated by hydroxyl radicals produced by iron-catalysed Haber-Weiss reaction. The oxygen radicals produced may directly damage the cell membrane, which contains a large amount of polyunsaturated fatty acids. Furthermore, the increased lipid peroxides in the blood would be deleterious to the blood vessels and to other intact organs and tissues. These processes enhance vascular permiability, histamine secretion, platelet aggregation and

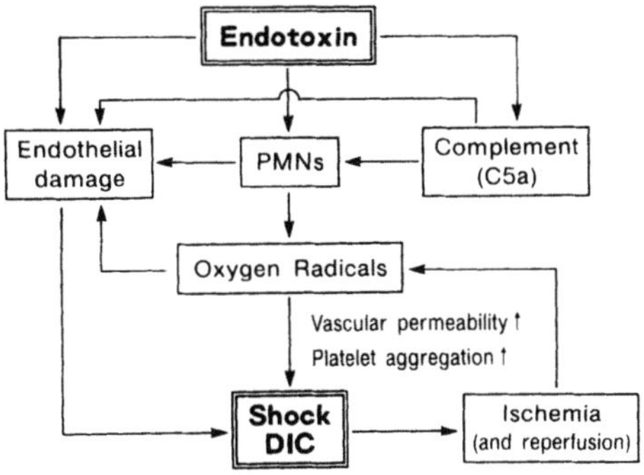

Figure 8. Postulated scheme of pathophysiologic events leading to shock and DIC induced by endotoxin.

generation of chemotactic factors. In shock and DIC, an insufficient defense system and excessive free radical generation may result in severe cell damage.

These observations provide evidence that complement activation products, PMNs and oxygen radicals are important factors in the pathogenesis of endotoxin-induced shock and DIC (Figure 8).

REFERENCES

1. T. Yoshikawa, M. Murakami, O. Seto, Y. Kakimi, T. Takemura, T. Tanigawa, S. Sugino, and M. Kondo, Effects of superoxide dismutase and catalase on endotoxin shock in rats, J. Clin. Biochem. Nutr. 1:165 (1986).
2. T. Yoshikawa, M. Murakami, N. Yoshida, O. Seto, and M. Kondo, Effects of superoxide dismutase and catalase on disseminated intravascular coagulation in rats, Thromb. Haemostas. 50:869 (1983).
3. T. Yoshikawa, Y. Furukawa, Y. Wakamatsu, S. Takemura, H. Tanaka, and M. Kondo, Experimental hypoxia and lipid peroxide in rats, Biochem. Med. 27:207 (1982).
4. R.F. Del Maestro, H.H. Thaw, J. Biork, M. Planker, and K.E. Arfors, Free radicals as mediators of tissue injury, Acta Physiol. Scand. Suppl. 492:43 (1980).
5. A.M. Lefer, H. Arai, and S. Okamatsu, Beneficial action of a free radical scavenger in traumatic shock and myocardial ischemia, Circ. Shock 8:273 (1981).
6. I.M. Goldstein, D. Roos, H.B. Kaplan, and G. Weissman, Complement and immunoglobulins stimulate superoxide production by human leukocytes independently of phagocytosis, J. Clin. Invest. 56:1155 (1975).
7. R.B.Jr. Johnston, L.A. Guthrie, and L.C. McPhail, Priming of neutrophils for enhanced oxidative metabolism by bacterial endotoxin, in:"Oxy Radicals and Their Scavenger Systems," Vol. II, R.A. Greenwald, and G. Cohen, eds., Elsevier Science Publishing Co., New York (1983).
8. B. Gray, Effects of superoxide dismutase on lipopolysaccharide-stressed mice and alteraction of lung enzyme levels by endotoxin, Toxicol. Appl. Pharmacol. 60:479 (1981).
9. I.M. Goldstein, M. Cerqueira, S. Lind, and H.B. Kaplan, Evidence that the superoxide-generating system of human leukocytes is associated with cell surface, J. Clin. Invest. 59:249 (1977).
10. K. Hong, T. Kinoshita, W. Miyazaki, T. Izawa, and K. Inoue, An anti-complementary agent, K-76 monocarboxylic acid: Its site and mechanism of inhibition of the complement activation cascade, J. Immunol. 122: 2418 (1979).

11. K. Abe, Y. Yuguchi, and G. Katsui, Quantitative determination of tocopherols by high-speed liquid chromatography, J. Nutr. Sci. Vitaminol. 21:183 (1975).
12. K. Yagi, A simple fluorometric assay for lipid peroxide in blood plasma, Biochem. Med. 15:212 (1976).
13. R. Bunag, Validation in awake rats of a tail-cuff method for measuring systolic pressure, J. Appl. Physiol. 34:279 (1973).
14. M.A. Andersch and A.J. Szczypinski, Use of p-nitrophenylphosphate as the substrate in determination of acid phosphatase, Am. J. Clin. Path. 17:571 (1947).
15. W.H. Fishman, K. Kato, C.L. Anstiss, and S. Green, Human serum β-glucuronidase; its measurement and some its properties, Clin. Chim. Acta, 15:435 (1967).
16. H.C. Ferreira and L.G. Murat, Immunological method for demonstrating fibrin degradation products in serum and its use the diagnosis of fibrinolytic states, Br. J. Haematol, 9:299 (1963).
17. J.F. Goodwin, Estimation of plasma fibrinogen using sodium sulfate fractionation, Am. J. Clin. Pathol. 35:277 (1961).
18. A.J. Quick. B.M. Stanly, and F.W. Bancroft, A Study of the coagulation defect in hemophilia and in jaundice, Am. J. Med. Sci. 190:501 (1935).
19. R.D. Langdell, R.H. Wager, and K.M. Brinkhous, Effects of antihemophilic factor on one-stage clotting time. J. Lab. Klin. Med. 47:637 (1953).
20. T. Yoshikawa, Y. Furukawa, M. Murakami, S. Takemura, and M. Kondo, Experimental model of disseminated intravascular coagulation induced by sustained infusion of endotoxin, Res. Exp. Med. 179:223 (1981).
21. T. Yoshikawa, M. Murakami, Y. Furukawa, H. Kato, S. Takemura, and M. Kondo, Lipid peroxidation and experimental disseminated intravascular coagulation in rats induced by endotoxin, Throm. Haemostas, 49:214 (1983).
22. T. Yoshikawa, O. Seto, K. Itani, Y. Kakimi, S. Sugino, and M. Kondo, Endotoxin-induced disseminated intravascular coagulation and shock, and their relationship with oxygen-derived free radicals, in: "Oxygen Free Radicals in Shock," G.P. Novelli and F. Ursini, eds., Kager, Basel (1986).
23. T. Yoshikawa, M. Murakami, and M. Kondo, Endotoxin-induced disseminated intravascular coagulation in vitamin E deficient rats, Toxicol, Appl. Pharmacol. 74:173 (1984).
24. H. Kato, H. Yaoita. T. Taki, and M. Nakano. LPS-induced hydrogen peroxide release from peritoneal macrophages of normal and immunodefective mice, in: "Immunopharmacology of Endotoxicosis," M.K. Agarwal and M. Yoshida eds., Walter de Gruyter, Berlin (1984).
25. R.I. Handin, R. Carabin, and G.J. Boxer, Enhancement of platelet function by superoxide anion, J. Clin. Invest. 59:959 (1977).
26. O. Higashi and Y. Kikuchi, Effects of vitamin E on the aggregation and the lipid peroxidation of platelets exposed to hydrogen peroxide, Tohoku J. Exp. Med. 112:271 (1974).
27. M. Steiner and J, Anastasi, Vitamin E. An inhibitor of the platelet release reaction, J. Clin. Invest. 57:732 (1976).
28. M. Greenberg, R.W. Grady, and C.M. Peterson, Inhibition of platelet function with 2,3-dihydroxybenzoic acid, Br. J. Haemat. 37:569 (1977).

PLASMA LIPID PEROXIDATION AND ATHEROSCLEROSIS

IN RESTRICTED OVULATOR CHICKENS

Terrance L. Smith and Fred A. Kummerow

Burnsides Research Laboratory
1208 W. Pennsylvania Avenue
Urbana, Illinois 61801, U.S.A.

INTRODUCTION

Restricted ovulator (RO, also called non-layer or NL) chickens carry a genetic mutation which results in the cessation of egg laying when under a prolonged photoperiod. As a consequence these chickens develop an extreme hyperlipidemia with blood lipid levels over 3,000 mg/triglyceride dl and 700 mg/dl cholesterol. Within 12 months these chickens develop atherosclerotic lesions in both the coronary artery[1] and aorta.[2] It was originally assumed the hyperlipidemia was responsible for the higher rate of atherosclerosis in these chickens. However, we also found a significant increase in plasma lipid peroxidation in RO chickens.[3] This increase in peroxidation correlates closely with the triglyceride levels.

METHODS

Since then we have been investigating the effects of plasma lipid peroxidation on atherogenesis. Normal laying hens and RO hens were kept in individual wire cages and allowed food and water _ad libitum_. The feed was a commercial cracked corn layer ration including a vitamin and mineral mix. Hens were randomly assigned to one of four groups: normal hens/normal diet, normal hen/vitamin E supplemented, RO hen/normal diet, and RO hen/vitamin E supplemented. Diets supplemented with vitamin E contained 1,000 I.U. of vitamin E per kilogram of feed. Each group contained 10 hens. After 12 weeks of treatment, blood was drawn from a wing vein into heparinized tubes and plasma was isolated by centrifugation. This was assayed for triglyceride using a clinical blood assay kit (Sigma, St. Louis, MO). Plasma lipid peroxidation was assayed by the fluorescent thiobartituric acid method of Yagi.[4] Intimal thickening of the aorta was used as an indicator of atherogenesis. Hens were sacrificed by lethal injection and the aortas removed. Samples were taken just below the aortic arch, fixed, embedded, and stained with alkaline toluidine blue.[5] Samples were thick sectioned and, mounted and intimal thickness were measured with an ocular micrometer. Data were analyzed using Tukey's method for multiple comparisons.[6]

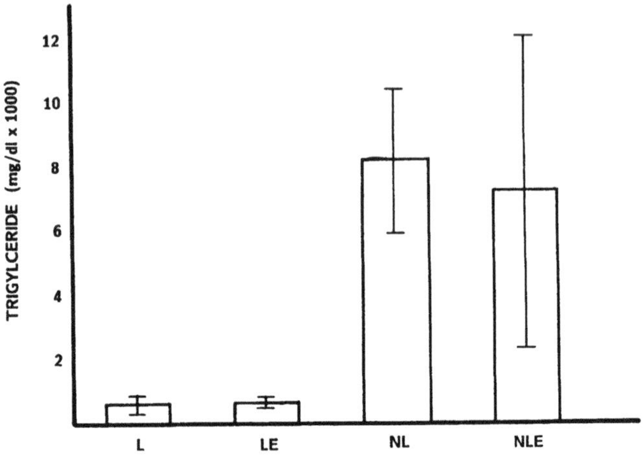

Figure 1. Plasma triglyceride L, layer; LE, layer with additional vitamin E; NL, non-layer, NLE, non-layer with additional vitamin E. L, LE and NL, NLE are statistically the same. L and LE are significantly different from NL and NLE. Tukey's test, $\alpha = 0.05$.

RESULTS AND DISCUSSION

The RO chickens had plasma triglyceride and cholesterol levels that were over tenfold greater than laying hens (Figure 1). Vitamin E had no effect on the hyperlipidemia of these chickens. However, vitamin E did reduce the plasma peroxidation of RO chickens from a more than fourfold increase above layers to the same as the layers (Figure 2). There was no further effect of vitamin E on the plasma peroxidation of layer hens. The intimal thickness of RO hens increased 1.6-fold but no

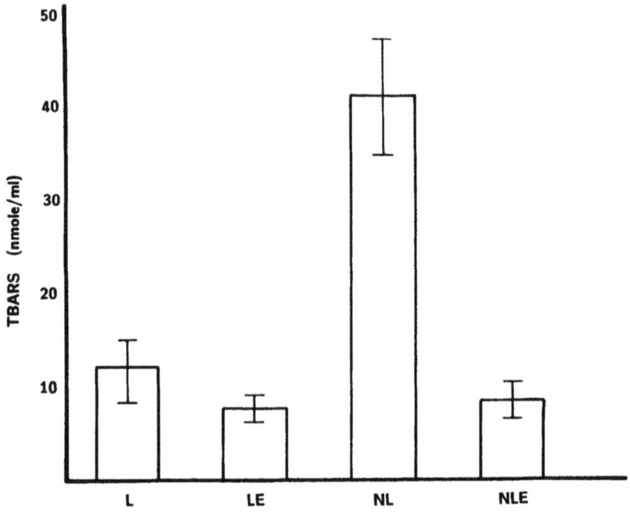

Figure 2. Plasma lipid peroxidation as malondialdehyde. L, layer; LE, layer with additional vitamin E; NL, non-layer; NLE, Non-layer with additional vitamin E. NL is significantly greater than L, LE and NLE. The latter are not different. Tukey's test $\alpha = 0.05$.

increase in intimal thickness was found in RO hens that were fed vitamin E (Figure 3).

The increase in plasma lipid of RO hens is due to inhibition of the turnover of blood lipids. Lipid synthesis is reduced due to feedback inhibition but still proceeds at a significantly greater pace than removal. The plasma peroxidation parallels this elevation in triglyceride, although the source of these peroxides is not known. That the RO vitamin E supplemented hens, which are hyperlipidemic but had basal levels of peroxidation, failed to demonstrate intimal thickening suggests that lipid peroxides may play a role in atherogenesis. Vitamin E does not appear to have had a direct effect on intimal thickness, as the layers with and without supplementation showed no differences in intimal thickening. It is not clear at what point lipid peroxides are affecting the tissue. There may be an effect directly on the endothelial cells, causing an increased uptake of lipid into the intima. Peroxidation-modified LDL is toxic to cells and is taken up by the "scavenger" pathway.[6] It is possible that low levels of peroxides obtained by endothelial cells in this manner could injure the cells, causing an alteration in their function without killing them. A second possibility is that the elevated peroxide level alters the balance between prostacyclin and thromboxin. Decreased prostacyclin synthesis has been reported in the presence of lipid peroxides.[8] A consequence of this would be greater platelet aggregation, and enhanced platelet aggregation has been advanced as a factor in atherogenesis through their potential damaging effect on the endothelium.

Lipid peroxidation of aortic tissue was first noted in association with heart disease by Glavind[9] in 1951. Since that time, blood lipid peroxidation has been found to be elevated in heart disease patients.[10,11] Elevated blood peroxidation has been found in diabetes[10] and smoking[12], both of which are risk factors for heart disease. When rabbits are injected with the peroxidation products of linoleic acid, there is a rapid alteration in the morphology of the aortic endothelial cell layer. This change in morphology is typical of those seen in the very early stages of atherosclerosis.[13] Lipid peroxidation products are known to be toxic. Secondary products such as malondyaldehyde are known to be capable of binding to and altering every class of cellular molecular constituent. Therefore, it should not be

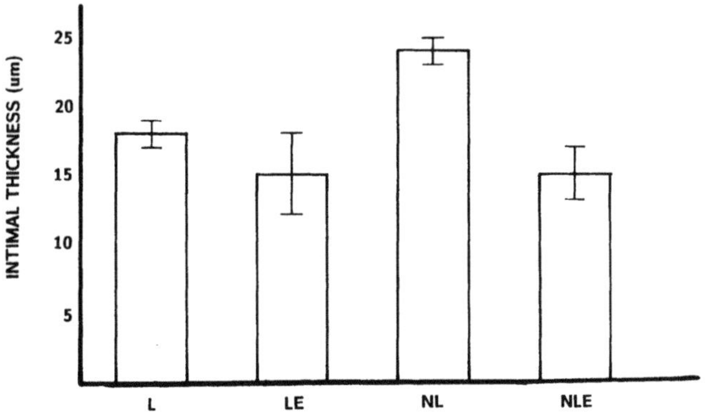

Figure 3. Aortic intimal thickness. L, layer; LE, layer with additional vitamin E. NL; Non-layer, NLE; Non-layer with additional vitamin E. L, LE and NLE are statistically the same while NL is greater than the other three. Tukey's test $\alpha = 0.05$.

surprising that lipid peroxidation products might play a role in atherosclerosis.

We conclude that serum lipid peroxidation plays a role in atherogenesis as measured by intimal thickening. It appears that peroxidation is of greater importance than hyperlipidemia as those chickens which were hyperlipidemic but had normal serum peroxide levels showed no intimal thickening.

ACKNOWLEDGEMENT

Dr. James Bitgood of the University of Wisconsin, Department of Poultry Science, supplied the RO chickens used in this study.

REFERENCES

1. T. Toda, D. Leszczynski, W. H. Gibbon and F. A. Kummerow, Coronary arterial lesions in sexually mature non-layers, layers, and roosters. Virchows Arach. A. Pathol. and Histol. 388:123 (1980).
2. K. Tokuyasu, H. Imai, S. Taura, B. H. S. Cho and F. A. Kummerow. Aortic lesions in nonlaying hens with endogenous hyperlipidemia. Arch. Pathol. Lab. Med. 104:41 (1980).
3. T. L. Smith, T. Toda and F. A. Kummerow. Lipid peroxidation in hyperlipidemic chickens. Atherosclerosis 57:119 (1985).
4. K. Yagi, A simple fluorometric assay for lipoperoxide in blood plasma. Biochem. Med. 15:212 (1976).
5. T. Smith, T. Toda, Y. Toda and F.A. Kummerow, The effect of elastase on chickens with endogenous hyperlipidemia. Biochem. Med. Metab. Biol. 37:96 (1987).
6. R. G. D. Steel and J. H. Torrie, Principles and Procedures of Statistics, McGraw-Hill Book Co., Inc., New York (1960).
7. O. Stein and Y. Stein, Bovine aortic endothelial cells display macrophage like properties toward acetylated ^{125}I-labeled low density lipoprotein. Biochim. Biophys. Acta 620:631 (1980).
8. S. Moncada, R.J. Gryglewski, S. Bunting and J.R. Vance, A lipid peroxide inhibits the enzyme in blood vessel microsomes that generates from prostaglandin endoperoxide the substance which prevents platelet aggregation. Prostaglandin 12:715 (1960).
9. J. Glavind and S. Hartmann, The occurrence of peroxidized lipids in atheromatous human aortas. Experientia 7:464 (1951).
10. Y. Goto, Lipid peroxides as a cause of vascular diseases. In: "Lipid Peroxides in Biology and Medicine," K. Yagi, ed. Academic Press, City, (1982).
11. Y. Soto, N. Hotta, N. Sakamoto, S. Matsuoka, N. Ohishi, and K. Yagi, Lipid peroxide level in plasma of diabetic patients. Biochem. Med. 21:104 (1979).
12. B. C. Blackman, P. White, W. Tsou, and D. Finkel, Peroxidation of plasma and platelet lipids in chronic cigarette smokers and insulin-dependent diabetics. Ann. N.Y. Acad. Sci. 435:385 (1984).
13. K. Yagi, H. Ohkawa, N. Ohisi, M. Yamashita, and T. Nakashima. Lesion of aortic intima caused by intravenous administration of linoleic and hydroperoxide. J. Appl. Biochem. 3:58 (1981).

ENZYMATIC LIPID PEROXIDATION: A STEP IN THE BREAKDOWN OF MITOCHONDRIA DURING THE MATURATION OF RED BLOOD CELLS

Hartmut Kühn, Tankred Schewe, Samuel M. Rapoport and Alan R. Brash*

Institute of Biochemistry, Humboldt University
Hessische St. 3-4, 1040, Berlin, GDR

*Department of Clinical Pharmacology
Vanderbilt University School of Medicine
Nashville, Tennessee 37232, U.S.A.

INTRODUCTION

The maturation of red blood cells is characterized by a loss of cellular organelles such as the ribosomes and mitochondria and by a partial breakdown of the cell membrane with a loss of various receptors and active transport systems. The lipoxygenase of reticulocytes was originally discovered as a protein in rabbit reticulocyte cytosol which could damage isolated rat liver mitochondria and cause an irreversible inactivation of respiratory enzymes. This protein was later shown to be a cell specific lipoxygenase; it was purified to electrophoretic homogeneity[1] and its chemical and enzymatic properties have been described.[2] In this paper we report recent results on the interaction of the pure reticulocyte lipoxygenase with esterified fatty acids and biological membranes in vitro, and on the intracellular action of the enzyme during the maturation process of reticulocytes.

RESULTS

Interaction with Phospholipids and Biological Membranes

The pure reticulocyte lipoxygenase oxygenates polyenoic fatty acids, in particular those with an n-8 doubly allylic methylene group, mainly to their corresponding n-6 hydroperoxy derivatives. Linoleic acid is specifically converted to $13L_S$-hydroperoxy-9Z,11E-octadecadienoic acid ($13L_S$-HPODE); arachidonic acid is oxygenated to its $15L_S$- and $12L_S$-hydroperoxy derivatives in a ratio of about 10:1.[3] In contrast to most lipoxygenases examined so far, the reticulocyte enzyme effectively oxygenates esterified fatty acids, phospholipids, and biological membranes. Incubation of the enzyme with a suspension of soybean phosphatidylcholine is characterized by a burst of oxygen uptake (Figure 1A), indicating the oxygenation of the phospholipids. The structures of the oxygenated phosphatidylcholine were described.[4]

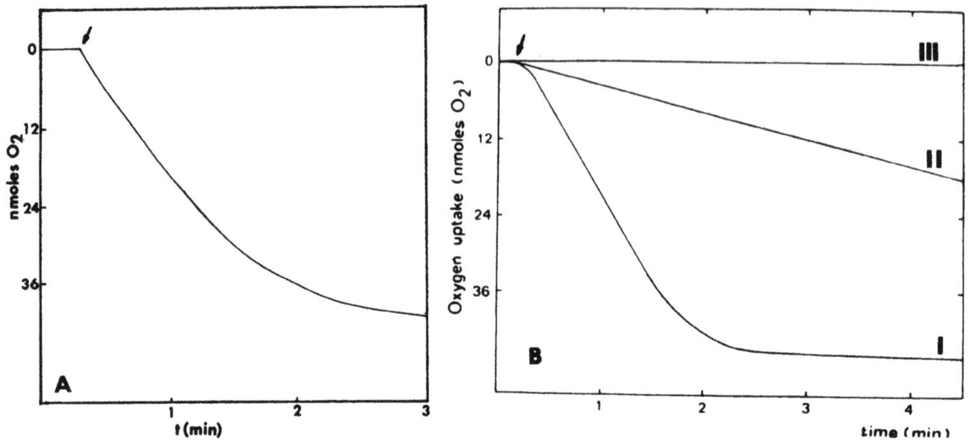

Figure 1. Reaction of the reticulocyte lipoxygenase with soybean phosphatidylcholine (A) and biological membranes. The pure reticulocyte lipoxygenase was incubated with (A) soybean phosphatidylcholine (0.5 mg/ml) and (B) preparations of biological membranes (1.7 mg protein/ml) in 0.1 M phosphate buffer, pH 7.4. I-rat liver mitochondria, II-rat liver microsomes, III-rabbit erythrocyte ghosts.

Among the biological membranes tested, rat liver mitochondrial membranes turned out to be the best substrate for this enzyme (Figure 1B, I). Their oxygenation rate is higher than that of rat liver microsomes (II). Erythrocyte ghosts (III), in contrast, are not oxygenated. These results are in line with earlier findings that the incubation of the reticulocyte lipoxygenase with isolated rat liver mitochondria leads to a severe damage of the organelles as shown by electron microscopy and by the release of matrix enzymes. Incubation of the enzyme with erythrocytes does not lead to hemolysis. The oxygenation of mitochondrial membranes is paralleled by a change in the physico-chemical properties of the membranes and by an inactivation of membrane-bound enzymes (Table 1).

The action of lipoxygenase leads to an inhibition of respiratory enzymes. Two types of lipoxygenase-mediated respiratory inhibitions can be distinguished: 1) Reversible inhibition of the cytochrome c oxidase (the activity could be restored by addition of unoxidized phospholipids to the reaction mixture). 2) A strong irreversible inactivation of NADH- and the

Table 1. Effect of Reticulocyte Lipoxygenase on the Properties of Mitochondrial Membranes

Membrane Property/ Enzymatic System	Effect of Lipoxygenase Treatment	Reference Number
Cytochrome c Oxidase	Reversible inhibition	6
NADH Oxidase	Strong irreversible inactivation	6
Succinate Oxidase	Irreversible inactivation	6
Monoamine Oxidase	Inactivation	6
Membrane Capacity	Fourfold increase	5
Membrane Conductivity	Twofold increase	5

succinate oxidase located in the respiratory chain between the proximal Fe-S centers and ubiquinone.[6] In addition to the respiratory enzymes, which are located in the inner mitochondrial membrane, monoamine oxidase, a marker enzyme for the outer mitochondrial membrane, is inactivated. Moreover, a complete destruction of the iron-sulfur center of the outer mitochondrial embranes was observed.[6] Oxygenation of the membranes is also characterized by marked changes in the passive electrical properties of the membranes (Table 1). These changes are due to an increase in the dielectric constant of the membranes, possibly a consequence of the formation of hydrophilic clusters consisting of hydroperoxy lipids or their degradation products. The changes in the electrical properties and the respiratory inhibition continued after the measuring chamber was depleted of oxygen. These observations indicate the involvement of secondary reactions, such as enzymatic and/or non-enzymatic degradation of hydroperoxy lipids, in the interaction of the lipoxygenase with mitochondrial membranes.[5]

In order to obtain information on the products formed during the reaction of the enzyme with mitochondrial membranes, the membrane bound hydroperoxy lipids were reduced with borohydride, and the lipids were extracted and hydrolyzed. The resulting free fatty acid derivatives were analyzed by reversed-phase HPLC. Two major compounds co-chromatographed with authentic standards; $15L_S$-hydroxy-5Z,8Z,11Z,13E-eicosatetraeneoic acid and $13L_S$-hydroxy 9Z-11E-octadecadienoic acid were detected. The chemical structure of both products has been confirmed by UV spectroscopy, co-chromatography on straight phase HPLC, gas chromatography/mass spectrometry and chiral phase HPLC. In addition to the main products, small anounts of 9-hydroxy-10E,12Z-octadecadienoic acid and of the all-trans isomers were detected.[7]

Intracellular Action of the Reticulocyte Lipoxygenase

The ability of the reticulocyte lipoxygenase to oxygenate mitochondrial membranes led to the hypothesis that this enzyme is involved in the degradation of mitochondria during the maturation of reticulocytes.[6]

Three lines of experimental evidence support the idea that the enzyme is actually functional in intact cells. Intact reticulocytes show an antimycin A-resistant oxygen uptake which corresponds to about 5% of the oxygen uptake in the absence of antimycin A. This remaining 5% is inhibited by the lipoxygenase inhibitors 5,8,11,14-eicosatetraenoic acid, 4-nitrocatechol, nordihydroguaiaretic acid, propyl gallate and salicyl hydroxamic acid.[6] Exogenously added and endogenously released linoleic acid and arachidonic acid are converted by intact reticulocytes to their corresponding n-6 hydroxy compounds.[6] The presence of specific lipoxygenase inhibitors during an *in vitro* maturation of rabbit reticulocytes suppressed the maturational decline of the cytochrome c oxidase.[6]

These results prompted us to look for lipoxygenase products in the cellular membranes of rabbit reticulocytes (H. Kühn and A.R. Brash, unpublished results). Lipid extracts of reticulocyte membranes were hydrolyzed and analyzed by RP-HPLC. The main lipoxygenase product(s) detected co-chromatographed with an authentic standard of $13L_S$-hydroxy-9Z, 11E-octadecadienoic acid (13-HODE). Erythrocyte membranes in contrast only contain small amounts (less than 1/10) of this product(s). A 4-hour *in vitro* maturation of the reticulocytes in the presence of calcium ionophore A23187, which stimulates the lipoxygenase path way in reticulocytes and other cells, led to a threefold increase in the formation of 13-HODE(s). Subcellular fractionation of the membranes and subsequent product analysis indicated that the mitochondrial membranes contained the highest amount of the typical lipoxygenase product 13-hydroxy-9Z,11E-octadecadienoic acids.

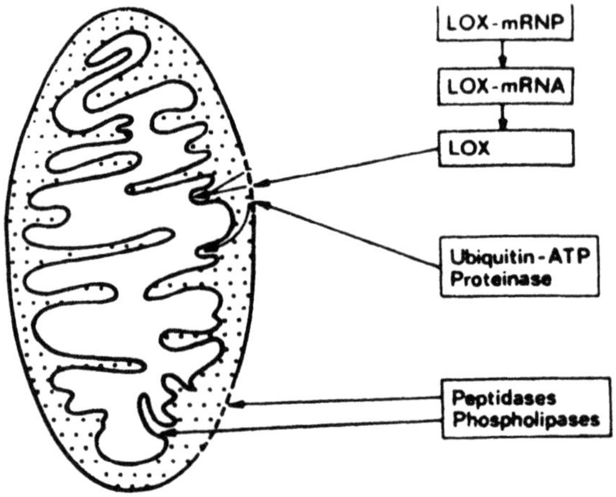

Figure 2. Scheme of the maturational degradation of mitochondria.

DISCUSSION

Several lines of evidence are presented here which support the hypothesis that the reticulocyte lipoxygenase is involved in the breakdown of mitochondria during the maturation of reticulocytes. It should be stressed that other enzymatic systems such as an ATP-dependent proteolysis system,[7,8] other proteases and other phospholipases are also involved in this process. The lipoxygenase appears to be involved in the early stages, making the mitochondrial membranes susceptible to attack by the ATP-dependent proteolysis system.

Figure 2 summarizes the proposed role of the lipoxygenase in the sequence of events leading to the breakdown of mitochondria. The first step is activation of the mRNP, probably effected by limited proteolysis. Unmasking of the mRNA is followed by synthesis of the lipoxygenase. This enzyme then oxygenates the mitochondrial membranes, making them susceptible to further degradation by the ATP-dependent proteolysis system and other hydrolytic enzymes.

ACKNOWLEDGMENTS

Structural analysis of the lipoxygenase products of phospholipids and membranes were conducted at Vanderbilt under NIH grant DK-35275. Part of this work was conducted during a fellowship in the Division of Clinical Pharmacology at Vanderbilt University, 1986-1987.

REFERENCES

1. T. Schewe, R. Wiesner, and S. M. Rapoport, Lipoxygenase from rabbit reticulocytes, Methods Enzymol. 71:430 (1981).
2. T. Schewe, S. M. Rapoport, and H. Kühn, Enzymology and physiology of reticulocyte lipoxygenase: Comparison with other lipoxygenases, Adv. Enzymol. 57:191 (1985).
3. R. W. Bryant, J. M. Bailey, T. Schewe, and S. M. Rapoport, Positional specificity of a reticulocyte lipoxygenase: Conversion of arachidonic acid to 15S-hydroperoxyeicosatetraenoic acid, J. Biol. Chem. 257:6050 (1982).

4. J. J. Murray and A. R. Brash, Rabbit reticulocyte lipoxygenase selectively oxygenates arachidonic acid esterified in phosphatidylcholine, Clin. Res. 33:350A (1985).
5. H. Kühn, F. Pliquett, S. Wunderlich, T. Schewe, and W. Krause, Reticulocyte lipoxygenase changes the passive electrical properties of bovine heart mitochondrial particles, Biochim. Biophys. Acta 735:283 (1983).
6. S. M. Rapoport and T. Schewe, The maturational breakdown of mitochondria in reticulocytes, Biochim. Biophys. Acta 864:471 (1986).
7. K. Tanaka, L. Waxman, and A. L. Goldberg, ATP serves two distinct functions in protein degradation in reticulocytes, one requiring and one independent of ubiquitin, J. Cell. Biol. 96:1580 (1983).
8. A. Hershko and C. Ciechanover, Mechanisms of intracellular protein breakdown, Annu. Rev. Biochem. 51:335 (1982).

MYOCARDIAL SALVAGE BY THE FREE RADICAL SCAVENGING ACTIONS OF XANTHINE OXIDASE INHIBITORS

Dipak K. Das, Richard M. Engelman*, Ronald Clement*,
Hajime Otani, M. Renuka Prasad, and Parinam S. Rao**

University of Connecticut School of Medicine
Farmington, CT 06032, U.S.A.
*Baystate Medical Center
Springfield, MA 01107, U.S.A.
**Long Island Jewish Medical Center
New Hyde Park, NY 11042, U.S.A.

INTRODUCTION

It is well established that oxygen-derived free radicals play a crucial role in the pathogenesis of ischemic-reperfusion injury.[1] However, the source of these radicals remains unknown. McCord[2] has proposed that oxygen radicals are generated during reoxygenation of ischemic myocardium by the action of xanthine oxidase on hypoxanthine. A number of investigators supported this hypothesis based on their observations of allopurinol-mediated reduction of reperfusion injury.[3] Recently, the beneficial action of allopurinol to salvage ischemic-reperfused myocardium has been challenged by Reimer et al.[4] In addition, most of the mammalian hearts, with the exception of rat heart, have been found to contain negligible or no xanthine oxidase activity at all.[5] The purpose of the present study was, therefore, to investigate the role of two xanthine oxidase inhibitors, allopurinol and oxypurinol, on the ischemic-reperfused heart and to explore their mechanisms of action. Our results suggest that these xanthine oxidase inhibitors salvage the ischemic-reperfused myocardium by scavenging the free radicals.

METHODS

Yorkshire pigs of either sex, weighing 15-20 kg, were properly anesthetized, placed on positive-pressure artificial respiration, and median sternotomy was performed. The animals were then placed on cardiopulmonary bypass, and hearts were isolated from the systemic circulation as described elsewhere[6]. Hearts were allowed to stabilize for 15 min and then were pre-perfused for a further period of 15 min with either oxypurinol or allopurinol (1 μmol/ml each). Control experiments were simultaneously performed in the absence of any xanthine oxidase inhibitors. Global hypothermic cardioplegic arrest was then induced with potassium crystalloid cardioplegia for 2 hr, followed by 1 hr of reperfusion. Functional parameters including left ventricular developed pressure (LVDP), left ventricular max. dp/dt (LV dp/dt), and left ventricular end-diastolic pressure (LVEDP) were measured as described previously.[6]

Biochemical assays included estimation of high-energy phosphate compounds, hypoxanthine and xanthine,[7] as well as assay for xanthine oxidase.[5] Left ventricular tissue biopsies were instantly frozen at liquid nitrogen temperature for subsequent free radical analysis by electron paramagnetic resonance spectrometer (EPR). Various free radicals were generated in vitro using known radical-generating systems. Superoxide radicals and hydroxyl radicals were generated by the action of $KO_2^{\cdot}$ in NaOH and (Fe^{2+} + DTPA + H_2O_2 + KTBA), respectively. Polymorphonuclear leukocytes were activated with FMLP or ZAS to generate free radicals. Hypohalite radical was generated with Alcide chemical. Scavenging actions of allopurinol and oxypurinol were examined using either EPR or Luminometer.

RESULTS AND DISCUSSION

Xanthine Oxidase-Hypoxanthine System In Pig Heart

Xanthine dehydrogenase, xanthine oxidase, hypoxanthine and xanthine were assayed from the biopsies obtained from control, ischemic, and reperfused hearts. Xanthine oxidase activity was not detected in any of the heart biopsies. Although measurable quantities of hypoxanthine were found in the biopsies, only a negligible amount of xanthine was detected. Thus, our results contradict the hypothesis that free radicals may be produced via xanthine oxidase reactions in pig heart. To the contrary, these results support the observations by Reimer and Jennings[4] that xanthine oxidase activity may not contribute to the myocardial ischemic-reperfusion injury. However, contrary to their results, we found significant protection of heart from reperfusion injury. This discrepancy may be explained by the species variation as well as protocol of the experiment, such as duration of ischemia, mode of ischemic insult, and dose of allopurinol.

Myocardial Salvage by Xanthine Oxidase Inhibitors

We examined the effects of allopurinol and its primary metabolite oxypurinol on the salvage of myocardial ischemic-reperfusion injury when added to the coronary perfusate prior to ischemia. Both oxypurinol and allopurinol were able to preserve the ATP levels at significantly higher values compared to control (Table 1). Creatine phosphate levels were decreased significantly during ischemia, but their values reached supranormal levels during reperfusion in the treated groups. Cardiac contractility and compliance were also better restored by allopurinol and oxypurinol. LVDP and LV max dp/dt were significantly higher in the treated groups. Thus, these results support the previous reports of myocardial protection by xanthine oxidase inhibitors.[3]

Identification of Free Radicals in Heart and Their Inhibition by Oxypurinol and Allopurinol

The left ventricular tissue biopsies, when examined at 77°K and scanned 6 times using EPR, showed multiple EPR signals after signal averaging (Figure 1) at g=1.985, 2.000, and 2.012. These EPR signals were significantly amplified in ischemic and reperfused heart, compared to those in pre-ischemic controls. Both oxypurinol and allopurinol were able to reduce these signals in ischemic and reperfused tissue, suggesting free radical scavenging abilities of these xanthine oxidase inhibitors. Our study, however, did not identify the EPR spectra generated in heart biopsies; it showed only enhancement of free radical signals during ischemia and reperfusion and their reductions by oxypurinol and allopurinol.

Table 1. Effects of Oxypurinol and Allopurinol on Myocardial High-Energy Phosphate Compounds, LVDP, and LV dp/dt[a]

	Control	Pre-Perfusion 15 min with XO-inhibitors	2 hr of Ischemia	60 min Reperfusion[b]
ATP (mM)				
Non-treated	28.0±2.8	26.1±1.9	23.5±2.0	23.5±1.7
Oxypurinol	27.6±3.1	29.8±2.3	26.3±2.5	29.6±1.5[c]
Allopurinol	26.9±4.0	28.0±2.7	27.7±1.8	28.9±1.4[c]
CP (mM)				
Non-treated	47.5±3.0	48.1±2.2	7.6±1.2	53.5±2.1
Oxypurinol	50.7±2.7	56.8±3.1	8.0±2.2	69.9±1.7[c]
Allopurinol	51.6±1.9	52.5±1.7	6.8±2.7	70.0±1.8[c]
LVDP (% control)				
Non-treated	100	99.7±1.64		75.0±5.56
Oxypurinol	100	99.3±3.84		94.4±6.91[c]
Allopurinol	100	99.5±2.66		95.5±5.48
LV dp/dt (% control)				
Non-treated	100	105.0±4.52		80.9±3.76
Oxypurinol	100	94.6±12.0		104.7±7.54[c]
Allopurinol	100	96.0±8.21		99.2±3.88[c]

[a]Results are expressed as Means ± SE of 6 animals in each group.
[b]ATP and CP values are given at 15 min of reperfusion.
[c]$p < 0.05$.

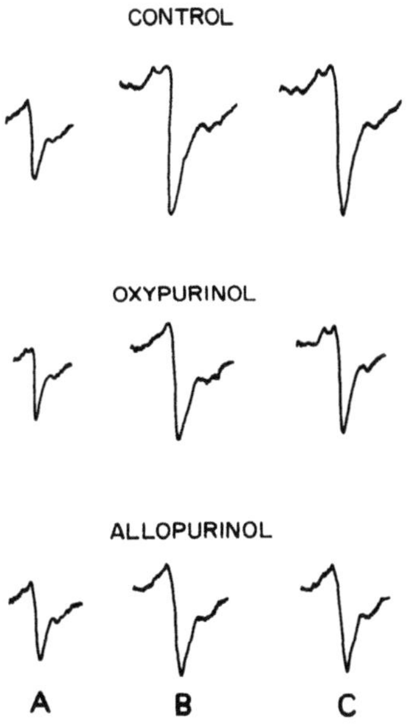

Figure 1. EPR spectra of (A) non-ischemic control, (B) ischemic, and (C) reperfused left ventricular tissue biopsies. Samples were analyzed at 77°K, microwave power 200 microwatts, and microwave frequency of 9.5 GHz.

Table 2. Effects of Oxypurinol and Allopurinol on Various Free-Radical Generating Systems

Free Radical Generating Systems	Radicals	% Inhibition of Chemiluminescent Response	
		Oxypurinol	Allopurinol
KO_2 + NaOH	O_2^-	0	0
Fe^{2+} + DTPA + H_2O_2 + KTBA	·OH	0	0
PMNs + FMLP	O_2^- + HOCL	50	50-60
PMNs + ZAS	O_2^- + HOCL	50-60	60-70
Alcide	Hypohalite	> 80	> 90

In Vitro Scavenging of Free Radicals by Oxypurinol and Allopurinol

Superoxide radicals and hydroxyl radicals were generated as described. Freshly isolated PMNs were activated with FMLP or ZAS to generate free radicals. Alcide has been found to generate hypohalite radicals in our laboratory. Scavenging actions of oxypurinol and allopurinol were studied by placing these xanthine oxidase inhibitors with the free-radical generating systems. Both EPR (data not shown) and Luminometer were used for the detection of free radicals. Table 2 shows the results, which indicate that both oxypurinol and allopurinol (1 μmol/ml) can scavenge free radicals, but allopurinol seems to be a better scavenger. We found that a lower concentration of these xanthine oxidase inhibitors do not function as free radical scavengers (data not shown). Recent work by Peterson and his coworkers[8] also indicated that allopurinol might function as a free radical scavenger and act as an electron-transfer agent from ferrous iron to ferric cytochrome C. Our results support this earlier report and further confirm that at a higher concentration (> 1 μmol/ml) these xanthine oxidase inhibitors can scavenge free radicals.

The results of our study clearly indicate that both oxypurinol and allopurinol salvage myocardial functions during ischemia and reperfusion by scavenging free radicals and not by inhibiting xanthine oxidase. Furthermore, a relatively higher concentration of these xanthine oxidase inhibitors must be used, because at lower doses they do not function as scavengers of free radicals.

ACKNOWLEDGEMENT

Alcide was a gift from Dr. Robert Cross, Alcide Corporation, Farmingdale, NY.

REFERENCES

1. D. K. Das, R. M. Engelman, J. A. Rousou, R. H. Breyer, H. Otani, and S. Lemeshow, Pathophysiology of superoxide radical as potential mediator of reperfusion injury in pig heart, Basic Res. Cardiol. 81:155 (1986).

2. J. M. McCord, Oxygen-derived free radicals in postischemic tissue injury, N. Engl. J. Med. 312:159 (1985).
3. S. W. Werns, M. J. Shea, S. E. Mitsos, R. C. Dysko, J. C. Fantone, M. A. Schork, G. D. Abrams, B. Pitt, and B. R. Lucchesi, Reduction of the size of infarction by allopurinol in the ischemic-reperfused canine heart, Circulation 73:518 (1986).
4. K. A. Reimer and R. B. Jennings, Failure of the xanthine oxidase inhibitor allopurinol to limit infarct size after ischemia and reperfusion, Circulation 71:1069 (1985).
5. D. A. Parks and D. N. Granger, Xanthine oxidase: biochemistry, distribution and physiology, Acta Physiol. Scand. (Suppl) 548:87 (1986).
6. H. Otani, R. M. Engelman, J. A. Rousou, R. H. Breyer, S. Lemeshow, and D. K. Das, Cardiac performance during reperfusion improved by pretreatment with oxygen free-radical scavengers, J. Thorac. Cardiovasc. Surg. 91:290 (1986).
7. D. K. Das, R. M. Engelman, H. Otani, J. A. Rousou, R. H. Breyer, and S. Lemeshow, Effect of superoxide dismutase and catalase on myocardial energy metabolism during ischemia and reperfusion, Clin. Physiol. Biochem. 4:187 (1986).
8. D. A. Peterson, B. Kelly, and J. M. Gerrard, Allopurinol can act as an electron transfer agent, Biochem. Biophys. Res. Commun. 137:76 (1986).

OXYGEN RADICAL SCAVENGERS AND MYOCARDIAL PRESERVATION DURING TRANSPLANTATION: ROLE OF XANTHINE OXIDASE

James R. Stewart, Walter H. Merrill, and William H. Frist

Department of Cardiac and Thoracic Surgery
Vanderbilt University Nashville, Tennessee 37212 U.S.A.

INTRODUCTION

Cardiac transplantation has emerged as an important therapeutic option in the treatment of end-stage heart disease. Advances in surgical techniques in the '50s and '60s and myocardial preservation and immunosuppressive therapy in the '70s have been responsible for the current success of cardiac transplantation in the '80s. Advances in myocardial preservation used in non-transplant heart surgery have been applied to the preservation of the donor heart during its procurement, transport, and implantation. In optimal situations the generally accepted limit of hypothermic preservation of the human heart for transplantation is approximately four hours.

It has been estimated that approximately 10,000 people in the U.S. would benefit from heart transplantation annually. In 1986, 1,400 transplants were performed, and the total is expected to plateau at 2,000 per year in the next few years due to the relatively constant supply of acceptable donors. Any manuever that would prolong viable cold ischemic time and possibly increase the potential donor population would be a welcomed advance in transplant surgery.

OXYRADICALS IN HEART TRANSPLANTATION

We have studied the possible benefit of superoxide dismutase (SOD) in preserving postischemic left ventricular (LV) function after 6 hours of hypothermic global ischemia and subsequent orthotopic transplantation.[1] Donor sheep were anesthetized and instrumented with an arterial pressure catheter, major and minor axis endocardial ultrasonic dimension transducers, and a micromanometer-tipped LV pressure catheter. Radioactive microspheres were used to measure preischemic coronary blood flow using standard reference organ methods, and left ventricular pressure-volume loops were generated. The ascending aorta was then clamped and standard cardioplegic solution administered. The heart was excised, wrapped in plastic bags, then stored for 6 hours at $4^{o}C$ in a solution of Ringer's lactate and 2,000 U heparin.

After the ischemic interval, the recipient animal was instrumented in an identical fashion, and placed on total cardiopulmonary bypass. The recipient's heart was removed, and the donor heart inserted in the form of an orthotopic transplant. With restoration of coronary blood flow, the hearts were rewarmed, defibrillated, and maintained in the beating, perfused, vented, and nonworking state for 45 minutes. Cardiopulmonary bypass was weaned, and functional measurements repeated. Platelet deposition was determined using In-111 labeled autologous platelets, and coronary blood flow measurements repeated after termination of bypass. At the end of the study, a sample of LV free wall was obtained for determination of lipid peroxidation products using thiobarbituric acid assay. Two groups were studied: the first served as experimental controls and received standard clinical hypothermic crystalloid cardioplegic solution. The second group received identical treatment with the exception that 10mg SOD was added to the cardioplegic solution, and 20 mg given into the ascending aorta immediately before reperfusion.

Our results demonstrate improved left ventricular function in the group receiving superoxide dismutase. Left ventricular contractility as determined by the slope of the end-systolic pressure-volume relationship, or Emax, was significantly better in Group 2 hearts (90 $\pm$ 23% versus 43 $\pm$ 32% control values). Similarly, the load-independent relationship of stroke-work versus left ventricular and diastolic volume showed improved postischemic recovery in hearts receiving SOD compared to Group 1 (0.42 $\pm$ 0.17 in Group 2, compared to 0.18 $\pm$ 13 in Group 1).

There was an overall increase in coronary blood flow with reperfusion in both groups due to hemodilution of cardiopulmonary bypass, but the redistribution of blood flow away from the endocardium (reflected in the endocardial versus epicardial ratios) was slightly but statistically less in Group 2 hearts.

The ratio of tissue-to-blood scintigraphic platelet counts was significantly greater in the Group 1 experimental controls (0.22 $\pm$ 0.15) compared to hearts receiving SOD (0.12 $\pm$ 0.04).

Finally the thiobarbituric acid reactive products were significantly greater in the control group (204.7 $\pm$ 52.1 nmol/gm) versus the Group 2 hearts (146.9 $\pm$ 24.6 nmol/gm).

Farid Gharagozloo and the cardiac surgical group at Brigham and Women's Hospital have reproduced these results using an ex vivo asanguinous perfusion system for extended preservation of the sheep heart for transplantation.[2] After cardiectomy and instrumentation with bilateral interventricular balloons connected to pressure transducers, hearts were placed on a perfusion apparatus. Control LV and right ventricular (RV) function, lactate production, and myocardial oxygen consumption were measured. Hearts were then arrested and perfused with 4°C oxygenated buffered crystalloid solution for 8 hours. After 8 hours of preservation, measurements of cardiac function and metabolism were repeated. Five controls and five animals receiving SOD + catalase (CAT) in the cold perfusate were studied.

Hearts receiving SOD and CAT had improved biventricular function after 8 hours of preservation. Isovolumic LV peak systolic pressures achieved 75% of preischemic value compared to 41% in the control group. The LV diastolic pressure and dP/dt in the group receiving SOD and catalase were 100%, and 85% of the preischemic value compared to 239% and 20% respectively in the group which did not receive oxygen radical scavengers. Right ventricular systolic and diastolic pressures in the group receiving OFR scavengers were 69% and 100% of preischemic value, compared to 30% and 179% respectively in the control group. Hearts receiving SOD + CAT had higher myocardial oxygen consumption than hearts in the control group, and lactate production was higher in the control group. Heart weight increased 67% in the control group and only 8% in the SOD + CAT group.

SOD + CAT clearly preserved systolic function and diastolic compliance after an extended period of ex vivo preservation in this study.

In both of these studies, an improvement of 40 to 50% of preischemic function was noted in the groups receiving oxygen radical scavengers, compared to controls. This would seem to imply that of the 60% reduction in function that occurs after 6 and 8 hours of ischemia, the majority (85% and 60% respectively) was due to oxygen radical-mediated injury.

ROLE OF XANTHINE OXIDASE IN MYOCARDIAL REPERFUSION INJURY

Xanthine oxidase has been proposed to be a source of oxygen radicals in the reperfused ischemic heart. Xanthine oxidase inhibitors allopurinol and oxypurinol have been demonstrated to improve postischemic myocardial function and reduce infarct size in canine models of hypothermic global and normothermic regional myocardial ischemia. We had demonstrated previously that pretreatment of the dog with allopurinol 50 mg/kg/d for 72 hours followed by 60 minutes of hypothermic global ischemia and reperfusion resulted in better recovery of LV function compared to controls receiving hypothermic ischemia alone. In addition, spontaneous recovery of normal sinus rhythm in the reperfusion period was also significantly more common in the experimental animals receiving SOD, mannitol, and allopurinol.[3] Miyamoto, at the University of Washington, has reproduced these results in dogs, demonstrating that allopurinol pretreatment or addition of allopurinol to the cardioplegic solution helps preserve myocardial function after cross-clamping and reperfusion.[4]

Using an isolated, perfused rabbit heart model, Myers and colleagues at the University of Michigan[5] noted that the addition of allopurinol to hypothermic crystalloid cardioplegic solution offered partial protection to LV systolic function and diastolic compliance compared to catalase when receiving global ischemia at $27^{o}C$ for 2 hours. This is one of only two studies to note a benefit using xanthine oxidase inhibitors in myocardial preservation in a xanthine oxidase-deficient species.[5,6] Allopurinol appears to work better than SOD to preserve LV function in this study.

Das et al[6] have demonstrated preservation of high-energy phosphates and postischemic left ventricular function in porcine hearts pretreated with xanthine oxidase inhibitors allopurinol and oxypurinol, subjected to 2 hours hypothermic global ischemia, followed by 60 minutes reperfusion. Because they were unable to measure any xanthine oxidase in porcine myocardium using the spectrophotometric assay and were able to confirm the oxyradical scavenging capabilities of allopurinol and oxypurinol reported by others,[20,21] they postulated that myocardial protection was achieved not by xanthine oxidase inhibition, but by non-specific scavenging of oxygen radicals by these compounds. Granger (personal communication), however, has demonstrated that in physiologic concentrations used to achieve xanthine oxidase inhibition, the "scavenging" capability of allopurinol and oxypurinol are negligible.

Several studies have demonstrated reduction of experimental infarct size with xanthine oxidase inhibition using allopurinol and oxypurinol in the ischemic reperfused canine heart. Chambers, using a model of reversible occlusion of the left anterior descending coronary artery (LAD) in dogs, demonstrated that 23% of the region at risk would infarct with no intervention. If allopurinol was administered, there was a marked and highly significant reduction in myocardial necrosis to about 9% of the area at risk. Incidentally, SOD had an even greater reduction to 6% of the area at risk.[7]

In two separate studies, Steven Werns and the group at the University of Michigan have demonstrated that allopurinol reduces experimental infarct size

(39% to 25% of area at risk). Allopurinol appeared to be partially protective, in that the addition of SOD allowed for a further reduction in infarct size to approximately 18% of area at risk.[9] In a follow-up study, administration of oxypurinol, and not allopurinol, started just 15 minutes before reperfusion, limited infarct size from 35% to 21% of the area at risk.[9] These data support the concept that the accumulation of oxypurinol from the metabolic conversion of its parent compound allopurinol is necessary to limit infarct size.

David Hearse, in collaboration with Jim Downey and Derek Yellon, has demonstrated striking effects of allopurinol in reduction of infarct size in the dog, and dramatic reduction of potentially lethal reperfusion induced arrhythmias in the rat.[10]

Jarasch has demonstrated, with an ultra-sensitive radioimmunoassay using bovine xanthine oxidase as a source of antibodies, that concentrations of xanthine oxidase were 1,000- to 10,000-fold higher in capillary endothelial cells than other mammalian cells, except mammary epithelial cells. Similar results were obtained with human tissues, specifically the human myocardium. They noted that in most human tissues sampled, XO and XD activities were usually below the limit of detection. Using antibodies against human milk lipid globules and Ab antibody against bovine xanthine oxidase that cross-reacts with the human polypeptide, they localized human xanthine oxidase to capillary endothelial and alveolar epithelial cells of the mammary gland.[11]

There is a conflicting body of evidence, however, on the potential benefit of xanthine oxidase inhibition in global and regional human myocardial ischemia.

In a study from Vanderbilt, Mervyn Forman has investigated the effect of oxypurinol on reperfusion injury in a region at risk model using 90 minutes of LAD occlusion. Oxypurinol failed to reduce infarct size at 24 hours, but radial shortening of the ischemic zone was significantly improved 24 hours after reperfusion. The author suggests that free radicals generated during reperfusion may contribute to myocardial stunning.[12]

Jim Downey has demonstrated that the rabbit heart has undetectable levels of xanthine oxidase. Using a region at risk model of coronary occlusion, free radicals clearly contribute to the ischemia-reperfusion injury in the rabbit, as evidenced by the protection afforded by SOD + CAT, but allopurinol failed to limit infarct size.[13]

Stanley, in collaboration with Jim Downey, measured cardiac uric acid production during reperfusion in 6 patients with evolving transmural myocardial infarcts. All patients had abrupt reperfusion of the occluded infarct-related artery by balloon angioplasty. No net uric acid production was detected in coronary sinus samples at any time following reperfusion.[14]

SOD and CAT have reduced experimental infarct size in rabbit and pig myocardium, confirming that oxygen radicals have an important role in the determination of infarct size. There is little evidence to date that xanthine oxidase inhibition preserves postischemic function or limits infarct size in xanthine oxidase-deficient species.

Xanthine oxidase activity is undetectable in the hearts of several species including the rabbit[13,15] and the pig.[16] There have been conflicting reports concerning the xanthine oxidase content in human heart. The three reports we could find in the literature yield values ranging in excess of those in the rat and dog, to trace amounts, to none detectable.[17-19] The applicability of the xanthine oxidase hypothesis to myocardial ischemia in humans may depend on whether or not the human heart contains xanthine oxidase. To answer this question, myocardial samples from 2 surgical patients and 2 organ donors were analyzed for xanthine oxidase activity by both the standard spectrophotometric assay and a more sensitive spectrofluorometric method. In all cases, samples were bisected, and one-half immediately frozen in liquid nitrogen. The other half was subjected to 30 minutes of bench ischemia at 37°C and then frozen in liquid nitrogen. After homogenization, centrifugation, and elution enzyme-rich fractions were pooled for assay. We used the standard spectrophotometric assay with a Perkin-Elmer Lambda 3 spectrophotometer. Rate of uric acid formation was measured by monitoring the absorbance at 295 nM for 10 minutes on a strip-chart recorder. Initial experiments using the spectrophotometric method showed no detectable xanthine oxidase or xanthine dehydrogenase activity. To increase sensitivity, samples were concentrated 20-fold in a rotovac. No activity was detected. Concentration via lyophilization and resuspension in reduced volume still yielded no activity. If enzyme were present, it was clearly below the limits of detectability for this assay (lower limit 0.2 mU/ml) of post-column sample.

We therefore examined specimens using a spectrofluorometric assay, which has been reported to be much more sensitive than the spectrophotometric method. The fluorometric assay measures the production of isoxanthopterin from pterine as an indication of xanthine oxidase activity. Addition of methylene blue as an electron acceptor allows for measurement of xanthine oxidase and xanthine dehydrogenase. Using dilutions of 1:30 to 1:1.5, no activity could be identified. Using the fluorometric method, xanthine oxidase could be detected as low as 2.1 nU/ml in post-column samples or 1.8 nU/gram wet weight of tissue.

In summary, any xanthine oxidase present in the human myocardium is present in trace amounts and cannot be a significant source of oxygen free radicals in the human heart, but may function as a "triggering" mechanism for leukocyte infiltration and further free radical formation. There is presently no convincing experimental nor clinical work involving heart transplantation in xanthine oxidase-deficient species that demonstrates a beneficial effect of xanthine oxidase inhibition. Further experimental, and eventually clinical, investigations are necessary to resolve this issue.

REFERENCES

1. J.R. Stewart, E.B. Gerhardt, C.J. Wehr, T. Shuman, W.H. Merrill, J.W. Hammon, and H.W. Bender, Free radical scavengers and myocardial preservation during transplantation, Ann. Thorac. Surg. 42:390 (1986).
2. F. Gharagozloo, F.S. Melendez, R.A. Heim, R.G. Laurence, M.S. Rosenzweig, R.J. Shemin, V.J. DiSesa, and L.H. Cohn, Superoxide dismutase and catalase preserve right and left ventricular function and compliance in the ex vivo sheep heart preserved for eight hours, Surg. Forum 37:243 (1986).
3. J.R. Stewart, S.L. Crute, V. Loughlin, M.L. Hess, and L.J. Greenfield, Prevention of free radical induced myocardial reperfusion injury with allopurinol, J. Thorac. Cardiovasc. Surg. 90:68 (1985).
4. M. Miyamoto, G.A. Misbach, K. Tabayashi, R. Thomas, A. Luedtke, and T.D. Ivey, Myocardial function preserved by allopurinol given systemically or in cardioplegic solution, J.A.C.C. 9:88A (1987).
5. C.L. Myers, S.J. Weiss, M.M. Kirsh, B.M. Shepard, and M. Shlafer, Effects of supplementing hypothermic crystalloid cardiplegic solution with catalase, superoxide dismutase, allopurinol, or deferoxamine on functional recovery of globally ischemic reperfused isolated hearts, J. Thorac. Cardiovasc. Surg. 91:281 (1986).

6. D.K. Das, R.M. Engleman, and R. Clement, H. Otani, M.R. Parasad, and P.S. Rao, Enhancement of myocardial salvage by the free radical scavenging actions of xanthine oxidase inhibitors, abstract, Fourth International Congress on Oxygen Radicals, La Jolla, CA 7/2/87, p. 149.
7. D.E. Chambers, D.A. Parks, G. Patterson, R. Roy, J.M. McCord, S. Yoshida, L.F. Parmley, and J.M. Downey, Xanthine oxidase as a source of free radical damage in myocardial ischemia, J. Mol. Cell Cardiol. 17:145 (1985).
8. S.W. Werns, M.J. Shea, S.E. Mitso, R.C. Dysko, J.C. Fantone, M.A. Schork, G.D. Abrams, B. Pitt, and B.R. Lucchesi, Reduction of the size of infarction by allopurinol in the ischemic-perfused canine heart, Circulation 73:518 (1986).
9. S.W. Werns, A. Ventura, and B.R Lucchesi, Effects of allopurinol or oxypurinol on myocardial reperfusion injury, Clin. Res. 35:79A (1987).
10. D.J. Hearse, A.S. Manning, J.M. Downey, and D.M. Yellon, Xanthine oxidase: A critical mediator of myocardial injury during ischemia and reperfusion? Acta. Physiol. Scand. Suppl. 548:65 (1986).
11. E.D. Jarasch, G. Bruder, and H.W. Heid, Significance of xanthine oxidase in capillary endothelial cells, Acta. Physiol. Scand. Suppl. 548:39 (1986).
12. D.W. Puett, M.B. Forman, C.U. Cates, B.H. Wilson, K.R. Hande, G.C. Friesinger, and R. Virmani, Oxypurinol limits myocardial stunning but does not reduce infarct size after reperfusion, Circulation 76:678 (1987).
13. J. Downey, D. Chambers, T. Miura, D. Yellon, and D. Jones, Allopurinol fails to limit infarct size in a xanthine oxidase - deficient species, Circulation 74 Suppl II-372 (1986).
14. A.W.H Stanley, C. Prickett, T. Simpson, J. Teague, J.G. Arciniegas, T.B. Cooper, R.G. Hess, W.A.H. MacLean, S.E. Papapietro, and J. Downey, Failure to detect xanthine oxidase activity in reperfused human myocardium, Clin. Res. 35:35A (1987).
15. L. Ketai, C. Grum, C. Meyers, and M. Shlafer, Purine efflux from ischemic rabbit hearts: Lack of xanthine oxidase activity relevance to allopurinol protection, Circulation 72 III-357 (1985).
16. T. Podzuweit, W. Brawn, A. Muller, and W. Schaper, Arrhythmias and infarction in the ischemic pig heart are not mediated by xanthine oxidase-derived free oxygen radicals, Circulation 74:II-346 (1986).
17. T.A. Krenitsky, J.V. Tuttle, E.L. Cattau, and P. Wang, Comparison of the distribution and electron specification of xanthine oxidase and aldehyde oxidase, Comp. Biochem. Physiol. 49(B):687 (1974).
18. R.W.E. Watts, J.E.M. Watts, and J.E. Seegmiller, Xanthine oxidase activity in human tissues and its inhibition by allopurinol, J. Lab. Clin. Med. 66:688 (1965).
19. C.R.H. Ramboer, A sensitive and non-radioactive assay for serum and tissu xanthine oxidase, J. Lab. Clin. Med. 74:828 (1969).
20. D.A. Peterson, B. Kelly, and J.M. Gerrard, Allopurinol can act as an electron transfer agent, Biochem. Biophys. Res. Commun. 137:76 (1986).
21. P.C. Moorhouse, M. GrootVeld, B. Halliwell, J.G. Quinlan, and J.M.C. Gutteridge, Allopurinol and oxypurinol are hydroxyl radical scavengers, F.E.B.S. 213:23 (1987).

ENDOTHELIAL CELLS: A SOURCE AND TARGET OF OXIDANT DAMAGE

Una S. Ryan and James M. Vann

Department of Medicine (R58)
University of Miami School of Medicine
Miami, Florida 33101, U.S.A.

INTRODUCTION

By virtue of their position interposed between the blood and the tissues, endothelial cells are ideally placed to receive both blood-borne and tissue-derived signals. By the same token, endothelial cells and their products are strategically situated to interact directly with cells and molecules of blood and with the underlying tissues. We now know that endothelial cells are highly metabolically active and responsive cells that control the passage of substances across the vascular wall and regulate the quality and quantity of circulating bioactive molecules.[1] Many of the interactions of endothelial cells with cell types such a neutrophils and macrophages may involve the endothelial cell as the target of oxidant injury[2,3] (see also P.A. Ward, this volume). However, endothelial cells themselves can respond to stimuli to release active products, including oxygen metabolites, into the surroundings. Some of these endothelial-derived products may serve a regulatory function as in the case of endothelium-derived relaxing factor (EDRF), and its actions on platelets and smooth muscle cells, while others are presumably designed as a cellular defense mechanism against, for example, microbial infection. Evidence is accumulating to indicate that endothelial cells can respond to a wide variety of stimuli with an equally wide range of responses. However, what is less clear is how the endothelium translates the signal into the response. We have recently set out to determine the mechanisms of signal transduction in endothelial cells. In this report we examine two systems: 1) agonist-induced release of EDRF from endothelial cells and 2) production of superoxide anion ($O_2^{\cdot -}$) in response to ingestion of bacteria.

RELEASE OF ENDOTHELIUM-DERIVED RELAXING FACTOR

We have used the release of EDRF (NO or a closely related compound) as a means of assaying the endothelial cells response to agonists. We have developed an on-line bioassay system (Figure 1) where endothelial cells are cultured on microcarriers[4] and suspended in a column above a rabbit aortic strip such that agonists can be applied directly to the smooth muscle strip or via the beads and the endothelial-derived products then are directed over the strip. Changes in tension are recorded on a tension transducer. Alternatively, the endothelial-derived products can

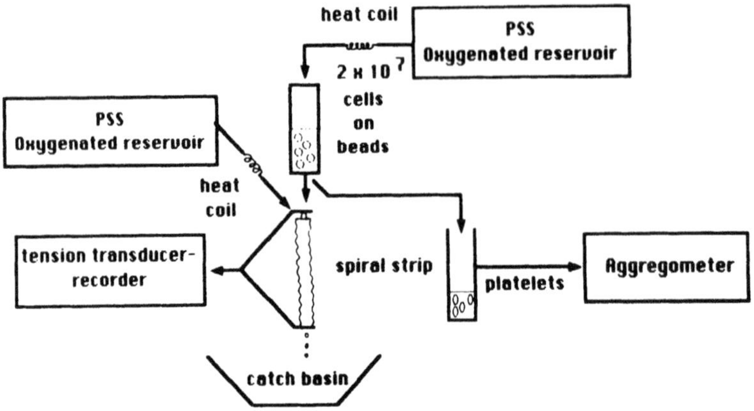

Figure 1. Release of EDRF from endothelial cells cultured on microcarriers (2 x 10^7 cells) was determined by bioassay using tension changes in a rabbit aortic strip from which the endothelium had been removed or by its effect on the aggregation of human platelets.

be directed to a platelet suspension in a platelet aggregometer. We have shown that endothelial cells respond to challenge with bradykinin, ATP, thrombin or calcium ionophore to release EDRF that reduces the platelet aggregating effects of thrombin and causes a vasodilator response in a denuded smooth muscle strip previously contracted with phenylephrine.[5] We have investigated the role of calcium in the receptor-mediated release of EDRF and have examined the mobilization of calcium from intracellular and extracellular stores, the role of second messengers, the

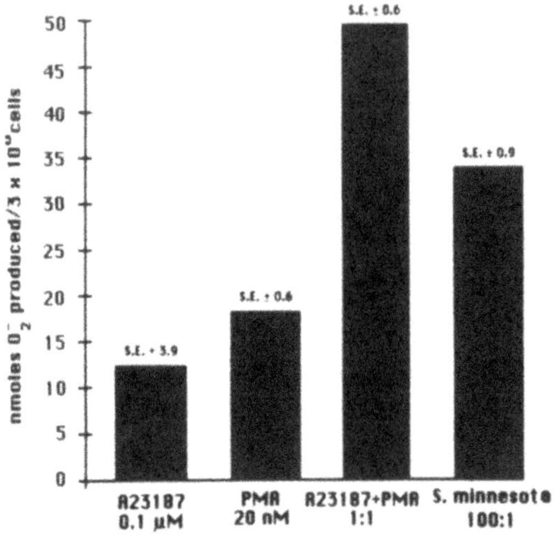

Figure 2. Superoxide anion released into the culture medium by endothelial cell monolayers (3 x 10^6 cells per well) was determined by assay of the superoxide dismutase inhibitable reduction of ferricytochrome C. The stimulation above control levels is shown for calcium ionophore (A23187), phorbol ester (PMA), both compounds acting synergistically, and after phagocytosis of the Re mutant of <u>S. minnesota</u> (100:1). All incubations were for 20 minutes.

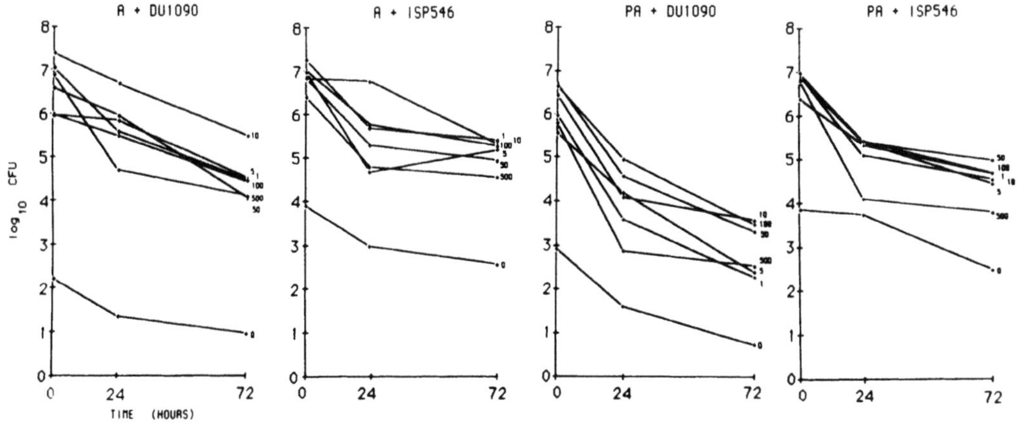

Figure 3. The effect of adenosine on bactericidal activity of bovine aortic (A) and bovine pulmonary artery (PA) endothelial cells is shown. Endothelial monolayers (3 x 10^5 cells) were incubated with S. aureus DU1090, a strain that lacks α-hemolysin activity, and ISP546, a strain that contains high catalase activity, for up to 72 hours in the presence of adenosine (1-500μg/ml) added concomitantly with the bacteria. Extracellular bacteria were removed by treatment with lysotaphin and the intracellular bacteria were estimated by lysis of the endothelial cells and counting of colony forming units (CFU). Controls were incubated in the absence of adenosine (0). In all cases adenosine reduced the bactericidal effectiveness of endothelial cells.

electrophysiological characteristics of the endothelial cell membrane and the ionic mechanisms underlying receptor operated influx of calcium.[5-8]

Our data indicate that

* Agonist-induced release of EDRF is controlled by intracellular free calcium $[Ca^{2+}]_i$
* Agonists exert their action on endothelial cells via activation of phosphoinositide turnover
* Ca^{2+} entry through receptor-operated ion channels contributes to excitation-secretion coupling in endothelial cells
* A negative feedback mechanism that blocks receptor-mediated $[Ca^{2+}]_i$ increase is triggered as a result of activation of protein kinase C

RELEASE OF SUPEROXIDE ANION

Endothelial cells not only respond to agonists but also respond to particulates. Although not normally considered "professional" phagocytes, endothelial cells can bind and ingest bacteria and can discriminate between bacterial strains.[9] We have shown that endothelial cells ingest the Re mutant of Salmonella minnesota but not the wild (S) type. Endothelial responses to phagocytosis of the Re mutant include unmasking of an Fc receptor, a respiratory burst and an increase in the rates of migration, division and further phagocytosis. The respiratory burst is not efficient at killing the Re mutant of S. minnesota[9, 10] or

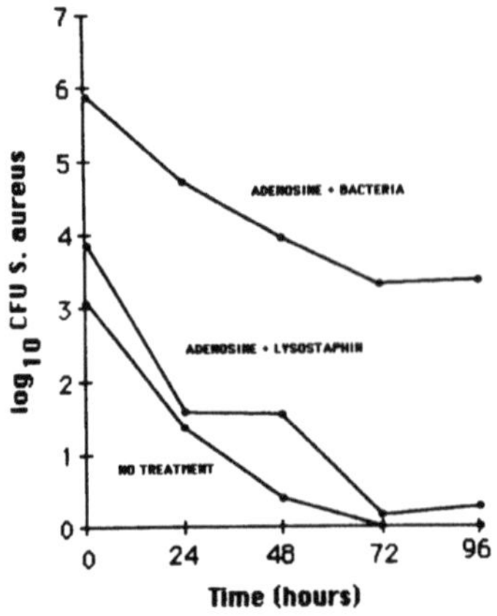

Figure 4. When adenosine was added to endothelial cultures at the same time as the bacteria, killing was inhibited. However, adenosine was not effective if added after phagocytosis (at the time lysostaphin was added to remove extracellular bacteria).

some strains of Staphylococcus aureus, especially strains having a high catalase activity. However, considerable release of $O_2^{\cdot-}$ into the extracellular medium can be observed after phagocytosis of bacteria (Figure 2). As shown in Figure 2, the release of $O_2^{\cdot-}$ appears to involve mobilization of Ca^{2+} and activation of protein kinase C since calcium ionophore A23187 and PMA, a powerful stimulator of protein kinase C, each appear to stimulate release. However, when the two systems act synergistically, the maximum physiological response is achieved.

Phagocytosis of S. minnesota Re mutant produces a greater release of $O_2^{\cdot-}$ than do either PMA or A23187 alone.

Effect of Purines on Survival of S. aureus within Endothelial Cells

In previous studies [11-13] we have shown that the adenine nucleotides (ATP, ADP and 5'-AMP) are rapidly metabolized by pulmonary endothelial cells and we were able to localize the ultrastructural sites of the nucleotidase enzymes to the luminal caveolae of endothelial cells.[11,12] The final products of the processing of adenine nucleotides are adenosine and free phosphate, and the adenosine so produced is rapidly taken up by the endothelial cells and reincorporated into cyclic nucleotides and ATP.[13,14] Recently, a series of separate studies have shown that neutrophils possess adenosine receptors,[15] that adenosine and its analogue 2-chloroadenosine, prevent neutrophils from generating superoxide anion in response to chemoattractants [16] and that adenine nucleotides may exert regulatory effects on $O_2^{\cdot-}$ generation by neutrophils.[17] Therefore, we set out to determine the effects of adenosine, adenine and the adenine nucleotides on the bactericidal effectiveness of endothelial cells.

Our data indicate that adenosine (5-50 μg/ml) markedly increases the numbers of S. aureus surviving phagocytosis by endothelial cells (Figure 3). This action of adenosine is apparently not due to an effect on binding or ingestion per se since the number of labelled bacteria phagocytized by endothelial cells is not altered by adenosine whether the bacteria are alive or dead. Experiments showing that the adenosine must

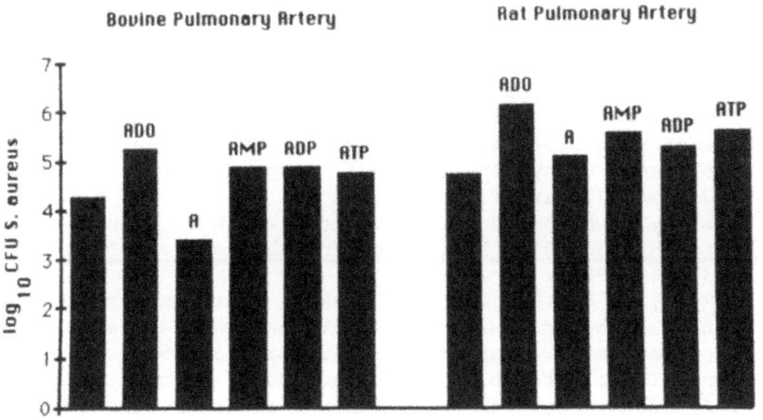

Figure 5. The effects of purines on endothelial cell bactericidal activity were tested on bovine or rat pulmonary artery endothelial monolayers. Compounds were added at the same time as S. aureus (DU1090 strain). CFUs were counted as described in Figure 3. In both species adenosine (ADO, $p < 0.01$), AMP, ADP and ATP ($p < 0.025$) increased survival of S. aureus, whereas adenine (A) appeared to increase bactericidal activity ($p < 0.025$) in bovine but not rat endothelial cells.

be incorporated along with the bacteria at the time of ingestion in order to be effective (Figure 4) have led us to postulate that the role of adenosine on endothelial bactericidal activity results from a reduction of endothelial $O_2^{\cdot-}$ production and that the regulatory effects occur at the level of the phagosome. The adenine nucleotides appear to inhibit killing in much the same way as adenosine, while in bovine cells adenine appears to stimulate the ability of endothelium to kill S. aureus. (Figure 5).

CONCLUSION

The mechanisms of signal transduction in the release of endothelial-derived $O_2^{\cdot-}$ are similar to those involved in the release of EDRF and involve elevation of intracellular calcium and activation of protein kinase C. Thus, endothelial cells form a dynamic interface providing a primary target for oxidant injury. However, far from being inert bystanders, endothelial cells can be activated in ways that include oxygen-dependent mechanisms leading to generation of a wealth of active substances that play both regulatory and protective roles. These inducible functions of endothelium, although stimulated by diverse stimuli and manifested in diverse ways, appear to be governed by relatively conservative mechanisms of signal transduction. Thus, endothelium provides a surface capable of maintaining homeostasis and of providing an early and critical host response to injury, but when inappropriately expressed or when compromised under conditions of endothelial injury, endothelial activation responses can underlie the pathogenesis of variety of vascular diseases.

ACKNOWLEDGEMENTS

This work was supported by grants HL 21568, HL 33064 from the N.I.H. and grant 814 from the Council for Tobacco Research U.S.A., Inc.

REFERENCES

1. U. S. Ryan, Metabolic activity of pulmonary endothelium: Modulation of structure and function, Ann. Rev. Physiol. 48:263-277 (1986).
2. J. Varani, S. E. G. Fligiel, G. O. Till, U. S. Ryan and P. A. Ward, Pulmonary endothelial cell killing by human neutrophils: Possible involvement of hydroxyl radical, Lab. Invest. 53:656-663 (1985).
3. D. E. Gannon, J. Varani, S. H. Phan, J. H. Ward, J. Kaplan, G. O. Till, R. H. Simon, U. S. Ryan and P. A. Ward, Source of Iron in Neutrophil-Mediated Killing of Endothelial Cells, Lab. Invest. 57:37-44 (1987).
4. U. S. Ryan, M. Mortara and C. Whitaker, Methods for microcarrier culture of bovine pulmonary artery endothelial cells avoiding the use of enzymes, Tissue & Cell 12:619-635 (1980).
5. A. Johns, T. W. Lategan, N. J. Lodge, U. S. Ryan, C. van Breemen and D. J. Adams, Calcium entry through receptor-operated channels in bovine pulmonary artery endothelial cells, Tissue & Cell 19:1-13 (1987).
6. U. S. Ryan, P. V. Avdonin, E. Ya. Posin, E. G. Popov, S. M. Danilov and V. A. Tkachuk, Influence of vasoactive agents on cytoplasmic free calcium concentration in Indo-1 loaded vascular endothelial cells, J. Appl. Physiol. in press (1988).
7. U. S. Ryan, G. Y. Grigorian, P. V. Avdonin and P. Bregestovski, Platelet-activating factor effects on endothelial cells, In: "New Trends in Lipid Mediators Res 1," P. Braquet, ed., Karger, Basel (1988).
8. G. Y. Grigorian and U. S. Ryan, Platelet-activating factor effects on bovine pulmonary artery endothelial cells, Circulation Research 61:389-395 (1987).
9. U. S. Ryan, Endothelial cell activation responses. In: "Pulmonary Endothelium in Health and Disease," U.S. Ryan, ed., Marcel Dekker Inc., New York Chap. 32 pp. 3-33 (1987).
10. J. M. Vann and U. S. Ryan, Adenosine decreases bactericidal activity of endothelial cells, J. Cell Biol. 105:326a Abstract 1840 (1987).
11. U. Smith and J. W. Ryan, Pinocytotic vesicles of the pulmonary endothelial cell, Chest 59:12s-15s (1971).
12. U. Smith and J. W. Ryan, Pulmonary endothelial cells and the metabolism of adenine nucleotides, kinins and angiotensin I, Adv. Exp. Med. Biol. N. Back and F. Sicuteri, ed., Plenum Press, New York (1972).
13. D. J. Crutchley, U. S. Ryan and J. W. Ryan, Effects of aspirin and dipyridamole on the degradation of adenosine diphosphate by cultured cells derived from bovine pulmonary artery, J. Clin. Invest. 66:29-35 (1980).
14. J. D. Pearson, J. S. Carlton, A. Hutchings, and J. L. Gordon, Uptake and metabolism of adenosine by pig aortic endothelial and smooth muscle cells in culture, Biochem. J. 170:265-271 (1978).
15. Y. Nishida, A. Takeuchi and T. Miyamoto, Modulation of Polymorphonuclear leukocyte function by Adenosine Analogues, Adv. Exp. Med. Biol. 195:487-490 (1986).
16. B. N. Cronstein, R. I. Levin, J. Belanoff, G. Weissmann, and R. Hirschhorn, Adenosine: An endogenous inhibitor of neutrophil-mediated injury to endothelial cells, J. Clin. Invest. 78:760-770 (1986).
17. T. W. Cunningham, P. A. Ward, and K. J. Johnson, Adenine nucleotide effects on oxygen radical responses of rat and human neutrophils, Fed. Proc. Vol. 46 No. 3 p. 986 Abstract 3902 (1987).

THE ROLE OF IRON IN INJURY OF ENDOTHELIAL CELLS IN VITRO AND IN VIVO

Peter A. Ward, Gerd O. Till, David E. Gannon*, James A. Varani, and Kent J. Johnson

Departments of Pathology and Internal Medicine*
The University of Michigan Medical School
1301 Catherine Street
Ann Arbor, Michigan 48109, U.S.A.

INTRODUCTION

Injury of cells and organs resulting from the generation of oxygen-derived free radicals is now a well-accepted observation (reviewed 1). Both the in vitro as well as the in vivo generation of oxygen radicals from activated phagocytic cells (neutrophils, eosinophils, monocytes and macrophages) via activation of NADPH oxidase results in injury of eukaryotic and prokaryotic cells (reviewed 1). All our studies have focused on acute injury of the rat lung and in vitro destruction of endothelial cells in an attempt to define the mechanisms of cell injury. What will be reviewed in this communication is the evidence for the critical role of iron in oxygen-radical-mediated cell injury and a linkage between these reactions and the appearance of products of lipid peroxidation.

IRON IN ACUTE LUNG INJURY

We have studied rather extensively the mechanism of acute lung injury in the rat following intravascular activation of the complement system induced by the injection of the cobra venom factor isolated from Naja naja venom.[2-4] This results in acute focal destruction of microvascular endothelial cells of the lung interstitium (as shown in Figure 1) when neutrophils come into intimate contact with endothelial cells within interstitial capillary beds. As described in earlier studies, the endothelial cell injury requires activation of the complement system and oxygen radical formation from neutrophils.[2] Catalase is especially potent with regard to its protective effects while superoxide dismutase (SOD) is of limited value in terms of protective effects. As shown in Figure 1, the resulting acute lung injury is iron dependent. The "index of lung injury" is the relative amount of radioactivity contained within lung parenchyma after washout of the pulmonary vasculature with saline. At the time of intravenous injection of 4 units of cobra venom factor (CVF) 30 min earlier, ^{125}I-bovine serum albumin was co-injected with the CVF in order to measure vascular permeability in the lung. At the time of sacrifice, lung parenchymal radioactivity (as well as the contents of the alveolar compartment) and blood radioactivity were measured, and lung injury expressed as ratio of radioactivity present in the lungs and in 1.0 ml peripheral blood of the same rat. As can be seen in Figure 1, infusion of CVF leads to an increase in lung vascular permeability of nearly fivefold. Intravenous treatment of rats at the time of CVF infusion with deferoxamine (desferal) in an amount of 15 mg/kg body weight led to a complete protection from the lung injury. The protective effects of desferal have been confirmed histologically. Animals protected in this manner show little or no evidence of endothelial cell injury, fibrin deposition, interstitial edema, or intraalveolar hemorrhage.[3] The specificity of the protective effects of desferal is shown by the finding that iron-saturated desferal does not protect against CVF-induced lung injury (Figure 1). In addition to the protective effects of desferal, we also have demonstrated that apolactoferrin (iron-free lactoferrin), which has been isolated from human breast milk, is highly protective. For these experiments, apolactoferrin was injected

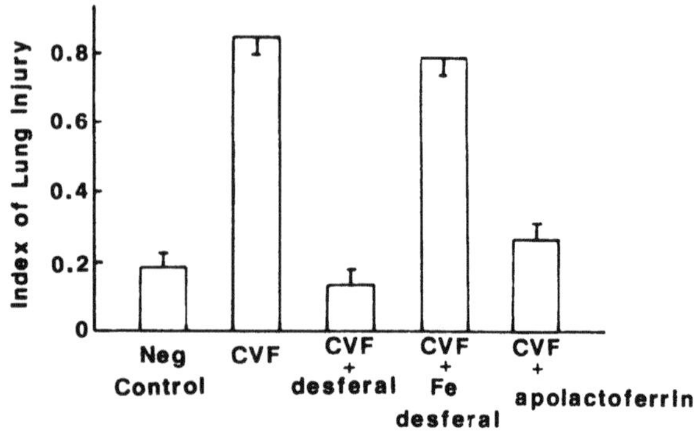

Figure 1. Effect of iron chelators on cobra venom factor (CVF)-induced acute lung injury in rats.

Table 1. Relationship Between Lung Injury and Plasma Conjugated Dienes*

	Reduction (% below control)	
Intervention	Lung Permeability	Conjugated Dienes
Neutrophil depletion	84	83
PEG-catalase	82	70
Heated PEG-catalase		-7.3
Deferoxamine	98	77
Iron saturated deferoxamine	24	28
Dimethyl sulfoxide	100	67
Dimethylthiourea	91	72

*Data obtained from Ward et al.[4]

intravenously along with CVF in amounts of 1.0 mg. In data not shown, iron-saturated lactoferrin is devoid of a protective effect against neutrophil-mediated acute lung injury.[3] Thus, iron appears to be an essential requirement for acute lung injury accompanying oxygen radical generation from neutrophils.

In Table 1, the relationship between the protective interventions in the CVF model of acute lung injury in the rat and the accompanying appearance of conjugated dienes (lipid peroxidation products) in plasma is shown. Details of the procedures employed in this study are described elsewhere.[4] Neutrophil depletion dramatically protects against lung injury as well as the appearance in plasma of conjugated dienes. The intravenous injection of appropriate amounts of polyethylene-glycol-modified catalase (PEG-catalase; 1,200 units/kg body weight) causes a similar pattern: dramatic reduction in both lung vascular permeability and accompanying levels of plasma conjugated dienes, whereas heat-inactivated PEG-catalase does not prevent the build up of conjugated dienes and, as shown elsewhere, does not affect increases in CVF-induced lung vascular permeability. As also shown in Table 1, deferoxamine (desferal) but not its iron-saturated counterpart protect against lung injury as well as the appearance of conjugated dienes. Finally, the hydroxyl radical ($HO^{\bullet}$) scavengers, dimethyl sulfoxide, and dimethyl thiourea, have similar protective effects on both parameters of injury. While these studies do not identify the sources of the conjugated dienes and do not reveal if the appearance of the dienes is primary or secondary to cell injury, the data do point to a relationship between the requirement for iron both for the development of acute lung injury as well as the appearance of conjugated dienes which appear to be linked to this type of oxygen radical-mediated injury. As discussed elsewhere,[4] monitoring of the levels of plasma conjugated dienes is a useful approach both for the intensity of tissue injury as well as for the assessment of the effects of protective interventions, including iron chelators.

IRON IN IN VITRO KILLING OF ENDOTHELIAL CELLS

Bovine or rat pulmonary artery endothelial cells have been incubated together with rat or human neutrophils stimulated with phorbol ester. The results of any of these combinations have been similar: killing of the endothelial cells as defined by the release of ^{51}Cr when phorbol-stimulated neutrophils are present but not when phorbol ester has been omitted.[5] As shown in Figure 2, $O_2^{\overline{\cdot}}$ responses of human neutrophils have been measured in the presence of various inhibitors, including soybean trypsin inhibitor (SBTI), catalase, dimethyl thiourea (DMTU), deferoxamine (desferal), and iron-saturated deferoxamine (Fe-desferal). It is demonstrated in the upper frame of Figure 2 that none of these interventions reduces the $O_2^{\overline{\cdot}}$ response, suggesting that the neutrophil activation process is not impaired. In the lower frame of Figure 2, the effects of the various agents on the killing of endothelial cells by phorbol ester (PMA)-activated human neutrophils are demonstrated. The protease inhibitor has no effect, while catalase is highly protective. The presence of superoxide dismutase leads to a slight attenuation in the killing process, while the $HO^{\bullet}$ scavenger DMTU and the iron chelator desferal are both highly protective. The specificity of the effects of iron binding by desferal is shown by the loss of the protective effects of iron-saturated desferal. In data not shown, inactivation of neutrophil-related myeloperoxidase by pretreatment with azide does not diminish the neutrophil-mediated killing of endothelial cells,[5] in confirmation of an earlier report.[6] These data suggest that in vitro the neutrophil-mediated cytotoxic effect on endothelial cells probably does not involve hypohalous products such as hypochlorous acid. The in vitro experiments together with the results of in vivo manipulations indicate that H_2O_2 or its derivatives are responsible for the cytotoxic events and that the $HO^{\bullet}$ may be the product directly responsible for the killing of endothelial cells.

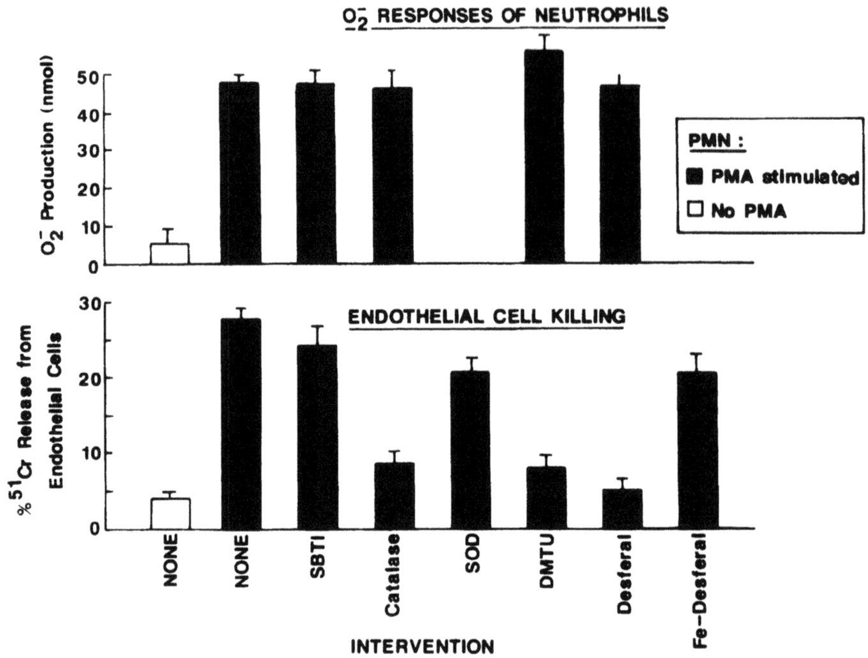

Figure 2. Effect of soybean trypsin inhibitor (SBTI), catalase, superoxide dismutase (SOD), dimethylthiourea (DMTU), and deferoxamine (desferal) on phorbol myristate acetate (PMA)-induced in vitro superoxide production by human neutrophils (upper frame) and in vitro ^{51}Cr release from rat pulmonary artery endothelial cells following exposure to PMA-stimulated neutrophils (lower frame).

SOURCE OF IRON IN NEUTROPHIL-MEDIATE KILLING OF ENDOTHELIAL CELLS

As indicated above, the killing of bovine pulmonary artery endothelial cells requires the availability of iron as defined by the protective effects of iron chelators. Experiments have been designed to assess the source of the iron required. Using ratios of human neutrophils:endothelial cells of 5:1 and employing phorbol ester as the agonist, human neutrophils or endothelial cells were pretreated with 0.5 mM desferal for 1/2 hr, washed, co-incubated, and release of ^{51}Cr from endothelial cells at 2 hr measured. The results of these experimental manipulations are shown in Figure 3. In the reference positive control, in which neither effector nor target cells have been pretreated with deferoxamine, 61% ^{51}Cr release from the endothelial cells occurs following addition of activated neutrophils. Desferal pretreatment of neutrophils produced the same response: full killing of endothelial cells as determined by ^{51}Cr release. Contrasting data are obtained when the endothelial cells are pretreated with deferoxamine: release of ^{51}Cr falls to levels (approximately 5%) that are virtually the same as those found when untreated endothelial cells are incubated with non-phorbol-stimulated neutrophils. These data provide decisive information suggesting that in oxygen radical-dependent killing of endothelial cells the source of the iron required for this event is the target cell.

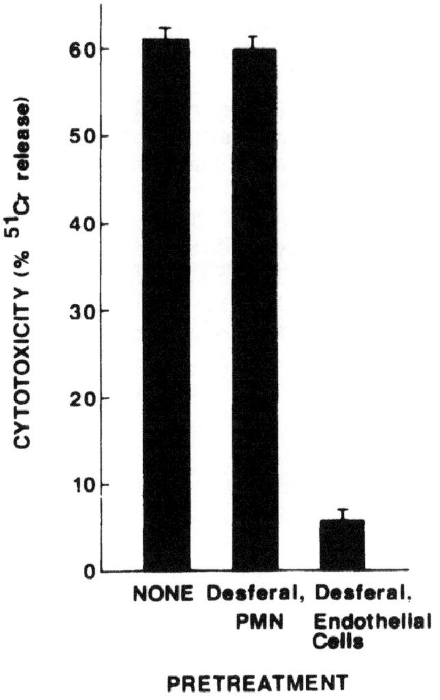

Figure 3. Pretreatment with desferal of endothelial cells protects against cytotoxic effects of PMA-stimulated neutrophils. Pretreatment of neutrophils has no effect.

In recent studies addressing the question of how deferoxamine interacts with bovine pulmonary endothelial cells to block the availability of iron, two different approaches have been used. In the first, incubation with deferoxamine of endothelial cells that had been preloaded with ^{59}Fe (using diferric transferrin) has failed to indicate the loss of any ^{59}Fe from the endothelial cells during the first 12 hrs. This suggests that the protective effects of deferoxamine are probably not related to the withdrawal of iron from the endothelial cells. HPLC analysis of extracts from endothelial cells that had been pretreated with deferoxamine has indicated the presence of small amounts (15 Fmol/cell) of cell-associated deferoxamine, all of which is present as an iron-deferoxamine complex. Thus, at least under the conditions employed, the protective effects of deferoxamine can be linked to the blocking of iron at some point on or within the endothelial cell. How deferoxamine is able to protect endothelial cells from killing by H_2O_2 and/or its derivatives awaits additional study.

REFERENCES

1. J.C. Fantone and P.A. Ward, Role of oxygen-derived free radicals and metabolites in leukocyte-dependent inflammatory reactions. Amer. J. Pathol. 107:395 (1982).
2. G.O. Till, K.J. Johnson, R. Kunkel, and P.A. Ward, Intravascular activation of complement and acute lung injury. Dependency on neutrophils and toxic oxygen metabolites. J. Clin. Invest. 69:1126-35 (1982).

3. P.A. Ward, G.O. Till, R. Kunkel, and C. Beauchamp, Evidence for role of hydroxyl radical in complement and neutrophil-dependent tissue injury. J. Clin. Invest. 72:789-801 (1983).
4. P.A. Ward, G.O. Till, J.R. Hatherill, T.M. Annesley, and R.G. Kunkel, Systemic complement activation, lung injury, and products of lipid peroxidation. J. Clin. Invest. 76:517-27 (1985).
5. J. Varani, S.E. Fligiel, G.O. Till, R.G. Kunkel, U.S. Ryan, and P.A. Ward, Pulmonary endothelial cell killing by human neutrophils. Possible involvement of hydroxyl radical. Lab. Invest. 53:656-63 (1985).
6. S.J. Weiss, J. Young, A.F. LoBuglio, A. Slivka, and N.F. Nimeh, Role of hydrogen peroxide in neutrophil-mediated destruction of cultured endothelial cells. J. Clin. Invest. 68:714 (1981).

ALTERATIONS IN CULTURED RAT PULMONARY ARTERY ENDOTHELIAL CELLS WITH HYPEROXIA; EFFECT OF AGE

R.S. Chuknyiska and R.E. Ratych

Department of Surgery, The Johns Hopkins Medical Institutions, Francis Scott Key Medical Center, 4940 Eastern Avenue, Baltimore, Maryland 21224 U.S.A.

INTRODUCTION

The integrity of the vascular endothelial cell layer is vital in separating the intravascular system from the parenchyma and thus in preserving whole-organ structure and function. As endothelial cells interface directly with the blood, they are exposed to relatively high concentrations of oxygen and its associated toxic metabolites. In order to adapt to this hyperoxic stress. the endothelial cells have evolved various antioxidant defense systems, including the enzymes superoxide dismutase (SOD) and catalase (CAT).[1]

Hyperoxia can result in the development of a severe form of pulmonary dysfunction referred to clinically as acute respiratory distress syndrome. Several studies have suggested that the susceptibility to hyperoxic stress may be age dependent, with senescent species being more susceptible to injury than younger or mature ones.[2] Given the critical role of the endothelial cells in maintaining tissue integrity, it is possible that these cells may be responsible for mediating this injury. The reason for the enhanced antioxidant defense capability in younger cells is unknown. Possible explanations accounting for this phenomenon could be higher endogenous basal levels of antioxidant enzymes or greater inductive capacity of these enzymes promoted by high oxygen level.

In order to determine which, if any, of these mechanisms are operative, rat pulmonary artery endothelial cells were isolated from either mautre or senescent rats and subjected to hyperoxia. The cells were assayed for SOD and CAT activity, levels of protein and DNA and cell injury and survival. Correlations between antioxidant enzyme activity and survival rate were then made.

METHODS

Rat pulmonary artery endothelial cells were isolated from mature (6-7 mo) and senescent (22 mo) female Wistar rats as described previously.[1] Conformation of the cells as endothelial cells was achieved by observing their cobblestone appearance under phase contrast microscopy and by factor VIII immunoassay.[1] In order to simulate hyperoxia, cells were transferred from the 95% air-5% CO_2 incubator to a chamber where the air was supplemented with 95% O_2-5% CO_2 for 24 hrs followed by 5 to 45 min incubation in 100% O_2. The cells were then transferred again in air/CO_2 and assayed at various times after being "rescued." All experimental determinations were performed in triplicate or quadruplicate.

SOD activity was determined by measuring the rate of reduction of cytochrome C spectrophotometrically at 550 nm at 25°C.[3] The reaction was performed in phosphate buffer, pH 7.8, containing 0.1 mM EDTA, 100 μM xanthine, 10 μM cytochrome C and 100 nM xanthine oxidase.

Catalase activity was determined spectrophotometrically at 240 nm by measuring the rate of utilization of 10 mM H_2O_2 using titration with $KMnO_4$ in 50 mM phosphate buffer, pH 7.8 at 23°C.[4] For this determination the cells were homogenized using 1% Triton X-100 and 0.2 M ETOH. DNA synthesis was determined by ^{3}H-thymidine. At the end of the incubation period the cells were washed with PBS and non-incorporated ^{3}H-thymidine was removed by wahsing with 0.2 M perchloric acid.[5]

Cell injury was assessed by both lactate dehydrogenase release and ^{3}H-adenine release. Lactate dehydrogenase released into the medium was assayed by measuring the appearance of NAD^+ from NADH, at 340 nm using pyruvate as a substrate.[6] One unit of lactate dehydrogenase activity was defined as the amount of enzyme producing one mol of NAD^+ per minute. Adenine release into the medium was assayed according to previously described methods.[7] Cells were incubated overnight with 1 μCi of ^{3}H-adenine and then washed to remove the unbound label.

Cells death was determined by trypan blue uptake. Following incubation with 0.1% trypan blue for 5 min at 37°C, cells were washed and the number of attached stained cells was counted in 3-5 randomly selected fields. Survival was assayed for up to 4 days following exposure to hyperoxia. Protein was measured according to Lowry[8] and DNA was measured according to the method of Labarca and Paigen.[9] All results are expressed as mean $\pm$ S.E. Differences between control and experimental samples were evaluated for statistical significance using the Student's t-test or repeated measurement analyses.

RESULTS AND DISCUSSION

Basal endogenous levels of the two major antioxidant enzymes SOD and CAT were not significantly different in endothelial cells derived from mature or senescent rat lungs (Table 1). However, after the cells have been exposed to 95/100% oxygen, there were apparent age differences in response to high oxygen treatement (Table 1). These differences appear afte rthe cells are transferred from the chamber with higher oxygen level to an incubator supplied with normal air/5% CO_2.

In both age groups, SOD and CAT estimated immediately after the endothelial cells were incubated in 95% and 100% oxygen or 100% oxygen alone, showed a slight but statstically significant decrease of the enzyme activaties. When cells treated with high oxygen are reseeded and transferred in 95% air-5% CO_2, a gradual increase in the SOD and CAT levels is observed (Table 2). The longer the treatment with 100% O_2, the higher increase in SOD and CAT activities measured. Senescent endothelial cells however are not stimulated to the same extent as mature cells

TABLE 1. SOD and CAT Activity in Rat Pulmonary Endothelial Cells at Different Times after the Cells Have Been Exposed to Hyperoxia (95/100% O_2) and Transferred in Air/CO_2

Treatment	Mature		Senescent	
	SOD	CAT	SOD	CAT
Control	0.42 $\pm$ 0.05	0.078 $\pm$ 0.09	0.36 $\pm$ 0.04	0.061 $\pm$ 0.05
0 hr	0.31 $\pm$ 0.03	0.058 $\pm$ 0.04	0.26 $\pm$ 0.04	0.047 $\pm$ 0.04
6 hr	0.45 $\pm$ 0.04	0.074 $\pm$ 0.06	0.33 $\pm$ 0.02	0.060 $\pm$ 0.05
12 hr	0.63 $\pm$ 0.07	0.106 $\pm$ 0.09	0.39 $\pm$ 0.05	0.072 $\pm$ 0.07
24 hr	0.78 $\pm$ 0.06	0.159 $\pm$ 0.08	0.50 $\pm$ 0.06	0.085 $\pm$ 0.05
48 hr	0.79 $\pm$ 0.08	0.162 $\pm$ 0.08	0.49 $\pm$ 0.04	0.084 $\pm$ 0.08

Results are expressed as Units/Well (10^5). Values represent means $\pm$ SE.

were. Thirty min of incubation in 100% O_2 stimulates SOD in mature cells by 89% above the control level, while in senescent cells the increase is only by 24%. Under the same conditons CAT activity increases by 132% in mature cells versus only 46% in senescent cells.

The greater inductive capacity of the antioxidant enzymes in mature endothelial cells coincide with the fact that protein synthesis in these cells increases in parallel to SOD and CAT activities (Table 3).

TABLE 2. SOD and CAT Activity in Rat Pulmonary Artery Endothelial Cells 48 Hours after Exposure to Hyperoxia.[a]

Duration of 100% O_2 Exposure (min)[2]				
5	107 $\pm$ 6	114 $\pm$ 5	98 $\pm$ 2	102 $\pm$ 3
10	134 $\pm$ 8	147 $\pm$ 8	103 $\pm$ 2	102 $\pm$ 2
15	191 $\pm$ 11	211 $\pm$ 15	105 $\pm$ 3	110 $\pm$ 2
30	189 $\pm$ 14	232 $\pm$ 14	124 $\pm$ 8	146 $\pm$ 9
45	204 $\pm$ 10	248 $\pm$ 13	129 $\pm$ 13	158 $\pm$ 14

[a]Results are expressed as % control. Values represent means $\pm$ SE.

That this greater enzyme activity was not due to enhancement in the proliferative rate by the mature cells is suggested by the fact that although mature cells have generally higher thymidine incorporation (Table 3), the differences in the increase of incorporated label in the two groups after hyperoxia were not great enough to explain the marked differences in the induced enzyme activity.

TABLE 3. Summary of Antioxidant Enzyme Activity, Protein Level, DNA Synthesis and Survival 48 hr Following Exposure to Hyperoxia[a]

	Mature	Senescent
SOD activity	189 ± 14	124 ± 8
CAT activity	232 ± 14	146 ± 9
Protein level	179 ± 15	121 ± 9
DNA synthesis	131 ± 9	122 ± 6
Survival	74 ± 3	53 ± 4

[a]Results are expressed as % control. Values represent means ± SE.

REFERENCES

1. R.E. Ratych, R.S. Chuknyiska, and G.B. Bulkley, The primary localization of free radical generation following anoxia/ reoxygenation in isolated endothelial cells, Surgery (in press).
2. L. Frank, J.R. Bucher and R.J. Roberts, Oxygen toxicity in neonatal and adult animals in various species, J. Appl. Physiol. 45:699 (1979).
3. J.M. McCord and I. Fridovich, The utility of superoxide dismutase in studying free radical reactions. II. The mechanism of cytochrome C reduction by a variety of electron carriers, J. Biol. Chem. 245:1374 (1970).
4. G. Cohen, D. Dembiec and J. Marcus, Measurement of catalase activity in tissue extracts, Annal. Biochem. 34:30 (1970).
5. P. Libby, P. Miao, J.M. Ordovas and E.J. Schafer, Lipoproteins increase growth of mitogen-stimulated arterial smooth muscle cells, J. Cell Physiol. 124:1 (1985).
6. M. Boehninger, Nicotic acid: Analogs and coenzymes, Methods Enzymol. 66:42 (1972).
7. S.P. Andreoli, R.L. Baehner and J.M. Bergstein, In vitro detection of endothelial cell damage using 2-deoxy-D-^{3}H-glucose: Comparison with chromium-51, ^{3}H-leucine, ^{3}H-adenine, and lactate dehydrogenase, J. Lab. Clin. Med. 106:253 (1985).
8. O.H. Lowry, N.J. Rosebrough, A.L. Farr and R.J. Randall, Protein measurement with the folin phenol reagent, J. Biol. Chem. 193:265 (1951).
9. C. Labarca and K. Paigen, A simple rapid and sensitive DNA assay procedure, Anal. Biochem. 102:344 (1980).

LUNG ENERGY METABOLISM FOLLOWING TWENTY-FOUR HOURS EXPOSURE TO ONE HUNDRED PERCENT OXYGEN

D.J.P. Bassett, E. Bowen-Kelly, and S.S. Reichenbaugh

The Johns Hopkins University School of Hygiene and Public Health 615 North Wolfe Street
Baltimore, MD 21205, U.S.A.

INTRODUCTION

The interstitial edema and endothelial cell damage observed in rat lungs following 48 hours of exposure to 100% oxygen is believed to result from oxygen-derived free-radical damage of cellular membranes, enzyme proteins, and possibly components of the extracellular matrix. Earlier physiological alterations of endothelial function have been demonstrated by depressed serotonin uptake in isolated perfused rat lungs following 18 to 24 hours of oxygen exposure,[1] at a time when tissue adenine nucleotide levels have been found not to be affected.[2] In vitro experiments using homogenates and isolated subcellular tissue fractions have demonstrated that key enzyme systems involved in energy generation are susceptible to oxygen inactivation. These include the dehydrogenases of glyceraldehyde-3-phosphate, pyruvate, succinate, and oxoglutarate.[3-5] The purpose of the present investigation was to use an intact organ preparation to determine whether lung glycolytic and mitochondrial enzyme systems are impaired during the first 24 hours of oxygen exposure when pathological alterations are not observed.

METHODS

In this study, the extent and pattern of glucose metabolism was assessed in lungs isolated from virus-free male Wistar rats (220-250 g)(Harlan-Sprague Dawley, Indianapolis) following 24 hours exposure to either air or 100% oxygen. Glycolytic rate was indicated by ^3H_2O production from the enolase and triose-phosphate isomerase reactions of glycolysis on perfusion with 5.6 mM [5-^{3}H] D-glucose (sp. act. 0.1 mCi/mmol) as previously described.[6,7] Glucose catabolism to carbon dioxide was measured as $^{14}CO_2$ produc-tion on perfusion with 5.6 mM [U-^{14}C] D-glucose (sp. act. 0.05 mCi/mmol). Perfused lungs were maintained under high oxygen conditions by ventilation with 95% O_2 and 5% CO_2. Maximal capacity of the lung to utilize glucose was determined by artificially lowering the tissue ATP to ADP ratio by perfusion with the uncoupler of oxidative phosphorylation 2,4-dinitrophenol (DNP).[6] As previously reported the maximal capacity of mitochondrial pyruvate dehydrogenase activity can be similarly assessed by measurements of $^{14}CO_2$ production on perfusion with [1-^{14}C] pyruvate in the absence and presence of DNP.[8] Determinations of changes in maximal metabolic rate by perfusion with uncoupler provided a more sensitive index of early impairment in the pathways of cellular energy generation than has been previously possible by measurements of adenine nucleotides,[2] and tissue homogenate enzyme activities.[4,5]

Data were analysed by standard two-way analyses of variance in combination with Hartley's test for variance homogeneity and Duncan's multiple comparison test of the means. The level of significance was taken as $P < 0.05$. It should be noted that metabolic rates were found to be linear in all cases giving regression coefficients of greater than 0.985.

RESULTS AND DISCUSSION

The effects of 24 hours of in vivo exposure to 100% oxygen on lung intermediary metabolism was indicated by a 53% increase in total glucose utilization, indicated by an increase in the linear appearance of ^3H_2O from control value of 18 $\pm$ 3 to an oxygen-exposed value of 28 $\pm$ 3 µmol [5-^{3}H] D-glucose per hour per g. dry wt. (SEM, n=4) (Figure 1). Although $^{14}CO_2$ production from [U-^{14}C] D-glucose was enhanced from 28 $\pm$ 4 to 35 $\pm$ 3 µmol $^{14}CO_2$/h/g. dry wt. (SEM, n=4), this increase was not statistically significant (Figure 1). The observed increase in glycolytic rate was therefore accounted for by a significant twofold increase in lactate plus pyruvate production from 16 $\pm$ 2 to 33 $\pm$ 4 µmol/h/g. dry wt.(SEM, n=4) (Figure 1), indicating enhanced cytosolic generation of ATP. This increased metabolism of oxygen-exposed lungs was associated with a mean tissue ATP to ADP ratio of 7.3 $\pm$ 0.8 that was not significantly different from an air-exposed control value of 8.5 $\pm$ 0.6 (SEM, n=4)($P > 0.05$). The ratio of perfusate lactate to pyruvate production was also unaltered from a control value of 9 $\pm$ 1, suggesting that oxygen exposure had not altered the ability of the mitochondria to oxidize reducing equivalents generated by the cytosolic glyceraldehyde-3-P dehydrogenase reaction of the Emden-Meyerhof glycolytic pathway.

In order to establish whether the maximal ability of the glycolytic pathway to generate ATP was adversely affected by oxygen exposure, lungs isolated from air- and oxygen-exposed rats were also perfused in the presence of 2,4-dinitrophenol (DNP). Previous studies using isolated rat lungs have demonstrated that a perfusate concentration of 0.8 mM DNP is most suitable for eliciting a maximal increase in glycolytic rate and subsequent

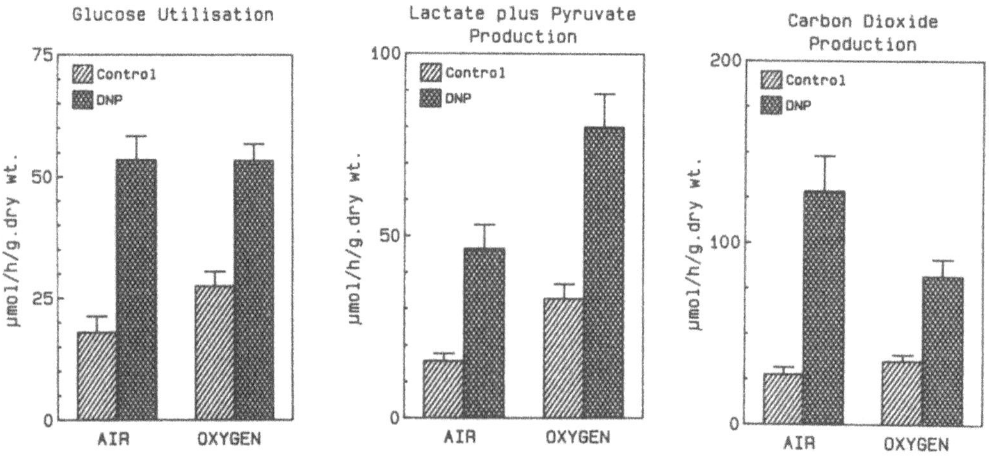

Figure 1. Results represent the means $\pm$ S.E.M. as µmol/h/g.dry wt. for four lungs perfused under each condition.

mitochondrial oxidation to carbon dioxide.[6,7] At this DNP concentration, perfusate lactate to pyruvate ratios were not affected, suggesting minimal damage to the mitochondrial respiratory chain. Tissue ATP to ADP ratios were decreased to 4.1 $\pm$ 0.1 and 2.4 $\pm$ 0.4 in air- and oxygen-exposed lungs, respectively. Perfusion of air-exposed lungs with DNP resulted in a three-fold increase in ^3H_2O production rate to 54 $\pm$ 5 μmol [5-^{3}H] D-glucose/h/g. dry wt. which was unaffected by oxygen exposure giving a value of 53 $\pm$ 3 μmol/h/g. dry wt.(Figure 1). This observation demonstrated that following 24 hours of exposure to 100% oxygen the ability of the lung to utilize glucose via the Embden-Meyerhof pathway was not limited. On the other hand, whereas $^{14}CO_2$ production from [U-^{14}C] D-glucose by air-exposed lungs was stimulated 4.5- fold to a linear rate of 128 $\pm$ 20 μmol $^{14}CO_2$ /h/g. dry wt. (SEM, n=4) by DNP, oxygen-exposed lungs exhibited a diminished increased rate of 82 $\pm$ 9 μmol/h/g. dry wt. (SEM, n=4) (Figure 1). Since previous studies have demonstrated that DNP perfusion does not alter the rate of cytosolic carbon dioxide generation in the pentose-phosphate cycle,[7] this failure to reach maximal DNP-stimulated rates of $^{14}CO_2$ production suggested possible mitochondrial impairment. This hypothesis was further supported by a larger rate of lactate plus pyruvate production of 80 $\pm$ 9 μmol/h/g. dry wt. (SEM, n=4) on DNP perfusion of oxygen-exposed lungs when compared with a mean value of 46 $\pm$ 7 μmol/h/g. dry wt (SEM, n=4) obtained on similar perfusion of lungs isolated from air-exposed animals (Figure 1). The pattern of glucose utilization in oxygen-exposed lungs when measured under uncoupler-stimulated conditions therefore demonstrated an altered pattern of metabolism that suggested a decrease in mitochondrial oxidation to carbon dioxide and a concomitant increase in lactate plus pyruvate production.

The effect of oxygen exposure on lung mitochondrial function was previously reported in a separate series of experiments,[8] that involved perfusion with a saturating 12.5 mM concentration of [1-^{14}C] pyruvate and subsequent measurements of $^{14}CO_2$ production in the presence and absence of 0.8 mM DNP.[8] Lungs were isolated from rats following 0, 3, 6, 12, and 24 hours of exposure to 100% oxygen. Unexposed lungs gave a linear rate of $^{14}CO_2$ production of 121 $\pm$ 16 μmol/h/g. dry wt. (n=5), which was maximally stimulated 84% by perfusion with 0.8mM DNP to 222 $\pm$ 32 μmol/h/g. dry wt. (SEM, n=5). The rates at which oxygen-exposed lungs metabolized pyruvate to CO_2 in the absence of DNP progressively decreased during 24 hours of oxygen exposure to 75 $\pm$ 17 μmol $^{14}CO_2$ /h/g.dry wt. (SEM, n=4). Although this trend suggested that oxygen exposure resulted in a 38% decrease in the rate of mitochondrial pyruvate metabolism and is consistent with previously observed decreases in lung homogenate oxygen utilization reported by Currie and coworkers,[4] the differences in $^{14}CO_2$ production were found not to be statistically significant. On the other hand, perfusion under stimulated conditions brought about by 0.8 mM DNP demonstrated 36% and 46% decreases in mean rates of $^{14}CO_2$ production from [1-^{14}C] pyruvate to 143 $\pm$ 32 and 121 $\pm$ 19 μmol/h/g. dry wt. (SEM, n=3 and 5) following 3 and 6 hours of in vivo oxygen exposure, respectively. Prolonging exposure to 12 and 24 hours of oxygen further diminished the rates of $^{14}CO_2$ production in the presence of DNP. Statistical comparisons between lungs exposed with and without DNP demonstrated that the uncoupler failed to stimulate pyruvate metabolism in all groups of lungs exposed in vivo to 100% oxygen. These data therefore suggested that lung mitochondria even after 3 hours of oxygen exposure were unable to respond to the increased demands for ATP generation asssociated with DNP uncoupling. It should be noted that the rates of lactate formation in the presence and absence of DNP were unaffected by oxygen exposure giving a mean production rate of 42 $\pm$ 2 μmol/h/g.dry wt. (SEM, n=18), suggesting that pyruvate distribution to all areas of the lung was not altered in these experiments.

These results demonstrated that lungs isolated from rats after 24 hours of exposure to 100% oxygen have an increased rate of exogenous glucose catabolism to lactate and pyruvate, with no observable alteration in the ability of the mitochondria to oxidize cytosolic-generated reducing equivalents. Since oxygen exposure ultimately leads to endothelial damage and edema formation, this observed increase in energy generation might reflect an early response of the cell to maintain homeostasis. Since glucose catabolism via the Emden-Meyerhof pathway in oxygen-exposed lungs was observed to proceed at a maximal rate in the presence of DNP, the cytosolic enzyme glyceraldehyde-3-P dehydrogenase most likely is not affected adversely by this level of oxygen exposure. On the other hand, it is possible that cells nearer to cell lysis and death might have inhibited glycolytic capacities due to possible NAD depletion as suggested by studies that have exposed isolated cells to lethal concentrations of hydrogen peroxide.[9,10] However, the failure to obtain maximal rates of $^{14}CO_2$ production from [1-^{14}C] pyruvate on perfusion with DNP in the present study suggested that lung mitochondrial metabolism via the pyruvate dehydrogenase enzyme complex might be impaired as early as 3 hours from the onset of 100% oxygen exposure. This study therefore suggests that dehydrogenases of the mitochondria are more likely to be the initial sites at which oxygen exposure adversely affects lung energy generation.

REFERENCES

1. E. R. Block and A. B. Fisher, Depression of serotonin clearance by rat lungs during oxygen exposure, J. Appl. Physiol. 42:33 (1977).
2. A. B. Fisher, Energy status of the rat lung after exposure to elevated pO_2, J. Appl. Physiol. 45:56 (1978).
3. N. Haugaard, Cellular mechanisms of oxygen toxicity, Ann. Rev. Physiol. 48:311 (1968).
4. W. D. Currie, P. C. Pratt, and A. P. Sanders, Hyperoxia and lung metabolism, Chest 66:19S (1974).
5. D. Jamieson and H. A. S. van den Brenk, Pulmonary damage due to high pressure oxygen breathing. 2. Changes in dehydrogenase activity of rat lung, Austral. J. Exp. Biol. 40:51 (1962).
6. D. J. P. Bassett and A. B. Fisher, Stimulation of rat lung metabolism with 2,4-dinitrophenol and phenazine methosulfate, Am. J. Physiol. 231:898 (1976).
7. D. J. P. Bassett and A. B. Fisher, Pentose cycle activity of the isolated perfused rat lung, Am. J. Physiol. 231:1527 (1976).
8. D. J. P. Bassett and E. Bowen-Kelly, Pyruvate metabolism of perfused rat lungs after exposure to 100% oxygen, J. Appl. Physiol. 60:1605 (1986).
9. R. G. Spragg, D. B. Hinshaw, P. A. Hyslop, I. U. Schraufstatter, and C. G. Cochrane, Alterations in adenosine triphosphate and energy charge in cultured endothelial and $P388D_1$ cells after oxidant injury, J. Clin. Invest. 76:1471 (1985).
10. I. U. Schraufstatter, D. B. Hinshaw, P. A. Hyslop, R. G. Spragg, and C. G. Cochrane, Glutathione cycle activity and pyridine nucleotide levels in oxidant-induced injury of cells, J. Clin. Invest. 76:1131 (1985).

ACKNOWLEDGMENTS

These investigations were mainly supported by the National Heart and Blood Institute grant number HL-34674, with additional support from the National Institute of Environmental Health grants ES-03505 and ES-03819.

PROSTAGLANDINS AND LEUKOTRIENES IN PULMONARY O_2 TOXICITY

Morris D. Faiman and Lavonne M. Patton

Department of Pharmacology and Toxicology
University of Kansas
Lawrence, Kansas 66045, U.S.A.

INTRODUCTION

O_2 radicals can initiate peroxidation of cellular membranes, resulting in the release of hydrolytic enzymes, decreased membrane fluidity, and compromised membrane integrity. Free radicals also may cause the release of vasoactive substances, including prostaglandins (PG) and leukotrienes (LT). These products of the arachidonic acid cascade have been shown to produce vasoconstriction, vasodilation, platelet aggregation, increased capillary permeability, plasma extravasation, and edema formation. Of interest is that a similar pathophysiology is seen after O_2 exposure. Employing the isolated perfused lung (IPL) prepared from lungs of rats exposed to 100% O_2, the role of PG and LT in O_2-induced lung damage was studied.

METHODS

Lung damage was assessed by increased protein and malondialdehyde (MDA) in lung perfusate, and changes in the lung dry weight/wet weight and wet weight/body weight ratios. Bioassay techniques for the determination of PG-like activity employed the rat stomach strip (RSS),[1] while LT activity was determined employing the guinea pig ileal smooth muscle (GPISM).[2] The RSS and GPISM were perfused with lung perfusate and the contractions recorded for 30 min. The area under the contraction curve was calculated which reflected the PG and LT activity respectively.

Adult respiratory disease-free male Sprague-Dawley rats were exposed to either one atmosphere of air, or 100% O_2 for 24 or 48 hours. After O_2 exposure, the animals were removed from the chamber, and an IPL prepared.[3] The lungs were perfused from a 17 ml constant-volume reservoir at a rate of 5 ml per minute. The lung perfusate was superfused over the RSS or GPISM and contractions recorded. Lung perfusate was used for the determination of lipid peroxide formation and protein content. After perfusion of the lung, the lungs were blotted, weighed, dried, and the lung wet weight/dry weight ratio determined. 6-Keto $PGF_{1\alpha}$ and thromboxane B_2 (TXB_2) which reflected prostacyclin (PGI_2) and thromboxane A_2 (TXA_2), respectively, were determined in lung perfusate by radioimmunoassay (RIA).

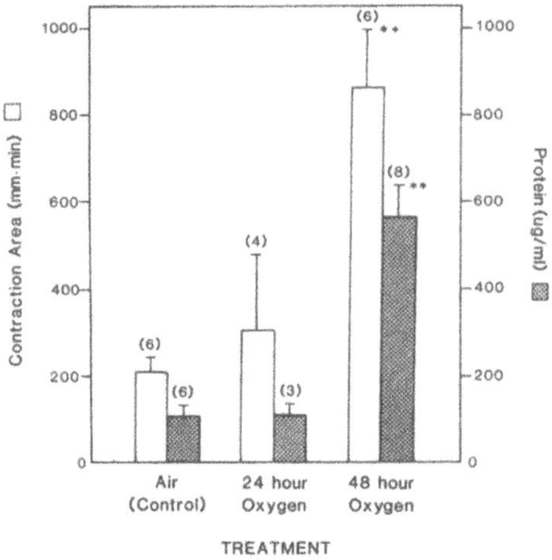

FIGURE 1. Male rats were exposed to 100% O_2 for 24 and 48 hours, killed, IPLS prepared, and PG-like activity (□) and perfusate protein (▩) determined. Values are mean ± SEM. Number of rats in parentheses. *p<0.01 when compared to control; **p<0.01 when compared to control and 24 hours; +p<0.05 when compared to 24 hours O_2.

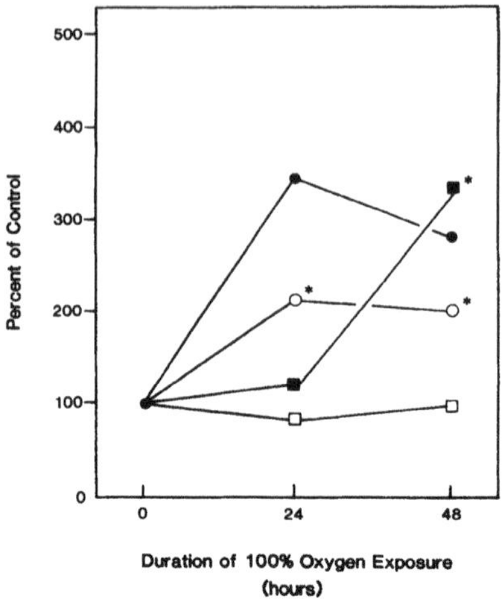

FIGURE 2. 6-Keto-$PGF_{1\alpha}$ (●), leukotriene-like activity (○), prostaglandin-like activity (■) and TXB_2 (□) in lung perfusate after exposure to 100% oxygen for 24 or 48 hours. *p<0.05

RESULTS AND DISCUSSION

Statistically significant increases in PG-like activity and protein were found in the perfusate from the IPL prepared from rats exposed to 48 hours of O_2 but not after 24 hours of exposure (Figure 1). The lung dry weight/wet weight ratio decreased, while the lung wet weight/body weight ratio increased in rats exposed to 48 hours of O_2 (data not shown). Perfusate PGI_2 exhibited a 3.5-fold increase after 24 hours of O_2 exposure when compared to controls, and remained at this concentration after 48 hours of O_2. TXA_2 did not change over the same 48-hour exposure period (Figure 2). Perfusate MDA was significantly elevated after 48 hours but not after 24 hours of O_2 exposure.

When rats were treated with indomethacin (1.8 mg/kg p.o. for 7 days) prior to O_2 exposure, PG-like activity in the lung perfusate decreased. However, the O_2-induced increase in perfusate protein was not blocked by indomethacin pretreatment (Figure 3). In addition, the lung dry weight/wet weight ratio decreased, while MDA again increased (data not shown) suggesting pulmonary damage. Disulfiram has been shown to be an inhibitor of superoxide dismutase (SOD). In disulfiram-treated rats, survival in O_2 decreased by 50% (66 hr vs 33 hr). Furthermore, when rats were pretreated with disulfiram and exposed to O_2, an increase in PG-like activity and protein in the perfusate from the IPL was found after only 28 hours of O_2 exposure (Figure 4). Also, the lung dry weight/wet weight ratio decreased, while the lung wet weight/body weight ratio increased, as did perfusate MDA (data not shown). Furthermore, in disulfiram-treated rats, 6-keto $PGF_{1\alpha}$ increased 3.5-fold within 24 hours and then decreased. No change in TXA_2 was found (Figure 5).

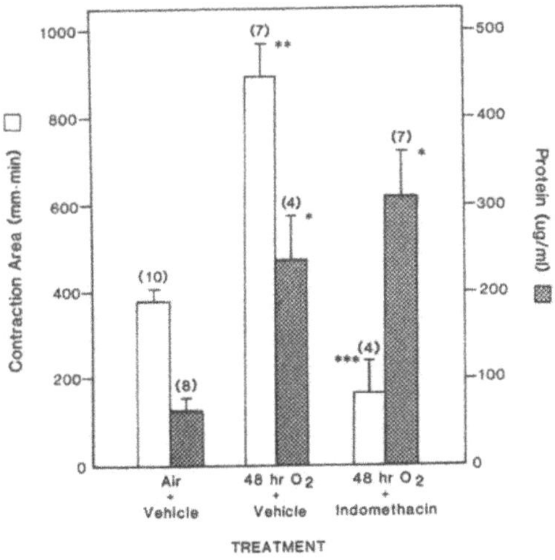

FIGURE 3. PG-like activity (□) and perfusate protein (▩) in O_2-exposed indomethacin (1.8 mg/kg p.o. for 7 days) treated rats. Mean ± SEM. Number of rats in parentheses. *p<0.05 when compared to control; **p<0.01 when compared to control; ***p<0.01 when compared to 48 hours O_2.

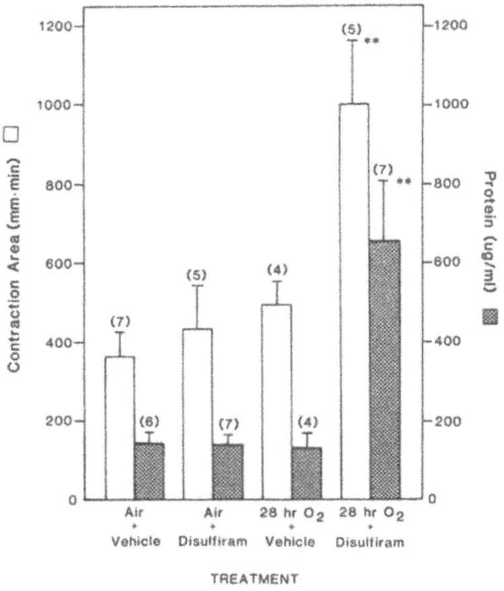

FIGURE 4. PG-like activity (□) and perfusate protein (▩) in O_2-exposed disulfiram (200 mg/kg i.p.) treated rats. Mean ± SEM. Number of rats in parentheses. **p<0.01 when compared with all groups.

In rats exposed to 100% O_2, an increase in LT-like activity in the perfusate of the isolated rat lung was found after 48 hours of O_2 but not after 24 hours of exposure (Figure 2). After 48 hours of O_2 exposure, the lung dry weight/wet weight ratio decreased, while the lung wet weight/body weight ratio, perfusate MDA, and perfusate protein increased (data not shown). Inhibition of lung SOD (disulfiram 200 mg/kg i.p.) markedly increased LT-like activity (Figure 5) and perfusate protein after only 28 hours of O_2. A decrease in the lung dry weight/wet weight ratio and an increase in lung wet weight/body weight ratio and perfusate MDA after 28 hours of O_2 also occurred (data not shown). BW755C at low doses is an inhibitor of lipoxygenase activity. BW755C administration (10 mg/kg i.p.) had no effect on perfusate PG-like activity (unpublished data). However, BW755C blocked the O_2-induced increase in perfusate LT activity and protein (Figure 6).

These studies showed that perfusate PGI_2 and LT increased before any significant change in PG-like activity was observed. After 48 hours of O_2 exposure, all of the vasoactive agents, except TXA_2, increased. Inhibition of SOD by disulfiram treatment decreased the survival of rats in O_2. Concomitant with this decreased survival, a significant increase in perfusate PG and LT activity occurred before any onset of O_2 toxicity. BW 755C, given at a dose in which only LT synthesis is inhibited, decreased lung perfusate protein, suggesting that capillary permeability had been inhibited. This appears to suggest that the LT may be responsible, at least in part, for the permeability associated with the toxic effects of O_2 in the lung, and to suggest a causative role for LT in pulmonary O_2

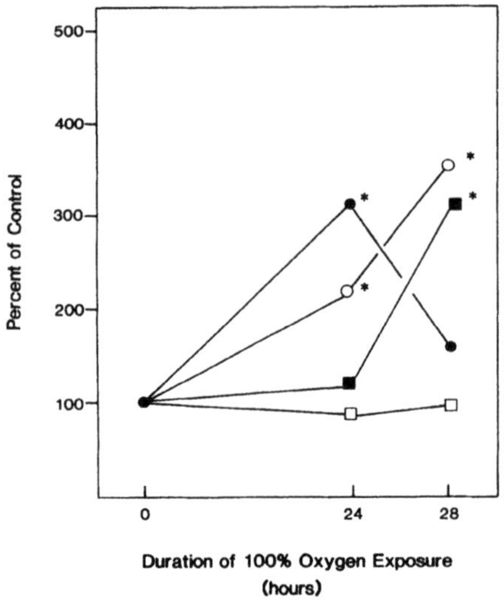

FIGURE 5. 6-Keto-PGF$_{1\alpha}$ (●), LT-like activity (○), PG-like activity (■) and TXB_2 (□) in lung perfusate of disulfiram (200 mg/kg, i.p.)-treated rats exposed to 100% O_2 for 24 or 48 hours. *p<0.05.

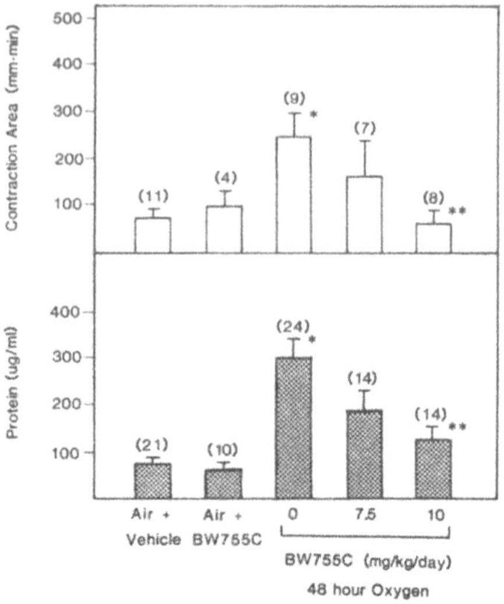

FIGURE 6. LT-like activity (□) and perfusate protein (▩) in O_2-exposed and BW775C-treated rats. Mean ± SEM. Number of rats in parentheses. *p<0.01 when compared with vehicle control, or BW775C (10.mg/kg/day i.p.); **p<0.05 when compared with 48 hours O_2.

toxicity. Lipoxygenase inhibitors also possess antioxidant activity which could account for some of the protection observed.[4] These data also suggest that the vasodilator PG may act to potentiate damage produced by the LT.

REFERENCES

1. J.R. Weeks, J.R. Schultz, and W.E. Brown, Evaluation of Smooth muscle bioassays for prostaglandins E_1 and F_1 , J. Appl. Physiol. 25: 783 (1968).
2. J.H. Fleisch, L.E. Rinkema, K.D. Haisch, D. Swanson-Bean, T. Goodson, P.P.K. Ho, and W.S. Marshall, LY171883, 1-<-2-Hydroxy-3-Propyl-4-L4-(1-H-Tetrazol-5-yl) butoxy > phenyl > ethanone, an orally active leukotriene D_4 antagonist, J. Pharmacol. Exp. Ther. 233: 148 (1985).
3. A.J. Januskiewicz and M.D. Faiman, The effect of in vivo hyperoxic exposure on the release of endogenous histamine from the rat isolated perfused lung, Toxicol. Appl. Pharmacol. 72: 134 (1984).
4. O.C. Burghuber, R.J. Strife, J. Zirrolli, P.M. Henson, J.E. Henson, M. Mathias, J.T. Reeves, R.C. Murphy, and N.F. Voelkel, Leukotriene inhibitors attenuate rat lung injury induced by hydrogen peroxide, Am. Rev. Respir. Dis. 131: 778 (1985).

ACKNOWLEDGMENTS

Supported in part by grants from the National Institute of Neurological and Communicative Disorders and Stroke, NS 07797, the National Institute of Alcohol Abuse and Alcoholism, AA 03577, and the Department of the Air Force, contract number F 33615-83-D-0601.

LIPID PEROXIDATION, FREE RADICALS AND EXPERIMENTAL INFLAMMATION

Andrew M. Symons, Ewa J. Dowling, and Dennis V. Parke

Department of Biochemistry
University of Surrey, Gildford
Surrey, GU2 5XH, England

INTRODUCTION

A variety of evidence has emphasised the significance of free radicals and products derived from lipid peroxidation in the development of inflammatory and arthritic diseases.[1] The damaging or injurious agent in inflammation causes a host of local reactions, including vasodilation, increased vascular permeability, platelet aggregation, fibrin accumulation and the release of numerous enzymes and small molecular mediators. Many of these enzymes and mediators are either integral components of lipid membranes or are products derived from lipid metabolism (i.e., the prostanoids). Local tissue damage and concomitant lipid peroxidation are therefore central to the promotive and contributing elements of an inflammatory reaction. These local events lead to systemic reactions such as increased body temperature, pain, increased fibrinogen, increased pituitary-adrenal function and increased gamma-globulin, which could all be viewed as protective and inhibitory components to the overall process. Whether cell death and necrosis or restitution results depends on the balance between the promotive and inhibitory components, and central to the overall process is the involvement of lipid membranes. The activated neutrophil, which plays an important defensive role in inflammation, produces the superoxide radical on activation of its membrane-bound NADPH oxidase, which can be converted to many other reactive oxygen species, such as hydrogen peroxide and the more damaging hydroxyl radical. The persistence of such free radical generation and its propagation of local membrane lipid peroxidation may determine the transition from an acute to a chronic inflammatory response.

METHODS

In this study a well-established model of acute foot-pad oedema was used.[2] Groups of male Wistar albino rats (150-200 g) were immunised in the scruff on Day 0 with Freund's Complete Adjuvant (FCA; 75 µl of 10 mg/ml suspension in liquid paraffin) and 6 days later inoculated (challenged) with 50 µl FCA s.c. in the right hind paw. Oedema was quantified as the increase in fresh weight (g) of inoculated versus uninoculated paws over the ensuing 96 hours. The presence of reactive oxygen species in 50 mg portions of inflamed tissue was determined using the luminol-amplified chemiluminescence (LAC) method.[2] Malondialdehyde (MDA), as index of lipid peroxidation, was measured in the inflamed paw, plasma and a 10,000 g hepatic supernatant, using the thiobarbituric acid method.[3] Different groups of animals were

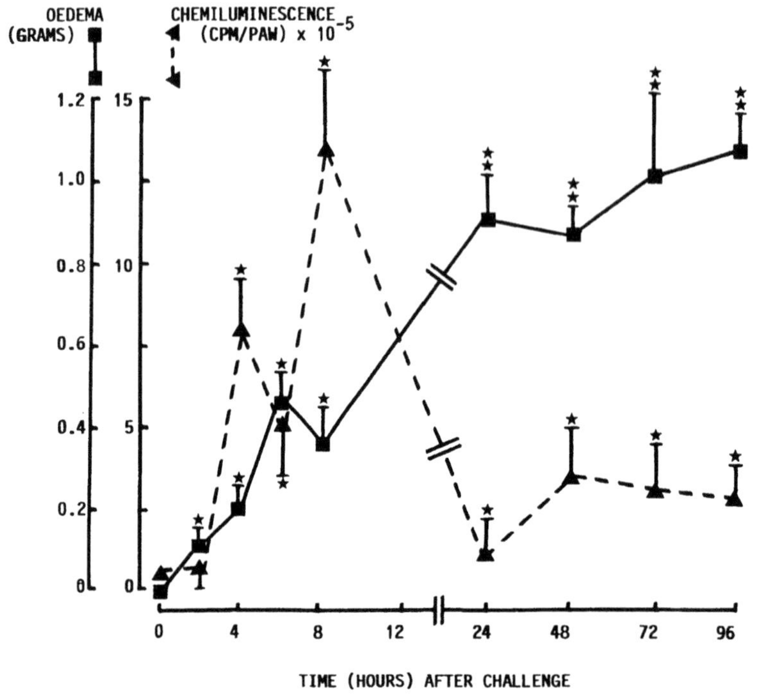

Figure 1. Oedema and luminol-amplified chemiluminescence in the rat foot-pad oedema model. Points are mean ± S.E.M. (n = 4). * $p < 0.05$, ** $p < 0.001$, versus zero time point for chemiluminescence; statistical analysis for oedema was performed using the actual weights of inflamed and non-inflamed control paws.

used for LAC and MDA experiments. In other experiments animals were dosed with diclofenac, indomethacin or piroxicam (all 3 mg/kg) or prednisolone (100 mg/kg) orally one hour before challenge with FCA, and the effects of the drugs on oedema, LAC and MDA determined.

RESULTS AND DISCUSSION

Figure 1 shows the oedema response and LAC in groups of animals up to 96 hours after challenge. Oedema developed rapidly over the initial 6 hours then more slowly over the ensuing 90 hours to reach maximal levels of approximately one gram. Associated with the initial phase of oedema, a significant increase in free radical activity, measured as LAC, was seen.

Figure 2 shows the changes in MDA seen over the initial 24 hours after challenge with FCA. Results have been expressed as total MDA in the plasma, liver and inflamed foot-pad. Maximal levels were reached between 2 and 8 hours after challenge, similar to the peak found with LAC, and well before maximum oedema.

Figure 3 shows the response to oedema, LAC and plasma MDA after treatment with the four anti-inflammatory drugs tested. With the exception of prednisolone, no drug showed any significant effect on oedema until 24 hours after challenge (prednisolone was effective after 12 hours), whereas significant decreases in LAC and plasma MDA were seen as soon as four hours after challenge with all drugs tested.

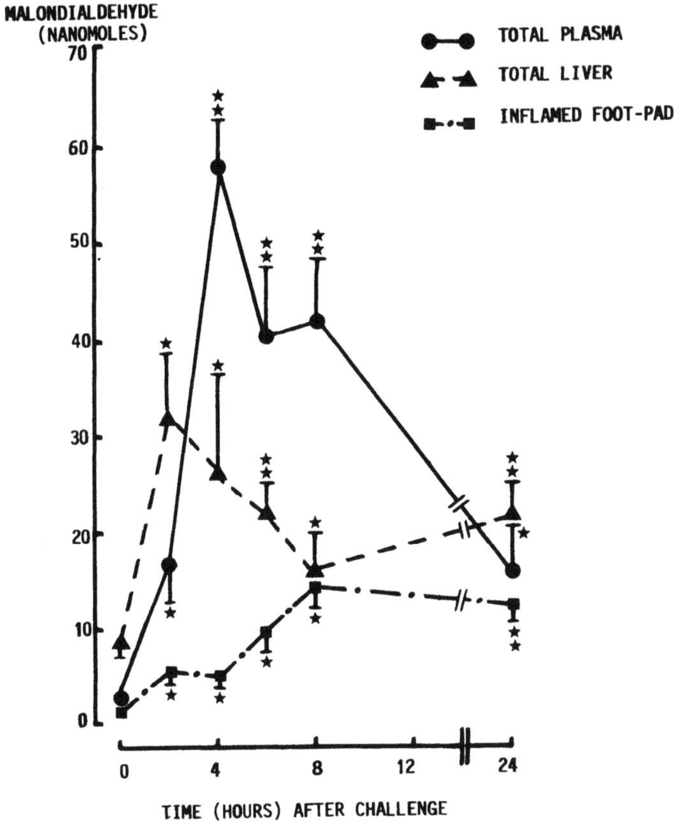

Figure 2. MDA levels in the rat foot-pad oedema model. Points are mean ± S.E.M. (n = 4). * $p < 0.05$, ** $p < 0.001$, versus zero time point.

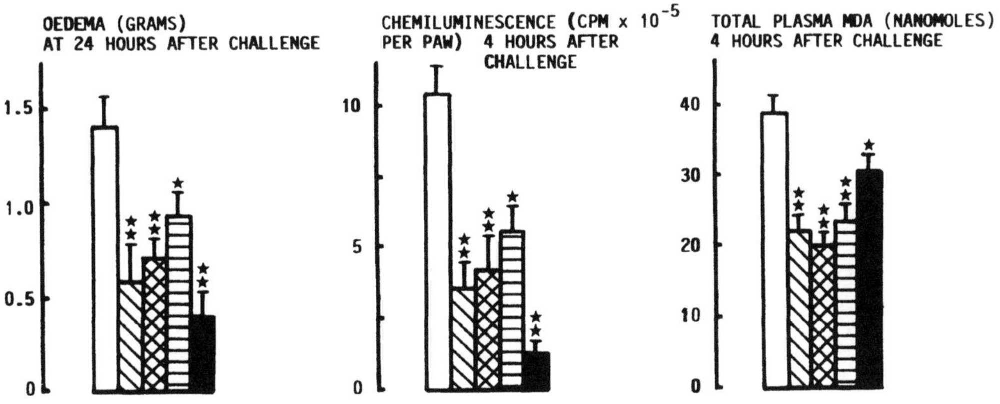

Figure 3. Effect of diclofenac (diagonal hatching), indomethacin (cross-hatching), piroxicam (horizontal hatching) and prednisolone (back shading) on oedema, LAC and plasma MDA. in the rat foot-pad oedema model. * $p < 0.05$, ** $p < 0.001$, versus untreated controls (open boxes). S.E.M. shown (n = between 4-16).

These results show that in the acute model of inflammation used, the presence of reactive oxygen species and evidence of lipid peroxidation (MDA) could be detected as early events in the development of the oedematous reaction. Maximum oedema was not apparent until 48 hours, whereas LAC and MDA were found at maximal levels within 8 hours.

The results with the anti-inflammatory drugs would suggest that the extent of lipid peroxidation and concomitant production of free radicals (including reactive oxygen species) is important in determining the later severity of an inflammatory response and indicates a possible mode of action for these drugs, which may prove useful in the testing of potential new compounds.

REFERENCES

1. D.R. Blake, R.E. Allen, and J. Lunec, Free radicals in biological systems — A review orientated to inflammatory processes, Brit. Med. Bull. 43:371 (1987).
2. E.J. Dowling, A.M. Symons, and D.V. Parke, Free radical production at the site of an acute inflammatory reaction as measured by chemiluminescence, Agents and Actions 19:203 (1986).
3. J.M.C Gutteridge, Free-radical damage to lipids, amino acids, carbohydrates and nucleic acids determined by thiobarbituric acid activity, Int. J. Biochem. 14:649 (1982).

INHIBITION BY PROSTACYCLIN ANALOGS OF $\cdot O_2^-$ PRODUCTION IN HUMAN POLYMORPHONUCLEAR LEUKOCYTES*

H. J. Stahlberg, G. Loschen, and L. Flohé

Grünenthal GmbH
Center of Research
D-5100 Aachen
Zieglerstrasse 6
Federal Republic of Germany

INTRODUCTION

While the interactions of prostacyclin with platelets have been intensively studied during the last decade[1,2] only very few and sometimes contradictory data point to an interaction of prostacyclin with stimulated leukocytes. The following events, which are inducible in polymorphonuclear leukocytes (PMNL) during inflammatory reactions, were reported to be inhibited by prostacyclin and PGE_1: leukotaxis,[3] degranulation of lytic enzymes,[4] adherence to endothelial cells,[5] oxygen radical formation.[6] An elevation of cAMP in stimulated leukocytes by prostacyclin has also been described.[7]

The chemical instability of prostacyclin, with a half life of only 3 minutes at pH 7.4,[8] proved to be a major obstacle in reliably evaluating its function. Therefore three different, chemically stabilized prostacyclin analogs (Figure 1) were investigated for their influence on $\cdot O_2^-$ radical formation and cAMP elevation induced by the chemoattractant N-formyl-L-methionyl-L-leneyl-phenylalanine (FMLP) in isolated human PMNL.

MATERIALS AND METHODS

The prostacyclin analogs CG 4203 (Taprosten), CG 4303 and CG 4305 (Figure 1) were synthesized in the chemistry department of Grünenthal GmbH.[9] Cytochalasin B, theophylline, dextran (MW 500 000) and FMLP were purchased from Sigma Chemical Company; bovine Cu/Zn superoxide dismutase (SOD) from Grünenthal GmbH and Percoll from Deutsche Pharmacia.

---

*This publication is part of the doctoral thesis of H.J. Stahlberg.

PMNL were isolated from the blood of healthy volunteers by using a discontinuous Percoll gradient.[10] Cell viability (> 95%) was controlled by the trypan blue exclusion test.[11] Intracellular cAMP of leukocytes was determined according to Gilman et al.[12] by means of a kit purchased from Amersham Buchler GmbH. 40 x 10^6 PMNL/ml were incubated in the presence of 2 mM theophylline, cytochalasin B (5 μg/ml) and various concentrations of the prostacyclin analogs for 5 min at 37°C before stimulation with FMLP (10^{-7} M). An aliquot of 75 μl (~3 x 10^6 PMNL) was taken exactly 1 minute later for determining cAMP. Inhibition of $\cdot O_2^-$ radical formation by prostacyclin analogs in FMLP-stimulated PMNL was determined spectrophotometrically in the presence of cytochalasin B (5 μg/ml) and cytochrome c (1.23 mg/ml) as described by Sedgwick et al..[13] $\cdot O_2^-$ formation was expressed as nmoles of reduced cytochrome c formed per 1 x 10^6 PMNL.

RESULTS

Addition of FMLP to isolated human PMNL in the presence of theophylline (2 mM) and cytochalasin B (0.5 μg/ml) and the prostacyclin analogs led to an intracellular increase of cAMP which peaks at about one minute after stimulation.

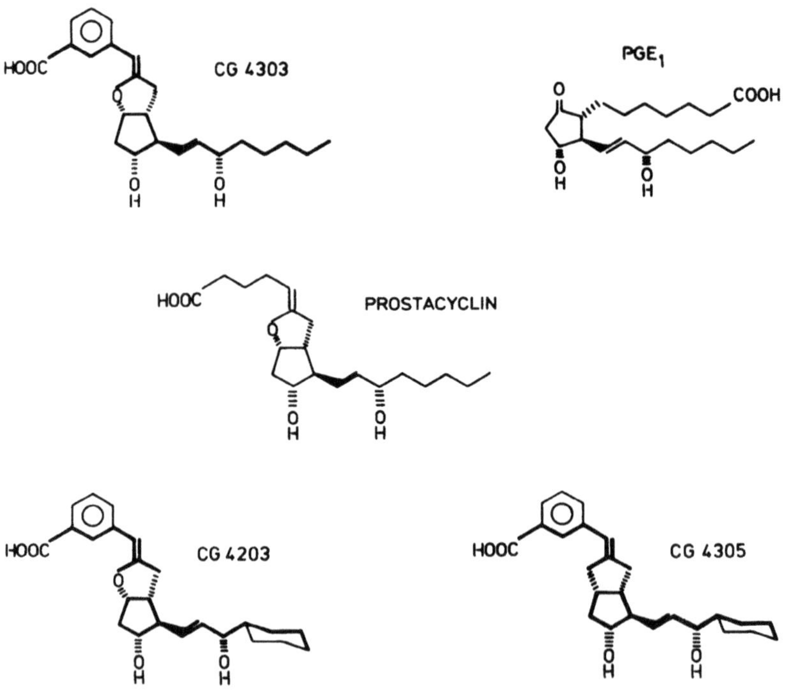

Figure 1. Structural formulae of the three chemically stable prostacyclin analogs used in our investigations. To demonstrate similarities with prostacyclin identical structural elements are stressed. Prostaglandin E_1 (PGE_1) is shown for comparison.

Table 1. Elevation of Intracellular cAMP by Prostacyclin Analogs in FMLP-stimulated Human PMNL.[a]

Concentration of analog	Stimulation factor compared to control CG 4203	CG 4303	CG 4305
10^{-7} M	3.0**	1.5*	1.4*
10^{-6} M	6.8**	2.1***	2.6***
10^{-5} M	9.7***	3.1***	3.0**
10^{-4} M	8.3***	2.8**	2.1**

[a] Statistical significance was calculated from the absolute values of cAMP determinations (n = 3). $*p < 0.05$, $**p < 0.01$, $***p < 0.001$.

The FMLP-dependent rise in cAMP was significantly increased by all three prostacyclin analogs in the dose range between 10^{-7} and 10^{-4} M (Table 1). Maximum cAMP levels were found at 10^{-5} M for all 3 analogs. Beyond this concentration the stimulation factor declined.

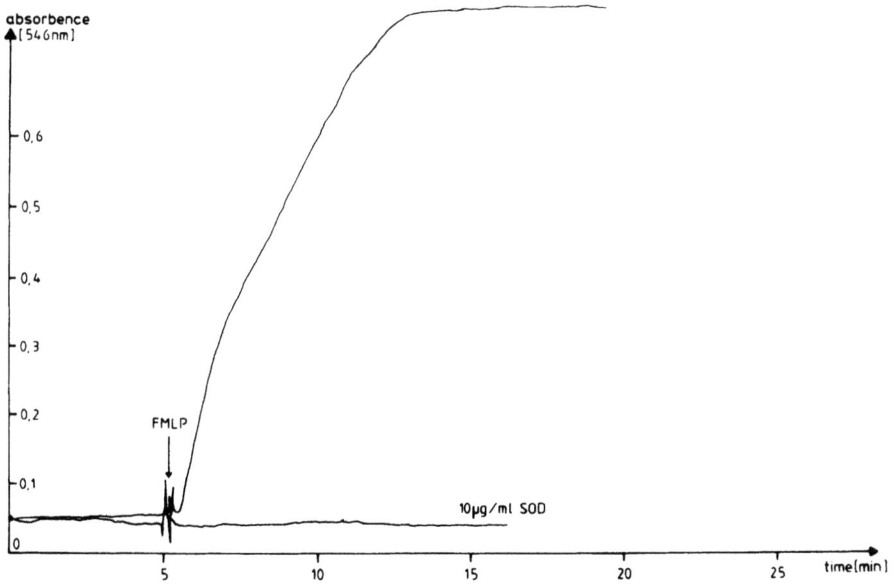

Figure 2. Induction of $\cdot O_2^-$ radical formation in isolated human PMNL by FMLP. Formation of $\cdot O_2^-$ radicals is detected by the SOD-sensitive reduction of cytochrome c (1.23 mg/ml) followed spectrophotometrically after addition of FMLP (10^{-7} M) to PMNL (2×10^6/ml).

Table 2. Inhibition of $\cdot O_2^-$ Formation by Prostacyclin Analogs in FMLP-stimulated Human PMNL[a]

Concentration of analog	CG 4203	% Inhibition CG 4303	CG 4305
10^{-7} M	35**	n.s.	n.s.
10^{-6} M	65**	17*	36*
10^{-5} M	73**	62**	66**
10^{-4} M	85**	87**	85**

[a] Statistical significance was calculated from the absolute values of cytochrome c reduced per 1 x 10^6 PMNL, (n = 3). *$p < 0.05$, **$p < 0.01$, ***$p < 0.001$.

Under similar conditions the addition of FMLP (10^{-7} M) to isolated human PMNL in the presence of cytochalasin B (0.5 μg/ml) and cytochrome c (1.23 mg/ml) induced a reduction of cytochrome c as measured spectrophotometrically at 546 nm (Figure 2). This reduction ceased within 10 minutes after stimulation.

The proof that this reduction of cytochrome c is due to an induction of $\cdot O_2^-$ radical formation was provided by the effect of superoxide dismutase (SOD; 10 μg/ml). The prostacyclin analogs significantly inhibited FMLP-stimulated $\cdot O_2^-$ radical formation (Table 2) at concentrations similar to those increasing the cAMP levels. At high concentrations of the analogs, however, the two effects induced by the analogs diverged: while the elevation of cAMP tended to decline, $\cdot O_2^-$ formation was further inhibited.

DISCUSSION

We have shown that all three prostacyclin analogs (CG 4203, CG 4303, CG 4305), in about the same range of concentration (10^{-7} M to 10^{-4} M), stimulate adenylate cyclase activity and inhibit $\cdot O_2^-$ radical formation in FMLP-stimulated human PMNL. Qualitatively the same effects have been described for prostacyclin and PGE_1.[4-6,13] The relative potencies of the prostacyclin analogs in suppressing stimulated $\cdot O_2^-$ release closely resemble those established for inhibition of platelet aggregation.[9] This suggests that the inhibitory effect is mediated by a real prostacyclin receptor of the PMNL. Although a causal connection between cAMP elevation and suppression of $\cdot O_2^-$ formation could be assumed, a significant correlation between these two phenomena was not verified. Other PMNL functions, such as FMLP-induced aggregation and chemotaxis (data not shown), do not correlate with $\cdot O_2^-$ or cAMP elevation in our experiments. However, from the leukocyte functions investigated, the inhibition of $\cdot O_2^-$ formation by all three prostacyclin analogs appeared to be the most pronounced effect in FMLP-stimulated PMNL. This dampening of the oxidative burst might be relevant in vivo and contribute to the beneficial effects observed with prostacyclin, prostacyclin analogs and PGE_1 in models of inflammation,[14] endotoxemia[15] and postischemic reperfusion injuries.[16]

REFERENCES

1. S. Moncada, Prostacyclin, from discovery to clinical application, J. Pharmacol. 16 (Suppl. I):71 (1985).
2. J.R. Vane, The road to prostacyclin, in: "Advances in Prostaglandin, Thromboxane, and Leukotriene Research," O. Hayaishi and S. Yamamoto, eds., Raven Press,New York 15:11 (1985).
3. B.B. Weksler, J.M. Knapp, and E.A. Jaffe, Prostacyclin (PGI_2) synthesized by cultural endothelial cells modulates polymorphonuclear leukocyte function, Blood 50:Suppl. 1, 287 (1977).
4. J.E. Smolen, H.M. Korchak, and G. Weissmann, Increased levels of cyclic adenosine-3',5'-monophosphate in human polymorphonuclear leukocytes after surface stimulation, J. Clin. Invest. 65:1077 (1980).
5. L.A. Boxer, J.M. Allen, M. Schmidt, M. Yoder, and R.L. Baehner, Inhibition of polymorphonuclear leukocyte adherence by prostacyclin, J. Lab. Clin. Med. 95:672 (1980).
6. J.C. Fantone and D.A. Kinnes, Prostaglandin E_1 and prostaglandin I_2 modulation of superoxide production by human neutrophils, Biochem. Biophys. Res. Commun. 113:506 (1983).
7. H.U. Keller, Granulocyte recruitment and its inhibition,in: "Handbook of Inflamm.," L.E. Glynn, J.C. Houk, and G.Weissman, eds., Elsevier, Amsterdam (1985).
8. G.J. Dusting, S. Moncada, and J.R. Vane, Disappearance of prostacyclin in the circulation of the dog, Br. J. Pharmacol. 62:414 (1977).
9. L. Flohé, H. Böhlke, E. Frankus, S.-M.A. Kim, W. Lintz, G. Loschen, G. Michel, B. Muller, J. Schneider, U. Seipp, W. Vollenberg, and K. Wilsmann, Designing prostacyclin analogues, Arzneim.-Forsch./Drug Res. 33:1240 (1983).
10. H. Pertoft, T.C. Laurent, T. Laas, and L. Kagedal, Density gradients prepared from colloidal silica particles coated by polyvinylpyrrolidone (Percoll), Anal. Biochem. 88:271 (1978).
11. E.A. Boyse, L.J. Old, and I. Chouroulinkov, VI. Cytotoxic test for demonstration of mouse antibody, Methods Med. Res. 10:39 (1964).
12. A.G. Gilman, A protein binding assay for adenosine 3', 5'-cyclic monophosphate, Proc. Natl. Acad. Sci. U.S.A. 67:305 (1970).
13. J.B. Sedgwick, M.L. Berube, and R.B. Zurier, Stimulus-dependent inhibition of superoxide generation by prostaglandins, Clin. Immun. Immunopath. 34:205 (1985).
14. E.M. Glenn and N. Rohloff, Antiarthritic and anti-inflammatory effects of certain protaglandins, Proc. Soc. Exper. Biol. Med. 139:290 (1972).
15. A. Wendel, G. Tiegs, and C. Werner, Evidence for the involvment of a reperfusion injury in galactosamine/endotoxin-induced hepatitis in mice, Biochem. Pharmacol. 36:2637 (1987).
16. P.J. Simpson, S.E. Mitsos, A. Ventura, K.P. Gallagher, J.C. Fantone, G.D. Abrams, M.A. Schork, and B.R.Lucchesi, Prostacyclin protects ischemic reperfused myocardium in the dog by inhibition of neutrophil activation, Am. Heart J. 113:129 (1987).

KAWASAKI DISEASE, OXYGEN RADICALS AND HIGH DOSE GAMMAGLOBULIN THERAPY

Tadatoshi Kuratsuji, Noriaki Shinomiya, Noboru Kobayashi, Kunihisa Takagi,* Shunichi Shimizu,* Tadashi Ojima,* Eisuke Takahashi,* and Hideji Hanabusa*

Division of Phagocytosis Research and Multidisciplinary Laboratories, National Children's Medical Research Center 3-35-31 Taishido, Setagayaku, Tokyo 154, Japan

*Department of Pediatrics, Keio University school of Medicine, 35 Shinanomachi, Shinjukuku, Tokyo 160, Japan

INTRODUCTION

Kawasaki disease is an acute febrile disease with severe inflammatory reactions. The disease, which occures predominantly in infants and young children, was first described in 1967 in Japan and subsequently reported worldwide. Despite its relatively frequent occurrence and more than a decade of research, the etiology of Kawasaki disease is currently unknown.[1]

The six primary signs of Kawasaki disease[2] are 1) high frever that lasts more than five days, 2) erythema and indurative edema of the hands and feet during the acute phase followed by desquamation in the subacute phase, 3) polymorphous skin rash, 4) bilateral nonexudative conjunctivitis, 5) pathological changes in the oropharynx such as red, dry, fissured lips, "strawberry" tongue and mucosal erythema, and 6) an enlargement of cervical lymph nodes. The most serious problem associated with this disease is complications involving cardiac coronary arteries. This condition is found in approximately 15-30% of children with Kawasaki disease[3] and has been reported to result from overproduction of reactive oxygen species by neutrophils during the early stage of the disease.[4] It has been reported that the treatment with a high dose of human gamma-globlin effectively prevents the development of coronary artery complications.[5]

The present study was initiated to examine the relationship between oxygen free radicals, lipid peroxidation and coronary complications in patients with Kawasaki disease, using free radical scavengers and gamma-globulin treatment.

MATERIALS AND METHODS

Informed consent was obtained from either patients or proxies. Kawasaki disease was diagnosed according to the method described in the 4th edition of manual for Kawasaki disease.[2]

Neutrophils and monocytes were separated by the density gradient and adherence method, respectively. In brief, blood obtained from patients with Kawasaki disease or healthy volunteers was subjected to a Ficoll-Hypaque gradient centrifugation. The bottom fraction including buffy coat was mixed with 2% methylcellulose. The mixture was kept at room temperature for 30 min, allowing the red cells to sediment. Subsequently, the upper phase was centrifuged at 450 x g for 10 min, and the resulting pellet was resuspended in a hypotonic solution to cause hemolysis of the residual erythrocytes. After a 25-sec exposure to the hypotonic solution, neutrophils were washed by a phosphate-buffered saline solution (PBS) using centrifugation and resuspension.

Mononuclear cells were harvested from the intermadiate layer of the Ficoll-Hypaque gradient centrifugation. The cells were resuspended in RPMI-1640 solution containing 10% fetal calf sreum, which was heat inactivated. The cell suspension was incubated at 37°C for 30 min in plastic Petri dishes. Subsequently, dishes were rinsed with an ice-cold PBS to remove floating cells, and cells adhering to dishes were collected using a rubber policeman. The purity of neutrophils and monocytes was 98% and 82.5%, respectively. Viability of either type of the cells was 99%.

The cell suspension was introduced to a platelet aggregometer, and aggregation was estimated from changes in transmittance caused by the addition of stimulants.

Suspension of 10^6 cells were stimulated either by opsonized zymosan (Op-Z; 1 mg/ml), formyl-methionyl-leucyl-phenylalanine (FMLP; 0.001-10 nM) or phorbol myristate acetate (PMA; 1-100 ng/ml). The production of superoxide anion ($\dot{O}_2^-$) by stimulated neutrophils was estimated by the method described by Nakano.[6] This method is based on the enhancement of the chemiluminescence of cypridina luciferin analogue (CLA) by oxygen free radicals. The chemiluminescence of CLA is completely inhibitable by superoxide dismutase, but not by catalase or sodium azide. Therefore, the method is specific to the superoxide anion.

The assay method for hydrogen peroxide (H_2O_2) was modified so that H_2O_2 -dependent formation of the oxidative product could be detectable using flow cytometry.[7] Isolated cells were incubated at 37°C for 15 min in PBS containing 5 µM 2',7'-dichlorofluorescin diacetate (DCFH-DA). This lipid-soluble compound readily diffuses into cells, where it is hydrolysed to 2',7'-dichlorofluorescin (DCFH), a nonfluorescent analogue of fluorescein which is trapped in the cells. During the oxidative metabolic burst of phagocytic cells, DCFH is rapidly oxidized to highly fluorescent 2',7'-dichlorofluorescein (DCF). Therefore, the degree of oxidative activity in the cells can be estimated by measuring the fluorescene of DCF using Epics-C fluorometric analyzer. The method has been shown to be capable of estimated H_2O_2 concentrations.[7]

Lipid peroxides (LP) in the cell membrane were asayed as thiobarbituric acid-reacting substances (TBA-RS), as described by Yagi et al.[8] The principle of this method is to isolate lipid by precipitating them with serum protein using phosphotungstic acid-sulfuric acid system. This method eliminates TBA-RS other than LP (e.g., bilirubin). The sediment was resuspended and products which react with thiobarbituric acid were measured fluorometrically using excitation wavelength at 515 nm and emission at 553 nm. LP are expressed as nmol/mg protein. Total protein concentration was estimated by the method of Lowry et al.[9]

Cu,Zn-superoxide dismutase (SOD) and Mn-SOD were separately estimated by radioimmunoassay.[10] Purified human Cu,Zn-SOD, Mn-SOD and specific radiolabeled antibodies to these enzymes were kindly supplied by Dr. K. Asayama of Yamanashi Medical college.

Monocytes were incubated at 37°C for 72 hours with lipopolysacchrides obtained from E. coli (LPS; 100 μg/ml) in the absence or presence of PMA (10 ng/ml). Subsequently, the supernatant solution was collected. Thymocytes obtained from C3H/He mice were incubated with this supernatant solution containing phytohemmagglutinin (1 μl/25 μl) and trace amounts of $[^3H]$thymidine. The uptake of $[^3H]$thymidine during an 8-hour incubation at 37°C was estimated. It has been shown previously that the incorporation of $[^3H]$thymidine by these cells in the presence of phytohemmagglutinin is strictly dependent on the concentration of interleukin-1 (IL-1), and therefore IL-1 concentration can estimated from the amount of $[^3H]$thymidine incorporated into the cells.[11]

Data were analyzed using Student's t-test.

RESULTS

Neutrophil Functions in Patients with Kawasaki Disease during the Acute Phase

Neutrophils were obtained from patients within five days of the onset of disease. Aggregation of the cells was increased twofold over the corresponding control value observed with neutrophils obtained from age-matched healthy control population (data not shown). Production of $\dot{O}_2^-$ by the neutrophils obtained from Kawasaki disease patients was also increased when stimulated by either Op-Z, FMLP or PMA. The production of $\dot{O}_2^-$ was above the normal range when neutrophls were stimulated by suboptimal concentration (e.g., 0.1 nM) of PMA (Figure 1, dotted column). In contrast, H_2O_2 production was within the normal limits in neutrophils obtained from patients with Kawasaki disease. LP in neutrophil membranes as well as LP in plasma were slightly elevated during the acute phase of the disease. Amounts of Cu,Zn-SOD in neutrophils were unchanged (118 ± 17 and 99 ± 21 ng/mg protein, respectively, in control and Kawasaki disease patients), whereas those of Mn-SOD were slightly elevated from 172 ± 44 in control to 223 ± 50 ng/mg protein in neutrophils obtained from the patients, but not significant.

Effects of the In Vivo Treatment of Patients with High Dose of Human Gamma-Globulin on Neutrophil Functions

Administration of human intact-type gamma-globulin (Glovenin-I, Nihon Pharmaceutical Co., Ltd.) to patients with acute Kawasaki disease at high doses of 200-400 mg/kg of body weight, for four days begining on day 5 of the disease caused the high fever to decline to normal on day 5-7, and reduced number of patients with coronary complications (data not shown),

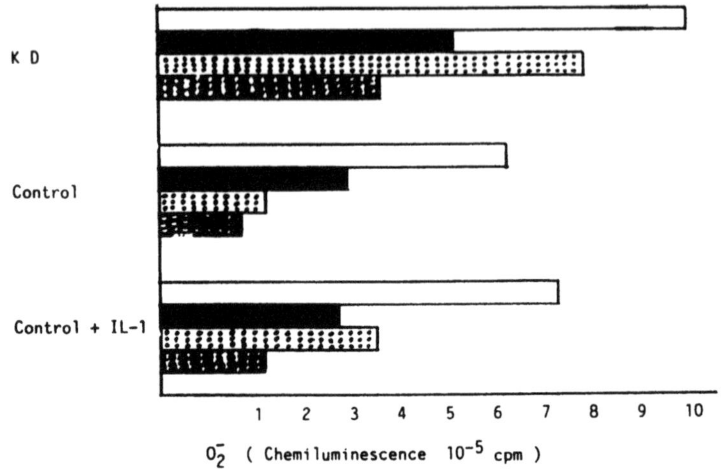

Fig.1. Superoxide anion production by neutrophils stimulated with FMLP. Open or solid columns and dotted columns indicate neutrophils stimulated with 10 and 0.1 nM of FMLP, respectively. Solid columns indicate neutrophils treated with IgG.

confirming an earlier observation.[5] Neutrophil functions were examined using blood samples obtained on the last day of the gamma-globulin treatment in order to test the hypothesis that the alteration of the clinical course of Kawasaki disease by the IgG therapy results from attenuation of the disorder of neutrophil functions. The production of $\cdot O_2^-$ by neutrophils obtained from IgG-treated patients was significantly lower than the corresponding value observed without IgG treatment, and was within the normal range even when neutrophils were stimulated by the optimal dose (10 nM) of FMLP. When neutrophils were stimulated with 0.1 nM FMLP, $\cdot O_2^-$ production by neutrphils obtained from patients after the treatment with high doses of the gamma-globulin was almost as low as that observed with control neutrophils obtained from normal subjects, whereas $\cdot O_2^-$ production was markedly elevated if the patients were not treated with the gamma-globulin (Figure 1, solid columns). The synthesis of H_2O_2 by neutrophils and Cu,Zn-SOD content were normal and Mn-SOD content was slightly increased in neutrophils obtained from IgG-treated patients. LP in neutrophil membranes and in the plasma were increased in patients treated with gamma-globulin; however, the increases were smaller compared to corresponding values observed in patients without the gamma-globulin treatment (data not shown).

The In Vitro Effect of IL-1 on Superoxide Anion and Hydrogen Peroxide Production by Neutrophils from Healthy Volunteers

We have reported at the Second International Conference on Kawasaki Disease that monocytes from patients with Kawasaki disease release IL-1 spontanously at high rates during the acute phase.[12] In order to test the hypothesis that IL-1 mediates changes in nutrophil functions, the effect of IL-1 on neutrophil functions was examined using neutrophils obtained from healthy volunteers. Preincubation of neutrophils with IL-1 (0.1-10 U/ml) for 10-120 min increased $\cdot O_2^-$ production compared to those which were not preincubated with IL-1 (Figure 1, four columns at the bottom). The synthesis of hydrogen peroxide was not influenced by the IL-1 pretreatment (data not shown). Neutrophils obtained from healthy volunteers aggregated strongly when preincubated with IL-1 and stimulated by FMLP. These results suggest that changes in neutrophil functions are brought about by an exposure of neutrophils to IL-1.

The In Vitro Effect of Gamma-Globulin on IL-1-Induced Hyperfunction of Neutrophils

In order to elucidate the mechanism by which IgG attenuates changes in neutrophil functions associated with the acute phase of Kawasaki disease, neutrophils obtained from patients in the acute phase of Kawasaki disease or IL-1-treated neutrophils obtained from healthy volunteers were incubated with IgG (50 mg/ml) for 30-120 min at 37°C. Subsequently, nuetrophils were washed and assayed for hyperfunctions. The production of $\cdot O_2^-$ by neutrophils obtained from Kawasaki disease patients was reduced by the IgG pretreatment to approximately 60% of the value observed without the IgG treatment. Similar results were observed with normal neutrophils treated with IL-1 when cells were exposed to IL-1, and then treated with IgG and subsequently tested for $\cdot O_2^-$ production (Figure 1, solid columns at the bottom). SOD contents in the neutrophils and membrane LP were not influenced by the IgG treatment (data not shown).

Effects of Gamma-Globulin on IL-1 Synthesis by Monocytes

The protective effect of IgG on high fever or coronary complications observed with Kawasaki disease[5] may result either from its direct action on

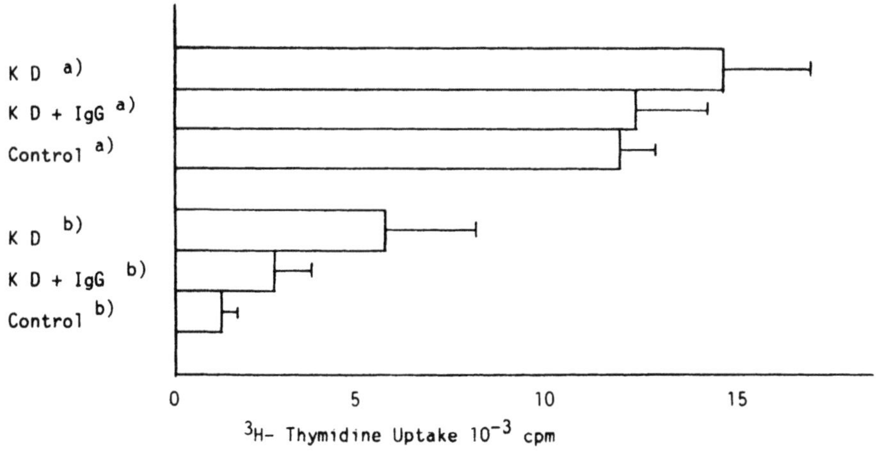

Fig.2. IL-1 Synthesis by monocytes. a) LPS 100 ug/ml + PMA 10 ng/ml; b) without mitogen.

neutrophils to inhibit enhanced functions or indirectly from an inhibition of IL-1 release from monocytes. To test the latter hypothesis, the effect of IgG on IL-1 synthesis by monocytes was examined. Monocytes obtained from patients with Kawasaki disease were capable of synthesizing IL-1 without any stimulantion, as estimated from stimulation of [^{3}H]thymidine uptake by thymocytes (Figure 2). This spontaneous IL-1 synthesis was significantly decreased when the monocytes were obtained from patients following the IgG administration. Subsequently, monocytes obtained from Kwasaki disease patients were incubated in vitro with IgG (50 mg/ml) for 120 min at 37°C, washed, and then assayed for IL-1 synthesis with or without stimulation by LPS or PMA. The results indicate that in vitro treatment with IgG reduces the spontaneous or stimulated IL-1 production by approximately 30 % (Figure 2).

DISCUSSION

The etiology of Kawasaki is unknown. Several investigators have proposed that either a retrovirus[13,14] or a propionibacterium[15] is one of the causative agents. We have reported a close relationship among Kawasaki disease, Streptococcus sanguis infection and IL-1 production.[12] Subsequently, several reports have been published supporting our data and indicated further that there is overproduction of $\cdot O_2^-$ in the early phase of the disease. Somiya et al administered liposomal SOD to patients with Kawasaki disease to test the hypothsis that the overproduction of $\cdot O_2^-$ is causelly related to etiology of Kawasaki disease,[16] however, the effects of the SOD treatment on coronary complications were variable. Circulating immune complex and gamma-interferon may damage vascular endothelial cells and might cause thrombosis or vasculitis in patients with Kawasaki disease.[17] Warner et al reported that IL-1 induced the expression of IL-1 gene in vascular smooth muscle cells.[18] These results suggest that IL-1 initiates a vicious cycle in Kawasaki disease. Because favorable effects of high dose of gamma-globulin have been reported,[5] we investigated the effects of gamma-globulin on IL-1 synthesis by monocytes and on $\cdot O_2^-$ production by neutrophils. The results suggest that gamma-globulin is capable of inhibiting IL-1 synthesis and inhibits neutrophil hyperfunctions observed in Kawasaki disease. The gamma-globulin treatment might be an effective means of preventing coronary complications.

ACKNOWLEDGMENTS

We thank Dr. Tai Akera, Director of the National Children's Hospital Medical Research Center, for the excellent advice and discussion, and Professor Minoru Nakano and Dr. Kohtaro Asayama of Yamanashi Medical College for the supply of CLA and for purified human SOD, respectively. This work was supported in part by a grant from the Ministry of Health and Welfare of Japan, Human Science Research Foundation, and Nihon Pharmaceutical Co.,Ltd.

REFERENCES

1. D.M. Bell, Kawasaki syndrome: Still a mystery after 20 years, JAMA 254:801 (1985).
2. T. Kawasaki, The new manual for diagnosis of Kawasaki disease, Shounika 26:985 (1985).
3. H. Fujiwara, C. Kawai, and Y. Hamashima, Clinicopathologic study of the conduction systems in 10 patients with Kawasaki disease, Am.Heart J. 96:744 (1978).
4. Y. Niwa and K. Somiya, Enhanced neutrophilic functions in mucocutaneous lymph node syndrome, with special reference to the possible role of increased oxygen intermediate generation in the pathogenesis of coronary thrombosis, J.Pediatr. 104:56 (1984).
5. K. Furusho, T.Kamiya, H. Nakano, N.Kiyosawa, K.Shinomiya, T.Hayashidera, T. Tamura, O. Hirose, Y. Manabe, Y. Yokoyama, M. Kawarano, K. Baba, and C. Mori, High-dose intravenous gammaglobulin for Kawasaki disease, Lancet 2:1055 (1984).
6. A. Nishida, K. Sugioka, M. Nakano, and K.Yagi, Use of 2-methyl-6-phenyl-3,7-dihydroimidazole[1,2-]-pyrazin-3-on for assay of superoxide anion in phagocytizing granulocytes, J.Clin.Biochem.Nutr. 1:5 (1986).
7. D.A. Bass, J.W. Parce, L.R. Dechatelet, P. Szejda, M.C. Seeds, and M. Thomas, Flow cytometric studies of oxidative product formation by neutrophils: A graded response to membrane stimulation, J.Immunol. 130:1910 (1983).
8. K. Yagi, A simple fluorometric assay for lipoperoxide in blood plasma, Biochem.Med. 15:212 (1976).
9. D.H. Lowry, N.J. Rosenbrough, A.L. Farr, and R.J. Randall, Protein measurement with the Folin phenol reagent, J.Biol.Chem. 193:265 (1951).
10. K. Asayama and I.M. Burr, Joint purification of mangano and cuprozinc superoxide dismutases froma single source, Analyt.Biochem. 136:336 (1984).
11. K.A. Smith, The functional relationship of the interleukins, J.Exp.Med. 151:1551 (1980).
12. N. Shinomiya, T. Takeda, T. Kuratsuji, K. Takagi, T.Kosaka, O.Tatsuzawa, T. Tsurumizu, T. Hashimoto, and N. Kobayashi, Variant Streptococcus sanguis as an etiological agent of Kawasaki disease, "Kawasaki Disease",571,(S.T. Shulman ed.), Alan R.Liss,Inc., New York (1987).
13. S.T. Shulman and A.H. Rowley, Does Kawasaki disease have a retroviral aetiology? Lancet 2:545 (1986).
14. J.C. Burns, R.S. Geha, E.E. Schneeberger, J.W. Newburger, F.S. Rosen. L.S. Glezen, A.S. Huang, J. Natale, and D.Y.M. Leung, Polymerase activity in lymphocyte culture supernatants from patients with Kawasaki disease, Nature 323:814 (1986).
15. H. Kato, T. Fujimoto, and O. Inoue, Variant strain of propionibacterium acnes: A clue to the aetiology of Kawasaki disease, Lancet 2:1383 (1983).
16. K. Somiya, S. Fukami, Y. Niwa, Y. Mizuno, and Y. Shimoda, Treatment with liposomal superoxide dismutase of patients with Kawasaki disease, Enshou 5:141 (1985).
17. D.Y.M. Leung, T. Collins, L.A. Lapierre, R.S. Geha, and J.S. Pober, Immunogloblin M antibodies present in the acute phase of Kawasaki syndrime lyse cultured vascular endothelial cells stimulated by gamma interferon, J.Clin.Invest. 77:1428 (1986).
18. S.J.C. Warner, K.R. Auger, and P. Libby, Human interleukin 1 induces interleukin 1 gene expression in human vascular smooth muscle cells, J.Exp.Med. 165:1316 (1987).

AN ENDOGENOUS SOURCE OF THE SUPEROXIDE ANION IN THE CENTRAL NERVOUS SYSTEM

Carol A. Colton and Daniel L. Gilbert*

Department of Physiology
Georgetown University
Washington, D.C. 20007 U.S.A.

*Laboratory of Biophysics, NINCDS
National Institutes of Health
Bethesda, Maryland 20892 U.S.A.

INTRODUCTION

While it is known that oxygen intermediates, i.e., the superoxide radical anion, hydrogen peroxide, and the hydroxyl radical are toxic[1-3], their role in pathological processes in the central nervous system (CNS) is not clear. To understand how these reactive oxygen metabolites are involved, two issues must be addressed. First, are there endogenous sources of oxygen intermediates in the CNS? Second, can oxygen intermediates produce functional neuronal changes?

Reactive oxygen intermediates can be generated from many different sources, some of which are found in the CNS.[3-7] For example, Betz[5,8] has shown the presence of xanthine oxidase in the capillary endothelium of cerebral blood vessels. In the heart, this enzyme has been shown to catalyze the production of superoxide when ATP levels are low.[6] Thus, activation of this pathway in a manner similar to that shown for the capillaries of the heart could lead to abnormally high local concentrations of superoxide. Brain mitochondria have been shown to generate hydrogen peroxide.[7] Kontos[9,10] has also provided evidence that superoxide levels rise in the CNS during episodes of hypertension and brain injury.

Another potentially significant source of oxygen intermediates in the CNS is the microglia. Microglia are one of the four major types of glial cells found in the CNS. Unlike astrocytes and oligodentrocytes, microglia are of hematogenous origin.[11-13] Evidence has now accumulated that these cells share morphological and functional characteristics with monocytes and tissue macrophages, including the presence of similar cell surface receptors, the presence of similar enzymatic processes, and the ability to phagocytize debris.[11-14] An electromicrograph of a typical cultured microglia is shown in Figure 1. Like a monocyte or macrophage, microglia have a large, prominent nucleus and many inclusion vesicles.

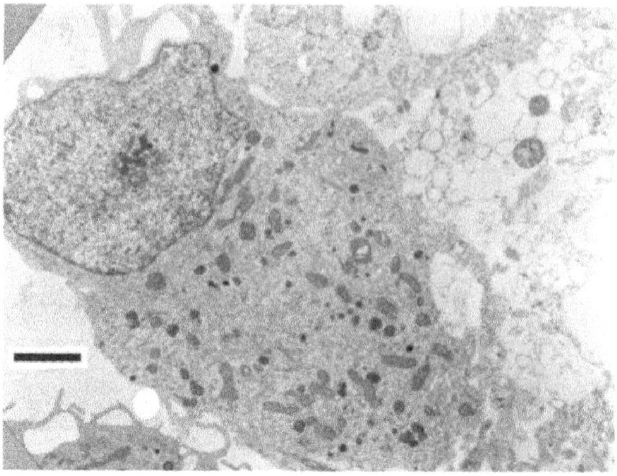

Figure 1. Electron micrograph of a cultured microglia. Length of line = 2 micrometers.

Among the well-described functions of monocytes and macrophages is the ability to kill invading organisms by the production and secretion of oxygen intermediates.[4] If microglia are the resident macrophages in the CNS, as has been suggested, then they might be expected to also secrete superoxide anion in response to inflammatory signals. We tested this hypothesis directly using cultured microglia and have found that these cells do secrete the superoxide anion. These studies were performed to assess the ability of different activating agents to produce superoxide anion secretion.

METHODS

A primary culture of microglia was prepared from 2-day-old rat cerebral cortices as described.[14] The cells were grown in a 10% CO_2, 90% air environment, using a modified Dulbecco's medium. After 10 to 14 days, the cells were separated by shaking and a microglia-enriched fraction was obtained. Purity of this fraction was assessed by histochemical means, using non-specific esterase as a marker for microglia.[12] Values ranged from 50 to 90% microglia, with fibrous astrocytes as the major contaminant cell type.[14]

The microglia enriched fraction was plated into 96 well culture dishes and allowed to adhere overnight. Cells were assayed for the superoxide anion using a cytchrome c reduction assay in the presence and absence of superoxide dismutase (SOD) as described.[14] Protein content of the wells was determined using a Lowry assay and the results expressed as nmoles superoxide/mg protein.

RESULTS

A typical time course for superoxide secretion by activated and resting rat microglia is shown in Figure 2. Activation was produced by the addition of either phorbol myristate acetate (PMA) or opsonized zymosan (OP Z) to the wells. These agents were chosen because of their known ability to produce superoxide anion secretion in monocytes and neutrophils.[4,15] Superoxide production was then followed over 4 hours. As shown, superoxide concentration rose rapidly in the presence of PMA or OP Z, reaching a maximum in approximately 2 hours. No significant change from baseline, however, was seen under non-stimulated, i.e., resting, conditions.

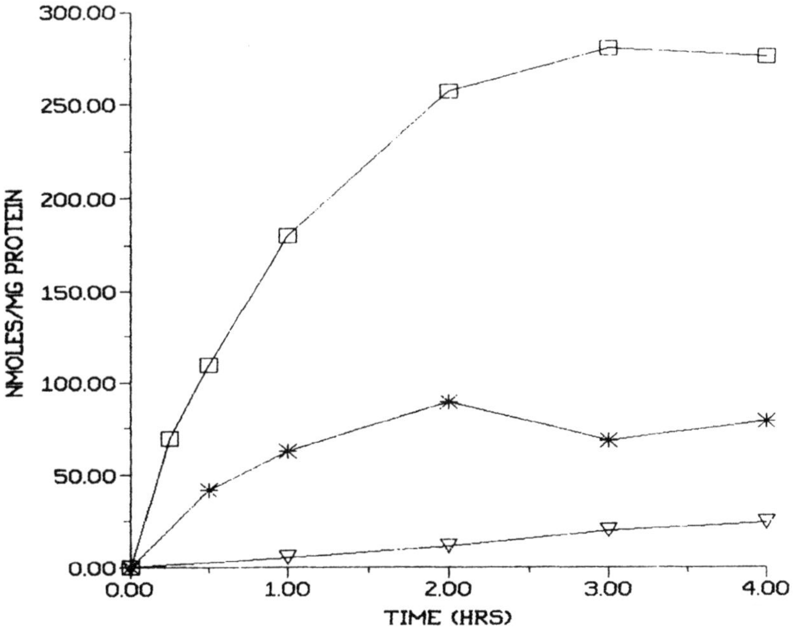

Figure 2. Time course of superoxide anion secretion by cultured rat microglia. Data points represent the average values of superoxide anion secretion over 4 hours with no stimulant (triangles), with 40 microgram/ml PMA (asterisks), and with 0.5 mg/ml OP Z (squares).

Various activating agents were examined for their ability to elicit superoxide anion secretion. These agents included increased extracellular potassium concentration (10 mM); platelet activating factor (PAF); PMA; OP Z; and a plant lectin, concanavalin A. Average values of superoxide secretion at 2 hours are compared in Figure 3. Each of the agents tested produced an increase in superoxide anion that was different from the unstimulated, i.e., resting, condition.

The level of superoxide anion secretion by cultured microglia was compared to that seen in monocytes. Human monocytes were obtained from fresh human blood by using a ficoll-hypaque gradient. These cells were then seeded at the same density as cultured rat microglia, allowed to adhere, and then assayed for superoxide secretion using PMA as the stimulant. As shown in Figure 4, superoxide production by rat microglia had an average value of 52.9 ± 8.5 nmoles/mg protein, while the average value for human monocytes was 48.7 ± 7.0 nmoles/mg protein.

DISCUSSION

The results from cultured rat microglia indicate that this cell is capable of producing the superoxide anion when activated and in amounts comparable to monocytes. If the same is true for endogenous microglia, local levels of oxygen intermediates would be expected to rise in the vicinity of these cells in the CNS. Although this action is potentially beneficial as described by Babior[4] for monocytes, it is also potentially deleterious. Fox[16] and others[17-19] have shown that part of the tissue

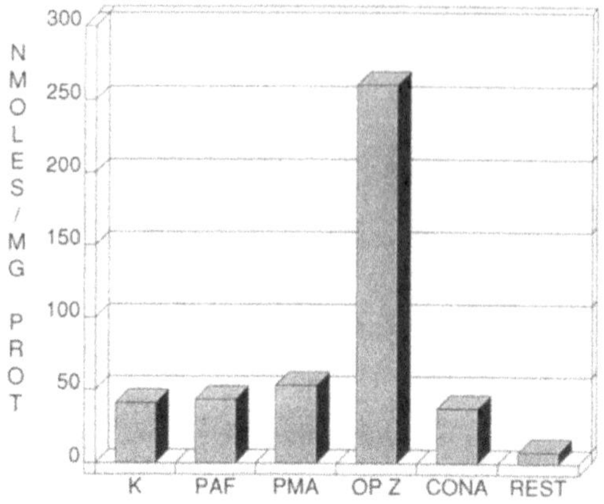

Figure 3. Comparison of various stimulants to superoxide anion secretion in cultured microglia. Values represent the average amount of superoxide anion secreted at 2 hours of incubation with either 10 mM potassium chloride, 0.4 microgram/ml platelet activating factor (PAF), 40 microgram/ml PMA, 0.5 mg/ml OP Z, 50 microgram/ml concanavilin, or no stimulant. Average values were 41.0 ± 7.0 for potassium chloride; 43.4 ± 29.1 for PAF; 52.9 ± 8.5 for PMA, 260.0 ± 18.8 for OP Z; 37.4 ± 6.1 for concanavalin; and 7.9 ± 1.4 for unstimulated (rest).

damage during an inflammatory response is due to oxygen intermediates produced by macrophages, monocytes, and neutrophils.

Is it possible that oxygen intermediates produce neuronal damage? The superoxide anion is especially harmful because it can be converted to other more reactive oxygen intermediates, such as hydrogen peroxide and the hydroxyl radical. Synaptic transmission is particularly susceptible.

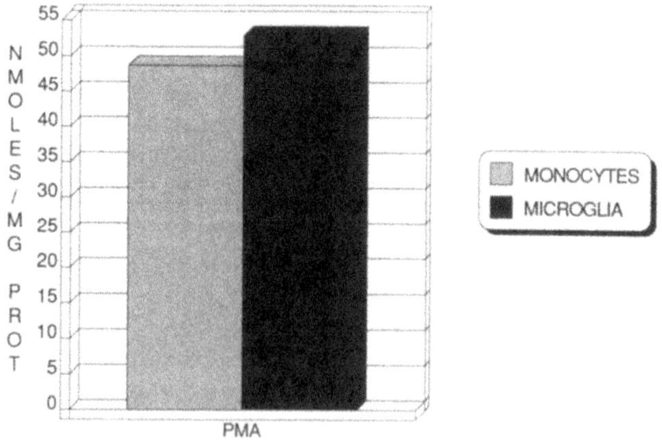

Figure 4. Comparison of the secretion of superoxide anion by microglia to monocytes. Values represent the average superoxide anion production at 3 hours of incubation using 40 microgram/ml PMA as stimulant.

Colton et al[20] in the lobster and Pellmar[21] in the hippocampal slice have shown that hydrogen peroxide depresses excitatory synaptic transmission by affecting pre-synaptic transmitter release. In the lobster, this action is seen at micromolar amounts of hydrogen peroxide and is not selective to excitatory transmission since inhibitory transmission is also depressed.[20] In the hippocampal slice, depression of excitatory transmission is seen when millimolar amounts of hydrogen peroxide are perfused over the slice.

Interestingly, modulation of synaptic transmission in the hippocampus is also affected by hydrogen peroxide. Long term potentiation (LTP), i.e., the long-lasting increase in amplitude of the synaptic potential and population action potential induced by high frequency stimulation of homosynaptic inputs is blocked by microinjection of hydrogen peroxide in the CA 1 region.[22] Since LTP has been described as a critical factor in the cellular mechanism of memory,[23] the action of hydrogen peroxide to block LTP could have wide ranging consequences on memory formation.

Thus, it is possible that the superoxide anion production by microglia can have functional consequences in the CNS. Activation of microglia results in locally high concentrations of oxygen intermediates in the vicinity of potentially susceptible regions, such as membranes. As indicated previously, microglia are activated by opsonized zymosan. Since opsonization involves mobilization of the complement cascade, particularly the C_3b component,[11] it is likely that other activators of the C_3b component (e.g., bacterial and viral products) would produce a similar action. Thus, one of the actions of microglia in the CNS might be similar to macrophages elsewhere, that is, to secrete superoxide in response to bacterial invasion. Other stimulants of microglial superoxide secretion such as increased extracellular K^+ levels or platelet activating factor, are also seen at sites of injury.[18] Again, if these conditions occur in the CNS, it might be expected that CNS microglia would respond with superoxide production. Microglia might also be involved in removing cellular debris and cells no longer functional. This possible function could be important during development.

CONCLUSIONS

How much damage to the surrounding tissue is done by this process is likely to depend on the balance between antioxidant and pro-oxidant factors. Like other tissues, neurons and glia have many different types of antioxidant defenses.[24-26] The type and level of antioxidant and pro-oxidant mechanisms will be determining factors in the ability of the neuron to function in face of oxidative attack.

REFERENCES

1. R. Gerschman, D. L. Gilbert, S. W. Nye, P. Dwyer, and W. O. Fenn, Oxygen poisoning and x-irradiation: A mechanism in common, Science 119:623 (1954).
2. D. L. Gilbert, ed., "Oxygen and Living Processes: An Interdisciplinary Approach," Springer-Verlag, New York (1981).
3. B. Halliwell and J. Gutteridge, Oxygen radicals and the nervous system, Trends in Neurosci. 8:22 (1985).
4. B. Babior, R. Kipnes, and J. Curnutte, Biological defense mechanisms: The production by leucocytes of superoxide, a potential bactericidal agent, J. Clin. Invest. 52:741 (1973).
5. W. Lo and L. Betz, Oxygen free radical reduction of brain capillary rubidium uptake, J. Neurochem. 46:394 (1986).

6. D. E. Parks, Reperfusion injury in ischemic tissues: Role of oxygen derived free radicals, in: "Oxygen: An Indepth Study of Its Pathophysiology," S. Gottlieb, I. Longmuir and J. Totter, eds., Undersea Medical Soc., Bethesda, MD (1983).
7. M. S. Patole, A. Swaroop, and T. Ramasarma, Generation of H_2O_2 in brain mitochondria, J. Neurochem. 47:1 (1986).
8. L. Betz, Identification of hypoxanthine transport and xanthine oxidase activity in brain capillaries, J. Neurochem. 44:574 (1985).
9. H. Kontos, Oxygen radicals from arachidonate metabolism in abnormal vascular responses, Am. Rev. Respir. Dis. 136:474 (1987).
10. H. A. Kontos and E. P. Wei, Superoxide production in experimental brain injury, J. Neurosurg. 64:803, 1986.
11. D. Giulian and T. Baker, Characterization of ameboid microglia isolated from developing mammalian brain, J. Neurosci. 6:2163 (1986).
12. E. Ling, C. Kaur, and W. Wong, Light and electron microscopic demonstration of non-specific esterase in amoeboid microglial cells in the corpus callosum in postnatal rats: A cytochemical link to monocytes, J. Anat. 135:385 (1982).
13. V. Perry, D. Hume, and S. Gordon, Immunocytochemical localization of macrophages and microglia in the adult and developing mouse brain, Neurosci. 15:313 (1985).
14. C. Colton and D. Gilbert, Production of superoxide anions by a CNS macrophage, the microglia, FEBS Lttrs. 223:284 (1987).
15. R. Johnston, Jr., Measurement of O_2^- secretion by monocytes and macrophages, Methods Enzymol. 105:365 (1984).
16. R. Fox, Prevention of granulocyte mediated oxidant lung injury in rats by a hydroxyl radical scavenger, dimethylurea, J. Clin. Invest. 74:1456 (1984).
17. W. Bannister, Superoxide dismutase and disease, in: "The Biology and Chemistry of Active Oxygen," J. Bannister and W. Bannister, eds., Elsevier, New York (1984).
18. J. Crapo, B. Freeman, B. Barry, J. Turrens, and S. Young, Mechanisms of hyperoxic injury to the pulmonary microcirculation, Physiologist, 26:170 (1983).
19. R. Del Maestro, An approach to free radicals in medicine and biology, Acta Physiol. Scand. Suppl. 492:153 (1980).
20. C. Colton, J. Colton, and D. Gilbert, Changes in synaptic transmission produced by hydrogen peroxide, J. Free Rad. Biol. Med. 2:141 (1986).
21. T. Pellmar, Electrophysiological correlates of peroxide damage in guinea pig hippocampus in vitro, Brain Res. 364:377 (1986).
22. C. Colton, L. Fagni, and D. Gilbert, The action of an oxygen intermediate, H_2O_2 on synaptic modulation in the hippocampus, in preparation.
23. L. Voronin, Long term potentiation in the hippocampus, Neurosci. 10:1051 (1983).
24. T. Brannan, H. Maker, and I. Raes, Regional distribution of catalase in the adult rat brain, J. Neurochem. 36:307 (1981).
25. P. H. Chan, M. Yurko, and R. Fishman, Phospholipid degradation and cellular edema induced by free radicals in brain cortical slices, J. Neurochem. 38:525 (1982).
26. R. Fried, Superoxide dismutase activity in the nervous system, J. Neurosci. Res. 4:435 (1979).

HYDROGEN PEROXIDE PRODUCTION IN DOPAMINE NEURONS

Mary Beth Spina and Gerald Cohen

Biomedical Sciences Doctoral Program (Neurobiology)
Department of Neurology, and Neurobiology Center
Mount Sinai School of Medicine, City University of New York
One Gustave Levy Place, New York, NY 10029, U.S.A.

INTRODUCTION

Hydrogen peroxide (H_2O_2) is an oxidant that is hazardous to neurons. A major cellular mechanism for the detoxification of hydrogen peroxide is through the glutathione peroxidase/glutathione reductase pathway. The oxidation of reduced glutathione (GSH) by glutathione peroxidase results in the formation of oxidized glutathione (GSSG). Increased steady-state levels of GSSG can provide an index of oxidative stress.

Dopamine (3,4-dihydroxyphenylethylamine), a prominent catecholamine neurotransmitter in brain, is metabolized by monoamine oxidase (a mitochondrial enzyme), and H_2O_2 is produced as a product (eq. 1). Mitochondrial production of H_2O_2 is generally coupled to the oxidation of GSH (eq. 2). Hence, metabolism of dopamine within dopamine nerve terminals in brain may be associated with an oxidative stress.

$$\text{Dopamine} + O_2 \xrightarrow[\text{oxidase}]{\text{monoamine}} H_2O_2 + \text{3,4-DiOH-phenylacetaldehyde} \quad (1)$$

$$H_2O_2 + 2\,\text{GSH} \xrightarrow[\text{peroxidase}]{\text{glutathione}} \text{GSSG} + 2\,H_2O \quad (2)$$

$$\text{GSSG} + \text{NADPH} + H^+ \xrightarrow[\text{reductase}]{\text{glutathione}} 2\,\text{GSH} + \text{NADP}^+ \quad (3)$$

The importance of this area lies in its potential relation to the degeneration of dopamine neurons that takes place during the aging process and becomes accelerated in Parkinson's disease. Dopamine neurons are also destroyed during chronic exposure to amphetamines or after exposure to neurotoxins, such as MPTP or 6-hydroxydopamine. It has been suggested that the production of hydrogen peroxide in dopamine neurons may accelerate the degeneration of dopamine neurons.[1]

We have attempted to detect an oxidation of glutathione associated with the turnover of dopamine by monoamine oxidase in nerve terminals in the striatum. The striatum is an output center that controls motor activity, and it is one of the richest sources of dopamine terminals in brain.

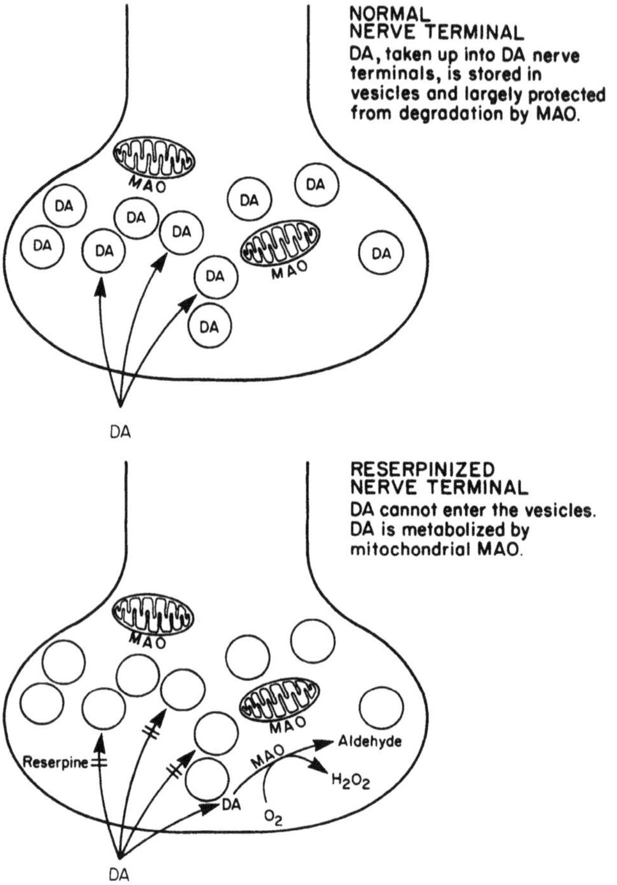

Figure 1. Schematic diagram of a dopamine nerve terminal before and after reserpinization.

Three problems present themselves in such studies:

1. Dopamine terminals in the striatum constitute less than 1% of the total weight of tissue. Therefore, even if all of the GSH were oxidized in an experiment, the fall in GSH would be extremely small, and probably not detectable. For this reason, we concentrated on the measurement of GSSG. GSSG constitutes roughly 1% of the total glutathione in tissues. Therefore, transformation of less than 1% of tissue GSH to GSSG could produce a significant change in GSSG.

2. When glucose is present, hexose shunt activity produces NADPH, which serves as the cofactor for the regeneration of GSH from GSSG by glutathione reductase (eq. 3). The reductase is very active. Therefore, only a fraction of the GSSG formed will be detected. Success in these experiments requires a significant elevation in the steady-state level of GSSG.

3. It is necessary to selectively activate the turnover of dopamine so that experimental changes will reflect the increment in monoamine oxidase activity and H_2O_2 production within dopamine nerve terminals. To achieve this, we added dopamine, which is selectively taken up into the dopamine neurons and, in addition, the synaptosomes were treated with reserpine, a drug that blocks the storage of dopamine in synaptic vesicles.

We report here the detection of significantly elevated GSSG during the oxidative deamination of dopamine by monoamine oxidase within isolated striatal nerve terminals.

METHODS

Striata were dissected from Sprague-Dawley rats and homogenized at 4°C in 0.32 M sucrose. A crude synaptosomal pellet (P2 fraction) was isolated by a standard centrifugation procedure. The P2 pellet was resuspended in Krebs-Ringer phosphate buffer (pH 7.4). Aliquots were incubated at 37°C with addition of appropriate agents (dopamine, L-dopa) and drugs. Subsequently, the aliquots were repelleted and homogenized in 0.4 M perchloric acid, followed by centrifugation at 11,000 x g for 15 minutes. The supernatant was analyzed for GSSG by a modification of the method of Tietze.[2] The supernatant was first incubated with N-ethylmaleimide to remove GSH, and the excess N-ethylmaleimide was removed by Sep-Pak chromatography. The assay procedure utilizes GSSG reductase and 5,5'-dithiobis-(2-nitrobenzoic acid) (DTNB). The rate of color formation was monitored at 412 nm. Data were evaluated with a paired t-test.

RESULTS AND DISCUSSION

When striatal synaptosomes were incubated with reserpine phosphate plus dopamine (10 μM), a significant rise in the steady-state level of GSSG was seen (22.9 ± 2.6% [SEM], Table 1). Similarly, incubation with reserpine phosphate plus L-dopa (1 mM) resulted in a significant rise in the steady-state level of GSSG (30.0% ± 2.5% [SEM], Table 1). These results signify an intracellular flux of H_2O_2 when exogenous dopamine is taken up into reserpinized nerve terminals or when dopamine is synthesized in situ from added L-dopa.

In other experiments, synaptosomes were preincubated with clorgyline, an inhibitor of monoamine oxidase. Clorgyline eliminated 85% of the increase in GSSG caused by reserpine plus dopamine and also eliminated 85% of that caused by reserpine and L-dopa. These results indicate that the H_2O_2 was generated by monoamine oxidase.

In additional experiments, synaptosomes were preincubated with 2 μg/ml catalase in order to eliminate the hydrogen peroxide formed by extraneuronal mitochondria, i.e., those mitochondria that were free in the medium. In the presence of catalase, the rise in GSSG was suppressed. Catalase eliminated 34% of the increase in GSSG caused by reserpine plus dopamine, and 50% of the increase seen with reserpine and L-dopa. These results indicate that approximately one-half of the dopamine is metabolized by free mitochondria in the medium. The remaining GSSG formed can be attributed to intraneuronal MAO activity.

Table 1. GSSG Levels in Rat Striatal Synaptosomes after Treatment with Reserpine + Dopamine or Reserpine + L-Dopa

Treatment	Increase in GSSG from Control Levels[a]	
	pmoles/mg original tissue	% Increase
Reserpine + Dopamine (n = 11)	4.07 ± 0.46[b]	22.9 ± 2.6
Reserpine + L-Dopa (n = 7)	5.89 ± 0.56[b]	30.0 ± 2.5

[a]Mean ± SEM. GSSG levels in control samples were 18.77 ± 1.79 in experiments with dopamine and 21.54 ± 1.65 in experiments with L-dopa.

[b]$p < 0.05$ compared to controls (paired t-test).

It is known that exposure of whole organs or tissues to H_2O_2 or organic peroxides produces a measurable elevation in the GSSG level.[3] The rise in GSSG is an indicator of oxidative stress. It has also been shown that the oxidation of amines by monoamine oxidase in whole liver is coupled to the oxidation of GSH to GSSG.[4] In our experiments, we increased the mitochondrial production of H_2O_2 in striatal nerve terminals by increasing the turnover of dopamine. Metabolism of dopamine by monoamine oxidase induced a substantial increase in the steady-state level of GSSG. Increased turnover of dopamine has been implicated in both Parkinson's disease and schizophrenia. Our experiments indicate that increased turnover of dopamine can evoke an oxidative stress. These observations provide insights into mechanisms that may be responsible for either alterations in neuronal function or neuronal senescence in CNS disorders that affect catecholamine neurons.

ACKNOWLEDGMENTS

Supported by USPHS Grant NS-23017.

REFERENCES

1. G. Cohen, The pathobiology of Parkinson's disease: Biochemical aspects of dopamine neuron senescence, J. Neural Trans. Suppl. 19:89 (1983).
2. F. Tietze, Enzymic method for quantitative determination of nanogram amounts of total and oxidized glutathione: Application to mammalian blood and other tissues, Anal. Biochem. 27:502 (1969).
3. H. Sies, Hydroperoxides and thiol oxidants in the study of oxidative stress in intact cells and organs, in: "Oxidative Stress," H. Sies, ed., Academic Press Inc., London, (1983).
4. N. Oshino and B. Chance, Properties of glutathione release observed during the reduction of organic hydroperoxide, demethylation of aminopyrine and oxidation of some substances in perfused rat liver, and their implications for the physiological function of catalase, Biochem. J. 162:509 (1977).

OXYGEN RADICAL-MEDIATED DAMAGE TO BRAIN TISSUE

Robert A. Floyd

Molecular Toxicology Research Group
Oklahoma Medical Research Foundation
825 N.E. 13th Street, Oklahoma City, Oklahoma 73104, USA

INTRODUCTION

The brain is the primary information-processing center of the body; its functioning in the most perfect state possible, within genetic constraints, is very important. Damage to brain most likely has more serious consequences than to other tissue because neuronal tissue is post-mitotic. Brain is a tissue having an abnormally high amount of 22:6 fatty acids,[1] which due to the large number of carbon-carbon double bonds are easily peroxidized. Brain is not well endowed with the antioxidant protective enzymes superoxide dismutase, catalase or glutathione peroxidase, nor does it have high levels of vitamin E.[1] It does, however, have higher levels of ascorbate,[2] which can in certain circumstances act as a prooxidant. Brain, especially human brain, does contain high levels of iron,[3] apparently stored as ferritin. Catalytic levels of iron, in combination with oxygen free radicals, appear to be extremely important in mediating oxidative damage to membranes.[4] The administration of free iron directly into the brain has been shown to cause death of neurons, resulting in permanent epiletiform centers.[5] The iron-induced damage to brain has been shown to be due to peroxidative events.[5] Thus, all of the background information clearly indicates the need to understand the basic aspects of oxidative damage to nervous tissue. Much of our research effort over the last few years is directed toward this aim, and the main results are summarized in the present report.

EXPERIMENTAL PROCEDURES

The experimental procedures have been described in detail previously (6,8,9,10). Briefly, female Sprague-Dawley rats, either 3 mo old in dopamine synthesis studies (8) or 15-18 mo old in the brain regional peroxidation studies (6), were used. Mongolian gerbils were used in the ischemia-reperfusion experiments (9). Rat brains were perfused _in situ_ with saline following light anesthesia, then rapidly removed and dissected into regions on a cold stage (6). Homogenization of brain regions in cold (4°C) phosphate buffer made up from Chelex-100 treated water did not cause peroxidation (6). Peroxidation was initiated by incubation of the homogenate in 0.05 M phosphate buffer (pH 7.0) at 37°C (6). The extent of lipid peroxidation was determined using the thiobarbituric acid procedure (6). The total iron content of various brain regions was determined by atomic absorption (6). Dopamine synthesis was determined on the striatum P_2 fractions by measurement of $^{14}CO_2$ liberation during the decarboxylation of L-(1-^{14}C) DOPA formed by the hydroxylation of L-(1-^{14}C) tyrosine (8). In the ischemia-reperfusion experiments with gerbils, surgical installation of dental floss around both carotid arteries allowed accurate control of ischemia and reperfusion for a finite period of time (9). Gerbils received salicylate (100 mg/kg, i.p.) 40 min prior to initiation of ischemia. After ischemia-reperfusion insults, gerbil brains were removed rapidly and plunged into cold 15% trichloroacetic acid and the amount of hydroxylation products of salicylate determined using HPLC-electrochemical detection as outlined previously (10).

RESEARCH OBSERVATIONS

Lipid Peroxidation of Brain

Rat brain readily peroxidizes when it is homogenized and the temperature is raised to 37°C, but rat brain homogenate does not peroxidize if the temperature is kept at 4°C.[6] Thus, if care is taken to homogenize brain at ice temperature, then peroxidation does not occur until the temperature is raised. Brain is thus more prone than most other tissue with regard to the ease with which it will peroxidize.[7] Since the homogenate of brain will peroxidize, in contrast to the intact tissue in its native state, this means that the components necessary are present but organization is a very important factor with regard to oxidative damage in brain.

Figure 1 illustrates the rate of peroxidation of rat brain cortex as a function of time.[6] Peroxidation at 37°C, as measured by the malondialdehyde (MDA) content, was nearly linear for about 1 hour for the homogenate. On the other hand, if ascorbate (250 μM) and iron (10 μM) were added to the homogenate, the initial rate of peroxidation was increased about fivefold; however, the rate slowed considerably after 30 minutes. Since MDA is used as an index of peroxidation, it is necessary to prove that MDA can be recovered from the homogenate. The data in Table 1 demonstrate that MDA added to the homogenate and either incubated for 0 time or for 60 minutes could be recovered completely.[6]

Peroxidation of the homogenate of rat brain cortex appears to be dependent on the endogenous iron content of the tissue. One series of experiments we have conducted which supports this conclusion is shown in Table 2.[6] The data demonstrate that chelators, which have high specificity for iron, inhibit peroxidation approximately in proportion to the degree of their iron-chelation strength. Compare, for instance, the ability of desferal to inhibit peroxidation with that of citrate or EDTA, both of the latter of which chelate iron more weakly than the former. It is interesting to note that the oxygen free radical scavengers had no effect on peroxidation under the conditions of these experiments. We conclude from these results that the availability of iron is an important rate-limiting step and that the oxygen free radical inhibitors did not scavenge the critical oxygen free radicals, possibly because they were not able to penetrate to a critical site where either they are formed or where they may be involved in a catalytic manner in the peroxidation process.

When the endogenous peroxidation rate of the homogenate of various regions of the brain are compared, a greater than twofold range is observed between regions as demonstrated by the data in Table 3.[6] Brain

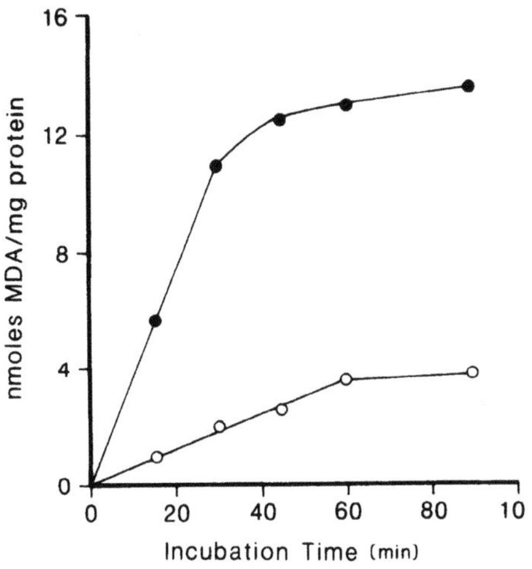

Figure 1. Accumulation of MDA in brain homogenate as a function of time.[6] Incubations were carried out in phosphate buffer, pH 7.0, at 37°C in air with (•) or without (o) $FeCl_2$ (10 μM) and ascorbate (250 μM). Brain cortex gray matter homogenate was used (about 1 mg protein/ml). Reproduced with permission of Plenum Publishing Corp.

stem has the lowest rate, whereas cerebellum and cortex have the highest rates. If the oxygen content of the atmosphere above the brain homogenate is increased from 21% to 100%, the rate of peroxidation increases nearly twofold for most regions of the brain. Also, the addition of Fe (10 μM) and ascorbate (250 μM) to the homogenate increases the peroxidation rate by at least tenfold for most regions of the brain.

If the rate of peroxidation of the homogenate is plotted versus total iron content from various regions of the brain, the data in Figure 2 demonstrate that there is a linear relationship.[6] The rate under 100% oxygen is approximately twofold greater than that under air. The one region which deviates from this general relationship is striatum, which

Table 1. Stability and Recovery of MDA in Brain Homogenates[6]

Incubation time (min)	Endogenous*	MDA (nmol) added	Found	% Recovery	(n)
0	0	1.5	1.48 ± 0.01	98.8 ± 0.8	(5)
0	0	3.0	3.07 ± 0.06	102.3 ± 1.4	(6)
60	2.14 ± 0.03	1.5	3.63 ± 0.07	99.8 ± 2.1	(5)
60	2.43 ± 0.26	3.0	5.58 ± 0.24	102.8 ± 4.4	(4)

*Zero time values were substracted from all experimental samples for calculation. Aliquots of brain cerebellum homogenates (about 0.8 mg protein) were added to phosphate buffer, pH 7.0, with or without exogenous MDA. Samples were taken immediately or following 60 min incubation at 37°C in air for the determination of MDA content as described under experimental procedure. Results are presented as mean ± SEM from the number of separate experiments shown in parentheses. Reproduced with permission of Plenum Publishing Corp.

Table 2. Effect of Iron Chelators and Oxygen Radical Scavengers on Lipid Peroxidation in Brain Cortex Homogenates under Hyperbabaric Oxygenation

Additions	Concentration	% Inhibition
None	0	0
α',α'-dipyridyl	0.1 mM	80 ± 0
1,10-phenanthroline	0.01 mM	75 ± 3
	0.1 mM	79 ± 4
Desferal	0.01 mM	90 ± 4
Citrate	0.01 mM	0
	0.1 mM	47 ± 4
	1.0 mM	75 ± 0
EDTA	0.01 mM	68 ± 6
	0.1 mM	95 ± 3
Apotransferin	0.01 mg/ml	39 ± 7
	0.1 mg/ml	72 ± 4
Catechol	0.1 mM	59 ± 3
	1.0 mM	75 ± 7
Thiourea	1 & 5 mM	0
Benzoate	1 & 5 mM	0
Mannitol	1 & 5 mM	0
1-histidine	5.0 mM	25 ± 7[a]
2,5-dimethylfuran	5.0 mM	20 ± 4[a]
Superoxide dismutase	1.0 mg/ml	7 ± 1[a]
Catalase	0.05 μg/ml	0

Homogenates of gray matter of cerebral cortex were incubated in 0.05 M phosphate buffer, pH 7.0, at 37°C under 5 atm of 100% oxygen for 60 min. MDA accumulation was determined as described under experimental procedure. Results are means ± SEM from 3-4 experiments, each using the tissue pooled from two rats.

[a]$P \geq 0.05$; all other values statistically different from control; $P <$ 0.01-0.001.

has less than one-half of the expected peroxidation rate, based upon the total iron content. That is, the cerebellum and striatum have approximately 22 μg iron per gm wet weight but endogenous peroxidation of cerebellum was about 2.4-fold higher than striatum under either 100% oxygen or air. Why this difference in endogenous peroxidation rate between these two brain regions occurs is an important question which we have attempted to understand.

Thus, even though the endogenous iron content of the two brain regions are equivalent, there are large differences in the dopamine content between them. Striatum is the primary dopamine synthesizing region of the brain whereas very little dopamine is synthesized in cerebellum. We had noted earlier that dopamine addition to cortex homogenate slowed down peroxidation. This capacity to slow down peroxidation could be related to the ability of dopamine to interact with iron in a manner such as to cause ferrous iron to be unable to catalyze hydrogen peroxide decomposition to hydroxyl free radicals. Based on these preliminary observations and the fact that cerebellum and striatum have the same iron content but widely dissimilar dopamine content, we have added dopamine at various levels to cerebellum to determine its effect on endogenous peroxidation of the homogenate of this brain region. These results are shown in Figure 3. The results clearly show that as increasing amounts of dopamine are added to the cerebellum homogenate,

Table 3. Regional Lipid Peroxidation in Rat Brain: Effect of Fe^{2+}/Ascorbate and Oxygen[6]

		MDA accumulation (nmol/mgp/30 min)	
Brain area	Fe^{2+}/Ascorbate	Air	Oxygen
Cortex	-	2.48 ± 0.14	4.13 ± 0.50[a]
	+	25.62 ± 1.43	49.25 ± 3.32[c]
Striatum	-	1.10 ± 0.20	2.32 ± 0.40[b]
	+	19.33 ± 1.82	35.80 ± 4.36[a]
Hippocampus	-	1.75 ± 0.12	3.66 ± 0.45[c]
	+	21.45 ± 1.34	39.72 ± 3.94[b]
Hypothalamus	-	1.04 ± 0.25	2.54 ± 0.50[b]
	+	21.03 ± 1.24	36.26 ± 2.01[c]
Stem	-	1.02 ± 0.17	2.02 ± 0.30[a]
	+	14.19 ± 2.65	21.20 ± 2.24[b]
Cerebellum	-	2.65 ± 0.28	5.51 ± 0.89[a]
	+	32.89 ± 4.24	42.96 ± 2.97

All values are means ± SEM from 4-6 separate experiments. Aliquots of brain homogenates (0.1-1.5 mg protein) were incubated in phosphate buffer, pH 7.0, at 37°C for 30 min, in the absence or presence of 10 μM $FeCl_2$ and 250 μM ascorbate. MDA content in the samples was determined as described under experimental procedure.
[a]$P < 0.02$, [b]$P < 0.05$, [c]$P < 0.005$, tested by Student's t-test against incubations in air. See the text for significance between brain areas.
Reproduced with permission of Plenum Publishing Corp.

peroxidation is slowed down. This was the case whether tranylcypromine, a monoamine oxidase inhibitor, was present or not. It is interesting that, when the dopamine content was raised to about 60 μM, then the peroxidation rate was decreased by about one-half of that observed when no exogenous dopamine was present.

Summarizing, our results clearly show that (a) peroxidation of brain homogenate occurs at 37°C but does not at 4°C, (b) the rate of peroxidation can be accelerated twofold by increasing the oxygen tension from 21% to 100%, (c) a 5- to 10-fold increase in peroxidation rate can be induced by the addition of Fe plus ascorbate, (d) chelators of iron inhibit brain peroxidation, and (e) different regions of rat brain peroxidize at different rates; the rates tend to correlate closely with total iron content, with the exception of striatum which peroxidizes more slowly, most likely due to the higher dopamine content.

Peroxidation Effects on Dopamine Synthesis

Dopamine synthesis is a vital brain function carried out with the most intensity in the striatum. We have conducted experiments to determine if the synthesis of dopamine from tyrosine by the P_2 fraction of striatum is affected by peroxidation of the membranes. Our results demonstrate that peroxidation of the membranes induced by incubation with ascorbate and increasing levels of ADP-chelated Fe caused a rather striking reduction of dopamine synthesis as shown in Table 4.[8] The extent of peroxidation was estimated by MDA formation. There was a striking inverse correlation between loss of dopamine synthesis and MDA formation as a function of added ADP-Fe in the presence of a constant level of ascorbate (250 μM). Malondialdehyde formation reached half-maximal amounts at about 5 μM Fe, whereas the half-maximal decrease in dopamine synthesis occurred at about 2.5 μM Fe. The malondialdehyde molecule per se was not directly responsible for the decrease in dopamine

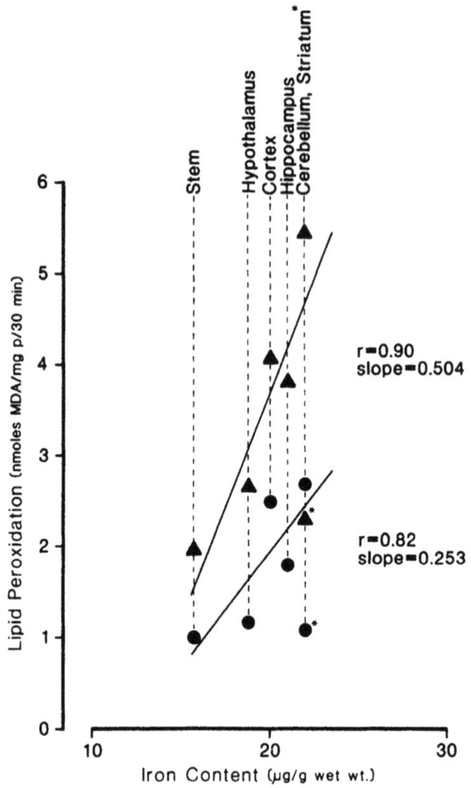

Figure 2. Relationship between endogenous (in the absence of exogenous iron and ascorbate) accumulation of lipid peroxides in vitro and iron content in various areas of rat brain.[6] Iron values for striatum and cerebellum were the same. Striatum values are designated by a superscript star. Lines were fitted by the least-squares method. The (o) represent values obtained under air and the (Δ) are the values obtained under 100% oxygen. Reproduced with permission of Plenum Publishing Corp.

synthesis.[8] Also, it is necessary that iron be reduced, by ascorbate in the present case, to initiate peroxidation. We have found that neither ascorbate only, ADP only, nor ADP-Fe(III) only (up to 10 μM) caused no significant peroxidation,[8] whereas ADP-Fe(III), at 25 μM or higher, caused a significant decrease in dopamine synthesis. Ascorbate present in the P_2 fraction, participating with the highest levels of ADP-Fe(III), may initiate peroxidation.

Ischemia-Reperfusion Injury to Brain

Several studies have implicated the involvement of oxygen free radicals in the injury which develops as a result of ischemia-reperfusion insult to brain.[9] The lack of a good method to monitor *in vivo* for the presence of oxygen free radicals has delayed a true evaluation of the importance of these toxic species in many biological processes. We have been developing a new approach for monitoring oxygen free radicals *in vivo* which is based upon the extreme sensitivity of high pressure liquid

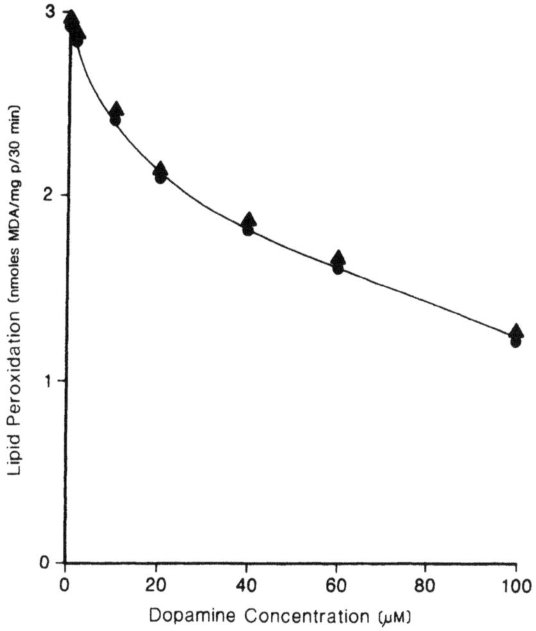

Figure 3. The inhibitory effect of dopamine on the accumulation of lipid peroxides in cerebellar homogenates.[6] Samples were incubated in phosphate buffer, pH 7.0, at 37°C for 30 min in air, in the absence (o) or presence (Δ) of tranylcyproanine (2×10^{-5} M). Results are mean triplicates of two separate experiments, agreeing within 3%. Reproduced with permission of Plenum Publishing Corp.

chromatography with electrochemical detection (LCED).[10,11] We have used LCED methodology to quantitate 2,3- and 2,5-dihydroxybenzoic acid (DHBA), products formed as a result of hydroxyl free radicals reacting with salicylate.[10] We have used salicylate as a possible trap for hydroxyl free radical formation *in vivo*.[10] Salicylate is relatively non-toxic and reacts essentially at a diffusion-controlled rate with hydroxyl free radicals. It should be noted that DHBA may be formed by other mechanisms; therefore, this must be kept in mind in the interpretation of the results. Our first study of a biological system using this approach involved adriamycin-treated rats.[10] Adriamycin causes tissue damage, especially to heart, in treated animals, apparently due to oxygen free radical-mediated processes.[10] The results obtained showed that adriamycin-treated rats had at least a 100-fold increase of DHBA in their hearts, lungs and gastrocnemius muscle as compared to control, i.e., non-adriamycin treated animals.[10] This data encouraged us that salicylate may be an effective trap for hydroxyl free radicals *in vivo*, and thus we decided to test the salicylate LCED methodological approach in ischemia-reperfusion damage to brain.

We have used the gerbil model in ischemia-reperfusion studies according to the protocols developed for inducing global ischemia.[9] The animals were followed by reperfusion results in a significant increase in

Table 4. Effect of Peroxidation on Dopamine Synthesis and Malondialdehyde Production by Crude Striatum Synaptosomes

ADP-Fe Added (μM)	Dopamine Synthesis[a] (pmole/mg protein)	Malondialdehyde (nmole/mg protein)
0	85.7	2.5
0.05	62.1	13.9
0.50	57.8	15.4
2.5	43.6	20.7
5.0	37.1	26.8
10.0	33.9	35.7
25.0	27.1	50.0
50.0	7.9	57.1

[a]Aliquots of the P_2 fraction (0.2-0.4 mg protein) from striatum were incubated for 10 min at 37°C followed by the addition L-(-^{14}C)-tyrosine, 15 μM, and then incubation for an additional 20 min. Iron was added as the ADP-Fe^{3+} chelate at a ratio of 33:1; ascorbate concentration was 250 μM. Values are mean ± S.E.M. from 3-5 separate experiments.[8]

DHBA levels in the brain. Ischemia without perfusion does not yield an increased level of DHBA. The level of DHBA correlates exactly with the damage to the brain, assessed anatomically as well as by behavioral studies. We found that salicylate administration did not protect the animals from ischemia-reperfusion-induced brain injury, as assessed behaviorally.[9]

DISCUSSION

It has become increasingly clear over the last few years that oxidative damage is very important in the etiology of several pathological states, most likely including stroke-related illnesses,[9] perhaps the initiation phases of epilepsy,[5] and trauma-related problems.[1] Our work summarized here on model systems clearly shows that the availability of low molecular weight complexes of iron, together with a reducing species such as ascorbate, most likely in cooperation with oxygen free radicals, certainly leads to oxidative damage of brain tissue. Drastic consequences, such as a decrease in dopamine synthesis as a result of oxidative damage, point out the possible seriousness of the damage in the overall functioning of the brain. New approaches for monitoring oxygen free radical formation <u>in vivo</u> have yielded data implicating oxygen free radical involvement in ischemia-reperfusion injury. It is clear that much is yet to be elucidated regarding the role of oxygen free radicals and damage to brain, but from all our results thus far and those of

Table 5. Changes in 2,5-DHBA Level Above Background in Gerbil Brain as a Result of an Ischemia-Reperfusion Insult

Ischemia[a] (min)	Reperfusion (min)	2,5-DHBA (ng/gm wet weight)	Significance
2	5	-2.5	NS
5	5	48.4	<0.001
15	5	50.1	<0.05
15	5	70.1	<0.01
15	0	3.3	NS

[a]The experiments have been conducted by the protocol outlined in Ref. 9.

others,[1] it is clear that the tissue damage must involve the participation of trace metals such as iron and/or possibly copper in the mechanism of action. The movement and availability of these metals[12], as well as their oxidation state[13], appear to be of prime importance in understanding oxidative damage to nervous tissue.

ACKNOWLEDGMENTS

The research work was supported in part by NIH grants AG02599, NS23307, American Heart Association Oklahoma Affiliate C613301, and funds from Johnson and Johnson.

REFERENCES

1. B. Halliwell and J. M. C. Gutteridge, Oxygen radicals and the nervous system, Trends in Neurosci. 8:22-26 (1985).
2. N. J. Chinoy, Ascorbic acid levels in mammalian tissues and its metabolic significance, Comp. Biochem. Physiol. 42A:945-952 (1972).
3. B. Hallgren and P. Sourander, The effect of age on the non-haemin iron in the human brain, J. Neurochem. 3:41-51 (1958).
4. C. E. Thomas, L. A. Morehouse and S. D. Aust, Ferritin and superoxide dependent lipid peroxidation, J. Biol. Chem. 260:3275-3280 (1985).
5. L. J. Willmore and J. J. Rubin, Formation of malondialdehyde and focal brain edema induced by subpial injection of $FeCl_2$ into rat isocortex, Brain Res. 246:113-119 (1982).
6. M. M. Zaleska and R. A. Floyd, Regional lipid peroxidation in rat brain in vitro: Possible role of endogenous iron, Neurochem. Res. 10:397-410 (1985).
7. H. Zalkin and A. L. Tappel, Studies of the mechanism of vitamin E action. IV. Lipid peroxidation in the vitamin E-deficient rabbit. Arch. Biochem. Biophys. 88:113-117 (1960).
8. M. M. Zaleska, K. Nagy and R. A. Floyd, Membrane peroxidation induced modification of dopamine synthesis in striatum synaptosomes, J. Neurochem., submitted.
9. W. Cao, J. M. Carney, A. Duchon, R. A. Floyd and M. Chevion, Evidence for the role of oxygen free radicals in ischemia-reperfusion induced brain damage, Neuroscience Letters, submitted.
10. R. A. Floyd, R. Henderson, J. J. Watson and P. K. Wong, Use of salicylate with high pressure liquid chromatography and electrochemical detection (LCED) as a sensitive measure of hydroxyl free radicals in adriamycin treated rats, J. Free Radicals Biol. Med. 2:13-18 (1986).
11. R. A. Floyd, J. J. Watson, J. Harris, M. West and P. K. Wong, Formation of 8-hydroxydeoxyguanosine hydroxyl free radical adduct of DNA in granulocytes exposed to the tumor promoter, tetra-deconylphorbolacetate, Biochem. Biophys. Res. Commun. 137:841-846 (1986).
12. N. R. Nayini, B. C. White, S. D. Aust, R. R. Huong, R. J. Indrieri, A. T. Evans, H. Bialek, W. A. Jacobs and J. Komara, Post resuscitation iron delocalization and malondialdehyde production in the brain following prolonged cardiac arrest, J. Free Radicals Biol. Med. 1:111-116 (1985).
13. J. M. Braughler, L. A. Duncan and R. L. Chase, The involvement of iron in lipid peroxidation. Importance of ferric to ferrous ratios in initiation, J. Biol. Chem. 261:10282-10289 (1986).

OXYGEN-INDUCED RETINOPATHY IN THE RAT

John S. Penn and Lisa A. Thum

Cullen Eye Institute
Baylor College of Medicine
Houston, Texas 77030, U.S.A.

INTRODUCTION

Through a decade of research, peroxidation has been implicated in human retinal degenerations. Ocular siderosis, or iron in the eye, has long been known to result in retinal degeneration. Certain retinotoxic drugs, such as adriamycin, are believed to act by peroxidative mechanisms. The retina can also be damaged by excess light, presumably through photo-oxidative processes. Senile macular degeneration is yet another candidate for peroxidation-induced degeneration. Finally, retinopathy of prematurity occurs only in premature infants housed in elevated levels of atmospheric oxygen. Although the agent in each of these insults may be different, cell destruction and death apparently proceed through pathways common to lipid or protein oxidation.

Retinopathy of Prematurity

Retinopathy of prematurity (ROP) is a condition evidenced by a degeneration of retinal blood vessels. It occurs in premature infants who, due to pulmonary complications, are housed in hyperoxic conditions. The development of the retinal vasculature is not complete in premature infants. The result of hyperoxia is the obliteration of a portion of the existing vessels and the abnormal development of new vessels, which eventually become fibrotic, contract, and cause retinal detachment.

The history of ROP in the United States shows two distinct "epidemics" (Table 1). The first occurred in the 1940s and 50s, after oxygen treatment was suggested for premature infants but before it was confirmed as a causal factor in the retinopathy. Once the oxygen risk was determined, the epidemic was eliminated by more careful dose administration and monitoring. A second epidemic is now underway and has been brought about, ironically, by our improved ability to care for premature infants. As survival, particularly of very low birthweight infants, has increased, so has the incidence of ROP. It should be emphasized that, although it has been 30 years since oxygen was recognized as a causal factor, the precise mechanism of cytotoxicity of oxygen on the retina still remains a mystery.

Since oxygen is the cause of ROP, it seems reasonable to suggest that peroxidation is involved in the pathophysiology of the degeneration. The retina is an extremely susceptible tissue to peroxidative reactions for several reasons: 1) it has a high oxygen flux across its cell layers even under normoxic conditions, 2) it has the highest levels of polyunsaturated fatty acids (effective peroxidation substrates) of any tissue, and 3) it has illumination focused on it a great deal of the time.

Table 1. History of ROP in the United States

YEAR	1945	1950	1960	1970	1982
Birthweight [a] <1500g (thousands of infants)	25	34	52	41	43
% Survival [a] <1500g	<8	10	30	39	62
% Incidence of ROP [b]	10	25	3	6	12

[a] Vital statistics of the United States, reference 1.
[b] Estimated from reference 2.

Direct evidence of the role of lipid peroxidation in oxygen-induced retinopathy has come from several laboratories that have measured different forms of lipid peroxidation products in the retinas of animals raised in high atmospheric oxygen.[3,4] The importance of peroxidation in retinal degenerations is also inferred from the retina's elaborate defense system. The retina is able to combat peroxidation by enzymatic detoxification of reactive oxygen molecules and lipid hydroperoxides or by termination of lipid-radical chain propagation by vitamin E. With respect to this latter mechanism, chain termination may be caused by the transfer of a hydrogen atom from vitamin E to the radical. Vitamin E then becomes a radical itself but is relatively stable. Packer et al.[5] proposed a scheme, which was later considered in detail by this lab,[6] whereby a coupled exchange of unpaired electrons brings vitamin C and the glutathione system into the battery of antioxidants, as shown in Figure 1. This scheme has been characterized in vitro by Winkler.[7]

Animal Model

We are currently developing a rat model of ROP so that we can make a careful and systemic analysis of biochemical factors that affect the retinal vasculature. The rat is an appropriate model because 1) the retinal vasculature develops postnatally, facilitating its study; 2) the rat retina is susceptible to oxygen-induced retinopathy; 3) as is the case in humans, retinal vessels arise from mesenchymal precursors; and 4) the newborn rat retina is vitamin E-deficient compared to adults, as it is in the premature human infant.

RESULTS AND DISCUSSION

Immediately after birth, pups and their mothers were placed in an atmosphere of 60% oxygen for 14 days. At that time, some animals were sacrificed for analysis of retinal morphology and/or biochemistry. The remainder were allowed to survive for longer periods of time in room air before being anesthetized for electroretinography and

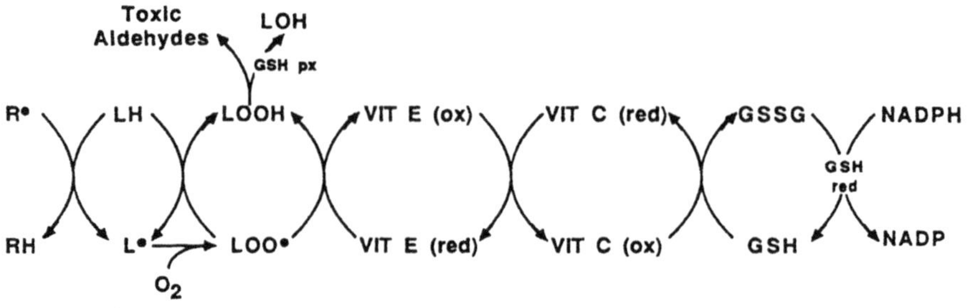

Figure 1. Proposed Mechanism for Retinal Antioxidants

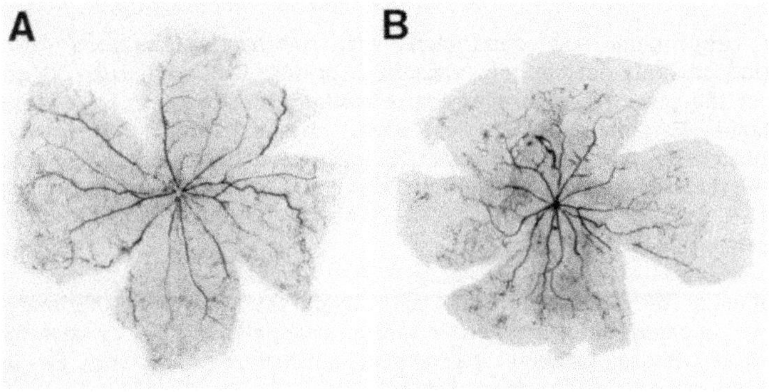

Figure 2. Flat-mounted retinas of rats perfused with India ink indicate normal vasculature in animals raised in room air (A). Rats subjected to 60% oxygen (B) display capillary-free zones in the central and peripheral retina.

sacrificed for tissue analysis. Figure 2A shows an ink-perfused, flat-mounted retina of a rat maintained from birth for 14 days on room air. Note the high degree of vascularization with capillary networks well formed and patent. Figure 2B shows the retina of an animal maintained on 60% oxygen for 14 days. There is a noticeable loss of retinal vasculature that consists of a retarded progression of vessels towards the periphery and the obliteration of central and peri-arterial capillaries. Morphometric evaluation of this loss showed that only 60% of the retina was vascularized in the hyperoxic rats, compared to 94% in room air animals.

Our biochemical analyses have yielded intriguing results as well. Animals raised in 60% oxygen had significantly lower levels of vitamins C (Figure 3A) and E (Figure 3B) in their retinas (20% and 34% reductions, respectively). The fact that these two antioxidants are compromised may, in itself, indicate that oxidative processes are involved in the retinal degeneration. No differences were seen in glutathione reductase, S-transferase, or peroxidase.

If one assumes the involvement of lipid peroxidation in the degeneration, then enhancement of the retina's antioxidants should provide some protection. In order to test this assumption, we have supplemented nursing mother rats with either vitamin C or vitamin E. It was possible to raise the serum level of vitamin C in mothers by daily injections of 0.5 mg/kg of sodium ascorbate, but this did not result in elevations of vitamin C in their retinas or the retinas of their pups.

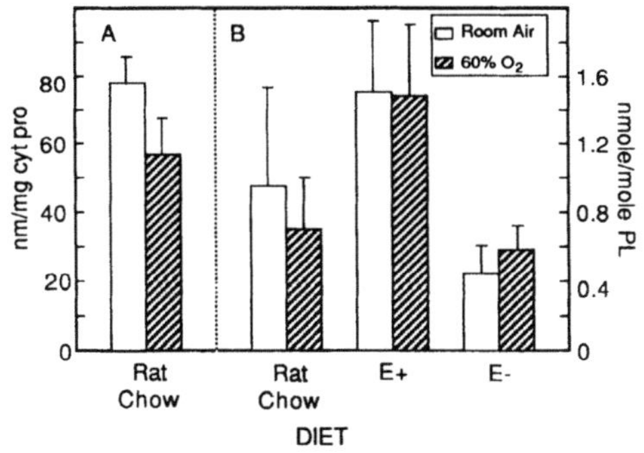

Figure 3. Retinal levels of vitamin C (A) and vitamin E (B) are reduced in hyperoxic rats, and the latter can be manipulated through the diet.

Another experiment was conducted with mothers maintained throughout the treatment period on diets deficient in vitamin E or supplemented with 1.0 g/kg food. In this experiment the pups from supplemented mothers had three to four times the retinal levels of vitamin E that those from depleted mothers had, both in hyperoxic and normoxic conditions (Figure 3B). Vitamin E supplementation of oxygen-exposed rats, in fact, improves retinal vascularization by 11% over exposed rats that are vitamin E-deprived. This exciting result aside, we feel the ability to manipulate the retinal antioxidant system is alone significant.

The challenge for the future will be to apply this ability to the understanding of the biochemical mechanisms of oxygen-induced retinopathy. This understanding must in turn be used to devise rational therapeutic approaches that will be applicable to retinopathy of prematurity in human infants.

ACKNOWLEDGMENTS

We are grateful to J. Jackson for preparing the manuscript and T. Oliver for producing the figures. This work was supported by grants from the Knights Templar Eye Foundation and the National Eye Institute (EY05820).

REFERENCES

1. J. W. Payne and A. Patz, Current status of retrolental fibroplasia: the retinopathy of prematurity, Ann. Clin. Res. 11:205 (1979).
2. The Vital Statistics of the United States. 1945, Vol I; 1950, Vols I and II; 1960, Vol II; 1970, Vols I and II; 1982, Vol II. Federal Security Agency (1945), U.S. Department of Health, Education, and Welfare, (1950-1970), and U.S. Department of Health and Human Services (1982). U. S. Govt. Print. Off.
3. K. Yagi, S. Matsuoka, H. Ohkawa, N. Ohishi, Y. K. Takauchi, and H. Sakai, Lipoperoxide level of the retina of chick embryo exposed to high concentration of oxygen, Clin. Chim. Acta 80:355 (1977).
4. T. Hiramitsu, Y. Hasegawa, K. Hirata, I. Nishigaki, and K. Yagi, Formation of lipoperoxide in the retina of rabbit exposed to high concentrations of oxygen, Experentia 32:622 (1976).
5. J. E. Packer, T. F. Slater, and R. L. Willson, Direct observation of a free radical interaction between vitamin E and vitamin C, Nature 278:737 (1979).
6. R. E. Anderson and J. S. Penn, The role of vitamin E in retinal degenerative diseases, Vitamin (Jap) 61:237 (1987).
7. B. S. Winkler. In vitro oxidation of ascorbic acid and its prevention by GSH, Biochim. Biophys. Acta 925:258 (1987).

SELECTIVE INHIBITION OF MEMBRANE ATPases BY HYDROGEN PEROXIDE IN THE LENS OF THE EYE

Douglas Borchman, Christopher Paterson and Nicholas Delamere

Kentucky Lions Eye Research Institute
301 E. Muhammad Ali Blvd.
Louisville, Kentucky 40202, U.S.A.

INTRODUCTION

Oxidative damage to the lens resulting in decreased Ca-ATPase activity may play a significant role in the formation of cataracts. Deterioration of membrane function is implicated in 70% of all human lens cataracts since the intracellular calcium and sodium content of these lenses are dramatically elevated.[1] Since oxidative damage to ion transport mechanisms might contribute to cataract formation,[2] we chose to study the effect of hydrogen peroxide on Na-K-ATPase and Ca-ATPase activity. In fact, significant concentrations of hydrogen peroxide has been reported in the vicinity of the lens in both the human and rabbit eye in vivo. We prepared a membrane-enriched sample from lens epithelium, cortex, and nucleus. The preparation is the same as that used routinely for preparing cardiac membranes enriched with Ca-ATPase.[3] This procedure was used because commonly used lens membrane preparations utilize chaotropic agents to solubilize cytosolic lens crystallins. These chaotropic agents abolish Ca-ATPase activity. Gel electrophoresis, lipid and protein quantitation, and ATPase activity show that our preparation is enriched with plasma membranes. The lipid:protein ratio was increased 22- and 23- fold in our final cortical and nuclear membrane preparations, respectively. Total phospholipid recovery was 98.6%.

RESULTS

Ca-ATPase activity in the lens cortex membrane may be more significant than previously believed. The activity and calcium stimulation curve are similar for both the epithelial and cortical membrane preparations (Figure 1). Cortical Ca-ATPase activity may have been masked in previous studies because of the high concentration of cytosolic membrane protein, disruption of membrane integrity by violent homogenization, or the inactivation of the Ca-ATPase by aggregation and cystine bonding to. cytosolic proteins. (This was prevented by teflon douncer homogenization, 2 mM dithiothreitol, and 0.3 M sucrose in our preparation.) Ca-ATPase is not observed in our nucleus preparation, perhaps

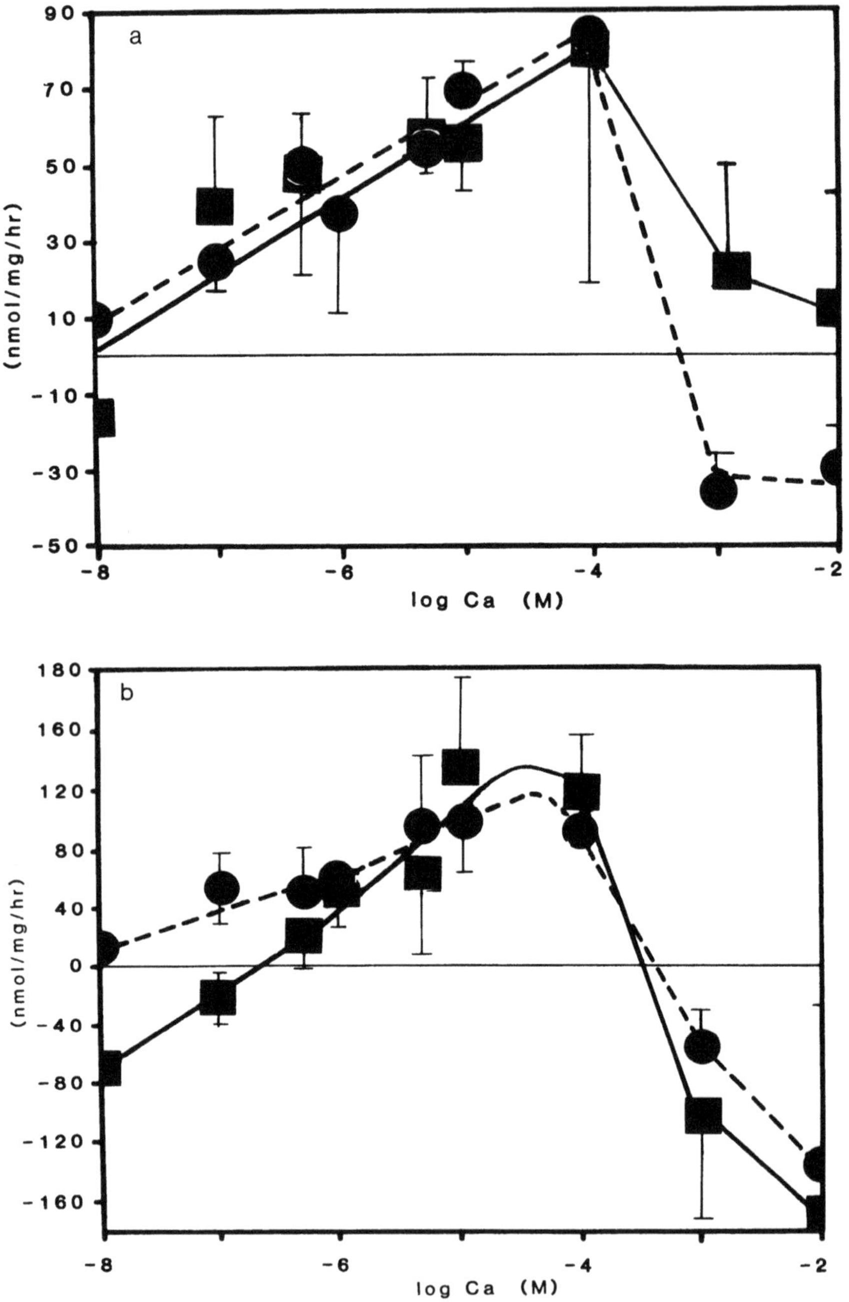

Figure 1. Calcium dependency of Ca-ATPase activity of membrane enriched bovine (a) and rabbit (b) lens preparations from --- ● Cortex and —— ■ Epithelium. Activity is expressed as nmoles phosphate released per mg per hour at 37°C, pH = 7.4. Calcium was buffered with EGIA, K_b 1X10^{6}.

because of the cholesterol content of the nuclear membrane which would be expected to inhibit Ca-ATPase activity.[4-6] The pattern of calcium stimulation of ATPase activity is similar in both rabbit and bovine lens (Figure 1). The distribution of Na-K-ATPase activity within both the rabbit and the bovine lens (Figure 2) is similar to that observed by other investigators;[7-9] the activity in the nucleus and cortex is less than in the surface epithelium. Evidence for inhibition of the Na-K-ATPase activity in the cataractous lens is inconclusive. Lenses challenged with hydrogen peroxide exhibit only moderate impairment of sodium transport capability. However, peroxide-treated lenses show a markedly deranged internal calcium level. This prompted us to examine the sensitivity of the lens calcium pump to oxidative damage. Preliminary findings indicate that the Ca-ATPase is more sensitive to

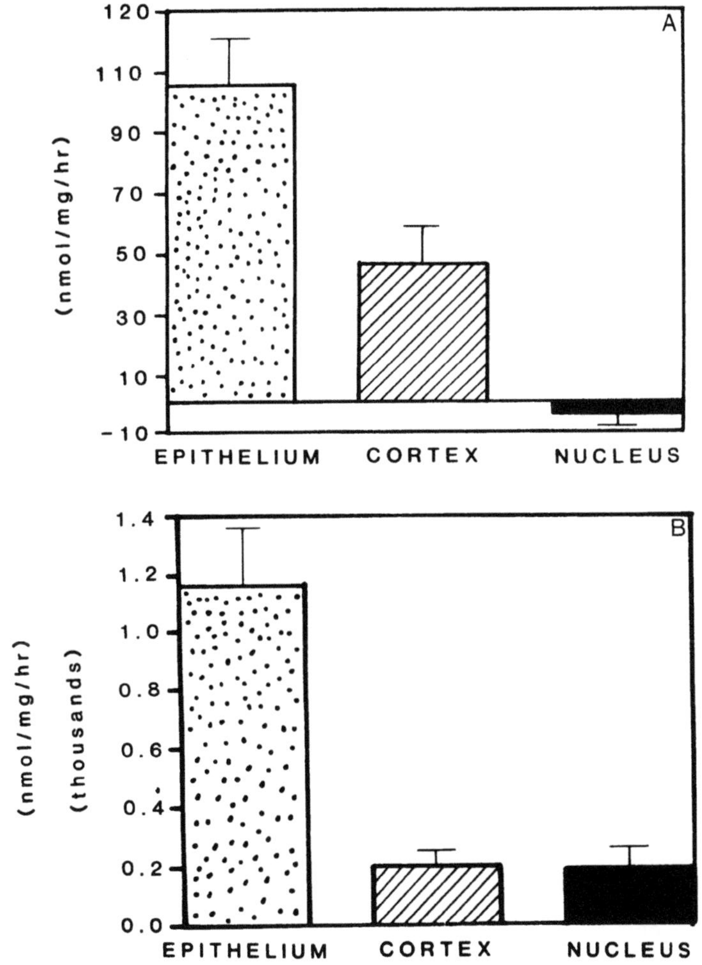

Figure 2. Distribution of ouabain-sensitive Na-K-ATPase activity in bovine (A) and rabbit (B) membrane enriched lens preparations. Activity expressed as nmoles phosphate released per mg protein per hour at 37°C, pH = 7.4. Mean ± S.E. n = 3-4.

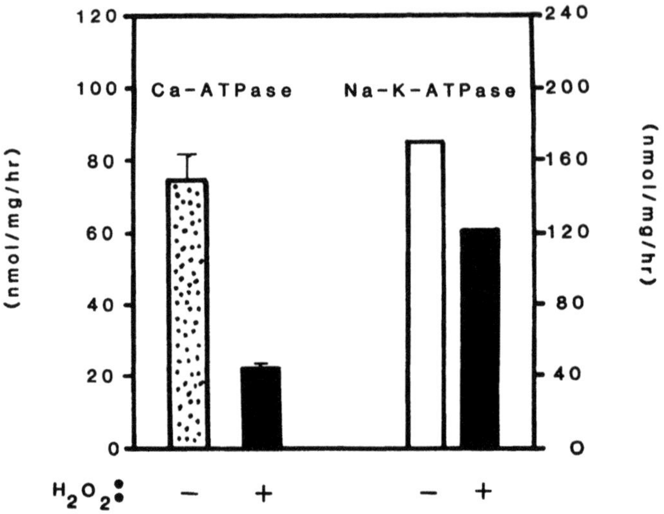

Figure 3. The inhibition of rabbit cortex Ca-ATPase at $[Ca^{2+}] = 5 \times 10^{-6}$ M, and ouabain sensitive Na-K-ATPase after a 15 minute exposure to 1×10^{-5} M H_2O_2. Temperature = 37°C, pH = 7.4.

oxidation. The Ca-ATPase is inhibited 69% after a brief 15 minute exposure to 10^{-5} M hydrogen peroxide (Figure 3). The Ca-ATPase may be inhibited by an oxidative metabolite, increased membrane lipid order, or direct inhibition of the Ca-ATPase pump. Ca-ATPase inhibition by hydrogen peroxide is greater than Na-K-ATPase inhibition under similar conditions.

DISCUSSION

Inhibition of the Ca-ATPase pump by oxidative damage would be expected to lead to elevation of calcium levels inside the lens. The immediate effect of elevated calcium and sodium levels is still under investigation. Calcium may cause lens opacity in a number of ways: 1) increased protein aggregation[10-12] and cystine bonding resulting in increased light scattering; 2) decreased Na-K-ATPase activity[13] indirectly causing an increase of intracellular sodium, osmotic swelling, and disruption of cellular structure increasing light scattering; 3) increased order of the lipid membrane causing a decrease in Ca-ATPase activity[14] and increased light scattering due to the aggregation of lipids; and 4) increased calmodulin-activated[15] pathways or inositol triphosphate release resulting in increased ion-membrane permeability.

REFERENCES

1. G. Duncan and A. R. Buschell, Ion analyses of human cataractous lenses, _Exp. Eye Res._ 20:223, 1975.
2. G. Duncan, "Mechanisms of cataract formation in the human lens", Academic Press, London, 1981.
3. M. A. Kirchberger and T. Antonetz, Calmodulin-mediated regulation of calcium transport and (Ca^{2+} + Mg^{2+})-activated ATPase activity in isolated cardiac sarcoplasmic reticulum, _J. Biol. Chem._ 257:5685, 1982.

4. G. B. Warren, M. D. Houslay, J. C. Metcalf and N. J. M. Birdsall, Cholesterol is excluded from the phospholipid annulus surrounding an active calcium transport protein, Nature (London) 255:684, 1975.
5. V. Madeira, M. C. Antunes-Maderia and A. Carvalho, Activation energies of the ATPase activity of sacroplasmic reticulum, Biochem. Biophys. Res. Commun. 58:897, 1974.
6. P. J. Quinn, R. Gomez and T. D. Madden, Biochem. Soc. Trans. 8:38, 1980.
7. M. C. Neville, C. A. Paterson and P. M. Hamilton, Evidence for two sodium pumps in the crystalline lens of the rabbit eye, Exp. Eye Res., 27:637, 1978.
8. S. Kobatashi, D. Roy and A. Spector, Sodium/Potassium ATPase in Normal and Cataractous Human Lenses, Curr. Eye Res., 2:327, 1983.
9. L. J. Alvarez, O. A. Candia and L. R. Grillone, Na^+-K^+ ATPase distribution in frog and bovine lenses, Curr. Eye Res. 4:143, 1985.
10. J. A. Jedziniak, J. H. Kinoshita, E. M. Yates, L. O. Hucker and G. B. Benedek, Calcium-induced aggregation of bovine lens alpha-crystallin, Invest. Ophthalmol. 11:905, 1972.
11. A. Spector and C. Rothschild, The effect of calcium upon the reaggregation of bovine alpha-crystallin, Invest. Ophthalmol. 12:225, 1973.
12. J. I. Clark, L. Mengel, A. Bagg and G. B. Benedek, Cortical opacity, calcium concentration and fiber membrane structure in the calf lens. Exp. Eye Res. 31:399, 1980.
13. P. M. Hamilton, N. A. Delamere and C. A. Paterson, The influence of calcium on the lens ATPase activity, Invest. Ophthalmol. Visual Sci. 18:434, 1979.
14. D. Borchman, R. Simon and E. Bicknell-Brown, Variation in the lipid composition of rabbit muscle sarcoplasmic reticulum membrane with muscle type, J. Biol. Chem. 257:14136, 1982.
15. S. Iwata, calcium-pump and its modulators in the lens: A review, Curr. Eye Res. 4:299, 1985.

ANTERIOR CHAMBER HYDROGEN PEROXIDE: EFFECTS OF 3-AMINOTRIAZOLE ON PEROXIDE KINETICS AND ON THE STATUS OF GLUTATHIONE

Stephen Csukas, Anastasios Costarides, Michael V. Riley*, and Keith Green

Department of Ophthalmology
Department of Physiology and Endocrinology
Medical College of Georgia
Augusta, Georgia 30912-3400, U.S.A.
*Eye Research Institute
Oakland University
Rochester, Michigan 48063, U.S.A.

INTRODUCTION

Previous studies using in vitro methods have shown that the anterior segment tissues, particularly the corneal endothelium, have the capacity to handle excess superoxide anion, but the capability to detoxify hydrogen peroxide can be exceeded. The data indicated that, of all the phototoxic by-products, hydrogen peroxide was the compound capable of inducing the most damage in the anterior segment of the eye.[1-3]

Hydrogen peroxide injected into the rabbit anterior chamber, to give an initial concentration of 3.6 mM in adults (from 3 to 6 months of age) and 3.6 mM in young (4 to 6 weeks of age), is cleared at a faster rate from young ($t_{½}$, 94 seconds) than from adult ($t_{½}$, 109 seconds) animals. This more rapid disappearance in young eyes may correlate with the 35 to 50% greater catalase levels measured in tissues bounding the anterior chamber in young compared to adult eyes.[4] Catalase levels in ocular and systemic tissues can be decreased using 3-amino-triazole (3AT), when 3AT is given either intravenously or orally.[5] In addition to catalase action, detoxification of hydrogen peroxide may also occur through interaction with glutathione, with conversion of the reduced to the oxidized form.

RESULTS AND DISCUSSION

A dose-response study of ocular tissues catalase to intravenous 3AT was performed. Tissues were harvested 4 hours after 3AT administration from anesthetized rabbits that were perfused intracardially with Ringer solution to remove erythrocytes. Ice-cold tissues were weighed, homogenized, sonicated and centrifuged. The superatant was analyzed spectrophotometrically for catalase at 240 nm in a quartz cuvette containing 16 mM hydrogen peroxide. Although this procedure also measures loss of hydrogen peroxide by glutathione reactions, the effect of a glutathione system at 16 mM hydrogen peroxide will be small, particularly when the catalase activity is high. The time course of hydrogen peroxide loss from the anterior chamber after bolus intracameral injections was determined in aqueous humor samples, using the dichlorophenol-indophenol method.[6] Tissues were also analyzed for total and oxidized glutathione.[7]

Catalase levels in the tissues bathed by aqueous humor (corneal endothelium, iris and ciliary processes) showed a dose-related decrease after 3AT treatment with 2, 4 and 6 ml of a 3M solution/kg body weight[8]; systemic tissues such as liver and lung also had markedly decreased catalase levels. Different tissues showed maximal decreases at different 3AT doses; iris and ciliary processes had an almost linear dose-related decrease, while corneal endothelium reached a near-maximum fall in catalase with 2 ml/kg. The maximal decreases averaged about 80% inhibition of catalase activity.

In the absence of 3AT, hydrogen peroxide loss from the anterior chamber occurred rapidly reaching base-line values in about 10 minutes. In adult rabbits the $t_{1/2}$ for hydrogen peroxide disappearance was 109 seconds (Table 1). After 2 ml of a 3M solution of 3AT/kg, the $t_{1/2}$ was 147 seconds; after 4 ml/kg the $t_{1/2}$ was 161 seconds, and after 6 ml/kg the $t_{1/2}$ was 184 seconds. The decrease in catalase activity caused by 3AT had a marked effect on the rate of disappearance of hydrogen peroxide from the anterior chamber. During the first 2 hours of measurement the $t_{1/2}$ for exogenous hydrogen peroxide added to aqueous humor in a cuvette was 35 minutes.

Table 1. Anterior Chamber Hydrogen Peroxide Concentrations *In Vivo* and $t_{1/2}$ as a Function of Age and Systemic 3AT Concentrations.[a]

Time	Anterior Chamber Hydrogen Peroxide Concentration (mM)				
	Young Control	Control	3AT-Treated Adults (3M solution) 2 ml/kg	4 ml/kg	6 ml/kg
18 sec	-	2.92± 0.48(5)	3.29(1)	-	-
1 min	2.10± 0.23(6)	1.81± 0.22(5)	2.65± 0.34(7)	2.74± 0.03(5)	2.68± 0.21(6)
3 min	0.69± 0.20(5)	0.81± 0.47(6)	1.93± 0.50(4)	1.89± 0.70(4)	1.80± 0.42(5)
5 min	0.36± 0.04(3)	0.47± 0.15(4)	0.72± 0.33(6)	0.93± 0.34(4)	1.03± 0.61(3)
	$t_{1/2}$=94.1± 6.8	$t_{1/2}$=108.8± 16.2	$t_{1/2}$=147.2± 27.6	$t_{1/2}$=160.8± 20.2	$t_{1/2}$=183.8±[b] 15.6

[a]All values are the mean ± SD. (N) indicates number of animals. All animals received a bolus intracameral injection of HP to give an initial aqueous humor concentration of approximately 3.6 mM. The $t_{1/2}$ (sec) was calculated as the best fit to an exponential decay curve utilizing a sum of least squares criterion. [b]P < 0.05 significantly different from adult control, calculated with z statistic.

The total glutathione (GSH + GSSG) of the corneal endothelium and iris was unaltered after intracameral hydrogen peroxide, but the oxidized fraction increased from control values of 5 - 10% to 80% at 18 seconds after peroxide injection. The oxidized fraction gradually decreased to near-normal levels over the next 20 - 30 minutes.

Catalase is one of the primary enzymes responsible for the breakdown of hydrogen peroxide. The present data show that catalase activity in different ocular tissues can be inhibited in a dose-related manner using 3AT and that the loss of exogenous intracameral hydrogen peroxide from the anterior chamber of the eye is correlated with the degree of catalase inhibition. Although different tissues bounding the anterior segment show different levels of inhibition of catalase activity by 3AT, the heavier tissues (viz. iris, ciliary processes) will predominate in determining the overall response of the anterior segment to injected hydrogen peroxide. Nevertheless, the breakdown of hydrogen peroxide is proportional to the available catalase activity. These experiments relate to the earlier findings of a decrease in catalase activity with increasing age.[4] If catalase activity progressively decreases with age, as do many other enzymes in ocular tissues,[9] then the capability of the anterior segment to detoxify either normal or excess loads of hydrogen peroxide normally found in the aqueous humor,[10] would be compromised. Such a change would render the eye tissues more susceptible to damage at lower peroxide concentrations. The dose response of catalase to 3AT indicates that catalase plays an important role in protection of the anterior segment tissues from the toxic effects of hydrogen peroxide. Detailed examination of the data indicates that the relationship between the $t_{1/2}$ for hydrogen peroxide clearance from the anterior chamber and the catalase levels is linear up to 2 ml/kg 3AT.[11] At 3AT concentrations of 4 and 6 ml/kg, the experimentally determined $t_{1/2}$ for hydrogen peroxide clearance from the anterior chamber differs from that calculated on the basis of catalase inhibition. The $t_{1/2}$ occurs more rapidly than that projected from tissue weights and percent inhibition of catalase. This is indicative of other processes (e.g., glutathione) assuming a proportionately more prominent role as the $t_{1/2}$ is extended and the influence of catalase reduced by 3AT-induced inhibition. These results show that beyond either a specific threshold of hydrogen peroxide, given a constant catalase level, or a specific threshold of catalase, given a constant hydrogen peroxide concentration, then relatively small changes in either modulator could lead to significant increases in tissue damage. This is especially true at the high hydrogen peroxide concentrations used in the present experiments.

The glutathione redox cycle can also play a role in the removal of oxyradical end products, yet the relationship to peroxide detoxification in tissues other than the lens is not clear. While the change in redox state of glutathione in the ocular tissues is significant after hydrogen peroxide injection into the anterior chamber, such responses could only account for a small portion of the total loss of hydrogen peroxide. The recycling of GSSG and GSH, however, could play a more important role that is dependent on the V_{max} of the reductase and peroxidase enzymes in the glutathione cycle.

CONCLUSIONS

In summary, modulation of catalase activity in anterior segment ocular tissues in a linear manner, using 3AT, results in increases in the $t_{1/2}$ for hydrogen peroxide disappearance from the anterior chamber up to 2 ml/kg 3AT. Catalase inhibition beyond those levels induced by 2 ml/kg (i.e., between 25 and 50% inhibition of normal catalase levels) reveals the involvement of other systems that decrease the projected $t_{1/2}$ for hydrogen peroxide clearance from the anterior chamber based on catalase inhibition alone. Protection of the anterior segment tissues from oxidative insult depends on the interplay between the concentration of the oxidant, the level of reduced glutathione, the activity of the GSH regenerating system, and catalase activity. Of the protective, or detoxification, mechanisms, catalase activity plays a major role under the experimental conditions utilized here in which a bolus of hydrogen peroxide is used. Under normal conditions the relative roles of catalase and the glutathione redox system await examination.

ACKNOWLEDGMENTS

Supported by NEI grants EY04558, EY04559 (KG) and EY00541 (MVR); by National Glaucoma Research, a program of the American Health Assistance Foundation; and by a Departmental Award from Research to Prevent Blindness, Inc. We thank Lisa Cheeks for technical help and Brenda Sheppard for secretarial assistance.

REFERENCES

1. D.S. Hull, S. Csukas, K. Green, and V. Livingston. Hydrogen peroxide and corneal endothelium. Acta Ophthalmol. 59:409 (1981).
2. D.S. Hull, K. Green, L. Thomas, and N. Alderman. Hydrogen peroxide mediated corneal endothelial damage: Induction by oxygen free radical. Invest. Ophthalmol. Vis. Sci. 25:1246 (1984).
3. D.S. Hull, K. Green, and R.D. Elijah. Effect of oxygen free radical products on iris vascular permeability. Acta Ophthalmol. 63:513 (1985).
4. S. Csukas and K. Green. Effects of intracameral hydrogen peroxide in the rabbit anterior chamber. Invest. Ophthalmol. Vis. Sci. 29:335 (1988).
5. K.C. Bhuyan and D.K. Bhuyan. Regulation of hydrogen peroxide in eye humors. Effect of 3-amino-1H-1,2,4-triazole on catalase and glutathione peroxidase of rabbit eye. Biochim. Biophys. Acta 497:641 (1977).
6. A. Pirie. Glutathione peroxidase in lens as a cause of hydrogen peroxide in aqueous humor. Biochem. J. 96:244 (1965).
7. M.C. Ng and M.V. Riley. Relation of intracellular levels and redox state of glutathione to endothelial function in the rabbit cornea. Exp. Eye Res. 30:511 (1980).
8. A. Costarides, D. Birnbaum, S. Csukas, E. Forbes, and K. Green. Morphological sequelae of anterior segment hydrogen peroxide in young and adult rabbits with or without 3-aminotriazole treatment, in: "Oxygen Radicals in Biology and Medicine," M. Simic, J. Ward and K. Taylor, eds., Plenum Press, New York (1988).
9. O. Hockwin and C. Ohrloff. The eye in the elderly: Lens, in: "Geriatrics 3," D. Platt, ed., Springer-Verlag, Berlin, Pp. 373 (1984).
10. A. Spector and W.H. Garner. Hydrogen peroxide and human cataracts. Exp. Eye. Res. 33:673 (1981).
11. S. Csukas, A. Costarides, M.V. Riley, and K. Green. Hydrogen peroxide in the rabbit anterior chamber: Effect on glutathione, and catalase effects on peroxide kinetics. Curr. Eye Res. 6:1395 (1987).

MORPHOLOGICAL SEQUELAE OF ANTERIOR SEGMENT HYDROGEN PEROXIDE IN YOUNG AND ADULT RABBITS WITH OR WITHOUT 3-AMINOTRIAZOLE TREATMENT

Anastasios Costarides[*,**], David Birnbaum[*], Stephen Csukas[*,**], Elizabeth Forbes[*], and Keith Green[*,**]

[*]Department of Ophthalmology
[**]Department of Physiology and Endocrinology
Medical College of Georgia
Augusta, Georgia 30912-3400, U.S.A.

INTRODUCTION

Both intravenous and oral 3-amino-triazole (3AT) are known to reduce catalase activity in ocular tissues;[1] a single intravenous injection of 3AT depresses catalase levels in iris, ciliary processes, retina, corneal endothelium and epithelium and lens capsule epithelium within 1 hour and is of several days duration. After intravenous 3AT we have shown that the $t_{½}$ of hydrogen peroxide clearance for rabbit anterior chamber is extended, dependent upon the dose of administered 3AT. This provided affirmation of the role of catalase in modulating the disappearance of anterior chamber hydrogen peroxide. No morphological studies have been made, however, of anterior segment tissues, save the lens,[2] after 3AT treatment. The present studies were designed to fill this void.

RESULTS AND DISCUSSION

Young (4 to 6 weeks old) and adult (from 3 to 6 months old) rabbits received 3AT either intravenously (4 ml/kg of a 3 M solution) or in drinking water (2% solution). Young and adult animals were divided into two main groups; each main group contained three subgroups. One subgroup received intravenous 3AT, one received intracameral hydrogen peroxide (to a final concentration of 3.6 mM), and the other received intracameral distilled water (the diluent for hydrogen peroxide). Those rabbits that drank 3AT for 30 days received intracameral hydrogen peroxide (final aqueous humor concentration of 3.6 mM).

Anterior segment tissues (corneal endothelium, iris and ciliary processes) were harvested at either 6 or 24 hours after hydrogen peroxide or distilled water injection. Tissues were fixed in a solution of 2% glutaraldehyde/1% paraformaldehyde in cacodylate buffer for 3 hours before transfer to buffer alone prior to fixation in 2% osmium. Specimens were examined in a JEOL 100CX electron microscope.

Oral 3AT alone had no effect on the morphology of the corneal endothelium, ciliary processes or iris despite almost 80% inhibition of catalase activity. After intracameral hydrogen peroxide the ciliary processes were edematous with dilated blood vessels containing thrombi; corneal endothelial cells were swollen.

Table 1. Summary of Morphological Changes Induced by Intracameral Hydrogen Peroxide in Normal and 3AT-treated Rabbits.

		Young Rabbits		Adult Rabbits	
Animal Status	Injection	Ciliary Processes	Corneal Endothelium	Ciliary Processes	Corneal Endothelium
Normal	Distilled water	No changes	No changes	No changes	No changes
	Hydrogen peroxide	Processes slightly swollen; pigment epithelium intact	No changes	Processes swollen; dilated vessels; thrombi present; attenuation of pigment epithelium	No changes
3-AT-Treated	Distilled water	No changes	No changes	No changes	No changes
	Hydrogen peroxide	Processes highly swollen; dilated vessels attenuation of pigment epithelium; damage to non-pigmented epithelium	Cells swollen; PMN's between cells	Processes very much swollen with cysts; dilated vessels; thrombi present; disruption of pigment epithelium; extensive damage to non-pigmented epithelium; PMN's in anterior chamber	Cells swollen; thicker endo-thelium; apical flaps disturbed; PMN's between cells

In young rabbits receiving intravenous 3AT, intracameral distilled water caused no morphological changes. Hydrogen peroxide injected into the anterior chamber of a normal animal caused only minor morphological changes of the ciliary processes, including edema of some processes (Table 1). Animals receiving 3AT and hydrogen peroxide showed highly swollen ciliary processes containing dilated vessels; some attenuation of the pigment epithelium was noted as was swelling of the corneal endothelium. The findings at 6 and 24 hours after hydrogen peroxide were essentially the same, except that some polymorphonuclear leucocyte (PMN) invasion was noted at 24 hours. In addition, PMNs were seen in the stroma near the basal cells of the corneal epithelium.

In adult animals, distilled water injection into normal animals caused no changes. Hydrogen peroxide injection into normal animals caused swelling of the ciliary process with dilation of blood vessels in which thrombi were seen (Table 1). When injected into 3AT-treated animals, hydrogen peroxide caused an exacerbation of the response with very highly swollen ciliary processes, with cyst formation between the pigmented and non-pigmented epithelia, vessel dilation, disruption of the pigmented epithelium and swelling of the corneal endothelium. PMN's were noted in all tissues, including the corneal epithelium that was disrupted, with enlarged extracellular spaces and cellular disorganization.

The marked difference in morphological response of normal young and adult rabbits to intracameral hydrogen peroxide injection correlates with the roughly 35% decrease in catalase levels of iris and corneal endothelium of adult relative to young rabbits.[3] Subsequent studies[4] showed that a graded reduction in catalase activity in ocular tissues can be achieved in a dose-related relationship to intravenous 3AT administered 4 hours previously (Table 2). Increasing intravenous 3AT doses cause an increase in the $t_{1/2}$ for hydrogen peroxide loss from the anterior chamber.[5] Oral 3AT has also been shown to markedly inhibit catalase activity[6] to levels less than those found after intravenous 3AT administration.

Table 2. Catalase Activity - Dose Response in Ocular Tissues to Single Systemic 3AT Injection.[a]

Dose of 3M 3AT (ml/kg)	Iris Catalase U/mg wet wt.	Iris Catalase U/mg soluble protein	Ciliary Process Catalase U/mg wet wt.	Ciliary Process Catalase U/mg soluble protein	Corneal Endothelium Catalase U/mg wet wt.	Corneal Endothelium Catalase U/mg soluble protein
0	0.93± 0.04(26)	19.66± 1.02(26)	1.45± 0.11(26)	17.49± 1.30(26)	0.231± 0.021(20)	5.48± 0.42(20)
2	0.66±[b] 0.09(6)	13.91±[b] 1.72(6)	1.12± 0.05(6)	17.17± 1.44(6)	0.12±[b] 0.01(4)	2.53±[b] 0.41(4)
4	0.43±[b] 0.11(9)	7.78±[b] 1.62(9)	0.85±[b] 0.21(8)	11.92± 2.92(8)	0.13± 0.07(6)	2.34±[b] 0.90(6)
6	0.28±[b] 0.03(4)	5.66±[b] 0.38(4)	0.49±[b] 0.11(3)	8.21±[b] 2.24(3)	0.15± 0.05(3)	2.32±[b] 0.71(3)

[a] Values are the mean catalase levels ± SEM; sample number given in parentheses. All tissues harvested between 3½ and 4 hours after intravenous 3AT administration. [b] Significantly different ($P < 0.05$) from controls (no 3AT) by student's unpaired t-test with SEM.

The present data indicate that morphological changes caused by intracameral hydrogen peroxide that are normally confined to normal adult animals are exacerbated in adults and occur in young animals that are treated with a catalase inhibitor, 3AT. Thus, young, normal animals that show only a marginal response to anterior chamber hydrogen peroxide respond vigorously to the same concentration of peroxide after 3AT pretreatment. The extension of the $t_{1/2}$ of hydrogen peroxide clearance after 3AT found earlier,[4] therefore, allows the toxic agent to cause greater damage. The catalase activity of young, normal animals is sufficient to protect anterior segment tissues against imposed levels of hydrogen peroxide up to 3.6 mM. It is evident that the level of catalase activity in anterior segment tissues is important in the regulation of aqueous humor hydrogen peroxide levels.

These findings relate to the decrease in catalase levels with age.[3] Hydrogen peroxide levels in the aqueous humor may not increase with age, except in certain cataractous patients,[7] but if catalase levels decrease with age (as do other enzymes[8]) then the anterior segment tissues should respond more vigorously with response to effectively greater amounts of hydrogen peroxide.

ACKNOWLEDGMENTS

Supported by NEI grant EY04559; by National Glaucoma Research, a program of the American Health Assistance Foundation; and by a departmental award from Research to Prevent Blindness, Inc. We thank Lisa Cheeks for her technical assistance and Brenda Sheppard for her secretarial assistance.

REFERENCES

1. K.C. Bhuyan and D.K. Bhuyan. Regulation of hydrogen peroxide in eye humors. Effect of 3-amino-1H-1,2,4-triazole on catalase and glutathione peroxidase of rabbit eye. Biochim. Biophys. Acta 497:641 (1977).
2. K.C. Bhuyan, D.K. Bhuyan, and H.M. Katzin. Amizol-induced cataract and inhibition of lens catalase in rabbit. Ophthalmic Res. 5:236 (1973).
3. S. Csukas and K. Green. Effects of intracameral hydrogen peroxide in the rabbit anterior chamber. Invest. Ophthalmol. Vis. Sci. 29:335 (1988).
4. S. Csukas, A. Costarides, M.V. Riley, and K. Green. Hydrogen peroxide in the rabbit anterior chamber: Effects on glutathione and catalase effects on peroxide kinetics. Curr. Eye Res. 6:1395 (1987).
5. S. Csukas, A. Costarides, M.V. Riley, and K. Green. Anterior chamber hydrogen peroxide: Effects of 3-aminotriazole on peroxide kinetics and effects on the status of glutathione, in: "Oxygen Radicals in Biology and Medicine," M. Simic, J. Ward and K. Taylor, eds., Plenum Press, New York (1988).
6. D. Birnbaum, S. Csukas, A. Costarides, E. Forbes, and K. Green. 3-amino-triazole effects on the eye of young and adult rabbits in the presence and absence of hydrogen peroxide. Curr. Eye Res. 6:1403 (1988).
7. A. Spector and W.H. Garner. Hydrogen peroxide and human cataract. Exp. Eye Res. 33:673 (1981).
8. O. Hockwin and C. Ohrloff. The eye in the elderly: Lens, in: "Geriatrics 3," D. Platt, ed., Pp. 373-424. Springer-Verlag, Berlin.

CORNEAL OXYGEN SCAVENGING SYSTEMS:

LYSIS OF CORNEAL EPITHELIAL CELLS BY SUPEROXIDE ANIONS

Rosalie Crouch, Zhu Ling, and Betty J. Hayden

Department of Ophthalmology
Medical University of South Carolina
Charleston, South Carolina 29425, USA

INTRODUCTION

The cornea is an unique tissue because it is exposed to insults from the external environment and the toxic effects of topically applied agents as well as the normal spectrum of physiological processes. Scarring can result from infection, trauma or numerous other causes which can compromise visual acuity. The cornea consists of three main layers: the epithelium, the stroma and the endothelium. Injury to either the epithelium or endothelium can lead to disruption of the underlying nonregenerative stroma resulting in permanent scarring.

Insults to the cornea, particularly the epithelium, may be in the form of oxygen radicals. We have shown that a major scavenging oxygen enzyme, superoxide dismutase,[1] is present in the corneal epithelium and endothelium but not in the stroma.[2] We report here the corneal distribution of two other major oxygen scavenging enzymes, catalase and glutathione peroxidase, and the effects of chemically generated superoxide radicals on corneal epithelial cells.

CORNEAL SCAVENGING ENZYMES

Tissue damage from oxygen radicals can arise from several species in addition to the superoxide radical. Hydrogen peroxide is perhaps the most damaging of these species, both in itself and as a precursor to the highly reactive hydroxyl radical. Therefore, in addition to superoxide dismutase, we have assayed the cornea for the enzymes catalase and glutathione peroxidase, both of which protect cells by metabolizing hydrogen peroxide. The levels of superoxide dismutase,[3] glutathione peroxidase[4] and catalase[5] in the rat cornea are summarized in Table 1. The levels of catalase and glutathione peroxidase are undetectable in the stromal layer.

In the epithelium, these enzymes are present but in low quantities. Catalase is particularly minimal, bordering on being undetectable. In the endothelium, no glutathione peroxidase was observed and again, catalase is minimal. These results are in agreement with the earlier reports by Bhuyan and Bhuyan[7] on levels of both enzymes in the rabbit and by Mayer on catalase levels in the cow and man.[8] In comparing old and young rats, no significant change on aging was observed for glutathione peroxidase or catalase levels, which parallels the lack of aging effects on corneal superoxide dismutase reported earlier.[9]

CYTOTOXIC EFFECTS OF SUPEROXIDE RADICALS

In order to determine if corneal cells are actually affected by oxygen radicals, superoxide radicals were generated in the presence of corneal epithelial cells and cell lysis determined by release of incorporated chromium-51.[10] The epithelial cells were selected for study as they are a highly vulnerable tissue in the eye and the cells can be routinely cultured. Because hydrogen peroxide may be the oxygen damaging radical species, catalase (which scavenges hydrogen peroxide) was examined for the potential to mediate this effect.

Table 1. Catalase, Glutathione Peroxidase, and Superoxide Dismutase Levels in the Rat[a] Corneal Epithelium, Stroma and Endothelium

Corneal Layer	Catalase[4] (mU/mg)	Glutathione Peroxidase[5] (U/mg)	Superoxide Dismutase[2,3] (U/mg)
Epithelium	0.2[b]	0.1[b]	101
Stroma	0.0	0	0
Endothelium	0.1[b]	0	93

[a] Wister males (3-4 months)
+ At the minimal limits of detection.
All values based on mg soluble protein.[6]

Epithelial cells cultured from both human and rat corneas were labeled with chromium 51. Superoxide radicals (and resulting products) were generated in a xanthine/ xanthine oxidase system and percent lysis of the corneal cells was based on chromium-51 release. This lysis could be reversed in the presence of catalase and is dependent on the catalase concentration (Figure 1A). Conversely, catalase inactivated with 3-amino-1,2,4-triazole[12] failed to protect against the oxygen radical species induced lysis (Figure 1B). Therefore the catalase effects are due to enzymatic activity rather than being a result of nonspecific protein interactions.

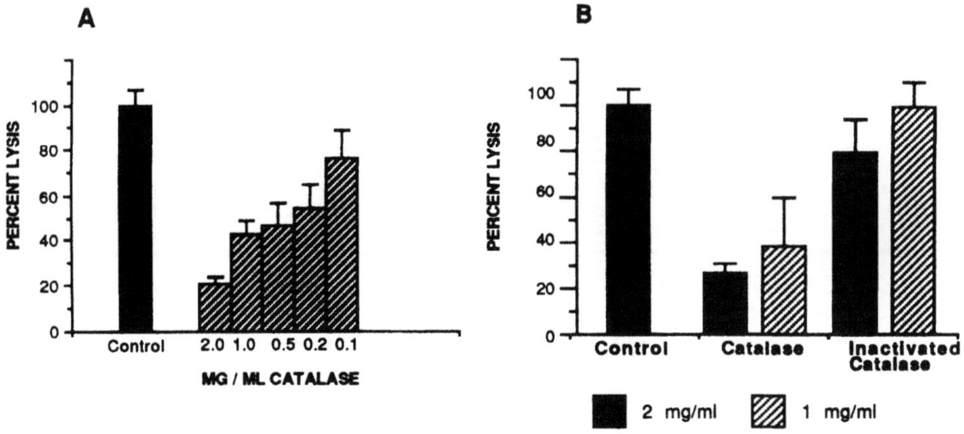

Figure 1. Protection of human corneal epithelial cells with catalase from cytolysis. Radicals generated by xanthine/xanthine oxidase. Cytolysis measured by chromium-51 release. Maximum lysis by in absence of catalase set at 100%. Incubation for 4 h at 37°C. (A) Catalase concentration varied between 0.1 and 2 mg/ml. (B) Catalase inactivated with 3-amino-1,2,4-triazole. Brackets represent SEM; n=3.

SUMMARY

The cornea has inherent protection from superoxide radicals from the presence of superoxide dismutase. However, the dismutation of these radicals results in the generation of hydrogen peroxide. The corneal scavenging systems for hydrogen peroxide is minimal as there are only low levels of catalase and glutathione peroxidase present. We have demonstrated the cytolytic capabilities of oxygen radicals in the cornea and reversed this lysis with catalase. These data indicate that hydrogen peroxide is the primary damaging agent in this system.

ACKNOWLEDGMENTS

This research was supported by NIH grants EY05757 and EY05981. We thank the South Carolina Eye Bank for donation of tissue.

REFERENCES

1. I. Fridovich, The biology of oxygen radicals, Science 201:875 (1978).
2. T. M. Redmond, E. J. Duke, W. H. Coles, J. A. V. Simson and R. K. Crouch, Localization of corneal superoxide dismutase by biochemical and histocytochemical techniques, Exp. Eye Res. 38:369 (1984).
3. C. Beauchamp and I. Fridovich, Superoxide dismutase: Improved assays and an assay applicable to acrylamide gels, Anal. Biochem. 44:276 (1971).

4. A. Clairborne, Catalase activity, in: "CRC Handbook of Methods for Oxygen Radical Research," R.A. Greenwald, ed., Boca Raton, FL, p. 283, (1985).
5. W. A. Gunzler, H. Kremers and L. Flohe, An improved coupled test procedure for glutathione peroxidase in blood, Z. Klin. Chem. Klin. Biochem. 12:444 (1974).
6. M. M. Bradford, A rapid and sensitive method for the quantitation of microgram quantities of protein utilizing the principle of protein-dye binding, Anal. Biochem. 72:248 (1976).
7. K. C. Bhuyan and D. K. Bhuyan, Regulation of hydrogen peroxide in eye humors. Effects of 3-amino-1H-1,2,4-triazole on catalase and glutathione peroxidase of rabbit eye. Biochim. Biophy. Acta, 497:641 (1977).
8. U. Mayer, Comparative investigations of catalase activity in different ocular tissues in cattle and man, Albrecht von Graefes Arch. Klin. Ophthal. 213:261 (1980).
9. R. Crouch, J. Patrick, J. Gossey, and W.H. Coles, The effect of aging corneal and lens superoxide dismutase, Curr. Eye Res. 3:1119 (1984).
10. K.T. Brunner, J. Mauel, J.C. Cerrottini, and B. Chapris, Quantitative assay of the lytic action of immune lymphoid cells on ^{51}Cr labeled target cells in vitro: Inhibition by isoantibody and by drugs, Immunol. 14:181 (1968).
11. R. Neville, P. Dennis, D. Sens, and R.K. Crouch, Preservative cytotoxicity to cultured corneal epithelial cells, Curr. Eye Res. 3:1119 (1984).
12. D. Darr and I. Fridovich, Irreversible inactivation of catalase by 3-amino-1,2,4-triazole, Biochem. Pharmacol. 35:36 (1986).

THE ROLE OF ALLOPURINOL AND OXYGEN FREE RADICAL SCAVENGERS IN LIVER PRESERVATION

Luis H. Toledo-Pereyra

Departments of Surgery and Surgical Research
Mount Carmel Mercy Hospital
Detroit, MI 48235 U.S.A.

INTRODUCTION

The increased use of long-distance procurement for donor livers has recently created a dire need for prolonged hepatic preservation. To date, a variety of experimental liver preservation techniques have been studied; however, consistent success for more than 12 hours has yet to be achieved.[1-3] Methods for testing the efficacy of preservation solutions that successfully and safely prolong the cold preservation time are needed and might offer a realistic solution to this problem. Toledo-Pereyra et al[4,5] and Lambotte et al[6] have suggested that the use of hyperosmolar perfusates may be beneficial in achieving prolonged hepatic cold-storage times. In addition, Toledo-Pereyra et al[4] and Fuller et al[7] have reported on enhanced hypothermic preservation of donor livers following the use of pulsatile perfusion with specially made perfusates. Nevertheless, the vulnerability of the liver to both warm and cold ischemic damage and reperfusion injury remains as the major obstacle to immediate post-transplantation graft function.

The primary focus of much of the experimental work in this area, therefore, has been based on the hypothesis that reperfusion of hypothermically preserved organs results in oxygen-derived free radical injury produced via the xanthine oxidase enzyme system.[8] One way to prevent an organ from reperfusion injury, however is to reduce the production of oxygen radicals. This can be achieved by the administration of allopurinol, a xanthine-oxidase inhibitor. In addition to the administration of allopurinol, oxygen free radical scavengers such as superoxide dismutase (SOD), catalase, and α-tocopherol can be given to protect the organ from reperfusion injury (Figure 1). However, the combined use of hyperosmolar perfusates and treatment with allopurinol and/or oxygen free radical scavengers for hepatic preservation has received only limited application in experimental liver transplantation and has yet to be used clinically. The scope of this paper, therefore, is to provide a selected review on the use of allopurinol and/or oxygen free radical scavengers in providing improved organ protection following hepatic transplantation.

ALLOPURINOL

Toledo-Pereyra et al[4] were the first to include allopurinol in the preservation solution itself. By conducting an extensive study, they were

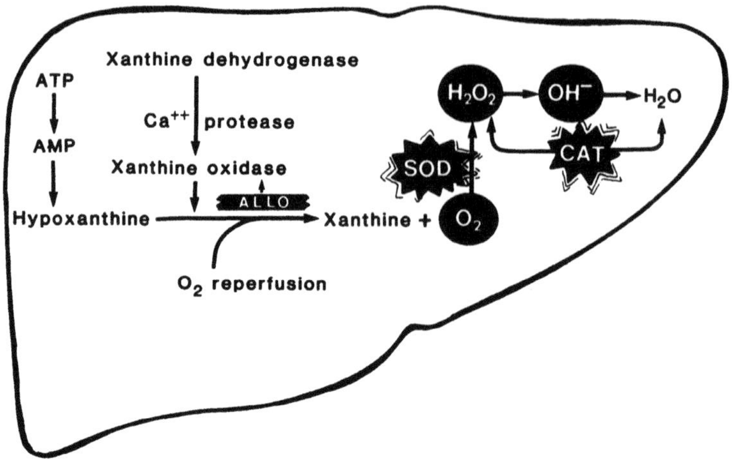

Figure 1. Proposed mechanism for the protective action of allopurinol (ALLO), superoxide dismutase (SOD), and catalase (CAT) on livers undergoing ischemia and reperfusion.

able to report that certain factors such as perfusate composition, type of flow, and flow rate were responsible for determining successful liver preservation for transplantation. Following 24 hours of hypothermic pulsatile perfusion with modified silica gel fraction, made hyperosmolar with glucose and containing 500 mg/L of allopurinol, the authors found that graft function after auxiliary liver transplantation was comparable to that seen among fresh-transplanted control animals (Table 1). Biopsy specimens of the 24-hour-perfused livers taken from recipient animals at the time of death revealed hepatocyte preservation with slightly central atrophy, mild bile stasis, and only minimal congestion.

Donor pretreatment with allopurinol prior to 24 hours of hypothermic pulsatile perfusion[9] or cold storage[10] has also been studied. Rizzuti et al[9] pretreated canines for three days with 50 mg/kg/day of allopurinol prior to 24 hours of hypothermic perfusion with a modified hyperosmolar Belzer perfusate containing sucrose and subsequent orthotopic transplantation. Survival among these animals was much improved when compared with recipients of livers that underwent no donor pretreatment and 24 hours of hypothermic pulsatile perfusion with an unmodified Belzer perfusate, but was significantly less in comparison to fresh-transplanted animals (Table 2). Hepatic ATP levels following reperfusion were consistently reduced from preharvest levels among the dogs receiving preserved livers pretreated

Table 1. Effect of the Modification of Perfusate Composition on the Survival of Dogs Following 24 Hours of Liver Preservation

Groups	Perfusate	Survivors >5 days	Mean Survival (M±SD)
I	None	5/5	14.5 ± 3.5
II	CPP	1/5	2.9 ± 1.1
III	SGF	2/5	4.7 ± 1.2
IV	MSGF	4/5	11.7 ± 2.9
V	MSGF*	5/5	13.6 + 4.1

CPP=Cryoprecipitated plasma; SGF=Silica gel fraction; MSGF=Modified silica gel fraction; MSGF*=Modified silica gel fraction with allopurinol.

Table 2. Survival and Cause of Death among the Experimental Groups

Group	Pretreatment	Perfusate	Survival	Cause of Death
I	None	None*	11 days	Peritonitis
			8 days	Unknown
			3 days	Liver necrosis
			44 days	Euthanasia
			25 days	Peritonitis
II	None	Belzer	2 hours	Bleeding
			1 hour	Bleeding
			0 hours	Bleeding
			4 hours	Bleeding
			2 hours	Bleeding
III	Allopurinol	Hyperosmolar Belzer	2 hours	Bleeding
			3 days	Unknown
			4 days	Peritonitis
			7 days	Euthanasia
			3 hours	Bleeding

*Dogs underwent immediate orthotopic liver transplantation.
Data from Rizzuti et al[9]

with allopurinol. However, no correlation between the percent reduction in ATP levels and future graft viability was evident.

Merion et al,[10] using the porcine model, were able to study the effects of allopurinol alone or in combination with SOD on hepatic histology, tissue xanthine-oxidase levels, survival, and post-transplant coagulopathy. Donor pretreatment with 100 mg/kg/day allopurinol for two days was implemented prior to storing the livers for 24 hours in Euro-Collins solution. Upon revascularization, allopurinol-pretreated animals revealed excellent preservation of hepatic architecture, with only minimal hepatocyte karyorrhexis, focal hepatic cord disorganization, and some extrasinusoidal hemorrhage. Xanthine-oxidase levels were reduced by 68% among these animals when compared with cold-stored, transplanted controls. In addition, survival was prolonged following this short course of therapy with allopurinol, and only one-third of the animals developed gross coagulopathy. The addition of recipient treatment with SOD, however, resulted in decreased survival. All animals developed severe coagulopathy, and died within two hours of revascularization. Despite low xanthine oxidase levels, hepatocellular necrosis, extrasinusoidal hemorrhage, and hepatic cord disorganization was most severe among this group.

Pretreatment of dogs with 5 mg/kg allopurinol 20 minutes prior to induction of warm hepatic ischemia by clamping the hepatic artery and portal vein for 30 minutes did not improve graft survival following auxiliary transplantation.[11] Toledo-Pereyra et al were able to achieve prolonged survival, however, when a combination of isoproterenol, allopurinol, and heparin was given to the animals prior to the induction of ischemia (Table 3). The exact mechanism by which the combination of these three drugs acts to improve post-transplant hepatic function and survival, however, is not clear.

Table 3. Dog Survival following Liver Ischemia and Transplantation

Treatment	Survival in Days (M±SD)	p Value
Control	6.0 ± 1.7	---
Allopurinol	11.2 ± 2.5	> 0.05
Allopurinol, isoproterenol, and heparin	13.2 ± 2.8	< 0.05

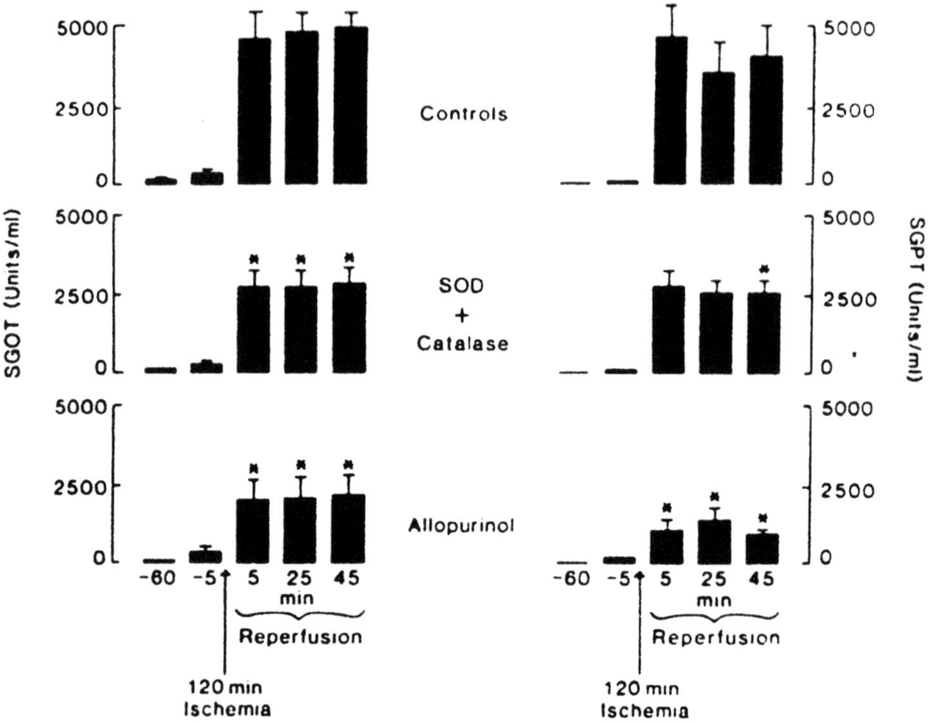

Figure 2. Effects of ischemia and reperfusion on SGOT and SGPT levels in normothermic rat livers. * indicates p< 0.05. Used with permission from Adkinson et al[12]

Adkinson et al[12] reported that in rats, pretreatment with decreasing doses of allopurinol (2.5-1.5 mg/kg), given at 36, 12, and 2 hours prior to surgery, or the combination of 1,400 U/ml SOD and 20,000 U/ml catalase prior to 120 minutes of warm ischemia, significantly decreased the depression in hepatic oxygen consumption induced by ischemia and reperfusion. In addition, significant decreases in both SGOT and SGPT were observed during reperfusion in those dogs given allopurinol. A significant decrease in SGOT levels was seen during reperfusion following the administration of SOD and catalase, whereas only a slight decrease in SGPT levels was noted (Figure 2).

SOD, CATALASE AND α-TOCOPHEROL

Atalla et al,[13] using the isolated hepatic-perfusion model (IHPM) for liver testing, studied the effects of donor pretreatment with SOD and/ or catalase on ischemic livers. Donor dogs were given either 5,000 U/kg SOD, 5,000 U/kg catalase, or both 5 minutes prior to liver ischemia, which was induced by cross-clamping of the portal vein and hepatic artery for 40 minutes. Following hepatectomy, livers were tested for 3 hours using the IHPM. Laboratory and physiological results from this study indicated that all three treatment modalities were able to provide partial protection against the oxygen-derived free radicals accumulated during liver ischemia and reperfusion, when compared with untreated controls (Tables 4 and 5). Histological specimens and 99-m Tc Hida scans confirmed these findings.

In a similar study using the IHPM, Toledo-Pereyra and Cederna[14] found that the combination of 5,000 U/kg of catalase and 1,000 U of Lente insulin

Table 4. Comparative Laboratory Values Following Donor Pretreatment With SOD and/or Catalase.

Group	Pretreatment	LDH	SGOT	AP	GGT	Lactic Acid
I	None	+++	+++	+++	+	++
II	Catalase	++	+	++	+	++
III	SOD	+	+	+++	+	++
IV	Catalase & SOD	++	+	++	+	++

Legend: +++ severe increase; ++ moderate increase; + minimal increase
LDH - +++(>1200 IU/L), ++(200-1200 IU/L), +(20-200 IU/L)
SGOT - +++(>300 IU/L), ++(150-300 IU/L), +(20-150 IU/L)
AP - +++(40-100 mM/ml), ++(10-40 mM/ml), +(<10 mM/ml)
GGT - +(1-6 U/L)
Lactic Acid - +++(>15 mM/L), ++(9-15 mM/L), +(3-8 mM/L)

given 5 minutes prior to 20 minutes of ischemia protected hepatic tissue and improved functional results.

Marubayashi et al[14] pretreated rats with the free radical scavenger α-tocopherol by either a single injection of 10 mg/kg one day pripr to warm ischemia or the same dose given for three days prior to surgery. Following 90 minutes of warm ischemia, percent survival among the animals given placebo pretreatment of a single dose of α-tocopherol, was zero. Pretreatment for 3 days resulted in an increase in the survival rate to 45.5%, which was accompanied by an increase in the content of α-tocopherol in both the liver tissue and mitochondria. The authors believed this restorative effect of α-tocopherol pretreatment on ischemic rat livers was due to its action as an antioxidant on lipid peroxidation.

These studies appear to indicate that in laboratory animals, oxygen free radical scavengers such as SOD, catalase, and α-tocopherol and xanthine oxidase inhibitors such as allopurinol are effective in limiting liver dysfunction produced during ischemia and reperfusion. Future studies incorporating the use of these or other oxygen free radical scavengers/inhibitors in combination with hyperosmolar perfusates are necessary if we are to improve liver preservation for transplantation.

Table 5. Mean Flow Rates Following Donor Pretreatment with SOD and/or Catalase

Group	Portal vein flow*		Hepatic artery flow*		Bile flow**	
	0 hr	3 hr	0 hr	3 hr	0 hr	3 hr
Control	234±51	127±50	138±10	90±11	3±1.7	.7±1.2
CAT	259±25	255±24³*	145±6	141±4³*	10±3.4³*	12±2.2³*
SOD	242±30	187±18³*	136±19	108±7³*	4±1	4±1³*
CAT&SOD	275±10³*	258±16³*	143±4	133±8³*	8±2³*	7±2.2³*

CAT = catalase
SOD = superoxide dismutase
* in cc/min
** in cc/hr
³* Student's t test vs. control, $p < 0.05$

REFERENCES

1. L.H. Toledo-Pereyra, J. Castellanos, and M. Chapman, Failure to preserve liver allografts for 24 hours: Experimental and theoretical considerations, Transplant Proc. 19(Suppl 3):68 (1987).

2. L. Lambotte, S. Pontegnie-Istace, and J.B. otte, The effect of isoproterenol and Collins' solution on the preservation of canine livers with simple cooling, Transplant Proc. 6:301 (1974).
3. K.N. Mallik, I.C. Pathak, S.K. Mitra, D.V. Datta, R. Bhatara, and S. Garg, Preservation of liver for 4 to 20 hours by the cold flushing-freeze storage technique, Indian J. Med. Res. 71:468 (1980).
4. L.H. Toledo-Pereyra, R.L. Simmons, and J.S. Najarian, Factors determining successful liver preservation for transplantation, Ann. Surg. 181:289 (1975).
5. L.H. Toledo-Pereyra, M. Chee, R.C. Lillehei, and R.M. Condie, Liver preservation by cold storage with hyperosmolar solutions for twenty four hours, Cryobiology 16:43 (1979).
6. L. Lambotte, S. Wojcik, and S. Pontegnie-Istace, Investigations on the mechanisms of action of hyperosmolar solutions used in liver preservation, in: "Organ Preservation II," P.E. Pegg and I.A. Jacobsen, eds., Churchill Livingstone, Edinburgh (1979).
7. B.J. Fuller, and V.D. Attenburrow, The effects of increasing the oncotic pressure of the perfusate on bloodless hypothermic perfusion of the liver in the rat, Cryobiology 15:545 (1978).
8. J.M. McCord, Oxygen derived free radicals in postischemic tissue injury, N. Engl. J. Med. 312:159 (1985).
9. R.P. Rizzuti, P.R. Cunningham, D. Easley, T. Fushiki, and F.T. Thomas, A new method of hepatic preservation using pulsatile perfusion and allopurinol, Current Surgery 43(6):492 (1986).
10. R.M. Merion, D.A. Campbell Jr., D.R. Averill Jr., D.C. Dafoe, J.C. Fantone, and J.G. Turcotte, Improved liver preservation with euro-collins and donor allopurinol, Transplant Proc. 18:582 (1986).
11. L.H. Toledo-Pereyra, R.L. Simmons, and J.S. Najarian, Protection of the ischemic liver by donor pretreatment before transplantation, Am. J. Surg. 129:513 (1975).
12. D. Adkinson, M.E. Hollwarth, J.N. Benoit, D.A. Parks, J.M. McCord, and D.N. Granger, Role of free radicals in ischemia-reperfusion injury to the liver, Acta Physiol. Scand. Suppl 548:101 (1986).
13. S.L. Attala, L.H. Toledo-Pereyra, G.H. MacKenzie, and J.P. Cederna, Influence of oxygen-derived free radical scavengers on ischemic livers, Transplantation 40:584 (1985).
14. L.H. Toledo-Pereyra, and J. Cederna, Protection of liver allografts from ischemic damage prior to transplantation using insulin and catalase, in: "Oxygen free radicals in shock," G.P. Novelli and F. Ursini eds., S. Karger, Basel (1986).
15. S. Marubayashi, K. Dohi, K. Ochi, and T. Kawasaki, Role of free radicals in ischemic rat liver cell injury: Prevention of damage by α -tocopherol, Surgery 90:184 (1986).

PREVENTION OF ACUTE RENAL FAILURE BY SOD AND SUCROSE

M. Wolgast, A. Bayati, O. Hellberg , Ö. Källskog and G. Öjteg

Department of Physiology and Medical Biophysics
Biomedical Center, Box 572, S-751 23 Uppsala, Sweden

INTRODUCTION

Acute renal failure is the most serious consequence resulting from renal ischemia and reperfusion. It is characterized by a greatly depressed glomerular filtration rate, isosthenuria and a reduced potassium secretion . It is generally agreed here that the depressed filtration rate is due to the obstruction of the tubules by cells and cell debris rejected from the tubular wall.

The damages to the urine concentration mechanism and the depressed potassium secretion are, on the other hand, probably the result of a medullary ischemia, which in previous works[1] has been found to be caused by the trapping of red cells in the early recirculation period into, primarily, the renal medulla. This process may also affect the formation of tubular obstructions in the sense that the dilated medullary capillaries would compress the tubules and also since the medullary ischemia leads to swelling of the tubular epithelial cells.

It is the aim of the present study to examine the factors behind both the formation of tubular obstruction and the trapping of red cells. Since oxygen-derived free radicals at least indirectly may lie behind the trapping of red cells, the effect of superoxide dismutase (SOD) was also investigated.

MATERIAL AND METHODS

The experiments were performed on Inactin-anaesthetized (120 mg/kg body weight) Sprague-Dawley rats weighing about 300 g. After tracheostomy, the rats were placed on a servo-controlled heating pad stabilizing the temperature at 37.5 ^{o}C. Acute renal failure was evoked by clamping the left renal artery for 45 min. During this time the kidney was placed in its natural position with the abdomen closed. After release of the clamp, the kidney was suspended into a lucite cup and prepared for micropuncture studies .

Regional Red Cell Volume (n=20)

In order to determine the post-ischemic regional renal red cell volume 5 uC of 51-Cr-labelled red cells were injected soon before the clamping of the kidney. Twenty minutes after the onset of circulation, the two kidneys were harvested and microdissected into pieces of the cortex, the outer and inner stripe of the outer medulla and the inner medulla. The percentage red cell volume was calculated from the quotient between the activity of the tissue specimens and that of the red cells in a blood sample withdrawn after the same 20 minutes.

Macromolecular Permeability (n=12)

The macromolecular permeability was assessed from the plasma-lymph transport of the plasma proteins. For this purpose a hilar lymph vessel was catheterized by a 100 um thick polyethylene tube. Lymph was collected (the flow was approximately 1 ul/min) for 15-30 min intervals both before and up to 1.5 h after the ischemia.

Micropuncture Studies (n= 15)

Standard micropuncture technique was employed to determine the pressures in both the vascular and tubular system of the kidney. Inulin clearance was used to assess the total and single nephron filtration rate.

Morphological Studies (n=5)

The morphology of the kidneys was investigated by scanning electronmicroscopy. In order to avoid tearing of the tissue, however, the sections were prepared by a Vibratom rather than by a razor blade.

RESULTS

The left panel of Figure 1 shows that 45 min of ischemia produced a fourfold rise in the average plasma-lymph transport of the plasma proteins in the time period up to 1.5 h after onset of circulation. This was also accompanied (Figure 1, right panel) by a trapping of red cells into, primarily, the inner stripe of the outer medulla. A single intravenous injection of 15 mg SOD or more as administered soon before recirculation was found to prevent both these unfortunate events.

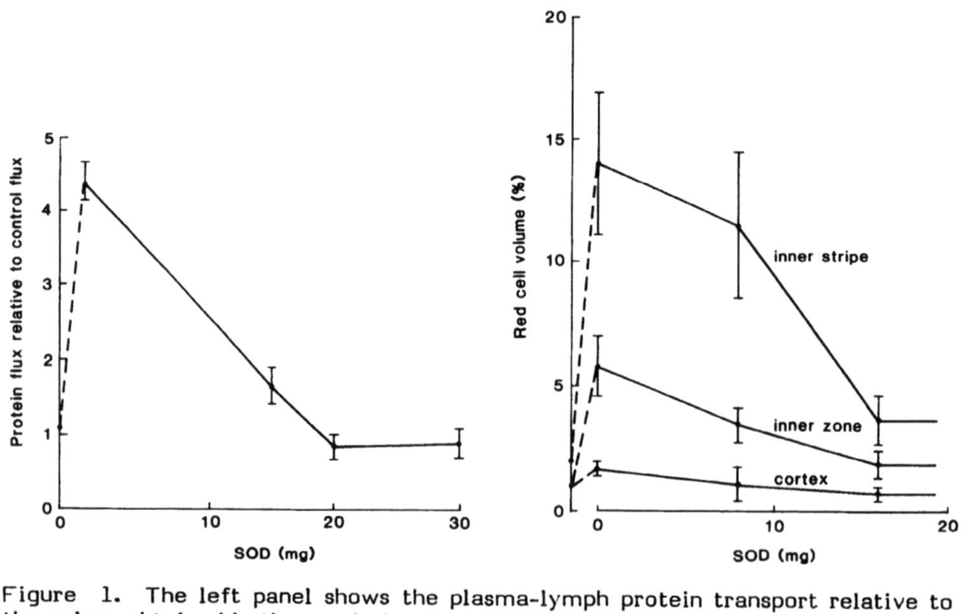

Figure 1. The left panel shows the plasma-lymph protein transport relative to the values obtained in the pre-ischemic control period. The right panel depicts the percentage post-ischemic red cell volume. The data on the ordinate refers to the the red cell volume obtained in the right control kidney.

The morphological studies showed that almost all the capillaries of the renal medulla contained aggregates of quasi-hexagonal-shaped red cells. The capillaries were also dilated and did thereby compress the adjacent tubules (figure 2).

In the micropuncture studies, the most striking finding was the steep rise of the proximal tubular pressure and the concommitant severe depression of the glomerular filtration rate to only a few percent of the normal. Although SOD was able to prevent the trapping of red cells, it did not prevent these serious abnormalities. However, if in addition to SOD, 2 ml of a 12% sucrose solution was infused soon before recirculation, the proximal tubular pressure and also the glomerular filtration rate was found to be largly restored. It was furthermore noted that the urine in the sucrose-treated animals contained an abundance of rejected cells and cell debris, i.e., which otherwise would have obstructed the tubules. It ought to be stressed, in this context, that treatment with sucrose alone, although capable of preventing obstruction of the tubules, leads to a massive urinary loss of fluid and proteins with concommitant uncontrollable dehydration. As SOD was found to restore the tubular reabsorption of fluid, the combined treatment would seem to represent a promising therapeutic approach.

DISCUSSION

The data suggest that the oxygen-derived free radicals produced during the early recirculation phase (directly or indirectly) lie behind the damage to the renal capillary membranes. We would also propose that the trapping of red cells is the result of this damage as follows: An increased macromolecular permeability (Figure 3) may thus, via the subsequent reduction in the transcapillary colloid osmotic pressure difference, lead to extravasation of plasma and hence to a hemoconcentration and rise in the vascular resistance. As shown in Figure 3, the consequent augmented capillary pressure would elicit further extravasation, further hemoconcentration and increase in the intracapillary hematocrit etc .

In capillaries with a short transit time, as in the cortex, the above process may only result in a congestion of the red cells in the late part of the capillary.As soon as the macromolecular permeability is restored, the intracapillary hematocrit and the blood flow will also be restored. In the renal medulla, where the transit times is five to ten times longer than in the cortex, this process may lead to a more or less complete loss of plasma. The red cells will thereby be forced to attain, in the ideal case, a hexagonal shape. The subsequent steep rise in the capillary hydrostatic pressure will then evoke a further trapping in the retrograde direction until the whole of the capillary loop becomes filled with aggregates of densly packed and trapped red cells. In accordance with this view, restoration of the capillary permeability would also lessen the risk for the red cell

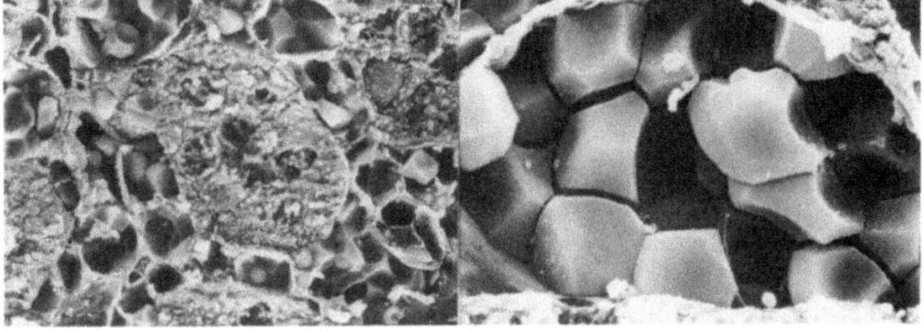

Figure 2. Scanning electronmicroscopy of the inner stripe of the outer medulla Note that all the capillaries seemingly contain aggregates of polygonal-shaped trapped red cells.

trapping, which also was established. However, SOD did also improve the tubular reabsorption. The density of the rejected cells in the lumen will then rise, which increases the risk for the formation of a tubular obstruction. For this reason, the tubular reabsorption was inhibited with sucrose, a diuretic which would seem advantageous in the sense that it affects all parts of the tubule and also since its effect gradually declines in parallel to its excretion.

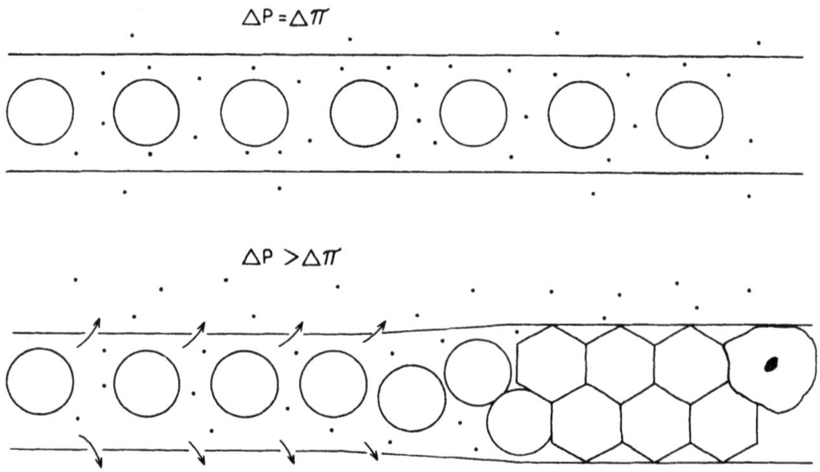

Figure 3. Upper panel depicts a capillary with no net trans-capillary turnover of fluid. The intravascular hematocrit will thereby remain unchanged. In the lower panel extravasation plasma has evoked a hemoconcentration, which if all plasma escapes the capillary leads to a dense packing of the red cells in the distal part of the capilllary.

REFERENCE

1. M. Wolgast, L.Karlberg , Ö. Källskog , B.J. Norlén ,K. Nygren ,G. Öjteg . Hemodynamic alterations in ischemic acute renal failure. Nephron. 31:301-303, 1982

THE AGING PROCESS

Denham Harman

Departments of Medicine and Biochemistry
University of Nebraska College of Medicine
Omaha, Nebraska 68105, U.S.A.

INTRODUCTION

Aging and death are universal phenomena. Aging is a progressive sequence of age-related, widespread, more-or-less common changes observed in every individual of a given species. These sequential changes are associated with an ever-increasing likelihood of disease and death, beginning at a minimum early in life and increasing until all members of the species are dead by a characteristic age, the maximum life span.

As causes of premature death are reduced the chance of death at any age is decreased, more so in the young than in the old, as is the age beyond which the chance of death increases in an exponential manner. Efforts to decrease premature deaths become progressively more difficult as such deaths are decreased so that the curve of the chance of dying as a function of age moves downward to the right toward a limiting curve; this is essentially exponential beyond the minimum. The foregoing is illustrated in Figure 1 by the semi-log plot of the chance of dying as a function of age for Swedish females for various periods from 1751 to 1950. Similar progressive changes have occurred in the United States; the dashed line curve superimposed on Figure 1 is the chance of dying (total population) in 1985; these data are plotted as an arithmetic curve in Figure 2. In developed countries these curves are now nearing the limiting curve for humans as life expectancies at birth approach a plateau value.

The limiting curve of the chance of dying as a function of age results from the irreducible contributions of the environment and disease to premature death and from an intrinsic process--the aging process--invoked to account for the progressive increase in the chance of dying with increasing age of the members of all species, even under optimal living conditions. Thus, in general, the chance that a combination of deleterious changes produced by the environment, disease, and the aging process capable of causing death will occur in a given individual increases continuously with time so that none live beyond a species-determined age.

All limiting curves have a similar shape. This suggests that 1) they are mainly determined by the aging process rather than by the environment and specific diseases; 2) the aging process is the same in all species,

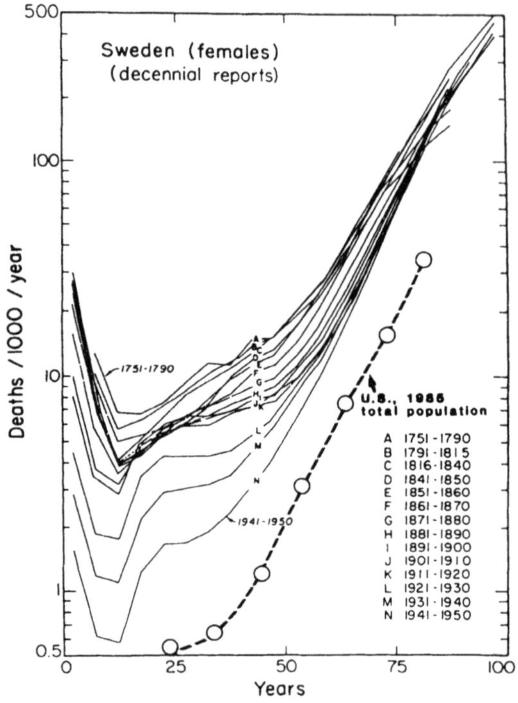

Figure 1. Age-specific death rates of Swedish females in various periods from 1751 to 1950. (Adapted from Jones, H. R.: The relation of human health to age, place and time, in Handbook of Aging and the Individual, edited by Birren, J. E., Chicago Univ., Chicago Press, 1959, pp. 336-363.) The superimposed dashed line is the 1985 data for the entire U. S. population.

the rate and the sequential changes produced by it being largely under genetic control; and 3) the aging process is responsible for the ever-increasing likelihood of disease and death with advancing age because the adverse changes produced by it per unit time increase exponentially with age.

The contributions of the aging process to disease and death, which are small in the early part of life, increase progressively with age, as shown by the dashed-line in Figure 2 (this is close to the limiting curve for humans). At some point it should become increasingly more difficult to increase the average life expectancy at birth (ALE) by further attempts to decrease premature death and should become impossible at a later age because of the exponentially increasing effect of the aging process. In the United States decreases in causes of premature death since 1900 (Figure 2) resulted, in a steady increase in average life expectancy (a rough measure of the healthy life span) from a value of 47.9 years in 1900 to 69.2 years in 1954-1955. After 1954-1955 the rate of increase began to slow, reaching 74.8 years in 1987. Apparently, at about 69 years of age, the ever-increasing widespread changes produced by the aging process begin to progressively thwart efforts to further reduce premature deaths by additional improvements in nutrition, housing, medical care, public health facilities, accident prevention, and other measures associated in the past

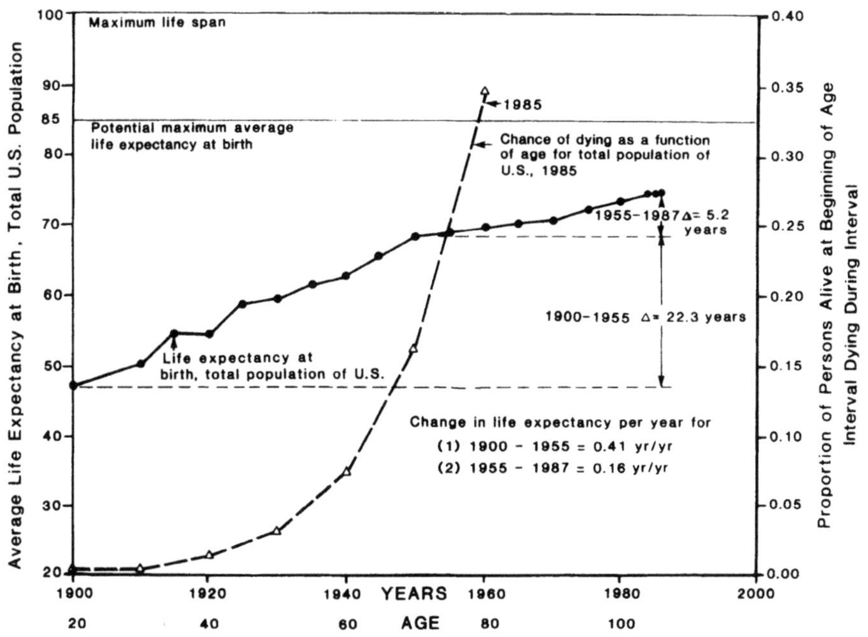

Figure. 2. Average life expectancy at birth since 1900 (solid line) and the chance of dying in 1985 as a function of age (dashed line) for the total United States population.

with advances in ALE. After about 75-76 years of age it is impossible, under normal circumstances, to reduce deaths; life expectancy in the United States and the other developed countries is near this plateau value. Thus, the aging process emerges as a major risk factor for disease and death at about age 69 and because the rate of the process increases continuously with age, it almost completely nullifies all present attempts to increase the average life span beyond the current value of 74.8 years.

The aging process is now the major cause of death in the developed countries. This fact is obscured by the protean nature of the contributions of this process to disease pathogenesis, i.e., cancer, atherosclerosis, and Alzheimer's disease, and to non-specific changes. In all probability, significant increases in average life expectancy in the future will be achieved only by slowing the aging process. Although the nature of the aging process is in dispute.[1] it may be due to free radical reactions.[2-5] These reactions, initiated by both enzymatic and non-enzymatic means, go on continuously through the cells and tissues. Because of the high chemical reactivity of the intermediates, all components of the body would be expected to constantly experience some degree of chemical change. It is these more-or-less random changes which may be the cause of aging.

ENDOGENOUS FREE RADICAL REACTIONS

Free radical reactions are ubiquitous in living things.[4,5] This observation probably reflects the seminal role of these unruly, irreversible reactions in the initiation and evolution of life, as depicted briefly in Figure 3. Life apparently arose spontaneously about 3.5

billion years ago from amino acids, nucleotides, and other basic chemicals of living things. These were produced from the simple, reduced components of the primitive atmosphere by free radical reactions, which were initiated mainly by ionizing radiation from the sun. Blue-green algae arose about 2.6 billion years ago and began to release O_2 into the atmosphere. As the atmospheric concentration of O_2 increased, so did the effectiveness of the ozone shield against ultraviolet light. By approximately 500 million years ago the intensity of radiations capable of initiating free radical reactions, from ultraviolet light and other sources, reaching the surface of the earth was reduced to a level compatible with life on land. The involvement of free radical reactions in metabolic processes and development of defenses against damage by such reactions probably evolved largely during the first 3 billion years after life arose (85 percent of the time life has existed), as living things sought to utilize these ever-present reactions and to protect themselves from free radical damage.

The original basic pattern of evolution does not appear to have changed except that the sun-initiated free radical reactions, which were essential for the origination and early evolution of the protocells, have largely been replaced in mammalian systems by those derived from enzymatic and non-enzymatic reactions.

Enzymatic reactions serving as sources of free radical reactions[4,5] include those involved in the respiratory chain, in phagocytosis, in prostaglandin synthesis, and in the cytochrome p-450 system. Free radicals also arise in the non-enzymatic reactions of oxygen with organic compounds as well as those reactions initiated by ionizing radiation.

<u>Years ago</u>	
3.5 Billion	Basic chemicals of life formed by free radical reactions, largely initiated by ionizing radiation from the sun. Life begins. Excision and recombinational process evolved to help maintain the functional integrity of DNA; meiosis evolved later. Ferredoxin RH or $H_2S + CO_2 \xrightarrow{(h\nu)} CH$
2.6 Billion	Blue-Green Algae $H_2O \xrightarrow{(h\nu)} 2H + O_2$
1.3 Billion	Atmospheric O_2 reaches one percent of present value. Anaerobic procaryotes disappear. Eucaryotes become dominant cells. Eucaryotes + blue-green algae ⟶ the green leaf plants. Eucaryotes + a procaryote able to reduce O_2 to H_2O ⟶ animal kingdom. Emergence of multicellular organisms and plants. Meiosis
500 Million	Atmospheric O_2 reaches 10 percent of present value. Ozone screen allows emergence of life from sea.
65 Million	Primates appear.
5 Million	Man

Figure 3. Overview of the origin and evolution of life.

Probably the vast majority of the free radical reactions are enzymatic ones involved in the maintenance and function of the organism, while the remainder, initiated by non-enzymatic means and by "leakage" of free radicals from enzymatic reactions, cause more-or-less random change.

Defenses have evolved[4,5] to protect against random free radical reaction-induced changes, including 1) tocopherols, 2) carotenes, 3) heme-containing peroxidases, 4) selenium-containing glutathione peroxidase, 5) superoxide dismutases, and 6) elevated serum uric acid levels. Protection against free radical reactions is not complete; therefore more-or-less random changes are produced continuously throughout living systems.[4,5]

FREE RADICAL THEORY OF AGING

The free radical theory assumes[2-5] that there is a single basic cause of aging, modified by genetic and environmental factors, and postulates that free radical reactions are involved in aging and disease. Although this theory is applicable to all life, the following discussion is largely directed to mammalian aging, in which O_2 is the main source of damaging free radical reactions, because of the importance attached to slowing the aging process in humans. Support for the theory, in addition to the studies on the origin of life and evolution discussed briefly above, includes[2-5], 1) the effect of ionizing radiation on living things, 2) life span experiments in which adverse endogenous free radical reactions were expected to be lowered by dietary manipulation, 3) the plausible explanations it provides for aging phenomena, and 4) the growing number of studies that implicate free radical reactions in the pathogenesis of specific diseases.

Ionizing Radiation

Whole-body exposure studies are particularly pertinent to aging. Such exposure produces changes which resemble those of normal aging[6] and shortens life span[7].

Life Span Studies

It should be possible to augment the normal defenses[4,5] against free radical reaction damage by at least three dietary changes: 1) caloric reduction; 2) minimizing dietary components, such as copper and polyunsaturated lipids or easily oxidized amino acids, which tend to increase free radical reaction levels; and 3) addition to the diet of one or more free radical reaction inhibitors, e.g., 2-mercaptoethylamine (2-MEA), α-tocopherol, butylated hydroxytoluene (BHT), and 1,2-dihydro-6-ethoxy-2,2,4-trimethylquinoline (ethoxyquin).

If free radical reactions do contribute to the degradation of biological systems and if the level of deleterious reactions can be lowered significantly by dietary changes, then such dietary alterations should have a beneficial effect. Studies are in accord with this possibility, for example:

Caloric reduction. The beneficial effects of reduced calories (with adequate essential nutrients) on life span and disease processes may be largely due[8] to a decrease in more-or-less random free radical damage throughout the cells and tissues and, in particular, in the mitochondria where over 90% of the oxygen consumed by a mammal is utilized.

Antioxidants. Many studies[4,5] have shown that the addition of one of a number of different antioxidants can increase the life span of mice, rats, fruit flies, nematodes, rotifers, and the "life span" of Neurospora. Thus, addition of 1.0%w 2-MEA to the diet of male LAF_1 mice increased the average life span by 30%. In this and other mouse studies, with the possible exception of two, there was no certain increase in maximum life span. Consideration of these results led to the suggestion that mitochondria[8] may serve as the biological clock that determines the maximum life span.

It is reasonable to expect, on the basis of present data, that lowering free radical reaction levels by a judicious selection of diets and antioxidant supplements will increase the healthy, active life span by 5-10 years or more.

Aging Phenomena

The free radical theory of aging provides plausible explanations for aging phenomena[4,5,8] which include 1) the inverse relationship between the average life spans of mammalian species and their basal metabolic rates; 2) the clustering of degenerative diseases in the terminal part of the life span; 3) the exponential nature of the mortality curve; 4) the beneficial effect of caloric restriction on life span and degenerative diseases, 5) the greater longevity of females, and 6) the increases in autoimmune manifestations with age.

For example, the greater longevity of females[9] may be due, at least in part, to the greater protection of female embryos from free radical reaction damage during a short period (about 48 hours in the mouse) of both high mitotic and metabolic activity. This period occurs just prior to the random inactivation of one of the two functioning X chromosomes in the late blastocyst stage of development.

The increase in autoimmune manifestations with age may be largely due to a disproportionate decrease in the radiosensitive T-suppressor cell function (compared with other cells involved in immunity)[10] due to the increasing levels of free radical reactions with age.

Table 1. "Free Radical" Diseases

Atherosclerosis	Fanconi´s anemia
Cancer	Bloom´s syndrome
Essential hypertension	Systemic lupus erythematosus
Senile dementia of the Alzheimer type	Diabetes mellitus, Type 1.
Amyloidosis	Preeclampsia
Osteoarthritis	Laennic´s cirrhosis
Senile macular degeneration	Acute pancreatitis
Senile cataract	Adult respiratory distress syndrome
Parkinson´s disease	Disseminated intravascular coagulation

The "Free Radical" Diseases

A disease is a combination of changes--usually forming a readily recognized pattern--that have detrimental effects on function, and in some cases may lead to death. Free radical reactions are implicated in the changes associated with a growing number of disorders (Table 1).

The ubiquitous free radical reactions would be expected to produce progressive adverse changes throughout the body. Those changes more-or-less common to all persons produce the normal aging pattern. Superimposed on this common pattern are changes that differ from individual to individual owing to genetic differences and to differences in the environment that modulate free radical reaction damage. The superimposed patterns of change[11] may become progressively more discernable with time, and in some individuals may eventually be recognized as diseases. Aging may also be viewed as a disease, differing from others in that the aging pattern is universal. The probability of developing any one of the "free radical" diseases should be decreased by lowering the free radical reaction level by any means, e.g., food restriction, antioxidants, increases in glutathione peroxidase or superoxide dismutases, etc., and in the case of a specific disease, lowered further by decreases in contributing environmental factors.

In addition to specific disorders, more subtle free radical-induced change may cause death. Thus, excessive cellular accumulation of lipofuscin, melanin, or the free radical reactions associated with pigment formation may kill and/or interfere with the function of cells critical to the organism as a whole, such as some groups in the central nervous system. The foregoing may account, at least in part, for the observation[12] that autopsies on individuals 85 years or older fail to find a cause of death in about one-third of the cases.

The above discussion indicates that the relationship between aging and diseases in which free radical reactions are involved is a direct one. The growing number of "free radical" diseases (Table 1) include[5,12] the two major causes of death, cancer and atherosclerosis, as well as other common degenerative diseases.

Some of the data[2-5] implicating free radical reactions in cancer and atherosclerosis are presented briefly below.

Cancer. Cancer initiation and promotion is associated with chromosomal defects and oncogene activation. Ionizing radiation is a "complete carcinogen" being both an initiator and promoter. It is a reasonable possibility that some endogenous free radical reactions, such as those initiated by ionizing radiation, will result in tumor formation by serving as a continuous source of tumor initiators and promoters. Numerous studies now support this possibility.

The increasing incidence of cancer with age is of particular interest. This increase is probably due, at least in part, to the increasing level of endogenous free radical reactions with advancing age.

Atherosclerosis. Atherosclerotic lesions can be initiated and enhanced by substances capable of irritating the arterial wall. A possible constant source of such compounds is the reaction of molecular oxygen with the polyunsaturated compounds present in serum and arterial wall lipids. The oxidation products, including peroxides and compounds of higher molecular weight formed through oxidative polymerization as well as substances arising from the reaction of intermediate lipid-free radicals with proteins and other substances, may be produced in amounts large enough to significantly contribute, directly or indirectly, to atherogenesis. In addition, oxyradicals arising from leucocytes may also serve as an important source of vessel wall irritants. Many studies now support the above possibility. Thus, the extensive atherosclerosis seen in patients with homocystinuria may be related to the formation of superoxide radicals during the ready oxidation by O_2 of homocysteine to homocystine. Xanthine oxidase has been implicated in atherogenesis; if so, it is probably related to superoxide production during the action of the enzyme. The enhanced atherosclerosis observed in chronic renal dialysis patients may also be due to free radical-induced endothelial cell injury; white cells tend to aggregate under these circumstances and to release H_2O_2, $O_2^{\bar{}}$ and HO•. In addition, the free radical reactions initiated by ionizing radiation also result in atherosclerosis. Further, the cardiovascular death rate of a group of health-conscious older persons, who supplemented their diets with relatively large amounts of vitamin E and vitamin C, was significantly below the expected value; this was also true for cancer.

CONCLUSIONS

The aging process is an intrinsic process invoked to account for the progressive increase in the chance of dying, even under optimal living conditions. This process, now the major cause of death in developed countries, is probably caused largely by free radical reactions arising in the course of normal metabolism.

The free radical theory of aging suggests practical means of increasing the healthy life span and provides reasonable explanations for aging phenomena, including the relationship between disease and aging.

REFERENCES

1. D. Harman, The aging process, Chem. and Engineering News, 65, No. 17:2 (1987).
2. D. Harman, Aging: A theory based on free radical and radiation chemistry, J. Gerontol. 11:298 (1956).
3. D. Harman, Role of free radicals in mutation, cancer, aging, and the maintenance of life, Rad. Res. 16:753 (1962).
4. D. Harman, The aging process, Proc. Natl. Acad. Sci. USA. 78:7124 (1981).
5. D. Harman, Free radical theory of aging: Role of free radicals in the origination and evolution of life, aging, and disease process, in "Free Radicals, Aging and Degenerative Diseases," J. E. Johnson, Jr., R. Walford, D. Harman, and J. Miquel, eds., Alan R. Liss, New York (1986).
6. G. W. Casarett, Similarities and contrasts between radiation and time pathology, in: "Advances in Gerontology Research," G. L. Strehler, ed., Claredon Press, Oxford (1964).
7. P. J. Lindrop and J. Rotblat, Long term effects of a single whole-body exposure of mice to ionizing radiation, Proc. Roy. Soc. (B) 154:332-349, 350-368 (1961).
8. D. Harman, Free radical theory of aging: Consequences of mitochondrial aging, Age 6:86 (1983).
9. D. Harman and D. E. Eddy, Free radical theory of aging: Beneficial effect of adding antioxidants to the maternal mouse diet on life span of offspring; possible explanation of the sex difference in longevity, Age 2:109 (1979).

10. D. Harman, Free radical theory of aging: Beneficial effect of antioxidants on the life span of male NZB mice: Role of free radical reactions in the deterioration of the immune system with age and in the pathogenesis of systemic lupus erythematosus, Age 3:64 (1980).
11. D. Harman, Free radical theory of aging: The "free radical" diseases, Age 7:111 (1984).
12. R. R. Kohn, Cause of death in very old people, JAMA 247:2793 (1982).

INFLUENCE OF LIFE-PROLONGING FOOD RESTRICTION ON MEMBRANE LIPOPEROXIDATION AND ANTIOXIDANT STATUS

Byung Pal Yu, Serge Laganiere, and Jae-Won Kim

Department of Physiology
University of Texas Health Science Center at San Antonio
San Antonio, Texas 78284-7756, U.S.A.

INTRODUCTION

The biological aging processes are time-dependent deteriorative changes occurring in all living cells. Several investigators have postulated that these age-related changes are due primarily to alterations in cellular membrane structure and function.[1,2] The free radical theory of aging proposed by Harman provides a molecular basis for the possible mechanisms for such ubiquitous phenomena.[3]

Among the many preventive interventions tested for anti-aging action, dietary calorie restriction is the most efficacious manipulation known to date[4]. In laboratory rodents, restricting caloric intake extends the life span, reduces the incidence of many age-related diseases and prevents age-related decline of many physiological functions.[4]

In search of the possible mechanisms underlying the action of food restriction, we have initiated an exploratory study[5] on lipid peroxidation (LPO) and age-related membrane alteration with the following hypothesis: food restriction protects against membrane lipid peroxidation and age-related impairment in cytosolic antioxidant defense systems.

METHODS

Animals and Diets

Rat maintenance and dietary procedures are the same as described.[6] Briefly, male specific pathogen-free Fischer 344 weanling rats were fed a diet of the following composition: 21% casein, 15% sucrose, 43.65% dextrin, 10% corn oil, 0.15% α-methionine, 0.2% choline chloride, 5% salt mix, 2% vitamin mix and 3% Solka-Floc. At 6 weeks of age, rats were divided into two groups: Group A continued to be fed the ad libitum diet, and group R rats were provided 60% of the mean amount of diet consumed by the group A rats.

Subcellular Membrane Preparation

Rat livers were homogenized in 10 mM HEPES buffer at pH 7.4 containing 10 mM KCl, 220 mM mannitol and 60 mM sucrose. The procedures for isolation and fractionation of mitochondria and microsomes by differential centrifugation were the same as described.[5]

Measurements of Lipid Peroxidation and Hydroperoxide

Lipid peroxidation was carried out by incubating the membranes with 37.5 mM Tris buffer (pH 7.4) containing 112 mM KCl, 10 mM $FeSO_4$ and 0.5 mM ascorbate at 30°C as described[5]. Lipid hydroperoxide was determined by the method of Buege and Aust.[7]

Fatty Acid and Other Biochemical Analyses

For determination of the fatty acid composition, membrane total lipid extracts were esterified by a methanolic-HCl method and quantified by gas chromatography.[5,8] The concentrations of several cytosolic antioxidants were determined in the following way. Reduced glutathione was measured with the 5, 5'-dithiobis (2-nitrobenzoic acid) method.[9] Ascorbate was assayed by measuring the reduction of Fe^{3+} to Fe^{2+} as described by Zannoni et. al.[10] GSH reductase activity was determined by the method of Racker.[11] GSH peroxidase was determined by the method of Paglia and Valentine.[12] GSH transferase was determined as described by Habig et. al.[13] Catalase activity was assayed by the method of Chance and Maehly.[14]

Measurement of Malondialdehyde (MDA) Oxidation

In vitro MDA oxidation was determined by measuring disappearance of MDA according to a modified method of Recknagel and Ghoshal.[15]

RESULTS

The evidence revealing that membrane lipid peroxidation (LPO) is age-dependent was obtained by the observation of increased generation of thiobarbituric-acid-positive material (malondialdehyde) during incubation of membrane fractions from older rats (Table 1). The production of malondialdehyde (MDA) from 6 to 24 months of Group A rats was approximately a two-fold increase without any lag phase. The qualitative and quantitative differences in the LPO process between the ad libitum fed and restricted rats were quite discernable; the typical characteristics of a rapid rate and the extent of LPO of ad libitum-fed rats were not observed in the restricted rats. Therefore, it is concluded that food restriction effectively suppressed in vitro membrane lipid peroxidation.

Lipid Hydroperoxide

The second line of evidence supporting age-related membrane modification by LPO was the quantitation of hydroperoxide content in membrane lipids. As shown in Table 2, the hydroperoxide contents in mitochondrial and microsomal membranes clearly increased with age. Although the progression of hydroperoxide formation with age seems to be similar in both groups, values were found approximately 1.5 to 3 times higher than those of the restricted rats in the ad libitum-fed rats. It is of interest to note that microsomal hydroperoxide content was consistently higher than the mitochondrial level, as was the case in the MDA-production study (Table 1).

Table 1. Membrane lipid peroxidation (nmole MDA produced/mg protein)

Group	Age (Mos)		Incubation Time (min) 5'	10'	40'
Group A	6	Mitochondria	2.61	5.60	5.93
		Microsomes	7.08	9.38	10.40
	24	Mitochondria	5.83	10.0	11.04
		Microsomes	9.38	15.52	19.48
Group R	6	Mitochondria	0.10	0.17	0.47
		Microsomes	0.23	0.23	0.94
	24	Mitochondria	0.65	0.83	2.54
		Microsomes	0.31	0.58	2.12

Table 2. Membrane Hydroperoxide (nmoles/mg protein)

	Group A		Group R	
Age (Mos)	Mitochondria	Microsomes	Mitochondria	Microsomes
6	1.31±0.38	1.47±0.33	0.44±0.09	0.74±0.12
12	1.57±0.10	2.81±0.15	0.99±0.18	1.64±0.17
24	1.98±0.29	2.42±0.29	1.24±0.23	1.54±0.10

Table 3. Fatty Acid Changes

Fatty Acids (%)	Age (Mos.)	Group A Mitochondria	Group A Microsomes	Group R Mitochondria	Group R Microsomes
18:2	6	11.67±0.5	11.30±0.2	12.98±0.6	12.78±0.7
22:5		3.05±0.3	3.17±0.2	3.90±0.6	3.34±0.4
unsat/sat. index		1.20	1.32	1.41	1.39
18:2	12	11.54±0.4	12.03±0.7	16.62±0.5	16.05±0.6
22:5		3.60±0.3	3.30±0.2	1.97±0.0	1.86±0.0
unsat/sat. index		1.26	1.31	1.41	1.43
18:2	24	8.72±0.9	9.70±0.6	15.98±1.0	16.44±1.4
22:5		5.57±0.8	5.96±0.8	2.96±0.5	2.63±0.4
unsat/sat. index		1.19	1.30	1.47	1.53

Fatty Acid Composition

Mitochondrial and microsomal membrane alteration by LPO was further evidenced by changes observed in membrane fatty acid composition (Table 3). Of the major fatty acids analyzed, linoleic (18:2) and docosapentaenoic (22:5) acids showed the most changes. Linoleic acid, consisting of 12% of the total fatty acids of both membranes, showed a substantial increase in group R rats, and, at the same time, a corresponding decrease in group A rats. The changes in docosapentaenoic acid, on the other hand, showed an opposite relationship: in the ad libitum-fed rats, the fatty acid increased, while in the restricted rats it decreased. The other important findings were changes in the total amount of unsaturated and saturated fatty acids. The index of the ratio of unsaturation/saturation, which is often used as an indicator for membrane fluidity, was clearly maintained by food restriction (Table 3).

Cytosolic Antioxidants and Related Enzyme Activities

In this study, glutathione (GSH), GSH reductase, GSH peroxidase, GSH transferase and catalase activities were measured in relation to age and diet (Table 4). The cytosolic GSH level was found to be fairly stable until 18 mos. of age; however, at 24 months a significant decrease was noticed in the control rats. GSH reductase, with very little age-related decline, was maintained at levels significantly higher than the control levels by food restriction. The age-related decline in GSH transferase activity of ad libitum fed-rats was effectively blunted by food restriction. Catalase activity showed a moderate age-related decrease in both dietary groups, but the enzyme activity in restricted rats was kept at higher levels than those of the controls. Changes in ascorbic acid content and glutathione peroxidase showed very little discernable dietary difference.

Table 4. Cytosolic Anti-oxidants and Related Enzymes

	6 Mos		18 Mos		24 Mos	
	Gp. A	Gp. R	Gp. A	Gp. R	Gp. A	Gp. R
Glutathione	59.8 ±7.7	50.9 ±2.9	56.4 ±13.1	51.5 ±2.0	46.0 ±5.0	62.3 ±5.2
GSH Reductase	41.6 ±2.8	49.8 ±4.0	57.4 ±2.0	67.5 ±2.4	41.0 ±2.7	52.3 ±3.3
GSH Transferase	527.9 ±24.1	565.0 ±50.6	439.2 ±56.0	533.2 ±166.4	473.6 ±22.6	653.2 ±40.6
GSH Peroxidase	262.5 ±14.9	279.6 ±30.4	347.2 ±217.1	249.1 ±87.7	273.4 ±26.4	292.6 ±20.7
Catalase	529.4 ±68.7	708.5 ±158.6	590.0 ±52.0	452.0 ±127.2	288.9 ±42.0	666.6 ±88.1
Ascorbic Acid	2249.7 ±221.3	2494.1 ±266.2	2866.8 ±273.5	2394.6 ±401.3	1618.2 ±94.6	1763.6 ±99.0

In Vitro Malondialdehyde Oxidation

Cellular protection systems against LPO include elimination of metabolites resulting from lipid peroxidation. In our study, malondialdehyde oxidation was used as a marker for assessing such cellular activity (Table 5). The highest oxidation of malondialdehyde was found in mitochondria, followed by cytosol and microsomes. It was found that malondialdehyde oxidation activity is age-related (Table 6), and the age-related loss was clearly prevented by food restriction, suggesting that food-restricted rats are more able to deal with potentially harmful lipid peroxidation byproducts.

Table 5. Subcellular Distribution of MDA Oxidation

	MDA Oxidation	
	(nmol/mg protein/10 min.)	Relative Activity
Homogenate	0.8915	1.0
Mitochondria	1.7857	2.0
Microsomes	0.2184	0.2
Cytosol	0.9803	1.1

Table 6. Mitochondrial MDA Oxidation (nmol MDA oxidized/mg protein/10 min)

Age (Mos.)	Group A	Group R
6	1.47	1.64
12	1.12	1.33
22	0.95	-
24	-	1.139

DISCUSSION

Our current data reaffirmed that hepatic mitochondrial and microsomal membranes undergo peroxidative changes with age.[5] The age-dependent changes were readily determined quantitatively by measuring malondialdehyde production and hydroperoxide contents of membrane lipid. Further data were obtained from the analysis of the membrane fatty acid composition.

It is of interest to speculate on how food restriction might elicit modulatory action on membranes and cytosolic components. Our data indicate that food restriction, apparently independent of the cytosolic influence, modulates membrane composition and membrane peroxidizability by balancing the amounts of 18:2 and 22:5 fatty acids. The mode of this modification may require a regulatory role of food restriction on the enzyme system involved in elongation and desaturation of fatty acid synthesis. On the other hand, for cytosolic modulation, food restriction seems to regulate only selected antioxidants and enzymes; ascorbic acid and GSH peroxidase are exceptions.

The action of food restriction in reducing lipid peroxidation has just begun to gain recognition as a plausible molecular mechanism for its anti-aging action.[5,16] Our study produces new evidence showing the influence of food restriction on the cytosolic defense system with modulation of key antioxidants, enzymes and malondialdehyde oxidation. The ramification and significance of the cytosolic modulation as part of the overall anti-lipid peroxidation by food restriction need further exploration in relation to its anti-aging process.

ACKNOWLEDGMENTS

This research was supported by NIH Grant AG01188. We thank Ms. Kathy LaRo and Anna Pittard for their expert technical assistance.

REFERENCES

1. I. Zs-Nagy, The role of membrane structure and function in cellular aging: A review, Mech. Age. Develop. 9:237 (1979).
2. D. Hegner, Age-dependence of molecular and functional changes in biological membrane properties, Mech. Age. Develop. 14:101 (1980).
3. D. Harman, Free radical theory of aging, in: "Free Radicals in Biology," W.A. Pryor, ed., Academic Press, New York, Vol. 5 (1982).
4. B. P. Yu, Recent advances in dietary restriction and aging, in: "Review of Biological Research in Aging", Vol. 2, M. Rothstein, Alan R. Liss, eds., New York (1985).

5. S. Laganiere and B. P. Yu, Anti-lipoperoxidation action of food restriction, Biochem. Biophy. Res. Comm. 145:1185 (1987).
6. B. P. Yu, E. J. Masoro, I. Murata, H. A. Bertrand, and F. T. Lynd, Life-span study of SPF Fischer 344 male rats fed ad libitum or restricted diets: longevity, growth, lean body mass and disease, J. Gerontol. 37:130 (1982).
7. J. A. Buege, and S. D. Aust, Microsomal lipid peroxidation, in: "Methods in Enzymology," Vol. 52, S. Fleischer and L. Packer, eds. (1978).
8. G. U. Liepa, E. J. Masoro, H. A. Bertrand, and B. P. Yu, Food restriction as a modulator of age-related changes in serum lipids, Am. J. Physiol. 238:E253 (1982).
9. M. S. Moron, J. W. Depierre, and B. Mannervik, Levels of glutathione, glutathione s-transferase activities in rat lung and liver, Biochim. Biophys. Acta 582:67 (1979).
10. V. Zannoni, M. Lynch, S. Goldstein, and P. Sato, A rapid micromethod for the determination of ascorbic acid in plasma and tissues, Biochem. Med. 11:41 (1974).
11. E. Racker, Glutathione reductase from baker's yeast and beef liver, J. Biol. Chem., 217:855 (1955).
12. D. E. Paglia and W. N. Valentine, Studies on the quantitative and qualitative characterization of erythrocyte glutathione peroxidase, J. Lab. Clin. Med. 70:158 (1967).
13. W. H. Habig, M. J. Pabst, and W. B. Jakoby, Glutathione s-transferase: The first enzymatic step in mercapturic acid formation, J. Biol. Chem. 249 (1974).
14. B. Chance, and A. C. Maehly, Assay of catalases and peroxidases, in: "Methods in Enzymology," Vol. 2, S.P.Colowick and N.O. Kaplan, eds. (1955).
15. R. O. Recknagel and A. K. Ghoshal, A. K., New data on the question of lipoperoxidation in carbon tetrachloride poisoning, Exp. Mol. Path. 5:108 (1966).
16. A. Koizumi, R. Weindruch, and R. L. Walford, Influence on dietary restriction on age on liver enzyme activities and lipid peroxidation in mice, J. Nutr. 117:361 (1987).

AUTHOR INDEX

SUBJECT INDEX

GPSR Compliance
The European Union's (EU) General Product Safety Regulation (GPSR) is a set of rules that requires consumer products to be safe and our obligations to ensure this.

If you have any concerns about our products, you can contact us on

ProductSafety@springernature.com

In case Publisher is established outside the EU, the EU authorized representative is:

Springer Nature Customer Service Center GmbH
Europaplatz 3
69115 Heidelberg, Germany

www.ingramcontent.com/pod-product-compliance
Ingram Content Group UK Ltd.
Pitfield, Milton Keynes, MK11 3LW, UK
UKHW022142141026
476618UK00005B/191
* 9 7 8 1 4 6 8 4 5 5 7 0 0 *